D1061084

Robust Control

OTHER IEEE PRESS BOOKS

System Design for Human Interaction, *Edited by A. P. Sage*
Adaptive Signal Processing, *Edited by L. H. Sibul*
Microcomputer Control of Power Electronics and Drives, *Edited by B. K. Bose*
Advances in Local Area Networks, *Edited by K. Kümmerle, J. O. Limb, and F. A. Tobagi*
Load Management, *Edited by S. Talukdar and C. W. Gellings*
Computers and Manufacturing Productivity, *Edited by R. K. Jurgen*
Being the Boss, *By L. K. Lineback*
Effective Meetings for Busy People, *By W. T. Carnes*
Modern Acoustical Imaging, *Edited by H. Lee and G. Wade*
Low-Temperature Electronics, *Edited by R. K. Kirschman*
Undersea Lightwave Communications, *Edited by P. K. Runge and P. R. Trischitta*
Multidimensional Digital Signal Processing, *Edited by the IEEE Multidimensional Signal Processing Committee*
Adaptive Methods for Control System Design, *Edited by M. M. Gupta*
Residue Number System Arithmetic, *Edited by M. A. Soderstrand, W. K. Jenkins, G. A. Jullien, and F. J. Taylor*
Singular Perturbations in Systems and Control, *Edited by P. V. Kokotovic and H. K. Khalil*
Getting the Picture, *By S. B. Weinstein*
Space Science and Applications, *Edited by J. H. McElroy*
Medical Applications of Microwave Imaging, *Edited by L. Larsen and J. H. Jacobi*
Modern Spectrum Analysis, *Edited by S. B. Kesler*
The Calculus Tutoring Book, *By C. Ash and R. Ash*
Imaging Technology, *Edited by H. Lee and G. Wade*
Phase-Locked Loops, *Edited by W. C. Lindsey and C. M. Chie*
VLSI Circuit Layout: Theory and Design, *Edited by T. C. Hu and E. S. Kuh*
Monolithic Microwave Integrated Circuits, *Edited by R. A. Pucel*
Next-Generation Computers, *Edited by E. A. Torrero*
Kalman Filtering: Theory and Application, *Edited by H. W. Sorenson*
Spectrum Management and Engineering, *Edited by F. Matos*
Digital VLSI Systems, *Edited by M. I. Elmasry*
Introduction to Magnetic Recording, *Edited by R. M. White*
Insights into Personal Computers, *Edited by A. Gupta and H. D. Toong*
Television Technology Today, *Edited by T. S. Rzeszewski*
The Space Station: An Idea Whose Time Has Come, *Edited by T. R. Simpson*
Marketing Technical Ideas and Products Successfully! *Edited by L. K. Moore and D. L. Plung*
The Making of a Profession: A Century of Electrical Engineering in America, *By A. M. McMahon*
Power Transistors: Device Design and Applications, *Edited by B. J. Baliga and D. Y. Chen*
VLSI: Technology and Design, *Edited by O. G. Folberth and W. D. Grobman*
General and Industrial Management, *By H. Fayol; revised by I. Gray*
A Century of Honors, *An IEEE Centennial Directory*
MOS Switched-Capacitor Filters: Analysis and Design, *Edited by G. S. Moschytz*
Distributed Computing: Concepts and Implementations, *Edited by P. L. McEntire, J. G. O'Reilly, and R. E. Larson*
Engineers and Electrons, *By J. D. Ryder and D. G. Fink*
Land-Mobile Communications Engineering, *Edited by D. Bodson, G. F. McClure, and S. R. McConoughey*
Frequency Stability: Fundamentals and Measurement, *Edited by V. F. Kroupa*
Electronic Displays, *Edited by H. I. Refioglu*
Spread-Spectrum Communications, *Edited by C. E. Cook, F. W. Ellersick, L. B. Milstein, and D. L. Schilling*
Color Television, *Edited by T. Rzeszewski*
Advanced Microprocessors, *Edited by A. Gupta and H. D. Toong*
Biological Effects of Electromagnetic Radiation, *Edited by J. M. Osepchuk*
Engineering Contributions to Biophysical Electrocardiography, *Edited by T. C. Pilkington and R. Plonsey*
The World of Large Scale Systems, *Edited by J. D. Palmer and R. Saeks*
Electronic Switching: Central Office Systems of the World, *Edited by A. E. Joel, Jr.*
A Guide for Writing Better Technical Papers, *Edited by C. Harkins and D. L. Plung*
Low-Noise Microwave Transistors and Amplifiers, *Edited by H. Fukui*
Digital MOS Integrated Circuits, *Edited by M. I. Elmasry*
Geometric Theory of Diffraction, *Edited by R. C. Hansen*
Modern Active Filter Design, *Edited by R. Schaumann, M. A. Soderstrand, and K. R. Laker*
Adjustable Speed AC Drive Systems, *Edited by B. K. Bose*

Robust Control

Edited by
Peter Dorato
Professor of Electrical and Computer Engineering
University of New Mexico

A volume in the IEEE PRESS Selected Reprint Series, prepared under the sponsorship of the IEEE Control Systems Society

The Institute of Electrical and Electronics Engineers, Inc., New York

PRINTED IN THE UNITED STATES OF AMERICA

IEEE Order Number: PC02204

Library of Congress Cataloging-in-Publication Data

Robust control.

(IEEE Press selected reprint series)
Bibliography: p.
Includes indexes.
1. Control theory—Collected works.
I. Dorato, Peter.
QA402.3.R63 1987 629.8′312 87-12448

ISBN 0-87942-233-5

Contents

Preface

The subject of this reprint volume deals with a fundamental problem in the design of control systems. This is the problem of accurately controlling the outputs of a system (plant) whose dynamics contain significant uncertainties. From the time of H. S. Black's patent in 1927, where feedback and large loop gains were first proposed for the solution of accurate design in the presence of uncertainty, much has been written on the subject. However, the term *robust control* did not appear in control literature until 1972. Prior to 1972, the problem was referred to as the *sensitivity design* problem or simply the problem of *control system design with uncertain plants*. The qualifier *robust*, having a similar technical meaning, had been used earlier in statistical analysis. The significance of the theory which developed around the term *robust control* from 1972 on was the focus on multivariable systems and synthesis. In contrast, most of the theory prior to 1972 focused on single-input-single-output systems and analysis (or at best cut-and-try design via iterated analysis).

In this reprint volume we have attempted to collect as many as possible of the significant papers that have been written in the period from approximately 1975 to 1985 on the subject of robust control. Unfortunately, because of the limited size of any volume of this type, many significant contributions also had to be omitted. In addition, it should be noted that the main thrust in this IEEE reprint volume is to survey the literature in IEEE publications and conference proceedings. Thus, only a limited number of non-IEEE papers are included here. Finally, no attempt was made to survey the literature in non-English journals, even where translations were available, again because of the constraint on size and the focus on IEEE publications. Of course, it is recognized that many significant contributions to robust control also exist in these other journals.

The problem of control system design in the presence of plant uncertainty is an extremely broad problem and many solutions have been proposed. Here we focus on those methods that lead to the *synthesis of fixed controllers for linear multivariable plants*, characterized by transfer function matrices. This excludes, for example, the so-called adaptive or self-tuning approach to the uncertainty problem. To survey the adaptive theory would require a separate volume, and indeed an IEEE reprint volume is available on this subject [11]*.

There is some question of a precise definition of the robust control problem. A particular question is whether one includes problems where the plant variations are arbitrarily small (differential sensitivity models) or one restricts the term *robust* to problems with finite variations. We select the former choice here, since some very interesting new theories of sensitivity optimization in multivariable systems have been developed in the past ten years. Of course, differential sensitivity designs are of limited practical interest since practical variations are always finite. However, like all first order differential methods, they can provide a first-cut at a suitable design.

The reprints are grouped into five parts. We summarize below the content of each part.

Part I: Robust Analysis

Here we include reprints of papers which provide the background for the analysis and synthesis of robust multivariable systems. A major focus in this part is the extension of single-input-single-output robustness concepts, i.e. stability margins, loop-gain, return-difference, etc., to multivariable systems.

Part II: Robust Stabilization

In this part we include reprints of papers which deal with the *synthesis* of controllers which yield closed-loop stable systems in the presence of *finite* plant variations.

Part III: H^2 and H^∞ Sensitivity Optimization

Here we include reprints which deal with with *synthesis* of controllers which yield closed-loop systems that minimize a *differential sensitivity measure*. Papers dealing with both H^2 and H^∞ sensitivity measures are included. While the H^2 theory (Wiener-Hopf theory) is classical, complete closed-loop design techniques for H^2 optimization have appeared only in the last ten years. On the other hand the H^∞ theory applied to control system design is entirely new.

Part IV: Other Approaches to the Control of Uncertain Systems

Here we include a collection of other approaches to the synthesis of robust systems, including: stochastic optimization, Lyapunov function methods, minimax optimization, guaranteed-cost control, robust servomechanism approach, and the quantitative-feedback-theory method. These methods represent deviations of the plant models and/or performance measures from those in Parts II and III. Yet they also deal with the problem of robust control synthesis.

Part V: Related Theory and Numerical Computation

In this part, reprints are included which do not directly address the robust control problem, but develop the underlying theory for the various synthesis approaches in Parts II and III. Also included in this part are some reprints on related numerical algorithms. We include this material here to make

* Bracketed numbers refer to references listed in the bibliography but not included in this reprint volume.

the volume as self-contained as possible. Most of this related theory, for example, coprime matrix-fraction approach to control system analysis and design, modern Wiener-Hopf theory, Hankel-norm approximation theory, etc., is theory that has been developed only over the past ten years.

In our introduction we review the literature in each of the above five areas, including significant results not included in this reprint volume. In addition, we include in the introduction sections on mathematical preliminaries, historical perspective, and applications. Finally, we include a fairly extensive bibliography of journal articles and conference papers.

This reprint volume should be most valuable for graduate studies in control system theory. Various combinations of reprints could be used for special topics courses in multivariable systems, robust analysis, robust control synthesis, H^2 and H^∞ system optimization, robust stablization, etc. It can also serve as a reference volume for textbooks on related subjects. Finally, while much of the theory presented here is new, and it is not clear how much of this theory will find serious applications, this volume should provide the practicing engineer easier access to recent results in robust control than would otherwise be possible.

The editor is grateful for the valuable inputs to this volume supplied by the following colleagues: J. J. Bongiorno, Jr., Polytechnic University; E. J. Davison, University of Toronto; J. W. Helton, University of California at San Diego; I. Horowitz, University of California at Davis; H. Kimura, Osaka University; H. Kwakernaak, Twente University; J. B. Pearson, Jr., Rice University; M. Vidyasagar, University of Waterloo; and G. Zames, McGill University.

The editor is also grateful to Gladys Ericksen for her patient typing of the volume manuscript. Finally the editor is indebted to L. Shaw, Polytechnic University, for his careful review of the original volume proposal, to M. G. Safonov, University of Southern California, for his many valuable inputs to the final manuscript, to H. P. Leander, Technical Editor of the IEEE PRESS for his help in expediting the publication of the volume, and to Yunzhi Li, from the Institute of Automation, Academia Sinica, for her valuable help in compiling and editing the bibliography.

Introduction

1. The Robust Control Problem

As noted in the preface, the robust control problem is roughly the problem of analyzing and designing accurate control systems given plants which contain significant uncertainty. To define the problem more precisely a number of problem elements must be carefully developed. In particular one must delineate the class of plant models, uncertainty models, and performance measures that are to be considered.

Plant Models—There are three common linear multivariable plant models used in the reprints in this volume; they are state-variable models, transfer-function matrix models, and matrix-fraction models. The state-variable models are generally written in the form

$$\dot{x}=Ax+Bu,\ y=Cx+Du \tag{1}$$

where x, y, and u are n, p and q dimensional state, output, and input vectors, respectively. The transfer-function matrix models are denoted $G(s)$, where $G(s)$ is a $p \times q$ matrix, with entries made up of rational functions in the complex Laplace transform variable s, and $G(s)$ relates the transform of the input vector u to the transform of the output vector y. Finally, matrix-fraction models are of the form, for example for a right-coprime fraction, $G(s) = N(s)D^{-1}(s)$, where $N(s)$ and $D(s)$ are matrices with either polynomial entries or stable-rational-function entries.

Plant-Uncertainty Models—Of course, each plant model requires its own type of uncertainty model. In addition, uncertainty is characterized in a number of different ways. For example, uncertainty is characterized as *parametric* versus *nonparametric, structured* versus *unstructured*, and *stochastic* versus *deterministic* (*set uncertainty*). To illustrate further the different types of uncertainty, consider the state-variable model of (1). The uncertainty in the term such as Ax can be modelled as follows:

(i) $Ax \to (A + \delta A)x$, where δA is constrained only in norm, i.e. $\|\delta A\| \le a$.
(ii) $Ax \to (A + \Sigma_i q_i A_i)x$, where $-1 \le q_i \le 1$.
(iii) $Ax \to Ax + \Sigma_i x_i G_i \dot{w}_2$ where $\dot{w}_2$ is a white noise random process.

In this case, (i) and (ii) represent *parametric set-uncertainty* models where in (i), the uncertainty, represented by δA, is *unstructured*, and where in (ii), the uncertainty, represented by q_i, is *structured*. Basically, structured uncertainty limits how the uncertainty enters into the matrix A. In (iii), the uncertainty is modelled *stochastically*, as a multiplicative white noise process $\dot{w}_2$. Examples of the use of the models in (i) and (ii) may be found in Chang and Peng (Reprint 25)*, and the model in (iii) in Wonham (Reprint 24).

* Reprints are identified in the table of contents.

For the transfer-function matrix model, uncertainty in G may be modelled as follows

(iv) $G \to (I + L)G$, where $\|L(j\omega)\| < l_m(\omega)$
(v) $G \to G(I + L)$, where $\|L(j\omega)\| < l_m(\omega)$
(vi) $G \to G + \delta G$, where $\|\delta G\| < l_a(\omega)$.

Here, $\|L\|$ denotes the norm of the matrix L. More will be said about norms in section 3. In (iv) and (v), L represents *multiplicative unstructured nonparametric* uncertainty, at the plant output and input, respectively. For multivariable systems, the location of multiplicative perturbations is critical, since matrix products do not in general commute. In (vi), δG represents *additive unstructured nonparametric* uncertainty. Uncertainties of these types are discussed in Doyle and Stein (Reprint 3). If only certain elements of L and δG are variable, the uncertainty would then be structured. An important advantage of nonparametric models is that they permit the treatment of uncertainty associated with unmodelled dynamics. Thus, for example, L in (iv) may be a rational matrix of degree higher than G. Perhaps some of the common causes of plant uncertainty should be recalled at this point, i.e., modelling errors, parameter variations, and parameter uncertainties.

Uncertainty in matrix fraction models is not considered extensively in this volume. Some discussion of their type of uncertainty, where $N(s)$ and $D(s)$ are additively perturbed, appears in Vidyasagar (Reprint 9).

Performance Measures—Of course, stability is often an implicit performance measure in control system design. In addition, however, other measures are often required, relating typically to tracking errors, disturbance rejection, and noise suppression. The most common mathematical forms for performance measures in this volume are

(i) $J = \int_0^\infty (x^T Q x + u^T R u)\, dt$
(ii) $J = \sup_\omega \|E(j\omega)\|$
(iii) $J = \int_{-\infty}^{\infty} \|E(j\omega)\|^2\, d\omega$.

The measures given by (i)–(iii) are referred to as the integral-quadratic measure, H^∞-norm measure, and H^2-norm measure, respectively. In (ii) and (iii), the vector E represents an "error" vector which is to be kept as small as possible. Because of the plant uncertainty one can at best minimize an upper bound on these performance measures, or in the stochastic case, an average value. Cases where $\|E(j\omega)\|$ is bounded at each frequency point, i.e., $\|E(j\omega)\| < l_p(\omega)$, can be reduced to a normalized design condition of the form

$$\sup_\omega \left\| \frac{E(j\omega)}{l_p(\omega)} \right\| < 1$$

which is of the H^∞-norm type measure. Designs which minimize an upper bound are referred to as *guaranteed-cost* designs, while those that minimize the least-upper-bound are referred to as *minimax* designs. See Chang and Peng (Reprint 25) and Bertsekas and Rhodes (Reprint 26) for examples of these kinds of designs. More about H^2 and H^∞ norms will be presented in section 2. Good introductory material for single-input single-output systems on design with these norms may be found in Youla, Bongiorno, and Jabr (Reprint 31) and Zames and Francis (Reprint 16).

In this volume, we focus on the design of fixed controllers which solve the robust control problem, i.e., controllers which result in control systems that meet performance measure bounds for given uncertainty bounds. This excludes from consideration adaptive or self-tuning solutions.

Finally, it should be noted that the robust control problem is generally associated with finite plant uncertainties. However, we include here some reprints, for examples those in Part III, that deal with infinitely small plant uncertainties (associated with differential sensitivity measures). We do so because the sensitivity theory associated with multivariable systems and H^∞-optimization is new and interesting, and also because the first-order design associated with sensitivity measures can sometimes yield useful results for finite uncertainties.

2. Robust Control: A Brief Historical Perspective*

Perhaps the earliest proposed solution to the "robust control problem" appears in the patent of H. S. Black [33] in 1927. In this now classical patent Black first proposed feedback and large-loop gains for the design of an accurate system (vacuum tube amplifier) given significant plant uncertainties (large variations in vacuum tube characteristics). Unfortunately, most "accurate" systems (Black used the term *stable* rather than *accurate*, but this should not be confused with dynamic stability) designed this way were dynamically unstable. It was not until the results of Nyquist [87] in 1932 that the trade-off between dynamic stability and large-loop gain was analytically understood. The Nyquist frequency domain stability criterion and Black's concept of large-loop gain for system accuracy formed the basis of robust control design developed in the classical book by Bode [5] published in 1945. Bode also introduced the differential sensitivity function to provide an analytical measure for system accuracy improvement, at least for sufficiently small plant variations. The Bode approach to the design of robust systems was extended to finite plant variations by Horowitz [12]. One could call the period from 1927 to 1960 the *classical sensitivity design period*. The focus during this period of time was on loop shaping of single-input single-output (SISO) systems for stability, sensitivity reduction, noise suppression, etc.

The next major period in control systems theory was the period between 1960 and 1975. We call this period the *state-variable period*. In the early 1960's R. E. Kalman introduced a number of key state-variable concepts, i.e., controllability, observability, optimal linear-quadratic state feedback (LQSF), optimal state estimation (Kalman filtering), etc. An exposition of the major results associated with this period may be found in the text of Anderson and Moore [3] published in 1971. Unfortunately, with a few notable exceptions, the problem of plant uncertainty was largely ignored during this period. One such notable exception was the introduction of the *sensitivity comparison matrix* in 1964 by Cruz and Perkins [40] for the analysis of multiple-input multiple-output (MIMO) systems. The sensitivity comparison matrix provided an analytic tool for the comparison of closed-loop versus open-loop accuracy improvement. It was an early attempt to extend SISO sensitivity results in MIMO systems. The problem of accurate control of uncertain systems was commonly referred to as the *sensitivity design problem* during this period of time. The major results in sensitivity theory developed during the state-variable period were summarized in a collection of reprints edited by Cruz [7] and published in 1973. In 1978 this material appeared in textbook form in Frank [10]. Further sensitivity results were reported in the special issue on sensitivity of the *Journal of the Franklin Institute* in 1981 (March/April issue). Included in the sensitivity design problems studied in this period of time are problems of trajectory sensitivity, performance sensitivity, and eigenvalue/eigenvector sensitivity.

In the late 1970's and early 1980's a renewed interest appeared in the problem of plant uncertainty. At about the same time, some significant results were being reported on the analysis of multivariable systems in the frequency domain. In particular the concept of *coprime matrix fraction* description of multivariable systems was introduced as a design tool by Youla, Jabr, and Bongiorno (Reprint 32), and Desoer, Liu, Murray, and Saeks (Reprint 33), in 1976 and 1980, respectively. Also, the Nyquist stability criterion was generalized to multivariable systems by Rosenbrock [21] and [100], and MacFarlane and Postlethwaite [79]. In Reprint 32, a parameterization of all stabilizing compensators was introduced. This parameterization has come to play a key role in the robust stabilization of multivariable systems, and is often referred to as the *Youla parameterization*. This confluence of interest in uncertainty and multivariable systems lead to the current period, which we refer to as the *modern robust control* period (1975–present). The actual term *robust control* first appeared in the text of a journal article by Davison [207] in 1972, and in the title of a journal article by Pearson and Staats [252] in 1974.

Actually the seeds for the modern robust control period were planted in two papers written in the early 1960's. One was a paper by Zames (Reprint 1) written in 1963 which introduced the concept of the "small gain" principle which plays such a key role in robust stability criteria. The other was a paper written by Kalman [68] in 1964 which demonstrated for SISO systems that optimal LQ state-feedback control laws had some very strong robustness properties, i.e., infinite gain margins and 60 degree phase margins. Safonov and Athans [105] demonstrated in 1977 that these gain and phase margins extended to MIMO systems for gain and phase variations in each input channel to the plant. These results are included in Lehtomaki, Sandell, and Athans (Reprint 6). Unfortunately, when state-estimate feedback is used instead of state feedback, these desirable robustness properties vanish (see Doyle [46]).

* The reader who is unfamiliar with some of the mathematical terms used in this section may wish to read Section 3 first.

Doyle and Stein [50] were able to show, however, that the desirable loop return-difference properties of the optimal LQSF control law could be recovered by suitable design of the Kalman filter in the feedback loop. Because of the considerable design experience with optimal LQ and LQG design during the state-variable period, there was strong interest in extending robustness results to this class of problems, and a multivariable robust design philosophy emerged, which was identified as the *LQG/LTR (linear-quadratic-Gaussian loop transfer recovery)* approach. See Doyle and Stein (Reprint 3) for a discussion of the LQG/LTR design approach.

In a research monograph published in 1980, Safonov [22] presented a generalized "sector" type stability criteria which was especially useful for the study of robustness in multivariable systems. This sector stability criterion was a generalization of the conic sector stability concepts developed earlier by Zames [121]. The above monograph also contains an excellent summary of LQG robustness and stability results. This monograph was also the first book on feedback systems to include the term *robust*.

In February 1981, the IEEE came out with a special issue of the TRANSACTIONS ON AUTOMATIC CONTROL on *Linear Multivariable Control Systems*. A number of papers on the robust control of multivariable systems appeared in that issue. The main focus of these papers was the use of singular values in the design of robust multivariable systems in the frequency domain. Basically, this represented an extension of the classical Bode design approach to multivariable systems. Reprints 3, 4, and 6 are taken from this special issue. The title of the lead paper in this issue (Reprint 3), *Multivariable feedback design: Concepts for a classical/modern synthesis*, is indicative of this theme.

Unfortunately, stability and performance measures based on singular values are generally too conservative for structured plant variations. In 1982, Doyle (Reprint 8) introduced the concept of *structured singular value* (SSV) to deal with this problem. The structured singular value of a matrix M is denoted $\mu(M)$ and reduces to the singular value $\bar{\sigma}(M)$ when perturbations are unstructured. The above paper by Doyle (Reprint 8) was included in a special issue of the IEE Proceedings, Part D, on *sensitivity and robustness* published in November, 1982. This special issue included a number of important papers on robust control. Reprints 7, 8, and 28 are taken from this special issue. The IEEE special issue in 1981 and the IEE special issue in 1982 provided the fundamental analysis tools for robust control.

The stage was set for the development of robust control synthesis procedures. One such procedure appeared in the paper by Zames and Francis (Reprint 16) in 1983. In this paper, the optimal H^∞ sensitivity design problem is posed for SISO systems and solved using the optimal Nevanlinna-Pick interpolation theory. The H^∞ approach to control system design was introduced in an earlier paper by Zames (Reprint 2). This paper also included the solution of the H^∞ sensitivity optimization problem for a special case, i.e., a single right-half s-plane plant zero. Solutions to the MIMO optimal H^∞ sensitivity problem were published the following year, 1984, simultaneously by a number of authors, i.e., Chang and Pearson (Reprint 19) and Francis, Helton, and Zames (Reprint 20).

In 1980, Tannenbaum [147] used Nevanlinna-Pick interpolation theory to synthesize *stable* controllers which maximized the allowable range of variation in plant-gain factor. In 1984, Kimura (Reprint 12) solved the SISO synthesis problem, also using Nevanlinna-Pick interpolation theory, for robust stabilization of plants with additive unstructured plant perturbations. Solutions to the MIMO robust stabilization problem were presented by several authors in 1986, i.e., Vidyasagar and Kimura (Reprint 14), and Glover [134].

In 1985, Youla and Bongiorno (Reprint 22) presented a solution to the H^2 optimal sensitivity design problem. These results were based on earlier work, in 1976, by Youla, Jabr, and Bongiorno (Reprint 32) on the Wiener-Hopf design of multivariable systems, and an early paper, in 1969, by Bongiorno (Reprint 15) on optimal H^2 sensitivity design.

This brief historical review summarizes the major developments that have led to the modern robust control period (1975–1985). In this reprint volume, we focus on the multivariable frequency domain theory approach to robust control. An up-to-date summary of the frequency domain approach may also be found in the recent text of Vidyasagar [26].

A number of other approaches to the control of uncertain systems were developed in parallel with the frequency domain approach. Some of these approaches are included in the reprints of Part IV. In particular, we include in Part IV a paper by Wonham (Reprint 24) published in 1967, on the optimal control of systems with state-dependent noise. The state-dependent noise serves as a model of random parameter variations. The probabilistic modelling of plant uncertainty represents an important class of uncertainty models.

Also included are reprints on state-variable minimax approaches to the robust control problem. Here the dynamics are given in state-space form and the performance measure is taken to be the maximum of an appropriate norm over all admissible parameter variations. See for example, Bertsekas and Rhodes (Reprint 26) and Chang and Peng (Reprint 25). This approach leads to a guaranteed level of performance (guaranteed-cost) for all possible parameter variation. The above paper by Chang and Peng, published in 1972, represents an early attempt to design for both *robust performance* and *robust stabilization*. The problem of simultaneous robust stability and performance design in the H^∞ setting is a continuing research problem. An approach to this design problem is embedded in Doyle's general μ-synthesis (Reprint 17). However, direct analytic solutions are not yet available. The special problem of nominal performance optimization subject to a robust-stability constraint has been studied by O'Young and Francis [179]. Bounds on performance subject to robust stabilization are developed by Hara and Katori [173], and a suboptimal to the simultaneous robust performance and stabilization problem is presented in Bird and Francis [153].

Lyapunov function techniques have yielded useful robust stabilization results, especially for nonlinear systems. We include a reprint by Barmish, Corless, and Leitmann (Reprint 29) to illustrate this approach to the control of uncertain systems.

Finally, it should be noted that some new mathematical results have been obtained recently which expedite the solution of multivariable H^∞ problems. In particular Ball and Helton, in a paper published in 1983 [284], used a Beurling-Lax theorem to solve a general optimal interpolation problem, and Glover, in a paper published in 1984 (Reprint 36), developed a theory of model reduction which has important computational implications for the optimal interpolation problem.

3. Some Mathematical Preliminaries

A great deal of mathematics, new to many engineers, has been introduced by the robust control theory developed over the past 10 years. In this section, we introduce a few of the key mathematical concepts found in many of the reprints included here. The mathematical concepts are grouped into three divisions: abstract algebra, matrix norms and singular values, and Hardy spaces. In order to motivate the application of these concepts to common engineering problems, the definitions we give here are not the most general. For example, in discussing Hardy spaces, we limit ourselves to rational functions. More general definitions may be found in the reprints and literature cited.

Abstract Algebra—In dealing with multivariable systems, the concept of a ring plays a key role. A *ring* is defined as a set R with two binary operations, denoted $+$ and $\cdot$ (called "addition" and "multiplication") with the following properties:

For every a, b, and c which belong to R,

- (i) $a + b$ belongs to R, (+closure),
- (ii) $a + b = b + a$, (+commutativity),
- (iii) $a + (b + c) = (a + b) + c$, (+associativity),
- (iv) There exists an element, denoted 0 (+ identity) and called the "zero" element, such that $a + 0 = 0 + a = a$,
- (v) there exists an element associated with a, denoted $-a$, such that $a + (-a) = 0$ (+inverse),
- (vi) $a \cdot b$ belongs to R ($\cdot$closure),
- (vii) $a \cdot (b \cdot c) = (a \cdot b) \cdot c$, ($\cdot$associativity),
- (viii) $a \cdot (b + c) = a \cdot b + a \cdot c$ and $(a + b) \cdot c = a \cdot c + b \cdot c$, ($\cdot$distributivity over +).

Important properties that a general ring does not have, compared to "ordinary" algebra, include: multiplication is not commutative, there is no multiplicative identity, and not every element has a multiplicative inverse. If a ring R has a multiplicative identity, denoted 1, and if an element a belonging to R has multiplicative inverse, i.e., there exists an element, denoted a^{-1}, such that $a \cdot a^{-1} = a^{-1} \cdot a = 1$, then a is said to be a *unit*. A *field* ("ordinary" algebra) is a ring where multiplication is commutative, a multiplicative identity exists, and every nonzero element is a unit.

Some important rings which arise in multivariable system theory include: the ring of square matrices under matrix addition and multiplication, the ring of polynomials in s under polynomial addition and multiplication, and the ring of proper rational functions in s with no poles in the closed right-half s-plane (stable transfer functions). Note that matrix multiplication is not commutative, that the reciprocal of a polynomial is not a polynomial, and that the reciprocal of a stable transfer function is not necessarily stable. Thus, in dealing algebraically with matrices, polynomials, and stable transfer functions, one is forced into the general structure of a ring. Excellent introductions to rings and related algebraic concepts may be found in Callier and Desoer [6] and Vidyasagar [26]. The application of the ring of stable transfer functions to the analysis and design of control systems is developed by Desoer, Liu, Murray, and Saeks (Reprint 33).

Another major algebraic concept that arises in multivariable system theory is that of coprime matrices. In particular related to coprime matrices are the concepts of right- and left-coprime matrix fractions, used to represent *transfer function matrices* $G(s)$ (matrices with entries given by rational functions of the Laplace transform variable s) as *matrix fractions* of the form $G = N_r D_r^{-1} = D_l^{-1} N_l$ where N_r, N_l, D_r, and D_l are *polynomial matrices* (matrices with polynomial entries in the variable s), and where N_r and D_r, $\det D_r \neq 0$, are right-coprime matrices, and N_l and D_l, $\det D_l \neq 0$, are left-coprime matrices. We define next the concepts of right- and left-coprime matrices. Following Vidyasagar and Kimura (Reprint 14), we denote $M(R)$ as the generic set of matrices with elements in the ring R. We assume here the ring R is commutative and has a multiplicative identity. Two matrices A and B having the same number of columns and belonging to $M(R)$ are said to be *right-coprime* if every greatest common right divisor (g.c.r.d.) of A and B is unimodular. A square matrix belonging to $M(R)$ is said to be *unimodular* if its determinant is a unit in R; and a square matrix D belonging to $M(R)$ is said to be a *greatest common right divisor* of A and B if $A = ED$ and $B = FD$ (i.e., D is a *common right divisor* of A and B), and if D' is any other common right divisor of A and B, then $D = GD'$ (i.e., D is a left multiple of D'). A similar definition holds for the *greatest common left divisor* for matrices with the same number of rows. A classical necessary and sufficient condition for two matrices A and B, belonging to $M(R)$, to be right-coprime is the existence of matrices X and Y, also belonging to $M(R)$ such that the so-called matrix *Diophantine* or *Bezout* equation,

$$XA + YB = I \tag{2}$$

is satisfied, where I denotes the identity matrix in $M(R)$.

Equation (2) plays a key role in the design of multivariable systems. See, for example, Youla, Jabr, and Bongiorno (Reprint 32) for the application of (2) over polynomial rings, and Desoer, Liu, Murray, and Saeks (Reprint 33) for the application of (2) over the ring of stable transfer functions.

Matrix Norms and Singular Values—Although most engineers are familiar with simple matrix norms and eigenvalues, a number of related concepts have arisen in robust control theory that the engineer is less likely to be familiar with, for example, H^∞-norms, singular values, Hankel-norms, etc. We review some of these concepts here, and we start with a review of the commonly used induced Euclidean norm. Let $\|x\| = (x^T x)^{1/2}$ denote the usual Euclidean norm of the vector x, where x^T denotes the transpose of x. Then the *induced Euclidean norm of the matrix A*, denoted $\|A\|$ is defined as

$$\|A\| = \max_x \|Ax\|, \ \|x\| = 1. \tag{3}$$

A special property of *induced* norms is that $\|AB\| \leq \|A\| \cdot \|B\|$.

Recall also that λ is called an *eigenvalue* and x the associated *eigenvector* of a matrix A, if $Ax = \lambda x$. We denote the ith eigenvalue of A by $\lambda_i(A)$.

The concept of singular value for a matrix A plays a very important role in robust control. The *singular values* of a square $n \times n$ complex matrix A, denoted $\sigma_i(A)$ are defined as

$$\sigma_i(A) = (\lambda_i(A^H A))^{1/2} \tag{4}$$

where A^H denotes the conjugate transpose of A. It is well known that any matrix A can be decomposed into the product (*singular value decomposition*)

$$A = U \Sigma V^H \tag{5}$$

where $U^H U = I$, $V^H V = I$, and $\Sigma = \text{diag}\,(\sigma_1, \sigma_2, \cdots \sigma_n)$. The singular values are real and non-negative, and $\bar{\sigma}$ is commonly used to denote the largest singular value and $\underline{\sigma}$ the smallest. It is easy to show that $\bar{\sigma}(A) = \|A\|$.

A recurrent problem that appears in robust analysis is the problem of determining conditions on the matrix M for nonsingularity of the matrix $I + M\Delta$, where I is the unit matrix and $\bar{\sigma}(\Delta) \leq 1$. A necessary and sufficient condition on M is that

$$\bar{\sigma}(M) < 1 \tag{6}$$

if no structure is assumed on the matrix Δ. See Lehtomaki, Sandell, and Athans (Reprint 6) for a brief discussion of singular values and singular-value decomposition. Unfortunately, for many problems Δ has some structure and the condition (6) is too conservative. To deal with this problem Doyle (Reprint 8) has introduced the concept of *structured singular value*, denoted $\mu(M)$ and defined as follows

$$\mu(M) = [\min_\Delta \{\bar{\sigma}(\Delta) | \det\,(I + M\Delta) = 0\}]^{-1} \tag{7}$$

where the minimization is done with respect to the set of Δ with *given structure* and $\bar{\sigma}(\Delta) < 1$. It can be shown that if Δ has no structure $\mu(M) = \bar{\sigma}(M)$, and also that if $\Delta = \delta I$, then $\mu(M) = \rho(M)$, where $\rho(M)$ is the *spectral radius* of the matrix M, i.e.

$$\rho(M) = \max_i |\lambda_i(M)|. \tag{8}$$

Some important norms are defined for matrices $G(s)$ whose entries are analytic functions of the complex variable s in the open right-half s-plane. We will have more to say about the space of such functions in the next section; however, we define here two key norms for such matrices. For ease of exposition we assume further that the matrix entries are real rational functions of s; thus, the only points where the entries can fail to be analytic are at their poles. The H^2-*norm* of $G(s)$ is denoted $\|G(s)\|_2$ and is defined to be

$$\|G(s)\|_2 = \left[\int_{-\infty}^{\infty} tr G^H(j\omega) G(j\omega)\, d\omega \right]^{1/2} \tag{9}$$

where $tr\,A$ denotes the trace of the matrix A, and the H^∞-*norm* of $G(s)$ is denoted $\|G(s)\|_\infty$ and is defined to be

$$\|G(s)\|_\infty = \sup_\omega \bar{\sigma}(G(j\omega)), \ \omega \text{ real}. \tag{10}$$

The final concepts we wish to introduce in this section are those of Hankel singular values and Hankel norms. Consider the transfer-function matrix associated with the state equations (1), i.e., $G(s) = C(sI - A)^{-1}B + D$, when all the eigenvalues of A have negative real parts, i.e., the state-variable system is asymptotically stable when $u = 0$. Assume the system is observable and controllable, and let P and Q be solutions, necessarily positive definite, of the Lyapunov equations

$$AP + PA^T = -BB^T \tag{11}$$

$$A^T Q + QA = -C^T C. \tag{12}$$

Then the ith Hankel singular value, denoted $\sigma_i(G(s))$, is defined as

$$\sigma_i(G(s)) = [\lambda_i(PQ)]^{1/2} \tag{13}$$

and the *Hankel norm* of $G(s)$, denoted $\|G(s)\|_H$ is defined as

$$\|G(s)\|_H = [\lambda_{\max}(PQ)]^{1/2}. \tag{14}$$

The Hankel norm is used by Glover (Reprint 36) to obtain efficient algorithms for the computation of reduced order models. However, the theory developed there can be applied to the efficient computation of multivariable H^∞ optimal-sensitivity problems. See Safonov and Verma (Reprint 21) for a discussion of the reduction of the H^∞ optimization problem to the model-reduction problem of Glover.

Hardy Spaces—As in the previous section, we limit our discussion here to matrices whose entries are real rational functions, and we focus on two Hardy spaces, H^2 and H^∞. The concept of a Hardy space is another mathematical concept that rarely appeared in the control engineering literature prior to 1975. Yet it should be a very natural space for control engineers since it basically corresponds to the space of transfer-function matrices of bounded-input bounded-output (BIBO) stable systems. In particular, a matrix $G(s)$ is said to belong to the matrix *Hardy space* H^2 if it is *analytic* in the open right-half s-plane and has *finite* H^2-*norm*, as defined in (9). Similarly, a matrix $G(s)$ is said to belong to the matrix *Hardy space* H^∞ if it is *analytic* in the open right-half s-plane and has *finite* H^∞-*norm*, as defined by (10). In most of the purely mathematical literature the region of analyticity is the open unit disk, i.e., $|s| < 1$, rather than the open right-half plane, i.e., Re $s > 0$, assumed here. We use the right-half plane here since it corresponds to the Laplace-transform transfer functions (continuous-time systems) that appear in most of the reprints. Of course, the unit disk would be natural for z-transform transfer functions (discrete-time systems). In any case, one can always map the right-half plane into the unit disk by the usual bilinear transformation, $s = (1 + z)/(1 - z)$ if necessary.

A key property of analytic functions is the so-called *maximum-modulus principle*, i.e., if the scalar function $u(s)$ is analytic in the open right-half s-plane, then the maximum value of its modulus, $|u(s)|$, is attained on the boundary, $s = j\omega$, unless $u(s)$ is identically constant. Although the Euclidean norm of an analytic matrix function $U(s)$ is not itself analytic, the maximum value of the norm is still attained on the boundary because the norm is a *subharmonic* function of the complex variable s (see Boyd and Desoer [288] for further details). Matrices $U(s)$ which satisfy the conditions

(i) $U(s)$ analytic for $\text{Re } s > 0$,
(ii) $\| U(s)\|_\infty \leq 1$,
(iii) $U(s)$ real for s real

are referred to as *bounded real matrices* (or *bounded real* functions in the scalar case). The maximum modulus property of analytic functions and matrices justifies replacing

$$\sup_{\text{Re } s>0} \| U(s)\|$$

by

$$\sup_{\omega} \| U(j\omega)\|$$

in the definition of H^∞-norms.

In some of the literature on robust control use is made of L^2 and L^∞ spaces rather than H^2 and H^∞ spaces. The basic difference is the L spaces do not require that $U(s)$ be analytic off the $s = j\omega$ axis.

The problem of finding H^∞ functions or matrices which interpolate to given points or matrices in the right-half s-plane plays an important role in a number of robust control problems. This problem is referred to as the *Nevanlinna-Pick interpolation problem*. An excellent introduction to the Nevanlinna-Pick interpolation problem and its application in circuit and system theory is given by Delsarte, Genin, and Kamp (Reprint 34). A classical algorithm for the solution of this problem and the optimal interpolation problem (where the interpolating function has minimal H^∞-norm) in the scalar case is given in Walsh [27]. This algorithm is exploited by Kimura (Reprint 12) for robust stabilization, and by Zames and Francis (Reprint 16) for sensitivity optimization.

We close this section with the definition of inner and outer matrices. These special Hardy space matrices play an important role in both multivariable robust stabilization and sensitivity optimization. A rational Hardy space matrix $F(s)$ is said to be *inner*, denoted $F_i(s)$, if $F_i^H(j\omega)F_i(j\omega) = I$ for all ω, and is said to be *outer*, denoted $F_0(s)$, if $F_0(s)$ has full row rank at all s in the open right-half plane. Every Hardy space matrix $F(s)$ can be factored into a product of an inner matrix times an outer matrix, e.g., $F(s) = F_i(s)F_0(s)$. Inner and outer matrices may be more familiar to engineers as "all-pass" and "minimum-plane" transfer matrices. A scalar inner function, necessarily of the form

$$F_i(s) = \prod_i \left(\frac{a_i - s}{\bar{a}_i + s}\right)$$

is referred to as a *Blaschke* product. See Vidyasagar and Kimura (Reprint 14) for a discussion and application of inner and outer matrices.

In the H^2 optimization theory, a critical step involves the *spectral factorization* of a matrix $\Phi(s)$ into a product $G^T(-s)G(s)$, where the matrix $G(s)$ is analytic and has constant rank r in the right-half s-plane, where r is the normal rank of $\Phi(s)$. The "spectral" matrix $\Phi(s)$ must also satisfy the conditions $\Phi^T(-s) = \Phi(s)$ and $\Phi(j\omega) \geq 0$ for all ω. Algorithms for spectral factorization may be found in Youla [343] and Davis [294]. Spectral factorization is applied to multivariable H^2 optimization problems in Reprints 15, 22, and 32.

4. Robust Analysis: Reprints—Part I

In Part I of this reprint volume we include a number of papers which lay the foundations for the robust analysis of multivariable systems in the frequency domain. Many of these results also provide background for the robust synthesis techniques presented in Parts II and III.

The first reprint paper, Zames (Reprint 1), introduces the concept of "contracting transformation" which leads to the stability requirement that a loop-gain have "amplification" less than unity. This result forms the foundation for the frequency domain "small gain theorems" subsequently developed for robust stabilization (see, for example, Doyle [47]). The functional analysis approach to stability analysis is further developed in Zames [121], where "conic sectors" are introduced for stability analysis of nonlinear systems. In Safonov [22], a general "sector" type stability theorem is derived which reduces to various well known stability criteria in special cases. The link between stability criteria and robustness criteria is well established in [22].

The next reprint paper, Zames (Reprint 2), deals with a number of topics in the design of "insensitive" systems, including the trade-off between right-half-plane poles and zeros and sensitivity, and the best choice of performance measure for sensitivity analysis. This paper introduces, for the first time in the control literature, the H^∞ performance measure for system optimization, and solves the H^∞ sensitivity optimization problem for the special case of a plant with one right-half-plane zero. H^∞ concepts had been applied at an earlier date by Helton [311] to electrical circuits problems.

In Doyle and Stein (Reprint 3), classical sensitivity reduction and stability margins concepts for SISO systems are extended to multivariable systems via singular values (see [47] for an early use of singular values in robust stability analysis). In Reprint 3, the authors show that for a system with a plant transfer matrix G and a controller transfer matrix K (see figure 1 in Reprint 3), sensitivity reduction requires that

$$ps(\omega) \leq \underline{\sigma}[I + G(j\omega)K(j\omega)], \text{ all } \omega \quad (15)$$

where $ps(\omega)$ represents a design level for sensitivity reduction, when sensitivity is measured by the sensitivity comparison matrix (see [40])

$$S = (I + GK)^{-1}. \quad (16)$$

The authors further show that for output multiplicative plant uncertainty of the form $G' = (I + L)G$, robust stability is assured if

$$l_m(\omega) < \underline{\sigma}[I + (G(j\omega)K(j\omega))^{-1}], \text{ all } \omega \tag{17}$$

where $l_m(\omega)$ denotes the uncertainty bound on $L(j\omega)$, i.e., $\bar{\sigma}[L(j\omega)] < l_m(\omega)$, all ω. Finally, the authors explore the robustness properties of the optimal LQ state-feedback solution, and the recovery of the full-state robustness properties by use of suitably designed Kalman-Bucy filters. Let $T(s)$ denote the optimal LQ loop transfer matrix, when $R = \rho I$, where R is the control cost matrix in the quadratic performance measure. The authors then show that the optimal LQ state-feedback system satisfies

$$\underline{\sigma}[I + T^{-1}(j\omega)] \geq 1/2, \text{ all } \omega \tag{18}$$

which guarantees robust stabilization for $l_m(\omega) < 0.5$. The concept of attempting to recover the full-state robustness properties by suitable choice of Kalman-Bucy filter (KBF) parameters was first introduced by Doyle and Stein [50] in 1979. In [50], it is shown that if the plant is minimum phase and if the term q^2BB^T is added to the dynamic noise matrix in the KBF model, the loop transmission of the state-estimate-feedback system approaches that of the full-state-feedback system as q approaches infinity. This approach to robust design, with some modifications, is commonly referred to as the LQG/LTR (linear-quadratic-Gaussian, loop transfer recovery) approach, and is currently one of the most widely used robust design techniques. Other publications on the loop recovery approach include References [70], [71], [80], [83], and [85]. Moore and Xia [83] investigate the problem of full-state loop recovery for non-minimum phase plants. See Athans [30], Stein and Athans [111], and Geering [363] for tutorial presentations of the LQG/LTR approach.

In the reprint paper of Safonov, Laub, and Hartmann (Reprint 4) the SISO return difference concept is generalized to multivariable systems. The return difference matrix $I + L$ and the inverse-return difference matrix $I + L^{-1}$ are defined and related to noise and disturbance attenuation, stability margins, and sensitivity design. Also, the trade-off between the sensitivity matrix $S = (I + L)^{-1}$ and the complementary sensitivity function $T = L(I + L)^{-1}$ is discussed in detail. It is noted that T represents noise response at the output, and from $S + T = I$, a trade-off is always necessary between sensitivity reduction and noise suppression (see also Reprint 10 and Reference [93] for discussions of sensitivity trade-offs). An early paper on the return-difference matrix and its application to multivariable design is the paper of MacFarlane [78] published in 1970. Special care is needed in loop-breaking of multivariable systems for the computation of return difference matrices. Freudenberg and Looze, [57] and [62], point out the differences that result when computing return-difference matrices with different loop-breaking points. In particular, the condition number of the plant enters into the relation of the return difference at one point relative to another. This could cause problems when a system is designed for robustness at a particular break point, but uncertainty enters in another point. See [62] for more details on this problem.

In Reprint 5, the circle criteria for nonlinear systems is generalized to multivariable systems and is used to obtain stability margins in the frequency domain. This reprint is included to illustrate the fact that frequency domain robustness results are available even for nonlinear plant variations. See also References [41] and [113] for the analysis of robustness in nonlinear systems.

In Lehtomaki, Sandell, and Athans (Reprint 6) the robustness properties of optimal LQ and LQG systems are explored further. It is shown, for example, that the optimal multivariable LQSF (linear quadratic state-feedback) system has, in each channel, an infinite increasing gain margin and a 60 degree phase margin. The singular value $\underline{\sigma}(I + G)$, where $I + G$ is the return difference matrix, is identified as a stability margin measure for multivariable systems. This is the direct generalization of the magnitude of $1 + G$ as a stability margin for SISO systems. Also noted in Reprint 6 is the deterioration, from full state-feedback, in stability margin for LQG design. Early papers on the robustness properties of optimal LQG designs include the 1977 paper of Safonov and Athans [105] and the 1978 paper of Doyle [46]. In Reference [46] an example is given that illustrates that, even with the strong gain and phase margins associated with optimal LQSF systems, arbitrarily small plant parameter variations can cause instability.

Unfortunately, all of the robustness results based on singular values require the assumption that the plant uncertainty is unstructured. If the uncertainty is actually structured, the singular-value results may be too conservative. The next two reprints deal with the problem of structured plant uncertainties. In Safonov (Reprint 7), the plant perturbations are assumed to be purely diagonal and stability margins are computed for this structured perturbation. Actually, it can be shown by "pulling out" parameters in a block diagram that it is possible to treat arbitrary structured perturbations as purely diagonal perturbations. In Doyle (Reprint 8), *structured singular values* are introduced to obtain nonconservative robustness conditions. Since the publication of Reprint 8 in 1982, a large number of papers have been published on the application of structured singular values (SSV) to analysis and design of robust control systems (see, for example, References [49], [52], [55], [84], [104], and [119]). Unfortunately, it is difficult to compute and optimize structured singular values. Each structure requires a separate computation. More research is required to develop a general synthesis theory for structured singular values.

The synthesis theory for robust stabilization in Part II requires the basic assumption that the plant perturbations do not change the number of unstable plant poles. Thus, for example, for additive perturbations $P' = P + \delta P$, δP must be stable. El-Sakkary and Zames, [53] and [124], exploit the *gap metric* to study robust feedback stabilization of unstable plants with rather arbitrary plant perturbations, including the case where the nominal and perturbed plant may not have the same number of right-half plane poles. In Vidyasagar (Reprint 9), the *graph metric* is introduced to deal with this problem in terms of co-prime factorizations of the plant transfer function. Graph topology and the graph metric are also discussed extensively in Vidyasagar [26].

The final reprint paper in this section, Freudenberg and Looze (Reprint 10), explores the limitations imposed on sensitivity and stability design for SISO systems by plant right-half plane poles and zeros. They show, for example, that if the loop transmission has an excess of two or more poles over zeros, and the plant has poles $p_1, \cdots, p_N$, in the right-half plane, the sensitivity function must satisfy the integral constraint

$$\pi \sum_{i=1}^{N} \mathrm{Re}(p_i) = \int_0^{\infty} \log |S(j\omega)| \, d\omega. \tag{19}$$

See also Horowitz and Liao [66] on the limitations of zeros in the right-half plane for SISO feedback systems, and Cheng and Desoer [36] for multivariable feedback systems. Poles and zeros in the right-half plane also place key constraints on the synthesis methods in the reprints in Parts II and III.

In Freudenberg and Looze [58], Bode's classical gain-phase relations [5] are generalized to multiloop systems.

See Section B of the bibliography for other papers on robust analysis.

5. Robust Stabilization: Reprints—Part II

In this part of the reprints we have collected some papers that present *synthesis* techniques for *robust stabilization* of unstable plants.

In the first reprint paper of Part II, Vidyasagar and Viswanadham (Reprint 11), the problem considered is that of designing a fixed controller which simultaneously stabilizes l different plants. This is an especially appropriate model for plants that suffer sensor or actuator failures, or plants that are linearized about a finite number of operating points. An interesting result obtained by the authors for this *simultaneous stabilization* problem is that the stabilization of l plants is equivalent to the stabilization of $l - 1$ plants using a stable compensator. Stabilization with a stable compensator is referred to as *strong stabilization*. Necessary and sufficient conditions for strong stabilizability are given in Youla, Bongiorno, and Lu (Reprint 30). In a slightly earlier paper than Vidyasagar and Viswanadham (Reprint 11), Saeks and Murray [144] also solve the simultaneous stabilization problem using an algebra-geometric approach.

Simultaneous stabilization represents an important, but very special, robust stabilization problem, i.e., a problem where the plant uncertainty is quantized into a finite number of possible plants.

In the second reprint of this part, Kimura (Reprint 12), a synthesis procedure is developed for general robust stabilization of SISO systems. To insure robust stability, a compensator must be found which guarantees nominal internal stability and satisfies the robustness condition

$$|r(j\omega)| \cdot \left| \frac{c(j\omega)}{1+p_0(j\omega)c(j\omega)} \right| < 1, \text{ all } \omega \tag{20}$$

where $r(j\omega)$ represents a bound on the additive plant uncertainty, i.e.,

$$|p(j\omega) - p_0(j\omega)| \le |r(j\omega)|, \text{ all } \omega \tag{21}$$

and $c(s)$ and $p_0(s)$ represent the compensator and nominal plant transfer functions, respectively. These conditions are then converted into the problem of finding a function $u(s)$ analytic for $\mathrm{Re}\, s > 0$ and with modulus bounded by one on the $j\omega$-axis, which interpolates to the points $u(\alpha_i) = \beta_i$, where α_i are the unstable plant poles, and β_i are given by

$$\beta_i = \frac{r(\alpha_i)}{\tilde{p}_0(\alpha_i)} \tag{22}$$

where

$$\tilde{p}_0(s) = p_0(s) \prod_i \left(\frac{\alpha_i - s}{\bar{\alpha}_i + s} \right). \tag{23}$$

This is basically the Nevanlinna-Pick interpolation problem, and Kimura uses the classical Nevanlinna-Pick algorithm (see, for example, Reprint 34) to solve the interpolation problem. Unfortunately, the classical Nevanlinna-Pick algorithm yields, in general, compensators with complex coefficients when the interpolation points are complex. A modification of the classical algorithm is given in [131]. This modification generates all *real* rational solutions to the interpolation problem without using excessively high order interpolation functions.

In an early paper on the application of Nevanlinna-Pick interpolation theory to robust stabilization, Tannenbaum [147] considers the problem of maximizing the admissible plant-gain factor variation in SISO feedback systems using *stable* compensators. To solve this problem, a modification of the classical Nevanlinna-Pick theory is also required [148]. In Reprint 13, Khargonekar and Tannenbaum introduce a non-Euclidean metric to design for the maximal allowable variations in a plant parameter for closed-loop stability.

In the last reprint in this section (Reprint 14), Vidyasagar and Kimura extend the SISO robust stabilization results in Reprint 12 to MIMO systems. An alternate procedure for robust stabilization of multivariable systems has been recently reported by Glover [134], based on the Hankel-norm model reduction procedures developed in Reprint 36. In Khargonekar, Georgiou, and Pascoal [136], it is shown that if there is no linear time-invariant robustly stabilizing controller, then there is no nonlinear time-varying robustly stabilizing controller.

It should be noted that the synthesis techniques developed in Reprints 12 and 14 for robust stabilization are based on unstructured plant perturbations, and may in some cases give results that are too conservative.

See Section C of the bibliography for other papers on robust stabilization.

6. H^2 and H^∞ Sensitivity Optimization: Reprints—Part III

In this part of the reprints we have collected papers that present *synthesis* techniques for the design of accurate control systems via *sensitivity minimization*. Papers dealing with both H^2 and H^∞ optimization techniques are included. It should be noted that since the sensitivity function and the transfer function between output disturbance and error are identical, the sensitivity optimization problem is identical mathematically to the disturbance rejection problem.

The first paper in this part, Bongiorno (Reprint 15), is an early paper on the application of Wiener-Hopf H^2-optimization theory to the sensitivity optimization of multivariable systems. Bongiorno assumes that the "small" plant parameter uncertainties are randomly distributed and uses the expected value of an integral-quadratic form as a performance measure. The problem is reduced to a Wiener-Hopf optimization problem and is solved by matrix spectral factorization (see Youla [343]). These results are extended by Youla and Bongiorno (Reprint 22) to multivariable feedback systems with two degrees of freedom. Further details on H^2 optimization theory are given in Reprints 31 and 32 of Part V.

In Zames and Francis (Reprint 16), an H^∞-norm is taken as a performance measure for the minimization of the weighted sensitivity, i.e.,

$$\|S(s)\,W(s)\|_\infty = \sup_\omega |S(j\omega)\,W(j\omega)|, \text{ all } \omega \tag{24}$$

and optimal Nevanlinna-Pick interpolation theory is used to obtain a solution of the SISO H^∞ sensitivity optimization problem. The optimal interpolation problem associated with the sensitivity problem is that of finding a $\tilde{X}(s)$, analytic for Re $s > 0$, with minimal H^∞-norm, which satisfies the interpolation condition

$$\tilde{X}(b_i) = \frac{W(b_i)}{B_p(b_i)}$$

where b_i are the right-help-plane plant zeros, and $B_p(s)$ is the Blaschke product associated with the unstable poles of the plant.

Some of the advantages of H^∞ optimization are:

(i) Signal spectra are not required.
(ii) Weighted errors are kept small at all frequencies.
(iii) Stability margins and other robust measures can be incorporated into the optimization criterion.

The case for H^∞ optimization for control system design is well made in Zames (Reprint 2). An analysis of the H^∞ approach to sensitivity design is given in recent paper of Freudenberg and Looze [303].

Several points are worth noting about the properties of the solution to the optimal H^∞ sensitivity design problem, i.e.,

(i) The optimal weighted sensitivity is all-pass, so that the amplitude spectrum of the optimal sensitivity function $S(j\omega)$ is the inverse of the spectrum of the weighting function $W(j\omega)$.
(ii) If there is more than one right-half plane plant zero, the optimal compensator is unstable.
(iii) The optimal performance value can only be approached asymptotically with proper compensators.
(iv) A considerable amount of pole-zero cancellation occurs in the computation of the optimal compensator.
(v) The sensitivity optimal solution may have poor robust stabilization properties.

In a follow-up paper Francis and Zames [167] solve the SISO optimal sensitivity problem using the operator theory of Sarason [335]. The Sarason theory reduces the computation for the optimal compensator to a standard quadratic minimization problem (see also Reprint 35 for numerical solutions of the SISO H^∞-optimization problem).

Solutions to the multivariable H^∞-sensitivity-optimization problem are presented in Doyle (Reprint 17), Chang and Pearson (Reprint 19), Francis, Helton, and Zames (Reprint 20), and Safonov and Verma (Reprint 21). Doyle (Reprint 17) presents a general framework for the synthesis of H^2, H^∞, or H^μ-optimal systems. He reduces the numerical computations required for a solution to the multivariable H^∞ problem to the solution of Riccati equations and the solution of the Hankel-norm model reduction problem, given in Glover (Reprint 36). In Francis, Helton, and Zames (Reprint 20) the multivariable H^∞-optimization problem is solved using Ball and Helton's operator-theoretic approach [284]. Chang and Pearson (Reprint 19) solve a slightly more general multivariable problem, where the measured outputs need not necessarily be the controlled outputs, and where the number of exogenous inputs may not necessarily be the same as the number of controlled outputs (as assumed in Reprint 20). Chang and Pearson use a combination of the Sarason theory and the matrix Nevanlinna-Pick theory (see Reference [295]) to solve the optimal H^∞ problem. In Safonov and Verma (Reprint 21), the relationship between the H^∞-optimization problem and the Hankel-norm approximation problem is developed for multivariable systems. Inner and outer matrices are used to convert the H^∞-optimization problem to the Hankel-norm approximation problem, and as previously noted, the latter problem is solved by Glover (Reprint 36). Since inner and outer matrix factorizations, co-prime factorizations, and Hankel-norm approximations can all be done in a state-space setting (see Reference [292], and Reprints 17, 36, and 37), the entire H^∞-optimization problem can be solved with state-space techniques, e.g., state-space realizations, solution by Lyapunov and Riccati equations, etc. In Safonov [182], the problem of multivariable optimal H^∞ synthesis with diagonally perturbed feedback systems is reduced to a finite-dimensional nonlinear programming problem.

In Reprints 18 and 23, and Reference [164], an optimal *mixed sensitivity* problem is considered, i.e., select a compensator which minimizes

$$\sup_\omega \left[|V(j\omega)S(j\omega)|^2 + |W(j\omega)T(j\omega)|^2\right] \tag{26}$$

where $S(j\omega)$ denotes the sensitivity function, and $T(j\omega) = 1 - S(j\omega)$ denotes the complementary sensitivity function. An important reason for adding the term which includes the function $T(j\omega)$ is that $T(j\omega) = (1 + (PC)^{-1})^{-1}$ is inversely proportional to the stability margin. Thus, a performance measure of this type tends to yield accurate systems which are also robustly stable. In Verma and Jonckheere (Reprint 18), the L^∞-optimization problem posed above is reduced to a broadband matching problem which has been solved by Helton [312]. Kwakernaak (Reprint 23) solves the same problem as Verma and Jonckheere by reducing the problem to a polynomial equation. The multivariable version of this problem is solved by Francis [165] using functional analysis techniques. Reprint 17 includes a solution to the MIMO mixed sensitivity problem as a special case.

Some related topics not covererd in the reprints of this volume include a study of the pole-zero cancellation phenomena in multivariable H^∞-optimal design found in Limebeer and Hung [178], a study of the relation between H^∞-optimization and LQG design found in Grimble [171], and a study of a conformal mapping approach to convert Horowitz templet design (see Reprint 28) to H^∞ design found in Sideris and Safonov [185].

Critiques of the H^∞ approach to robust design may be found in Horowitz (Reprint 28) and Freudenberg and Looze [303].

In Feintuck, Khargonekar, and Tannenbaum [161], and Georgiou and Khargonekar [170], the H^∞ optimal sensitivity problem is solved for periodically varying plants.

In tracking and disturbance rejection problems where "persistent" inputs, with unbounded energy, are present, H^2 and H^∞ optimization techniques are not appropriate. Dahleh and Pearson introduce L^1-optimization theory to deal with this class of problems in Reference [158].

See Section D of the bibliography for other papers on H^2 and H^∞ sensitivity optimization.

7. Other Approaches to the Control of Uncertain Systems: Reprints—Part IV

In this part of the reprint volume we include a miscellaneous collection of papers that deal with the robust synthesis problem, but which differ considerably from the linear multivariable synthesis techniques represented in Parts II and III.

The first reprint paper in this part, Wonham (Reprint 24), considers a stochastic model for the uncertain coefficients of a state-variable plant model. This reprint illustrates the structured *stochastic approach* to plant uncertainty. In particular, Wonham assumes plant dynamics of the form

$$\dot{x} = Ax - Bu + C\dot{w}_1 + G(x)\dot{w}_2 \tag{27}$$

where

$$G(x) = \sum_{k=1}^{n} x_k G_k \tag{28}$$

with a performance measure.

$$\lim_{t\to\infty} E\{x^T(t)Mx(t) + u(t)N(t)\}. \tag{29}$$

The vectors x, u, $\dot{w}_1$, and $\dot{w}_2$ are state, input, additive noise, multiplicative noise vectors with dimensions n, m, d_1, and d_2, respectively. Also, $\dot{w}_1$ and $\dot{w}_2$ are assumed to be uncorrelated "white noise" vector processes with unit noise intensity. The multiplicative noise term $G(x)\dot{w}_2$ represents the plant uncertainty as randomly varying coefficients of a differential equation. Equation (27) may be given a rigorous interpretation as a stochastic differential equation (eq. 2.1 in Reprint 24). The *structured* way in which the uncertainty enters the dynamical equations is specified by the matrices G_k. Wonham shows that the optimal control law is given by $u = N^{-1}B^TP$ where P satisfies the modified (stochastic) algebraic Riccati equation

$$A^TP + PA + M - PBN^{-1}B^TP + \Gamma(P) = 0 \tag{30}$$

where $\Gamma(P)$ is a matrix with kl-entry equal to tr $(G_k^TPG_l)$. Stochastic uncertainty has introduced an additional term $\Gamma(P)$ in the standard Riccati equation. Wonham further shows that a solution to the optimal stochastic problem exists, and the stochastic system is stable if

$$\sum_{k=1}^{n} \|G_k\|^2 < (K_1 d_2)^{-1} \tag{31}$$

where

$$K_1 = \inf_K \left\| \int_0^\infty e^{t(A-BK)^T} e^{t(A-BK)}\, dt \right\|. \tag{32}$$

Condition (31) is essentially a robust stability test, comparable to the frequency domain test of (17). An extension of this stochastic theory to include dynamic output feedback may be found in Bernstein and Greely [196]. Robust stabilization of uncertain systems modelled as stochastic equations, i.e., (27), is also discussed in Willems and Willems [267].

In the second paper in Part IV, Chang and Peng (Reprint 25), uncertainty is modelled as unknown, but structured, parameter variables in state-variable model, i.e.,

$$\dot{x} = A(q)x + B(q)u \tag{33}$$

where the uncertainty structure is given by

$$A(q) = A_0 + \sum_{i=1}^{n'} q_i(t)A_i \tag{34}$$

$$B(q) = q_b(t)B \tag{35}$$

where $-1 \le q_i(t) \le 1$ and $1 \le q_b(t) \le b$. The vector q, made up of q_i plus q_b, represents the unknown parameter vector. The performance measure considered by Chang and Peng is of the form

$$J(q) = 1/2x^THx + 1/2 \int_0^T (x^TPx + u^TQu)\, dt. \tag{36}$$

They then demonstrate that the control law $u = Q^{-1}B^TR(t)x$, where $R(t)$ satisfies the *modified* Riccati equation

$$A_0^TR + RA_0 + P - RBQ^{-1}B^T + U(R, A) = \frac{-dR}{dt},$$

$$R(T) = H \tag{37}$$

with

$$U(R, A) = \sum_{i=1}^{n'} S_i|\Lambda_i|S_i^T \tag{38}$$

where S_i is the orthogonal transformation which diagonalizes

$(RA_i + A_i^T R)$, i.e., $S_i^T(RA_i + A_i^T R)S_i = \Lambda_i$, and where $|\Lambda_i|$ denotes a diagonal matrix with elements equal to the absolute values of the diagonal elements of Λ_i, guarantees that the cost $J(q)$ is bounded by $x^T(0)R(0)x(0)$ for all admissible q. The term *guaranteed-cost control* derives from the above property of the given control law. It should be noted that like the stochastic approach of Wonham, a modified Riccati equation results for the "robust" control system, if there is no uncertainty $U(R, A) \equiv 0$, and the Riccati equation reduces once more to the standard Riccati equation, and the control law to the optimal LQ state-feedback control law. If $q_B(t) \equiv 1$ and the pair (A_0, B) is controllable, a sufficient condition given by Chang and Peng for a steady-state solution to exist for $R(t)$ is that

$$\sum_{i-1}^{n'} \|A_i\| < 1/2 \left\{ \inf_K \left\| \int_0^\infty e^{(A_0+BK)^T t} e^{(A_0+BK)t}\, dt \right\| \right\}^{-1}. \tag{39}$$

Further, they show that if the steady-state solution for $R(t)$ is positive definite, the closed-loop system is robustly stable. Thus, (39) is part of a state-variable robust stability condition. Unfortunately, (39) alone is not sufficient to insure stability, so that issue of robust stability is not directly verifiable in this theory. Nevertheless, it is of interest to note that this paper, which predates most of the modern robust control theory, provides for *robust performance*, a research issue of current interest. Goldstein [227], and Dorato and Menga, [220] and [248], have extended the guaranteed-cost approach to state-estimate-feedback systems. A comparison of the stabilization problem for uncertain dynamics as characterized in Reprints 24 and 25 may be found in Bernstein and Greely [197].

The *guaranteed-cost* approach of Chang and Peng is based on the minimization of an upper bound on the cost function. What makes the computation tractable in this case is that the upper bound is independent of the plant variations. However, this solution may be too conservative. To obtain a sharper upper bound, one must solve the exact minimax problem. A number of authors have investigated the game-theoretic approach to uncertainty design, where the plant uncertainty is viewed as an antagonist that attempts to maximize the performance measure, while the control input is selected to minimize the performance measure (minimax control). Some early papers on the subject include the papers of Ragade and Sarma [256], and Dorato and Kestenbaum [219] published in 1967; and the papers of Salmon [257] and Witsenhausen [268] published in 1968. In the paper included in this volume, Bersekas and Rhodes (Reprint 26), dynamic programming is used to solve the discrete-time minimax problem. A significant feature of the solution given by Bertsekas and Rhodes is the introduction of *sufficient-information functions*. These functions define how much information the control function must accumulate for optimal control action. They are the counterpart of *sufficient statistics* in stochastic formulations of the uncertainty problem. Unfortunately, the recursive equations generated by the dynamic programming approach are extremely complicated and can be solved in only very special cases, for example, linear-quadratic problems with additive uncertainty. The theory presented by Bertsekas and Rhodes is probably the most complete theory possible for optimal control of uncertain systems. Adaptive control theory could be considered a subset of this minimax theory, where a separation assumption is made on estimation and control, i.e., the state or parameters are estimated first, and then controller parameters are modified according to the estimated values. Robust control could be considered a further subset, where the controller is constrained to be a function of only the current state or output.

In Vinkler and Wood [264], a comparison is made of these three approaches, i.e., expected-value cost (stochastic model, Reprint 24), guaranteed cost (Reprint 25), and minimax (Reprint 26). They use, as an example for comparison, a fifth order lateral autopilot problem. The expected-cost and guaranteed-cost approaches are singled out as "methods of choice," at least for this particular example.

In Part IV, we include Davison and Ferguson (Reprint 27), a paper which uses parameter optimization techniques to solve the multivariable *robust servomechanism problem*. The robust servomechanism problem, as defined by Davison and colleagues [207]–[214], is basically the problem of designing stable tracking systems in the presence of finite plant uncertainties. Although analytic solutions to the problem are given in the references cited above, we include here a paper based on parameter optimization, since, in many engineering problems, parameter optimization is the most feasible way to obtain numerical solutions. Also, design via parameter optimization contrasts sharply with the open structure synthesis procedures defined in Parts II and III. Numerical examples treated by Davison and Ferguson include a two-input two-output batch reactor and a four-input four-output boiler furnace. In Davison [211], a design method is developed for the robust servomechanism problem when the plant is open-loop stable. A very practical aspect of the design procedure in [211] is that the compensator can be designed "on line" directly from some simple experiments on the open-loop plant. In Davison and Copland [212], a synthesis procedure is developed to meet specifications on gain margin and time-lag tolerance in the multivariable robust servomechanism problem.

Horowitz and his co-workers have developed a design approach to the robust control problem which is referred to as the *qualitative-feedback-theory* (QFT) approach (see, for example, References [232]–[239], and [270]). We include here a survey paper, Horowitz (Reprint 28), on the subject which summarizes the theory developed up to 1982. In the QFT approach, plant uncertainty is represented as a templet of possible complex values of the plant transfer function $P(j\omega)$ at a given frequency ω. We will refer to this templet as the *Horowitz templet*. The actual loop-gain at any given frequency is a set of values given by the nominal gain plus the Horowitz templet of possible plants. The theory is then based on shaping the nominal loop gain so that the feedback system is robustly stable and satisfies various design objectives, such as input-output accuracy, noise rejection, etc. It should be noted that for SISO systems the Horowitz templets, in combination with the Nyquist stability criterion, provide nonconservative stability results for structured perturbations. This is in contrast to the stability results in Reprint 12. The distinction is that the templets contain "exact" information on the shape of plant uncertainty, while the results of Reprint 12 contain the uncertainty in a ball of possible values, i.e., $|\delta P(j\omega)| < |r(j\omega)|$. Of course, there is a problem in computing the templets exactly at all frequencies. In [185], conformal mappings are used to convert Horowitz templet design to H^∞ design. The QFT has been extended to multivariable systems (see, for example, References [239] and [270]). The multivariable QFT approach involves reducing the robust MIMO problem to a sequence of robust SISO problems. Unfortunately, in the process, some design conservatism is introduced. See Doyle [221] for a general critique of the QFT approach to robust design, and Yaniv and Horowitz [271] for a response to this critique.

In the last reprint of Part IV, Barmish, Corless, and Leitmann (Reprint 29), Lyapunov-function theory is used to design robustly stable controllers for both linear and nonlinear systems. Classically, Lyapunov functions have been used to establish asymptotic stability of nonlinear systems. Roughly speaking, a nonlinear system $\dot{x} = f(x)$ is asymptotically stable if there exists a function (*Lyapunov function*) $V(x)$ which is positive for all non-zero x and which has a time-derivative $\dot{V}(x)$ which is negative for all non-zero x. Obviously, if the inequalities in question are true for admissible variations in $f(x)$, the Lyapunov stability theory also provides conditions for robust stabilization. An early paper on the application of Lyapunov function to robust stabilization is the paper of Gutman [229], published in 1979. The monograph of Safonov [22] also contains an extensive discussion of the application of Lyapunov functions to robust stabilization. In Reprint 29, the authors show for linear systems that if the plant dynamics are given by

$$\dot{x} = (A + \Delta A)x + (B + \Delta B)u \tag{40}$$

and the "*matching conditions*"

$$\Delta A = BD, \ \Delta B = BE \tag{41}$$

hold for the plant uncertainty, with $\|D\| \leq \rho_D$ and $\|E\| < 1$, then the state feedback $u = -2\gamma_0 B^T P x$ will always robustly stabilize the system (40), provided that the positive scalar γ_0 is chosen sufficiently large. The matrix P in the state-feedback control law is given as a solution of the Lyapunov matrix equation, $A^T P + PA = -H$, where H is any positive definite matrix. The matching conditions, (41), do impose a special structure on the permissible plant variations. However, robust stabilization conditions can be derived from Lyapunov theory with these conditions removed (see, for example, References [195] and [259]). In Galimidi and Barmish [225], Hollot [230], and Hollot and Galimidi [231], the *Lyapunov function* approach is used to study the robust stabilization of closed-loop systems with observer feedback. Patel, Toda and Sridhar [94], and Yedavalli and colleagues [114]–[118], exploit Lyapunov function theory to obtain conditions on the A matrix in the state-variable plant description (eq. (1)) for robust stabilization.

We conclude this section with a discussion of a few other approaches to the robust control problem which are not represented in the reprints in this volume.

One is the so-called *norm-uncertain* approach to the control of uncertain systems developed by Donati and colleagues, [201] and [216]. In the norm-uncertain approach, the plant uncertainty is characterized by a bound on the norm of the output variation, i.e.,

$$\|y - \hat{y}\| \leq E^* \|u\| + D^* \tag{42}$$

where $\hat{y}$ represents the nominal system and u the system input, and where the norms in (42) are norms on the Banach space of input and output signals. The design data for robust control is then the nominal operator which relates u to $\hat{y}$, and the positive constants E^* and D^*. This uncertainty model is used to design guaranteed performance controllers in [216].

Another approach we wish to discuss is the so-called *Hurwitz-condition* approach. The basic idea behind this approach is to use Hurwitz stability conditions on the coefficients of a polynomial to insure robust stabilization (see, for example, Keel and Bhattacharyya [242], Wei and Barmish [266], and Bose, Jury, and Zeheb [200]). An advantage of this approach is that structured parameter variation can be dealt with directly, and Hurwitz conditions are necessary and sufficient conditions for stability. Of course, the difficulty in this approach is in computing the stability boundaries in parameter space defined by the Hurwitz conditions.

Finally, we would like to mention an approach to robust control of time-invariant plants based on the use of periodically time-varying compensators. This *periodic compensation* approach to robust control is developed in Khargonekar, Poolla, and Tannenbaum [244]. The authors show for discrete-time systems that periodic compensation offers the following advantages over time-invariant compensation:

(i) Improved gain and phase margins.
(ii) Stabilization of *arbitrary* LTI plants with state compensators (normally the parity interlacing property is required of the plant for strong stabilization with an LTI compensator; see Reprint 30 for details).
(iii) Simultaneous stabilization of *any* finite collection of plants (see Reprint 11 for the usual restriction on the collection of plants for simultaneous stabilization with an LTI compensator).

The authors also show in [244] the very interesting result that, while periodic compensation can be used to obtain a number of improved robustness results, it cannot yield better results than the LTI compensator for the H^∞ optimal sensitivity problem. Khargonekar and Poolla [318] show further that even nonlinear time-varying compensators cannot yield improvements over the LTI compensators in the case of H^∞ optimal sensitivity.

8. Related Theory and Numerical Computation: Reprints—Part V

In this final reprint part we include papers that do not deal directly with the robust control problem, but present some of the underlying theory and computations required for robust control.

In the first reprint paper of Part V, Youla, Bongiorno, and Lu (Reprint 30), conditions are given for the existence of a *stable compensator* which stabilizes an unstable plant in a closed-loop configuration. A plant which can be so stabilized is said to be *strongly stabilizable*. Both SISO and multivariable systems are considered by the authors. It is shown, for example, that for strictly proper SISO plants a necessary and sufficient condition for strong stabilizability is that between real zeros of the plant in the right-half plane there lie an *even* number of real plant poles, counted according to their multiplicity. This property is referred to as the *parity interlacing property*. Strong stabilizability plays a key role in simultaneous stabilization. For example, if plant P_0 is stable and P_1 is arbitrary, then P_0 and P_1 can be simultaneously stabilized if and only if $P_1 - P_0$ is strongly stabilizable (for further details see Vidyasagar [26] and Reprint 11).

In the next two reprints (Reprints 31 and 32), the theory for optimal H^2 design is developed for both SISO and multivariable systems. Matrix spectral factorization techniques are used to compute the controller which minimizes an H^2-norm on given error vectors in the frequency domain. In the above reprints, Youla, Bongiorno, and Jabr introduce the parameterization of all stabilizing compensators, now often referred to as the *Youla parameterization*, which plays such a key role in much of modern robust control theory. Further discussion of the H^2 optimization problem may be found in Francis [301] and Vidyasagar [26]. The theory of matrix spectral factorization is developed in various references including [281], [294], and [343], and computational algorithms in [283], [292], [315], and [338].

In the reprint of Desoer, Liu, Murray, and Saeks (Reprint 33), an algebraic theory is developed for the analysis and synthesis of general linear systems, including both lumped and distributed parameter systems. This theory provides the unified approach to the synthesis of robust control systems developed in Vidyasagar's text [26]. The basic idea is to represent transfer functions as co-prime fractions, e.g., $G = n_r d_r^{-1}$ for a right fractional representation, over the ring of stable transfer functions, rather than the ring of polynomials, as exploited, for example, by Youla, Jabr, and Bongiorno (Reprint 32).

The next two reprints in this part deal with the classical Nevanlinna-Pick interpolation problem. The basic problem is that of finding a function $u(s)$, analytic in the right-half plane, with H^∞ norm bounded by one, which interpolates to a given set of complex values in the right-half plane. This problem was originally defined by Nevanlinna [326] and Pick [329]. As previously noted, this problem plays a key role in the design of robustly stable systems and optimally insensitive systems (see, for example, Reprints 12 and 16). In Delsarte, Genin, and Kamp (Reprint 34), the Nevanlinna-Pick interpolation theory is reviewed, and the theory is applied to a number of problems in circuit and system design, including stability analysis, Darlington filter synthesis, and model reduction. In this reprint, the region of analyticity is the interior of the unit disk, rather than the right-half plane, but as noted in Section 3 of this introduction, a simple mapping can be used to convert one region into the other. A generalization of the classical Nevanlinna-Pick algorithm for the solution of the interpolation problem for matrix functions is given in Delsarte, Genin, and Kamp [295]. In Allison and Young (Reprint 35), the Sarason [335] and AAK [278] operator-theoretic approach to H^∞ interpolation is applied to the numerical solution of the scalar Nevanlinna-Pick interpolation problem. Allison and Young reduce the computations to the computation of singular values and eigenvalues, and corresponding singular vectors and eigenvectors. This computational approach is also used in Francis and Zames [167] to solve the H^∞-optimal sensitivity problem for SISO systems. In Chiang and Safonov [291], detailed program listings are given for multivariable L^∞ controller design for implementation on a VAX/UMS 11/750 computer or an IBM PC, using the PC-MATLAB language. Reference to other computational efforts may be found on page 1156 of Helton [314]. Zames and Tannenbaum [349] present a new approach to the H^∞ optimization problem which is computationally attractive when the weighting function is of low order.

In the next reprint, Glover (Reprint 36), a theory is developed for model reduction which relates directly to the problem of multivariable H^∞-optimization (see Doyle and Chu [300] for the detailed linking of the two problems). Particularly noteworthy of the results in Reprint 36 are the reduction of the required computation to certain standard state-space computations, i.e., balanced realizations, solution of Lyapunov matrix equations, etc., and the use of standard numerically robust transformations and canonic forms, i.e., Householder transformations, real upper Schur forms, etc. The key link between the H^∞-optimization problem and Hankel-norms is given by theorem 6.1 in Reprint 36, which basically states that

$$\inf_F \sup_\omega \bar{\sigma}(G(j\omega) - F(j\omega)) = \| G(s) \|_H \tag{43}$$

where $G(s)$ is a given matrix function which is analytic in the closed right-half plane (causal), and $F(s)$ is constrained to be analytic in the open left-half plane (anticausal). This result was first obtained by Nehari [324] for scalar functions, and by AAK [279] for multivariable systems. Earlier work on the model reduction problem is included in Silverman and Bettayeb [336], Bettayeb, Silverman, and Safonov [286], and Kung and Lin [320].

Much of modern robust control theory centers around co-prime matrix fractions and solutions of the Bezout, or Diophantine, equation. In Reprint 37, Nett, Jacobson, and Balas exhibit an explicit solution for the Bezout identity, and explicit solutions for right- and left-co-prime matrix fractions in terms of state-space realization of the given transfer matrix. It should be noted that results for state-space realizations for co-prime factorization are also embedded in an earlier paper by Khargonekar and Sontag [319] (in particular, see eq. 5.9 and 5.10 of Reference [319]). The computer-aided design of multivariable systems based on co-prime matrix fractions is discussed in [340], and algorithms for the solution of the Diophantine equations are presented in [350].

In Poolla and Khargonekar [330], the equivalent of co-prime stable-proper factorizations are developed for time-varying systems.

See Section F of the bibliography for other papers on related mathematics and computational methods.

9. Applications

Although no reprints specifically devoted to applications are included in this volume, Section G of the bibliography includes a number of recently reported applications. It should be noted that, in most cases, these applications are at best feasibility studies and do not represent final operating systems. At the present time, the two approaches to robust control that have the most reported applications are the *LQG/LTR design technique* and the *robust servomechanism* approach. Both have been applied to a wide range of control problems. The LQG/LTR technique has been applied to control of submersible vehicles [368], engine speed [352], [363], [371], helicopters [373], ship steering [364], and large flexible space structures [381]. The robust servomechanism approach has been applied to control of heat exchangers [359], spacecraft altitude [358], binary distillation columns [355], and large flexible space structures [384].

One application area where a diverse combination of robust approaches have been applied is the area of *flight control*. Reported applications of robust control theory in this area include QFT theory (Reprint 28), L^∞ mixed-sensitivity theory [374], μ-synthesis theory [366], guaranteed-cost theory [383], Lyapunov-function theory [376], and LQSF theory [353]. Other applications to flight control may be found in [361] and [385]. Another significant application area is *robotics*. Application papers in this area include [356], [362], [367], [370], [375], [377]–[380].

10. A Short List of Acronyms

We conclude this introduction with a short list of acronyms that are found in the reprints included here.

AAK—Adamjan/Arov/Krein
BR—Bounded real
CL—Characteristic loci
CLHP—Closed left half plane
CRHP—Closed right half plane
FDLTI—Finite dimensional linear time-invariant
INA—Inverse Nyquist array
KBF—Kalman-Bucy filter
LCF—Left-co-prime fraction
LFSS—Large flexible space structure
LHP—Left-half plane
LQ—Linear quadratic
LQG—Linear quadratic Gaussian
LQR—Linear quadratic regulator
LQSF—Linear quadratic state feedback
LTI—Linear time invariant
LTR—Loop transfer recovery
MIMO—Multiple-input multiple-output
NMP—Non-minimum phase
NTV—Nonlinear time varying
OLHP—Open left half plane
ORHP—Open right half plane
PR—Positive real
QFT—Quantitative feedback theory
RCF—Right co-prime fraction
RHP—Right-half plane
RMV—Robust multivariable
SBR—Strictly bounded real
SISO—Single-input single-output
SPR—Strictly positive real
SSV—Structured singular value
SV—Singular value
SVD—Singular value decomposition
WHK—Wiener/Hopf/Kalman.

Part I
Robust Analysis

Functional Analysis Applied to Nonlinear Feedback Systems*

GEORGE ZAMES†, MEMBER, IEEE

Summary—**An operator theory is outlined for the general, nonlinear, feedback loop. Methods for bounding system responses and investigating stability are introduced. An iterative expansion of the feedback loop, valid for large nonlinearities and unstable systems, is derived.**

The theory is applied to the study of nonlinear distortion in a class of amplifiers; it is shown that feedback reduces distortion for band-limited inputs. A model of the distortion is obtained, shown to be stable, and an iteration whose rate of convergence is optimized is derived.

I. Introduction

Objectives

FUNCTIONAL analysis provides the mathematical foundation for system theory. In this paper some of its basic ideas are used to construct, heuristically, the elements of an operational theory of the nonlinear feedback loop (Fig. 1), for systems whose output is determined by the history of the input, without the postulation of state. The theory is applied to the study of nonlinear distortion in the amplifier in Fig. 2, whose open loop consists of a linear time-invariant system $\underline{L}$ (Fig. 3), a nonlinear no memory device $\underline{N}$ (Fig. 4), and a variable, pure gain $\underline{K}$ in cascade. The following topics are discussed:

1) An operator algebra suitable for system theory is outlined. Feedback equations are written in a general form, and necessary and sufficient conditions for the existence of solutions are derived.

2) A calculus of bounds on the responses of nonlinear systems is presented, in terms of a system "gain," which is an operator norm.

3) A method is introduced for investigating the stability of certain nonlinear systems by means of "contracting transformations" (not to be confused with contraction mappings).

* Received November 3, 1962; revised manuscript received May 10, 1963. This work was made possible in part by support extended the Massachusetts Institute of Technology, Research Laboratory of Electronics, jointly by the U. S. Army (Signal Corps), the U. S. Navy (Office of Naval Research), and the U. S. Air Force (Office of Scientific Research), under Signal Corps Contract DA 36-039-sc-78108, Department of the Army Task 3-99-20-001 and Project 3-99-00-000 and in part by support under U. S. Navy Contract Nonr 1866(16) at Harvard University. Portions of this paper have been taken from a thesis submitted in partial fulfillment of the Sc.D. at the Massachusetts Institute of Technology.

† Department of Electrical Engineering, Massachusetts Institute of Technology, and Division of Applied Physics, Harvard University, Cambridge, Mass.

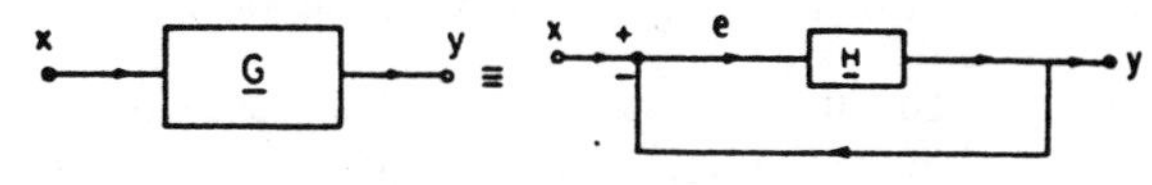

Fig. 1—A feedback system.

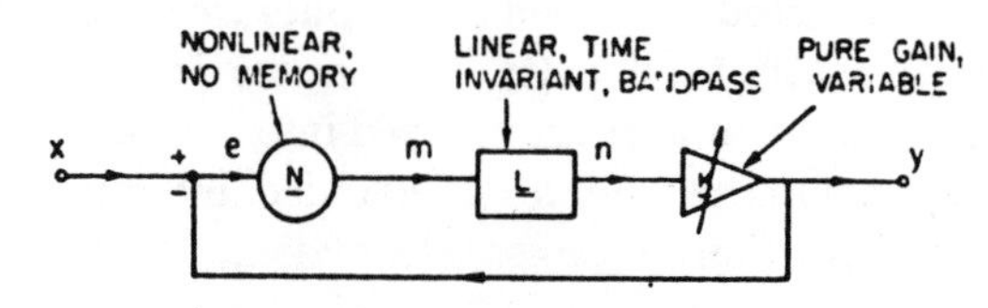

Fig. 2—A feedback amplifier.

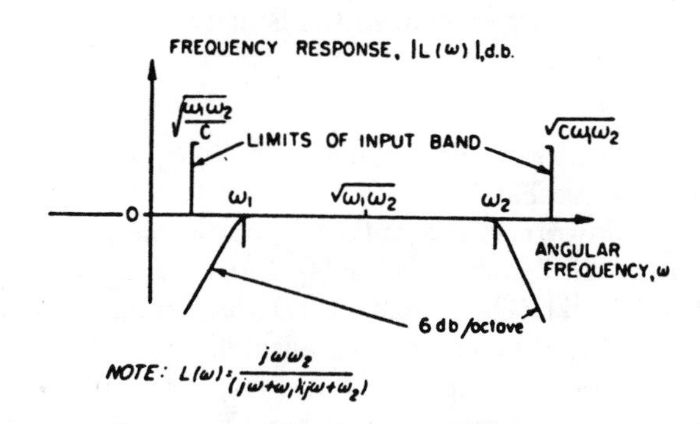

Fig. 3—Frequency response of $\underline{L}$ and limits of input band.

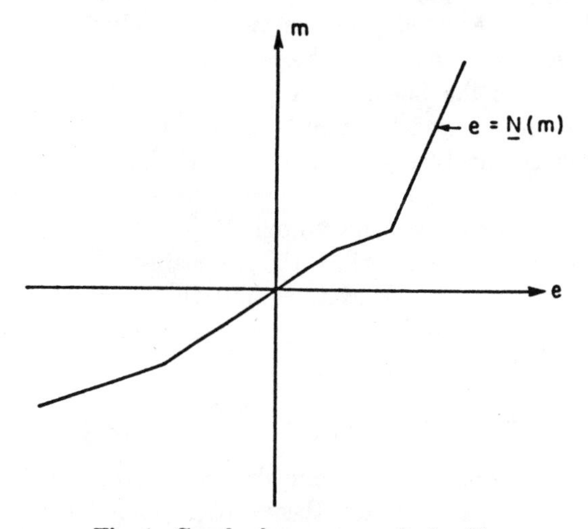

Fig. 4—Graph of no memory device $\underline{N}$.

4) An iterative expansion of the feedback loop, whose validity is not restricted to stable systems or small nonlinearities, is described. The convergence of this iteration results from delay or smoothing, which are present in all physical systems, unlike that of the contraction iteration, which is limited to feedback loops having small amplifications.

Reprinted from *IEEE Trans. Circuit Theory*, pp. 392–404, Sept. 1963.

The theory is applied to the problem of nonlinear distortion as follows:

5) The nonlinear distortion of the system in Fig. 2 is bounded using perturbation, and the system is shown to be stable. It is shown that the method is useful for monotonic nonlinearities having upper and lower bounds to their slope; the distortion bound as well as the rate of convergence are optimized by splitting the nonlinearity along the mean of the largest and smallest slopes.

6) One of the important properties of feedback is its ability to reduce distortion; the nonlinear distortion of the amplifier in Fig. 2 is shown to go to zero as the loop gain $\underline{K}$ is increased, provided that inputs are band-limited. The mathematical difficulty that arises from the fact that band-limited inputs are mapped into nonband-limited errors by nonlinear systems is avoided by relating the nonlinear distortion to a linear distortion.

7) An iterative model of the distortion which is valid for arbitrarily large nonlinearities is derived.

Background

Functionals were first studied by Volterra [1]. Functional analysis has been applied to system theory by a number of writers; the author is particularly indebted to Wiener [2], Zadeh [3], Bose [4], Barret [5], Brilliant [6] and George [7]. The author [8]–[11] has studied applications to the formulation of realizability conditions for nonlinear systems, power series representations, the effects of nonlinear systems on band-limited inputs, and stability; this paper is a summary of those concepts that occur repeatedly in the various applications. Two good, recent references on the general subject of functional analysis are the books of Kolmogoroff and Fomin [12], and Lusternik and Sobolev [13], and many of the ideas on which our theory is based are explored, in greater depth, in the first three chapters of each. A source paper on iteration theory is Kantorovic [14].

The author has described the application of contraction iteration to feedback systems and to the derivation of distortion models [8]. A very similar analysis has been given by Desoer [15].

II. An Operator Theory of Systems

Our approach to system theory emerges from the desire to have a unified theory with unified definitions of properties such as stability, independent of the nature of the equations—differential, integral, etc.—used in describing systems. Such a theory is predicated on having an axiomatic definition of a system. The content of a suitable definition is a subject for discussion.

On a microscopic scale the concept of a physical device independent of observations made on it is excluded in modern physics. On a macroscopic scale the fiction of an R.L.C. network or a finite state machine is a most useful one. However, a system in the communication and control sense is an input-output relation distinct from any particular physical realization. Thus a predictor, compensator, integrator, etc., are well-defined systems having many physical realizations—the usual object is to find the cheapest one. Conversely a dc electric motor is (ideally) an integrator for the purpose of controlling position, but a different system—a pure gain—for the purpose of controlling velocity; the former is unstable with respect to input noise while the latter is not.

An input-output system is distinct also from the equations describing it; many equations can describe the same system.

The operator theory originates in the view that observed input and output time functions constitute the ultimate reality in any system—whether it be a control system or a physical device. Furthermore, the definitions of a system and of system properties should be operational—*i.e.*, directly in terms of observable quantities, in order to be most readily verifiable, unambiguous and meaningful. A system, according to this view, is a causal relation between observed quantities, *i.e.*, a mapping which assigns one and only one output function of time to each (vector-valued) input function of time. Output components which cannot be causally related to inputs are accounted for by noise sources in the input. Definitions of a system and of stability are made without reference to the concept of state; state is regarded as a transformed variable useful in the formulation of certain problems, which is not postulated axiomatically but is defined in terms of inputs and outputs. A system is thus a "black-box" divested of all features that leave outputs unaffected. The mathematical counterpart of a system is an operator.

Definition of an Operator

A "time interval" $[t_1, t_2]$ is any set of all real numbers t with $t_1 \leq t \leq t_2$; t_1 and t_2 may be infinite. A "time function" is any *vector*-valued function on a time interval. Operators will be denoted by underlined capitals such as $\underline{H}$. We conceive of two sets of time functions on some fixed time interval, called the domain, $Do(\underline{H})$, and the range, $Ra(\underline{H})$, respectively. $\underline{H}$ assigns one and only one element y in $Ra(\underline{H})$ to each element x in $Do(\underline{H})$ and this is denoted by the equation $y = \underline{H}(x)$. Note that "y" refers to the totality of y as opposed to the particular value $y(t)$ at some time t.

The specification of any operator requires three statements concerning 1) the domain, 2) the range, and 3) the operational relation. For example, a "square root operator" might be specified as follows: 1) domain = set of all postitive, real-valued time functions on the infinite interval; 2) range = set of all real-valued time functions on the infinite interval; 3) for each x in the domain, $\underline{U}(x) = +\sqrt{x}$.

The words "operator" and "system" will be used interchangeably, and x and y will be referred to as input and output time functions, depending on their intuitive role.

Interpretations

In modeling a physical device *all* independent variables known to affect the output are included as components of the input. Any part of the output that cannot be related to the input is accounted for by a general noise component in the input. Conversely, an input-output statement is assumed to be the complete specification of some situation. For example, the amplifier model shown in Figs. 2–4, in which no information concerning the amplifier's internal structure is given, specifies a situation in which the open-loop output is completely determined by only one input. It is irrelevant to ask about the effects of internal noises (or initial conditions); it is assumed that there are no effects that are not explicitly shown. Of course it is relevant to ask whether this is a good model of an acutal physical device.

The possibility of internally excited states that are inaccessible to the input but may affect the output is not admitted in operator theory. Either the internal structure of the system that causes such states is known—in which case the excitations are included as input components—or else their contribution is attributed to the general noise component in the input.

The operator's input-output relation is always deterministic; random perturbations are attributed to input noise components. The input-output relation thus embodies our total knowledge of causality in the system.

Oscillators—usually specified as interconnections—are systems having certain "rest" inputs with unstable neighborhoods. Noise at these inputs disturbs the system, and outputs tend towards "oscillations" having stable neighborhoods.

Initial Conditions and State

Initial conditions have an obvious significance in differential equation theory. However, in general there is no clear distinction between one system with two initial conditions and between two different systems. For example, the initial conditions in a time-varying, nonlinear network may depend on the initial charges on the capacitors—on the initial capacities themselves—in fact, on the initial construction of the network. Therefore no special provision is made for initial conditions in operatory theory. When initial conditions are themselves independent variables—for example, in ballistic problems—they are treated as additional input components. For example, in a linear, R.L.C. network the effect of initial conditions is to add output components which are viewed as linear filterings of impulse input components.

The concept of state simplifies the formulation of certain problems, and is postulated axiomatically in some theories (16). It is, potentially at least, a useful concept in operator theory as well, although one that will not be used here. In operator theory state is defined as any transformation of the input domain having certain properties; in particular, the state at any time, together with the subsequent input completely determine the subsequent output. The state is a classifier of input pasts.

Operator Algebra

Interconnections of systems are described by means of an operator algebra, in which the sum $\underline{H} + \underline{K}$ of two operators $\underline{H}$ and $\underline{K}$, and the cascade $\underline{H} * \underline{K}$ of $\underline{H}$ following K, are defined, and correspond to the system sum and cascade shown in Fig. 5. For the sake of simplicity an abridged algebra, limited to operators having identical domains and ranges which are linear spaces (see Section IV), will be employed. The properties of such an algebra are conventional (see, for example, Kolmogoroff [12] and Lusternik and Sobolev [13]) and are therefore relegated to Appendix I. Note, however, that since we are dealing with nonlinear operators cascading is not distributive after addition and not commutative.

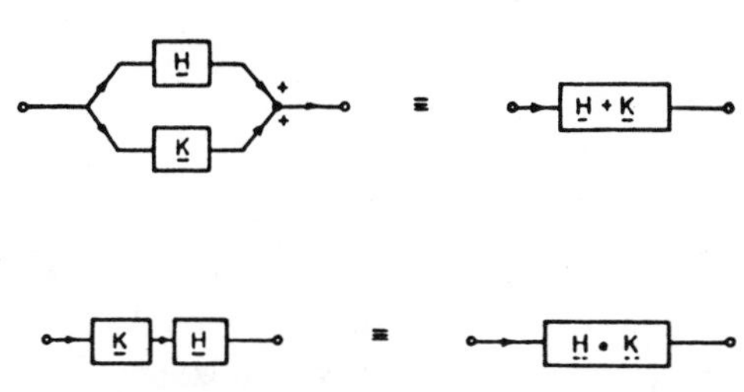

Fig. 5—Operator sum and cascade.

The Feedback Equation Derived

Consider the system in Fig. 1. Let $\underline{X}$ denote a linear space and assume that all operators map $\underline{X}$ into itself. Assume that the open-loop system $\underline{H}$ is given. If x in $\underline{X}$ is any input to the closed-loop system then the error e and output y must simultaneously satisfy the two equations,

$$y = \underline{H}(e) \tag{1}$$

$$e = x - y. \tag{2}$$

These are equations relating *time functions*. We prefer to replace them by equations relating *systems—i.e.*, the closed-loop system $\underline{G}$, and the error system $\underline{E}$. This is possible provided that a unique solution exists for e and for y for each x in the given space $\underline{X}$ of (actual) input functions; G is then defined to be the operator which assigns $y = \underline{G}(x)$ to each x in $\underline{X}$; E is defined to be the operator which assigns $e = \underline{E}(x)$ to each x in $\underline{X}$. If $\underline{E}$ and $\underline{G}$ exist in this sense they can be substituted into (1) and (2) to give

$$\underline{G}(x) = \underline{H}(\underline{E}(x))$$

$$\underline{E}(x) = x - \underline{G}(x)$$

and since these hold for all possible inputs x they imply the following operator forms of the simultaneous feedback equations:

$$\underline{G} = \underline{H} * \underline{E} \tag{3}$$

$$\underline{E} = \underline{I} - \underline{G} = \underline{I} - \underline{H} * \underline{E} \tag{4}$$

(I being the identity), which can be combined to give the single feedback equation

$$G = H * (I - G). \tag{5}$$

Eq. (5) defines G implicitly when H is given.

Explicit Expressions for E and G

Consider (4) for E; after transposing $H*E$ to the left-hand side and factoring out E we obtain

$$(I + H) * E = I$$

which, on being cascaded with $(I + H)^{-1}$ (provided that inverse exists) gives

$$E = (I + H)^{-1} \tag{6}$$

whence (3) gives

$$G = H * (I + H)^{-1}. \tag{7}$$

Eqs. (6) and (7) are the fundamental expressions for the error and closed-loop feedback systems in terms of the open loop. (Note the resemblance of (7) to the well-known expression $G(\omega) = H(\omega)/1 + H(\omega)$ relating the open- and closed-loop frequency responses of a linear, time-invariant system.)

Existence of Solutions

The existence of the inverse $(I + H)^{-1}$ is the necessary and sufficient condition for the existence of solutions for E and G to the feedback equations. If this condition is satisfied then the solutions are unique and are given by (6) and (7). This general, set-theoretical, result is independent of other properties of the feedback system and is valid for unstable systems. In the more general case when $Do(H)$ and $Ra(H)$ are not identical linear spaces a similar result can be obtained by defining the inverse suitably.

To show the sufficiency of this condition note that the existence of $(I + H)^{-1}$ implies the existence and uniqueness of the expressions for E and G in (6) and (7); direct substitution of these expressions into the feedback equations (4) and (5) shows that they are indeed solutions. They must be unique since it has been shown that every solution that exists is given by (6) and (7).

To show the necessity of the condition suppose that $(I + H)^{-1}$ fails to exist; this can happen in two ways: (See Appendix I.)

1) $(I + H)$ is not onto.
2) $(I + H)$ is not one to one.

In the first case there is at least one x in X for which there is no e in E satisfying $x = e + H(e)$, that is, for some input there is no solution for the error. In the second case there is an input for which there are at least two solutions for the error. In either case E fails to exist; similarly G fails to exist.

Realizability

If $(I + H)^{-1}$ fails to exist then there is no operator solution to the feedback problem, *i.e.*, for some input there is no solution for the error, or there are multiple solutions. For example, if H is a perfect squarer then there are two solutions for positive inputs and there is no solution for negative inputs. Clearly this does not correspond to the physical situation, in which each input produces a well defined, unique output. The squarer model, then, lacks certain essential properties of a physical system, *i.e.*, it is not realizable. The following questions arise now: What property of physical systems causes them to have unique solutions? What is an adequate model of a nonlinear system? These questions are explored in Zames [10].[1] (A paper on this subject is now in preparation.) The essential property is a generalization of inevitable delay around the loop (or of attenuation of high frequencies).

Determination of Inverses

Although the expression $G = H * (I + H)^{-1}$ defines G explicitly in terms of H in a form whose meaning is easy to grasp, in general it is difficult to evaluate $(I+H)^{-1}$. Special cases in which this is possible are those in which 1) H is linear and time invariant in which case the spectra $G(\omega)$ and $H(\omega)$ are related by the expression $G(\omega) = H(\omega)\ /1+H(\omega)$; 2) H is linear but time varying, in which case solutions in terms of eigenfunction expansions are possible (though difficult to obtain); and 3) H is nonlinear and has no memory where inversion is merely a reflection of the graph of H about the 45° line.

In the more general nonlinear case a closed form for the solution of $(I + H)^{-1}$ is not possible. Instead $(I + H)^{-1}$ must be evaluated by one of a variety of iteration schemes, some of which are discussed in Section VIII. However, those properties of feedback systems that can be obtained without actually evaluating inverses will be discussed first.

More General Interconnections

Our primary concern here is the simple feedback loop of Fig. 1. However, more general interconnections may be described by equations of the form of (3) and (4), with the result that most of our statements, bounds and iteration schemes are generally valid for them. Consider, for example, the n-node interconnection shown in Fig. 6; if x, e and y now denote n-component vectors having components x_i, e_i and y_i, respectively, the equations

$$e_i = x_i - \sum_{j=1}^{n} H_{ij}(e_j)$$

$$y_i = \sum_{j=1}^{n} H_{ij}(e_j), \qquad i = 1, 2, \cdots, n$$

are satisfied; these equations may be written in the form of (3) and (4) with I, H, E and G having domains and ranges consisting of n-component vectors.

[1] Zames [11], section 6.

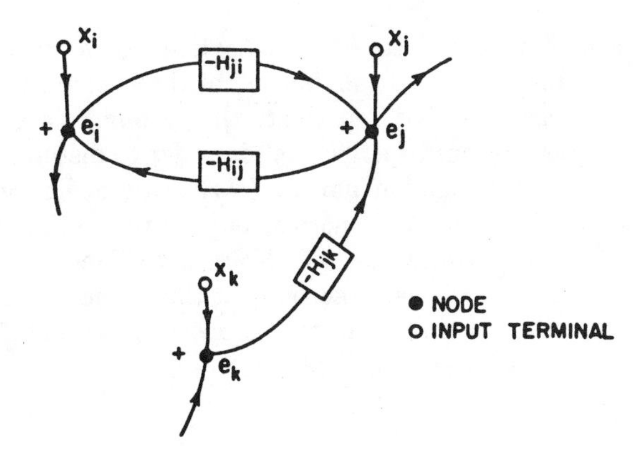

Fig. 6—An interconnection.

III. Loop Transformation for Distortion

The main development of the theory will now be suspended to demonstrate an application to the manipulation of nonlinear flow graphs, for which the theory supplies a rigorous justification. The configuration of a feedback loop can be manipulated into a variety of forms; a form will be derived which exhibits the effects on the closed loop of an open-loop distortion, as an additive component in the closed loop.

General Distortion Model

Consider the linear, time-invariant system H_l, which on being placed in a feedback loop produces the closed-loop system (Fig. 7),

$$G_l = H_l * (I + H_l)^{-1}. \tag{8}$$

The open loop is now distorted by the addition of a nonlinear distortion operator ΔH to give a new open-loop system H, *i.e.*,

$$H = H_l + \Delta H. \tag{9}$$

The new closed-loop system is G and we seek that distortion ΔG which must be added to the original G_l in order to produce G, *i.e.*,

$$G = G_l + \Delta G. \tag{10}$$

To find ΔG we first write the feedback equation (5) for the new system G, *i.e.*,

$$G = (H_l + \Delta H) * (I - G),$$

invoke the left distributive law,

$$G = H_l * (I - G) + \Delta H * (I - G),$$

followed by the right distributive law (since H_l is linear),

$$G = H_l - H_l * G + \Delta H * (I - G),$$

bring the term $H_l * G$ to the left-hand side, and factor out G,

$$(I + H_l) * G = H_l + \Delta H * (I - G)$$

and cascade with the inverse of $(I + H_l)$, (which must exist whenever H_l is an adequate model of an actual physical system),

$$G = (I + H_l)^{-1} * H_l + (I + H_l)^{-1} * \Delta H * (I - G).$$

The first term on the right-hand side is recognized to be G_l. [Compare with (8) and note that $(I + H_l)^{-1}$ and H_l are time invariant and hence commutative.] Therefore the second term must equal ΔG, *i.e.*,

$$\Delta G = (I + H_l)^{-1} * \Delta H * (I - G)$$

which, after replacing G by $G_l + \Delta G$ can also be written,

$$\Delta G = (I + H_l)^{-1} * \Delta H * (I - G_l - \Delta G).$$

This can be made easier to interpret by introducing the linear error operator E_l, defined by

$$E_l = I - G_l \tag{11}$$

which generates the error signal in the linear feedback system G_l. (E_l is also given by $E_l = (I + H_l)^{-1}$). Substituting (11) into the expression for ΔG we arrive at the following result for the closed-loop distortion ΔG:

$$\Delta G = E_l * \Delta H * (E_l - \Delta G). \tag{12}$$

Eq. (12) is implicit in ΔG; it is in fact the equation for a feedback system which, by inspection, appears as in Fig. 8. It involves only two kinds of elements; the original open-loop distortion term ΔH and the linear error operator E_l.

Note that this result is valid for any representation for H and ΔH, say by differential or integral equations. It may be shown to be valid for a time-varying H_l too.

Distortion Model for Amplifier

The general distortion model can be applied to the amplifier problem by splitting the no memory operator N (Fig. 2) into linear and nonlinear distortion terms, *i.e.*,

$$N = N_l + \Delta N \tag{13}$$

identified by the subscript l and the prefix Δ, respectively. The over-all open-loop linear term and distortion are now identified as $H_l = K * L * N_l$ and $\Delta H = K * L * \Delta N = H * N_l^{-1} * \Delta N$, respectively, whereupon (12) gives

$$\Delta G = E_l * H_l * N_l^{-1} * \Delta N_l * (E_l - \Delta G).$$

This can be simplified when it is realized that $E_l * H_l = G_l$, whence we obtain the following model:

$$\Delta G = G_l * N_l^{-1} * \Delta N * (E_l - \Delta G) \tag{14}$$

which is shown in Fig. 9. Note that N_l^{-1} is a constant.

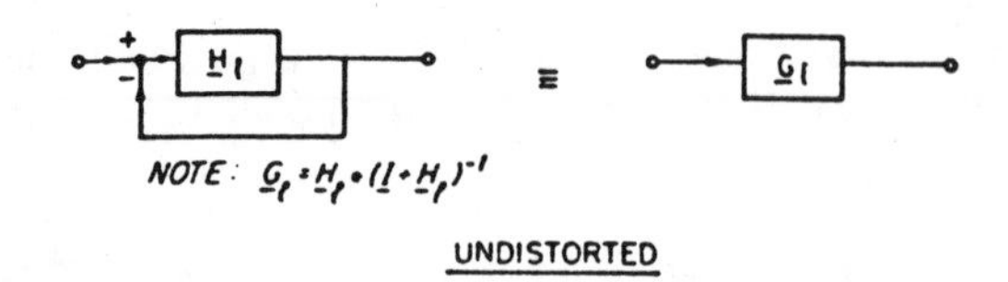

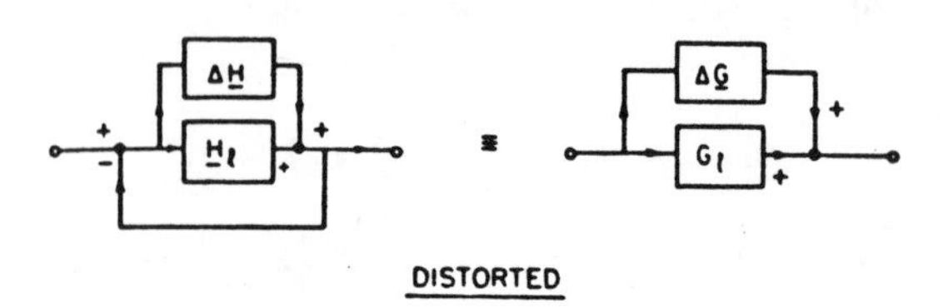

Fig. 7—Related open- and closed-loop distortions.

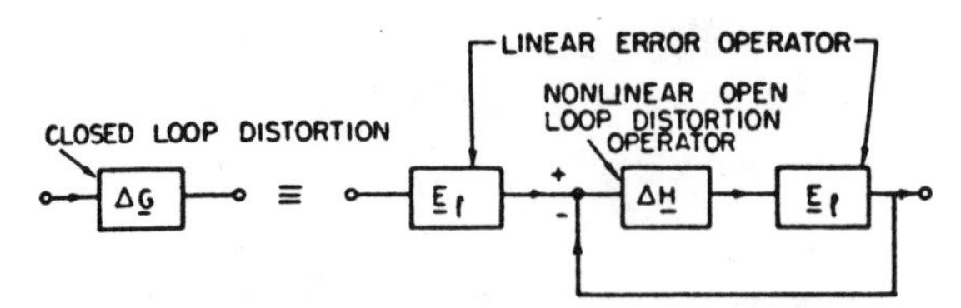

Fig. 8—General model for closed-loop distortion.

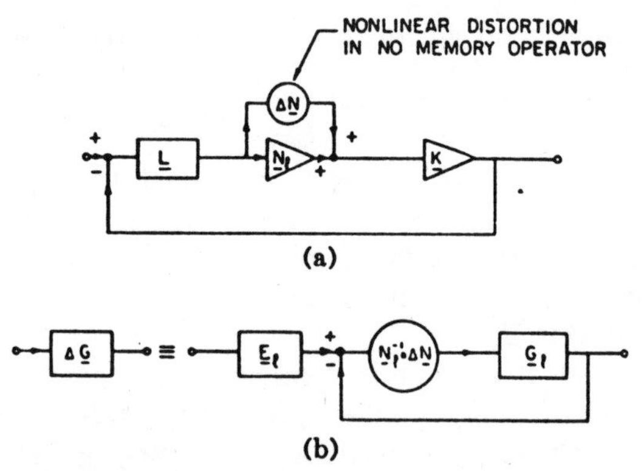

Fig. 9—Model for nonlinear feedback amplifier distortion. (a) Amplifier showing distorted open loop. (b) Closed-loop distortion. *Note*: $E_l(w) = 1/[1 + H_l(w)]$, $G_l(w) = H_l(w)/[1 + H_l(w)]$, where $H_l(w)$, $E_l(w)$ and $H_l(w)$ are frequency responses

IV. Bounds on System Responses

The theory will now be resumed and a calculus of bounds on system responses will be described. It is desirable to be able to bound system responses for a variety of purposes, for example: 1) A bound on the distortion or error in a system is a tolerance on system quality. 2) The stability of a system can be verified by bounding its responses. In particular, a feedback system is stable if a "contracting transformation" can be found which manipulates the loop into a form in which the open-loop "amplification" is less than unity. 3) In the solution of feedback equations by iterations the convergence of iterations is established by bounding differences between true and approximate solutions.

In order to have a calculus of system bounds, it is necessary to specify, by a single number, the "size" of a time function or the "amplification" of a system. Such a number is called the "norm" of the time function or of the operator that represents the system. We shall refer to operator norms as "gains" in order to distinguish them from time-function norms. (*Note*: Our operational notation and definitions of norms conform to the conventions of linear operator theory; Kolmogoroff and Fomin [12] have a comprehensive account of this subject. However, we shall not require any properties beyond the elementary ones described here.)

Norms

The measure of the size of a time function x is its norm, denoted $\| x \|$. Two commonly used norms are: 1) the largest height attained by the function,

$$\| x \|_1 = \underset{t}{\text{l.u.b.}} \, | x(t) | .$$

l.u.b.$_t$ denotes least upper bound with respect to t—it equals the maximum if a maximum is attained.) 2) The root integral of the square

$$\| x \|_2 = \left\{ \int_{t_1}^{t_2} | x(t) |^2 \, dt \right\}^{1/2}$$

where x has been assumed to be a real-valued function on $[t_1, t_2]$. If x is vector-valued with n components then $\| x \|$ equals the largest of the component norms. More generally, however, a norm can be defined axiomatically as follows.

Norms in Linear Spaces

Consider a space $\mathfrak{X}$ of time functions. $\mathfrak{X}$ is said to be a *linear space* if it is closed under addition and scalar multiplication, *i.e.*, if x_1 belongs to $\mathfrak{X}$, and x_2 belongs to $\mathfrak{X}$, then $(x_1 \pm x_2)$ belongs to $\mathfrak{X}$; if x belongs to $\mathfrak{X}$ and a is a real number, then ax belongs to $\mathfrak{X}$. For example, the space of all time functions is linear, but the space of positive functions is not.

The space $\mathfrak{X}$ is said to be *normed* if each element x in X has a norm $\| x \|$ which can be defined in any way at all so long as the following three properties are fulfilled: 1) $\| x \|$ is a real, positive number and is different from zero unless x is identically zero. 2) $\| ax \| = | a | \cdot \| x \|$. 3) $\| x_1 + x_2 \| \leq \| x_1 \| + \| x_2 \|$. The last inequality is called the "triangle inequality" since it is analogous to the rule that the sum of the lengths of any two sides of a triangle is less than the third.

The preceding definition of a norm is motivated by the need for a calculus of bounds which is independent of how the size of a function is specified, but which permits the relation of the size of a time function formed under any one of the operations postulated for a linear space (*i.e.*, addition and scalar multiplication) to the component sizes. Furthermore, the definition imposes on a norm the metric properties, *i.e.*, the minimum properties of a measure of distance between approximations (but not necessarily of cost or performance).

Gains

An operator norm—or "gain"—must have all the properties of norms that we have cited. In addition, it must have the following property, relating the gain of a cascade to the component gains:

$$g(H * K) \geq g(H) \cdot g(K) \tag{15}$$

where $g(H)$ denotes the gain of H, and H and K are arbitrary gained operators.

We shall adopt a particular gain, namely, the *maximum incremental amplification*; *i.e.*, if H is any system and x_1 and x_2 are any two inputs then the incremental amplification is

$$\frac{\| H(x_1) - H(x_2) \|_i}{\| x_1 - x_2 \|_i}$$

and the gain, $g_i(H)$, is the least upper bound to it,

$$g_i(H) = \underset{x_1, x_2 \text{ in } D_0(H)}{\text{l.u.b.}} \frac{\| H(x_1) - H(x_2) \|_i}{\| x_1 - x_2 \|_i}.$$

The subscript i refers to the fact that the gain depends on the particular norm being used. In order to use this definition of gain it is necessary to limit our attention to operators whose domains and ranges consist of elements taken from a normed, linear space.

Some Particular Gains

The gain of a no memory system is the maximum absolute value of the slope in *either* of the two norms. It may be shown (proofs are given in Zames[2]) that the gains of a stable, linear, time-invariant system, on the infinite interval $[-\infty, \infty]$, are $g_1(H) = \int_{-\infty}^{\infty} | h(t) | dt$, $g_2(H) = \text{l.u.b.}_\omega | H(\omega) |$. The first is the area under the absolute impulse response, while the second is the peak magnitude of the frequency response $H(\omega)$. Gains of open-loop systems are usually straightforward to compute.

Lipschitz Conditions

The definition of gain as a ratio of norms implies the following inequality, known as a "Lipschitz condition," for any pair x_1, x_2, of time functions,

$$\| H(x_1) - H(x_2) \|_i \geq \alpha \| x_1 - x_2 \|_i, \tag{16}$$

where $\alpha = g_i(H)$. Conversely, the gain of any operator H that satisfies a Lipschitz condition exists and is bounded, $g_i(H) \leq \alpha$; thus, in order to bound gains Lipschitz conditions are sought.

Bounds on Feedback System Gains

Consider any feedback system G satisfying the feedback equations (3) and (4). Suppose E exists. To bound $g(E)$ we shall obtain a Lipschitz condition for E. Consider any pair of inputs x_1 and x_2; (4) and the triangle inequality imply the following:

$$\begin{aligned} \| E(x_1) - E(x_2) \| \\ &= \| (I - H * E)(x_1) - (I - H * E)(x_2) \| \\ &\geq \| x_1 - x_2 \| + g(H) \cdot \| E(x_1) - E(x_2) \| \end{aligned}$$

where (16) has been used. Provided that $g(H) < 1$ the last inequality may be rearranged to give the Lipschitz condition

$$\| E(x_1) - E(x_2) \| \leq \frac{1}{1 - g(H)} \cdot \| x_1 - x_2 \|$$

which implies that $g(E)$ exists and $g(E) \leq 1/1 - g(H)$. Finally, since the existence of E implies that of G with $G = H * E$, it may similarly be deduced that $g(G)$ exists and

$$g(G) \leq g(H) \cdot g(E) \leq \frac{g(H)}{1 - g(H)} \tag{17}$$

provided that $g(H) < 1$ and E exists.

Contractions and Fixed Point Theorems

In order that inequality (17) may be used, the open-loop operator must have a gain $g(H)$ that is less than unity. An operator having this property is called a "contraction," since it reduces the distances between all pairs of functions.

It must also be established, by means of a "fixed point theorem," that E exists. It will appear in Section VIII, however, that under certain conditions—which are fulfilled in the following applications—the existence of E is automatically ensured if H is a contraction.

The Method of Contracting Transformations

This is a method for bounding the gain of a feedback system, which consists of finding a transformation that renders the open-loop contracting, so that (17) is valid. Since the original and transformed systems are externally equivalent, the bound given by (17) applies to the original system as well. In order that this method may be employed it must be established that either the original or the transformed solution for E exists. It will appear in Section VIII that this prerequisite is automatically fulfilled in our applications.

V. Applications: Amplifier Distortion

The theory of gains will now be applied to the system in Fig. 2, for which a contracting transformation will be found; the result will be used to study the effect of feedback on nonlinear distortion. The ability to reduce distortion is one of the chief reasons for using feedback. However, distortion is reduced only if inputs are essentially band-limited so that errors lie within the open loop passband. This creates a mathematical obstacle, since nonlinearities map band-limited inputs into errors that are not band-limited, making it difficult to introduce

[2] Zames [11], section 2.6.1.

band-limiting constraints. The difficulty is avoided by bounding the nonband-limited nonlinear distortion by one occurring within a linear loop, which is bandlimited.

It will be shown that the transformation of Section III is contracting provided that the slope of the nonlinearity has an upper bound and a positive lower bound. The loop gain and the distortion bound are minimized by splitting the nonlinearity along the average of the least-upper and greatest-lower slope bounds.

The domains and ranges of all operators are defined to consist of real-valued time functions on the infinite interval, having the second norm (L_2), in which it is easiest to evaluate gains.

Consider the system in Fig. 2. It is transformed as in Section III into the sum of a linear system $\underline{G}_l$ and a distortion $\Delta\underline{G}$, illustrated in Fig. 9. The distortion is bounded by applying the properties of gains to (14) which defines the model, giving

$$g(\Delta\underline{G}) \leq \frac{g(\underline{G}_l * \underline{N}_l^{-1} * \Delta\underline{N})g(\underline{E}_l)}{1 - g(\underline{G}_l * \underline{N}_l^{-1} * \Delta\underline{N})} \tag{18}$$

provided that the open loop operator in Fig. 9,

$$\underline{G}_l * \underline{N}_l^{-1} * \Delta\underline{N},$$

is a contraction. In order to establish the circumstances under which this condition is fulfilled, the open-loop gain will be evaluated.

Evaluation of Open-Loop Gain

The gain $g_2(\underline{G}_l * \underline{N}_l^{-1} * \Delta\underline{N})$ may be shown to *equal* the product of the gains $g_2(\underline{G}_l)$ and $g_2(\underline{N}_l^{-1} * \Delta\underline{N})$ of the linear feedback operator $\underline{G}_l$ and the no memory operator $\underline{N}_l^{-1} * \Delta\underline{N}$, respectively. The no memory gain is determined by the manner in which $\underline{N}$ is split into linear and nonlinear components; if β and γ denote the least-upper and greatest-lower bounds to the slope of $\underline{N}$, and c denotes the slope of $\underline{N}_l$, we have,

$$g_2(\underline{N}_l^{-1} * \Delta\underline{N}) = \max\left(\left|\frac{\gamma - c}{c}\right|, \left|\frac{\beta - c}{c}\right|\right).$$

The linear gain is the maximum magnitude of the frequency response of $\underline{G}_l$. Assuming that

$$L(\omega) = j\omega\omega_2/(j\omega + \omega_1)(j\omega + \omega_2),$$

the maximum response of $\underline{G}_l$ occurs at the center of the pass band, $\sqrt{\omega_1\omega_2}$, and is

$$g_2(\underline{G}_l) = \frac{Kc}{1 + Kc + \frac{\omega_1}{\omega_2}}$$

which is always less than one and approximately equals one. The over-all open-loop gain is the product of the linear and no memory gains, and is minimum when $c = \frac{1}{2}(\beta + \gamma)$, which gives

$$g_2(\underline{G} *_l \underline{N}_l^{-1} * \Delta\underline{N}) = \left(\frac{\gamma - \beta}{\gamma + \beta}\right)\left[\frac{1}{1 + \frac{2(\omega_1 + \omega_2)}{K\omega_2(\gamma + \beta)}}\right] \lessapprox \frac{\gamma - \beta}{\gamma + \beta}. \tag{19}$$

This is less than one, *i.e.*, a contraction results—provided that we have $\gamma \geq \beta > 0$, *i.e.*, the minimum slope of $\underline{N}$ is greater than zero. This method of splitting $\underline{N}$ is illustrated in Fig. 10.

Distortion Bound—Arbitrary Signals

The distortion bound (18) is smallest when the open-loop gain is smallest because $g_2(\underline{E}_l)$ is independent of c and in fact equals two. Substituting (19) into (18) we obtain

$$g_2(\Delta\underline{G}) \leq \left(\frac{\gamma}{\beta} - 1\right). \tag{20}$$

This bound fails to show that increasing K reduces distortion. In fact, of course, distortion is not reduced for signals having significant components outside the pass-band, where there is no feedback effect.

Distortion Bound for Band-limited Signals

Suppose next that the input is band-limited between frequencies $\sqrt{B\omega_1\omega_2}$ and $\sqrt{\omega_1\omega_2/B}$, *i.e.*, on a log scale the bandwidth is log B units symmetrically placed with respect to the amplifier's pass band. The error gain,

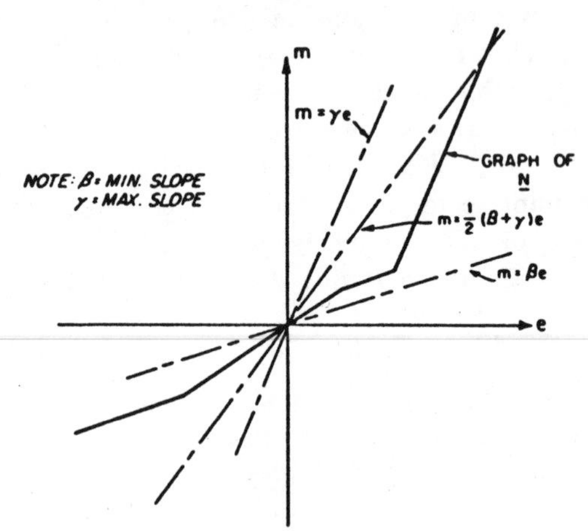

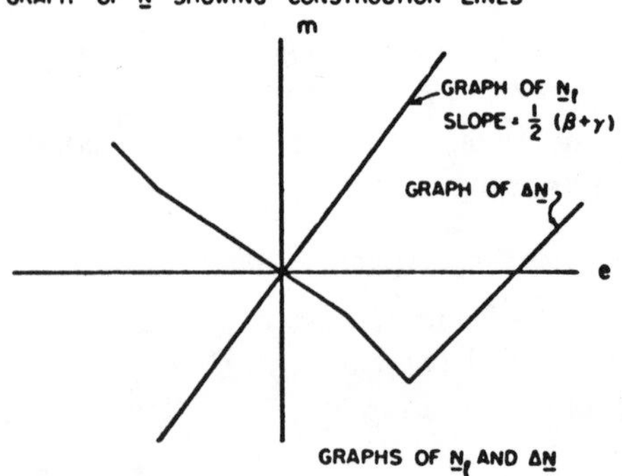

Fig. 10—Optimum splitting of $\underline{N}$.

$g_2(E_l)$, is the maximum magnitude of the error response $E_l(\omega)$, which occurs at the edges of the pass band, and is

$$g_2(E_l) = E_l(\sqrt{B\omega_1\omega_2}) = E_l\left(\sqrt{\frac{\omega_1\omega_2}{B}}\right)$$

$$= \frac{1}{\left\{1 + \frac{K(\gamma + \beta)B\omega_2[(\omega_1 + \omega_2) + \frac{1}{4}K(\gamma+\beta)\omega_2]}{\left[1 + \frac{B\omega_2}{\omega_1}\right]\left[1 + \frac{B\omega_1}{\omega_2}\right]}\right\}^{1/2}}.$$

Hence we have

$$g_2(\Delta G) \leq \frac{\frac{1}{2}\left(\frac{\gamma}{\beta} - 1\right)}{\left\{1 + \frac{K(\gamma + \beta)B\omega_2[(\omega_1 + \omega_2) + \frac{1}{4}K(\gamma+\beta)\omega_2]}{\left[1 + \frac{B\omega_2}{\omega_1}\right]\left[1 + \frac{B\omega_1}{\omega_2}\right]}\right\}^{1/2}} \tag{21}$$

which indeed goes to zero as K is increased. This result has been obtained without recomputing the nonlinear loop gain for band-limited signals.

Comments

The nonlinear distortion has been bounded without solving the feedback equation or having an analytical expression for H. The bound is a function of loop gain K for band-limited signals, and goes to zero as K is increased. It is a good upper bound (though not a least upper bound) for unfavorably shaped nonlinearities, but may be weak in favorable cases; much poorer results would be obtained if the linear system had resonant peaks.

The procedure is valid for arbitrary nonlinearities, provided that they have maximum slopes and minimum slopes greater than zero. It can be carried through in the peak value norm $\| \ \|_1$ leading to similar results except that then it may be shown[3] that $g_1(G_l) \approx 2$, and that the procedure is limited as a result to nonlinearities whose slope ratios γ/B are less than 3. The nonlinearity must precede the linear system or the assumption that inputs are band-limited must be replaced by a more complicated one.

VI. Stability

An operational definition of stability entirely in terms of inputs and outputs is desired. The definition adopted here is simply that of continuity of the operator: A system H is uniformly stable in the large provided that for every real $\epsilon > 0$ there exists a real $\delta(\epsilon) > 0$ with the property that whenever two inputs x_1 and x_2 are less than $\delta(\epsilon)$ apart, *i.e.*, $\| x_1 - x_2 \| < \delta(\epsilon)$, the corresponding outputs are less than ϵ apart, *i.e.*,

$$\| H(x_1) - H(x_2) \| < \epsilon.$$

[3] Zames [11], section 3.6.

This definition, which reduces to the commonly accepted "bounded inputs produce bounded outputs" for linear systems, is not equivalent to Lyapunov's. although it usually implies Lyapunov's in well-behaved situations. The principal difference between the former, in terms of inputs and outputs, and the latter, in terms of state, lies in the possibility of having states that are not accessible from the input and not observable from the output (17)—a possibility which is excluded in operator theory.

It follows from the Lipschitz condition (16) that finite gain implies uniform stability in the large. The gain of a stable system is a useful measure of its degree of stability (in the sense of sensitivity to noise). Thus the gain g_2 of a linear time-invariant system is the maximum magnitude of the frequency response, a quantity which has been used in the past as an empirical stability measure in servomechanism design [18]. It may be more meaningful, for this purpose, if the gain is normalized with respect to a desired amplification.

Feedback Stability by Contracting Transformations

A theorem will now be stated which leads to a method for establishing the stability of certain nonlinear systems: Any closed-loop feedback system whose solution exists is uniformly stable in the large if a contracting transformation can be found. For example, the distortion model is stable. To prove this note that any feedback system whose solution exists and for which a contracting transformation exists is equivalent, in the closed loop, to a feedback system having a contracting open loop. The closed-loop gain must therefore be finite by virtue of (17). Finally, finite gain implies uniform stability in the large; our theorem follows.

VII. Transformation

Some effort has been made[4] to classify transformations according to their effects on loop gain. So far, however, they have been used on a trial and error basis, since no general scheme has been found for predicting their ability to produce contractions.

Perturbation is the best-known transformation and is illustrated in Fig. 11. An operator T is added and subtracted to the reverse path, and a new configuration results. Two conditions must be fulfilled: the solution to the resulting minor feedback loop, $H * (I + T * H)^{-1}$, must be known and the resulting major loop gain, $g([K - T] * [H * (I + T * H)^{-1}])$, must be less than one. The second condition requires that $(K - T)$, which is known as the "perturbation about the known solution" be small. The amplifier example in Section V is a modification of perturbation for the case in which T is linear.

Inversion is illustrated in Fig. 12, and rotations are shown in Fig. 13. Their usefulness when they are used separately is limited, but they can be alternated to generate an infinite variety of useful compound transformations.

[4] Zames [11], section 3.5.

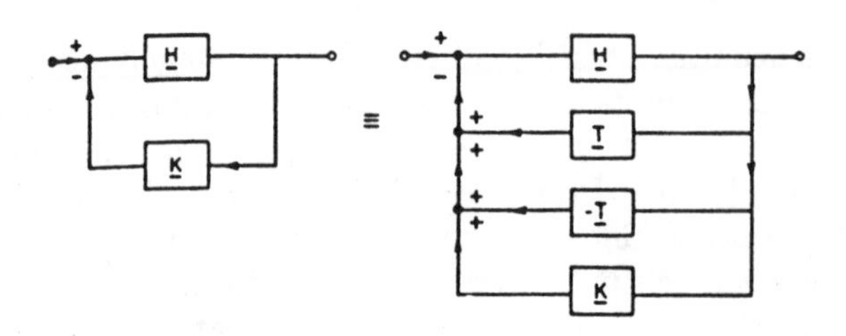

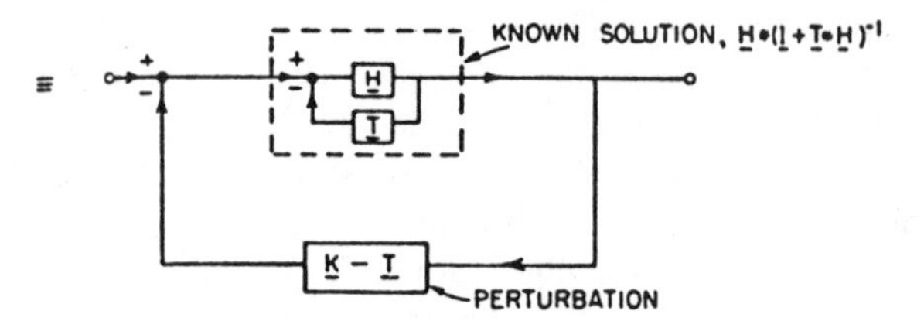

Fig. 11—Perturbation.

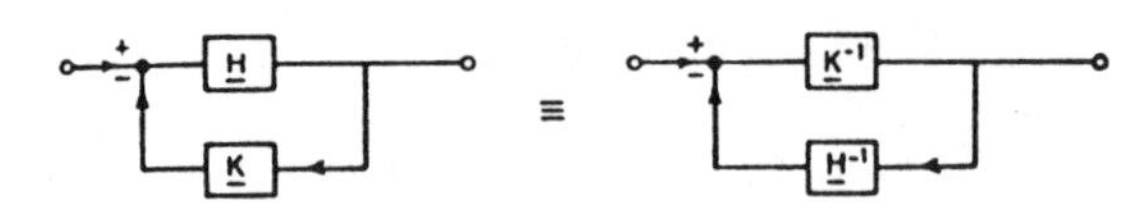

Fig. 12—Inversion.

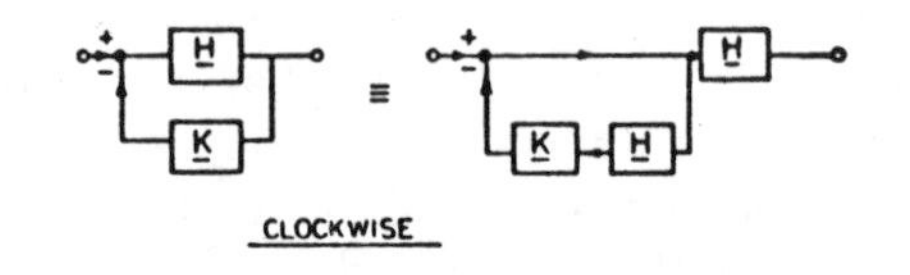

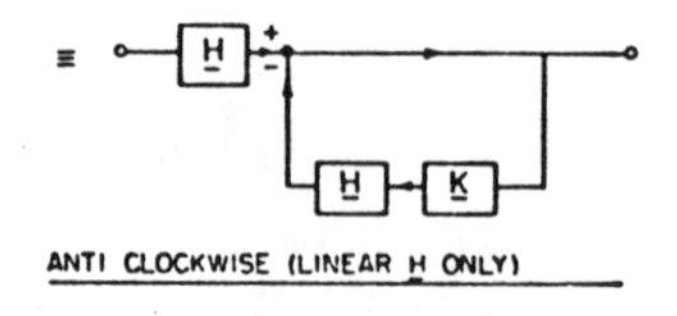

Fig. 13—Rotations.

VIII. Iteration Theory

Except in a very few special cases in which a closed solution is possible, the response of a nonlinear feedback system is computed by iteration, which replaces the implicit loop by an explicit structure. Two iterations will be described; one depends on the contraction principle, the other on having an integration in the loop. The former is useful in a limited class of problems involving stable systems and small nonlinearities; the latter suffers from no such restrictions, and is valid for most physical systems.

The simplest iteration for the equation $\underline{G} = \underline{H} * (\underline{I} - \underline{G})$ is described by the following equations:

$$\underline{G}_1 = \underline{H}$$

$$\underline{G}_n = \underline{H} * (\underline{I} - \underline{G}_{n-1}), \qquad n = 2, 3, \cdots \tag{22}$$

$$\underline{G}(x) = \lim_{n\to\infty} \underline{G}_n(x)$$

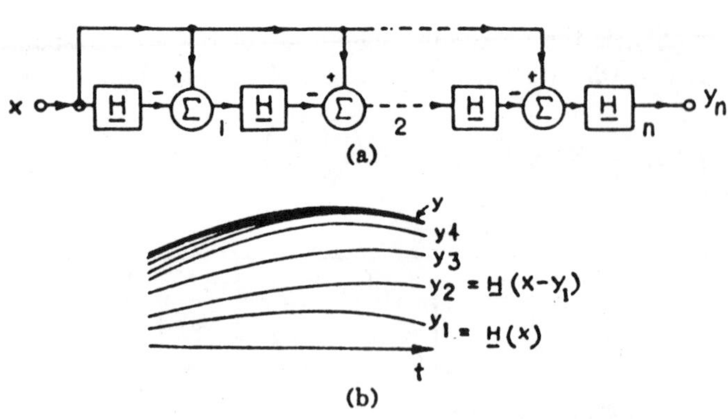

Fig. 14—(a) Iterative expansion of feedback loop. (b) Convergence of sequence of approximations to the output.

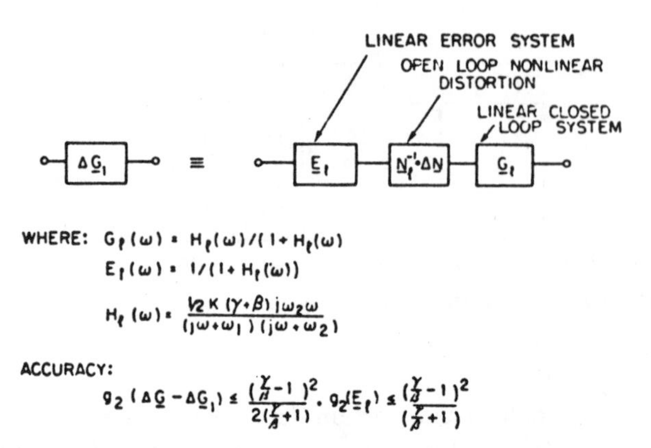

Fig. 15—First approximation to amplifier distortion.

and is illustrated in Fig. 14. In the first approximation the closed loop $\underline{G}$ is approximated by the open loop $\underline{H}$ and feedback is disregarded. The approximate output obtained thus leads to a second approximation to the error, and so on round and round the loop. Provided that the iteration converges, a finite number of stages provide an arbitrarily good approximation to $\underline{G}$. In certain cases the first stage provides an adequate approximation, so that iteration can be avoided altogether. Thus for small nonlinearities the distortion model in Fig. 9 can be replaced by that in Fig. 15, in which the feedback loop is absent.

Convergence Conditions; Banach Space

Three conditions must be fulfilled if the iteration in (22) is to solve the feedback equation: 1) the sequence $\underline{G}_n$ must converge; 2) an operator $\underline{G}$ must exist to which $\underline{G}_n$ converges; 3) it must be shown that $\underline{G}$ obtained thus is a solution to the feedback equation.

A sequence of time functions y_n in some space $\mathfrak{X}$ can conceivably fail to meet the second condition by converging in the Cauchy sense (*i.e.*, given any $\epsilon > 0$ an integer $N(\epsilon)$ can be found for which $\| y_m - y_n \| < \epsilon$ for all $m, n > N$) without finding any element in $\mathfrak{X}$ to converge to. This never happens if $\mathfrak{X}$ is a "complete" space, *i.e.*, one in which by definition every Cauchy sequence converges. Thus the simplest setting for operator iteration is a space which, in addition to being normed and linear, is also complete. Such a space is called a Banach space.

The spaces of bounded time functions and of square integrable (L_2) time functions are Banach spaces. (The completeness of the former is a trivial consequence of the completeness of reals; that of the latter is established by the Riesz-Fisher theorem.)

Geometrically Converging Iteration

A sufficient condition for the feedback equation $\underset{\sim}{G} = \underset{\sim}{H} * (\underset{\sim}{I} - \underset{\sim}{G})$ to have a unique solution for $\underset{\sim}{G}$ that can be found by iteration, (22), is for $\underset{\sim}{H}$ to be a contraction, *i.e.*, $g(\underset{\sim}{H}) < 1$, provided that all operators map a Banach space into itself. (This is a form of the Banach-Cacciopolli theorem, also known as the contraction principle. Its proof will be omitted since it is readily available in Kolmogoroff and Fomin;[5] it is also given in Kantorovic [14], which is a source paper on functional iteration, and in Zames [8], [9], [11], which are concerned with applications to feedback.) The resulting convergence rate is faster than that of the geometric series $\alpha + \alpha^2 + \alpha^3 + \cdots$, where α is the loop gain, $\alpha = g(\underset{\sim}{H})$. The truncation error at the nth stage is

$$g(\underset{\sim}{G} - \underset{\sim}{G}_n) \leq \frac{\alpha^{n+1}}{1 - \alpha}.$$

For example, the first approximation to the nonlinear distortion in Fig. 15 is accurate to within

$$\frac{\left(\frac{\gamma}{B} - 1\right)^2}{2\left(\frac{\gamma}{B} + 1\right)} \cdot g_2(E_l) \leq \frac{\left(\frac{\gamma}{B} - 1\right)^2}{\left(\frac{\gamma}{B} + 1\right)}.$$

An accuracy of 5 per cent of the bound on the amplifier's true output (*i.e.*, 2) is attained for slope ratios $\gamma/B < 1.5$, *i.e.*, for small nonlinearities.

Note that the rate of convergence depends solely on the loop gain, which can be affected by transformations, and has been minimized here by choosing $c = \frac{1}{2}(\beta + \gamma)$.

Exponentially Converging Iteration[6]

An iteration will now be described that requires an integration in the loop instead of a small loop gain, and which is useful for most physical systems; in particular, it extends the validity of the iterative amplifier model in Fig. 14 to arbitrarily large nonlinearities.

Iteration with Pure Delay

If a pure delay is present in the loop, iteration converges even with large loop gains. A delay of length T in the loop eliminates all feedback for the first T seconds after the signal is applied; the system is then effectively open loop, and the first iteration in (22) is accurate. Similarly the nth iteration is accurate until after nT seconds.

[5] Kolmogoroff and S. V. Fomin [12], ch. 3.
[6] The possibilities of exponential iteration were brought to the author's attention by George [7].

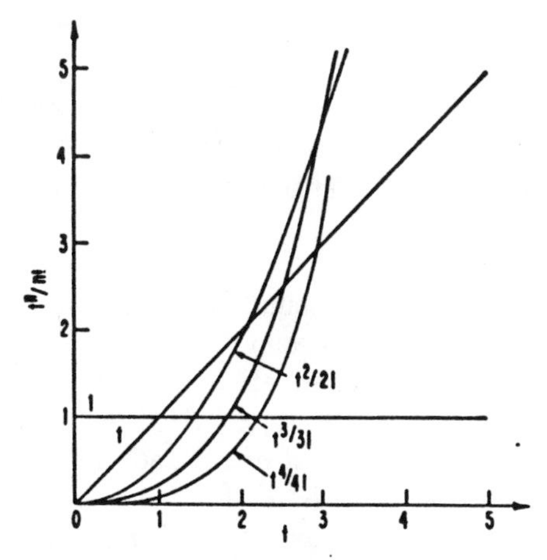

Fig. 16—The functions t^n/n.

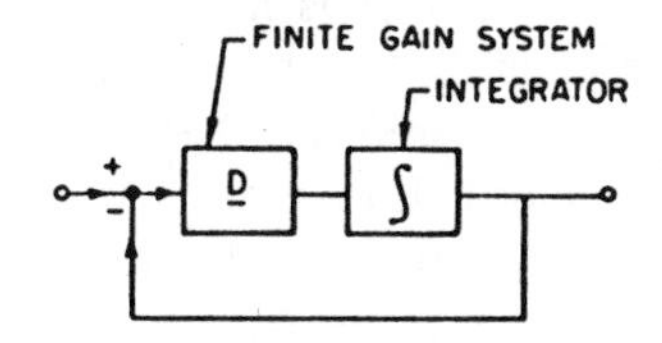

Fig. 17—Feedback loop with integrator.

The rate of convergence is "pointwise," *i.e.*, it depends on the length of time after the first application of input, and is less satisfactory than the uniform convergence of geometric iteration.

Iteration with an Integral in the Loop

The effect of an integration in the loop is similar to that of a delay. Consider, for example, the effect of successive integrations on a unit step; the sequence of time functions $1, t, t^2/2!, \cdots$, shown in Fig. 16, results. Successive functions have their bulk delayed at each integration. At any particular time t the nth integration, $t^n/n!$, eventually goes to zero with n increasing. (After an initial increase for $t > 1$) the series $1 + t + t^2/2! + \cdots$ converges to e^t, and the mechanism of convergence is very similar to that where a pure delay is present.

It is shown in Appendix II that if the open loop $\underset{\sim}{H}$ of a feedback system consists of an integrator in cascade with a system $\underset{\sim}{D}$ (Fig. 17) then the iteration equation (22) for $y(t)$ converges faster than the exponential series $(\alpha t + (\alpha t)^2/2! + \cdots + (\alpha t)^n/n! + \cdots)$ where $\alpha = g_1(\underset{\sim}{D})$. The output is bounded by

$$|\, y(t)\, | \leq \|\, x\, \|_1\, (\epsilon^{\alpha t} - 1). \qquad (22')$$

Application to Amplifier

If the nonlinearity in the amplifier is not monotonic then transformation cannot produce a contraction and the iteration does not converge geometrically. However, it converges exponentially for nonlineaities having arbitrary upper and lower bounds to their slope, γ and β, respectively, where β can be negative, because the open loop can be represented as an integrator and a finite gain system $\underline{D}$ in cascade. $\underline{D}$ is the derivative of the original open-loop system $\underline{G}_l * \underline{N}_l^{-1} * \Delta\underline{N}$ and its gain may be shown to be $\alpha \leqq K(\gamma - \beta)$ for wide-band amplifiers. Convergence is exponential and satisfies (22').

Existence of Solutions

A prerequisite to the bounding procedures for gain and stability (Sections IV and VI) is the existence of a unique solution to the equation $\underline{G} = \underline{H} * (\underline{I} - \underline{G})$. The contraction principle is a "fixed-point theorem" which ensures that $\underline{G}$ indeed exists and that its gain is bounded by the series $\alpha + \alpha^2 + \alpha^3 + \cdots$. The same bound that was derived by using (17) is obtained by summing the series. However, the former procedure is more general than the latter, for the series may fail to converge—for example, if $\underline{H}$ is a contraction on only a part of its domain, or if the gain is defined to be the maximum (but not incremental) amplification—while (17) is valid in conjunction with *any* gain and fixed point theorem. (See, for example, Benes [19].)

Appendix I

Operator Algebra

The most important algebraic properties will be stated without proof. (The algebraic properties are often assumed in conventional operator theory, but must be derived from the set properties of functions in our theory.)

Consider the class C of operators that map the linear space $\underline{X}$ into itself.

The sum $(\underline{H} + \underline{K})$ of any two operators $\underline{H}$ and $\underline{K}$ corresponds to the system sum in Fig. 5, and is that operator in C for which $(\underline{H} + \underline{K})(x) = \underline{H}(x) + (\underline{K}x)$ for each x in $\underline{X}$.

Sums may be shown to be unique, commutative, *ie.*, $(\underline{H} + \underline{K}) = (\underline{K} + \underline{H})$, and associative, *i.e.*, $(\underline{G} + \underline{H}) + \underline{K} = \underline{G} + (\underline{H} + \underline{K})$ for arbitrary operators $\underline{G}, \underline{H}, \underline{K}$. Each operator $\underline{H}$ has a unique negative $(-\underline{H})$ with the property that $\underline{H} + (-\underline{H}) = \underline{0}$ in which $\underline{0}$ is the zero operator. The $\underline{0}$ operator is that operator in C which assigns $\underline{0}(x) =$ zero to each x in $\underline{X}$. $\underline{0}$ is unique.

The operator cascade $\underline{K} * \underline{H}$ of $\underline{K}$ following $\underline{H}$ corresponds to the system cascade of Fig. 5, and is defined to be that operator in C for which $(\underline{K} * \underline{H})(x) = \underline{K}(\underline{H}(x))$ for each x in $\underline{X}$. Cascades are unique, associative, but in general *not* commutative unless, for example, $\underline{K}$ and $\underline{H}$ are linear and time invariant or homogeneous and without memory.

Sums and cascades always have the right distributive property $(\underline{G} + \underline{H}) * \underline{K} = \underline{G} * \underline{K} + \underline{H} * \underline{K}$ but *not the left distributive property*.

$$\underline{K} * (\underline{G} + \underline{H}) = \underline{K} * \underline{G} + \underline{K} * \underline{H},$$

unless $\underline{K}$ is linear.

The identity operator $\underline{I}$ is that operator in C for which $\underline{I}(x) = x$ for all x in $\underline{X}$. $\underline{I}$ is unique.

Inverses

The inverse $\underline{H}^{-1}$ of any operator $\underline{H}$ is that operator in C which has the properties $\underline{H}^{-1} * \underline{H} = \underline{I}$, $\underline{H} * \underline{H}^{-1} = \underline{I}$.

The necessary and sufficient condition for $\underline{H}^{-1}$ to exist and be unique are that $\underline{H}$ be one to one and onto, *i.e.*, first, if $x_1 \neq x_2$ then $\underline{H}(x_1) \neq \underline{H}(x_2)$ for each x_1 and x_2 in $Do(\underline{H})$, second, for each y in $Ra(\underline{H})$ there exists an x in $Do(\underline{H})$ with the property that $y = \underline{H}(x)$.

Algebraic Manipulations

Our algebra has the convenient property that expressions derived from each other, by finite combinations of additions and of cascadings with invertible operators, imply each other. Thus implicit equations may be rendered explicit with concern only for the existence of inverses.

Appendix II

Exponential Iteration Theorem

The exponential iteration is general enough to be useful for unstable systems. Since these tend to "blow up" towards infinity, attention must necessarily be restricted to a finite time interval, such as $[0, T]$. The linear space of normed, real-valued functions of this interval will be denoted by $\underline{X}_T$, and the norm of x restricted to any interval $[0, t]$ will be denoted by $\| x, t \|$. The norm is left unspecified, but $\underline{X}_T$ is assumed to be a Banach space; this assumption is fulfilled in the first and second norms.

Theorem

The feedback equation $\underline{G} = \underline{H} * (\underline{I} - \underline{G})$, in which all the operators map the Banach space $\underline{X}_T$ into itself, has a unique solution for $\underline{G}$ that may be found as the limit of the iteration (22), which converges uniformly on $[0, T]$, provided that $\underline{H}$ satisfies the inequality

$$\| \underline{H}(x) - \underline{H}(y), t \| \leq h \int_0^t \| x - y, t_1 \| \, dt_1 \qquad (23)$$

for all x, y, in $\underline{X}_T$ and for all t in $[0, T]$; h is any constant; it is assumed for simplicity that $\underline{H}(0) = 0$. [Eq. (23) is satisfied by the system in Fig. 17 with $h = g_1(\underline{D})$ in the first norm.]

Proof

The sequence is shown to be Cauchy. The distance between any two terms is bounded by means of the triangle inequality,

$$\| \underline{G}_m(x) - \underline{G}_n(x), t \| \leq \sum_{i=0}^{m-n-1} \| \underline{G}_{n+i+1}(x) - \underline{G}_{n+i}(x), t \| \qquad m > n = 0, 1, 2, \cdots \qquad (24)$$

where $\mathcal{G}_0 = 0$. Each term of the series in (24) can be bounded by the preceding term;

$$\| \mathcal{G}_l(x) - \mathcal{G}_{l-1}(x), t \|$$

$$= \| \mathcal{H} * (I - \mathcal{G}_{l-1})(x) - \mathcal{H} * (I - \mathcal{G}_{l-2})(x), t \|$$

$$\geq h \int_0^t \| \mathcal{G}_{l-1}(x) - \mathcal{G}_{l-2}(x), t_1 \| \, dt_1, \; l = 2, 3, \cdots . \quad (25)$$

Here (22) and (23) have been used. Substituting (25) into itself $l - 2$ times, then using (23), and recalling that $\mathcal{H}(0) = 0$, we get

$$\| \mathcal{G}_l(x) - \mathcal{G}_{l-1}(x) \|$$

$$\leq h^{l-1} \int_0^t \cdots \int_0^{t_s} \| H(x), t_1 \| \, dt_1 \cdots dt_{l-1}$$

$$\leq \frac{(ht)^l}{l!} \| x, t \|, \quad l = 2, 3, \cdots . \quad (26)$$

Applying (26) or (23) to each term of the series in (24), and choosing $m > n \geq N \geq 0$ we get

$$\| \mathcal{G}_m(x) - \mathcal{G}_n(x), t \|$$

$$\leq \sum_{l=n+1}^{m} \frac{(ht)^l}{l!} \| x, t \| \leq \frac{(hT)^n}{n!} (\epsilon^{hT} - 1) \| x, T \|. \quad (27)$$

Since the right-hand side of (27) can be made arbitrarily small by making N large enough, the sequence is Cauchy. Since $\mathcal{X}_T$ is complete the sequence is uniformly convergent on $[0, T]$ and we may write $\mathcal{G}(x) = \lim_{n\to\infty} \mathcal{G}_n(x)$. It may be established that $\mathcal{G}(x)$ is a solution, *i.e.*, that

$$\mathcal{G} - \mathcal{H} * (I - \mathcal{G}) = \underline{0}$$

and that it is unique, *i.e.*, if $\mathcal{G}'$ is any other solution then $\mathcal{G} - \mathcal{G}' = 0$, by bounding the differences and showing that the bounds can be made arbitrarily small. A bound on $\mathcal{G}(x)$ is obtained by setting $n = N = 0$.

Bibliography

[1] V. Volterra, "Sopra le Funzioni che Dipendose de altre funzioni, R.C." *Accad. Linci*, vol. 3, pp. 97–105, 141–146, 153–158; 1887.

[2] N. Wiener, "Nonlinear Problems in Random Theory," The Technology Press of the Massachusetts Institute of Technology, Cambridge, Mass., and John Wiley and Sons, Inc., New York, N. Y.; 1958.

[3] L. A. Zadeh, "A contribution to the theory of nonlinear systems," *J. Franklin Inst.*, vol. 255, pp. 387–408; 1953.

[4] A. G. Bose, "A Theory of Nonlinear Systems," M.I.T. Research Lab. of Electronics, Cambridge, Mass., Tech. Rept. No. 309; May 15, 1956.

[5] J. F. Barrett, "The Use of Functionals in the Analysis of Non-linear Physical Systems," Ministry of Supply, Great Britain, Statistical Advisory Unit Rept. 1/57; 1957.

[6] M. B. Brilliant, "Theory of the Analysis of Nonlinear Systems, M.I.T. Research Lab. of Electronics, Cambridge, Mass., Tech. Rept. No. 345; March 3, 1958.

[7] D. A. George, Continuous Nonlinear Systems, M.I.T. Research Lab. of Electronics, Cambridge, Mass., Tech. Rept. No. 355; July 24, 1959.

[8] G. Zames, "Nonlinear Operators—Cascading, Inversion and Feedback." M.I.T. Research Lab. of Electronics, Cambridge, Mass., Quarterly Progress Rept. No. 53, April 15, 1959, pp. 93–107.

[9] ——, "Conservation of Bandwidth in Nonlinear Operations," M.I.T. Research Lab. of Electronics, Cambridge, Mass., Quarterly Progress Rept. No. 55, pp. 98–109; October 15, 1959.

[10] ——, "Realizability of Nonlinear Filters and Feedback Systems," M.I.T. Research Lab. of Electronics, Cambridge, Mass., Quarterly Progress Rept. No. 56, pp. 137–143; January 15, 1960.

[11] ——, "Nonlinear Operators for System Analysis," M.I.T. Research Laboratory of Electronics, Cambridge, Mass., Tech. Rept. No. 370; August 25, 1960.

[12] A. N. Kolmogoroff and S. V. Fomin, "Elements of the Theory of Functionals," Graylock Press, Rochester, N. Y., vol. 1; 1957.

[13] L. A. Lusternik and V. J. Sobolev, "Elements of Functional Analysis," Gordon and Breach, New York, N. Y.; 1961.

[14] L. V. Kantorovic, "The method of successive approximations for functional equations," *Acta. Math.*, vol. 71, pp. 63–97; 1939.

[15] C. A. Desoer, "Nonlinear distortion in feedback amplifiers," IRE Trans. on Circuit Theory, vol. CT-9, pp. 2–6; March, 1962.

[16] V. V. Nemytskii and V. V. Stepanov, "Qualitative Theory of Differential Equations," Princeton University Press, Princeton, N. J.; 1960.

[17] R. E. Kalman, Y. C. Ho, and K. S. Narendra, "Controllability in Linear Dynamical Systems," in "Contributions to Differential Equations," 1961.

[18] G. C. Newton, Jr., L. A. Gould, and J. F. Kaiser, "Analytical Design of Linear Feedback Controls," John Wiley and Sons, Inc., New York, N. Y., p. 352; 1957.

[19] V. E. Benes, "A nonlinear integral equation from the theory of servomechanisms," *Bell Sys. Tech. J.*, pp. 1309–1321; September, 1961.

Feedback and Optimal Sensitivity: Model Reference Transformations, Multiplicative Seminorms, and Approximate Inverses

G. ZAMES, FELLOW, IEEE

Abstract—**In this paper, the problem of sensitivity reduction by feedback is formulated as an optimization problem and separated from the problem of stabilization. Stable feedback schemes obtainable from a given plant are parameterized. Salient properties of sensitivity reducing schemes are derived, and it is shown that plant uncertainty reduces the ability of feedback to reduce sensitivity.**

The theory is developed for input–output systems in a general setting of Banach algebras, and then specialized to a class of multivariable, time-invariant systems characterized by $n \times n$ matrices of H^∞ frequency response functions, either with or without zeros in the right half-plane.

The approach is based on the use of a *weighted seminorm* on the algebra of operators to measure sensitivity, and on the concept of an *approximate inverse*. Approximate invertibility of the plant is shown to be a necessary and sufficient condition for sensitivity reduction. An indicator of approximate invertibility, called a *measure of singularity*, is introduced.

The measure of singularity of a linear time-invariant plant is shown to be determined by the location of its right half-plane zeros. In the absence of plant uncertainty, the sensitivity to output disturbances can be reduced to an optimal value approaching the singularity measure. In particular, if there are no right half-plane zeros, sensitivity can be made arbitrarily small.

The feedback schemes used in the optimization of sensitivity resemble the lead–lag networks of classical control design. Some of their properties, and methods of constructing them in special cases are presented.

I. Introduction

In this paper we shall be concerned with the effects of feedback on uncertainty, where uncertainty occurs either in the form of an additive disturbance d at the output of a linear plant P (Fig. 1), or an additive perturbation in P representing "plant uncertainty." We shall approach this subject from the point of view of classical sensitivity theory, with the difference that feedbacks will not only reduce but actually optimize sensitivity in an appropriate sense.

The theory will be developed at two levels of generality. At the higher level, a framework will be sought in which the essence of the classical ideas can be captured. To this end, systems will be represented by mappings belonging to a normed algebra. The object here is to obtain general answers to such questions as: how does the usefulness of feedback depend on plant invertibility? are there measures

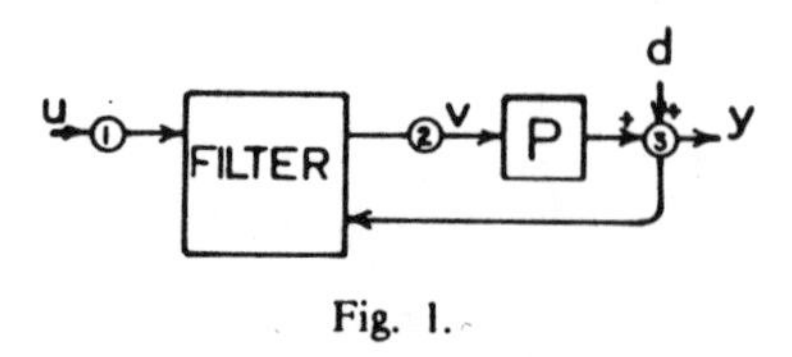

Fig. 1.

of sensitivity or plant uncertainty that are natural for optimization? how does plant uncertainty affect the possibility of designing a feedback scheme to reduce plant uncertainty?

At a more practical level, the theory will be illustrated by simple examples involving single variable and multivariable frequency responses. The questions here are: can the classical "lead–lag" controllers be derived from an optimization problem? How do RHP (right half-plane) zeros restrict sensitivity? in multivariable systems without RHP zeros, can sensitivity be made arbitrarily small, and if so how?

A. *Motivation*

A few observations might serve to motivate this reexamination of feedback theory.

One way of attenuating disturbances is to introduce a filter of the WHK (Wiener–Hopf–Kalman) type in the feedback path. Despite the unquestioned success of the WHK and state-space approaches, the classical methods, which rely on lead–lag "compensators" to reduce sensitivity, have continued to dominate many areas of design. On and off, there have been attempts to develop analogous methods for multivariable systems. However, the classical techniques have been difficult to pin down in a mathematical theory, partly because the purpose of compensation has not been clearly stated. One of our objectives is to formulate the compensation problem as the solution to a well defined optimization problem.

Another motivating factor is the gradual realization that classical theory is not just an old-fashioned way of doing WHK, but is concerned with a different category of mathematical problems. In a typical WHK problem, the quadratic norm of the response to a disturbance d is minimized by a projection method (see Sections III′-A′ and IV-C); in a deterministic version, the power spectrum

Manuscript received October 8, 1979; revised December 4, 1980 and December 17, 1980. Paper recommended by A. Z. Manitius, Past Chairman of the Optimal Systems Committee. An earlier version of this paper [23] was presented the 17th Allerton Conference, October 1979.

The author is with the Department of Electrical Engineering, McGill University, Montreal, P.Q., Canada.

Reprinted from *IEEE Trans. Automat. Contr.*, vol. AC-26, no. 2, pp. 301–320, Apr. 1981.

$|\hat{d}(j\omega)|$ is a *single*, *known* vector in, e.g., the space $L_2(-\infty,\infty)$; in stochastic versions, d belongs to a *single* random process of *known* covariance properties. However, there are many practical problems in which $|\hat{d}(j\omega)|$ is unknown but belongs to a prescribed set, or d belongs to a class of random processes whose covariances are uncertain but belong to a prescribed set. For example, in audio design, d is often one of a set of narrow-band signals in the 20–20K Hz interval, as opposed to a single, wide-band signal in the same interval. Problems involving such more general disturbance sets are not tractable by WHK or projection techniques. In a feedback context, they are now usually handled by empirical methods resembling those of classical sensitivity. One objective here is to find a systematic approach to problems involving such sets of disturbances.

Another observation is that many problems of plant uncertainty can be stated easily in the classical theory, e.g., in terms of a tolerance-band on a frequency response as in [2], but are difficult to express in a linear-quadratic-state-space framework. One reason for this is that frequency-response descriptions and, more generally, input–output descriptions preserve the operations of system addition and multiplication, whereas state-space descriptions do not. Another reason is that the quadratic norm is hard to estimate for system products (see Sections III′-A′ and IV-B1), whereas the induced norm (or "gain") that is implicit in the classical theory is easier to estimate. We would like to exploit these advantages in the study of plant uncertainty.

Finally, sensitivity theory is one of the few tools available for the study of organization structure: feedback versus open-loop, aggregated versus disaggregated, etc. For example, feedback reduces complexity of identification roughly for the same reason that it reduces sensitivity [12], [13]. However, it is hard to draw definitive conclusions about the effects of organization without some notion of optimality, and such a notion is missing in the old theory.

B. *Weighted Seminorms and Approximate Inverses*

One way of defining the optimal sensitivity of a feedback system, and of addressing some of the issues mentioned in Section I-A, is in terms of an induced norm of the sensitivity operator. However, it will be shown in Section III′-B′ that the primary norm of an operator in a normed algebra is useless for this purpose. Perhaps that is why operator norm optimization has not been pursued extensively in the past.

Instead, we shall introduce an auxiliary "weighted" seminorm, which retains some of the multiplicative properties of the induced norm, but is amenable to optimization. Plant uncertainty will be described in terms of belonging to a sphere in the weighted seminorm.

Approximate invertibility of the plant is one of the features which distinguishes control from, say, communication problems. We shall define the concept of an approximate inverse under a weighted seminorm, and show that sensitivity reduction is possible if there is such an inverse.

C. *Background*

Many of the ideas in this paper are foreshadowed in the classical theory [1], [2] of single-input single-output convolution systems, especially as presented by Horowitz [2], who derived various limits on sensitivity imposed by the plant, and stressed the need to consider plant uncertainty in design. The author posed the feedback problem in a normed algebra of operators on a Banach space, and introduced [4], [5] perturbation formulas of the type

$$(I-P)^{-1}-(I-P_0)^{-1}=(I-P)^{-1}(P-P_0)(I-P_0)^{-1} \tag{1.1}$$

which were used to show that high-gain feedback reduces the sensitivity of linear amplifiers to large nonlinear perturbations [3]–[5]. Desoer studied a related problem in [6], and recently [7] has obtained results for the case of P and P_0 both nonlinear (also see footnote 8). Perkins and Cruz [8] used perturbation formulas similar to (1.1) to calculate the sensitivity of linear multivariable systems. Porter [9] posed various sensitivity problems in Hilbert space, and in a paper with Desantis [10] obtained circle type conditions for sensitivity reduction. Willems [11] has stressed the Banach algebraic aspects of feedback theory.

In [1]–[10], the disturbance is either a fixed vector, or lies in some band of frequencies, and sensitivity is measured in terms of an output norm, as opposed to an induced operator norm. The approach of using weighted operator norms, and relating optimal sensitivity to weighted invertibility via a fractional transformation was used in [12], but has since been reworked and expanded.

D. *Two Problems*

We shall be concerned with the system of Fig. 1. Here, P is a given *plant* with a single (possibly multivariable) input v accessible to control, and an output y to which a disturbance d, not accessible to control, has been added. The plant input v is generated by a filter whose only inputs consist of observations on the plant output y and a reference input u. Two types of problems will be considered.

Problem 1—Disturbance Attenuation: This problem will be the subject of Sections V–VII. Suppose that $u=0$. The input–output behavior of the system between the nodes (2,3) can be modeled by the flowgraph of Fig. 2, which consists of the plant P and a single additional operator F in the feedback path. The disturbance d is uncertain in the sense that it can be any one of a set of disturbances. Initially (through Section VI) P is assumed to be known exactly, but later (Section VII) to be uncertain. We would like to *characterize the feedback operators F which attenuate the response y to d in some appropriately optimal sense, and examine the effects of uncertainty about P on disturbance attenuation.*

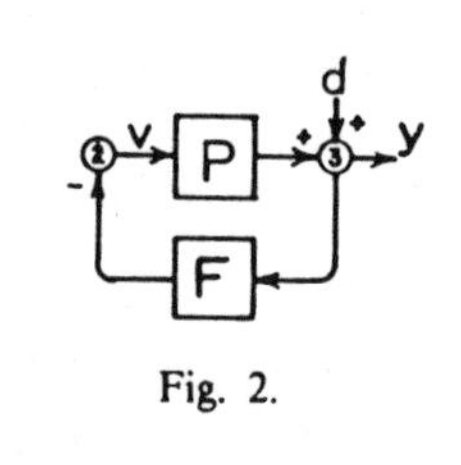

Fig. 2.

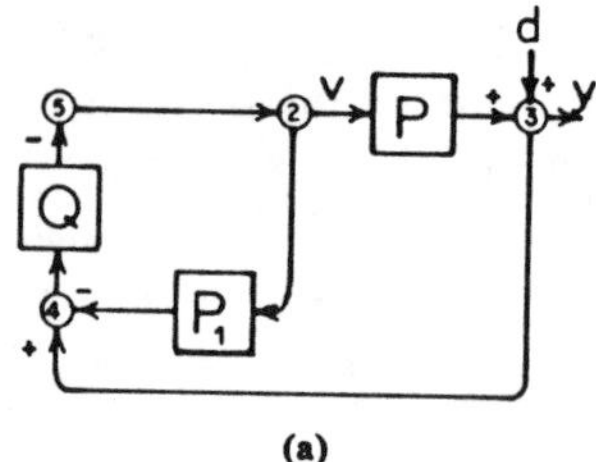

(a)

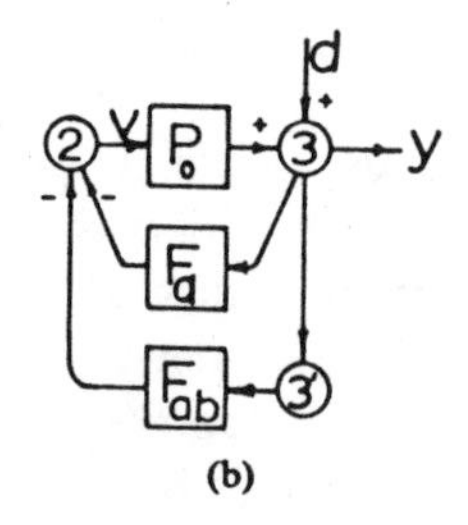

(b)

Fig. 3.

Problem 2—Plant Uncertainty Attenuation: (Problem 2 is the subject of Section VIII.) Suppose that $d=0$, and the plant P is uncertain to the extent that it can be any one of a "ball" of possible plants centered around some nominal value P_1. If the filter is linear, the behavior of the system between nodes (1,2,3) can be modeled by the flowgraph of Fig. 4. The filter can be characterized by a pair of operators (U, F). We would like to *find operators* (U, F) *which shrink the ball of uncertainty but leave the nominal plant invariant; to find bounds on the optimal shrinkage and to look at its dependence on plant uncertainty.*

E. Outline of the Paper

See Synopsis following Appendixes.

II. Spaces and Algebras of Systems

The purpose of this section is to specify the meaning which will be attached to the terms "frequency-response" and "linear system," and to summarize their properties for later use.

A feature of the input–output approach is that systems can be added, multiplied by other systems or by scalars, and the sums or products obtained are still systems, i.e., they form an *algebra*. Frequently, it will be assumed that the largest amplification produced by a system can be measured by a norm, typically the maximum frequency response amplitude over some region of analyticity; under this assumption the algebra of systems becomes a *normed algebra*. Normed algebras provide the natural setting for the study of system interconnections such as feedback. Their elementary properties will be used freely here, and

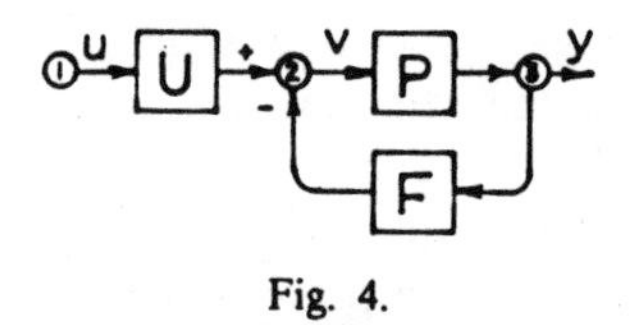

Fig. 4.

may be found in such texts as Naimark [19]. Occasionally, it will be assumed that a normed algebra is a *Banach algebra*, i.e., has the property that every convergent sequence of elements of the algebra has a limit in the algebra.

It will be assumed that all linear spaces and algebras are over the real field.

A. Algebras of Frequency Response Systems

The frequency response of a stable, causal, linear time-invariant system is a function analytic in the right-half of the complex plane. An accepted setting for such functions involves the H^p Hardy spaces [15], which we shall employ with some modifications to accommodate unstable systems.

The algebra H_e^∞ consists of functions $\hat{p}(\cdot)$ of a complex variable $s=\sigma+j\omega$, each of which is analytic in some open half-plane $\mathrm{Re}(s)>\sigma_p$ possibly depending on $\hat{p}(\cdot)$, and is bounded there, i.e., $\hat{p}(s)\leqslant$const. for $\mathrm{Re}(s)>\sigma_p$. The functions in H_e^∞ will be referred to as *causal* frequency responses. If $\hat{p}$ is in H_e^∞, then the domain of definition of $\hat{p}$ can be extended by analytic continuation to a unique, maximal, open half-plane of analyticity $\mathrm{Re}(s)>\sigma_{pm}$, where $\sigma_{pm}\leqslant\sigma_p$. In general, $\hat{p}$ need not be bounded on this maximal open RHP, but if it is, then it can further be extended to the boundary by the limit $\hat{p}(\sigma_{pm}+j\omega \triangleq \lim_{\sigma\to\sigma_{pm}}\hat{p}(\sigma+j\omega)$, which exists for almost all ω provided $\sigma+j\omega$ approaches the boundary nontangentially from the right. Assume that all functions in H_e^∞ have been so extended.

The algebra H^∞ (of *stable causal* frequency responses) consists of functions $\hat{p}$ of H_e^∞ for which $\sigma_p\leqslant 0$, i.e., the region of bounded analyticity includes the RHP. The norm $\|\hat{p}\|=\sup\{|\hat{p}(s)|: \mathrm{Re}(s)>0\}$ is defined on H^∞, making H^∞ a normed and, indeed, Banach algebra.

A *strictly proper* function in H_e^∞ satisfies the condition $\hat{p}(s)\to 0$ as $|s|\to\infty$ in $\mathrm{Re}(s)\geqslant\sigma_p$. The symbols H_{e0}^∞ and H_0^∞ will denote the algebras of strictly proper frequency responses in H_e^∞ and H^∞, respectively. By a straightforward application of the maximum modulus principle, the normed algebra H_0^∞ of strictly proper stable frequency responses has the property that $\|\hat{p}\|=\mathrm{ess\,sup}\{|\hat{p}(j\omega)|: \omega \text{ real}\}$ for any $\hat{p}$ in H_0^∞, i.e., the norm can be computed from $j\omega$-axis measurements.

Spaces of Inputs and Outputs: For any integer $1\leqslant q<\infty$, the linear space H_e^q consists of functions $\hat{u}(\cdot)$ of a complex variable, each $\hat{u}$ being analytic in some open half-plane $\mathrm{Re}(s)>\sigma_u$ in which the restriction of $\hat{u}$ to any vertical line is in L^q and $\int_{-\infty}^{\infty}|\hat{u}(\sigma+j\omega)|^q d\omega\leqslant$const. for all $\sigma>\sigma_u$. Again, the domain of definition of each $\hat{u}$ in H_e^q is extended by analytic continuation to a maximal open RHP of analyticity and then, if $\hat{u}$ is L^q-bounded in this RHP, to its boundary by a nontangential limit. The space H^q consists of functions $\hat{u}$ in H_e^q for which $\sigma_u \leq 0$, and is a Banach space under the norm $\|\hat{u}\| = \sup_{\sigma>0}\{\int_{-\infty}^{\infty}|\hat{u}(\sigma+j\omega)|^q d\omega\}^{1/q}$.

Inputs and outputs will belong either to H_e^∞ or one of the H_e^q spaces, $1 \leq q < \infty$. In the special case $q=2$ it follows from the Paley–Wiener theory that every function of H^2 is a Laplace transform of a time function in $L^2(0,\infty)$ and vice versa. In the general case of H^q, some functions can be viewed as transforms of time functions, and the others as frequency functions that do not appear in physical applications and can be disregarded.

Frequency Response Operators: Let q be any integer, $1 \leq q \leq \infty$, which will be held fixed. For any causal frequency response $\hat{p}(\cdot)$ in H_e^∞ an operator P: $H_e^q \to H_e^q$ is defined by the multiplication $\widehat{Pu}(s) = \hat{p}(s)\hat{u}(s)$. P will be called a causal frequency response operator, and the algebra of all such operators will be denoted by the bordered capital $\mathbb{H}_e^\infty$. Similarly, for each of the algebras of frequency responses H^∞, H_{e0}^∞, and H_0^∞, an algebra of operators mapping H_e^q into itself is defined and denoted by the corresponding bordered capital, i.e., $\mathbb{H}^\infty$, $\mathbb{H}_{e0}^\infty$, or $\mathbb{H}_0^\infty$. In the case of normed frequency response algebras, the corresponding operator algebras are similarly normed, e.g., $\|P\|_{\mathbb{H}^\infty} = \|\hat{p}\|_{H^\infty}$.

The stable operators in $\mathbb{H}^\infty$ map H^q, which is a proper subspace of H_e^q, into H^q, and are in fact completely determined by their behavior on H^q. They can therefore be represented by their restrictions of the form P: $H^q \to H^q$. In sections devoted entirely to stable systems, we shall concentrate on operators of the form P: $H^q \to H^q$ without distinguishing them as restrictions of operators on H_e^q.

B. *More General Algebras of Systems*

We would like to take an axiomatic approach to the problem of sensitivity reduction by feedback, i.e., to single out the relevant properties of linear systems and postulate them as axioms. For example, the related properties of causality, realizability, and strong or strict causality have definitions [see, e.g., [14] and [11]) reflecting the fact that the response to a sudden input to a physical system can not anticipate the input, and cannot occur instantaneously. These properties of physical systems preclude the pathological phenomena associated with instantaneous response around a feedback loop, and ensure that the feedback operator $(I+P)^{-1}$ is well defined. However, these details are not relevant here. The only items of interest are that causal systems form an algebra of mappings, and that strictly causal systems form a subalgebra whose salient feature is the existence of the inverse $(I+P)^{-1}$ for all of its members, i.e., a "radical." Accordingly we postulate the following.

$\mathcal{X}$ is a linear space whose elements will be called *inputs* or *outputs*. $\mathbf{A}$ is a linear algebra of linear mappings P: $\mathcal{X} \to \mathcal{X}$ with identity I, whose elements will be called *causal operators*. $\mathbf{A}_s$ is a radical of $\mathbf{A}$, i.e.,[1] a proper nontrivial subalgebra of $\mathbf{A}$ with the property that for any P in $\mathbf{A}_s$, the inverse $(I+P)^{-1}$ exists in $\mathbf{A}$, and for any F in $\mathbf{A}$ the products PF and FP are in $\mathbf{A}_s$. The elements of $\mathbf{A}_s$ will be called *strictly causal operators.* (The concept of an algebra of "realizable" systems was introduced by the author in [14]; the related notions of strong causality of Willems [11], and later strict causality of Porter, Saeks, and Desantis are compared in a paper of Feintuch [20].)

An example of the space $\mathcal{X}$ of inputs and the algebra $\mathbf{A}$ of causal operators is provided by the space H_e^q of (transforms of) inputs and the algebra $\mathbb{H}_e^\infty$ of causal frequency response operators. In this context the algebra $\mathbb{H}_{e0}^\infty$ of strictly proper frequency response operators is an example of an algebra $\mathbf{A}_s$ of strictly causal operators. Henceforth, we shall refer to the strictly proper operators as strictly causal.

In the case of stability, we shall need the fact that a stable input–output system produces a finite amplification of inputs that can be measured by a suitable norm, and that stable systems form an algebra (see, e.g. [5]). Accordingly, we postulate the following.

$\mathcal{B}$ is a Banach subspace of $\mathcal{X}$ whose elements will be called *bounded* inputs or outputs [for example, H^2 or $L_2(0,\infty)$]. $\mathbf{B}$ is a normed subalgebra of $\mathbf{A}$ containing the identity I, whose elements will be called *stable* causal operators, under the following assumption: the norm of any P in $\mathbf{B}$ is the $\mathcal{B}$-induced norm, that is, $\|P\| \triangleq \sup\{\|Pu\|/\|u\|: u \text{ in } \mathcal{B},\ u \neq 0\}$, the sup being finite.[2]

If $\mathbb{H}_e^\infty$ is taken as an example of $\mathbf{A}$, then $\mathbb{H}^\infty$ is an example of $\mathbf{B}$.

$\mathbf{B}_s$ is the subalgebra of $\mathbf{B}$ obtained by intersecting $\mathbf{A}_s$ and $\mathbf{B}$, consisting of *strictly-causal stable* operators. It should be noted that $\mathbf{B}_s$ is not a radical of $\mathbf{B}$, as P stable does not imply that $(I+P)^{-1}$ is stable.

For the purpose of estimating the effects of small perturbations, it will be assumed that $\mathbf{B}_s$ has the *small-gain property*, i.e., for any P in $\mathbf{B}_s$, if $\|P\| < 1$ then $(I+P)^{-1}$ is in $\mathbf{B}$. If $\mathbf{B}$ is complete, i.e., a Banach space, this assumption is redundant for then the series $I - P + P^2 - \cdots$ converges to the inverse in $\mathbf{B}$ of $(I+P)$. However, we have applications in mind in which completeness of $\mathbf{B}$ is replaced by other assumptions.

A frequency response $\hat{p} \in H^\infty$ which is not strictly proper can not be realized exactly, but can be approximated by a sequence of strictly proper responses of the form $n(s+n)^{-1}\hat{p}(s)$, $n = 1, 2, \cdots$. The sequence $n(s+n)^{-1}$ is an example of an "identity sequence." More generally, identity sequences will be used to construct strictly proper approximations to improper responses, and are defined as follows: an *identity sequence* $\{I_n\}_{n=1}^\infty$ for $\mathbf{B}$ is a sequence of operators in $\mathbf{B}$ with the property that for any F in $\mathbf{B}$ the sequences $\|I_n F - F\|$ and $\|F I_n - F\|$ approach 0 as $n \to \infty$. It is assumed that $\mathbf{B}_s$ contains an identity sequence for $\mathbf{B}$.

The following well-known (cf. Naimark [19, p. 162]) properties of a normed algebra will be crucial in many

[1] The properties of radicals are discussed in Naimark [19, p. 162].

[2] Whenever the norm of x is not identified by a subscript, it should be taken to be the principal norm of the space to which x belongs.

parts of the paper. For convenience they are proved in Appendix I.

Let P and Q be in B.

Proposition 2.1: a) If $(I+PQ)^{-1}$ is in B, then $(I+QP)^{-1}$ is in B, and the formula $P(I+QP)^{-1}=(I+PQ)^{-1}P$ is valid. b) If R is a radical in B and P is in R, then P has no inverse in B. (Strictly causal operators have no inverses in $\mathbf{B}$.) c) If P and $(I+P)^{-1}$ are in B and $\|P\|<1$, then $\|(I+P)^{-1}\|\leqslant(1-\|P\|)^{-1}$.

III. Feedback Decomposition: Stabilizing and Stabilized Stages

We proceed to derive a decomposition principle to be employed in disturbance attenuation. Suppose that there is no plant uncertainty, and that the plant and feedback are constrained not to be simultaneously unstable. Under these hypotheses, any closed-loop stable feedback design can be decomposed into two stages: a first stage involving plant stabilization (which can be omitted for stable plants); and a second stage, involving a model reference scheme in which only stable elements are used, and which is automatically closed-loop stable. The choice of a stabilizing stage is independent of, and does not prejudice the choice of the second stage. Having established this fact, we shall be free to concentrate on the second stage of the disturbance attenuation problem under the condition that the plant is stable (or has been stabilized), without loss of generality.

Consider the system of Fig. 2. The plant input v, output y, and disturbance d are all in $\mathfrak{X}$ and satisfy the equations

$$y=Pv+d \tag{3.1a}$$

$$v=-Fy \tag{3.1b}$$

in which P and F are operators in $\mathbf{A}_s$ [see Remark 3.1c)]. We shall refer to (3.1) as a *feedback scheme* with plant P and feedback F. Since P is strictly causal, the inverse $(I+PF)^{-1}$ exists in $\mathbf{A}$. Therefore, for each d in $\mathfrak{X}$, (3.1) have unique solutions for v and y in $\mathfrak{X}$, given by the formulas

$$y=(I+PF)^{-1}d \tag{3.2a}$$

$$v=-F(I+PF)^{-1}d. \tag{3.2b}$$

Let $K_{32}: \mathfrak{X}\to\mathfrak{X}$ denote the "closed-loop" operator mapping d to v. K_{32} is an operator in $\mathbf{A}$ given by $K_{32}=-F(I+PF)^{-1}$.

The flowgraphs in this paper are simple, and will be approached informally in order to avoid lengthy definitions. Expressions for some of the subsidiary c.l. (closed-loop) operators, which can be found by inspection, will be listed without derivation as needed.

For a system to be physically realizable on an infinite time interval, it is usual to postulate [14] that all c.l. input–output operators must be stable, though "open-loop" operators such as the plant P and feedback F may be unstable. The set of c.l. operators for (3.1) consists of: $K_{22}=(I+FP)^{-1}$, $K_{23}=PK_{22}$, $K_{33}=(I+PF)^{-1}$, and K_{32} specified above. Accordingly, the feedback scheme (3.1) will be called *c.l. stable* if K_{ij} is in $\mathbf{B}$ for $i,j=2$ or 3.

We shall be interested in situations in which P is at or near some nominal value P_1, and the feedback F appears as an operator variable in an optimization problem whose object is to minimize response to d. Unstable operator variables are difficult to handle, and so our first step will be to show that F has an equivalent realization in terms of a stable operator.

A. *The Model Reference Transformation*

The flowgraph of Fig. 3 is described by the equations

$$y=Pv+d \tag{3.3a}$$

$$v=-Q(y-P_1v) \tag{3.3b}$$

in which y, v, and d are in $\mathfrak{X}$, and P, P_1, Q are in $\mathbf{A}_s$. Equation (3.3) will be called a *model reference scheme* with *comparator* Q, as the output of the plant P with disturbance d added is compared to the output of a model P_1 of the plant without disturbance, and the difference actuates Q.

The two sets of equations (3.1) and (3.3) are called *equivalent* iff every input–output triple (d,v,y) in $\mathfrak{X}^3$ satisfying (3.1) satisfies (3.3), and vice versa. Their equivalence will be established under the assumption that the equations

$$Q=F(I+P_1F)^{-1} \tag{3.4a}$$

$$F=Q(I-P_1Q)^{-1} \tag{3.4b}$$

hold. If either equation in (3.4) is valid, then so is the other, and

$$(I-P_1Q)=(I+P_1F)^{-1}. \tag{3.5}$$

To derive (3.5), suppose (3.4a) is valid. Therefore,

$$\begin{aligned}I-P_1Q&=I-P_1F(I+P_1F)^{-1}\\&=(I+P_1F)(I+P_1F)^{-1}-P_1F(I+P_1F)^{-1}\\&=(I+P_1F)^{-1}\end{aligned}$$

and (3.5) is true; here the expression $I=(I+P_1F)(I+P_1F)^{-1}$ and the distributive law for multiplication on the right was used. Equations (3.4a) and (3.5) can now be used to give the identities

$$Q(I-P_1Q)^{-1}=F(I+P_1F)^{-1}(I+P_1F)=F,$$

so (3.4b) is true as claimed. The converse proposition is proved similarly.

Assumption: For the present, and until the end of Section VI, assume that $P=P_1$, i.e., there is no plant uncertainty.

Theorem 1:

a) Any closed-loop stable feedback scheme (3.1) with stable plant $P\in\mathbf{B}_s$ and (not necessarily stable) feedback $F\in\mathbf{A}_s$ is equivalent to a model reference scheme whose branches are all stable, i.e., $Q\in\mathbf{B}_s$ and $P_1=P$, where F and Q are related by (3.4). Conversely, any model reference scheme with stable branches is closed-loop stable, and equivalent to a closed-loop stable feedback scheme subject to (3.4).

b) If (3.4) holds, and d and y satisfy either the feedback or model reference equations they satisfy the equation

$$y=(I-P_1Q)d. \tag{3.6}$$

Proof:

a) For any feedback scheme (3.1), if $(d, v, y)\in\mathfrak{X}^3$ satisfies (3.1) and F is given by (3.4b), then (d, v, y) satisfies (3.3), and conversely. Therefore, (3.1) is equivalent to (3.3). If the feedback scheme is c.l. stable, then Q must be stable as it equals $-K_{32}$. If, in addition, P is assumed stable, then all branches in (3.3) are stable, as claimed.

Conversely, by a similar argument, any model reference scheme (3.1) is equivalent to a feedback scheme (3.3). Suppose that the branches of (3.3), namely, P_1, Q, and $P=P_1$ are all stable. Then, all the c.l. operators of (3.3), namely, $\{K_{ij}\}_{i,j=2,3,4,5}$, must be stable because they can be expressed in terms of sums and products of the stable operators P, P_1, Q, and I. The last assertion follows from the expressions for the diagonal c.l. operators K_{ii} of (3.3), namely,

$$K_{22}=I-QP_1,\ K_{33}=I-P_1Q,\ K_{44}=K_{55}=I$$

and the fact, easily checked by inspection, that the remaining c.l. operators K_{ij}, $i\neq j$, are products of the K_{ii} by P, P_1, Q, or I. It follows that (3.3) is c.l. stable.

b) If $P=P_1$ and (d, y) satisfies (3.1) or (3.3), and (3.4) holds, then (3.6) is obtained by substitution of (3.5) into (3.2a). Q.E.D.

The operator $(I-P_1Q)$ appearing in (3.6) will reappear as a factor in most expressions for sensitivity. It will be called the *sensitivity operator* and denoted by E. For equivalent schemes $E=(I+P_1F)^{-1}$.

Remarks 3.1:

a) The model-reference scheme has some remarkable features. Unlike most feedback arrangements, it is a realization which cannot be made unstable by any choice of Q, at least for stable plants in the absence of plant uncertainty. Under these assumptions,[3] any allowable feedback law can be realized in the form of an equivalent model reference scheme, with the guarantee that all branches will be stable, and the closed-loop system automatically stable. *The design of Q, whether for small sensitivity or other purposes, can be accomplished without concern for closed-loop stability.*

In engineering applications, model reference schemes are realizable in principle, but may have undesirable features. For example, they may have high sensitivity to errors in the realization of Q. Unstable inner loops, obtained whenever $(I-P_1Q)^{-1}$ is unstable, may present reliability problems. Even then, the fractional transformation remains advantageous from the viewpoint of theory, as potentially unstable feedbacks F are replaced by stable operators Q. In later sections on plant uncertainty, the flowgraph interpretation of the model reference scheme will provide a convenient guide to perturbation analysis. It will also appear that model-reference schemes have a useful plant-invariance property.

b) Implicit in our notion of an allowable feedback is the view that each feedback realization involves a graph, and that although most of the internal details of the realization may be unimportant, closed-loop stability at *all* internal nodes is essential.

c) Theorem 1 holds even if F and Q are in $\mathbf{A}$ but not strictly causal. However, strict causality is a prerequisite for physical realizability, and will therefore have to be assumed in subsequent theorems.

B. Unstable Plants

The assumption in Theorem 1 that the plant P is stable will now be relaxed. Consider a plant $P_0\in\mathbf{A}_s$ with disturbance d at the output, which is unstable but for which there exists a *stabilizing feedback*, i.e., an operator $F_0\in\mathbf{A}_s$ which gives a c.l. stable feedback scheme on being fed back around P_0. The stabilized system can be incorporated in a model reference scheme, by letting P be the stabilized c.l. operator $P_0(I+F_0P_0)^{-1}$ and d be the stabilized disturbance $(I+P_0F_0)^{-1}d_0$, and Q (or F) can be selected as for a stable plant. At this point the question arises: "can F_0 be selected independently of Q (or F), or could the prior choice of F_0 prejudice the class of achievable systems?"

In general, the choices are not independent, even for stable plants, because the application of two unstable feedbacks in succession may give a result different from the application of a single feedback equal to their sum. Consider the following frequency response example in $\mathbf{H}_e^\infty$. Let $\hat{p}(s)=1$ and $\hat{f}_1(s)=\hat{f}_2(s)=s^{-1}$. The application of a single feedback $\hat{f}(s)$ equal to $\hat{f}_1(s)+\hat{f}_2(s)$ gives a c.l. stable feedback scheme, with c.l. responses 1, $(s+2)^{-1}$, and $s(s+2)^{-1}$. However, if the feedback is split into two branches, the c.l. response across either one of these branches is $(s+1)/s(s+2)$, i.e., the system is not c.l. stable. Popular belief notwithstanding, *c.l. stable systems do not form an additive group under feedback if the complete set of c.l. operators is considered.*

However, if feedbacks are constrained to be stable then choices are independent, as the following construction shows.

Let $P_0\in\mathbf{A}_s$ be an unstable plant which can be stabilized by either one of two feedbacks, F_a and F_b in $\mathbf{A}_s$, and label the resulting feedback schemes (a) and (b), respectively. We would like to find an operator $F_{ab}\in\mathbf{A}_s$ which on being fed back around scheme (a), as shown in Fig. 3(a), produces a *two-stage feedback scheme* equivalent to scheme (b). (Observe that the two-stage feedback scheme has extra nodes in the feedback branches to allow for the possibility of noise sources there.)

Proposition 3.2: If F_a and F_b are stable, then the stable feedback $F_{ab}\triangleq F_b-F_a$ makes the two-stage feedback scheme c.l. stable and equivalent to scheme (b).

Proof: The two-stage scheme is obviously equivalent to scheme (b), and is c.l. stable because its c.l. operators consist of: i) the c.l. operators K_{22}, K_{23}, K_{32}, and K_{33} of scheme (a) or scheme (b), which are stable by hypothesis, or ii) sums and products of the operators listed in i), and

[3] If P is unstable, the parameterization of c.l. stable schemes by a single operator Q is obviously still possible. However, some of our other conclusions, concerning existence of a feedback realization with stable elements, structural stability, or decomposition properties, may no longer be valid.

the operators F_a or F_b which are stable by hypotheses. Q.E.D.

The following *decomposition principle* can be obtained immediately from Theorem 1 and Proposition 3.2. Let $P_0 \in \mathbf{A}_s$ be any unstable plant stabilizable by a set of stable feedbacks in $\mathbf{B}_s$. *Any closed-loop stable feedback scheme employing a stable feedback around the plant P_0 is equivalent to a closed-loop stable scheme consisting of: i) a stabilizing feedback $F_0 \in \mathbf{B}_s$ which can be selected arbitrarily, followed by ii) a model reference scheme with stable operators Q and P_1.*

It follows that under our hypotheses,[3] and in particular under the assumption that plant and feedback are not simultaneously unstable, the problem of sensitivity reduction can be decomposed into two independent problems: stabilization followed by desensitization of a stable system. Henceforth, we shall confine ourselves to the second problem.

III′. Approaches to Feedback-Sensitivity Minimization

A′. Quadratic versus Induced Norms

The main properties of feedback cannot be deduced without some notion of uncertainty. Suppose that the disturbance d is uncertain but belongs to some subset $\mathcal{D}$ of possible disturbances in $\mathcal{X}$. From (3.6) it is clear that for disturbances to be attenuated, $(I-P_1Q)$ must be small on $\mathcal{D}$, i.e., Q must act as an approximate inverse of P_1 on $\mathcal{D}$. The various approaches to the disturbance attenuation problem are differentiated by the way in which uncertainty is described, and this approximate inversion is metricized and calculated.

A typical WHK approach in a deterministic version could be viewed as follows: $\mathcal{D}$ consists of the set of disturbances d in $L_2(0,\infty)$ possessing a single, fixed, known power spectrum $|\hat{d}(j\omega)|$ in $L_2(-\infty,\infty)$, and the object of design is to find a filter Q that minimizes the quadratic distance $\|d-P_1Qd\|_{L_2}$, where in general Q depends on $|\hat{d}(j\omega)|$. In the stochastic analog of this problem $\mathcal{D}$ is a random process characterized by probability-covariance functions and metricized by a quadratic norm. This description of uncertainty has certain limitations that we would like to circumvent, namely, the following.

1) The covariance properties of the random process must be known. In practice, they are often unknown elements of prescribed sets.

This is merely a limitation on the class of random processes for which WHK is valid. More serious from the point of view of feedback theory is the following observation.

2) The quadratic norm on plants employed in the WHK method lacks the multiplicative property $\|PQ\|_{L_2} \leq \|P\|_{L_2}\|Q\|_{L_2}$, and in general it may be difficult or impossible to estimate the norm of a product PQ from the norms of P and Q. The product norm $\|PQ\|_{L_2}$ may be large even though $\|P\|_{L_2}$ and $\|Q\|_{L_2}$ are small.

Consequently, if plant uncertainty is metricized by the quadratic norm, its propagation through products and inverses is hard to study. This is a serious limitation in feedback problems in which expressions such as $(I+PF)^{-1}$ play a major role.

By contrast, the "M_m" spec which is widely used in classical design measures the maximum frequency response magnitude, and is essentially the induced operator norm $\|P\| \triangleq \sup\{\|Px\|_{L_2}/\|x\|_{L_2}: x\in L_2\}$, which has the multiplicative property, and is therefore convenient to estimate in cascaded systems. By describing plant uncertainty in terms of a sphere of specified radius in a norm having such a multiplicative property, it is possible to obtain a general approach to problems involving disturbances/random processes which are unknown but belong to prescribed sets, as we shall see in Section IV.

Minimization of an induced norm in effect amounts to a minimax solution. Minimax methods do not necessarily represent uncertainty with greater fidelity than quadratic methods. However, the concern here is less with fidelity than with the ability to handle product systems.

B′. Constraints on the Norm of a Sensitivity Operator

It is natural then to try to pose sensitivity reduction problems in terms of the minimization of norm of the sensitivity operator, and to employ a norm having multiplicative properties. The primary norm of a Banach algebra has such properties, but the following propositions show it to be useless for this purpose.

Proposition 3.3: If P and Q are in a Banach algebra $\mathbf{B}$ and $\|I-PQ\|<1$, then PQ has an inverse in $\mathbf{B}$.

Proof: Denote $(I-PQ)$ by E. As $\|E\|<1$, the power series $I-E+E^2-\cdots$ converges to an operator which inverts $(I-E)$. As $(I-E)=PQ$, $(PQ)^{-1}$ exists. Q.E.D.

Proposition 3.3 has occasionally been interpreted as showing that invertibility is necessary for sensitivity reduction. This interpretation is empty. In fact, since strongly causal operators never have inverses in $\mathbf{B}$ [see Proposition 2.1b)], we have the following.

Corollary 3.4: If PQ is in $\mathbf{B}_s$, then $\|I-PQ\|\geq 1$.

It is impossible to make the sensitivity operator less than 1 in the original $\mathbf{B}$ norm. In H_0^∞ this simply means that the frequency response of PQ approaches 0 at infinite frequencies and $(I-PQ)$ approaches 1.

An obvious idea at this point is to make $(I-P_1Q)$ small in norm over some finite frequency band, i.e., over an invariant subspace. The next proposition shows that norms over invariant subspaces usually are not useful measures of sensitivity for optimization purposes. Let $\mathcal{B}_1$ be a subspace of $\mathcal{B}$, Π a projection operator onto $\mathcal{B}_1$, and suppose that $\mathcal{B}_1$ is invariant[4] under $\mathbf{B}$, i.e., $R\Pi=\Pi R\Pi$ for each R in $\mathbf{B}$. Let α denote the norm of $(I-P_1Q)$ restricted to the subspace $\mathcal{B}_1$ and optimized over all Q, i.e., $\alpha \triangleq \inf_{Q\in\mathbf{B}_s}\sup\{\|(I-P_1Q)\Pi d\|: d\in\mathcal{B} \text{ and } \|d\|=1\}$.

Proposition 3.5: For any P in $\mathbf{B}_s$, $\alpha=1$ or $\alpha=0$.

The proof is in Appendix I. If $\alpha=0$, sensitivity can be made arbitrarily small over the subspace $\mathcal{B}_1$. In practice there are special cases involving "minimum phase" systems

[4]More generally, Proposition 3.5 holds if $\mathbf{B}$ is replaced by any of its norm preserving extensions. Note that Π is not necessarily in the algebra $\mathbf{B}$ or causal.

in which solutions that approach $\alpha=0$ may be useful. More typically, this result is achieved at the expense of increasing the sensitivity without bound on complements of $\mathfrak{B}_1$; in such cases, the norm of a restriction of $(I-P_1Q)$ is not a candidate for minimization.

Corollary 3.4 and Proposition 3.5 delineate some of the peculiarities of the sensitivity optimization problem. In one form or another these peculiarities were recognized in the classical theory, and are probably the reason why it stopped short of optimization. We shall try to circumvent them by introducing an auxiliary (semi) norm to which they do not apply.

IV. Multiplicative Seminorms and Approximate Inverses

Uncertainty in a disturbance (or plant) in a linear space can be specified in terms of belonging to a ball of disturbances (or plants) centered at some nominal value, and of radius specified in some norm. Such a description of uncertainty may be cruder than a probabilistic description, but is usually more tractable in feedback problems.

One of the axioms of a norm asserts that only the zero element has zero norm. This axiom is often not needed, and with its elimination a norm is replaced by the slightly more general concept of a seminorm.

A ball in any seminorm can be shown to be a convex set.[4] Conversely, any convex set[5] in a linear space generates a seminorm (see Rudin [22, p. 24]) known as the Minkowski functional of that set. In linear spaces convex sets[4] of uncertainty can therefore always be described in terms of seminorms. We shall employ seminorms to obtain a systematic approach to such sets of uncertainty (cf. the objectives outlined in Section I-A).

In the next section, we shall define classes of left and right seminorms. To motivate the definitions, let us find seminorm descriptions for two disturbance sets which can be generated by the interaction of filters and certain "flat" disturbance sets.

Henceforth, W will denote a stable causal operator of unit norm, which will play the role of a weighting filter. For concreteness, W can be thought of as an operator in $\mathbf{H}_0^\infty$, $W: H^\infty \to H^\infty$, with response $\hat{w}(s)=k(s+k)^{-1}$. $\mathfrak{D}$ will denote a flat disturbance set (analogous to white noise) consisting of the unit ball in the space of inputs, in this case in H^∞:

$$\mathfrak{D}_1=\{d_1 \in H^\infty: |\hat{d}_1(j\omega)| \leq 1\}$$

whose elements are frequency functions of unknown but bounded magnitude. Consider two situations.

1) Let $\mathfrak{D}$ be the set mapped by W into flat disturbances, i.e., $W\mathfrak{D}=\mathfrak{D}_1$. $\mathfrak{D}$ can be described as the unit ball in the seminorm $\|\cdot\|_\ell$, defined on H^∞ by the equation $\|d\|_\ell \triangleq \|Wd\|$.

2) Let $\mathfrak{D}$ be the set into which W maps flat disturbances, i.e., $W\mathfrak{D}_1=\mathfrak{D}$. $\mathfrak{D}$ is also the set

$$\mathfrak{D}=\{d \in \text{range}(W): |\hat{d}(j\omega)| \leq |\hat{w}(j\omega)|\}.$$

$\mathfrak{D}$ can be described as the unit ball in the seminorm $\|\cdot\|_r$ defined on the range of W (which is a proper subspace of H_0^∞) by the equation $\|d\|_r \triangleq \|W^{-1}d\|$.

There are many engineering problems in which the apriori information about disturbances is in the form of an upper bound to the magnitudes of their possible frequency responses. The seminorm description 2) is natural for such problems, and 1) occurs in inverse problems.

The seminorm $\|\cdot\|_r$ employs an up-weighting, and $\|\cdot\|_\ell$ employs a down-weighting. These two examples generalize into the notion of left and right seminorms, defined as follows.

A. *Seminorms for Inputs and Outputs*[6]

Let $\mathfrak{Y}$ be a $\|\cdot\|$-normed linear space. Let $\|\cdot\|_\ell$ be any seminorm defined on all of $\mathfrak{Y}$, and $\|\cdot\|_r$ a seminorm defined on some nontrivial subspace $\mathfrak{Y}_r$ of $\mathfrak{Y}$. The seminorm $\|\cdot\|_r$ is said to *dominate* $\|\cdot\|_\ell$ iff $\|y\|_\ell \leq \|y\|_r$ for all y in $\mathfrak{Y}_r$; this dominance is denoted by $\|\cdot\|_\ell \leq \|\cdot\|_r$.

Definition: A *left seminorm* is any seminorm defined on all of $\mathfrak{Y}$ with the property that $\|\cdot\|_\ell \leq \|\cdot\|$. A *right seminorm* is any seminorm defined on a nontrivial subspace $\mathfrak{Y}_r$ of $\mathfrak{Y}$ with the property that $\|\cdot\| \leq \|\cdot\|_r$.

For example, if W is any $\mathbf{H}^\infty$ filter of unit norm, the expressions $\|y\|_\ell \triangleq \|Wy\|$ and $\|y\|_r \triangleq \|W^{-1}y\|$ define left and right seminorms, on H^∞ and the range of W, respectively. The range of W is a subspace of H^∞, proper whenever $\hat{w}(s)$ has zeros in the right half-plane or at ∞.

B. *Weighted Seminorms for Plants*

Definition: A weighted seminorm is any seminorm $\|\cdot\|_W$ on the $\mathbf{B}$ with the property that $\|\cdot\|_W \leq \|\cdot\|$.

The terms "weighted" and "left" are synonymous. We shall use the term "weighted" to distinguish the left seminorms on $\mathbf{B}$ used as measures of plant sensitivity from the others.

A weighted seminorm on $\mathbf{B}$ is *induced* by a pair $(\|\cdot\|_\ell, \|\cdot\|_r)$, where $\|\cdot\|_\ell$ is a left seminorm on the space $\mathfrak{B}$ (of outputs), and $\|\cdot\|_r$ is a right seminorm on a subspace $\mathfrak{B}_r \subset \mathfrak{B}$ (of inputs), iff $\|\cdot\|_W$ is defined for $A \in \mathbf{B}$ by the equation

$$\|A\|_W = \sup\{\|Au\|_\ell / \|u\|_r: u \in \mathfrak{B}_r \text{ and } \|u\|_r \neq 0\}.$$

It follows that $\|I\|_W \leq 1$.

In control problems, weightings are often introduced by filters, which act on disturbances either before entering a plant or after leaving it. For example, let W_ℓ and W_r be linear mappings in $\mathfrak{X} \times \mathfrak{X}$, each of unit $\mathfrak{B}$-induced norm. A

[5]Satisfying the following additional assumptions: 1) if x is in the linear space, then αx is in the set for some real α; 2) if y is in the set then so is $-y$.

[6]*Convention*: Whenever x belongs to a space on which several norms are defined, the unsubscripted norm $\|x\|$ denotes the principal norm. Weighted norms will be designated by subscripts.

left seminorm is defined on $\mathfrak{B}$ by the equation $\|y\|_t \triangleq \|W_t y\|$. Let $\mathfrak{B}_r$ be the range of W_r; if W is 1:1, a right seminorm is defined on $\mathfrak{B}_r$ by the equation $\|u\|_r \triangleq \|W_r^{-1}u\|$. The pair $(\|\cdot\|_r, \|\cdot\|_t)$ induces the weighted seminorm $\|A\|_W = \|W_t A W_r\|$ on the space $\mathbf{B}$.

Although weightings produced by filters will be emphasized in this paper, they can be produced by other means. For example, a weighted seminorm on $\mathbf{H}_0^\infty$ is given by the supremum over a shifted half-plane,

$$\|P\|_W = \sup\{|\hat{p}(s)|: \operatorname{Re}(s) \geq \alpha\}, \alpha > 0.$$

1) Multiplicative Seminorms—Symmetric Case: In general, weighted seminorms lack the multiplicative property $\|CD\|_W \leq \|C\|_W \|D\|_W$ of algebra norms. In problems involving the attenuation of a single disturbance (or single random process) this need not matter, as multiplications can be avoided. However, in problems involving plant uncertainty, closed-loop perturbations have the product form $(I-PQ)\Delta P$. We shall employ seminorms with weaker multiplicative properties suitable for such products.

Definition: A *symmetric (weighted) seminorm* on the algebra $\mathbf{B}$ is a weighted seminorm $\|\cdot\|_W$ on the space $\mathbf{B}$ which satisfies the multiplicative inequalities

$$\|CD\|_W \leq \|C\|_W \|D\|, \quad \|CD\|_W \leq \|C\| \cdot \|D\|_W. \quad (4.1)$$

Any operator $W: \mathfrak{X} \to \mathfrak{X}$ of unit $\mathfrak{B}$-induced norm which commutes with all operators of $\mathbf{B}$ defines a symmetric (weighted) seminorm by the equation $\|A\|_W \triangleq \|WA\| = \|AW\|$.

Symmetric seminorms have the property that $\|I\|_W = 1$.

2) Multiplicative Seminorms—General Case: In multivariable systems the plant perturbation ΔP always appears on the right of the product $(I-PQ)\Delta P$, and often does not commute with $(I-PQ)$. In such cases a more general class of multiplicative seminorms will be used. If ΔP is strictly causal, then the product $(I-PQ)\Delta P$ lies in a proper subspace of $\mathbf{B}$ (which is a left ideal, although we have no immediate use for this fact). With such products in mind, we make the following definition.

Definition: Let $\mathbf{B}_r$ be any subspace of $\mathbf{B}$, and $\mathbf{B} \cdot \mathbf{B}_r$ denote the space of products $\{CD: C \text{ in } \mathbf{B} \text{ and } D \text{ in } \mathbf{B}_r\}$. A *multiplicative seminorm* on the space of products $\mathbf{B} \cdot \mathbf{B}_r$ is any weighted seminorm on the space $\mathbf{B} \cdot \mathbf{B}_r$ with the following additional property: there is a left seminorm $\|\cdot\|_{Wt}$ defined on $\mathbf{B}$, a seminorm $\|\cdot\|_{W1}$ defined on $\mathbf{B}_r$, and the inequality $\|CD\|_W \leq \|C\|_{Wt} \|D\|_{W1}$ holds[7] for all C in $\mathbf{B}$ and D in $\mathbf{B}_r$.

Note that $\|\cdot\|_{W1}$ is not necessarily a right seminorm.

An example of a multiplicative seminorm is obtained as follows. Let $\|\cdot\|_W$ be the seminorm produced by operators W_t and W_r in the example of Section IV-B [preceding Section IV-B1)]; $V: \mathfrak{X} \to \mathfrak{X}$ be any 1:1 map of unit norm; and $\mathbf{B}_r \triangleq \{VD_1: D_1 \in \mathbf{B}\}$. Define the seminorms $\|C\|_{Wt} \triangleq \|W_t C V\|$ for $C \in \mathbf{B}$, and $\|D\|_{W1} \triangleq \|V^{-1} D W_r\|$ for $D \in \mathbf{B}_r$. Then, $\|\cdot\|_W$ has the multiplicative property claimed, as

$$\|CD\|_W = \|W_t C D W_r\| \leq \|W_t C V\| \cdot \|V^{-1} D W_r\|$$
$$= \|C\|_{Wt} \|D\|_{W1}.$$

A *symmetric* seminorm can be viewed as a special case of a multiplicative seminorm on the space of products $\mathbf{B} \cdot \mathbf{B}_r$, in which $\mathbf{B}_r = \mathbf{B}$, and the multiplicative inequality holds for each of two pairs of seminorms, namely, $(\|\cdot\|_W, \|\cdot\|)$ and $(\|\cdot\|, \|\cdot\|_W)$.

C. Approximate Inverses and Singularity Measures

Many problems of feedback theory, both classical and modern, can be reduced to the construction of an approximate inverse. Let $\|\cdot\|_W$ be a fixed, weighted seminorm on the space $\mathbf{B}$.

Definition: For any operator P in $\mathbf{B}$, an *approximate right inverse* (in $\mathbf{B}_s$) of P is any operator Q in $\mathbf{B}_s$ for which[8] $\|I-PQ\|_W < \|I\|_W$; the *right singularity measure* (in $\mathbf{B}_s$) of P (under $\|\cdot\|$), denoted by $\mu(P)$, is

$$\mu(P) = \inf\{\|I-PQ\|_W: Q \text{ in } \mathbf{B}_s\}.$$

In general, $\mu(P)$ is a number in the interval $0 \leq \mu(P) \leq 1$. The last inequality follows from the observation that $Q=0$ gives $\|I-PQ\|_W = \|I\|_W$.

In all of the following $\mathbf{H}^\infty$ examples, $\|\cdot\|_W$ will be the symmetric weighted norm defined by $\|A\|_W \triangleq \|WA\|$ for $A \in \mathbf{H}^\infty$, where $W \in \mathbf{H}_0^\infty$ is a fixed (strictly causal) operator of unit norm. For example, W can be the "low-pass" frequency response $\hat{w}(s) = k(s+k)^{-1}$, $k>0$. As $\|\cdot\|_W$ is symmetric, it has the multiplicative properties defined in Section IV-B1).

Example 4.1: P_a is a plant in $\mathbf{H}_0^\infty$ with frequency response $\hat{p}_a(s) = \alpha(s+\alpha)^{-1}$, $\alpha>0$. The sequence of operators Q_n in $\mathbf{H}_0^\infty$ with frequency responses $\hat{q}_n(s) = \alpha^{-1}(s+\alpha) \cdot n^2(s+n)^{-2}$, $n=1,2,\cdots$, satisfies the equation

$$\|\hat{w}(1-\hat{p}\hat{q}_n)\|_{H^\infty} = \sup_{\operatorname{Re}(s)\geq 0} \left|\hat{w}(s)\left(1-\frac{n^2}{(s+n)^2}\right)\right|. \quad (4.1)$$

The right-hand side (RHS) of (4.1) approaches 0 as $n \to \infty$. Therefore, the singularity measure in $\mathbf{H}_0^\infty$ of P_a under $\|\cdot\|_W$ is $\mu(P_a)=0$. The operators Q_n are approximate inverses, and the sequence Q_n is an example of what will be called an inverting sequence.

Example 4.2: P_a is the operator of Example 4.1; P_b is the "nonminimum phase" operator in $\mathbf{H}^\infty$ with frequency response $\hat{p}_b(s) = (\beta-s)(\beta+s)^{-1}$, $\beta>0$; P_1 is the product, $P_1 \triangleq P_b P_a$. For any Q in $\mathbf{H}_0^\infty$, $\hat{w}(s)[1-\hat{p}_1(s)\hat{q}(s)]$ has the value $\hat{w}(\beta)$ at the zero of $\hat{p}_b(s)$. Therefore, $\|I-P_1 Q\|_W \geq |\hat{w}(\beta)|$, and we get the lower bound $\mu(P_1) \geq |\hat{w}(\beta)|$. In

[7] The definition can obviously be generalized for the case of perturbations appearing on the left.

[8] Recall that $\|I\|_W \leq 1$, and $\|I\|_W = 1$ if $\|\cdot\|_W$ is symmetric.

Section V-A and Corollary 6.1 it will be established that in fact $\mu(P_1)=|\hat{w}(\beta)|$.

In these examples we have emphasized approximate inverses under a multiplicative seminorm. In passing, it may be worth mentioning that WHK problems can be viewed as approximate inversion problems in which the weighted seminorm of $(I-PQ)$ is obtained by weighting by a fixed vector $d\in L^2(0,\infty)$, to obtain $\|I-PQ\|_W=\|(I-PQ)d\|_{L^2}$. Here $\mu(P)$ is the irreducible error. However, $\|\cdot\|_W$ lacks the multiplicative properties.

No matter which seminorm is used $\mu(\cdot)$ has the following property.

Proposition 4.3: For any P_a and P_b in $\mathbf{B}$, $\mu(P_bP_a)\geqslant\mu(P_b)$.

Proof: If the contrary is assumed to be true, there is a Q in $\mathbf{B}_s$ for which $\|I-P_bP_aQ\|_W<\mu(P_b)$; P_aQ now acts as an approximate inverse for P_b, and there is a contradiction.

V. Sensitivity to Disturbances and Approximate Invertibility

We shall show that approximate invertibility of the plant is a necessary and sufficient condition for the existence of a feedback to attenuate disturbances, and the optimal sensitivity depends on the measure of singularity of the plant.

Consider the feedback scheme, (3.1), and suppose the plant P in $\mathbf{B}$ equals the nominal P_1. Let $\|\cdot\|_r$ be a fixed right seminorm defined on some subspace $\mathfrak{B}_r$ of (inputs), $\|\cdot\|_\ell$ a fixed left seminorm on (outputs) $\mathfrak{B}$, and $\|\cdot\|_W$ the resulting weighted seminorm induced on $\mathbf{B}$. The *sensitivity to disturbances* η_1 is defined by the equation $\eta_1\triangleq\sup\{\|y\|_\ell/\|d\|_r:\ d \text{ in } \mathbf{B}_r,\ \|d\|_r\neq 0\}$. From (3.2a) it follows that $\eta_1=\|(I+P_1F)^{-1}\|_W$. Whenever the equivalence equations (3.4) hold, η_1 also equals $\|I-P_1Q\|_W$.

The sensitivity η_0 obtained when $F=0$ will be called the *open-loop sensitivity*. As $\eta_0=\|I\|_W$, $\eta_0\leqslant 1$, and $\eta_0=1$ whenever $\|\cdot\|_W$ is symmetric.

Theorem 2:

a) A necessary and sufficient condition for the existence of a feedback F in $\mathbf{A}_s$ for which (3.1) is closed-loop stable and the sensitivity η_1 is less than the open-loop value η_0, is that P_1 have an approximate (stable) right inverse Q in the algebra[9] $\mathbf{B}_s$.

b) For any $\epsilon>0$, there is a feedback F in $\mathbf{A}_s$ for which (3.1) is closed-loop stable and for which $\eta_1<\mu(P_1)+\epsilon$, but no F for which (3.1) is closed-loop stable and $\eta_1<\mu(P_1)$, where $\mu(P_1)$ is the measure of right singularity in $\mathbf{B}_s$ of P_1 under $\|\cdot\|_W$.

Proof: By Theorem 1, any closed-loop stable feedback scheme (3.1) with F in $\mathbf{A}_s$ is equivalent to a model reference scheme (3.3) with Q in $\mathbf{B}_s$. Let (3.3) be equivalent to (3.1). η_1 can be expressed by $\eta_1=\|I-P_1Q\|_W$. Therefore, a) $\eta_1<\eta_0$ iff Q is an approximate right inverse of P_1. Also, b) by definition of $\mu(P_1)$, there is a Q in $\mathbf{B}_s$ for which $\|I-P_1Q\|_W<\mu(P_1)+\epsilon$, but none for which $\|I-P_1Q\|_W<\mu(P_1)$. The conclusion concerning F follows from the equivalence of (3.1) and (3.3). Q.E.D.

In general, it may be impossible to attain the sensitivity $\mu(P_1)$.

Definition: A sequence $Q_n\in\mathbf{A}$, $n=1,2,\cdots$, will be called *optimal* for P_1 iff the sequence of sensitivities $\|I-P_1Q_n\|_W$ approaches $\mu(P_1)$ as $n\to\infty$. $\mu(P_1)$ will be called the *optimal sensitivity* for P_1.

A sequence of feedbacks $F_n\in\mathbf{A}_s$ will be called optimal iff the equivalent comparators Q_n are optimal.

Remark: If the disturbance d lies in a balanced set $\mathfrak{D}$ (i.e., d in $\mathfrak{D}$ implies $-d$ in $\mathfrak{D}$), it is simple to show that no open-loop control, obtained by letting $Q=0$ and applying an input at node 2, can make $\|y\|_\ell$ less than $\|d\|_\ell$ for all d in $\mathfrak{D}$. It follows from Theorem 2 that, for right invertible plants, optimal sensitivity achievable with feedback is smaller than without; in other words, ability of control schemes to cope with unknown disturbances depends on their configuration. This can be viewed as a continuation of the internal model principle [16] to seminormed disturbances.

A. An Example of Sensitivity Optimization

We would like to show that feedback optimization is feasible under a seminorm that has the multiplicative properties (4.1), unlike the quadratic norm of WHK methods. If weighting is obtained from a filter, the optimal feedback in $\mathbf{H}^\infty$ resembles a classical lead–lag network. We shall try to demonstrate these points by an example. A more comprehensive theory of H^∞ optimization would take too long to present here.

Let P_1 in $\mathbf{H}_0^\infty$ be any plant with a single $\mathrm{Re}(s)>0$ zero at $s=\beta$, subject to the high frequency restriction $|\omega\hat{p}_1(j\omega)|^{-1}\leqslant$const. for $|\omega|>1$. Let $\|\cdot\|_W$ be the weighted seminorm, $\|P\|_W\triangleq\|WP\|$, where $\hat{w}$ in H_0^∞ satisfies the condition $|\omega^\ell\hat{w}(j\omega)|^{-1}\leqslant$const. for some integer ℓ. An optimal sequence of comparators Q_n in $\mathbf{H}_0^\infty$ is sought.

It will be shown that *a sequence with frequency responses*

$$\hat{q}_m(s)=c_mP_b^{-1}(s)\left[1-\hat{w}(\beta)\hat{w}^{-1}(s)\right](s+m)^{-\ell}(s+n_m)^{-2},\qquad m,n_m=1,2,\cdots\quad(5.1)$$

in which: P_b is the operator of Example 4.2; $c_m=m^\ell n_m^2$; and for each m, n_m is a sufficiently large integer, is such an optimal sequence. In fact, $\mu(P_1)=|\hat{w}(\beta)|$, and the sequence of sensitivities $\|I-P_1Q_m\|_W$ approaches $\mu(P_1)$ as $m\to\infty$.

Q_m will be constructed here, but some of the proof details will be postponed until Section VI. P_1 can be factored into the product[10] $P_1=P_aP_b$, in which $\hat{p}_b(s)=(\beta-s)(\beta+s)^{-1}$, and $|\hat{p}_b(j\omega)|=1$; and $\hat{p}_a$ is in H_0^∞, $\hat{p}_a(j\omega)=$

[9]The fact that given $P\in\mathbf{B}$, the closed-loop system is stable iff $F(I+P_1F)^{-1}$ is stable is pointed out by Desoer–Chan [21, Theorem 3], who scrutinize the relationship between open- and closed-loop stability in the context of convolution algebras.

[10]P_a/P_b are often referred to as all-pass/minimum phase or inner/outer factors.

$\hat{p}_1(j\omega)/\hat{p}_b(j\omega)$ for all real ω, and $\hat{p}_a$ has no zeros in $\mathrm{Re}(s)\geq 0$. Optimal sequences will be constructed for P_a and P_b separately.

P_b coincides with the operator of Example 4.2. By the reasoning of that example we get the lower bound $\mu(P_1)\geq |\hat{w}(\beta)|$; in fact, for the $\hat{p}_b$ factor alone, $\|\hat{w}(1-\hat{p}_b\hat{q})\|_{H^\infty}\geq |\hat{w}(\beta)|$ whenever $\hat{q}$ preserves the analyticity of $\hat{w}\hat{p}_b\hat{q}$ in $\mathrm{Re}(s)\geq 0$.

We observe next that the function $\hat{q}_b$: $\mathbb{C}\to\mathbb{C}$, $\hat{q}_b \triangleq \hat{p}_b^{-1}(s)[1-\hat{w}(\beta)\hat{w}^{-1}(s)]$ exactly minimizes $\|\hat{w}(1-\hat{p}_b\hat{q}_b)\|$, since $\hat{w}(s)[1-\hat{p}_b(s)\hat{q}_b(s)]$ equals the lower bound $\hat{w}(\beta)$ for all s in $\mathbb{C}$. Also, $\hat{q}_b$ is analytic in $\mathrm{Re}(s)\geq 0$, as $[1-\hat{w}(\beta)\hat{w}^{-1}(s)]$ has a zero at β to cancel the pole $\hat{p}_b^{-1}$. We now combine $\hat{q}_b$ with enough high frequency attenuation to obtain an H^∞ sequence:

$$\hat{q}_{bm}(s)=\hat{p}_b^{-1}(s)\left[1-\hat{w}(\beta)\hat{w}^{-1}(s)\right]m^\ell(s+m)^{-\ell}$$

$$m=1,2,\cdots.$$

(It will become apparent that the attenuation $m\ell(s+m)^{-\ell}$ will be at a high enough frequency to have arbitrarily little effect on weighted sensitivity.) The factor $P_a(s)$ can be inverted by the sequence in H_0^∞,

$$\hat{q}_{an}(s)=\hat{p}_a^{-1}(s)n^2(s+n)^{-2}, \qquad n=1,2,\cdots.$$

The sequence $\hat{q}_m$ of (5.1) is constructed from the product of $\hat{q}_{an}$ and $\hat{q}_{bm}$. The validity of this construction is established in Corollary 6.1.

Remarks: It may be worth looking at a special case of (5.1) to get a better feel for the kind of feedbacks our approach generates. Let $\hat{p}_a(s)=\alpha(s+\alpha)^{-1}$ and $\hat{w}(s)=k(s+k)^{-1}$, $\alpha>0$, $k>0$. Equation (5.1) gives

$$\hat{q}_m(s)=c_m(s+\beta)(s+m)^{-1}(s+n_m)^{-2}. \qquad (5.2)$$

$c_m \triangleq m^2n_m(k+\beta)^{-1}$. This optimal $\hat{q}_m$ consists of the "lead" factor $c(s+\beta)$, and high frequency poles whose purpose is to make $\hat{q}_m(s)$ strictly proper. The feedback law $\hat{f}(s)$ produced by $\hat{q}_m(s)$ [via (3.4b)] is a lead–lag network typical of classical control.

If the high frequency poles are neglected, the sensitivity operator $E \triangleq I-P_1Q_m$ has the frequency response $\hat{e}(j\omega)=\hat{w}(\beta)\hat{w}^{-1}(s)$; $\hat{e}$ is small at highly weighted frequencies and vice versa. If $\hat{e}(j\omega)$ is compared to the value that would be obtained by letting $\hat{q}(s)=1$, it appears that $\hat{q}_m(s)$ has the effect of trading undesirable low-frequency phase lags $\arg\hat{p}(j\omega)$, for undesirably large magnitudes $|\hat{p}(j\omega)\hat{q}(j\omega)|$ at high frequencies, where they matter less, at least if the specified weighting is correct. This, too, is a typical strategy of classical design.

The growth of $|\hat{e}(j\omega)|$ as $\omega\to\infty$ depends on the decay of $|\hat{w}(j\omega)|$; if the former is too high, the choice of the latter was inappropriate, if $\|\hat{w}(j\omega)|$ is bounded from below, then $|\hat{e}(j\omega)|$ is bounded from above. The dependence of the optimal filter on the weighting W is not surprising, as W describes the convex set of disturbances to be attenuated. Filters optimal in this sense are known to be very sensitive to plant uncertainty, and are practical only where accurate plant models are available.

B. *Unstable Plants*

If P_0 is an unstable plant in $\mathbf{A}_s$ with output disturbance d_0 in $\mathfrak{B}$, and there is a *stabilizing feedback* F_0 with the property that $P_0(I+F_0P_0)^{-1}$ and $(I+P_0F_0)^{-1}$ are in $\mathbf{B}$, then we can let P_1 in Theorem 2 be the stabilized system $P_0(I+F_0P_0)^{-1}$, and d be the stabilized disturbance $(I+P_0F_0)^{-1}d_0$. In that case, (3.6) takes the form

$$y=(I-P_1Q)(I+P_0F_0)^{-1}d_0. \qquad (5.3)$$

The term $(I+P_0F_0)^{-1}$ contributed by stabilization appears in (5.3) as an extra right weighting on $(I-P_1Q)$. Provided $\|\cdot\|_W$ is modified to inlude the extra weighting, Theorem 2 remains valid.

Remark: We have preferred to separate feedback synthesis into two consecutive stages: 1) stabilization and 2) desensitization of (input–output) stable systems. From the point of view of input–output sensitivity theory, the separation is in a certain sense unavoidable, as the perturbations allowed for robust stabilization are radically different from those for desensitization. This point is elaborated in Appendix II.

C. *Optimal Sequences: Symmetric Case*

Usually, optimal filters for sensitivity reduction are not strictly proper and can not be attained, although they can be approached by sequences of filters of increasing bandwidth. The behavior of such sequences is conveniently described in terms of the concept of an "identity sequence," or "approximate identity" drawn from Banach algebras. Here we shall summarize those properties of identity sequences which are employed in our filter construction.

Let $\|\cdot\|_W$ be a fixed, symmetric[11] weighted seminorm on the algebra $\mathbf{B}$. A *weighted identity sequence* (widseq) is any sequence I_m, $m=1,2,\cdots$ in $\mathbf{B}$ with the property that $\|I_m\|=1$ and, for any A in $\mathbf{B}$, $\|A-I_nA\|_W$ and $\|A-AI_n\|_W$ approach 0 as $n\to\infty$. For any P in $\mathbf{A}$, a sequence Q_n, $n=1,2,\cdots$, is a *right weighted inverting sequence* (winvseq) iff PQ_n is a widseq.

Whenever $\|I-PQ_n\|_W\to 0$ as $n\to\infty$, Q_n must be a right winvseq for P; for then, the inequalities

$$\|A-PQ_nA\|_W\leq\|I-PQ_n\|_W\|A\|,$$
$$\|A-APQ_n\|_W\leq\|A\|\cdot\|I-PQ_n\|_W,$$

obtained using the multiplicative property of symmetric seminorms, have left-hand sides which converge to 0 as $n\to\infty$.

For example, in $\mathbf{H}^\infty$ the sequence of operators with frequency responses $n^r(s+n)^{-r}$, $n=1,2,\cdots$ and $r>0$ any constant integer, is a widseq for the weighted norm of

[11] Without symmetry a distinction between left and right identity sequences, etc., must be made.

Example 4.1. The sequence Q_n of that example is a winvseq for P_a.

It will be shown that in certain cases an optimal sequence of approximate inverses for a product $P_b P_a$ can be obtained from separate sequences for the factors P_a and P_b, and consequently some optimization problems can be decomposed into simpler ones. Let $\|\cdot\|_a \geq \|\cdot\|_b$ be a pair of symmetric, weighted seminorms on the algebra $\mathbf{B}$. Let P_a in $\mathbf{B}_s$ and P_b in $\mathbf{B}$ have singularity measures $\mu_a(P_a)$ and $\mu_b(P_b)$ under $\|\cdot\|_a$ and $\|\cdot\|_b$, respectively, and optimal sequences $\{Q_{am}\}_{m=1}^{\infty}$ in $\mathbf{B}_s$, $\{Q_{bn}\}_{n=1}^{\infty}$ in $\mathbf{B}$, i.e., $\|I-P_iQ_{in}\|_i \to \mu_i(P_i)$ as $n\to\infty$ for $i=a$ or b.

Lemma 5.1: If $\mu_a(P_a)=0$, then $\mu_b(P_bP_a)=\mu_b(P_b)$, and P_bP_a has an optimal sequence of $\|\cdot\|_b$-weighted right approximate inverses in $\mathbf{B}_s$ of the form $Q_{an_m}Q_{bm}$, where for any $m=1,2,\cdots,n_m$ is a sufficiently large integer.

In fact, for any $\epsilon>0$ and $0<\alpha<1$, if we pick integers m and n_m such that $\|I-P_bQ_{bm}\|_b \leq \mu_b(P_b)+\alpha\epsilon$, and $\|I-P_aQ_{an_m}\|_a \leq (1-\alpha)\epsilon \|P_b\|^{-1}\|Q_{bm}\|^{-1}$, then

$$\|I-P_bP_aQ_{an_m}Q_{bm}\|_b \leq \mu_b(P_b)+\epsilon. \tag{5.4}$$

Proof:

a) $\mu_b(P_bP_a)\geq \mu_b(P_b)$ by Proposition 4.3. Let us establish the opposite inequality. For any $m>0$, $n>0$, we have

$$\begin{aligned}\|I-P_bP_aQ_{an}Q_{bm}\|_b &= \|I-P_bQ_{bm}-P_n(P_aQ_{an}-I)Q_{bm}\|_b\\ &\leq \|I-P_bQ_{bm}\|_b+\|P_b\|\cdot\|P_aQ_{an}-I\|_a\|Q_{bm}\|\end{aligned} \tag{5.5}$$

where the triangle inequality, the multiplicative property of symmetric seminorms, and the dominance $\|\cdot\|_b \leq \|\cdot\|_a$ have been used. The integers m and n specified in the hypothesis exist by definition of $\mu_b(P_b)$ and by the hypothesis that $\mu_a(P_a)=0$. (5.4) now follows from (5.5). As ϵ was arbitrary, (5.4) implies that $\mu_b(P_bP_a)\leq \mu_b(P_b)$. Therefore, $\mu_b(P_bP_a) = \mu_b(P_b)$, and $Q_{an_m}Q_m$ is an optimal sequence. Q.E.D.

VI. Multivariable Systems in $\mathbf{H}^{\infty N}$

At present there is much interest in the design of multivariable systems along classical lines, but there is little formal theory. Our next objective is to show that the theory outlined in Sections III–V provides a suitable framework for multivariable design.

Multivariable frequency responses will be viewed as elements of an n-dimensional version of the Hardy space H^∞. Our approach will be to decompose the frequency response matrix into a product of nearly invertible and noninvertible parts, and to design a feedback for each part separately. The validity of approaches based on decompositions will be established in Theorem 3. The optimal sensitivity scheme of Section V-A will be obtained as a corollary. Another corollary will validate the common hypothesis that sensitivity can be made arbitrarily small if the plant is nearly invertible. Finally, a lower bound to sensitivity in terms of the location of RHP zeros will be derived.

A. *Terminology*

N is a fixed integer. H^{2N} is the Banach space consisting of the N-fold product of H^2, on which the norm of any vector $\hat{u}=(\hat{u}_1,\hat{u}_2,\cdots,\hat{u}_N)$ is $\|\hat{u}\| \triangleq (\sum_{i=1}^N \|\hat{u}_i\|^2)^{1/2}$. The elements of H^{2N} can be represented by columns of input or output Laplace transforms.

$H^{\infty N}$ is the algebra of $N\times N$ matrices, $\hat{P}=[\hat{p}_{ij}]$, whose elements $\hat{p}_{ij}$ are causal, stable, frequency responses in H^∞. For any $\hat{P}$ in $H^{\infty N}$ and s in $\mathbb{C}$, $\sigma[\hat{P}(s)]$ denotes the square root of the magnitude of the largest eigenvalue of $[\hat{P}(s)]^*\hat{P}(s)$, where $*$ denotes the conjugate transpose of a matrix. $H^{\infty N}$ is a Banach algebra under the norm $\|\hat{P}\| = \sup_\omega \sigma[\hat{P}(j\omega)]$.

Each frequency response matrix $\hat{P}$ in $H^{\infty N}$ determines an operator P: $H^{2N}\to H^{2N}$, $\widehat{Pu}=\hat{P}\hat{u}$. The algebra of such operators is denoted by $\mathbf{H}^{\infty N}$, and is a Banach algebra under the norm $\|P\|_{\mathbf{H}^{\infty N}} = \|\hat{P}\|_{H^{\infty N}}$ The norm defined in this way coincides with the norm induced on $\mathbf{H}^{\infty N}$ by H^{2N}, i.e., $\sup\{\|Pu\|$: u in H^{2N} and $\|u\|=1\}$.

A frequency response $\hat{P}$ in $H^{\infty N}$ is called *strictly proper* iff each of its components $\hat{p}_{ij}$ is strictly proper; the corresponding operator P in $H^{\infty N}$ is called *strictly causal.* The subalgebras of strictly proper or strictly causal elements in $H^{\infty N}$ or $\mathbf{H}^{\infty N}$ are denoted by $H_0^{\infty N}$ or $\mathbf{H}_0^{\infty N}$, respectively.

Let $\hat{w}$ be a (scalar) weighting function in H_0^∞. The weighted seminorm $\|P\|_W \triangleq \|\hat{w}\hat{P}\|_{H^{\infty N}}$ is defined on $\mathbf{H}^\infty$. Observe that $\hat{w}$ commutes with all operators in $\mathbf{H}^{\infty N}$, so that $\|\cdot\|_W$ is symmetric, and the multiplicative inequalities (4.1) are satisfied.

B. *A Decomposition Theorem*

We shall be interested in frequency response matrices whose rate of approach to zero as $s\to\infty$ is comparable to that of a power of $(1+s)$, and employ the following.

Notation: For any integer n, $(1+s)^nH^{\infty N}$ denotes the set of functions $\hat{P}(s)$ with the property that $(1+s)^{-n}\hat{P}(s)$ is in $H^{\infty N}$.

In the rest of Section VI, plants will be decomposed into products P_aP_b of strictly causal and approximately invertible parts under the following *assumption*:

a) P_a is a strictly causal operator in $\mathbf{H}_0^{\infty N}$; and

b) P_b is in $\mathbf{H}^{\infty N}$. The frequency response $\hat{P}_b$ has a *minimal* approximate right inverse $\hat{Q}_b$ in $(1+s)^lH^{\infty N}$; i.e., $\hat{w}\hat{P}_b\hat{Q}_b$ is in $H^{\infty N}$ and $\|\hat{w}(1-\hat{P}_b\hat{Q}_b)\|_{H^{\infty N}}=\mu(P_b)$.

Theorem 3: If i) $\det \hat{P}_a(s)\neq 0$ for $\mathrm{Re}(s)\geq 0$ and ii) there are constants $c>0$, $\rho>0$, and an integer $k>0$ for which the inequality $|s|^k\sigma[\hat{P}_a(s)]\geq c$ is valid in the region $|s|>\rho$, $\mathrm{Re}(s)\geq 0$, then

a) $\mu(P_bP_a)=\mu(P_b)$, and

b) P_bP_a has an optimal sequence of approximate right inverses Q_m in $\mathbf{H}_0^{\infty N}$, with frequency responses

$$\hat{Q}_m(s) \triangleq \hat{P}_a^{-1}(s)\hat{Q}_b(s)\left[m(s+m)^{-1}\right]^l\left[n_m(s+n_m)^{-1}\right]^{k+1}$$

where for any $m=1,2,\cdots$, n_m is a sufficiently large integer (given explicitly in Lemma 5.1).

Theorem 3 is an application of Lemma 5.1. Q_m will be constructed out of separate optimal sequences for P_a and P_b, both under the symmetric $\|\cdot\|_W$ norm. First, however, a lemma will be proved.

For any integer $r>0$, let J_n^r denote the sequence in $\mathbf{H}^{\infty N}$ with "low-pass" frequency responses

$$\hat{J}_n^r(s)=n^r(s+n)^{-r}\hat{I}, \qquad n=1,2,\cdots.$$

Lemma 6.1: $\|J_n^r\|=1$; J_n^r is a weighted identity sequence,[12] i.e., for any $r>0$, $\|I-J_n^r\|_W \to 0$ as $n\to\infty$.

Proof: First of all, $\|J_n^r\|=\|I\|\sup_\omega|n(j\omega+n)^{-1}|^r=1$. Second, we have

$$\|I-J_n^r\|_W=\sup_\omega|\hat{w}(j\omega)|\cdot\left|1-\left(\frac{n}{j\omega+n}\right)^r\right| \tag{6.1}$$

$$\leq 2\sup_{|\omega|\geq\delta}|\hat{w}(j\omega)|+\|\hat{w}\|\sup_{|\omega|<\delta}\left|1-\left(\frac{n}{j\omega+n}\right)^r\right| \tag{6.2}$$

$\delta>0$ being any number. Let $\epsilon>0$ be given. As $\hat{w}$ is in H_0^∞, there is a $\delta>0$ for which the first term in (6.2) is less than $\epsilon/2$; for fixed δ there is an integer n for which the second term is less than $\epsilon/2$; i.e., $\lim_{n\to\infty}\|I-J_n^r\|_W=0$. Q.E.D.

Proof of Theorem 3: First, let us show that the sequence Q_{an} with frequency responses

$$\hat{Q}_{an}\triangleq\hat{P}_a^{-1}\hat{J}_n^{k+1}, \qquad n=1,2,\cdots \tag{6.3}$$

is a right weighted inverting sequence in $\mathbf{H}_0^{\infty N}$ for P_a. As $\det(\hat{P}_a(s))\neq 0$, $\hat{P}_a^{-1}(s)$ is analytic in $\mathrm{Re}(s)\geq 0$. The inequality $|s|^k\sigma[\hat{P}_a(s)]\geq c$ ensures that the functions defined by (6.3) are in $H_0^{\infty N}$. Now, $\|I-P_aQ_{an}\|_W=\|I-J_n^{k+1}\|_W \to 0$ as $n\to\infty$, by Lemma 6.1. Therefore, $\mu(P_a)=0$ and Q_{an} is a right winvseq in $\mathbf{H}_0^{\infty N}$ for P_a.

Next, consider P_b. By hypothesis, $\hat{P}_b$ has a minimal approximate right inverse $\hat{Q}_b$ in $(1+s)^tH^{\infty N}$. The frequency responses

$$\hat{Q}_{bm}\triangleq\hat{Q}_b\hat{J}_m^t, \qquad m=1,2,\cdots \tag{6.4}$$

lie in $H^{\infty N}$, and determine operators Q_{bm} in $\mathbf{H}^{\infty N}$. Let us show that Q_{bm} is optimal. Certainly, $\|I-P_bQ_{bm}\|_W\geq\mu(P_b)$. Now,

$$\|I-P_bQ_{bm}\|_W=\|(I-P_bQ_b)J_m^t+I-J_m^t\|_W$$

$$\leq\|I-P_bQ_b\|_W\|J_m^{t+1}\|+\|I-J_m^t\|_W$$

by the triangle inequality and the multiplicative property of symmetric seminorms. As $\|I-P_bQ_b\|_W=\mu(P_b)$ by hypothesis, and by Lemma 6.1 $\|J_m^{t+1}\|=1$ and $\lim_{m\to\infty}\|I-J_m^t\|_W\to 0$, we obtain $\lim_{m\to\infty}\|I-P_bQ_{bm}\|_W\leq\mu(P_b)$. As this upper bound coincides with the lower bound obtained above, $\lim_{m\to\infty}\|I-P_bQ_{bm}\|_W=\mu(P_b)$, i.e., Q_{bm} is a right winvseq in $\mathbf{H}^{\infty N}$ for P_b.

[12] For a symmetric weighting, $\|A-J_n^rA\|_W\leq\|I-J_n^r\|_W\|A\|$, so $\|A-J_n^rA\|_W\to 0$ whenever $\|I-J_n^r\|_W$ does; similarly for $\|A-AJ_n^r\|_W$. Therefore, J_n^r is a widseq iff $\lim_{n\to\infty}\|I-J_n^r\|_W=0$.

The conclusions of Theorem 3 are now true by Lemma 5.1. Here $\|\cdot\|_a=\|\cdot\|_b=\|\cdot\|_W$, and Q_m is the product $P_a^{-1}J_{n_m}^{k+1}Q_bJ_m^t$. Q.E.D.

Corollary 6.1: The conclusion of the example of Section V-A is true.[13] That example is simply a special case of Theorem 3 for $N=k=1$, and $\hat{q}_b=\hat{p}_b^{-1}(1-\hat{w}(\beta)\hat{w}^{-1})$ in $(1+s)^tH^\infty$.

C. *Sensitivity Reduction in Nearly Invertible Multivariable Systems*

Consider the feedback scheme of equations (3.1)–(3.3). Suppose that the plant $P_1(s)$ is in $\mathbf{H}_0^{\infty N}$ and satisfies the restrictions: $\det P_1(s)\neq 0$ for $\mathrm{Re}(s)\geq 0$, and $|s|^k\sigma[\hat{P}_1(s)]>c$ in some region $|s|>p$, $\mathrm{Re}(s)\geq 0$, where c, p are constants and k is an integer. The sensitivity $\eta\triangleq\|I-P_1Q\|_W$ is defined[14] as in Section V.

Corollary 6.2:[15] The sensitivity $\|I-P_1Q_n\|_W$ can be made smaller than any $\epsilon>0$ by a comparator Q_n in $\mathbf{H}_0^{\infty N}$ with frequency response

$$\hat{Q}_n(s)=\hat{p}_1^{-1}(s)\left[n(s+n)^{-1}\right]^{k+1}$$

n being a (sufficiently large) integer.

Proof: The hypotheses of Theorem 3 are satisfied, with $P_a=P_1$ and $P_b=I$. Therefore, $\mu(P_1)=0$ and $\|I-P_1Q_n\|_W\to 0$ as $n\to\infty$. Q.E.D.

D. *Lower Bounds to Sensitivity*

It might be expected that the singularity measure of an $H^{\infty N}$ frequency response matrix would be limited by the location of its RHP zeros, and that the optimal sensitivity would be similarly limited. The following theorem shows this to be true.

Let $\|\cdot\|_W$ be any weighted seminorm on $\mathbf{H}^{\infty N}$ of the form $\|P\|_W=\|\hat{w}\hat{P}\|$, $\hat{w}$ being (a scalar function) in H_0^∞. For any plant P in $\mathbf{H}^{\infty N}$, let the *RHP zeros* of P be the points s_i in $\mathrm{Re}(s)\geq 0$ $i=1,2,\cdots$, at which $\det[\hat{P}(s_i)]=0$.

Theorem 4:

$$\mu(P)\geq\sup\{|\hat{w}(s_i)|: i=1,2,\cdots\}.$$

Proof: For any Q in $\mathbf{H}^{\infty N}$ let $E\triangleq(I-PQ)$. For any RHP zero s_i, let ξ be a unit vector in the nullspace of the matrix $\hat{P}(s_i)\hat{Q}(s_i)$. Let f: $C\to C$ be the function $f(s)=\xi^*\hat{w}(s)\hat{E}(s)\xi$. Since $\hat{P}(s_i)\hat{Q}(s_i)=0$, $f(s_i)=\xi^*\xi\hat{w}(s_i)=\hat{w}(s_i)$. Now $f(s)$ is an H_0^∞ function, and by the maximum modulus principle attains its maximum on the $j\omega$ axis, i.e., $|f(j\omega)|\geq|\hat{w}(s_i)|$ for some ω. But for any transformation A in Euclidean N-space, and any unit N-vector ξ, $\sigma(A)\geq\xi^*A\xi$. The last assertion is established by the inequalities

[13] For simplicity we have considered inputs in H^2 here, but the conclusion is easily extended to H^P, $1\leq p\leq\infty$.

[14] Recall that sensitivity of a feedback scheme equals $\|I-P_1Q\|_W$, by Theorem 1.

[15] This corollary is based on results obtained with D. Bensoussan [17].

$$|\xi^* A\xi|^2 \leqslant |\xi^*|^2 |A\xi|^2 = |A\xi|^2$$
$$= (A^*A\xi)^*\xi \leqslant |A^*A\xi| \cdot |\xi|$$
$$\leqslant \sigma^2(A)|\xi|^2 = \sigma^2(A).$$

We employ this fact to obtain

$$\mu(P) \geqslant \|E\| = \sup_\omega \sigma\left[\hat{w}(j\omega)\hat{E}(j\omega)\right] \geqslant \sup_\omega |f(j\omega)| \geqslant |\hat{w}(s_i)|.$$

The theorem follows. Q.E.D.

Remark: Theorem 4 implies that no feedback can produce small sensitivity if a plant zero is present in any heavily weighted part of the right half-plane. For "low-pass" weightings such as $k(s+k)^{-1}$, $k>0$, this means that if sensitivity is to be reduced, the only RHP zeros possibly allowed are those at very high complex frequencies, $|s_i| \gg k$.

VII. Effects of Plant Uncertainty on Disturbance Attenuation

Two opposing tendencies can be found in most feedback systems. On the one hand, to the extent that feedback reduces sensitivity it reduces the need for plant identification. On the other hand, the less information is available about the plant, the less possible it is to select a feedback to reduce sensitivity. The balance between these tendencies establishes a maximum to the amount of tolerable plant uncertainty and, equivalently, a minimum to the amount of identification needed.

It can be argued that the search for such a minimum should be basic to the theory of adaptive systems. Actually, even the existence of such a minimum appears not to have been stated, perhaps because plant uncertainty is so difficult to study in the WHK framework in the absence of the multiplicative properties (4.1), and because there is no notion of optimality in the classical setup.

Here, we would like to take a step in the direction of articulating these issues, by defining the tradeoff between minimal sensitivity and plant uncertainty and deducing its simpler properties. Sensitivity to disturbances will be considered in this section, and to plant uncertainty in the next.

Let $\mathbb{B}_r$ be a subspace of the causal operators $\mathbb{B}_s$. Consider the feedback (3.1) and model reference (3.3) schemes. Suppose that some nominal plant P_1 in $\mathbb{B}_r$ is specified and (3.4) hold, so the two schemes are equivalent. The true plant P in $\mathbb{B}_r$ will differ from P_1 in general. Let K_{33}: $\mathcal{X} \to \mathcal{X}$, $K_{33}(d) = y$ be the operator in $\mathbb{A}$ mapping disturbances into outputs. K_{33} can be expressed in terms of F [using (3.2a)],

$$K_{33} = (I+PF)^{-1}$$

or in terms of P_1 and Q [using 3.4b)],

$$K_{33} = \left[I + PQ(I-P_1Q)^{-1}\right]^{-1}$$
$$= \left[(I - P_1Q + PQ)(I-P_1Q)^{-1}\right]^{-1}$$
$$= (I-P_1Q)\left[I + (P-P_1)Q\right]^{-1}. \tag{7.1}$$

We shall be interested in the way in which closed-loop operators such as K_{33} behave as functions of the open-loop operators P, P_1, Q, and F. In particular, let us define two functions mapping open-loop into closed-loop operators. Let $\mathbb{E}$: $\mathbb{A}_s^3 \to \mathbb{A}$, $\mathbb{E}(P, P_1, Q) = K_{33}$ be the function relating K_{33} to the model reference variables, and $\mathbb{E}_f: \mathbb{A}_s^2 \to \mathbb{A}$, $\mathbb{E}_f(P, F) = K_{33}$ the function for the feedback variables. Any pair (d, y) in $\mathcal{X}^2$ satisfying (3.1) or (3.3) also satisfies the equations

$$y = \mathbb{E}_f(P, F)d = \mathbb{E}(P, P_1, Q)d. \tag{7.2}$$

The c.l. (closed-loop) operators K_{ij}, $i, j = 2, 3$, of the feedback scheme (3.1) were introduced in Section III, and are well defined for the model reference scheme (3.3) under the assumed equivalence. The model reference scheme shown in Fig. 3 has two extra nodes labeled 4 and 5. We shall avoid the lengthly but straightforward calculation of the remaining c.l. operators, and instead assume the following elementary properties of the model reference scheme: define the operators $K_{44} \triangleq [I+(P-P_1)Q]^{-1}$ and $K_{55} \triangleq [I+Q(P-P_1)]^{-1}$ and let K be the sextuplet of operators $\{K_{44}, K_{55}, P, P_1, Q, I\}$; the set $\{K_{ij}\}_{i,j=2}^5$ of all c.l. operators consists of algebraic combinations (i.e., involving sums and products only of) the operators in K; furthermore $K_{32} = -QK_{44}$.

A. Stabilizing Feedbacks

It is well known that any stable plant which is stabilized by feedback is surrounded by a ball of "admissible" perturbations which preserve closed-loop stability. Here, the radius of such a ball will be calculated, and sensitivity will be defined with respect to plant uncertainty within the ball.

Suppose a right seminorm $\|\cdot\|_r$ to be defined on $\mathbb{B}_r$, and recall that $\|\cdot\|_r \geqslant \|\cdot\|$. The true plant P will be supposed to lie in a *ball of uncertainty of radius* $\delta \geqslant 0$ in $\mathbb{B}_r$ *around the nominal* P_1, $b(P_1, \delta) \triangleq \{P \text{ in } \mathbb{B}_r: \|P-P_1\|_r \leqslant \delta\}$. Since $\|P-P_1\| \leqslant \|P-P_1\|_r$, $b(P_1, \delta)$ is a subset of the ball of radius δ in $\mathbb{B}$.

An operator K in $\mathbb{A}$ which depends on $P \in \mathbb{B}_r$ will be called *bounded* over a ball $b(P_1, \delta)$ iff K is in $\mathbb{B}$ and there is a constant $c \geqslant 0$ with the property that $\|K\| \leqslant c$ for all P in $b(P_1, \delta)$. A feedback or model reference scheme will be *c.l. (closed-loop) bounded* on $b(P_1, \delta)$ iff all its c.l. operators are bounded on $b(P_1, \delta)$.

Proposition 7.1: The following statements are equivalent on any ball $b(P_1, \delta)$. i) The feedback scheme is closed-loop bounded; ii) Q is in $\mathbb{B}$ and $K_{44} \triangleq [I+(P-P_1)Q]^{-1}$ is bounded; iii) the model reference scheme is closed-loop bounded.

Proof: We shall prove the sequence of implications, i)⇒ii)⇒iii)⇒i). If i) is true, then K_{44} is bounded by definition. Also, K_{32} is bounded by definition, and therefore in $\mathbb{B}$ for all P in $b(P_1, \delta)$. But P_1 is in $b(P_1, \delta)$, and $K_{32} = -Q$ when $P = P_1$; therefore, Q is in $\mathbb{B}$, and ii) is true. If ii) is true then K_{55} is also bounded, as the equations $K_{55} = [I + Q(P-P_1)]^{-1} = I - QK_{44}(P-P_1)$ imply that

$\|K_{55}\| \leq 1+\|Q\| \cdot \|K_{44}\| \cdot \delta$, for any P in $b(P_1, \delta)$. Therefore, and by the assumed property of the K_{ij}, each c.l. operator K_{ij}, $i, j=2, \cdots, 5$, is an algebraic combination of the operators of K, which are bounded on $b(P_1, \delta)$. It follows that each K_{ij} is bounded there, and that (3.3) is c.l. bounded, i.e., iii) is true. If iii) is true then i) is true, as (3.1) and (3.3) are equivalent. Q.E.D.

Again, only stable Q need be considered. The operators F in $\mathbf{A}_s$ and Q in $\mathbf{B}_s$ will be called *stabilizing* for $b(P_1, \delta)$ iff (3.1) and (3.3), respectively, are c.l. bounded on $b(P_1, \delta)$. The set of all Q in $\mathbf{B}_s$ stabilizing for $b(P_1, \delta)$ will be denoted by $\mathbf{B}_s(P_1, \delta)$. For any P_1 and Q in $\mathbf{B}_s$, the set of real points δ for which Q is stabilizing for $b(P_1, \delta)$ will be denoted by $\Delta(P_1, Q)$, and abbreviated to Δ when dependence on (P_1, Q) is not of interest.

The next lemma shows that any stable Q gives a c.l. bounded scheme (3.1), and is therefore stabilizing, for δ small enough. Let P_1 and Q be in $\mathbf{B}_s$.

Proposition 7.2: If δ satisfies $0 \leq \delta < \|Q\|^{-1}$, then δ is in Δ and for any P in $b(P_1, \delta)$ the inequality

$$\|\{I+(P-P_1)Q\}^{-1}\| < (1-\delta\|Q\|)^{-1} \tag{7.3}$$

holds. Δ is a half-open interval, $[0, \delta_1)$.

Proof: Under the hypothesis, $\|(P-P_1)Q\| < 1$. Therefore, the small gain property ensures that $[I+(P-P_1)Q]^{-1}$ is in $\mathbf{B}$, and (7.3) is true by Proposition 2.1c. Δ is an interval beginning at 0, because $\delta \in \Delta$ and $0 \leq \delta' \leq \delta$ implies that $b(P_1, \delta)$ contains $b(P_1, \delta')$, which implies that $\delta' \in \Delta$. Δ is half open by a standard perturbation argument for the openness of resolvent sets which will be omitted, as this property is not important here. Δ contains $[0, \|Q\|^{-1}]$ because, for any δ in $[0, \|Q\|^{-1}]$, (7.3) implies that K_{44} is bounded on $b(P_1, \delta)$, so Q is stabilizing by Proposition 7.1.

B. Sensitivity

Suppose that right and left seminorms are defined on $\mathfrak{B}$ as in Section V, and induce a weighted seminorm $\|\cdot\|_W$ on $\mathbf{B}$. For the plant P in any ball of uncertainty $b(P_1, \delta)$, and any feedback F in $\mathbf{A}_s$ or comparator Q in $\mathbf{B}_s$, the *sensitivity to disturbances under plant uncertainty* of the equivalent schemes (3.1), (3.3) is defined to be

$$\eta(P_1, Q; \delta) = \sup_{P \in b(P_1, \delta)} \sup \{\|y\|_l / \|d\|_r : d \text{ in } \mathfrak{B}_r, \|d\|_r \neq 0\}$$

if Q is (stabilizing) in $\mathbf{B}_s(P_1, \delta)$, and $\eta(P_1, Q; \delta) = \infty$ otherwise. For any stabilizing Q,

$$\eta(P_1, Q; \delta) = \sup_{P \in b(P_1, \delta)} \|\mathbf{E}(P, P_1, Q)\|_W$$

by (7.2). (Recall, also, that $\mathbf{E}(P, P_1, Q) = \mathbf{E}_f(P, F)$.) $\eta(P_1, Q, \delta)$ is the smallest assured sensitivity for P in the ball. We are interested in finding Q to minimize this sensitivity. Accordingly, we define the *minimal sensitivity to plant uncertainty* to be

$$\eta(P_1, \delta) = \inf_{Q \in \mathbf{B}_s} \sup_{P \in b(P_1, \delta)} \|\mathbf{E}(P, P_1, Q)\|_W$$

with the proviso that the sup is replaced by ∞ if Q is not stabilizing. For any fixed nominal plant[16] P_1 in $\mathbf{B}_s$, $\eta(P_1, \delta)$ is a positive-real valued function of $\delta > 0$.

Theorem 5: For any nominal plant P_1 in $\mathbf{B}_s$, the minimal sensitivity to plant uncertainty $\eta(P_1, \delta)$ is a monotone nondecreasing function of δ for $\delta \geq 0$; $\eta(P_1, \delta)$ approaches the singularity measure $\mu(P_1)$ as $\delta \to 0$; and $\eta(P_1, \delta) = \|I\|_W$ for $\delta = \|P_1\|_r$.

Lemma 7.3: For any P_1 and Q in $\mathbf{B}_s$, $\eta(P_1, Q; \delta)$ is a monotone nondecreasing function of $\delta \geq 0$, finite for δ in Δ. Δ contains the interval $[0, \|Q\|^{-1}]$. For any δ in $[0, \|Q\|^{-1}]$, the inequalities

$$\|I-P_1Q\|_W \leq \eta(P_1, Q; \delta) \leq \|I-P_1Q\|_W + \|I-P_1Q\|_W \delta \|Q\| (1-\delta\|Q\|)^{-1} \tag{7.4}$$

and, if $\|\cdot\|_W$ is symmetric, also the inequality

$$\eta(P_1, Q; \delta) \leq \|I-P_1Q\|_W (1-\delta\|Q\|)^{-1} \tag{7.5}$$

are satisfied.

Proof of Lemma 7.3: The symbol $\uparrow$ will denote a monotone nondecreasing function of $\delta \geq 0$. By definition of Δ, Q is stabilizing for any $b(P_1, \delta)$ with δ in Δ, so $\eta(P_1, Q; \delta)$ is finite on Δ. By Proposition 7.2, Δ is an interval containing $[0, \|Q\|^{-1}]$. δ is a $\uparrow$ function on Δ since it is a sup over sets $b(P_1, \delta)$ which are nested and nondecreasing as δ increases.

From (7.1) we get

$$\mathbf{E}(P, P_1, Q) = (I-P_1Q)\{I-(P_1-P)Q[I+(P-P_1)Q]^{-1}\}. \tag{7.6}$$

The triangle inequality and dominance condition $\|\cdot\|_W \leq \|\cdot\|$ give

$$\|\mathbf{E}(P, P_1, Q)\|_W \leq \|I-P_1Q\|_W + \|(I-P_1Q)(P_1-P)Q[I+(P-P_1)Q]^{-1}. \tag{7.7}$$

For δ in $[0, \|Q\|^{-1}]$, the upper bound in (7.4) is obtained from (7.7) by (7.3) and the multiplicative property of norms. The lower bound is valid as P_1 is in $b(P_1 \delta)$, and $E(P_1, P_1, Q) = I-P_1Q$. If $\|\cdot\|_W$ is symmetric, the multiplicative property of *symmetric* seminorms applied to (7.1) gives

$$\|\mathbf{E}(P, P_1, Q)\|_W \leq \|I-P_1Q\|_W \|\{I+(P-P_1)Q\}^{-1}\|$$

and (7.4) follows. Q.E.D.

Proof of Theorem 5: From Lemma 7.3 and the inequality $\|I-P_1Q\|_W \geq \mu(P_1)$, it can be concluded that for any Q in $\mathbf{B}_s$, $\eta(P_1, Q; \delta)$ is a $\uparrow$ function with values in $[\mu(P_1), \infty]$. As $\eta(P_1, \delta) = \inf_{Q \in \mathbf{B}} \eta(P_1, Q; \delta)$ by definition, $\eta(P_1, \delta)$ is the inf of a set of $\uparrow$ functions with values in $[\mu(P_1, \infty]$. Therefore, $\eta(P_1, \delta)$ is $\uparrow$ and $\eta(P_1, \delta) \geq \mu(P_1)$. Since 0 is in

[16]Again, the assumption that plant and feedback are both strictly causal can be relaxed in Theorem 5, Lemma 5.3, etc.

$\mathbf{B}_s$, $\eta(P_1,0;\delta)$ is in the set, and $\eta(P_1,\delta)\leq\eta(P_1,0;\delta)=\|I\|_W$, where the last identity is true by (7.1).

Let us show that $\lim_{\delta\to 0}\eta(P_1,\delta)=\mu(P_1)$. By definition of $\mu(P_1)$, for any $\epsilon>0$ there is a Q in $\mathbf{B}_s$ satisfying $\|I-P_1Q\|_W\leq\mu(P_1)+\epsilon/2$; as the RHS of (7.4) approaches $\|I-P_1Q\|_W$ as $\delta\to 0$, there is a $\delta>0$ for which $\eta(P,Q;\delta)\leq\mu(P_1)+\epsilon$. Since ϵ was arbitrary, and it has been shown that $\eta(P_1,\delta)\geq\mu(P_1)$, the conclusion follows.

Finally, let us show that $\eta(P_1,\|P_1\|_r)=\|I\|_W$. Suppose the contrary to be true. It has been shown that $\eta(P_1,\|P_1\|_r)\leq\|I\|_W$, so $\eta(P_1,\|P_1\|_r)\leq\|I\|_W$. Therefore, there is a Q in $\mathbf{B}_s$ for which $\|\mathbf{E}(P,P_1,Q)\|_W<\|I\|_W$. Now, 0 is in $b(P_1,\|P_1\|_r)$ and, by (7.1), $\eta(P_1,\|P_1\|_r)\geq\|\mathbf{E}(0,P_1,Q)\|_W=\|I\|_W$ which is a contradiction. Q.E.D.

Remark: Theorem 5 implies that whenever feedback reduces sensitivity for the nominal plant P_1, there are points in the interval $[0,\|P_1\|]$ at which $\eta(P_1,\delta)$ increases, i.e., where "the less we know about P, the less able we are to construct a feedback to attenuate disturbances."

VIII. Filtering of Plant Uncertainty: Invariant Schemes

We now turn to the second problem outlined in the introduction, Section I-D. The plant P lies in a ball of uncertainty[17] around some nominal value P_1, and one object of using feedback is to shrink the size of the uncertainty. Of course, uncertainty can be reduced to zero by disconnecting the input (u in Fig. 4) from the system, but then P_1 is also transformed into zero. Clearly, the problem is trivial unless there is a normalization or constraint on the control law that transforms P_1 into a closed-loop system.

We would prefer as far as possible to separate the reduction of uncertainty from the transformation of P_1, and therefore seek a definition of uncertainty which is independent of the eventual closed-loop system.

If P_1 were a real number, uncertainty could be normalized by specifying it as a percentage of the nominal value. This possibility is not open for noninvertible plants. Instead, we shall achieve a normalized definition of uncertainty by employing the device of a plant-invariant scheme, which leaves the nominal plant invariant while shrinking the ball of uncertainty. Such a scheme will be shown always to be realizable in the form of a model reference scheme.

This device will also enable us to separate the design process into two consecutive stages: 1) reduction of uncertainty and 2) transformation of the nominal plant into a nominal closed-loop system (cf. the separation into estimation and control stages in Kalman filtering).

A possible disadvantage to this approach is that the two stages may be dependent, and yield a suboptimal sensitivity. We shall try to get the best of both worlds, and simultaneously formulate normalized and unnormalized

[17] Departures of the true plant from the nominal can be interpreted as uncertain in some applications; in others, they may simply represent known perturbations to be attenuated.

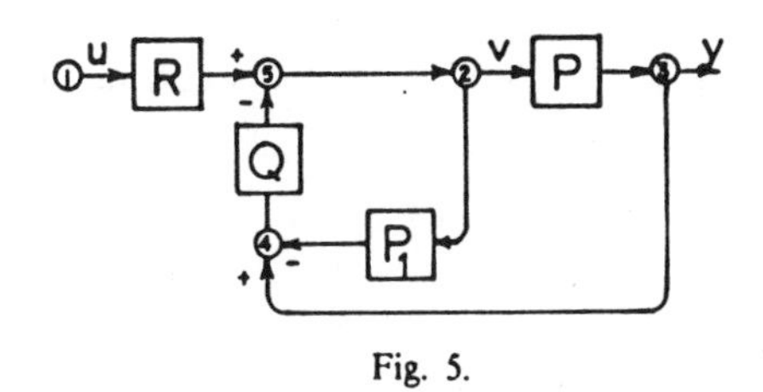

Fig. 5.

versions of the problem, by allowing the plant invariant scheme to be multiplied by a constant R_1, representing a desired nominal control law.

A. Plant Invariant Schemes

Any feedback arrangement of the type shown in Fig. 1, incorporating a plant P with a single accessible input, has an equivalent description in terms of the flowgraph of Fig. 4. This flowgraph is completely specified by the two additional operators U and F, which may be unstable even though the original arrangement is closed-loop stable. Note that Fig. 4 is an enlargement of Fig. 2 by a new branch, U, connected to a new node labeled 1. For simplicity, it will be assumed that $d=0$. The new *feedback scheme* equations are

$$y=Pv \tag{8.1a}$$

$$v=Uu-Fy \tag{8.1b}$$

in which $P\in\mathbf{A}_s$, $F\in\mathbf{A}_s$, and[18] $U\in\mathbf{A}$ are operators, and u, v, and y are in $\mathcal{X}$. (Equations (8.1) represent an enlargement of (3.1) by the term Uu, under the constraint that $d=0$.) As P is strictly causal, $(I+FP)^{-1}$ exists in $\mathbf{A}$, and for each u in $\mathcal{X}$, (8.1) have unique solutions for v and y in $\mathcal{X}$, given by the equations

$$v=(I+FP)^{-1}Uu \tag{8.2a}$$

$$y=P(I+FP)^{-1}Uu. \tag{8.2b}$$

Let K_{12} and K_{13} be the operators mapping $\mathcal{X}$ into $\mathcal{X}$ which satisfy, $K_{12}(u)=v$, $K_{13}(u)=y$. By (8.2), $K_{12}=(I+FP)^{-1}U$ and $K_{13}=P(I+FP)^{-1}U$.

The full set of closed-loop operators of (8.1) consists of K_{12}, K_{13}, $K_{11}\triangleq I$, $K_{i1}\triangleq 0$, and the operators K_{ij} defined in Section III, where $i,j=2,3$.

A variant of the model-reference transformation of Section III-A will be used. The flowgraph of Fig. 5 is described by the new *model reference scheme* equations

$$y=Pv \tag{8.3a}$$

$$v=Ru-Q(y-P_1v) \tag{8.3b}$$

in which $P, P_1\in\mathbf{B}_s$, $R\in\mathbf{A}$, and $u,v,y\in\mathcal{X}$. For any $u\in\mathcal{X}$, (8.3) have unique solutions for v and y in $\mathcal{X}$, as P, P_1, and Q are strictly causal. (Equations (8.3) can be viewed as an enlargement of (3.3) by the term Ru, and subject to the constraint that $d=0$.)

The schemes (8.1) and (8.3) will be called *equivalent* iff

[18] In order to be physically realized, U would have to be approximated by a strictly causal operator. We prefer not to assume that $U\in\mathbf{A}_s$ as this simplifies the presentation. As U is followed by a strictly causal element, and has no feedback around it, there is no loss of generality.

every triple (u, v, y) in $\mathcal{X}^3$ satisfying (8.1) satisfies (8.3) and vice versa. Equivalence will be established subject to the equations

$$Q=F(I+P_1F)^{-1} \quad (8.4a)$$

$$F=Q(I-P_1Q)^{-1} \quad (8.4b)$$

$$R=(I+FP_1)^{-1}U \quad (8.5a)$$

$$U=(I+FP_1)R. \quad (8.5b)$$

Equation (8.4) is a repetition of (3.4). If either of (8.5) holds then, clearly, both are true.

For equivalent schemes, let $\mathbf{E}(P, P_1, Q)$ be defined as in Section VII; Let $\mathbf{K}$ be the function $\mathbf{K}$: $\mathbf{A}_s^3 x \mathbf{A} \rightarrow \mathbf{A}$, $\mathbf{K}(P, P_1, Q, R)=K_{13}$, which maps operators appearing in (8.3) into the c.l. (closed-loop) operator K_{13}. Whenever all operators except P are regarded as fixed, $\mathbf{K}(P, P_1, Q, R)$ will be denoted by $\mathbf{K}(P)$. $\mathbf{K}(P)$ can be expressed in terms of the model-reference scheme operators, by the formula

$$\mathbf{K}(P)=P[I+Q(P-P_1)]^{-1}R \quad (8.6a)$$

which is obtained from the following sequence of equations. For any (u, v, y) in $\mathcal{X}^3$ satisfying (8.3), we have

$$v=Ru-Q(P-P_1)v$$

$$[I+Q(P-P_1)]v=Ru$$

$$y=Pv=P[I+Q(P-P_1)]^{-1}Ru \quad (8.6b)$$

in which the inverse exists in $\mathbf{A}$ as $Q(P-P_1)$ is strictly causal. Since K_{13} maps u into y, K_{13} must coincide with the last operator of (8.6b), and (8.6a) is true.

The set of c.l. operators of the model reference scheme (8.3) is defined as in the preceding Section VII, except that K is augmented by the operator R.

For any P_1 in $\mathbf{A}_s$, $\mathbf{K}$ will be called (nominal) *plant invariant* iff $\mathbf{K}(P_1)=P_1$; and, for any R_1 in $\mathbf{A}$, *plant invariant* $(\times R_1)$ iff $\mathbf{K}(P_1)=P_1R_1$. From (8.4) it is clear that $\mathbf{K}$ has these properties whenever $R=I$ or $R=R_1$, respectively.

An operator $0_r \in \mathbf{B}$ is a *right zero*[19] of P_1 iff $P_1 0_r=0$.

Theorem 7:

a) Any feedback scheme (8.1) with (P, F, U) in $\mathbf{A}_s^2 \times \mathbf{A}$ is equivalent to a model reference scheme (8.3) with (P, P_1, Q, R) in $\mathbf{A}_s^3 \times \mathbf{A}$, in which (F, U) and (P_1, Q, R) are related by (8.4)–(8.5); and vice versa. If (8.4)–(5) hold, then we have the following.

b) If P assumes the (nominal) value $P_1 \in \mathbf{A}_s$, then $\mathbf{K}(P_1)=P_1R$, and if the feedback scheme is closed-loop stable, then Q and R of the model reference scheme are stable (although F and U may be unstable).

c) The differences $\mathbf{K}(P)-\mathbf{K}(P_1)$ and $P-P_1$ are related by the formulas

$$\mathbf{K}(P)-\mathbf{K}(P_1)=(I-P_1Q)(P-P_1)\{I+Q(P-P_1)\}^{-1}R \quad (8.7a)$$

$$=\mathbf{E}(P, P_1, Q)(P-P_1)R. \quad (8.7b)$$

d) For any $R_1 \in \mathbf{B}$, $\mathbf{K}$ is plant invariant $(\times R_1)$ iff $R=R_1+0_r$, where $0_r \in \mathbf{B}$ is any right zero of P_1.

Proof:

a) If (u, v, y) satisfies (8.1), then we have

$$(I-QP_1)v=(I-QP_1)(Uu-Fy) \quad [\text{by } (8.1b)]$$

$$=(I-QP_1)[(I+FP_1)Ru-Q(I-P_1Q)^{-1}Y] \quad [\text{by } (8.5b) \text{ and } (8.4b)]$$

$$=(I-QP_1)[(I-QP_1)^{-1}Ru-(I-QP_1)^{-1}Qy]$$

$$=Ru-Qu$$

where the second last equation was obtained using the identity $(I+FP_1)=(I-QP_1)^{-1}$ [see (3.5)] and Property 2.1a. Hence, (u, v, y) satisfies (8.3). The reverse assertion is proved similarly. It follows that (8.1) and (8.3) are equivalent.

b) From (8.4), $\mathbf{K}(P_1)=P_1R$; if (8.1) is closed-loop stable, then the c.l. operators $K_{32}=F(I+P_1F)^{-1}$ and $K_{12}=(I+FP)^{-1}U$ are stable, and equal Q and R.

c) From (8.4) the following sequence of equations is obtained:

$$\mathbf{K}(P)-\mathbf{K}(P_1)=P[I+Q(P-P_1)]^{-1}R-P_1R$$

$$=\{P-P_1[I+Q(P-P_1)]\}[I+Q(P-P_1)]^{-1}R$$

from which (8.7a) follows; (8.7b) is obtained by Proposition 2.1a.

d) $\mathbf{K}$ is plant invariant $(\times R_1)$ iff $\mathbf{K}(P_1)-P_1R_1=P_1(R-R_1)=0$, i.e., $R-R_1=0_r$. Q.E.D.

Remark: Plant invariant schemes allow us to divide any control-law synthesis into two stages: 1) filtering of plant uncertainty $P-P_1$ and 2) design of a control law for a nominal plant P_1, with the assurance at least that the filtering stage will improve the design. In general, there is no "separation principle" to guarantee optimality of the division.

Either stage may come first. Therefore, in our theorems, P can be interpreted either as a plant without controller, for which a controller will be designed eventually; or, as a plant with controller attached, which requires additional filtering only to the extent that P differs from P_1.

B. Stabilizing Feedbacks

In the rest of Section VIII it will be assumed that: P and P_1 belong to a subspace $\mathbf{B}_r$ of the strongly causal operators $\mathbf{B}_s$, on which a right seminorm $\|\cdot\|_r$ is defined; P lies in a ball $b(P_1, \delta)$, defined as in Section VII-A, of what can be interpreted either as uncertainty or perturbations; and the equivalence conditions (8.4), (8.5) hold, so that the feedback and model reference schemes (8.1), (8.3) are equivalent.

Q in $\mathbf{A}_s$ and R in $\mathbf{A}$ are sought which give low sensitivity and maintain c.l. boundedness. In view of Theorem 7b), the assumption that Q is in $\mathbf{B}_s$ and R in $\mathbf{B}$, i.e., that both are stable, can be made without loss of generality.

[19]More simply, 0_r is an operator whose range is in the nullspace of P_1. The term "zero" is appropriate for a normed algebra.

For any $b(P_1,\delta)$, the definitions of a bounded operator, c.l. bounded scheme, and stabilizing Q were given in Section VII-A. The set $\mathbf{B}_s(P_1,\delta)$ of Q in $\mathbf{B}_s$ stabilizing for $b(P_1,\delta)$ was introduced.

Proposition 8.1: Under the present hypotheses, Propositions 7.1 and 7.2, are valid for the feedback scheme (8.1) and model reference scheme (8.3), and the set of Q in $\mathbf{B}_s$ stabilizing for $b(P_1,\delta)$ coincides with $\mathbf{B}_s(P_1,\delta)$.

Proof: As R is in $\mathbf{B}$, the set K augmented by R consists of operators bounded on $b(P_1,\delta)$ iff the unaugmented set K consists of operators bounded on $b(P_1,\delta)$. Therefore, $\{K_{ij}\}_{i,j=1}^{6}$ are bounded iff $\{K_{ij}\}_{i,j=1}^{5}$ are bounded. The conclusion follows. Q.E.D.

C. Sensitivity to Plant Perturbations or Uncertainty

Suppose that a pair of seminorms $(\|\cdot\|_W,\|\cdot\|_1)$ is defined on the spaces $(\mathbf{B},\mathbf{B}_r)$, respectively. We assume $\|\cdot\|_W$ to be a left seminorm, but leave open the possibility that $\|\cdot\|_1$ is not a right seminorm, and assume instead that $\|\cdot\|_1\leqslant\|\cdot\|_r$, i.e., $\|\cdot\|_1$ is dominated by the right seminorm $\|\cdot\|_r$. A weighted seminorm $\|\cdot\|_v$ is assumed to be defined on the set of products $\mathbf{B}\cdot\mathbf{B}_r$, and to have the multiplicative property, $\|CD\|_v\leqslant\|C\|_W\|D\|_1$ for all C in $\mathbf{B}$ and D in $\mathbf{B}_r$. The pair $(\|\cdot\|_W,\|\cdot\|_1)$ will be called *aligned* iff there is a $D\in\mathbf{B}_r$ for which the preceding inequality is an equality for all C in $\mathbf{B}$.

For the plant P in any ball of uncertainty $b(P_1,\delta)$, and any ($F\in\mathbf{A}_s$, $U\in\mathbf{A}$), or ($Q\in\mathbf{B}_s$, $R\in\mathbf{B}$), the *sensitivity to plant perturbations* (or uncertainty) of the equivalent schemes (8.1)–(8.3) is defined to be

$$\nu(P_1,Q,R;\delta)=\sup_{P\in b(P_1,\delta)}\left\{\|\mathbf{K}(P)-\mathbf{K}(P_1)\|_v\delta^{-1}\right\} \tag{8.8}$$

for Q stabilizing (i.e., in $\mathbf{B}_s(P_1,\delta)$), and $\nu(P_1,Q,R,;\delta)=\infty$ otherwise. For any stabilizing Q, (8.7) gives the equation

$$\nu(P_1,Q,R;\delta)=\sup_{P\in b(P_1,\delta)}\left\{\|\mathbf{E}(P,P_1,Q)(P-P_1)R\|_v\delta^{-1}\right\}. \tag{8.9}$$

The following lemma relates the disturbance and plant perturbation sensitivities, η and ν, to each other, and to the $\|\cdot\|_W$-singularity measure $\mu(P_1)$, when $R=I$.

Lemma 8.2: $\nu(P_1,Q,I;\delta)$ is a monotone nondecreasing function of $\delta\geqslant 0$ satisfying the inequality

$$\nu(P_1,Q,I;\delta)\leqslant\eta(P_1,Q;\delta); \tag{8.10}$$

and if $\|\cdot\|_1=\|\cdot\|_r$ and the pair $(\|\cdot\|_W,\|\cdot\|_r)$ is aligned then

$$\nu(P_1,Q,I;\delta)\geqslant\mu(P_1). \tag{8.11}$$

Remark: $\|\cdot\|_v$ can coincide with the principal norm $\|\cdot\|$. For example, if W is in $\mathbf{B}_r$, $\|W\|=1$, $\|C\|_W\triangleq\|CW\|$, and $\|D\|_1\triangleq\|W^{-1}D\|$, then the principal norm has the multiplicative property $\|CD\|_v\leqslant\|C\|_W\|D\|_1$ and may be used as the $\|\cdot\|_v$ norm. Even though $\|\cdot\|$ provides no weighting, (8.10) shows that sensitivities ν smaller than 1 can be achieved whenever $\nu(P_1)<1$. In effect, the plant perturbations supply their own weighting.

This example has the alignment property, as $\|CW\|_v=\|C\|_W\|W\|_1$.

Proof of Lemma 8.2: $\nu(P_1,Q,I;\delta)$ is a $\uparrow$ function of δ, as it is a sup over sets $b(P_1,\delta)$ which are nested and nondecreasing with $\delta\in\Delta$, and is ∞ for $\delta\notin\Delta$. For any P in $b(P_1,\delta)$ and stabilizing Q, we have the inequalities

$$\begin{aligned}\|\mathbf{E}(P,P_1,Q)(P-P_1)\|_v\delta^{-1}&\\ &\leqslant\|\mathbf{E}(P,P_1,Q)\|_W\|P-P_1\|_1\delta^{-1}\\ &\leqslant\|\mathbf{E}(P,P_1,Q)\|_W\end{aligned} \tag{8.12}$$

by the multiplicative property of $\|\cdot\|_v$, and as

$$\|P-P_1\|_1\leqslant\|P-P_1\|_r\leqslant\delta$$

is obtained by taking the sup of both sides of (8.12) over all P in $b(P_1,\delta)$. If $\|\cdot\|_1=\|\cdot\|_r$ and $(\|\cdot\|_W,\|\cdot\|_r)$ is aligned there is an operator $D\in\mathbf{B}_r$ for which $\|\mathbf{E}(P,P_1,Q)D\|_v=\|\mathbf{E}(P,P_1,Q)\|_W\|D\|_r$. The operator $P_0\triangleq P_1+\delta D\|D\|_r^{-1}$ is in $b(P_1,\delta)$ and $\|\mathbf{E}(P_0,P_1,\delta)(P-P_1)\|_v\delta^{-1}=\|\mathbf{E}(P_1,P_1,Q)\|_W\geqslant\mu(P_1)$. The sup in (8.9) must have the lower bound $\mu(P_1)$, and (8.11) follows. Q.E.D.

D. Assumptions on R

As $R=0$ gives $\nu=0$, the attainment of a small sensitivity is trivial unless R is constrained. In many problems a target value of the c.l. operator K_{13} is specified for some nominal value of the plant. By Theorem 7b), the target value can be attained iff it has the form P_1R. We therefore make the following assumption.

Assumption 1: An $R_1\in\mathbf{B}$ is given for which the equation $\mathbf{K}(P_1)=P_1R_1$ must be satisfied, i.e., $\mathbf{K}$ is plant invariant $(\times R_1)$. By Theorem 7d), $R=R_1+O_r$, where $O_r\in\mathbf{B}$ is any right zero of P_1. There are now two variables to be optimized, Q and O_r. Their simultaneous optimization can be difficult, and we make the following simplifying assumption.

Assumption 2: $O_r=0$. Assumption 2 may constrain the class of allowable feedbacks and thereby give suboptimal sensitivities. However, it is clear from (8.7b) that there is no such constraint if the condition

$$(P-P_1)O_r=0 \tag{8.13}$$

is fulfilled. Equation (8.13) is fulfilled whenever $\mathsf{N}(P_1)$ (nullspace of P_1) is contained in $\mathsf{N}(P)$ for all P in $b(P_1,\delta)$; for then, for any $u\in\mathfrak{X}$, either $O_ru=0$, or $O_ru\in\mathsf{N}(P_1)$ and so $O_ru\in\mathsf{N}(P)$. In either case, $(P-P_1)O_ru=0$, i.e., (8.13) is true. If $\mathsf{N}(P_1)$ is trivial, then, of course, (8.13) is fulfilled. For example, $\mathsf{N}(P_1)$ is trivial for any nonzero P_1 in $\mathbf{H}_0^\infty$, because functions analytic and not identically zero in $\mathrm{Re}(s)\geqslant 0$ have at most a countable number of $\mathrm{Re}(s)\geqslant 0$ zeros.

As R in (8.6) is now fixed, cR can be absorbed into the weighting for some constant λ in (0, 1] by replacing $\|\cdot\|_v$ with the new seminorm $\|C\|_v\triangleq\lambda\|CR\|_v$, and similarly for $\|\cdot\|_1$. Without loss of generality, we can therefore make the following assumption.

Assumption 3: $R=I$. We prefer to absorb R rather than to show it explicitly, as this approach reduces the number of variables and allows sensitivity to be related to weighted invertibility.

E. *Minimal Sensitivity*

For any P_1 in $\mathbf{B}_s$ and $\delta>0$, the *minimal sensitivity to plant perturbations* is defined to be

$$\nu(P_1,\delta)=\inf_{Q\in\mathbf{B}}\{\nu(P_1,Q,I;\delta)\}$$

$$=\inf_{Q\in\mathbf{B}}\sup_{P\in b(P_1,\delta)}\{\|\mathbf{E}(P,P_1,Q)(P-P_1)\|_v\delta^{-1}\}$$

provided the last sup is replaced by ∞ for nonstabilizing Q. For fixed P_1, $\nu(P_1,\delta)$ is a function of $\delta\geq 0$ which represents the dependence of perturbation sensitivity on plant uncertainty.

Theorem 8: For any nominal plant P_1 in $\mathbf{B}_s$, the minimal sensitivity to plant perturbations $\nu(P_1,\delta)$ is a monotone nondecreasing function of $\delta\geq 0$.

$\nu(P_1,\delta)$ satisfies the upper bound conditions $\nu(P_1,\delta)\leq\eta(P_1,\delta)$ and $\lim_{\delta\to 0}\nu(P_1,\delta)\leq\mu(P_1)$. If $\|\cdot\|_1=\|\cdot\|_r$, and $(\|\cdot\|_W,\|\cdot\|_1)$ are aligned, then $\lim_{\delta\to 0}\nu(P_1,\delta)=\mu(P_1)$ and, for $\delta=\|P_1\|_r$, $\|P_1\|_v\|P_1\|_r^{-1}\leq\nu(P_1,\delta)\leq\|I\|_W$.

Proof: $\nu(P_1,\delta)$ is a $\uparrow$ function of $\delta\geq 0$, as by definition it is a sup over Q in $\mathbf{B}_s$ of functions which are $\uparrow$ by Lemma 8.2. The inequality $\nu(P_1,\delta)\leq\eta(P_1,\delta)$ is obtained by taking $\sup_{Q\in\mathbf{B}}$ of both sides of (8.10). It follows by Theorem 5 that $\lim_{\delta\to 0}u(P_1,\delta)\leq\mu(P_1)$, and that $\nu(P_1,\delta)\leq\|I\|_W$ for $\delta=\|P_1\|_r$. If alignment and $\|\cdot\|_1=\|\cdot\|_r$ are assumed, then taking the $\inf_{Q\in\mathbf{B}}$ of both sides of (8.11) gives the inequality $\nu(P_1,\delta)\geq\mu(P_1)$, and the conclusion that $\lim_{\delta\to 0}\nu(P_1,\delta)=\mu(P_1)$ follows.

As the operator $P=0$ lies in $b(P_1,\|P_1\|_r)$ and, for each Q in $\mathbf{B}_s$, $\|E(0,P_1,Q)(-P_1)\|_v\|P_1\|_r^{-1}=\|P_1\|_v\|P_1\|_r^{-1}$; therefore $\|P_1\|_v\|P_1\|_r^{-1}\leq\nu(P_1,\|P_1\|_r)$. Q.E.D.

Remark: The ratio $\|P_1\|_v\|P_1\|_r^{-1}$ is a measure of the amount by which the $\|\cdot\|_v$ norm weighs down the nominal plant. Provided norms are used for which the weighting is not excessive, in fact whenever the ratio exceeds $\mu(P_1)$, there must be points in the interval $[0,\|P_1\|_r]$ at which any increase in plant uncertainty causes a worsening of minimal sensitivity.

APPENDIX I

Proof of Proposition 2.1:

a) If $(I+PQ)^{-1}$ exists in B, let $R\triangleq I-Q(I+PQ)^{-1}P$. Now, $(I+QP)R=I$, because

$$\begin{aligned}(I+QP)R&=I-Q(I+PQ)^{-1}P+QP-QPQ(I+PQ)^{-1}P\\&=I-Q(I+PQ)^{-1}P+Q\left[I-PQ(I+PQ)^{-1}\right]P\\&=I-Q(I+PQ)^{-1}P+Q(I+PQ)^{-1}P=I.\end{aligned}$$

Similarly, $R(I+QP)=I$, so R is the inverse of $(I+QP)$ in B. The required formula is obtained by multiplying both sides of the equation $(I+PQ)P=P(I+QP)$ by $(I+PQ)^{-1}$ on the left, and by $(I+QP)^{-1}$ on the right.

b) If P is in R and P^{-1} in B, there is a contradiction: any F in B belongs to R as $F=PP^{-1}F$, so R is not a proper subspace and not a radical.

c) Let $R\triangleq(I+P)^{-1}$ and observe that $R=I-PR$. By the triangle inequality, $\|R\|\leq 1+\|P\|\|R\|$ and, as $\|P\|<1$, $\|R\|\leq(1-\|P\|)^{-1}$. Q.E.D.

Proof of Proposition 3.5: Since $\|(I-PQ)\Pi\|=1$ when $Q=0$, the infimum α satisfies $\alpha\leq 1$. Suppose $\alpha\neq 1$; then there is a Q in $\mathbf{B}_s$ for which $\|(I-PQ)\Pi\|\triangleq\alpha_1<1$. It will now be shown that given any $\epsilon>0$, there is a Q_1 in $\mathbf{B}_s$ for which $\|(I-PQ_1)\Pi\|<\epsilon$, and so the proposition is true.

Let n be any integer for which $\alpha^n<\epsilon$. Now, $(I-PQ)^n=I+a_1PQ+\cdots+a_n(PQ)^n$ is a polynomial in PQ, in which every term except the first has P as a left factor, and Q in $\mathbf{B}_s$ as a factor. Therefore, there is a Q_1 in $\mathbf{B}_s$ for which the equation $(I-PQ)^n=(I-PQ_1)$ holds. Now as $\Pi\mathbf{B}$ is invariant under all operators in $\mathbf{B}$, we get $\|(I-PQ_1)\Pi\|=\|\{(I-PQ)\Pi\}^n\|\leq\|(I-PQ)\Pi\|^n\leq\alpha_1^n<\epsilon$, as claimed. Q.E.D.

APPENDIX II
UNSTABLE PLANTS: REMARKS

Consider the problem of input–output stabilizing an unstable plant P_0 in $\mathbf{H}^\infty_{e0}$ by identity feedback, given that the *a priori* information about P consists of $\hat{p}(j\omega)$ and the number N_+ of $\mathrm{Re}(s)\geq 0$ poles of P. By Nyquist's criterion, uncertainty as to N_+ translates into uncertainty[20] as to the number of unstable poles of $(I+P_0)^{-1}$. N_+ must not be underestimated if stability is to be assured. However, it is impossible to be sure that N_+ has not been underestimated from purely input–output measurements. For example, let s_i, $i=1,2,\cdots n$, be any finite set of frequencies, and $\hat{p}_m(s_i)$ a set of n measurements of $\hat{p}_0(s)$ of tolerance ϵ, i.e., $|\hat{p}_m(s_i)-\hat{p}_0(s)|<\epsilon$. It is impossible to deduce N_+ from the measurements, as it is always possible to find a dipole, $(s+\alpha)(s+\alpha+\delta)^{-1}$, with $\delta>0$ so small that multiplication of $\hat{p}_0(s)$ by the dipole changes $\hat{p}_0(s_i)$ by less than ϵ, but increases N_+ by 1. In our particular setting, it appears that input–output stabilization cannot be accomplished using solely input–output data. The implication is that some information concerning internal structure is essential, and that stabilization is not a legitimate problem for a purely input–output theory.

On the other hand, desensitization can be achieved using input–output data. If P is in H^∞_0, and $p(t)$ satisfies appropriate restrictions on smoothness and convergence to 0 as $t\to\infty$, then $\hat{p}(j\omega)$ has a finite modulus of continuity, and $\hat{p}$ can be located in a ϵ-ball of H^∞ by a finite set of measurements [14]. This is sufficient for the purpose of desensitization, as sensitivity depends continuously on $\hat{p}$ in H^∞ (see Theorems 5 and 7).

[20] A more complete discussion is in Zames and El-Sakkary [18].

Synopsis

The main results of this paper are the Theorems 1–7.

Section II: Spaces of frequency responses and algebras of input–output mappings are defined, and their relevant properties summarized.

Section III: A decomposition principle is derived. The disturbance attenuation problem is separated into two independent stages: stabilization, followed by desensitization of a stable system. The second stage is a model reference scheme.

Section III′: Some constraints on sensitivity norms are displayed. It is shown that the induced operator norm is useless as a measure of sensitivity.

Section IV: The concepts of a weighted seminorm, approximate inverse, and measure of singularity are introduced and illustrated by examples.

Section V: The plant is assumed to be stable and known precisely. Sensitivity is defined. Approximate invertibility is shown to be a necessary and sufficient condition for sensitivity reduction in Theorem 2, and optimal sensitivity is shown to be equal to the measure of singularity. An example of sensitivity minimization is solved in Section V-A. Unstable plants are discussed in Section V-B. Identity and inverting sequences are introduced in Section V-C, and a lemma on products of inverting sequences is proved.

Section VI: Devoted to multivariable systems. In Theorem 3, an optimal sequence of compensators is derived for a plant factorable into a product of nearly invertible and noninvertible factors. Corollary 6.2 specializes this result to a single-input single-output plant with a RHP zero. Corollary 6.2 gives conditions under which the sensitivity of multivariable systems without RHP zeros can be made arbitrarily small. Theorem 4 shows that sensitivity can never be made small if there are zeros in any heavily weighted part of the RHP.

Section VII: The plant is assumed to lie in a ball of uncertainty around a nominal value. Feedbacks stabilizing over a ball are defined. Optimal sensitivity is shown to be a monotone nondecreasing function of uncertainty in Theorem 5, and various bounds are obtained.

Section VIII: Problem 2 is formulated, again in terms of an equivalence between feedback and model reference schemes, this time subject to a plant invariance property, in Theorem 6. Sensitivity to plant perturbations is defined, bounded, and optimal sensitivity is shown to be a monotone nondecreasing function of plant uncertainty in Theorem 7.

Acknowledgment

The author thanks C. A. Desoer for many useful suggestions concerning the draft of this paper.

References

[1] H. W. Bode, *Network Analysis and Feedback Amplifier Design.* Princeton, NJ: Van Nostrand, 1945.

[2] M. Horowitz, *Synthesis of Feedback Systems.* New York: Academic, 1963.

[3] G. Zames, "Nonlinear operators—Cascading inversion, and feedback" M.I.T., *R.L.E., Quart. P.R.*, vol. 53, pp. 93–107, Apr. 1959.

[4] ——, "Nonlinear operators for system analysis," Res. Lab. Electron., M.I.T., Tech. Rep. 370, Sept. 1960.

[5] ——, "functional analysis applied to nonlinear feedback systems," *IEEE Trans. Circuit Theory*, vol. CT-10, pp. 392–404, Sept. 1963.

[6] C. A. Desoer, "Nonlinear distortion in feedback amplifiers," *IRE Trans. Circuit Theory*, vol. CT-9, pp. 2–6, Mar. 1962.

[7] ——, "Perturbation in the I/O map of a nonlinear feedback system caused by large plant perturbation," *J. Franklin Inst.*, vol. 306, pp. 225–235, Sept. 1978.

[8] J. B. Cruz, Jr. and W. R. Perkins, "A new approach to the sensitivity problem in multivariable feedback system design," *IEEE Trans. Automat. Contr.*, vol. AC-9, pp. 216–223, 1964.

[9] W. A. Porter and C. L. Zahm, "Basic concepts in system theory," System Eng. Lab., Univ. of Michigan, Tech. Rep. 33, 1969.

[10] R. M. Desantis and W. A. Porter, "A generalized Nyquist plot and its use in sensitivity analysis," *Int. J. Syst. Sci.*, vol. 5–12, pp. 1143–1153, 1974.

[11] J. C. Willems, *The Analysis of Feedback Systems.* Cambridge, MA: M.I.T., 1971.

[12] G. Zames, "Feedback hierarchies and complexity," in *Proc. IEEE Conf. Decision Contr.*, Dec. 1976, addenda.

[13] ——, "On the metric complexity of causal linear systems, ϵ-entropy and ϵ-dimension for continuous time," *IEEE Trans. Automat. Contr.*, vol. AC-12, pp. 222–230, Apr. 1979.

[14] ——, "Realizability conditions for nonlinear feedback systems," *IEEE Trans. Circuit Theory*, vol. CT-11, pp. 186–194, June 1964.

[15] R. G. Douglas, *Banach Algebra Techniques in Operator Theory.* New York: Academic, 1972.

[16] B. A. Francis and W. M. Wonham, "The internal model principle for linear multivariable regulators," *J. Appl. Math. Optimiz.*, vol. 2, pp. 170–194, 1975.

[17] D. M. Bensoussan, "High-gain feedback and sensitivity," Ph.D. dissertation, Dep. Elec. Eng., McGill Univ., to appear.

[18] A. K. El-Sakkary, "The gap metric for unstable systems," Ph.D. dissertation, Dep. Elec. Eng., McGill Univ.; also in G. Zames and A. K. El-Sakkary, *Proc. 18th Allerton Conf.*, Oct. 1980.

[19] M. A. Naimark, *Normed Rings.* Groningen: Noordhooff, 1964.

[20] A. Feintuch, "Strong Causality Conditions and Causal Invertibility," *SIAM J. Contr. Optimiz.*, vol. 18, pp. 317–324, May 1980.

[21] C. A. Desoer and W. S. Chan, "Feedback interconnection of lumped linear time invariant systems," *J. Franklin Inst.*, vol. 300, pp. 335–351, 1975.

[22] W. Rudin, *Functional Analysis.* New York: McGraw Hill, 1973.

[23] G. Zames, "Feedback and optimal sensitivity: Model reference transformations, weighted seminorms, and approximate inverses," in *Proc. 17th Allerton Conf.*, Oct. 1979, pp. 744–752.

Multivariable Feedback Design: Concepts for a Classical/Modern Synthesis

JOHN C. DOYLE AND GUNTER STEIN, MEMBER, IEEE

Abstract—This paper presents a practical design perspective on multivariable feedback control problems. It reviews the basic issue—feedback design in the face of uncertainties—and generalizes known single-input, single-output (SISO) statements and constraints of the design problem to multiinput, multioutput (MIMO) cases. Two major MIMO design approaches are then evaluated in the context of these results.

I. Introduction

THE last two decades have brought major developments in the mathematical theory of multivariable linear time invariant feedback systems. These include the celebrated state space concept for system description and the notions of mathematical optimization for controller synthesis [1], [2]. Various time-domain-based analytical and computational tools have been made possible by these ideas. The developments also include certain generalizations of frequency-domain concepts which offer analysis and synthesis tools in the classical single-input, single-output (SISO) tradition [3], [4]. Unfortunately, however, the two decades have also brought a growing schism between practitioners of feedback control design and its theoreticians. The theory has increasingly concentrated on analytical issues and has placed little emphasis on issues which are important and interesting from the perspective of design.

This paper is an attempt to express the latter perspective and to examine the extent to which modern results are meaningful to it. The paper begins with a review of the fundamental practical issue in feedback design—namely, how to achieve the benefits of feedback in the face of uncertainties. Various types of uncertainties which arise in physical systems are briefly described and so-called "unstructured uncertainties" are singled out as generic errors which are associated with all design models. The paper then shows how classical SISO statements of the feedback design problem in the face of unstructured uncertainties can be reliably generalized to multiinput, multioutput (MIMO) systems, and it develops MIMO generalizations of the classical Bode gain/phase constraints [5], [6] which limit ultimate performance of feedback in the face of such uncertainties. Several proposed MIMO design procedures are examined next in the context of the fundamental feedback design issue. These include the recent frequency domain inverse Nyquist array (INA) and characteristic loci (CL) methods and the well-known linear-quadratic Gaussian (LQG) procedure. The INA and CL methods are found to be effective, but only in special cases, while LQG methods, if used properly, have desirable general features. The latter are fortunate consequences of quadratic optimization, not explicitly sought after or tested for by the theoretical developers of the procedure. Practitioners should find them valuable for design.

Manuscript received March 13, 1980; revised October 6, 1980. This research was supported by the ONR under Contract N00014-75-C-0144, by the DOE under Contract ET-78-C-01-3391, and by NASA under Grant NGL-22-009-124.

J. C. Doyle is with the Systems and Research Center, Honeywell, Inc., Minneapolis, MN 55413 and the University of California, Berkeley, CA 94720.

G. Stein is with the Systems and Research Center, Honeywell, Inc., Minneapolis, MN 55413 and the Department of Electrical Engineering and Computer Sciences, Massachusetts Institute of Technology, Cambridge, MA 02139.

II. Feedback Fundamentals

We will deal with the standard feedback configuration illustrated in Fig. 1. It consists of the interconnected plant (G) and controller (K) forced by commands (r), measurement noise (η), and disturbances (d). The dashed precompensator (P) is an optional element used to achieve deliberate command shaping or to represent a nonunity feedback system in equivalent unity feedback form. All disturbances are assumed to be reflected to the measured outputs (y), all signals are multivariable, in general, and both nominal mathematical models for G and K are finite dimensional linear time invariant (FDLTI) systems with transfer function matrices $G(s)$ and $K(s)$. Then it is well known that the configuration, if it is stable, has the following major properties:

1) Input–Output Behavior:

$$y = GK(I+GK)^{-1}(r-\eta)+(I+GK)^{-1}d \tag{1}$$

$$\begin{aligned} e &\triangleq r-y \\ &= (I+GK)^{-1}(r-d)+GK(I+GK)^{-1}\eta. \end{aligned} \tag{2}$$

2) System Sensitivity [7]:

$$\Delta H_{cl} = (I+G'K)^{-1}\Delta H_{ol}. \tag{3}$$

In (3), ΔH_{cl} and ΔH_{ol} denote changes in the closed-loop system and changes in a nominally equivalent open-loop system, respectively, caused by changes in the plant G, i.e., $G' = G + \Delta G$.

Equations (1)–(3) summarize the fundamental benefits and design objectives inherent in feedback loops. Specifically, (2) shows that the loop's errors in the presence of

Reprinted from *IEEE Trans. Automat. Contr.*, vol. AC-26, no. 1, pp. 4–16, Feb. 1981.

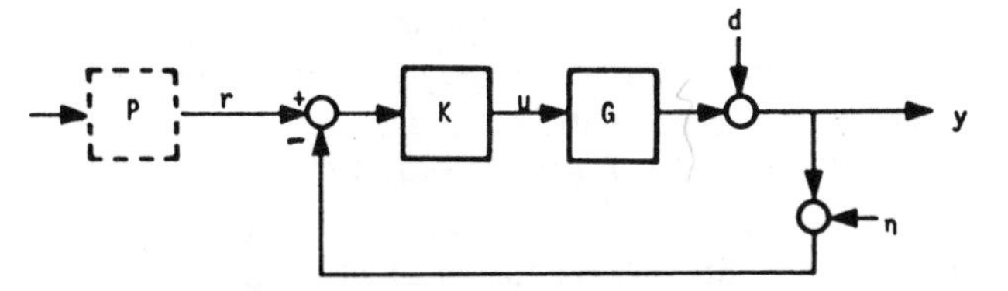

Fig. 1. Standard feedback configuration.

commands and disturbances can be made "small" by making the sensitivity operator, or inverse return difference operator, $(I+GK)^{-1}$, "small," and (3) shows that loop sensitivity is improved under these same conditions, provided G' does not stray too far from G.

For SISO systems, the appropriate notion of smallness for the sensitivity operator is well-understood—namely, we require that the complex scalar $[1+g(j\omega)k(j\omega)]^{-1}$ have small magnitude, or conversely that $1+g(j\omega)k(j\omega)$ have large magnitude, for all real frequencies ω where the commands, disturbances and/or plant changes, ΔG, are significant. In fact, the performance objectives of SISO feedback systems are commonly stipulated in terms of explicit inequalities of the form

$$ps(\omega) < |1+g(j\omega)k(j\omega)| \qquad \forall\omega < \omega_0, \tag{4}$$

where $ps(\omega)$ is a (large) positive function and ω_0 specifies the active frequency range.

This basic idea can be readily extended to MIMO problems through the use of matrix norms. Selecting the spectral norms as our measure of matrix size, for example, the corresponding feedback requirements become

$$\bar{\sigma}\left[(I+G(j\omega)K(j\omega))^{-1}\right] \text{ small}$$

or conversely

$$ps(\omega) < \underline{\sigma}[I+G(j\omega)K(j\omega)] \tag{5}$$

for the necessary range of frequencies. The symbols $\bar{\sigma}$ and $\underline{\sigma}$ in these expressions are defined as follows:

$$\bar{\sigma}[A] \triangleq \max_{\|x\|=1} \|Ax\| \equiv \sqrt{\lambda_{\max}[A^*A]} \tag{6}$$

$$\underline{\sigma}[A] \triangleq \min_{\|x\|=1} \|Ax\| \equiv \sqrt{\lambda_{\min}[A^*A]} \tag{7}$$

where $\|\cdot\|$ is the usual Euclidean norm, $\lambda[\cdot]$ denotes eigenvalues, and $[\cdot]^*$ denotes conjugate transpose. The two σ's are called maximum and minimum singular values of A (or principal gains [4]), respectively, and can be calculated with available linear system software [8]. More discussion of singular values and their properties can be found in various texts [9].

Condition (5) on the return difference $I+GK$ can be interpreted as merely a restatement of the common intuition that large loop gains or "tight" loops yield good performance. This follows from the inequalities

$$\underline{\sigma}[GK]-1 < \underline{\sigma}[I+GK] < \underline{\sigma}[GK]+1 \tag{8}$$

which show that return difference magnitudes approximate the loop gains, $\underline{\sigma}[GK]$, whenever these are large compared with unity. Evidently, good multivariable feedback loop design boils down to achieving high loop gains in the necessary frequency range.

Despite the simplicity of this last statement, it is clear from years of research and design activity that feedback design is not trivial. This is true because loop gains cannot be made arbitrarily high over arbitrarily large frequency ranges. Rather, they must satisfy certain performance tradeoffs and design limitations. A major performance tradeoff, for example, concerns command and disturbance error reduction versus sensor noise error reduction [10]. The conflict between these two objectives is evident in (2). Large $\underline{\sigma}[GK(j\omega)]$ values over a large frequency range make errors due to r and d small. However, they also make errors due to η large because this noise is "passed through" over the same frequency range, i.e.,

$$y = GK(j\omega)[I+GK(j\omega)]^{-1}\eta \approx I\eta. \tag{9}$$

Worse still, large loop gains can make the control activity (variable u in Fig. 1) quite unacceptable. This follows from

$$\begin{aligned} u &= K[I+GK]^{-1}(r-\eta-d) \\ &\approx G^{-1}(j\omega)(r-\eta-d). \end{aligned} \tag{10}$$

Here we have assumed G to be square and invertible for convenience. The resulting equation shows that commands, disturbances, and sensor noise are actually amplified at u whenever the frequency range significantly exceeds the bandwidth of G, i.e., for ω such that $\bar{\sigma}[G(j\omega)] \ll 1$ we get

$$\underline{\sigma}[G^{-1}(j\omega)] \equiv \frac{1}{\bar{\sigma}[G(j\omega)]} \gg 1. \tag{11}$$

One of the major contributions of modern feedback theory is the development of systematic procedures for conducting the above performance tradeoffs. We are referring, of course, to the LQG theory [11] and to its modern Wiener–Hopf frequency domain counterpart [12]. Under reasonable assumptions on plant, disturbances, and performance criteria, these procedures yield efficient design compromises. In fact, if the tradeoff between command/disturbance error reduction and sensor noise error reduction were the only constraint on feedback design, practitioners would have little to complain about with respect to the relevance of modern theory. The problem is that these performance trades are often overshadowed by a second limitation on high loop gains—namely, the requirement for tolerance to uncertainties. Although a controller may be designed using FDLTI models, the design must be implemented and operate with a real physical plant. The properties of physical systems, in particular the ways in which they deviate from finite-dimensional linear models, put strict limitations on the frequency range over which the loop gains may be large. In order to properly motivate these restrictions, we digress in Section III to a brief description of the types of system uncertainties most frequently encountered. The manner in which these uncertainties can be accounted for in MIMO design then forms the basis for the rest of the paper.

III. Uncertainties

While no nominal design model $G(s)$ can emulate a physical plant perfectly, it is clear that some models do so with greater fidelity than others. Hence, no nominal model should be considered complete without some assessment of its errors. We will call these errors the "model uncertainties," and whatever mechanism is used to express them will be called a "representation of uncertainty."

Representations of uncertainty vary primarily in terms of the amount of structure they contain. This reflects both our knowledge of the physical mechanisms which cause differences between model and plant and our ability to represent these mechanisms in a way that facilitates convenient manipulation. For example, a set membership statement for the parameters of an otherwise known FDLTI model is a highly structured representation of uncertainty. It typically arises from the use of linear incremental models at various operating points, e.g., aerodynamic coefficients in flight control vary with flight environment and aircraft configurations, and equation coefficients in power plant control vary with aging, slag buildup, coal composition, etc. In each case, the amounts of variation and any known relationships between parameters can be expressed by confining the parameters to appropriately defined subsets of parameter space. A specific example of such a parameterization for the F-8C aircraft is given in [13]. Examples of less-structured representations of uncertainty are direct set membership statements for the transfer function matrix of the model. For instance, the statement

$$G'(j\omega)=G(j\omega)+\Delta G(j\omega)$$

with

$$\bar{\sigma}[\Delta G(j\omega)]<l_a(\omega) \qquad \forall\omega>0 \tag{12}$$

where $l_a(\cdot)$ is a positive scalar function, confines the matrix G' to a neighborhood of G with magnitude $l_a(\omega)$. The statement does not imply a mechanism or structure which gives rise to ΔG. The uncertainty may be caused by parameter changes, as above, or by neglected dynamics, or by a host of other unspecified effects. An alternative statement for (12) is the so-called multiplicative form:

$$G'(j\omega)=[I+L(j\omega)]G(j\omega)$$

with

$$\bar{\sigma}[L(j\omega)]<l_m(\omega) \qquad \forall\omega>0. \tag{13}$$

This statement confines G' to a normalized neighborhood of G. It is preferable over (12) because compensated transfer functions have the same uncertainty representation as the raw model (i.e., the bound (13) applies to GK as well as to G). Still other alternative set membership statements are the inverse forms of (12) and (13) which confine $(G')^{-1}$ to direct or normalized neighborhoods about G^{-1}.

The best choice of uncertainty representation for a specific FDLTI model depends, of course, on the errors the model makes. In practice, it is generally possible to represent some of these errors in a highly structured parameterized form. These are usually the low frequency error components. There are always remaining higher frequency errors, however, which cannot be covered this way. These are caused by such effects as infinite-dimensional electromechanical resonances [16], [17], time delays, diffusion processes, etc. Fortunately, the less-structured representations, (12) or (13), are well suited to represent this latter class of errors. Consequently, (12) and (13) have become widely used "generic" uncertainty representations for FDLTI models.

Motivated by these observations, we will focus throughout the rest of this paper exclusively on the effects of uncertainties as represented by (13). For lack of a better name, we will refer to these uncertainties simply as "unstructured." We will assume that G' in (13) remains a strictly proper FDLTI system and that G' has the same number of unstable modes as G. The unstable modes of G' and G do not need to be identical, however, and hence $L(s)$ may be an unstable operator. These restricted assumptions on G' make exposition easy. More general perturbations (e.g., time varying, infinite dimensional, nonlinear) can also be covered by the bounds in (13) provided they are given appropriate "conic sector" interpretations via Parseval's theorem. This connection is developed in [14], [15] and will not be pursued here.

When used to represent the various high frequency mechanisms mentioned above, the bounding functions $l_m(\omega)$ in (13) commonly have the properties illustrated in Fig. 2. They are small ($\ll 1$) at low frequencies and increase to unity and above at higher frequencies. The growth with frequency inevitably occurs because phase uncertainties eventually exceed ± 180 degrees and magnitude deviations eventually exceed the nominal transfer function magnitudes. Readers who are skeptical about this reality are encouraged to try a few experiments with physical devices.

It should also be noted that the representation of uncertainty in (13) can be used to include perturbation effects that are in fact not at all uncertain. A nonlinear element, for example, may be quite accurately modeled, but because our design techniques cannot deal with the nonlinearity effectively, it is treated as a conic linearity [14], [15]. As another example, we may deliberately choose to ignore various known dynamic characteristics in order to achieve a simpler nominal design model.

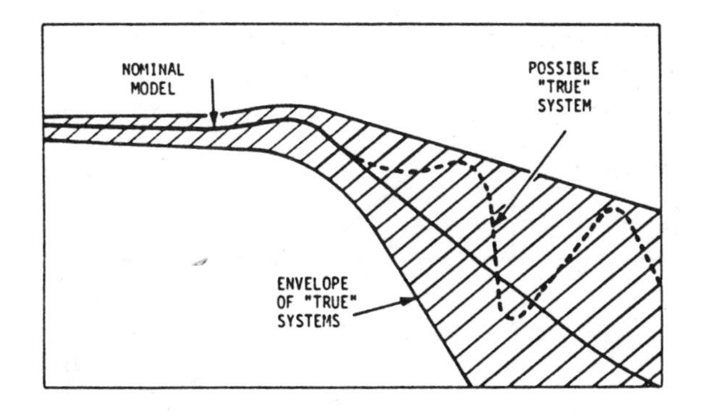

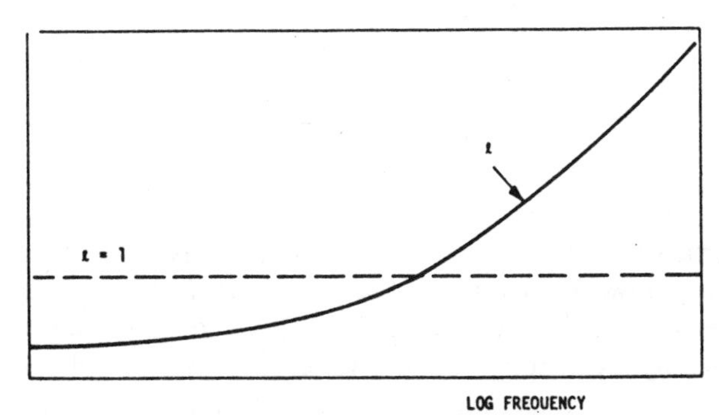

Fig. 2. Typical behavior of multiplicative perturbations.

Another important point is that the construction of $l_m(\omega)$ for multivariable systems is not trivial. The bound assumes a single worst case uncertainty magnitude applicable to all channels. If substantially different levels of uncertainty exist in various channels, it may be necessary to scale the input–output variables and/or apply frequency-dependent transformations [15] in such a way that l_m becomes more uniformly tight. These scale factors and transformations are here assumed to be part of the nominal model $G(s)$.

IV. Feedback Design in the Face of Unstructured Uncertainties

Once we specify a design model, $G(s)$, and accept the existence of unstructured uncertainties in the form (13), the feedback design problem becomes one of finding a compensator $K(s)$ such that

1) the nominal feedback system, $GK[I+GK]^{-1}$, is stable;

2) the perturbed system, $G'K[I+G'K]^{-1}$, is stable for *all possible* G' *allowed by (13)*; and

3) performance objectives are satisfied *for all possible* G' *allowed by (13)*.

All three of these requirements can be interpreted as frequency domain conditions on the nominal loop transfer matrix, $GK(s)$, which the designer must attempt to satisfy.

Stability Conditions

The frequency domain conditions for requirement 1) are, of course, well known. In SISO cases, they take the form of the standard Nyquist criterion,[1] and in MIMO cases, they involve its multivariable generalization [18]. Namely, we require that the encirclement count of the map $\det[I+GK(s)]$, evaluated on the standard Nyquist D-contour, be equal to the (negative) number of unstable open loop modes of GK.

Similarly, for requirement 2) the number of encirclements of the map $\det[I+G'K(s)]$ must equal the (negative) number of unstable modes of $G'K$. Under our assumptions on G', however, this number is the same as that of GK. Hence, requirement 2) is satisfied if and only if the number of encirclements of $\det[I+G'K(s)]$ remains unchanged for all G' allowed by (13). This is assured iff $\det[I+G'K]$ remains nonzero as G is warped continuously toward G', or equivalently, iff

$$0<\underline{\sigma}\left[I+\left[I+\epsilon L(\bar{s})\right]G(\bar{s})K(\bar{s})\right] \tag{14}$$

for all $0<\epsilon<1$, all $\bar{s}$ on the D-contour, and all $L(s)$ satisfying (13). Since G' vanishes on the infinite radius segment of the D-contour, and assuming, for simplicity, that the contour requires no indentations along the $j\omega$-axis,[2] (14) reduces to the following equivalent conditions:

$$0<\underline{\sigma}\left[I+G(j\omega)K(j\omega)+\epsilon L(j\omega)G(j\omega)K(j\omega)\right] \tag{15}$$

for all $0<\epsilon<1$, $0<\omega<\infty$, and all L

$$\Leftrightarrow\ 0<\underline{\sigma}\left[I+LGK(I+GK)^{-1}\right] \tag{16}$$

for all $0<\omega<\infty$, and all L

$$\Leftrightarrow\ \bar{\sigma}\left[GK(I+GK)^{-1}\right]<1/l_m(\omega) \tag{17}$$

for all $0<\omega<\infty$.

The last of these equations is the MIMO generalization of the familiar SISO requirement that loop gains be small whenever the magnitude of unstructured uncertainties is large. In fact, whenever $l_m(\omega)\gg 1$, we get the following constraint on GK:

$$\bar{\sigma}\left[GK(j\omega)\right]<1/l_m(\omega) \tag{18}$$

for all ω such that $l_m(\omega)\gg 1$.

We emphasize that these are *not* conservative stability conditions. On the contrary, if the uncertainties are truly unstructured and (17) is violated, then there exists a perturbation $L(s)$ within the set allowed by (13) for which the system is unstable. Hence, these stability conditions impose hard limits on the permissible loop gains of practical feedback systems.

Performance Conditions

Frequency domain conditions for requirement 3) have already been described in (5) in Section II. The only

[1] See any classical control text.

[2] If indentations are required, (14) and (17) must hold in the limit for all $\bar{s}$ on the indented path as the radius of indentation is taken to zero.

modification needed to account for unstructured uncertainties is to apply (5) to G' instead of G, i.e.,

$$ps < \underline{\sigma}[I+(I+L)GK]$$
$$\Leftarrow ps < \underline{\sigma}[I+LGK(I+GK)^{-1}]\underline{\sigma}[I+GK]$$
$$\Leftarrow \frac{ps(\omega)}{1-l_m(\omega)} < \underline{\sigma}[GK(j\omega)] \tag{19}$$

for all ω such that $l_m(\omega)<1$ and $\underline{\sigma}[GK(j\omega)]\gg 1$.

This is the MIMO generalization of another familiar SISO design rule—namely that performance objectives can be met in the face of unstructured uncertainties if the nominal loop gains are made sufficiently large to compensate for model variations. Note, however, that finite solutions exist only in the frequency range where $l_m(\omega)<1$.

The stability and performance conditions derived above illustrate that MIMO feedback design problems do not differ fundamentally from their SISO counterparts. In both cases, stability must be achieved nominally and assured for all perturbations by satisfying conditions (17) and (18). Performance may then be optimized by satisfying condition (19) as well as possible. What distinguishes MIMO from SISO design conditions are the functions used to express transfer function "size." Singular values replace absolute values. The underlying concepts remain the same.

We note that the singular value functions used in our statements of design conditions play a design role much like classical Bode plots. The $\bar{\sigma}[I+GK]$ function in (5) is the minimum return difference magnitude of the closed-loop system; $\underline{\sigma}[GK]$ in (8) and $\bar{\sigma}[GK]$ in (18) are minimum and maximum loop gains, and $\bar{\sigma}[GK(I+GK)^{-1}]$ in (17) is the maximum closed-loop frequency response. These can all be plotted as ordinary frequency dependent functions in order to display and analyze the features of a multivariable design. Such plots will here be called σ-plots.

One of the σ-plots which is particularly significant with regard to design for uncertainties is obtained by inverting condition (17), i.e.,

$$l_m(\omega) < \frac{1}{\bar{\sigma}[GK(I+GK)^{-1}]} = \underline{\sigma}[I+GK(j\omega)^{-1}] \tag{20}$$

for all $0 \le \omega < \infty$.

The function on the right-hand side of this expression is an explicit measure of the degree of stability (or stability robustness) of the feedback system. Stability is guaranteed for all perturbations $L(s)$ whose maximum singular values fall below it. This can include gain or phase changes in individual output channels, simultaneous changes in several channels, and various other kinds of perturbations. In effect, $\underline{\sigma}[I+(GK)^{-1}]$ is a reliable multivariable generalization of SISO stability margin concepts (e.g., frequency dependent gain and phase margins). Unlike the SISO case, however, it is important to note that $\underline{\sigma}[I+(GK)^{-1}]$ measures tolerances for uncertainties at the plant outputs only. Tolerances for uncertainties at the input are

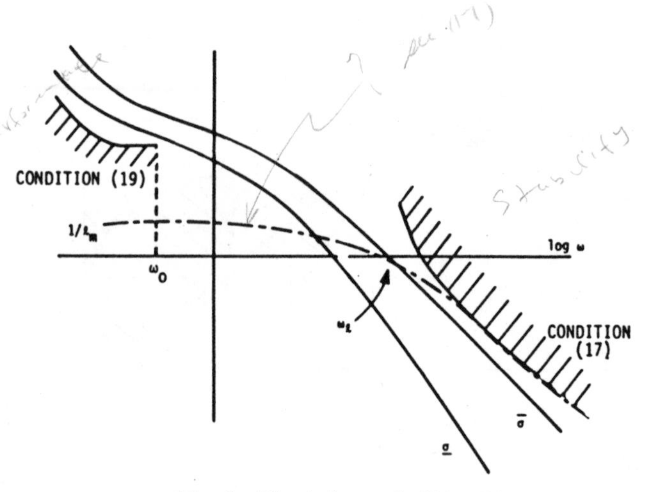

Fig. 3. The design tradeoff for GK.

generally not the same. They can be analyzed with equal ease, however, by using the function $\underline{\sigma}[I+(KG)^{-1}]$ instead of $\underline{\sigma}[I+(GK)^{-1}]$ in (20). This can be readily verified by evaluating the encirclement count of the map $\det(I+KG)$ under perturbations of the form $G'=G(I+L)$ (i.e., uncertainties reflected to the input). The mathematical steps are directly analogous to (15)–(18) above.

Classical designers will recognize, of course, that the difference between these two stability robustness measures is simply that each uses a loop transfer function appropriate for the loop breaking point at which robustness is being tested.

V. Transfer Function Limitations

The feedback design conditions derived above are pictured graphically in Fig. 3. The designer must find a loop transfer function matrix, GK, for which the loop is nominally stable and whose maximum and minimum singular values clear the high and low frequency "design boundaries" given by conditions (17) and (19). The high frequency boundary is mandatory, while the low frequency one is desirable for good performance. Both are influenced by the uncertainty bound, $l_m(\omega)$.

The σ-plots of a representative loop transfer matrix are also sketched in the figure. As shown, the effective bandwidth of the loop cannot fall much beyond the frequency ω_l for which $l_m(\omega_l)=1$. As a result, the frequency range over which performance objectives can be met is explicitly constrained by the uncertainties. It is also evident from the sketch that the severity of this constraint depends on the rate at which $\underline{\sigma}[GK]$ and $\bar{\sigma}[GK]$ are attenuated. The steeper these functions drop off, the wider the frequency range over which condition (19) can be satisfied. Unfortunately, however, FDLTI transfer functions behave in such a way that steep attenuation comes only at the expense of small $\underline{\sigma}[I+GK]$ values and small $\underline{\sigma}[I+(GK)^{-1}]$ values when $\underline{\sigma}[GK]$ and $\bar{\sigma}[GK]\approx 1$. This means that while performance is good at lower frequencies and stability robustness is good at higher frequencies, both are poor near crossover. The behavior of FDLTI transfer functions, therefore, imposes a second major limitation on the achievable performance of feedback systems.

SISO Transfer Function Limitation

For SISO cases, the conflict between attenuation rates and loop quality at crossover is again well understood. We know that any rational, stable, proper, minimum phase loop transfer function satisfies fixed integral relations between its gain and phase components. Hence, its phase angle near crossover (i.e., at values of ω such that $|gk(j\omega)| \approx 1$) is determined uniquely by the gain plot in Fig. 3 (for $\bar{\sigma} = \underline{\sigma} = |gk|$). Various expressions for this angle were derived by Bode using contour integration around closed contours encompassing the right half plane [5, ch. 13, 14]. One expression is

$$\phi_{gkc} \triangleq \arg[gk(j\omega_c)]$$

$$= \frac{1}{\pi} \int_{-\infty}^{\infty} \frac{\ln|gk(j\omega(\nu))| - \ln|gk(j\omega_c)|}{\sinh \nu} d\nu \quad (21)$$

where $\nu = \ln(\omega/\omega_c)$, $\omega(\nu) = \omega_c \exp \nu$. Since the sign of $\sinh(\nu)$ is the same as the sign of ν, it follows that ϕ_{gkc} will be large if the gain $|gk|$ attenuates slowly and small if it attenuates rapidly. In fact, ϕ_{gkc} is given explicitly in terms of weighted average attenuation rate by the following alternate form of (21) (also from [5]):

$$\phi_{gkc} = \frac{1}{\pi} \int_{-\infty}^{\infty} \frac{d \ln|gk|}{d\nu} \left(\ln \coth \frac{|\nu|}{2} \right) d\nu. \quad (22)$$

The behavior of ϕ_{gkc} is significant because it defines the magnitudes of our two SISO design conditions (17) and (19) at crossover. Specially, when $|gk| = 1$, we have

$$|1 + gk| \equiv |1 + (gk)^{-1}| = 2 \left| \sin\left(\frac{\pi + \phi_{gkc}}{2} \right) \right|. \quad (23)$$

The quantity $\pi + \phi_{gkc}$ is the phase margin of the feedback system. Assuming gk stable, this margin must be positive for nominal stability and, according to (23), it must be reasonably large (≈ 1 rad) for good return difference and stability robustness properties. If $\pi + \phi_{gkc}$ is forced to be very small by rapid gain attenuation, the feedback system will amplify disturbances ($|1 + gk| \ll 1$) and exhibit little uncertainty tolerance at and near ω_c. The conflict between attenuation rate and loop quality near crossover is thus clearly evident.

It is also known that more general nonminimum phase and/or unstable loop transfer functions do not alleviate this conflict. If the plant has right half-plane zeros, for example, it may be factored as

$$g(s) = m(s)p(s) \quad (24)$$

where $m(s)$ is minimum phase and $p(s)$ is an all-pass (i.e., $|p(j\omega)| = 1 \ \forall \omega$.) The (negative) phase angle of $p(s)$ reduces total phase at crossover, i.e.,

$$\phi_{gkc} = \phi_{mkc} + \phi_{pc} < \phi_{mkc} \quad (25)$$

and therefore aggravates the tradeoff problem. In fact, if $|\phi_{pc}|$ is too large, we will be forced to reduce the crossover frequency. Thus, RHP zeros limit loop gain (and thus performance) in a way similar to the unstructured uncertainty. A measure of severity of this added limitation is $|1 - p(j\omega)|$, which can be used just like $l_m(\omega)$ to constrain a nominal minimum phase design.

If $g(s)$ has right half-plane poles, the extra phase lead contributed by these poles compared with their mirror images in the left half-plane is needed to provide encirclements for stability. Unstable plants thus also do not offer any inherent advantage over stable plants in alleviating the crossover conflict.

Multivariable Generalization

The above transfer limitations for SISO systems have multivariable generalizations; with some additional complications as would be expected. The major complication is that singular values of rational transfer matrices, viewed as functions of the complex variable s, are not analytic and therefore cannot be used for contour integration to derive relation such as (21). Eigenvalues of rational matrices, on the other hand, have the necessary mathematical properties. Unfortunately, they do not in general relate directly to the quality of the feedback design. (More is said about this in Section VI.) Thus, we must combine the properties of eigenvalues and singular values through the bounding relations

$$\underline{\sigma}[A] < |\lambda[A]| < \bar{\sigma}[A] \quad (26)$$

which holds for any eigenvalue, λ_i, of the (square) matrix A. The approach will be to derive gain/phase relations as in (21) for the eigenvalues of $I + GK$ and $I + (GK)^{-1}$ and to use these to bound their minimum singular values. Since good performance and stability robustness requires singular values of both of these matrices to be sufficiently large near crossover, the multivariable system's properties can then be no better than the properties of their eigenvalue bounds.

Equations for the eigenvalues themselves are straightforward. There is a one-to-one correspondence between eigenvalues of GK and eigenvalues of $I + GK$ such that

$$\lambda_i[I + GK] = 1 + \lambda_i[GK]. \quad (27)$$

Likewise for $I + (GK)^{-1}$;

$$\lambda_i[I + (GK)^{-1}] = 1 + \frac{1}{\lambda_i[GK]}. \quad (28)$$

Thus, when $|\lambda_i[GK]| = 1$ for some λ_i and $\omega = \omega_c$, we have

$$|\lambda_i[I + GK]| \equiv |\lambda_i[I + (GK)^{-1}]|$$

$$= 2 \left| \sin\left(\frac{\pi + \phi_{\lambda_i c}}{2} \right) \right|. \quad (29)$$

Since this equation is exactly analogous to (23) for the scalar case, and since $|\lambda_i|$ bounds $\underline{\sigma}$, it follows that the loop will exhibit poor properties whenever the phase angle $(\pi+\phi_{\lambda,c})$ is small.

In order to derive expressions for the angle $\phi_{\lambda_i c}$ itself, we require certain results from the theory of algebraic functions [20]–[26]. The key concepts needed from these references are that the eigenvalues λ_i of a rational, proper transfer function matrix, viewed as a function of the complex variable s constitute one mathematical entity, $\lambda(s)$, called an algebraic function. Each eigenvalue λ_i is a branch of this function and is defined on one sheet of an extended Riemann surface domain. On its extended domain an algebraic function can be treated as an ordinary meromorphic function whose poles and zeros are the system poles and transmission zeros of the transfer function matrix. It also has additional critical points, called branch points, which correspond to multiple eigenvalues. Contour integration is valid on the Riemann surface domain provided that contours are properly closed.

In the contour integral leading to (21), $gk(s)$ may therefore be replaced by the algebraic function, $\lambda(s)$, with contour taken on its Riemann domain. Carrying out this integral yields several partial sums:

$$\sum_i \phi_{\lambda_i c} = \frac{1}{\pi}\int_{-\infty}^{\infty} \sum_i \frac{[\ln|\lambda_i(j\omega(\nu))| - \ln|\lambda_i(j\omega_c)|]}{\sinh \nu} d\nu \tag{30}$$

where each sum is over all branches of $\lambda(s)$ whose sheets are connected by right half-plane branch points. Thus the eigenvalues $\{\lambda_i\}$ are restricted in a way similar to scalar transfer functions but in summation form. The summation, however, does not alter the fundamental tradeoff between attenuation rate and loop quality at crossover. In fact, if we deliberately choose to maximize the bound (29) by making ω_c and $\phi_{\lambda_i c}$ identical for all i, then (30) imposes the same restrictions on multivariable loops as (21) imposes on SISO loops. Hence, multivariable systems do not escape the fundamental transfer function limitations.

As in the scalar case, expression (30) is again valid for minimum phase systems only. That is, GK can have no transmission zeros[3] in the right half-plane. If this is not true, the tradeoffs governed by (29) and (30) are aggravated because every right half-plane transmission zero adds the same phase lag as in (25) to one of the partial sums in (30). The matrix GK may also be factored, as in (24), to get

$$GK(s) = M(s)P(s) \tag{31}$$

where $M(s)$ has no right half-plane zeros and $P(s)$ is an all-pass matrix $P^T(-s)P(s)=I$. Analogous to the scalar case, $\bar{\sigma}(I-P(s))$ can be taken as a measure of the degree of multivariable nonminimum phaseness and used like $l_m(\omega)$ to constrain a nominal minimum phase design.

[3]For our purposes, transmission zeros [41] are values $\bar{s}$ such that $\det[G(\bar{s})K(\bar{s})]=0$. Degenerate systems with $\det[GK]\equiv 0$ for all s are not of interest because they cannot meet condition (19) in Fig. 3.

VI. Multivariable Design by Modern Frequency Domain Methods

So far, we have described the FDLTI feedback design problem as a design tradeoff involving performance objectives [condition (19)], stability requirements in the face of unstructured uncertainties [condition (17)], and certain performance limitations imposed by gain/phase relations which must be satisfied by realizable loop transfer functions. This tradeoff is essentially the same for SISO and MIMO problems. Design methods to carry it out, of course, are not.

For scalar design problems, a large body of well-developed tools exists (e.g., "classical control") which permits designers to construct good transfer functions for Fig. 3 with relatively little difficulty. Various attempts have been made to extend these methods to multivariable design problems. Probably the most successful of these are the inverse Nyquist array (INA) [3] and the characteristic loci (CL) methodologies [4]. Both are based on the idea of reducing the multivariable design problem to a sequence of scalar problems. This is done by constructing a set of scalar transfer functions which may be manipulated more or less independently with classical techniques. In the INA methodology, the scalar functions are the diagonal elements of a loop transfer function matrix which has been pre- and post-compensated to be diagonally dominant. In the CL methodology, the functions are the eigenvalues of the loop transfer matrix.

Based on the design perspective developed in the previous sections, these multiple single-loop methods turn out to be reliable design tools only for special types of plants. Their restrictions are associated with the fact that the selected set of scalar design functions are not necessarily related to the system's actual feedback properties. That is, the feedback system may be designed so that the scalar functions have good feedback properties if interpreted as SISO systems, but the resulting multivariable system may still have poor feedback properties. This possibility is easy to demonstrate for the CL method and, by implication, for the INA method with perfect diagonalization. For these cases, we attempt to achieve stability robustness by satisfying

$$l_m(\omega) < |\lambda_i[I+(GK)^{-1}]| = |1+1/\lambda_i(GK)| \tag{32}$$

for all i and $0<\omega<\infty$ and similarly, we attempt to achieve performance objectives by making

$$\frac{ps(\omega)}{1-l_m(\omega)} < |\lambda_i[I+GK]| = |1+\lambda_i(GK)| \tag{33}$$

for all i and $0<\omega<\infty$.

As discussed in Section V, however, the eigenvalues on the right-hand sides of these expressions are only upper

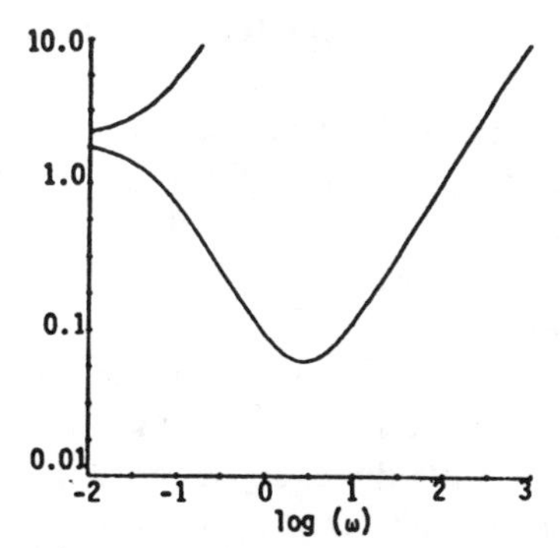

Fig. 4. σ-plots for $I+G^{-1}$. [G from (34).]

bounds for the true stability robustness and performance conditions (20) and (19). Hence, $\underline{\sigma}[I+(GK)^{-1}]$ and/or $\underline{\sigma}[I+GK]$ may actually be quite small even when (32) and (33) are satisfied.

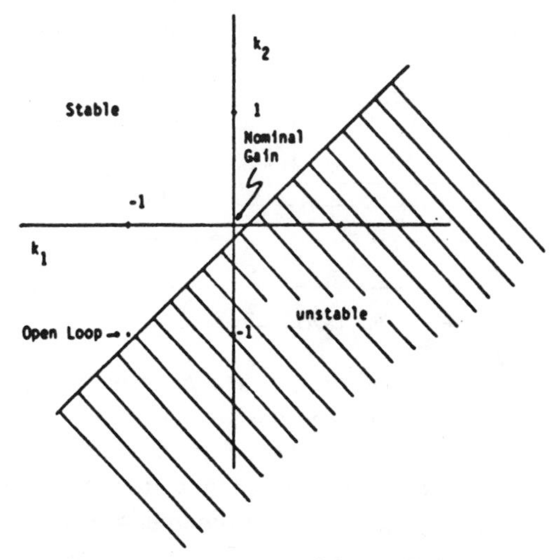

Fig. 5. Stability regions for $\left\{I+\begin{bmatrix} k_1 & 0 \\ 0 & k_2 \end{bmatrix}\right\} G$. [$G$ from (34).]

An Example

These potential inadequacies in the INA and CL methods are readily illustrated with a simple example selected specifically to highlight the limitations. Consider

$$G(s)=\frac{1}{(s+1)(s+2)}\begin{bmatrix} -47s+2 & 56s \\ -42s & 50s+2 \end{bmatrix}. \quad (34)$$

This system may be diagonalized exactly by introducing constant compensation. Let

$$U=\begin{bmatrix} 7 & 8 \\ 6 & 7 \end{bmatrix} \quad U^{-1}=\begin{bmatrix} 7 & -8 \\ -6 & 7 \end{bmatrix}. \quad (35)$$

Then

$$\hat{G}=UGU^{-1}=\begin{bmatrix} \frac{1}{s+1} & 0 \\ 0 & \frac{2}{s+2} \end{bmatrix}. \quad (36)$$

If the diagonal elements of this $\hat{G}$ are interpreted as independent SISO systems, as in the INA approach, we could readily conclude that no further compensation is necessary to achieve desirable feedback properties. For example, unity feedback yields stability margins at crossover of $\pm\infty$ dB in gain and greater than 90 degrees in phase. Thus, an INA design could reasonably stop at this point with compensator $K(s)=UU^{-1}=I$. Since the diagonal elements of $\hat{G}$ are also the eigenvalues of G, we could also be reasonably satisfied with this design from the CL point of view.

Singular value analysis, however, leads to an entirely different conclusion. The σ-plots for $(I+G^{-1})$ are shown in Fig. 4. These clearly display a serious lack of robustness with respect to unstructured uncertainties. The smallest value of σ is approximately 0.1 near $\omega=2$ rad/s. This means that multiplicative uncertainties as small as $l_m(2)=0.1$ (≈10% gain changes, ≈6 deg phase changes) could produce instability. An interpretation of this lack of stability robustness is given in Fig. 5. This figure shows stability regions in "gain space" for the compensator $K(s)=\text{diag}(1+k_1, 1+k_2)$. The figure reveals an unstable region in close proximity to the nominal design point. The INA and CL methods are not reliable design tools because they fail to alert the designer to its presence.

This example and the discussion which precedes it should not be misunderstood as a universal indictment of the INA and CL methods. Rather, it represents a caution regarding their use. There are various types of systems for which the methods prove effective and reliable. Conditions which these systems satisfy can be deduced from (32) and (33)—namely they must have tight singular value/eigenvalue bounds. This includes naturally diagonal systems, of course, and also the class of "normal" systems [28]. The limitations which arise when the bounds are not tight have also been recognized in [4].

We note in passing that the problem of reliability is not unique to the INA and CL methods. Various examples can be constructured to show that other design approaches such as the "single-loop-at-a-time" methods common in engineering practice and tridiagonalization approaches suffer similarly.

VII. Multivariable Design Via LQG

A second major approach to multivariable feedback design is the modern LQG procedure [11], [12]. We have already introduced this method in connection with the tradeoff between command/disturbance error reduction and sensor noise error reduction. The method requires that we select stochastic models for sensor noise, commands and disturbances and define a weighted mean square error criterion as the standard of goodness for the design. The rest is automatic. We get an FDLTI compensator $K(s)$ which stabilizes the nominal model $G(s)$ (under mild assumptions) and optimizes the criterion of goodness. All too often, of course, the resulting loop

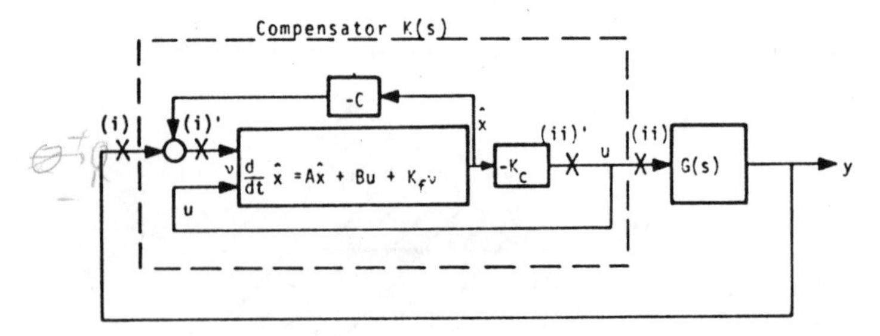

Fig. 6. LQG feedback loop.

transfer functions, GK or KG, are entirely unacceptable when examined against the design constraints of Fig. 4. We are then forced to iterate the design—adjust weights in the performance criterion, change the stochastic disturbance and noise models, add dynamics, etc. There are so many parameters to manipulate that frustration sets in quickly and the schism between practitioners and theoreticians becomes easier to understand.

Fortunately, such design iterations of LQG controllers have become easier to carry out in the last few years because the frequency domain properties of these controllers are better understood. Some of the key new results are summarized below and their significance with respect to Fig. 3 are discussed. For our purposes, LQG controllers are ordinary FDLTI compensators with a special internal structure. This structure is shown in Fig. 6 and is well known. It consists of a Kalman–Bucy filter (KBF) designed for a state space realization of the nominal model $G(s)$, including all appended dynamics for disturbance processes, commands, integral action, etc. The model is

$$\dot{x}=Ax+Bu+\xi; \qquad x\in R^n, \quad u\in R^m \qquad (37)$$
$$y=Cx+\eta; \qquad y\in R^r$$

and satisfies

$$G(s)=C\Phi(s)B \qquad (38)$$
$$d=C\Phi(s)\xi$$

with

$$\Phi(s)=(sI_n-A)^{-1} \qquad (39)$$

The symbols ξ and η denote the usual white noise processes. The filter's gains are denoted by K_f and its state estimates by $\hat{x}$. The state estimates are multiplied by full-state linear-quadratic regulator (LQR) gains, K_c, to produce the control commands which drive the plant and are also fed back internally to the KBF. The usual conditions for well-posedness of the LQG problem are assumed.

In terms of previous discussions, the functions of interest in Fig. 6 are the loop transfer, return difference, and stability robustness functions

$$GK, \quad I_r+GK, \quad I_r+(GK)^{-1},$$

and also their counterparts

$$KG, \quad I_m+KG, \quad I_m+(KG)^{-1}.$$

As noted earlier, the first three functions measure performance and stability robustness with respect to uncertainties at the plant outputs (loop-breaking point (i) in Fig. 6), and the second three measure performance and robustness with respect to uncertainties at the plant input (loop-breaking point (ii) in Fig. 6). Both points are generally significant in design.

Two other loop-breaking points, $(i)'$ and $(ii)'$, are also shown in the figure. These are internal to the compensator and therefore have little direct significance. However, they have desirable loop transfer properties which can be related to the properties of points (i) and (ii). The properties and connections are these.

Fact 1: The loop transfer function obtained by breaking the LQG loop at point $(i)'$ is the KBF loop transfer function $C\Phi K_f$.

Fact 2: The loop transfer function obtained by breaking the LQG loop at point (i) is GK. It can be made to approach $C\Phi K_f$ pointwise in s by designing the LQR in accordance with a "sensitivity recovery" procedure due to Kwakenaak [29].

Fact 3: The loop transfer function obtained by breaking the LQG loop at point $(ii)'$ is the LQR loop transfer function $K_c\Phi B$.

Fact 4: The loop transfer function obtained by breaking the LQG loop at point (ii) is KG. It can be made to approach $K_c\Phi B$ pointwise in s by designing the KBF in accordance with a "robustness recovery" procedure due to Doyle and Stein [30].

Facts 1 and 3 can be readily verified by explicit evaluation of the transfer functions involved. Facts 2 and 4 take more elaboration and are taken up in a later section. They also require more assumptions. Specifically, $G(s)$ must be minimum phase with $m>r$ for Fact 2, $m<r$ for Fact 4, and hence, $G(s)$ must be square for both. Also, the names "sensitivity recovery" and "robustness recovery" are overly restrictive. "Full-state loop transfer recovery" is perhaps a better name for both procedures, with the distinction that one applies to points (i), $(i)'$ and the other to points (ii), $(ii)'$.

The significance of these four facts is that we can design LQG loop transfer functions on a full-state feedback basis and then approximate them adequately with a recovery procedure. For point (i), the full state design must be done with the KBF design equations (i.e., its Riccati equation) and recovery with the LQR equations, while for point (ii), full-state design must be done with the LQR equations and recovery with the KBF. The mathematics of these two options are, in fact, dual. Hence, we will describe only one option [for point (ii)] in further detail. Results for the other are stated and used later in our example.

Full-State Loop Transfer Design

The intermediate full-state design step is worthwhile because LQR and KBF loops have good classical properties which have been rediscovered over the last few years [31]–[33]. The basic result for the LQR case is that LQR loop transfer matrices

$$T(s) \triangleq K_c\Phi(s)B \tag{40}$$

satisfy the following return difference identity [32];

$$[I_m + T(j\omega)]^* R [I_m + T(j\omega)] = R + [H\Phi(j\omega)B]^*[H\Phi(j\omega)B] \quad \forall 0<\omega<\infty \tag{41}$$

where $R=R^T>0$ is the standard control weighting matrix, and $H^TH=Q>0$ is the corresponding state weighting matrix. Without loss of generality, H can be of size $(m\times n)$ [34]. Using the definitions (6) and (7), (41) with $R=\rho I$ implies that

$$\sigma_i[I_m+T(j\omega)] = \sqrt{\lambda_i\left[I+\frac{1}{\rho}(H\Phi B)^* H\Phi B\right]}$$

$$= \sqrt{1+\frac{1}{\rho}\lambda_i[(H\Phi B)^* H\Phi B]}$$

$$= \sqrt{1+\frac{1}{\rho}\sigma_i^2[H\Phi(j\omega)B]}\ . \tag{42}$$

This expression applies to all singular values σ_i of $T(s)$ and, hence, specifically to $\underline{\sigma}$ and $\bar{\sigma}$. It governs the performance and stability robustness properties of LQR loops.

Performance Properties (Condition 19)

Whenever $\underline{\sigma}[T]\gg 1$, the following approximation of (42) shows explicitly how the parameters ρ and H influence $T(s)$:

$$\sigma_i[T(j\omega)] \approx \sigma_i[H\Phi(j\omega)B]/\sqrt{\rho}\ . \tag{43}$$

We can thus choose ρ and H explicitly to satisfy condition (19) and also to "balance" the multivariable loop such that $\underline{\sigma}[T]$ and $\bar{\sigma}[T]$ are reasonably close together.[4] This second objective is consistent with our assumption in Section III that the transfer function $G(s)$ has been scaled and/or transformed such at $l_m(\omega)$ applies more or less uniformly in all directions. This is also the justification for considering control weighting matrices in the form $R=\rho I$ only. Nonidentity R's are subsumed in G as $GR^{1/2}$.

Robustness Properties [Conditions (17) and (20)]

It also follows from (42) that the LQR return difference always exceeds unity, i.e.,

$$\underline{\sigma}[I_m+T(j\omega)] \geq 1 \qquad \forall 0<\omega<\infty. \tag{44}$$

[4]It may also be necessary to append additional dynamics. In order to achieve zero steady state errors, for example, $\sigma[H\Phi B]$ must tend to ∞ as $\omega\to 0$. This may require additional integrations in the plant.

This implies [35] that

$$\underline{\sigma}[I_m+T^{-1}(j\omega)] \geq 1/2 \qquad \forall 0<\omega<\infty. \tag{45}$$

Hence, LQR loops are guaranteed to remain stable for all unstructured uncertainties (reflected to the input) which satisfy $l_m(\omega)<0.5$. Without further knowledge of the types of uncertainties present in the plant, this bound is the greatest robustness guarantee which can be ascribed to the regulator.[5]

While it is reassuring to have a guarantee at all, the $l_m<0.5$ bound is clearly inadequate for the requirements of condition (20) with realistic $l_m(\omega)$'s. In order to satisfy condition (20) in LQR designs, therefore, it becomes necessary to directly manipulate the high-frequency behavior of $T(s)$. This behavior can be derived from known asymptotic properties of the regulator as the scalar ρ tends to zero [29], [47], [48], [34], [36]. The result needed here is that under minimum phase assumptions on $H\Phi B$, the LQR gains K_c behave asymptotically as [29]

$$\sqrt{\rho}\,K_c \to WH \tag{46}$$

where W is an orthonormal matrix. The LQR loop transfer function, $T(s)$, evaluated at high frequencies, $s=jc/\sqrt{\rho}$ with c constant, is then given by[6]

$$T(jc/\sqrt{\rho}) = \sqrt{\rho}\,K_c(jcI-\sqrt{\rho}\,A)^{-1}B \to WHB/jc. \tag{47}$$

Since crossovers occur at $\sigma_i[T]=1$, this means that the maximum (asymptotic) crossover frequency of the loop is

$$\omega_{c\,\max} = \bar{\sigma}[HB]/\sqrt{\rho}\ . \tag{48}$$

As shown in Fig 3, this frequency cannot fall much beyond ω_l, where unstructured uncertainty magnitudes approach unity. Hence, our choice of H and ρ to achieve the performance objectives via (43) are constrained by the stability robustness requirement via (48).

Note also from (47) that the asymptotic loop transfer function in the vicinity of crossover is proportional to $1/\omega$ (-1 slope on log-log plots). This is a relatively slow attenuation rate which, in view of Section V, is the price the regulator pays for its excellent return difference properties. If $l_m^{-1}(\omega)$ attenuates faster than this rate, further reduction of ω_c may be required. It is also true, of course, that no physical system can actually maintain a $1/\omega$ characteristic indefinitely [6]. This is not a concern here since $T(s)$ is a nominal (design) function only and will

[5]The $l_m<0.5$ bound turns out to be tight for pure gain changes, i.e., $0.5\Leftrightarrow 6$ dB, which is identical to regulator's celebrated guaranteed gain margin [33]. The bound is conservative if the uncertainties are known to be pure phase changes, i.e., $0.5\Leftrightarrow\approx 30$ deg, which is less than the known ± 60 deg guarantee [33].

[6]This specific limiting process is appropriate for the so-called generic case [36] with full rank HB. More general versions of (55) with rank $[HB]<m$ are derived in [44].

later be approximated by one of the full-state loop transfer recovery procedures.

Full-State Loop Transfer Recovery

As described earlier, the full-state loop transfer function designed above for point $(ii)'$ can be recovered at point (ii) by a modified KBF design procedure. The required assumptions are that $r > m$ and that $C\Phi B$ is minimum phase. The procedure then consists of two steps.

1) Append additional dummy columns to B and zero row to K_c to make $C\Phi B$ and $K_c\Phi B$ square $(r \times r)$. $C\Phi B$ must remain minimum phase.

2) Design the KBF with modified noise intensity matrices.

$$E(\xi\xi^T) = \left[M_0 + q^2 BB^T\right]\delta(t-\tau)$$
$$E(\eta\eta^T) = N_0\delta(t-\tau)$$

where M_0, N_0 are the nominal noise intensity matrices obtained from stochastic models of the plant and q is a scalar parameter. Under these conditions, it is known that the filter gains K_f have the following asymptotic behavior as $q \to \infty$ [30]:

$$\frac{1}{q}K_f \to BWN_0^{-1/2}. \tag{49}$$

Here W is another orthonormal matrix, as in (46). When this K_f is used in the loop transfer expression for point (ii), we get pointwise loop transfer recovery as $q \to \infty$, i.e.,

$$K(s)G(s) = K_c\left[\Phi^{-1} + BK_c + K_f C\right]^{-1} K_f C\Phi B \tag{50}$$

$$= K_c\left[\bar{\Phi} - \bar{\Phi}K_f\left(I_r + C\bar{\Phi}K_f\right)^{-1} C\bar{\Phi}\right] K_f C\Phi B \tag{51}$$

$$= K_c\bar{\Phi}K_f\left(I_r + C\bar{\Phi}K_f\right)^{-1} C\Phi B \tag{52}$$

$$\to K_c\bar{\Phi}B(C\bar{\Phi}B)^{-1}C\Phi B \tag{53}$$

$$= K_c\Phi B(I_r + K_c\Phi B)^{-1} \cdot \left[C\Phi B(I_r + K_c\Phi B)^{-1}\right]^{-1} C\Phi B \tag{54}$$

$$= \left\{K_c\Phi B(C\Phi B)^{-1}\right\} C\Phi B \tag{55}$$

$$= K_c\Phi B. \tag{56}$$

In this series of expressions, $\bar{\Phi}$ was used to represent the matrix $(sI_n - A + BK_c)^{-1}$, (49) was used to get from (52) to (53), and the identity $\bar{\Phi}B = \Phi B(I + K_c\Phi B)^{-1}$ was used to get from there to (54). The final step shows explicitly that the asymptotic compensator $K(s)$ [the bracketed term in (55)] inverts the nominal plant (from the left) and substitutes the desired LQR dynamics. The need for minimum phase is thus clear, and it is also evident that the entire recovery procedure is only appropriate as long as the target LQR dynamics satisfy Fig. 3's constraints (i.e., as long as we do not attempt inversion in frequency ranges where uncertainties do not permit it). Closer inspection of (50)–(56) further shows that there is no dependence on LQR or KBF optimality of the gains K_c or K_f. The procedure requires only that K_f be stabilizing and have the asymptotic characteristic (49). Thus, more general state feedback laws can be recovered (e.g., pole placement), and more general filters can be used for the process (e.g., observers).[7]

An Example

The behavior of LQG design iterations with full-state loop transfer recovery is illustrated by the following abstracted longitudinal control design example for a CH-47 tandem rotor helicopter. Our objective is to control two measured outputs—vertical velocity and pitch attitude—by manipulating collective and differential collective rotor thrust commands. A nominal model for the dynamics relating these variables at 40 knot airspeed is [45]

$$\frac{d}{dt}x = \begin{bmatrix} -0.02 & 0.005 & 2.4 & -32 \\ -0.14 & 0.44 & -1.3 & -30 \\ 0 & 0.018 & -1.6 & 1.2 \\ 0 & 0 & 1 & 0 \end{bmatrix} x + \begin{bmatrix} 0.14 & -0.12 \\ 0.36 & -8.6 \\ 0.35 & 0.009 \\ 0 & 0 \end{bmatrix} u$$

$$y = \begin{bmatrix} 0 & 1 & 0 & 0 \\ 0 & 0 & 0 & 57.3 \end{bmatrix} x.$$

Major unstructured uncertainties associated with this model are due to neglected rotor dynamics and unmodeled rate limit nonlinearities. These are discussed at greater length in [46]. For our present purposes, it, suffices to note that they are uniform in both control channels and that $l_m(\omega) > 1$ for all $\omega > 10$ rad/s. Hence, the controller bandwidth should be constrained as in Fig. 3 to $\omega_{c\max} < 10$.

Since our objective is to control two measured outputs at point (i), the design iterations utilize the duals of (40)–(56). They begin with a full state KBF design whose noise intensity matrices, $E(\xi\xi^T) = \Gamma\Gamma^T\delta(t-\tau)$ and $E(\eta\eta^T) = \rho I\delta(t-\tau)$, are selected to meet performance objectives at low frequencies, i.e.,

$$\underline{\sigma}[T] \approx \underline{\sigma}[C\Phi\Gamma]/\sqrt{\rho} > ps, \tag{57}$$

while satisfying stability robustness constraints at high frequencies,

$$\omega_{c\max} = \bar{\sigma}[C\Gamma]/\sqrt{\rho} < 10 \text{ r/s}. \tag{58}$$

[7]Still more generally, the modified KBF procedure will actually recover full-state feedback loop transfer functions at any point, u^1, in the system for which $C\Phi B^1$ is minimum phase [30].

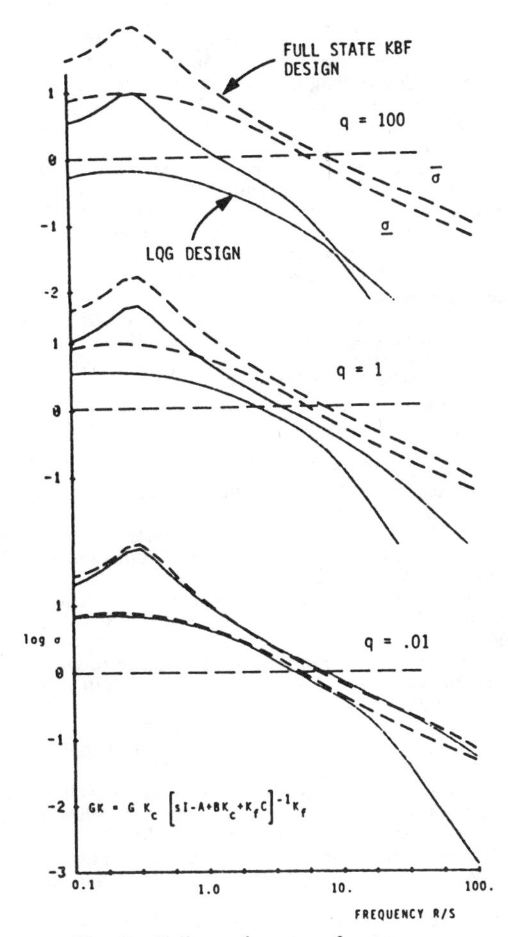

Fig. 7. Full-state loop transfer recovery.

For the choice $\Gamma = B$, (58) constrains ρ to be greater than or equal to unity.[8] The resulting KBF loop transfer for $\rho = 1$ is shown in Fig. 7. For purposes of illustration, this function will be considered to have the desired high gain properties for condition (19), with low gains beyond $\omega = 10$ for condition (20).[9] It then remains to recover this function by means of the full-state recovery procedure for point (i). This calls for LQR design with $Q = Q_0 + q^2 CC^T$ and $R = R_0$. Letting $Q_0 = 0$, $R_0 = I$, the resulting LQG transfer functions for several values of q are also shown in Fig. 7. They clearly display the pointwise convergence properties of the procedure.

VIII. Conclusion

This paper has attempted to present a practical design perspective on MIMO linear time invariant feedback control problems. It has focused on the fundamental issue—feedback in the face of uncertainties. It has shown how classical SISO approaches to this issue can be reliably generalized to MIMO systems, and has defined the extent to which MIMO systems are subject to the same uncertainty constraints and transfer function gain/phase limitations as SISO ones. Two categories of design procedures, were then examined in the context of these results.

There are numerous other topics and many other proposed design procedures which were not addressed, of course. Modal control, [42] eigenvalue-vector assignments, [43] and the entire field of geometric methods [19] are prime examples. These deal with internal structural properties of systems which, though important theoretically, cease to have central importance in the face of the input–output nature of unstructured incertainties. Hence, they were omitted. We also did not treat certain performance objectives in MIMO systems which are distinct from SISO systems. These include perfect noninteraction and integrity. Noninteraction is again a structural property which loses meaning in the face of unstructured uncertainties. [It is achieved as well as possible by condition (19).] Integrity, on the other hand, cannot be dismissed as lightly. It concerns the ability of MIMO systems to maintain stability in the face of actuator and/or sensor failures. The singular value concepts described here are indeed useful for integrity analysis. For example, a design has integrity with respect to actuator failures whenever

$$\underline{\sigma}\left[I + (KG)^{-1}\right] > 1 \qquad \forall \omega. \tag{59}$$

This follows because failures satisfy $l_m < 1$. Moreover it can be shown [37] that full-state control laws designed via Lyapunov equations, as opposed to Riccati equations, as in Section VII, satisfy (59). It is also worth noting that integrity properties claimed for design methods such as INA and CL suffer from the reliability problem discussed in Section VI and, hence, may not be valid in the system's natural (nondiagonal) coordinate system.

The major limitations on what has been said in the paper are associated with the representation chosen in Section III for unstructured uncertainty. A single magnitude bound on matrix perturbations is a worst case representation which is often much too conservative (i.e., it may admit perturbations which are structurally known not to occur). The use of weighted norms in (8) and (9) or selective transformations applied to G (as in [39]) can alleviate this conservatism somewhat, but seldom completely. For this reason, the problem of representing more structured uncertainties in simple ways analogous to (13) is receiving renewed research attention [38].

A second major drawback is our implicit assumption that all loops (all directions) of the MIMO system should have equal bandwidth ($\underline{\sigma}$ close to $\bar{\sigma}$ in Fig. 3). This assumption is consistent with a uniform uncertainty bound but will no longer be appropriate as we learn to represent more complex uncertainty structures. Research along these lines is also proceeding.

[8] If $C\Gamma$ (or HB) is singular, (58) or (48) are still valid in the nonzero directions.

[9] The function should not be considered final, or course. Better balance between $\bar{\sigma}$ and $\underline{\sigma}$ and greater gain at low frequencies via appended integrators would be desirable in a serious design.

References

[1] M. Athans and P. L. Falk, *Optimal Control*. New York: McGraw-Hill, 1966.

[2] A. E. Bryson and Y. C. Ho, *Applied Optimal Control*. Waltham, MA: Ginn, 1969.

[3] H. H. Rosenbrock, *Computer-Aided Control System Design*. New York: Academic, 1974.

[4] A. G. J. MacFarlane and B. Kouvaritakis, "A design technique for linear multivariable feedback systems," *Int. J. Contr.*, vol. 25, pp. 837–879, 1977.

[5] H. W. Bode, *Network Analysis and Feedback Amplifier Design*. Princeton, NJ: Van Nostrand, 1945.

[6] I. M. Horowitz, *Synthesis of Feedback Systems*. New York: Academic, 1963.

[7] J. B. Cruz, Ed., *System Sensitivity Analysis*. Stroudsburg, PA: Dowden, Hutchinson & Ross, 1973.

[8] B. S. Garbow *et al.*, *Matrix Eigensystem Routines—EISPACK Guide Extension*, Lecture Notes in Computer Science, vol. 51. New York: Springer-Verlag, 1977.

[9] G. W. Stewart, *Introduction to Matrix Computations*. New York: Academic, 1973.

[10] I. M. Horowitz and U. Shaked, "Superiority of transfer function over state variable methods," *IEEE Trans. Automat. Contr.*, pp. 84–97, Feb. 1975.

[11] Special Issue on the LQG Problems, *IEEE Trans. Automat. Contr.*, Dec. 1971.

[12] D. C. Youla, H. A. Jabr, and J. J. Bongiorno, Jr., "Modern Wiener-Hopf design of optimal controllers—Part II, The multivariable case," *IEEE Trans. Automat. Contr.*, June 1976.

[13] G. Stein, G. L. Hartmann, and R. C. Hendrick, "Adaptive control laws for F-8C flight test," *IEEE Trans. Automat. Contr.*, Oct. 1977.

[14] M. G. Safonov, "Robustness and stability aspects of stochastic multivariable feedback system design," Ph.D. dissertation, Massachusetts Institute of Technology, 1977.

[15] G. Zames, "On the input-output stability of time-varying nonlinear feedback systems—Part I," *IEEE Trans. Automat. Contr.*, vol. AC-11, no. 2, pp. 228–238, Apr. 1966.
——, "On the input-output stability of time-varying nonlinear feedback systems—Part II," *IEEE Trans. Automat. Contr.*, vol. AC-11, no. 3 pp. 465–476, July 1966.

[16] R. L. Bisplinhoff, H. Ashley and R. L. Halfman, *Aeroelasticity*. Reading, MA: Addison-Wesley, 1957.

[17] C. E. J. Bowler, D. N. Ewart, and C. Concordia, "Self-excited torsional frequency oscillations with series capacitors," *IEEE Trans. Power App. Syst.*, Sept./Oct. 1973.

[18] H. H. Rosenbrock, "The stability of multivariable systems," *IEEE Trans. Automat. Contr.*, vol. AC-17, pp. 105–107, Feb. 1972.

[19] W. M. Wonham, *Linear Multivariable Control: A Geometric Approach*. New York: Springer-Verlag, 1974.

[20] K. Knopp, *Theory of Functions*, vol. II. New York: Dover, 1947.

[21] G. Springer, *Introduction to Riemann Surfaces*. Reading, MA: Addison-Wesley, 1957.

[22] G. A. Bliss, *Algebraic Functions*. New York: Dover, 1966 (reprint of 1933 original).

[23] A. G. J. MacFarlane and I. Postlethwaite, "Characteristic frequency functions and characteristic gain functions," *IJC*, vol. 26, pp. 265–278, 1977.

[24] A. G. J. MacFarlane, and I. Postlethwaite, "The generalized Nyquist stability criterion and multivariable root loci," *IJC*, vol. 25, pp. 81–127, 1977.

[25] A. G. J. MacFarlane, B. Kouvaritakis, and J. M. Edmunds, "Complex variable methods for multivariable feedback systems analysis and design," *Alternatives for Linear Multivariable Control*, National Engineering Consortium, Chicago, IL, pp. 189–228, 1977.

[26] I. Postlethwaite, and A. G. J. MacFarlane, *A Complex Variable Approach to the Analysis of Linear Multivariable Feedback Systems*. Berlin: Springer-Verlag, 1979.

[27] C. A. Desoer, and M. Vidyasagar *Feedback Systems: Input–Output Properties*. New York: Academic, 1975.

[28] Gantmacher, *The Theory of Matrices*. New York: Chelsea, 1960.

[29] H. Kwakernaak and R. Sivan, *Linear Optimal Control Systems*. New York: Wiley-Interscience, 1972.

[30] J. C. Doyle and G. Stein, "Robustness with observers," *IEEE Trans. Automat. Contr.*, Aug. 1979.

[31] R. E. Kalman, "When is a linear system optimal?" *Trans. ASME Ser. D: J. Basic Eng.*, vol. 86, pp. 51–60, 1964.

[32] B. D. O. Anderson and J. B. Moore, *Linear Optimal Control*. Englewood Cliffs, NJ: Prentice-Hall, 1971.

[33] M. G. Safonov and M. Athans, "Gain and phase margin of multiloop LQG regulators," *IEEE Trans. Automat. Contr.*, Apr. 1977.

[34] C. A. Harvey and G. Stein, "Quadratic weights for asymptotic regulator properties," *IEEE Trans. Automat. Contr.*, vol. AC-23, pp. 378–387, 1978.

[35] A. J. Laub, "Robust stability of linear systems—Some computational considerations," Laboratory for Information and Decision Systems, M.I.T., LIDS-R-904, Feb. 1979.

[36] G. Stein, "Generalized quadratic weights for asymptotic regulator properties," *IEEE Trans. Automat. Contr.*, vol. AC-24, pp. 559–566, 1979.

[37] C. A. Harvey, "On feedback systems possessing integrity with respect to actuator outages," in *Proc. MIT/ONR Workshop on Recent Developments in the Robustness Theory of Multivariable Systems*, MIT Rep. LIDS-R-954, Apr. 1979.

[38] M. F. Barrett, "Conservatism with sector-based robustness tests," Ph.D. dissertation, Univ. of Minnesota, 1980.

[39] M. G. Safonov, "Frequency-domain design of multivariable control systems for insensitivity to large plant modeling errors," presented at the18th CDC, Fort Lauderdale, FL, Dec. 12–14, 1979.

[40] H. H. Rosenbrock, "Design of multivariable control systems using the inverse Nyquist array," *Proc. IEEE*, vol. 110, no. 11, Nov. 1969.

[41] A. G. J. MacFarlane and N. Karcanias, "Poles and zeros of linear multivariable systems: A survey of algebraic, geometric, and complex variable theory," *Int. J. Contr.*, pp. 33–74, July 1976.

[42] J. D. Simon and S. K. Mitter, "A theory of modal control," *Inform. Contr.*, vol. 13, pp. 316–353, 1968.

[43] B. C. Moore, "On the flexibility offered by state feedback in multivariable systems beyond closed loop eigenvalue assignment," *IEEE Trans. Automat. Contr.*, vol. AC-21, pp. 685–692, 1976.

[44] G. Stein, "Asymptotic loop transfer functions of LQ regulators and filters," in preparation.

[45] G. L. Hartmann, G. Stein, and S. G. Pratt, "Digital adaptive controllers for VTOL vehicles," NASA Contractor Rep. 159 NASA Langley Research Center, Nov. 1979.

[46] G. Stein and J. C. Doyle, "Singular values and feedback: Design examples," in *Proc. 1978 Allerton Conf.*, Urbana, IL, 1978.

[47] Shaked, "The asymptotic behavior of the root loci of multivariable optimal regulators," *IEEE Trans. Automat. Contr.*, pp. 425–430, June 1978.

[48] B. Kouvaritakis, "The optimal root loci of linear multivariable systems," *Int. J. Contr.*, pp. 33–62, July 1978.

Feedback Properties of Multivariable Systems: The Role and Use of the Return Difference Matrix

MICHAEL G. SAFONOV, MEMBER, IEEE, ALAN J. LAUB, MEMBER, IEEE, AND GARY L. HARTMANN, MEMBER, IEEE

Abstract—For linear time-invariant multivariable feedback systems, the *feedback properties* of plant disturbance attenuation, sensor noise response, stability margins, and sensitivity to plant and sensor variation are quantitatively related to the Bode magnitude versus frequency plots of the singular values of the return difference matrix $I+L$ and of the associated inverse-return difference matrix $I+L^{-1}$. Implied fundamental limits of feedback performance are quantitatively described and design tradeoffs are discussed.

The penalty function in the stochastic linear quadratic Gaussian (LQG) optimal control problem is found to be a weighted-sum of the singular values, with the weights determined by the quadratic cost and noise intensity matrices. This enables systematic "tuning" of LQG cost and noise matrices so that the resulting optimal return difference and inverse-return difference meet inequality constraints derived from design specifications on feedback properties.

The theory has been used to synthesize a multivariable automatic controller for the longitudinal dynamics of an advanced fighter aircraft.

I. INTRODUCTION

THE TERM *feedback properties* is used here to describe those linear system properties such as stability (including stability margins), sensitivity, and disturbance attenuation that can be altered *only* through the use of feedback. Systems with large stability margins, good disturbance attenuation, and/or low sensitivity have also been described as being robust and having good robustness properties. The usual method for improving the robustness properties of a system involves the use of feedback.[1]

Properties of a control system that are *not* feedback properties include, for example, its impulse response matrix, its transfer function matrix, and other characterizations of its response to commands. This is because by appropriate prefiltering of the command signal, command response can be readily altered without feedback via an open-loop control scheme—see Fig. 1. Indeed, it has been proven by Pernebo [24] that there is a complete and rigorous separation between command response and one's choice of feedback loop transfer matrix when prefiltering of commands is permitted.

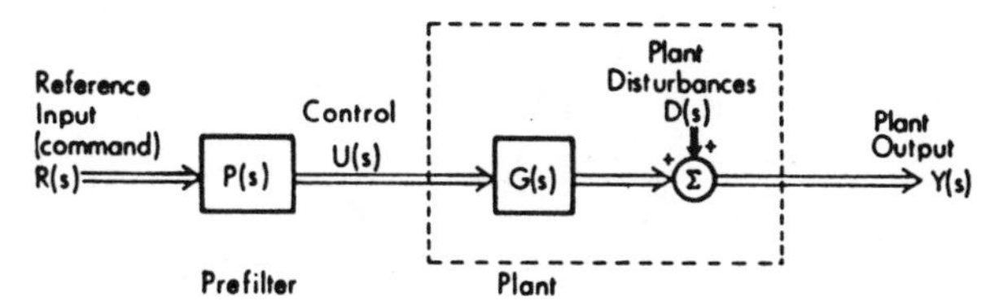

Fig. 1. Open-loop control system.

Consequently, the feedback properties of a control system have little, if any, relation to the system's response to commands. It is thus unfortunate that the main design chapters of most classical control textbooks tend to stress the use of feedback to modify the plant's response to commands. This has led many casual students of servomechanism and control theory to naively suppose that a feedback controller with good transient response will also have acceptably good feedback properties: the fact is that even "optimal" feedback control designs may have some feedback properties that are very poor [3], [4].

We hasten to note that classical frequency response and root-locus design tools are quite well-suited to the purpose of synthesizing control systems with good feedback properties. The book of Horowitz [5], for example, focuses on root-locus based and frequency-response based methodologies for synthesizing feedback control systems to meet explicit specifications on noise and disturbance attenuation, sensitivity to large but bounded parameter variations, and (of course) stability margins. However, these methods are, for practical purposes, limited to single-loop feedback systems or to multiloop systems that are designed one loop at a time.

Multivariable feedback analysis and design techniques have proliferated in the last two decades (e.g., [6]–[19], but, as with the classical single-input single-output literature, the majority of the multivariable literature has either not stressed feedback properties at all or has tended to stress only certain feedback properties, e.g., sensitivity (e.g., [14), stability margins [15]–[16], or disturbance rejection [17]–[18]. The purpose of the present paper is to provide a more unified view of the analysis and synthesis of linear time-invariant multivariable control systems with good feedback properties. To the extent possible, the

Manuscript received October 21, 1980. This work was supported in part by AFOSR Grant 80-0013 and in part by the Honeywell Systems and Research Center, Minneapolis, MN 55413.

M. G. Safonov and A. J. Laub are with the Department of Electrical Engineering, University of Southern California, Los Angeles, CA 9007.

G. L. Hartmann is with the Honeywell Systems and Research Center, Minneapolis, MN 55413.

[1]For some nonlinear systems, robustness can also be improved via open-loop control, e.g., using dither [1], [2].

Reprinted from *IEEE Trans. Automat. Contr.*, vol. AC-26, no. 1, pp. 47–65, Feb. 1981.

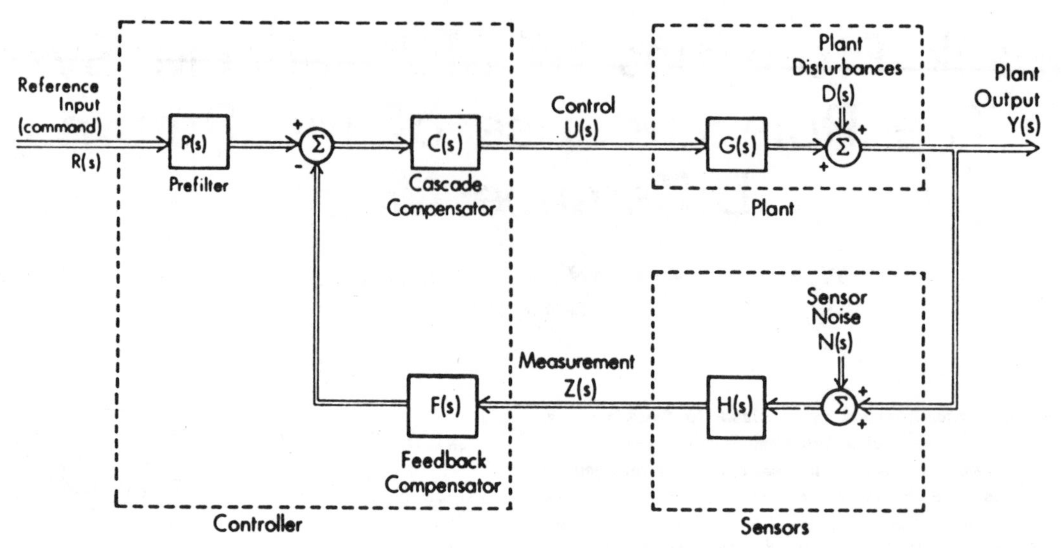

Fig. 2. Feedback control system.

results for the analysis of multivariable feedback systems are developed so as to form a natural extension to multivariable systems of the proven classical methods based on root locus and frequency response. For the actual synthesis of multivariable feedback controllers with specified frequency domain properties, a methodology based on stochastic linear quadratic Gaussian (LQG) theory is suggested.

The paper is organized as follows. The remainder of the first section is devoted to defining the system and associated notation. Connections between return difference and multivariable feedback properties are developed in Section II. In Section III we describe the roles of Bode magnitude plots of return-difference matrix norms and singular values in providing useful quantitative measures of feedback qualities. In Section IV some fundamental limits on achievable feedback performance are discussed. Connections between an engineer's choice of LQG quadratic cost and noise matrices and the resulting optimal system's feedback properties are developed in Section V, and an LQG-based aircraft autopilot design synthesis example is described in Section VI. Conclusions are in Section VII.

A. *The System*

We consider the following linear time-invariant multivariable control system (see Fig. 2):

plant

$$Y(s)=G(s)U(s)+D(s) \tag{1.1}$$

controller

$$U(s)=C(s)(P(s)R(s)-F(s)Z(s)) \tag{1.2}$$

sensors

$$Z(s)=H(s)(Y(s)+N(s)) \tag{1.3}$$

where $Y(s)$, $U(s)$, $D(s)$, $R(s)$, $Z(s)$, and $N(s)$ are vectors and $G(s)$, $C(s)$, $P(s)$, $F(s)$, and $H(s)$ are compatibly dimensioned matrices. The controller-sensor portion of the feedback loop has transfer matrix

$$M(s) \triangleq C(s)F(s)H(s). \tag{1.4}$$

The matrices

$$L_2(s) \triangleq G(s)M(s) \tag{1.5}$$

$$L_1(s) \triangleq M(s)G(s) \tag{1.6}$$

are called the *return ratio* [10] matrices at the $Y(s)$ node and the $U(s)$ node, respectively. The matrices

$$I+L_2(s) \tag{1.7}$$

and

$$I+L_1(s) \tag{1.8}$$

are called, respectively, the $Y(s)$ node and $U(s)$ node *return difference* [10] matrices. It is also convenient for us to introduce the new term, *inverse-return difference*, to describe the return difference matrix associated with a hypothetical system whose return ratio is the inverse of the original system's return ratio. The $Y(s)$ node and the $U(s)$ node inverse-return difference matrices are, respectively,

$$I+L_2^{-1}(s) \tag{1.9}$$

$$I+L_1^{-1}(s). \tag{1.10}$$

II. Feedback Properties

The feedback properties of the system (1.1)–(1.3) are intimately related to the associated return difference and inverse-return difference matrices. This section is devoted to establishing these relations for the feedback properties of noise and disturbance attenuation, stability margin, and sensitivity to modeling errors.

A. Command, Disturbance, and Noise Response

The measurement signal $Z(s)$, plant output $Y(s)$, and control $U(s)$ may be represented as sums of three components, generated by $R(s)$, $D(s)$ and $N(s)$, respectively,

$$\begin{bmatrix} Z(s) \\ Y(s) \\ U(s) \end{bmatrix} = \begin{bmatrix} Z_R(s) \\ Y_R(s) \\ U_R(s) \end{bmatrix} + \begin{bmatrix} Z_D(s) \\ Y_D(s) \\ U_D(s) \end{bmatrix} + \begin{bmatrix} Z_N(s) \\ Y_N(s) \\ U_N(s) \end{bmatrix}. \quad (2.1)$$

The components of $Z(s)$, $Y(s)$, and $U(s)$ are given in terms of $R(s)$, $D(s)$, and $N(s)$ as follows:

$$\begin{bmatrix} Z_R \\ Y_R \\ U_R \end{bmatrix} \triangleq \begin{bmatrix} T_{ZR} \\ T_{YR} \\ T_{UR} \end{bmatrix} R = \begin{bmatrix} HGS_1CP \\ GS_1CP \\ S_1CP \end{bmatrix} R = \begin{bmatrix} HS_2GCP \\ S_2GCP \\ S_1CP \end{bmatrix} R \quad (2.2)$$

$$\begin{bmatrix} Z_D \\ Y_D \\ U_D \end{bmatrix} \triangleq \begin{bmatrix} T_{ZD} \\ T_{YD} \\ T_{UD} \end{bmatrix} D = \begin{bmatrix} HS_2 \\ S_2 \\ -W \end{bmatrix} D \quad (2.3)$$

$$\begin{bmatrix} Z_N \\ Y_N \\ U_N \end{bmatrix} \triangleq \begin{bmatrix} T_{ZN} \\ T_{YN} \\ T_{UN} \end{bmatrix} D = \begin{bmatrix} HS_2 \\ -T_2 \\ -W \end{bmatrix} N \quad (2.4)$$

where

$$S_i \triangleq (I+L_i)^{-1}, \qquad (i=1,2) \quad (2.5)$$

$$T_i \triangleq I - S_i \equiv L_i(I+L_i)^{-1} \equiv (I+L_i)^{-1}L_i, \qquad (i=1,2) \quad (2.6)$$

$$W \triangleq S_1 M \equiv MS_2. \quad (2.7)$$

One also has the identities

$$T_2 \equiv GW \quad (2.8)$$

$$T_1 \equiv WG. \quad (2.9)$$

The connection between noise response and disturbance response on the one hand and inverse-return difference and return-difference on the other hand can now be clarified. From (2.5)–(2.9) it is clear that the following four identities hold, subject only to the existence of the indicated inverses:

$$S_i^{-1}(s) = (I+L_i(s)), \qquad (i=1,2) \quad (2.10)$$

$$T_i^{-1}(s) = (I+L_i^{-1}(s)), \qquad (i=1,2) \quad (2.11)$$

$$W_l(s) = (I+L_2^{-1}(s))G(s) \quad (2.12)$$

$$W_r(s) = G(s)(I+L_1^{-1}(s)) \quad (2.13)$$

where A_r and A_l denote, respectively, right and left inverses of a not necessarily square, matrix A (when such inverses exist), i.e.,

$$A_l A \triangleq I \quad (2.14)$$

$$AA_r \triangleq I; \quad (2.15)$$

note that $G_r(s)$ and $W_l(s)$ exist if rank$(L_2(s))$=dimension $(Y(s))$ and that $G_l(s)$ and $W_r(s)$ exist if rank$(L_1(s))$= dimension$(U(s))$. Thus, the response $Y_D(s)$ of the plant output to plant disturbances is determined by the return-difference $I+L_2(s)$. The response $Y_N(s)$ to sensor noise is determined by the inverse-return difference $I+L_2^{-1}(s)$. For any given plant $G(s)$, the responses of the control inputs $U_D(s)$ and $U_R(s)$ are both determined by the inverse return difference $I+L_2^{-1}(s)$, or $I+L_1^{-1}(s)$, depending on whether $W(s)$ has a left inverse or a right inverse.

B. Stability and Stability Margin

The issue in characterizing stability margin for the feedback system (1.1)–(1.3) is to determine a lower bound on the "size"—any physically meaningful measure of size will suffice—of the smallest perturbation to the return ratio matrix that will destabilize the system. The stability of the system (1.1)–(1.3), like its response to noise and disturbances, is directly related to the matrices W, S_i, and T_i. Subject to mild stabilizability and detectability assumptions [13] the feedback system (1.1)–(1.3) is asymptotically stable if and only if $W(s)$, $S_i(s)$, and $T_i(s)$ (equivalently) have no poles in the closed right-half of the complex plane.

One reasonable way to measure the stability margin of the system (1.1)–(1.3) is to suppose that the true value of $L_i(s)$ is related to its nominal value, say, $L_{0i}(s)$, by the equality

$$L_i(s) = (I+\Delta L_i(s))L_{0i}(s) \quad (2.16)$$

where $\Delta L_i(s)$ is a (not necessarily small) perturbation; $i=1$ or 2. Then the stability margin of the system could be defined as the "size" of the "smallest" perturbation $\Delta L_i(s)$ such that $S_i(s) \triangleq (I+L_i(s))^{-1}$ is unstable, i.e., such that the feedback system (1.1)–(1.3) is destabilized.[2] Note that the condition that $S_i(s)$ be unstable is equivalent to the condition that, for some s_0 with $\mathrm{Re}(s_0) > 0$ and some nonzero $x_0 \in \mathbf{C}^n$,

$$0 = (I+L_i(s_0))x_0 e^{s_0 t}. \quad (2.17)$$

If one assumes that $S_{0i}(s) \triangleq (I+L_{0i}(s))^{-1}$ is stable then, taking $x_1 = (I+L_{0i}(s_0))x_0$ and simplifying, one may obtain from (2.17) that $S_i(s)$ is not asymptotically stable if

[2]We say a proper rational transfer function matrix is stable if it has no poles in the closed right-half of the complex plane; otherwise it is unstable.

and only if there exist an s_0 with $\mathrm{Re}(s_0)>0$ and a nonzero $x_1 \in \mathbf{C}^n$ such that

$$x_1 = -\Delta L_i(s_0) T_{0i}(s_0) x_1 \tag{2.18}$$

where $T_{0i}(s) \triangleq L_{0i}(s)(I+L_{0i}(s))^{-1}$.

From (2.18) one may deduce that the stability margin of (1.1)–(1.3), i.e., the size of the smallest $\Delta L_i(s)$ such that (1.1)–(1.3) is not asymptotically stable, is in some sense inversely proportional to the size of $T_i(s)$ for values of s in the closed right-half complex plane.

Indeed, classical single-loop gain and phase margin may be computed from (2.18) by considering the special case of scalar $L_i(s)$. In this case (2.18) becomes

$$\Delta L_i(s_0) = \frac{-1}{T_{0i}(s_0)}. \tag{2.19}$$

The classical gain margin is readily computed by considering $\Delta L_i = g-1$, leading to the conclusion that the gain margin g_M is the least constant $g>1$ such that

$$g = -L_{0i}^{-1}(s_0) \equiv \frac{-1}{T_{0i}(s_0)} + 1 \tag{2.20}$$

for some s_0 in $\mathrm{Re}(s_0)>0$. Phase margin may be similarly computed by considering $\Delta L_i(s) = e^{-j\theta}-1$ leading to the conclusion that the phase margin θ_M is the smallest $\theta>0$ such that

$$e^{-j\theta} - 1 = -L_{0i}^{-1}(s_0) - 1 \equiv \frac{1}{T_{0i}(s_0)}. \tag{2.21}$$

It is a well-known classical control result that the value of s_0 satisfying the gain margin condition (2.20) is a $-180°$ phase crossover frequency $s_0 = j\omega_{180}$ satisfying

$$\angle L_{0i}(j\omega_{180}) = -180° \tag{2.22}$$

and the value of S_0 satisfying the phase margin condition (2.21) is a 0 dB magnitude crossover frequency $s_0 = j\omega_c$ satisfying

$$|L_{0i}(j\omega_c)| = 1. \tag{2.23}$$

For the more general case of multivariable feedback systems the meaning of gain and phase becomes clouded by the fact that there are infinitely many different input and output vector "directions."

C. Sensitivity

The sensitivity of a feedback control system to variations in the plant or sensor transfer function matrices has to do with the effects that these variations have on the plant input and output signals, viz., $U(s)$ and $Y(s)$.[3] In order to be able to meaningfully compare the sensitivities of various feedback systems it is customary in the single-loop case to define sensitivity in terms of percentage variation in $Y_R(s)$ or $U_R(s)$ resulting from a given percentage change in $G(s)$ or in $H(s)$ with the input $R(s)$ held fixed. The classical Bode sensitivity function [20] may be defined this way. A similar result has been developed for multivariable systems by Cruz and Perkins [14], who define the sensitivity of $Y_R(s)$ to variations in $G(s)$ to be a matrix that multiplies the $Y_R(s)$-variation from an "equivalent open-loop system" to yield the $Y_R(s)$-variation that results in the actual feedback control system (1.1)–(1.3). This approach to multivariable sensitivity analysis, known as the *comparison sensitivity* approach because of the way closed-loop and open-loop systems are compared, has the important limitation that it cannot be used to evaluate sensitivity when variations are confined to the "backward" path in the feedback loop [e.g., variations in the sensor transfer function matrix $H(s)$] because the comparison sensitivity matrix becomes singular in this case. Also, although it yields a sensitivity relation that, in principle, holds for large plant variations, the Cruz and Perkins comparison sensitivity matrix is difficult to interpret when all that is known about the plant variations is a bound on their size, since the sensitivity matrix depends on the actual values of the plant variations. There has been some progress in resolving this latter difficulty (e.g., [43, Ch. 3], [44]).

The following theorems provide an alternative for multivariable sensitivity characterization that overcomes the aforementioned limitations of the comparison sensitivity approach. These results were originally presented in [45].

Theorem 1 (Plant Output Sensitivity): Suppose that in (1.1)–(1.3)[4]

$$F(s)H(s) = F_0(s)H_0(s)(I+\Delta_{FH}(s)) \tag{2.24}$$

$$G(s)C(s) = (I+\Delta_{GC}(s))^{-1} G_0(s)C_0(s) \tag{2.25}$$

for some specified matrices $F_0(s)$, $H_0(s)$, $G_0(s)$, $C_0(s)$, $\Delta_{FH}(s)$, and $\Delta_{GC}(s)$. Define

$$S_{02}(s) \triangleq S_2(s)|_{F=F_0,H=H_0,G=G_0,C=C_0}$$
$$\equiv (I+G_0(s)C_0(s)F_0(s)H_0(s))^{-1}, \tag{2.26}$$

$$T_{02}(s) \triangleq I - S_{02}(s), \tag{2.27}$$

and

$$T_{0YR}(s) \triangleq T_{YR}(s)|_{F=F_0,H=H_0,G=G_0,C=C_0}$$
$$\equiv S_{02}(s)G_0(s)C_0(s)P(s). \tag{2.28}$$

[3]Much of the literature on feedback sensitivity deals only with the narrower issue of the effects of plant variation on plant output signal. Obviously, a control engineer must, in practice, consider sensor variations too, and the effects on the control signal $U(s)$ as well as on $Y(s)$.

[4]The quantities $G_0(s)C_0(s)$ and $F_0(s)H_0(s)$ may be regarded as the nominal "forward path" and "backward path" matrices with $\Delta_{GC}(s)$ and $\Delta_{FH}(s)$ being perturbations whose respective "sizes" may be regarded as proportional to the percentage variation about these nominals, at least when Δ_{GC} and Δ_{FH} are small.

Then

$$T_{YR}(s)=(I+\Delta_{YR}(s))^{-1}T_{OYR}(s) \qquad (2.29a)$$

where

$$\Delta_{YR}(s) \triangleq S_{02}(s)\Delta_{GC}(s)+T_{02}(s)\Delta_{FH}(s). \qquad (2.29b)$$

Proof: Define $L_{02} \triangleq G_0C_0F_0H_0$. Then

$$(I+L_2)Y_R=GCPR. \qquad (2.30)$$

or, equivalently,

$$\left(I+(I+\Delta_{GC})^{-1}L_{02}(I+\Delta_{FH})\right)Y_R=(I+\Delta_{GC})^{-1}G_0C_0PR.$$

Multiplying on the left by $(I+L_{02})^{-1}(I+\Delta_{GC})$ yields

$$\left(I+(I+L_{02})^{-1}\Delta_{GC}+(I+L_{02})^{-1}L_{02}\Delta_{FH}\right)Y_R$$
$$=(I+L_{02})^{-1}G_0C_0PR.$$

Equivalently,

$$(I+\Delta_{YR})Y_R=T_{OYR}R.$$

Thus,

$$Y_R=(I+\Delta_{YR})^{-1}T_{OYR}R.$$

But from (2.2) $Y_R \triangleq T_{YR}R$; hence $T_{YR}=(I+\Delta_{YR})^{-1}T_{OYR}$. □

Theorem 2 (Control Sensitivity): Suppose that in (1.1)–(1.3)

$$C(s)=(I+\Delta_C(s))^{-1}C_0(s) \qquad (2.31)$$

$$F(s)H(s)G(s)=F_0(s)H_0(s)G_0(s)(I+\Delta_{FHG}(s)) \qquad (2.32)$$

for some specified matrices $F_0(s)$, $H_0(s)$, $C_0(s)$, $G_0(s)$, $\Delta_C(s)$, and $\Delta_{FHG}(s)$. Define

$$S_{01}(s) \triangleq S_1(s)|_{F=F_0,H=H_0,C=C_0,G=G_0}$$
$$\equiv(I+C_0(s)F_0(s)H_0(s)G_0(s))^{-1}, \qquad (2.33)$$

$$T_{01}(s) \triangleq I-S_{01}(s), \qquad (2.34)$$

and

$$T_{OUR}(s) \triangleq T_{UR}(s)|_{F=F_0,H=H_0,C=C_0,G=G_0}$$
$$\equiv S_{01}(s)C_0(s)P(s). \qquad (2.35)$$

Then,

$$T_{UR}(s)=(I+\Delta_{UR}(s))^{-1}T_{OUR}(s) \qquad (2.36a)$$

where

$$\Delta_{UR}(s) \triangleq S_{01}(s)\Delta_C(s)+T_{01}(s)\Delta_{FHG}(s). \qquad (2.36b)$$

Proof: Let $L_{01}=C_0F_0H_0G_0$. Then proceeding analogously to the proof of Theorem 1, we have

$$(I+L_1)U_R=CPR$$

or, equivalently,

$$\left(I+(I+\Delta_C)^{-1}L_{01}(I+\Delta_{FHG})\right)U_R=(I+\Delta_C)^{-1}C_0PR.$$

Multiplying on the left by $(I+L_{01})^{-1}(I+\Delta_{UR})$ and simplifying yields

$$(I+\Delta_{UR})U_R=T_{OUR}R.$$

Since by (2.2) $U_R \triangleq T_{UR}R$, it follows that

$$G_{UR}=(I+\Delta_{UR})^{-1}T_{OUR}. \qquad \square$$

The connections between sensitivity and return-difference and inverse-return difference are immediately evident in (2.29) and (2.36) if one recalls the identities (2.10) and (2.11).

Like the comparison sensitivity result of Cruz and Perkins [14], Theorems 1 and 2 hold for arbitrarily large variations, not just for vanishingly small variations. Also, the sensitivity matrices S_{01} and S_{02} for variations in the forward loop of the system are the exact same sensitivity matrices that one would obtain by the approach of Cruz and Perkins [14], *except that our sensitivity matrices (S_{0i} and T_{0i}) are evaluated at the known nominal matrix values (G_0, C_0, F_0, H_0) rather than the unknown true values (G, C, F, H).*[5] Additionally, our result enables evaluation of the sensitivity to system variations (Δ_{FH} and Δ_{FHG}) which are confined to the "backward" path of the feedback loop, which is not possible using the Cruz and Perkins approach.

It is also interesting to consider what Theorems 1 and 2 imply about the deviation of the signal $Y(s)$ from its nominal value, say $Y_0(s)$. Equation (2.25) implies that for any given reference input $R(s)$,

$$Y(s)=(I+\Delta_{YR})^{-1}Y_0(s)$$

where $Y_0(s) \triangleq (T_{OYR}(s)R(s)$ is the nominal value of $Y_0(s)$ resulting when $\Delta_{GC}\equiv\Delta_{FH}\equiv 0$. Thus, the deviation $Y_0(s) - Y(s)$ is linearly related to $Y(s)$ by

$$Y_0-Y=\Delta_{YR}Y$$

and Δ_{YR} is linearly related to the open-loop variations Δ_{GC} and Δ_{FH} by (2.29b). This may be loosely interpreted as saying that Δ_{YR} is the "percentage change" in the nominal output Y_0 relative to the true-but-unknown output Y. If one prefers to express percentage deviation relative to the known nominal Y_0, the preceding formula yields

[5]Note that one may also interpret Theorems 1 and 2 by regarding (G_0, C_0, F_0, H_0) as the true values and (G, C, F, H) as the nominal values. In this case, the sensitivity matrices S_{0i} and T_{0i} would be evaluated at unknown true values, like the Cruz and Perkins results in [14].

$$Y - Y_0 = -\Delta_{YR}(I+\Delta_{YR})^{-1}Y_0,$$

which, regrettably, is no longer linear in Δ_{YR}. Loosely speaking, $-\Delta_{YR}(I+\Delta_{YR})^{-1}$ is the percentage deviation in Y relative to the nominal output Y_0. Observe that when $\|\Delta_{YR}\| \ll 1$ then the closed-loop sensitivity is small and the percentage deviations relative to Y and relative to Y_0 are about equal; this may be the case even when the open-loop forward path variation Δ_{GC} is very large. Note, however, that careful interpretation is required when $\|\Delta_{YR}\| > 1$ since in this case $\Delta_{YR}(I+\Delta_{YR})^{-1}$ may have poles in the right-half complex plane, leading to instability and infinite closed-loop "percentage deviation" in Y relative to the nominal Y_0.

We believe that our approach is close in spirit to the classical Bode sensitivity function [20], which defines sensitivity as a means for relating percentage change in component response to percentage change in feedback system response. Only the percentage plant and sensor variation is of interest in practical control problems because it is always possible to select the $C(s)$ and $F(s)$ matrices to manipulate the nominal values $G_0(s)C_0(s)$ and $F_0(s)H_0(s)$, subject only to the constraint that the right-half plane transmission zeros must remain fixed (lest unstable pole-zero "cancellations" be introduced). However, the percentage variation in the "forward" and "backward" paths of the feedback loop are normally fixed *a priori* by the precision of the models available for the plant $G(s)$ and sensors $H(s)$. Likewise, only the percentage variation in closed-loop response is of practical interest because the nominal closed-loop response can be manipulated as needed by appropriate choice of the prefilter $P(s)$ independent of S_{0i}, again subject only to the constraint that right-half plane transmission zeros must remain fixed [24]. In the multivariable case considered in Theorems 1 and 2 the matrices Δ_{GC}, Δ_{FH}, Δ_C, and Δ_{FHG} constitute multivariable analogs of the percentage changes in open-loop system components and the matrices Δ_{YR} and Δ_{UR} constitute multivariable analogs of resulting percentage changes in closed-loop system response to commands. In fact, assuming the indicated inverses exist, the equations (2.24) and (2.25), (2.31) and (2.32), (2.29a), and (2.36a) may be rearranged to yield direct multivariable generalizations of Bode's representation of percentage change, viz.,

$$\Delta_{GC} = (G_0C_0 - GC)(GC)^{-1} \tag{2.37}$$

$$\Delta_C = (C_0 - C)C^{-1} \tag{2.38}$$

$$\Delta_{FH} = (F_0H_0)^{-1}(FH - F_0H_0) \tag{2.39}$$

$$\Delta_{FHG} = (F_0H_0G_0)^{-1}(FHG - F_0H_0G_0) \tag{2.40}$$

$$\Delta_{YR} = (T_{OYR} - T_{YR})T_{YR}^{-1} \tag{2.41}$$

$$\Delta_{UR} = (T_{OUR} - T_{UR})T_{UR}^{-1}. \tag{2.42}$$

D. Summary of Feedback Properties

The relations (2.3), (2.4), (2.18), (2.29), and (2.36) establish unequivocally that the "smallness" of the matrices W, S_1, S_2, T_1, and T_2 is directly related to the quality of the feedback properties of noise and disturbance attenuation, stability margin, and sensitivity. If $W(s)$ vanishes, $T_1(s)$ and $T_2(s)$ both vanish, too [see (2.8) and (2.9)], and the following desirable feedback properties result:

1) $U_D(s)$, $Y_N(s)$, and $U_N(s)$ vanish [see (2.3) and (2.4)];

2) stability margins approach infinity, i.e., arbitrarily large $\Delta L_i(s)$ cannot destabilize the system [see (2.18)];

3) sensitivity of $Y_R(s)$ to variations in $F(s)H(s)$ vanishes and sensitivity of $U_R(s)$ to variations in $F(s)H(s)G(s)$ vanishes [see (2.29) and (2.36)].

On the other hand, if $S_1(s)$ and $S_2(s)$ vanish, one gets other desirable properties, viz.,

1) $Y_D(s)$ vanishes [see (2.3)];

2) sensitivity of $Y_R(s)$ to variations in $G(s)C(s)$ vanishes and sensitivity of $U_R(s)$ to variations in $C(s)$ vanishes.

Unfortunately both $S_i(s)$ and $T_i(s)$ cannot be made to vanish simultaneously because of the identity $S_i(s)+T_i(s) = I$ [see (2.6)]. Consequently, the feedback design engineer is faced with the task of attempting to optimally tradeoff the "smallness" of $(T_1(s), T_2(s))$ against the smallness of $(S_1(s), S_2(s))$. Equivalently, he must tradeoff the "largeness" of the return-difference matrices $S_i^{-1}(s) \equiv I + L_i(s)$, $(i=1,2)$ against the "largeness" of the inverse-return difference matrices $T_i^{-1}(s) \equiv I + L_i^{-1}(s)$, $(i=1,2)$. In either case, the question of how to meaningfully quantify the "size" (i.e., the "largeness' and "smallness") of frequency dependent matrices arises.

III. Quantitative Measures of Feedback Quality

To usefully quantify the feedback properties of the multivariable system (1.1)–(1.3), simple scalar (but perhaps frequency-dependent) measures of these qualities are required so that alternative designs can be directly compared and so that the problem of choosing an "optimal" design can be meaningfully posed. The relations described in Section II suggest that the "sizes" of the matrices $W(s)$, $S_1(s)$, $S_2(s)$, $T_1(s)$ and $T_2(s)$ may provide the required scalar measures.

A. Matrix Norms and Singular Values

A wide choice of measures of the "size" of a matrix is available. For example, the magnitudes of angles of the frequency-dependent eigenvalues of $T_1(s)$, $T_2(s)$, $S_1(s)$, $S_2(s)$, and $W(s)$ offer one tempting possibility. This seems to be superficially attractive because of a number of elegantly simple relations between the eigenvalues and eigenvectors of these matrices. For example,[6]

[6]The notation $\lambda_j(A)$ denotes the jth eigenvalue of the square matrix A.

$$1+\lambda_j(L(s))=\lambda_j(I+L(s))\equiv\lambda_j(S^{-1}(s))$$
$$=\lambda_j^{-1}(S(s)) \quad (3.1)$$
$$1+\lambda_j(L^{-1}(s))=\lambda_j(I+L^{-1}(s))\equiv\lambda_j(T^{-1}(s))$$
$$=\lambda_j^{-1}(T(s)). \quad (3.2)$$

Further, the left and right eigenvectors, say $\xi_j(s)$ and $\eta_j(s)$, respectively, are shared by $T(s)$, $S(s)$, and $L(s)$. These facts have been extensively exploited by MacFarlane, Kouvaritakis, and their colleagues who have advocated the use of the Nyquist loci and root loci of the frequency-dependent functions $\lambda_j(L(s))$ to extend classical single-loop methods to multivariable feedback systems [10], [19].

However, despite the fact that eigenvalues have the nice properties (3.1) and (3.2), they may be very poor indicators of the "size" of a matrix. This may be readily seen by considering the 2×2 matrix

$$A=\begin{bmatrix} 0.1 & 10^9 \\ 0 & 0.1 \end{bmatrix} \quad (3.3)$$

whose eigenvalues are both 0.1 but which is seemingly "large" in most norms. Certainly, it is large in the input-output sense that the "gain" $\|Au\|/\|u\|$ can be very large for inputs such as $u=(0,1)^T$. It might be argued, however, that the problem here is only a matter of scale in the sense that if u is scaled say, by dividing all components by 10^9), the gain no longer seems large. But there are actually deeper problems than just scale in trying to use eigenvalues to measure the "size" of a matrix, as the following example illustrates. Consider the matrix $A:\mathbf{R}^n\to\mathbf{R}^n$

$$A=\begin{bmatrix} -1 & +1 & \cdots & +1 \\ \cdot & & \ddots & \vdots \\ 0 & \cdot & & +1 \\ & & \cdot & -1 \end{bmatrix} \quad (3.4)$$

which appears quite "large," having eigenvalues $\lambda_j(A)=-1$. Yet, the input $u_0=(1,\frac{1}{2},\frac{1}{4},\cdots,1/2^{n-1})^T$ yields an output $Au_0=-(2^{-n+1},2^{-n+1},\cdots,2^{-n+1})^T$. In other words, for inputs in the direction u_0 the matrix A can have a vanishingly small gain as $n\to\infty$ despite the fact that its eigenvalues all have magnitude one. See [21] for further details.

It is our feeling that in the control system context a physically meaningful measure of the "size" of a system described by a transfer function matrix ought to bear some relation to the input–output gain ratio of the system. Consider a transfer function matrix $A(s)$ taking input signals $U(s)$ into output signals $Y(s)$. Suppose that it has been determined that the physically relevant measures of input and output signal sizes at each frequency s are the norms $\|U(s)\|_{\mathcal{U}(s)}$ and $\|Y(s)\|_{\mathcal{Y}(s)}$, respectively, where $\|\cdot\|_{\mathcal{U}(s)}$ and $\|\cdot\|_{\mathcal{Y}(s)}$ are norms, possibly s-dependent, on $\mathbf{C}^n$. For example, $\|\cdot\|^2_{\mathcal{U}(s)}$ and $\|\cdot\|^2_{\mathcal{Y}(s)}$ might be signal power. Given an input signal, say $U(s)$, the gain ratio would then be $\|A(s)U(s)\|_{\mathcal{Y}(s)}/\|U(s)\|_{\mathcal{U}(s)}$.

Definition: The maximum gain of $A(s)$, denoted by $\sigma_{\max}(A(s);\mathcal{U}(s),\mathcal{Y}(s))$, is the subordinant matrix norm

$$\sigma_{\max}(A(s);\mathcal{U}(s),\mathcal{Y}(s))\triangleq\|A(s)\|_{\mathcal{U}(s),\mathcal{Y}(s)}$$
$$\triangleq\max_{U(s)\neq 0}\frac{\|A(s)U(s)\|_{\mathcal{Y}(s)}}{\|U(s)\|_{\mathcal{U}(s)}}.$$

Similarly, the minimum gain of $A(s)$ is

$$\sigma_{\min}(A(s);\mathcal{U}(s),\mathcal{Y}(s))\triangleq\min_{U(s)\neq 0}\frac{\|A(s)U(s)\|_{\mathcal{Y}(s)}}{\|U(s)\|_{\mathcal{U}(s)}}. \quad \square$$

Norms on $\mathbf{C}^n$ that have been found to be particularly useful for control system analysis are the Euclidean norms, so we shall specialize our discussion for the rest of this section to these norms.

Definition: Given a positive definite matrix $Q\in\mathbf{C}^{n\times n}$, we define the unitary space $\mathbf{C}^n_Q$ as the set of complex n-vectors $x=(x_1,\cdots,x_n)^T$ together with the inner product $\langle\cdot,\cdot\rangle_Q$ and induced Euclidean norm $\|\cdot\|_Q$ defined by[7]

$$\langle x,y\rangle_Q\triangleq\bar{y}^TQx$$
$$\|x\|_Q\triangleq(\bar{x}^TQx)^{1/2}. \quad \square$$

Let us now reconsider our two examples in (3.3) and (3.4) with $\mathcal{U}(s)=I,\mathcal{Y}(s)=I$ for convenience. Then for (3.3) we have that $\|A\|$ must be at least of the order of 10^9 (take $U(s)=(1,1)^T$, for example) while for (3.4) we have that $\sigma_{\min}\lesssim\sqrt{n}/2^{n-1}$ [take $U(s)=(1,\frac{1}{2},\frac{1}{4},\cdots,1/2^{n-1})^T$], i.e., A has "small" minimum gain at least for moderately large n.

For the case of a matrix A mapping one unitary space into another, say, $A:\mathbf{C}^n_{Q_1}\to\mathbf{C}^m_{Q_2}$, the quantities $\sigma_{\max}(A;Q_1,Q_2)$ and $\sigma_{\min}(A;Q_1,Q_2)$ are, in fact, equal to certain singular values of A. We shall review a few properties of singular values here. Further details and applications can be found in [39].

Definition: The singular values[8] of a matrix $A:\mathbf{C}^n_{Q_1}\to\mathbf{C}^m_{Q_2}$, denoted

$$\sigma_j(A;Q_1,Q_2),\qquad j=1,\cdots,n$$

are the nonnegative square roots of the eigenvalues of A^*A, where $A^*\triangleq Q_1^{-1}\bar{A}^TQ_2$.[9]

When A has rank r, exactly r of the singular values will be positive and it will be convenient to order them as follows.

$$\sigma_1\geqslant\cdots\geqslant\sigma_r>0=\sigma_{r+1}=\cdots=\sigma_n.$$

Of course, it is also possible to define the singular values

[7]An overbar denotes complex conjugate, e.g., $\bar{x}$ and $\bar{A}$.

[8]Singular values have also been called s numbers [40], and principal gains [41].

[9]$A^*:\mathbf{C}^m_{Q_2}\to\mathbf{C}^n_{Q_1}$ is the *adjoint* of the linear mapping $A:\mathbf{C}^n_{Q_1}\to\mathbf{C}^m_{Q_2}$ defined by the relation $\langle Ax,y\rangle_{Q_2}\triangleq\langle x,A^*y\rangle_{Q_1}\forall x,y$.

the nonnegative square roots of the m eigenvalues of AA^*. The number of zero singular values is different if $m \neq n$ but the r positive singular values are identical in either case. In the special case $Q_1 = I$ and $Q_2 = I$, we use the abbreviated notation

$$\sigma_j(A) \triangleq \sigma_j(A; I, I).$$

Note also that

$$\sigma_j(A; Q_1, Q_2) = \sigma_j(Q_2^{1/2} A Q_1^{-1/2}).$$

The following well-known results establish the connections between the singular values of A and the maximum and minimum gains of A.

Lemma 1:

$$\sigma_{\max}(A; Q_1, Q_2) \triangleq \max_{u \neq 0} \left(\frac{\|Au\|_{Q_2}}{\|u\|_{Q_1}} \right) = \max_j \{\sigma_j(A; Q_1, Q_2)\} \tag{3.5}$$

$$\sigma_{\min}(A; Q_1, Q_2) \triangleq \min_{u \neq 0} \left(\frac{\|Au\|_{Q_2}}{\|u\|_{Q_1}} \right) = \min_j \{\sigma_j(A; Q_1, Q_2)\} \tag{3.6}$$

$$\operatorname{tr}(AQ_1^{-1}\bar{A}^T Q_2) \equiv \operatorname{tr}(AA^*) = \sum_{j=1}^{r} \sigma_j^2(A; Q_1, Q_2) \tag{3.7}$$

Proof: See, e.g., Stewart [22].

Although the singular values of a matrix do not have such convenient properties as the eigenvalues [e.g., (3.1) and (3.2)], they do possess certain other useful properties. Associated with each singular value $\sigma_j(A; Q_1, Q_2)$ are a left singular vector $u_j \in \mathbf{C}^m_{Q_2}$ satisfying

$$AA^* u_j = \sigma_j^2(A; Q_1, Q_2) u_j \tag{3.8}$$

and a right singular vector $v_j \in \mathbf{C}^n_{Q_1}$ satisfying

$$A^* A v_j = \sigma_j^2(A; Q_1, Q_2) v_j. \tag{3.9}$$

The singular vectors v_j and u_j may be selected so that the sets $\{u_j | j = 1, \cdots, n\}$ and $\{v_j | j = 1, \cdots, m\}$ are orthonormal sets in $\mathbf{C}^n_{Q_2}$ and $\mathbf{C}^m_{Q_1}$, respectively. The matrix A admits the singular value decomposition (SVD)

$$A = \sum_{j=1}^{n} \sigma_j(A; Q_1, Q_2) u_j v_j^* = \sum_{j=1}^{\operatorname{rank}(A)} \sigma_j(A; Q_1, Q_2) u_j v_j^* \tag{3.10}$$

where $v_j^* \triangleq \bar{v}_j^T Q_1$. One readily verifies from (3.10) that if $A: \mathbf{C}^n_{Q_1} \to \mathbf{C}^n_{Q_2}$ and A is invertible then the SVD of $A^{-1}: \mathbf{C}^n_{Q_2} \to \mathbf{C}^n_{Q_1}$ is given by

$$A^{-1} = \sum_{j=1}^{n} \sigma_j^{-1}(A; Q_1, Q_2) v_j u_j^* \tag{3.11}$$

where $u_j^* \triangleq \bar{u}_j^T Q_2$; so

$$\sigma_j(A^{-1}; Q_2, Q_1) = 1/\sigma_{n-j+1}(A; Q_1, Q_2), \qquad j = 1, \cdots, n. \tag{3.12}$$

Assuming $S^{-1}(s) = I + L(s)$ and $T^{-1} = I + L^{-1}(s)$ exist, this leads to the useful identities

$$\sigma_j(S(s); Q_1, Q_2) = 1/\sigma_{n-j+1}(I + L(s); Q_2, Q_1) \tag{3.13}$$

$$\sigma_j(T(s); Q_1, Q_2) = 1/\sigma_{n-j+1}(I + L^{-1}(s); Q_2, Q_1). \tag{3.14}$$

Thus, the singular values of the return difference and inverse-return difference matrices are simply the inverses of the singular values of the $S_i(s)$ and $T_i(s)$ matrices; $i = 1, 2$.

Let us now consider the implications of using singular values to quantitatively evaluate the feedback properties of (1.1)–(1.3).

B. Noise and Disturbance Attenuation

From (2.3) and (2.4) it follows via (3.13) that for any positive definite Q_1, Q_2 and for any ω

$$\sup_{N(j\omega) \neq 0} \left(\frac{\|Y_D(j\omega)\|_{Q_2}}{\|D(j\omega)\|_{Q_1}} \right) = 1/\sigma_{\min}(I + L_2(j\omega); Q_2, Q_1) \tag{3.15}$$

$$\sup_{N(j\omega) \neq 0} \left(\frac{\|Y_N(j\omega)\|_{Q_2}}{\|N(j\omega)\|_{Q_1}} \right) = 1/\sigma_{\min}(I + L_2^{-1}(j\omega); Q_2, Q_1). \tag{3.16}$$

If, say, rank$(G(s)) >$ dimension$(Ys))$, then similar identifies may be obtained for $U_D(j\omega)$ and $U_N(j\omega)$, viz.,

$$\sup_{D(j\omega) \neq 0} \left(\frac{\|U_D(j\omega)\|_{Q_2}}{\|D(j\omega)\|_{Q_1}} \right) = 1/\sigma_{\min}(I + L_1^{-1}(j\omega); Q_2, \bar{G}^T(j\omega) Q_1 G(j\omega)) \tag{3.17}$$

$$= \sup_{N(j\omega) \neq 0} \left(\frac{\|U_N(j\omega)\|_{Q_2}}{\|N(j\omega)\|_{Q_1}} \right). \tag{3.18}$$

Alternatively, suppose in (1.1)–(1.3) that the signals $D(s)$ and $N(s)$ are generated by random processes having the respective power spectra $\Sigma_D(j\omega)$ and $\Sigma_N(j\omega)$. Then it follows that for any frequency dependent positive definite Hermitian matrices, say, $Q_2(j\omega)$ and $Q_1(j\omega)$,

$$\|Y_D(j\omega)\|^2_{Q_2(j\omega)} \triangleq \bar{Y}_D^T(j\omega) Q_2(j\omega) Y_D(j\omega)$$

$$= D(j\omega) \bar{S}_2^T(j\omega) Q_2(j\omega) S_2(j\omega) D(j\omega)$$

$$= \operatorname{tr}(S_2(j\omega) D(j\omega) \bar{D}^T(j\omega) \bar{S}_2^T(j\omega) Q_2(j\omega))$$

$$= \mathrm{tr}\left(S_2(j\omega)\Sigma_D(j\omega)\bar{S}_2^T(j\omega)Q_2(j\omega)\right)$$
$$= \sum_{j=1}^{\mathrm{rank}(S(j\omega))} \sigma_i^2\left(S_2(j\omega);\ \Sigma_D^{-1}(j\omega), Q_2(j\omega)\right) \tag{3.19}$$

and similarly,

$$\|Y_N(j\omega)\|^2_{Q_2(j\omega)} = \mathrm{tr}\left(T_2(j\omega)\Sigma_N(j\omega)\bar{T}_2^T(j\omega)Q_2(j\omega)\right)$$
$$= \sum_{j=1}^{\mathrm{rank}(T_2(j\omega))} \sigma_i^2\left(T_2(j\omega);\ \Sigma_N^{-1}(j\omega), Q_2(j\omega)\right) \tag{3.20}$$

$$\|U_D(j\omega)\|^2_{Q_1(j\omega)} = \mathrm{tr}\left(W(j\omega)\Sigma_D(j\omega)\bar{W}^T(j\omega)Q_1(j\omega)\right)$$
$$= \sum_{j=1}^{\mathrm{rank}(W(j\omega))} \sigma_j^2\left(W(j\omega);\ \Sigma_D^{-1}(j\omega), Q_1(j\omega)\right) \tag{3.21}$$

$$\|U_N(j\omega)\|^2_{Q_1(j\omega)} = \mathrm{tr}\left(W(j\omega)\Sigma_N(j\omega)\bar{W}^T(j\omega)Q_1(j\omega)\right)$$
$$= \sum_{j=1}^{\mathrm{rank}(W(j\omega))} \sigma_j^2\left(W(j\omega);\ \Sigma_N^{-1}(j\omega), Q_1(j\omega)\right). \tag{3.22}$$

From (3.19)–(3.22) the connections between the singular values of $T_2(s)$, $S_2(s)$ and $W(s)$ and the noise and disturbance responses of (1.1)–(1.3) are clear.

C. Stability Margins

The "size" of the stability margins of the system (1.1)–(1.3) is directly related to the largest singular value of $T(j\omega)$ by the following theorem.

Theorem 3: Let $Q_1(s)$ and $Q_2(s)$ be appropriately dimensioned invertible para-Hermitian [23] matrices and let the elements of the matrices $Q_1(s)$, $Q_2(s)$, $Q_1^{-1}(s)$ and $Q_2^{-1}(s)$ be proper[10] rational transfer functions with no poles on the $j\omega$-axis. As before, let $L_i(s) = (I + \Delta L_i(s))L_{0i}(s)$ $(i=1,2)$. Let $\Delta L_i(s)$ be rational, proper, and stable. If

$$\|\Delta L_i(j\omega)\|_{Q_1(j\omega), Q_2(j\omega)} < \sigma_{\min}\left(I + L_{0i}^{-1}(j\omega);\ Q_1(j\omega), Q_2(j\omega)\right) \forall\omega \in [0,\infty], \tag{3.23}$$

then both $S_i(s) \triangleq (I+L_i(s))^{-1}$ and $T_i(s) \triangleq I - S_i(s)$ are stable whenever $S_{0i}(s) \triangleq (I+L_{0i}(s))^{-1}$ and $T_{0i}(s) \triangleq I - S_{0i}(s)$ are stable, $i=1$ or 2.[11]

[10]A rational transfer function $A(s)$ is proper if it has at least as many poles as zeros, i.e., $\|A(\infty)\| < \infty$. It is strictly proper if it has more poles than zeros, i.e., $A(\infty)=0$.

[11]We say a proper rational matrix $A(s)$ is stable if it has no poles in the closed right-half of the complex plane.

Proof: The matrices $Q_1(s)$ and $Q_2(s)$ have stable spectral factors $Q_1^{1/2}(s)$ and $Q_2^{1/2}(s)$ with stable inverses $Q_1^{-1/2}(s)$ and $Q_2^{-1/2}(s)$ such that [23]

$$(Q_i^{1/2}(-s))^T(Q_i^{1/2}(s)) = Q_i(s), \qquad i=1,2. \tag{3.24}$$

Thus, the condition (3.23) implies (suppressing s) that for all $s = j\omega$

$$\begin{aligned}1 &> (\|\Delta L\|_{Q_1,Q_2})\cdot(\sigma_{\min}(I+L_{0i}^{-1};\ Q_1,Q_2))\\ &= \|\Delta L_i\|_{Q_1,Q_2}\cdot\|T_{0i}\|_{Q_2,Q_1}\\ &> \|\Delta L_i T_{0i}\|_{Q_1,Q_1}\\ &= \|Q_1^{1/2}\Delta L_i T_{0i}(Q_1^{1/2})^{-1}\|_{I,I}.\end{aligned} \tag{3.25}$$

Let $\tilde{Q}(s) \triangleq (\bar{Q}_1^{1/2}(s))^T(Q_1^{1/2}(s))$. Note that since $Q_1^{1/2}$ and $Q_1^{1/2}\Delta L_i T_{0i}$ are asymptotically stable, they are analytic in $\mathrm{Re}(s) > 0$. Using (3.25), it follows via the Cauchy residue theorem that for every s_0 with $\mathrm{Re}(s_0) > 0$ and every nonzero $x_1 \in \mathbb{C}^n$

$$\begin{aligned}\|x_1\|_{\tilde{Q}(s_0)} &\equiv \|Q_1^{1/2}(s_0)x_1\|_I\\ &= \left\|\frac{1}{2\pi j}\int_{-\infty}^{\infty}\left[\frac{1}{s-s_0}Q_1^{1/2}(s)x_1\right]_{s=j\omega}d\omega\right\|_I\\ &> \left\|\frac{1}{2\pi j}\int_{-\infty}^{\infty}\left[\frac{1}{s-s_0}\left(Q_1^{1/2}(s)\Delta L_i(s)T_{0i}(s)\cdot(Q_1^{1/2}(s))^{-1}\right)Q_1^{1/2}(s)x_1\right]_{s=j\omega}d\omega\right\|_I\\ &= \|Q_1^{1/2}(s_0)\Delta L_i(s_0)T_{0i}(s_0)x_1\|_I\\ &= \|\Delta L_i(s_0)T_{0i}(s_0)x_1\|_{\tilde{Q}(s_0)}.\end{aligned} \tag{3.26}$$

From the inequality (3.26) it follows that equation (2.18) has no nonzero solution $x_1 \in \mathbb{C}^n$ for any s_0 with $\mathrm{Re}(s_0) > 0$. Therefore, $S_i(s)$ and $T_i(s)$ are stable. □

The key implication of Theorem 3.1[12] regarding the stability margin of (1.1)–(1.3) is that any stable perturbation $\Delta L_i(s)$ satisfying (3.23) cannot destabilize the system (1.1)–(1.3) if the unperturbed system having return ratio $L_{0i}(s)$ is stable; $i=1$ or 2. Among the more specific implications of Theorem 3.1 is that the system has a gain margin in each of the plant's control input channels of at least (see [26])

$$g_M = \inf_\omega\left(20\log_{10}\left(1+\sigma_{\min}\left(I+L_{01}^{-1}(j\omega)\right)\right)\right)\ \mathrm{dB}, \tag{3.27}$$

even when the gain variations occur simultaneously in several or all input channels. Also, the system has a phase

[12]Results similar to Theorem 3 abound in the literature, even extending to the case of nonlinear ΔL_i—see, for example, [25], [26] and the references therein. The proof provided here is offered primarily for the sake of completeness. Note that our present proof makes use of standard complex variable theory arguments in contrast to previous proofs which make use of extended normed spaces.

margin in each control channel of at least [26]

$$\theta_M = 2\inf_{\omega} \arcsin\left(\tfrac{1}{2}\sigma_{\min}\left(I + L_{01}^{-1}(j\omega)\right)\right).$$

Other results in [26] establish a lower bound on the amount of unmodeled nonlinearity the system (1.1)–(1.3) can tolerate without instability.

D. Sensitivity

From (2.29) and (2.36) one sees that the variations in Δ_{YR} and Δ_{UR} resulting from variations in $G(s)$, $C(s)$, $H(s)$ and $F(s)$ can each be represented as the sum of two components:

$$\Delta_{YR} = S_{01}\Delta_{YE} + T_{02}\Delta_{EY}$$
$$\Delta_{UR} = S_{01}\Delta_{UE} + T_{01}\Delta_{EU}.$$

Using the inequality

$$\|AB\|_{Q_1,Q_3} < \|A\|_{Q_2,Q_3} \cdot \|B\|_{Q_1,Q_2}$$

together with the identity (3.12) it follows, for any appropriately dimensioned positive definite Q_i $(i=1,2,3)$, that

$$\|S_{02}\Delta_{GC}\|_{Q_1,Q_3} < \|S_{20}\|_{Q_2,Q_3} \cdot \|\Delta_{GC}\|_{Q_1,Q_2}$$
$$< \|\Delta_{GC}\|_{Q_1,Q_2} / \sigma_{\min}(I + L_{02};\, Q_3, Q_2)$$
$$\|S_{01}\Delta_C\|_{Q_1,Q_3} < \|\Delta_C\|_{Q_1,Q_2} / \sigma_{\min}(I + L_{01};\, Q_3, Q_2)$$
$$\|T_{02}\Delta_{FH}\|_{Q_1,Q_3} < \|\Delta_{FH}\|_{Q_1,Q_2} / \sigma_{\min}(I + L_{02}^{-1};\, Q_3, Q_2)$$
$$\|T_{01}\Delta_{FHG}\|_{Q_1,Q_3} < \|\Delta_{FHG}\|_{Q_1,Q_2} / \sigma_{\min}(I + L_{01}^{-1};\, Q_3, Q_2).$$

These sensitivity bounds have the simple interpretation as bounds on the percentage changes in T_{YR} and T_{UR} due to open-loop system percentage variations Δ_{G_C}, Δ_C, Δ_{FH}, and Δ_{FHG}. The bounds are inversely proportional to the smallest singular values of the nominal return-differences $I + L_{0i}$ and nominal inverse return differences $I + L_{0i}^{-1}$, $i = 1,2$. The bounds hold for arbitrarily large open-loop variations Δ_{GC}, Δ_C, Δ_{FH}, and Δ_{FGH}.

IV. Fundamental Limits of Feedback and Tradeoffs Imposed by the Equality $S+T=I$

In this section we shall derive a number of inequalities (some of which were discussed in [42]) which quantify some of the tradeoffs implicit in the fundamental relationship

$$S(s) + T(s) \equiv I \tag{4.1}$$

where $S(s) \triangleq (I + L(s))^{-1}$ and $T(s) \triangleq L(s)(I + L(s))^{-1}$.

In what follows $\|\cdot\|$ will denote any complex matrix norm such as those induced on unitary spaces of the type considered in Section III. Note that, in general, $\|I\| \neq 1$. Results for the particular matrix norm $\|\cdot\|_{Q_1,Q_2}$ (subordinate to the weighted Euclidean vector norms) can, of course, be interpreted in terms of singular values. Finally, all results will be stated with the frequency variable s or $j\omega$ suppressed and with the understanding that the result is valid for all admissible values of the frequency variable.

We shall make frequent use of the following result.

Lemma 2: Let $B, C \in \mathbb{C}^{m\times m}$ with C invertible. Furthermore, suppose $A = B + C$ is invertible. Then, in addition to the usual triangle inequality

$$\|A\| < \|B\| + \|C\| \tag{4.2}$$

we have

$$\frac{1}{\|A^{-1}\|} < \|B\| + \frac{1}{\|C^{-1}\|}. \tag{4.3}$$

Proof: Clearly $C^{-1} = A^{-1}BC^{-1} + A^{-1}$, whence

$$\|C^{-1}\| < \|A^{-1}\| \cdot \|B\| \cdot \|C^{-1}\| + \|A^{-1}\|$$

from which (4.3) follows. □

Using (4.2) and (4.3) and various forms of (4.1), the following useful inequalities are easily derived.

Theorem 4:

$$\max\left\{ \big|\|I\| - \|S\|\big|, \left|\|I\| - \frac{1}{\|S^{-1}\|}\right| \right\} < \|T\| < \|I\| + \|S\| \tag{4.4}$$

$$\max\left\{ \big|\|I\| - \|T\|\big|, \left|\|I\| - \frac{1}{\|T^{-1}\|}\right| \right\} < \|S\| < \|I\| + \|T\| \tag{4.5}$$

$$\left| \frac{1}{\|S^{-1}\|} - \frac{1}{\|T^{-1}\|} \right| < \|I\|. \tag{4.6}$$

□

Using the definitions of S and T in terms of L it is then easy to derive the following further sets of inequalities.

Theorem 5:

$$\frac{1}{\|L^{-1}\|} < \frac{\|T\|}{\|S\|} < \|L\| \tag{4.7}$$

$$\left| \frac{1}{\|S\|} - \frac{1}{\|L^{-1}\|} \right| < \|I\| \tag{4.8}$$

$$\left| \frac{1}{\|T\|} - \frac{1}{\|L\|} \right| < \|I\|. \tag{4.9}$$

□

For reference, we also include the above inequalities specialized to the case of the norm $\|\cdot\|_{Q_1,Q_2}$ in which case, for a nonsingular $A: \mathbb{C}^m_{Q_1} \to \mathbb{C}^m_{Q_2}$ with singular values $\sigma_1 > \cdots > \sigma_m > 0$, we have $\|A\|_{Q_1,Q_2} = \sigma_1$, $\|A^{-1}\|_{Q_1,Q_2} = 1/\sigma_m$. For convenience, weighting matrices on the underlying unitary spaces $\mathbb{C}^m_{Q_1}$ and $\mathbb{C}^m_{Q_2}$ will not be denoted explicitly while $\|I\|_{Q_1,Q_2}$ will be denoted by γ. Equations (4.4)–(4.9) then become

$$\max\{|\gamma-\sigma_1(S)|,|\gamma-\sigma_m(S)|\}<\sigma_1(T)<\gamma+\sigma_1(S) \quad (4.4')$$

$$\max\{|\gamma-\sigma_1(T)|,|\gamma-\sigma_m(T)|\}<\sigma_1(S)<\gamma+\sigma_1(T) \quad (4.5')$$

$$|\sigma_m(S)-\sigma_m(T)|<\gamma \quad (4.6')$$

$$\sigma_m(L)<\frac{\sigma_1(T)}{\sigma_1(S)}<\sigma_1(L) \quad (4.7')$$

$$\left|\frac{1}{\sigma_1(S)}-\sigma_m(L)\right|<\gamma \quad (4.8')$$

$$\left|\frac{1}{\sigma_1(T)}-\frac{1}{\sigma_1(L)}\right|<\gamma. \quad (4.9')$$

Note also that $\sigma_1(S)=1/\sigma_m(I+L)$, $\sigma_m(S)=1/\sigma_1(I+L)$ and, if L^{-1} exists,

$$\sigma_1(T)=\frac{1}{\sigma_m(I+L^{-1})},\ \sigma_m(T)=\frac{1}{\sigma_1(I+L^{-1})}.$$

We have seen in Section III that the quantities $\sigma_m(I+L)$ and $\sigma_m(I+L^{-1})$ play key roles in analyzing stability margins and sensitivity of linear systems for various frequency ranges. In fact, the Bode plot graphs of these quantities [or of their reciprocals, i.e., $\sigma_1(S)$ and $\sigma_1(T)$] are related to graphs of $\sigma_1(L)$ and $\sigma_m(L)$ by (4.7′). Since magnitudes are usually more conveniently displayed with a logarithmic scale, the various σ_j are commonly measured in dB. Thus, from (4.7′) we have

$$\sigma_m(L)\,\mathrm{dB}<\sigma_1(T)\,\mathrm{dB}-\sigma_1(S)\,\mathrm{dB}<\sigma_1(L)\,\mathrm{dB}. \quad (4.10)$$

Notice that for very small ω, L becomes very large so that T is near I while S is near 0 (assuming L is strictly proper). Thus, (4.10) becomes

$$\sigma_m(L)\,\mathrm{dB}\lesssim-\sigma_1(S)\,\mathrm{dB}\lesssim\sigma_1(L)\,\mathrm{dB}. \quad (4.11)$$

Furthermore, for very high ω, L tends to 0 so that S is near I while T is near 0 and (4.10) becomes

$$\sigma_m(L)\,\mathrm{dB}\lesssim\sigma_1(T)\,\mathrm{dB}\lesssim\sigma_1(L)\,\mathrm{dB}. \quad (4.12)$$

Other interpretations of the inequalities (4.4)–(4.9) or (4.4′)–(4.9′) follow easily. For example, a stability margin specification of the form $\sigma_m(I+L^{-1})>\alpha$ for some $\alpha>1/\gamma$ becomes

$$\sigma_1(T)<\frac{1}{\alpha}<\gamma$$

which implies, by (4.5′), that

$$|\sigma_1(S)-\gamma|<\frac{1}{\alpha}$$

or

$$\frac{\alpha}{\alpha\gamma+1}<\sigma_m(I+L)<\frac{\alpha}{\alpha\gamma-1}.$$

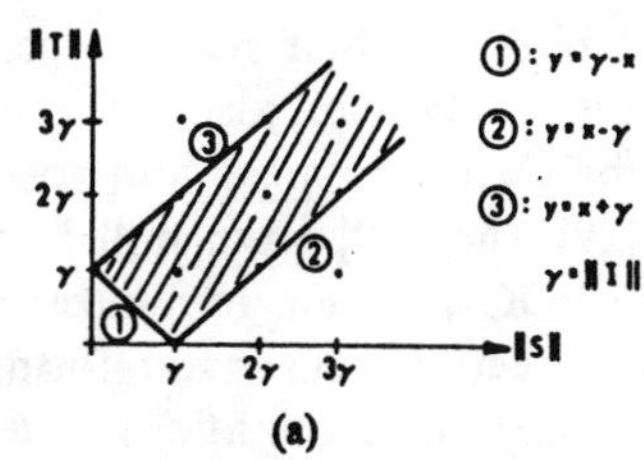

(a)

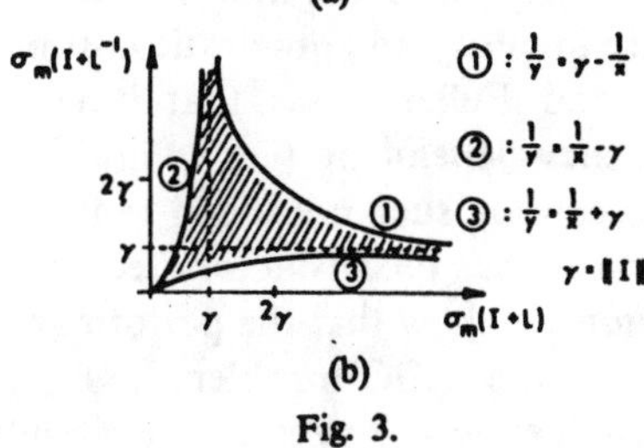

(b)

Fig. 3.

Further, if

$$0<\alpha<\sigma_m(I+L^{-1})<\beta<\frac{1}{\gamma}$$

then

$$\gamma<\frac{1}{\beta}<\sigma_1(T)<\frac{1}{\alpha}$$

which implies, by (4.5′), that

$$\frac{1}{\beta}-\gamma<\sigma_1(S)<\frac{1}{\alpha}+\gamma$$

or

$$\frac{\alpha}{\alpha\gamma+1}<\sigma_m(I+L)<\frac{\beta}{1-\beta\gamma}.$$

The relationships between $\|S\|$ and $\|T\|$ in (4.4) and (4.5) are more easily seen in Fig. 3(a). The shaded area indicates permissible values for $\|S\|$ and $\|T\|$. Similarly, permissible values for tradeoffs between $\sigma_m(I+L^{-1})$ and $\sigma_m(I+L)$ are also easily seen in the shaded area of Fig. 3(b)). The fact that the smallest singular values of the return difference $I+L$ and inverse-return difference $I+L^{-1}$ must always lie in the shaded region in Fig. 3(b) establishes fundamental limits on achievable performance in feedback control systems in view of the relations developed in Section III.

V. LQG Synthesis for Good Feedback Properties[13]

It has long been known that the choice of quadratic noise and cost matrices in stochastic linear quadratic Gaussian (LQG) optimal synthesis is intimately related to the feedback properties of the resultant optimal system. Indeed, the original motivation for LQG optimal synthesis was to minimize the root-mean-square noise and disturbance response of a control system. Based on explicit

[13] Some of the results in this section were originally presented in [27].

solutions to the LQG problem for some simple examples, Fuller [28] has observed that such properties as stability margin and sensitivity can also be related to certain properties of the quadratic noise and cost matrices. Also, Youla *et al.* [29] and [30] have noted that the optimal solution to the LQG problem minimizes a weighted sum of the traces of certain quadratic forms involving the $S_2(j\omega)$ matrix, with the weightings determined by the Gaussian noise intensity and quadratic cost matrices. Both Youla *et al.* and Fuller stress that it is important to employ frequency-dependent (i.e., nonwhite) noise and that failure to employ such noise can result in an optimal control design that has unsatisfactory feedback properties.

In this section we show that the performance index that is optimized in the LQG problem assigns frequency-dependent penalties to the sizes of the singular values of the $S_i(j\omega)$, $T_i(j\omega)$, and $W(j\omega)$ matrices, with the magnitudes of the penalties at each frequency determined by the quadratic cost and Gaussian noise intensity matrices. In view of the connections described in Section III of this paper between singular values of (S_i, T_i, W) and the feedback properties of the system, this can be applied directly to the problem of "tuning" the quadratic cost and noise matrices to obtain a feedback design that meets inequality specifications on its feedback properties.

Consider the following stochastic linear quadratic optimal control problem of minimizing the expected value of the cost

$$J=\frac{1}{2\pi}\int_{-\infty}^{\infty}\|Z(j\omega)\|^2_{Q_3(j\omega)}+\|Y(j\omega)\|^2_{Q_2(j\omega)}+\|U(j\omega)\|^2_{Q_1(j\omega)}\,d\omega \tag{5.1}$$

subject to (1.1)–(1.3), with $R(s)\equiv 0$, and with $D(s)$ and $N(s)$ generated by Gaussian random processes having respective power spectrum matrices $\Sigma_D(s)$ and $\Sigma_N(s)$. This problem may in general be solved via the spectral factorization approach (cf. [29]–[32]). Alternatively, one can apply the standard state-space LQG approach involving Riccati equations and an optimal Kalman–Bucy filter (cf. [17]) to a suitably augmented Gaussian white noise driven equivalent system in state-space form. In any case, there are well established methods for computing the optimal feedback solution for this problem. Our concern here is not with how to solve this problem, but with the feedback properties of the resulting optimal feedback system.

Substituting the identities (2.1)–(2.4) into (5.1) and rearranging, the expected value of the lost J in (5.1) is found to be

$$E(J)=\frac{1}{2\pi}\int_{-\infty}^{\infty}\operatorname{tr}\left[\begin{bmatrix}Q_1 & 0 & 0\\ 0 & Q_2 & 0\\ 0 & 0 & H^*Q_3H\end{bmatrix}\begin{bmatrix}W & W\\ S_2 & T_2\\ S_2 & S_2\end{bmatrix}\cdot\begin{bmatrix}\Sigma_D & 0\\ 0 & \Sigma_N\end{bmatrix}\begin{bmatrix}W^* & S_2^* & S_2^*\\ W^* & T_2^* & S_2^*\end{bmatrix}\right]d\omega$$

$$=\frac{1}{2\pi}\int_{-\infty}^{\infty}\operatorname{tr}((Q_1W(\Sigma_D+\Sigma_N)W^*)+\operatorname{tr}Q_2S_2\Sigma_DS_2^*+\operatorname{tr}(Q_2T_2\Sigma_NT_2^*)+\operatorname{tr}((H^*Q_3H)S_2(\Sigma_D+\Sigma_N)S_2^*)\,d\omega. \tag{5.2}$$

where for any $A(s)$ the notation A^* denotes $A^T(-s)$. Thus, the foregoing LQG optimization problem is equivalent to the minimization of (5.2) over the set of all stabilizing feedback matrices $M(s)$.

Recalling the identity

$$\operatorname{tr}(QA\Sigma\bar{A}^T)\equiv\sum_{i=1}^{\text{rank}}\sigma_i^2(A;\Sigma^{-1},Q)$$

for any A and for any positive definite Q,Σ, and the fact that $A^*(j\omega)=\bar{A}^T(j\omega)$ whenever $A(s)$ is a matrix whose elements are Laplace transforms or real-valued signals, it is immediately evident that (5.2) is simply a weighted sum of terms penalizing for large singular values of $S_2(j\omega)$, $T_2(j\omega)$, and $W(j\omega)$. Moreover, if $G(s)$ has a left inverse $G_l(s)$, then the terms penalizing W may be replaced by terms penalizing T_2 using the identity $W=G_lT_2$:

$$E(J)\equiv\frac{1}{2\pi}\int_{-\infty}^{\infty}\operatorname{tr}((G_l^*Q_1G_l)T_2(\Sigma_D+\Sigma_N)T_2^*)+\operatorname{tr}(Q_2S_2\Sigma_DS_2^*)+\operatorname{tr}(Q_2T_2\Sigma_NT_2^*)+\operatorname{tr}((H^*Q_3H)S_2(\Sigma_D+\Sigma_N)S_2^*)\,d\omega. \tag{5.3}$$

Thus, the cost function $E(J)$ may be regarded as the sum of two terms penalizing for large singular values of $S_2(j\omega)$ and two other terms penalizing for large singular values of $T_2(j\omega)$. Equivalently, there are two terms penalizing for small singular values of the return difference $(I+L_2)$ and two other terms penalizing for small singular values of the inverse-return difference $(I+L_2^{-1})$.

The representations (5.2) and (5.3) of the stochastic LQG cost functional $E(J)$ have obvious implications regarding the influence of one's choice of the matrices $(Q_1, Q_2, Q_3, \Sigma_D, \Sigma_N)$ on the singular values of S_2, T_2, and W and hence on the singular values of the return differences and inverse-return difference matrices associated with the cost-minimizing feedback matrix $M(s)$. The possibility of using frequency-dependent matrices $(Q_1(s), Q_2(s), Q_3(s), \Sigma_D(s), \Sigma_N(s))$ (see [28]–[32]) provides additional flexibility that can be used to assign frequency-dependent penalties to S_2, T_2, and W, albeit at the expense of a more complicated feedback matrix $M(s)$. This approach is used in the design example considered in the following section.

VI. An Aircraft Autopilot Design Example

In this section the preceding methodology is used to design a set of control laws for an advanced fighter aircraft. This example is used to explore and illustrate the shaping of (analysis of) the S_2 and T_2 matrices in an LQG synthesis. An LQG synthesis procedure is used to design

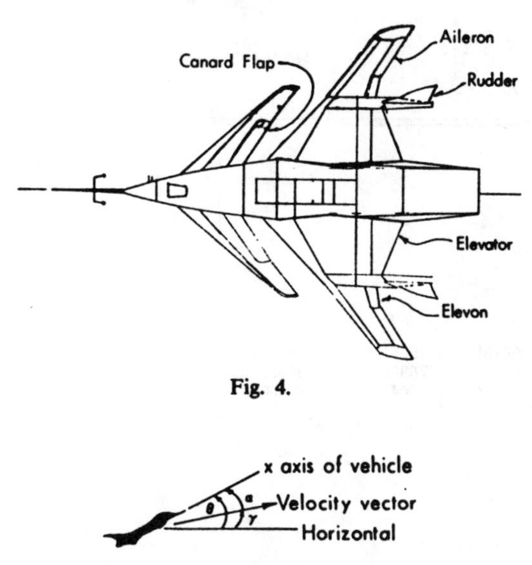

Fig. 4.

Fig. 5. Vertical plane geometry.

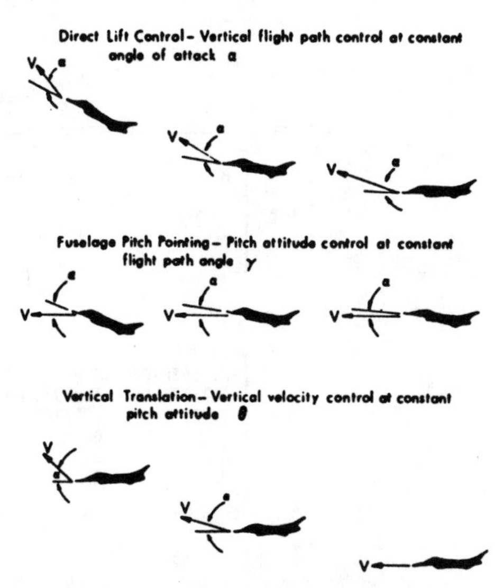

Fig. 6. Longitudinal CCV modes.

linear time invariant multiple feedback loops satisfying S_2 and T_2 constraints. We begin the example with a discussion of the vehicle and our desired control objectives. The feedback design goals are stated as inequality constraints on the singular values of the S_2 and T_2 matrices. Several design iterations are used to illustrate the role of the cost and noise matrices in shaping the Bode plots of the singular values of S_2 and T_2 to satisfy our objectives. Finally, a prefilter $P(s)$ is designed to ensure acceptable response to commands.

A. Vehicle Model

The vehicle used for illustration is a scaled version of an advanced (1990's) fighter presently being flight tested as an RPV [33]. This vehicle was selected primarily because its multiple, independently controlled surfaces make it an ideal candidate for multiloop synthesis of advanced control modes (see Fig. 4).

In our example, we will design a set of direct force modes for the longitudinal axis. Linearized (small perturbation) models for a collection of flight conditions are given in [33]. For design purposes the longitudinal dynamics are uncoupled from the lateral-directional dynamics. The state vector consists of the vehicle's basic rigid body variables $x^T=(\delta V, \alpha, q, \theta)$ (forward velocity, angle of attack, pitch rate, and pitch attitude angle). The usual definitions apply where the flight path angle (γ) is defined as $\gamma=\theta-\alpha$. The state variables used to describe motions in the vertical plane are shown in Fig. 5:

δV—perturbations along velocity vector;

α—angle between velocity vector and aircraft's longitudinal axis;

q—rate-of-change of aircraft attitude angle;

θ—aircraft attitude angle.

The angles (α, γ, θ) will be used in describing the control modes in the next section.

The control inputs used in our design are the elevon (δ_e) and the canard (δ_c). Two first-order lags are appended to the state to represent actuator dynamics. For our control design we will measure two of the states α and θ. Thus, the problem posed has 2 inputs, 2 outputs and 6 states, which is of realistic, yet easily manageable size.

B. Control Objectives

With the advent of digital flight control and the availability of multiple surfaces on new fighters there is interest in designing control modes for precision flight path control. These modes have the objective of decoupling attitude from flight path control. These nonconventional modes offer precise control for certain tasks as well as the possibility of new tactics for the advanced fighters.

Further discussions are given in [34] and [35], for example. Consider three precision longitudinal modes [34] (see Fig. 6).

Vertical Translation: Control vertical velocity at constant θ (α varies). Note that attitude holds constant as the velocity vector rotates.

Pitch Pointing: Control attitude at a constant flight path angle (i.e., $\theta=\alpha$). Note velocity vector does not rotate. The angle of attack is varied at a constant normal acceleration.

Direct Lift: Control of flight path at constant angle of attack (i.e., $\gamma=\theta$). This mode produces a normal acceleration response without changing angle-of-attack.

C. Design Specs

For good performance we would like each singular value of $T_2(j\omega)$ to have roughly equal bandwidth (speed of response). Because of uncertainties in the model at high frequencies, the bandwidth must be limited. Let us specify that each singular value be attenuated by 20 dB (one order of magnitude) at frequencies beyond $\omega=100$ rad/s (15.9 Hz). From Theorem 3, this ensures that the system will have sufficient stability margin to tolerate variations in loop transfer function of magnitude as large as a factor of ten (i.e., 1000 percent) at frequency $\omega=100$ rad/s. We

TABLE I
THE PLANT MATRICES (C, A, B)

$$C = \begin{bmatrix} 0. & 1.0000E\ 00 & 0. & 0. & 0. & 0. \\ 0. & 0. & 0. & 1.0000E\ 00 & 0. & 0. \end{bmatrix}$$

$$A = \begin{bmatrix} -2.2567E-02 & -3.6617E\ 01 & -1.8897E\ 01 & -3.2090E\ 01 & 3.2509E\ 00 & -7.6257E-01 \\ 9.2572E-05 & -1.8997E\ 00 & 9.8312E-01 & -7.2562E-04 & -1.7080E-01 & -4.9652E-03 \\ 1.2338E-02 & 1.1720E\ 01 & -2.6316E\ 00 & 8.7582E-04 & -3.1604E\ 01 & 2.2396E\ 01 \\ 0. & 0. & 1.0000E\ 00 & 0. & 0. & 0. \\ 0. & 0. & 0. & 0. & -3.0000E\ 01 & 0. \\ 0. & 0. & 0. & 0. & 0. & -3.0000E\ 01 \end{bmatrix}$$

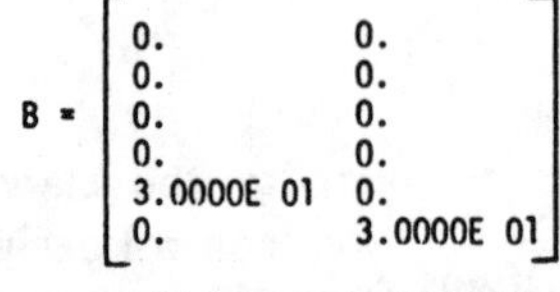

$$B = \begin{bmatrix} 0. & 0. \\ 0. & 0. \\ 0. & 0. \\ 0. & 0. \\ 3.0000E\ 01 & 0. \\ 0. & 3.0000E\ 01 \end{bmatrix}$$

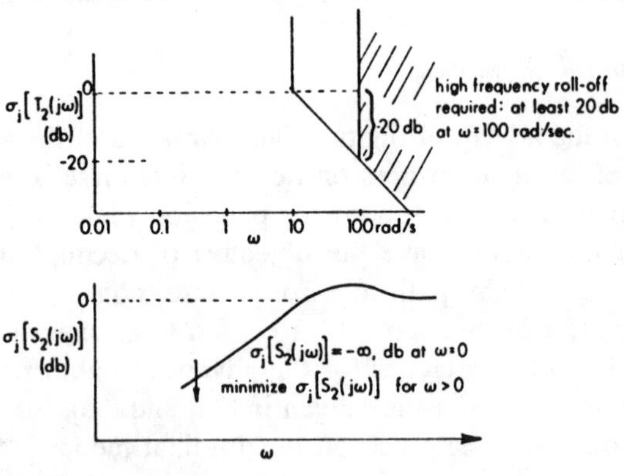

Fig. 7. Design specifications as S_2 and T_2 inequalities.

would also like the loops to exhibit approximately first order roll-off beyond $\omega = 100$ rad/s. We would like the maximum bandwidth possible subject to this constraint. This corresponds to minimizing the sensitivity to disturbances and low frequency plant variations. The primary disturbances in our example are vertical gusts. They may be modeled as white noise passed through a low-pass filter $1/(\tau s+1)$ where the bandwidth $1/\tau$ is determined by the vehicles' velocity and a "gust scale length" [36]. In the present application gust energy bandwidth $1/\tau$ is always less than 1 rad/s.

Since the control modes are intended to be precision control modes, we require command tracking with no steady-state errors. This is equivalent to specifying zero sensitivity to errors at zero frequency.[14]

Fig. 7 contains a sketch of the constraints on the singular values of S_2 and T_2 versus frequency based on the above considerations. Note that we have expressed these in dB to exploit familiarity with classical Bode magnitude plots.

[14]Obviously this specification is satisfied by using integral control.

D. Trial Designs

The design presented below is for an operating point of mach $= 0.9$, altitude $= 25{,}000$ ft. The general feedback is shown in Fig. 2. The controller-sensor portion of the feedback loop, $M(s) \triangleq C(s)F(s)H(s)$, was designed using LQG synthesis. The transfer function matrix of the plant model for this flight condition is of the form

$$G(s) = C(Is - A)^{-1}B$$

where the matrices (C, A, B) are given in Table I. The plant output vector is $y \equiv (y_1, y_2)^T$ where $y_1(t) \triangleq \alpha(t)$ is the angle of attack and $y_2(t) = \theta(t)$ is the pitch attitude angle. The sensor matrix is

$$H(s) = I \in \mathbb{C}^{2\times 2}.$$

Controller Design: The controller was designed as the usual LQG feedback law (cf. Section V) minimizing a quadratic performance index. The cost matrix $Q_2(s) = Q_2^T(-s)$ is a 2×2 matrix that weights the two outputs α and θ. Similarly, $Q_1(s) = Q_1^T(-s)$ is a 2×2 matrix that weights inputs to the two surfaces—elevon and canard. The matrix $Q_3(s) = Q_3^T(-s)$ weights the measurements of (α, θ) corrupted by the additive sensor noise $N(s)$.

The plant disturbance intensity matrix $\tilde{\Sigma}_D(s)$ was selected to be of the form $G^T(-s)\,\tilde{\Sigma}_D G(s)$ where $\tilde{\Sigma}_D$ is a constant matrix. This is equivalent to the plant being driven by white noise of intensity Σ_D entering at the plant input (cf. Doyle [37]). The measurement noise matrix $\Sigma_N(s)$ is a 2×2 matrix describing the sensor noise levels; Σ_N was chosen to be a constant matrix.

Loop Properties: Four trial designs are presented to illustrate shaping S_2 and T_2 by adjusting $Q_3(s)$, $Q_2(s)$, $Q_1(s)$, $\tilde{\Sigma}_D$ and Σ_N in the LQG synthesis procedure.

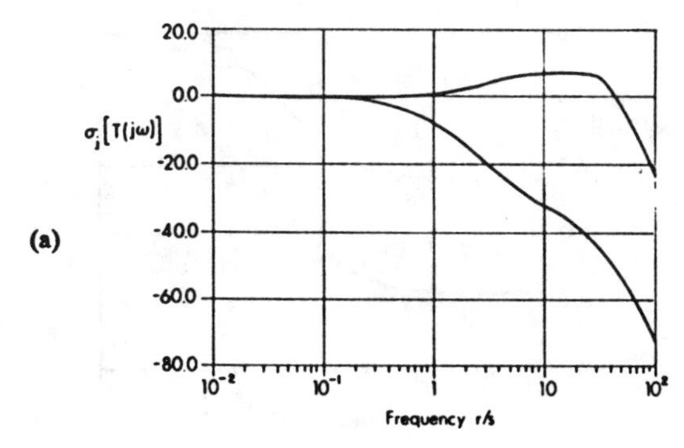

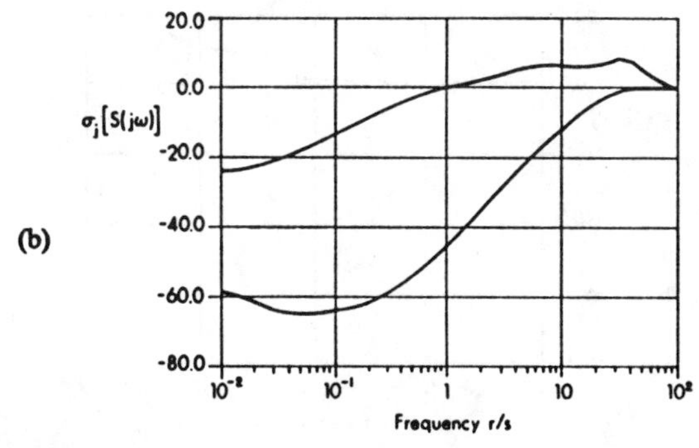

Fig. 8. (a) Design 1. (b) Design 1.

Any design procedure is iterative and requires some insight (usually gained by experience) on the part of the designer. The present procedure uses the connection between the singular values of S_2 and T_2 and the quadratic cost and noise matrices to obtain a design satisfying our feedback objectives. For Design 1 the cost/noise matrices were selected to have a simple diagonal structure. The weighting $Q_3(s)$ on α and θ was chosen to include a term proportional to $-1/s^2$. (This is equivalent to weighting the integral of α and θ in the cost function and automatically ensures that the resultant optimal LQG feedback will be type 1, i.e., proportional integral.)[15] Several initial designs were performed by scaling the diagonal elements using the observation [see (5.2)] that reducing Q_1 or Σ_N tends to increase $\|T_2\|$ and decrease $\|S_2\|$. Similarly reducing Q_2, Q_2, Σ_D tends to decrease $\|T_2\|$ and increase $\|S_2\|$. Based on the foregoing considerations, we selected the following $(Q_3, Q_2, Q_1, \tilde{\Sigma}_D, \Sigma_N)$ for Design 1:

$$Q_3 = \begin{bmatrix} -\dfrac{5000}{s^2} & 0 \\ 0 & -\dfrac{5000}{s^2} \end{bmatrix};$$

$$Q_1 = \begin{bmatrix} 50 & 0 \\ 0 & 50 \end{bmatrix}; \; Q_2 = \begin{bmatrix} 0 & 0 \\ 0 & 0 \end{bmatrix};$$

$$\tilde{\Sigma}_D = \begin{bmatrix} 11 & 0 \\ 0 & 11 \end{bmatrix}; \; \Sigma_N = \begin{bmatrix} 0.01 & 0 \\ 0 & 0.01 \end{bmatrix}.$$

Plots of $\sigma_j[S_2(j\omega)]$ and $\sigma_j[T_2(j\omega)]$ versus ω for Design 1 are shown in Fig. 8. The "T" plot shows that the larger singular value just meets the bandwidth limitation specification (i.e., maximum bandwidth limited by 20 dB attenuation at $\omega = 100$ rad/s). The other singular value has somewhat less bandwidth and meets the specification with room to spare. For this design, T_2 is approximately the identity matrix for $\omega < 0.1$ rad/s. For ω in the midfrequency range (near $\omega = 10$) the left singular vector associated with the larger singular value of $T_2(j\omega)$ tended to align approximately with the "θ-direction," viz., the $y = [0, 1]^T$ direction. The other left singular vector aligned approximately with the "α-direction" $[1, 0]^T$. Accordingly, it seemed that it might be possible to further reduce the singular values of $S_2(j\omega)$ in the midfrequency range—trading a larger $\sigma_2(T_2(j\omega))$ for smaller $\sigma_i(S_2(j\omega))$—by altering the weighting matrices Q_1 and Q_2 to penalize the output of $T(j\omega)$ more in the θ-direction and less in the α-direction.

For the second design the Q_1 matrix was modified but Q_2 was not changed. The cost function (5.3) shows that the penalty on the left side of T_2 is $G_l^{-T} Q_1 G_l$ where the matrix G_l is the left inverse of the plant $G(s)$. To accommodate this effect, Q_1 was expressed as a diagonal matrix weighted by $\tilde{G}$ where $\tilde{G}$ is an approximation to the plant transfer function

$$G(s) = C(sI - A)^{-1} B$$

over the midfrequency range. The approximation used is

$$\tilde{G} = \begin{bmatrix} a_{25} & a_{26} \\ a_{35} & a_{36} \end{bmatrix} = \begin{bmatrix} -1.7080E\text{-}01 & -4.9652E\text{-}03 \\ -3.1604E\ 01 & 2.2396E\ 01 \end{bmatrix}$$

which ignores the frequency dependence and is satisfactory over a midrange of frequencies. The matrix $\tilde{G}$ basically contains terms indicating how effective the inputs (u_1, u_2) are in controlling the two outputs (α, θ). Thus, we took

$$Q_1 = \tilde{G}^T \begin{bmatrix} S_1 & 0 \\ 0 & S_2 \end{bmatrix} \tilde{G}$$

with $S_1 = 10$, and $S_2 = 0.1$

$$Q_1 = \begin{bmatrix} 103. & -70. \\ -70. & 48. \end{bmatrix}$$

(Note that this adds some off-diagonal terms to Q_1). Plots of $\sigma_j[T_2(j\omega)]$ and $\sigma_j[S_2(j\omega)]$ are shown in Fig. 9. Note that the revised Q_1 has improved the loop properties. The cross-over of the low bandwidth ("slow") singular-vector direction has been extended in frequency without changing the bandwidth ("fast") singular vector direction. The

[15] Using an s-dependent $Q_3(s)$ of the form $Q_3(s) = (1/-s^2)\bar{Q}_3$ is also equivalent to augmenting the system by adding an integrating sensor (i.e., replacing $H(s) = I$ by $H(s) = \begin{bmatrix} I \\ \frac{1}{s} I \end{bmatrix}$ and replacing $Q_3(s)$ by the *constant* matrix $\begin{bmatrix} 0 & 0 \\ 0 & \bar{Q}_3 \end{bmatrix}$.

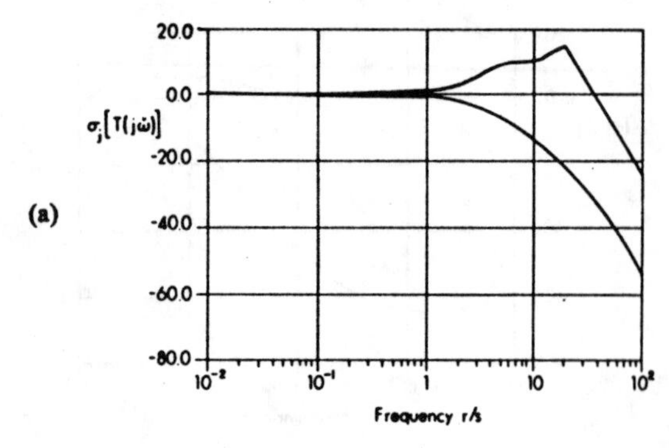

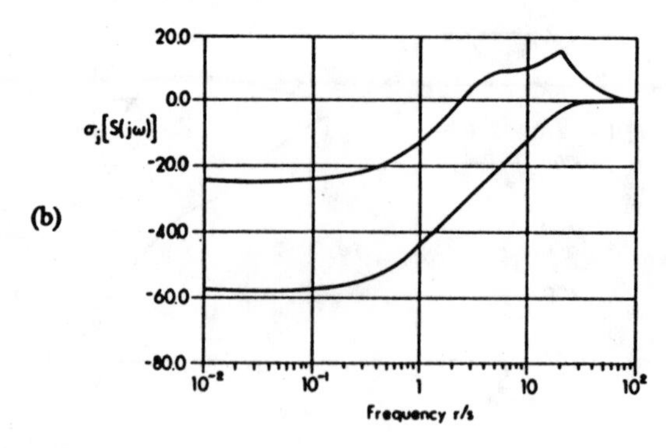

Fig. 9. (a) Design 2. (b) Design 2.

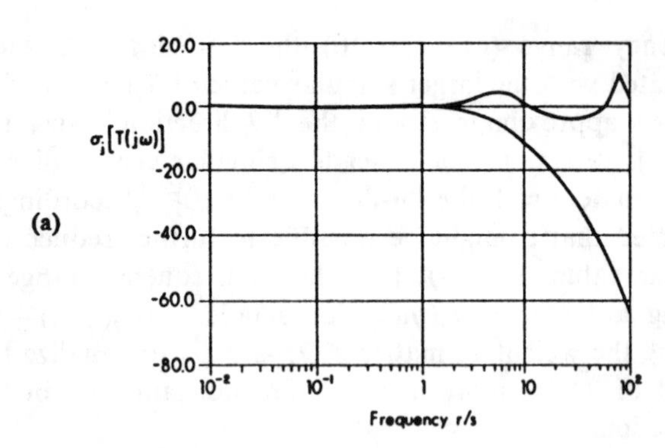

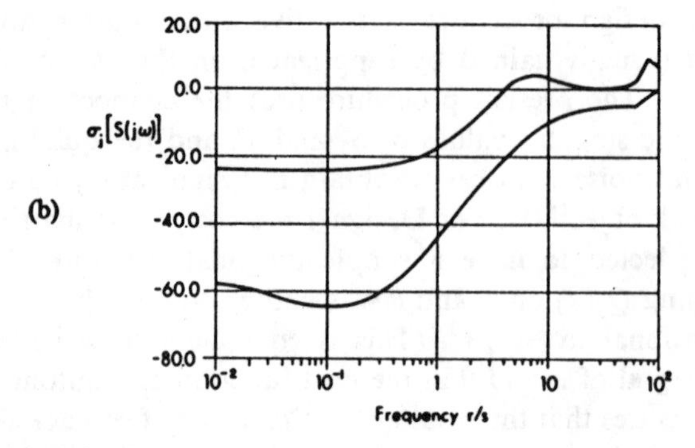

Fig. 10. (a) Design 3. (b) Design 3.

"small direction" of T_2 still tended to be along the α direction. For Design 3 the control weight was reduced in the α direction by changing S_1 from 10. to 1. This resulted in a slight modification to Q_1 as shown below.

$$Q_1 = \begin{bmatrix} 99.86 & -70.78 \\ -70.78 & 50.18 \end{bmatrix}.$$

Plots of $\sigma_j[T(j\omega)]$ and $\sigma_j[S(j\omega)]$ are presented in Fig. 10. Note that S_2 is smaller in both directions over a wider frequency range compared to Design 2. Crossover of the "slow" singular vector direction of S_2 moved from 1 rad/s to 4 rad/s. However, in checking the T_2 plot note the "fast" singular vector direction has extended its bandwidth past $\omega = 100$ r/s which violates the specification. Rather than adding frequency dependence to Q_1, the use of the other weighting matrices was investigated.

For the next iteration the Q_1 from Design 2 was used, but Q_2 and Q_3 were modified. As noted in Design 2, the singular vectors of T_2 indicated the "slow" direction was close to the (1,0) direction (i.e., the α direction). Therefore, the weight on α was increased by modifying Q_2 and Q_3 to be

$$Q_2 = \begin{bmatrix} 5000 & 0 \\ 0 & 5000 \end{bmatrix}; \quad Q_3 = \begin{bmatrix} \dfrac{15{,}000}{-s^2} & 0 \\ 0 & \dfrac{15{,}000}{-s^2} \end{bmatrix}.$$

Plots of $\sigma_j[T_2(j\omega)]$ and $\sigma_j[S_2(j\omega)]$ for the final design are shown in Fig. 11.

A comparison of Designs 2 and 4 shows that in Design 4 the bandwidth of the "slow" direction has been pushed out in frequency and the fast direction is still satisfactory. Comparing the S_2 plots also indicates that the crossover frequencies in Design 4 are higher than in Design 2.

Thus for illustrative purposes we will accept Design 4. If it were desired to "balance" the bandwidth more closely in both singular vector directions it would probably be necessary to make additional frequency dependent weighting adjustments, which would lead to a more complicated optimal LQG controller having more poles (i.e., more states in its internal Kalman–Bucy filter). For this design the extra complexity is not warranted.

Recall that one of the objectives of the design is to obtain a good response to pilot commands $r(t) = (\alpha_{\text{desired}}(t), \theta_{\text{desired}}(t))^T \in \mathbf{R}^2$. We want equally fast response to α and θ commands, zero steady-state tracking error for step inputs, and good decoupling between response to α commands and to θ commands. Up to this point the issue of command response was ignored while attention was focused on specifying the controller portion of the loop transfer function matrix $M(s) \triangleq C(s)F(s)H(s)$ in order to ensure that the system has good feedback properties. The LQG theory has been used to determine $M(s)$. Sufficient freedom to specify the response to pilot commands is provided by the possibility of using any $P(s)$, $C(s)$, and $F(s)$ matrices so long as the product $C(s)F(s)H(s)$ equals $M(s)$. To ensure asymptotic tracking of step commands, we elected to exploit the fact that the optimal $M(s)$ has a pole at $s=0$ and is of the form[16]

[16]The fact that $M(s)$ is of this form is a consequence of the $1/-s^2$ term in the quadratic weight $Q_3(s)$. In general, all the open-loop poles of $Q_3(s)$ or their LHP mirror images are also poles of $M(s)$, as may be seen by careful examination of the solution to the LQG problem [29]–[32].

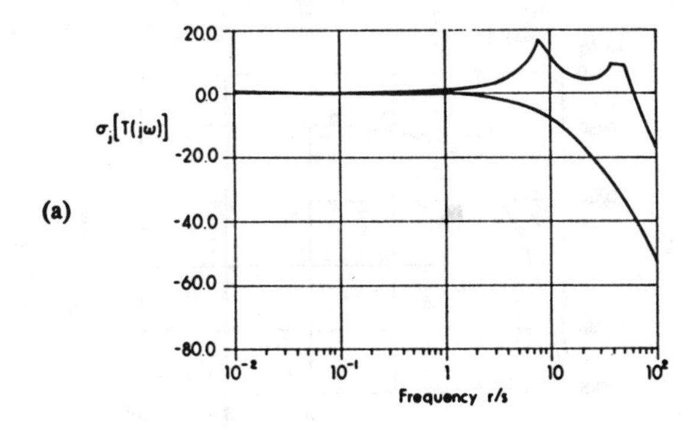

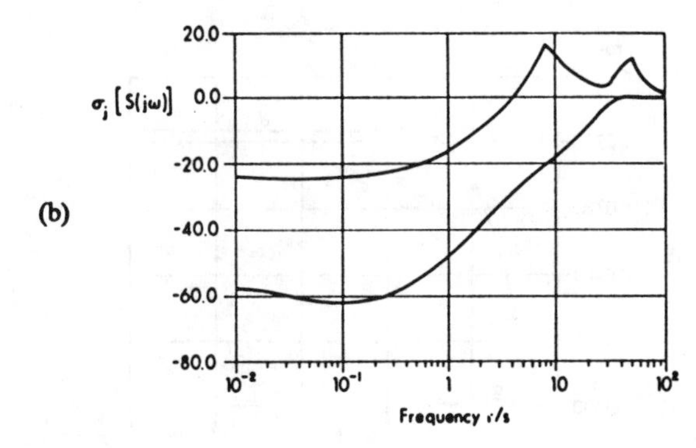

Fig. 11. (a) Design 4. (b) Design 4.

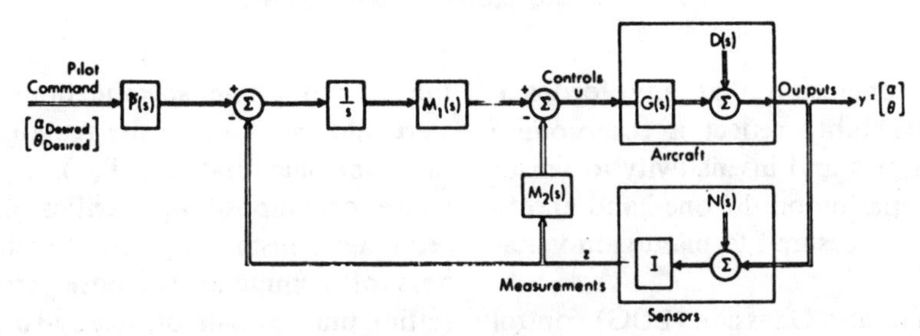

Fig. 12. LQG-designed aircraft autopilot.

$$M(s)=\frac{1}{s}M_1(s)+M_2(s)$$

to obtain the type 1 [38] control law (see Fig.12):

$$C(s)=\left[\frac{1}{s}M_1(s),\ M_2(s)\right]\in\mathbb{C}^{2\times4}$$

$$F(s)=\begin{bmatrix} I \\ I \end{bmatrix}\in\mathbb{C}^{4\times2}$$

$$P(s)=\begin{bmatrix} \tilde{P}(s) \\ 0_{2\times2} \end{bmatrix}\in\mathbb{C}^{4\times2}.$$

The type 1 control configuration assures zero steady-state tracking error for step commands (provided that $\tilde{P}(0)=I$). In order to achieve equally fast response to both α and θ commands we tried the prefilter

$$\tilde{P}(s)=\begin{bmatrix} 1 & 0 \\ 0 & \dfrac{3}{s+3} \end{bmatrix}.$$

As an independent check on the command response of Design 4 we evaluated the time history response for separate α and θ step commands. Responses for a 0.1 rad α step command and a 0.1 rad θ step command are shown in Fig 13. We found these time histories to be entirely satisfactory, so no further adjustment of the prefilter was deemed necessary.

Recall that one of the goals of the design was to decouple α and θ to permit flying the three modes previously described. This decoupling has been achieved with our prefilter as evidenced by the minimal cross-axis response to commands. The response times on the order of 1 s can be considered very satisfactory.

In summary, the preceding design has successfully illustrated the procedure of shaping S_2 and T_2 matrices in multivariable design. Of course, additional analyses would have to be performed before this design can be considered flightworthy. The design would have to be evaluated on a full six degree of freedom nonlinear simulation. The design would also be extended to the aircraft's operational flight envelope; this could be done by repeating the design procedure at a collection of operating points spanning the flight envelope. The resulting controllers would then be scheduled as a function of measured flight condition parameters (mach number, dynamic pressure, etc., as derived from air data).

VII. Conclusions

The results described in this paper establish precise connections between return difference and inverse-return difference matrices and the feedback properties of noise/disturbance response, stability margins, and sensitivity to large plant and sensor modeling errors. The frequency-dependent singular values of these matrices have been found to provide one possible physically meaningful quantitative measure of the quality of these properties, and it appears to be relatively easy to translate specifications on feedback system performance into inequality constraints on the magnitude Bode plots of the singular values.

Certain fundamental limits on achievable feedback performance have been found to result from the identity $S+T=I$, from which it has been shown in Section IV that the singular values of the return-difference and inverse-return difference matrices are not completely indepen-

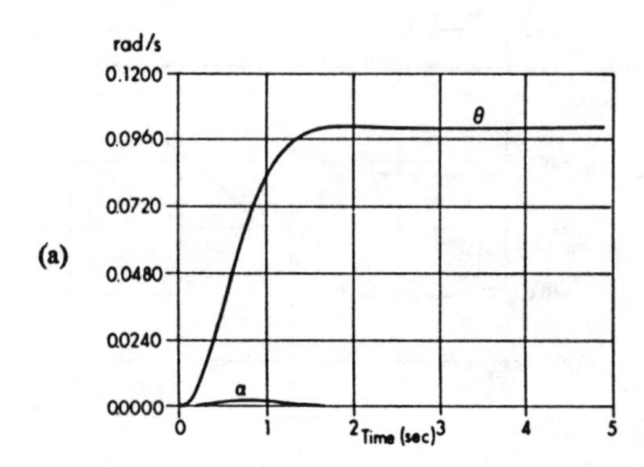

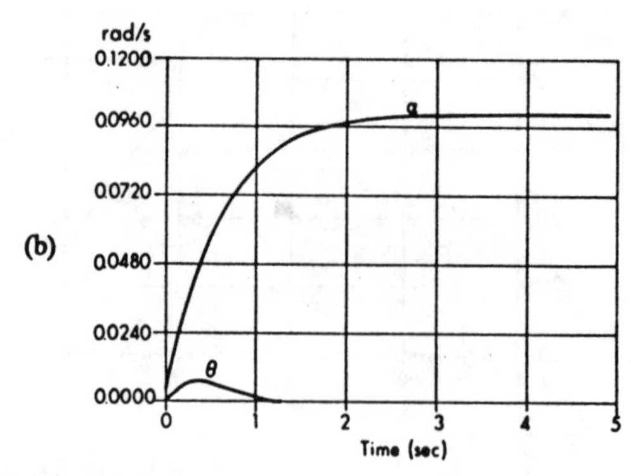

Fig. 13. (a) θ step process. (b) α step process.

dent. This leads to the conclusion that any feedback controller design must inevitably reflect a compromise between good stability margins and insensitivity to sensor noise and to sensor gain variation on the one hand and to insensitivity to plant disturbances and to plant gain variations on the other hand.

The stochastic linear quadratic Gaussian (LQG) control theory has been found to minimize a weighted sum of terms involving the singular values of the return difference and inverse-return difference matrices. This fact establishes a connection between one's choice of quadratic cost and noise intensity matrices and the shapes of the Bode magnitude plots of the resulting optimal system's return and inverse-return difference singular values. It also suggests a method for iteratively adjusting the LQG quadratic weights and noise matrices as an aid in shaping the singular value Bode plots to meet inequality specifications on noise/disturbance response, stability margins, and sensitivity. Frequency dependent cost and noise matrices provide added flexibility in manipulating the singular values, enabling more precise shaping of the singular value Bode plots and enabling the creation of feedback laws with special structures for disturbance rejection (e.g., type 1 proportional-integral feedback).

The advanced aircraft design example considered in Section VII demonstrates the power of the theory in coping with a true multivariable design problem involving highly cross-coupled plant input-output channels. Practical performance criteria were readily translated into inequality constraints on the magnitude Bode plots of singular values. Iterative adjustment of frequency-dependent LQG cost and noise matrices via the approach outlined in Section V proved to be a systematic and effective procedure for synthesizing a dynamical output feedback controller satisfying the constraints.

The aircraft control design example also illustrates how the problem of designing for good command response is essentially separate from the problem of designing for good feedback properties. The design was approached in two distinct phases. In the first phase specifications on feedback properties were addressed, leading to the selection of a particular LQG design for $M(s)$ (the controller/sensor portion of the loop transfer matrix). In the second phase specifications on command response were addressed, using the remaining freedom in choosing the controller matrices $P(s)$, $C(s)$ and $F(s)$. This two-phase decomposition clarifies the underlying issues in servomechanism design by stressing the inherent separateness of command response properties (e.g., rise time, setting time, overshoot, steady-state error, decoupling, etc.) and feedback properties (e.g., stability, stability margin, disturbance attenuation, sensor noise response, sensitivity to plant and sensor variation, etc.). It can also simplify the actual design computations.

Acknowledgment

We are grateful to Dr. G. Stein and J. Doyle of the Honeywell Systems and Research Center, Minneapolis, MN for numerous thought-provoking discussions. We also thank Dr. R. Pope and Dr. C. A. Harvey, also of the Honeywell Systems and Research Center, for their encouragement and support. The comments of several anonymous reviewers contributed significantly to this paper.

References

[1] G. Zames, "Structural stabilization and quenching by dither in nonlinear systems," *IEEE Trans. Automat. Contr.*, vol. AC-22 pp. 352–360, 1977.

[2] S. M. Meerkov, "Vibrational control theory," *J. Franklin Inst.*, vol. 305, pp. 117–128, 1977.

[3] H. H. Rosenbrock and Peter D. McMorran, "Good, bad, or optimal," *IEEE Trans. Automat. Contr.*, vol. AC-16, pp. 552–554, Dec. 1971.

[4] J. C. Doyle, "Guaranteed margins for LQG regulators," *IEEE Trans. Automat. Contr.*, vol. AC-23, pp. 756–757, 1978.

[5] I. Horowitz, *Synthesis of Feedback Systems*. New York: Academic, 1963.

[6] H. H. Rosenbrock, *Computer-Aided Control System Design*. New York: Academic, 1974.

[7] L. S. Pontryagin, V. Boltyanskii, R. Gamkrelidze, and E. Mishchenko, *The Mathematical Theory of Optimal Processes*. New York: Interscience, 1962.

[8] R. E. Bellman, *Dynamic Programming*. Princeton, NJ.: Princeton Univ. Press, 1957.

[9] R. E. Kalman, "Contributions to the theory of optimal control," *Boletin de la Sociedad Matematica Mexicana*, pp. 101–119, 1960.

[10] A. G. J. MacFarlane, "Return-difference and return-ratio matrices and their use in the analysis and design of multivariable feedback control systems," *Proc. IEE*, vol. 117, no. 10, pp. 2037–2049, Oct. 1970.

[11] D. Q. Mayne, "Design of linear multivariable systems," *Automatica*, vol. 9, no. 2, pp. 201–208, Mar. 1973.

[12] N. T. Hung and B. D. O. Anderson, "Triangularization technique for the design of multivariable control systems," *IEEE Trans. Automat. Contr.*, vol. AC-24, pp. 455–460, 1979.

[13] W. M. Wonham, *Linear Multivariable Control*, 2nd ed. New York: Springer-Verlag, 1979.

[14] J. B. Cruz and W. R. Perkins, "A new approach to the sensitivity problem in multivariable feedback system design," *IEEE Trans. Automat. Contr.*, vol. AC-9, pp. 216–223, 1964.

[15] H. H. Rosenbrock, "Progress in the design of multivariable control systems," *Trans. Inst. Measur. Contr.*, vol. 4, pp. 9–11, 1971.

[16] M. G. Safonov, *Stability and Robustness of Multivariable Feedback Systems*. Cambridge, MA: MIT Press, 1980.

[17] M. Athans, "The role and use of the stochastic linear-quadratic Gaussian problem in control system design," *IEEE Trans. Automat. Contr.*, vol. AC-16, pp. 529–552, 1971.

[18] B. Francis and W. M. Wonham, "The internal model principle for linear multivariable regulators," *J. Appl. Math. Optimiz.*, vol. 2, 1975.

[19] A. G. J. MacFarlane and B. Kouvaritakis, "A design technique for linear multivariable feedback systems," *Int. J. Contr.*, vol. 25, pp. 837–874, 1977.

[20] H. W. Bode, *Network Analysis and Feedback Amplifier Design*. New York: Van Nostrand, 1945.

[21] A. J. Laub, "Linear multivariable control: Numerical considerations," Lab. Inform. Decision Syst., Massachusetts Inst. Technol., Cambridge, MA, Rep. ESL-P-833, July 1978.

[22] G. W. Stewart, *Introduction to Matrix Computations*. New York: Academic, 1973.

[23] D. C. Youla, "On the factorization of rational matrices," *IRE Trans. Inform. Theory*, vol. IT-7, pp. 172–189, July 1961.

[24] L. Pernebo, "Algebraic control theory for linear multivariable systems," Ph.D. dissertation, Dept. Automat. Contr., Lund Inst Technol., Lund, Sweden, May 1978.

[25] M. G. Safonov and M. Athans, "A multiloop generalization of the circle stability criterion," in *Proc. Asilomar Conf. Circuits, Syst., Comput.*, Pacific Grove, CA, Nov. 6–8, 1978.

[26] M. G. Safonov and M. Athans, "A multiloop generalization of the Circle criterion for stability margin analysis," *IEEE Trans. Automat. Contr.*, to be published.

[27] M. G. Safonov, "Choice of quadratic cost and noise matrices and the feedback properties of multiloop LQG regulators," in *Proc. Asilomar Conf. Circ., Syst., Comput.*, Pacific Grove, CA, Nov. 5–7, 1979.

[28] A. T. Fuller, "Feedback control systems with low frequency stochastic disturbances," *Int. J. Contr.*, vol. 24, pp. 165–207, 1976.

[29] D. C. Youla, J. J. Bongiorno, and H. A. Jabr, "Modern Weiner-Hopf design of optimal controllers—Part I: The single-input-output case," *IEEE Trans. Automat. Contr.*, vol. AC-21, pp. 3–13, 1976.

[30] D. C. Youla, H. A. Jabr, and J. J. Bongiorno, "Modern Weiner-Hopf design of optimal controllers—Part II: The multivariable case," *IEEE Trans. Automat. Contr.*, vol. AC-21, pp. 319–338, 1976.

[31] U. Shaked, "A general transfer function approach to linear stationary filtering and steady-state optimal control problems," *Int. J. Contr.*, vol. 24, pp. 741–770, 1976.

[32] U. Shaked, "A general transfer function approach to the steady-state linear quadratic Gaussian stochastic control problem," *Int. J. Contr.*, vol. 24, pp. 771–800, 1976.

[33] G. L. Hartmann, M. F. Barrett, and C. S. Greene, "Control design for an unstable vehicle," NASA Dryden Flight Research Center, Contract Rep. NAS 4-2578, Dec. 1979.

[34] P. A. Merkel R. A. Whitmoyer, "Development and evaluation of precision control modes for fighter aircraft," AIAA Guidance and Contr. Conf., San Diego, CA, Paper 76-1950, 1976.

[35] A. R. Mitchell, "Direct force mode flight control for a vectored lift fighter," AIAA Guidance and Contr. Conf., Boulder, Colorado, Paper 79-1744, Aug. 1979.

[36] C. R. Chalk *et al.*, "Background information and users' guide for MIL-F-8785B (ASG) military specification—Flying qualities of piloted airplanes," Wright-Patterson Air Force Base, OH, Rep. AFFDL-TR-69-72, Aug. 1979.

[37] J. C. Doyle and G. Stein, "Robustness with observers," *IEEE Trans. Automat. Contr.*, vol. AC-24, pp. 607–611, 1979.

[38] N. Sandell and M. Athans, "On 'type-L' multivariable linear systems," *Automatica*, vol. 9, pp. 131–136, 1973.

[39] V. C. Klema and A. J. Laub, "The singular value decomposition: Its computation and some applications," *IEEE Trans. Automat. Contr.*, vol. 25, Apr. 1980.

[40] I. C. Gohberg and M. G. Krein, *Introduction to the Theory of Linear Nonselfadjoint Operators*. Providence, RI: American Math. Society, 1969.

[41] A. G. J. MacFarlane and D. F. A. Scott-Jones, "Vector gain," *Int. J. Contr.*, vol. 29, pp. 65–91, 1979.

[42] A. J. Laub, "Robust stability of linear systems—Some computational considerations," in *Information Linkages between Applied Mathematics and Industry*, C. Wilde *et al.*, Eds. New York: Academic, 1980.

[43] J. B. Cruz, Ed., *Feedback Systems*. New York: McGraw-Hill, 1972.

[44] M. G. Safonov, "Frequency domain design of multivariable control systems for insensitivty to large plant modeling errors," in *Proc. IEEE Conf. Decision Contr.*, Ft. Lauderdale, FL, Dec. 12–14, 1979.

[45] M. G. Safonov, "The sensitivity problem in multivariable control, Revisited," in *Proc. 14th Asilomar Conf. Circ., Syst., Comput.*, Pacific Grove, CA, Nov. 17–19, 1980.

A Multiloop Generalization of the Circle Criterion for Stability Margin Analysis

MICHAEL G. SAFONOV, MEMBER, IEEE, AND
MICHAEL ATHANS, FELLOW, IEEE

***Abstract*—In order to provide a theoretical tool well suited for use in characterizing the stability margins (e.g., gain and phase margins) of multiloop feedback systems, multiloop input–output stability results generalizing the circle stability criterion are considered. Generalized conic sectors with "centers" and "radii" determined by linear dynamical operators are employed to enable an engineer to specify the stability margins which he desires as a frequency-dependent convex set of modeling errors–including nonlinearities, gain variations, and phase variations–which the system must be able to tolerate in each feedback loop without instability. The resulting stability criterion gives sufficient conditions for closed-loop stability in the presence of such frequency-dependent modeling errors, *even when the modeling errors occur simultaneously in all loops*; so, for example, stability is assured as loop gains and phases vary throughout a "set of nonzero measure" whose boundaries are frequency-dependent. The stability conditions yield an easily interpreted scalar measure of the amount by which a multiloop system exceeds, or falls short of, its stability margin specifications.**

I. Introduction

A key step in the synthesis of robustly stable feedback systems is the characterization of a *set* of feedback laws that are stabilizing for every element of the set of possible plant dynamics. This type of information is precisely what is provided for single-loop feedback systems by such input–output stability criteria as the Nyquist, Popov, and circle theorems. Indeed, the practical merit of classical feedback design procedures involving Nyquist loci, Bode plots, and Nichols charts is in a large measure directly attributable to the fact that these design procedures provide the designer with easily interpretable characterizations of such sets of robustly stable feedback laws. For single-loop feedback systems, these stability theorems enable engineers to meaningfully characterize the tolerable amount of gain and phase variation in the loop at each frequency, and even the tolerable amount of unmodeled nonlinearity. These tolerances of modeling error are in broad terms what we call *stability margin*, classical gain and phase margin being two familiar measures of stability margin for single feedback loops. Although multiloop generalizations of the Nyquist stability criterion have been developed (e.g., [1]–[3]), it has been difficult to meaningfully relate the conditions of these multiloop criteria to tolerance of open-loop modeling error except in special cases such as diagonally dominant systems, normal systems, and systems in which feedback loop gains vary only over certain "sets of zero measure" [4]. The results of the present paper are intended to address the need for an improved method for characterizing the stability margins of multiloop feedback systems.

In broad and imprecise terms what seems to be necessary to meaningfully characterize multiloop stability margins is a stability criterion that guarantees stability for every multiloop feedback operator within a given "frequency-dependent ball" in an appropriate space of input–output relations, this ball being centered at the system's nominal "open-loop gain" operator. It is important that the size of this ball be permitted to be frequency-dependent so that one can account for frequency-dependent variations in the precision of mathematical models such as result from such unbiquitous effects as singular perturbations, hysteresis, imprecisely known time delays, or any sort of unmodeled dynamics. Also, since in general one may expect modeling imprecision in certain feedback loops to be large relative to other loops, it should be possible to specify that this ball be somewhat egg-shaped, having different diameters in the various "directions" corresponding to the "gains" of individual feedback loops. So, perhaps the necessary ball of stable multiloop feedback operators could be better described as a "frequency-dependent egg." All of this is of course too vague and imprecise to be of immediate use—what is needed is a stability criterion dealing with a precise mathematical description of this frequency-dependent ball (or "egg") and of the space of operators in which it is embedded.

Stability results in this general spirit are provided by the input–output stability theory of Zames [5], [27]–[29] and of Sandberg [6], [7]. Sandberg's frequency-domain stability criterion [6] for systems with multiple nonlinearities can be interpreted as guaranteeing stability for a collection of nonlinear feedback operators inside a spherical (i.e., not "egg-shaped") nonfrequency-dependent ball centered at the identity operator times a scalar; the now well-known circle stability criterion emerged in [7] as a special case of this result. Zames' conic sector stability theorem [5, theorem 2] is an abstract generalization of Sandberg's criterion that makes the connection with balls of stable multiloop feedback operators even more transparent: the conditions of Zames' theorem involve conic sectors which it happens are simply spherical balls, centered at the identity operator times a scalar in an extended normed space of input–output relations.

In the past 15 years frequency domain stability criteria based on the theory of Zames and Sandberg have been improved in many significant ways. Reference [8] provides a good overview of much of this work. Reference [9] discusses similar results in a Lyapunov setting. References [10] to [12] and the additional references cited therein describe many stability results developed specifically for interconnected (i.e., multiloop) nonlinear systems. However, the previous literature in this area has focused primarily on *nonlinear* stability; although Zames [5] makes some key suggestive remarks about the broader implications of conic sector results regarding imprecisely modeled systems. The stability margin implications of the results have not been stressed and no results based on this theory have been published which address the need for a *frequency-dependent* characterization of multiloop stability margins.

The main objective of the present paper is to present a multiloop input–output stability criterion that is tailored to the task of multiloop feedback stability margin analysis. Our main result (Theorem 1) shows that multiloop stability margins–including tolerance of unmodeled nonlinearity and of dynamical modeling errors of *frequency-dependent* magnitude–can be directly related to open-loop system frequency-response quantities. The results, expressed in terms of the "singular values" of certain matrices, are observed to yield an easily interpretable scalar measure of a system's "excess stability margin", i.e., of the amount by which a multiloop feedback design exceeds its stability margin specifications. A related result described in [21] makes use of Theorem 1 in generating generalized conic sector bounds for characterizing the sensitivity of multiloop systems to large dynamical modeling errors of frequency-dependent magnitudes; in effect the result of [21] provides a nonlinear multiloop generalization of classical M-circle ideas.

The role of singular values in connection with stability was first established by Sandberg [6] who developed an early stability result of the type reported in this paper. However, Sandberg does not specifically use the term singular value. Earlier versions of the results in present paper and their connections with stability margin analysis were first reported by the authors in [13] and [14]. Stressing the use of singular values, Doyle [15] establishes important geometric connections between multivariable Nyquist criteria and the stability margin results of [13], [14], and the present paper; additionally, Doyle [15] and Stein and Doyle [16] cite a number of illustrative examples that present a compelling case for the use of results of this type in the analysis of multivariable feedback stability

Manuscript received January 8, 1979; revised August 29, 1979 and August 25, 1980. Paper recommended by M. Vidyasagar, Past Chairman of the Stability, Nonlinear, and Distributed Systems Committee. This work was supported in part by NASA/Ames under Grant NGL-22-009-124, by NASA/Langley under Grant NSG-1312, by the Joint Services Electronics Program under Contract F44620-76C-0061 monitored by AFOSR, by the National Science Foundation unde Grant ENG78-05628, and by Honeywell Systems and Research Center, Minneapolis, MN with support from the Office of Naval Research under Contract N00014-75-C-0061.

M. G. Safonov is with the Department of Electrical Engineering, University of Southern California, Los Angeles, CA 90007.

M. Athans is with the Laboratory for Information and Decision Systems, Massachusetts Institute of Technology, Cambridge, MA 02139.

Reprinted from *IEEE Trans. Automat. Contr.*, vol. AC-26, no. 2, pp. 415–422, Apr. 1981.

margins. MacFarlane and Scott-Jones [25] discuss at length the relationships between the eigenvalues and the "principal gains" (i.e., singular values) of a multiloop system's transfer matrix. Nuzman and Sandell [23] establish some inequalities relating these results to the singular values of the return difference matrix for multiloop systems and discuss the connection with the guaranteed stability margins of full-state feedback linear optimal regulators (see [24]). The paper by Sandell [17] discusses in broad and simple terms the role of singular values in coping with modeling imprecision in a wide range of engineering and numerical problems. Numerical aspects of singular value computation are surveyed in [18], wherein sophisticated and widely available computer routines for singular value computation are also referenced.

II. Notation

The following notation is used: A^T and x^T denote, respectively, the transpose of the matrix A and the vector x; A^* and x^* denote the complex conjugate of the matrix A^T and the vector x^T, respectively; the determinant of a matrix A is denoted $\det(A)$; the Euclidian norm of a vector x is

$$\|x\|_E \triangleq \sqrt{x^* x}. \tag{1}$$

The extended inner product space L_{2e} is the set of functions $x\colon R_+ \to R^n$ for which $\|x(t)\|_E^2$ is integrable over every closed bounded interval $[0,\tau] \subset R_+$. For every $\tau \in R_+$ and every $x, x_1, x_2 \in L_{2e}$, the norm $\|x\|_\tau$ and inner product $\langle x_1, x_2\rangle_\tau$ are defined as

$$\|x\|_\tau \triangleq \sqrt{\langle x, x\rangle_\tau} \tag{2}$$

where

$$\langle x_1, x_2\rangle_\tau \triangleq \int_0^\tau x_1^T(t) x_2(t)\, dt. \tag{3}$$

Laplace transforms are denoted by capital letters, e.g., $X(s)$ denotes the Laplace transform of $x(t)$.

Given any matrix A, the square roots of the eigenvalues of A^*A are called the *singular values* of A. For any matrix A, we use the notation $\sigma_{\max}(A)$ to denote the largest singular value of A and $\sigma_{\min}(A)$ to denote the smallest singular value of A. Singular values are always nonnegative real numbers since A^*A is always positive semidefinite.

A functional relation is a mapping taking functions into *sets* of functions; for example, a dynamical system mapping inputs in L_{2e} into outputs in L_{2e} defines a relation (e.g., [5]). An *operator* is a special type of relation which takes each function in its domain into exactly one function, i.e., into a set with exactly one element. All functional relations considered in this paper are mappings of L_{2e} into L_{2e}. A relation $\underline{H}$ is said to be nonanticipative if, for all t_0, the value of $(\underline{H}x)(t_0)$ is independent of the values of $x(t)$ for $t>t_0$. We say that a relation $\underline{H}$ is L_{2e}*-stable*[1] if there exists a constant $k<\infty$ such that for all $x \in L_{2e}$ and all $\tau \in R_+$

$$\|\underline{H}x\|_\tau \le k\|x\|_\tau. \tag{4}$$

Generalizing some of the L_{2e} conic sector conditions of Zames [5], we employ the following definitions which are a special case of the generalized sector conditions of [13], [19], and [20]; the sector conditions of [13], [19], and [20] are similar to the quadratic "dissipativeness" conditions of Moylan and Hill [12]. Given an operator $\underline{H}$, if there exist operators $\underline{C}$, $\underline{R}$, and $\underline{S}$ and a scalar $\epsilon>0$ such that

$$\|\underline{S}(y-\underline{C}x)\|_\tau^2 \le \|\underline{R}x\|_\tau^2 - \epsilon\left(\|x\|_\tau^2 + \|y\|_\tau^2\right) \tag{5}$$

for all $y=\underline{H}x$, all x, and all τ, then we say "$\underline{H}$ is strictly inside the L_{2e} conic sector with center $\underline{C}$ and radius $(\underline{R}, \underline{S})$"; equivalently, we write

$$\underline{H} \text{ strictly inside } L_{2e}\text{-Cone}(\underline{C}, \underline{R}, \underline{S}).^2 \tag{6}$$

Given a relation $\underline{G}$, if

$$\|\underline{S}(y-\underline{C}x)\|_\tau^2 \ge \|\underline{R}x\|_\tau^2 \tag{7}$$

for all $x=-\underline{G}y$, all y, and all τ, then we say "the inverse relation of $-\underline{G}$ is outside the L_{2e} conic sector with center $\underline{C}$ and radius $(\underline{R}, \underline{S})$";[3] equivalently, we write

$$(-\underline{G})^I \text{ outside } L_{2e}\text{-Cone}(\underline{C}, \underline{R}, \underline{S}). \tag{8}$$

The notation $\mathrm{col}(x_1,\cdots,x_N)$ denotes the column vector

$$\mathrm{col}(x_1,\cdots,x_N) \triangleq \begin{bmatrix} x_1 \\ \vdots \\ x_N \end{bmatrix}. \tag{9}$$

The relation $\mathrm{diag}(\underline{H}_1,\cdots,\underline{H}_N)$ is defined by

$$\mathrm{diag}(\underline{H}_1,\cdots,\underline{H}_N)\mathrm{col}(x_1,\cdots,x_N) = \mathrm{col}(\underline{H}_1x_1,\ldots,\underline{H}_Nx_N). \tag{10}$$

III. Problem Formulation

Our results concern the input–output stability of systems consisting of a dynamical *linear time-invariant* (LTI) interconnection of N imprecisely modeled components, including imprecisely modeled LTI components and *nonlinear time-varying* (NTV) components, as well as dynamical nonlinear components comprised of interconnections of LTI and NTV subcomponents. The system equations thus take the following form (see Fig. 1):

components

$$y_i = \underline{H}_i x_i, \qquad (i=1,\cdots,N) \tag{11}$$

dynamical LTI interconnection

$$X(s) = -G(s)(Y(s)+V(s)) + U(s) \tag{12}$$

where

$$Y(s) = \mathrm{col}(Y_1(s),\cdots,Y_N(s)) \tag{13}$$

$$X(s) = \mathrm{col}(X_1(s),\cdots,X_N(s)) \tag{14}$$

$$U(s) = \mathrm{col}(U_1(s),\cdots,U_N(s)) \tag{15}$$

$$V(s) = \mathrm{col}(V_1(s),\cdots,V_N(s)) \tag{16}$$

$$G(s) = \begin{bmatrix} G_{11}(s) & \cdots & G_{1N}(s) \\ & \vdots & \\ G_{N1}(s) & \cdots & G_{NN}(s) \end{bmatrix}. \tag{17}$$

The endogenous variables $y_i(t)$ and $x_i(t)$ are the system "outputs" and the exogenous variables $u_i(t)$ and $v_i(t)$ are the system "inputs." Each of the "components" $\underline{H}_i$ may itself be a multiinput–multioutput (MIMO) system in general, although our results are most easily used and interpreted when the components are single-input–single-output (SISO).

We assume that for each of the imprecisely modeled components $\underline{H}_i$ we have a crude approximate LTI model $\underline{C}_i$ and that LTI operators $\underline{R}_i$ and $\underline{S}_i$

[1] For nonanticipative operators, L_{2e}-stability as defined here is equivalent to the usual notion of L_2-stability, e.g., [8].

[2] The "strictly inside" conic sector condition of [5] can be demonstrated to be a special case of (6) resulting when $S=I$, $R=rI$, and $C=cI$, where c, r are scalars; however, the term $\epsilon\|y\|_\tau^2$ is only implicit in [5].

[3] Following [13], [19], and [20], the notation $(-\underline{G})^I$ is used for the inverse of the relation $-\underline{G}$; i.e., $(-\underline{G})^I$ is the relation which takes each $y \in L_{2e}$ into the *set* of functions $x \in L_{2e}$ such that $y=-\underline{G}x$. The inverse *relation* $(-\underline{G})^I$ always exists even for operators $\underline{G}$ for which the inverse operator, denoted $(-\underline{G})^{-1}$, does not exist.

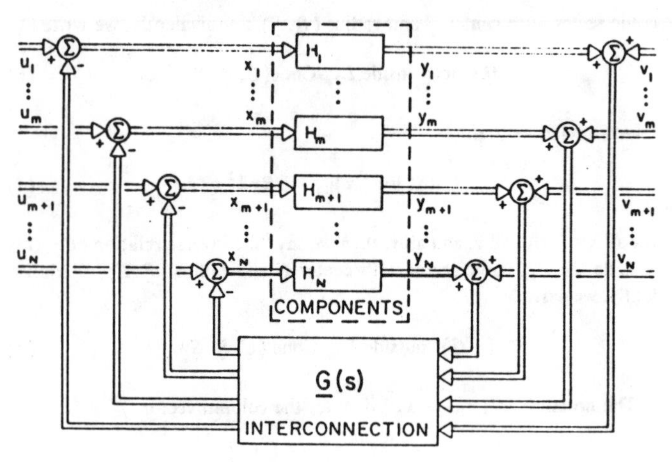

Fig. 1. The system.

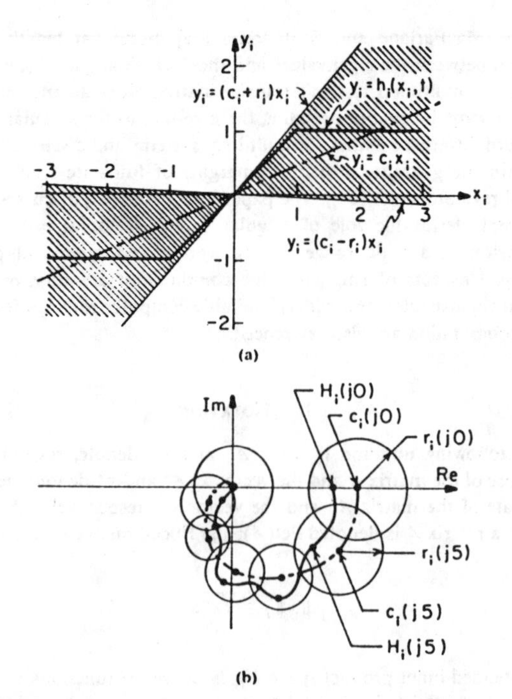

Fig. 2. SISO components. (a) Nonlinear component satisfying (26). (b) Nyquist locus of LTI component satisfying (22).

can be found such that the modeling error in each $\underset{\sim}{H}_i$ is bounded by a generalized conic sector condition such as (6), i.e.,

$$\underset{\sim}{H}_i \text{ strictly inside } L_{2e}\text{-Cone}\,(\underset{\sim}{C}_i, \underset{\sim}{R}_i, \underset{\sim}{S}_i) \tag{18}$$

for $i=1,\cdots,N$. For notational convenience we define

$$\underset{\sim}{C} \triangleq \text{diag}(\underset{\sim}{C}_1,\cdots,\underset{\sim}{C}_N) \tag{19}$$

$$\underset{\sim}{R} \triangleq \text{diag}(\underset{\sim}{R}_1,\cdots,\underset{\sim}{R}_N) \tag{20}$$

$$\underset{\sim}{S} \triangleq \text{diag}(\underset{\sim}{S}_1,\cdots,\underset{\sim}{S}_N). \tag{21}$$

Comments: The generalized conic sector error bound (18), although somewhat abstract, is fairly easily related to meaningful quantities. Lemmas A4 and A5 in the Appendix relate condition (18) to simple Euclidean norm bounds for multiinput–multioutput (MIMO) NTV and LTI $\underset{\sim}{H}_i$. For example, if $\underset{\sim}{H}_i$ is a stable SISO LTI element, then it follows from Lemma A4 that the simple frequency domain condition

$$|H_i(j\omega)-c_i(j\omega)|^2 \leq |r_i(j\omega)|^2-\epsilon \tag{22}$$

for some $\epsilon>0$ and all ω [see Fig. 2(b)], implies that (18) is satisfied for any stable SISO LTI nonanticipative $\underset{\sim}{S}$ and $\underset{\sim}{R}$ satisfying

$$C_i(j\omega)=c_i(j\omega) \tag{23}$$

$$R_i(j\omega)=r_i(j\omega)S_i(j\omega). \tag{24}$$

If $\underset{\sim}{H}_i$ is a SISO memoryless NTV element defined by

$$y_i(t)=h_i(x_i(t),t) \tag{25}$$

and if for some c_i, r_i and some $\epsilon>0$

$$\left|\frac{h_i(\alpha,t)-c_i\alpha}{\alpha}\right|^2 < r_i^2-\epsilon \qquad \forall\alpha\neq 0 \tag{26}$$

then from Lemma A5 it follows that (18) is satisfied for any *constant* $C_i(s)$, $R_i(s)$, $S_i(s)$ satisfying (23) and (24); this is the usual sector nonlinearity condition (e.g., [5]–[8])–see Fig. 2(a). In more complicated situations where a component $\underset{\sim}{H}_i$ consists of an interconnection of several LTI elements and/or memoryless NTV elements, the result of [21] frequently may be invoked to determine suitable $\underset{\sim}{C}_i$, $\underset{\sim}{R}_i$, and $\underset{\sim}{S}_i$.

IV. Main Result

Our main result is now stated.

Theorem 1 (Multiloop Circle Criterion): Let $\underset{\sim}{G}$, $\underset{\sim}{C}$, $\underset{\sim}{R}$, and $\underset{\sim}{S}$ have respective proper rational transfer function matrices $G(s)$, $C(s)$, $R(s)$, and $S(s)$; let $\underset{\sim}{R}^{-1}$ and $\underset{\sim}{S}^{-1}$ exist; and let $\underset{\sim}{C}$, $\underset{\sim}{R}$, $\underset{\sim}{R}^{-1}$, $\underset{\sim}{S}$, and $\underset{\sim}{S}^{-1}$ be nonanticipative and L_{2e}-stable;[4,5] let $\underset{\sim}{H}_i$ $(i=1,\cdots,N)$ satisfy the condition (18). Suppose the feedback system (11), (12) is nonanticipative and L_{2e}-stable in the special case where $\underset{\sim}{H}_i=\underset{\sim}{C}_i$ $\forall i=1,\cdots,N$. Then a sufficient condition for the system (11) and (12) to be L_{2e}-stable for every collection $\underset{\sim}{H}_i$ $(i=1,\cdots,N)$ satisfying (18) is

$$\sigma_{\max}\Big(R(j\omega)G(j\omega)\big(I+C(j\omega)G(j\omega)\big)^{-1}S^{-1}(j\omega)\Big) \quad \leq 1 \;\forall\omega. \tag{27}$$

Further, when $G(s)$ is a square matrix and is invertible almost everywhere on the $j\omega$-axis, then the above condition (27) can be expressed as

$$\sigma_{\min}\Big(S(j\omega)\big(C(j\omega)+G^{-1}(j\omega)\big)R^{-1}(j\omega)\Big)\geq 1 \tag{28}$$

for all ω at which $G^{-1}(j\omega)$ exists. □

Proof: We apply the results in the Appendix. From Lemma A3 and (27), it follows that

$$\underset{\sim}{H} \triangleq \text{diag}(\underset{\sim}{H}_1,\cdots,\underset{\sim}{H}_N) \text{ strictly inside } L_{2e}\text{-Cone}\,(\underset{\sim}{C},\underset{\sim}{R},\underset{\sim}{S}). \tag{29}$$

From Lemma A2, it follows that (27) and (28) are equivalent when $G^{-1}(j\omega)$ exists almost everywhere and that

$$(-\underset{\sim}{G})^{I} \text{ outside } L_{2e}\text{-Cone}\,(\underset{\sim}{C},\underset{\sim}{R},\underset{\sim}{S}). \tag{30}$$

L_{2e}-stability of (11) and (12) follows from the conic sector stability theorem, Theorem A1. □

Remarks: It is also possible to prove Theorem 1 by applying Parseval's theorem to verify that the conditions of the well-known "small gain theorem" (e.g., [8]) are satisfied by the transformed system defined by (A5)–(A10) of the Appendix. We consider the present proof more appealing because it stresses the direct connection between the conditions of Theorem 1 and the simple conic sector conditions of Theorem A1; just as Zames' proof in [22] of the well-known circle criterion stresses the direct connection between circle theorem conditions and the simple, but less

[4] A rational transfer function matrix, say $A(s)$, is *proper* if $\sigma_{\max}(A(\infty))<\infty$. If, additionally $\sigma_{\max}(A(\infty))=0$ then $A(s)$ is *strictly proper*–see [26].

[5] Note that a system described by a proper rational transfer function is L_{2e}-stable if and only if its poles all have negative real parts; so for this class of systems, L_{2e}-stability is equivalent to exponential stability [8].

general, conic sector conditions of [5, theorem 2a]. This approach has also been used more recently by Moylan and Hill (e.g., [12]) in proving certain generalized sector stability results, although they do not specifically mention the term sector.

V. Discussion

There are essentially two main conditions which must be satisfied to conclude stability from Theorem 1: 1) the system must be stable when the uncertain components $\underline{H}_i$ are replaced by the respective LTI approximations $\underline{C}_i$; and 2) the frequency-domain condition (27) [or (28)] must be satisfied. The former condition can be verified a variety of ways; for example, one may check that the roots of the characteristic equation

$$\det(I+C(s)G(s))=0 \tag{31}$$

all have negative real parts;[6] alternatively, one may apply the multivariable Nyquist criterion, checking that the polar plot of the locus of $\det(C(j\omega)G(j\omega))$ encircles the point $-1+j0$ exactly once counterclockwise for each unstable open-loop pole of $C(s)G(s)$ (multiplicities counted) [3]. The latter condition (27) [or (28)] requires that one plot the variable $\sigma_{\max}(\cdot)$ [or $\sigma_{\min}(\cdot)$] versus ω and verify that the appropriate inequality holds for all ω.

In the special case in which there is a single SISO nonlinearity $h(x,t)$ both of the conditions of Theorem 1 can be verified by inspection of the polar plot of $G(j\omega)$ $\forall\omega$. In this case the conditions of Theorem 1 become precisely the conditions of the well-known circle stability criterion ([7],[8]). It is this which motivates us to refer to Theorem 1 as a "multiloop circle stability criterion"—despite the fact that in general no circles are employed in verifying its conditions.

One can interpret the uncertainty bounds (R_i, S_i) as specifications for the gain margins and phase margins of the system (11) and (12). For example, if the $\underline{H}_i$ are LTI and SISO and if $H_i(s)\equiv C_i(s)$ $(i=1,\cdots,N)$, then it follows from (22) that under the conditions of Theorem 1 the system will remain stable despite variations in the individual component gains of magnitudes as great as $|r_i(j\omega)| \triangleq |R_i(j\omega)/S_i(j\omega)|$, *even when the variations occur simultaneously in all components.* So, for example, the system can tolerate simultaneous gain variations or phase variations of at least

$$G_{M_i} \triangleq \inf_{\omega} 20\log\left(1+\left|\frac{r_i(j\omega)}{C_i(j\omega)}\right|\right), \text{dB} \tag{32}$$

or

$$\Theta_{M_i} \triangleq 2\inf_{\omega} \arcsin\left|\frac{r_i(j\omega)}{2C_i(j\omega)}\right| \tag{33}$$

in each of the respective component feedback loops; i.e., the system has gain margins of *at least* G_{M_i} and phase margins of *at least* Θ_{M_i} at the inputs to the respective components $C_i(s)$ $(i=1,\cdots,N)$. The quantity

$$k_m(j\omega) \triangleq \sigma_{\min}\left(S(j\omega)\left(C(j\omega)+G^{-1}(j\omega)\right)R^{-1}(j\omega)\right) \tag{34}$$

is the amount by which the uncertainty bounding matrices $R_i(j\omega)$ can be simultaneously increased without violating the stability conditions of Theorem 1—$k_m(j\omega)$ can be viewed as a lower bound on the factor by which the system (11) and (12) *exceeds* the stability margin specifications (18) at each frequency ω.

In general, the stability conditions of Theorem 1—and the estimate (34) of excess stability margin k_m—will be conservative. This conservativeness can usually be reduced by substituting weighted uncertainty bounding matrices $(\alpha_i R_i(s),\ \alpha_i S_i(s))$ for the original matrices $(R_i(s), S_i(s))$. Further, if it happens that $\underline{H}_i$ is linear time-invariant for some i, then as a consequence of Lemma A4, the corresponding α_i may be replaced by a frequency-dependent $\alpha_i(j\omega)$ (provided that for some $k<\infty$ and some $\epsilon>0$, $k>|\alpha_i(j\omega)|^2>\epsilon$ for all ω). Iterative numerical methods would be required to enable one to efficiently compute the "optimal" weightings (i.e., the weightings leading to the least conservative stability conclusions). We hasten to add that the idea of using constant weightings to reduce conservativeness in multiloop nonlinear input–output stability results is not new: M-matrix tests provide a simple but conservative method to implicitly ensure the existence of constant weightings (see [10] and [11]); other results have been stated in which constant weightings appear explicitly (e.g., Moylan and Hill [12, theorems 5 and 6]).

The results of [12] involving explicit weightings may be viewed as a special case in which the matrices $(C(j\omega), R(j\omega), S(j\omega))$, the interconnection matrix $G(j\omega)$, and the weightings $\alpha_i(j\omega)$ are not permitted to be frequency-dependent. We emphasize that the advantages offered by frequency-dependent $C(j\omega)$, $R(j\omega)$, $S(j\omega)$ and $\alpha_i(j\omega)$ are crucial in stability margin analysis where it usually is necessary to be able to characterize tolerance of dynamical modeling errors of frequency-dependent magnitude. Allowing the matrix $G(j\omega)$ to be frequency dependent eliminates the need for incorporating the dynamics of $\underline{G}$ in additional dynamical $\underline{H}_i$'s, thereby reducing the dimension N of $(\underline{G}, \underline{C}, \underline{R}, \underline{S})$; this in general leads to less conservative stability conclusions from Theorem 1 and also broadens its scope of applicability (since $\underline{G}$ need not be L_{2e}-stable under the conditions of Theorem 1, whereas each $\underline{H}_i$ must).

We note that Theorem 1 is fairly broad in its scope of applicability. The transfer matrix $G(s)$ may be nonsquare and need not be open-loop stable (though the $\underline{H}_i$ must be). The $\underline{H}_i$ operators may be multiinput–multioutput and need not have equal numbers of inputs and outputs. Unlike some previous interconnected system results, no condition is imposed requiring either $c_i^2-r_i^2>0$ $\forall i$ or $c_i^2-r_i^2<0$ $\forall i$. The operators $(\underline{C}_i, \underline{R}_i, \underline{S}_i)$ defining the conic sector condition (18) may be dynamical. Further, with the aid of Lemmas A4 and A5 and the aid of results such as in [21], it is practical to verify the conic sector condition (18), even for multiinput–multioutput dynamical nonlinearities.

VI. Conclusions

With a view towards developing a stability criterion well suited to the problem of multiloop feedback stability margin analysis, nonlinear input–output stability techniques generalizing the circle criterion have been re-examined. The stability margin implications of existing results have been stressed and an improved result has been generated allowing one to take account of the frequency-dependence of the magnitude of system modeling errors which commonly occurs in situations involving imprecisely modeled dynamics–e.g., singular perturbations, hysteresis, etc. Theorem 1 together with the related Lemmas A1 and A2 provide verifiable sufficient conditions for the stability of multiloop feedback systems using only crude conic sector bounds on system parameters, subsystem frequency responses, and nonlinearities. Applications include the testing of system integrity in the presence of actuator and/or sensor failures ([2]) and the characterization of frequency-dependent gain and phase margins for multiloop feedback designs subject to multiple singular perturbations and dynamical nonlinearities leading to simultaneous frequency-dependent variations in gains and phases in the feedback loops.

The main result, Theorem 1, also plays a key role in a result described in [21] for generating conic sector bounds to characterize the sensitivity of multiloop systems to large dynamical modeling errors of frequency-dependent magnitudes in a manner similar to the way classical M-circle and Nichols diagram techniques enable one to quantitatively gauge the effect of open-loop gain variations in single-loop, unity-feedback systems. The result of [21] also can be useful in determining the conic sector bounds $(\underline{C}, \underline{R}, \underline{S})$ required by Theorem 1.

Several examples demonstrating the merits of the singular value approach to multiloop stability margin analysis are described in [15]. The results are applied to the design of a controller for a twin rotor helicopter with uncertain rotor blade dynamics in [16].

[6] If $C(s)G(s)$ has any "decoupling zeros" (i.e., uncontrollable or unobservable poles), then these will not be roots of (31) and one must check separately that these poles have negative real parts [3].

APPENDIX

In this appendix several results are stated which are needed in connection with Theorem 1. Theorem A1 and Lemmas A2 and A3 are used in the proof of Theorem 1. Lemmas A4 and A5 are useful in verifying the conic sector condition (18) for memoryless nonlinear H_i and for linear time-invariant dynamical H_i.

We note that while results similar to Lemmas A2, A4, and A5 have been presented in various forms elsewhere (e.g., [6]–[8],[13],[19],[20]), the very general case considered here (admitting, for example, dynamical and multiinput–multioutput $\underset{\sim}{C}, \underset{\sim}{R}, \underset{\sim}{S}$) is new, as is the explicit appearance of the term $\epsilon\|y\|_\tau^2$ in the "strict" conicity condition (5),(6). The differences are sufficient to mandate the inclusion here of proofs for these lemmas.

Theorem A1 (Conic Sector Stability Theorem): Consider the feedback system

$$y=\underset{\sim}{H}x \tag{A1}$$

$$x=-\underset{\sim}{G}(y+v)+u \tag{A2}$$

where $x, y, u, v \in L_{2e}$ and $\underset{\sim}{G}, \underset{\sim}{H}: L_{2e} \to L_{2e}$; (u, v) is the "input" and (x, y) is the "output." If L_{2e}-stable linear operators $\underset{\sim}{C}, \underset{\sim}{R}$, and $\underset{\sim}{S}$ can be found such that

$$\underset{\sim}{H} \text{ strictly inside } L_{2e}\text{-Cone}(\underset{\sim}{C}, \underset{\sim}{R}, \underset{\sim}{S}) \tag{A3}$$

and

$$(-\underset{\sim}{G})' \text{ outside } L_{2e}\text{-Cone}(\underset{\sim}{C}, \underset{\sim}{R}, \underset{\sim}{S}) \tag{A4}$$

then the feedback system (A1) and (A2) is L_{2e}-stable.

Proof: This result is a special case of the "sector stability theorem" [13, p. 65 and 19, theorem 6.1]. □

Remark: Theorem A1 also can be proved by applying the small gain theorem (e.g., [8]) to the "transformed" system

$$\tilde{y}=\underset{\sim}{S}(\underset{\sim}{H}-\underset{\sim}{C})\underset{\sim}{R}^{-1}\tilde{x}+\tilde{v} \tag{A5}$$

$$\tilde{x}=-\underset{\sim}{R}\underset{\sim}{G}(\underset{\sim}{I}+\underset{\sim}{C}\underset{\sim}{G})^{-1}\underset{\sim}{S}^{-1}\tilde{y}+\tilde{u} \tag{A6}$$

where

$$\tilde{y}=\underset{\sim}{S}(y-\underset{\sim}{C}(x-u)+v) \tag{A7}$$

$$\tilde{x}=\underset{\sim}{R}x \tag{A8}$$

$$\tilde{u}=\underset{\sim}{R}u \tag{A9}$$

$$\tilde{v}=\underset{\sim}{S}(v+\underset{\sim}{C}u) \tag{A10}$$

provided $\underset{\sim}{S}^{-1}$, $\underset{\sim}{G}(\underset{\sim}{I}+\underset{\sim}{C}\underset{\sim}{G})^{-1}$, and $\underset{\sim}{R}^{-1}$ exist and are L_{2e}-stable and nonanticipative. It can be shown that under the conditions of Theorem A1, the two operators $\underset{\sim}{S}(\underset{\sim}{H}-\underset{\sim}{C})\underset{\sim}{R}^{-1}$ and $\underset{\sim}{R}\underset{\sim}{G}(\underset{\sim}{I}+\underset{\sim}{C}\underset{\sim}{G})^{-1}\underset{\sim}{S}^{-1}$ each have L_{2e}-gain less than one. Thus, Theorem A1 may be viewed as a characterization of the improvements on the small gain theorem obtainable by use of the linear dynamical transformation (A5)–(A10); such transformations have been described as "loop-shifting" and "multiplier" transformations (e.g., [8]).

Lemma A2 (LTI Outside Conicity): Let $\underset{\sim}{G}, \underset{\sim}{C}, \underset{\sim}{R}, \underset{\sim}{S}$ be linear time-invariant operators with respective proper rational transfer functions $G(s)$, $C(s)$, $R(s)$, $S(s)$. Suppose that $S^{-1}(s)$ exists and has a proper rational transfer function matrix with no poles in $\mathrm{Re}(s)\geq 0$. Suppose that $\underset{\sim}{R}$, $\underset{\sim}{G}(\underset{\sim}{I}+\underset{\sim}{C}\underset{\sim}{G})^{-1}$, $(\underset{\sim}{I}+\underset{\sim}{C}\underset{\sim}{G})^{-1}$, and $\underset{\sim}{S}^{-1}$ are L_{2e}-stable and nonanticipative. Then

$$(-\underset{\sim}{G})' \text{ outside } L_{2e}\text{-Cone}(\underset{\sim}{C}, \underset{\sim}{R}, \underset{\sim}{S}) \tag{A11}$$

if and only if the following condition holds for all real ω

$$\sigma_{\max}\Big(R(j\omega)G(j\omega)(I+C(j\omega)G(j\omega))^{-1}S^{-1}(j\omega)\Big)\leq 1. \tag{A12}$$

almost everywhere.

When $G^{-1}(s)$ and $R^{-1}(s)$ exist almost everywhere on the $j\omega$-axis, then condition (A12) is equivalent to

$$\sigma_{\min}\Big(S(j\omega)\big(C(j\omega)+G^{-1}(j\omega)\big)R^{-1}(j\omega)\Big)\geq 1 \tag{A13}$$

Proof: It is trivial to see that (A12) and (A13) are equivalent when $G^{-1}(j\omega)$ and $R^{-1}(j\omega)$ exist, since for any invertible matrix A

$$\sigma_{\min}(A^{-1})=1/\sigma_{\max}(A).$$

Suppose that (A12) holds. Let (x, y) be any input–output pair satisfying $x=-\underset{\sim}{G}y$; let

$$\tilde{y}=\underset{\sim}{S}(y-\underset{\sim}{C}x) \tag{A14}$$

and let

$$\tilde{y}_\tau=\begin{cases}\tilde{y}(t), & \text{if } 0\leq t\leq\tau\\ 0, & \text{otherwise.}\end{cases} \tag{A15}$$

Let $\tilde{Y}_\tau(j\omega)$ denote the Fourier transform of $\tilde{y}_\tau$. Note that from (A12) it follows that for all $\tilde{Y}_\tau(j\omega)$

$$\|\tilde{Y}_\tau(j\omega)\|^2-\|R(j\omega)G(j\omega)(\underset{\sim}{I}+\underset{\sim}{C}(j\omega)\underset{\sim}{G}(j\omega))^{-1}\underset{\sim}{S}^{-1}(j\omega)\tilde{Y}_\tau(j\omega)\|^2\geq 0.$$

Now

$$\begin{aligned}
\|\underset{\sim}{R}x\|_\tau^2&=\|\underset{\sim}{R}\underset{\sim}{G}(\underset{\sim}{I}+\underset{\sim}{C}\underset{\sim}{G})^{-1}\underset{\sim}{S}^{-1}\tilde{y}\|_\tau^2\\
&=\|\underset{\sim}{R}\underset{\sim}{G}(\underset{\sim}{I}+\underset{\sim}{C}\underset{\sim}{G})^{-1}\underset{\sim}{S}^{-1}\tilde{y}_\tau\|_\tau^2
\end{aligned}$$

by the nonanticipativeness of $\underset{\sim}{R}, \underset{\sim}{G}(\underset{\sim}{I}+\underset{\sim}{C}\underset{\sim}{G})^{-1}, \underset{\sim}{S}^{-1}$

$$\begin{aligned}
&\leq\int_0^\infty\|\underset{\sim}{R}\underset{\sim}{G}(\underset{\sim}{I}+\underset{\sim}{C}\underset{\sim}{G})^{-1}\underset{\sim}{S}^{-1}\tilde{y}_\tau\big)(t)\|^2\,dt\\
&=\frac{1}{2\pi}\int_{-\infty}^{\infty}\|R(j\omega)G(j\omega)(I+C(j\omega)G(j\omega))^{-1}\\
&\quad\cdot S^{-1}(j\omega)\tilde{Y}_\tau(j\omega)\|^2\,d\omega
\end{aligned}$$

by Parseval's Theorem and the hypotheses that $\underset{\sim}{R}, \underset{\sim}{G}(\underset{\sim}{I}+\underset{\sim}{C}\underset{\sim}{G})^{-1}$, and $\underset{\sim}{S}^{-1}$ and L_{2e}-stable

$$\begin{aligned}
&\leq\frac{1}{2\pi}\int_{-\infty}^{\infty}\|\tilde{Y}_\tau(j\omega)\|^2\,d\omega\\
&=\|\tilde{y}_\tau\|_\tau^2=\|\tilde{y}\|_\tau^2\\
&=\|\underset{\sim}{S}(y-\underset{\sim}{C}x)\|_\tau^2
\end{aligned} \tag{A17}$$

which proves (A11) is implied by (A12).

Conversely, suppose that (A11) holds. Let Y_0 and ω_0 be arbitrary. Consider the L_{2e}-stable feedback system

$$x=-\underset{\sim}{G}y \tag{A18}$$

$$y=\underset{\sim}{C}x+\underset{\sim}{S}^{-1}\tilde{y}. \tag{A19}$$

Let $\tilde{y}(t)\to Y_0e^{j\omega_0 t}$. Then [letting $\tau\to\infty$ in (7)] it follows from (A11) that

$$\|R(j\omega)G(j\omega)(I+C(j\omega)G(j\omega))^{-1}S^{-1}(j\omega)Y_0\|_E^2\leq\|Y_0\|_E^2 \tag{A20}$$

and hence (A12) holds. □

Lemma A3 (Composite Operator Conicity): Let

$$\underset{\sim}{H}=\operatorname{diag}(\underset{\sim}{H}_1,\cdots,\underset{\sim}{H}_N) \tag{A21}$$

$$\underset{\sim}{C}=\operatorname{diag}(\underset{\sim}{C}_1,\cdots,\underset{\sim}{C}_N) \tag{A22}$$

$$\underset{\sim}{R}=\operatorname{diag}(\underset{\sim}{R}_1,\cdots,\underset{\sim}{R}_N) \tag{A23}$$

$$\underset{\sim}{S}=\operatorname{diag}(\underset{\sim}{S}_1,\cdots,\underset{\sim}{S}_N). \tag{A24}$$

If for all $i=,\cdots,N$

$$\underset{\sim}{H}_i \text{ strictly inside } L_{2e}\text{-Cone}(\underset{\sim}{C}_i,\underset{\sim}{R}_i,\underset{\sim}{S}_i) \tag{A25}$$

then

$$\underset{\sim}{H} \text{ strictly inside } L_{2e}\text{-Cone}(\underset{\sim}{C},\underset{\sim}{R},\underset{\sim}{S}). \tag{A26}$$

Proof: This is a special case of the results in [13, p. 70] and [19, Lemma 6.2 (vi)]. □

Lemma A4 (LTI Conicity): Let $\underset{\sim}{H}$, $\underset{\sim}{C}$, $\underset{\sim}{R}$, $\underset{\sim}{S}$ be nonanticipative L_{2e}-stable linear-time-invariant operators with respective rational transfer function matrices $H(s)$, $C(s)$, $R(s)$, $S(s)$. Suppose that $R^{-1}(s)$ exists, is proper and has no poles in $\mathrm{Re}(s)\geq 0$. Then

$$\underset{\sim}{H} \text{ strictly inside } L_{2e}\text{-Cone}(\underset{\sim}{C},\underset{\sim}{R},\underset{\sim}{S}) \tag{A27}$$

if and only if

$$\|S(j\omega)(H(j\omega)-C(j\omega))X(j\omega)\|_E^2\leq\|R(j\omega)X(j\omega)\|_E^2-\epsilon\|X(j\omega)\|_E^2 \tag{A28}$$

for all $X(j\omega)$, all ω, and some $\epsilon>0$.

Proof: Let $\underset{\sim}{R}^{-1}$ denote the stable nonanticipative LTI operator having transfer function matrix $R^{-1}(s)$. Suppose that (A28) holds and let

$$\tilde{x}_\tau(t)=\begin{cases}(\underset{\sim}{R}x)(t), & \text{if } t\leq\tau\\ 0, & \text{if } t>\tau\end{cases} \tag{A29}$$

and let $\tilde{X}_\tau(j\omega)$ denote the Fourier transform of $\tilde{x}_\tau(t)$. Then for all $y=\underset{\sim}{H}x$ we have

$$\|\underset{\sim}{S}(y-\underset{\sim}{C}x)\|_\tau^2=\|\underset{\sim}{S}(\underset{\sim}{H}x-\underset{\sim}{C}x)\|_\tau^2$$

$$=\|\underset{\sim}{S}(\underset{\sim}{H}-\underset{\sim}{C})x\|_\tau^2 \quad \text{(by linearity)}$$

$$=\|\underset{\sim}{S}(\underset{\sim}{H}-\underset{\sim}{C})\underset{\sim}{R}^{-1}\underset{\sim}{R}x\|_\tau^2 \quad (\text{since } \underset{\sim}{R}^{-1} \text{ exists})$$

$$=\|\underset{\sim}{S}(\underset{\sim}{H}-\underset{\sim}{C})\underset{\sim}{R}^{-1}\tilde{x}_\tau\|_\tau^2 \quad \text{(by nonanticipativeness)}$$

$$\leq\int_0^\infty\|(\underset{\sim}{S}(\underset{\sim}{H}-\underset{\sim}{C})\underset{\sim}{R}^{-1}x_\tau)(t)\|_E^2\,dt$$

(the integral exists since $\underset{\sim}{S}$, $\underset{\sim}{H}$, $\underset{\sim}{C}$, and $\underset{\sim}{R}^{-1}$ are stable)

$$=\frac{1}{2\pi}\int_{-\infty}^{\infty}\|S(j\omega)(H(j\omega)-C(j\omega))R^{-1}(j\omega)\tilde{X}_\tau(j\omega)\|_E^2\,d\omega$$

(by Parseval's Theorem)

$$\leq\frac{1}{2\pi}\int_{-\infty}^{\infty}\|\tilde{X}_\tau(j\omega)\|_E^2-\epsilon\|R^{-1}(j\omega)\tilde{x}_\tau(j\omega)\|_E^2\,d\omega \quad [\text{by A28}]$$

$$=\int_0^\infty\|\tilde{x}_\tau(t)\|_E^2\,dt-\epsilon\int_0^\infty\|(\underset{\sim}{R}^{-1}\tilde{x}_\tau)(t)\|_E^2\,dt \quad \text{(by Parseval's Theorem)}$$

$$\leq\int_0^\tau\|\tilde{x}_\tau(t)\|_E^2\,dt-\epsilon\int_0^\tau\|(\underset{\sim}{R}^{-1}\tilde{x}_\tau)(t)\|_E^2\,dt$$

$$=\|\tilde{x}_\tau\|_\tau^2-\epsilon\|\underset{\sim}{R}^{-1}\tilde{x}_\tau\|_\tau^2$$

$$=\|\underset{\sim}{R}x\|_\tau^2-\epsilon\|\underset{\sim}{R}^{-1}\underset{\sim}{R}x\|_\tau^2 \quad (\text{by nonanticipativeness of } \underset{\sim}{R}^{-1})$$

$$=\|\underset{\sim}{R}x\|_\tau^2-\epsilon\|x\|_\tau^2$$

$$\leq\|\underset{\sim}{R}x\|_\tau^2-\epsilon'(\|x\|_\tau^2+\|\underset{\sim}{H}x\|_\tau^2). \tag{A30}$$

where

$$\epsilon'=\frac{\epsilon}{1+\alpha^2} \tag{A31}$$

and

$$\alpha=\sup_{x,\tau}\left(\frac{\|\underset{\sim}{H}x\|_\tau}{\|x\|_\tau}\right)<\infty \quad (\text{since } \underset{\sim}{H} \text{ is stable}). \tag{A32}$$

Conversely, suppose (A27) holds. Let X_0 and ω_0 be arbitrary. Then, letting $x(t)\to X_0\cos(\omega_0 t+\phi)$ and $\tau\to\infty$, it follows that

$$\|S(j\omega_0)(H(j\omega_0)-C(j\omega_0))X_0\|$$

$$\leq\|R(j\omega_0)X_0\|_E^2-\epsilon(\|X_0\|_E^2+\|H(j\omega_0)X_0\|_E^2)$$

$$\leq\|R(j\omega_0)X_0\|_E^2-\epsilon\|X_0\|_E^2. \tag{A33}$$

□

Lemma A5 (Nonlinear Time-Varying Conicity): Let $h(x(t),t)$ be any function of $x(t)$ and t and let $\underset{\sim}{H}$ be given by

$$(\underset{\sim}{H}x)(t)=h(x(t),t). \tag{A34}$$

Let C, R, and S be constant matrices and let $\underset{\sim}{C}$, $\underset{\sim}{R}$, and $\underset{\sim}{S}$ be the operators defined by

$$(\underset{\sim}{C}x)(t)=Cx(t) \quad \forall x \tag{A35}$$

$$(\underset{\sim}{R}x)(t)=Rx(t) \quad \forall x \tag{A36}$$

$$(\underset{\sim}{S}y)(t)=Sy(t) \quad \forall y. \tag{A37}$$

Suppose S^{-1} exists; then

$$\underset{\sim}{H} \text{ strictly inside } L_{2e}\text{-Cone}(\underset{\sim}{C},\underset{\sim}{R},\underset{\sim}{S}) \tag{A38}$$

if and only if

$$\|S(h(x(t),t)-Cx(t))\|_E^2\leq\|Rx(t)\|_E^2-\epsilon\|x(t)\|_E^2 \quad \forall x(t),t. \tag{A39}$$

Proof: Let $y(t)=h(x(t),t)$.

Suppose (A39) holds. Then,

$$\|y(t)\|\leq\alpha\|x(t)\| \tag{A40}$$

where

$$\alpha=\frac{\sigma_{\max}(R)}{\sigma_{\min}(S)}+\sigma_{\max}(C). \tag{A41}$$

Thus, taking

$$\epsilon'=\frac{\epsilon}{1+\alpha^2} \tag{A42}$$

we have that

$$\|\underset{\sim}{S}(y-\underset{\sim}{C}x)\|_\tau^2=\int_0^\tau\|S(h(x(t),t)-Cx(t)\|_E^2\,dt$$

$$\leq\int_0^\tau\|Rx(t)\|_E^2-\epsilon\|x(t)\|_E^2\,dt$$

$$\leq\int_0^\tau\|Rx(t)\|_E^2-\epsilon'(\|x(t)\|_E^2+\|y(t)\|_E^2\,dt$$

$$=\|\underset{\sim}{R}x\|_\tau^2-\epsilon(\|x\|_\tau^2+\|y\|_\tau^2). \tag{A43}$$

Conversely, when (A38) holds, then taking $x(t)$ to be the constant function $x(t) \equiv x_0$ we have that for some $\epsilon > 0$

$$\begin{aligned} \| S(h(x_0, t) - Cx_0) \|_E^2 &= \frac{1}{\tau} \cdot \| \underset{\sim}{S}(\underset{\sim}{H}x - \underset{\sim}{C}x) \|_\tau^2 \\ &\leq \frac{1}{\tau} \left(\| \underset{\sim}{R}x \|_\tau^2 - \epsilon \left(\| x \|_\tau^2 + \| y \|_\tau^2 \right) \right. \\ &\leq \frac{1}{\tau} \left(\| \underset{\sim}{R}x \|_\tau^2 - \epsilon \| x \|_\tau^2 \right) \\ &= \| Rx_0 \|_E^2 - \epsilon \| x_0 \|_E^2. \end{aligned} \tag{A44}$$

□

REFERENCES

[1] A. G. J. MacFarlane and I. Postlethwaite, "The generalized Nyquist criterion and multivariable root loci," *Int. J. Contr.* vol. 25, no. 1, pp. 81-127, 1977.

[2] A. G. J. MacFarlane and J. J. Belletrutti, "The characteristic locus design method," *Automatica*, vol. 9, pp. 575-588, 1973.

[3] H. H. Rosenbrock, *Computer-Aided Control System Design*. New York: Academic, 1974.

[4] H. H. Rosenbrock and P. A. Cook, "Stability and the eigenvalues of $G(s)$," *Int. J. Contr.*, vol. 21, no. 1, pp. 99-104, 1975.

[5] G. Zames, "On the input-output stability of time-varying nonlinear feedback systems—Part I: Conditions using concepts of loop gain conicity, and positivity," *IEEE Trans. Automat. Contr.*, vol. AC-11, pp. 228-238, 1966.

[6] I. W. Sandberg, "On the L_2-boundedness of solutions of nonlinear functional equations," *Bell Syst. Tech. J.*, vol. 43, pp. 1581-1599, 1964.

[7] —, "A frequency-domain condition for the stability of feedback systems containing a single time-varying nonlinear element," *Bell Syst. Tech. J.*, vol. 43, pp. 1601-1608, 1964.

[8] C. A. Desoer and M. Vidyasagar, *Feedback Systems: Input-Output Properties*. New York: Academic, 1975.

[9] K. S. Narendra and J. H. Taylor, *Frequency-Domain Cirteria for Absolute Stability*. New York: Academic, 1973.

[10] A. N. Michel and R. K. Miller, *Qualitative Analysis of Large Scale Systems*. New York: Academic, 1977.

[11] M. Araki, "Stability of large-scale nonlinear systems—quadratic order theory of composite system method using M-matrices," *IEEE Trans. Automat. Contr.*, vol. AC-23, pp. 129-142, 1978.

[12] P. J. Moylan and D. J. Hill, "Stability criteria for large-scale systems," *IEEE Trans. Automat. Contr.*, vol. AC-23, pp. 143-149, 1978.

[13] M. G. Safonov, "Robustness and stability aspects of stochastic multivariable feedback system design," Ph.D. dissertation, Mass. Inst. Technol., Cambridge, MA, Sept. 1977; also, Electron. Syst. Lab., Mass. Inst. Technol., Cambridge, MA, Rep. ESL-R-763, Sept. 1977.

[14] M. G. Safonov and M. Athans, "A multiloop generalization of the circle stability criterion," in *Proc. Asilomar Conf. on Circuits, Syst. and Comput.*, Pacific Grove, CA, Nov. 6-8, 1978.

[15] J. C. Doyle, "Robustness of multiloop linear feedback systems," in *Proc. 1978 IEEE Conf. on Decision and Contr.*, San Diego, CA, Jan. 10-12, 1979.

[16] G. Stein and J. C. Doyle, "Singular values and feedback: Design examples," in *Proc. Allerton Conf. on Commun. Contr., and Comput.*, Monticello, IL, Oct. 4-6, 1978.

[17] N. R. Sandell, "Robust stability of multivariable feedback systems," in *Proc. Allerton Conf. on Commun. Contr., and Comput.*, Monticello, IL, Oct. 4-6, 1978.

[18] A. J. Laub, "Computational aspects of singular value decomposition and some applications," in *Proc. Allerton Conf. on Commun., Contr., and Comput.*, Monticello, IL, Oct. 4-6, 1978.

[19] M. G. Safonov and M. Athans, "On stability theory," in *Proc. IEEE Conf. on Decision and Contr.*, San Diego, CA, Jan. 10-12, 1979.

[20] M. G. Safonov, *Stability and Robustness of Multivariable Feedback Systems*. Cambridge, MA: MIT Press, to be published.

[21] ——"Tight bounds on the response of multivariable systems with component uncertainty," in *Proc. Allerton Conf. on Commun. Contr., and Comput.*, Monticello, IL, Oct. 4-6, 1978.

[22] G. Zames, "On the input-output stability of time-varying nonlinear feedback systems—Part II: Conditions involving circles in the frequency plane and sector nonlinearities," *IEEE Trans. Automat. Contr.*, vol. AC-11, pp. 465-476, 1966.

[23] D. W. Nuzman and N. R. Sandell, Jr., "An inequality arising in robustness analysis of multivariable systems," *IEEE Trans. Automat. Contr.*, vol. AC-24, pp. 492-493, 1979.

[24] M. G. Safonov and M. Athans, "Gain and phase margin for multiloop LQG regulators," *IEEE Trans. Automat. Contr.*, AC-22, pp. 173-178, 1977.

[25] A. G. J. MacFarlane and D. F. A. Scott-Jones, "Vector gain," *Int. J. Contr.*, vol. 29, pp. 65-91, 1979.

[26] L. Zadeh and C. A. Desoer, *Linear System Theory*. New York: McGraw-Hill, 1963.

[27] G. Zames, "Nonlinear operators for system analysis," Res. Lab. for Electron., Mass. Inst. Technol., Cambridge, MA, Tech. Rep. 370, Sept. 1960.

[28] —, "Functional analysis applied to nonlinear feedback systems," *IEEE Trans. Circuit Theory*, vol. CT-10, pp. 392-404, 1963.

[29] —, "On the stability of nonlinear, time-varying feedback systems," *Proc. NEC*, vol. 20, pp. 725-730, Oct. 1964.

Robustness Results in Linear-Quadratic Gaussian Based Multivariable Control Designs

NORMAN A. LEHTOMAKI, STUDENT MEMBER, IEEE, NILS R. SANDELL, JR., MEMBER, IEEE, AND MICHAEL ATHANS, FELLOW, IEEE

***Abstract*—The robustness of control systems with respect to model uncertainty is considered using simple frequency domain criteria. Available and new results are derived under a common framework in which the minimum singular value of the return difference transfer matrix is the key quantity. In particular, robustness results associated with multivariable control systems designed on the basis of linear-quadratic (LQ) and the linear-quadratic Gaussian (LQG) design methodologies are presented.**

I. INTRODUCTION

THE importance of obtaining robustly stable feedback control systems has long been recognized by designers. Indeed, a principal reason for using feedback rather than open-loop control is the presence of model uncertainties. Any model is at best an approximation of reality, and the relatively low-order linear time-invariant models most often used for controller synthesis are bound to be rather crude approximations.

More specifically, a given system model can usually be characterized as follows. There is a certain range of inputs typically bounded in amplitude and in rate of change for which the model gives a reasonable approximation to the system. Outside of this range, due to neglected nonlinearities and dynamic effects, the model and system may behave in grossly different ways. Unfortunately, this range of permissible inputs is rarely spelled out explicitly along with the model, but is rather implicit in the technology that the model came from—there is no "truth in modeling" law in systems theory.

In classical frequency domain techniques for single-input single-output (SISO) control system design, the robustness issue is naturally handled.[1] These techniques employ various graphical means (e.g., Bode, Nyquist, inverse Nyquist, Nichols plots) of displaying the system model in terms of its frequency response. From these plots, it is automatic to determine by inspection the minimum change in the model frequency response that leads to instability.

Manuscript received March 4, 1980; revised October 6, 1980. This work was supported by DOE under Contract DE-AC01-78RA03395, by NASA Ames under Grant NGL-22-009-124, and by ONR under Contract ONR/N0014-79-C-0377.

N. A. Lehtomaki and M. Athans are with the Laboratory for Information and Decision Systems, Massachusetts Institute of Technology, Cambridge, MA 02139.

N. R. Sandell, Jr. is with Alphatech Inc., Burlington, MA 01803.

[1]See the fundamental work of Bode [6], and any good classical textbook, but especially [9].

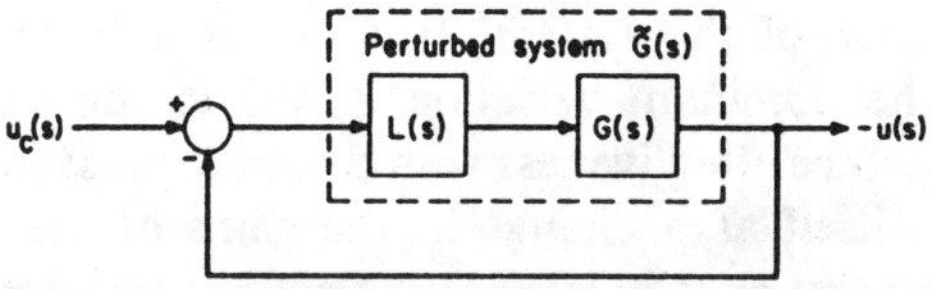

Fig. 1. Feedback system with multiplicative representation of uncertainty in $G(s)$.

Commonly used measures of the closeness of an SISO feedback system to instability are its gain and phase margins. The margins are defined with reference to Fig. 1. Here the nominal feedback system (with $L(s)\equiv 1$) is assumed stable. The positive phase margin is the smallest value of ϕ greater than 0 such that the system of Fig. 1 with $L(j\omega)=e^{j\phi}$ is unstable. The negative phase margin is defined in an analogous fashion. The upward gain margin is the smallest value of $L(s)=\text{constant}>1$ for which the system is unstable (usually expressed in decibels with respect to $L(s)=+1$), and the downward gain margin is similarly defined. The notions of gain and phase margins have gained such widespread acceptance that they are commonly incorporated into the specifications for a control system design.

To give a concrete and explicit example, consider the military specifications on the design of flight control systems for piloted aircraft. Among other requirements, a feedback control systems must have certain gain and phase margins in order to be acceptable. To quote from the military specifications:

> "Stability margins are required for FCS to allow for variations in system dynamics. Three basic types of variations exist:
> - Math modeling and data errors in defining the nominal system and plant.
> - Variations in dynamic characteristics caused by changes in environmental conditions, manufacturing tolerances, aging, wear, noncritical material failures, and off-nominal power supplies.
> - Maintenance induced errors in calibration, installation and adjustment."

It would seem from the above quotation that the robustness issue is well-understood from the viewpoint of classical frequency domain techniques, at least for flight control systems. Indeed, this is the case, *for single-loop*

Reprinted from *IEEE Trans. Automat. Contr.*, vol. AC-26, no. 1, pp. 75–92, Feb. 1981.

systems. However, the situation is quite different for multiple-loop systems. To quote again from the military specifications:

> "In multiple-loop systems, variations shall be made with all gain and phase values in the feedback paths held at nominal values except for the path under investigation."

The fundamental difficulty with this approach is that it fails to check the effect of *simultaneous* gain and phase variations in several paths. Of course, real-world model uncertainty cannot be expected to nicely confine itself to a single loop of the system! In fact, for a flight control system, the dominant variation is due to the change in control surface effectiveness with dynamic pressure, which manifests itself as a change in the gains of the transfer functions from control surface deflections to the response variables of interest. The dynamic pressure variation is due to changes in aircraft altitude and speed, and clearly affects all loops simultaneously.

From this discussion, it is clear that a satisfactory notion of stability margins for a multivariable feedback system must be able to characterize the ability of the system to tolerate gain and phase variations in all its loops simultaneously. It is only very recently, in the context of studying the feedback properties of controllers derived using linear-quadratic Gaussian (LQG) techniques, that an appropriate formulation has emerged. The purpose of this paper is to develop the ideas of this formulation, and to derive in a relatively simple way the robustness properties of LQG controllers.

The paper begins in Section II by developing a characterization of the robustness of a multivariable feedback system in terms of the minimum singular value of its return difference transfer function matrix. This quantity is a natural one to consider as it directly generalizes the classical notion of the distance of the Nyquist locus of a single-input feedback system to the critical (-1) point. The characterization is original, and is derived rather simply from arguments based on the multivariable Nyquist theorem. Multivariable gain, phase, and crossfeed margins are derived in terms of the minimum singular value.

The development in Section II permits an efficient derivation of the robustness properties of linear-quadratic (LQ) controllers in Section III since the Kalman frequency domain inequality provides a bound on the minimum singular value of an optimal return difference transfer function matrix. The dependence of these robustness properties on the form of the control weighting matrix R is illustrated by introducing a counterexample of an LQ (*not* LQG) regulator with vanishingly small gain margins. The robustness degradation associated with introducing a Kalman filter into a feedback system are briefly discussed, and robustness recovery procedures are mentioned.

The paper closes in Section IV with a summary and discussion of the results of the paper. Also, since singular values are used throughout the paper, a brief discussion of them is given in the Appendix. The Appendix also contains the proofs of the theorems and corollaries.

NOTATION

(A, B, C)	realization of linear system specified by the time domain description $\dot{x}=Ax+Bu$, $y=Cx$,
$G(s)$	loop-transfer matrix,
$\phi_{OL}(s)$	open-loop characteristic polynomial,
$\phi_{CL}(s)$	closed-loop characteristic polynomial,
$L(s)$	multiplicative perturbation of $G(s)$,
$\lambda(A)$	an eigenvalue of A,
A^H	complex conjugate transpose of A,
$\bar{\sigma}(A)$	maximum singular value of $A \triangleq \lambda^{1/2}_{\max}(A^H A)$,
$\underline{\sigma}(A)$	minimum singular value of $A \triangleq \lambda^{1/2}_{\min}(A^H A)$,
$\mathcal{N}(\Omega, f(s), \mathcal{C})$	number of clockwise encirclements of the point Ω by the locus of $f(s)$ as s traverses the closed contour $\mathcal{C}$ in the complex plane in a clockwise sense,
D_R	Nyquist contour of radius R given in Fig. 2 with $(1/R)$-radius indentations,
Ω_R	segment of D_R for which $\mathrm{Re}[s]<0$.
SISO	single-input single-output,
MIMO	multiple-input multiple-output,
ORHP(CRHP)	open (closed) right half plane,
OLHP(CLHP)	open (closed) left half plane,
$A>B$	$A-B$ is positive definite matrix,
$A \geq B$	$A-B$ is positive semidefinite matrix,
$\Leftrightarrow$	if and only if,
$\exists$	there exists,
$\ni$	such that,
$\Rightarrow$	implies,
$\triangleq$	defined as,
LQ	linear-quadratic,
LQG	linear-quadratic Gaussian
KF	Kalman filter,
z^*	complex conjugate of z
$\forall$	for all.

II. MULTIVARIABLE STABILITY MARGINS

In this section we will develop the basic characterizations of the robustness of linear multivariable feedback systems, i.e., multivariable stability margins, to be used in the remainder of the paper. The feedback system to be discussed is depicted in Fig. 1 with $L(s)\equiv I$, where the loop transfer matrix $G(s)$ is assumed to incorporate both the plant dynamics and any compensation employed, and has the state space realization

$$G(s)=C(sI-A)^{-1}B. \tag{2.1}$$

The basic issue of concern is to characterize the robustness of the feedback system, i.e., the extent to which the elements of the loop transfer function matrix $G(s)$ can vary from their nominal design values without comprom-

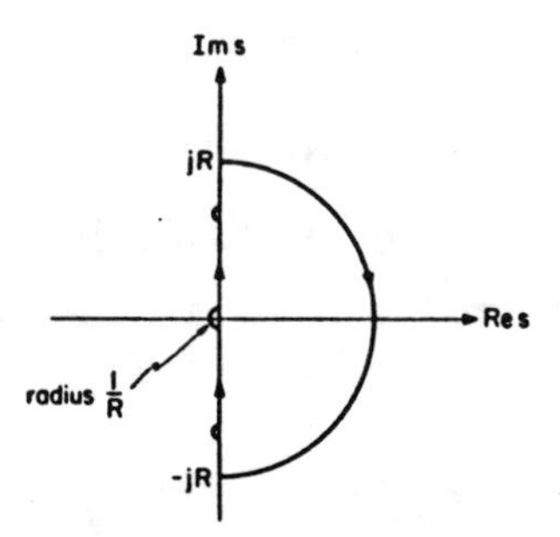

Fig. 2. Nyquist contour D_R which encloses all zeros of $\phi_{OL}(s)$ in the CRHP, avoiding imaginary zeros by indentations of radius of $1/R$.

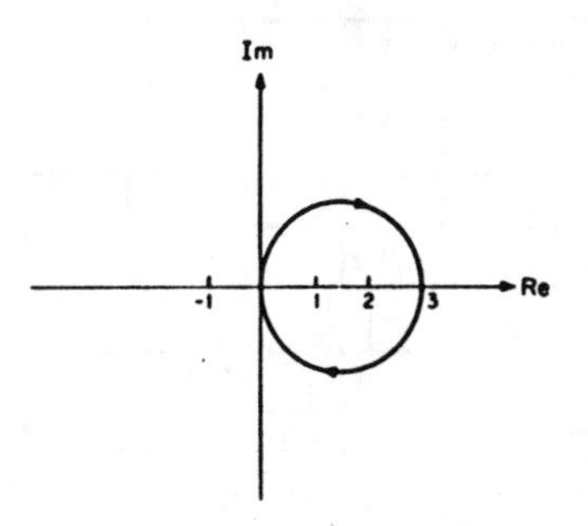

Fig. 3. Nyquist diagram of $2s+3/(s+1)^2$.

ising the stability of the system. The analysis is based on the multivariable Nyquist theorem.

A. Multivariable Nyquist Theorem

The multivariable Nyquist theorem is derived from the relationship

$$\det(I+G(s))=\frac{\phi_{CL}(s)}{\phi_{OL}(s)} \tag{2.2}$$

where

$$\phi_{OL}(s)=\det(sI-A) \tag{2.3}$$

$$\phi_{CL}(s)=\det(sI-A+BC) \tag{2.4}$$

and from the Principle of the Argument of complex variable theory. The following statement of the Nyquist theorem is a variation of a version given by Rosenbrock in [1].

Theorem 2.1: The system of Fig. 1 with $L(s)\equiv I$, is closed-loop asymptotically stable (in the sense that $\phi_{CL}(s)$ has no CRHP zeros) if and only if for all R sufficiently large

$$\mathfrak{N}(0,\det[I+G(s)],D_R)=-P \tag{2.5}$$

or equivalently,

$$\mathfrak{N}(-1,-1+\det[I+G(s)],D_R)=-P \tag{2.6}$$

where D_R is the contour of Fig. 2 which encloses all CRHP zeros of $\phi_{OL}(s)$, P is the number of CRHP zeros of $\phi_{OL}(s)$ and where for convenience, we define $\mathfrak{N}(\Omega,f(s),\mathcal{C})\triangleq\frac{1}{2}$ when $f(s_0)=\Omega$, for some $s_0\in\mathcal{C}$. ▲

Notice that no controllability or observability assumptions have been made. If $[A,B,C]$ is a nonminimal realization then pole-zero cancellations will occur when $G(s)$ is formed, eliminating uncontrollable or unobservable modes. The stability of these modes cannot be tested in terms of $G(s)$. However, by using the zeros of $\phi_{OL}(s)$ instead of the poles of the loop transfer matrix $G(s)$, this version of the Nyquist theorem allows one to test for the internal stability of the closed-loop system.

When compared with the classical Nyquist theorem for the SISO case, the multivariable Nyquist theorem is much more difficult to use, for two reasons. First, the dependence of $\det(I+G(s))$ on the compensation implicit in $G(s)$ is complicated, and cannot be easily depicted with a Nyquist, Bode, or related plot. This fact has motivated a considerable amount of research on synthesis methods, e.g., [2]–[5], which need not concern us here. Second, and this is the key observation, one cannot get a satisfactory notion of multivariable stability margins *directly* from the multivariable Nyquist theorem. The following extremely simple example illustrates this fact.

Example 2.1: Consider the linear system specified by

$$\begin{bmatrix}\dot{x}_1\\ \dot{x}_2\end{bmatrix}=\begin{bmatrix}-1 & 0\\ 0 & -1\end{bmatrix}\begin{bmatrix}x_1\\ x_2\end{bmatrix}+\begin{bmatrix}1 & b_{12}\\ 0 & 1\end{bmatrix}\begin{bmatrix}u_1\\ u_2\end{bmatrix} \tag{2.7}$$

$$\begin{bmatrix}y_1\\ y_2\end{bmatrix}=\begin{bmatrix}x_1\\ x_2\end{bmatrix} \tag{2.8}$$

which is illustrated in Fig. 4 by the nominal plant.

If the feedback compensation

$$\begin{bmatrix}u_1\\ u_2\end{bmatrix}=-\begin{bmatrix}x_1\\ x_2\end{bmatrix}+\begin{bmatrix}u_{C1}\\ u_{C2}\end{bmatrix} \tag{2.9}$$

is used the closed-loop system is given by

$$\begin{bmatrix}\dot{x}_1\\ \dot{x}_2\end{bmatrix}=\begin{bmatrix}-2 & -b_{12}\\ 0 & -2\end{bmatrix}\begin{bmatrix}x_1\\ x_2\end{bmatrix}+\begin{bmatrix}1 & b_{12}\\ 0 & 1\end{bmatrix}\begin{bmatrix}u_{C1}\\ u_{C2}\end{bmatrix}. \tag{2.10}$$

The eigenvalues of this system are -2, -2 and therefore it is stable. The return difference matrix $I+G(s)$ is given by

$$I+G(s)=\begin{bmatrix}\frac{s+2}{s+1} & \frac{b_{12}}{s+1}\\ 0 & \frac{s+2}{s+1}\end{bmatrix} \tag{2.11}$$

and thus

$$\det(I+G(s))-1=\frac{2s+3}{(s+1)^2}. \tag{2.12}$$

The multivariable Nyquist diagram is just the usual Nyquist diagram of $2s+3/(s+1)^2$ and is shown in Fig. 3. If this is interpreted as for a SISO system, one would conclude that the system has an infinite upward gain margin, a gain reduction margin of $-\frac{1}{3}$ and a phase

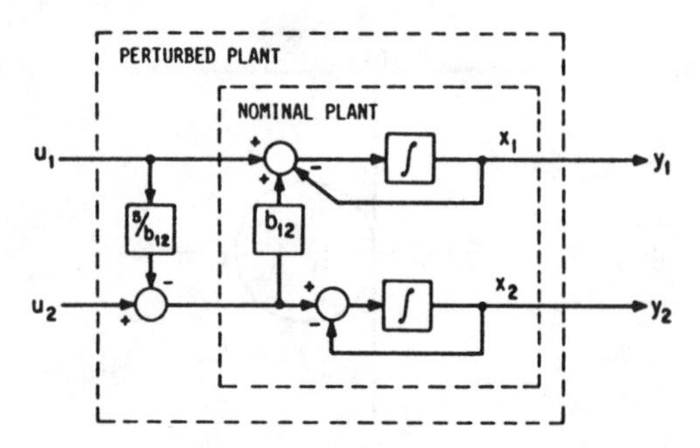

Fig. 4. Perturbation in nominal open-loop system which makes closed-loop system unstable.

margin of $\pm 106°$ margins indicative of a highly robust system. Note, however, that the Nyquist diagram does not depend on the value of b_{12} and as b_{12} becomes large the closed-loop system is close to instability in the following sense. If the nominal open-loop system of Fig. 4 is perturbed slightly to obtain the perturbed system of Fig. 4, the closed-loop system obtained by negative identity feedback is unstable (as is easily verified). This situation *cannot* be detected by inspection of the multivariable Nyquist theorem.[2]

The difficulty we have uncovered can be explained as follows. A multivariable system will not be robust to modeling errors if its return difference transfer function matrix $I+G(j\omega)$ is nearly singular at some frequency ω_c, since then a small change in $G(j\omega_c)$ will make $I+G(j\omega_c)$ exactly singular. When this happens, $\det(I+G(j\omega_c))=0$ and the number of encirclements counted in the multivariable Nyquist theorem changes. In this example a small change in $I+G(j\omega_c)$ produces a large change in $\det(I+G(j\omega_c))$ showing that the near singularity of a matrix cannot always be detected in terms of its determinant. Instead, tests such as those developed in the remainder of this section must be employed.

B. Robustness Characterization

From the above example, we can see that the problem of determining characterizations of the robustness of a multivariable feedback system, i.e., its distance from instability, is of fundamental importance. The basic work in this area is due to Safonov [7], [8], who generalized an approach of Zames [10], [11]. Safonov's work heavily utilizes concepts of functional analysis, as is standard in the modern input–output formulation of stability theory.[3] However, in the finite-dimensional linear time-invariant case considered in this paper, powerful robustness characterization can be derived more simply in terms of the multivariable Nyquist theorem.

The feedback system under consideration in the following theorems is shown in Fig. 1 where $G(s)$ again represents the loop-transfer matrix (open loop plant plus controller) and $L(s)$ is a multiplicative perturbation applied to $G(s)$ to account for uncertainty in the open-loop plant model.

We denote the perturbed system as $\tilde{G}(s)$ given by

$$\tilde{G}(s)=G(s)L(s) \tag{2.13}$$

which is assumed to have a state-space realization $(\tilde{A}, \tilde{B}, \tilde{C})$, open-loop characteristic polynomial $\tilde{\phi}_{OL}(s)$ given by

$$\tilde{\phi}_{OL}(s)=\det(sI-\tilde{A}) \tag{2.14}$$

and similarly a closed-loop characteristic polynomial $\tilde{\phi}_{CL}(s)$ given by

$$\tilde{\phi}_{CL}(s)=\det(sI-\tilde{A}+\tilde{B}\tilde{C}). \tag{2.15}$$

All the results of this section are based on the following theorem.

Theorem 2.2: The polynomial $\tilde{\phi}_{CL}(s)$ of (2.14) has no CRHP zeros if the following conditions hold:

a) $\phi_{OL}(s)$ and $\tilde{\phi}_{OL}(s)$ have the same number of CRHP zeros, (2.16)

b) if $\tilde{\phi}_{OL}(j\omega_0)=0$ then $\phi_{OL}(j\omega_0)=0$, (2.17)

c) $\phi_{CL}(s)$ has no CRHP zeros, (2.18)

d) $\det[I+(I-\epsilon)G(s)+\epsilon\tilde{G}(s)]\neq 0$, $s\in D_R$ for all R sufficiently large and for all ϵ on the interval [0, 1]. (2.19)

▲

The basic idea behind this theorem is that of continuously deforming the Nyquist diagram of the nominal system $G(s)$ to one corresponding to the Nyquist diagram of the perturbed system $\tilde{G}(s)$ without changing the number of encirclements of the critical point. If this can be done and the number of encirclements of the critical point required for $\tilde{G}(s)$ and $G(s)$ are the same, then no CRHP zeros of $\tilde{\phi}_{CL}(s)$ will result from this perturbation. Imbedding arguments of this type have been previously used, implicitly by Rosenbrock [1] and explicitly by Doyle [14], in connection with linear systems and in the more general context of nonlinear or multidimensional systems by DeCarlo, Saeks, and Murray [15]–[17], utilizing homotopy theory from algebraic topology.

The significance of Theorem 2.2 is that various multivariable robustness characterizations can be stated in terms of conditions that guarantee that (2.19) is satisfied. In checking condition (2.19), it is unnecessary to consider all $s\in D_R$ because $\bar{\sigma}[G(s)]$ and $\bar{\sigma}[\tilde{G}(s)]\to 0$ as $|s|\to\infty$. This is due to the assumption that their respective state-space realizations have no direct feedthrough from input to output so that there are more poles than zeros in both $\tilde{G}(s)$ and $G(s)$. It is therefore convenient to define Ω_R as

$$\Omega_R \triangleq \{s \mid s\in D_R \text{ and } \mathrm{Re}(s)<0\} \tag{2.20}$$

which is the only part of the Nyquist contour on which (2.19) need be verified. Using this simplification, one characterization that guarantees (2.19) is based on the return difference matrix $I+G(s)$. This result is new, although related to conicity conditions discussed by Zames [10] and Safonov [7], and is fundamental to the derivation of the LQ state-feedback and LQG stability margins.

[2] This situation cannot be detected by classical, single-loop tests or by characteristic loci plots [5] either.

[3] See, e.g., [12] or [13].

We emphasize that the following theorem is distinct from the main results in [14], [18]. These papers work with $\underline{\sigma}(I+G^{-1}(s))$ (for a multiplicative representation of model uncertainty), which is complementary to the quantity $\underline{\sigma}(I+G(s))$, which measures the distance between the Nyquist locus and the critical (-1) point in the SISO case. Moreover, these papers work with $\bar{\sigma}(L(s)-I)$ (in our notation) rather than $\bar{\sigma}(L^{-1}(s)-I)$. As a consequence, it is *not* possible to derive the gain and phase margin properties of LQ and LQG controllers using the results stated in [14], [18].

Theorem 2.3: The polynomial $\tilde{\phi}_{CL}(s)$ has no CRHP zeros if the following conditions hold for all R sufficiently large:

a) conditions (2.16)–(2.18) of Theorem 2.2 hold (2.21)

b) $\bar{\sigma}(L^{-1}(s)-I)<\alpha \triangleq \underline{\sigma}(I+G(s)), s\in\Omega_R$ (2.22)

c) any one of the following is satisfied at each $s\in\Omega_R$

i) $\alpha\leq 1$ (2.23)

ii) $L^H(s)+L(s)\geq 0$ (2.24)

iii) $4(\alpha^2-1)\underline{\sigma}^2(L(s)-I)>\alpha^2\bar{\sigma}^2(L(s)+L^H(s)-2I)$. (2.25)

▲

An analytical proof of Theorem 2.3 and a discussion of singular values are provided in the Appendix. It is, however, instructive to given a simple graphical proof for the SISO case.

By the embedding argument of Theorem 2.2, our nominal feedback system and its perturbed version will have the same number of encirclements of the critical (-1) point if we can continuously deform the nominal into the perturbed Nyquist locus. This will be the case if, for $0\leq\epsilon\leq 1$, we have

$$1+[\epsilon l(s)+(1-\epsilon)]g(s)\neq 0 \qquad s\in\Omega_R \tag{2.26}$$

or equivalently

$$-\frac{1}{g(s)}\neq\epsilon l(s)+(1-\epsilon), \qquad s\in\Omega_R. \tag{2.27}$$

In the scalar case, the inequality

$$|1+g(s)|\geq\alpha, \tag{2.28}$$

shown in Fig. 5(d), simply states that $g(s)$ lies outside or on a circle of radius α about the -1 point. Consequently, when $|1+g(s)|=\alpha$, $-1/g(s)$ will lie on a circle of radius α/α^2-1 (infinite when $\alpha=1$) centered at $1/1-\alpha^2$ as illustrated in Fig. 5(a), (b), and (c) for various α. Then, the allowable values of $l(s)$ are those that can be connected to the $+1$ point by a straight line not intersecting the circle of possible boundary values of $-[1/g(s)]$.

We emphasize that α varies as a function of $s\in\Omega_R$, so that Fig. 5 only represents the situation at a single (complex) frequency. Similarly, the condition (2.22) of Theorem 2.3 must be tested for all $s\in\Omega_R$. This is most readily accomplished by computing and plotting $\bar{\sigma}(L^{-1}(s)-I)$ and $\underline{\sigma}(I+G(s))$. The situation is roughly analogous to looking at a magnitude Bode plot. Computational techniques are discussed in [19].

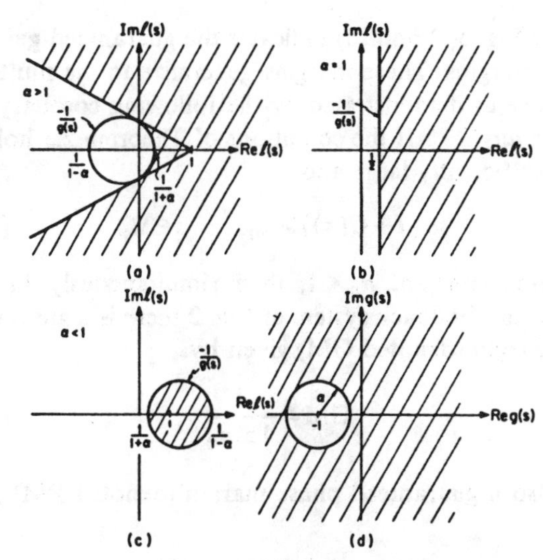

Fig. 5. Set of allowable values of $l(s)$ and $g(s)$ for various α in Theorem 2.3.

To show that condition (2.22), in Theorem 2.3, alone is not enough to guarantee stability of the perturbed closed-loop system of Fig. 1, a simple counterexample is given.

Example 2.2: Let

$$g(s)=\frac{9}{s+1} \tag{2.29}$$

$$\phi_{OL}(s)=s+1 \tag{2.30}$$

$$\phi_{CL}(s)=s+10 \tag{2.31}$$

$$l(s)=\frac{s-1}{s+1} \tag{2.32}$$

$$\tilde{\phi}_{OL}(s)=(s+1)^2, \tag{2.33}$$

then

$$|l^{-1}(j\omega)-1|<|1+g(j\omega)| \qquad \forall\omega \tag{2.34}$$

but

$$\tilde{\phi}_{CL}(s)=s^2+11s-8 \tag{2.35}$$

has a CRHP zero.

From Theorem 2.3, it is clear that the quantity $\underline{\sigma}(I+G(s))$ is a multivariable stability margin for the feedback system at the complex frequency s. However, this quantity is unconventional, and possibly somewhat difficult to interpret. Therefore, we will explore further some of the consequences of the theorem.

Note from Figs. 5(a)–(d) that in the SISO case a system satisfying the conditions of Theorem 2.3 will automatically have certain guaranteed minimum gain and phase margins if α (i.e., $|1+g(s)|$) as a function of $s\in\Omega_R$ has a constant nonzero lower bound, say α_0. If $\alpha_0=1$, then Fig. 5(d) shows that $g(s)$ must have exactly a one-pole roll-off. It is well known (see, e.g., [20]) that physical systems always exhibit at least a two pole roll-off which again from Fig. 5(d) indicates that $\alpha_0<1$.

The case $\alpha_0>1$ is inconsistent with the assumption that the state-space realization of $g(s)$ does not have a feedthrough term. Thus for $\alpha_0\leq 1$ examination of Fig.

5(b) or Fig. 5(c) and (d) indicates the guaranteed gain and phase margins. These margins generalize to the multivariable case as demonstrated by the following corollary.

Corollary 2.1: If the conditions of Theorem 2.3 hold for all R sufficiently large and

$$\underline{\sigma}(I+G(s))>\alpha_0, \qquad s\in\Omega_R \tag{2.36}$$

for some constant $\alpha_0<1$, then simultaneously in each loop of the feedback system of Fig. 2 there is a guaranteed gain margin (denoted GM) given by

$$\mathrm{GM}=\frac{1}{1\pm\alpha_0} \tag{2.37}$$

and also a guaranteed phase margin (denoted PM) given by

$$\mathrm{PM}=\pm\cos^{-1}\left[1-\frac{\alpha_0^2}{2}\right]. \tag{2.38}$$

▲

The interpretation of the gain and phase margin quantities specified in Corollary 2.1 require some explanation. First of all, the word "simultaneously" in Corollary 2.1, means that the gains or the phases of all the feedback loops may be changed at the same time within the limits prescribed by (2.37) and (2.38) without destabilizing the closed-loop system. It does *not* mean, just as it does not in SISO case, that the gains *and* phases may be changed simultaneously. Secondly, (2.37) is to be interpreted as meaning any gains, γ_i, inserted in the feedback loops of the system of Fig. 1 satisfying

$$\frac{1}{1+\alpha_0}<\gamma_i<\frac{1}{1-\alpha_0} \tag{2.39}$$

will not destabilize the closed-loop system. Similarly for (2.38), every loop may have a phase factor $e^{j\phi_i}$ inserted provided

$$|\phi_i|<\cos^{-1}\left[1-\frac{\alpha_0^2}{2}\right] \tag{2.40}$$

and the system will remain closed-loop stable.

The ability to consider simultaneous gain or phase variations in *all* the loops of a multivariable feedback system is physically very appealing. A typical example is a flight control system, in which the effectiveness of all control surfaces varies simultaneously as a function of altitude and airspeed. Another common model uncertainty in flight control applications is a crossfeed arising from a neglected interaxis coupling. Bounds on the ability of a multivariable system to tolerate crossfeed uncertainty are given in the following result.

Corollary 2.2: Provided the conditions of Theorem 2.3 are satisfied for all R sufficiently large and

$$\underline{\sigma}(I+G(s))>\alpha_0, \qquad s\in\Omega_R \tag{2.41}$$

for some constant $\alpha_0<1$, then the feedback system of Fig. 1 will tolerate a crossfeed perturbation of the form

$$L(s)=\begin{bmatrix} I_k & X(s)\\ 0 & I_m\end{bmatrix} \quad\text{or}\quad \begin{bmatrix} I_k & 0\\ X(s) & I_m\end{bmatrix} \tag{2.42}$$

where I_k is the $k\times k$ identity and

$$\bar{\sigma}(X(s))<\alpha_0, \qquad s\in\Omega_R \tag{2.43}$$

without becoming unstable, that is $\tilde{\phi}_{CL}(s)$ will have no CRHP zeros. ▲

There is one final result that will be of use to us in the next section. This theorem involves the well-known passivity condition [12], [13].

Theorem 2.4: The polynomial $\tilde{\phi}_{CL}(s)$ has no CRHP zeros if the following conditions hold for all R sufficiently large:

$$\text{a) conditions (2.16)–(2.18) of Theorem 2.2 hold} \tag{2.44}$$

$$\text{b) } G(s)+G^H(s)>0,\ s\in\Omega_R \tag{2.45}$$

$$\text{c) } L(s)+L^H(s)>0,\ s\in\Omega_R. \tag{2.46}$$

▲

Just as for Theorem 2.3, specializing to the SISO case illustrates the types of $G(s)$ and $L(s)$ that are required as well as the associated guaranteed minimum margins. The constraints (2.45) and (2.46) in the SISO case keep $g(s)$ and $l(s)$ from entering the OLHP and show that since $g(s)$ cannot encircle the "-1" point it must be open-loop stable in order to apply the theorem. It is fairly obvious that the phase of $g(s)l(s)$ is less than 180° and thus $g(s)l(s)\neq-1$ is assured. The gain and phase margins are also immediately apparent since $g(s)$ and $l(s)$ never enter the OLHP and again generalize to the multivariable case given in Corollary 2.3.

Corollary 2.3: If the conditions of Theorem 2.4 hold for all R sufficiently large then simultaneously in each feedback loop of the system of Fig. 1 there is a guaranteed gain margin given by

$$\mathrm{GM}=0,+\infty \tag{2.47}$$

and also a guaranteed phase margin given by

$$\mathrm{PM}=\pm 90^\circ. \tag{2.48}$$

▲

Corollary 2.4: If the conditions of Theorem 2.4 are satisfied, then the feedback system of Fig. 1 will tolerate a crossfeed perturbation of the form given in (2.42) where

$$\bar{\sigma}(X(s))<2, \qquad s\in\Omega_R \tag{2.49}$$

without becoming unstable, that is $\tilde{\phi}_{CL}(s)$ will have no CRHP zeros. ▲

Returning to Example 2.1, the nearness to instability can be easily detected using Theorem 2.3. Fig. 6 shows $\underline{\sigma}(I+G(j\omega))$ as a function of ω, where b_{12} has been selected as 50. At low frequencies $\underline{\sigma}(I+G(j\omega))$ is very small giving (see Corollary 2.1)

$$\mathrm{GM}=0.93, 1.08 \tag{2.50}$$

and

$$\mathrm{PM}=\pm 4.1^\circ \tag{2.51}$$

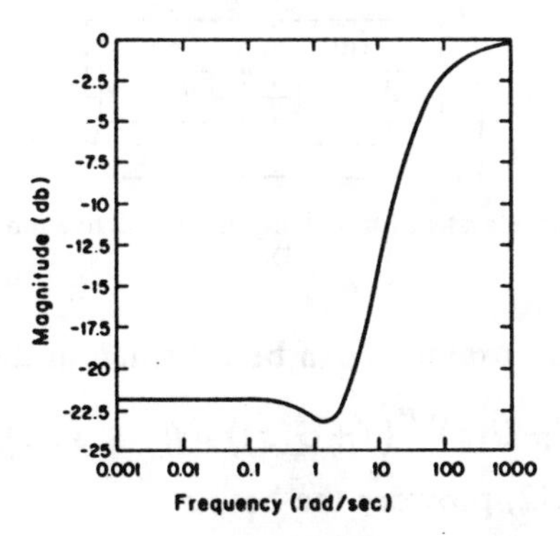

Fig. 6. Singular value plot of $\underline{\sigma}(I+G(j\omega))$ for Example 2.1 when $b_{12}=50$.

which give worst case bounds on the actual gain and phase margins. However, we know by inspection that each feedback loop in the system has an actual infinite upward gain margin and a -1 gain reduction margin which indicates that (2.50) and (2.51) are very conservative estimates of gain and phase margins. Nevertheless, they indicate a robustness problem which is exhibited by the very small crossfeed tolerance (see Corollary 2.2)

$$\bar{\sigma}(X(s))<\min_{s\in\Omega_R}\underline{\sigma}(I+G(s))\simeq 0.071=-23\text{ dB}. \quad (2.52)$$

This again is a worst case bound on the allowable amount of crossfeed but corresponds precisely to the magnitude of the perturbation given previously in Fig. 4 to show nearness to instability. In general, the type of perturbation that the feedback system is most sensitive to at each s is given by

$$L(s)=\left[I-\underline{\sigma}(I+G(s))u(s)v^H(s)\right]^{-1} \quad (2.53)$$

where $u(s)$ and $v(s)$ are, respectively, the left and right singular vectors[4] of $I+G(s)$ corresponding to $\underline{\sigma}(I+G(s))$. This is the perturbation which makes $I+G(s)L(s)$ exactly singular with minimum $\bar{\sigma}[(\tilde{G}^{-1}(s)-G^{-1}(s))G(s)]$.

Thus in Theorem 2.3 and its corollaries, if a global minimum of $\underline{\sigma}(I+G(s))$ occurs at $s_0\in\Omega_R$ then there exists a perturbation $L(s)$ that satisfies

$$\bar{\sigma}(L^{-1}(s)-I)\leq\underline{\sigma}(I+G(s_0)) \qquad \forall s\in\Omega_R \quad (2.54)$$

such that the closed-loop system is unstable. Note that $L(s)$ in (2.54) need correspond to $L(s)$ in (2.53) only at $s=s_0$ since $I+G(s)L(s)$ need only be singular at a single $s\in\Omega_R$ to induce instability of the closed-loop system.

III. LQ and LQG Stability Margins

The subject of qualitative feedback properties of LQ control systems is not a new one. An early and fundamental paper by Kalman [21] detailed properties shared by all LQ regulators in the single-input case. Kalman showed that the SISO return difference transfer function of a single-input LQ state feedback regulator satisfies the inequality

$$|1+g(j\omega)|\geq 1, \qquad \forall\omega. \quad (3.1)$$

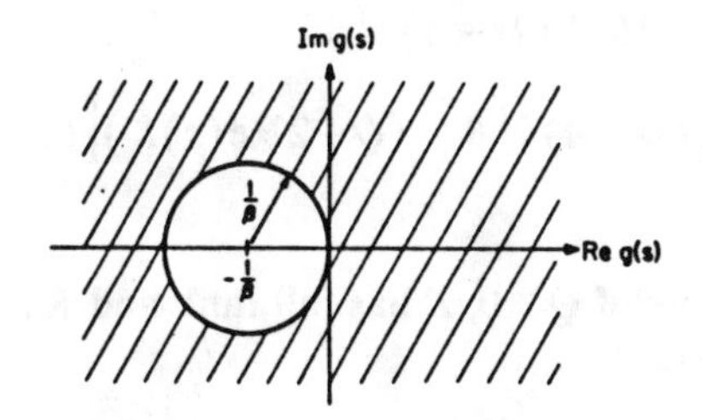

Fig. 7. Set of allowable values of $g(s)$ when $|1+\beta g(s)|>1$.

This is both a classical condition for the reduction of sensitivity at the feedback input to the system (see, e.g., [19]) as well as necessary and sufficient for a (stable) state feedback regulator to be optimal with respect to some quadratic cost index. By inspection of the Nyquist diagram corresponding to (3.1), (Fig. 7 with $\beta=1$), it is straightforward to observe [22, pp. 70–76] that a SISO LQ state feedback regulator has a guaranteed infinite upward gain margin, at least a 50 percent gain reduction margin and also a guaranteed minimum phase margin of $\pm 60°$.

Anderson [23] developed a multivariable version of condition (3.1) as a property of LQ state-feedback regulators; a similar generalized condition arises in sensitivity theory[5] (see, e.g., Cruz and Perkins [24]). In the remainder of this paper, we will exploit the multivariable form of (3.1) together with the results of the preceding section to establish the stability margin properties of LQ and LQG optimal regulators.

A. Multivariable Kalman Inequality

We will need a precise statement of the multivariable version of condition (3.1) in the sequel, and this is provided by the following theorem. The proof is by straightforward manipulation of the algebraic Riccati equation and is included, for completeness, in the Appendix.

For convenience we will assume that in all remaining theorems and corollaries that the Nyquist contour D_R is chosen with R sufficiently large so that the theorems of Section II may be applied.

Theorem 3.1: If the matrix K satisfies the matrix algebraic Riccati equation

$$A^TK+KA+Q-KBR^{-1}B^TK=0 \quad (3.2)$$

with $R>0$ and $Q\geq 0$ then

$$(I+G(s))^HR(I+G(s))=R+H(s) \quad (3.3)$$

where

[4]The left and right singular vectors $u(s)$ and $v(s)$ are, respectively, unit length eigenvectors of $(I+G(s))(I+G(s))^H$ and $(I+G(s))^H(I+G(s))$ such that $(I+G(s))v(s)=\underline{\sigma}(I+G(s))u(s)$.

[5]Sensitivity refers to the variation in system responses due to infinitesimal changes in the nominal system parameters. Robustness refers to the delineation of finite regions of allowable variation in nominal system parameters that preserve stability.

$$G(s)=R^{-1}B^TK(Is-A)^{-1}B \tag{3.4}$$

$$H(s)=\left[(Is-A)^{-1}B\right]^H(Q+2\mathrm{Re}(s)K)\left[(Is-A)^{-1}B\right]. \tag{3.5}$$

Furthermore, if $Q>0$, B has full rank and $K>0$ then (3.3) implies that

$$(I+G(s))^HR(I+G(s))>R, \qquad s\in D_R. \tag{3.6}$$

Alternatively, if $\det(j\omega I-A)\neq 0\ \forall\omega$ and $K>0$ then (3.3) implies that

$$(I+G(s))^HR(I+G(s))\geq R, \qquad s\in D_R. \tag{3.7}$$

▲

It is important to point out that this theorem uses $G^H(s)$ rather than $G^T(-s)$ as in [23]. These two quantities are the same when $s=j\omega$, but are different when $\mathrm{Re}(s)\neq 0$. This is the case when s is evaluated along the Nyquist D_R contour and this contour is indented along the imaginary axis. It is necessary to use $G^H(s)$ in order to apply the theorems of Section II. Note, however, that when $\det(j\omega I-A)\neq 0$, $\forall\omega$, that Ω_R is just the imaginary axis from $-jR$ to jR. In this case (3.7) could be written as

$$(I+G(j\omega))^HR(I+G(j\omega))\geq R, \qquad \forall\omega \tag{3.8}$$

which is the previously mentioned multivariable generalization of condition (3.1).

Of course, the Riccati equation (3.2) arises in connection with the LQ regulator problem defined by the optimization problem

$$\min_{u(t)} J(u)=\int_0^\infty\left[x^T(t)Qx(t)+u^T(t)Ru(t)\right]dt \tag{3.9}$$

such that

$$\dot{x}(t)=Ax(t)+Bu(t) \tag{3.10}$$

which has the usual solution

$$u(t)=-R^{-1}B^TKx(t) \tag{3.11}$$

with $K>0$ satisfying (3.2), provided $(A,B,Q^{1/2})$ is minimal.

B. Stability Margins of LQ Regulators

We can now employ Theorem 3.1 in conjunction with the results of Section II to establish the robustness properties of multivariable LQ regulators. Recall from Section II that the key quantity for multivariable robustness analysis is the minimum singular value $\underline{\sigma}(I+G(s))$, where $G(s)$ is the loop transfer matrix. Unfortunately, the inequalities (3.6) and (3.7) of Theorem 3.1 do *not* provide a bound on $\underline{\sigma}(I+G(s))$, where $G(s)$ is the LQ regulator loop transfer matrix defined by (3.4). However, if we define

$$\hat{G}(s)=R^{1/2}G(s)R^{-1/2}$$

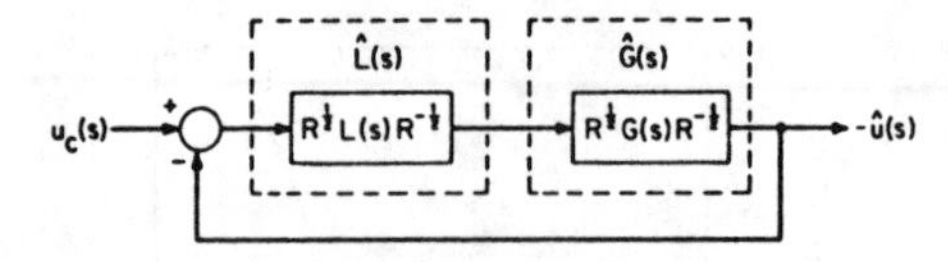

Fig. 8. Feedback system for stability margin derivation (compare Fig. 1).

then (3.7) (for example) can be rewritten in the form

$$(I+\hat{G}(s))^H(I+\hat{G}(s))\geq I, \qquad s\in D_R. \tag{3.12}$$

Equation (3.12) provides the bound

$$\underline{\sigma}(I+\hat{G}(s))\geq 1, \qquad s\in D_R \tag{3.13}$$

on the minimum singular value of $I+\hat{G}(s)$.

To work with $\hat{G}(s)$ instead of $G(s)$, it is necessary to manipulate Fig. 1 into the equivalent (for stability analysis) form depicted in Fig. 8. Then using (3.6) and (3.7) together with Theorem 2.3 leads directly to the following result.

Theorem 3.2: The polynomial $\tilde{\phi}_{CL}(s)$ has no CRHP zeros provided the following conditions are satisfied:

a) conditions (2.16) and (2.17) hold (3.14)

b) $G(s)$ is specified by (3.4) where $K>0$ satisfies (3.2) and $[A,B]$ is stabilizable, $[A,Q^{1/2}]$ is detectable and B has full rank (3.15)

c) with $\gamma(s)\triangleq\bar{\sigma}(R^{1/2}L^{-1}(s)R^{-1/2}-I)$ either of the following hold

i) $Q>0$ and $\gamma(s)\leq 1$, $s\in\Omega_R$ (3.16)

ii) $\phi_{OL}(j\omega)\neq 0\ \forall\omega$ and $\gamma(s)<1$, $s\in\Omega_R$. (3.17)

Note that the condition $\bar{\sigma}(R^{1/2}L^{-1}(s)R^{-1/2}-I)\leq 1$ in (3.16) can be rewritten as

$$RL(s)+L^H(s)R-R\geq 0, \qquad s\in\Omega_R \tag{3.18}$$

or with $s=j\omega$

$$L(j\omega)R^{-1}+R^{-1}L^H(j\omega)-R^{-1}\geq 0 \qquad \forall\omega. \tag{3.19}$$

▲

The inequality (3.19) is used by Safonov and Athans [25] to prove the LQ state feedback guaranteed gain and phase margins although their method, of proof is quite different. They implicitly assume that $L(j\omega)$ is stable, something which we do not require.

Theorem 3.2 can now be employed to establish the guaranteed minimum multivariable gain and phase margins associated with LQ regulators. We emphasize that these margins are guaranteed *only* if R is chosen to be a diagonal matrix; we will subsequently present an example showing that the margins can be made arbitrarily small for an appropriately chosen nondiagonal R matrix.

Corollary 3.1: The LQ regulator with loop transfer matrix $G(s)$ satisfying (3.13) has simultaneously in each feedback loop a guaranteed minimum gain margin (GM) given by

$$\mathrm{GM}=\frac{1}{2},\infty \tag{3.20}$$

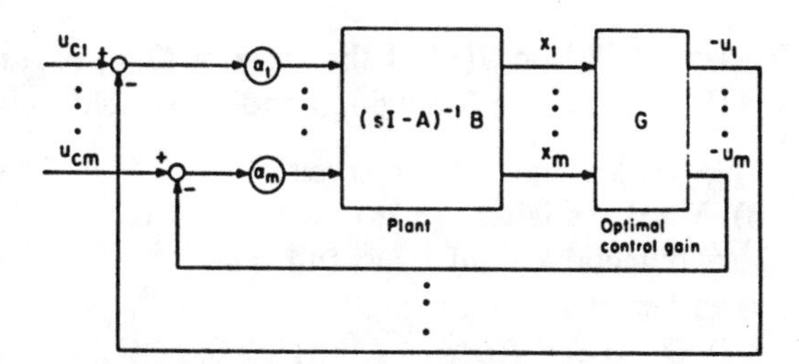

Fig. 9. Configuration for definition of multiloop LQ stability margins.

and also a guaranteed minimum phase margin (PM)

$$\text{PM} = \pm 60° \tag{3.21}$$

if R is diagonal and either $Q>0$ in (3.2) or $\phi_{OL}(j\omega)\neq 0$, $\forall\omega$. ▲

If α_i represents a pure gain change or $\alpha_i = e^{j\phi_i}$ represents a pure phase change, then Fig. 9 illustrates the placement of the α_i in the feedback loop. Note that the interpretation of the gain and phase margins specified in (3.20) and (3.21) changes slightly depending on whether $Q>0$ or $\phi_{OL}(j\omega)\neq 0$, $\forall\omega$ holds. If $Q>0$, then the pure gain α_i must satisfy $\alpha_i > \frac{1}{2}$. Similarly, for $\alpha_i = e^{j\phi_i}$ we must have $|\phi_i| < 60°$. If only $\phi_{OL}(j\omega)\neq 0$, $\forall\omega$ holds then pure gains α_i must satisfy $\alpha_i > \frac{1}{2}$ while pure phase factors $\alpha_i = e^{j\phi_i}$ must satisfy $|\phi_i| < 60°$.

Results related to Corollary 3.1 have been derived by various authors [26]–[29]; but the definitive treatment including the multivariable phase margin result is due to Safonov and Athans [25]. The approach of this paper, based on relatively simple frequency domain arguments, is new.

If R is not diagonal then the guarantees of Corollary 3.1 do not apply. The following example illustrates that the gain margins may become arbitrarily small.

Example 3.1: Consider the LQ regulator specified by (3.15) with

$$(A, B, Q^{1/2}) = \left(I_2, \begin{bmatrix} 1 & \beta \\ 0 & 1 \end{bmatrix}, I_2\right) \tag{3.22}$$

where I_2 is the 2×2 identity matrix and $R>0$ is a nondiagonal control-weighting matrix given by

$$R = B^T[K^{-2} + 2K^{-1}]^{-1}B \tag{3.23}$$

where $K>0$ is arbitrary. By selection of R in (3.23), K satisfies (3.2). Now let the multiplicative perturbation $L(s)$ be given by the constant matrix L where

$$L = \begin{bmatrix} 1 & 0 \\ 0 & 1+\epsilon \end{bmatrix} \tag{3.24}$$

and $\epsilon \neq 0$ is arbitrary. The zeros of $\tilde{\phi}_{CL}(s)$ are the eigenvalues of the perturbed closed-loop system matrix $\tilde{A}_{CL}$ where

$$\tilde{A}_{CL} = A - BLB^{-1}(2I + K^{-1}) \tag{3.25}$$

or

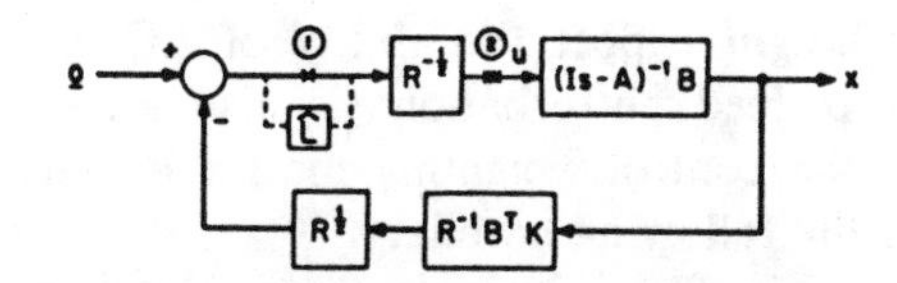

Fig. 10. LQ regulator with margins guaranteed at point ① for an $R>0$ and at both ① and ② for diagonal $R>0$.

$$\tilde{A}_{CL} = -\begin{bmatrix} (p_1+2)+\beta\epsilon p_2 - 1 & p_2 + \beta\epsilon(p_3+2) \\ (1+\epsilon)p_2 & (1+\epsilon)(p_3+2)-1 \end{bmatrix} \tag{3.26}$$

where we have let K^{-1} be denoted by

$$K^{-1} = \begin{bmatrix} p_1 & p_2 \\ p_2 & p_3 \end{bmatrix}. \tag{3.27}$$

For $\tilde{A}_{CL}$ to have no CRHP eigenvalues it is necessary for $\operatorname{tr}\tilde{A}_{CL} < 0$. However, by inspection of (3.26), if $p_2 \neq 0$ then for any $\epsilon \neq 0$ there exists a β that will make $\operatorname{tr}\tilde{A}_{CL} > 0$ and therefore for arbitrarily small ϵ, the perturbed closed-loop system will be unstable.

The basic problem exposed here is that the margins are really guaranteed at a different point in the loop than where we would like. This is illustrated in Fig. 10 where the perturbation $\hat{L}(s)$ is inserted at point ①. When $\hat{L}(s)$ is diagonal, as when calculating gain and phase margins, and R is also diagonal then $R^{-1/2}$ and $\hat{L}(s)$ commute and points ① and ② have identical guaranteed gain and phase margins. Point ② is where it is important to have margins (i.e., at the input to the *physical* plant), not inside the compensator at point ①.

Returning to Example 2.1 of the previous section once more, an LQ feedback control law is given that has the same closed-loop poles as before, but avoids the near instability associated with the negative identity feedback.

Example 3.2: With $b_{12} = 50$ in (2.10) as in the plot of $\underline{\sigma}(I + G(j\omega))$, an LQ design using $R = 1$ and

$$Q = 3\begin{bmatrix} 2601 & -50 \\ -50 & 1 \end{bmatrix} \tag{3.28}$$

gives a feedback gain of

$$R^{-1}B^TK = \begin{bmatrix} 1 & -50 \\ 0 & 1 \end{bmatrix} \tag{3.29}$$

and a closed-loop system matrix A_{CL} of

$$A_{CL} = A - BR^{-1}B^TK = -2I. \tag{3.30}$$

This makes $I + G(s)$

$$I + G(s) = \begin{bmatrix} \dfrac{s+2}{s+1} & 0 \\ 0 & \dfrac{s+2}{s+1} \end{bmatrix} \tag{3.31}$$

and thus

$$\underline{\sigma}(I + G(j\omega)) = \left[\frac{\omega^2+4}{\omega^2+1}\right]^{1/2} > 1. \tag{3.32}$$

As one might expect the ability of LQ regulators to tolerate crossfeed perturbations is also affected by the choice of the control weighting matrix R. This is made precise in the following corollary.

Corollary 3.2: The LQ regulator with loop transfer matrix $G(s)$ satisfying (3.15) will tolerate (i.e., $\tilde{\phi}_{CL}(s)$ will have no CRHP zeros) a crossfeed perturbation of the form (2.42) satisfying (2.16) and (2.17) provided

$$\bar{\sigma}^2(X(s))<\min\left\{\frac{\lambda_{\min}(R_1)}{\lambda_{\max}(R_2)},\frac{\lambda_{\min}(R_2)}{\lambda_{\max}(R_1)}\right\},\qquad s\in\Omega_R \tag{3.33}$$

where R is given by

$$R=\begin{bmatrix} R_1 & 0\\ 0 & R_2\end{bmatrix} \tag{3.34}$$

and is conformably partitioned with $L(s)$ in (2.42) and either $Q>0$ or $\phi_{OL}(j\omega)\neq 0\ \forall\omega$ holds. ▲

Note that

$$\frac{\lambda\min(R)}{\lambda\max(R)}<\min\left\{\frac{\lambda\min(R_1)}{\lambda\max(R_2)},\frac{\lambda\min(R_2)}{\lambda\max(R_1)}\right\}<\left[\frac{\lambda\min(R)}{\lambda\max(R)}\right]^{1/2} \tag{3.35}$$

which indicates that if the ratio of $\lambda\min(R)/\lambda\max(R)$ is very small that the ability to tolerate crossfeed perturbations is drastically reduced. As illustrated in Fig. 10 the use of R scales the inputs and outputs such that the stability margins are obtained in the scaled system rather than the original system. This means that if our original model has the coordinate system in which we would like to guarantee margins, that R should be selected as $R=\rho I$ for some positive scalar ρ.

Since LQ designs have inherently good margins provided R is selected appropriately, it is natural to search for variations of this method. One such variation, proposed by Wong and Athans [27], is to solve a Lyapunov rather than a Riccati equation to compute K in (3.11).

The Lyapunov equation with $Q\geq 0$ given by

$$A^TK+KA+Q=0 \tag{3.36}$$

guarantees that the eigenvalues of A lie in the CLHP if $K>0$ and $[A,Q^{1/2}]$ is detectable. The corresponding Kalman type inequality for loop transfer matrices $G(s)$ specified by (3.4) where $K>0$ satisfies (3.36) is given by

$$RG(j\omega)+G^H(j\omega)R\geq 0,\qquad \forall\omega \tag{3.37}$$

and is the fundamental inequality used to derive stability margins. When $Q>0$ the inequality (3.37) may be changed to strictly greater than. The stability margins for this type of feedback are given in the next theorem and its corollaries.

Theorem 3.3: For $G(s)$ of the form of (3.4), $\tilde{\phi}_{CL}(s)$ has no CRHP zeros if the following conditions hold:

a) $\tilde{\phi}_{OL}(s)$ has no CRHP zeros (3.38)

b) $K>0$ satisfies (3.36) with $Q\geq 0$, $R>0$ and $[A,Q^{1/2}]$ detectable, and B has full rank (3.39)

c) either of the following holds

i) $Q>0$ and $RL(s)+L^H(s)R\geq 0, s\in\Omega_R$ (3.40)

ii) $RL(s)+L^H(s)R>0, s\in\Omega_R$. (3.41)

▲

Corollary 3.3: For $G(s)$ as in Theorem 3.3 with R diagonal the guaranteed gain and phase margins are given by

$$\text{GM}=0,\infty \tag{3.42}$$

and

$$\text{PM}=\pm 90°. \tag{3.43}$$

Proof: Similar to Corollary 2.3.

The importance of Corollary 3.3 is that the standard LQ guaranteed gain reduction margin of $\frac{1}{2}$ can be reduced to 0 by using K satisfying the Lyapunov equation (3.36) with $Q>0$ rather than the Riccati equation (3.2). Of course, it is possible to have a zero gain reduction margin only for open-loop stable systems. However, standard LQ state feedback does *not* guarantee a zero gain reduction margin even in the open-loop stable case, and has been criticized on these grounds [20]. Having a zero gain reduction margin is important in situations where actuators may fail or saturate, and there is no opportunity to reconfigure the control system. In fact, the motivation for the thesis [26] (which in turn lead to most of the developments reported in this paper) was a study supporting the design of the automatic depth-keeping controller for the Trident submarine, in which saturation of one of the two hydrodynamic control surfaces produced an unstable closed-loop system.

Corollary 3.4: For $G(s)$ as in Theorem 3.3 the crossfeed tolerance is given by

$$\bar{\sigma}^2(X(s))<4\min\left[\frac{\lambda\min(R_1)}{\lambda\max(R_2)},\frac{\lambda\min(R_2)}{\lambda\max(R_1)}\right],\qquad s\in\Omega_R \tag{3.44}$$

where $L(s)$ is given by (2.42), $R>0$ is given by (3.34) and (3.38) holds ensuring $\tilde{\phi}_{CL}(s)$ has no CRHP zeros.

Proof: Analogous to Corollary 3.2.

Another way to modify the LQ design procedure that is a compromise between Theorem 3.2 and Theorem 3.3 involves using a parameterized Riccati equation given by

$$A^TK+KA+Q-\beta KBR^{-1}B^TK=0 \tag{3.45}$$

where β is an adjustable parameter and $0<\beta<2$. The feedback law is still given by (3.11) and $G(s)$ is still given by (3.4) with $K>0$. Since the β in (3.45) may be lumped together with the R matrix, (3.45) is just a standard Riccati equation and therefore has a unique solution $K>0$ under the appropriate assumptions (3.15). The standard

LQ optimal feedback law associated with (3.45) is given by

$$u(t) = -\beta R^{-1}B^TKx(t). \qquad (3.46)$$

Instead of (3.46) we will use $u(t) = -R^{-1}B^TKx(t)$ as in (3.11). Thus depending on whether $\beta > 1$ or $\beta < 1$ we are merely decreasing or increasing, respectively, the optimal feedback gain by a scalar factor of $1/\beta$. Also with $G(s)$ given by (3.4) the standard LQ loop transfer matrix is simply $\beta G(s)$. From Theorem 3.1 we know that if $Q > 0$

$$[I+G(s)]^H \frac{1}{\beta} R[I+\beta G(s)] > \frac{1}{\beta} R, \qquad s \in \Omega_R \qquad (3.47)$$

which in the SISO case becomes

$$\left|\frac{1}{\beta} + g(s)\right| > \frac{1}{\beta}, \qquad s \in \Omega_R \qquad (3.48)$$

and is illustrated in Fig. 7. To obtain bounds on $L(s)$ to ensure stability we merely work with $1/\beta L(s)$ and $\beta G(s)$ and apply Theorem 3.2 for the standard LQ regulator problem. Doing this we obtain, in the SISO case, the inequality

$$|\beta l^{-1}(s) - 1| < 1, \qquad s \in \Omega_R \qquad (3.49)$$

illustrated in Fig. 11. Note that from Fig. 7 that the critical "-1" point is no longer contained inside the circle if $\beta > 2$ and thus there are no guaranteed margins. If $\beta \to 0$ the guaranteed minimum margins approach those of the Lyapunov feedback case. In general, for the multivariable case the guaranteed minimum margins, again if R is diagonal and Q is positive definite, are given by

$$\text{GM} = \beta/2, \infty, \qquad 0 < \beta < 2 \qquad (3.50)$$

and

$$\text{PM} = \pm\cos^{-1}\frac{\beta}{2}, \qquad 0 < \beta < 2. \qquad (3.51)$$

These guaranteed margins (when $\beta \leq 1$) can also be obtained by similar but distinctly different procedures reported in [30] and [31] which utilize standard LQ regulators with vanishingly small control weights.

C. Stability Margins of LQG Regulators

A basic limitation associated with the LQ guaranteed stability margins is that they are obtained only under the assumption of full state feedback. State feedback can never be exactly realized, and often it is impossible or too expensive to provide enough sensors to achieve even an approximate realization. Thus one is motivated to investigate what guaranteed stability margins might be associated with LQG controllers, in which a Kalman filter (KF) is used to provide state estimates for feedback.

Since the Kalman filter is the dual of the LQ regulator, dual robustness results are obtainable. They ensure a

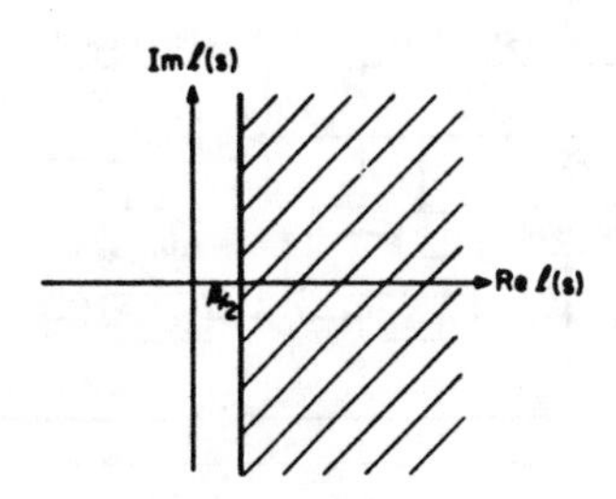

Fig. 11. Set of allowable values of $l(s)$ when $|\beta l^{-1}(s) - 1| < 1$ and $0 < \beta < 2$.

nondivergent Kalman filter under variations in the nominal model parameters of the plant whose state is to be estimated. To make the precise connection between the regulator and filter problems, consider the linear system

$$\dot{x}(t) = Ax(t) + \xi(t) \qquad (3.52)$$

$$y(t) = Cx(t) + \theta(t) \qquad (3.53)$$

where $\xi(t)$ and $\theta(t)$ are zero mean white noise sources with spectral intensity matrices Ξ and Θ, respectively. We wish to estimate $x(t)$ given $y(\tau)$, $-\infty < \tau < t$, such that the mean square error is minimized. Under the assumption the $[A, C]$ is detectable, it is well-known that the state estimate is specified by

$$\dot{\hat{x}}(t) = A\hat{x}(t) + \Sigma C^T\Theta^{-1}\nu(t) \qquad (3.54)$$

$$\nu(t) = y(t) - C\hat{x}(t) \qquad (3.55)$$

where

$$A\Sigma + \Sigma A^T + \Xi - \Sigma C^T\Theta^{-1}C\Sigma = 0, \qquad \Sigma > 0. \qquad (3.56)$$

If we calculate the transfer matrix from $\nu(s)$ to $\hat{y}(s) = C\hat{x}(s)$, we find that

$$\hat{y}(s) = \left[C(Is-A)^{-1}\Sigma C^T\Theta^{-1}\right]\nu(s) \triangleq F(s)\nu(s). \qquad (3.57)$$

Then, if $\Xi > 0$, $F(s)$ satisfies the dual of (3.6) given by

$$(I+F(s))\Theta(I+F(s))^H > \Theta, \qquad s \in \Omega_R \qquad (3.58)$$

which guarantees the stability of the error dynamics under a range of perturbations in $F(s)$. Thus, if $F(s)$ is perturbed to $\tilde{F}(s) = F(s)L(s)$, where usual assumptions about $G(s)$ are applied to $F(s)$, the Kalman filter will remain nondivergent if

$$\bar{\sigma}(\Theta^{-1/2}L^{-1}(s)\Theta^{1/2} - I) < 1, \qquad s \in \Omega_R \qquad (3.59)$$

or equivalently,

$$\Theta L^H(s) + L(s)\Theta - \Theta \geq 0. \qquad (3.60)$$

It is now readily apparent that $F(s)$, the loop transfer matrix of the error dynamics loop of the Kalman filter, is the dual of $G(s)$ in the LQ regulator and has the same guaranteed margins at its input, $\nu(s)$, for diagonal Θ.

Safonov and Athans [32] have developed these dual results for the nondivergence of the extended Kalman

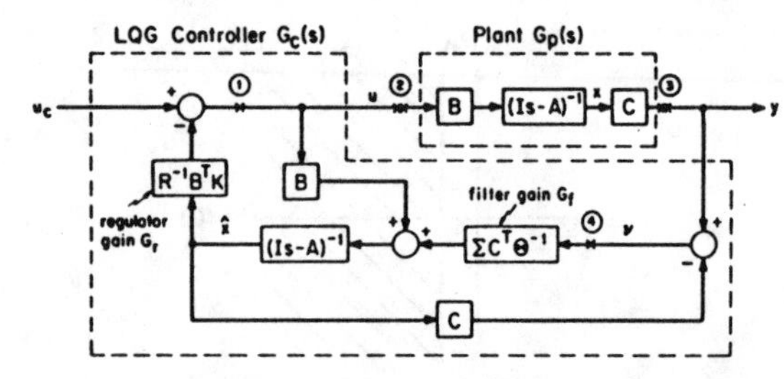

Fig. 12. LQG control system.

filter. Furthermore, they have considered the robustness properties of a nonlinear LQG control system formed by the cascade of a constant gain extended Kalman filter and the LQ state feedback gain. The LQ state feedback gain and the constant gain of the extended Kalman filter are computed from the linearized model parameters. However, the extended Kalman filter must have the true nonlinear model of the plant. In the completely linear case the LQG stability margins are much easier to obtain.

The standard LQG control system block diagram is shown in Fig. 12 with various points of the loop marked. To determine the robustness of the LQG control system we insert perturbations at points ② and ③ (the input and output of the physical plant) and find out how large they can be made without destabilizing the closed-loop system. It is therefore convenient to calculate the loop transfer matrices at points ① to ④. The loop transfer matrix at point Ⓚ will be denoted $T_K(s)$ and is calculated breaking the loop at point Ⓚ and using it as the input as well as the output. For the four points indicated in Fig. 12 we have

$$T_1(s)=G_r\phi(s)B \tag{3.61}$$

$$T_2(s)=G_r\left(\phi^{-1}(s)+BG_r+G_fC\right)^{-1}G_fC\phi(s)B \tag{3.62}$$

$$T_3(s)=C\phi(s)BG_r\left(\phi^{-1}(s)+BG_r+G_fC\right)^{-1}G_f \tag{3.63}$$

$$T_4(s)=C\phi(s)G_f \tag{3.64}$$

where

$$G_r \triangleq R^{-1}B^TK = \text{regulator gain} \tag{3.65}$$

$$G_f \triangleq \Sigma C^T\Theta^{-1} = \text{filter gain} \tag{3.66}$$

$$\phi(s) \triangleq (Is-A)^{-1}. \tag{3.67}$$

Note that points ① and ④ have the standard LQ regulator and Kalman filter loop transfer matrices, respectively. Thus at points ① and ④ (inside the LQG controller) the LQ and KF minimum guaranteed stability margins apply. The following theorem is a much simplified version of a theorem proved in [32] and gives LQG stability margins at points ② and ③ (the input and output of the physical plant).

Theorem 3.4: The LQG feedback control system of Fig. 12 is asymptotically stable under variations in the open-loop plant $G_p(s) \triangleq C(Is-A)^{-1}B$ if the following conditions hold.

a) The perturbed open-loop plant $\tilde{G}_p(s) \triangleq \tilde{C}(Is-\tilde{A})^{-1}\tilde{B}$ is such that the $\det(sI-\tilde{A})$ and $\det(sI-A)$ have the same number of CRHP zeros and if $\det(j\omega_0 I-\tilde{A})=0$ then $\det(j\omega_0 I-A)=0$. (3.69)

b) $[A, B]$ is stabilizable, $Q>0$, $R>0$ and $K>0$ satisfies (3.2) and B has full rank. (3.70)

c) $\tilde{G}_p(s)=G_p(s)L(s)=N(s)G_p(s)$ (3.71)

and either

$$\bar{\sigma}\left(R^{1/2}L^{-1}(s)R^{-1/2}-I\right)<1 \tag{3.72}$$

or

$$\bar{\sigma}\left(\Theta^{-1/2}N^{-1}(s)\Theta^{1/2}-I\right)<1 \tag{3.73}$$

hold for all $s\in\Omega_R$.

d) The LQG controller transfer matrix $G_c(s)$ from the plant output to the plant input is given by

$$G_c(s)=G_r\left(Is-\tilde{A}+\tilde{B}G_r+G_f\tilde{C}\right)^{-1}G_f \tag{3.74}$$

where G_r and G_f, respectively satisfy (3.65) and (3.66). ▲

Notice that in (3.71) $L(s)$ represents the same perturbation in $G_p(s)$ at the input to the plant as $N(s)$ represents at the output of the plant and that $\tilde{G}_p(s)$ is the same in both cases. Now the basic idea of the proof is quite simple. At point ① we have an LQ state feedback regulator loop transfer matrix and the LQ guaranteed margins apply. By moving $L(s)$, the perturbation, to point ② we simply change B in the Kalman filter to $BL(s)$ leaving $G_r=R^{-1}B^TK$ fixed. This, however, is the same as giving the Kalman filter the correct dynamic model of the perturbed open-loop system without changing either the filter or the regulator gains. The same result follows if we start with a perturbation, $N(s)$, at point ④, where the KF guarantees apply and move it to point ③ changing C to $CN(s)$.

Thus the LQ and KF guaranteed stability margins will apply to LQG controllers at the input and output of the physical plant but under the restrictive assumption that the system model embedded within the Kalman filter is always the same as the true system (i.e., the perturbed system). For the more realistic case in which the internal model of the Kalman filter remains unchanged, there are unfortunately *no* guarantees, as Doyle has demonstrated with a simple counterexample [33]. This counterexample is extreme, but it is possible to obtain LQG controllers with inadequate stability margins that look quite reasonable in the time domain. Fig. 13 shows the Nyquist plot of a single-input design reported in the literature [34]; note that the phase margin is less than 10°.

Fortunately, there are two dual procedures that do not require the Kalman filter to have the true system model and that still recover the LQ and KF guaranteed minimum margins. These procedures use the asymptotic properties of the Kalman filter and LQ regulator (see [43] this issue) and can be used only if the plant is minimum phase.

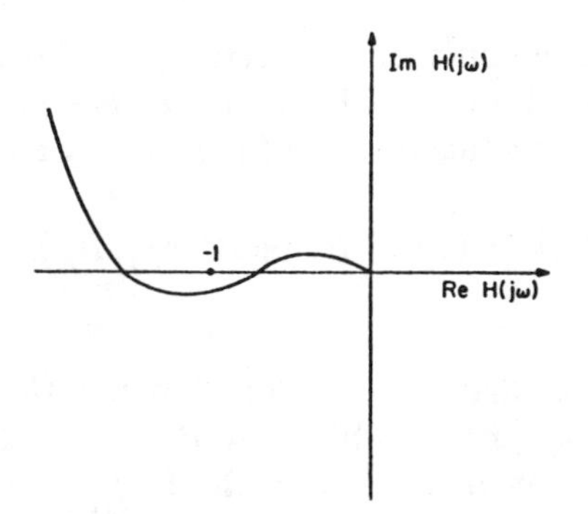

Fig. 13. Nyquist diagram for LQG design in [34]. ($H(j\omega)$ = loop transfer function).

If W is a nonsingular arbitrary matrix, then by selecting Ξ in (3.56) as $\rho\ BWW^TB^T$ and letting $\rho \to \infty$ the loop transfer matrix $T_2(s)$ in (3.62) approaches $T_1(s)$ of (3.61) if the minimum phase assumption holds [35]. Thus the LQ regulator guaranteed margins will be recovered at the *input* to the plant. Kwakernaak [36] proposed the dual of the above procedure to obtain low sensitivity feedback systems. His procedure makes $T_3(s)$ of (3.63) approach $T_4(s)$ of (3.64) and thus the KF guaranteed minimum margins will be recovered at the *output* of the plant.[6] However, it is not always the case that an LQG controller needs to be robustified by these procedures since in some cases the LQG control system will have better stability margins than its full state feedback counterpart [42].

Even when these procedures are used, the guaranteed stability margins apply at the input *or* output of the physical plant but not necessarily at both input *and* output. It is desirable to have margins at both these locations since the perturbations in $G_p(s)$ are represented as either $G_p(s)L(s)$ or $N(s)G_p(s)$ and we should not like small perturbations in either input or output to destabilize the system. Margins at both input and output can be ensured if the inequalities

$$\underline{\sigma}(I+G_c(s)G_p(s))>1 \tag{3.75}$$

and

$$\underline{\sigma}(I+G_p(s)G_c(s))>1 \tag{3.76}$$

both hold. The relationship between these two quantities when $G_p(s)$ and $G_c(s)$ are square matrices is given by

$$\frac{1}{k}\underline{\sigma}(I+G_p(s)G_c(s))<\underline{\sigma}(I+G_c(s)G_p(s))$$
$$<k\underline{\sigma}(I+G_p(s)G_c(s)) \tag{3.77}$$

where

$$k=\min\left[\frac{\bar{\sigma}(G_p(s))}{\underline{\sigma}(G_p(s))},\frac{\bar{\sigma}(G_c(s))}{\underline{\sigma}(G_c(s))}\right]>1. \tag{3.78}$$

[6]Dowdle [37] has adapted these procedures for use with minimal order observer based compensators and their duals.

The quantity k is the minimum of the condition numbers of $G_p(s)$ and $G_c(s)$ with respect to inversion. The proof is accomplished by a simple calculation and is omitted. From (3.77) we conclude that if k is close to unity then approximately the same robustness guarantees will apply at both input and output. Note that we have no control over $G_p(s)$ so that if $G_p(s)$ is nearly singular we must design our compensator so that $\bar{\sigma}(G_c(s))\simeq\underline{\sigma}(G_c(s))$. On the other hand, if our plant is well-conditioned with respect to inversion our compensator $G_c(s)$ need not be so severely constrained, allowing more flexibility in achieving performance objectives.

IV. Summary and Conclusions

A. Summary

In this paper we have stressed the importance of the robustness of feedback systems with respect to uncertainty in the nominal plant model. Arguing from the multivariable Nyquist theorem, it was shown that if the return difference matrix $I+G(s)$ is nearly singular, then there exists a small perturbation in $G(s)$ that will destabilize the closed-loop system. To detect this condition, singular values, familiar from numerical analysis, were introduced. In particular, $\underline{\sigma}(I+G(s))$ measures the nearness to singularity of $I+G(s)$. Interpreting $\underline{\sigma}(I+G(s))$ lead to a direct multivariable generalization of the classical notions of gain and phase margins. These margins were shown to hold *simultaneously* in all loops of the feedback system. Also a channel crossfeed margin was derived from $\underline{\sigma}(I+G(s))$ which is the key quantity in the determination of all the stability margin results.

Direct synthesis procedures involving $\underline{\sigma}(I+G(s))$ are unknown at present. This is due in part to the complicated dependence of $\underline{\sigma}(I+G(s))$ on the compensator implicit in $G(s)$. However, the LQ state feedback regulator was shown to provide an indirect synthesis procedure which automatically ensures a degree of robustness in the coordinate system specified by the control weighting matrix R. Thus LQ state feedback is preferable to state feedback specified by pole placement techniques since there are no robustness guarantees provided by this latter method. Next, using the Lyapunov equation to compute the feedback gains, a regulator was specified that was stable in spite of the failure of any of its feedback channels. For LQG control systems there are no automatic guarantees. However, in the case of minimum phase plants, the LQ guaranteed margins are asymptotically recoverable if necessary.

Even so, these guarantees will simply be inadequate if excessively large feedback gains are used. This is most clear in the SISO case when the gain crossover frequency occurs at a frequency at which the phase of the open-loop transfer function is completely uncertain. In this case the phase margin necessary for stability is $\pm 180°$ requiring the loop gain to be less than 1. This illustrates the more generally valid point that stability margins are completely useless in themselves without further information about the size of the expected perturbations in the nominal system.

B. Conclusions and Future Research

Although the singular value approach is useful in detecting the near instability of a control system, it is sometimes unnecessarily conservative. This is due to the fact that some of the small perturbations that would theoretically destabilize the closed-loop system will never occur in the physical system. Nevertheless they are still detected by a small $\underline{\sigma}(I+G(s))$. One direction for further research is characterization of the robustness of a feedback system in which perturbations in certain directions are ruled out as impossibilities on physical grounds. This points out the fundamental problem of obtaining the characterization of the uncertainty associated with a given model. It seems that this knowledge can be acquired only by experience with real applications. One of the advantages of the singular value approach is that it singles out the worst type of perturbations for scrutiny. Another avenue for research is, of course, LQG controller design procedures that ensure a minimum size for $\underline{\sigma}(I+G(s))$ or its complementary quantity $\underline{\sigma}(I+G^{-1}(s))$ and yet maintain a satisfactory degree of performance.

In the decade since the promulgation of the LQG methodology [41], much has been learned about the pitfalls of its application to practical problems. It has been criticized on the grounds that the optimal control problem is over specified to the extent that there is only one solution merely to be found by the computer [1]. Practice has shown however, that the LQG method is not merely a "cookbook" procedure to be used blindly, but one in which a fair amount of iteration on the design of a controller is necessary to obtain satisfactory results. Unlike, however, many of the frequency domain techniques which reduce a multiloop problem to a series of single loop designs, the LQG procedure is inherently a multiloop procedure. It provides a reasonable place to start a control system design and when used intelligently should provide the designer a good chance of success.

Appendix

A. Proofs for Sections II and III

Proof of Theorem 2.2: For all R sufficiently large, D_R will enclose all ORHP zeros of both $\phi_{OL}(s)$ and $\tilde{\phi}_{OL}(s)$ and by virtue of (2.17) and the indentation construction of D_R this can be extended to all CRHP zeros of both $\phi_{OL}(s)$ and $\tilde{\phi}_{OL}(s)$. Also, for R sufficiently large, D_R avoids all OLHP zeros of $\phi_{OL}(s)$, $\tilde{\phi}_{OL}(s)$, and $\phi_{CL}(s)$. From Theorem 2.1 and (2.18) we conclude that

$$\mathfrak{N}(0,\det[I+G(s)],D_R)=-P \quad \text{(A.1)}$$

where P is the number of CRHP zeros of $\phi_{OL}(s)$ and, by (2.16), also of $\tilde{\phi}_{OL}(s)$. Clearly $\det[I+(1-\epsilon)G(s)+\epsilon\tilde{G}(s)]$ is a continuous function of ϵ for all $s\in D_R$ and from (A.1)

$$\mathfrak{N}(0,\det[I+(1-\epsilon)G(s)+\epsilon\tilde{G}(s)],D_R)|_{\epsilon=0}=-P. \quad \text{(A.2)}$$

Now suppose that ϵ is varied continuously from 0 to 1 and that $\mathfrak{N}(0,\det[I+(1-\epsilon)G(s)+\epsilon G(s)],D_R)$ does *not* remain constant at $-P$. From the Principle of the Argument we know that for some ϵ on [0,1] the number of zeros minus the number of poles of $\det[I+(1-\epsilon)G(s)+\epsilon G(s)]$ enclosed in D_R must change. However, since the poles and zeros are also continuous functions of ϵ, it must be that for some ϵ_0 on [0,1] that they lie on D_R and thus

$$\det[I+(1-\epsilon_0)G(s)+\epsilon_0\tilde{G}(s)]=0 \text{ or } \infty. \quad \text{(A.3)}$$

Condition (2.19) eliminates the possibility that $\det[I+(1-\epsilon_0)G(s)+\epsilon_0\tilde{G}(s)]$ evaluates to zero. Since R is chosen sufficiently large, D_R must avoid all zeros of $\tilde{\phi}_{OL}(s)$ and $\phi_{OL}(s)$ which include the poles of $\det[I+(1-\epsilon_0)G(s)+\epsilon_0\tilde{G}(s)]$ and thus the possibility that $\det[I+(1-\epsilon_0)G(s)+\epsilon_0\tilde{G}(s)]$ is infinite is also ruled out. This contradicts the assumption that $\mathfrak{N}(0,\det[I+(1-\epsilon)G(s)+\epsilon G(s),D_R)$ changes as ϵ is varied on [0,1] and therefore it must be that it remains constant for all ϵ on [0,1]. However, this implies that for $\epsilon=1$ that

$$\mathfrak{N}(0,\det[I+\tilde{G}(s)],D_R)=-P \quad \text{(A.4)}$$

and thus by (2.16) and Theorem 2.1, $\tilde{\phi}_{CL}(s)$ has no CRHP zeros. Q.E.D.

Lemma A.1: For square matrices G and L, $\det(I+GL)\neq 0$ if any of the following conditions hold:

a) $\bar{\sigma}(L^{-1}-I)<\underline{\sigma}(I+G)$ (A.5)
b) $G+G^H\geq 0$ and $L+L^H>0$ (A.6)
c) $G+G^H>0$ and $L+L^H\geq 0$. (A.7)

Proof: To prove a) rewrite $I+GL$ as

$$I+GL=[(L^{-1}-I)(I+G)^{-1}+I](I+G)L \quad \text{(A.8)}$$

since L and $I+G$ are, by (A.5), assumed to be nonsingular, $I+GL$ is nonsingular if and only if $[(L^{-1}-I)(I+G)^{-1}+I]$ is nonsingular. Condition (A.5) guarantees that

$$\|(L^{-1}-I)(I+G)^{-1}\|_2<1 \quad \text{(A.9)}$$

which ensures that $[(L^{-1}-I)(I+G)^{-1}+I]$ is invertible. To prove b), suppose, contrary to what we wish to prove that

$$\det(I+GL)=0. \quad \text{(A.10)}$$

Thus $\exists$ a vector $x\neq 0 \ni (I+GL)x=0$ and hence

$$x=-GLx. \quad \text{(A.11)}$$

Defining $z = Lx$, (A.11) implies $z \neq 0$ and

$$z = -LGz. \tag{A.12}$$

Condition (A.6) and (A.12) imply

$$z^H Gz + z^H G^H z = -z^H G^H [L + L^H] Gz > 0 \tag{A.13}$$

and since $Gz \neq 0$ a contradiction to (A.6) and (A.10) is obtained. The proof for c) is analogous to that of b). Q.E.D.

Lemma A.2: If L is a square matrix and $P(\epsilon) = (1-\epsilon)I + \epsilon L$, then for all ϵ on $[0,1]$ the following implications hold:

a) $\bar{\sigma}(L^{-1} - I) < \alpha < 1 \Rightarrow \bar{\sigma}(P^{-1}(\epsilon) - I) < \alpha$ (A.14)

b) $\bar{\sigma}(L^{-1} - I) < \alpha$ and $L + L^H > 0 \Rightarrow \bar{\sigma}(P^{-1}(\epsilon) - I) < \alpha$ (A.15)

c) $\bar{\sigma}(L^{-1} - I) < \alpha$ and $4(\alpha^2 - 1)\underline{\sigma}^2(L - I) > \alpha^2 \bar{\sigma}^2(L + L^H - 2I) \Rightarrow \bar{\sigma}(P^{-1}(\epsilon) - I) < \alpha$ (A.16)

d) $L + L^H > 0 \Rightarrow P(\epsilon) + P^H(\epsilon) > 0.$ (A.17)

Proof: For a) and b) rewrite $\bar{\sigma}(P^{-1}(\epsilon) - I) < \alpha$ as

$$\alpha^2 P^H(\epsilon) P(\epsilon) - (P(\epsilon) - I)^H (P(\epsilon) - I) > 0. \tag{A.18}$$

Expanding the left-hand side of (A.18) gives

$$\begin{aligned} \alpha^2 P^H(\epsilon) P(\epsilon) &- (P(\epsilon) - I)^H (P(\epsilon) - I) \\ &= \epsilon^2 \left[\alpha^2 L^H L - (L - I)^H (L - I) \right] \\ &\quad + \alpha^2 (1-\epsilon) \left[(1-\epsilon) I + \epsilon (L + L^H) \right]. \end{aligned} \tag{A.19}$$

By (A.14) or (A.15) the first term on the right-hand side of (A.19) is assumed to be positive definite. Now with $\alpha < 1$, (A.14) also gives

$$L^H + L > I + (1 - \alpha^2) L^H L > 0 \tag{A.20}$$

so that (A.14) or (A.15) guarantee $L + L^H > 0$ which makes the second term on the right-hand side of (A.19) positive semidefinite. Thus (A.18) is true for all ϵ on $[0,1]$ which is equivalent to $\bar{\sigma}(P^{-1}(\epsilon) - I) < \alpha$. To prove c) rewrite (A.19) as

$$\begin{aligned} \alpha^2 P^H(\epsilon) P(\epsilon) &- (P(\epsilon) - I)^H (P(\epsilon) - I) \\ &= (\alpha^2 - 1)\epsilon^2 (L - I)^H (L - I) + \alpha^2 \epsilon \left[L + L^H - 2I \right] + \alpha^2 I. \end{aligned} \tag{A.21}$$

For the right-hand side of (A.21) to positive definite, it is sufficient to show that $f(\epsilon) > 0$ where $f(\epsilon)$ is given by

$$f(\epsilon) = \beta^2 \epsilon^2 - 2\phi\epsilon + \alpha^2 \tag{A.22}$$

and where

$$\beta^2 = (\alpha^2 - 1)\underline{\sigma}^2(L - I) \tag{A.23}$$

$$\phi = \alpha^2 \bar{\sigma}\left(\frac{L + L^H}{2} - I \right). \tag{A.24}$$

Differentiating $f(\epsilon)$ to find an ϵ^* such that $f(\epsilon^*)$ is minimum results in

$$\epsilon^* = \frac{\phi}{\beta^2} \tag{A.25}$$

and

$$f(\epsilon^*) = \alpha^2 - \frac{\phi^2}{\beta^2}. \tag{A.26}$$

Thus $f(\epsilon^*)$ will be positive if

$$\alpha^2 \beta^2 > \phi^2 \tag{A.27}$$

or

$$4(\alpha^2 - 1)\underline{\sigma}^2(L - I) > \alpha^2 \bar{\sigma}^2(L + L^H - 2I) \tag{A.28}$$

which proves c).

Implication d) follows trivially by direct substitution. Q.E.D.

Proof of Theorems 2.3 and 2.4: Conditions (2.21) and (2.44) ensure that only condition (2.19) of Theorem 2.2 need be verified. Since $\bar{\sigma}[G(s)]$ and $\bar{\sigma}[\tilde{G}(s)] \to 0$ as $|s| \to \infty$, $\det[I + (1-\epsilon)G(s) + \epsilon\tilde{G}(s)] \neq 0$ for all $|s| = R$ for R sufficiently large. Thus (2.19) need only be checked for $s \in \Omega_R$. Lemmas A.1 and A.2 applied at every $s \in \Omega_R$, guarantee that (2.19) holds for $s \in \Omega_R$ and thus for $s \in D_R$. The conditions of Theorem 2.2 are satisfied and thus $\tilde{\phi}_{CL}(s)$ has no CRHP zeros. Q.E.D.

Proof of Corollaries 2.1 and 2.3: In Theorems 2.3 and 2.4 take $L(s)$ to be a diagonal matrix given by

$$L(s) = \operatorname{diag}\left[l_1(s), l_2(s), \cdots, l_n(s) \right] \tag{A.29}$$

which simplifies conditions (2.22) and (2.23) to

$$|l_i^{-1}(s) - 1| < \alpha_0, \qquad \forall i \tag{A.30}$$

and condition (2.46) to

$$\operatorname{Re}\left[l_i(s) \right] > 0, \qquad \forall i. \tag{A.31}$$

To obtain gain margins, let

$$l_i(s) = l_i, \qquad l_i \text{ real}$$

then (A.30) becomes

$$\frac{1}{1 + \alpha_0} < l_i < \frac{1}{1 - \alpha_0} \tag{A.32}$$

and (A.31) becomes

$$l_i > 0. \tag{A.33}$$

Similarly, to obtain phase margins, let

$$l_i(s) = e^{j\phi_i(s)}, \qquad \phi_i(s) \text{ real},$$

then (A.30) becomes

$$1 - \frac{\alpha_0^2}{2} < \cos\phi_i(s) \tag{A.34}$$

or

$$|\phi_i(s)|<\cos^{-1}\left[1-\frac{\alpha_0^2}{2}\right] \quad \text{(A.35)}$$

while (A.31) becomes

$$\cos\phi_i(s)>0 \quad \text{(A.36)}$$

or

$$|\phi_i(s)|<\pm 90°. \quad \text{Q.E.D.} \quad \text{(A.37)}$$

Proof of Corollaries 2.2 and 2.4: Only conditions (2.22) and (2.46) need verification. With $L(s)$ given by (2.42) we have

$$L^{-1}(s)-I=\begin{bmatrix} 0 & -X(s) \\ 0 & 0 \end{bmatrix} \text{ or } \begin{bmatrix} 0 & 0 \\ -X(s) & 0 \end{bmatrix} \quad \text{(A.38)}$$

and thus

$$\bar{\sigma}(L^{-1}(s)-I)=\bar{\sigma}(X(s))<\alpha_0 \leq \underline{\sigma}(I+G(s)). \quad \text{(A.39)}$$

Also,

$$L(s)+L^H(s)=\begin{bmatrix} 2I & X(s) \\ X^H(s) & 2I \end{bmatrix} \text{ or } \begin{bmatrix} 2I & X^H(s) \\ X(s) & 2I \end{bmatrix} \quad \text{(A.40)}$$

so that

$$\begin{aligned}\lambda_{\min}(L(s)+L^H(s)) &\geq 2-\lambda_{\max}\left(\begin{bmatrix} 0 & X(s) \\ X^H(s) & 0 \end{bmatrix}\right) \\ &\geq 2-\bar{\sigma}\left(\begin{bmatrix} 0 & X(s) \\ X^H(s) & 0 \end{bmatrix}\right) \\ &=2-\bar{\sigma}(X(s))>0. \end{aligned} \quad \text{(A.41)}$$

Q.E.D.

Proof of Theorem 3.1: Direct manipulation of (3.2) gives

$$(s^*I-A^T)K+K(sI-A)+KBR^{-1}B^TK = (Q+2\mathrm{Re}(s)K) \quad \text{(A.42)}$$

where s^* denotes the complex conjugate of s. Premultiplying and postmultiplying (A.42) by $[(sI-A)^{-1}B]^H$ and $[(sI-A)^{-1}B]$, respectively, we obtain

$$RG(s)+G^H(s)R+G^H(s)RG(s)=H(s). \quad \text{(A.43)}$$

Adding R to both sides of (A.43) gives (3.3). Now $Q+2\mathrm{Re}(s)K$ will be positive semidefinite for $s\in D_R$ if $Q>0$ and the indentations of Ω_R are sufficiently small or if $\mathrm{Re}(s)\geq 0$, $s\in D_R$ which happens if $\det(j\omega I-A)\neq 0\ \forall\omega$. Thus under these conditions $H(s)>0$ or $H(s)\geq 0$ respectively for all $s\in D_R$. Q.E.D.

Proof of Theorem 3.2: It is well known that condition (3.15) ensures that $\phi_{CL}(s)$ has no CRHP zeros. Defining $\hat{G}(s) \triangleq R^{1/2}G(s)R^{-1/2}$, we see that $G(s)$ has a statc-space realization $(A, BR^{-1/2}, R^{-1/2}B^TK)$ and thus its open- and closed-loop characteristic polynomials $\hat{\phi}_{OL}(s)$ and $\hat{\phi}_{CL}(s)$ are identical to those of $(A, B, R^{-1}B^TK)$. Thus any assumptions about $\phi_{OL}(s)$ and $\phi_{CL}(s)$ obviously apply to $\hat{\phi}_{OL}(s)$ and $\hat{\phi}_{CL}(s)$. Similarly, by defining $\hat{L}(s) \triangleq R^{1/2}L(s)R^{-1/2}$, we may work with $\hat{G}(s)$ and $\hat{L}(s)$ instead of $G(s)$ and $L(s)$. The conditions (3.6) and (3.7) of Theorem 3.1 are equivalent to $\underline{\sigma}(I+\hat{G}(s))>1$ and $\underline{\sigma}(I+\hat{G}(s))\geq 1$, respectively. Condition (3.16) and Theorem 3.1 require that

$$\bar{\sigma}(\hat{L}^{-1}(s)-I)\leq 1<\underline{\sigma}(I+\hat{G}(s)), \quad s\in\Omega_R \quad \text{(A.44)}$$

and by Theorem 2.3 we conclude that $\tilde{\phi}_{CL}(s)$ has no CRHP zeros. Alternatively condition (3.17) and Theorem 3.1 require that

$$\bar{\sigma}(\hat{L}^{-1}(s)-I)<1\leq\underline{\sigma}(I+\hat{G}(s)) \quad \text{(A.45)}$$

which again by Theorem 2.3 means $\tilde{\phi}_{CL}(s)$ has no CRHP zeros. Q.E.D.

Proof of Corollary 3.1: From Theorems 3.1 and 3.2 we know if $Q>0$ then

$$\bar{\sigma}(R^{1/2}L^{-1}(s)R^{-1/2}-I)=\bar{\sigma}(L^{-1}(s)-I)\leq 1, \quad s\in\Omega_R \quad \text{(A.46)}$$

to satisfy (3.16) when $L(s)$ and R are diagonal. If $\phi_{CL}(j\omega)\neq 0\ \forall\omega$ then

$$\bar{\sigma}(L^{-1}(s)-I)<1, \quad s\in\Omega_R \quad \text{(A.47)}$$

to satisfy (3.17) when $L(s)$ and R are diagonal. The remainder of the proof of completely analogous to Corollary 2.1. Q.E.D.

Proof of Corollary 3.2: Only conditions (3.16) and (3.17) of Theorem 3.2 need to be verified for the $L(s)$ of (2.42) the rest are satisfied by assumption. Note that for $s\in\Omega_R$

$$\begin{aligned}&\bar{\sigma}(R^{1/2}L^{-1}(s)R^{-1/2}-I) \\ &=\bar{\sigma}\left(\begin{bmatrix} 0 & -R_1^{1/2}X(s)R_2^{-1/2} \\ 0 & 0 \end{bmatrix}\right) \text{ or } \\ &\bar{\sigma}\left(\begin{bmatrix} 0 & 0 \\ -R_2^{1/2}X(s)R_1^{-1/2} & 0 \end{bmatrix}\right) \\ &\leq\bar{\sigma}(X(s))\max\{\bar{\sigma}(R_1^{1/2})\bar{\sigma}(R_2^{-1/2}),\bar{\sigma}(R_2^{1/2})\bar{\sigma}(R_1^{-1/2})\}\end{aligned} \quad \text{(A.48)}$$

and hence if

$$\bar{\sigma}(X(s))\max\left[\frac{\lambda_{\max}(R_1^{1/2})}{\lambda_{\min}(R_2^{1/2})},\frac{\lambda_{\max}(R_2^{1/2})}{\lambda_{\min}(R_1^{1/2})}\right]<1, \quad \text{(A.49)}$$

then conditions (3.16) and (3.17) are both satisfied. However, (A.49) is equivalent to (3.33). Q.E.D.

Proof of Theorem 3.3: Conditions (3.38) and (3.39) and the Lyapunov stability criterion guarantee that condition (2.44) of Theorem 2.4 is satisfied. As in the proof of Theorem 3.2 we may work with $\hat{G}(s)=R^{1/2}G(s)R^{-1/2}$ and $\hat{L}(s)=R^{1/2}L(s)R^{-1/2}$ instead of $G(s)$ and $L(s)$. Condition (3.37) is simply condition (2.45) of Theorem 2.4 with $\hat{G}(s)$ replacing $G(s)$ and $\hat{L}(s)$ replacing $L(s)$ in (2.46) is simply (3.41). Thus by Theorem 2.4 the theorem is proved when (3.41) holds. When $Q>0$ and (3.40) is satisfied, the strictness of the inequality (2.46) of Theorem 2.4 may be changed to $\geq$ and the $\geq$ of (2.45) to $>$ and Theorem 2.4 remains valid. Thus when (3.40) holds Theorem 3.1 is again proved. Q.E.D.

Proof of Theorem 3.4: Breaking the loop at point ① of Fig. 12 we have a loop transfer function matrix of

$$G_r(sI-A+G_fC)^{-1}\left[G_fC(Is-A)^{-1}B+B\right]$$
$$=G_r(sI-A)^{-1}B \triangleq G(s) \quad \text{(A.50)}$$

so that

$$\phi_{OL}(s)=\det\left[sI-A+G_fC\right]\det\left[sI-A\right] \quad \text{(A.51)}$$

and

$$\tilde{\phi}_{OL}(s)=\det\left[sI-A+G_fC\right]\det\left[sI-\tilde{A}\right]. \quad \text{(A.52)}$$

Since the Kalman filter error dynamics are stable given (3.70) and since (3.68) holds, conditions (2.16) and (2.17) of Theorem 2.2 hold. Now by direct application of Theorem 3.2 we conclude that the system of Fig. 12 is stable if $L(s)$ is inserted at point ①. However, this is not the location we desire to have the margins guaranteed. Nevertheless, by manipulation of the block diagram of Fig. 12 we may place $L(s)$ at point ② if we change B to $BL(s)$ inside the controller leaving $G_r=R^{-1}B^TK$ fixed. This, however, is equivalent to changing (A,B,C) to $(\tilde{A},\tilde{B},\tilde{C})$ inside the controller leaving G_f and G_r fixed which is the desired result for perturbation $L(s)$. The proof for a perturbation $N(s)$ at the output is analogous. Q.E.D.

B. Singular Values [19], [38]–[40]

The singular values of a square $n\times n$ complex matrix A, denoted $\sigma_i(A)$, are defined as

$$\sigma_i(A)\triangleq\lambda_i^{1/2}(A^HA)=\lambda_i^{1/2}(AA^H) \quad \text{(B.1)}$$

where A^H denotes the complex conjugate transpose of A and $\lambda_i(A^HA)$ the ith largest eigenvalue of A^HA. A way of representing the matrix A, known as the singular value decomposition (SVD) is given by

$$A=U\Sigma V^H=\sum_{i=1}^{n}\sigma_i(A)u_iv_i^H \quad \text{(B.2)}$$

where

$$U\triangleq\left[u_1,u_2,\cdots,u_n\right]; \qquad U^HU=I \quad \text{(B.3)}$$

$$V\triangleq\left[v_1,v_2,\cdots,v_n\right]; \qquad V^HV=I \quad \text{(B.4)}$$

$$\Sigma=\text{diag}\left[\sigma_1,\sigma_2,\cdots,\sigma_n\right] \quad \text{(B.5)}$$

and the columns of V and U are eigenvectors of A^HA and AA^H, respectively. The minimum and maximum singular values denoted $\underline{\sigma}$ and $\bar{\sigma}$, respectively, are sometimes equivalently defined in terms of the spectral matrix norm $\|\cdot\|_2$ as

$$\bar{\sigma}(A)=\max_{\|x\|_2\neq 0}\frac{\|Ax\|_2}{\|x\|_2}=\|A\|_2 \quad \text{(B.6)}$$

and

$$\underline{\sigma}(A)=\min_{\|x\|_2\neq 0}\frac{\|Ax\|_2}{\|x\|_2}$$
$$=\begin{cases}\|A^{-1}\|_2^{-1}, & \text{if } \det A\neq 0\\ 0, & \text{if } \det A=0.\end{cases} \quad \text{(B.7)}$$

The minimum singular value $\underline{\sigma}(A)$ provides a measure of the nearness to singularity of the matrix A in the following sense. If $A+E$ is singular then

$$\|E\|_2=\bar{\sigma}(E)\geq\underline{\sigma}(A). \quad \text{(B.8)}$$

Other facts involving singular values that are useful for manipulation follow. The inequality

$$\underline{\sigma}(A)\geq\bar{\sigma}(B) \quad \text{(B.9)}$$

implies that

$$A^HA\geq B^HB. \quad \text{(B.10)}$$

Also, since $\bar{\sigma}(\cdot)$ is the same as $\|\cdot\|_2$, the triangle inequality

$$\bar{\sigma}(A+B)\leq\bar{\sigma}(A)+\bar{\sigma}(B) \quad \text{(B.11)}$$

holds. Finally, if A^{-1} exists then (B.6) and (B.7) give

$$\underline{\sigma}(A)=\frac{1}{\bar{\sigma}(A^{-1})}. \quad \text{(B.12)}$$

References

[1] H. H. Rosenbrock, *Computer-Aided Control System Design*. London: Academic Press, 1974.

[2] ——, "Design of multivariable control systems using the inverse Nyquist array," *Proc. IEEE*, vol. 116, pp. 1929–1936, Nov. 1969.

[3] ——, "Multivariable circle theorem," in *Recent Mathematical Developments in Control*, D. B. Bell, Ed. London: Academic Press, 1974.

[4] A. G. J. MacFarlane and I. Postlethwaite, "The generalized Nyquist stability criterion and multivariable root loci," *Int. J. Contr.*, vol. 25, pp. 81–127, Jan. 1977.

[5] ——, "Characteristic frequencies functions and characteristic gain functions," *Int. J. Contr.*, vol. 26, pp. 265–278, Aug. 1977.

[6] H. W. Bode, *Network Analysis and Feedback Amplifier Design*. New York: Van Nostrand, 1945.

[7] M. G. Safonov, "Robustness and stability aspects of stochastic multivariable feedback system design," Ph.D. dissertation, MIT, Rep. ESL-R-763, Cambridge, MA, Sept. 1977; also, MIT Press, 1980.

[8] ——, "Tight bounds on the response of multivariable systems with component uncertainty," in *Proc. Allerton Conf. on Communication, and Computing*, Monticello, IL, Oct. 4–6, 1978.

[9] I. M. Horowitz, *Synthesis of Feedback Systems*. New York: Academic Press, 1963.

[10] G. Zames, "On the input-output stability of time-varying nonlinear feedback systems—Part I: Condition using concepts of loop gain conicity and positivity," *IEEE Trans. Autom. Contr.*, vol. AC-11, pp. 228–238, Apr. 1966.

[11] G. Zames, "On the input-output stability of time-varying nonlinear feedback systems—Part II: Conditions involving circles in the frequency plane and sector nonlinearities," *IEEE Trans. Autom. Contr.*, vol. AC-11, pp. 465–476, July 1966.

[12] C. A. Desoer and M. Vidysager, *Feedback Systems: Input-Output Properties*. New York: Academic Press, 1975.

[13] J. C. Willems, *The Analysis of Feedback Systems*. Cambridge, MA: MIT Press, 1971.

[14] J. C. Doyle, "Robustness of multiloop linear feedback systems," in *Proc. 1978 IEEE Conf. on Decision and Control*, San Diego, CA, Jan. 10–12, 1979.

[15] R. DeCarlo and R. Saeks, "The encirclement condition an approach using algebraic topology," *Int. J. Contr.*, vol. 26, pp. 279–287, 1977.

[16] R. DeCarlo, J. Murray, and R. Saeks, "Multivariable Nyquist theory," *Int. J. Contr.*, vol. 25, pp. 657–675, 1977.

[17] R. Saeks and R. DeCarlo, "Stability and homotopy," in *Alternatives for Linear Multivariable Control*, M. Sain, J. Peczkowski, and J. Melsa, Eds. Chicago, IL: Nat. Eng. Consortium, 1978.

[18] M. G. Safonov and M. Athans, "A multiloop generalization of the circle stability criterion," in *Proc. 12th Annu. Asilomar Conf. on Circuits, Systems and Computers*, Pacific Grove, CA, Nov. 6–8, 1978.

[19] A. J. Laub, "Computational aspects of singular value decomposition and some applications," in *Proc. Allerton Conf. on Communication, Control and Computing*, Monticello, IL, Oct. 4–6, 1978.

[20] H. H. Rosenbrock and P. D. McMorran, "Good, bad, or optimal?" *IEEE Trans. Automat. Control*, vol. AC-16, pp. 552–553, Dec. 1971.

[21] R. E. Kalman, "When is a linear control system optimal?," *Trans. ASME Ser. D: J. Basic Eng.*, vol. 86, pp. 51–60, Mar. 1964.

[22] B. D. O. Anderson and J. B. Moore, *Linear Optimal Control*. Englewood Cliffs, NJ: Prentice-Hall, 1971.

[23] B. D. O. Anderson, "The inverse problem of optimal control," Stanford Electronics Laboratories, Tech. Rep. SEL-66-038 (T.R. No. 6560-3), Stanford, CA, Apr. 1966.

[24] J. B. Cruz, Jr., and W. R. Perkins, "A new approach to the sensitivity problem in multivariable feedback system design," *IEEE Trans. Automat. Contr.*, vol. AC-9, pp. 216–233, July 1964.

[25] M. G. Safonov and M. Athans, "Gain and phase margin for multiloop LQG regulators," *IEEE Trans. Automat. Contr.*, vol. AC-22, pp. 173–179, Apr. 1977.

[26] P. K. Wong, "On the interaction structure of multi-input feedback control systems," M.S. thesis, MIT, Cambridge, MA, Sept. 1975.

[27] P. K. Wong and M. Athans, "Closed-loop structural stability for linear-quadratic optimal systems," *IEEE Trans. Automat. Contr.*, vol. AC-22, pp. 94–99, Feb. 1977.

[28] B. D. O. Anderson, "Stability results for optimal systems," *Electron. Lett.*, vol. 5, p. 545, Oct. 1969.

[29] S. Barnett and C. Storey, "Insensitivity of optimal linear control systems to persistent changes in parameters," *Int. J. Contr.*, vol. 4, pp. 179–184, 1966.

[30] C. A. Harvey, "On feedback systems possessing integrity with respect to actuator outages," in *Recent Developments in the Robustness Theory of Multivariable Systems*, N. R. Sandell, Jr. Ed., LIDS, Rep. LIDS-R-954, MIT, Cambridge, MA, Aug. 1979.

[31] P. Molander and J. C. Willems, "Robustness designs by state feedback," in *Recent Developments in the Robustness Theory of Multivariable Systems*, N. R. Sandell, Jr., Ed. Laboratory for Information and Decision Systems, Rep. LIDS-R-954, MIT, Cambridge, MA, Aug. 1979.

[32] M. G. Safonov and M. Athans, "Robustness and computational aspects of nonlinear stochastic estimators and regulators," *IEEE Trans. Automat. Cont.*, vol. AC-23, pp. 717–725, Aug. 1978.

[33] J. C. Doyle, "Guaranteed margins for LQG regulators," *IEEE Trans. Automat. Contr.*, vol. AC-23, pp. 756–757, Aug. 1978.

[34] A. E. Bryson, "Random problems in control theory," Stanford University, Rep. SUDAAR 447, Sept. 1972.

[35] J. C. Doyle and G. Stein, "Robustness with observers," *IEEE Trans. Automat. Contr.*, vol. AC-24, Aug. 1979.

[36] H. Kwakernaak, "Optimal low-sensitivity linear feedback systems," *Automatic*, vol. 5, May 1969.

[37] J. R. Dowdle, "Robust observer based compensator," Ph.D. dissertation, MIT, Cambridge, MA, Aug. 1979.

[38] B. Noble, *Applied Linear Algebra*. Englewood Cliffs, NJ: Prentice-Hall, 1969.

[39] C. L. Lawson and R. J. Hanson, *Solving Least Squares Problems*. Englewood Cliffs, NJ: Prentice-Hall, 1974.

[40] J. H. Wilkinson and C. Reinsch, *Handbook for Automatic Computation, II, Linear Algebra*, F. L. Bauer *et al.*, Ed. New York: Springer-Verlag, 1971.

[41] M. Athans, "The role and use of the stochastic linear-quadratic-Gaussian problem in control system design," *IEEE Trans. Automat. Contr.*, vol. AC-16, pp. 529–552, Dec. 1971.

[42] J. B. Lewis, "Automotive engine control: As linear-quadratic approach," S.M. thesis, Laboratory for Information and Decision Systems, MIT, Cambridge, MA, Mar. 1980.

[43] J. C. Doyle and G. Stein, "Multivariable feedback design: Concepts for a classical/modern synthesis," *IEEE Trans. Automat. Contr.*, vol. AC-26, no. 1, Feb. 1981.

Stability margins of diagonally perturbed multivariable feedback systems

M.G. Safonov, B.Sc., M.Sc., Ph. D., Sen. Mem.I.E.E.E.

Indexing terms: *Control theory, Feedback, Multivariable control systems, Stability, Sensitivity*

Abstract: For diagonally perturbed linear time-invariant multivariable feedback systems, the problem of finding an improved characterisation of the stability margin is examined. A readily computable lower bound for diagonally perturbed systems is developed using Perron-Frobenius non-negative matrix results. The present theory improves upon the existing singular-value stability-margin theory, providing a simple constructive method for determining previously unspecified norm weighting parameters (i.e. scaling factors) so as to minimise the conservativeness of stability-margin bounds.

1 Introduction

In characterising the stability of a multivariable linear time-invariant system, what is usually desired is a lower bound on the 'size' of the smallest perturbation to the transfer matrix that can induce the system to become unstable. Because the nominal value of the transfer function can be manipulated via compensator matrices, the relevant measure of the size of a perturbation should be independent of such compensation. Hence, for a feedback system having a *loop transfer-function matrix* $L(s) \in C^{n \times n}$ with nominal value $L_0(s)$, the relevant measure of size must depend only on the matrix

$$\Delta(s) \triangleq (L(s) - L_0(s))L_0^{-1}(s) \tag{1}$$

The true loop transfer-function matrix $L(s)$ is related to its nominal value $L_0(s)$ by

$$L(s) = (I + \Delta(s))L_0(s) \tag{2}$$

where $\Delta(s)$ is the perturbation matrix whose size is of interest, see Fig. 1. The representation $\Delta(s)$ of the perturbation in the loop transfer-function matrix $L(s)$ is called a *multiplicative perturbation* representation; loosely, the matrix $\Delta(s)$ reflects the percentage variation of $L(s)$ relative to $L_0(s)$. When the *only* information available about $\Delta(s)$ is a bound on the norm of $\Delta(s)$ induced† by some norm on C^n, then the condition [1, 2]

$$\|\Delta(j\omega)\| < 1/\|(I + L_0^{-1}(j\omega))^{-1}\| \tag{3}$$

provides the least conservative possible characterisation of the stability margin. Singular-value stability conditions are a special case of expr. 3 that results when the matrix norm induced by the l_2-norm (i.e. the Euclidean norm) on C^n is employed in expr. 3. For this norm, the stability condition given in expr. 3 becomes the standard singular-value inequality

$$\sigma_{max}(\Delta(j\omega)) < \sigma_{min}(I + L_0^{-1}(j\omega))$$

where $\sigma_{max}(\cdot)$ and $\sigma_{min}(\cdot)$ denote, respectively, the greatest and least singular values of $(\cdot)$ [1–11].

Unfortunately, it is usually the case that much more is known about $\Delta(j\omega)$ than just a simple bound on its norm or its singular values; thus the condition given in expr. 3 usually yields a *conservative* bound on stability margins. Specifically, it is commonly the case that the perturbation $\Delta(j\omega)$ is a *diagonal* matrix, i.e.

$$\Delta(s) = \text{diag}(\Delta_1(s), \ldots, \Delta_n(s)) \tag{4}$$

where $\Delta_i(s) \in C$ for all i,s. The reason for the special interest in diagonal perturbation matrices is that they naturally represent the case of simultaneous variations in the gains and phases in several distinct system components or feedback paths. The problem of multiloop feedback system 'integrity' in the face of simultaneous variations in loop *gains* throughout a set of nonzero measure is an example of a type of diagonal perturbation stability problem; see Reference 12 for an overview of this problem, including a brief survey of some early theoretical studies and connections with the eigenvalues of $L_0(s)$. The multiloop LQG gain and phase margin results of Safonov and Athans [13] introduced the concept of multiloop robustness in the face of simultaneous *phase* perturbations and showed that full state feedback linear optimal regulators have nice properties in this regard. The multivariable stability-margin problem was treated in depth in Reference 14, where connections between stability robustness and generalised conic sectors, positive real conditions and Lyapunov functions are developed. The singular value was introduced as a method for bounding the stability margins of diagonally perturbed systems by Doyle [4]. Barrett [1] has made an extensive study of the conservativeness of singular-value conditions and other conditions related to expr. 3 for bounding the stability margins of diagonally perturbed systems. Despite its conservativeness, the singular-value condition has since gained broad acceptance as a stability-margin measure for multivariable feedback systems (see, for example, References 1–3 and 6–11), mainly because no adequately simple straightforward technique for reducing this conservativeness has yet emerged. The result in this paper addresses this problem.

The use of diagonal multiplier matrices (also known as norm weights or scaling factors) has been recognised as one possibility for conservativeness reduction. This old idea has been suggested for use in the multivariable stability-margin context by several authors, for example, Barrett [1], Safonov and Athans [5] and Postlethwaite *et al.* [7] to name a few. The use of a variety of unusual norms and even more exotic robustness measures has also been suggested by Barrett [1]. However, the choice of multipliers or norms for any given feedback system to minimise the conservativeness of the resultant stability-margin bounds has until now remained largely trial and error.

The principal contribution of this paper is a simple constructive technique for determining the multipliers which minimise the conservativeness of stability-margin bounds obtained from expr. 3 when the matrix norm in expr. 3 is the

†Given any norm $\|\cdot\|$ on C^n, this induces a norm on the set of matrices $A \in C^{n \times n}$: namely $\|A\| \triangleq \sup_{x \neq 0} \|Ax\|/\|x\|$. For example, the largest singular value of A is the norm of A induced by the Euclidean norm on C^n.

Paper 2201D, first received 6th May and in revised form 15th September 1982

The author is with the Department of Electrical Engineering Systems, University of Southern California, University Park, Los Angeles, CA 90089, USA

Reprinted with permission from *IEE Proc.*, vol. 129, Pt. D, no. 6, pp. 251-256, Nov. 1982.

one induced by the Hölder l_p-norm on C^n for $p = 1$ and for $p = \infty$. Good suboptimal mulitpliers are also found for $1 < p < \infty$, including the $p = 2$ case corresponding to singular-value results.

In this paper, it is assumed that the perturbation $\Delta(s)$ is diagonal (compare with eqn. 4) and, following References 3 and 5, we suppose that bounds on the individual perturbations $\Delta_i(s)$ have been determined beforehand, namely

$$|\Delta_i(j\omega)| < |r_i(j\omega)| \quad \text{for all } i,\omega \tag{5a}$$

As in Reference 5, we characterise the stability margin as follows.

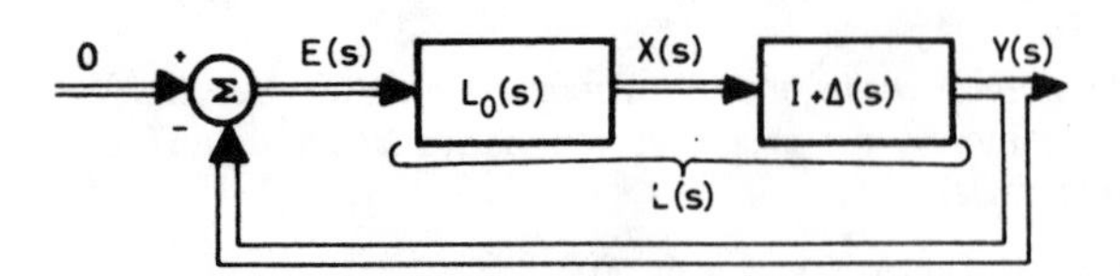

Fig. 1 *Perturbed multivariable feedback*

Definition
The stability margin of the system in Fig. 1 is the largest positive $k_m(j\omega)$ such that this system is stable for every stable diagonal $\Delta(s)$ for which

$$|\Delta_i(j\omega)| < k_m(j\omega)|r_i(j\omega)| \quad \text{for all } i,\omega \tag{5b}$$

The number $k_m(j\omega)$ is a measure of the amount, at each frequency ω, by which the system exceeds [if $k_m(j\omega) \geq 1$], or falls short of [if $k_m(j\omega) < 1$], having ample stability margin to accommodate simultaneous perturbations $\Delta_i(s)$, $i = 1, \ldots, n$, bounded by expr. 5a. For notational convenience, we define

$$R(s) \triangleq \text{diag}(r_1(s), \ldots, r_n(s)) \tag{6}$$

Theorem 1 of the following Section gives a simple formula for the optimal l_1-norm and l_∞-norm weights for minimising the conservativeness of the condition given in expr. 3. Suboptimal weights are given for other l_p-norms, including the case $p = 2$, thus providing a method for suboptimally choosing good weights to reduce the conservativeness of conventional singular-value stability-margin bounds, which are based on the Euclidean l_2-norm on C^n.

2 Main result

In this Section, the main result (theorem 1) is stated. Readers unfamiliar with the Perron-Frobenius theory of nonnegative matrices should refer to Appendix 8.1 before reading this Section.

Definition
Given a number $p \in [1,\infty]$ and a diagonal matrix $D \triangleq \text{diag}(d_1, \ldots, d_n) \in C^{n \times n}$ with $d_i \neq 0$ for all $i = 1, \ldots, n$, the D-weighted Hölder l_p-norm on C^n is

$$\|x\|_{pD} \triangleq \|Dx\|_p \triangleq \left(\sum_{i=1}^{n} |d_i x_i|^p\right)^{1/p} \quad \text{for all } x \in C^n \tag{7}$$

The subordinate bound norm induced on $C^{n \times n}$ by the l_p-norm on C^n is

$$\|A\|_{pD} \triangleq \max_{x \neq 0} \frac{\|Ax\|_{pD}}{\|x\|_{pD}} \quad \text{for all } x \in C^{n \times n} \tag{8}$$

Definition
For every matrix $A = \{a_{ij}\}_{i,j=1}^n \in C^{n \times n}$, we define the nonnegative matrix

$$|A| \triangleq \{|a_{ij}|\}_{i,j=1}^n \in R^{n \times n} \tag{9a}$$

Likewise, for every vector $x = \{x_j\}_{j=1}^n$, we define the nonnegative vector

$$|x| \triangleq \{|x_j|\}_{j=1}^n \in R^n \tag{9b}$$

Define

$$B(j\omega) \triangleq (I + L_0(j\omega))^{-1} L_0(j\omega) R(j\omega)$$

and let $\pi(j\omega)$ denote the Perron eigenvalue of $|B(j\omega)|$. Denote by $x(j\omega)$ and $y(j\omega)$ the respective right and left Perron eigenvectors of $|B(j\omega)|$ satisfying $\pi x = |B|x$ and $\pi y^T = y^T|B|$ for all ω, normalised so that $\|x\|_\infty = \|y\|_\infty = 1$ for all ω.

The importance of D-weighted l_p-norms in the context of diagonally perturbed system stability margins is that the quantity $\|\Delta(j\omega)\|$ on the left-hand side of expr. 3 is independent of the weight D whenever $\Delta(s)$ is a diagonal matrix. Specifically, one has the following lemma, which is readily derived from the stability condition given in expr. 3.

Lemma 1
For every $p \in [1,\infty]$ and every $D(s) \triangleq \text{diag}(d_1(s), \ldots, d_n(s))$ with $d_i(j\omega) \neq 0$ for all $i = 1, \ldots, n$ and for all ω, the stability margin $k_m(j\omega)$ of the diagonally perturbed system in Fig. 1 is bounded below by

$$k_m(j\omega) \geq 1/\|B(j\omega)\|_{pD(j\omega)} \tag{10}$$

i.e. the system of Fig. 1 is stable for every diagonal $\Delta(s) = \text{diag}(\Delta_1(s), \ldots, \Delta_n(s))$ satisfying $|\Delta_i(j\omega)| < |r_i(j\omega)|/\|B(j\omega)\|_{pD(j\omega)}$

Proof
The proof is given in Appendix 8.2.

In view of lemma 1, it is apparent that, to minimise the conservativeness of the stability condition expr. 3, one should choose the norm and weight D so as to minimise the right-hand side of the expr. 3. Existing stability-margin conditions based on expr. 3, including singular-value conditions, either arbitrarily suppose $D = I$ (for example, References 2, 4 and 6), leading to unduly conservative results, or else leave the choice of D unspecified (for example, References 5, 7 and 9), leading to a trial-and-error search for a suitable D. The following lemma provides a means circumventing this trial-and-error search for a good D, giving the optimal D for $p = 1$ and $p = \infty$ and giving a good suboptimal D for $1 < p < \infty$.

Lemma 2
Define

$$D_p^* \triangleq \text{diag}(y_1^{1/p} x_1^{-1/q}, \ldots, y_n^{1/p} x_n^{-1/q}) \tag{11}$$

where q is determined by

$$\frac{1}{q} + \frac{1}{p} = 1$$

The p-norm weight D_p^* is optimal for $p = 1$ and $p = \infty$ in the sense that

$$\|B(j\omega)\|_{pD_p^*(j\omega)} = \| \, |B(j\omega)| \, \|_{pD_p^*(j\omega)} = \inf_D \| \, |B(j\omega)| \, \|_{pD} = \pi(j\omega) \quad \text{for all } \omega \tag{12}$$

For $1 < p < \infty$, the weight D_p^* is suboptimal in that the equality given in eqn. 12 must be replaced by the following inequality:

$$\|B(j\omega)\|_{pD_p^*(j\omega)} \leq \| \, |B(j\omega)| \, \|_{pD_p^*(j\omega)} = \inf_D \| \, |B(j\omega)| \, \|_{pD} = \pi(j\omega) \quad \text{for all } \omega \tag{13}$$

Proof
The proof is given in Appendix 8.3.

The implications of lemmas 1 and 2 for diagonal $\Delta(s)$ are readily seen from the following theorem, which is an immediate consequence of these lemmas.

Theorem 1 (main result)
Suppose that the unperturbed system $L_0(s)(I + L_0(s))^{-1}$ is stable, that $\Delta(s)$ is a stable diagonal matrix and that $|B(j\omega)|$ is irreducible except at a few values of ω.‡ Then the stability margin $k_m(j\omega)$ of the system of Fig. 1 is bounded below by

$$k_m(j\omega) \geqslant 1/\|B(j\omega)\|_{pD_p^*(j\omega)} \tag{14}$$

This stability-margin bound is *optimal* for $p = 1$ and $p = \infty$ in the sense that the equality given in eqn. 12 holds. For $1 < p < \infty$, the stability-margin bound of expr. 14 is suboptimal in the sense that the inequality given in expr. 13 holds.

Proof
The proof follows directly from lemmas 1 and 2.

There is really no need to compute the weight $D^*(j\omega)$ explicitly; it suffices to compute $\pi(j\omega)$, the Perron eigenvalue of $|B(j\omega)|$, since from eqn. 12 and expr. 13 we have

$$\|B(j\omega)\|_{pD_p^*(j\omega)} \leqslant \pi(j\omega) \tag{15}$$

(with equality holding for $p = 1$ and $p = \infty$) where the weighting matrix D_p^* is given by eqn. 11 of lemma 2. It follows from theorem 1 that

$$k_m(j\omega) \geqslant 1/\pi(j\omega) \tag{16}$$

Thus one has that the *optimal* D-weighted l_1-norm and l_∞-norm stability-margin bound is simply $1/\pi(j\omega)$.

3 Discussion

The question arises: 'Are there instances besides $p = \infty$ and 1 for which D_p^* is optimal in the sense that expr. 15 holds with equality?' From expr. 13, it is immediately evident that equality always holds if $B(j\omega)$ is itself a nonnegative matrix, i.e. if $|B(j\omega)| = B(j\omega)$. Furthermore, since for every pair of unitary diagonal matrices $\Theta, \Psi \in C^{n\times n}$ [i.e. every $\Theta = \text{diag}(\exp j\theta_1, \ldots, \exp j\theta_n)$ and $\Psi = \text{diag}(\exp j\psi_1, \ldots, j\psi_n)$] $\|B\|_{pD} = \|\Psi B^{-1}\theta\|_{pD}$, it follows that expr. 15 holds with equality irrespective of the value of $p \in [1,\infty]$ if

$$|B(j\omega)| = \Theta(j\omega)B(j\omega)\Psi^{-1}(j\omega) \tag{17}$$

for some unitary diagonal Θ and Ψ.

We note that it is relatively straightforward to modify the proof and conditions of theorem 1 so that theorem 1 and its corollaries will accommodate *multi-input multi-output nonlinear* Δ_is satisfying conic sector conditions of the type [3,5]

$$\Delta_i \text{ inside cone } (0; S_i, R_i)$$

i.e. we just replace $|\cdot|$ by a suitable $\|\cdot\|$ throughout. In fact, certain existing nonlinear interconnected-system stability results by Michel, Lasley, Cook, Porter, Siljak, Araki and others essentially accomplish this for the case $p = 2$ (see the survey paper of Reference 16 for references). However, these nonlinear results require $|B(j\omega)|$ to be replaced to the larger constant matrix $\tilde{B} \triangleq \{\sup|b_{ij}(j\omega)|\}_{i,j=1}^n$, leading to more conservative results than theorem 1 in the case of linear time-invariant systems. The best result obtainable via their results would be that the system of Fig. 1 is stable for all diagonal $\Delta(s)$ satisfying expr. 5*a* if I-$\tilde{B}$ is an 'M-matrix', but no bound on $k_m(j\omega)$ would be provided, except of course the bound $k_m(j\omega) \geqslant 1$ which holds trivially whenever stability is assured. This M-matrix condition is equivalent to the condition that the Perron eigenvalue of $\tilde{B}$ be less than one.§ This is more conservative than our results since the Perron eigenvalue of $\tilde{B}$ is always greater than or equal to the Perron eigenvalue of $|B(j\omega)|$ (see Reference 17, p. 57 of vol. 2).

Although diagonal dominance is not directly related to diagonal perturbations it has been shown by Mees [18] that the problem of finding an optimal diagonal postcompensator $D = \text{diag}(d_1, \ldots, d_n)$ for an $n \times n$ plant $G(s_0)$ so that $DG(s_0)$ will be row dominant is solved by taking $(d_1, \ldots, d_n)^T$ to be the left Perron eigenvector of the $n \times n$ nonnegative matrix $T \triangleq \{|G_{ij}(s_0)/G_{ii}(s_0)|\}_{i,j=1}^n$. Noting that lemma 5 implies that $DG(s_0)$ is row dominant if, and only if, $\|T-I\|_{1D} < 1$, we can see that Mees' result can also be obtained as a corollary to our lemma 1, corresponding the special case $p = 1$. Similarly, the optimal diagonal precompensator D^{-1} to make GD^{-1} column dominant is obtained by taking $p = \infty$ in lemma 1. Of course, the result in lemma 1 is more general in that it addresses values of p other than 1 and ∞, including the important case of $p = 2$ that arises in selecting suitable scaling factors for computing stability margins via singular values.

As a final note, we mention that the matrix scaling algorithm of E.E. Osborne [19] also has the potential to be used for suboptimally choosing norm weight D when $p = 2$. Osborne's algorithm is an efficient iterative routine that converges to a diagonal D which minimises $\sum_{i=1}^n \sigma_i^2\,(BDB^{-1})$ for any given matrix B. In view of the inequality

$$\sigma_{max}(DBD^{-1}) \leqslant \frac{1}{\sqrt{n}}\left\{\sum_{i=1}^n \sigma_i^2(DBD^{-1})\right\}^{1/2} \tag{18}$$

a stability-margin estimate

$$|r_i^{-1}\,\Delta_i| < 1/\sigma_{max}(DBD^{-1}) \tag{19}$$

obtained using the Osborne D may be as much as a factor $\sqrt{n}$ more conservative than an estimate obtained using the true optimal D. It is difficult to compare the Osborne D with the D_p^* given for $p = 2$ by eqn. 11, since both D_p^* for $p = 2$ and the Osborne D are suboptimal. We note, however, that our D_p^* for $p = 2$ actually achieves the optimal stability-margin bound for the relatively broad class of matrices satisfying eqn. 17, whereas no comparable optimality properties are known for the Osborne D.

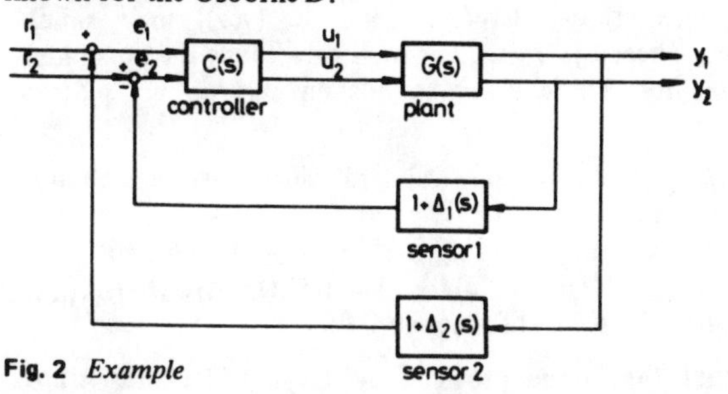

Fig. 2 *Example*

‡ For reducible $|B(j\omega)|$ one may separately consider each of the irreducible submatrices of $|B(j\omega)|$, since these irreducible submatrices correspond precisely to the *strongly connected subsystems* (see Reference 15). In cases where $|B(j\omega)|$ is reducible at only a few points $\omega \in [0, \infty]$ theoretical rigour demands that one avoid these points by making infinitesimal indentations in the Nyquist D-contour. However, because $\|B(s)\|_{D^*(s)}$ is continuous at points in the complex s-plane where it exists, such indentations have no pratical effect on the conditions of theorem 1.

§ This equivalence with I-$\tilde{B}$ being an M-matrix follows directly from the standard result (for example theorem A1 of Reference 9) that I-$\tilde{B}$ is an M-matrix if, and only if, all its real eigenvalues $\lambda_i(I\text{-}\tilde{B})$ are positive. But $\lambda_i(I\text{-}B) = 1 - \lambda_i(B)$; hence the Perron eigenvalue $\tilde{\pi} \triangleq \max_i \lambda_i(\tilde{B})$ is less than one if, and only if, I-$\tilde{B}$ is an M-matrix.

4 Example

To illustrate the application of theorem 1 and to demonstrate its merit as a conservativeness-reducing technique, we consider the simple system depicted in Fig. 2 having two equally uncertain sensors. The two sensors have precision ± 50% for $\omega \ll 1$, and accuracy decreases roughly in proportion to ω for $\omega > 1$; more precisely, we have

$$|\Delta_i(j\omega)| < |r_i(j\omega)| \qquad \text{for all } \omega \qquad i = 1, 2 \tag{20}$$

where $r_1(s) = r_2(s) = (1 + s)/2$, i.e.

$$R(s) = \begin{bmatrix} \frac{1+s}{2} & 0 \\ 0 & \frac{1+s}{2} \end{bmatrix} \tag{21}$$

The loop transfer matrix $L(s) \triangleq G(s)C(s)$ is such that

$$B(s) \triangleq (I + L_0(s))^{-1} L_0(s) R(s) = \begin{bmatrix} \frac{0.5}{s+1} & \frac{0.5}{s+0.1} \\ \frac{0.5}{s+10} & \frac{0.5}{s+1} \end{bmatrix} \tag{22}$$

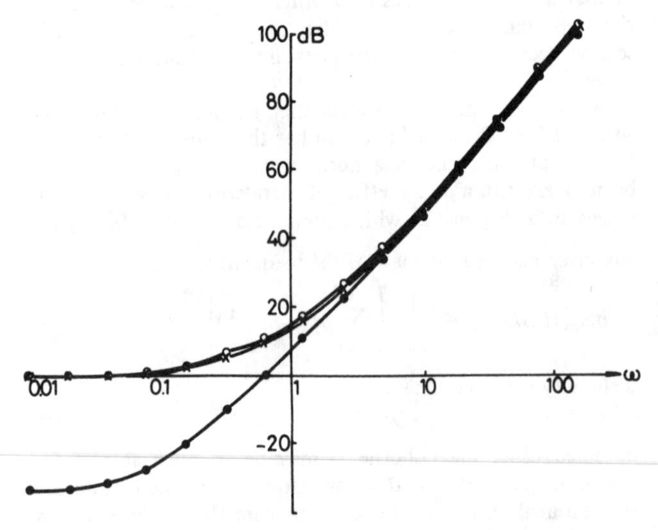

Fig. 3 *Comparison of results for example*

o $1/\sigma\,(D^*BD^{*-1})$
× $1/\pi\,(B)$
• $1/\sigma\,(B)$

The stability margin $k_m(j\omega)$ is the amount by which the sensor uncertainties $|\Delta_1(j\omega)|$ and $|\Delta_2(j\omega)|$ may simultaneously exceed $|1 + j\omega|/2$ at each ω without the possibility of instability. Three lower bounds on $k_m(j\omega)$ are plotted in Fig. 3:

(a) $1/\sigma_{max}(B(j\omega))$, the 'standard' singular-value stability-margin bound

(b) $1/\pi(j\omega)$, the Perron eigenvalue bound of expr. 16

(c) $1/\sigma_{max}(D_2^*(j\omega)B(j\omega)D_2^{*-1}(j\omega)) \equiv 1/\|B(j\omega)\|_{2D_2^*(j\omega)}$, the bound of expr. 14 of theorem 1 for $p = 2$.

Note that the bound provided by $1/\sigma_{max}(B(j\omega))$ does not even ensure stability for $|\Delta_i(j\omega)| < |r_i(j\omega)|$, since $1/\sigma_{max}(B) < 1$ for some frequencies ω. However, the improved bound $1/\pi(j\omega)$ and the even less conservative bound $1/\sigma_{max}(D_2^*BD_2^{*-1})$ both show ample stability margin [i.e. $k_m(j\omega) \geq 1$] at all frequencies. Fig. 3 shows a more than thirty-fold reduction in conservativeness relative to the standard singular-value bound $k_m \geq 1/\sigma_{max}(B)$ at low frequencies, where the $B(j\omega)$ given by eqn. 22 is relatively 'skew', i.e. it has non-orthogonal eigenvectors. At higher frequencies where $B(j\omega)$ is 'non-skew', the scaling provided by D_2^* produces no benefit. This is as one would expect, since for matrices with orthogonal eigenvectors the singular values cannot be reduced by scaling or other similarity transformations. Note that the difference between $1/\pi(j\omega)$ and $1/\sigma_{max}(D_2^*BD_2^{*-1})$ is small, and vanishes (as expected) at frequencies where eqn. 17 holds, in this case at low frequencies and high frequencies.

5 Conclusions

The problem of minimising the conservativeness of stability-margin bounds for multivariable feedback systems has been addressed. As is customary, it has been assumed that the system perturbations against which stability must be assured are uncorrelated and arise *simultaneously* in several distinct system components or feedback paths, thus giving rise to a *diagonal* perturbation matrix. This should not be confused with old-fashioned one-loop-at-a-time stability-margin analysis techniques, which do not admit consideration of simultaneous loop perturbations in several loops; nor should it be confused with results such as arise from the multivariable root locus theory, which require completely correlated perturbations of the form $k(s)I$, where $k(s)$ is scalar.

The main result (theorem 1) gives the *optimal* diagonal norm-weighting matrix D_p^* (or scaling factor matrix) so as to minimise the conservativeness of the stability condition given in expr. 3 when the norm in question is the bound norm of eqn. 8 induced by the l_p-norm on C^n for $p = 1$ and $p = \infty$. Theorem 1 also gives a good suboptimal diagonal weighting matrix D_p^* for other values of $p \in (1, \infty)$, and shows that, *with the weighting* D_p^* given by eqn. 11, the stability-margin bounds obtained for any $p \in (1, \infty)$ will be at least as good as (i.e. no more conservative than) the optimum bound $k_m(j\omega) \geq 1/\pi(j\omega)$ which is attained by D_p^* for $p = 1$ and $p = \infty$. Of course, if one uses some weight D other than the D_p^* of eqn. 11, then in general one may get very conservative results, as the example illustrates.

In practice, there is no need to compute D_p^* or even the Perron eigenvectors x and y. Only the Perron eigenvalue $\pi(j\omega)$ need be computed since one has the stability-margin bound of expr. 15, which is optimal for $p = 1$ and $p = \infty$. The Perron eigenvalue itself is relatively easy to compute. Being real, positive and the eigenvalue of largest magnitude of the real non-negative matrix $|B(j\omega)| \in R^{n \times n}$, it can be readily computed even by simplistic algorithms such as direct iteration on a single vector (see. p. 571 of Reference 20). In contrast, singular-value computations require that one work with an $n \times n$ *complex* matrix and normally require that one invoke more complex computational algorithms than direct iteration. Furthermore, singular-value stability-margin bounds obtained without using the $p = 2$ suboptimal weight D_2^* obtained from eqn. 11 are typically, but not always, more conservative than the Perron eigenvalue bound of expr. 15. The standard singular-value bound is merely a special case of the conservative result of lemma 1, in which the weight D is arbitrarily set at $D = I$. If by trial and error, some weight other than the $p = 2$ value D_2^* is found to produce even less conservative singular-value stability-margin bounds than theorem 1 over some frequency ranges, one may of course use this other weight over these frequency ranges.

The contribution of this paper to the existing singular-value stability-margin is that it provides a simple constructive, but suboptimal, procedure for choosing the l_2-norm weight D, which previously would have been arbitrarily set at $D = I$

or, perhaps, determined by trial and error. More importantly, the stability-margin bound

$$k_m(j\omega) \geqslant 1/\pi(j\omega)$$

provides a new simpler-to-compute alternative to standard singular-value stability-margin bounds. Moreover, this Perron eigenvalue stability-margin bound is typically less conservative than using singular values with arbitrarily chosen weights such as $D = I$.

6 Acknowledgment

This research was supported in part by Honeywell Systems & Research Center, Minneapolis, USA, and in part by Joint Services Electronics Program contract F4462–76C–0061 monitored by AFOSR.

7 References

1 BARRETT, M.F.: 'Conservatism with robustness tests for linear feedback control systems'. Ph.D. thesis, University of Minnesota, USA, June 1980; report 80SRC35, Honeywell Systems Research Center, Minneapolis, Minnesota, USA

2 STEIN, G., and SANDELL, N.R., Jr.: 'Classical and modern methods for control system design'. Notes for MIT course 6.291, Department of Electrical Engineering & Computer Sciences, MIT, Cambridge, Massachusetts, USA, 1979

3 SAFONOV, M.G., and ATHANS, M.: 'A multiloop generalisation of the circle stability criterion'. Proceedings of twelfth annual Asilomar conference on circuits, systems and computers, Pacific Grove, California, USA, Nov. 1978

4 DOYLE, J.C.: 'Robustness of multiloop linear feedback systems'. Proceedings of IEEE conference on decision and control, Fort Lauderdale, Florida, USA, Dec. 1979

5 SAFONOV, M.G., and ATHANS, M.: 'A multiloop generalisation of the circle criterion for stability margin analysis', *IEEE Trans.*, 1981, **AC-26**, pp. 415–422

6 SANDELL, N.R.: 'Robust stability of systems with application to singular perturbations', *Automatica*, 1979, **15**, pp. 467–470

7 POSTLETHWAITE, I., EDMUNDS, J.M., and MacFARLANE, A.G.J.: 'Principal gains and phases in the analysis of linear multivariable feedback systems', *IEEE Trans.*, 1981, **AC-26**, pp. 32–46

8 SAFONOV, M.G., LAUB, A.J., and HARTMANN, G.L.: 'Feedback properties of multivariable systems: The role and use of the return difference matrix', *ibid.*, 1981, **AC-26**, pp. 75–92

9 LEHTOMAKI, N.A., SANDELL, N.R., and ATHANS, M.: 'Robustness results in linear-quadratic Gaussian based multivariable control designs', *ibid.*, 1981, **AC-26**, pp. 75–92

10 BARRETT, M.F.: 'Conservatism with robustness tests for linear feedback control systems', Proceedings of IEEE conference on decision and control, Albuquerque, New Mexico, USA, Dec. 1980

11 DOYLE, J.C., and STEIN, G.: 'Multivariable feedback system design: Concepts for a classical modern synthesis', *IEEE Trans.*, 1981, **AC-26**, pp. 4–16

12 ROSENBROCK, H.H., and COOK, P.A.: 'Stability and the eigenvalues of $G(s)$', *Int. J. Control*, 1975, **21**, pp. 99–104

13 SAFONOV, M.G., and ATHANS, M.: 'Gain and phase margin for multiloop LQG regulators', *IEEE Trans.*, 1977, **AC-22**, pp. 173–178

14 SAFONOV, M.G.: 'Robustness and stability aspects of stochastic multivariable feedback system design', Ph.D. thesis, MIT, Cambridge, Massachusetts, USA, Aug. 1977, report ESL-R-763, Electronics Systems Laboratory, MIT, Cambridge, Massachusetts, USA, Sept. 1977

15 CALLIER, F.M., CHAN, W.S., and DESOER, C.A.: 'Input-output stability of interconnected systems using decompositions: An improved formulation', *IEEE Trans.*, 1978, **AC-23**, pp. 150–163

16 ARAKI, M.: 'Input-output stability of composite feedback systems', *ibid.*, 1976, **AC-21**, pp. 254–259

17 GANTMACHER, F.R.: 'Theory of matrices' (Chelsea, 1960)

18 MEES, A.J.: 'Achieving diagonal dominance', *Syst. Control Lett.*, 1981, pp. 155–158

19 OSBORNE, E.E.: 'On preconditioning of matrices', *J. Assoc. Comput. Mach.*, 1960, **7**, pp. 338–345

20 WILKENSON, J.H.: 'Algebraic eigenvalue problem' (Clarendon Press, 1965)

21 STOER, J., and WITZGALL, C.: 'Transformations by diagonal matrices in a normed space', *Numer. Math.*, 1962, **4**, pp. 458–471

22 BAUER, F.L.' 'Optimally scaled matrices', *ibid.*, 1963, **5**, pp. 73–87

8 Appendixes

8.1 Theory of non-negative matrices

Non-negative matrices are matrices whose individual elements are non-negative. They have a number of remarkable properties. A good Reference is chapter 13 of Reference 17.

Definition (Reference 17, vol. 2., p. 50)

A square matrix is called *redicible* if there is a permutation (i.e. relabelling of indices or, equivalently, interchanging of rows and/or columns) that puts it in the form

$$\begin{bmatrix} A & 0 \\ B & C \end{bmatrix}$$

where A and C are square.

Obviously every positive matrix (i.e. every matrix whose elements are all positive) is irreducible. Since every non-negative matrix can be made positive by a vanishingly small perturbation, it is also clear that reducibility of a non-negative matrix is a sort of pathological condition of limited practical significance. Reducible matrices can often be treated as degenerate limiting cases of irreducible matrices, although we do not use this property in this paper.

The key results used in this paper concerning non-negative matrices are summarised in the following theorem attributed to Perron and Frobenius by Gantmacher (Reference 17, vol. 2, p. 53).

Theorem 2

An irreducible non-negative matrix $A \in R^{n \times n}$ always has a real non-negative eigenvalue $\lambda_{max} \geqslant 0$ such that

$$\lambda_{max} \geqslant |\lambda_i|$$

for all eigenvalues $\lambda_i (i = 1, \ldots, n)$ of A. To this maximal eigenvalue λ_{max}, there correspond respective right and left eigenvectors $x, y \in R^n$ with positive co-ordinates $x_i, y_i > 0$ for all $i = 1, \ldots, n$.

Definition

The real non-negative maximal eigenvalue λ_{max} in theorem 2 is called the *Perron eigenvalue* of A, and the corrsponding positive eigenvectors $x, y \in R^n$ are called *Perron eigenvectors*.

8.2 Proof of lemma 1

The system of Fig. 1 is described by the equations

$$x(s) = -L_0(s)y(s) \tag{23}$$

$$y(s) = (I + \Delta(s))x(s) \tag{24}$$

Under the change of variable

$$\tilde{y}(s) = y(s) - x(s) \tag{25}$$

$$\tilde{x}(s) = x(s) \tag{26}$$

this system becomes

$$\tilde{x}(s) = -L_0(s)(I + L_0(s))^{-1}\tilde{y}(s) \tag{27}$$

$$\tilde{y}(s) = \Delta(s)\tilde{x}(s) \tag{28}$$

Following the approach of Stein and Sandell [2] and Barrett [1], one has that a sufficient condition for stability of eqns. 27 and 28 (and hence for eqns. 23 and 24) is the existence of two norms on C^n, say $\|\cdot\|_{x(j\omega)}$ and $\|\cdot\|_{y(j\omega)}$ such that

$$\|\Delta(j\omega)\|_{x,y} \leqslant 1/\|L_0(j\omega)(I + L_0(j\omega))^{-1}\|_{y,x} \quad \text{for all } \omega \tag{29}$$

where $||\cdot||_{x,y}$ and $||\cdot||_{y,x}$ are the subordinate bound norms

$$||\Delta(j\omega)||_{x,y} \triangleq \sup_{x \in C^n} \frac{||\Delta(j\omega)x||_{y(j\omega)}}{||x||_{x(j\omega)}} \tag{30}$$

$$||L_0(j\omega)(I + L_0(j\omega))^{-1}||_{y,x} \triangleq \sup_{y \in C^n} \left(\frac{||L_0(j\omega)(I + L_0(j\omega))^{-1}y||_{x(j\omega)}}{||y||_{y(j\omega)}} \right) \tag{31}$$

Evidently, the condition given in expr. 10 is merely a special case of expr. 29 in which

$$||\cdot||_{x(j\omega)} = ||\cdot||_{pD}\Big|_{D = D^*(j\omega)} \tag{32}$$

and

$$||\cdot||_{y(j\omega)} = ||\cdot||_{pD}\Big|_{D = R^{-1}(j\omega)D^*(j\omega)} \tag{33}$$

Thus lemma 1 is essentially a trivial corollary to existing results. This completes the proof of lemma 1.

8.3 Proof of lemma 2

The equality

$$\inf_D || \, |B(j\omega)| \, ||_{pD} = \pi(j\omega)$$

is shown to hold for any matrix $B(j\omega)$ by Stoer and Witzgall (theorem 2 of Reference 21) for a much broader class of norms than $||\cdot||_{pD}$. Stoer and Wtizgall [21] also show that the minimising weighting D is the D^* given by eqn. 11 for the case of the diagonally weighted Hölder norm $||\cdot||_{pD}$, thus establishing the equality

$$|| \, |B(j\omega)| \, ||_{pD^*} = \inf_D || \, |B(j\omega)| \, ||_{pD} \tag{34}$$

The inequality

$$|| \, B(j\omega) \, ||_{pD^*} \leqslant || \, |B(j\omega)| \, ||_{pD^*} \tag{35}$$

is shown to hold for a much broader class of norms than $||\cdot||_{pD^*}$ by Bauer (lemma 2 of Reference 22). That equality holds in the cases $p = \infty$, and $p = 1$ follows directly with the aid of the following lemmas.

Lemma 3

For every matrix $B = \{b_{ij}\}_{i,j=1}^n \in C^{n \times n}$ and every invertible diagonal matrix $D = \text{diag}(d_1, \ldots, d_n)$, the following holds:

$$||B||_{\infty D} = || \, |B| \, ||_{\infty D} = \max_i \sum_j |d_i b_{ij} d_j^{-1}| \tag{36}$$

Proof

$$\begin{aligned} ||B||_{\infty D} &\triangleq \sup_x \frac{||Bx||_{\infty D}}{||x||_{\infty D}} \\ &= \sup_{||x||_\infty = 1} ||DBD^{-1}x||_\infty \\ &= \max_{\substack{i \\ |x_j| \leqslant 1}} |\sum_j d_i b_{ij} d_j^{-1} x_j| \\ &= \max_i \sum_j |d_i b_{ij} d_j^{-1}| \\ &= \max_i \sum_j |(d_i |b_{ij}| d_j^{-1})| \\ &= || \, |B| \, ||_{\infty D} \end{aligned} \tag{37}$$

Lemma 4

For every $p \in [1, \infty]$, every $B = \{b_{ij}\}_{i,j=1}^n \in C^{n \times n}$ and every invertible diagonal $D = \text{diag}(d_1, \ldots, d_n)$. The following identity holds:

$$||B||_{pD} = ||\bar{B}^T||_{q\bar{D}^{-1}} \tag{38}$$

where $1/p + 1/q = 1$ and $\bar{B}^T$ denotes the complex-conjugate transpose of B.

Proof

Let $C = DBD^{-1}$. Then

$$\begin{aligned} ||B||_{pD} &= ||C||_p \\ &= \sup_x \frac{||Cx||_p}{||x||_p} \\ &= \sup_{x,y} \frac{\text{Re}(\bar{y}^T Cx)}{||y||_q \, ||x||_p} \\ &= \sup_{x,y} \frac{\text{Re}(\bar{x}^T \bar{C}^T y)}{||x||_p \, ||y||_q} \\ &= \sup_y \frac{||\bar{C}^T y||_q}{||y||_q} \\ &= ||\bar{C}^T||_q \\ &= ||\bar{B}^T||_{q\bar{D}^{-1}} \end{aligned} \tag{39}$$

where the third and fifth equalities in eqn. 39 follow from the Hölder inequality, namely

$$\text{Re}(\bar{y}^T x) \leqslant ||y||_q \, ||x||_p \quad \text{for all } x,y \tag{40}$$

where equality holds if, and only if, $y_i^q = x_i^p$ for all i.

Lemma 5

For every matrix $B \in C^{n \times n}$ and every invertible diagonal weight $D = \text{diag}(d_1, \ldots, d_n)$, the following holds:

$$||B||_{1D} = || \, |B| \, ||_{1D} = \max_j (\sum_i |d_i b_{ij} d_j^{-1}|) \tag{41}$$

Proof

Let $C = DBD^{-1}$. Then from lemmas 3 and 4 we have

$$\begin{aligned} ||B||_{1D} &= ||C||_1 = ||\bar{C}^T||_\infty \\ &= || \, |\bar{C}^T| \, ||_\infty = || \, |C|^T ||_\infty \\ &= \max_j \sum_i |d_i b_{ij} d_j^{-1}| \\ &= \max_j \sum_i |(\bar{d}_i |b_{ij}| \bar{d}_j^{-1})| \\ &= || \, |B|^T ||_{\infty \bar{D}^{-1}} \\ &= || \, |B| \, ||_{1D} \end{aligned} \tag{42}$$

This completes the proof of lemma 2.

Analysis of feedback systems with structured uncertainties

John Doyle

Indexing terms: *Control theory, Feedback, Robustness, Sensitivity*

Abstract: The paper introduces a general approach for analysing linear systems with structured uncertainty based on a new generalised spectral theory for matrices. The results of the paper naturally extend techniques based on singular values and eliminate their most serious difficulties.

1 Introduction

The last several years has seen something of a revolution in multivariable control theory. Central to this revolution is a renewed focus on the fundamental issue in feedback design: providing performance in the face of uncertainty. This development has been supported by a renewed interest in the frequency domain. An indication of the growing acceptance of these points of view can be found in the recent Special Issue on 'Linear multivariable control systems' in the February 1981 *IEEE Transactions on Automatic Control*, where five of the first six articles [1–5] deal with uncertainty, robustness and sensitivity using frequency-domain tools.

There are many results of both practical and theoretical importance which have come out of this research. As a consequence, it is now well known that multivariable sensitivity and robustness cannot be reliably evaluated one loop at a time [1, 6]. Singular-value methods have proved useful in providing both a reliable multivariable measure of sensitivity and robustness [1, 3, 7] and, more generally, a framework within which to develop reliable multiloop versions of the main tools of classical control theory [1–9].

While these results are reasonably well understood within the control theory community, many of the more subtle aspects of multivariable feedback systems are less well understood. Multiloop systems have important properties that have no analogues in the single-loop case, For example, there are problems caused by the fact that the signals and responses in multiloop systems vary not only with frequency, but also with direction. While some published results [8–10] have given preliminary indications of this, the issue of directionality in multiloop systems is largely unexplored.

The most serious limitation associated with the methods based on singular values is that they treat a limited, albeit important, class of uncertainties. Roughly speaking, most of these results give, at best, exact characterisation (i.e. involving necessary and sufficient conditions) of the feedback properties of a system model relative to perturbations which are norm bounded but otherwise unconstrained. There is no systematic mechanism for exploiting information about the structure of a perturbation.

The aim of this paper is to develop techniques for analysing systems with arbitrary constraints on their structure. Section 2 discusses some particularly important examples of structured uncertainties in an attempt to motivate the general results in later Sections.

Sections 3–5 present the main theoretical results. In Section 3, the block-diagonal perturbation problem is formulated. This problem is quite general since any norm-bounded perturbation problem, regardless of structure, can be trivially rewritten as a block-diagonal perturbation problem. This problem leads to the definition of a function μ that provides necessary and sufficient conditions for structured matrix perturbation problems. Section 4 examines the properties of the function μ and expresses it in familiar matrix algebraic terms. Section 5 develops the tools necessary to find gradients for singular values, which are used in Section 6 to provide techniques to compute μ in important special cases. Section 7 has some examples and a discussion of recent experience in computing μ, and Section 8 gives a summary and conclusions.

Paper 2205D, first received 4th May and in revised form 14th September 1982

The author is with Honeywell Inc., MN 17–2375, 2600 Ridgway Parkway, PO Box 312, Minneapolis, MN 55440, USA

2 Importance of the block-diagonal perturbation problem

Analysis methods based on singular values have been successful in providing a framework within which to develop multiloop generalisations of classical single-loop techniques. While this has been an important development, singular-value methods have serious limitations. Consider, for example, the problem of analysing a linear multivariable feedback system with two multiplicative perturbations appearing simultaneously at the inputs and outputs [8–10].

There are essentially two direct applications of singular-value techniques to this problem. First, the system may be rewritten to isolate the two perturbations as a single perturbation with a two-block-diagonal structure. This structure can then be ignored by replacing the block-diagonal perturbation with one full matrix perturbation. In this configuration, standard singular-value analysis can be applied. Of course, the results may be, in general, arbitrarily conservative. A second approach is simply to treat the two perturbations one at a time, leading possibly to arbitrarily optimistic answers.

If these first two approaches to the simultaneous perturbation problem happen to give the same results, then the 'true' answer is obviously determined. In general, this will not be the case. It should be emphasised again that these two approaches can yield arbitrarily bad estimates; i.e., depending on the problem, the gap between the upper and lower bounds may be arbitrarily large. No other information can be obtained from these bounds other than that the 'true' answer lies between them. This will become evident in the course of this paper.

An extension of the direct application of singular values is to analyse the differential sensitivity of the singular values evaluated at one point relative to perturbations at the other [10]. Reference 10 is extremely well written and would provide excellent background for the present paper. However, this approach suffers from the fundamental difficulty associated with differential sensitivity techniques in not applying to large perturbations. In addition, the results in Reference 10 provided only directional sensitivity information (i.e. directional derivatives) when singular values are not clustered. If differential sensitivity methods are to be effective, it is often important that worst-case directions (as with gradients) be obtained. The results of this paper can be modified for computing worst-case differential directions for clustered singular values. Although not developed further here, this may make it

Reprinted with permission from *IEE Proc.*, vol. 129, Pt.D, no. 6, pp. 242–250, Nov. 1982.

possible to complete the results in Reference 10 for clustered singular values.

What is really desired, of course, is an exact analysis of the original simultaneous independent perturbation problem, which applies to large perturbations. One application of the results in this paper is just such an analysis method. This will be illustrated in Section 7 by a simple 'textbook' example.

The problem of evaluating robustness with respect to simultaneous input and output perturbations is a special case of the more general problem of evaluating robustness with respect to perturbations with arbitrary constraints on their structure. Consider the general problem of an interconnected linear system with multiple independent norm-bounded perturbations occurring throughout. By rearranging the system, it is always possible to isolate the perturbations as a single large block-diagonal perturbation.

Stability analysis, for example, then essentially boils down to ensuring that $I + M\Delta$ remains nonsingular at all frequencies and for all Δ under consideration. Here, Δ is a block-diagonal perturbation and M is the transfer function from the collective outputs of the perturbations to their inputs. Readers unfamiliar with manipulating systems to isolate specific components as a diagonal matrix are encouraged to try some examples and convince themselves that it can be done. This notion is essential to understanding how the results in this paper may be applied to analyse control systems.

The key problem then involves the matrix problem of determining necessary and sufficient conditions such that $\det(I + M\Delta) \neq 0$. All norm-bounded linear perturbation problems reduce almost trivially to this. The main results in this paper are a partial solution to the general block-diagonal perturbation problem and a complete solution for the case of three or fewer blocks.

3 Definitions and problem formulation

The following notation will be used throughout:

$\mathcal{M}(k)$ = algebra of complex $k \times k$ matrices
$\mathcal{U}(k)$ = unitary matrices in $\mathcal{M}(k)$
$\sigma_{max}(M)$ = maximum singular value of M
$\rho(M)$ = spectral radius = magnitude of largest eigenvalue
$\lambda_{min}(H)$ = least (most negative) eigenvalue for Hermitian H
M^* = conjugate transpose of M
$\text{diag}(M_1, M_2, M_3, \ldots, M_n)$ = block-diagonal matrix with M_j (not necessarily square) on the diagonal

To provide a description of block-diagonal perturbations, let $\mathcal{K} = (m_1, m_2, \ldots, m_n, k_1, k_2, \ldots, k_n)$ be a $2n$-tuple of positive integers. All the definitions that follow depend on $\mathcal{K}$, but, to simplify notation, this dependency will not be explicitly represented. Unless specifically noted otherwise, $\mathcal{K}$ will be assumed to be an arbitrary but fixed $2n$-tuple of positive integers. Let

$$K = \sum_{j=1}^{n} m_j k_j \quad \text{and} \quad m = \sum_{j=1}^{n} m_j$$

For each $\delta \in R$, $\delta \geq 0$, let $X_\delta \subset \mathcal{M}(K)$ be

$$X_\delta = \{\text{diag}(\overbrace{\Delta_1, \Delta_1, \ldots, \Delta_1}^{m_1}, \overbrace{\Delta_2, \Delta_2, \ldots, \Delta_2}^{m_2}, \Delta_3, \ldots \ldots, \Delta_{n-1}, \overbrace{\Delta_n, \Delta_n, \ldots, \Delta_n}^{m_n}) \mid \Delta_j \in \mathcal{M}(k_j)$$
$$\text{and } \sigma_{max}(\Delta_j) \leq \delta \text{ for each } j = 1, 2, \ldots, n\}$$

Let

$$X_\infty = \bigcup_{j=1}^{\infty} X_j$$

Thus X_δ is the set of block-diagonal matrices with structure determined by $\mathcal{K}$ whose norm is not greater than δ, and X_∞ is all such matrices with no restriction on the norm.

What is desired is a function (depending, of course, on $\mathcal{K}$)

$$\mu : \mathcal{M}(K) \to [0, \infty)$$

with the property that for all $M \in \mathcal{M}(K)$

$$\det(I + M\Delta) \neq 0 \qquad \text{for all } \Delta \in X_\delta \tag{1}$$

if, and only if,

$$\delta\mu(M) < 1$$

This could be taken as a definition of μ. Alternatively, μ could be defined as

$$\mu(M) = \begin{cases} 0 & \text{if no } \Delta \in X_\infty \text{ solves } \det(I + M\Delta) = 0 \\ \left(\min\limits_{\Delta \in X_\infty} \left\{\sigma_{max}(\Delta) \mid \det(I + M\Delta) = 0\right\}\right)^{-1} & \text{otherwise} \end{cases}$$

This definition of μ shows that a well defined function satisfies eqn. 1. It probably has little additional value since the optimisation problem involved does not appear to have any useful properties.

In order to proceed further, some additional definitions are needed. Again, assume the $2n$-tuple $\mathcal{K}$ is given.

Then let

$$\mathcal{U} = U(K) \cap X_1 \qquad \text{(block-diagonal unitary matrices)}$$

$$\mathcal{D} = \{\text{diag}(d_1 I_{k_1}, d_2 I_{k_1}, \ldots, d_{m_1} I_{k_1}, d_{m_1+1} I_{k_2}, \ldots, d_m I_{k_n}) \mid d_i \in R^+ = (0, \infty)\}$$

$$\mathcal{S} = \{\text{diag}(d_1, d_2, \ldots, d_K) \mid d_i \in [0, \infty)\}$$

$$\mathcal{V} = \{\text{diag}(\overbrace{v_1, v_1, \ldots, v_1}^{m_1}, \overbrace{v_2, \ldots, v_2}^{m_2}, v_3, \ldots \ldots, v_{n-1}, \overbrace{v_n, \ldots, v_n}^{m_n}) \mid v_j \in \mathbf{C}^{k_j \times 1}, v_j^* v_j = 1, j = 1, 2, \ldots, n\}$$

$$\mathcal{V} \subset \mathbf{C}^{K \times m}$$

Note that these definitions all depend on $\mathcal{K}$, but to explicitly represent the dependency (for example, $\mu_{\mathcal{K}}$, $X_{\mathcal{K},\delta}$, $\mathcal{U}_{\mathcal{K}}$ etc.) would be unnecessarily cumbersome. Using these definitions, several useful properties of μ are stated in the following Section, leading up to the first main result of the paper.

The problem formulation taken here is certainly not the most general possible. For example, the entire development in this paper could be done with nonsquare perturbations with no conceptual change in the results or proofs. The notation, however, would become even more cumbersome than it already is. The particular approach taken is the simplest that allows both ρ and σ_{max} to emerge as special cases of μ. Hopefully, this will aid the reader by maintaining continuity with standard linear algebra as well as with some of the more popular existing methods in multivariable control.

4 Properties of μ

The following properties of μ are easily proven from eqn. 1. As always, $\mathscr{K} = (m_1, \ldots, m_n, k_1, \ldots, k_n)$ is an arbitrary $2n$-tuple for some $n \in \mathbf{Z}^+$, unless specified otherwise, and

$$K = \sum_{j=1}^{n} m_j k_j$$

(*a*) $\mu(\alpha M) = |\alpha|\mu(M) \quad$ for all $M \in \mathscr{M}(K)$

(*b*) $\mu(I) = 1$

(*c*) $\mu(AB) \leqslant \sigma_{max}(A)\mu(B) \quad$ for all $A, B \in \mathscr{M}(K)$

(*d*) $\mu(\Delta) = \sigma_{max}(\Delta) \quad$ for all $\Delta \in X_\delta$

(*e*) If $n = 1$ and $m_1 = 1$, then

$$\mu(M) = \sigma_{max}(M) \quad \text{for all } M \in \mathscr{M}(K)$$

(*f*) If $n = 1, k_1 = 1$, then

$$K = m_1$$

$$X_\delta = \{\lambda I \mid \lambda \in \mathbf{C}, |\lambda| \leqslant \delta\}$$

$$\mu(M) = \rho(M) \quad \text{for all } M \in \mathscr{M}(K)$$

(*g*) If $\Delta \in X_\delta$, $U \in \mathscr{U}$, then $U\Delta \in X_\delta$ and $\Delta U \in X_\delta$

(*h*) For all $\Delta \in X_\infty$ and for all $D \in \mathscr{D}$,

$$D\Delta D^{-1} = \Delta$$

(*i*) For all $U \in \mathscr{U}$ and $M \in \mathscr{M}(K)$,

$$\mu(MU) = \mu(UM) = \mu(M)$$

(*j*) For all $D \in \mathscr{D}$ and $M \in \mathscr{M}(K)$,

$$\mu(DMD^{-1}) = \mu(M)$$

(*k*) $\max_{U \in \mathscr{U}} \rho(UM) \leqslant \mu(M) \leqslant \inf_{D \in \mathscr{D}} \sigma_{max}(DMD^{-1})$

for all $M \in \mathscr{M}(K)$

(*l*) For all $\Delta \in X_\infty$, there exist $U, V \in \mathscr{U}$ and $\Sigma \in \mathscr{S}$ such that

$$\Delta = U \Sigma V^*$$

(*m*) μ is continuous in the usual metric topology on $\mathscr{M}(K)$.

Property (*l*) is just a block version of the singular-value decomposition (SVD). Properties (*e*) and (*f*) show that μ has as special cases both the spectral radius and maximum singular value or spectral norm. Property (*i*) means that μ is $\mathscr{U}$-invariant. The main theorem of this Section will show that the left-hand inequality in property (*k*) is actually an equality and thus that μ is, in some sense, ***the*** fundamental $\mathscr{U}$-invariant. In order to prove this theorem, some simple lemmas involving polynomial equations are needed.

Suppose $p: \mathbf{C}^k \rightarrow \mathbf{C}$ is a polynomial in k complex variables of degree no more that $q \in \mathbf{Z}$ for each variable. For $z = (z_1, \ldots, z_k) \in \mathbf{C}^k$, let $\|z\|_\infty = \max |z_j|$ be the usual $\|\cdot\|_\infty$ norm on $\mathbf{C}^k$. Let $\hat{z} \in \mathbf{C}^k$ be such that $p(\hat{z}) = 0$ and $|\hat{z}| = \min_{z \in \mathbf{C}^k} (\|z\|_\infty | p(z) = 0)$; i.e. $\hat{z}$ is a solution of $p(z) = 0$ with minimum $\|\cdot\|_\infty$ norm.

Lemma 1
There exists $x \in \mathbf{C}^k$ such that $p(x) = 0$ and $|x_j| = \|\hat{z}\|_\infty$ for all $j = 1, \ldots, k$.

Proof
Let $x_j = \hat{z}_j$ for each $\hat{z}_j$ such that $|\hat{z}_j| = \|\hat{z}\|_\infty$. If $|\hat{z}_j| = \|\hat{z}\|_\infty$ for all j, then $x = \hat{z}$ satisfies the lemma. So suppose that one of the components of $\hat{z}$, say for convenience $\hat{z}_k$, has $|\hat{z}_k| < \|z\|_\infty$. Then $p(z) = \Sigma p_r(z_1, \ldots, z_{k-1})z_k$, where each p_r is a polynomial in $k-1$ variables. Suppose that for some r, $p_r(\hat{z}_1, \ldots, \hat{z}_{k-1}) \neq 0$. This will lead to a contradiction. By continuity of the roots of a polynomial, for any $\epsilon > 0$ there exists $\delta > 0$ such that, for each $y \in \mathbf{C}^{k-1}$ such that $\|y - (\hat{z}_1, \ldots, \hat{z}_{k-1})\| < \delta$, there exists a $w \in \mathbf{C}$ such that $|w - \hat{z}_k| < \epsilon$ and $\Sigma p_r(y)w^r = 0$. Then, by choosing ϵ sufficiently small, there exists a $\hat{y} \in \mathbf{C}^{k-1}$ and $\hat{w} \in \mathbf{C}$ such that $p(\hat{y}_1, \hat{y}_2, \ldots, \hat{y}_{k-1}, \hat{w}) = 0$ and $\|(\hat{y}_1, \ldots, \hat{y}_{k-1}, \hat{w})\|_\infty < \|\hat{z}\|_\infty$. This contradicts the minimising property of $\hat{z}$. Thus $p_r(\hat{z}_1, \ldots, \hat{z}_{k-1}) = 0$ for all r, $0 \leqslant r \leqslant q$, and $p(\hat{z}_1, \ldots, \hat{z}_{k-1}, w) \equiv 0$ independent of w. Let $x_k = \|\hat{z}\|_\infty$. This argument may be repeated for each $j = 1, 2, \ldots, k$ so that either $x_j = \hat{z}_j$ or $x_j = \|\hat{z}\|_\infty$ and $p(x) = 0$ for $x = (x_1, \ldots, x_k)$. This completes the proof.

This lemma may seem a bit obscure at first glance. What it basically says is that if $\|z\|_\infty$ is to be minimised subject to the constraint that $p(z) = 0$, then at least one minimising solution x lies on the polydisc where each $|x_j| = \|x\|_\infty$. This will allow a great simplification in the characterisation of μ. First though, note that the proof of lemma 1 immediately implies the following two additional lemmas.

Lemma 2
If $\hat{z}$ is real $[\text{Im}(\hat{z}_j) = 0$ for all $j]$, then there exists $x \in \mathbf{R}^k$ such that $p(x) = 0$ and $|x_j| = \|\hat{z}\|_\infty$ for all $j = 1, 2, \ldots, k$.

Lemma 3
If $\hat{z}$ is real and non-negative $[\text{Im}(\hat{z}_j) = 0$ and $\text{Re}(\hat{z}_j) \geqslant 0$ for all $j]$, then there exists $x \in \mathbf{R}^k$ non-negative such that $p(x) = 0$ and $x_j = \|\hat{z}\|_\infty$.

Theorem 1
For all $M \in \mathscr{M}(K)$,

$$\mu(M) = \max_{U \in \mathscr{U}} \rho(MU)$$

Proof
Suppose $M \in \mathscr{M}(K)$. If $\mu(M) = 0$, then the result follows immediately from property (*k*), and so assume $\mu(M) = 1/\delta > 0$. Then there exists $\Delta \in X_\delta$ (not necessarily unique) such that $\sigma_{max}(\Delta) = \delta$ and $\det(I + M\Delta) = 0$. By property (*l*), there exist $U, V \in \mathscr{U}$ and $\Sigma \in \mathscr{S}$ such that

$$\Delta = U \Sigma V^*$$

Then

$$\det(I + M\Delta) = 0$$

if, and only if,

$$\det(I + MU\Sigma V^*) = 0$$

This last equation may be viewed as a polynomial in the diagonal elements of Σ. By assumption, Σ is a minimum norm (now viewed as $\|\cdot\|_\infty$ on the diagonal elements) solution to this polynomial. This satisfies the conditions of lemma 3 and therefore Σ may be replaced by a scalar δI. Thus

$$\det(I + MU\Sigma V^*) = 0$$

implies

$$\det(I + \delta MUV^*) = 0$$

which implies

$$\rho(MUV^*) \geqslant 1/\delta = \mu(M)$$

By property (k), the reverse inequality also holds, and so

$$\max_{U \in \mathscr{U}} \rho(MU) = \mu(M)$$

Corollary

$$\mu(M) = \max_{V, W \in \mathscr{V}} \rho(V^*MW) \qquad \text{for all } M \in \mathscr{M}(K)$$

Proof

$\max_{U \in \mathscr{U}} \rho(MU) = \lambda \Rightarrow$ there exist $U \in \mathscr{U}$ and $x \in \mathbf{C}^k$ such that $MUx = \lambda x$. This implies there exist

$V, W \in \mathscr{V}$ such that $MWV^*x = \lambda x$

which implies

$$\rho(MWV^*) = \rho(V^*MW) \geqslant \lambda$$

which in turn implies

$$\max_{V, W \in \mathscr{V}} \rho(V^*MW) \geqslant \lambda$$

Similarly,

$$\max_{V, W \in \mathscr{V}} \rho(V^*MW) = \lambda$$

implies

$$\max_{U \in \mathscr{U}} \rho(MU) \geqslant \lambda$$

Thus

$$\mu(M) = \max_{V, W \in \mathscr{V}} \rho(V^*MW) \qquad \text{for all } M \in \mathscr{M}(K)$$

This theorem and corollary express μ in terms of familiar linear algebraic quantities. Note that it is now possible to use the corollary to define a general matrix decomposition that would have both the singular-value decomposition and Jordan form as special cases. While such a unification may prove to be of a great theoretical interest, it is not essential to the aims of this paper and will not be pursued further here.

Instead, the remainder of this paper will be devoted to developing methods to compute $\mu(M)$. Unfortunately, the optimisation problem expressed in the corollary cannot be solved in general by simple gradient or local techniques. Examples have been generated which have multiple local maxima. While recent computational experience with some alternative algorithms is very promising, no efficient algorithms have been developed which have proven guaranteed convergence to the global maximum.

An alternative approach will now be taken by considering the right-hand side inequality of property (k). In important special cases it will be shown that this inequality is actually an equality, and that $\inf \sigma_{max}(DMD^{-1})$ has no local minimum which is not global. In fact, a tedious but straightforward calculation shows that $\sigma_{max}(DMD^{-1})$ is convex in D. The proof of this is omitted since it is not actually needed for any of the main results, but it makes this optimisation problem an attractive alternative. The first step needed for solving $\inf \sigma_{max}(DMD^{-1})$ is a technique for computing descent directions. This problem is taken up in the following Section.

5 Differentiability properties of singular values

This Section will develop the necessary tools for computing 'gradients' for singular values. Although singular values are not in general differentiable functions of the matrix elements, it is always possible to compute a generalised gradient which serves the same purpose as would a gradient. This Section will be brief and not at all self-contained and only original results will be proved. An excellent background for this Section is Reference 10, as well as standard texts [11–13].

Suppose $M: \Omega \to \mathscr{M}(k)$ is a matrix-valued function whose elements are real analytic in some neighbourhood Ω of $0 \in \mathbf{R}^q$. Denote by $\sigma_{max}(x)$ [for $\sigma_{max}(M(x)), x \in \Omega$] the maximum singular value of M as a function of $x \in \Omega \subset \mathbf{R}^q$.

The directional derivatives of $\sigma_{max}(x)$ at $x = 0$ depend only on $M(x)$ to first order, and so without loss of generality assume

$$M(x) = M_0 + \sum_{j=1}^{q} x_j M_j$$

for $x = (x_1, x_2, \ldots, x_q)$. Using the SVD, $M(0)$ may be written as

$$M(0) = M_0 = \sigma_{max}(0) U_1 V_1^* + U_2 \Sigma_2 V_2^*$$

Here $U_1, V_1, \in \mathbf{C}^{k \times r}$ with r the multiplicity of $\sigma_{max}(0)$ and

$$[U_1 U_2], [V_1 V_2] \in U(k)$$

Suppose $\sigma_{max}(0) > 0$ [i.e. $M(0) \neq 0$]. Then, letting $(\sigma_{max})_0 = \sigma_{max}(0)$, $\sigma_{max}(x)$ may be written as

$$\sigma_{max}(x) = |(\sigma_{max})_0 + \lambda_{max}(\Sigma x_j H_j) + 0(x)| \tag{2}$$

where

$$H_j = H_j^* = \mathrm{Re}(U_1^* M_j V_1) = \tfrac{1}{2}(U_1^* M_j V_1 + V_1^* M_j^* U_1)$$

Define $\nabla_1 \subset \mathbf{R}^q$ by

$$\nabla_1 = \{x \in \mathbf{R}^q, x_j = v^* H_j v \,|\, v^* v = 1 \text{ and } v \text{ is an eigenvector of } \Sigma y_j H_j \text{ for some } 0 \neq y \in \mathbf{R}^q\}$$

and denote by $\mathrm{co}\nabla_1$ the convex hull of ∇_1. Denote by $\hat{x} = \min(\mathrm{co}\nabla_1)$ the unique point $\hat{x} \in \mathrm{co}\nabla_1$ that minimises

$$\min_{x \in \mathrm{co}\nabla_1} \|x\|_2$$

A well known property of convex sets that will be used in the following theorem is that $\langle \hat{x}, x \rangle \geqslant \|x\|^2$ for all $x \in \mathrm{co}\nabla_1$ [14]. The significance of ∇_1 is expressed in the following theorem.

Theorem 2

If $\hat{x} \neq 0$, then there exists $\epsilon_0 > 0$ such that

$$\sigma_{max}(-\epsilon\hat{x}) < (\sigma_{max})_0 \qquad \text{for all } 0 < \epsilon < \epsilon_0$$

Proof

Let v be a unit eigenvector associated with the eigenvalue $\lambda_{min}(\Sigma \hat{x}_j H_j)$ and let $y \in \nabla_1$ be defined by $y_j = v^* H_j v$ for $j = 1, 2, \ldots, q$. Then

$$\lambda_{min}(\Sigma \hat{x}_j H_j) = v^*(\Sigma \hat{x}_j H_j) v = \langle \hat{x}, y \rangle > 0$$

Then from eqn. 2 there exists $\epsilon_0 > 0$ such that $\sigma_{max}(-\epsilon\hat{x}) < (\sigma_{max})_0$ for all $0 < \epsilon < \epsilon_0$.

This theorem implies that, although $\sigma_{max}(x)$ is not necessarily differentiable, $\hat{x} = \min(\mathrm{co}\nabla_1)$ serve one of the purposes of a gradient in providing a descent direction. If $x \neq 0$, then a direction (for example, $-x$) can be found which reduces all the singular values in the cluster. In the special case where $\sigma_{max}(0)$ has multiplicity one, then ∇_1 trivially reduces to the ordinary gradient. It may be thought of as the direct limit of the sets of possible gradients in neighbourhoods of 0. Unfortunately, ∇_1 is a rather awkward set to work with since it depends, in part, on singular vectors of $M(x)$ for $x \neq 0$.

As an alternative, consider the set

$$\nabla_2 = \{x \in \mathbf{R}^q, x_j = v^* H_j v \,|\, v \in \mathbf{C}^q, v^* v = 1\}$$

Clearly, $\nabla_1 \subset \nabla_2$. While ∇_2 has a much simpler description than ∇_1, it serves the same purpose, as is shown by the following lemma and theorem.

Lemma 4

$\min(\mathrm{co}\nabla_1) = \min(\mathrm{co}\nabla_2)$

Proof

Let $y \in \nabla_2$ and $\hat{x} = \min(\mathrm{co}\nabla_1)$. Then $y_j = v^* H_j v$ for some v such that $v^* v = 1$. Then

$$\begin{aligned} \langle \hat{x}, y \rangle &= \Sigma \hat{x}_j (v^* H_j v) \\ &\geqslant \lambda_{min}(\Sigma \hat{x}_j H_j) \\ &= \langle \hat{x}, x \rangle \quad \text{for some } x \in \nabla_1 \\ &\geqslant \|\hat{x}\|^2 \quad \text{since } \hat{x} = \min(\mathrm{co}\nabla_1) \end{aligned}$$

Therefore $\|\hat{x}\| \leqslant \|\min(\mathrm{co}\nabla_2)\|$ and since $\nabla_1 \subset \nabla_2$, $\hat{x} = \min(\mathrm{co}\nabla_2)$.

Theorem 3

$$\mathrm{co}\nabla_1 = \mathrm{co}\nabla_2$$

Proof

It suffices to prove that $\mathrm{dist}(x, \mathrm{co}\nabla_1) = \mathrm{dist}(x, \mathrm{co}\nabla_2)$ for all $x \in \mathbf{R}^q$. Fix $x = (x_1, \ldots, x_q) \in \mathbf{R}^q$. Define $\hat{\nabla}_1$ and $\hat{\nabla}_2$ by replacing H_j by $H_j - x_j I$ in the definitions of ∇_1 and ∇_2, respectively. Then, by lemma 4,

$$\begin{aligned} \mathrm{dist}(x, \mathrm{co}\nabla_1) &= \|\min(\mathrm{co}\hat{\nabla}_1)\|_2 = \|\min(\mathrm{co}\hat{\nabla}_2)\|_2 \\ &= \mathrm{dist}(x, \mathrm{co}\nabla_2) \end{aligned}$$

This holds for any $x \in \mathbf{R}^q$, and so the proof is complete.

In order to make effective use of ∇_2 in, say, an optimisation algorithm or a sensitivity calculation, there must be some reliable algorithm for computing $\min(\mathrm{co}\nabla_2)$ [or maybe $\max(\mathrm{co}\nabla_2)$ for sensitivity] given the $\{M_j\}$. The Appendix briefly describes an approach to this. The rest of this Section is concerned with properties of ∇_2 that will prove useful in the following Section.

Suppose that $H_j = H_j^* \in \mathscr{M}(r)$, $j = 1, 2, \ldots, q$, and let $f\colon \mathbf{C}^r \to \mathbf{R}^q$ be defined by $f_j(x) = x^* H_j x$ for $x \in \mathbf{C}^r$. Let

$$P^n = \{x \in \mathbf{C}^n \,|\, x^* x = 1\} \subset \mathbf{C}^n$$

and

$$S^n = \{x \in \mathbf{R}^{n+1} \,|\, x^T x = 1\} \subset \mathbf{R}^{n+1}$$

Note that this notation is not standard.

Theorem 4

For $r = 2$, there exists an affine map $g\colon \mathbf{R}^3 \to \mathbf{R}^q$ such that $f(P^2) = g(S^2)$.

Proof

Suppose

$$H_j = \begin{bmatrix} a & b \\ \bar{b} & c \end{bmatrix}$$

for some j. Then

$$[\cos\theta \quad \sin\theta(\cos\psi - i\sin\psi)] \begin{bmatrix} a & b \\ \bar{b} & c \end{bmatrix} \times \begin{bmatrix} \cos\theta \\ \sin\theta(\cos\psi + i\sin\psi) \end{bmatrix}$$

$$\begin{aligned} &= a\cos^2\theta + c\sin^2\theta + 2\mathrm{Re}(b\cos\theta\sin\theta(\cos\psi + i\sin\psi)) \\ &= (1/2)(a+c) + (1/2)(a-c)\cos 2\theta \\ &\quad + \sin 2\theta\,\mathrm{Re}(b(\cos\psi + i\sin\psi)) \\ &= (1/2)(a+c) \\ &\quad + [(1/2)(a-c) \quad \mathrm{Re}(b) \quad \mathrm{Im}(b)] \begin{bmatrix} \cos 2\theta \\ \sin 2\theta\cos\psi \\ \sin 2\theta\sin\psi \end{bmatrix} \end{aligned}$$

Let $g_1\colon \mathbf{R}^3 \to \mathbf{R}$ be the affine map defined by the scalar and vector in the last equation so that $f_1(P^2) = g_1(S^2)$ as shown. Define g_j for $j = 2, 3, \ldots, q$ similarly so that the resultant $g\colon \mathbf{R}^3 \to \mathbf{R}^q$ is affine. Then $f(P^2) = g(S^2)$ as desired.

Corollary

For $f\colon \mathbf{C}^2 \to \mathbf{R}^q$ defined above, if $q = 1$ or 2, then $f(P^2)$ is convex.

Proof

If $q = 1$ or 2, then for any affine map $g\colon \mathbf{R}^3 \to \mathbf{R}^q$, $g(S^2)$ is convex. Apply the theorem.

Corollary

Let $f\colon \mathbf{C}^r \to \mathbf{R}^q$ be defined as above for arbitrary $r \in \mathbf{Z}^+$. If $q = 1$ or 2, then $f(P^r)$ is convex.

Proof

Let $x, y \in \mathbf{C}^r$, $x \neq y$ and $\|x\|_2 = \|y\|_2 = 1$. It suffices to show that $z(t) = tf(x) + (1-t)f(y) \in f(P^r)$ for all $t \in [0, 1]$. Let

$$\hat{H}_j = \begin{bmatrix} x^* \\ y^* \end{bmatrix} H_j [x \quad y]$$

for all $j = 1, 2, \ldots, q$, and let $\hat{f}\colon \mathbf{C}^2 \to \mathbf{R}^q$ be defined as before in terms of the $\hat{H}_j$. Then $f(x)$, $f(y) \in \hat{f}(P^2) \subset f(P^r)$ and $\hat{f}(P^2)$ is convex by the previous corollary. Thus $z(t) \in \hat{f}(P^2)$ $f(P^r)$ for all $t \in [0, 1]$, and so $f(P^r)$ is convex.

This last corollary implies immediately that, if $M\colon \Omega \to \mathscr{M}(K)$ for $\Omega \in \mathbf{R}^q$ and $q = 1$ or 2, then the ∇_2 defined as before is convex. This will prove quite useful in the following Section.

Note that the results of this Section and Appendix could be used to study the sensitivity of clustered singular values to infinitesimal parameter variations. The aim of this paper, however, is to develop techniques for handling large perturbations, and so the following Section will make use of these results to compute the function μ.

6 Computation of μ

This Section will combine the results of the preceding two Sections and concentrate on the right-hand side inequality in property (*k*) of Section 4. Suppose in this Section that in the $2n$-tuple $\mathscr{K}$, $m_j = 1$ for each j (there are no repeated blocks). Then reduce $\mathscr{K}$ to an n-tuple $\mathscr{K} = (k_1, \ldots, k_n)$ and let $K = \Sigma k_j$. Define $D: \mathbf{R}^n \to \mathscr{M}(K)$ by

$$D(x) = \operatorname{diag}(\exp(x_1)I, \exp(x_2)I, \ldots, \exp(x_n)I)$$

For $M_0 \in \mathscr{M}(K)$, define $M: \mathbf{R}^n \to \mathscr{M}(K)$ by

$$M(x) = D(x)M_0(D(x))^{-1} = D(x)M_0D(-x)$$

and let $\sigma_{max}(x)$, U_1, V_1, ∇_1, ∇_2 etc. be defined as in Section 4 for $M(x)$. Let $A = U_1$ and $B = U_2$ and assume that M is normalised so that $\sigma_{max}(0) = 1$. Then write

$$A = [a_1 \quad a_2 \ldots a_r] = \begin{bmatrix} \alpha_1^* \\ \alpha_2^* \\ \cdot \\ \cdot \\ \alpha_n^* \end{bmatrix}$$

and similarly for B in terms of $\{b_j\}$ and $\{\beta_j\}$. Here $\alpha_j^* \in \mathbf{C}^{K \times 1}$, $\alpha_j^* \in \mathbf{C}^{kj \times r}$ (similarly for B). It is then easy to verify that $H_j = \alpha_j\alpha_j^* - \beta_j\beta_j^*$.

Theorem 5
$\mu(M) = 1$ if, and only if, $0 \in \nabla_2(M)$.

Proof
$0 \in \nabla_2(M)$ if, and only if:

(i) there exists $y \in p^r$ such that

$$y^* H_j y = 0 \qquad \text{for all } j$$

if, and only if,

(ii) $\|\beta_j^* y\| = \|\alpha_j^* y\| \qquad$ for all j

if, and only if,

(iii) for all j, there exists $U_j \in U(k_i)$ such that

$$\beta_j^* y = U_j \alpha_j^* y$$

if, and only if,

(iv) there exists $U \in \mathscr{U}$ such that $By = UAy$
$(U = \operatorname{diag}(U_1, U_2, \ldots, U_n))$

if, and only if,

(v) $B^*UAy = y$

if, and only if,

(vi) there exists $0 \neq x \in \mathbf{C}^K$ such that

$$AB^*Ux = x$$

if, and only if,

(vii) $MUx = x \qquad$ (since $\sigma_{max}(M) = 1$)

if, and only if,

(viii) $\rho(MU) = 1$

Therefore, using property (*k*) from Section 4, $\mu(M) = 1$ if, and only if, $0 \in \nabla_2(M)$.

Theorem 6
For $n \leqslant 3$, $\sigma_{max}(M) = \mu(M)$ if, and only if, $0 \in \operatorname{co}\nabla_2$.

Proof
For any n, ∇_2 depends only on $n-1$ variables, since ∇_2 lies in the $(n-1)$-dimensional subspace of $\mathbf{R}^n$ orthogonal to $(1, 1, \ldots, 1)$; i.e. $D(x)M_0D(-x)$ is constant along any line parallel to the line through 0 and $(1, 1, \ldots, 1)$. For $n \leqslant 3$, by the second corollary to theorem 3, $\nabla_2 = \operatorname{co}\nabla_2$, and the result then follows from Theorem 1.

These two theorems have important implications for computing μ. Suppose that, conceptually, a gradient search is to be used to compute

$$\inf_{D \in \mathscr{D}} \sigma_{max}(DMD^{-1})$$

with $\min(\operatorname{co}\nabla_2)$ serving as a gradient when there are clustered singular values. This search can proceed until a local minimum is found, i.e. $0 \in \operatorname{co}\nabla_2$. [As was mentioned earlier, $\sigma_{max}(DMD^{-1})$ is convex in D, and so such a local minimum would be global.] If, in fact, $0 \in \nabla_2$ at this point, then, by theorem 5, μ has also been found. Unfortunately, it is not generally true that $0 \in \operatorname{co}\nabla_2$ implies $0 \in \nabla_2$.

Theorem 6 implies that, for the case $n \leqslant 3$ (three or fewer blocks), $\nabla_2 = \operatorname{co}\nabla_2$, and so such a gradient search would (at least conceptually) always yield μ. This result is rather remarkable since the optimisation involves only $n-1$ variables, regardless of the size of the blocks. Furthermore, in view of theorem 3 and its corollaries, the computation of the generalised gradients for these cases should be quite straightforward. This provides the desired technique for exact analysis of multivariable systems with simultaneous input/output perturbations (i.e. two blocks), as well as more general situations.

7 Examples and computational experience

Two numerical examples and a discussion of computational experience are contained in this Section. The examples are intended to be purely illustrative, and no particular significance should be attributed to them.

The first example uses a 3×3 matrix for which μ will be computed for five different structured perturbations. The nominal matrix is

$$M = \begin{bmatrix} 4-2i & -\frac{1}{2}-\frac{1}{2}i & -10 \\ -24+6i & 3i & 60-80i \\ -\frac{6}{5} & -\frac{1}{5}-\frac{1}{5}i & 2+2i \end{bmatrix}$$

and the five uncertainty structures are shown in Table 1. The perturbation in case 1 is block diagonal, and so μ can be computed directly from M. In cases 2–5, however, the matrix must be rearranged to make the perturbation block diagonal. This can always be done by pre- and postmultiplying M by appropriate matrices, denoted here by L and R, respectively. The last two columns in Table 1 give $\mathscr{K}$ for that case and the corresponding μ (LMR).

Cases 1–3 have three or fewer blocks, and μ was computed using the approach suggested in Section 6. Cases 4 and 5 have more than three blocks, and μ was computed using an algorithm based on theorems 1 and 2. The global maximum was found for case 4, as was verified by computing $\inf \sigma_{max}(DLMRD^{-1})$ and obtaining agreement to within the accuracy of the algorithms' termination conditions (in this case, four digits for single precision).

This example illustrates that μ depends heavily on the assumed structure of the uncertainty. This should not be interpreted as a limitation on its usefulness. On the contrary, it emphasises the importance of having a method for analysing matrix perturbation problems that maintains structure. The next example illustrates how a particular uncertainty structure could naturally arise in a feedback system.

The following example was originally constructed by the author to illustrate that loop-at-a-time analysis was inadequate for studying sin ultaneous variations in multiloop systems and to introduce the use of singular values. An example was published by the author [6]. The actual design analysed here was proposed by Sain *et al.* [15]. A thorough and lucid discussion of this example, including Sain's design, can be found in Reference 10.

Consider the feedback configuration in Fig. 1, where the nominal plant P has a transfer function

$$P(s) = \begin{bmatrix} \frac{9}{s+1} & \frac{-10}{s+1} \\ \frac{-8}{s+2} & \frac{9}{s+2} \end{bmatrix}$$

and

$$K(s) = \frac{1}{0.0159s} \begin{bmatrix} 9(s+1) & 10(s+2) \\ 8(s+1) & 9(s+2) \end{bmatrix}$$

The feedback compensation K is the product of the two elements of the compensator in References 15 and 10. Only the product K affects the system's feedback properties.

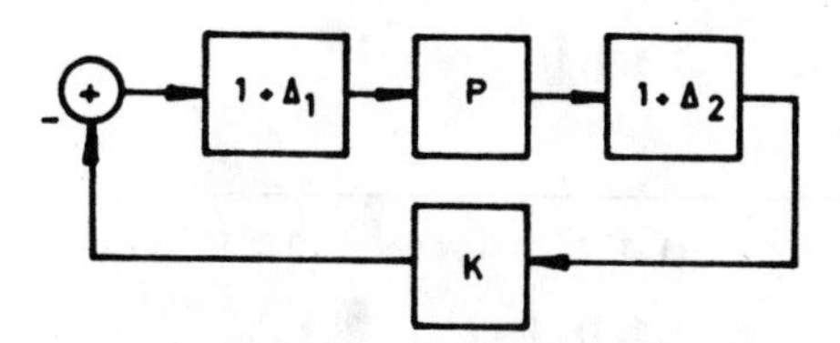

Fig. 1 *Block diagram for example design*

For this example, the nominal plant is considered to have simultaneous multiplicative perturbations [1] at the input and output. In applications, such perturbations might arise from actuators, sensors and unmodelled dynamics. The perturbations could then be weighted to reflect the fact that the level of uncertainty varies in frequency and direction. This example has no physical motivation, and so for simplicity this system will be evaluated for robustness with respect to unweighted size of the norm-bounded perturbations. It is desired to compute the smallest norm-bounded perturbation (as a function of frequency) which produces instability.

If only one perturbation is considered at a time, then μ is simply the maximum singular value of the transfer function $(I + PK)^{-1}PK$ or $(I + KP)^{-1}KP$, for an output or input perturbation, respectively. These two transfer functions happen to be the same for this example, and their maximum singular value is plotted as the lower curve in Fig. 2. This implies, for example, that there exists a destabilising perturbation of norm 1 and that all smaller perturbations can be tolerated without instability. Unfortunately, this provides very little information (other than the upper bound) on the tolerable level for simultaneous perturbations; i.e. these maximum singular values are a lower bound for μ for simultaneous perturbations.

In order to analyse the system for simultaneous variations, it is rearranged to isolate the Δ_is as a block-diagonal perturbation in standard feedback configuration, as in Fig. 3. For this example, one choice is

$$M = \begin{bmatrix} (I+KP)^{-1}KP & (I+KP)^{-1}K \\ -(I+PK)^{-1}P & (I+PK)^{-1}PK \end{bmatrix}$$

with

$$\Delta = \begin{bmatrix} \Delta_1 & 0 \\ 0 & \Delta_2 \end{bmatrix}$$

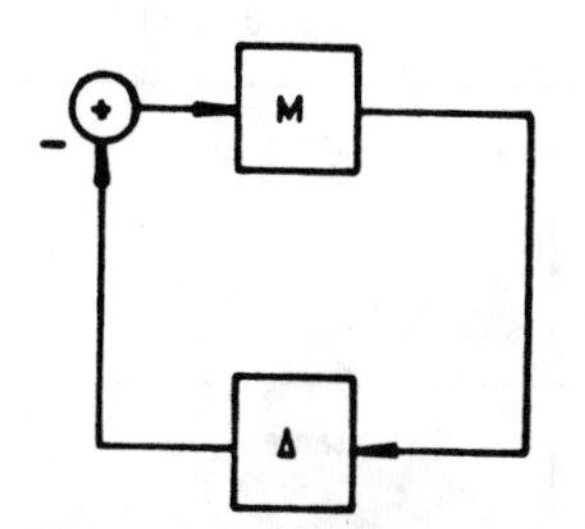

Fig. 3 *Standard feedback configuration*

If the block structure were ignored at this point, then $\sigma_{max}(M)$ (the upper curve in Fig. 2) would provide a tight bound on tolerable perturbations Δ. Unfortunately, $\sigma_{max}(M)$ provides only a conservative bound for the block-diagonal case.

For the structure as given, the corresponding μ is plotted in Fig. 4. This plot has several interpretations. The simplest is that the system can be stabilised by simultaneous perturbations with norms approximately equal to 0.1, and smaller perturbations may be tolerated without instability. The bounds in Fig. 2 were almost useless for this example. More generally, it is quite easy to construct examples where the gap for both bounds is arbitrarily large.

The relatively large value of μ in this analysis should not be considered as an indictment of this design. It merely indicates that the design has poor margins relative to this particular uncertainty structure; with respect to another structure they may be much better (for example, the margins for uncertainty on just one side may be good). A design that tolerated larger simultaneous variations of the type considered here

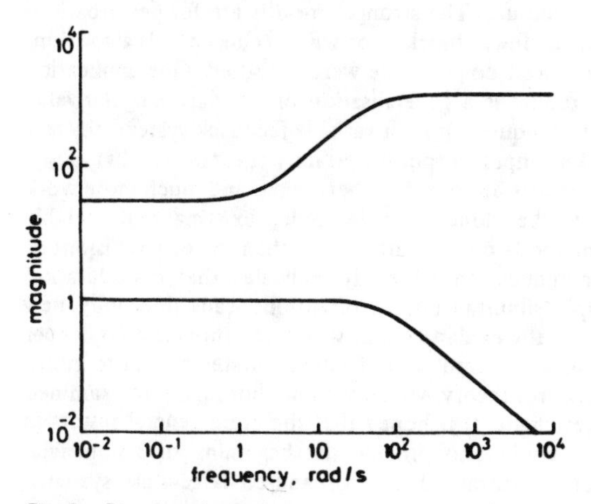

Fig. 2 *Singular-value bounds*

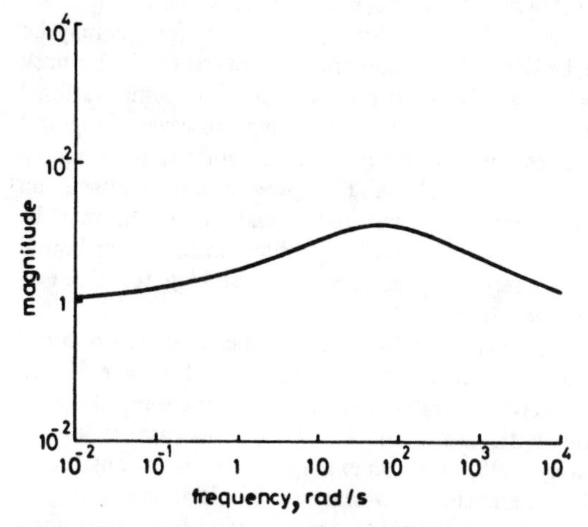

Fig. 4 *Plot of μ for example design*

Table 1: Uncertainty structures

Case	Δ	L	R	$\mathscr{K}$	$\underline{\mu}$
1	$\begin{bmatrix} \Delta_1 & 0 & 0 \\ 0 & \Delta_2 & 0 \\ 0 & 0 & \Delta_3 \end{bmatrix}$	I_3	I_3	(1, 1, 1, 1, 1, 1)	10.24
2	$\begin{bmatrix} \Delta_1 & 0 & 0 \\ 0 & 0 & \Delta_2 \\ 0 & 0 & 0 \end{bmatrix}$	$\begin{bmatrix} 1 & 0 & 0 \\ 0 & 0 & 1 \end{bmatrix}$	$\begin{bmatrix} 1 & 0 \\ 0 & 1 \\ 0 & 0 \end{bmatrix}$	(1, 1, 1, 1)	4.65
3	$\begin{bmatrix} \Delta_1 & 0 & 0 \\ \Delta_2 & \Delta_3 & \\ 0 & & \end{bmatrix}$	$\begin{bmatrix} 1 & 0 & 0 \\ 1 & 0 & 0 \\ 0 & 1 & 0 \\ 0 & 0 & 1 \end{bmatrix}$	$\begin{bmatrix} 1 & 0 & 0 & 0 \\ 0 & 1 & 1 & 0 \\ 0 & 0 & 0 & 1 \end{bmatrix}$	(1, 1, 1, 1, 1, 2)	102.8
4	$\begin{bmatrix} \Delta_1 & 0 & 0 \\ \Delta_2 & \Delta_3 & 0 \\ 0 & 0 & \Delta_4 \end{bmatrix}$	same	same	(1, 1, 1, 1, 1, 1, 1, 1)	10.8
5	$\begin{bmatrix} \Delta_1 & 0 & 0 \\ \Delta_2 & \Delta_3 & 0 \\ 0 & 0 & \Delta_3 \end{bmatrix}$	same	same	(1, 1, 2, 1, 1, 1)	10.6

would have to sacrifice some other aspect of performance. There was no physical basis for the uncertainty: it was chosen to be illustrative. Thus the results should not be interpreted too broadly.

This example reiterates the point that vastly different numbers are obtained, depending on the assumed structure of the problem. For practical problems, the structure is dictated to a large degree by physical reality and engineering constraints. The extent to which a design engineer can capture and handle the natural structure determines the extent to which the conclusions based on any analysis are relevant to the practical problem. It is hoped that the ideas introduced in this paper and illustrated in the examples will make a contribution towards better techniques for handling structured uncertainty in feedback systems.

An important consideration in the application of μ is a numerical software. In as much as the ideas in this paper are relatively new and have not been published before, it may be some time before reliable numerical software (as in Linpack or Eispack) is available to compute μ. The computational experience to date is most encouraging, however. Programs have been developed to compute both bounds in property (k) of Section 4. Recall that the lower bound is always an equality (see Section 4), but the global maximum may be difficult to find. The optimisation problem in the upper bound is convex, but it is only guaranteed to yield μ for three or fewer blocks (see Section 6).

The ratio between the lower and upper bounds produced by these new programs has been computed for over 50 000 pseudo-randomly generated matrices of dimension three to ten, mostly with scalar blocks. The worst-case ratio was approximately 0.95, although examples have been constructed analytically where the ratio was $\simeq 0.85$. It is interesting to note that the ratio seems not to decrease after four dimensions. This is suggestive but, of course, not conclusive. As expected, for three dimensions it is always 1 to within reasonable numerical error.

Because the computer programs are experimental, i.e. containing many diagnostics and obviously inefficient code, it is impossible to draw any meaningful conclusions about computational speed. As a single data point, it took approximately three times longer to compute the curve in Fig. 4 as it did to compute the bounds in Fig. 2.

8 Summary and conclusions

This paper has introduced a general approach for analysing linear systems with structured uncertainties based on a new generalised spectral theory for matrices. This basic theory addresses the norm-bounded perturbation problem with arbitrary structure. The strongest results are for perturbations with three or fewer blocks, for which (conceptual) algorithms with guaranteed convergence were proposed. One application of these results is a generalisation of standard singular-value analysis techniques for multivariable feedback systems to treat simultaneous input/output uncertainty (i.e. two blocks).

These results are merely a beginning, and much more work remains to be done. For example, existing multivariable control methods provide little more than minor extensions of SISO techniques. Initial study indicates that consideration of multiple simultaneous perturbations leads to wholly new phenomena, the explanation of which will provide a far deeper understanding of multiloop feedback systems. Linear multivariable control theory will need to be thoroughly re-examined in this new light. It is hoped that the more general results in this paper could also provide the beginning of a nontrivial theory of decentralised control and/or large-scale systems, where use of structural information is essential.

9 Acknowledgments

Many people contributed to this paper, but I would particularly like to thank Dr. Joe Wall, of Honeywell Systems and Research Center, for his help throughout the research and writing. Jim Freudenberg of the University of Illinois and Honeywell SRC made a major contribution to the implementation of the algorithms used in the examples and offered many useful suggestions regarding the paper.

This work has been supported by Honeywell internal research and development funding, the US Office of Naval Reserach under ONR research grant N00014-82-C-0157, and the US Air Force Office of Scientific Research grant F49620-82-C-0090.

10 References

1 DOYLE, J.C., and STEIN, G.: 'Multivariable feedback design: Concepts for a classical/modern synthesis', *IEEE Trans.*, 1981, AC-26, pp. 4–16

2 POSTELTHWAITE, I., EDMUNDS, J.M., and MacFARLANE, A.G.J.: 'Principal gains and principal phases in the analysis of linear multivariable feedback systems', *ibid.*, 1981, AC-26, pp. 32–46

3 SAFONOV, M.G., LAUB, A.J., and HARTMANN, 'Feedback properties of multivariable systems: The role and use of the return difference matrix', *ibid.*, 1981, AC-26, pp. 47–65

4 CRUZ, J.B., FREUDENBERG, J.S., and LOOZE, D.P.: 'A relationship between sensitivity and stability of multivariable feedback systems', *ibid.*, 1981, AC-26, pp. 66–74

5 LEHTOMAKI, N.A., SANDELL, N.R., Jr., and ATHANS, M.: 'Robustness results in linear-quadratic Gaussian based multivariable control designs', *ibid.*, 1981, AC-26, pp. 75–92

6 DOYLE, J.C.: 'Robustness of multiloop linear feedback systems' 17th IEEE conference on decision and control, San Diego, USA, Jan. 1979

7 DOYLE, J.C.: 'Multivariable design techniques based on singular value generalisation of classical control' AGARD lecture series 117 on multivariable analysis and design techniques. Sept. 1981

8 WALL, J.E., DOYLE, J.C., and HARVEY, C.A.: 'Tradeoffs in the design of multivariable feedback systems'. Proceedings of 18th Allerton conference on communication control and computing, Oct. 1980, pp. 715–725

9 DOYLE, J.C.: 'Limitations on achievable performance of multivariable feedback systems'. AGARD lecture series 117 on multivariable analysis and design techniques, September. 1981

10 FREUDENBERG, J.S., LOOZE, D.P., and CRUZ, J.B.: 'Robustness analysis using singular value sensitivities', *Int. J. Control*, 1982, pp. 95–116

11 KATO, T.: 'Perturbation theory for linear operators' (Springer-Verlag, 1976)

12 WILKENSON, J.H.: 'The algebraic eigenvalue problem' (Clarendon Press, 1965)

13 RELLICH, F.: 'Perturbation theory of eigenvalue problems' (Gordon & Breach, 1969)

14 LUENBERGER, D.G.: 'Introduction to linear and nonlinear programming' (Addison-Wesley, 1973)

15 SAIN, M.K., MA, A., and PERKINS, D.: 'Sensitivity issues in decoupled control system design'. Proceedings of Southeast symposium on system theory

11 Appendix

Let $\{H_j\}_{j=1}^{q}$, f, and P^r be defined as in the paragraph preceding theorem 3, and let $\nabla_2 = f(P^r)$. To compute $\min(\text{co}\nabla_2)$, consider the following algorithm to produce a sequence $\{x_n\}$, where $x_n = [x_n^1, x_n^2, \ldots, x_n^q] \in \mathbb{R}^q$:

(a) Pick any $v_1 \in P^r$ and let $x_1 = f(v_1)$.
(b) Inductively define $x_n = \min(\text{co}[x_{n-1}, f(v_{n-1})])$.
(c) Then v_n = any unit eigenvector for $\lambda_{min}(\Sigma x_n^j H_j)$.

This produces a sequence $\{x_n\}$ that has nonincreasing norm, and thus $\{\|x_n\|\}$ converges. Therefore there exists a subsequence $\{x_{n_k}\}$ such that both $\{x_{n_k}\}$ and $\{v_{n_k}\}$ converge to, say, x_0 and v_0, respectively. Note that $y = f(v_n)$ minimises the $\min\langle x_n, y\rangle$ and $\langle x_n f(v_n)\rangle = \lambda_{min}(\Sigma x_n^j H_j)$. Then, by taking limits, $\langle x_0 f(v_0)\rangle = \lambda_{min}(\Sigma x_0^j H_j)$ and $y = f(v_0)$ minimises $\min\langle x_0, y\rangle$.

Now suppose $x_0 \neq \min(\text{co}\nabla_2)$ (leading to a contradiction). Then by definition of the convex hull, $\langle x_0, f(v_0)\rangle < \|x_0\|^2$ and $\|x_0\| - \|\min(\text{co}[x_0, f(v_0)])\| > \epsilon$ for some $\epsilon > 0$. By existance of a convergent subsequence, there exists N sufficiently large such that $\|x_N - x_0\| < \epsilon/2$ and $\|f(v_N) - f(v_0)\| < \epsilon/2$. Then

$$\|x_0\| > \epsilon + \|\min(\text{co}[x_0, f(v_0)])\|$$
$$\geqslant \epsilon + [\min(\text{co}[x_N, f(v_N)]) - \epsilon/2]$$
$$= \epsilon/2 + \|x_{N+1}\|$$

This contradicts the fact that $\{\|x_n\|\}$ is nonincreasing, and thus $\{x_n\}$ converges to $\min(\text{co}\nabla_2)$.

In this algorithm $f(v_n)$ is an intersection point of ∇_2 and the support hyperplane perpendicular to x_n. There are, of course, a number of other ways that this point could be used to obtain a convergent algorithm. The particular algorithm outlined here is rather crude in not taking full advantage of the structure of ∇_2.

The Graph Metric for Unstable Plants and Robustness Estimates for Feedback Stability

MATHUKUMALLI VIDYASAGAR, FELLOW, IEEE

***Abstract*—In this paper, a "graph metric" is defined that provides a measure of the distance between unstable multivariable plants. The graph metric induces a "graph topology" on unstable plants, which is the weakest possible topology in which feedback stability is robust. Using the graph metric, it is possible to derive estimates for the robustness of feedback stability *without* assuming that the perturbed and unperturbed plants have the same number of RHP poles. If the perturbed and unperturbed systems have the same RHP poles, then it is possible to obtain necessary and sufficient conditions for robustness with respect to a given class of perturbations. As an application of these results, the design of stabilizing controllers for unstable singularly perturbed systems is studied. Finally, the relationship of the graph metric to the "gap metric" introduced by Zames and El-Sakkary is studied in detail. In particular, it is shown that the robustness results of Zames and El-Sakkary do not enable one to conclude the *causality* of the perturbed system, whereas the present results do.**

Manuscript received July 18, 1982; revised February 2, 1983. Paper recommended by E. W. Kamen, Past Chairman of the Linear Systems Committee. This work was supported by the Natural Sciences and Engineering Research Council of Canada under Grant A-1240.

The author is with the Department of Electrical Engineering, University of Waterloo, Waterloo, Ont., Canada.

I. Introduction

THE objective of this paper is to study the robustness of feedback stability. Consider the standard feedback configuration shown in Fig. 1, where P represents the plant and C the compensator. Suppose this system is stable. Much research has been devoted to deriving conditions under which the system remains stable as P, C are replaced by some perturbed systems P_1, C_1 (see, e.g., the special issue [1]). Almost all of the research to date has been concentrated on the case where $P_1(C_1)$ has the same number of RHP poles as $P(C)$. Yet this is an artificial restriction that arises from the methods of analysis used. Suppose, for example, that $P = 1/s$; then it is intuitively clear that the system will remain stable if P is replaced by $P_1 = 1/(s+\epsilon)$, provided ϵ is sufficiently small, and *it is immaterial whether ϵ is positive or negative*. The point is that in some sense both $1/(s-\epsilon)$ and $1/(s+\epsilon)$ are "close" to $1/s$, even though one system is stable and the other unstable.

Accordingly, the approach adopted in this paper is to define a notion of distance on the set of (possibly) unstable plants, and to obtain robustness margins for the feedback system in Fig. 1 in terms of the "distances" between P and P_1 and between C and

Reprinted from *IEEE Trans. Automat. Contr.*, vol. AC-29, no. 5, pp. 403–417, May 1984.

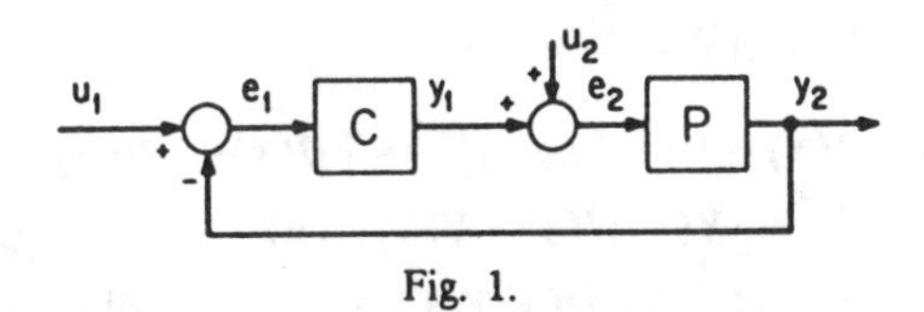

Fig. 1.

C_1. This would include, but would not be restricted to, the case where P_1 has the same number of RHP poles as P.

This paper is organized as follows. The rest of the introduction is devoted to demonstrating, by means of examples, some of the difficulties in defining a notion of proximity for unstable multivariable plants. In Section II we define a topology on the set of unstable plants, i.e., a notion of convergence of sequences. In this topology, a perturbation of the plant consists of perturbations of the "numerator" and "denominator" of the plant transfer matrix when it is expressed as a ratio of stable transfer matrices. We show that this is the weakest topology in which feedback stability is robust; that is, one can robustly stabilize against plant uncertainty if and only if the uncertainty can be expressed as perturbations in the stable numerator and denominator of the plant. In Section III we define a metric, i.e., a measure of distance, for unstable plants, and show that this metric induces the topology defined in Section II. Using this metric, we derive quantitative estimates for the robustness of feedback stability in Section IV. These results do *not* require the perturbed and unperturbed plants to have the same number of RHP poles, and as a consequence, they give only sufficient conditions for robustness. In case the perturbed and unperturbed plants have the same RHP poles, it is possible to obtain necessary and sufficient conditions for robustness, and this is done in Section V. In Section VI the results derived in the preceding sections are used to study the design of controllers for unstable singularly perturbed systems. The extension of the results given here to the case of distributed systems is discussed in Section VII.

In some earlier work [14], [15], a so-called "gap metric" is defined, and some robustness results are proved. In the Appendix we show that these robustness results do not enable one to conclude the *causality* of the perturbed system, whereas the present results do.

We now give some motivating discussion concerning the problem of defining a metric on the set of (possibly unstable) multivariable plants. While there are readily available and reasonable notions of distance on the set of *stable* plants, the issue is much more tricky in the case of *unstable* plants. The best way to illustrate this is by means of several examples.

As is customary, let $R(s)$ denote the set of rational functions in s with real coefficients. We use $\mathscr{S}$ (to suggest "stable") for the subset of $R(s)$ consisting of proper rational functions whose poles are all in the open left half-plane. Thus, $\mathscr{S}$ is the set of transfer functions of BIBO stable systems. For every f in $\mathscr{S}$, define

$$\|f\| = \sup_{\omega} |f(j\omega)| = \sup_{\operatorname{Re} s \geq 0} |f(s)|. \tag{1.1}$$

Then $\|\cdot\|$ defines a norm on $\mathscr{S}$, and the distance between two functions f and g in $\mathscr{S}$ is simply $\|f-g\|$. Thus, two BIBO stable systems are "close" if (and only if) their frequency responses are close at all frequencies. More generally, a sequence of functions $\{f_i\}$ in $\mathscr{S}$ converges to f in $\mathscr{S}$ if and only if $f_i(j\omega)$ converges to $f(j\omega)$ uniformly for all real ω (or equivalently, $f_i(s)$ converges to $f(s)$ uniformly for all s in the closed RHP). With this norm, $\mathscr{S}$ becomes a topological ring, i.e., addition and multiplication $\mathscr{S}$ are continuous. Thus, if $f_i \to f$ and $g_i \to g$, then $f_i + g_i \to f + g$ and $f_i g_i \to fg$.

The extension of the norm in (1.1) to stable multivariable plants is straightforward. Let $\operatorname{mat}(R(s))$ (resp. $\operatorname{mat}(\mathscr{S})$) denote the set of matrices with elements in $R(s)$ (resp. $\mathscr{S}$). For every $F \in \operatorname{mat}(\mathscr{S})$, define

$$\|F\| = \sup_{\omega} \bar{\sigma}(F(j\omega)) = \sup_{\omega} \left\{ \lambda_{\max}\left[F^{\dagger}(j\omega) F(j\omega) \right] \right\}^{1/2} \tag{1.2}$$

where $\bar{\sigma}$ denotes the largest singular value [2] and $\dagger$ denotes the conjugate transpose. This is a norm on $\operatorname{mat}(\mathscr{S})$ and this defines a metric on $\operatorname{mat}(\mathscr{S})$ in the obvious way. Moreover, a sequence of matrices $\{F^{(k)}\}$ in $\operatorname{mat}(\mathscr{S})$ converges to F in $\operatorname{mat}(\mathscr{S})$ [in the sense of the norm (1.2)] if and only if each of the component sequences $\{f_{ij}^{(k)}\}$ converges to f_{ij} [in the sense of the norm (1.1)].[1] Once again, addition and multiplication on $\operatorname{mat}(\mathscr{S})$ are continuous.

In the case of unstable plants, one can ask: what *should* proximity and convergence mean? Consider again the standard feedback configuration shown in Fig. 1, where P is a (possibly unstable) plant and C is the compensator. In other words, both P and C belong to $\operatorname{mat}(R(s))$. Let

$$H(P,C) = \begin{bmatrix} (I+PC)^{-1} & -P(I+CP)^{-1} \\ C(I+PC)^{-1} & (I+CP)^{-1} \end{bmatrix} \tag{1.3}$$

denote the transfer matrix relating (u_1, u_2) to (e_1, e_2).[2] The system in Fig. 1 is *stable* if $H(P,C) \in \operatorname{mat}(\mathscr{S})$, in which case we say that the pair (P,C) *is stable*, or that C *stabilizes* P. Now a reasonable notion of convergence in $\operatorname{mat}(R(s))$ is the following: A sequence of plants $\{P_i\}$ converges to P if there is a compensator C that stabilizes P as well as P_i for all large enough i, and in addition $H(P_i, C)$ converges to $H(P,C)$ in the sense of the norm (1.2). If one thinks of P as a nominal plant and of P_i as perturbations of P, then the above definition states that $P_i \to P$ if one can find a stabilizing compensator C for the nominal plant P that also stabilizes the perturbed plant P_i for large enough i, and in addition the perturbed *stable* closed-loop response $H(P_i, C)$ approaches the nominal *stable* closed-loop response $H(P,C)$. For reasons explained in Section II, we refer to the topology on $\operatorname{mat}(R(s))$ induced by the above notion of convergence as the *graph topology*.

This notion of convergence for unstable plants is very weak. There is no requirement that P_i and P should have the same McMillan degree, or that they should have the same number of RHP poles. The only requirement is that one can find a compensator C that stabilizes P as well as P_i, in such a way that $H(P_i, C)$ approaches $H(P,C)$. The contrapositive of this is that if P_i does *not* approach P, then either P_i and P cannot be simultaneously stabilized, or else the resulting closed-loop responses $H(P_i, C)$ and $H(P,C)$ will be widely different. Thus, the foregoing concept of convergence gives rise to the *weakest topology* in which feedback stability is a robust property, i.e., the weakest topology in which the function $P \to H(P,C)$ is continuous for some C. Note that we *do not* demand that *every* C that stabilizes P must also stabilize P_i for large enough i. As a result, in the case where P, P_i are all stable, the convergence of $\{P_i\}$ to P in the sense of the norm (1.2) implies that $P_i \to P$ in the graph topology as well—just take $C = 0$. (The converse is also true; see Lemma 2.2.)

To illustrate the above notion of convergence, consider the scalar case and let

$$P(s) = \frac{1}{(s+1)}. \tag{1.4}$$

[1] Mathematically, this means that the topology on $\operatorname{mat}(\mathscr{S})$ is the product topology obtained from $\mathscr{S}$.

[2] It is assumed that the system is well-posed so that the indicated inverses exist.

Let $\{\epsilon_i\}$ be any sequence, converging to zero, and let

$$P_i(s)=\frac{s-\epsilon_i}{(s+1)(s+\epsilon_i)}. \tag{1.5}$$

With a bit of "hand-waving," the reader can convince himself that, in order for C to stabilize both P as well as P_i for all large enough i, $C(0)$ must equal 0; but in this case, $H(P_i,C)|_{s=0}$ will not approach $H(P,C)|_{s=0}$. Thus, the family of plants in (1.5) does not represent a "valid perturbation" of P. As another example, suppose

$$P(s)=\frac{1}{s(s+1)},\qquad P_i(s)=\frac{s-\epsilon_i}{s(s+1)(s+\epsilon_i)}. \tag{1.6}$$

Then *no* compensator can be found that stabilizes P as well as all P_i for large enough i.

In the above examples, the nonconvergence of $\{P_i\}$ to P can be explained by the "illegal" RHP pole-zero cancellation at $s=0$ as $i\to\infty$. But this "intuition" can lead one astray in the case of multivariable systems. Consider

$$P(s)=\begin{bmatrix}0 & 1\\ \dfrac{2s^2-1}{s^2-1} & \dfrac{-s^2+s+1}{s^2-1}\end{bmatrix} \tag{1.7}$$

$$P_\epsilon(s)=\begin{bmatrix}\dfrac{-\epsilon s}{s-1} & \dfrac{s-1-\epsilon}{s-1}\\ \dfrac{2s^2-1}{s^2-1} & \dfrac{-s^2+s+1}{s^2-1}\end{bmatrix}. \tag{1.8}$$

Then as $\epsilon\to 0$ there is a pole-zero cancellation at $s=1$ in the (1,2) component of P_ϵ. Nevertheless, P_ϵ can be shown to converge to P as $\epsilon\to 0$. In fact, the following stronger statement is true. Let C be *any* compensator that stabilizes P; then C also stabilizes P_ϵ for small enough ϵ, and in addition $H(P_\epsilon,C)\to H(P,C)$ as $\epsilon\to 0$.

As a final example, consider

$$P_\epsilon(s)=\begin{bmatrix}\dfrac{s+1}{s-1+\epsilon} & \dfrac{s+1}{s-1-\epsilon}\end{bmatrix} \tag{1.9}$$

$$P(s)=\begin{bmatrix}\dfrac{s+1}{s-1} & \dfrac{s+1}{s-1}\end{bmatrix}. \tag{1.10}$$

Then each component of P_ϵ converges to the corresponding component of P as $\epsilon\to 0$. Nevertheless, P_ϵ does *not* converge to P as $\epsilon\to 0$.

The preceding examples illustrate the difficulties involved in determining what is and what is not a "valid" perturbation of an unstable multivariable plant. In the next two sections we define a topology and a metric, respectively, that can be used to determine unambiguously whether or not a sequence $\{P_i\}$ converges to a limit candidate P.

II. The Graph Topology

In this section we define a topology on the set $\mathrm{mat}(R(s))$ of (possibly) unstable plants, and study some of its properties. The most important of these is that the topology presented here is the weakest one on $\mathrm{mat}(R(s))$ in which feedback stability is a robust property.

Recall [3] that every matrix in $\mathrm{mat}(R(s))$ has both a right-coprime factorization (RCF) as well as a left-coprime factorization (LCF) over the ring $\mathscr{S}$ of proper stable rational functions. Thus, if $P\in\mathrm{mat}(R(s))$, then there exist N, D, $\tilde{N}$, $\tilde{D}$, X, Y, $\tilde{X}$, $\tilde{Y}\in\mathrm{mat}(\mathscr{S})$ such that

$$P(s)=N(s)[D(s)]^{-1}=[\tilde{D}(s)]^{-1}\tilde{N}(s) \tag{2.1}$$

$$X(s)N(s)+Y(s)D(s)=I \tag{2.2}$$

$$\tilde{N}(s)\tilde{X}(s)+\tilde{D}(s)\tilde{Y}(s)=I,\qquad \text{for all } s. \tag{2.3}$$

An easy way to find an RCF and an LCF is the following [3]. Given $P\in\mathrm{mat}(R(s))$, define $P_1\in\mathrm{mat}(R(\lambda))$ by $P_1(\lambda)=P((1-\lambda)/\lambda)$, and then find an RCF and an LCF of P_1 over the ring of *polynomials* $R[\lambda]$ using standard methods [4], [5]. If $(N_1(\lambda),D_1(\lambda))$ is an RCF of $P_1(\lambda)$ over $R[\lambda]$, then $N(s)=N_1(1/(s+1))$, $D(s)=D_1(1/(s+1))$ gives an RCF of P over the ring $\mathscr{S}$. Similar remarks apply to LCF's.[3]

In [3] it is also shown that an RCF of an unstable plant parametrizes the graph of the plant in a simple way. Suppose $P\in\mathrm{mat}(R(s))$ is of order $n\times m$. Then the *graph* of P, denoted by $\mathscr{G}(P)$, consists of the subspace of L_2^{n+m} defined by[4]

$$\mathscr{G}(P)=\{(u,y)\in L_2^{n+m}: y=Pu\}. \tag{2.4}$$

Thus, the graph of P simply consists of the bounded (in the L_2-sense) input–output pairs corresponding to the possibly unstable plant P. If in particular $P\in\mathrm{mat}(\mathscr{S})$, then P maps *every* $u\in L_2^m$ into L_2^n, and

$$\mathscr{G}(P)=\{(u,Pu): u\in L_2^m\}. \tag{2.5}$$

If P does not belong to $\mathrm{mat}(\mathscr{S})$, then not every u in L_2^m gets mapped into L_2^n by P. In this case [3, Theorem 2]

$$\mathscr{G}(P)=\{(Dz,Nz): z\in L_2^m\} \tag{2.6}$$

where (N,D) is any RCF over $\mathscr{S}$ of P.[5] Note that if $P\in\mathrm{mat}(\mathscr{S})$, then (P,I) is an RCF of P, so that (2.6) reduces to (2.5).

With this background, we are ready to define a notion of convergence in $\mathrm{mat}(R(s))$.

Definition 2.1: A sequence of plants $\{P_i\}$ in $\mathrm{mat}(R(s))$ converges to P in the *graph topology* if there exist RCF's (N_i,D_i) of P_i and (N,D) of P such that $N_i\to N$, $D_i\to D$ in $\mathrm{mat}(\mathscr{S})$.[6]

Thus, $P_i\to P$ if one can factorize P_i and P in such a way that the right numerator and denominator of P_i, respectively, approach the right numerator and denominator of P. In this way, convergence of unstable plant sequences can be examined in terms of the convergence of stable plant sequences, which is described in Section I. It is immediate from (2.6) that, if $P_i\to P$ in the sense of Definition 2.1, then the *graph* of P_i converges to the *graph* of P (in a sense to be made precise in Section III). This is why the topology on $\mathrm{mat}(R(s))$ introduced in Definition 2.1 is referred to as the graph topology.

At this stage one can ask: i) does it matter which RCF of P is used, and ii) can LCF's be used instead of RCF's? These questions are answered in the following result from [7, Lemma 4.7].

Lemma 2.1: Suppose $\{P_i\}$ is a sequence in $\mathrm{mat}(R(s))$ and that $P\in\mathrm{mat}(R(s))$. Then the following are equivalent.

[3]From a numerical viewpoint, finding factorizations over $\mathscr{S}$ is a more stable operation than finding polynomial factorizations. This is because the set of units in $\mathscr{S}$ is open, whereas the set of units in $R[s]$ is not. This issue will be discussed in more detail elsewhere.

[4]Here P is used to denote a rational matrix in s as well as the corresponding operator mapping an input space of functions into an output space.

[5]Hereafter all factorizations are over $\mathscr{S}$ unless stated otherwise.

[6]Actually, in order to fully characterize the topology on $\mathrm{mat}(R(s))$, Definition 2.1 would have to be broadened to discuss the convergence of *nets* rather than sequences [6]. The reader who is familiar with point set topology can easily do this. In Section III it is shown that the graph topology of Definition 2.1 is actually induced by a *graph metric*. As a result, $\mathrm{mat}(R(s))$ is a first-countable topological space, and the graph topology is fully characterized by giving conditions for the convergence of *sequences*.

i) P_i converges to P in the graph topology.

ii) there exist RCF's (N_i, D_i) of P_i and (N, D) of P such that $N_i \to N$, $D_i \to D$ in $\mathrm{mat}(\mathscr{S})$.

iii) For *every* RCF (N, D) of P, there is a corresponding sequence $\{(N_i, D_i)\}$ of RCF's of P_i such that $N_i \to N$, $D_i \to D$ in $\mathrm{mat}(\mathscr{S})$.

iv) There exist LCF's $(\tilde{D}_i, \tilde{N}_i)$ of P_i and $(\tilde{D}, \tilde{N})$ of P such that $\tilde{D}_i \to \tilde{D}$, $\tilde{N}_i \to \tilde{N}$ in $\mathrm{mat}(\mathscr{S})$.

v) For *every* LCF $(\tilde{D}, \tilde{N})$ of P, there is a corresponding sequence $\{(\tilde{D}_i, \tilde{N}_i)\}$ of LCF's of P_i such that $\tilde{D}_i \to \tilde{D}$, $\tilde{N}_i \to \tilde{N}$ in $\mathrm{mat}(\mathscr{S})$.

Note that the equivalence of i) and ii) is just a restatement of Definition 2.1. Condition iii) states that it does not matter which RCF of P is used. Conditions iv) and v) are the "left" analogs of ii) and iii).

Now we come to the main results of this section. To make the theorem statements more compact, we use the following notation. If $P \in \mathrm{mat}(R(s))$, then $S(P)$ denotes the set of compensators that stabilize P. In other words,

$$S(P) = \{C \in \mathrm{mat}(R(s)) \colon H(P,C) \in \mathrm{mat}(\mathscr{S})\} \tag{2.7}$$

where $H(P,C)$ is defined in (1.3).

Theorem 2.1: Suppose $\{P_i\}$ converges to P in the graph topology. Then every C in $S(P)$ also belongs to $S(P_i)$ for large enough i; moreover, for every $C \in S(P)$, $H(P_i, C) \to H(P, C)$ in $\mathrm{mat}(\mathscr{S})$. Conversely, suppose $\{P_i\}$ is a sequence in $\mathrm{mat}(R(s))$, that $P \in \mathrm{mat}(R(s))$, and suppose there exists a $C \in S(P)$ such that $C \in S(P_i)$ for large enough i and $H(P_i, C) \to H(P, C)$ in $\mathrm{mat}(\mathscr{S})$; then $\{P_i\}$ converges to P in the graph topology.

Corollary 2.1.1: Suppose $\{P_i\}$ is a sequence in $\mathrm{mat}(R(s))$ and that $P \in \mathrm{mat}(R(s))$. Suppose there exists *one* $C \in S(P)$ such that $C \in S(P_i)$ for large enough i and $H(P_i, C) \to H(P, C)$ in $\mathrm{mat}(\mathscr{S})$. Then *every* compensator in $S(P)$ has these properties.

Theorem 2.1 shows that the graph topology is the weakest one on $\mathrm{mat}(R(s))$ in which the function $P \to H(P, C)$ is continuous for some C, i.e., the weakest one in which feedback stability is robust against plant perturbations. A related result, given in [7, Theorem 4.1], shows that the graph topology is the weakest one in which the function $(P, C) \to H(P, C)$ is continuous, i.e., the weakest one in which feedback stability is robust against *simultaneous* plant and compensator perturbations. The present theorem statement is much more direct, and its proof much less technical, than in [7]. The result corresponding to the one in [7] is given next.

Theorem 2.2: Suppose $\{P_i\}$, $\{C_i\}$ are sequences in $\mathrm{mat}(R(s))$, that $P, C \in \mathrm{mat}(R(s))$, and that (P, C) is stable. Suppose $P_i \to P$, $C_i \to C$ in the graph topology; then (P_i, C_i) is stable for large enough i and $H(P_i, C_i) \to H(P, C)$ in $\mathrm{mat}(\mathscr{S})$. Conversely, suppose (P_i, C_i) is stable and $H(P_i, C_i) \to H(P, C)$ in $\mathrm{mat}(\mathscr{S})$; then $P_i \to P$, $C_i \to C$ in the graph topology.

The proof of Theorem 2.1 is not very difficult, and is based on a few well-known facts that are summarized below [7], [8]: A function $u \in \mathscr{S}$ is called a *unit* of $\mathscr{S}$ if $1/u$ also belongs to $\mathscr{S}$. Clearly, u is a unit if and only if it has neither poles nor zeros in the closed RHP, and does not vanish at infinity (such functions are called *miniphase* in [9]). Let $\mathscr{U}$ denote the set of units of $\mathscr{S}$. Then $\mathscr{U}$ is an *open* subset of $\mathscr{S}$; in other words, if u is a unit, and if $\{u_i\}$ is a sequence in $\mathscr{S}$ converging to u, then u_i is also a unit for large enough i. Moreover, the map $u \to u^{-1}$ from $\mathscr{U}$ into itself is continuous; thus if u is a unit and $\{u_i\}$ is a sequence in $\mathscr{S}$ converging to u, then $\{u_i^{-1}\}$ converges to u^{-1}. Similar results hold in the multivariable case as well. A matrix $U \in \mathrm{mat}(\mathscr{S})$ is *unimodular* if its determinant is a unit of $\mathscr{S}$; in this case U^{-1} also belongs to $\mathrm{mat}(\mathscr{S})$. The set of unimodular matrices is open and inversion is continuous; thus if U is unimodular and $\{U_i\}$ is a sequence in $\mathrm{mat}(\mathscr{S})$ converging to U, then U_i is also unimodular for large enough i and $\{U_i^{-1}\}$ converges to U^{-1}.

Proof of Theorem 2.1: To prove the first part of the theorem, suppose $\{P_i\}$ converges to P in the graph topology, and select RCF's (N_i, D_i) of P_i and (N, D) of P such that $N_i \to N$, $D_i \to D$ in $\mathrm{mat}(\mathscr{S})$. Let C be *any* stabilizing compensator for P. Then from [7], [10] there exists an LCF $(\tilde{D}_c, \tilde{N}_c)$ of C such that $\tilde{D}_c D + \tilde{N}_c N = I$. Moreover,

$$H(P,C) = \begin{bmatrix} I - N\tilde{N}_c & -N\tilde{D}_c \\ D\tilde{N}_c & D\tilde{D}_c \end{bmatrix}. \tag{2.8}$$

Now define $\Delta_i = \tilde{D}_c D_i + \tilde{N}_c N_i$. Then, as $\Delta_i \to I$, it follows that Δ_i is unimodular for large enough i and that $\Delta_i^{-1} \to I$. Hence, C stabilizes P_i for large enough i. Moreover, since

$$H(P_i,C) = \begin{bmatrix} I - N_i\Delta_i^{-1}\tilde{N}_c & -N_i\Delta_i^{-1}\tilde{D}_c \\ D_i\Delta_i^{-1}\tilde{N}_c & D_i\Delta_i^{-1}\tilde{D}_c \end{bmatrix} \tag{2.9}$$

from (1.3), and since addition and multiplication are continuous on $\mathrm{mat}(\mathscr{S})$, it follows that $H(P_i, C) \to H(P, C)$ in $\mathrm{mat}(\mathscr{S})$.

To prove the second part of the theorem, suppose $\{P_i\}$ is a sequence in $\mathrm{mat}(R(s))$, that $P \in \mathrm{mat}(R(s))$, and suppose there is a $C \in S(P)$ such that $C \in S(P_i)$ for large enough i and $H(P_i, C) \to H(P, C)$. Let (N, D) be any RCF of P. Since C stabilizes P, there is an LCF $(\tilde{D}_c, \tilde{N}_c)$ of C such that $\tilde{D}_c D + \tilde{N}_c N = I$. Moreover, $H(P, C)$ is given by (2.8). By discarding a finite number of terms from the sequence $\{P_i\}$ we can assume that C stabilizes P_i for all i. This implies that there is an RCF (N_i, D_i) of P_i such that $\tilde{D}_c D_i + \tilde{N}_c N_i = I$. Further,

$$H(P_i,C) = \begin{bmatrix} I - N_i\tilde{N}_c & -N_i\tilde{D}_c \\ D_i\tilde{N}_c & D_i\tilde{D}_c \end{bmatrix} \tag{2.10}$$

from (1.3). As $H(P_i, C) \to H(P, C)$ in $\mathrm{mat}(\mathscr{S})$, (2.8) and (2.10) imply that

$$\begin{bmatrix} N_i\tilde{N}_c & N_i\tilde{D}_c \\ D_i\tilde{N}_c & D_i\tilde{D}_c \end{bmatrix} \to \begin{bmatrix} N\tilde{N}_c & N\tilde{D}_c \\ D\tilde{N}_c & D\tilde{D}_c \end{bmatrix} \quad \text{in } \mathrm{mat}(\mathscr{S}). \tag{2.11}$$

Multiplying both sides of (2.11) by $\begin{bmatrix} N \\ D \end{bmatrix}$ and noting that $\tilde{N}_c N + \tilde{D}_c D = I$ gives

$$\begin{bmatrix} N_i \\ D_i \end{bmatrix} \to \begin{bmatrix} N \\ D \end{bmatrix} \quad \text{in } \mathrm{mat}(\mathscr{S}). \tag{2.12}$$

Hence, $\{P_i\}$ converges to P in the graph topology. □

The proof of Theorem 2.2 is deferred to Section IV, as it makes use of some further concepts that are developed therein.

The graph topology gives a notion of convergence in $\mathrm{mat}(R(s))$. As $\mathrm{mat}(\mathscr{S})$ is a subset of $\mathrm{mat}(R(s))$, the graph topology also gives a notion of convergence on $\mathrm{mat}(\mathscr{S})$. We now show that this is the same as convergence in the sense of the norm (1.2). This shows that the graph topology on $\mathrm{mat}(R(s))$ is a genuine extension to unstable plants of the familiar topology for stable plants.

Lemma 2.2: Suppose $\{P_i\}$ is a sequence in $\mathrm{mat}(\mathscr{S})$, $P \in \mathrm{mat}(\mathscr{S})$, and $\|P_i - P\| \to 0$; then $P_i \to P$ in the graph topology. Conversely, suppose $\{P_i\}$ is a sequence in $\mathrm{mat}(R(s))$, $P \in \mathrm{mat}(\mathscr{S})$, and $P_i \to P$ in the graph topology; then $P_i \in \mathrm{mat}(\mathscr{S})$ for large enough i, and $\|P_i - P\| \to 0$.

Proof: To prove the first part of the lemma, note that $C = 0$ stabilizes P as well as all P_i, since $P, P_i \in \mathrm{mat}(\mathscr{S})$. Further, since $\|P_i - P\| \to 0$, it follows from (1.3) that $H(P_i, C) \to H(P, C)$ in $\mathrm{mat}(\mathscr{S})$. Hence, by Theorem 2.1, $P_i \to P$ in the graph topology.

To prove the second part of the lemma, note that (P, I) is an RCF of P, since $P \in \mathrm{mat}(\mathscr{S})$. By assumption, there exist RCF's (N_i, D_i) of P_i such that $N_i \to P$, $D_i \to I$. Since the set of unimodular matrices is open and inversion is continuous, this implies that D_i is unimodular for large enough i and that $D_i^{-1} \to I$ in $\mathrm{mat}(\mathscr{S})$. Hence, $P_i = N_i D_i^{-1}$ is in $\mathrm{mat}(\mathscr{S})$ for large enough i, and approaches P in $\mathrm{mat}(\mathscr{S})$, i.e., $\|P_i - P\| \to 0$. □

The second part of Lemma 2.2 shows that, in the graph topology, mat($\mathscr{S}$) is an open subset of mat($R(s)$). In words, this means that the set of all stable (unstable) plants is an open (closed) subset of the set of all plants.

We now develop some further details concerning the convergence of sequences in the graph topology. These results are often helpful in concluding nonconvergence.

Lemma 2.3: Suppose $\{P_i\}$ converges to P in the graph topology. Let $s_1,\cdots,s_l$ denote the poles of P in the closed RHP. Then we have the following.

i) Let $\mathscr{R}$ be any compact subset of the open RHP such that none of the s_i lies on the boundary of $\mathscr{R}$, and let ν denote the number of poles of P inside $\mathscr{R}$, counted according to their McMillan degrees. Then for large enough i, each P_i has exactly ν poles inside $\mathscr{R}$.

ii) Let $\mathscr{R}$ be any closed subset of the closed RHP that does not contain $s_1,\cdots,s_l$; if P has a pole at infinity, suppose in addition that $\mathscr{R}$ is bounded. Then $P_i(s)\to P(s)$ as $i\to\infty$, uniformly for all s in $\mathscr{R}$.

Remarks: In general, the number of *closed* RHP poles of P_i and of P need not be equal; for example, consider $P_i(s)=1/(s+\epsilon_i)$, $P(s)=1/s$, where $\{\epsilon_i\}$ is any sequence of positive numbers converging to zero.

Proof: Since $P_i\to P$ in the graph topology, there exist RCF's (N_i,D_i) of P_i and (N,D) of P such that $N_i\to N$, $D_i\to D$ in mat($\mathscr{S}$). To prove i), observe that $\det D_i\to\det D$ in $\mathscr{S}$.That is, $\det D_i(s)\to\det D(s)$ uniformly over the closed RHP. Recall [20, Appendix I] that the RHP poles of P are precisely the RHP zeros of $\det D$, and the McMillan degree of an RHP pole of P is equal to its multiplicity as a zero of $\det D$; similar remarks apply to P_i and $\det D_i$. Since $\det D$ and $\det D_i$ are both analytic in the open RHP, and since $\det D$ has no zeros on the boundary of $\mathscr{R}$, i) follows from applying the principle of argument to $\det D$ and $\det D_i$. To prove ii), observe that since $D_i(s)\to D(s)$ uniformly over the closed RHP, it follows that $[D_i(s)]^{-1}\to[D(s)]^{-1}$ wherever the latter is well-defined, i.e., wherever $\det D(s)$ is nonzero. Moreover, if we exclude a neighborhood of every closed RHP zero of $\det D$, the convergence is uniform with respect to s. Since $P_i(s)=N_i(s)[D_i(s)]^{-1}$, $P(s)=N(s)[D(s)]^{-1}$, the same is true of P_i and P. □

One of the main difficulties with the graph topology is that the convergence of a sequence of matrices cannot be related in a simple way to the convergence of the component sequences. In other words, the graph topology on mat($R(s)$) is *not* the same as the product topology on mat($R(s)$) obtained from the graph topology on $R(s)$; in fact, neither contains the other. This was illustrated in Section I by means of two examples. The justification for the statements made in those examples is given next.

First consider the plant $P_\epsilon(s)$ of (1.8). This has the RCF

$$N_\epsilon=\begin{bmatrix}\lambda+\epsilon & \lambda-1\\ \lambda-2 & 1\end{bmatrix},$$

$$D_\epsilon=D_0=\begin{bmatrix}\lambda-1 & \lambda\\ \lambda & \lambda-1\end{bmatrix},$$

$$\lambda=\frac{1}{s+1}. \tag{2.13}$$

As $\epsilon\to 0$, N_ϵ approaches N_0, where

$$N_0=\begin{bmatrix}\lambda & \lambda-1\\ \lambda-2 & 1\end{bmatrix}. \tag{2.14}$$

Further, D_ϵ is independent of ϵ, so call it D_0. So, in the graph topology, $P_\epsilon=N_\epsilon D_\epsilon^{-1}$ converges to $P_0=N_0D_0^{-1}$, where the latter is given by (1.7). However, $(P_\epsilon)_{12}$ does *not* converge to $(P_0)_{12}$ [see (1.8)].

Now consider the plant P_ϵ of (1.9). This has the LCF

$$\tilde{d}_\epsilon=\frac{(s-1-\epsilon)(s-1+\epsilon)}{(s+1)^2},\qquad \tilde{N}_\epsilon=\begin{bmatrix}\frac{s-1-\epsilon}{s+1} & \frac{s-1+\epsilon}{s+1}\end{bmatrix} \tag{2.15}$$

whereas P_0 of (1.10) has the LCF

$$\tilde{d}_0=\frac{s-1}{s+1},\qquad \tilde{N}_0=[1\quad 1]. \tag{2.16}$$

From (1.9), we see that each component of P_ϵ converges in the graph topology to the corresponding component of P_0. Nevertheless, P_ϵ does not converge to P_0. Consider the set $\mathscr{R}=\{s\colon |s-1|\leqslant 1/2\}$. Then P_0 has one pole inside $\mathscr{R}$, whereas P_ϵ has two poles inside $\mathscr{R}$ whenever $\epsilon\neq 0$. Hence from i) of Lemma 2.3, it follows that P_ϵ does not approach P_0.

Even though the graph topology is *not* a product topology in general, it is a product topology in two important special cases: for stable plants and for block-diagonal plants. The case of stable plants is covered by Lemma 2.2, which states that the graph topology, when restricted to the set mat($\mathscr{S}$), is the same as the topology induced by the norm (1.2), which is of course a product topology. For block-diagonal plants, we have the following result.

Lemma 2.4: Suppose $\{P_i\}$ is a sequence in mat($R(s)$), $P\in$ mat($\mathscr{R}(\mathscr{S})$), and suppose P_i, P are of the form

$$P_i=\begin{bmatrix}P_{1i} & & 0\\ & \ddots & \\ 0 & & P_{ni}\end{bmatrix}\quad\text{for all } i,\qquad P=\begin{bmatrix}P_1 & & 0\\ & \ddots & \\ 0 & & P_n\end{bmatrix} \tag{2.17}$$

where all partitions are of commensurate size. Then $P_i\to P$ in the graph topology if and only if $P_{li}\to P_l$ in the graph topology for each 1 in $\{1,\cdots,n\}$.

The proof of Lemma 2.4 is also given in Section IV.

To summarize, in this section we have defined a graph topology on the set mat($R(s)$) of (possibly) unstable plants, and have shown that convergence in the graph topology has a nice interpretation in terms of the ability to design stabilizing compensators and the continuity of the closed-loop response. In fact, the graph topology is the weakest topology on unstable plants such that feedback stability is robust. On the other hand, convergence of a sequence of matrices cannot be related in a simple way to the convergence of the component sequences, except in special cases. It is therefore desirable to have a simple test for convergence in the graph topology. This is provided by the graph metric introduced in Section III.

Finally, in some applications one may not wish to use the norms (1.1) and (1.2) to measure distances in $\mathscr{S}$ and mat($\mathscr{S}$). A careful examination of the proof of Theorem 2.1 reveals that any topology on mat($\mathscr{S}$) can be used [and not necessarily the one induced by the norm (1.2)], and the theorem remains valid, provided only that i) addition and multiplication in $\mathscr{S}$ are continuous, and ii) the set of units is open and inversion is continuous.

III. The Graph Metric

In Section II, we defined a topology on the set of unstable plants, and derived several qualitative properties (see Theorems 2.1 and 2.2). However, in order to obtain *quantitative* estimates of stability margins, it is desirable to have a quantitative measure of the disparity between two unstable plants. Towards this end, in this section we define a metric on the set mat($R(s)$) of possibly unstable plants, and show that the topology on mat($R(s)$) induced by this metric is the same as the graph topology of

Definition 2.1. For this reason, we refer to this metric as the *graph metric*. Robustness estimates for feedback stability based on the graph metric are given in Section IV.

A preliminary concept is needed to define the graph metric. Suppose $M(s)$ is a square rational matrix with the properties: i) $M(s)=M'(-s)$ for all s, where $'$ denotes the transpose; ii) $M(j\omega)$ is uniformly positive definite for all ω and is bounded as a function of ω; and iii) $M(s)$ is nonsingular for almost all s. Under these conditions, it is well known [12] that there exists a matrix A in $\mathrm{mat}(\mathscr{S})$ such that $M(s)=A'(-s)A(s)$ and such that A is actually a *unit* of $\mathrm{mat}(\mathscr{S})$(i.e., $A^{-1}\in \mathrm{mat}(\mathscr{S})$). Such a matrix A is called a *spectral factor* of M and is unique to within left multiplication by an orthogonal matrix; that is, if A and B are both spectral factors of M, then there is an orthogonal matrix U such that $B=UA$.

Definition 3.1: A pair (N, D) is called a *normalized RCF* of a plant $P\in \mathrm{mat}(R(s))$ if (N, D) is an RCF of P, and in addition

$$D'(-s)D(s)+N'(-s)N(s)=I, \qquad \text{for all } s. \tag{3.1}$$

Lemma 3.1: Every plant has a normalized RCF, which is unique to within right multiplication by an orthogonal matrix.

Proof: Suppose $P\in \mathrm{mat}(R(s))$, and let (N_1, D_1) be any RCF of P. Define

$$M(s)=\begin{bmatrix} D_1'(-s) & N_1'(-s)\end{bmatrix}\begin{bmatrix} D_1(s) \\ N_1(s)\end{bmatrix}. \tag{3.2}$$

Then the coprimeness of N_1 and D_1 assures that M has a spectral factorization. Let A be a spectral factor of M and define $N=N_1A^{-1}$, $D=D_1A^{-1}$. Then (N, D) is a normalized RCF of P.

To show that (N, D) is unique except for the possibility of right multiplication by an orthogonal matrix, let (N_2, D_2) be any other RCF of P. Then $N_2=N_1V$, $D_2=D_1V$ for some unimodular matrix $V\in \mathrm{mat}(\mathscr{S})$, and

$$M_2(s)=\begin{bmatrix} D_2'(-s) & N_2'(-s)\end{bmatrix}\begin{bmatrix} D_2(s) \\ N_2(s)\end{bmatrix}=V'(-s)M(s)V(s). \tag{3.3}$$

Thus, AV is a spectral factor of M_2; moreover, if B is any other spectral factor of M_2, then $B=UAV$ for some orthogonal matrix U. Hence, $(N_2B^{-1}, D_2B^{-1})=(N_1A^{-1}U^{-1}, D_1A^{-1}U^{-1})=(NU^{-1}, DU^{-1})$. □

Suppose (N, D) is a normalized RCF of P and let $A=\begin{bmatrix} D \\ N\end{bmatrix}$. If $R\in \mathrm{mat}(\mathscr{S})$, then it is a ready consequence of (1.2) and (3.1) that $\|AR\|=\|R\|$. Thus, the map $R\to AR$ is an isometry on $\mathrm{mat}(\mathscr{S})$. Similarly, if U is an orthogonal matrix, then $\|UR\|=\|R\|$ for all R in $\mathrm{mat}(\mathscr{S})$.

We now define the graph metric.

Definition 3.2: Suppose $P_1, P_2\in \mathrm{mat}(R(s))$ have the same dimensions, and let (N_i, D_i) be a normalized RCF of P_i for $i=1,2$. Define

$$A_i=\begin{bmatrix} D_i \\ N_i\end{bmatrix}, \qquad i=1,2 \tag{3.4}$$

$$\delta(P_1, P_2)=\inf_{\|U\|\leqslant 1,\, U\in \mathrm{mat}(\mathscr{S})}\|A_1-A_2U\| \tag{3.5}$$

$$d(P_1, P_2)=\max\{\delta(P_1, P_2), \delta(P_2, P_1)\}. \tag{3.6}$$

Then d is called the *graph metric* on $\mathrm{mat}(R(s))$.

It is left to the reader to verify that $d(P_1, P_2)$ is a well-defined quantity, even though A_1, A_2 are only unique to within right multiplication by an orthogonal matrix. This is because multiplication by an orthogonal matrix does not change the norm.

Lemma 3.2: d is a metric on $\mathrm{mat}(R(s))$ taking values in the interval $[0,1]$.

Proof: If $U=0$, then $\|A_1-A_2U\|=\|A_1\|=1$. Hence, $\delta(P_1, P_2)\leqslant 1$ and $d(P_1, P_2)\leqslant 1$ for all P_1, P_2. In proving that d is a metric, the only nonobvious part is the proof of the triangle inequality. Accordingly, let $P_1, P_2, P_3\in \mathrm{mat}(R(s))$, and select $U, V\in \mathrm{mat}(\mathscr{S})$ such that $\|U\|\leqslant 1$, $\|V\|\leqslant 1$, and

$$\|A_1-A_2U\|\leqslant \delta(P_1, P_2)+\epsilon \tag{3.7}$$

$$\|A_2-A_3V\|\leqslant \delta(P_2, P_3)+\epsilon \tag{3.8}$$

where ϵ is some positive number. Then $VU\in \mathrm{mat}(\mathscr{S})$ and $\|VU\|\leqslant \|V\|\cdot\|U\|\leqslant 1$; moreover,

$$\begin{aligned}\|A_1-A_3VU\| &\leqslant \|A_1-A_2U\|+\|A_2U-A_3VU\| \\ &\leqslant \|A_1-A_2U\|+\|A_2-A_3V\|\cdot\|U\| \\ &\leqslant \delta(P_1, P_2)+\delta(P_2, P_3)+2\epsilon \end{aligned} \tag{3.9}$$

where we use the fact that $\|U\|\leqslant 1$. Now (3.9) shows that

$$\delta(P_1, P_3)\leqslant \delta(P_1, P_2)+\delta(P_2, P_3)+2\epsilon. \tag{3.10}$$

Since this is true for *every* $\epsilon>0$, it follows that

$$\delta(P_1, P_3)\leqslant \delta(P_1, P_2)+\delta(P_2, P_3). \tag{3.11}$$

By symmetry, we get the triangle inequality for $\delta(P_3, P_1)$, and hence for d. □

The main result of this section is Theorem 3.1 which states that the topology on $\mathrm{mat}(R(s))$ induced by the graph metric is the same as the graph topology of Definition 2.1. To prove this theorem, we need the following technical fact.

Lemma 3.3: Suppose $P_1, P_2\in \mathrm{mat}(R(s))$ have the same dimensions, and let A_1, A_2 be associated matrices in $\mathrm{mat}(\mathscr{S})$ obtained from their normalized RCF's. Suppose $U, V\in \mathrm{mat}(\mathscr{S})$ satisfy $\|U\|\leqslant 1$, $\|V\|\leqslant 1$, and $\|A_1-A_2U\|+\|A_2-A_1V\|<1$. Then U, V are unimodular.

Proof: We have

$$\begin{aligned}\|I-VU\|=\|A_1-A_1VU\| &\leqslant \|A_1-A_2U\|+\|A_2U-A_1VU\| \\ &\leqslant \|A_1-A_2U\|+\|A_2-A_1V\|<1. \end{aligned} \tag{3.12}$$

Hence, VU is unimodular, i.e., $\det(VU)$ is a unit of $\mathscr{S}$. Since $\det(VU)=\det V\cdot\det U$, it follows that both $\det V$ and $\det U$ are units of $\mathscr{S}$, i.e., that both U and V are unimodular in $\mathrm{mat}(\mathscr{S})$. □

Theorem 3.1: A sequence $\{P_i\}$ in $\mathrm{mat}(R(s))$ converges to P in the graph topology of Definition 2.1 if and only if $d(P_i, P)\to 0$.

Proof:

"If": Suppose $d(P_i, P)\to 0$. Then for each ϵ in the interval $(0,1/4)$ there is an integer i_0 such that $d(P_i, P)\leqslant\epsilon$ whenever $i\geqslant i_0$. For each i, there exist U_i, V_i in $\mathrm{mat}(\mathscr{S})$ with norm at most one such that $\|A-A_iU_i\|\leqslant d(P_i, P)+\epsilon$, $\|A_i-AV_i\|\leqslant d(P_i, P)+\epsilon$. Now, if $i\geqslant i_0$, then

$$\|A-A_iU_i\|+\|A_i-AV_i\|\leqslant 4\epsilon<1. \tag{3.13}$$

Hence, by Lemma 3.3, $U_i, V_i\in\mathscr{U}$. Thus, if

$$A=\begin{bmatrix} D \\ N\end{bmatrix}, \qquad A_i=\begin{bmatrix} D_i \\ N_i\end{bmatrix},$$

then (N_iU_i, D_iU_i) is an RCF of P_i, and $\|A-A_iU_i\|\leqslant 2\epsilon$ whenever $i\geqslant i_0$. It follows from Definition 2.1 that $P_i\to P$ in the graph topology.

"Only If": Suppose $P_i\to P$ in the graph topology, and let (N, D) be a normalized RCF of P. Then, by Lemma 2.1, there exist RCF's (N_i, D_i) of P_i such that

$$\begin{bmatrix} D_i \\ N_i\end{bmatrix}\to\begin{bmatrix} D \\ N\end{bmatrix}.$$

However, (N_i, D_i) need not be normalized. Let

$$M_i = \begin{bmatrix} D_i \\ N_i \end{bmatrix},$$

and suppose $M_i = A_i R_i$ where A_i corresponds to a normalized RCF and $R_i \in \mathscr{U}$. Now $\|M_i - A\| \to 0$, and $\|M_i\| = \|A_i R_i\| = \|R_i\|$ since A_i is an isometry. Hence, $\|R_i\| \to 1$. Define $U_i = R_i / \|R_i\|$. Then $\|U_i\| = 1$; moreover,

$$\begin{aligned} \|A - A_i U_i\| &\leq \|A - A_i R_i\| + \|A_i R_i - A_i U_i\| \\ &= \|A - M_i\| + \|R_i - U_i\| \qquad \text{since } A_i \text{ is an isometry} \\ &= \|A - M_i\| + |1 - \|R_i\|| \\ &\to 0 \qquad \text{as } i \to \infty. \end{aligned} \tag{3.14}$$

This shows that $\delta(P, P_i) \to 0$. To prove that $\delta(P_i, P) \to 0$, it is necessary to estimate $\|R_i^{-1}\|$. Let $\nu_i = \|A - M_i\|$. Then for all $x \in L_2$, we have

$$\|M_i x\| \geq \|Ax\| - \|(A - M_i)x\| \geq (1 - \nu_i)\|x\|.$$

Since

$$\|x\| = \|A_i x\| = \|M_i R_i^{-1} x\| \geq (1 - \nu_i)\|R_i^{-1} x\|,$$

it follows that $\|R_i^{-1}\| \leq 1/(1 - \nu_i)$ whenever $\nu_i < 1$. In particular, $\|R_i^{-1}\| \to 1$ as $i \to \infty$. Let $V_i = R_i^{-1} / \|R_i^{-1}\|$. Then $\|V_i\| = 1$, and

$$\begin{aligned} \|A_i - A V_i\| &= \|M_i R_i^{-1} - A V_i\| \\ &= \|M_i R_i^{-1} - A R_i^{-1} + A R_i^{-1} - A V_i\| \\ &\leq \|M_i - A\| \cdot \|R_i^{-1}\| + \|R_i^{-1} - V_i\| \\ &\to 0 \qquad \text{as } i \to \infty. \end{aligned} \tag{3.15}$$

Thus, $\delta(P_i, P) \to 0$ and the proof is complete. □

At present, the problem of computing the infimum in (3.5) is an unsolved problem in interpolation theory. Clearly,

$$\delta(P_i, P_2) \geq \inf_{U \in \mathrm{mat}(\mathscr{S})} \|A_1 - A_2 U\|. \tag{3.16}$$

The infimum in (3.16) can be computed using the results in [13], and provides a lower bound on $d(P_1, P_2)$ to go with the upper bound given next.

Lemma 3.4: Suppose (N, D) is a normalized RCF of P, that (N_1, D_1) is a (not necessarily normalized) RCF of P_1, and let

$$A = \begin{bmatrix} D \\ N \end{bmatrix}, \qquad M_1 = \begin{bmatrix} D_1 \\ N_1 \end{bmatrix}.$$

Suppose $\|A - M_1\| \triangleq \nu < 1$. Then

$$d(P, P_1) \leq 2\nu/(1 - \nu). \tag{3.17}$$

Proof: Suppose $M_1 = A_1 R_1$, where A_1 is a normalized RCF of P_1. Now $\|M_1\| \leq \|A\| + \|M_1 - A\| = 1 + \nu$. Hence $\|R_1\| = \|M_1\| \leq 1 + \nu$ and $\|R_1^{-1}\| \leq 1/(1 - \nu)$, as in the proof of Theorem 3.1. From (3.14) and (3.15) we get, successively,

$$\delta(P, P_1) \leq 2\nu \tag{3.18}$$

$$\delta(P_1, P) \leq 2\nu/(1 - \nu) \tag{3.19}$$

$$d(P, P_1) = \max\{\delta(P, P_1), \delta(P_1, P)\} \leq 2\nu/(1 - \nu). \tag{3.20}$$

This completes the proof. □

As mentioned earlier, $\mathrm{mat}(\mathscr{S})$ is an open subset of $\mathrm{mat}(R(s))$ in the graph topology (see the paragraph after the proof of Lemma 2.2). Thus, if $P \in \mathrm{mat}(\mathscr{S})$, then there is a number $c > 0$ such that $P_1 \in \mathrm{mat}(\mathscr{S})$ whenever $d(P, P_1) < c$. Estimating this number c leads to some robustness results, as discussed in the next section.

If P is an unstable plant, and particularly if P has $j\omega$-axis poles, then it is possible for *every* neighborhood of P to contain plants with a different number of RHP poles from P. For example, the reader can verify using Lemma 3.4 that

$$d\left(\frac{1}{s+\epsilon}, \frac{1}{s}\right) = 0(\epsilon), \qquad d\left(\frac{1}{s-\epsilon}, \frac{1}{s}\right) = 0(\epsilon). \tag{3.21}$$

Thus, every neighborhood of the unstable plant $1/s$ contains stable plants.

IV. Robustness Estimates for Feedback Stability

In this section, the graph metric introduced in Section III is used to derive some estimates for the robustness of feedback stability. We present at once the main result of this section, and defer the proof.

Theorem 4.1: Suppose the plant–compensator pair (P, C) is stable, and let $H(P, C)$ be the associated stable closed-loop transfer matrix defined in (1.3). Let $r = (1 + \|H(P, C)\|^2)^{1/2}$. Then the pair (P_1, C_1) is also stable whenever

$$\max\{d(P, P_1), d(C, C_1)\} \triangleq d < 1/(2 + 4r). \tag{4.1}$$

Moreover, if (4.1) holds, then

$$\|H(P_1, C_1) - H(P, C)\| \leq \frac{r\gamma(1 + r)}{1 - \gamma r} \tag{4.2}$$

where $\gamma = 4d/(1 - 2d)$.

Thus, Theorem 4.1 shows that if P_1, C_1 are sufficiently close to (P, C), then (P_1, C_1) is also stable, and $H(P_1, C_1)$ is close to $H(P, C)$. The measure of proximity of the perturbed pair (P_1, C_1) to the unperturbed pair (P, C) is provided by the larger of the graph metrics $d(P, P_1), d(C, C_1)$. The noteworthy features of Theorem 4.1 are the following: 1) $P_1(C_1)$ need not have the same number of RHP poles as $P(C)$; 2) $P_1(C_1)$ need not have the same dynamic order as $P(C)$; in fact, P_1 can be infinite-dimensional (representing a distributed parameter system), whereas P can be finite-dimensional (see Section VII for a discussion of distributed systems).

If P_1 has the same RHP poles as P, then it is in fact possible to derive *necessary and sufficient* conditions for robustness. This is done in Section V.

We now present a series of lemmas, culminating in the proof of Theorem 4.1.

Lemma 4.1: Suppose $P \in \mathrm{mat}(\mathscr{S})$, and let $r = (1 + \|P\|^2)^{1/2}$. Then $P_1 \in \mathrm{mat}(\mathscr{S})$ whenever $d(P, P_1) < 1/r$; moreover,

$$\|P_1 - P\| \leq [rd(1 + r)]/(1 - rd) \tag{4.3}$$

where d denotes $d(P, P_1)$.

Proof: Since P is stable, (P, I) is an RCF for P, although it may not be normalized. Let R be a spectral factor of $I + P'(-s)P(s)$. Then

$$R^{\dagger}(j\omega)R(j\omega) = I + P^{\dagger}(j\omega)P(j\omega) \tag{4.4}$$

which shows that $\|R\|^2 \leq 1 + \|P\|^2 = r^2$. Now (PR^{-1}, R^{-1}) is a normalized RCF of P. Further, since $d(P, P_1) < 1/r$, there exists a normalized RCF $(\bar{N}_1, \bar{D}_1)$ of P_1 and a matrix $U \in \mathrm{mat}(\mathscr{S})$ such that $\|U\| \leq 1$ and

$$\left\| \begin{bmatrix} \bar{D}_1 U \\ \bar{N}_1 U \end{bmatrix} - \begin{bmatrix} R^{-1} \\ PR^{-1} \end{bmatrix} \right\| \leq d(P, P_1) + \epsilon < 1/r. \tag{4.5}$$

In particular, $\|\bar{D}_1 U - R^{-1}\| < 1/r \leqslant 1/\|R\|$. This shows that $\bar{D}_1 U$ is a unimodular matrix, and by an easy calculation, that

$$\|(\bar{D}_1 U)^{-1} - R\| \leqslant \frac{\|R\|^2 \cdot \|\bar{D}_1 U - R^{-1}\|}{1 - \|R\| \cdot \|\bar{D}_1 U - R^{-1}\|} \leqslant \frac{r^2 \gamma}{1 - r\gamma} \qquad (4.6)$$

where $\gamma = d(P, P_1) + \epsilon$. Now, since $\bar{D}_1 U$ is unimodular, it follows that U is also unimodular, so that $P_1 = \bar{N}_1 U \cdot (\bar{D}_1 U)^{-1}$ is stable.

It only remains to estimate $\|P_1 - P\|$. For notational convenience, let $N = PR^{-1}$, $D = R^{-1}$, $N_1 = \bar{N}_1 U$, $D_1 = \bar{D}_1 U$. Then

$$\begin{aligned} \|P_1 - P\| &= \|N_1 D_1^{-1} - ND^{-1}\| \\ &\leqslant \|N_1 D_1^{-1} - ND_1^{-1}\| + \|ND_1^{-1} - ND^{-1}\| \\ &\leqslant \|N_1 - N\| \cdot \|D_1^{-1}\| + \|N\| \cdot \|D_1^{-1} - D^{-1}\|. \end{aligned} \qquad (4.7)$$

From (4.6), we get

$$\|D_1^{-1}\| \leqslant \|D_1^{-1} - R\| + \|R\| \leqslant \frac{r^2\gamma}{1 - r\gamma} + r = \frac{r}{1 - r\gamma}. \qquad (4.8)$$

Also, $\|N_1 - N\| \leqslant \gamma$ and $\|N\| \leqslant 1$, since $\left\| \begin{bmatrix} D \\ N \end{bmatrix} \right\| = 1$. Substituting all these into (4.7) gives

$$\|P_1 - P\| \leqslant \frac{r\gamma}{1 - r\gamma} + \frac{r^2\gamma}{1 - r\gamma} = \frac{r\gamma(1 + r)}{1 - r\gamma}. \qquad (4.9)$$

Since $\gamma = d(P, P_1) + \epsilon$, (4.3) follows by letting ϵ approach 0. □

Lemma 4.2: Suppose S is a unimodular matrix. If P is a plant with an RCF (N, D), let SP denote the plant with RCF (N_s, D_s), where

$$\begin{bmatrix} D_s \\ N_s \end{bmatrix} = S \begin{bmatrix} D \\ N \end{bmatrix}. \qquad (4.10)$$

Let P_1 be another plant with $d(P, P_1) < 1/2$. Then

$$d(SP, SP_1) \leqslant \frac{2\|S\| \cdot \|S^{-1}\| \cdot d(P, P_1)}{1 - \|S\| \cdot \|S^{-1}\| d(P, P_1)} \qquad (4.11)$$

provided $\|S\| \cdot \|S^{-1}\| \cdot d(P, P_1) < 1$.

Proof: First we show that N_s, D_s are indeed right-coprime. Suppose X, Y satisfy

$$[Y \quad X] \begin{bmatrix} D \\ N \end{bmatrix} = I. \qquad (4.12)$$

Then

$$[Y \quad X] \cdot S^{-1} \cdot \begin{bmatrix} D_s \\ N_s \end{bmatrix} = I. \qquad (4.13)$$

Now suppose (N, D), (N_1, D_1) are normalized RCF's of P, P_1, and let

$$A = \begin{bmatrix} D \\ N \end{bmatrix}, \qquad A_1 = \begin{bmatrix} D_1 \\ N_1 \end{bmatrix}.$$

Select a U with $\|U\| \leqslant 1$ such that $\|A_1 U - A\| \leqslant d(P, P_1) + \epsilon < 1/2$. Then by Lemma 3.3, U is unimodular. Let R be a spectral factor of $A'(-s)S'(-s)S(s)A(s)$, and define

$$\begin{bmatrix} \bar{D}_s \\ \bar{N}_s \end{bmatrix} = S \begin{bmatrix} D \\ N \end{bmatrix} R^{-1} = SAR^{-1}. \qquad (4.14)$$

Then $(\bar{N}_s, \bar{D}_s)$ is a normalized RCF of the plant SP. Now $SA_1 UR^{-1}$ is of the form $\begin{bmatrix} \bar{D}_1 \\ \bar{N}_1 \end{bmatrix}$, where $(\bar{N}_1, \bar{D}_1)$ is an RCF of the plant SP_1 (observe that both U and R^{-1} are unimodular). Moreover,

$$\begin{aligned} \left\| SA_1 UR^{-1} - \begin{bmatrix} \bar{D}_s \\ \bar{N}_s \end{bmatrix} \right\| &= \|SA_1 UR^{-1} - SAR^{-1}\| \\ &\leqslant \|S\| \cdot \|A_1 U - A\| \cdot \|R^{-1}\| \\ &\leqslant \|S\| \cdot \|R^{-1}\| \cdot [d(P, P_1) + \epsilon]. \end{aligned} \qquad (4.15)$$

Now note that

$$\|R^{-1}\| = \|AR^{-1}\| = \|S^{-1} \cdot SAR^{-1}\| = \|S^{-1}\| \qquad (4.16)$$

where we use the fact that both A and SAR^{-1} are isometries. Thus, (4.15) and (4.16) show that

$$\|SA_1 UR^{-1} - SAR^{-1}\| \leqslant \|S\| \cdot \|S^{-1}\| \cdot [d(P, P_1) + \epsilon]. \qquad (4.17)$$

Now, SAR^{-1} gives a normalized RCF of the plant SP, while $SA_1 UR^{-1}$ gives a (not necessarily normalized) RCF of the plant SP_1. If we apply Lemma 3.4 together with (4.17) and let ϵ approach zero, we get (4.11). □

Proof of Theorem 4.1: Consider the feedback system of Fig. 1, and let

$$G = \begin{bmatrix} C & 0 \\ 0 & P \end{bmatrix}, \qquad G_1 = \begin{bmatrix} C_1 & 0 \\ 0 & P_1 \end{bmatrix} \qquad (4.18)$$

where P_1, C_1 are perturbed versions of P, C. Clearly, if (N_p, D_p), (N_c, D_c) are normalized RCF's of P and C, respectively, then

$$\left(\begin{bmatrix} N_c & 0 \\ 0 & N_p \end{bmatrix}, \begin{bmatrix} D_c & 0 \\ 0 & D_p \end{bmatrix} \right) \qquad (4.19)$$

is a normalized RCF of G. Thus, $d(G, G_1) \leqslant \max\{d(P, P_1), d(C, C_1)\}$.

Now note that

$$H(P, C) = (I + FG)^{-1} \qquad (4.20)$$

where

$$F = \begin{bmatrix} 0 & I \\ -I & 0 \end{bmatrix}. \qquad (4.21)$$

Thus, if (N_g, D_g) is an RCF of G, then

$$\begin{bmatrix} D_h \\ N_h \end{bmatrix} = \begin{bmatrix} I & F \\ I & 0 \end{bmatrix} \begin{bmatrix} D_g \\ N_g \end{bmatrix} \qquad (4.22)$$

is an RCF of H (in other words, $H = (I + FG)^{-1} = D_g(D_g + FN_g)^{-1}$). Hence, the plant H is of the form SG, where S is the unimodular matrix

$$S = \begin{bmatrix} I & F \\ I & 0 \end{bmatrix}. \qquad (4.23)$$

An easy calculation shows that $\|S\| = \|S^{-1}\| = \sqrt{2}$. Hence, from Lemma 4.2,

$$\begin{aligned} d(H(P_1, C_1), H(P, C)) &= d\left[(I + FG_1)^{-1}, (I + FG)^{-1}\right] \\ &\leqslant 4d(G, G_1)/[1 - 2d(G, G_1)]. \end{aligned} \qquad (4.24)$$

Now suppose the pair (P,C) is stable, and let $r=(1+\|H(P,C)\|^2)^{1/2}$. By Lemma 4.1, if the quantity on the right side of (4.24) is less than $1/r$, then $H(P_1,C_1)$ is also stable. Since $d(G,G_1)\leqslant \max\{d(P,P_1),d(C,C_1)\}\triangleq d$, the right side of (4.24) is no larger than

$$\gamma \triangleq 4d/(1-2d). \tag{4.25}$$

Thus, from Lemma 4.1, a sufficient condition for (P_1,C_1) to be stable is $\gamma r<1$, or equivalently

$$d<1/(2+4r). \tag{4.26}$$

This is the same as (4.1). If (4.26) holds, then (4.3) with γ replacing d leads to (4.2). □

We now examine the case where a nominal pair (P,C) is not necessarily stable.

Lemma 4.3: The map $(P,C)\to H(P,C)$ is continuous in the graph metric. Specifically, whenever $d(P,P_1)<1/2, d(C,C_1)<1/2$, we have

$$d(H(P,C),H(P_1,C_1))\leqslant 4d/[1-2d] \tag{4.27}$$

where $d=\max\{d(P,P_1),d(C,C_1)\}$.

Proof: See (4.24). □

We are now in a position to give a proof of Theorem 2.2. For convenience, the theorem is restated here.

Theorem 2.2: Suppose $\{P_i\},\{C_i\}$ are sequences in $\mathrm{mat}(R(s))$, that $P,C\in \mathrm{mat}(R(s))$, and that (P,C) is stable. Suppose $P_i\to P$, $C_i\to C$ in the graph topology; then (P_i,C_i) is stable for large enough i and $H(P_i,C_i)\to H(P,C)$ in $\mathrm{mat}(\mathscr{S})$. Conversely, suppose (P_i,C_i) is stable and $H(P_i,C_i)\to H(P,C)$; then $P_i\to P$, $C_i\to C$ in the graph topology.

Proof: The second sentence is already proved in Theorem 4.1. To prove the last sentence, define

$$G=\begin{bmatrix} C & 0\\ 0 & P\end{bmatrix},\quad G_i=\begin{bmatrix} C_i & 0\\ 0 & P_i\end{bmatrix} \tag{4.28}$$

and let F be as in (4.21). The hypothesis is that $H(P_i,C_i)\to H(P,C)$, or that $(I+FG_i)^{-1}\to(I+FG)^{-1}$ in $\mathrm{mat}(\mathscr{S})$. Now

$$H(G_i,F)=\begin{bmatrix} (I+G_iF)^{-1} & -G_i(I+FG_i)^{-1}\\ F(I+G_iF)^{-1} & (I+FG_i)^{-1}\end{bmatrix} \tag{4.29}$$

from (1.3). Since $(I+FG_i)^{-1}\to(I+FG)^{-1}$, it follows that

$$(I+FG_i)^{-1}F=F(I+G_iF)^{-1}\to(I+FG)^{-1}F = F(I+GF)^{-1}.$$

Next, since

$$G_i(I+FG_i)^{-1}=F^{-1}\left[I-(I+FG_i)^{-1}\right] \tag{4.30}$$

it follows that $G_i(I+FG_i)^{-1}\to G(I+FG)^{-1}$. Finally, since

$$(I+G_iF)^{-1}=I-G_i(I+FG_i)^{-1}F \tag{4.31}$$

it follows that $(I+G_iF)^{-1}\to(I+GF)^{-1}$. In other words, $H(G_i,F)\in\mathrm{mat}(\mathscr{S})$, and $\|H(G_i,F)-H(G,F)\|\to 0$. If we now apply Theorem 2.1 with P,P_i,C replaced by G,G_i,F, respectively, we can conclude that $G_i\to G$ in the graph topology.

The trick now is to prove from this that $P_i\to P$, $C_i\to C$ in the graph topology. As the example in (1.7) and (1.8) shows, this is not automatic. Let (N_{1i},D_{1i}), (N_{2i},D_{2i}) be normalized RCF's of C_i,P_i, and let (N_1,D_1), (N_2,D_2) be normalized RCF's of C,P, respectively. Then normalized RCF's for G_i and G can be formed as in (4.19). Since $G_i\to G$ in the graph topology, there exists a sequence $\{U_i\}$ of unimodular matrices such that

$$\begin{bmatrix} N_{1i} & 0\\ 0 & N_{2i}\\ D_{1i} & 0\\ 0 & D_{2i}\end{bmatrix}\begin{bmatrix} U_{1i} & U_{2i}\\ U_{3i} & U_{4i}\end{bmatrix}\to\begin{bmatrix} N_1 & 0\\ 0 & N_2\\ D_1 & 0\\ 0 & D_2\end{bmatrix}\quad \text{in } \mathrm{mat}(\mathscr{S}) \tag{4.32}$$

where U_i is partitioned in the obvious way. Since the topology on $\mathrm{mat}(\mathscr{S})$ is the product topology, each block in the partitioned matrix on the left side of (4.32) converges to the corresponding block on the right side of (4.32). Thus,

$$\begin{bmatrix} N_{1i}U_{2i}\\ D_{1i}U_{2i}\end{bmatrix}\to\begin{bmatrix}0\\0\end{bmatrix},\quad \begin{bmatrix} N_{2i}U_{3i}\\ D_{2i}U_{3i}\end{bmatrix}\to\begin{bmatrix}0\\0\end{bmatrix}. \tag{4.33}$$

Since $\begin{bmatrix} N_{1i}\\ D_{1i}\end{bmatrix},\begin{bmatrix} N_{2i}\\ D_{2i}\end{bmatrix}$ are all isometries, it follows that $U_{2i}\to 0$, $U_{3i}\to 0$. Now let A_i (resp. A) denote the rectangular matrix on the left (resp. right) side of (4.32), and define

$$B_i=A-A_iU_i \tag{4.34a}$$

$$W_i=\begin{bmatrix} 0 & U_{2i}\\ U_{3i} & 0\end{bmatrix} \tag{4.34b}$$

$$c_i=\|U_i-W_i\|=\left\|\begin{bmatrix} U_{1i} & 0\\ 0 & U_{2i}\end{bmatrix}\right\| \tag{4.34c}$$

$$\bar{U}_i=\frac{1}{c_i}\begin{bmatrix} U_{1i} & 0\\ 0 & U_{2i}\end{bmatrix}. \tag{4.34d}$$

Then $c_i\to 1$ since $\|W_i\|\to 0$ and $\|U_i\|\to 1$. Therefore,

$$\begin{aligned}\|A-A_i\bar{U}_i\| &= \|A-A_i(U_i-W_i)/c_i\|\\ &\leqslant \|A-A_iU_i/c_i\|+\|A_iW_i/c_i\|\\ &\leqslant \|A-A_iU_i\|+\left\|A_iU_i\left(1-\frac{1}{c_i}\right)\right\|+\|A_iW_i/c_i\|\\ &\to 0\end{aligned}$$

since A_i is an isometry, $\|U_i\|$ is bounded, $c_i\to 1$, and $\|W_i\|\to 0$. Observe also that $\|\bar{U}_i\|=1$. Since $G_i\to G$ in the graph metric, there exists a sequence V_i of matrices such that $\|V_i\|\leqslant 1$ and $\|AV_i-A_i\|\to 0$. Since $\|A-A_i\bar{U}_i\|+\|AV_i-A_i\|\to 0$, it is less than 1 for large enough i. By Lemma 3.3, it now follows that $\bar{U}_i, V_i$ are unimodular for large enough i. This implies that U_{1i}, U_{4i} are unimodular for large enough i, and that

$$\begin{bmatrix} N_{1i} & U_{1i}\\ D_{1i} & U_{1i}\end{bmatrix}\to\begin{bmatrix} N_1\\ D_1\end{bmatrix},\quad \begin{bmatrix} N_{2i} & U_{4i}\\ D_{2i} & U_{4i}\end{bmatrix}\to\begin{bmatrix} N_2\\ D_2\end{bmatrix} \tag{4.35}$$

which shows that $P_i\to P$, $C_i\to C$ in the graph topology. □

Proof of Lemma 2.4: In the preceding proof of Theorem 2.2 it was shown that

$$\begin{bmatrix} C_i & 0\\ 0 & P_i\end{bmatrix}\to\begin{bmatrix} C & 0\\ 0 & P\end{bmatrix}\Rightarrow C_i\to C,\quad P_i\to P. \tag{4.36}$$

The converse of the above implication is easy to establish: Suppose $C_i\to C$, $P_i\to P$. Let (N_c,D_c), (N_p,D_p) be RCF's of C and P, respectively, and select sequences of RCF's (N_{ci},D_{ci}) of C_i, (N_{pi},D_{pi}) of P_i such that

$$\begin{bmatrix} D_{ci}\\ N_{ci}\end{bmatrix}\to\begin{bmatrix} D_c\\ N_c\end{bmatrix},\quad \begin{bmatrix} D_{pi}\\ N_{pi}\end{bmatrix}\to\begin{bmatrix} D_p\\ N_p\end{bmatrix}. \tag{4.37}$$

Then

$$\begin{bmatrix} D_{ci} & 0 \\ 0 & D_{pi} \end{bmatrix} \to \begin{bmatrix} D_c & 0 \\ 0 & D_p \end{bmatrix}, \quad \begin{bmatrix} N_{ci} & 0 \\ 0 & N_{pi} \end{bmatrix} \to \begin{bmatrix} N_c & 0 \\ 0 & N_p \end{bmatrix} \tag{4.38}$$

which shows that

$$\begin{bmatrix} C_i & 0 \\ 0 & P_i \end{bmatrix} \to \begin{bmatrix} C & 0 \\ 0 & P \end{bmatrix}. \tag{4.39}$$

This completes the proof of Lemma 2.4 in the case of block-diagonal matrices with two blocks. The case of more than two blocks now follows readily by induction. □

The robustness estimates of Theorem 4.1 represent a good beginning, but are quite conservative. For instance, in Lemma 4.2, if we substitute $S = I$, the identity operator, then the right side of (4.11) becomes

$$\frac{2d(P, P_i)}{1 - d(P, P_1)} > 2d(P, P_1). \tag{4.40}$$

Compared to the true value of $d(SP, SP_1)$ (which is $d(P, P_1)$ since $S = I$), this bound is not very tight. This looseness is carried over into (4.1). Another failure of Theorem 4.1 is in not accommodating some sort of frequency-dependent weighting in computing the various distances. These problems are left for future research.

V. Specialized Robustness Results

In this section, we study the robustness of feedback systems when the perturbed plant has the same RHP poles as the unperturbed plant. Both additive as well as multiplicative perturbations are studied. By restricting to this special class of perturbations, we are able to derive *necessary and sufficient conditions* for robustness. The current results are weaker than existing ones in the case of multiplicative perturbations, since other authors [23] give necessary and sufficient conditions for robust stability when the perturbed plant has the same *number* of RHP poles as the unperturbed plant, although possibly at different locations. However, the results in the case of additive perturbations are new.

The main tool used in this section is the following.

Lemma 5.1: Suppose $F \in \mathrm{mat}(\mathscr{S})$. Then $I + RF$ is unimodular for all $R \in \mathrm{mat}(\mathscr{S})$ with $\|R\| \leqslant r$ if and only if $\|F\| < 1/r$.

Proof:

"If": Suppose $\|F\| < 1/r$. Then, whenever $\|R\| \leqslant r$, we have $\|FR\| < 1$, which implies that $I + FR$ is unimodular.

"Only If": This part is constructive. Suppose $\|F\| \geqslant 1/r$; we construct an R with $\|R\| \leqslant r$ such that $I + RF$ is not unimodular. Let $\|A\|_2$ denote the matrix norm defined by

$$\|A\|_2 = \left[\lambda_{\max}(A^\dagger A)\right]^{1/2}. \tag{5.1}$$

Then

$$\|F\| = \sup_\omega \|F(j\omega)\|_2. \tag{5.2}$$

Hence, if $\|F\| \geqslant r^{-1}$, then either $\|F(\infty)\|_2 \geqslant r^{-1}$ or else $\|F(j\omega)\|_2 \geqslant r^{-1}$ for some ω. Suppose first that $\|F(\infty)\|_2 \geqslant r^{-1}$. Then, since $F(\infty)$ is a real matrix, there exist *real* vectors u, v such that $\|v\|_2 = 1$, $\|u\|_2 \geqslant r^{-1}$, and $Fv = u$. Let $c = \|u\|_2$ and let R equal the constant matrix $-vu'/c^2$. Then $\|R\| = c^{-1} \leqslant r$; moreover $I + RF(\infty)$ is singular, since $RF(\infty)v = -v$. Thus, $I + RF$ is not unimodular. Next, suppose $\|F(\infty)\|_2 < r^{-1}$, but $\|F(j\omega)\|_2 \geqslant r^{-1}$ for some finite ω. The only additional complication is that $F(j\omega)$ might be a complex matrix. If $\omega = 0$, then $F(j0)$ is real and the above proof applies. If $\omega > 0$, select (possibly complex) vectors v, u such that $\|v\|_2 = 1$, $\|u\|_2 \geqslant r^{-1}$, and $Fv = u$. Suppose

$$v = \begin{bmatrix} v_1 e^{j\theta_1} \\ \vdots \\ v_m e^{j\theta_m} \end{bmatrix}, \quad u = \begin{bmatrix} u_1 e^{j\phi_1} \\ \vdots \\ u_n e^{j\phi_n} \end{bmatrix} \tag{5.3}$$

where the v_i, u_i are real, and the θ_i, ϕ_i lie in $[-\pi, 0)$. Choose

$$R(s) = -\begin{bmatrix} \frac{s-\alpha_1}{s+\alpha_1} & & 0 \\ & \ddots & \\ 0 & & \frac{s-\alpha_m}{s+\alpha_m} \end{bmatrix} \begin{bmatrix} v_1 \\ \vdots \\ v_m \end{bmatrix} [u_1 \cdots u_n] \begin{bmatrix} \frac{s-\beta_1}{s+\beta_1} & & 0 \\ & \ddots & \\ 0 & & \frac{s-\beta_n}{s+\beta_n} \end{bmatrix} \frac{1}{\|u\|_2^2} \tag{5.4}$$

where the constants α_i, β_i are adjusted so that

$$\arg\left(\frac{j\omega - \alpha_i}{j\omega + \alpha_i}\right) = \theta_i, \quad \arg\left(\frac{j\omega - \beta_i}{j\omega + \beta_i}\right) = \phi_i, \qquad \text{for all } i. \tag{5.5}$$

Then $\|R\| \leqslant r$ and $(I + RF)(j\omega)$ is singular, so that $I + RF$ is not unimodular. □

In what follows, we consider the robustness of the feedback system of Fig. 1 when P is perturbed to either $(I + L)P$ where $L \in \mathrm{mat}(\mathscr{S})$ and $\|L\| < 1$ or to $P + L$, $L \in \mathrm{mat}(\mathscr{S})$. The former is referred to as a multiplicative perturbation and the latter as an additive perturbation. In both cases, both Q and P have the same RHP poles.

Theorem 5.1: Suppose the pair (P, C) is stable. Then C stabilizes every Q of the form $(I + L)P$ for all $L \in \mathrm{mat}(\mathscr{S})$ with $\|L\| \leqslant r < 1$, if and only if $\|PC(I + PC)^{-1}\| < r^{-1}$.

Proof: Let (N_p, D_p) be any RCF of P. Then $((I + L)N_p, D_p)$ is an RCF of Q. Since (P, C) is stable, it follows from [10] that there exists an LCF $(\tilde{D}_c, \tilde{N}_c)$ of C such that $\tilde{D}_c D_p + \tilde{N}_c N_p = I$. If P is replaced by Q, this "return difference" matrix becomes $\tilde{D}_c D_p + \tilde{N}_c(I + L)N_p = I + \tilde{N}_c L N_p$. Thus, C stabilizes every Q of the form $(I + L)P$ if and only if $I + \tilde{N}_c L N_p$ is unimodular for all L in $\mathrm{mat}(\mathscr{S})$ with $\|L\| \leqslant r$. Now $\det(I + \tilde{N}_c L N_p) = \det(I + L N_p \tilde{N}_c)$; hence $I + \tilde{N}_c L N_p$ is unimodular if and only if $I + L N_p \tilde{N}_c$ is. Now from Lemma 5.1, it follows that $I + L N_p \tilde{N}_c$ is unimodular whenever $\|L\| \leqslant r$ if and only if $\|N_p \tilde{N}_c\| < r^{-1}$. It is easily verified that $N_p \tilde{N}_c = PC(I + PC)^{-1}$. □

Theorem 5.2: Suppose (P, C) is stable. Then C stabilizes every Q of the form $P + L$ for all $L \in \mathrm{mat}(\mathscr{S})$ with $\|L\| \leqslant r$, if and only if $\|C(I + PC)^{-1}\| < r^{-1}$.

Sketch of Proof: $P + L$ has the RCF $(N_p + LD_p, D_p)$. So the return difference matrix is $\tilde{D}_c D_p + \tilde{N}_c(N_p + LD_p) = I + \tilde{N}_c L D_p$. Now $\det(I + \tilde{N}_c L D_p) = \det(I + L D_p \tilde{N}_c)$. So we require $\|D_p \tilde{N}_c\| < r^{-1}$. Finally, $D_p \tilde{N}_c = C(I + PC)^{-1}$. □

VI. Application to Singularly Perturbed Systems

In this section, we apply the theory developed in Section II to analyze the continuity of singularly perturbed systems in the graph topology. Consider the system

$$\begin{bmatrix} \dot{x} \\ \epsilon \dot{z} \end{bmatrix} = \begin{bmatrix} A_{11} & A_{12} \\ A_{21} & A_{22} \end{bmatrix} \begin{bmatrix} x \\ z \end{bmatrix} + \begin{bmatrix} B_1 \\ B_2 \end{bmatrix} u \tag{6.1}$$

$$(P_\epsilon)$$

$$y = [C_1 \quad C_2] \begin{bmatrix} x \\ z \end{bmatrix} + D_1 u \tag{6.2}$$

where $\epsilon > 0$ and A_{22} is a Hurwitz matrix (i.e., all eigenvalues of A_{22} have negative real parts). Let P_0 denote the system obtained from (6.1), (6.2) by substituting $\epsilon = 0$, namely

$$\dot{x} = Ax + Bu, \quad y = Cx + Du \qquad (P_0) \tag{6.3}$$

where

$$\begin{aligned} A &= A_{11} - A_{12}A_{22}^{-1}A_{21}, & B &= B_1 - A_{12}A_{22}^{-1}B_2 \\ C &= C_1 - C_2A_{22}^{-1}A_{21}, & D &= D_1 - C_2A_{22}^{-1}B_2. \end{aligned} \tag{6.4}$$

The question studied here is: does P_ϵ approach P_0 in the graph topology as $\epsilon \to 0$?[7] The motivation for this study is the well-known fact (see, e.g., [16]) that, if P_0 is unstable and a controller C stabilizes P_0, then C may not in general stabilize P_ϵ for small enough ϵ. In this context, Theorem 2.1 is significant. Applied to the problem at hand, it implies that if $P_\epsilon \to P_0$ in the graph topology, then *every* controller C that stabilizes P_0 also stabilizes P_ϵ for small enough ϵ, and the resulting closed-loop transfer matrix $H(P_\epsilon, C)$ approaches $H(P_0, C)$. On the other hand, if P_ϵ does not approach P_0 and C stabilizes P_0, then one of two things happens: either C does not stabilize P_ϵ for small enough ϵ, or else C does stabilize P_ϵ for small enough ϵ, but $H(P_\epsilon, C)$ does not approach $H(P_0, C)$.

One could also explore the possibility of using an ϵ-dependent family of controllers C_ϵ such that C_ϵ stabilizes P_ϵ for small enough ϵ, and $H(P_\epsilon, C_\epsilon) \to H(P_0, C_0)$ as $\epsilon \to 0$. This situation is addressed by Theorem 2.2. In the present situation, this theorem implies that if such a family $\{C_\epsilon\}$ exists, then $P_\epsilon \to P_0$, $C_\epsilon \to C_0$ as $\epsilon \to 0$. Thus, if P_ϵ does not converge to P_0, then one cannot even find an ϵ-dependent family C_ϵ such that (P_0, C_0) is stable and $H(P_\epsilon, C_\epsilon) \to H(P_0, C_0)$. On the other hand, if $P_\epsilon \to P_0$, then one can get by with an "ϵ-independent" controller.[8]

We state at once the main result of this section and devote the rest of the section to its proof.[9]

Theorem 6.1: Suppose the system (6.3) is stabilizable and detectable, and that A_{22} is Hurwitz. Then P_ϵ approaches P_0 in the graph topology as $\epsilon \to 0$ if $C_2(sI - A_{22})^{-1}B_2 \equiv 0$.

Many previously known results can be obtained as ready consequences of Theorem 6.1. For instance, it is shown in [17] that if the system (6.3) is stabilized by state feedback, then the same state feedback stabilizes the system (6.1), (6.2) for small enough ϵ. In the present setup, this corresponds to the case where $y = x$ and the stabilizing controller for P_0 is just a static gain (call it K). Since $y = x$, we have $C_1 = I$, $C_2 = 0$, hence $C_2(sI - A_{22})^{-1}B_2 \equiv 0$. Thus, Theorem 6.1 implies that $P_\epsilon \to P_0$, and Theorem 2.1 now implies that K also stabilizes P_ϵ for small enough ϵ. We also get an added bit of information from Theorem 2.1 that is not discussed in [17], namely that $H(P_\epsilon, K)$ converges to $H(P_0, K)$. Similarly, it is shown in [16] that if $B_2 = 0$ or $C_2 = 0$, and if P_0 is stabilized using an observer-controller scheme, the same scheme stabilizes P_ϵ for small enough ϵ. This too can be deduced from Theorems 6.1 and 2.1. More generally, Theorem 6.1 shows that if $C_2(sI - A_{22})^{-1}B_2 \equiv 0$, then *every* controller that stabilizes P_0 also stabilizes P_ϵ for small enough ϵ. No details need be known about the configuration of the controller. Thus, Theorem 6.1 (together with Theorem 2.1) gives means of unifying several known results.

Now we move toward a proof of Theorem 6.1. Lemma 6.1 below is of independent interest, as it provides a method for obtaining a left-coprime factorization of a system described in state-space form.[10] Since finding an RCF of a plant $P(s)$ is equivalent to finding an LCF for $P'(s)$, the same lemma can also be used for finding RCF's.

[7] I thank Prof. M. Suzuki of Nagoya University for drawing my attention to the possibility of applying this topological approach to singularly perturbed systems.

[8] I thank Prof. A. Willsky of M.I.T. for raising this issue.

[9] I thank Prof. H. Khalil and P. Kokotovic for pointing out an error in an earlier version of this theorem.

[10] The expression for the left-coprime factorization is due to my colleague D. Aplevich, while the main idea of the proof is due to my student M. McIntyre.

Lemma 6.1: Consider a system of the form (6.3), and suppose it is stabilizable and detectable. Select a matrix F such that $A - FC \triangleq \bar{A}$ is Hurwitz, and define

$$\tilde{N}(s) = C(sI - \bar{A})^{-1}(B - FD) + D \tag{6.5a}$$

$$\tilde{D}(s) = I - C(sI - \bar{A})^{-1}F. \tag{6.5b}$$

Then $(\tilde{D}, \tilde{N})$ is a left-coprime factorization of $P_0(s) = C(sI - A)^{-1}B + D$.

Proof: First we show that $P_0(s) = [\tilde{D}(s)]^{-1}\tilde{N}(s)$ by establishing that $\tilde{D}(s)y(s) = \tilde{N}(s)u(s)$. From (6.3), we get

$$\dot{x} = Ax + Bu = \bar{A}x + Fy + (B - FD)u \tag{6.6}$$

$$x(s) = (sI - \bar{A})^{-1}[Fy(s) + (B - FD)u(s)] \tag{6.7}$$

$$y(s) = C(sI - \bar{A})^{-1}[Fy(s) + (B - FD)u(s)] + Du(s). \tag{6.8}$$

It only remains to show that $\tilde{D}, \tilde{N}$ are left-coprime. This is equivalent to showing that the matrix $[\tilde{D}(s) \quad \tilde{N}(s)]$ has full row rank whenever $\operatorname{Re} s \geq 0$ and at $s = \infty$. At $s = \infty$, we have $[\tilde{D}(\infty) \quad \tilde{N}(\infty)] = [I \quad D]$, so let us examine the case where $s \in C_+$. If $|\tilde{D}(s)| \neq 0$ for some s, then $[\tilde{D}(s) \quad \tilde{N}(s)]$ has full rank; so it is only necessary to examine those $s \in C_+$ at which $|\tilde{D}(s)| = 0$ to ensure that $[\tilde{D}(s) \quad \tilde{N}(s)]$ has full rank at these values of s as well.

In order to do this, it is first established that any C_+-zero of $|\tilde{D}(\cdot)|$ must be an eigenvalue of A. This is done by showing that

$$\tilde{D}(s) = \left[I + C(sI - A)^{-1}F\right]^{-1}. \tag{6.9}$$

To prove (6.9) let E denote $sI - A$. Then from (6.5b)

$$\tilde{D} = I - C(E + FC)^{-1}F. \tag{6.10}$$

We want to show that

$$\tilde{D} = [I + CE^{-1}F]^{-1}. \tag{6.11}$$

Thus, the claim is that

$$\left[I - C(E + FC)^{-1}F\right](I + CE^{-1}F) = I$$

or, equivalently

$$I + CE^{-1}F - C(E + FC)^{-1}F - C(E + FC)^{-1}FCE^{-1}F = I \tag{6.12}$$

$$C\left[E^{-1} - (E + FC)^{-1} - (E + FC)^{-1}FCE^{-1}\right]F = 0. \tag{6.13}$$

The quantity inside the brackets equals

$$\begin{aligned} &\left[I - (E + FC)^{-1}E - (E + FC)^{-1}FC\right]E^{-1} \\ &= \left[I - (E + FC)^{-1}(E + FC)\right]E^{-1} = 0. \end{aligned} \tag{6.14}$$

Thus, (6.9) holds. Now let $m(s) = |I + C(sI - A)^{-1}F|$. Then (6.9) shows that every (not necessarily C_+) zero of $|\tilde{D}(\cdot)|$ must be a pole of $m(\cdot)$. But $m(\cdot)$ is analytic except at the eigenvalues of A. Thus, every (not necessarily C_+) zero of $|\tilde{D}(\cdot)|$ must be an eigenvalue of A. Further, if $|\tilde{D}(s_0)| = 0$, then the multiplicity of s_0 as a zero of $|\tilde{D}(\cdot)|$ is no larger than its multiplicity as an eigenvalue of A.

To conclude the proof, suppose $s_0 \epsilon C_+$ and that $|\tilde{D}(s_0)| = 0$. By the preceding discussion, s_0 is an eigenvalue of A, say of multiplicity μ. Then the multiplicity of s_0 as a zero of $|\tilde{D}(\cdot)|$ is no larger than μ, and by the stabilizability and detectability assumptions, the McMillan degree of s_0 as a pole of $P(\cdot)$ equals μ; that is, there exists a minor $p\binom{J}{K}$ of P such that $p\binom{J}{K}(\cdot)$ has a pole of order μ at s_0. By [24, p. 50], there is a one-to-one correspondence between the minors of P and the minors of $[\tilde{D} \;\; \tilde{N}]$. In fact,

$$p\binom{J}{K} = \left|\tilde{D}\left[M\backslash J\right] \;\vdots\; \tilde{N}\left[\begin{matrix}\cdot\\ K\end{matrix}\right]\right| \cdot |\tilde{D}|^{-1} \tag{6.15}$$

where $M = \{1,\cdots,m\}$ is the set of all columns of D. Let $F = [\tilde{D} \;\; \tilde{N}]$ and let $f(J,K)$ denote the minor in (6.15), so that $p\binom{J}{K} = f(J,K)\;|\tilde{D}|$. Now $p\binom{J}{K}$ has a pole at s_0 of order μ, while $\tilde{D}^{-1}$ has a pole at s_0 or order no larger than μ. Hence, $f(J,K)(s_0) \neq 0$, i.e., $F(s_0)$ has full row rank. Since this argument can be repeated at *all* C_+-zeros of $|D(\cdot)|$, it follows that $\tilde{D}, \tilde{N}$ are left-coprime. □

Proof of Theorem 6.1: Suppose $C_2(sI - A_{22})^{-1}B_2 \equiv 0$ select F such that $A - FC$ is Hurwitz, and define $\tilde{N}, \tilde{D}$ by (6.4) and (6.5). Then $(\tilde{D}, \tilde{N})$ is an LCF of P_0, by Lemma 6.1. To obtain an LCF of P_ϵ, define $F_0 = \begin{bmatrix} F \\ 0 \end{bmatrix}$, and consider

$$A_\epsilon = \begin{bmatrix} A_{11} - FC_1 & A_{12} - FC_2 \\ A_{21}/\epsilon & A_{22}/\epsilon \end{bmatrix}. \tag{6.16}$$

From [17], the eigenvalues of $\bar{A}_\epsilon$ are asymptotically equal to those of A_{22}/ϵ, plus those of

$$\begin{aligned} A_{11} - FC_1 - (A_{12} - FC_2)A_{22}^{-1}A_{21} &\\ = \left(A_{11} - A_{12}A_{22}^{-1}A_{21}\right) - F\left(C_1 - C_2A_{22}^{-1}A_{21}\right)&\\ = A - FC. & \end{aligned} \tag{6.17}$$

Hence $\bar{A}_\epsilon$ is Hurwitz for sufficiently small ϵ. Thus, from Lemma 6.1, an LCF of P_ϵ is given by $(\tilde{D}_\epsilon, \tilde{N}_\epsilon)$, where

$$\tilde{N}_\epsilon = C_0\left(sI - \bar{A}_\epsilon\right)^{-1}\left(B_\epsilon - F_0D_1\right) + D_1 \tag{6.18}$$

$$\tilde{D}_\epsilon = I - C_0\left(sI - \bar{A}_\epsilon\right)^{-1}F_0 \tag{6.19}$$

with

$$B_\epsilon = \begin{bmatrix} B_1 \\ B_2/\epsilon \end{bmatrix}, \qquad C_0 = [C_1 \;\; C_2]. \tag{6.20}$$

The remainder of the proof consists of studying $\tilde{N}_\epsilon$ and $\tilde{D}_\epsilon$ in detail, and showing that $\tilde{N}_\epsilon \to \tilde{N}$, $\tilde{D}_\epsilon \to \tilde{D}$. Recall that the inverse of a partitioned matrix is given by

$$\begin{bmatrix} X & Y \\ W & V \end{bmatrix}^{-1} = \begin{bmatrix} \Delta^{-1} & -\Delta^{-1}YV^{-1} \\ -V^{-1}W\Delta^{-1} & V^{-1} + V^{-1}W\Delta^{-1}YV^{-1} \end{bmatrix} \tag{6.21}$$

where

$$\Delta = X - YV^{-1}W. \tag{6.22}$$

Hence,

$$\begin{aligned} \tilde{D}_\epsilon &= I - [C_1 \;\; C_2]\begin{bmatrix} sI - A_{11} + FC_1 & -A_{12} + FC_2 \\ -A_{21}/\epsilon & sI - A_{22}/\epsilon \end{bmatrix}^{-1}\begin{bmatrix} F \\ 0 \end{bmatrix} \\ &= I - C_1\Delta^{-1}F + C_2V^{-1}W\Delta^{-1}F \end{aligned} \tag{6.23}$$

where

$$\begin{aligned} \Delta &= (sI - A_{11} + FC_1) - (A_{12} - FC_2)\left(sI - \frac{A_{22}}{\epsilon}\right)^{-1} \cdot \frac{A_{21}}{\epsilon} \\ &= (sI - A_{11} + FC_1) - (A_{12} - FC_2)(\epsilon sI - A_{22})^{-1}A_{21} \end{aligned} \tag{6.24}$$

$$V = sI - A_{22}/\epsilon, \qquad W = -A_{21}/\epsilon. \tag{6.25}$$

Simplification of (6.23) yields

$$\tilde{D}_\epsilon = I - M\Delta^{-1}F \tag{6.26}$$

where

$$M = C_1 - C_2V^{-1}W = C_1 + C_2(\epsilon sI - A_{22})^{-1}A_{21}. \tag{6.27}$$

Now

$$\begin{aligned} \tilde{D} &= I - C(sI - \bar{A})^{-1}F \\ &= I - C_1(sI - \bar{A})^{-1}F + C_2A_{22}^{-1}A_{21}(sI - \bar{A})^{-1}F. \end{aligned} \tag{6.28}$$

Hence, we can conclude that $\tilde{D}_\epsilon \to \tilde{D}$ if we can show that

$$\Delta^{-1} \to (sI - \bar{A})^{-1} \tag{6.29}$$

$$(\epsilon sI - A_{22})^{-1}A_{21}\Delta^{-1} \to -A_{22}^{-1}A_{21}(sI - \bar{A})^{-1}. \tag{6.30}$$

Note that $(\epsilon sI - A_{22})^{-1}$ does *not* approach $-A_{22}^{-1}$.[11] Hence, (6.30) does not automatically follow from (6.29).

To prove (6.29), let $\bar{A}_{11}$ denote $A_{11} - FC_1$, and note that

$$\begin{aligned} \Delta^{-1} &= \left[sI - \bar{A}_{11} - (A_{12} - FC_2)(\epsilon sI - A_{22})^{-1}A_{21}\right]^{-1} \\ &= \Big\{sI - \bar{A}_{11} - (A_{12} - FC_2)A_{22}^{-1}A_{21} \\ &\qquad - (A_{12} - FC_2)\left[(\epsilon sI - A_{22})^{-1} - A_{22}^{-1}\right]A_{21}\Big\}^{-1} \\ &= (I - R)^{-1}(sI - \bar{A})^{-1} \end{aligned} \tag{6.31}$$

where

$$R = (A_{12} - FC_2)\left[(\epsilon sI - A_{22})^{-1} - A_{22}^{-1}\right](sI - \bar{A})^{-1}. \tag{6.32}$$

Hence,

$$\Delta^{-1} - (sI - \bar{A})^{-1} = \left[(I - R)^{-1} - I\right](sI - \bar{A})^{-1}. \tag{6.33}$$

Now $(I - R)^{-1}(j\omega)$ approaches I uniformly on every *finite* interval $[-\omega_0, \omega_0]$ as $\epsilon \to 0$, and $(j\omega I - \bar{A})^{-1} \to 0$ as $|\omega| \to \infty$. So $\Delta^{-1} \to (sI - \bar{A})^{-1}$ and (6.29) is proved. To establish (6.30), observe that $(j\epsilon\omega I - A_{22})^{-1}$ approaches A_{22}^{-1} uniformly on every finite interval $[-\omega_0, -\omega_0]$ as $\epsilon \to 0$, and that both Δ^{-1}, $(sI - \bar{A})^{-1}$ are strictly proper. Thus, (6.30) follows from (6.29) plus the strict properness property of Δ^{-1} and its limit. Now (6.29) and (6.30) together show that $\tilde{D}_\epsilon \to D$ as $\epsilon \to 0$.

Now let us look at the "numerator" matrix $\tilde{N}_\epsilon$. From (6.18),

$$\tilde{N}_\epsilon = [C_1 \;\; C_2]\begin{bmatrix} sI - A_{11} + FC_1 & -A_{12} + FC_2 \\ -A_{21}/\epsilon & sI - A_{22}/\epsilon \end{bmatrix}^{-1} \cdot \begin{bmatrix} B_1 - FD_1 \\ B_2/\epsilon \end{bmatrix} + D_1. \tag{6.34}$$

[11]Observe that $(\epsilon s - A_{22})^{-1} \to 0$ as $|s| \to \infty$, whenever $\epsilon > 0$. Hence, $\|(\epsilon s - A_{22})^{-1} - A_{22}^{-1}\| \geq \|A_{22}^{-1}\|$ no matter how small ϵ is.

Ignoring the D_1 term which is simply added on, we can write $\tilde{N}_\epsilon$ as a sum of four terms obtained by expanding (6.34). Using (6.21), the first term can be expressed as $C_1\Delta^{-1}(B_1 - FD_1)$. From (6.29), it follows that this converges to $C_1(sI - \bar{A})^{-1}(B_1 - FD_1)$. The second term is

$$-C_2V^{-1}W\Delta^{-1}(B_1 - FD_1)$$
$$= C_2\left(sI - \frac{A_{22}}{\epsilon}\right)^{-1}\frac{A_{21}}{\epsilon}\Delta^{-1}(B_1 - FD_1)$$
$$= C_2(\epsilon sI - A_{22})^{-1}A_{21}\Delta^{-1}(B_1 - FD_1). \quad (6.35)$$

From reasoning analogous to that used to establish (6.30), it follows that this term converges to $-C_2A_{22}^{-1}A_{21}(sI - \bar{A})^{-1}(B_1 - FD_1)$. The third term is

$$-C_1\Delta^{-1}YV^{-1}B_2/\epsilon = C_1\Delta^{-1}(A_{12} - F_1C_2)\cdot\left(sI - \frac{A_{22}}{\epsilon}\right)^{-1}\frac{B_2}{\epsilon}$$
$$= C_1\Delta^{-1}(A_{12} - F_1C_2)(\epsilon sI - A_{22})^{-1}B_2. \quad (6.36)$$

As before, this converges to

$$-C_1(sI - \bar{A})^{-1}(A_{12} - F_1C_2)A_{22}^{-1}B_2$$
$$= -C_1(sI - \bar{A})^{-1}A_{12}A_{22}^{-1}B_2 \quad (6.37)$$

since $C_2A_{22}^{-1}B_2 = 0$. The fourth and final term is

$$C_2V^{-1}(I + W\Delta^{-1}YV^{-1})B_2/\epsilon$$
$$= C_2\left(sI - \frac{A_{22}}{\epsilon}\right)^{-1}$$
$$\cdot\left[I + \frac{A_{21}}{\epsilon}\cdot\Delta^{-1}(A_{12} - FC_2)\left(sI - \frac{A_{22}}{\epsilon}\right)^{-1}\right]\cdot\frac{B_2}{\epsilon}$$
$$= C_2(\epsilon sI - A_{22})^{-1}\left[I + A_{22}\Delta^{-1}(A_{12} - FC_2)(\epsilon sI - A_{22})^{-1}\right]B_2$$
$$= C_2(\epsilon sI - A_{22})^{-1}B_2 + C_2(\epsilon sI - A_{22})^{-1}A_{21}\Delta^{-1}$$
$$\cdot(A_{12} - FC_2)(\epsilon sI - A_{22})^{-1}B_2. \quad (6.38)$$

Now the second term on the right side of (6.38) is strictly proper and therefore converges without any difficulty to $C_2A_{22}^{-1}A_{21}(sI - \bar{A})^{-1}(A_{12} - FC_2)A_{22}^{-1}B_2$. Further, the first term is identically zero by assumption. Moreover, the limit of the second term simplifies to $C_2A_{22}^{-1}A_{21}(sI - \bar{A})^{-1}A_{12}A_{22}^{-1}B_2$. Putting everything together, we see that N_ϵ approaches

$$(C_1 - C_2A_{22}^{-1}A_{21})(sI - \bar{A})^{-1}(B_1 - FD_1 - A_{12}A_{22}^{-1}B_2) + D_1$$
$$= C(sI - \bar{A})^{-1}(B - FD_1) + D_1 = \tilde{N} \quad \text{since } D_1 = D. \quad (6.39)$$

Hence $P_\epsilon \to P_0$ in the graph topology. □

In conclusion, observe that if all matrices in (6.1), (6.2) are known, then Theorem 4.1 can be used to make the qualitative phrase "for small enough ϵ" more precise in a quantitative sense.

VII. Conclusions

In this paper, we have defined a "graph metric" that provides a measure of the distance between unstable plants. The graph metric induces a "graph topology" on unstable plants, which is the weakest possible topology in which feedback stability is robust. Using the graph metric, it is possible to derive estimates for the robustness of feedback stability *without* assuming that the perturbed and unperturbed plants have the same number of RHP poles. If the perturbed and unperturbed plants have the same RHP poles, then one can derive necessary and sufficient conditions for robustness with respect to a given class of perturbations.

The generalization of the results of this paper to linear distributed systems is straightforward. One simply replaces $\mathscr{S}$ by the set of transfer functions of all BIBO stable systems (lumped as well as distributed). This is the set $\hat{\mathscr{A}}$ defined in [18]. It is now known [7] that not all transfer functions of the form a/b, where $a, b \in \hat{\mathscr{A}}$, have coprime factorizations over $\hat{\mathscr{A}}$. Hence, the graph metric can only be defined for those plants that have an RCF over $\hat{\mathscr{A}}$. Fortunately, this class includes all lumped systems, as well as all the class mat$(\hat{B})$; see [11] and the references therein. The existence of the spectral factors needed to define the graph metric follows from [19]. In order to do full justice to the technicalities in the case of distributed systems, a detailed treatment will be given elsewhere.

In defining the graph metric between two plants, we have used their right-coprime factorizations. We could have also used their LCF's to define an equivalent, although not equal, metric.

There are some open problems for future research. The first is to find a formula for explicitly computing the graph metric (we give only an upper and a lower bound). The second is to improve the bounds given in Theorem 4.1 by making them less conservative.

Appendix
Comparison to the Gap Metric

In this Appendix, we analyze the gap metric defined in [14], [15] and analyze its relationship to the graph metric defined in the present paper. A few preliminaries are required for this purpose.

Recall [21] that the Hardy space H^2 consists of analytic functions f of the complex variable s with the property that

$$\sup_{\sigma > 0}\int_{-\infty}^{\infty}|f(\sigma + j\omega)|^2\,d\omega < \infty. \quad (A1)$$

The space H^2 is a Hilbert space, with inner product defined by

$$\langle f, g\rangle = \frac{1}{2\pi}\int_{-\infty}^{\infty}f^*(j\omega)g(j\omega)\,d\omega. \quad (A2)$$

Moreover, it is well known [21, p. 471] that a "time" function $f(t)$ belongs to $L_2[0,\infty)$ if and only if its Laplace transform $\hat{f}(s)$ belongs to H^2. Further, Laplace transformation is a linear isometry from L_2 into H^2.

Let $\mathscr{B}(L_2)$ denote the set of linear continuous operators mapping L_2 into itself. Then $\mathscr{B}(L_2)$ is a Banach algebra if each operator is equipped with the standard supremum norm. It is important to note that not all operators in $\mathscr{B}(L_2)$ are causal in the sense of [18, p. 39]. Since L_2 and H^2 are isomorphic, we let $\mathscr{B}(H^2)$ denote the set of continuous linear operators on H^2 corresponding to those in $\mathscr{B}(L_2)$.

The notion of stability employed in [14], [15] is that an input–output map is stable if it belongs to mat$(\mathscr{B}(L_2))$. Thus, a system is deemed to be stable if it maps L_2-inputs into L_2-outputs in a continuous (although not necessarily causal) manner. Thus, the above notion of stability is equivalent to requiring the system transfer matrix to lie in mat$(\mathscr{B}(H^2))$. This is a weaker notion of stability than the one employed in this paper, which (at this level of generality) corresponds to requiring the system transfer matrix to lie in mat$(\mathscr{A})$ (see Section VII). Thus, we require a stable input–output map to be causal as well as bounded, whereas in [14], [15] a stable input–output map is only required to be bounded.

In [14], [15] a gap metric is defined on the set of closed operators (i.e., operators whose graphs are closed subspaces) mapping some subset of L_2^m into itself, for some integer m. If δ denotes the gap metric, then the principal result of [14], [15] can be stated as follows. Suppose $P \in \text{mat}(\mathscr{B}(L_2))$ is *square*, and that $\delta(P, P_1) < (1+\|P\|^2)^{-1/2}$. Then $P_1 \in \text{mat}(\mathscr{B}(L_2))$, and

$$\|P-P_1\| \leq \frac{(1+\|P\|^2)\delta(P,P_1)}{1-(1+\|P\|^2)^{1/2} \cdot \delta(P,P_1)}. \tag{A.3}$$

Even if P is causal, the above result does not guarantee that P_1 is *causal*, only that it is *bounded*. In fact causality is not addressed in [14], [15]. Using (A.3), it is shown that if $(I+P)^{-1} \in \text{mat}(\mathscr{B}(L_2))$ and $\delta(P, P_1)$ is sufficiently small, then $(I+P_1)^{-1} \in \text{mat}(\mathscr{B}(L_2))$, and $\|(I+P_1)^{-1}-(I+P)^{-1}\|$ is $0(\delta(P, P_1))$; the actual formulas can be found in [14], [15]. This result means that if P is stabilized by unit feedback and $\delta(P, P_1)$ is sufficiently small, then P_1 is also stabilized by unit feedback and $(I+P_1)^{-1}$ is close to $(I+P)^{-1}$.

Thus, in summary, a comparison of the contents of this paper with those of [14], [15] results in the following observations.

1) In the present paper, causality as well as boundedness of the input–output map are requirements for stability; in [14], [15] only boundedness is required. It is not known at present whether the results of [14], [15] can be modified to conclude the causality as well as boundedness of the perturbed system, if the original system is causal as well as bounded.

2) The present analysis is carried out for nonsquare plants and general (not necessarily unit or even stable) feedback. The analysis in [14], [15] is for square plants under unit feedback.

3) The scope of the results in [14], [15] can be extended by treating the plant–compensator combo as a square system. That is, given the system of Fig. 1, define

$$G = \begin{bmatrix} 0 & -C \\ P & 0 \end{bmatrix} \tag{A.4}$$

and observe that $H(P, C) = (I+G)^{-1}$. But in this case, $\delta(G, G_1)$ is not related in [14], [15] to individual variations in P or C. This is done in the present paper.

In [17], a topology is defined for unstable plants in a very general setting. One begins with a set $\mathscr{H}$ of "stable" plants, which is assumed to have two properties: i) $\mathscr{H}$ is a topological ring with no zero divisors, and ii) the set $\mathscr{U}$ of units in $\mathscr{H}$ is open, and the map $u \to u^{-1}$ mapping $\mathscr{U}$ into itself is continuous. The universe of unstable plants is then taken to be the set $\text{mat}(\mathscr{F})$, where $\mathscr{F}$ is the field of fractions associated with $\mathscr{H}$. Let $\mathscr{C}$ denote the subset of $\text{mat}(\mathscr{F})$ consisting of those matrices that have both an RCF as well as an LCF over $\mathscr{H}$; then a natural topology can be defined over $\mathscr{C}$ in a manner entirely analogous with Definition 2.1, by just replacing the set $\mathscr{S}$ by $\mathscr{H}$. We may refer to this as the *graph topology induced by* $\mathscr{H}$. In the remainder of this Appendix, we show that the topology induced by the gap metric is the same as the graph topology induced by the set $\mathscr{B}(L_2)$.

This comment in no way implies that the graph metric of Section III is equivalent to the gap metric. The topology of Section II (which is the one induced by the graph metric) is the graph topology induced by the set $\mathscr{S}$, whereas the topology induced by the gap metric is the graph topology induced by the set $\mathscr{B}(L_2)$. They are both graph topologies, corresponding to different choices for the set of "stable" systems. Since $\mathscr{S}$ (or $\hat{\mathscr{A}}$) is a *proper* subset of $\mathscr{B}(H^2)$, the topology on rational functions induced by the gap metric is certainly no stronger than the one induced by the graph metric. Whether it is actually weaker is not yet known.

Before presenting the main result of the Appendix, we state and prove a few lemmas. In what follows, P is always of dimension $m \times m$, and $\mathscr{C}$ denotes the set of plants that possess both an RCF and an LCF over $\mathscr{B}(L_2)$.

Lemma A.1: Suppose $P \in \mathscr{C}$. Then P is closed.

Proof: We will actually show that if P has an RCF, then the graph of P is closed. It will be seen that this result also holds for nonsquare plants. Let (N, D) be an RCF of P over $\mathscr{B}(L_2)$. Then, by an easy modification of [3, Theorem 2], it follows that the graph of P is described by

$$G(P) = \{(Dz, Nz) : z \in L_2^m\}. \tag{A.5}$$

Select X, Y in $\text{mat}(\mathscr{B}(L_2))$ such that $XN + YD = I$, and let $\{v_i\}$ be any sequence in $\mathscr{G}(P)$ converging to $v \in L_2^{2m}$; we will show that $v \in \mathscr{G}(P)$. Suppose $v_i = (Dz_i, Nz_i)$ where $z_i \in L_2^m$. Then $z_i = [Y \;\; X]v_i$. Since $v_i \to v$, it follows that $\{z_i\}$ is convergent, with $[Y \;\; X]v$ as its limit. Define $z = [Y \;\; X]v \in L_2^m$. Then $(Dz, Nz) \in \mathscr{G}(P)$. Moreover, $(Dz, Nz) = \lim(Dz_i, Nz_i) = \lim v_i = v$. □

Since $\mathscr{G}(P)$ is a closed subspace of L_2^{2m}, there is a well-defined orthogonal projection mapping L_2^{2m} onto $\mathscr{G}(P)$. Let $\Pi(P)$ denote this projection. Then, for any $x \in L_2^{2m}$, $\Pi(P)x$ is the unique element v in $\mathscr{G}(P)$ that minimizes $\|x - v\|$. Since every $v \in \mathscr{G}(P)$ is of the form (Dz, Nz) for some $z \in L_2^m$, it is an easy exercise to show that

$$\Pi(P) = \begin{bmatrix} D \\ N \end{bmatrix} (D^*D + N^*N)^{-1} [D^* \;\; N^*] \tag{A.6}$$

where * denotes the adjoint operator. Note that $\Pi(P)$ is self-adjoint, which means that it is noncausal except in the most trivial cases.

Lemma A.2: Suppose $(I+P)^{-1} \in \text{mat}(\mathscr{B}(L_2))$. Then $P \in \mathscr{C}$.

Proof: Let $(I+P)^{-1} = R$. Then $P = R^{-1} - I = (I-R)R^{-1} = R^{-1}(I-R)$. Since $R + (I-R) = I$, R and $I-R$ are both left- and right-coprime.

Lemma A.2 shows that $\mathscr{C}$ is large enough to include all plants stabilized by unit feedback.

Now we present the main result of this section. Note that the plants in Theorem A.1 need not be square.

Theorem A.1: A sequence $\{P_i\}$ in $\mathscr{C}$ converges to $P \in \mathscr{C}$ in the graph topology induced by $\mathscr{B}(L_2)$ if and only if $\delta(P, P_i) \to 0$.

Proof:

"Only If": Recall [15, p. 88] that

$$\delta(P, P_i) = \|\Pi(P) - \Pi(P_i)\|. \tag{A.7}$$

Suppose $P_i \to P$ in the graph topology induced by $\mathscr{B}(L_2)$. Then there exist RCF's (N_i, D_i) of P_i and (N, D) of P such that $D_i \to D$, $N_i \to N$ in $\mathscr{B}(L_2)$. Since formation of adjoints and inversion are both continuous on $\mathscr{B}(L2)$, it follows that

$$\Pi(P_i) = \begin{bmatrix} D_i \\ N_i \end{bmatrix} (D_i^*D_i + N_i^*N_i)^{-1} [D_i^* \;\; N_i^*] \to \Pi(P). \tag{A.8}$$

"If": Suppose $\delta(P, P_i) \to 0$, and let (N, D) be an RCF of P over $\mathscr{B}(L_2)$. We will construct RCF's (N_i, D_i) of P_i such that $N_i \to N$, $D_i \to D$. In fact, define

$$\begin{bmatrix} D_i \\ N_i \end{bmatrix} = \Pi(P_i) \begin{bmatrix} D \\ N \end{bmatrix}. \tag{A.9}$$

Since $\Pi(P_i) \to \Pi(P)$,

$$\begin{bmatrix} D_i \\ N_i \end{bmatrix} \to \Pi(P) \begin{bmatrix} D \\ N \end{bmatrix} = \begin{bmatrix} D \\ N \end{bmatrix} \tag{A.10}$$

and all that remains to be shown is that (N_i, D_i) is actually an RCF of P_i. We state this as a separate lemma, since it might be of independent interest.

Lemma A.3: Suppose (N, D) is an RCF of P, that $\delta(P,P_1)<1$, and define (N_1, D_1) by

$$\begin{bmatrix} D_1 \\ N_1 \end{bmatrix} = \Pi(P_1)\begin{bmatrix} D \\ N \end{bmatrix}. \tag{A.10}$$

Then (N_1, D_1) is an RCF of P_1.

Proof: The proof is divided into four steps.

Step 1: The set $\{(D_1 z, N_1 z)\colon z \in L_2^m\}$ is a closed subspace of L_2^{2m}. We show that there exists a constant α such that

$$\|z\| \leqslant \alpha \left\| \begin{bmatrix} D_1 z \\ N_1 z \end{bmatrix} \right\| \qquad \text{for all } z \in L_2^m. \tag{A.11}$$

The above claim will then follow readily from [22, p. 513, Problem 15(ii)].

Let δ denote $\delta(P, P_1)$ and recall that $\delta<1$. Now, if $v \in \mathcal{G}(P)$, then

$$\|v - \Pi(P_1)v\| = \|\Pi(P)v - \Pi(P_1)v\| \leqslant \delta\|v\|. \tag{A.12}$$

Hence

$$\|\Pi(P_1)v\| \geqslant (1-\delta)\|v\| \qquad \text{for all } v \in \mathcal{G}(P). \tag{A.13}$$

So for any $z \in L_2^m$, we have

$$\left\| \begin{bmatrix} D_1 z \\ N_1 z \end{bmatrix} \right\| \geqslant (1-\delta) \left\| \begin{bmatrix} Dz \\ Nz \end{bmatrix} \right\|. \tag{A.14}$$

Select $X, Y \in \text{mat}(\mathcal{B}(L_2))$ such that $XN + YD = I$. Then

$$z = [Y \quad X]\begin{bmatrix} Dz \\ Nz \end{bmatrix},$$

which implies that

$$\|z\| \leqslant \|[Y \quad X]\| \cdot \left\| \begin{bmatrix} Dz \\ Nz \end{bmatrix} \right\|. \tag{A.15}$$

The inequality (A.11) readily follows from (A.14) and (A.15).

Step 2: The set $\{(D_1 z, N_1 z)\colon z \in L_2^m\}$ equals $\mathcal{G}(P_1)$. In view of Step 1, it is enough to show that the range of $\begin{bmatrix} D_1 \\ N_1 \end{bmatrix}\colon L_2^m \to \mathcal{G}(P_1)$ is dense in $\mathcal{G}(P_1)$. Since $\mathcal{G}(P_1)$ is a closed subspace of L_2^{2m}, it is a Hilbert space in its own right. Hence, to establish the claim, it is enough to show that if $v \in \mathcal{G}(P_1)$ and $\left\langle v, \begin{bmatrix} D_1 \\ N_1 \end{bmatrix} z \right\rangle = 0$ for all $z \in L_2^m$, then $v = 0$. Suppose $v \in \mathcal{G}(P_1)$ has this property. Then

$$0 = \left\langle v, \begin{bmatrix} D_1 \\ N_1 \end{bmatrix} z \right\rangle = \langle [D_1^* \quad N_1^*]v, z \rangle \qquad \text{for all } z \in L_2^m \tag{A.16}$$

which implies that $[D_1^* \quad N_1^*]v = 0$, i.e., that $[D^* \quad N^*]\Pi(P_1)^* v = 0$ because of (A.10). Now $\Pi(P_1)$ is self-adjoint, and $v \in \mathcal{G}(P_1)$; so $\Pi(P_1)^* v = v$. Hence, $[D^* \quad N^*]v = 0$, which implies that $\pi(P)v = 0$, from (A.6). Now, interchanging P and P_1 in (A.13) gives

$$\|\Pi(P)v\| \geqslant (1-\delta)\|v\| \qquad \text{for all } v \in \mathcal{G}(P_1). \tag{A.17}$$

Since $\delta<1$, $\Pi(P)v = 0$ implies $v = 0$.

Step 3: The map $z \to (D_1 z, N_1 z)$ is a one-to-one map of L_2^m onto $\mathcal{G}(P_1)$. The "onto" part is established in Step 2, and the one-to-one part follows from (A.11).

Step 4: Let $(\bar{N}_1, \bar{D}_1)$ be any RCF of P_1, and select $X_1, Y_1 \in \text{mat}(\mathcal{B}(L_2))$ such that $X_1\bar{N}_1 + Y_1\bar{D}_1 = I$; then $X_1 N_1 + Y_1 D_1$ is a unit of $\text{mat}(\mathcal{B}(L_2))$. Clearly this step completes the proof. Let $U = X_1 N_1 + Y_1 D_1$. The action of U can be simply explained as follows. Every $v \in \mathcal{G}(P_1)$ can be uniquely represented as $(D_1 z, N_1 z)$ for some $z \in L_2^m$, and as $(\bar{D}_1\bar{z}, \bar{N}_1\bar{z})$ for some $\bar{z} \in L_2^m$, U merely maps z into $\bar{z}$. As such, U is a continuous one-to-one map of L_2^m onto itself, and, by the open mapping theorem [22, p. 57], U^{-1} is also continuous. □

In closing, we observe that all of the above reasoning breaks down completely if we insist on using *causal* bounded operators instead of $\mathcal{B}(L_2)$.

References

[1] *IEEE Trans. Automat. Contr.*, Special Issue on Linear Multivariable Control Systems, vol. AC-26, Feb. 1981.

[2] J. H. Wilkinson, *The Algebraic Eigenvalue Problem*. Oxford: Clarendon, 1965, p. 57.

[3] M. Vidyasagar, "On the use of right-coprime factorizations in distributed feedback systems containing unstable subsystems," *IEEE Trans. Circuits Syst.*, vol. CAS-25, pp. 916–921, Nov. 1978.

[4] W. A. Wolovich, *Linear Multivariable Systems*. New York: Springer-Verlag, 1974.

[5] T. Kailath, *Linear Systems*. Englewood Cliffs, NJ: Prentice-Hall, 1980.

[6] J. L. Kelley, *General Topology*. New York: Van Nostrand, 1955.

[7] M. Vidyasagar, H. Schneider, and B. A. Francis, "Algebraic and topological aspects of feedback stabilization," *IEEE Trans. Automat. Contr.*, vol. AC-27, pp. 880–894, Aug. 1982.

[8] B. A. Francis and M. Vidyasagar, "Algebraic and topological aspects of the regulator problem for lumped linear systems," *Automatica*, vol. 19, pp. 87–90, Jan. 1983.

[9] R. Saeks and J. Murray, "Feedback system design: The tracking and disturbance rejection problems," *IEEE Trans. Automat. Contr.*, vol. AC-26, pp. 203–208, Feb. 1981.

[10] C. A. Desoer, R. W. Liu, J. Murray, and R. Saeks, "Feedback system design: The fractional approach to analysis and synthesis," *IEEE Trans. Automat. Contr.*, vol. AC-25, pp. 399–412, June 1980.

[11] F. M. Callier and C. A. Desoer, "Stabilization, tracking, and disturbance rejection in multivariable convolution systems," *Ann. Societe Scientifique de Bruxelles*, tome 94, I, pp. 7–51, 1980.

[12] D. C. Youla, "On the factorization of rational matrices," *IRE Trans. Inform. Theory*, vol. IT-7, pp. 172–189, July 1961.

[13] J. A. Ball and J. W. Helton, "A Beurling–Lax theorem for the lie group $U(m, n)$ which contains most classical interpolation theory," *J. Opt. Theory*, vol. 9, pp. 107–142, 1983.

[14] G. Zames and A. El-Sakkary, "Unstable systems and feedback: The gap metric," in *Proc. Allerton Conf.*, 1980, pp. 380–385.

[15] A. El-Sakkary, "The gap metric for unstable systems," Ph.D. dissertation, McGill University, Montreal, P.Q., Canada, Mar. 1981.

[16] H. K. Khalil, "On the robustness of output feedback control methods to modeling errors," *IEEE Trans. Automat. Contr.*, vol. AC-26, pp. 524–526, Apr. 1981.

[17] J. H. Chow and P. V. Kokotovic, "Eigenvalue placement in two-time-scale systems," in *Proc. IFAC Symp. on Large Scale Syst.*, Udine, Italy, 1976, pp. 321–326.

[18] C. A. Desoer and M. Vidyasagar, *Feedback Systems: Input-Output Properties*. New York: Academic, 1975.

[19] B. Sz-Nagy and C. Foias, *Harmonic Analysis of Operators on Hilbert Space*. New York: Elsevier, 1970, p. 205.

[20] F. M. Callier and C. A. Desoer, "Open-loop unstable convolution feedback systems with dynamical feedback," *Automatica*, vol. 12, pp. 507–512, 1976.

[21] E. Hille, *Analytic Function Theory*, Volume 2. New York: Chelsea, 1962.

[22] N. Dunford and J. T. Schwartz, *Linear Operators*, Part I. New York: Interscience, 1959.

[23] J. C. Doyle and G. Stein, "Multivariable feedback design: Concepts for a classical/modern synthesis," *IEEE Trans. Automat. Contr.*, vol. AC-26, pp. 4-16, Feb. 1981.

[24] H. H. Rosenbrock, *State Space and Multivariable Systems*. New York: Wiley, 1970.

Right Half Plane Poles and Zeros and Design Tradeoffs in Feedback Systems

JAMES S. FREUDENBERG, STUDENT MEMBER, IEEE, AND DOUGLAS P. LOOZE, MEMBER, IEEE

Abstract—**This paper expresses limitations imposed by right half plane poles and zeros of the open-loop system directly in terms of the sensitivity and complementary sensitivity functions of the closed-loop system. The limitations are determined by integral relationships which must be satisfied by these functions. The integral relationships are interpreted in the context of feedback design.**

I. Introduction

A CENTRAL issue in the design of feedback systems is that of sensitivity of the closed-loop system to uncertainty in the plant model and to disturbance inputs. The system sensitivity function, denoted $S(s)$, has played a key role in the classical design and theory of feedback systems. The importance of the sensitivity function has been discussed by many authors [1]–[3], [9]–[18], [20], [21]. Briefly, the magnitude of the sensitivity function evaluated along the $j\omega$-axis directly quantifies such feedback properties as output disturbance rejection and sensitivity to small parameter variations.

Another function which expresses important feedback properties is the complementary sensitivity function [17], defined as $T(s) \triangleq 1 - S(s)$. The magnitude of $T(s)$ along the $j\omega$-axis quantifies the response of the feedback system to sensor noise. In addition, this quantity has recently been used as a measure of stability margin [2], [13]–[16], [19].

The importance of $|T(j\omega)|$ and $|S(j\omega)|$ to design properties motivates the expression of design limitations imposed by the open-loop transfer function directly in terms of these quantities. For example, a well-known theorem of Bode [1], [3] states that for stable open-loop transfer functions with greater than one pole rolloff, the integral over all frequencies of the log magnitude of the sensitivity function must equal zero. In the presence of bandwidth limitations this imposes a design tradeoff among system sensitivity properties in different frequency ranges ([1] and Section III below).

It has not been common, however, to formulate other limitations explicitly in terms of $|S(j\omega)|$ and $|T(j\omega)|$. For example, it has long been recognized that the presence of right half plane poles and zeros in the open-loop transfer function imposes limitations upon the design of feedback systems. These limitations are frequently expressed in terms of the effect on the phase of the open-loop transfer function. Suppose the plant is nonminimum phase. Then using classical analysis techniques it can be seen qualitatively that requiring $|S(j\omega)|$ to be less than one over some frequency interval implies that $|S(j\omega)|$ is greater than one elsewhere. This fact is proven by Francis and Zames [6, Theorem 3]. These authors show that if the plant has a right half plane zero, then requiring $|S(j\omega)|$ to be arbitrarily small over some interval forces $|S(j\omega)|$ to be arbitrarily large elsewhere.

Manuscript received October 3, 1983; revised May 7, 1984. This paper is based on a prior submission of March 29, 1983. Paper recommended by Past Associate Editor, W. A. Wolovich. This work was supported in part by the Joint Services Electronics Program under Contract N00014-79-C-0424, in part by the U.S. Air Force under Grant AFOSR 78-3633, and in part by the National Science Foundation under Grant ECS-82-12080.

J. S. Freudenberg is with the Department of Electrical Engineering and Computer Science, University of Michigan, Ann Arbor, MI 48109-1109.

D. P. Looze is with Alphatech, Inc., Burlington, MA 01803.

Despite the importance of $|S(j\omega)|$ and $|T(j\omega)|$ as measure of design quality, it has been more common to express limitations due to open right half plane poles and zeros as constraining the values of $S(s)$ and $T(s)$ at isolated points away from the $j\omega$-axis [6], [11], [17]–[19], [22]. The purpose of this paper is to present equivalent statements of the right half plane pole and zero constraints in terms of integral relations which must be satisfied by $|S(\omega)|$ and $|T(j\omega)|$. These constraints show that desirable properties of the sensitivity and complementary sensitivity functions in one frequency range must be traded off against undesirable properties at other frequencies. These tradeoffs are a direct consequence of properties of linear time-invariant systems. Thus, the limitations discussed in this paper are independent of any particular choice of design method.

The remainder of this paper is organized as follows. Section II is devoted to the derivation of the integral relations from the right half plane pole and zero constraints. In Section III Bode's integral theorem, referred to above, is extended to open-loop unstable plants and consequences for feedback design are discussed. Section IV contains a discussion of the limitations imposed upon system sensitivity properties by the integral constraints due to open right half plane zeros. Section V contains a similar discussion of the limitations imposed upon the complementary sensitivity function by unstable open-loop poles. The effect of the relative location of right half plane poles and zeros to frequency ranges of interest is discussed in Section VI. Some brief remarks on limitations in multivariable systems are found in Section VII. The paper is summarized in Section VIII.

II. Right Half Plane Pole and Zero Constraints

Consider the linear time-invariant feedback system of Fig. 1. Let the transfer functions of the plant model and the feedback compensator be denoted $P(s)$ and $F(s)$, respectively. The open-loop transfer function is given by

$$L(s) \triangleq P(s)F(s). \tag{2.1}$$

The sensitivity function of this system is

$$S(s) = \frac{1}{1+L(s)} \tag{2.2}$$

and the complementary sensitivity function [17] is

$$T(s) \triangleq 1 - S(s)$$

$$= \frac{L(s)}{1+L(s)}. \tag{2.3}$$

The response of the system of Fig. 1 to disturbance inputs is given by

$$y_d(s) = S(s)d(s)$$

and the response to sensor noise is given by

$$y_n(s) = -T(s)n(s).$$

Reprinted from *IEEE Trans. Automat. Contr.*, vol. AC-30, no. 6, pp. 555–565, June 1985.

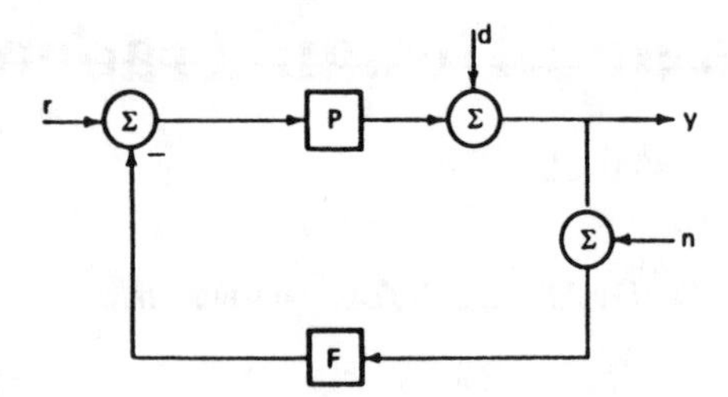

Fig. 1. Feedback system.

From these equations it is seen that, at a particular frequency, the effect of disturbances can be reduced by requiring that $|S(j\omega)| < 1$ at that frequency. Similarly, requiring $|T(j\omega)| < 1$ leads to a reduction in the effects of sensor noise at that frequency. Since $S(j\omega) + T(j\omega) = 1$, there is a well-known tradeoff between the two types of response at a given frequency. The integral relations to be derived in this section and Section III reveal that there also exist tradeoffs among feedback properties at *different* frequencies.

Assume that $L(s)$ is free of unstable hidden modes. Then the feedback system is stable if $S(s)$ is bounded in the closed right half plane. Note this assumption on $L(s)$ implies that the closed right half plane poles and zeros of the plant and compensator must appear with at least the same multiplicity in $L(s)$.

Assume also that $L(s)$ can be factored as

$$L(s) = \tilde{L}(s)B_p^{-1}(s)B_z(s)e^{-s\tau}. \tag{2.4}$$

The term $e^{-s\tau}$, $\tau \geq 0$, represents a time delay if $\tau > 0$. The term

$$B_z(s) = \prod_{i=1}^{N_z} \frac{z_i - s}{\bar{z}_i + s} \tag{2.5}$$

is the Blaschke product of open right half plane zeros, including multiplicities

$$\mathcal{Z} = \{z_i;\ i = 1, \cdots, N_z\}. \tag{2.6}$$

Similarly,

$$B_p(s) = \prod_{i=1}^{N_p} \frac{p_i - s}{\bar{p}_i + s} \tag{2.7}$$

is the Blaschke product of open right half plane poles, including multiplicities

$$\mathcal{P} = \{p_i;\ i = 1, \cdots, N_p\}. \tag{2.8}$$

Finally, $\tilde{L}(s)$ is proper and has no poles or zeros in the open right half plane.

From (2.2) and (2.3) it is clear that right half plane poles and zeros of $L(s)$ constrain the values of $S(s)$ and $T(s)$ at these points in the right half plane [11], [18], [19], [22]. One way of expressing these constraints, following immediately from (2.2) and (2.3), is now given.

At each closed right half plane zero z of multiplicity m, it follows that

A)
$$S(z) = 1$$
$$\vdots$$
$$\left.\frac{d^i}{ds^i} S\right|_{s=z} = 0 \qquad i = 1, \cdots, m-1.$$

Similarly, at each closed right half plane pole p of multiplicity n, it follows that

B)
$$T(p) = 1$$
$$\vdots$$
$$\left.\frac{d^i}{ds^i} T\right|_{s=p} = 0 \qquad i = 1, \cdots, n-1.$$

Note that the above constraints can be expressed in terms of either S or T via the identity

$$S(s) + T(s) = 1. \tag{2.9}$$

Assume that the feedback system is stable. Then $S(s)$ and $T(s)$ have no poles in the closed right half plane. In order to express the constraints A) due to open right half plane zeros in terms of $|S(j\omega)|$, it is necessary to remove the zeros of $S(s)$ at the open right half plane poles of $L(s)$. Note that the sensitivity function can be factored as

$$S(s) = \tilde{S}(s)B_p(s) \tag{2.10}$$

where $\tilde{S}(s)$ has no poles or zeros in the open right half plane. Since Blaschke products are all-pass of unit magnitude ($|B_p(j\omega)| = 1$ $\forall\omega$), it follows that $|S(j\omega)| = |\tilde{S}(j\omega)|$ $\forall\omega$.

It is necessary to constrain the behavior of $\log S(s)$ and $d^i/ds^i \log S(s)$ at infinity. Consider the following class of functions. Given $F(s)$, define

$$M(R) = \sup_\theta |F(Re^{j\theta})|, \qquad \theta \in [-\pi/2, \pi/2].$$

Then $F(s)$ is said to be in class $\mathcal{R}$ provided

$$\lim_{R\to\infty} \frac{1}{R} M(R) = 0. \tag{2.11}$$

Class $\mathcal{R}$ includes many functions of interest. If $L_o(s)$ is a proper rational function, then $\log L_o(s)$ and $d^i/ds^i \log L_o(s)$ are in class $\mathcal{R}$. Functions of the form $\log L(s)$, with $L(s) = L_o(s)e^{-s\tau}$, $\tau > 0$ are not. If, however, the feedback system with sensitivity function (2.2) is stable, then $\log S(s)$ and $d^i/ds^i \log S(s)$ are in class $\mathcal{R}$ despite any time delay in $L(s)$.

The constraints upon the sensitivity function at open right half plane zeros A) can be expressed in terms of the sensitivity function on the $j\omega$-axis as follows.

Theorem 1: Let $z = x + jy$ be an open right half plane zero, with multiplicity m, of the open-loop transfer function $L(s)$. Assume that $d^i/ds^i \log \tilde{S}(s)$ is in class $\mathcal{R}$, $i = 0, 1, \cdots, n-1$. Then, if the corresponding feedback system is stable, the sensitivity function must satisfy the following integral constraints:[1]

$$\pi \log |B_p^{-1}(z)| = \int_{-\infty}^{\infty} \log |S(j\omega)|\ d\theta_z(\omega) \tag{2.12}$$

$$\pi \measuredangle B_p^{-1}(z) = \int_{-\infty}^{\infty} \measuredangle \tilde{S}(j, \omega)\ d\theta_z(\omega) \tag{2.13}$$

$$\pi \frac{d^i}{ds^i} \log B_p^{-1}(s)\bigg|_{s=z} = \int_{-\infty}^{\infty} \frac{d^i}{ds^i} \log \tilde{S}(s)\bigg|_{s=j\omega} d\theta_z(\omega)$$

$$(i = 1, \cdots, m-1). \tag{2.14}$$

[1] The results of Theorems 1 and 3 remain valid even if $L(s)$ has $j\omega$-axis poles. Similarly, Theorem 2 remains valid if $L(s)$ has $j\omega$-axis zeros. This is discussed further in the Appendix.

The function $\theta_z(\omega)$ is given by

$$\theta_z(\omega) = \arctan\left[\frac{\omega - y}{x}\right]. \qquad \blacksquare \quad (2.15)$$

Proof: A simple application of Poisson's integral formulas [4], [5]. See the Appendix for details. ■

In order to express the constraints B) due to open right half plane poles in terms of $|T(j\omega)|$, it is necessary to remove any zeros of $T(s)$ at the open right half plane zeros of $L(s)$ as well as any time delay. Assume again that the feedback system is stable. Then the complementary sensitivity function can be factored as

$$T(s) = \tilde{T}(s)B_z(s)e^{-s\tau} \qquad (2.16)$$

where $\tilde{T}(s)$ has no poles or zeros in the open right half plane. The fact that $B_z(s)$ and $e^{-s\tau}$ are all-pass functions of unit magnitude implies $|T(j\omega)| = |\tilde{T}(j\omega)|$. The constraints upon the complementary sensitivity function at open right half plane poles B) can be expressed in terms of this function on the $j\omega$-axis as follows.

Theorem 2: Let $p = x + jy$ be an open right half plane pole, with multiplicity n, of the open-loop transfer function $L(s)$. Assume that $d^i/ds^i \log \tilde{T}(s)$ is class $\mathcal{R}$, $i = 0, 1, \cdots, m - 1$. Then, if the corresponding feedback system is stable, the complementary sensitivity function must satisfy the following integral constraints:

$$\pi \log |B_z^{-1}(p)| + \pi x\tau = \int_{-\infty}^{\infty} \log |T(j\omega)| \, d\theta_p(\omega) \quad (2.17)$$

$$\pi \measuredangle B_z^{-1}(p) + \pi y\tau = \int_{-\infty}^{\infty} \measuredangle \tilde{T}(j\omega) \, d\theta_p(\omega) \quad (2.18)$$

$$\pi \frac{d^i}{ds^i}\left(\log B_z^{-1}(s)e^{s\tau}\right)\bigg|_{s=p}$$

$$= \int_{-\infty}^{\infty} \frac{d^i}{ds^i} \log \tilde{T}(s)\bigg|_{s=j\omega} d\theta_p(\omega) \quad (i=1, \cdots, n-1). \quad (2.19)$$

The function $\theta_p(\omega)$ is given by

$$\theta_p(\omega) = \arctan\left[\frac{\omega - y}{x}\right]. \qquad \blacksquare \quad (2.20)$$

Proof: See the Appendix.

Each of the integral relations of Theorems 1 and 2 places a constraint upon the sensitivity or complementary sensitivity function. For the purposes of this paper the constraints which are most insightful are those given by (2.12) and (2.17). These constraints give the area under the $\log |S(j\omega)|$ and $\log |T(j\omega)|$ curves; the area is calculated using the $j\omega$-axis weighted by the location of a right half plane zero or pole, respectively. The weighting function

$$\theta_s(\omega) = \arctan\left[\frac{\omega - y}{x}\right], \qquad s = x + jy$$

is shown in Fig. 2.

It is of particular significance that (2.12) and (2.17) constrain the *integrals* of $\log |S(j\omega)|$ and $\log |T(j\omega)|$. This fact implies that feedback properties in different frequency ranges are *not* independent. To see this, note that since

$$\frac{d\theta_s(\omega)}{d\omega} = \frac{x}{x^2 + (y-\omega)^2} > 0$$

it follows that $\theta_s(\omega)$ is an increasing function of ω. Moreover, the terms on the left-hand sides of (2.12) and (2.17) are nonnegative. These facts reveal that systems which reduce the response due to disturbances or sensor noise ($|S(j\omega)| < 1$ or $|T(j\omega)| < 1$) in some frequency range *necessarily* increase this response at other frequencies. Thus, feedback properties at different frequencies must be traded off against one another to achieve a satisfactory design. These tradeoffs will be discussed further in Sections IV and V.

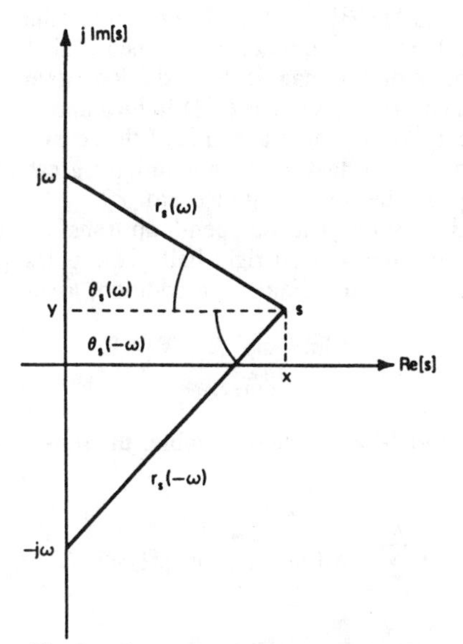

Fig. 2. Geometry of weighting functions.

It seems intuitive that right half plane poles and zeros which are close to frequency ranges over which design objectives are given constitute a greater obstacle to the achievement of these objectives than if these poles and zeros were far away. The weighting function appearing in the integral relations verifies this intuition and yields a precise notion of the proximity of a zero or pole. The relation between the locations of zeros and poles and the corresponding weightings is discussed further in Section VI.

Finally, note that the weighted length of the $j\omega$-axis is finite (and equal to π). This implies, for example, that it is *not* possible to trade off a given amount of sensitivity reduction by allowing $|S(j\omega)|$ to exceed one by an arbitrarily small amount over an arbitrarily large frequency range. The amount by which $|S(j\omega)|$ exceeds one cannot be made arbitrarily small. The significance of this observation will become clear in Section III, where an extension of the well-known Bode integral theorem is presented.

III. Generalization of Bode's Integral Theorem

The purpose of this section is to extend a well-known theorem of Bode [3] to open-loop unstable systems. Bode's original result was valid only for open-loop stable systems, despite a claim of Horowitz [1, p. 307] to the contrary. The implications of this result for feedback design have been discussed by Horowitz [1] and others [9], [20].

Bandwidth constraints in feedback design typically require that the loop gain be small above a specified frequency. In addition, it is frequently required that the loop gain possess greater than a one pole rolloff above that frequency. These constraints commonly arise due to the need to provide for stability robustness despite uncertainty in the plant model at high frequencies. Bandwidth constraints also arise due to limitations imposed by actuators and sensors. One way of quantifying such constraints is by requiring that

$$|L(j\omega)| \le \frac{M}{\omega^{1+k}}, \qquad \omega > \omega_o \qquad (3.1)$$

where $k > 0$ and $M/(\omega_0^{1+k}) \leq m$. The positive value of k ensures that a greater than one pole rolloff is obtained while the value of m imposes a bound on the magnitude of the loop gain.

Bandwidth constraints such as (3.1) in turn impose a constraint upon the integral of the log magnitude of the sensitivity function. First, the requirement that the loop gain have greater than a one pole rolloff yields the following theorem.

Theorem 3: Assume that the open-loop transfer function $L(s)$ possesses finitely many open right half plane poles $\{p_i: i = 1, \cdots, N_p\}$ including multiplicities. In addition, assume that

$$\lim_{R\to\infty} \sup_{\substack{|s|\geq R \\ \mathrm{Re}[s]\geq 0.}} R|L(s)| = 0. \tag{3.2}$$

Then, if the closed-loop system is stable, the sensitivity function must satisfy

$$\pi \sum_{i=1}^{N_p} \mathrm{Re}\ [p_i] = \int_0^\infty \log\ |S(j\omega)|\ d\omega. \tag{3.3}$$

■

Proof: See the Appendix. ■

Note that Theorem 3 is valid for systems which include right half plane zeros and time delays in $L(s)$.

If $N_p = 0$, then Theorem 3 reduces to Bode's theorem. This theorem states that on a plot of log $|S(j\omega)|$ versus ω the sensitivity reduction area (log $|S(j\omega)| < 0$) must equal the area of sensitivity increase (log $|S(j\omega)| > 0$) in units of decibels × (radians/second).

If $N_p > 0$, then the area of sensitivity reduction is less than the area of sensitivity increase by an amount proportional to the sum of the distances from the unstable poles to the imaginary axis. This indicates that a portion of the loop gain which could otherwise contribute to sensitivity reduction must instead be used to pull the unstable poles into the left half plane.

By itself, Theorem 3 does not impose a meaningful design limitation since the necessary area of sensitivity increase can be obtained by allowing $|S(j\omega)|$ to exceed one by an arbitrarily small amount over an arbitrarily large frequency range. (In this respect Theorem 3 differs from Theorems 1 and 2.) However, only part of the bandwidth constraint (3.1) was used to obtain Theorem 3; namely, the fact that $k > 0$ implies (3.2) is satisfied. Practical bandwidth constraints also specify the value of m in (3.1). Thus, (3.1) implies that there exists a frequency ω_c such that $M/(\omega_c^{1+k}) = \epsilon < 1$. Thus, it follows that

$$|L(j\omega)| \leq \frac{M}{\omega^{1+k}} \leq \epsilon, \qquad \omega \geq \omega_c \tag{3.4}$$

where $k > 0$. This property of the open-loop transfer function yields the following bound.

Corollary: Assume that, in addition to (3.2), the transfer function $L(s)$ satisfies the bound (3.4). Then

$$\int_{\omega_c}^\infty \log\ |S(j\omega)|\ d\omega \leq \frac{\log\left[\dfrac{1}{1-\epsilon}\right] \cdot \omega_c}{k}. \tag{3.5}$$

■

Proof: See the Appendix. ■

The bound (3.5) is crude and, in fact, is an optimistic estimate of the integral in question. Nonetheless, it indicates how bandwidth constraints which limit the loop gain as a function of frequency impose a tradeoff upon system sensitivity properties. Suppose that a given level of sensitivity reduction is desired over some low frequency range. Then (3.5) places an upper bound on the area of sensitivity increase which can be obtained at frequencies greater than ω_c and, therefore, a lower bound on the area of sensitivity increase which must be present at lower frequencies. This fact can be used to obtain a lower bound (greater than one) on the maximum value of sensitivity increase below ω_c. Note that the bound (3.5) can be increased only by relaxing the bandwidth specification. In practice, this may not be possible due to the necessity of ensuring stability robustness. Thus, a tradeoff is imposed among system sensitivity properties in different frequency ranges. The benefits of sensitivity reduction in one frequency range must be obtained at the cost of increased sensitivity at other frequencies whenever bandwidth constraints are imposed. The generalization of Bode's theorem presented here shows that this cost is greater for open-loop unstable systems.

IV. Limitations on the Sensitivity Function Due to Open Right Half Plane Zeros

In Section II it was shown that the presence of open right half plane zeros places constraints upon the system sensitivity function. These constraints show that if sensitivity reduction ($|S(j\omega)| < 1$) is present in some frequency range, then there *necessarily* exist other frequencies at which the use of feedback increases sensitivity ($|S(j\omega)| > 1$). The purpose of this section is to illustrate this requirement by deriving some lower bounds on the maximum amount of sensitivity increase given that a certain level of sensitivity reduction has been achieved over some frequency range.

For a given plant model $P(s)$ suppose it is desired to design a feedback compensator $F(s)$ such that a specified level of sensitivity reduction is obtained over a conjugate symmetric[2] range of frequencies Ω. Let the desired level of sensitivity reduction be given by

$$|S(j\omega)| \leq \alpha < 1 \qquad \forall\omega \in \Omega. \tag{4.1}$$

Let z be an open right half plane zero of $L(s)$ and let the weighted length of the frequency range Ω be denoted

$$\Theta_z(\Omega) \triangleq \int_\Omega d\theta_z(\omega). \tag{4.2}$$

The weighted length of the complementary frequency range $\Omega^c = \{\omega : \omega \notin \Omega\}$ is given by

$$\Theta_z(\Omega^c) = \pi - \Theta_z(\Omega). \tag{4.3}$$

From Fig. 3 it is clear that $\pi > \Theta_z(\Omega) > 0$ provided that $\Omega \neq \phi$ and $\Omega \neq \mathbb{R}$. It is also clear that if $\Omega_1 \subseteq \Omega_2$, then $\Theta_z(\Omega_2) \geq \Theta_z(\Omega_1)$.

Define the maximum sensitivity

$$\|S\|_\infty \triangleq \sup_\omega |S(j\omega)|. \tag{4.4}$$

The following theorem gives a lower bound on the maximum sensitivity due to achievement of the sensitivity reduction level (4.1) for a nonminimum phase system. As this lower bound is greater than one, it follows that the closed-loop system exhibits a sensitivity increase over some frequency range.

Theorem 4: Let the open-loop transfer function $L(s)$ have open right half plane poles and zeros given by (2.8) and (2.6). Suppose that the closed-loop system is stable and that the level of sensitivity reduction (4.1) has been achieved. Then for each $z \in \mathcal{Z}$ the following bound must satisfied:

$$\|S\|_\infty \geq \left(\frac{1}{\alpha}\right)^{\frac{\Theta_z(\Omega)}{\pi-\Theta_z(\Omega)}} |B_p^{-1}(z)|^{\frac{\pi}{\pi-\Theta_z(\Omega)}} \tag{4.5}$$

where $\Theta_z(\Omega)$ is given by (4.2). ■

[2] Conjugate symmetry implies that if $\omega \in \Omega$, then $-\omega \in \Omega$.

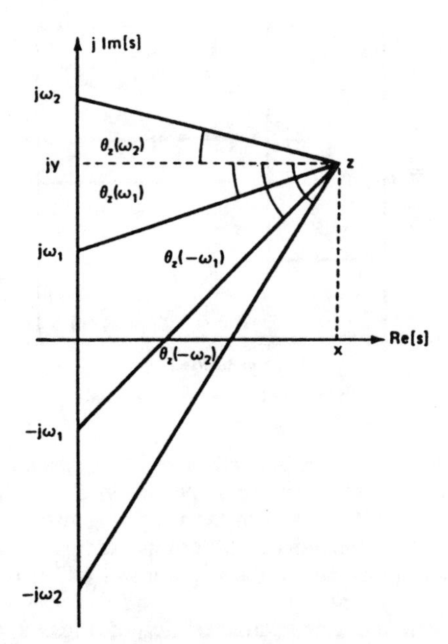

Fig. 3. Geometry of weighted frequency interval.

Proof: From (2.12)

$$\pi \log |B_p^{-1}(z)| = \int_{\Omega} \log |S(j\omega)| \, d\theta_z(\omega) + \int_{\Omega^c} \log |S(j\omega)| \, d\theta_z(\omega). \tag{4.6}$$

Since $\sup_{\Omega} |S(j\omega)| \le \alpha$ by design, and since $\sup_{\Omega^c} |S(j\omega)| \le \|S\|_\infty$ by definition, it follows that

$$\pi \log |B_p^{-1}(z)| \le \log (\alpha)\Theta_z(\Omega) + \log \|S\|_\infty \Theta_z(\Omega^c). \tag{4.7}$$

Exponentiating both sides of (4.7) yields the result. ■

In general, a different lower bound is obtained for each zero (an exception being pairs of complex conjugate zeros). Note that a similar bound has been derived by Francis and Zames [6, Theorem 3]. The bound of Theorem 4 is less crude, and the proof and results are more insightful. In addition, although (4.5) is derived for the simple specification (4.1) the method readily extends to more general specifications. This is clear from the proof.

Inequalities (4.5) and (4.7) must be satisfied for each open right half plane zero of $L(s)$. Before discussing the tightness of these inequalities the significance of each term will be briefly explained. First, however, note the facts that $\alpha < 1$, $|B_p^{-1}(z)| > 1$, and $\Theta_z(\Omega) < \pi$ imply that the right-hand side of (4.5) is strictly greater than one. This verifies that the maximum sensitivity is indeed greater than one.

The term $\Theta_z(\Omega)$ given by (4.2) is the weighted length of the frequency range over which sensitivity reduction is desired; the term $\Theta_z(\Omega^c)$ given by (4.3) is the weighted length of the complementary frequency range. The relation between these weighted lengths and the location of the zero is discussed in Section VI. To illustrate, Fig. 4 shows how the weighted length of the frequency interval $\Omega = [0, \omega_2]$ varies as a function of ω_2 for a zero at $s = 1$ and at $s = 1 + j$. From Fig. 4 and inequalities (4.5) and (4.7) it is clear that a significant level of sensitivity reduction at frequencies near a right half plane zero is necessarily accompanied by a large sensitivity increase at other frequencies. Moreover, suppose that sensitivity reduction is desired over all but a lightly weighted portion of the $j\omega$-axis. Then Theorem 4 shows that the accompanying sensitivity increase must be greater than if sensitivity is permitted to exceed one at more heavily weighted frequencies. The above comments are illustrated in Fig. 5 by plotting the lower bound on $\log \|S\|_\infty$ for the frequency interval $\Omega = [0, 1 \text{ rad/s}]$, zeros at $s = 1$ and $s = 1 + j$, and

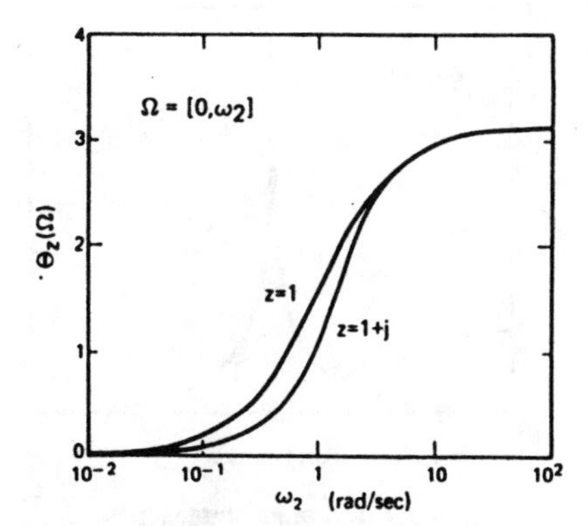

Fig. 4. Weighted length of frequency interval.

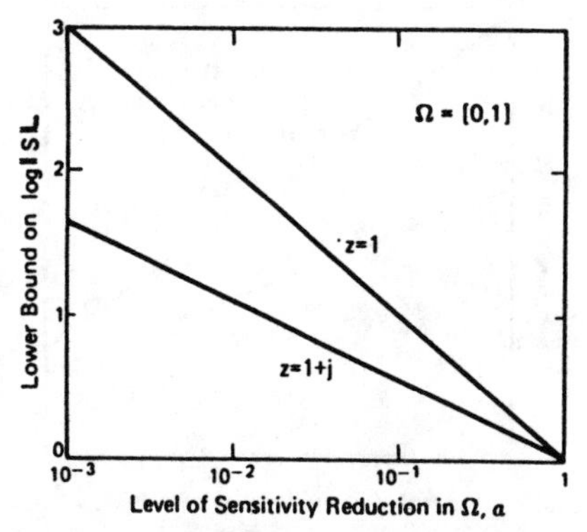

Fig. 5. Lower bound on maximum sensitivity: open loop stable.

various levels of sensitivity reduction. The system in this example is assumed to be open-loop stable.

If the system is open-loop unstable, then the lower bound (4.5) on $\|S\|_\infty$ is increased as a function of the proximity of the unstable poles to the zero in question. This is a consequence of the fact that the weighted area under the $\log |S(j\omega)|$ curve is positive for open-loop unstable systems. Thus, the presence of unstable poles in the open-loop transfer function tends to worsen the sensitivity performance of the closed-loop system. Since

$$\pi \log |B_p^{-1}(z)| = \pi \sum_{i=1}^{N_p} \log \left| \frac{\bar{p}_i + z}{p_i - z} \right| \tag{4.8}$$

it follows (unsurprisingly) that systems with approximate right half plane pole-zero cancellations can have especially bad sensitivity properties. As an example the magnitude of one term of (4.8) is plotted in Fig. 6 versus various locations of a real pole for a zero at $s = 1$. The effect of an unstable pole at $s = 2$ upon the lower bound (4.5) for zeros at $s = 1$ and $s = 1 + j$ and various levels of sensitivity reduction over the frequency interval $\Omega = [0, 1]$ is illustrated in Fig. 7. This bound should be compared to that for an open-loop stable system plotted in Fig. 5.

In general, the bounds of Theorem 4 will not be tight for a variety of practical as well as theoretical reasons. Consider the bound (4.5) for a single zero. From (4.6) it follows that this bound is satisfied with equality by a function $\hat{S}(s)$ for which

$$|\hat{S}(j\omega)| = \begin{cases} \alpha & \omega \in \Omega \\ \|\hat{S}\|_\infty & \omega \in \Omega^c \end{cases} \tag{4.9}$$

where $\|\hat{S}\|_\infty$ is given by the right-hand side of (4.5). The function $\hat{S}(s)$ is illustrated in Fig. 8 for an open-loop stable system with a zero at $s = 1 + j$ and a level of sensitivity reduction $\alpha = 0.1$ over the frequency interval [0, 1 rad/s]. This function has the minimum possible value of maximum sensitivity increase of all functions satisfying the sensitivity specification (4.1).

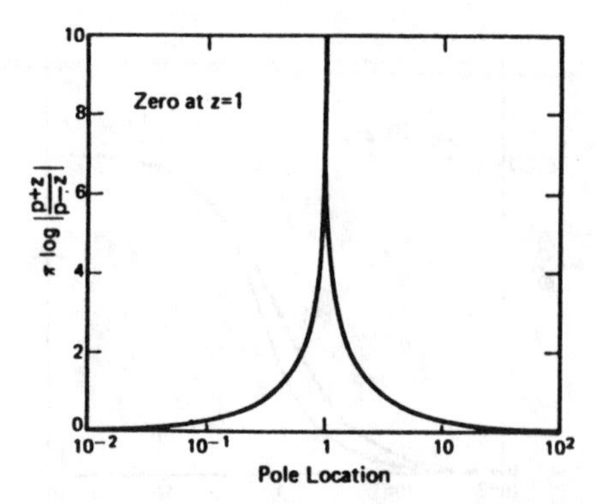

Fig. 6. Effect of an unstable pole.

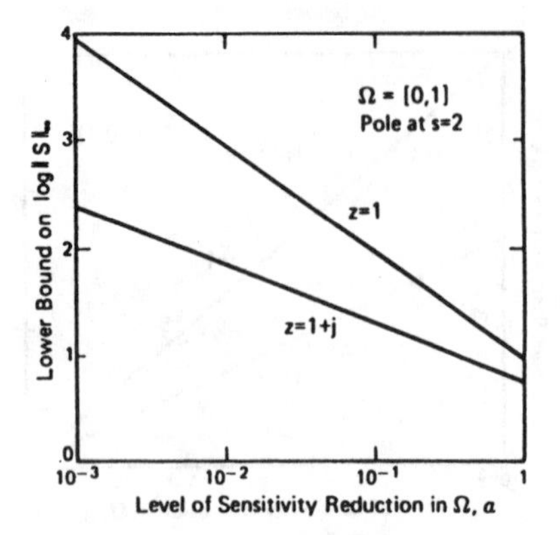

Fig. 7. Effect of unstable pole on maximum sensitivity bound.

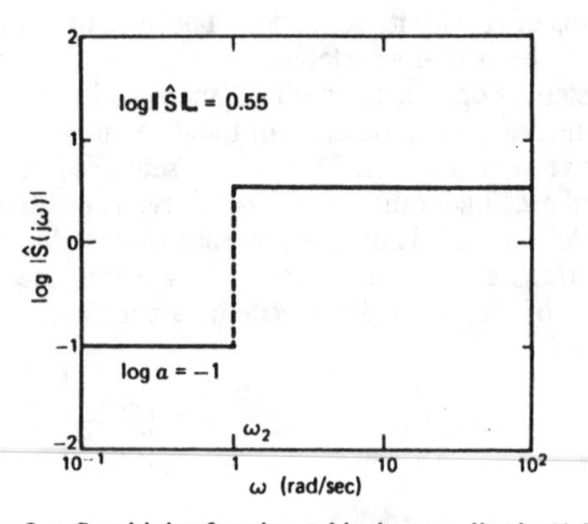

Fig. 8. Sensitivity function achieving equality in (4.5).

As mentioned earlier, a sensitivity function with a gain characteristic as in (4.9) and Fig. 8 is in practice neither achievable nor desirable. It is not achievable due to the corresponding requirement of an infinite bandwidth open-loop transfer function nor desirable due to robustness considerations. In most cases robustness constraints will, at a minimum, require that

$$|S(j\omega)-1|<\epsilon \qquad \text{for } \omega \in \Omega_3 \triangleq [\omega_3, \infty]. \tag{4.10}$$

The effect of the constraint (4.10) may be analyzed by a straightforward modification of Theorem 4. As $\epsilon \to 0$, the bound (4.5) is replaced by

$$\|S\|_\infty \geq \left(\frac{1}{\alpha}\right)^{\frac{\Theta_z(\Omega)}{\Theta_z(\Omega')}} |B_p^{-1}(z)|^{\frac{\pi}{\Theta_z(\Omega')}} \tag{4.11}$$

where $\Omega' \triangleq \Omega^c \cap [-\omega_3, \omega_3]$. As $\Theta_z(\Omega')$ is an increasing function of ω_3, it follows that for fixed values of z, α, ω_1, and ω_2 the minimum possible value of $\|S\|_\infty$ increases as ω_3 decreases. This minimum value would be attained by a design for which equality

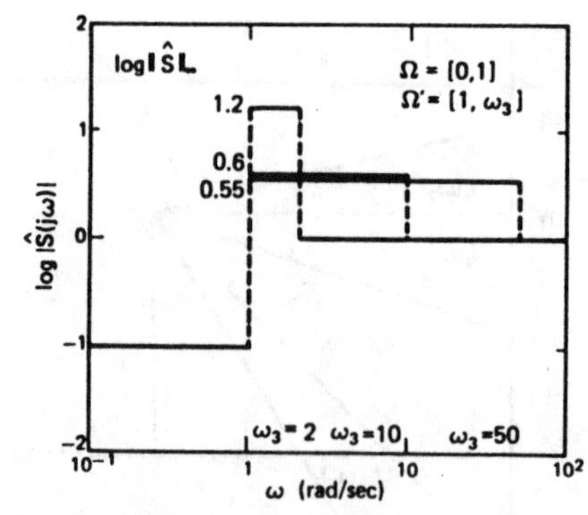

Fig. 9. Sensitivity functions achieving equality in (4.11).

is achieved in (4.11). The gain characteristic of such a sensitivity function for the data of Fig. 8 is plotted in Fig. 9 for various values of ω_3. Note the effect of the weighting function upon the value of $\|S\|_\infty$ as a function of ω_3. For this example the increase in maximum sensitivity due to the requirement (4.10) is negligible for ω_3 sufficiently large.

Even though the gain characteristics of Figs. 8 and 9 are discontinuous, they may be approximated arbitrarily closely by stable rational functions. Thus, the bounds (4.5) or (4.11) would be tight if no other constraints on sensitivity were present. In order to realize these sensitivity functions by applying feedback around a given plant, however, integral relations (2.12)–(2.14) due to each open right half plane zero of $L(s)$ must be satisfied. Thus, a sensitivity function constructed as in Fig. 8 or 9 to satisfy (2.12) for one zero will not, in general, satisfy integral gain relations for other zeros. Recall that the sensitivity functions of Figs. 8 and 9 are constructed to yield the minimum possible value of $\|S\|_\infty$ for the given sensitivity specification under the constraint imposed by a single zero. Thus, the implication of the preceding discussion is that this minimum possible value is optimistic and the lower bounds (4.5) and (4.11) cannot be tight. Finally, the practical need to limit complexity of the compensator will limit the ability to realize piecewise constant functions for which the bounds are tight.

The results of Theorem 4 can be useful in applications by allowing an estimate of the minimum price, in terms of sensitivity increase, which must be paid for a given level of sensitivity reduction over an interval. For example, given Ω, Ω', and α as in Fig. 9, the minimum possible value of $\|S\|_\infty$ can be computed using the above procedure for each zero. If this value is too large for any zero, then it may be necessary to reduce the level of sensitivity reduction α. Alternately, the locations of the frequency intervals Ω and Ω' could be modified. There are, of course, other tradeoffs involved; for example, increasing the system bandwidth may not be permissible due to robustness considerations.

V. Limitations on the Complementary Sensitivity Function Imposed by Unstable Poles

In Section II it was shown that the presence of open right half plane poles places constraints upon the complementary system sensitivity function. These constraints show that there exists a tradeoff among system sensor noise rejection properties in different frequency ranges. This tradeoff can be thought of as dual to that imposed upon the sensitivity function by right half plane zeros.

Lower bounds on the maximum value of $|T(j\omega)|$ can be derived which are similar to those of Section IV. Specifications on sensor noise response analogous to (4.1) are usually imposed at high frequencies. Assumption of such a specification leads to a bound similar to (4.5). At low frequencies, $|T(j\omega)|$ is usually constrained to be near unity by the requirement of small sensitivity. This fact can be used to construct a lower bound similar to (4.11).

One difference between the results of Sections IV and V is that time delays worsen the tradeoff upon sensor noise reduction imposed by unstable poles. This is plausible for the following reasons. Use of feedback around an open-loop unstable system is necessary to achieve stability. Time delays, as well as right half plane zeros, impede the processing of information around a feedback loop. Thus, it is reasonable to expect that limitations due to unstable poles are worse when time delays and/or right half plane zeros are present. Note, in particular, that the term due to the time delay in (2.17) is proportional to the product of the length of the time delay and the distance from the unstable pole in question to the $j\omega$-axis. This is consistent with the above interpretation.

It should also be noted that the reciprocal of the complementary sensitivity function has been interpreted as a measure of system stability margin against unstructured multiplicative uncertainty [2], [13]-[16], [19]. Under this interpretation, Theorem 2 shows that unstable poles also impose a tradeoff upon the size of this measure of stability margin in different frequency ranges. Thus, this stability margin cannot be large at all frequencies.

VI. Dependence of Weightings Upon Pole/Zero Location

In the previous sections it was shown that the weighted length of various frequency intervals is important in determining the tradeoffs imposed by right half plane poles and zeros via the integral relations of Theorems 1 and 2. Intuitively, the difficulty in achieving the benefits of feedback is a function of the proximity of such zeros and poles to frequency ranges over which design specifications are imposed. The fact that weighting functions appear in the integral relations justifies this intuition and allows the notion of proximity to be made precise.

The purpose of this section is to discuss the dependence of the weighting assigned to a frequency interval upon the relative location of the pole or zero and the interval. This dependence can be seen qualitatively from Fig. 3. The quantitative analysis in this section should prove useful in constructing design specifications which reflect the tradeoff between benefits and cost of feedback imposed by right half plane poles and zeros. The discussion of this section uses weightings imposed by right half plane zeros, but identical results hold for weightings imposed by right half plane poles.

For purposes of illustration, consider the frequency interval

$$\Omega=[\omega_1,\ \omega_2]\cup[-\omega_2,\ -\omega_1]. \tag{6.1}$$

Then (4.2) yields

$$\Theta_z(\Omega)=\theta_z(\omega_2)-\theta_z(\omega_1)+\theta_z(-\omega_1)-\theta_z(-\omega_2). \tag{6.2}$$

From Fig. 3 and (6.2) it is obvious that $\Theta_z(\Omega)$ is a monotonically increasing function of $\Delta\omega \triangleq \omega_2-\omega_1$. For fixed values of ω_1 and z, a simple calculation reveals that

$$\frac{\partial\Theta_z(\Omega)}{\partial\Delta\omega}=x\left\{\left(\frac{1}{r_z(\omega_1+\Delta\omega)}\right)^2+\left(\frac{1}{r_z(-(\omega_1+\Delta\omega))}\right)^2\right\} \tag{6.3}$$

where (see Fig. 2)

$$r_s(\omega)\triangleq\sqrt{x^2+(y-\omega)^2},\qquad s=x+jy. \tag{6.4}$$

Equation (6.3) confirms that the severity of the tradeoff imposed by the integral relation (2.12) and estimated by the bound (4.5) becomes increasingly worse as the length of the frequency interval over which a given level of sensitivity reduction is desired is increased. For a real zero, the two terms on the right-hand side of (6.3) are equal in magnitude and are monotonically decreasing functions of $\Delta\omega$. For a complex zero in the upper half plane, the first term dominates, is an increasing function of $\Delta\omega$ until $r_z(\omega_1+\Delta\omega)=\mathrm{Im}\,[z]$, and decreases thereafter (Fig. 3). The second term

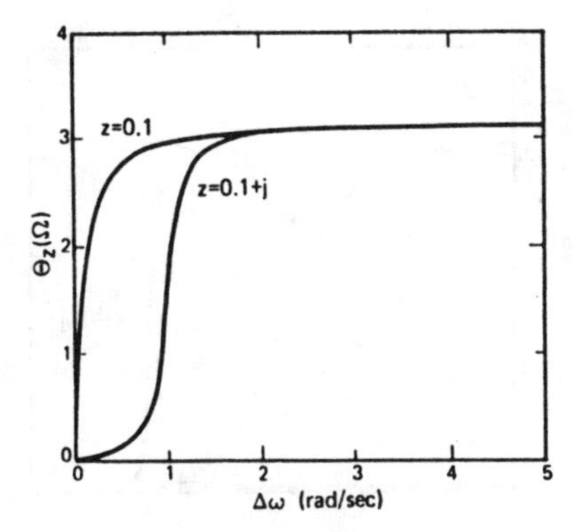

Fig. 10. Weighted length of $\Omega=[0,\ \Delta\omega]$.

is monotonically decreasing. These observations indicate that the greatest incremental degradation in performance due to an incremental increase in the length of an interval of sensitivity reduction occurs for values of $\Delta\omega$ such that ω_2 is in the vicinity of $y=\mathrm{Im}\,[z]$. This is verified for the frequency interval $\omega_1=0$, $\Delta\omega=\omega_2$ and zeros at $s=0.1$ and $s=0.1+j$ in Fig. 10.

Another interesting result is obtained by fixing the length of the frequency interval and varying the location of the interval relative to the zero. If $\Delta\omega<|z|$, then from Fig. 3 it follows that as ω_1 is increased from 0 to ∞ the weighted length of the frequency interval increases and then decreases. This is verified quantitatively from

$$\frac{\partial\Theta_z(\Omega)}{\partial\omega_1}=x\left\{\left(\frac{1}{r_z(\omega_1+\Delta\omega)}\right)^2-\left(\frac{1}{r_z(\omega_1)}\right)^2\right\}$$
$$+x\left\{\left(\frac{1}{r_z(-(\omega_1+\Delta\omega))}\right)^2-\left(\frac{1}{r_z(-\omega_1)}\right)^2\right\}. \tag{6.5}$$

Again, for a real zero the two bracketed terms in (6.5) are equal and negative. For a complex zero in the upper half plane the second term is negative. The first term is monotonically decreasing and equal to zero for $(\omega_1+\omega_2)/2=y$. Together, the contributions of the two terms indicate that the weighted length of the frequency interval reaches a maximum when the midpoint of the frequency interval is somewhat less than the imaginary component of the zero, with corresponding effect on the difficulty of achieving suitable sensitivity performance. This is illustrated in Fig. 11 for a frequency interval $\Omega=[\omega_1,\ \omega_1+1]$ and zeros at $s=1$ and $s=1+10j$.

It is also interesting to consider the effect of varying the location of the zero relative to a fixed frequency interval. Of course, plant zeros cannot be varied in practice; however, this analysis can provide information as to which of several zeros causes the most difficulty in design.

A simple calculation shows that, for $z=x+jy$,

$$\frac{\partial\Theta_z(\Omega)}{\partial x}=\left\{\frac{(y-\omega_2)}{r_z(\omega_2)^2}-\frac{(y-\omega_1)}{r_z(\omega_1)^2}\right\}-\left\{\frac{(y+\omega_2)}{r_z(-\omega_2)^2}-\frac{(y+\omega_1)}{r_z(-\omega_1)^2}\right\}. \tag{6.6}$$

For a real zero (6.6) reduces to

$$\frac{\partial\Theta_z(\Omega)}{\partial x}=\frac{2(\omega_2-\omega_1)(-x^2+\omega_1\omega_2)}{r_x(\omega_2)^2r_x(\omega_1)^2}. \tag{6.7}$$

From Fig. 3 it is clear that as x is increased from zero, $\Theta_z(\Omega)$ first increases and then decreases. Equation (6.7) reveals that the maximum value of $\Theta_z(\Omega)$ is achieved for $x=\sqrt{\omega_1\omega_2}$. For a complex zero in the upper half plane, each of the two terms in (6.6) increases and then decreases as x is increased from zero. The first term is zero at $x=\sqrt{(y-\omega_1)(y-\omega_2)}$; the second term is

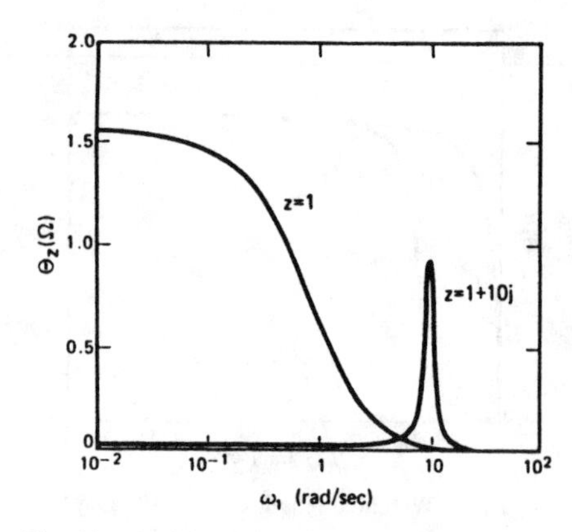

Fig. 11. Weighted length of $\Omega = [\omega_1, \omega_1 + 1]$.

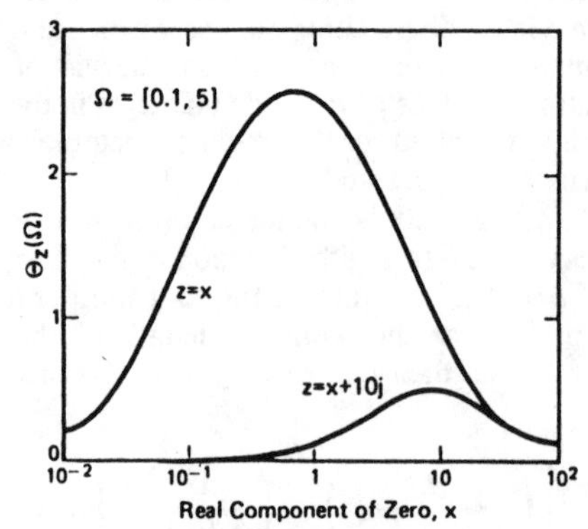

Fig. 12. Weighted length of frequency interval as a function of real component of zero.

zero at $x = \sqrt{(y+\omega_1)(y+\omega_2)}$. Thus the maximum value of $\Theta_z(\Omega)$ is achieved at some intermediate value of x, which could be determined explicitly from (6.6). These results are illustrated in Fig. 12 for the frequency interval $\Omega = [0.1, 5]$ and zeros at $s = x$ and $s = x + 10j$.

Another simple calculation shows that

$$\frac{\partial\Theta_z(\Omega)}{\partial y} = x\left\{\frac{-1}{r_z(\omega_2)^2} + \frac{1}{r_z(\omega_1)^2}\right\} + x\left\{\frac{1}{r_z(-\omega_2)^2} - \frac{1}{r_z(-\omega_1)^2}\right\}$$

$$= \frac{x(\omega_2-\omega_1)(\omega_2+\omega_1-2y)}{r_z(\omega_2)^2 r_z(\omega_1)^2} - \frac{x(\omega_2-\omega_1)(\omega_2+\omega_1+2y)}{r_z(-\omega_2)^2 r_z(-\omega_1)^2}. \tag{6.8}$$

As y is increased from 0, the second term in (6.8) is negative; the first term is monotonically decreasing and is zero at $y = (\omega_1 + \omega_2)/2$. Note that this result is consistent with (6.5).

VII. Remarks on Multivariable Systems

The purpose of this section is to briefly comment on the constraints that right half plane poles and zeros of a matrix open-loop transfer function impose upon the corresponding matrix sensitivity and complementary sensitivity functions. Although the situation is more complicated than for single-loop systems, some useful results can be obtained using the results of this paper. Again, the results are illustrated for the sensitivity function although analogous statements can be made about the complementary sensitivity function.

Let $L(s) \in \mathbb{C}^{n\times n}$ be a matrix of transfer functions and assume that the feedback system whose sensitivity function is the matrix $S(s) = [I + L(s)]^{-1}$ is stable. A commonly used measure of system sensitivity reduction [2], [13]–[16] is the largest singular value, or matrix two-norm, of the sensitivity matrix. The results of this paper cannot, in general, be applied to singular values. Consider instead functions of the form

$$S_{vu}(s) = v^H S(s) u \tag{7.1}$$

where u and v are *constant* unit vectors in $\mathbb{C}^n$. The response of the system to disturbances entering in the direction spanned by u is given by $S(s)u$. The function $S_{vu}(s)$ is the component of this disturbance response appearing in the output direction spanned by v. Each function of the form (7.1) has the property that

$$|v^H S(j\omega)u| \le \bar{\sigma}[S(j\omega)] = \|S(j\omega)\|_2. \tag{7.2}$$

Thus, to ensure good sensitivity reduction properties at a given frequency, it is necessary (but not sufficient) to ensure that $|v^H S(j\omega)u|$ is small at that frequency for a particular choice of u and v. For stable feedback systems, functions of the form (7.1) are analytic and bounded in the closed right half plane. Hence, the results of this paper can be used to study the effect of right half plane transmission zeros upon the magnitude of $|v^H S(j\omega)u|$ and thus to indirectly study $\bar{\sigma}[S(j\omega)]$.

If $S_{vu}(s)$ has zeros in the right half plane, then these are either isolated or $S_{vu}(s) \equiv 0$. The latter case is trivial. If the zeros are isolated then they can be factored out using a Blaschke product as in (2.10) to form a function

$$\tilde{S}_{vu}(s) = S_{vu}(s) B^{-1}(s) \tag{7.3}$$

which has no right half plane zeros and for which $|\tilde{S}_{vu}(j\omega)| = |S_{vu}(j\omega)|$.

One difference between the functions $S_{vu}(s)$ and a scalar sensitivity function is that $S_{vu}(s)$ can possess right half plane zeros which are *not* due to unstable poles of $L(s)$. For example, let

$$L(s) = \begin{bmatrix} \frac{1}{s+3} & \frac{2(s-2)}{s+3} \\ 0 & \frac{s-2}{s+3} \end{bmatrix}$$

$$u = \begin{bmatrix} 0 \\ 1 \end{bmatrix} \quad v = \begin{bmatrix} \frac{\sqrt{2}}{2} \\ \frac{\sqrt{2}}{2} \end{bmatrix}. \tag{7.4}$$

Then the function $S_{vu}(s)$ corresponding to (7.4) has a zero at $s = 8$. This type of zero does not necessarily appear as a consequence of internal stability as do zeros at the poles of $P(s)$ and $F(s)$. Nonetheless, zeros of this type can be present in a given design and will be seen to worsen tradeoffs due to the transmission zeros of $L(s)$.

If $L(s)$ has a transmission zero at $s = z$, then it is easily verified that $S(z)$ has an eigenvalue equal to one. Let w be a unit magnitude right eigenvector of $S(z)$ corresponding to this eigenvalue. Then the unit vector u may be written as

$$u = \alpha w + \beta w_\perp \tag{7.5}$$

where α and β are complex scalars $|\alpha|^2 + |\beta|^2 = 1$ and $w_\perp$ is a unit vector orthogonal to w. Then the magnitude of (7.1) at z is

$$|S_{vu}(z)| = |\alpha v^H w + \beta v^H S(z) w_\perp|. \tag{7.6}$$

Lemma A.1 can be applied to show that

$$\pi \log |\alpha v^H w + \beta v^H S(s) w_\perp| + \pi \log |B^{-1}(z)|$$

$$= \int_{-\infty}^{\infty} \log |S_{vu}(j\omega)| \, d\theta_z(\omega). \tag{7.7}$$

If the input direction u is equal to the right eigenvector w, then (7.7) reduces to

$$\pi \log |v^H w| + \pi \log |B^{-1}(z)| = \int_{-\infty}^{\infty} \log |S_{vu}(j\omega)| \, d\theta_z(\omega). \tag{7.8}$$

This equation shows how the presence of a right half plane zero constrains the magnitude of $v^H S(j\omega) w$ provided that v, the output direction of interest, has a nonzero component in the direction of the eigenvector w. If, in fact, $v = w$, then $v^H w = 1$ and (7.8) is similar to the constraint (2.12). If v and w are orthogonal, however, then $v^H S(s) w$ has a zero at $s = z$. If this zero is isolated then it can be removed via the Blaschke product; the value of the resulting function $\tilde{S}_{vu}(s)$ is not constrained at $s = z$. Otherwise, $v^H S(s) w \equiv 0$. In either case there is no constraint imposed upon the weighted integral of $\log |S_{vu}(j\omega)|$. If $0 < |v^H w| \leq 1$, then a constraint is imposed whose severity decreases as $|v^H w|$ decreases.

The purpose of the above discussion is merely to show how some information about the effects of transmission zeros on closed-loop sensitivity can be obtained and is not intended to be complete. For example, results can also be obtained using left eigenvectors of $S(z)$. It should be pointed out that the results of this section are consistent with those of Wall, Doyle, and Harvey [21], who discuss multivariable right half plane zeros in terms of the extra phase lag they produce in certain directions.

VIII. Summary

In this paper we have discussed limitations on feedback design due to right half plane poles and zeros of the open-loop transfer function. These limitations are expressed directly in terms of the magnitude of the sensitivity and complementary sensitivity functions evaluated along the $j\omega$-axis. This form of expressing the limitations should prove useful in that $|S(j\omega)|$ and $|T(j\omega)|$ are directly related to the quality of a feedback design. We have also extended the Bode integral theorem to the case of open-loop unstable systems.

The limitations of Theorems 1-3 were interpreted as imposing tradeoffs among system properties in *different* frequency ranges. It should be pointed out that the origin of these tradeoffs is physical realizability. The property of realizability relevant to the present context is the fact that the Laplace transform of the impulse response of a physical system is a locally analytic function of the complex frequency variable. Thus, the origin, as well as the implications, of the tradeoffs discussed above are significantly different from the well-known tradeoff between performance and robustness at a *single* frequency imposed by the algebraic identity $S(s) + T(s) \equiv 1$.

Appendix

The following lemma follows from a minor modification of the well-known Poisson integral formulas [4], [5] for the recovery of a function analytic in the right half plane from its values on the imaginary axis. The proof found in [5] is modified to show that singularities of $\log f(s)$ at $j\omega$-axis zeros of $f(s)$ do not contribute to the values of the integrals. As the required modification is slight, the proof will only be sketched.

Lemma A.1: Let $f(s)$ be analytic and nonzero in the closed right half plane except for possible zeros on the imaginary axis. Assume that $d^i/ds^i \log f(s)$ is in class $\mathcal{R}$, $i = 0, 1, \cdots$.

Then at each point $s_o = x_o + jy_o$, $x_o > 0$ it follows that

$$\log |f(s_o)| = \frac{1}{\pi} \int_{-\infty}^{\infty} \log |f(j\omega)| \frac{x_o}{x_o^2 + (y_o - \omega)^2} \, d\omega \tag{A1}$$

$$\measuredangle f(s_o) = \frac{1}{\pi} \int_{-\infty}^{\infty} \measuredangle f(j\omega) \frac{x_o}{x_o^2 + (y_o - \omega)^2} \, d\omega \tag{A2}$$

$$\left. \frac{d^i}{ds^i} \log f(s) \right|_{s_o} = \frac{1}{\pi} \int_{-\infty}^{\infty} \left. \frac{d^i}{ds^i} \log f(s) \right|_{s = j\omega} \frac{x_o}{x_o^2 + (y_o - \omega)^2} \, d\omega. \tag{A3}$$

Proof: Define the contour C_δ to be the imaginary axis traversed from $+j\infty$ to $-j\infty$ with semicircular indentations of radius δ into the right half plane at the $j\omega$-axis zeros of $f(s)$. Then $f(s)$ is analytic and nonzero to the right of C_δ. It follows that $\log f(s)$ and its derivatives are analytic in this region also.

The proof of the Poisson integral formulas in [5] may be followed to show that for each point s_o in the open right half plane there exists δ sufficiently small so that

$$\log f(s_o) = \frac{1}{j\pi} \int_{C_\delta} \log f(\zeta) \frac{x_o}{(\zeta - s_o)(\zeta + \bar{s}_o)} \, d\zeta. \tag{A4}$$

A straightforward calculation shows that in the limit as $\delta \to 0$ the integrals taken around the semicircular indentations vanish. Thus, setting $\zeta = j\omega$ in (A4) yields

$$\log f(s_o) = \frac{1}{\pi} \int_{-\infty}^{\infty} \log f(j\omega) \frac{x_o}{x_o^2 + (y_o - \omega)^2} \, d\omega. \tag{A5}$$

Similarly,

$$\left. \frac{d^i}{ds^i} \log f(s) \right|_{s_o} = \frac{1}{\pi} \int_{-\infty}^{\infty} \left. \frac{d^i}{ds^i} \log f(s) \right|_{s = j\omega} \frac{x_o}{x_o^2 + (y_o - \omega)^2} \, d\omega \tag{A6}$$

where the improper integrals are defined using Cauchy principal values [5, pp. 203-204]. The result follows by taking real and imaginary parts of (A5). ■

Proof of Theorem 1: Follows from Lemma A.1 setting $f(s) = \tilde{S}(s)$ and noting $\tilde{S}(z) = B_p^{-1}(z)$. ■

Proof of Theorem 2: Follows from Lemma A.1 by setting $f(s) = \tilde{T}(s)$ and noting $\tilde{T}(p) = B_z^{-1}(p) e^{p\tau}$. ■

Theorem 3 could be verified directly using contour integration in the right half plane with branch cuts extending from the zeros of $S(s)$ to infinity deleted. Care must be taken to deal correctly with the fact that $\measuredangle S(s)$ jumps by a multiple of 2π across these cuts. (This led to the incorrect result in [1].) An alternative procedure is to prove Theorem 3 as a limiting case of the integral relation (A1).

Proof of Theorem 3: Evaluating (A1) for $f(s) = \tilde{S}(s)$ and $s = x > 0$ yields [noting conjugate symmetry of $S(j\omega)$ implies $S(j\omega) = S(-j\omega)$]

$$\log |\tilde{S}(x)| = \frac{2}{\pi} \int_0^{\infty} \log |S(j\omega)| \frac{x}{x^2 + \omega^2} \, d\omega. \tag{A7}$$

Observing that $\log |S(j\omega)| x^2/(x^2 + \omega^2)$ converges to $\log |S(j\omega)|$ pointwise as x approaches infinity suggests that

$$\lim_{x \to +\infty} x \log |\tilde{S}(x)| = \lim_{x \to +\infty} \frac{2}{\pi} \int_0^{\infty} \log |S(j\omega)| \frac{x^2}{x^2 + \omega^2} \, d\omega \tag{A8}$$

might be used to evaluate

$$\int_0^{\infty} \log |S(j\omega)| \, d\omega. \tag{A9}$$

The proof of this conjecture consists of three parts. First it is shown that (A9) may be approximated arbitrarily closely by $\int_0^{\bar{\omega}} \log |S(j\omega)| d\omega$ for $\bar{\omega}$ sufficiently large. Next it is shown that for $\omega \in$

$[0, \bar{\omega}]$ the sequence $n^2/(n^2 + \omega^2)$ converges uniformly to 1 as n approaches infinity. Together these facts show that the limit on the right-hand side of (A8) is finite and equal to (A9). Finally, the limit on the left-hand side of (A8) is evaluated.

The power series expansion of $\log S(s) = -\log[1 + L(s)]$ for $|L(s)| < 1$ is given by [7, p. 158]

$$\log S(s) = -L(s) + \frac{L^2(s)}{2} + \text{higher order terms.}$$

Thus, by (3.2) $\omega \log |S(j\omega)|$ approaches zero at infinity. This implies that there exists a frequency ω_o and positive constants M_o and δ such that $|\log|S(j\omega)|| \leq M_o/\omega^{1+\delta}$ for $\omega > \omega_o$. For $\bar{\omega} > \omega_o$,

$$\int_{\bar{\omega}}^{\infty} |\log| S(j\omega)|| \, d\omega \leq \int_{\bar{\omega}}^{\infty} \frac{M_o}{\omega^{1+\delta}} \, d\omega = \frac{M_o}{\delta \bar{\omega}^{\delta}}. \tag{A10}$$

From (A10) it follows that for any $\epsilon > 0$ there exists a frequency $\bar{\omega}$ such that

$$\int_{\bar{\omega}}^{\infty} \log |S(j\omega)| \, d\omega < \epsilon. \tag{A11}$$

On the interval $\omega \in [0, \bar{\omega}]$ it is easy to verify that the sequence $n^2/(n^2 + \omega^2)$ converges uniformly to one as n approaches infinity. If $L(s)$ has no $j\omega$-axis poles, then uniform convergence of the integrand suffices to show that [8, p. 71]

$$\lim_{x\to\infty} \int_0^{\bar{\omega}} \log |S(j\omega)| \frac{x^2}{x^2+\omega^2} \, d\omega = \int_0^{\bar{\omega}} \log |S(j\omega)| \, d\omega. \tag{A12}$$

(If $L(s)$ has poles in $[0, \bar{\omega}]$, then it is necessary to consider indentations into the right half plane as in the proof of the lemma; these details are omitted.)

Together, (A11) and (A12) imply that

$$\lim_{x\to\infty} \int_0^{\infty} \log |S(j\omega)| \frac{x^2}{x^2+\omega^2} \, d\omega = \int_0^{\infty} \log |S(j\omega)| \, d\omega. \tag{A13}$$

It remains to evaluate $\lim_{x\to\infty} x \log |\tilde{S}(x)|$. By definition

$$\log |\tilde{S}(x)| = \log |B_p^{-1}(x)| + \log |S(x)|$$

$$= \sum_{i=1}^{N_p} \log \left|\frac{\bar{p}+x}{p-x}\right| + \log |S(x)|. \tag{A14}$$

Since $\lim_{x\to\infty} x \log |S(x)| = 0$, it follows that

$$\lim_{x\to\infty} x \log |\tilde{S}(x)| = \sum_{i=1}^{N_p} \lim_{x\to\infty} x \log \left|\frac{\bar{p}_i+x}{p_i-x}\right|. \tag{A15}$$

From the power series expansions [7, p. 158]

$$\log \left(1+\frac{\bar{p}}{x}\right) = \frac{\bar{p}}{x} - \frac{1}{2}\left(\frac{\bar{p}}{x}\right)^2 + \cdots \quad \left|\frac{\bar{p}}{x}\right| < 1$$

$$\log \left(1-\frac{p}{x}\right) = -\frac{p}{x} - \frac{1}{2}\left(\frac{p}{x}\right)^2 + \cdots \quad \left|\frac{p}{x}\right| < 1$$

it follows that for $x > |p|$

$$\log \frac{\bar{p}_i+x}{x-p_i} = \log \left(\frac{1+\frac{\bar{p}_i}{x}}{1-\frac{p_i}{x}}\right)$$

$$= \frac{\bar{p}_i+p_i}{x} + \text{higher order terms.}$$

Thus,

$$\lim_{x\to\infty} x \log \left|\frac{\bar{p}_i+x}{p_i-x}\right| = 2 \operatorname{Re}[p_i]. \tag{A16}$$

Substituting (A16) into (A15) yields

$$\lim_{x\to\infty} x \log |\tilde{S}(x)| = \sum_{i=1}^{N_p} 2 \operatorname{Re}[p_i]. \tag{A17}$$

Substituting (A17) into (A8) and using (A13) yields

$$\pi \sum_{i=1}^{N_p} \operatorname{Re}[p_i] = \int_0^{\infty} \log |S(j\omega)| \, d\omega. \quad \blacksquare$$

Proof of Corollary: At each frequency $\omega \geq \omega_c$

$$|S(j\omega)| \leq \left|\frac{1}{1+L(j\omega)}\right| \leq \frac{1}{1-|L(j\omega)|}.$$

This follows since $|L(j\omega)| < 1$ for $\omega \geq \omega_c$ by (3.4). Moreover, since

$$|L(j\omega)| \leq \frac{M}{\omega^{1+k}} < 1 \quad \text{for } \omega > \omega_c$$

$$|S(j\omega)| \leq \frac{1}{1-\frac{M}{\omega^{1+k}}}.$$

Expanding

$$\log \left(\frac{1}{1-\frac{M}{\omega^{1+k}}}\right)$$

in a power series yields (for $\omega > \omega_c$)

$$\log \left[\frac{1}{1-\frac{M}{\omega^{1+k}}}\right] = \sum_{n=1}^{\infty} \frac{\left(\frac{M}{\omega^{1+k}}\right)^n}{n} = \frac{1}{\omega^{1+k}} \sum_{n=1}^{\infty} \frac{\left(\frac{M}{\omega^{1+k}}\right)^n \omega^{1+k}}{n}.$$

The infinite sum attains its maximum for $\omega = \omega_c$. Define

$$M_o \triangleq \sum_{n=1}^{\infty} \frac{\left(\frac{M}{\omega^{1+k}}\right)^n \omega^{1+k}}{n} = \log \left[\frac{1}{1-\frac{M}{\omega_c^{1+k}}}\right] \omega_c^{1+k}.$$

Thus, $\omega > \omega_c$ implies that

$$\log |S(j\omega)| \leq \frac{M_o}{\omega^{1+k}}.$$

Thus,

$$\int_{\omega_c}^{\infty} \log |S| \, d\omega \leq \int_{\omega_c}^{\infty} \frac{M_o}{\omega^{1+k}} \, d\omega$$

$$= \frac{M_o}{k\omega_c^k}.$$

Substituting M_o:

$$\frac{M_o}{k\omega_c^k} = \frac{\log\left[\dfrac{1}{1-\dfrac{M}{\omega_c^{1+k}}}\right]\omega_c^{1+k}}{k\omega_c^k}.$$

Since $M/\omega_c^{1+k} = \epsilon$, this reduces to

$$\int_{\omega_c}^{\infty} \log |S(j\omega)| \, d\omega \leq \frac{\log\left[\dfrac{1}{1-\epsilon}\right]\omega_c}{k}. \qquad \blacksquare$$

References

[1] I. M. Horowitz, *Synthesis of Feedback Systems.* New York: Academic, 1963.

[2] J. C. Doyle and G. Stein, "Multivariable feedback design: Concepts for a classical/modern synthesis," *IEEE Trans. Automat. Contr.*, vol. AC-26, Feb. 1981.

[3] H. W. Bode, *Network Analysis and Feedback Amplifier Design.* Princeton, NJ: Van Nostrand, 1945.

[4] A. E. Guillemin, *The Mathematics of Circuit Analysis.* New York: Wiley, 1949.

[5] N. Levinson and R. M. Redheffer, *Complex Variables.* San Francisco, CA: Holden-Day, 1970.

[6] B. A. Francis and G. Zames, "On optimal sensitivity theory for SISO feedback systems," *IEEE Trans. Automat. Contr.*, vol. AC-29, Jan. 1984.

[7] K. Knopp, *Infinite Sequences and Series.* New York: Dover, 1956.

[8] J. B. Conway, *Functions of One Complex Variable,* 2nd ed. New York: Springer-Verlag, 1978.

[9] J. B. Cruz, Jr. and W. R. Perkins, "A new approach to the sensitivity problem in multivariable feedback system design," *IEEE Trans. Automat. Contr.*, vol. AC-9, 1964.

[10] G. Zames, "Feedback and optimal sensitivity: Model reference transformations, Multiplicative seminorms, and approximate inverses," *IEEE Trans. Automat. Contr.*, vol. AC-26, 1981.

[11] G. Zames and B. A. Francis, "Feedback, minimax sensitivity, and optimal robustness," *IEEE Trans. Automat. Contr.*, vol. AC-28, May 1983.

[12] J. S. Freudenberg, D. P. Looze, and J. B. Cruz, Jr., "Robustness analysis using singular value sensitivities," *Int. J. Contr.*, vol. 35, no. 1, 1982.

[13] J. B. Cruz, Jr., J. S. Freudenberg, and D. P. Looze, "A relationship between sensitivity and stability of multivariable feedback systems," *IEEE Trans. Automat. Contr.*, vol. AC-26, Feb. 1981.

[14] M. B. Safonov, A. J. Laub, and G. L. Hartmann, "Feedback properties of multivariable systems: The role and use of the return difference matrix," *IEEE Trans. Automat. Contr.*, vol. AC-26, Feb. 1981.

[15] N. A. Lehtomaki, N. R. Sandell, Jr., and M. Athans, "Robustness results in linear-quadratic Gaussian based multivariable control design," *IEEE Trans. Automat. Contr.*, vol. AC-26, Feb. 1981.

[16] I. Postlethwaite, J. M. Edmunds, and A. G. J. MacFarlane, "Principal gains and principal phases in the analysis of linear multivariable feedback systems," *IEEE Trans. Automat. Contr.*, vol. AC-26, Feb. 1981.

[17] H. Kwakernaak, "Robustness optimization of linear feedback systems," in *Proc. 22nd IEEE Conf. Decision Contr.*, Dec. 1983, pp. 618–624.

[18] D. C. Youla, H. A. Jabr, and J. J. Bongiorno, "Modern Wiener–Hopf design of optimal controllers—Parts I and II," *IEEE Trans. Automat. Contr.*, vol. AC-21, 1976.

[19] M. G. Safonov and B. S. Chen, "Multivariable stability margin optimization with decoupling and output regulation," in *Proc. 21st IEEE Conf. Decision Contr.*, Dec. 1982, pp. 616–622.

[20] P. M. Frank, *Introduction to System Sensitivity Theory.* New York: Academic, 1978.

[21] J. E. Wall, Jr., J. C. Doyle, and C. A. Harvey, "Tradeoffs in the design of multivariable feedback systems," in *Proc. 18th Allerton Conf.*, 1980.

[22] V. H. L. Cheng and C. A. Desoer, "Limitations on the closed-loop transfer function due to right-half plane transmission zeros of the plant," *IEEE Trans. Automat. Contr.*, vol. AC-25, Dec. 1980.

Part II
Robust Stabilization

Algebraic Design Techniques for Reliable Stabilization

M. VIDYASAGAR, SENIOR MEMBER, IEEE, AND N. VISWANADHAM

Abstract—In this paper we study two problems in feedback stabilization. The first is the simultaneous stabilization problem, which can be stated as follows. Given plants $G_0, G_1, \cdots, G_l$, does there exist a single compensator C that stabilizes all of them? The second is that of stabilization by a stable compensator, or more generally, a "least unstable" compensator. Given a plant G, we would like to know whether or not there exists a stable compensator C that stabilizes G; if not, what is the smallest number of right half-place poles (counted according to their McMillan degree) that any stabilizing compensator must have? We show that the two problems are equivalent in the following sense. The problem of simultaneously stabilizing $l+1$ plants can be reduced to the problem of simultaneously stabilizing l plants using a stable compensator, which in turn can be stated as the following purely algebraic problem. Given $2l$ matrices $A_1, \cdots, A_l, B_1, \cdots, B_l$, where A_i, B_i are right-coprime for all i, does there exist a matrix M such that $A_i + MB_i$ is unimodular for all i? Conversely, the problem of simultaneously stabilizing l plants using a stable compensator can be formulated as one of simultaneously stabilizing $l+1$ plants.

The problem of determining whether or not there exists an M such that $A + BM$ is unimodular, given a right-coprime pair (A, B), turns out to be a special case of a question concerning a matrix division algorithm in a proper Euclidean domain. We give an answer to this question, and we believe this result might be of some independent interest. We show that, given two $n \times m$ plants G_0 and G_1, we can generically stabilize them simultaneously provided *either* n or m is greater than one. In contrast, simultaneous stabilizability of two single-input–single-output plants, g_0 and g_1, is *not* generic.

I. Introduction

IN THIS paper we study two problems in feedback stabilization. The first is the simultaneous stabilization problem, which can be stated as follows. Given plants $G_0, G_1, \cdots, G_l$, does there exist a single compensator C that stabilizes all of them? This can be viewed as a problem of reliable stabilization, where G_0 is the nominal description of a particular plant, which changes to $G_1, \cdots, G_l$ in the case of some failures (e.g., failure of sensors, severance of loops or software breakdown). The second problem tackled in this paper is that of stabilization by a stable compensator or, more generally, a "least unstable" compensator. Given a plant G, we would like to know whether or not there exists a stable compensator C that stabilizes G; if not, what is the smallest number of right half-plane poles (counted according to their McMillan degree) that any stabilizing compensator must have?

We show that the two problems are equivalent in the following sense. The problem of simultaneously stabilizing $l+1$ plants can be reduced to the problem of simultaneously stabilizing l plants using a stable compensator, which in turn can be stated as the following purely algebraic problem. Given $2l$ matrices $A_1, \cdots, A_l, B_1, \cdots, B_l$, where A_i, B_i are right coprime for all i, does there exist a matrix M such that $A_i + MB_i$ is unimodular for all i? Conversely, the problem of simultaneously stabilizing l plants using a stable compensator can be formulated as one of simultaneously stabilizing $l+1$ plants. All this is done in Section III.

The problem of determining whether or not there exists an M such that $A + BM$ is unimodular, given a right-coprime pair (A, B), turns out to be a special case of a question concerning a matrix division algorithm in a proper Euclidean domain. We give an answer to this question in Section IV, and we believe this result might be of some independent interest. Using this result, in Section V we study the problem of stabilization using a "least unstable" compensator.

One of the surprising aspects of the problems studied here is the generic nature of the solutions. We show that, given two $n \times m$ plants G_0 and G_1, we can generically stabilize them simultaneously provided *either* n or m is greater than one. In other words, even if the given G_0 and G_1 cannot be simultaneously stabilized, there exist plants $\overline{G}_1$ arbitrarily close to G_1 such that $\overline{G}_1, G_0$ can be simultaneously stabilized. In contrast, simultaneous stabilizability

Manuscript received March 11, 1981; revised December 15, 1981. Paper recommended by E. W. Kamen, Chairman of the Linear Systems Committee. This work was supported by the Natural Sciences and Engineering Research Council of Canada under Grant A-1240.

M. Vidyasagar is with the Department of Electrical Engineering, University of Waterloo, Waterloo, Ont., Canada N2L 3G1.

N. Viswanadham was on leave at the Department of Electrical Engineering, University of Waterloo, Waterloo, Ont., Canada N2L 3G1. He is with the School of Automation, Indian Institute of Science, Bangalore 560 012 India.

Reprinted from *IEEE Trans. Automat. Contr.*, vol. AC-27, no. 5, pp. 1085–1095, Oct. 1982.

of two single-input–single-output plants g_0 and g_1 is *not* generic. This is shown in Section VI.

It turns out that the results given here for simultaneous stabilizability can be readily extended to the case of an arbitrary (i.e., not necessarily finite) family of plants. However, the resulting necessary and sufficient conditions are not computationally verifiable, except in special cases (e.g., the case where $l=1$, i.e., there are only two plants, or, where all plants are "close" to a nominal plant). As far as we are aware, the problem of simultaneous stabilization has been studied only in [1], [18]. Our results were derived independently of [1], [18] and generalize those of [1] to the case of multiinput–multioutput systems.

The problem of stabilizing a single plant using a stable compensator is studied in [2], where necessary and sufficient conditions are given for the existence of a stable stabilizing compensator for a given plant. These conditions are very elegant and easily verifiable, involving some interlacing of plant poles and zeros, and they figure in an important way in our proofs. The question of a "least unstable" stabilizing compensator has not previously been addressed in the literature, but is resolved here.

To keep the exposition simple, we deal for the most part with plants whose transfer matrices contain only rational functions of s, and define a "stable" transfer function to be a proper rational function whose poles are in the open left half-plane. However, the generalizations to the case of distributed systems and/or systems whose transfer functions have poles in a prescribed region of the complex plane (not necessarily the open left half-plane) are straightforward and are indicated in Section VII.

II. Background and Notation

Throughout the paper we let $R(s)$ denote the set of rational functions in s with real coefficients, and we let $\mathcal{H}$ denote the subset of $R(s)$ consisting of proper rational functions whose poles lie in the open left half-plane. In the case of lumped linear time-invariant systems, $\mathcal{H}$ consists of precisely the transfer functions of BIBO stable systems. The set $\mathcal{H}$ is a ring; thus, if two functions f_1 and f_2 belong to $\mathcal{H}$, so do their difference and product.[1] The ring $\mathcal{H}$ is clearly commutative ($f_1f_2=f_2f_1$) and is an integral domain ($f_1f_2=0$ implies $f_1=0$ or $f_2=0$). The set $R(s)$ is the quotient field generated by $\mathcal{H}$; i.e., every $g\in R(s)$ can be written as $g=f_1/f_2$, $f_1, f_2\in\mathcal{H}$, $f_2\neq 0$, and conversely, every ratio f_1/f_2 where $f_1, f_2\in\mathcal{H}, f_2\neq 0$, belongs to $R(s)$.

A function f in $\mathcal{H}$ is called a *unit* if its reciprocal belongs to $\mathcal{H}$. Clearly the units in $\mathcal{H}$ are the properly invertible minimum phase transfer functions.

Given any rational function h, we can find two functions f and g in $\mathcal{H}$ such that $h=f/g$, and such that f and g are relatively prime (i.e., 1 is a greatest common divisor of f and g). Such a pair (f, g) is called a *coprime factorization* of h. It is essential to recognize that we are doing factorizations in the ring $\mathcal{H}$, and not in the ring of polynomials. In other words, we are expressing a given rational function h as a ratio of proper stable transfer functions with no common factors, rather than as a ratio of polynomials with no common zeros.

We let $\mathcal{H}^{n\times m}$ denote the set of $n\times m$ matrices whose elements all belong to $\mathcal{H}$. Thus $\mathcal{H}^{n\times m}$ is the set of transfer functions of BIBO stable lumped linear time-invariant systems with m inputs and n outputs. A matrix $F\in\mathcal{H}^{n\times n}$ is *unimodular* if its inverse belongs to $\mathcal{H}^{n\times n}$. Clearly F is unimodular if and only if $\det F$ is a unit.

Given any $H\in R^{n\times m}(s)$ (which means H is an $n\times m$ matrix whose elements are rational functions of s), we can find matrices $N\in\mathcal{H}^{n\times m}$, $D\in\mathcal{H}^{m\times m}$ such that $H(s)=N(s)[D(s)]^{-1}$ and the matrices N, D are *right coprime*, i.e., there exist $P\in\mathcal{H}^{m\times n}$, $Q\in\mathcal{H}^{m\times m}$ such that

$$P(s)N(s)+Q(s)D(s)=I_m, \qquad \forall s. \tag{2.1}$$

Similarly, we can find $\tilde{N}\in\mathcal{H}^{n\times m}$, $\tilde{D}\in\mathcal{H}^{n\times n}$, $\tilde{P}\in\mathcal{H}^{m\times n}$, $\tilde{Q}\in\mathcal{H}^{n\times n}$ such that $H(s)=[\tilde{D}(s)]^{-1}\tilde{N}(s)$, and

$$\tilde{N}(s)\tilde{P}(s)+\tilde{D}(s)\tilde{Q}(s)=I_n, \qquad \forall s. \tag{2.2}$$

We refer to (N, D) as a *right-coprime factorization* (*r.c.f.*) of H and to $(\tilde{D}, \tilde{N})$ as a *left-coprime factorization* (*l.c.f.*) of H.

If (N, D) is an r.c.f. of $H\in R^{n\times m}(s)$, so is (NU, DU) whenever U is an $m\times m$ unimodular matrix. The converse is also true, i.e., if $(N_1, D_1), (N_2, D_2)$ are two r.c.f.'s of H, then there exists a unimodular matrix U such that $N_1=N_2U$, $D_1=D_2U$. Similar statements apply to l.c.f.'s.

Given an $n\times m$ rational matrix H, it is possible to select an r.c.f. (N, D), an l.c.f. $(\tilde{D}, \tilde{N})$, and matrices $P, Q, \tilde{P}, \tilde{Q}$ such that

$$\begin{bmatrix} Q & P \\ -\tilde{N} & \tilde{D} \end{bmatrix}\begin{bmatrix} D & -\tilde{P} \\ N & \tilde{Q} \end{bmatrix}=\begin{bmatrix} I_m & 0 \\ 0 & I_n \end{bmatrix}. \tag{2.3}$$

For further discussion of these topics, see [5]–[8].

Next, we state and prove a result concerning coprime factorizations of a strictly proper matrix.

Lemma 2.1: Let H be strictly proper, let (N, D) be any r.c.f. of H, and let P, Q be any matrices over $\mathcal{H}$ such that $PN+QD=I_m$. Then, i) N is strictly proper, and ii) $\det D(\infty)\neq 0$, $\det Q(\infty)\neq 0$ (i.e., both $\det D$ and $\det Q$ have relative degree zero).

Proof: Since H is strictly proper, we can write every element of H as $h_{ij}=\bar{n}_{ij}/\bar{d}_{ij}$, where $\bar{n}_{ij}, \bar{d}_{ij}\in\mathcal{H}$, $\bar{n}_{ij}$ is strictly proper, and $\bar{d}_{ij}(\infty)\neq 0$. One way of doing this is as follows. If

$$h_{ij}(s)=\frac{\alpha_{ij}(s)}{\beta_{ij}(s)}, \tag{2.4}$$

let

$$\bar{n}_{ij}(s)=\frac{\alpha_{ij}(s)}{(s+1)^{\gamma_{ij}}},$$

$$\bar{d}_{ij}(s)=\frac{\beta_{ij}(s)}{(s+1)^{\gamma_{ij}}},$$

$$\gamma_{ij}=\deg\beta_{ij}. \tag{2.5}$$

[1]See [3], [4] for the requisite background in abstract algebra.

Now let $\bar{d} = \Pi_i \Pi_j \bar{d}_{ij}$; then $\bar{d} \in \mathcal{H}$ and $\bar{d}(\infty) \neq 0$. Moreover, we can write $H(s) = N_1(s)(\bar{d}(s))^{-1}$, where $N_1 \in \mathcal{H}^{n \times m}$ is strictly proper. Of course, the matrices N_1 and $\bar{d}I_m$ need not be right coprime. However, we can extract a greatest common right divisor $R \in \mathcal{H}^{m \times m}$ using standard methods [8, Theorem 2.1], [11, pp. 30–35]. Let $N_1 = \bar{N}R$, $\bar{d}I_m = \bar{D}R$; then $(\bar{N}, \bar{D})$ is an r.c.f. of H. Since $\det \bar{D}(\infty) \cdot \det R(\infty) = (\bar{d}(\infty))^m \neq 0$, it follows that $\det R(\infty) \neq 0$. Hence $\bar{N}(\infty) = N_1(\infty)[R(\infty)]^{-1} = 0$, since N_1 is strictly proper; this shows that $\bar{N}$ is also strictly proper. In this way we have constructed *one* r.c.f. $(\bar{N}, \bar{D})$ of H such that $\bar{N}$ is strictly proper. Now let (N, D) be *any* r.c.f. of H. Then there exists a unimodular matrix U such that $N = \bar{N}U$, $D = \bar{D}U$. Hence $N(\infty) = \bar{N}(\infty)U(\infty) = 0$, which shows that N is strictly proper. This completes the proof of i).

To prove ii), let P, Q be any matrices over $\mathcal{H}$ such that

$$P(s)N(s) + Q(s)D(s) = I_m, \qquad \forall s. \tag{2.6}$$

Letting $s \to \infty$ in (2.6) gives $Q(\infty)D(\infty) = I_m$, which shows that $\det D(\infty) \neq 0$, $\det Q(\infty) \neq 0$. □

A similar result holds for l.c.f.'s.

Now we briefly summarize some results on feedback stability, taken from [8]. Consider the feedback system shown in Fig. 1, where G and C are rational matrices of order $n \times m$ and $m \times n$, respectively, and assume that $\det(I_n + GC) \not\equiv 0$ (otherwise the system is not well-defined). Then it is easy to verify that

$$\begin{bmatrix} e_1 \\ e_2 \end{bmatrix} \begin{bmatrix} (I_m + CG)^{-1} & -C(I_n + GC)^{-1} \\ G(I_m + CG)^{-1} & (I_n + GC)^{-1} \end{bmatrix} \begin{bmatrix} u_1 \\ u_2 \end{bmatrix} \tag{2.7}$$

or, more concisely,

$$e = Hu. \tag{2.8}$$

We will say that the pair (G, C) *is stable* if $H \in \mathcal{H}^{(n+m) \times (n+m)}$. (It is necessary to consider all four transfer functions in (2.4) because any three of these can be BIBO stable while the fourth is not; see [9] for some examples.) We say that C *stabilizes* G if (G, C) is stable. Note that (2.7) is essentially symmetric in G and C, so that C stabilizes G if and only if G stabilizes C.

Next, we state without proof the necessary and sufficient conditions for a pair (G, C) to be stable, which are taken from [8], [10].

Lemma 2.2: Let $(N, D), (\tilde{D}, \tilde{N})$ be any r.c.f. and l.c.f. of $G \in R^{n \times m}(s)$, and let $(N_C, D_C), (\tilde{D}_C, \tilde{N}_C)$ be any r.c.f. and l.c.f. of $C \in R^{n \times m}(s)$. Suppose $\det(I_n + GC) = \det(I_m + CG) \neq 0$. Then the following conditions are equivalent.

i) The pair (G, C) is stable.

ii) The matrix $\tilde{D}_C D + \tilde{N}_C N$ is unimodular.

iii) The matrix $\tilde{D} D_C + \tilde{N} N_C$ is unimodular.

Corollary 2.2.1: Let $(N, D), (\tilde{D}, \tilde{N})$ be any r.c.f. and l.c.f. of $G \in R^{n \times m}(s)$, and suppose $C \in \mathcal{H}^{m \times n}$. Then the following conditions are equivalent.

i) The pair (G, C) is stable.

ii) $D + CN$ is unimodular.

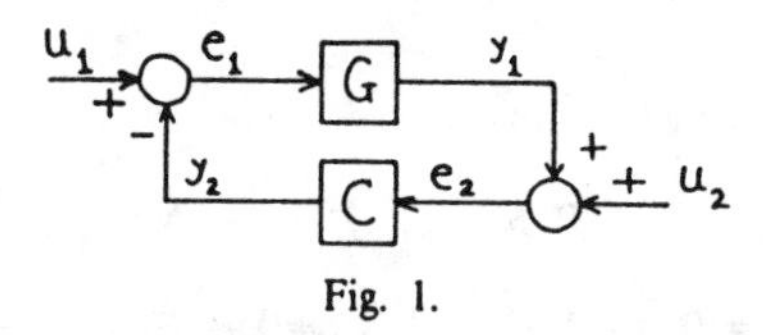

Fig. 1.

iii) $\tilde{D} + \tilde{N}C$ is unimodular.

The next result characterizes all compensators that stabilize a given strictly proper plant. The proof is given in [8, Theorem 3.1].

Lemma 2.3: Let $G \in R^{n \times m}(s)$ be strictly proper, and let $(N, D), (\tilde{D}, \tilde{N})$ be any r.c.f. and l.c.f. of G. Select matrices $P, Q, \tilde{P}, \tilde{Q}$ such that

$$PN + QD = I_m, \qquad \tilde{N}\tilde{P} + \tilde{D}\tilde{Q} = I_n. \tag{2.9}$$

Then

i) every C such that (G, C) is stable is proper;

ii) the set of C such that (G, C) is stable is given by

$$\begin{aligned} \mathcal{C}(G) &= \{(Q - R\tilde{N})^{-1}(P + R\tilde{D}), R \in \mathcal{H}^{m \times n}\} \\ &= \{(\tilde{P} + DS)(\tilde{Q} - NS)^{-1}, S \in \mathcal{H}^{m \times n}\}. \end{aligned} \tag{2.10}$$

Corollary 2.3.1: Suppose $G \in \mathcal{H}^{n \times m}$ is strictly proper. Then

$$\begin{aligned} \mathcal{C}(G) &= \{(I - RG)^{-1}R, R \in \mathcal{H}^{m \times n}\}. \\ &= \{S(I - GS)^{-1}, S \in \mathcal{H}^{m \times n}\}. \end{aligned} \tag{2.11}$$

III. Simultaneous Stabilization

In this section we study the problem of simultaneously stabilizing $l+1$ plants $G_0, G_1, \cdots, G_l$ using the *same* compensator C. We begin by studying the case of two plants G_0 and G_1, and show that G_0, G_1 can be simultaneously stabilized if and only if an associated system can be stabilized using a stable compensator. Since necessary and sufficient conditions for this are known [2], our test for simultaneous stabilizability of two plants is computationally verifiable. We then show that the problem of simultaneously stabilizing $l+1$ plants can be reduced to one of simultaneously stabilizing l plants using a stable compensator. At present, no computable tests are available for the latter, so that this result can only be viewed as a starting point for further work.

We begin with the problem of simultaneously stabilizing two given $n \times m$ strictly proper plants G_0 and G_1.[2] Without loss of generality we assume that for $i = 0, 1$, we have available an r.c.f. (N_i, D_i) and an l.c.f. $(\tilde{D}_i, \tilde{N}_i)$ of G_i, together with matrices $P_i, Q_i, \tilde{P}_i, \tilde{Q}_i$, such that

$$\begin{bmatrix} Q_i & P_i \\ -\tilde{N}_i & \tilde{D}_i \end{bmatrix} \begin{bmatrix} D_i & -\tilde{P}_i \\ N_i & \tilde{Q}_i \end{bmatrix} = I_{n+m}. \tag{3.1}$$

[2] The assumption that G_0 and G_1 are strictly proper is not essential to the theory and is only made to simplify a few expressions. The general case is discussed in Section VII.

Theorem 3.1: Define

$$A_1 = Q_0 D_1 + P_0 N_1, \qquad B_1 = -\tilde{N}_0 D_1 + \tilde{D}_0 N_1. \quad (3.2)$$

Then $\det A_1 \not\equiv 0$, so that A_1^{-1} is well-defined, and A_1, B_1 are right coprime. Moreover, there exists a C that stabilizes both G_0 and G_1 if and only if there exists an $M \in \mathcal{H}^{m\times n}$ that stabilizes $B_1 A_1^{-1}$.

Proof: First, from Lemma 2.1 it follows that $A_1(\infty) = Q_0(\infty)D_1(\infty)$ (since $N_1(\infty)=0$), and that $\det A_1(\infty) \neq 0$ (since $\det Q_0(\infty)\neq 0$ and $\det D_1(\infty)\neq 0$). Hence, A_1^{-1} is well-defined. Next, since

$$\begin{bmatrix} A_1 \\ B_1 \end{bmatrix} = \begin{bmatrix} Q_0 & P_0 \\ -\tilde{N}_0 & \tilde{D}_0 \end{bmatrix} \begin{bmatrix} D_1 \\ N_1 \end{bmatrix}, \quad (3.3)$$

it follows from (3.1) that A_1, B_1 are right coprime.

From Lemma 2.3, there exists a C that stabilizes both G_0 and G_1 if and only if there exist R_0, R_1 in $\mathcal{H}^{m\times n}$ such that

$$(Q_0 - R_0\tilde{N}_0)^{-1}(P_0 + R_0\tilde{D}_0) = (Q_1 - R_1\tilde{N}_1)^{-1}(P_1 + R_1\tilde{D}_1). \quad (3.4)$$

Observe now that $Q_0 - R_0\tilde{N}_0, P_0 + R_0\tilde{D}_0$ are right coprime, and that $Q_1 - R_1\tilde{N}_1, P_1 + R_1\tilde{D}_1$ are right coprime. Hence, (3.4) holds if and only if there exists a unimodular matrix U such that

$$Q_0 - R_0\tilde{N}_0 = U(Q_1 - R_1\tilde{N}_1);$$
$$P_0 + R_0\tilde{D}_0 = U(P_1 + R_1\tilde{D}_1). \quad (3.5)$$

Thus, we have shown that G_0 and G_1 can be simultaneously stabilized if and only if there exist *stable* R_0 *and* R_1 and a *unimodular* U such that (3.5) holds. We now show that this is the case if and only if there exists a stable M such that $A_1 + MB_1$ is unimodular.

To do this, rewrite the two equations in (3.5) as

$$[I \;\; R_0]\begin{bmatrix} Q_0 & P_0 \\ -\tilde{N}_0 & \tilde{D}_0 \end{bmatrix} = U[I \;\; R_1]\begin{bmatrix} Q_1 & P_1 \\ -\tilde{N}_1 & \tilde{D}_1 \end{bmatrix} \quad (3.6)$$

and recall from (3.1) that

$$\begin{bmatrix} Q_1 & P_1 \\ -\tilde{N}_1 & \tilde{D}_1 \end{bmatrix}^{-1} = \begin{bmatrix} D_1 & -\tilde{P}_1 \\ N_1 & \tilde{Q}_1 \end{bmatrix}. \quad (3.7)$$

Multiplying both sides of (3.6) by the matrix in (3.7) gives

$$[I \;\; R_0]\begin{bmatrix} Q_0 & P_0 \\ -\tilde{N}_0 & \tilde{D}_0 \end{bmatrix}\begin{bmatrix} D_1 & -\tilde{P}_1 \\ N_1 & \tilde{Q}_1 \end{bmatrix} = U[I \;\; R_1] \quad (3.8)$$

$$[I \;\; R_0]\begin{bmatrix} A_1 & X_1 \\ B_1 & Y_1 \end{bmatrix} = U[I \;\; R_1] \quad (3.9)$$

as the equations to be satisfied, where the definitions of X_1 and Y_1 are self-evident. We will now show that there exist stable R_0, R_1 and a unimodular U satisfying (3.9) if and only if there exists a stable M such that $A_1 + MB_1$ is unimodular. To prove the "if" part, select M such that $A_1 + MB_1$ is unimodular and let $R_0 = M$, $U = A_1 + MB_1$, and $R_1 = U^{-1}(X_1 + MY_1)$. To prove the "only if" part, select R_0, R_1, U such that (3.9) holds, and let $M = R_0$; then $A_1 + MB_1 = U$ is unimodular.

To complete the proof observe that $M \in \mathcal{H}^{m\times n}$ stabilizes $B_1A_1^{-1}$ if and only if $A_1 + MB_1$ is unimodular. □

During the course of the above proof we have actually characterized *all* compensators that simultaneously stabilize G_0 and G_1. Let $\mathfrak{M}$ denote the set of all $M \in \mathcal{H}^{m\times n}$ such that $A_1 + MB_1$ is unimodular; thus $\mathfrak{M}$ is the set of all stable compensators that stabilize $B_1A_1^{-1}$. Then the set of all compensators that simultaneously stabilize G_0 and G_1 is given by

$$\{(Q_0 - R\tilde{N}_0)^{-1}(P_0 + R\tilde{D}_0), R \in \mathfrak{M}\}. \quad (3.10)$$

In the multivariable case, an explicit expression for the set $\mathfrak{M}$ is not available, but $\mathfrak{M}$ can be explicitly described in the case $m = n = 1$ [17].

To clarify the result contained in Theorem 3.1 we now study the case where one of the systems (say G_0) is stable. Actually, there is no loss of generality in making this assumption. Suppose that we are given two plants $\overline{G}_0$ and $\overline{G}_1$, and we would like to know whether or not they can be simultaneously stabilized. First, we select a compensator $\overline{C}$ that stabilizes $\overline{G}_0$, and define $G_0 = \overline{G}_0(I + \overline{C}\overline{G}_0)^{-1}$, $G_1 = \overline{G}_1(I + \overline{C}\overline{G}_1)^{-1}$. A little reflection will show that i) $\overline{G}_0, \overline{G}_1$ can be simultaneously stabilized if and only if G_0, G_1 can be simultaneously stabilized, and ii) if C stabilizes both G_0 and G_1, then $C + \overline{C}$ stabilizes both $\overline{G}_0$ and $\overline{G}_1$.

Corollary 3.1.1: Suppose G_0 is strictly proper and stable, and G_1 is strictly proper. Then G_0 and G_1 can be simultaneously stabilized if and only if $G_1 - G_0$ can be stabilized by a stable compensator.

Proof: Since G_0 is stable, we can apply Theorem 3.1 with $N_0 = \tilde{N}_0 = G_0$, $\tilde{D}_0 = I$, $D_0 = I$, $P_0 = 0$, $\tilde{P}_0 = 0$, $Q_0 = I$, $\tilde{Q}_0 = I$. This gives $B_1A_1^{-1} = G_1 - G_0$. However, we give a proof independent of Theorem 3.1.

From Corollary 2.3.1 the set of compensators that stabilize G_0 is given by

$$\mathcal{C}(G_0) = \{(I - RG_0)^{-1}R, R \in \mathcal{H}^{m\times n}\}. \quad (3.11)$$

Now, from Fig. 2 we see that a compensator of the form $R(I - G_0R)^{-1}$ stabilizes G_1 if and only if R stabilizes $G_1 - G_0$. □

To aid in the application of Theorem 3.1 we quote below a result from [2] on stabilization using a stable compensator, which we shall encounter again in Section V.

Lemma 3.1 [2]: Let $\sigma_1, \cdots, \sigma_{l-1}, \sigma_l = \infty$ denote the extended real nonnegative zeros of the largest invariant factor of B_1.[3] Then there exists an $M \in \mathcal{H}^{m\times n}$ that stabilizes

[3]The *blocking zeros* of the rational matrix $B_1A_1^{-1}$ are the values of s such that $B_1(s)[A_1(s)]^{-1} = 0$. Since B_1 and A_1 are right coprime, the blocking zeros of $B_1A_1^{-1}$ in the right half-plane are the same as the zeros of the largest invariant factor of B_1 (in the right half-plane). Note that $B_1(\infty) = 0$.

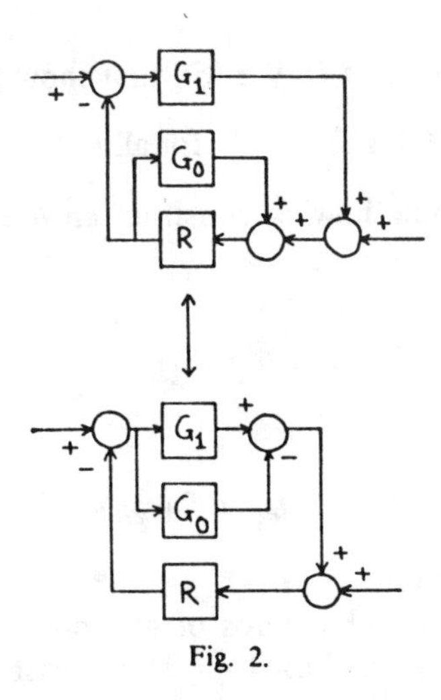

Fig. 2.

$B_1 A_1^{-1}$ if and only if the number of (real) zeros of $\det A_1$ in the interval (σ_i, σ_j), counted according to multiplicity, is even for every i, j.

In the case where G_0 is strictly proper and stable, we have seen that $B_1 A_1^{-1}$ equals $G_1 - G_0$. Thus $\sigma_1, \cdots, \sigma_{l-1}$ are precisely the real nonnegative blocking zeros of $G_1 - G_0$, i.e., the real nonnegative values of s such that $G_1(s) = G_0(s)$. The zeros of $\det A_1$ in the right half-plane are precisely the poles of $G_1 - G_0$, which in turn are the poles of G_1 (since G_0 is stable). In the general case, where neither G_0 nor G_1 is assumed to be stable, $\sigma_1, \cdots, \sigma_{l-1}$ are still the real nonnegative blocking zeros of $G_1 - G_0$. This can be seen by observing that

$$B_1(s) = 0 \Leftrightarrow [-\tilde{N}_0 D_1 + \tilde{D}_0 N_1](s) = 0 \Leftrightarrow [\tilde{D}_0^{-1} \tilde{N}_0](s) = [N_1 D_1^{-1}](s). \quad (3.12)$$

However, the interpretation of the zeros of $\det A_0$ is no longer simple.

Next, we consider the problem of simultaneously stabilizing several strictly proper plants $G_0, G_1, \cdots, G_l$. Without loss of generality we assume that we have available matrices satisfying (3.1) for $i = 0, \cdots, l$. By proceeding as in the proof of Theorem 3.1, we can derive the following result.

Theorem 3.2: Define

$$A_i = Q_0 D_i + P_0 N_i, \quad B_i = -\tilde{N}_0 D_i + \tilde{D}_0 N_i, \qquad i = 1, \cdots, l. \quad (3.13)$$

Then $\det A_i \not\equiv 0$ for all i, and B_i, A_i are right coprime for all i. Moreover, there exists a C that stabilizes G_i for $i = 0, \cdots, l$ if and only if there exists an $M \in \mathcal{H}^{m \times n}$ that stabilizes $B_i A_i^{-1}$ for $i = 1, \cdots, l$.

Outline of Proof: We leave it to the reader to verify that $\det A_i \not\equiv 0$ and that B_i, A_i are right coprime for all i. Now, there exists a C that stabilizes G_i for all i if and only if there exist $R_0, \cdots, R_l$ in $\mathcal{H}^{m \times n}$ such that

$$(Q_0 - R_0 \tilde{N}_0)^{-1}(P_0 + R_0 \tilde{D}_0) = (Q_i - R_i \tilde{N}_i)^{-1}(P_i + R_i \tilde{D}_i) \qquad \text{for } i = 1, \cdots, l. \quad (3.14)$$

Next, (3.14) is true if and only if there exist unimodular matrices $U_1, \cdots, U_l$ such that

$$Q_0 - R_0 \tilde{N}_0 = U_i (Q_i - R_i \tilde{N}_i),$$
$$P_0 + R_0 \tilde{D}_0 = U_i (P_i + R_i \tilde{D}_i) \qquad \text{for } i = 1, \cdots, l. \quad (3.15)$$

The rest of the steps are as in the proof of Theorem 3.1. □

Theorem 3.2 shows that the simultaneous stabilization of $l+1$ plants can be reduced to the simultaneous stabilization of l plants using a stable compensator. The converse is also true. Given l plants $H_1, \cdots, H_l$, there exists a stable compensator stabilizing all the plants $H_1, \cdots, H_l$ if and only if there exists a compensator that simultaneously stabilizes $H_0 = 0, H_1, \cdots, H_l$ [the fact that the compensator stabilizes the zero plant implies that the compensator must be stable; see (2.7)]. However, at present the criterion of Theorem 3.2 is not computationally verifiable except when $l = 1$.

If we wish to study the simultaneous stabilization of an arbitrary (i.e., not necessarily finite) family of plants $\{G_0, G_\alpha, \alpha \in \mathcal{C}\}$, the required generalization of Theorem 3.2 is readily apparent. For an interpretation of this problem in a differential-geometric setting, see [18]. Single-input–single-output versions of Theorems 3.1, 3.2 can be found in [1].

IV. Euclidean Division in the Ring $\mathcal{H}$

In this section we derive two results concerning Euclidean division in the ring $\mathcal{H}$. These results prove to be useful when we study the problem of stabilization using a "least unstable" compensator in Section V. The simple statements of these results (namely, Lemmas 4.1 and 4.2) are in sharp contrast with the tediousness and technical nature of their proofs. Thus, for clarity of exposition, we state the two lemmas in succession, and then give the proof of each.

Given a nonzero function $f \in \mathcal{H}$, define its *gauge* $\gamma(f)$ as

$$\gamma(f) = \text{relative degree of } f + \#\text{ zeros of } f \text{ in closed right half-plane}$$
$$= \#\text{zeros of } f \text{ in closed RHP, including } \infty. \quad (4.1)$$

Thus $\gamma(f)$ is a well-defined nonnegative integer for all nonzero $f \in \mathcal{H}$. Moreover, it can be shown [13] that $\mathcal{H}$ is a Euclidean domain; that is, given any f in $\mathcal{H}$ and any $g \neq 0$ in $\mathcal{H}$, there exists an $h \in \mathcal{H}$ such that either $f + gh = 0$, or else $\gamma(f + gh) < \gamma(g)$. In other words, a division algorithm can be performed in $\mathcal{H}$, and a greatest common divisor of a given pair of functions f and g can be found in a finite number of steps using the familiar Euclidean algorithm.

With the above definition of gauge, $\mathcal{H}$ is actually a *proper* Euclidean domain, i.e., we have $\gamma(fg) = \gamma(f) + \gamma(g)$ for all nonzero f, g. However, it is possible to have $\gamma(f+g) > \max\{\gamma(f), \gamma(g)\}$.[4] For example, consider

$$f = \frac{s+4}{s+1}, \quad g = -\frac{2s+3}{s+1}, \quad f + g = \frac{-s+1}{s+1}. \quad (4.2)$$

[4]Contrast this with the case of polynomials, where $\deg(f+g) \geq \max\{\deg f, \deg g\}$.

Then $\gamma(f)=\gamma(g)=0$, but $\gamma(f+g)=1$. This has an important consequence; namely, given f and $g \neq 0$, there may exist more than one h such that $\gamma(f+gh)<\gamma(g)$. [Contrast this with the case of polynomials. Given polynomials f, g with $g \neq 0$, there exists a *unique* polynomial h such that $\deg(f-gh)<\deg(g)$]. With this in mind, we define

$$I(f,g)=\min_{h \in \mathcal{H}} \gamma(f+gh). \tag{4.3}$$

If f is a multiple of g, so that $f=gh$ for some h, we set $I(f,g)=-\infty$.

Observe now that f is a unit in $\mathcal{H}$ if and only if $\gamma(f)=0$. Also, given f and g in $\mathcal{H}$, their greatest common divisor is well-defined to within a unit. Thus, if we let $\langle f, g\rangle$ denote a g.c.d. of f and g, then $\gamma(\langle f,g\rangle)$ is a well-defined integer[5] even though $\langle f,g\rangle$ is only defined to within a unit factor. Now suppose f and g are not relatively prime, and let w be a g.c.d. of f and g. If $w \neq 0$, it is easy to see that

$$I(f,g)=\gamma(w)+I(f/w, g/w). \tag{4.4}$$

Thus, for computational purposes, it is enough if we can calculate $I(f,g)$ when f, g are relatively prime.

Lemma 4.1: Let f, g be two elements of $\mathcal{H}$ with a greatest common divisor of 1. Let $\sigma_1,\cdots,\sigma_k$ denote the distinct nonnegative real zeros of the function g, including ∞ as appropriate, arranged in ascending order. Then

$$I(f,g)=\#\text{ sign changes in the sequence } \{f(\sigma_i), i=1,\cdots,k\} \triangleq \nu. \tag{4.5}$$

Remarks: Since f, g are relatively prime, $f(\sigma_i)\neq 0$, so that $f(\sigma_i)$ has a definite sign for all i.

The next lemma presents a result on matrix Euclidean division, which might be of some independent interest.

Lemma 4.2: Let $\mathfrak{R}$ be a proper Euclidean domain, with gauge γ. Given f, g in $\mathfrak{R}$, define

$$I(f,g)=\min_{h\in\mathfrak{R}} \gamma(f+gh) \tag{4.6}$$

where we take $\gamma(0)=-\infty$. Let $A\in\mathfrak{R}^{m\times m}$, $B\in\mathfrak{R}^{n\times m}$ be right coprime, with $\det A\neq 0$. Then

$$\min_{M\in\mathfrak{R}^{m\times n}} \gamma(\det(A+MB))=I(a,b_1) \tag{4.7}$$

where $a=\det A$ and b_1 is the greatest common divisor of all elements of B.

Remarks: Suppose that A and B are *not* right coprime; then we can write $A=A_1F$, $B=B_1F$, where A_1, B_1 are right coprime and F is a greatest common right divisor of A and B. Since $\gamma(\det(A+MB))=\gamma(\det(A_1+MB_1)+\gamma(\det F)$, Lemma 4.2 can be used to compute the minimum value of $\gamma(\det(A+MB))$ even when A and B are not right coprime.

We now give the proofs of Lemmas 4.1 and 4.2; these borrow heavily from [2].

[5] We can accommodate the case $\langle f,g\rangle=0$ by defining $\gamma(0)=-\infty$.

Proof of Lemma 4.1: We will first show that

$$\gamma(f+gh)\geqslant\nu \qquad \text{for all } h\in\mathcal{H}. \tag{4.8}$$

Then we will show how to construct an $h\in\mathcal{H}$ such that $\gamma(f+gh)=\nu$.

First, define

$$r=f/(f+gh). \tag{4.9}$$

Then

$$1-r=hg/(f+gh). \tag{4.10}$$

We now observe that

i) $r(s)=1$ at all RHP zeros of g; moreover, the multiplicity of s as a zero of $1-r$ is at least equal to its multiplicity as a zero of g.

ii) Every zero of r in the closed right half-plane is also a zero of f; moreover, the multiplicity of any such zero of r is less than or equal to its multiplicity as a zero of f.

iii) Conversely, let r be any function satisfying i) and ii) above, and define

$$h=f(1-r)/gr. \tag{4.11}$$

Then $h\in\mathcal{H}$.

To show that $\gamma(f+gh)\geqslant\nu$ for all h, let h be selected arbitrarily; then the resulting function r satisfies i) and ii). Now, write f and $f+gh$ as

$$f(s)=u_1(s)\phi_1(s)/[s^{l_1}\psi_1(s)] \tag{4.12}$$

$$f(s)+g(s)h(s)=u_2(s)\phi_2(s)/[s^{l_2}\psi_2(s)] \tag{4.13}$$

where u_1, u_2 are units of $\mathcal{H}$, ϕ_1, ϕ_2 are monic polynomials whose zeros are all in the closed right half-plane, and ψ_1, ψ_2 are monic strictly Hurwitz polynomials. Since i) holds, we have

$$\frac{\phi_1(\sigma_i)}{\phi_2(\sigma_i)}\cdot\frac{u_1(\sigma_i)s^{l_2}\psi_2(\sigma_i)}{u_2(\sigma_i)s^{l_1}\psi_1(\sigma_i)}=1, \qquad \text{for all } i. \tag{4.14}$$

Since the second term in (4.14) does not change sign, $\phi_1(\sigma_i)/\phi_2(\sigma_i)$ must always be of the same sign; i.e., the sequence $\{\phi_2(\sigma_i)\}$ must contain exactly as many sign changes as the sequence $\{\phi_1(\sigma_i)\}$. It is easy to see that the latter number of sign changes is ν, so that the sequence $\{\phi_2(\sigma_i)\}$ must have exactly ν sign changes. From this, it follows that $f+gh$ has at least ν zeros in the closed RHP, i.e., $\gamma(f+gh)\geqslant\nu$.[6]

We now give an iterative procedure for constructing an $h\in\mathcal{H}$ such that $\gamma(f+gh)=\nu$. Given f and g, find a polynomial $\xi(s)$ which has only ν nonnegative real zeros, such that the sequence $\{f(\sigma_i)/\xi(\sigma_i)\}$ does not change sign. This is clearly possible; indeed, if $f(\sigma_i)$ has a different sign from $f(\sigma_{i+1})$, we select any real number in the interval (σ_i, σ_{i+1}) to be a zero of $\xi(\cdot)$. Define

[6] Some care is needed in applying this argument when ∞ is a zero of $1-r$. But the required modifications are minor and are left to the reader.

$$r_0(s)=\frac{f(s)(s+1)^{\nu}}{\xi(s)}. \qquad (4.15)$$

Then r_0 satisfies ii) above, but not necessarily i). We show, following [2], that given any r_i, we can construct an r_{i+1} such that r_{i+1} satisfies ii).

iv) Given any s_0 in the closed RHP, $1-r_{i+1}$ can be made to have a zero at s_0; moreover, the multiplicity of the zero of $1-r_{i+1}$ at s_0 exceeds that of the zero of $1-r_i$ at s_0, if any. Finally, the RHP poles of r_{i+1} are exactly the same as the RHP poles of r_i, multiplicities included.

If we can indeed accomplish the construction of such an r_{i+1}, we can eventually find an r_j with exactly ν RHP poles satisfying i) and ii). For such a choice of $r=r_j$, we see that h given by (4.11) belongs to $\mathcal{K}$, and $\gamma(f+gh)=\#$ RHP poles of $r=\nu$.

The iterative procedure is

$$r_{i+1}=u_{i+1}r_i \qquad (4.16)$$

where u_{i+1} is a unit. The details of the procedure are exactly as in [2], with very minor differences. Reference [2, eq. (25)] is replaced by

$$\epsilon d_p^+(\sigma_i)/\xi(\sigma_i)>0 \qquad \forall i \qquad (4.17)$$

while [2, eq. (26)] is replaced by

$$S(s)=\frac{\epsilon d_p^+(s)h(s)}{\xi(s)g(s)}. \qquad (4.18)$$

This completes the proof of the lemma. □

Proof of Lemma 4.2:[7] In the first part of the proof we show that

$$\gamma(\det(A+MB))\geqslant I(a,b_1) \qquad \text{for all } M\in R^{m\times m}. \qquad (4.19)$$

Since $a\neq 0$, BA^{-1} is well-defined. Put BA^{-1} in Smith–McMillan form, and suppose U,V are unimodular matrices such that

$$U^{-1}(BA^{-1})V=\operatorname{diag}[s_1/t_1,\cdots,s_k/t_k,0,\cdots,0]\triangleq D \qquad (4.20)$$

where k is the rank of B, s_i divides s_{i+1}, t_{i+1} divides t_i, and s_i, t_i are relatively prime. Then clearly the ordered pair

$$(U\operatorname{diag}[s_1,\cdots,s_k,0,\cdots,0],\ V\operatorname{diag}[t_1,\cdots,t_k,1,\cdots,1]) \qquad (4.21)$$

is also an r.c.f. of BA^{-1}. Thus there exists a unimodular matrix W such that

$$B=U\operatorname{diag}[s_1,\cdots,s_k,0,\cdots,0]W \qquad (4.22)$$

$$A=V\operatorname{diag}[t_1,\cdots,t_k,1,\cdots,1]W. \qquad (4.23)$$

In particular, we see that

$$s_1=b_1=\text{g.c.d. of all elements of } B \qquad (4.24)$$

$$a\sim\prod_{i=1}^{k}t_i \qquad (4.25)$$

where "$\sim$" denotes "is equivalent to."

Next, let A^{adj} denote the adjoint matrix of A. Then $A^{-1}=A^{\text{adj}}a^{-1}$, and we have

$$BA^{\text{adj}}=BA^{-1}a\sim Da. \qquad (4.26)$$

Thus, if

$$C=\operatorname{diag}[c_1,\cdots,c_k,0,\cdots,0] \qquad (4.27)$$

is a Smith form for BA^{adj}, we have

$$c_i\sim\frac{s_i}{t_i}a. \qquad (4.28)$$

Hereafter, we suppose, without loss of generality, that

$$a=\prod_{i=1}^{k}t_i, \qquad c_i=\frac{s_i}{t_i}a=s_i\prod_{j\neq i}t_j. \qquad (4.29)$$

Next, observe that

$$\begin{aligned}\gamma(\det(A+MB))&+(n-1)\gamma(a)\\&=\gamma(\det(A+MB))+\gamma(\det A^{\text{adj}})\\&=\gamma(\det(AA^{\text{adj}}+MBA^{\text{adj}}))\\&=\gamma(\det(aI_m+MBA^{\text{adj}})) \qquad (4.30)\end{aligned}$$

so that

$$\begin{aligned}\min_M\gamma(\det(A+MB))&\\&=\min_M\gamma(\det(aI_m+MBA^{\text{adj}}))-(n-1)\gamma(a). \qquad (4.31)\end{aligned}$$

Suppose that

$$UBA^{\text{adj}}V=C; \qquad (4.32)$$

then

$$\begin{aligned}\gamma(\det(aI_m&+MBA^{\text{adj}}))\\&=\gamma(\det(AA^{\text{adj}}+MBA^{\text{adj}}))\\&=\gamma(\det(V^{-1}AA^{\text{adj}}V+V^{-1}MBA^{\text{adj}}V))\\&\qquad\text{since } \gamma(\det V)=\gamma(\det V^{-1})=0\\&=\gamma(\det(aI_m+M_1UBA^{\text{adj}}V))\\&\qquad\text{where } M_1=V^{-1}MU^{-1}\\&=\gamma(\det(aI_m+M_1C)). \qquad (4.33)\end{aligned}$$

Hence minimizing $\gamma(\det(aI_m+MBA^{\text{adj}}))$ is equivalent to minimizing $\gamma(\det(aI_m+M_1C))$.

By a well-known expansion formula [11, p. 9], we have

$$\begin{aligned}\det(aI_m+M_1C)&=a^m+\text{a multiple of } a^{m-1}c_1\\&+\text{a multiple of } a^{m-2}c_1c_2+\cdots\\&+\text{a multiple of } a^{m-k}c_1c_2\cdots c_k \qquad (4.34)\end{aligned}$$

[7]For notational simplicity we assume that $m=n$. If $m\neq n$, the only change is that all Smith–McMillan forms are bordered by zero matrices.

where we use the fact that c_i divides c_{i+1}. Now, (4.34) implies that

$$\det(aI_m + M_1C) = a^{m-k}\big[a^k + \text{a multiple of g.c.d. of } \{a^{k-1}c_1, a^{k-2}c_1c_2, \cdots, ac_1c_2\cdots c_{k-1}, c_1c_2\cdots c_k\}\big]. \quad (4.35)$$

Now, note that, from (4.29), we have

$$\begin{aligned} a^{k-1}c_1 &= a^k s_1/t_1 = a^{k-1}s_1t_2t_3\cdots t_k \\ a^{k-2}c_1c_2 &= a^k\frac{s_1s_2}{t_1t_2} = a^{k-1}s_1s_2t_3\cdots t_k \\ c_1\cdots c_k &= a^{k-1}s_1s_2s_3\cdots s_k. \end{aligned} \quad (4.36)$$

Since g.c.d. $\{t_2t_3\cdots t_k, \cdots, s_2s_3\cdots s_k\} = 1$,[8] we have that g.c.d. $\{a^{k-1}c_1, \cdots, c_1\cdots c_k\} = a^{k-1}s_1$. Hence

$$\begin{aligned} \det(aI_m + M_1C) &= a^{m-k}(a^k + \text{a multiple of } a^{k-1}s_1) \\ &= a^{m-1}(a + \text{a multiple of } s_1). \end{aligned} \quad (4.37)$$

Hence

$$\gamma(\det(aI_m + M_1C)) \geqslant \gamma(a^{m-1}) + I(a, s_1). \quad (4.38)$$

From (4.31), (4.33), and (4.38), we get

$$\gamma(\det(A + MB)) \geqslant I(a, s_1) = I(a, b_1) \quad \text{since } s_1 = b_1. \quad (4.39)$$

This proves (4.18).

To prove that the bound (4.18) is exact, let U, V be as in (4.32), and select $\theta_1, \cdots, \theta_k$ such that

$$\theta_1(t_2\cdots t_k) + \cdots + \theta_k(s_2\cdots s_k) = 1 \quad (4.40)$$

(this is possible because g.c.d. $(t_2\cdots t_k, \cdots, s_2\cdots s_k) = 1$); let r be chosen so that $\gamma(a + rb_1) = I(a, b_1)$; let

$$E = \begin{bmatrix} 0 & & 1 \\ & 1 & \\ 1 & & 0 \end{bmatrix} \in \Re^{m\times m} \quad (4.41)$$

and define

$$M = EV\begin{bmatrix} r\theta_m & 0 & 0 & \cdots & 0 \\ r\theta_{m-1} & 0 & 0 & & 1 \\ \vdots & & & & \vdots \\ r\theta_2 & 0 & 1 & & 0 \\ r\theta_1 & 1 & 0 & & 0 \end{bmatrix} U \triangleq EVRU, \text{ say} \quad (4.42)$$

where $\theta_i = 0$ if $i > k$. Then (noting that E is unimodular), we have

$$\begin{aligned} &\gamma(\det(A + MB))\cdot\gamma(\det A^{\text{adj}}) \\ &= \gamma(\det E^{-1}V^{-1}A + E^{-1}V^{-1}MB)\cdot\gamma(\det(A^{\text{adj}}VE)) \\ &= \gamma(\det(aI_m + E^{-1}V^{-1}MBA^{\text{adj}}VE)) \\ &= \gamma(\det(aI_m + RUBVE)) = \gamma(\det(aI_m + RCE)). \end{aligned} \quad (4.43)$$

However,

$$\begin{aligned} \det(aI_m + RCE) &= a^m\det(I_m + RCa^{-1}E) \\ &= a^m\left(1 + r\frac{b_1}{a}\right) = a^{m-1}(a + rb_1) \end{aligned} \quad (4.44)$$

where the last calculation is tedious, but straightforward (see [2, eqs. (79), (80)]. Hence, for this choice of M,

$$\begin{aligned} &\gamma(\det(A + MB)) + \gamma(\det A^{\text{adj}}) \\ &= \gamma(\det(A + MB)) + \gamma(a^{m-1}) \\ &= \gamma\big[a^{m-1}(a + rb_1)\big] = \gamma(a^{m-1}) + \gamma(a + rb_1) \end{aligned} \quad (4.45)$$

which shows that

$$\gamma(\det(A + MB)) = \gamma(a + rb_1) = I(a, b_1). \quad (4.46)$$

□

V. Stabilization Using a "Least Unstable" Compensator

In this section we study the following problems. Given a strictly proper plant G, what is the smallest number of right half-plane poles (counted according to their McMillan degree) that any stabilizing compensator for G can have? This question is of interest for two reasons: i) it generalizes the question of stabilization using a stable compensator, and ii) if the answer to this question is known, it is possible to obtain a lower bound on the dynamic orders of all stabilizing compensators for G. The main result of this section is given next.

Theorem 5.1: Let $G \in R^{n\times m}(s)$ be strictly proper, and let $\sigma_1, \cdots, \sigma_{l-1}, \sigma_l = \infty$ denote the nonnegative real blocking zeros of G (i.e., the nonnegative real values of s such that $G(s) = 0$), arranged in ascending order. Define

$$n_i = \begin{cases} 1 & \text{if the number of poles of } G \text{ in } (\sigma_i, \sigma_{i+1}) \text{ is even} \\ -1 & \text{if the number of poles of } G \text{ in } (\sigma_1, \sigma_{i+1}) \text{ is odd} \end{cases} \quad (5.1)$$

where the poles of G are counted according to their McMillan degree. Let ν denote the number of -1's in the

[8] This follows from the up and down divisibility properties of the s_i's, t_i's, and the primeness of (s_i, t_i).

sequence $n_1, \cdots, n_{l-1}$. Then every compensator C that stabilizes G has at least ν poles (counted according to McMillan degree) in the closed right half-plane. Moreover, there exists a compensator C_0 with exactly ν poles in the closed right half-plane that stabilizes G.

Proof: Let $(N, D), (\tilde{D}, \tilde{N})$ be any r.c.f. and l.c.f. of G, and select P, Q such that

$$PN + QD = I. \tag{5.2}$$

Then, by Lemma 2.3, every C that stabilizes G must be of the form

$$C = (Q - R\tilde{N})^{-1}(P + R\tilde{D}) \qquad \text{for some } R \in \mathcal{H}^{m \times n}. \tag{5.3}$$

By a result in [14], the number of RHP poles of C, counted according to their McMillan degree, is equal to the number of RHP zeros of $\det(Q - R\tilde{N})$. Since $\det[Q(\infty) - R(\infty)\tilde{N}(\infty)] = \det Q(\infty) \neq 0$, the number of RHP zeros of $\det(Q - R\tilde{N})$ equals $\gamma(\det(Q - R\tilde{N}))$. By Lemma 4.2, the minimum value of $\gamma(\det(Q - R\tilde{N}))$ as a function of R is $I(q, n_1)$, where $q = \det Q$ and n_1 is the smallest invariant factor of $\tilde{N}$. Now observe that the RHP zeros of n_1 are precisely the RHP blocking zeros of G. Thus, from Lemma 4.1, $I(q, n_1)$ equals the number of sign changes in the sequence $\{q(\sigma_i)\}_{i=1}^{l}$. Next, from (5.2) we get

$$Q(\sigma_i)D(\sigma_i) = I, \qquad \forall i \tag{5.4}$$

since $N(\sigma_i) = 0$. This shows that the signs of $\det Q(\sigma_i)$ and $\det D(\sigma_i)$ are the same. To complete the proof, note that, from [14],

$$\det D(s) \sim \prod_{i=1}^{k} \left[\frac{s - p_i}{s + 1}\right]^{m_i} \tag{5.5}$$

where $p_1, \cdots, p_k$ are the RHP poles of G, with McMillan degrees $m_1, \cdots, m_k$, respectively. It is easy to verify that the number of sign changes in the sequence $\{\det D(\sigma_i)\}_{i=1}^{l}$ equals the number of -1's in the sequence $\{n_i\}_{i=1}^{l-1}$. □

Note that the previous result on stabilization using a stable compensator, which is proved in [2] and stated here as Lemma 3.1, is a corollary of Theorem 5.1.

VI. Genericity of Simultaneous Stabilizability of Two Plants

In this section we study the genericity of simultaneous stabilizability of two plants, and of stabilizability of a single plant using a stable compensator. We show that both properties are generic in the case of multivariable systems, but not in the case of single-input–single-output systems.

We begin by reviewing the conditions for two functions f and g in $\mathcal{H}$ to be relatively prime (i.e., the common divisors of f and g are units).

Lemma 6.1: Two functions f and g in $\mathcal{H}$ are relatively prime if and only if i) at least one of them is nonzero at infinity, and ii) they have no common zeros in the closed right half-plane.

The proof is easily deduced from [16].

Next, we define a norm on $\mathcal{H}$, so that we then have a natural notion of neighborhoods. For any $f \in \mathcal{H}$, we define

$$\|f\| = \sup_{\omega \in R} |f(j\omega)|. \tag{6.1}$$

Thus, a *ball* $\mathcal{B}(f; \epsilon)$ is defined by

$$\mathcal{B}(f; \epsilon) = \{g: \|f - g\| < \epsilon\}. \tag{6.2}$$

A set $\mathcal{F}$ in $\mathcal{H}$ is *open* if, for every $f \in \mathcal{H}$, there is an $\epsilon > 0$ such that $\mathcal{B}(f; \epsilon) \subset \mathcal{F}$. A *neighborhood* of f is any open set containing f. A sequence $\{f_i\}$ in $\mathcal{H}$ *converges* to f if every neighborhood of f contains all but a finite number of terms in the sequence $\{f_i\}$. Finally, a set $\mathcal{F}$ in $\mathcal{H}$ is *dense* if every $f \in \mathcal{H}$ is the limit of a sequence in $\mathcal{F}$.

Recall that a *binary relation* on $\mathcal{H}$ is a subset of $\mathcal{H} \times \mathcal{H}$. If $\mathcal{R} \subset \mathcal{H} \times \mathcal{H}$ is a binary relation (or a *relation* for short), we write $a\mathcal{R}b$ to denote $(a, b) \in \mathcal{R}$. In other words, we say that a is related to b via $\mathcal{R}$ if $(a, b) \in R$. Given a relation $\mathcal{R}$ on $\mathcal{H}$, we define, for every $a \in \mathcal{H}$,

$$\mathcal{S}(a; \mathcal{R}) = \{b \in \mathcal{H}: (a, b) \in \mathcal{R}\}. \tag{6.3}$$

Finally, we say that the relation $\mathcal{R}$ is *generic* if $\mathcal{S}(a; \mathcal{R})$ is open and dense in $\mathcal{H}$ for every a in $\mathcal{H}$. Thus, if $\mathcal{R}$ is a generic relation on $\mathcal{H}$, then for any a, b in $\mathcal{H}$ one of two things is true: i) $a\mathcal{R}b$, and moreover, there is a ball $\mathcal{B}(b; \epsilon)$ such that $a\mathcal{R}\bar{b}$ for every $\bar{b}$ in $\mathcal{B}(b; \epsilon)$; or ii) a is not related to b via $\mathcal{R}$, in which case there is a sequence $\{b_i\}$ converging to b such that $a\mathcal{R}b_i$. Roughly speaking, if $\mathcal{R}$ is a generic relation and if $a\mathcal{R}b$, then small perturbations in b will not destroy the relationship, while if a is not related to b, then arbitrarily small perturbations in b will cause the relationship to hold true.

Lemma 6.2: Define a relation $\mathcal{C}$ on $\mathcal{H}$ by

$$\mathcal{C} = \{(a, b): a, b \in \mathcal{H} \text{ and are relatively prime}\}. \tag{6.4}$$

Then $\mathcal{C}$ is generic.

The proof using Lemma 6.1 is easy and therefore omitted.

In order to consider multivariable systems, we define a norm on $\mathcal{H}^{n \times m}$. Given $F \in \mathcal{H}^{n \times m}$, we define

$$\|F\| = \max_{1 \leq i \leq n} \sum_{j=1}^{m} \|f_{ij}\|. \tag{6.5}$$

This norm defines a topology on $\mathcal{H}^{n \times m}$, as before. It is easy to verify that a sequence $\{F^k\}$ in $\mathcal{H}^{n \times m}$ converges to $F \in \mathcal{H}^{n \times m}$ if and only if each of the component sequences $\{f_{ij}^k\}$ converges to f_{ij}.

Lemma 6.3: let $F \in \mathcal{H}^{n \times m}$ and suppose either n or m is greater than one. Then either 1 is the smallest invariant factor of F, or else there is a sequence $\{F_i\}$ converging to F such that 1 is the smallest invariant factor of F_i for all i.

Proof: Recall that the smallest invariant factor of a matrix is the greatest common divisor of all of its elements. The result now follows readily from Lemma 6.2. □

Up to now, we have only defined notions of neighborhood and convergence on $\mathcal{H}^{n\times m}$, which is the set of *stable* $n\times m$ transfer functions. We now extend these notions to $R_s^{n\times m}(s)$, which is the set of $n\times m$ matrices whose elements are strictly proper rational functions of s. Given a $G\in R^{n\times m}(s)$, a *neighborhood* of G consists of all ratios $N_1(s)[D_1(s)]^{-1}$, where (N, D) is any r.c.f. of G, and N_1, D_1 belong to some neighborhoods of N, D in $\mathcal{H}_s^{n\times m}$, $\mathcal{H}^{m\times m}$, respectively. A sequence $\{G_i\}$ in $R_s^{n\times m}(s)$ *converges* to $G\in R_s^{n\times m}(s)$ if there are r.c.f.'s (N_i, D_i) of G_i and (N, D) of G such that $N_i\to N$ in $\mathcal{H}_s^{n\times m}$ and $D_i\to D$ in $\mathcal{H}^{m\times m}$, respectively. The reader is referred to [8, sect. 4] for further details of the above topology. Once we have a topology on $R_s^{n\times m}(s)$, it is clear what is meant by a relation on $R_s^{n\times m}(s)$ being generic.

Theorem 6.1: Define a relation $\mathcal{SS}$ on $R_s^{n\times m}(s)$ as follows:[9]

$$\mathcal{SS}=\{(G_0,G_1)\colon G_0,G_1\in R_s^{n\times m}(s),\text{ and }G_0,G_1\text{ can be simultaneously stabilized}\}. \tag{6.6}$$

If either n or m is greater than one, then $\mathcal{SS}$ is generic.

Proof: Suppose that either n or m is greater than one. In order to show that $\mathcal{SS}$ is generic, we must establish two things: 1) whenever $(G_0, G_1)\in\mathcal{SS}$, there is a neighborhood $\mathfrak{N}_1$ of G_1 such that $(G_0,\bar{G}_1)\in\mathcal{SS}$ for all $\bar{G}_1\in\mathfrak{N}_1$, and 2) for any G_1, there is a sequence $\{G_1^{(i)}\}$ converging to G_1 such that $(G_0, G_1^{(i)})\in\mathcal{SS}$.

To prove the first statement, we prove first that the set of units in $\mathcal{H}$ is open. Let f be a unit in $\mathcal{H}$; then

$$\inf_{\operatorname{re} s\geq 0}|f(s)|\triangleq\epsilon>0 \tag{6.7}$$

i.e., f has no zeros in the closed RHP, including infinity. Now suppose $g\in\mathcal{B}(f;\epsilon/2)$. Then

$$\inf_\omega|g(j\omega)|\geq\inf_\omega|f(j\omega)|-\sup_\omega|f(j\omega)-g(j\omega)|>\epsilon/2>0. \tag{6.8}$$

Hence, by the Nyquist criterion, $g(\cdot)$ has no zeros in the closed RHP and is thus a unit. Since the mapping $F\to\det F$ from $\mathcal{H}^{m\times m}$ to $\mathcal{H}$ is continuous, and since F is unimodular if and only if $\det F$ is a unit, it follows that the set of unimodular matrices in $\mathcal{H}^{m\times m}$ is open.

Now suppose $(G_0, G_1)\in\mathcal{SS}$, and let $(N_0, D_0),(N_1, D_1)$ be any r.c.f.'s of G_0 and G_1, respectively. Define $A_1=Q_0D_1+P_0N_1$, $B_1=-\tilde{N}_0D_1+\tilde{D}_0N_1$, as in (3.2). Then, by Theorem 3.1, there is a matrix $M\in\mathcal{H}^{m\times n}$ such that A_1+MB_1 is unimodular. Since the set of unimodular matrices is open, there is a neighborhood $\mathfrak{N}$ of B_1 such that A_1+MB is unimodular for every $B\in\mathfrak{N}$. It is easy to see also that there exist neighborhoods $\mathfrak{N}_1$ of N_1 and $\mathfrak{N}_2$ of D_1 such that $-\tilde{N}_0D+\tilde{D}_0N\in\mathfrak{N}$ whenever $N\in\mathfrak{N}_1$, $D\in\mathfrak{N}_2$. To summarize, whenever $N\in\mathfrak{N}_1$, $D\in\mathfrak{N}_2$, the matrix $A_1+MB=A_1+M(-\tilde{N}_0D+\tilde{D}_0N)$ is unimodular, so that $(G_0, G)\in\mathcal{SS}$. Since the set of ratios $\{ND^{-1}, N\in\mathfrak{N}_1, D\in\mathfrak{N}_2\}$ is a neighborhood of $G_1=N_1D_1^{-1}$, we have shown that $\mathcal{S}(G_0;\mathcal{SS})$ is open for every G_0.[10]

To complete the proof that $\mathcal{SS}$ is generic, let $G_0\in R_s^{n\times m}(s)$ be given, let $G_1\in R_s^{n\times m}(s)$ be arbitrary, and define $A_1=Q_0D_1+P_0N_1$, $B_1=-\tilde{N}_0D_1+\tilde{D}_0N_1$ as before. Clearly B_1 is strictly proper, so let $B_1=F/(s+1)^\alpha$, where $F\in\mathcal{H}^{n\times m}$ and $F(\infty)\neq 0$. By Lemma 6.3, either 1 is the smallest invariant factor of F, or else there is a sequence $\{F_i\}$ converging to F such that 1 is the smallest invariant factor of G_i for all i. In the first case, $b_1=1/(s+1)^\alpha$ is the smallest invariant factor of B_1 and b_1 vanishes only at infinity. Hence, by Lemma 3.1, $(G_0, G_1)\in\mathcal{SS}$. In the second case, select a sequence $\{F_i\}$ converging to F such that 1 is the smallest invariant factor of F_i for all i. Now let

$$N_i=N_1+\tilde{Q}_0(F_i-F)/(s+1)^\alpha,$$

$$D_i=D_1-\tilde{P}_0(F_i-F)/(s+1)^\alpha \tag{6.9}$$

$$G_i=N_iD_i^{-1} \tag{6.10}$$

where $\tilde{P}_0$, $\tilde{Q}_0$ are selected such that

$$\tilde{N}_0\tilde{P}_0+\tilde{D}_0\tilde{Q}_0=I_n. \tag{6.11}$$

Clearly, $N_i\in\mathcal{H}_s^{n\times m}$ for all i and $N_i\to N_1$ in $\mathcal{H}_s^{n\times m}$ as $i\to\infty$.[11] Similarly, $D_i\to D_1$ in $\mathcal{H}^{m\times m}$ as $i\to\infty$. Hence $G_i\in R_s^{n\times m}(s)$ and $G_i\to G$ in $R_s^{n\times m}(s)$ as $i\to\infty$ (recall our definition of convergence in $R_s^{n\times m}(s)$). Moreover,

$$B_i\triangleq-\tilde{N}_0D_i+\tilde{D}_0N_i=B_1+(F_i-F)/(s+1)^\alpha=F_i/(s+1)^\alpha \tag{6.12}$$

has $1/(s+1)^\alpha$ as its smallest invariant factor. Hence, by Lemma 3.1, $(G_0, G_i)\in\mathcal{SS}$ for all i. □

Using exactly the same reasoning, one can show that the set of G in $R^{n\times m}(s)$ that can be stabilized by a stable compensator is dense in $R^{n\times m}(s)$, provided either n or m is greater than one. This is a formalization of an observation made in [2].

VII. Conclusions

In this paper, we have studied the problems of simultaneous stabilization and stabilization using a stable compensator. We have shown that the simultaneous stabilization of $l+1$ plants is equivalent to the stabilization of l plants using a stable compensator. We have given computationally verifiable tests for the simultaneous stabilizability of two plants, and have shown that this property is generic in the case of multivariable systems. Finally, we have derived an expression for the least unstable compensator that stabilizes a given plant.

In order to simplify the presentation, we have only studied the case of strictly proper plants. An examination

[9] $\mathcal{SS}$ is the set of pairs that can be simultaneously stabilized. See (3.1) for the definition of $g(G_0)$.

[10] This statement is true even if $n=m=1$; i.e., G_0 is a single-input–single-output system.

[11] Here we use $\mathcal{H}_s$ to denote the subset of $\mathcal{H}$ consisting of strictly proper functions.

of our proofs reveals that this assumption is quite unnecessary; it is only made so that various inverses are guaranteed to exist.

In some applications, we may wish to place the poles of the closed-loop system not just in the open left half-plane, but in some subset thereof. This would be the case, for example, if we wish the closed-loop system to have a certain maximum settling time and minimum damping factor. The generalization of our results to this case is extremely straightforward. Suppose $\mathcal{S}$ is a region in the complex plane which is symmetric about the real axis, and let $\mathcal{H}_{\mathcal{S}}$ denote the set of proper rational functions with real coefficients whose poles are all in $\mathcal{S}$. Then [13] $\mathcal{H}_{\mathcal{S}}$ is also a proper Euclidean domain, and the gauge $\gamma_{\mathcal{S}}(f)$ of a function in $\mathcal{H}_{\mathcal{S}}$ is defined by

$$\gamma_{\mathcal{S}}(f) = \text{relative degree of } f + \# \text{ zeros of } f \text{ outside } \mathcal{S}.$$

We can also study distributed systems by letting $\mathcal{H}$ be a ring and $\mathcal{H}_s$ a prime ideal in $\mathcal{H}$; see [6], [8] for details.

Acknowledgment

M. Vidyasagar thanks R. Saeks and J. J. Murray for drawing his attention to the simultaneous stabilization problem. Both authors thank B. A. Francis for helpful discussions.

References

[1] R. Saeks, J. Murray, O. Chua, and C. Karmokolias, "Feedback system design: The single-variate case," Dep. Elec. Eng., Texas Tech. Univ., Lubbock, TX, Tech. Rep., Jan. 1981.

[2] D. C. Youla, J. J. Bongiorno, Jr., and C. N. Lu, "Single-loop feedback stabilization of linear multivariable plants," *Automatica*, vol. 10, pp. 159–173, 1974.

[3] N. Jacobson, *Lectures in Abstract Algebra*, vol. I. Berlin, Germany: Springer-Verlag, 1964.

[4] O. Zariski and P. Samuel, *Commutative Algebra*, vol. I. Berlin, Germany: Springer-Verlag, 1958.

[5] M. Vidyasagar, "On the use of right-coprime factorizations in distributed feedback systems containing unstable subsystems," *IEEE Trans. Circuits Syst.*, vol. CAS-25, pp. 916–921, Nov. 1978.

[6] F. M. Callier and C. A. Desoer, "Stabilization, tracking and disturbance rejection in multivariable convolution systems," *Annales de la Societé Scientifique de Bruxelles*, vol. 94, no. I, pp. 7–51, 1980.

[7] R. Saeks and J. Murray, "Feedback system design: The tracking and disturbance rejection problems," *IEEE Trans. Automat. Contr.*, vol. AC-26, pp. 203–217, Feb. 1981.

[8] M. Vidyasagar, H. Schneider, and B. A. Francis, "Algebraic and topological aspects of feedback stabilization," Dep. Elec. Eng., Univ. of Waterloo, Waterloo, Ont., Canada, Tech. Rep. 80-09, Sept. 1980; see also *IEEE Trans. Automat. Contr.*, vol. AC-27, pp. 880–894, Aug. 1982.

[9] C. A. Desoer and W. S. Chan, "The feedback interconnection of lumped linear time-invariant systems," *J. Franklin Inst.*, vol. 300, pp. 335–351, 1975.

[10] C. A. Desoer, R. W. Liu, J. Murray, and R. Saeks, "Feedback system design: The fractional representation approach to analysis and synthesis," *IEEE Trans. Automat. Contr.*, vol. AC-25, pp. 399–412, June 1980.

[11] C. C. MacDuffee, *Theory of Matrices*. New York: Chelsea, 1956.

[12] M. Vidyasagar, "Coprime factorization and stability of multivariable distributed feedback systems," *SIAM J. Contr.*, vol. 13, pp. 1144–1155, Nov. 1975.

[13] N. T. Hung and B. D. O. Anderson, "Triangularization technique for the design of multivariable control systems," *IEEE Trans. Automat. Contr.*, vol. AC-24, pp. 455–460, June 1979.

[14] F. M. Callier and C. A. Desoer, "Open-loop unstable convolution feedback systems with dynamical feedback," *Automatica*, vol. 12, pp. 507–512, 1976.

[15] T. Kailath, *Linear Systems*. Englewood Cliffs, NJ: Prentice-Hall, 1980.

[16] A. S. Morse, "System invariants under feedback and cascade control," in *Mathematical System Theory*, G. Marchesini and S. K. Mitter, Eds. Heidelberg, Germany: Springer-Verlag, 1976.

[17] M. Vidyasagar, "A characterization of all stable stabilizing compensators for single-input-output systems," in preparation.

[18] R. Saeks and J. Murray, "Fractional representations, algebraic geometry and the simultaneous stabilization problem," *IEEE Trans. Automat. Contr.*, vol. AC-27, pp. 895–903, Aug. 1982.

Robust Stabilizability for a Class of Transfer Functions

HIDENORI KIMURA, MEMBER, IEEE

Abstract—This paper is concerned with the robust stabilizability for single-input single-output plants. Robust stabilizability means that a fixed controller can stabilize simultaneously all the plants in a given class which is characterized by a frequency-dependent uncertainty band function around the transfer function of a nominal model. A necessary and sufficient condition for robust stabilizability is derived based on the well-known Nevanlinna–Pick theory in classical analysis. It is shown that the values of the uncertainty band function should be restricted within a certain range at the unstable poles of the nominal model, in order for the class to be robustly stabilizable. A procedure of synthesizing a robust stabilizer is given and the parametrization of all the robust stabilizers is also shown.

I. Introduction

In the design of control systems, we are faced with the uncertainty of the model of the plant originated from various sources such as identification error, the nonlinearity of the plant dynamics, the simplification of the model for the purpose of controller design, the variation of the plant parameters during the operation, and so on. These uncertainties are almost inevitable for any model of real systems and should be regarded as an important feature of control system synthesis in order for the synthesis to be workable and reliable from the practical point of view. The robust control theory, which has received increasing attention in the last several years, is expected to give a design philosophy incorporating the model uncertainty.

One of the most important issues in the robust control theory is the robust stability. Apparently, the robustness estimation of the optimal LQ regulator derived by Safonov and Athans [1] was the first significant contribution in this field. More general cases have been considered in the frequency domain based on the singular value analysis of the closed-loop system, which was extensively used by Doyle [2]. This methodology which is based on the results of modern stability theory [23], is intuitively appealing and has produced some substantial results (Doyle and Stein [3], Lehtomaki *et al.* [4], and Cruz *et al.* [5]). Its connection to the nonlinear stability theory was discussed by Safonov [6]. In these results, the emphasis has been placed on the analysis aspects of the problem. The estimation of the robustness for a given control system has been the central issue. Very little has been exposed concerning the synthesis aspects.

To design robust control systems, it is essential to develop the synthesis aspects of the robust stability. The problems are as follows. Is it possible to design a stable control system for a given uncertainty of the model? If it is, how can such a system be designed? These problems are concerned with the problem of *robust stabilizability*, which was initially discussed by Kwakernaak [21]. It is in a sense a generalization of the simultaneous stabilization problem posed by Vidyasagar and Viswanadham [7].

Manuscript received April 5, 1983; revised October 7, 1983. Paper recommended by W. A. Wolovich, Associate Editor for Linear Systems.

The author is with the Department of Control Engineering, Faculty of Engineering Science, Osaka University, Osaka, Japan.

This paper deals with the problem of robust stabilization for single-input single-output plants with a prescribed uncertainty band in the frequency domain. A crucial result obtained by the singular value approach is now a starting point for further development. Use is made of the well-known Nevanlinna–Pick theory in classical analysis, to derive a condition for robust stabilizability. The result shows that, in order for the plant to be robustly stabilizable, the value of the uncertainty band function of the plant should be restricted within a certain range at the unstable poles of the nominal model. A procedure of synthesizing the robust stabilizer is derived based on the linear fractional transformations, which, at the same time, gives a parametrization of all the robust stabilizers. These results are closely related to the recently established optimization technique in the frequency domain [8], [9], [19].

Section II is devoted to the formulation of robust stabilizability. In Section III the Nevanlinna–Pick theory is briefly reviewed and some of its generalizations are discussed. In Section IV we reduce the problem of robust stabilization to an interpolation problem of bounded real functions, and derive a necessary and sufficient condition for the robust stabilizability using the Nevanlinna–Pick theory. Some illustrative examples are given. In Section V the results of Section IV are extended to a more general case, where the plant nominal model includes an integrator.

II. Robust Stabilizability

Consider the closed-loop system of Fig. 1, where $p(s)$ denotes the transfer function of the single-input single-output plant and $c(s)$ the transfer function of the controller. To consider the robustness issues of the closed-loop system, we assume that $p(s)$ belongs to a class of transfer functions which represents the uncertainty of the plant dynamics. A natural way of describing the uncertainty is given in terms of the variation of the frequency response curve [1]–[5].

Definition 1: A transfer function $p(s)$ is said to be in the class $C(p_0(s), r(s))$ if

i) $p(s)$ has the same number of the unstable poles as that of $p_0(s)$,

ii) $|p(j\omega)-p_0(j\omega)| \leq |r(j\omega)|, \quad |r(j\omega)|>0, \quad \forall \omega. \quad (1)$

In the above definition, $p_0(s)$ denotes the nominal model of the plant dynamics and $r(s)$ characterizes the uncertainty of the nominal plant model. We assume that $r(s)$ is a stable proper rational function. It should be noted that, in the above definition, only the values of the $r(s)$ on the imaginary axis $s=j\omega$ are relevant because we are only concerned with the uncertainty of the frequency response curve. Therefore, any $r(s)$ having the same values on the imaginary axis $s=j\omega$ represents the same class of $p(s)$.

Reprinted from *IEEE Trans. Automat. Contr.*, vol. AC-29, no. 9, pp. 788–793, Sept. 1984.

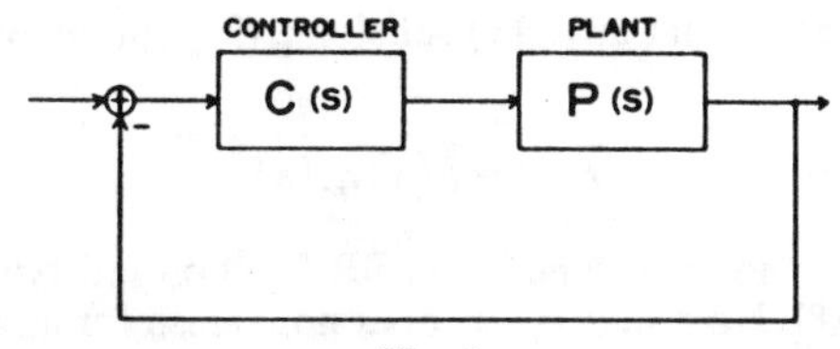

Fig. 1.

Now we formulate the robust stablizability of the class $C(p_0(s), r(s))$.

Definition 2: A class $C(p_0(s), r(s))$ is said to be *robustly stabilizable* if there exists a controller $c(s)$ such that the closed-loop system of Fig. 1 is stable for each $p(s) \in C(p_0(s), r(s))$. Such a controller is called a *robust stabilizer* for $C(p_0(s), r(s))$.

The characterization of the robust stabilizer has been the principal issue in the literature of robust control [1]–[6]. The following is the standard result due to Doyle [2], Doyle and Stein [3], Mannerfelt [10], and in a more sophisticated form, due to Chen and Desoer [11].

Lemma 1: A controller $c(s)$ is a robust stabilizer for $C(p_0(s), r(s))$, if and only if the closed-loop system of Fig. 1 is stable for $p(s) = p_0(s)$ and

$$|r(j\omega)c(j\omega)| < |1 + p_0(j\omega)c(j\omega)|, \quad \forall\omega. \tag{2}$$

In the later development, the function $q(s)$ defined as

$$q(s) = \frac{c(s)}{1 + p_0(s)c(s)} \tag{3}$$

plays an essential role. This function was introduced by Zames and Francis [8] to characterize the class of stabilizing controller. The controller $c(s)$ is conversely characterized by $q(s)$ as

$$c(s) = \frac{q(s)}{1 - p_0(s)q(s)}. \tag{4}$$

Using $q(s)$, we can establish the following stability criterion for the closed-loop system of Fig. 1 [8].

Lemma 2: The closed-loop system of Fig. 1 is stable for $p(s) = p_0(s)$, if and only if

i) $q(s)$ is stable,

ii) $1 - p_0(s)q(s)$ has the zeros at the unstable poles of $p_0(s)$, multiplicity included.

An extension of Lemma 2 to multivariable cases was derived by Safonov and Chen [12].

III. The Nevanlinna–Pick Theory

In this section, we briefly review the Nevanlinna–Pick theory. This is concerned with an interpolation problem for some classes of analytic functions [13], [14]. The relevance of this theory to circuit theory and signal processing was reported in the recent literature [15], [16]. It also plays an essential role in a model approximation problem [22].

We start with defining a class of functions.

Definition 3: A function $u(s)$ analytic in $\mathrm{Re}[s] \geq 0$ satisfying

$$|u(j\omega)| \leq 1, \quad \forall\omega \tag{5}$$

is called *bounded real* (BR). If $\leq$ is replaced by $<$ in (5), we call it *strongly bounded real* (SBR).

The BR or SBR function is used to describe the reflectance of passive networks [17]. In classical analysis, it is sometimes called *the function of class S* (Schur) [14].

Now we formulate an interpolation problem for BR or SBR functions which is usually referred to as the Nevanlinna–Pick problem.

The Nevanlinna–Pick Problem

Assume that $2l$ complex numbers α_i, β_i, $i = 1, \cdots, l$, are given satisfying

$$\mathrm{Re}[\alpha_i] > 0, \quad |\beta_i| < 1, \quad i = 1, \cdots, l. \tag{6}$$

Find a BR or SBR function $u(s)$ which satisfies

$$u(\alpha_i) = \beta_i, \quad i = 1, \cdots, l. \tag{7}$$

The solvability condition for the Nevanlinna–Pick problem is well known [13], [14]. Let us define a Hermite matrix, which is called the *Pick matrix*, associated with α_i and β_i as follows:

$$P = \begin{bmatrix} \frac{1-\beta_1\bar{\beta}_1}{\alpha_1+\bar{\alpha}_1} & \cdots & \frac{1-\beta_1\bar{\beta}_l}{\alpha_1+\bar{\alpha}_l} \\ \cdots & \cdots & \cdots \\ \frac{1-\beta_l\bar{\beta}_1}{\alpha_l+\bar{\alpha}_1} & \cdots & \frac{1-\beta_l\bar{\beta}_l}{\alpha_l+\bar{\alpha}_l} \end{bmatrix}. \tag{8}$$

We only state the solvability condition for SBR functions which is relevant to the later discussions.

Lemma 3: The Nevanlinna–Pick problem is solvable for SBR functions, if and only if the matrix P is positive definite.

The proof of the above lemma can be found in [13], [14]. A procedure of finding a solution $u(s)$ under the condition of Lemma 3 is also well known. It enables us to parametrize all the solutions. Although it can be found in the literature, we briefly discuss it for later use.

First, we form the so-called Fenyves array $\beta_{i,j}$ as follows:

$$\beta_{i,1} = \beta_i, \quad i = 1, \cdots, l, \tag{9}$$

$$\beta_{i,j+1} = \frac{(\alpha_i + \bar{\alpha}_j)(\beta_{i,j} - \beta_{j,j})}{(\alpha_i - \alpha_j)(1 - \bar{\beta}_{j,j}\beta_{i,j})}, \quad 1 \leq j \leq i-1 \leq l-1. \tag{10}$$

It is known that P is positive definite if and only if $|\beta_{i,j}| < 1$. Let

$$\rho_j = \beta_{j,j}.$$

Then, a solution $u(s)$ is given by repeating the linear fractional transformations

$$u_j(s) = \frac{(s - \alpha_j)u_{j+1}(s) + \rho_j(s + \bar{\alpha}_j)}{s + \bar{\alpha}_j + \bar{\rho}_j(s - \alpha_j)u_{j+1}(s)}, \quad j = l, l-1, \cdots, 1, \tag{11}$$

$$u(s) = u_1(s) \tag{12}$$

where $u_{l+1}(s)$ is an arbitrary SBR function. The essential fact is that all the solutions are obtained by the procedure (11), (12). This implies that the procedure (11), (12) parametrizes all the solutions in terms of a "free" SBR function $u_{l+1}(s)$.

For later use, we need a slight generalization of Lemma 3. We impose the two additional interpolation conditions at $s = 0$ and $s = \infty$

$$u(0) = \beta_0 \tag{13}$$

$$u(\infty) = \beta_{l+1}. \tag{14}$$

The solvability condition for the Nevanlinna–Pick problem with these two additional constraints are given as follows.

Lemma 4: The Nevanlinna–Pick problem with the additional interpolation conditions (13), (14) is solvable if and only if P in (8) is positive definite and

$$|\beta_0|<1, \qquad |\beta_{l+1}|<1. \tag{15}$$

(Proof is found in the Appendix.)

IV. The Condition for Robust Stabilizability

In this section, we reduce the robust stabilization problem to the Nevanlinna–Pick problem, and derive a necessary and sufficient condition for the robust stabilizability. We make the following assumptions on the class $C(p_0(s), r(s))$.

A1: All the unstable poles of $p_0(s)$ are simple and their real parts are positive.

A2: The uncertainty band function $r(s)$ is rational and satisfies either $\lim_{s\to\infty} r(s) = \text{const.} \neq 0$ or $\lim_{s\to\infty} sr(s) = \text{const.} \neq 0$. In other words, the relative degree of $r(s)$ is either 0 or 1.

The assumption A1 enables us to apply the Nevanlinna–Pick theory directly. Its relaxation will be discussed in the next section. The assumption A2 is for assuring the properness of the controller transfer function $c(s)$. It should be noted that we can always choose $r(s)$ satisfying the assumption A2 by estimating the uncertainty of the plant conservatively in the high frequency band.

Let $\alpha_1, \alpha_2, \cdots, \alpha_l$ be the unstable poles of $p_0(s)$. The corresponding *Blaschke product* $B(s)$ is defined as

$$B(s) = \frac{(\alpha_1 - s)\cdots(\alpha_l - s)}{(\bar{\alpha}_1 + s)\cdots(\bar{\alpha}_l + s)}. \tag{16}$$

Since $\mathrm{Re}[\alpha_i] > 0$ from the assumption A1, $B(s)$ is a BR function satisfying

$$|B(j\omega)| = 1, \qquad \forall\omega. \tag{17}$$

Let

$$\tilde{p}_0(s) = p_0(s)B(s). \tag{18}$$

According to the assumption A1, $\tilde{p}_0(s)$ is a stable proper transfer function. Assume that the closed-loop system of Fig. 1 is stable for some $c(s)$. Then, due to Lemma 2 ii), all the unstable poles of $p_0(s)$ must be cancelled out by the zeros of $q(s)$. Therefore, the function

$$\tilde{q}(s) = \frac{q(s)}{B(s)} \tag{19}$$

is analytic in $\mathrm{Re}[s] \geq 0$. Since $p_0(s)q(s) = \tilde{p}_0(s)\tilde{q}(s)$, the condition ii) of Lemma 2 is equivalent to

$$\tilde{p}_0(\alpha_i)\tilde{q}(\alpha_i) = 1, \qquad i = 1, \cdots, l, \tag{20}$$

where the simplicity of α_i assumed in A1 is taken into account. Thus, the stabilization of Fig. 1 for $p(s) = p_0(s)$ is reduced to finding a stable proper rational function $\tilde{q}(s)$ satisfying (20).

To extend the stability to the robust stability, Lemma 1 must be considered. The relation (2) is equivalent to

$$|r(j\omega)q(j\omega)| < 1, \qquad \forall\omega. \tag{21}$$

Due to (17), this inequality is equivalent to

$$|r(j\omega)\tilde{q}(j\omega)| < 1, \qquad \forall\omega. \tag{22}$$

Since both $r(s)$ and $\tilde{q}(s)$ are stable, the relation (22) implies that $r(s)\tilde{q}(s)$ is an SBR function. As was remarked in Section II $r(s)$ may be chosen freely, subject to the condition that its modulus takes a given value on the $j\omega$-axis. One of the selections is the *minimal phase factor* $r_m(s)$ of $r(s)r(-s)$, i.e., the function with no zero in $\mathrm{Re}[s] \geq 0$ satisfying

$$|r(j\omega)| = |r_m(j\omega)|. \tag{23}$$

It is well known that any $r(s)$ satisfying (23) can be written as a product

$$r(s) = b(s)r_m(s) \tag{24}$$

where $b(s)$ is an inner function (a BR function satisfying $|b(j\omega)| = 1, \forall\omega$) [18]. Note that $r(j\omega)$ does not vanish for any ω due to (1).

Let

$$u(s) = r_m(s)\tilde{q}(s) \tag{25}$$

$$\beta_i = r_m(\alpha_i)/\tilde{p}_0(\alpha_i), \qquad i = 1, \cdots, l. \tag{26}$$

Due to (22) and (23), $u(s)$ is an SBR function. Also, from (20), it should satisfy

$$u(\alpha_i) = \beta_i, \qquad i = 1, \cdots, l. \tag{27}$$

Thus, the robust stabilizability implies the existence of an SBR function $u(s)$ satisfying the interpolation condition (27). This is exactly the Nevanlinna–Pick problem which we discussed in the previous section.

Conversely, if such a $u(s)$ exists, then, from (25) and (19), we have

$$q(s) = \frac{B(s)}{r_m(s)}u(s). \tag{28}$$

The function $q(s)$ is stable from the minimal phase property of $r_m(s)$ and obviously satisfies the conditions of Lemma 2 and (21). Thus, the controller $c(s)$ given by (4) is a robust stabilizer. Therefore, we have reduced the problem to the Nevanlinna–Pick problem with the interpolation condition (27).

Before we use the results of the previous section, we must assure that the robust stabilizer obtained above is dynamic, i.e., $c(s)$ is proper. From (4), $c(s)$ is proper if and only if $q(s)$ is. Due to (28), the properness of $q(s)$ depends on the relative degree of $r_m(s)$. If $\lim_{s\to\infty} r_m(s) \neq 0$, any proper $u(s)$ yields a proper $q(s)$. Since any SBR function is proper, we can directly use Lemma 3. If $\lim_{s\to\infty} r_m(s) = 0$, the assumption A2 implies that $\lim_{s\to\infty} sr_m(s) \neq 0$. By writing $q(s) = B(s)su(s)/sr_m(s)$, we see that the properness of $q(s)$ is assured, if we choose $u(s)$ satisfying $\lim_{s\to\infty} su(s) < \infty$. Since $u(s)$ is a rational function, this is equivalent to $\lim_{s\to\infty} u(s) = 0$, which imposes another interpolation condition on $u(s)$ in addition to (27). This condition is exactly the same as the condition (14) with $\beta_{l+1} = 0$. Lemma 4 has already worked out the solvability of such an interpolation problem (we can neglect the condition (13) here). Thus, we have established the main result of the paper.

Theorem 1: Under the assumptions A1 and A2, $C(p_0(s), r(s))$ is robustly stabilizable, if and only if the matrix P in (8) with β_i being given by (26) is positive definite.

The theorem imposes a strong restriction on the magnitude of the uncertainty band function at the unstable poles of the nominal plant model $p_0(s)$, in order for the given class to be robustly stabilizable. In Theorem 1, the restriction is expressed in terms of the minimal phase factor $r_m(s)$. In fact, this expression is the strongest one in the sense that, if the condition of Theorem 1 holds, then, for any $r(s)$ satisfying (23), Theorem 1 also holds. More precisely, if P with β_i being given by (26) is positive definite, then the matrix P' defined by

$$(P')_{ij} = \frac{1 - \beta_i'\bar{\beta}_j'}{\alpha_i + \bar{\alpha}_j}, \qquad \beta_i' = r(\alpha_i)/\tilde{p}_0(\alpha_i) \tag{29}$$

is positive definite for any stable $r(s)$ satisfying (23). Indeed, since $r(s)$ is represented in the form of (24), $\beta_i' = \beta_i b(\alpha_i)$ for some BR function $b(s)$. A version of Lemma 3 for BR functions asserts that the matrix Q given by

$$(Q)_{ij}=\frac{1-b(\alpha_i)\bar{b}(\alpha_j)}{\alpha_i+\bar{\alpha}_j}$$

is nonnegative definite. Note that $b(\alpha_i)$ is a "natural" interpolation value of $b(s)$ at $s=\alpha_i$. Now, we can write

$$(P')_{ij}=\frac{1-\beta_i\bar{\beta}_j}{\alpha_i+\bar{\alpha}_j}+\beta_i(Q)_{ij}\bar{\beta}_j$$

which establishes that P' is positive definite if P is.

Example 1: Let $p_0(s)$ be given by

$$p_0(s)=\frac{s+2}{(s-1)(s+1)}.$$

Since $\alpha_1=1$ is the only unstable pole of $p_0(s)$, we have, according to (18) and (26),

$$\tilde{p}_0(s)=-\frac{s+2}{(s+1)^2}$$

$$\beta_1=\frac{r_m^{(1)}}{\tilde{p}_0(1)}=-\frac{4}{3}r_m(1).$$

Since $P>0$ is equivalent to $|\beta_1|<1$ for $l=1$, the inequality

$$|r_m(1)|<3/4 \tag{30}$$

is the necessary and sufficient condition for the robust stabilizability.

We shall derive the parametrization of all the robust stabilizers under the condition (30). Due to (11), the solution of the Nevanlinna–Pick problem in this case is expressed as

$$u_1(s)=\frac{(s-1)u_2(s)+\beta_1(s+1)}{s+1+\bar{\beta}_1(s-1)u_2(s)}$$

where $u_2(u)$ is any SBR function. Simple but lengthy calculations using (28) and (4) yield

$$c(s)=-\frac{l_1(s)u_2(s)+l_2(s)}{m_1(s)u_2(s)+m_2(s)} \tag{31}$$

where $l_i(s)$, $i=1,2$ and $m_i(s)$, $i=1,2$ are given by

$$l_1(s)=s^2-1, \qquad l_2(s)=\beta_1(s+1)^2$$

$$m_1(s)=\bar{\beta}_1(s+1)^2r_m(s)+(s+2)$$

$$m_2(s)=(s+1)[r_m(1)(s+5/3)+(s+1)(r_m(s)-r_m(1))/(s-1)].$$

The formula (31) gives a parametrization of all the robust stabilizers in terms of SBR function $u_2(s)$. If $r_m(s)=b/(s+1)(b>0)$, then (30) becomes $b<3/2$ and the representation (31) is simplified as

$$c(s)=-\frac{(s+1)[(s-1)u_2(s)-2b(s+1)/3]}{[(1-2b^2/3)s+2(1-b^2/3)]u_2(s)+b(s+1)/3}.$$

In this case, $u_2(s)$ has to be chosen as satisfying

$$u_2(\infty)=2b/3,$$

in order to guarantee the properness of $c(s)$.

Example 2: Let $p_0(s)$ be given by

$$p_0(s)=\frac{1}{s^2-2s+2}, \tag{32}$$

which has the two unstable poles $\alpha_1=1+j$ and $\alpha_2=1-j$. According to (18), we have

$$\tilde{p}_0(s)=\frac{1}{s^2+2s+2}.$$

Therefore, from (26),

$$\beta_1=4(1+j)r_m(\alpha_1) \qquad \beta_2=\bar{\beta}_1. \tag{33}$$

The robust stabilizability condition is represented as

$$\begin{bmatrix}\frac{1-|\beta_1|^2}{2} & \frac{1-\beta_1^2}{2(1+j)}\\ \frac{1-\bar{\beta}_1^2}{2(1-j)} & \frac{1-|\beta_1|^2}{2}\end{bmatrix}>0$$

which yields

$$1-|\beta_1|^2>2|\mathrm{Im}\,\beta_1|.$$

From (33), this can be written as

$$\frac{1}{32}-|r_m(1+j)|^2>\frac{1}{4}|\mathrm{Re}\,r_m(1+j)+\mathrm{Im}\,r_m(1+j)|.$$

If, for instance, $r_m(s)=b/(s+1)(b>0)$, then the above inequality holds if and only if

$$0<b<\frac{1}{8}(\sqrt{11}-1)=0.289578.$$

V. Extension to the Plant with an Integrator

In the preceding section, we restricted our attention to the plant whose unstable poles α_i are in the open right half plane, i.e., $\mathrm{Re}[\alpha_i]>0$. In this section, we shall extend the result to the plant with a simple pole at $s=0$ in addition to the unstable poles α_i, $i=1,\cdots,l,\mathrm{Re}[\alpha_i]>0$. This extension is essential when we consider the servo problem which requires the stabilization of the plant with an integrator.

If we consider (1) as a defining property of $C(p_0(s),r(s))$ in this case, we must assume that $|r(j\omega)|\to\infty$ as $\omega\to 0$ because of the presence of the pole at $s=0$ in $p_0(s)$. Therefore, without loss of generality, $r(s)$ is assumed to be written as

$$r(s)=\frac{r_m'(s)}{s} \tag{34}$$

where $r_m'(s)$ is a minimal phase function. From the assumption,

$$\tilde{p}_0(s)=sB(s)p_0(s) \tag{35}$$

is stable, where $B(s)$ is the Blaschke product defined by (16). From the same reasoning as in the previous section, the function

$$\tilde{q}(s)=\frac{q(s)}{sB(s)} \tag{36}$$

is stable, if $c(s)$ stabilizes the nominal plant $p_0(s)$, where $q(s)$ is given by (3). Due to Lemma 2 and $p_0(s)q(s)=\tilde{p}_0(s)\tilde{q}(s)$, we conclude that

$$\tilde{p}_0(\alpha_i)\tilde{q}(\alpha_i)=1, \qquad i=1,\cdots,l \tag{37}$$

$$\tilde{p}_0(0)\tilde{q}(0)=1 \tag{38}$$

must be satisfied if $c(s)$ stabilizes the nominal plant. Also, from (21), (17), and (36),

$$u(s)=r_m'(s)\tilde{q}(s)$$

must be an SBR function in order for $c(s)$ to be a robust stabilizer. Therefore, we have reduced the problem again to the interpolation problem

$$u(\alpha_i)=\beta_i,\quad \beta_i=r_m'(\alpha_i)/\tilde{p}_0(\alpha_i),\qquad i=1,\cdots,l \tag{39}$$

with the additional interpolation condition at $s=0$,

$$u(0)=\beta_0,\quad \beta_0=r_m'(0)/\tilde{p}_0(0). \tag{40}$$

Lemma 4 is now applied to derive the robust stabilization condition (take $\beta_{l+1}=0$, if necessary).

Theorem 2: The class of plants $C(p_0(s),r(s))$ which satisfies the assumptions A1 and A2, with the exception that $p_0(s)$ has a simple pole at $s=0$ and $r(s)$ is written in the form of (34), is robustly stabilizable, if and only if the matrix P in (8) with β_i being given by (39) is positive definite and $|\beta_0|<1$.

If the conditions of Theorem 2 are satisfied, we can calculate a robust stabilizer $c(s)$ using a solution $u(s)$ of the Nevanlinna–Pick problem through (4) and

$$q(s)=\frac{sB(s)u(s)}{r_m'(s)}.$$

In the usual servo problem, the integrator is regarded as part of a controller. Hence, all the plants can be written in the form of $p(s)=p'(s)/s$. Then, the inequality (1) can be written as

$$|p'(j\omega)-p_0'(j\omega)|\leqslant|r_m'(j\omega)|$$

where $p_0(s)=p_0'(s)/s$. Since $p_0'(s)$ has the unstable poles only in the open right half plane, the problem is reduced to the one discussed in the previous section to which the interpolation conditions (39) correspond, except that the additional condition $|\beta_0|<1$ is imposed, to which the interpolation condition (40) corresponds. This additional condition restricts the uncertainty of the static gain of the plant.

Example 3: Consider a nominal plant

$$p_0(s)=\frac{K_0}{s(s+1)}. \tag{41}$$

In this case, $B(s)=1$ and $\tilde{p}_0(s)=K_0/(s+1)$. The condition of robust stabilizability is given by

$$|r_m'(0)|<|\tilde{p}_0(0)|=|K_0|. \tag{42}$$

Thus, the robust stabilizability is assured for large uncertainty at $s=0$, if the static gain K_0 of the nominal plant is large.

The condition (42) can be interpreted as follows. Consider the class of plant with the form $p(s)=K/s(s+1)$, where the gain K is assumed to lie in the range $[K_1,K_2]$. If we take the nominal gain at the middle of this range $K_0=(K_1+K_2)/2$, then we can take $r_m'(s)=(K_2-K_1)/2(s+1)$ and the inequality (42) becomes equivalent to $K_1K_2>0$. This implies that the gain K should not change its sign in order for a robust stabilizer to exist. This result is clear by examining the characteristic equation $s(s+1)d(s)+Kn(s)=0$, where $c(s)=n(s)/d(s)$. Indeed, the constant term $Kn(0)$ of this equation should be kept positive for all values $K_1\leqslant K\leqslant K_2$. This is possible only when the sign of K is fixed.

VI. Conclusion

We have derived a condition for the robust stabilizability of single-input single-output systems, based on the Nevanlinna–Pick theory in classical analysis. According to this result, the robust stabilizability requires that the uncertainty of the plant is rather severely restricted within a certain range at the unstable poles of the nominal plant model. A procedure of synthesizing and parametrizing the robust stabilizer has been discussed. An extension of the results has been derived.

The results in this paper are closely related to the optimization technique in the frequency domain [8], [9], [19]. The extension of the present results to multivariable cases using these techniques will be reported [20].

Appendix
Proof of Lemma 4

The necessity part is obvious. In order to prove the sufficiency, consider the solution procedure (11). If $u_{l+1}(s)$ is chosen to be a rational function, all the $u_j(s)$ in (11) are rational. Then, writing

$$u_j(s)=\frac{\pi_j(s)}{\chi_j(s)}$$

we have, from (11),

$$\begin{bmatrix}\pi_j(s)\\ \chi_j(s)\end{bmatrix}=\left(1-|\rho_j|^2\right)^{1/2}(s+\bar{\alpha}_j)L_j(s)\begin{bmatrix}\pi_{j+1}(s)\\ \chi_{j+1}(s)\end{bmatrix} \tag{A1}$$

where

$$L_j(s)=\frac{1}{\left(1-|\rho_j|^2\right)^{1/2}}\begin{bmatrix}-B_j(s) & \rho_j\\ -\bar{\rho}_jB_j(s) & 1\end{bmatrix}$$

$$B_j(s)=\frac{\alpha_j-s}{\bar{\alpha}_j+s}.$$

It is easily seen that

$$L_j^*(j\omega)JL_j(j\omega)=J \tag{A2}$$

where the asterisk denotes the Hermitian conjugate and $J=\operatorname{diag}[1,\ -1]$. The relation (A2) represents the *J-lossless property* of $L_j(s)$ (see [16], for details).

Let

$$T(s)=L_1(s)L_2(s)\cdots L_l(s)=\begin{bmatrix}t_{11}(s) & t_{12}(s)\\ t_{21}(s) & t_{22}(s)\end{bmatrix}. \tag{A3}$$

From (A2) and (A3), it follows that $T^*(j\omega)JT(j\omega)=J$, which implies that

$$|t_{11}(j\omega)|^2-|t_{21}(j\omega)|^2=1 \tag{A4}$$

$$\bar{t}_{11}(j\omega)t_{12}(j\omega)-\bar{t}_{21}(j\omega)t_{22}(j\omega)=0 \tag{A5}$$

$$|t_{12}(j\omega)|^2-|t_{22}(j\omega)|^2=-1. \tag{A6}$$

By eliminating $t_{12}(j\omega)$ and $t_{21}(j\omega)$ from the above relations, we have

$$|t_{11}(j\omega)|=|t_{22}(j\omega)|. \tag{A7}$$

According to (A1) and (A3), $u_1(s)$ is represented as

$$u_1(s)=\frac{t_{11}(s)u_{l+1}(s)+t_{12}(s)}{t_{21}(s)u_{l+1}(s)+t_{22}(s)}. \tag{A8}$$

As was remarked in Section III $u_1(s)$ is a solution of the original Nevanlinna–Pick problem for any SBR function $u_{l+1}(s)$. Then, the proof will be completed if we can find an SBR function $u_{l+1}(s)$ for which

$$u_1(0) = \beta_0, \quad u_1(\infty) = \beta_{l+1} \tag{A9}$$

under the condition (15). Solving (A8) with respect to $u_{l+1}(s)$ yields

$$u_{l+1}(s) = f_1(s)\frac{u_1(s) + f_3(s)}{1 + f_2(s)u_1(s)} \tag{A10}$$

where $f_1(s) = t_{22}(s)/t_{11}(s)$, $f_2(s) = -t_{21}(s)/t_{11}(s)$, and $f_3(s) = -t_{12}(s)/t_{22}(s)$. Due to (A4)–(A7), we conclude that, for each ω

$$|f_1(j\omega)| = 1, \qquad |f_2(j\omega)| = |f_3(j\omega)| < 1,$$
$$f_2(j\omega) = \bar{f}_3(j\omega). \tag{A11}$$

Let

$$\delta_0 = f_1(0)\frac{\beta_0 + f_3(0)}{1 + f_2(0)\beta_0}, \qquad \delta_\infty = f_1(\infty)\frac{\beta_{l+1} + f_3(\infty)}{1 + f_2(\infty)\beta_{l+1}}. \tag{A12}$$

In view of (A11), we can easily show that $|\delta_0| < 1, |\delta_\infty| < 1$ under the condition (15). If we choose an SBR function $u_{l+1}(s)$ as satisfying

$$u_{l+1}(0) = \delta_0, \qquad u_{l+1}(\infty) = \delta_\infty,$$

then the corresponding solution $u_1(s)$ satisfies (A9) due to (A10). An example of such an SBR function is given by

$$u_{l+1}(s) = \frac{\delta_\infty s + \delta_0}{s + 1}. \qquad \square$$

References

[1] M. G. Safonov and M. Athans, "Gain and phase margin for multiloop LQG regulators," *IEEE Trans. Automat. Contr.*, vol. AC-22, pp. 173–179, Apr. 1977.

[2] J. C. Doyle, "Robustness of multiloop linear feedback systems," in *Proc. IEEE Conf. Decision and Contr.*, Jan. 1978.

[3] J. C. Doyle and G. Stein, "Multivariable feedback design: Concepts for classical/modern synthesis," *IEEE Trans. Automat. Contr.*, vol. AC-26, pp. 4–16, Feb. 1981.

[4] N. A. Lehtomaki, N. R. Sandell, and M. Athans, "Robustness results in linear-quadratic-Gaussian based multivariable control design," *IEEE Trans. Automat. Contr.*, vol. AC-26, pp. 75–93, Feb. 1981.

[5] J. B. Cruz, J. S. Freudenberg, and D. P. Looze, "A relationship between sensitivity and stability of multivariable feedback systems," *IEEE Trans. Automat. Contr.*, vol. AC-26, pp. 66–74, Feb. 1981.

[6] M. G. Safonov, *Stability and Robustness of Multivariable Feedback Systems*. Cambridge, MA: M.I.T. Press, 1980.

[7] M. Vidyasagar and N. Viswanadham, "Algebraic techniques for reliable stabilization," *IEEE Trans. Automat. Contr.*, vol. AC-27, pp. 1085–1095, Oct. 1982.

[8] G. Zames and B. A. Francis, "Feedback, minimax sensitivity, and optimal robustness," *IEEE Trans. Automat. Contr.*, vol. AC-28, pp. 585–601, May 1983.

[9] B. A. Francis, J. W. Helton, and G. Zames, "H^∞-optimal feedback controllers for linear multivariable systems," presented at the MTNS Symp., Beer Sheva, Israel, June 1983.

[10] C. F. Mannerfelt, "Robust control design with simplified models," *Dep. Contr., Lund Inst. Technol.*, Rep. LUTFD2/(TFRT-1021), 1981.

[11] M. J. Chen and C. A. Desoer, "Necessary and sufficient condition for robust stability of linear distributed feedback systems," Electron. Research Lab., Univ. Calif., Berkeley, Memo. UCB/ERL, Aug. 1981.

[12] M. G. Safonov and B. S. Chen, "Multivariable stability margin optimization with decoupling and output feedback," in *Proc. IEEE Conf. Decision and Contr.*, Dec. 1982.

[13] J. L. Walsh, *Interpolation and Approximation by Rational Function in the Complex Domain*. American Math. Society, 1935.

[14] N. I. Akhizer, *The Classical Moment Problem and Some Related Questions in Analysis*. Kemmer, Oliver & Boyd, (trans.), 1965.

[15] Ph. Delsarte, Y. Genin, and Y. Kamp, "On the role of the Nevanlinna–Pick problem in circuit and system theory," *Circuit Theory and Appl.*, vol. 9, pp. 177–187, 1981.

[16] P. Dewilde, A. Vieira, and T. Kailath, "On a generalized Szegö–Levinson realization algorithm for optimal linear predictors based on a network synthesis approach," *IEEE Trans. Circuits Syst.*, vol. CAS-25, pp. 663–675, Sept. 1978.

[17] V. Belevitch, *Classical Network Theory*. San Francisco, CA: Holden-Day, 1968.

[18] W. L. Duren, *Theory of H^p Spaces*. New York: Academic, 1970.

[19] B. C. Chang and J. B. Pearson, "Optimal disturbance reduction in linear multivariable system," Dep. Elec. Eng., Rice Univ., Houston, TX, Tech. Rep. 8214, 1982.

[20] M. Vidyasagar and H. Kimura, "Robust controlers for uncertain linear multivariable systems," presented at the 9th IFAC World Congress, Budapest, 1984.

[21] H. Kwakernaak, "A condition for robust stabilizability," *Syst. Contr. Lett.*, vol. 2, pp. 1–4, July 1982.

[22] H. Kimura, "Optimal L_2-approximation with fixed poles," *Syst. Contr. Lett.*, vol. 2, pp. 257–261, Feb. 1983.

[23] G. Zames, "On the input-output stability of time-varying nonlinear feedback systems: Part II," *IEEE Trans. Automat. Contr.*, vol. AC-11, pp. 465–476, July 1966.

Non-Euclidian Metrics and the Robust Stabilization of Systems with Parameter Uncertainty

PRAMOD P. KHARGONEKAR, MEMBER, IEEE, AND ALLEN TANNENBAUM

Abstract—**This paper considers, from a complex function theoretic point of view, certain kinds of robust synthesis problems. In particular, we use a certain kind of metric on the disk (the "hyperbolic" metric) which allows us to reduce the problem of robust stabilization of systems with many types of real and complex parameter variations to an easily solvable problem in non-Euclidian geometry. It is shown that several apparently different problems can be treated in a unified general framework. A new result on the gain margin problem for multivariable plants is also given. Finally, we apply our methods to systems with real zero or pole variations.**

NOTATION

$\mathbb{C}$	{complex numbers}
$\mathbb{R}$	{real numbers}
$\mathbb{P}^1$	$\mathbb{C} \cup \{\infty\}$
H	open right half plane $= \{s \in \mathbb{C}: \text{Re } s > 0\}$
$\bar{H}$	closed right half plane $= \{s \in \mathbb{C}: \text{Re } s \geq 0\}$
$\hat{H}$	$\bar{H} \cup \{\infty\}$
D	open unit disk $= \{s \in \mathbb{C}: \|s\| < 1\}$
$\bar{D}$	closed unit disk $= \{s \in \mathbb{C}: \|s\| \leq 1\}$
T	unit circle $= \{s \in \mathbb{C}: \|s\| = 1\}$

$\hat{H}$ and $\bar{D}$ are well known to be conformally equivalent.

INTRODUCTION

THIS paper is devoted to solving certain kinds of robust stabilization problems using techniques from complex analysis, and, in particular, interpolation theory. Particular cases of these problems have been considered by Tannenbaum [26]–[28]. In this paper, we continue the investigation of these robust design problems.

In general terms, the problem may be formulated as follows. Let $P_k(s)$ be a parametrized family of (linear, continuous-time, finite-dimensional, time-invariant, proper) plants, where the parameter vector k takes values in some compact set K. Then we want to design a controller $C(s)$ such that for each k in K, the closed-loop system as seen in Fig. 1 is (internally) asymptotically stable.

The problem stated above, in its complete generality, is very hard and no general solution is known. However, for certain special cases of importance in practical design, one can give a complete algorithmic solution. For example, consider the following family of SISO plants:

$$P_k(s) = kP_o(s) \tag{0.1}$$

Manuscript received March 12, 1984; revised July 3, 1984, November 20, 1984, and January 25, 1985. Paper recommended by Past Associate Editor, B. R. Barmish. This work was supported in part by the U.S. Army Research Office under Grant DAAG29-81-K-0136, in part by the U.S. Air Force under Grant AFOSR81-0238, and in part by the National Science Foundation under Grant ECS-8200607.

P. P. Khargonekar is with the Department of Electrical Engineering, University of Minnesota, Minneapolis, MN 55455.

A. Tannenbaum is with the Department of Mathematics, Ben Gurion University of Negev, Beer Sheva, Israel.

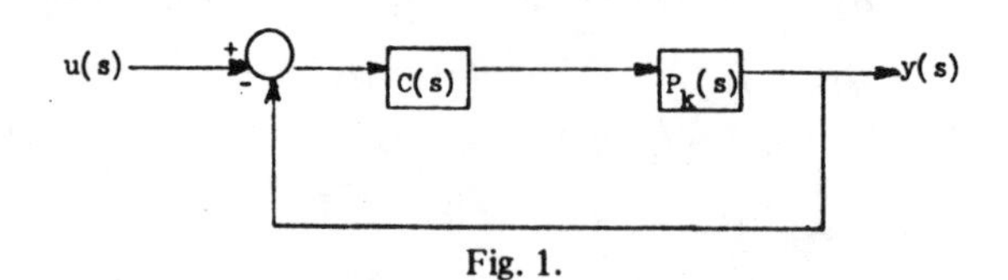

Fig. 1.

where $P_o(s)$ is the (fixed) nominal plant and k is a variable parameter taking values in $[a, b]$, $b > 1 > a > 0$. Then the above problem becomes one of finding (if possible) a proper compensator $C(s)$ which stabilizes the closed-loop system for all k in $[a, b]$. (If such a compensator exists, then by definition, $C(s)$ guarantees a gain margin of at least 20 log b/a dB for the nominal plant $P_o(s)$. Even though the gain margin only depends on the ratio b/a, the solution $C(s)$ depends on the interval $[a, b]$. However, given intervals $[a_1, b_1]$, $[a_2, b_2]$, such that $b_1/a_1 = b_2/a_2$, and $C_1(s)$ which stabilizes $kP_o(s)$ for all $k \in [a_1, b_1]$, clearly $(b_1/b_2)C_1(s)$ stabilizes $kP_o(s)$ for all $k \in [a_2, b_2]$.) It turns out that given the nominal model $P_o(s)$, one can compute a number β such that this problem is solvable if and only if

$$\frac{b}{a} < \beta. \tag{0.2}$$

Indeed, it is easy to see that 20 log β is the maximal attainable gain margin for the nominal plant $P_o(s)$ by suitable design of $C(s)$. Thus, this special problem may be viewed as the problem of maximization of gain margin by feedback. It will be seen that this new invariant β depends only on the zeros and poles of $P_o(s)$ in the open right half plane. Given a, b such that (0.2) holds, we give an explicit parametrization of all controllers that solve this design problem. The above problem (which was considered in [14] and [15], and solved by Tannenbaum [26]) is a very special case of a whole class of design problems for which our techniques work. In point of fact, we will argue that some of the standard robustness and H^∞-sensitivity minimization problems can be embedded in a unified framework and solved using essentially the same techniques.

Our techniques are complex analytic going back to some of the ideas of Nevanlinna and Pick [22], [1]. In particular, we make strong use of Pick's formulation of the Schwarz lemma in terms of a certain non-Euclidian (hyperbolic) metric. *This approach enables us to treat real as well as complex variations in the same framework*. We feel that this is an important contribution of this paper.

The paper is organized as follows. In Section I, we discuss some general results on Nevanlinna–Pick interpolation which we will need in the subsequent sections. Most of the results in this section are standard. However, we observe some important (from a control-theoretic point of view) facts about boundary interpolation. Moreover, we introduce a new invariant $\alpha_{\max}$ in terms of which many bounds on robust design can be expressed. In Section II, for SISO systems, using the concept of hyperbolic metric, we formulate and solve a general problem under which most problems involving multiplicative uncertainty and sensitivity minimization can be considered. This general formulation allows

Reprinted from *IEEE Trans. Automat. Contr.*, vol. AC-30, no. 10, pp. 1005–1013, Oct. 1985.

us to consider real as well as complex variations in the same framework. In Section III, we present an interesting new result on the multivariable analog of the gain margin maximization problem. Essentially, the result says that for plants with no blocking zeros in the open right half plane, one can obtain an arbitrarily large gain margin by suitable design of $C(s)$. In Section IV, we apply our methods for certain types of pole-zero variations. In Section V, we draw some general conclusions.

I. Interpolation Theory

It is well known that interpolation theory plays a major role in certain feedback design problems. See, for example, [6]–[8], [17], [26]–[28], [33], [34], and [36], [37], and the references cited therein. (We should mention that interpolation theory has also been used in the circuit theory literature. See, e.g., [32] and [10].) In this section, we will describe those aspects of the classical interpolation theory which are relevant to the design problems treated in the subsequent sections. See [11] for a thorough treatment of interpolation theory and related subjects.

Let $a_i \in D$, $\bar{b}_i \in \bar{D}$, $i = 1, 2, \cdots, q$ with $a_i \neq a_j$, $i \neq j$. The classical Nevanlinna-Pick interpolation problem is to find (if one exists) an analytic function $f: D \to \bar{D}$ such that $f(a_i) = \bar{b}_i$, $i = 1, 2, \cdots, q$. As is well known [23], [21] an interpolating function f exists if and only if the following *Nevanlinna-Pick matrix*

$$N := \left[\frac{1-\bar{b}_i b_j}{1-a_i\bar{a}_j}\right]_{i,j=1,2,\cdots,q}$$

is positive semidefinite.

Our work depends on the following slight variation of the above problem. Let $a_i \in D$, $b_i \in \mathbb{C}$, $i = 1, 2, \cdots, q$ with the a_i distinct as above. Let $\alpha \geq 0$ be in $\mathbb{R}$. Then we are interested in finding an analytic $f_\alpha: D \to \bar{D}$ such that $f_\alpha(a_i) = \alpha b_i$, $i = 1, 2, \cdots, q$. Clearly, for $\alpha = 0$, one can find such a function, namely $f_\alpha \equiv 0$. Therefore, by continuity, one can do this for α sufficiently small. Indeed, it is an easy exercise to compute the maximal α, $\hat{\alpha}_{max}$, such that for each $\alpha \leq \hat{\alpha}_{max}$, f_α exists. Explicitly, $\hat{\alpha}_{max}$ can be computed as follows. Define

$$A := \left[\frac{1}{1-a_i\bar{a}_j}\right]_{i,j=1,2,\cdots,q}$$

$$B := \left[\frac{b_i\bar{b}_j}{1-a_i\bar{a}_j}\right]_{i,j=1,2,\cdots,q}$$

Clearly, in order for the above problem to be solvable we must require that $A - \alpha^2 B \geq 0$. If $b_i = 0$, $i = 1, 2, \cdots, q$, then $A - \alpha^2 B > 0$ for all α in $\mathbb{R}$. In this case, we set $\hat{\alpha}_{max} := \infty$. On the other hand if at least one of the $b_i \neq 0$, then

$$\hat{\alpha}_{max} = 1/\sqrt{\lambda_{max}}$$

where λ_{max} is the largest eigenvalue of $A^{-1}B$. (It is not difficult to see that $\lambda_{max} > 0$ if $B \neq 0$.) Note that $\hat{\alpha}_{max} := \hat{\alpha}_{max}(a_i, b_i)$ only depends on the interpolation data a_i, b_i, $i = 1, 2, \cdots, q$. We will see in Section II that $\hat{\alpha}_{max}$ plays a central role in robust stabilization problems.

We should also note that the assumption of the distinctness of the a_i's is only done for simplicity. Indeed, if one wants to interpolate with multiplicities, i.e., put interpolation conditions on the derivatives of f_α at the points a_i, one also has a corresponding Nevanlinna-Pick matrix from which $\hat{\alpha}_{max}$ may be derived. See [10], [24], and [2].

In Section II, we will show that the construction of solutions to certain kinds of robust stabilization problems amounts to finding solutions to Nevanlinna-Pick interpolation problems. Therefore, we would now like to sketch an explicit parameterization of all holomorphic functions $f_\alpha(a_i) = \alpha b_i$, $i = 1, 2, \cdots, q$. We will assume for simplicity that all the a_i's are distinct. By using an appropriate conformal equivalence $D \to D$, if necessary, we may clearly assume without loss of generality, that all of the a_i's are nonzero. Let $\alpha < \hat{\alpha}_{max}$. Then

$$N_\alpha := A - \alpha^2 B > 0$$

is the appropriate Nevanlinna-Pick matrix. We now follow the development in [18] to describe all the interpolating functions f_α. In order to do this, let us first set up some notation

$$B(z) := \prod_{i=1}^{q} \frac{a_i - z}{1-\bar{a}_i z} \cdot \frac{\bar{a}_i}{|a_i|},$$

$$y_i := \bar{B}(0)/\bar{a}_i, \quad y' := [y_1\, y_2 \cdots y_q].$$

Let $x' = [x_1 x_2 \cdots x_q]$ be the (unique) vector such that

$$N_\alpha x = y. \tag{1.1}$$

Now let

$$P(z) := \overline{B(0)}B(z) - \sum_{i=1}^{q} \frac{B(z)}{(z-a_i)} x_i$$

$$Q(z) := (-1)^q z\left(\sum_{i=1}^{q} \frac{\alpha \bar{b}_i}{(1-\bar{a}_i z)} \bar{x}_i\right)$$

$$\tilde{P}(z) := B(z)\bar{P}(1/z), \quad \tilde{Q}(z) := B(z)\bar{Q}(1/z).$$

Then all solutions to our interpolation problem are given by

$$f_\alpha = \frac{\tilde{P}(z)g(z) + \tilde{Q}(z)}{P(z) + Q(z)g(z)} \tag{1.2}$$

where $g(z)$ is any arbitrary analytic function $g: D \to \bar{D}$. Note that the only nontrivial computation involves solving the linear equations (1.1). As N_α is Hermitian, this is easily done.

Finally, consider the degenerate case, when $\alpha = \hat{\alpha}_{max}$. Then N_α is singular. In this case, there is a unique function $f_\alpha: D \to \bar{D}$ such that $f_\alpha(a_i) = \alpha b_i$. This function is an "all pass," i.e., has constant modulus on the unit circle T. This is precisely the case which occurs in the work of Zames and Francis [37]. It is easy to use the parameterization given above to find this unique all-pass function. Indeed, let l be the rank of N_α, $l < q$. After a suitable reordering of the a_i's, we may without loss of generality, assume that the top left $l \times l$ principal minor M of N_α is nonsingular. Now consider the restricted interpolation problem of finding all the holomorphic functions $h: D \to \bar{D}$ such that $h(a_j) = \hat{\alpha}_{max} b_j$, $j = 1, 2, \cdots, l$. Then M is the *corresponding Nevanlinna-Pick matrix which is nonsingular.* Then we can find, as above, $P(z)$, $Q(z)$, $\tilde{P}(z)$, $\tilde{Q}(z)$ for this restricted problem such that all solutions h are given by

$$h = \frac{\tilde{P}(z)g(z) + \tilde{Q}(z)}{P(z) + Q(z)g(z)}$$

where $g(z)$ is an analytic function from $D \to \bar{D}$. Now, to solve the original problem, we must choose $g(z)$ such that h satisfies the rest of the interpolation conditions, i.e., $h(a_i) = \hat{\alpha}_{max} b_i$, $i = l+1, l+2, \cdots, q$. Therefore, $g(z)$ must satisfy

$$\hat{\alpha}_{max} b_i = \frac{\tilde{P}(a_i)g(a_i) + \tilde{Q}(a_i)}{P(a_i) + Q(a_i)g(a_i)}, \qquad i = l+1,\ l+2, \cdots, q.$$

Since rank $N_\alpha = l < q$, it is a standard fact from Nevanlinna-Pick interpolation theory that there is a unique constant g_o with $|g_o| = 1$ such that $g(z) \equiv g_o$ is the only function which satisfies the above requirements. Hence, the unique solution to the degenerate interpolation problem for $\alpha = \hat{\alpha}_{max}$ is given by

$$f = \frac{\tilde{P}(z)g_o + \tilde{Q}(z)}{P(z) + Q(z)g_o}.$$

This is the required all-pass function.

We refer the interested reader to the excellent recent paper of Helton [11] for a comprehensive treatment of interpolation theory and an extensive list of references on the above topic.

Remark 1.3: The formula (1.2) is the standard linear fractional representation of all solutions to the Nevanlinna-Pick interpolation problem. This formula occurs in various different forms in the system theory and mathematics literature. Specifically, we would like to note that such formulas arise in certain types of spectral estimation problems in signal processing. In these problems, orthogonal polynomials on the unit circle and Toeplitz matrices play a major role. (See [9] and the references cited therein.) It is interesting to note that Delsarte, Denin, and Kamp [4] show that a general Nevanlinna-Pick matrix can be transformed into a Toeplitz matrix by certain matrix operations. Thus, it seems, that P, Q, $\tilde{P}$, $\tilde{Q}$ are, in some sense, "orthogonal polynomials." We feel that the various computational techniques developed in the signal processing literature to deal with Toeplitz matrices may prove very useful in computation aspects of the Nevanlinna-Pick interpolation problems.

Finally, it will be seen that we need to consider certain kinds of interpolation problems with some of the points lying on the boundary T of the unit disk D. Contrary to the seemingly popular impression, for the problems which arise in robust stabilization theory, boundary interpolation is easily treated. Here we extend our notation of $\hat{\alpha}_{max}$ to cover boundary interpolation. Let $a_j \in D$, $j = 1, \cdots, l$, $a_{l+r} \in T (r = 1, \cdots, q - l)$, and $b_i \in \mathbb{C}$, $i = 1, \cdots, q$. Given a real number $\alpha \geq 0$, we are required to find an analytic function $f_\alpha : \bar{D} \to D$ such that $f_\alpha(a_i) = \alpha b_i$ for $i = 1, \cdots, q$. Let α_1 be the $\hat{\alpha}_{max}$ for the "interior" interpolation data a_j, b_j, $j = 1, 2, \cdots, l$. Define

$$\alpha_{max}(a_j, b_i) := \min\left(\alpha_1, \frac{1}{|b_{l+1}|}, \frac{1}{|b_{l+2}|}, \cdots, \frac{1}{|b_q|}\right) \quad \text{for}$$

$$j = 1, \cdots, l \text{ and } i = 1, \cdots, q. \tag{1.4}$$

We can now state the general theorem.

Theorem 1.5: Let a_i in $\bar{D}$ and b_i in $\mathbb{C}$, $i = 1, 2, \cdots, q$ be as above. Then there exists an analytic function $f_\alpha : \bar{D} \to D$ such that $f_\alpha(a_i) = \alpha b_i$ if and only if $\alpha < \alpha_{max}(a_j, b_i)$.

Proof: Let $h_\alpha : D \to D$ be an analytic function such that $h_\alpha(a_j) = \alpha b_j$, $j = 1, 2, \cdots, l$. This exists since $\alpha < \hat{\alpha}_{max}$. Then from (1.2) there exist rational functions, completely determined by the interpolation data, P, $\tilde{P}$, Q, $\tilde{Q}$ such that

$$h_\alpha = \frac{\tilde{P}g + \tilde{Q}}{P + Qg} \tag{1.6}$$

where $g : D \to D$ is an *arbitrary* holomorphic function. We need, therefore, to find g such that $h_\alpha(a_{l+r}) = b_{l+r}$, $r = 1, \cdots, q - l$. But from (1.6) we have

$$g = \frac{\tilde{Q} - Ph_\alpha}{-\tilde{P} + Qh_\alpha}$$

and, therefore, $h_\alpha(a_{l+r}) = b_{l+r}$, $r = 1, \cdots, q - l$ if and only if

$$g(a_{l+r}) = \frac{\tilde{Q}(a_{l+r}) - P(a_{l+r})b_{l+r}}{\tilde{P}(a_{l+r}) + Q(a_{l+r})b_{l+r}} =: \gamma_r$$

$r = 1, \cdots, q - l$. Consequently, we need $g : \bar{D} \to D$ such that $g(a_{l+r}) = \gamma_r$, $a_{l+r} \in T$, $\gamma_r \in D$, $r = 1, \cdots, q - l$. Such a g always exists. Indeed, for $\epsilon > 0$, set $D_{1+\epsilon} := (|z| < 1 + \epsilon\}$. Then computing the corresponding Nevanlinna-Pick matrix for functions $g : D_{1+\epsilon} \to D$, it is trivial to check that for ϵ sufficiently small, with the given interpolation data, the matrix will be positive definite. □

Remark 1.7: An identical result holds if we consider interpolation with multiplicities, i.e., we impose interpolation conditions on the function h_α and its derivatives. (This fact was first seen by the authors, and later verified in a personal communication with J. Ball.) Indeed, this is an immediate corollary of the generalized Nevanlinna-Pick matrices for interpolation with multiplicities due to Helton [10], Rosenblum and Rovnyak [24]. To avoid a proliferation of multiindexes, we consider the case $q = 2$. Then we are interested in the following problem. Find an analytic function $\psi : \bar{D} \to D$ such that $\psi(\lambda_1) = w_{11}$, $\psi'(\lambda_1) = w_{12}$, $\psi(\lambda_2) = w_{21}$, $\psi'(\lambda_2) = w_{22}$. The main point of Theorem 1.7 is that this problem is *always* solvable if $\lambda_1, \lambda_2, \in T$ (= boundary of the disk). Note we are assuming (and this assumption is crucial) that $w_{ii} \in D$, $i = 1, 2$. Indeed if we mimic the proof of Theorem 1.7, and consider the functions $\psi : D_{1+\epsilon} - D$, as $\epsilon \to 0$ the Helton [10] generalization of the Nevanlinna-Pick matrix will approach

$$\begin{bmatrix} \infty & \bar{w}_{11}w_{12} & \alpha & \bar{\lambda}_1\alpha \\ \bar{w}_{12}w_{11} & \infty & \lambda_1\bar{\alpha} & 0 \\ \bar{\alpha} & \bar{\lambda}_1\alpha & \infty & w_{22}\bar{w}_{21} \\ \lambda_1\bar{\alpha} & 0 & \bar{w}_{22}w_{21} & \infty \end{bmatrix}$$

where $\alpha := (1 - \bar{w}_{11}w_{21})/(1 - \bar{\lambda}_1\lambda_2)$.

In other words for ϵ small, one gets a positive definite matrix. The same argument (using the full Helton matrix) shows that Remark 1.7 extends for interpolation with multiplicities as required.

II. Robust Stabilization and Other Problems

In this section, we will consider certain types of robust stabilization and related problems which were alluded to in the Introduction. To motivate our approach, let us begin by reviewing precisely how the problem of internal stabilization by feedback amounts to an interpolation problem. Let $P_o(s)$ be a fixed SISO nominal plant with closed right half plane zeros $z_1, z_2, \cdots, z_m$, and closed right half plane poles $p_1, p_2, \cdots, p_n$. (Note that some of the z_i's will be ∞ since we are dealing with a strictly proper plant.) For a given compensator $C(s)$ define the *sensitivity function*

$$S(s) = (1 + P_o(s)C(s))^{-1}. \tag{2.1}$$

As is well known (see, e.g., [31]) in order for the closed-loop system to be internally asymptotically stable, it is necessary and sufficient that $S(s)$ have the following properties:

i) $S(s)$ is real rational and analytic in $\bar{H}$;

ii) the zeros of $S(s)$ contain $\{p_1, p_2, \cdots, p_n\}$ multiplicities included; and

iii) the zeros of $S(s) - 1$ contain $\{z_1, z_2, \cdots, z_m\}$ multiplicities included. (2.2)

Given any such $S(s)$, one can find the corresponding (proper) compensator $C(s)$ using (2.1).

Let us begin by considering the problem of internal stabilization for plants with parameter uncertainty as discussed in the Introduction. Consider the family of SISO plants $P_k(s) = kP_o(s)$ as given by (0.1) where $P_o(s)$ is the nominal model and k belongs to the interval $[a, b]$, $b > 1 > a > 0$. Let $C(s)$ be a proper compensator. We can now state the following.

Lemma 2.3: The feedback system (Fig. 1) is internally asymptotically stable for all k in $[a, b]$ if and only if the sensitivity function $S(s)$ satisfies (2.2) and

$$S(s) \notin \left(-\infty, \frac{a}{a-1}\right] \cup \left[\frac{b}{b-1}, \infty\right).$$

Proof: Clearly (2.2) must hold. For internal stability, we must have

$$1+kP_o(s)C(s)\neq 0, \qquad \text{for all } s \text{ in } \bar{H}.$$

From the definition of $S(s)$, and since $1 \in [a, b]$, we have that $S(s)$ is not contained in $(-\infty, a/(a-1)] \cup [b/(b-1), \infty)$.□

Gain Margin Problem 2.4: Lemma 2.3 shows that the gain margin problem of the Introduction is equivalent to the following interpolation problem. For given $P_o(s)$ and interval $[a, b]$, $0 < a < 1 < b$, find a real rational function $S(s)$ such that

i) $S(s) : \bar{H} \to \mathbb{C} \setminus \left\{ \left(-\infty, \frac{a}{a-1}\right] \cup \left[\frac{b}{b-1}, \infty\right) \right\}$,

ii) $S(s)$ satisfies (2.2).

Next, let us consider the problem of sensitivity minimization of Zames [34], Zames and Francis [37], and Francis and Zames [8]. First we will consider the unweighted sensitivity function and then, a bit later, consider the weighted sensitivity function. Let $P_o(s)$ be the fixed SISO plant. Then we are required to find

$$\inf \{\sup_{s \in \bar{H}} |S(s)| : C(s) \text{ internally stabilizes } P_o(s)\}.$$

We can reformulate this problem in the following way.

Minimal Sensitivity Problem 2.5: Let $r > 0$ be a real number such that there exists

$$S(s) : \bar{H} \to D_r := \{s \text{ in } \mathbb{C} : |s| < r\}$$

satisfying (2.2). Clearly, the Francis–Zames problem stated above is to find the infimum r_o, of all such real numbers r.

Next we would like to consider a kind of parameter variation which is motivated by the work of Doyle, Wall, and Stein [5] and Lehtomaki [19]. These authors consider various types of uncertainties in modeling dynamics. Their work shows that in several cases these uncertainties are equivalent to complex uncertainties in the multiplicative factor. We will therefore consider the following family of plants. Let $r > 0$ be given. Define

$$K_r := \{k : k=(1+s)^{-1} \quad \text{where } s \in \mathbb{C} \text{ and } |s| \leq r\}. \tag{2.6}$$

Now consider the family of plants

$$P_k(s) = kP_o(s)$$

where k belongs to K_r, and $P_o(s)$ is the nominal plant. (Doyle, Wall, and Stein [5] consider other types of modeling uncertainties as well. Each of these cases can also be translated into interpolation problems with different data and interpolating functions.) For this family of plants we consider the corresponding robust stabilization problem. Using the same method as in Lemma 2.3, it is easy to see that this problem can be formulated as follows.

Complex Parameter Variations 2.7: Let $D'_{1/r} := \{s \in \mathbb{C} : |s| \geq 1/r\}$, and $D_{1/r} = \mathbb{C} \setminus D'_{1/r} = \{s \in \mathbb{C} : |s| < 1/r\}$. Then for given $P_o(s)$ and $r > 0$, find

i) $S(s) : \bar{H} \to D_{1/r}$, and

ii) $S(s)$ satisfies (2.2).

We will now solve problems 2.4, 2.5, and 2.7 (and their weighted analogs) in a unified way. Let us first note that the conditions 2.4-i), 2.5-i), and 2.7-i) require the sensitivity function $S(s)$ to have range in a domain which is simply connected and not all of $\mathbb{C}$. But by the Riemann mapping theorem [25] these domains are all conformally equivalent to the unit disk D. In point of fact, in all these cases it is trivial to write explicit conformal equivalences between these domains and D which we will do shortly. But first, let us abstract the problem.

General Problem 2.8: Let $G \subsetneq C$ be given simply connected domain containing 0, 1. *Find (if possible) a rational analytic function*

$$S(s) : \bar{H} \to G$$

satisfying (2.2).

[It is clear that the general problem 2.8 includes problems 2.5, 2.6, and 2.7, and other problems such as gain-phase margin, etc., as special cases. As far as sensitivity optimization is concerned, 2.8 includes the unweighted sensitivity minimization problem but does not include the weighted sensitivity minimization problem. For the weighted case, see [37], [8], and (2.19).]

We will now give a simple procedure to solve this general problem which will lead to explicit solutions of problems 2.4, 2.5, and 2.7. In order to do this, we will have to describe, briefly, a certain notion from complex function theory, namely the *hyperbolic* or *Poincare metric*. For complete details, see the classic work of Nevanlinna [22]. We should note that in Helton [11] non-Euclidian metrics and their relations to problems in system theory have been discussed.

Hyperbolic Metrics 2.9: It is a classical fact that Nevanlinna–Pick interpolation is a generalization of the Schwarz lemma, and that the Schwarz lemma is a statement about the relationship between the properties of analyticity and a certain non-Euclidian metric on the disk called *hyperbolic* or *Poincare metric*. Since this notion will be so important to us in the sequel we would like to briefly review some of the basic properties of this metric. We follow the treatment of Ahlfors [1] to which we refer the reader for proofs of all the facts which we state below.

Let z_1, z_2 be in D. Define

$$\delta(z_1\, z_2) := \left| \frac{z_1 - z_2}{1 - \bar{z}_1 z_2} \right|. \tag{2.10}$$

The quantity $\delta(z_1, z_2)$ is a conformal invariant in the following sense. Given $\gamma : D \to D$ is a conformal equivalence, $\delta(z_1, z_2) = \delta(\gamma(z_1), (\gamma(z_2))$. Moreover, it is easy to check that $\delta(z_1, z_2) < 1$. Letting z_1 approach z_2, we get a metric on D, $(|dz|)/(1 - |z|^2)$. The *hyperbolic metric* on D is given infinitesimally by $(2|dz|)/(1 - |z|^2)$. Explicitly, the *hyperbolic distance* between two points z_1, z_2 in D is given by

$$d_D(z_1,\ z_2) = \log \frac{1 + \left| \dfrac{z_1 - z_2}{1 - \bar{z}_1 z_2} \right|}{1 - \left| \dfrac{z_1 - z_2}{1 - \bar{z}_1 z_2} \right|}. \tag{2.11}$$

In particular, for $r > 0$,

$$d_D(0,\ r) = \log \frac{1+r}{1-r}.$$

Next, let $G \subseteq \mathbb{C} \cup \{\infty\}$ be a simply connected domain with at least two boundary points. Then by the Riemann mapping theorem there exists $\lambda : G \to D$ a conformal equivalence. We define the *hyperbolic distance* on G by

$$d_G(z_1,\ z_2) = d_D(\lambda(z_1),\ \lambda(z_2)). \tag{2.12}$$

It is a fact that this definition is independent of the choice of conformal equivalence λ. (In [22], there are some variational formulas for d_G. There are also methods of finding this hyperbolic distance using a Green's function and the kernel functions for the domain G.)

The key fact which we need is the following version of the Schwarz lemma. See [1] for a proof.

Theorem 2.13: Let G_1, $G_2 \subseteq \mathbb{C} \cup \{\infty\}$ be simply connected domains with at least two boundary points. Let $f : G_1 \to G_2$ be *an analytic map. Then for all z_1, z_2 in G_1,*

$$d_{G_1}(z_1, z_2) \geq d_{G_2}(f(z_1), f(z_2)).$$

Moreover, one has equality if and only if f is a conformal equivalence.

This result will be the key in our treatment of robust stabilization. Before stating our solutions to the general problem 2.8, we need to set up some notation. Let $P_o(s)$ be the nominal plant as above with z_i in $\bar{H}$ the zeros and p_j in $\bar{H}$ the poles. Let $r: \bar{H} - \bar{D}$ be a fixed conformal equivalence. Let $\xi_i := \varphi(z_i)$ and $\psi_j := r(p_j)$. In the notation of Section I, define the interpolation data

$$a_i = \xi_i, \qquad i = 1, 2, \cdots, m$$

$$a_{j+m} = \psi_j, \qquad j = 1, 2, \cdots, n$$

$$b_i = 1, \qquad i = 1, 2, \cdots, m$$

$$b_{j+m} = 0, \qquad j = 1, 2, \cdots, n.$$

As in (1.4) consider now the α_{max} defined relative to this interpolation data. Note that if $P_o(s)$ has at least one open right half plane zero and one open right half pole, then $\alpha_{max} = \hat{\alpha}_{max}$ (see Section I). Moreover, since $P_o(s)$ has a zero at ∞, $\alpha_{max} \leq 1$. Indeed, if there is at least one zero of $P_o(s)$ in the open right half plane, then α_{max} becomes independent of the zeros and poles of $P_o(s)$ on the $j\omega$-axis and ∞. If $P_o(s)$ has no zeros in the open right half plane, then $\alpha_{max} = 1$. We can now state the following key result.

Theorem 2.14: The general problem 2.8 is solvable if and only if

$$d_G(0, 1) < d_D(0, \alpha_{max}) = \log \frac{1 + \alpha_{max}}{1 - \alpha_{max}}.$$

Proof: Since $G \subsetneq \mathbb{C}$ is a simply connected domain, we can find a conformal equivalence $\theta: G \to D$ such that $\theta(0) = 0$. Now consider the following commutative diagram

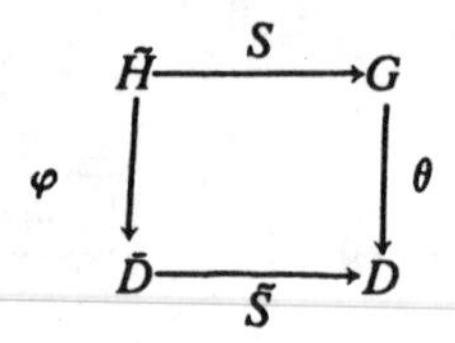

where $\tilde{S} := \theta \circ S \circ \varphi^{-1}$. Clearly, we can find S satisfying (2.2) if and only if we can find $\tilde{S}$ such that

$$\tilde{S}(\xi_i) = 0, \qquad i = 1, 2, \cdots, n \text{ and}$$

$$\tilde{S}(\psi_j) = \theta(1), \qquad j = 1, 2, \cdots, m.$$

By Theorem 1.5 and the definition (1.4) of α_{max}, it follows that we can find such an $\tilde{S}$ if and only if

$$|\theta(1)| < \alpha_{max},$$

or equivalently,

$$d_D(0, \theta(1)) < d_D(0, \alpha_{max}).$$

(Note that the inequality must be strict. Indeed, suppose we could find $\tilde{S}: \bar{D} \to D$ such that (2.15) holds with $|\theta(1)| = \alpha_{max}$. But $\tilde{S}(\bar{D})$ is compact, and therefore there exists $\epsilon > 0$ such that $\tilde{S}(\bar{D}) \subseteq D_{1-\epsilon} := \{|z| < 1 - \epsilon\}$. Therefore, we can find an $r > 1$ such that $r\tilde{S}: \bar{D} \to D$. Clearly, $r\tilde{S}$ has zeros at the ξ_i's, $r\tilde{S}(\psi_j) = r\theta(1)$, and $|r\theta(1)| > \alpha_{max}$. This contradicts the definition of α_{max}.) Thus, by Theorem 2.13 we can find $S: \bar{H} \to G$ with the required properties if and only if

$$d_G(0, 1) = d_D(0, \theta(1)) < d_D(0, \alpha_{max}) = \log \frac{1 + \alpha_{max}}{1 - \alpha_{max}}. \qquad \square$$

Remark: Theorem 2.14 essentially solves problems 2.4, 2.5, and 2.7. Indeed, we see that solving these problems can be divided into two parts. The first part requires computation of α_{max} which depends only on the zeros and poles of the nominal plant in the open right half plane when the plant has at least one open right half plane zero, and is 1 otherwise.

The second part of the solution of these problems is the computation of $d_G(0, 1)$. Certainly, this depends on the choice of G which in turn depends on the kind of uncertainty in the given problem. Given the domain, G, $d_G(0, 1)$ can be computed as explained in 2.9.

We shall now give explicit solutions to the above three problems.

2.4′ Solution to 2.4: We need to find

$$\theta: \mathbb{C} \setminus \left\{\left(-\infty, \frac{a}{a-1}\right] \cup \left[\frac{b}{b-1}, \infty\right)\right\} \to D$$

a conformal equivalence, such that $\theta(0) = 0$. Following standard procedures in conformal mapping theory (see, e.g., [26]), we find

$$\theta(s) = \frac{1 - \left[\left(1 - \left(\frac{b-1}{b}\right)s\right) \Big/ \left(1 - \left(\frac{a-1}{a}\right)s\right)\right]^{1/2}}{1 + \left[\left(1 - \left(\frac{b-1}{b}\right)s\right) \Big/ \left(1 - \left(\frac{a-1}{a}\right)s\right)\right]^{1/2}}.$$

It is easy to compute that

$$\theta(1) = \frac{1 - \sqrt{a/b}}{1 + \sqrt{a/b}}.$$

Theorem 2.14 implies that the gain margin problem is solvable if and only if

$$d_G(0, 1) = d_D(0, \theta(1)) < d_D(0, \alpha_{max})$$

which can be rewritten as

$$\frac{1 - \sqrt{a/b}}{1 + \sqrt{a/b}} < \alpha_{max}$$

or equivalently,

$$\frac{b}{a} < \left(\frac{1 + \alpha_{max}}{1 - \alpha_{max}}\right)^2 =: \beta_{max}. \qquad (2.17)$$

From this expression, certain interesting control theoretic implications can be drawn. For example, as α_{max} approaches 1, the maximal attainable gain margin goes to ∞. If the nominal plant $P_o(s)$ has no zeros in the *open* right half plane, i.e., we have a minimum phase plant, then it is immediate that $\alpha_{max} = 1$. Thus, for such plants given $b > 1 > a > 0$, one can always solve 2.4. In Section III we shall prove a similar result for multivariable plants.

On the other hand as α_{max} approaches zero, the maximal b/a approaches 1. In Theorem 2.21, we shall give a very simple useful upper bound for α_{max}.

2.5′ Solution to 2.5: In this case we need to find $\theta: D_r \to D$ such that $\theta(0) = 0$. Trivially $\theta(s) = s/r$, and

$$d_{D_r}(0, 1) = d_D(0, 1/r) = \log \frac{1 + 1/r}{1 - 1/r}.$$

Applying Theorem 2.14, problem 2.5 is solvable if and only if

$$\log \frac{1+1/r}{1-1/r}=d_{D_r}(0,\ 1)<d_D(0,\ \alpha_{\max})=\log \frac{1+\alpha_{\max}}{1-\alpha_{\max}}.$$

That is,

$$r>1/\alpha_{\max}.$$

Therefore, by definition, the minimal sensitivity

$$\inf_C \sup_{s\in \bar{H}} |S(s)|=1/\alpha_{\max} \tag{2.18}$$

where the infimum is taken over all internally stabilizing compensators.

This result reveals a basic connection between the sensitivity minimization problem and the gain margin problem. From this new general viewpoint, it is clear that Tannenbaum [26] and Zames and Francis [37] have solved two aspects of the same general problem.

2.7′ Solution to 2.7: In this case

$$G=D_{1/r}=\{s \in \mathbb{C} : |s|<1/r\}.$$

Thus, this is precisely the Zames–Francis [37] problem and for each $r < \alpha_{\max}$, the problem is solvable.

Theorem 2.14 gives a necessary and sufficient condition for the solvability of the general problem 2.8. Moreover, the proof of Theorem 2.14 shows that the construction of a stabilizing compensator $C(s)$ to solve problem 2.8 amounts to an interpolation problem from the unit disk to itself. Since in Section I, following classical interpolation theory, we have reviewed a parameterization of all solutions to any given interpolation problem, we can therefore explicitly write down all solutions to 2.8. For the gain margin problem, see [26] for explicit examples.

It is not difficult to incorporate the question of weighted sensitivity minimization into our general framework. This is the general problem considered by Zames and Francis [37]. Specifically, as above, let $P_o(s)$ be the given nominal plant transfer function. Let $W(s)$ be a proper stable rational function with no zeros in $\bar{H}$. Given a compensator $C(s)$, define the *weighted sensitivity function* to be

$$T(s) := W(s)(1+P_o(s)C(s))^{-1}.$$

Then the problem is to find

$$\mu(W) := \inf \{\sup_{s\in H} |T(s)| : C(s) \text{ is a stabilizing controller}\}.$$

Zames and Francis [37] showed that a compensator $C(s)$ internally stabilizes the plant $P_o(s)$ if and only if

i) $T(s)$ is analytic in $\bar{H}$;

ii) the zeros of $T(s)$ contain the set $\{p_j : i=1,\ 2,\ \cdots,\ n\}$; and

iii) $T(z_i)=W(z_i)$, $i=1,\ 2,\ \cdots,\ m$, multiplicities included.

In view of this result, we can define the following "interpolation data" following the notation of Section II:

$$a_i=\xi_i, \qquad i=1,\ 2,\ \cdots,\ m$$

$$a_{j+m}=\psi_j, \qquad j=1,\ 2,\ \cdots,\ n$$

$$b_i=W(z_i), \qquad i=1,\ 2,\ \cdots,\ m$$

$$b_{j+m}=0, \qquad j=1,\ 2,\ \cdots,\ n.$$

(Recall that ξ_i, ψ_j are the images in $\bar{D}$ via $\varphi:\bar{H} \to \bar{D}$ of the zeros and poles of $P_o(s)$ in the closed right half plane.) Let $\alpha_{\max}(W)$ denote the $\alpha_{\max}$ defined relative to this interpolation data (see (1.4) above). Then as in the case of unweighted sensitivity minimization, it is easy to see that

$$\mu(W)=1/\alpha_{\max}(W). \tag{2.19}$$

Remark 2.20: In the work of Doyle, Wall, Stein [5], Kimura [17], Lehtomaki [19], several synthesis problems arise which are very similar to the problems considered above. The analysis tests for robust stability lead to the question of finding holomorphic functions $f:\bar{H} \to D_r$ subject to certain interpolation conditions which arise from the internal stability constraint. Each of these problems can be easily treated using the interpolation theory discussed in Section I. Indeed, our observation on boundary interpolation (see Theorem 1.5) allows us in certain cases, to extend previous results, and to consider poles and zeros on the $j\omega$-axis and ∞. For example, using our techniques, we can relax the assumption (A_2 in Section IV) of [17] on the relative degree of the *uncertainty band function* by multiple interpolation at ∞.

It is of course useful to have an explicit formula for $\alpha_{\max}$. Using the Nevanlinna–Pick matrix, or the theory of Walsh [30, pp. 290–291], one can write down an exact expression for $\alpha_{\max}$ which is quite complicated. However, following some ideas of Nevanlinna [22, p. 52], it is easy to write down some very useful upper bounds for $\alpha_{\max}$. The exact result is the following.

Theorem 2.21: Let $f:D \to D$ be an analytic function such that $f(a_i) = 0$, $i = 1, \cdots, k$, $f(a_{j+k}) = \alpha$, $j = 1, \cdots, l$. Define

$$\lambda := \min_{1\le i\le k} \left\{\prod_{j=1}^{l} \left|\frac{a_{j+k}-a_i}{1-\bar{a}_i a_{j+k}}\right|\right\}$$

and

$$\mu := \min_{1\le j\le l} \left\{\prod_{i=1}^{k} \left|\frac{a_i-a_{j+k}}{1-a_i\bar{a}_{j+k}}\right|\right\}.$$

Then

$$|\alpha| \le \min\ (\lambda,\ \mu).$$

Proof: First via the conformal equivalence $\gamma:D \to D$ defined by

$$\gamma(z) := \frac{z-\alpha}{1-\bar{\alpha}z}$$

which sends 0 to α and α to 0, it is clearly enough to prove that $|\alpha| \le \lambda$ (to show $|\alpha| \le u$, we merely consider $\gamma \circ f$ and apply the previous inequality). Moreover, via the equivalence

$$e_i(z) := \frac{a_i-z}{1-\bar{a}_i z}$$

which sends 0 to a_i (for each $i = 1, \cdots, k$) and such that

$$e_i^{-1}(a_{j+k})=\frac{a_{j+k}-a_i}{1-\bar{a}_i a_{j+k}} =: \beta_{ji} \qquad j=1,\ \cdots,\ l$$

we are clearly reduced to proving the following assertion. Given $g_i:D \to D$ analytic such that $g(0) = 0$ and $g_i(\beta_{ji}) = \alpha$, $j = 1, \cdots, l$, then

$$|\alpha| \le \prod_{j=1}^{l} |\beta_{ji}|.$$

(Just take $g_i = f \circ e_i$, $i = 1, \cdots, k$.) Since this will be true for each i, we will be done. But the proof of this assertion is trivial using the following argument of Nevanlinna [22, p. 52]. Indeed, define

$$u_i(z) := \frac{g_i(z)-\alpha}{1-\bar{\alpha} g_i(z)} \Big/ \prod_{j=1}^{l} \left(\frac{z-\beta_{ji}}{1-\bar{\beta}_{ji} z} \right).$$

Clearly, u_i is analytic in D, and $\sup_{|z|=1} |u_i(z)| \leq 1$. Therefore, by the maximum modulus principle $|u_i(z)| \leq 1$ for all $z \in D$. For $z = 0$, we get then

$$|\alpha| \leq \prod_{j=1}^{l} |\beta_{ji}|$$

as required. □

Remark 2.22: Theorem 2.21 gives us a nice and useful upper bound on α_{max}. From this upper bound, we see that generally as the number of right half plane poles or zeros of $P_o(s)$ increases, α_{max} decreases, and hence the minimal sensitivity increases, and the maximal obtainable gain margin decreases. This bound provides a justification for some of the classical observations of Horowitz [13].

We conclude this section with some illustrative examples.

Examples 2.23: i) Consider a nominal model $P_o(s)$ which has one open right half plane zero at z_o, and one open right half plane pole at p_o. In this case, it is easy to compute that

$$\alpha_{max} = \left| \frac{z_o - p_o}{z_o + p_o} \right|.$$

Note from this formula that as the distance between z_o and p_o increases, α_{max} approaches 1. For the gain margin problem, this means that as $|z_o - p_o| \to \infty$, the maximal obtainable gain margin goes to ∞ as well; and similarly for the minimal sensitivity problem, the minimal sensitivity goes to 1. Conversely, as $|z_o - p_o| \to 0$, the maximal gain margin goes to 0 (dB), and the minimal sensitivity approaches ∞. [See formulas (2.17) and (2.18).]

ii) In [26] for the nominal plant $P_o(s) = (s-1)(s-2)/(s-3)(s-4)$, α_{max} was computed to be 0.027. It is interesting to compute a corresponding "α_{max}" in case we restrict our internally stabilizing compensators to be stable themselves. A procedure for doing this was given in [28] where a generalization of the famous result of Youla, Bongiorno, and Lu [31] was derived for variations in the gain factor. (Moreover, using an argument involving the logarithm it is possible to give an explanation of the parity interlacing property of that paper.) The value of "α_{max}" taken over stable compensators turned out to be 0.0146.

III. Remarks on the Multivariable Case

In this section, we present a simple result on the multivariable version of the gain margin problem. Let us consider the family of $p \times m$ real rational proper transfer matrices

$$P(s) = kP_o(s), \; k \in [a, b], \; b > 1 > a > 0.$$

We want to find a real rational compensator transfer matrix $C(s)$ such that the feedback system shown in Fig. 1 is internally asymptotically stable for all k in $[a, b]$. Let R denote the ring of stable proper rational functions. It is well known that R is a Euclidian domain (see [20] and [16]). Let $P_o(s) = N(s)D^{-1}(s)$ be a coprime factorization of $P_o(s)$, where $N(s)$, $D(s)$ have their entries in R. (See [29].) Let $\alpha(s)$ be the g.c.d. (over R) of all entries of $N(s)$. Then the zeros of $\alpha(s)$ in H are the blocking zeros of $P_o(s)$ in the open right half plane. We now can state the following.

Theorem 3.1: Suppose $P_o(s)$ has no blocking zeros in the open right half plane. Suppose that the roots of det $D(s)$ *in the open right half plane have multiplicity no greater than one. Then given any $b > 1 > a > 0$, there exists a compensator $C(s)$ such that the closed-loop system is internally asymptotically stable for each k in $[a, b]$.*

Proof: Let us consider the Smith-McMillan form of $P_o(s)$ over R. As is well known, there exist unimodular matrices U, V over R such that

$$UP_oV = \begin{bmatrix} \text{diag}\,(n_1/d_1, \; n_2/d_2, \; \cdots, \; n_l/d_l) & 0 \\ 0 & 0 \end{bmatrix},$$

where l is the nominal rank of P_o, 0 represents the zero matrix of appropriate size, n_i divides n_{i+1}, d_i divides d_{i-1}. It is a standard fact that $n_1(s)$ is the g.c.d. of all the entries of $N(s)$. Hence, by our assumption on blocking zeros of $P_o(s)$, $n_1(s)$ has no zeros in the open right half plane. Further,

$$\det D(s) = d_1 d_2 \cdots d_l.$$

By our assumption on the open right half plane zeros of det $D(s)$, and the divisibility properties of d_i's, it follows that for $i \geq 2$, $d_i(s)$ has no roots in the open right half plane. Consequently, for each $i = 1, \cdots, l$

$$P_i(s) := n_i(s)/d_i(s)$$

has either no zeros in the open right half plane or no poles in the open right half plane. It follows from Section II that there exist $c_i(s)$, $i = 1, 2, \cdots, l$ such that $c_i(s)$ internally stabilizes $kP_i(s)$ for each k in $[a, b]$, $i = 1, 2, \cdots, l$. Now define

$$C(s) = V(s) \begin{bmatrix} \text{diag}\,(c_1, \; c_2, \; \cdots, \; c_l) & 0 \\ 0 & 0 \end{bmatrix} U(s).$$

It is now easy to check that $C(s)$ internally stabilizes $kP_o(s)$ for each k in $[a, b]$. □

The above result shows that if the nominal plant $P_o(s)$ has no *blocking zeros* and has distinct right half plane poles, then there is *no upper bound* on the achievable gain margin by suitable design of stabilizing compensators. (At this point, we do not know if it is possible to remove the hypothesis of distinct right half plane poles in Theorem 3.1. It is easy to see, however, that our proof goes over under the slightly weaker hypothesis that the denominator matrix $D(s)$ has only one nontrivial invariant factor over the ring R. Of course, the condition of distinct poles holds generically.) This result is similar to the known results on systems with no right half plane *transmission zeros*. Zames [34], Zames and Bensoussan [35], Francis and Zames [8], and Helton [12] show that for systems with no right half plane transmission zeros, perfect tracking is possible.

IV. Pole-Zero Variations

It is also possible to consider robust stabilization problems involving variations in poles and zeros in our general framework. As an illustration of our methods, we shall treat in this section the case of a variations of a real pole. (Analogous considerations apply in case of variations of a real zero.)

Consider the following family of plants:

$$P_a(s) := \frac{\hat{P}(s)}{s-a}$$

where $a \in [a_o - \alpha, a_o + \beta]\alpha, \beta > 0$, and

$$P_o(s) := P_{a_o}(s) = \frac{\hat{P}(s)}{s-a_o}$$

is the nominal plant. In this case, we are required to find a proper compensator $C(s)$ such that

$$1 + C(s)\frac{\hat{P}(s)}{s-a} \neq 0 \tag{4.1}$$

for all $s \in \tilde{H}$, $a \in [a_o - \alpha, a_o + \beta]$, and, of course, we require that there are no unstable pole-zero cancellations between $P_a(s)$ and $C(s)$.

Now clearly we can rewrite (4.1) as

$$(s-a_o)+\hat{P}(s)C(s)=(s-a_o)(1+P_o(s)C(s)) \notin [-\alpha, \beta] \tag{4.2}$$

for all $s \in \bar{H}$. Define

$$T(s) := \frac{1}{(s-a_o)(1+P_o(s)C(s))}. \tag{4.3}$$

Then it is easy to see that we are required to find a real rational holomorphic function

$$T(s) : \bar{H} \rightarrow \mathbb{C} \setminus \{(-\infty, -1/\alpha] \cup [1/\beta, \infty)\} \tag{4.4}$$

such that

i) $T(z_i)=\dfrac{1}{z_i-a_o} \quad i=1, \cdots, m$

ii) $T(p_j)=0 \quad j=1, \cdots, n$ (4.5)

where $z_1, z_2, \cdots, z_m$ are the zeros of $P_o(s)$ in $\bar{H}$, and $p_1, \cdots, p_n$ are the poles of $\hat{P}(s)$ in $\bar{H}$ (multiplicities included).

From our discussion in Section II it is clear that in order to solve this problem, we can choose conformal equivalences

$$\varphi_1 : \mathbb{C} \setminus \{(-\infty, -1/\alpha] \cup [1/\beta, \infty)\} \rightarrow D$$

$$\varphi_2 : \bar{H} \rightarrow \bar{D}$$

and via φ_1, φ_2 derive a standard Nevanlinna–Pick interpolation problem from $\bar{D} \rightarrow D$ which can be solved using the techniques described in Section I. Since the images $\varphi_1(1/(z_i - a_o))$ $i = 1, \cdots, m$ depend on the given uncertainty (i.e., α, β), the relationship of the "maximal" α, β to the poles and zeros of the plant may be quite complicated. However, it is possible to check the solvability in any given instance and compute a compensator $C(s)$ (if one exists).

We now conclude this section with some illustrative examples.

Examples 4.6: i) Let

$$P_a(s) := \frac{(s-1)}{(s-a)} \frac{1}{(s+1)}$$

where $a \in [a_o - \alpha, a_o + \beta]$, $a_o > 1$, be a family of plants with a zero at 1, and an uncertain pole a. We wish to compute the maximal interval $[a_o - \alpha, a_o + \beta]$ for which it is possible to find a proper compensator $C(s)$ which satisfies (4.1).

Using the interpolating conditions (4.4), (4.5i), (4.5ii), and the conformal equivalences φ_1, φ_2 it is easy to show that for all α, $\beta > 0$ such that $1 \notin [a_o - \alpha, a_o + \beta]$, one can find an internally stabilizing compensator $C(s)$. This means of course that we can internally stabilize the uncertain family $P_a(s)$ on any interval around the nominal a_o, as long as there are no unstable pole-zero cancellations.

ii) In this example, we would like to consider some of the complications that can arise if one considers simultaneous pole-zero variations. In point of fact, we would like to show that even if one has "small" simultaneous pole-zero variations in which an open right half plane pole circles around an open right half plane zero, or vice versa, one may not be able to find an internally stabilizing compensator. The argument which follows below is taken from Tannenbaum [27, pp. 136–137] but because of its obvious relevance to our present discussion, we wish to reproduce it in part.

Consider a family of proper plants $P_k(s)$ continuously parameterized by a compact set K (i.e., $k \in K$), and with no unstable pole-zero cancellations. Set

$$P_k(s) = \frac{P_{1k}(s)}{P_{2k}(s)}$$

where $P_{ik}(s) \in \mathbb{R}[s]$ for each $k \in K$, $i = 1, 2$. Then to find an internally stabilizing compensator $C(s)$ for this family (as described in the Introduction), we are required in particular to find fixed polynomials $C_1(s)$, $C_2(s)$ such that

$$P_{1k}(s)C_1(s)+P_{2k}(s)C_2(s) \neq 0 \tag{4.7}$$

for all $s \in \bar{H}$, $k \in K$.

Now suppose that there exist points $z_1, z_2 \in H$ ($z_1 \neq z_2$) such that

a) $P_{1k}(z_1)=P_{1k}(z_2)=0$ for all $k \in K$;

b) $P_{2k}(z_1)$ circles around $0 \in c$ as k varies in K;

c) $P_{2k}(z_2)$ is a fixed nonzero constant for all $k \in K$.

Under these hypothesis, we claim that (4.7) has no solution even if we require C_1 and C_2 to be only continuous. To see this, suppose to the contrary that we could find complex continuous functions C_1, C_2 such that

$$F_k := P_{1k}C_1 + P_{2k}C_2$$

has no right half plane zeros. Note that $F_k(z_1) = C_2(z_1)P_{2k}(z_1) \neq 0$ (since otherwise z_1 would be a right half plane zero), and hence at z_1, the function F_k circles around 0 as k varies in K. Similarly, $F_k(z_2) = C_2(z_2)P_{2k}(z_2)$ is a fixed nonzero constant for all $k \in K$. By continuity, since $z_1, z_2 \in H$, and the line connecting z_1 and z_2 lies in H, for some point on this line F_k must vanish, contradicting our supposition that F_k had no right half plane zeros. Indeed, to see this, just note that as we move along the line from z_1 to z_2, the closed loop which $F_k(z_1)$ describes about the origin as k varies in K is deformed to the point $F_k(z_2) \neq 0$, and consequently must cross the origin.

In [27], this failure of the possibility of robust stabilization in such cases of simultaneous pole-zero variations in the plant is related to some results of [3] on the topology of rational transfer functions.

V. Conclusions

In this paper we have used certain classical techniques from complex function theory to solve problems in robust control system synthesis. One of our main contributions is to show that real parameter uncertainties, complex parameter uncertainties (arising from errors in modeling dynamics), and sensitivity minimization problems are essentially the same. We were able to decompose these problems into two parts: calculation of the invariant α_{max} which depends on right-half plane poles and zeros of the nominal plant and calculation of the hyperbolic distance. The first part can be easily approached via the Nevanlinna–Pick method. The second part depends crucially on the kind of uncertainty being considered. For the kinds of uncertainty we considered in this paper, this computation of the hyperbolic distance is relatively straightforward. It is possible to imagine parameter uncertainties which can lead to regions G in the general problem 2.8 which can be quite complicated. In this case, one may be interested in obtaining upper and lower bounds on $d_G(0, 1)$. For getting these bounds the following fact is often useful. If $G_1 \subseteq G_2 \subsetneq \mathbb{C}$ are simply connected regions containing 0 and 1, then

$$d_{G_2}(0, 1) \leq d_{G_1}(0, 1).$$

Thus, by finding suitable regions inside and outside G such as disks, one may be able to get good upper and lower bounds on the hyperbolic distance. It remains to be seen whether classical tools such as Green's function are useful in this regard.

In the multivariable case, we have given a result which is generically applicable (since generic multivariable systems do not have blocking zeros). However, a full investigation of the multivariable problem is still an open area for future research.

Finally, one would like to be able to consider simultaneous variations in poles and zeros of the plant. This appears to be a difficult problem which we are currently investigating.

REFERENCES

[1] L. Ahlfors, *Conformal Invariants*. New York: McGraw-Hill, 1973.

[2] J. Ball, "Interpolation problems of Pick-Nevanlinna and Loewner types for meromorphic matrix functions," *Integral Equations and Operator Theory*, vol. 6, pp. 804-840, 1983.

[3] R. W. Brockett, "Some geometric questions in the theory of linear systems," *IEEE Trans. Automat. Contr.*, vol. AC-21, pp. 449-464, 1976.

[4] P. Delsarte, Y. Genin, and Y. Kamp, "The Nevanlinna-Pick problem for matrix valued functions," *SIAM J. Appl. Math.*, vol. 36, pp. 47-61, 1979.

[5] J. C. Doyle, J. E. Wall, and G. Stein, "Performance and robustness analysis for unstructured uncertainty," in *Proc. 21st IEEE Conf. Decision Contr.*, Orlando, FL, 1982, pp. 629-636.

[6] B. A. Francis, J. W. Helton, and G. Zames, "H^∞-optimal feedback controllers for linear multivariable systems," *IEEE Trans. Automat. Contr.*, vol. AC-29, pp. 888-900, 1984.

[7] B. A. Francis and G. Zames, "On optimal sensitivity theory for SISO feedback systems," in *Proc. 21st IEEE Conf. Decision Contr.*, Dec. 1982, pp. 623-628.

[8] ——, "On H^∞-optimal sensitivity theory for SISO feedback systems," *IEEE Trans. Automat. Contr.*, vol. AC-29, pp. 9-16, 1984.

[9] S. Haykin, *Nonlinear Methods of Spectral Analysis*. New York: Springer-Verlag, 1979.

[10] J. W. Helton, "Orbit structure of the mobins transformation semigroup action on H^∞ (broadband matching)," in *Advances in Math. Suppl. Studies*, vol. 3. New York: Academic, 1978.

[11] ——, "Non-Euclidian functional analysis and electronics," *Bull. Amer. Math. Soc.*, vol. 7, pp. 1-64, 1982.

[12] ——, "Worst case analysis in the frequency domain: The H^∞-approach to control," Dep. Math., Univ. California, San Diego, Tech. Rep., 1983.

[13] I. Horowitz, *Synthesis of Feedback Systems*. New York: Academic, 1963.

[14] I. Horowitz and A. Gera, "Blending of uncertain nonminimum phase plants for elimination or reduction of nonminimum-phase property," *Int. J. Syst. Sci.*, vol. 10, pp. 1007-1024, 1976.

[15] I. Horowitz and M. Sidi, "Optimum synthesis of nonminimum-phase feedback systems with parameter uncertainty," *Int. J. Contr.*, vol. 27, pp. 361-386, 1978.

[16] N. T. Hung and B. D. O. Anderson, "Triangularization technique for the design of multivariable control systems," *IEEE Trans. Automat. Contr.*, vol. AC-24, pp. 455-460, 1979.

[17] H. Kimura, "Robust stabilizability for a class of transfer functions," *IEEE Trans. Automat. Contr.*, vol. AC-29, pp. 788-793, 1984.

[18] M. G. Krein and A. A. Nudelman, *The Markov Movement Problem and Extremal Problems*. Providence, RI: AMS, 1977.

[19] N. Lehtomaki, "Practical robustness measures in multivariable control system analysis," Ph.D. dissertation, Mass. Inst. Technol., Cambridge, MA, 1981.

[20] A. S. Morse, "System invariants under feedback and cascade control," in *Proc. Int. Symp. Math. Syst. Theory*, Udine, Italy, 1976.

[21] R. Nevanlinna, "Über beschränkte Funktionen die in gegebenen Punkten vorgeschriebene Werte annehmen," *Ann. Acad. Sci. Fenn.*, vol. 13, 1919.

[22] ——, *Analytic Functions*. New York: Springer-Verlag, 1953.

[23] G. Pick, "Über die Beschränkungen analytischer Funktionen, welche durch vorgegebenen Werte annehmen," *Math. Ann.*, vol. 77, pp. 7-23, 1916.

[24] M. Rosenblum and M. Rovnyak, "An operator theoretic approach to theorems of the Pick-Nevanlinna and Loewner types. I," *Integral Equations Operator Theory*, vol. 3, pp. 408-436, 1980.

[25] W. Rudin, *Real and Complex Analysis*. New York: McGraw-Hill, 1966.

[26] A. Tannenbaum, "Feedback stabilization of plants with uncertainty in the gain factor," *Int. J. Contr.*, vol. 32, pp. 1-16, 1980.

[27] A. Tannenbaum, *Invariance and System Theory: Algebraic and Geometric Aspects*. New York: Springer-Verlag, 1981.

[28] ——, "Modified Nevanlinna-Pick interpolation and feedback in the gain factor," *Int. J. Contr.*, vol. 36, pp. 331-336, 1982.

[29] M. Vidyasagar, "On the use of right-coprime factorization in distributed feedback systems containing unstable systems," *IEEE Trans. Circuits Syst.*, vol. CAS-26, pp. 916-921, 1978.

[30] J. Walsh, *Interpolation and Approximation by Rational Functions in the Complex Domain*. AMS Colloquium Publications, vol. 20, Fourth Edition, 1965.

[31] D. C. Youla, J. Bongiorno, and Y. Lu, "Single loop feedback stabilization of linear multivariable dynamic plants," *Automatica*, vol. 10, pp. 159-173, 1974.

[32] D. C. Youla and M. Saito, "Interpolation with positive real functions," *J. Franklin Inst.*, vol. 284, pp. 77-108, 1967.

[33] G. Zames, "Feedback and optimal sensitivity: Model reference transformations, multiplicative seminorms, and approximate inverses," in *Proc. 17th Allerton Conf.*, pp. 744-751, 1979.

[34] ——, "Feedback and optimal sensitivity: Model reference transformations, multiplicative seminorms, and approximate inverses," *IEEE Trans. Automat. Contr.*, vol. AC-26, pp. 301-320, 1981.

[35] G. Zames and D. Bensoussan, "Multivariable sensitivity reduction by feedback," *IEEE Trans. Automat. Contr.*, vol. AC-27, pp. 1030-1035, 1983.

[36] G. Zames and B. A. Francis, "A new approach to classical frequency methods: Feedback and minimax sensitivity," in *Proc. 20th IEEE Conf. Decision Contr.*, pp. 867-874, Dec. 1981.

[37] G. Zames and B. A. Francis, "Feedback, minimax sensitivity, and optimal feedback," *IEEE Trans. Automat. Contr.*, vol. AC-28, pp. 585-601, 1983.

Robust Controllers for Uncertain Linear Multivariable Systems*

M. VIDYASAGAR† and H. KIMURA‡

Necessary and sufficient conditions are derived for a particular controller to achieve robust stabilization in the face of various types of plant uncertainty; in addition, necessary and sufficient conditions are derived for the existence of such a controller.

Key Words—Control system design; feedback control; robust control; mathematical system theory; stability criteria.

Abstract—This paper is addressed to three distinct yet related topics in the design of controllers for imprecisely known linear multivariable systems. In the first part, it is supposed that the plant to be stabilized is subject to additive or multiplicative uncertainties, and necessary and sufficient conditions are derived for the existence of a controller that stabilizes *all* plants within this band of uncertainty. In the second part, in contrast with the first part, it is supposed that the number of unstable poles of the plant to be stabilized is not precisely known. The type of plant uncertainty is the so-called "stable-factor" uncertainty, and necessary and sufficient conditions are given for robust stabilization. In the third part, the model of uncertainty is a ball in the space of rational matrices metrized by the so-called graph metric, and sufficient conditions for robust stabilization are derived.

INTRODUCTION

THIS PAPER is addressed to three distinct yet related problems in the design of stabilizing controllers for imprecisely known linear multivariable systems. To state each of these problems formally, suppose the imprecisely known plant which is to be stabilized is nominally modeled by a rational transfer matrix $P_0(s)$; the "true" plant is not necessarily P_0, but lies within some "domain of uncertainty" containing P_0. The three problems differ in the representation of this domain of uncertainty.

In the first problem, the true plant P is supposed to have the same number of right half-plane (RHP) poles as P_0, though not necessarily at the same locations. In the case of additive uncertainty, P is assumed to satisfy

$$\|P(j\omega) - P_0(j\omega)\| < |r(j\omega)|, \forall\omega. \qquad (1.1)$$

where r is a prespecified stable rational function. In the case of multiplicative uncertainty, P is assumed to satisfy

$$P(s) = (I + L(s))P_0(s). \qquad (1.2)$$

where

$$\|L(j\omega)\| < |r(j\omega)| \forall\omega. \qquad (1.3)$$

Let $A(P_0, r)$ (resp. $M(P_0, r)$) denote the class of all plants P that have the same number of RHP poles as P_0 and satisfy (1.1) [resp. (1.2)]. In Doyle and Stein (1981), Chen and Desoer (1982), necessary and sufficient conditions are presented that a controller must satisfy in order to stabilize all plants in the class $A(P_0, r)$ or $M(P_0, r)$. These papers leave open the question of whether such a controller actually exists. This is the question tackled in Section 3. Specifically, necessary and sufficient conditions are given for the existence of robustly stabilizing controllers in the case of $A(P_0, r)$ and $M(P_0, r)$.

The assumption that the number of unstable poles of the plant to be stabilized is exactly known is rather restrictive. In some applications, such as large flexible spacecraft, the number of unstable poles can and does change as the configuration of the spacecraft is changed. The objective of Section 4 is to put forward a model of plant uncertainty wherein the various plants within the domain of uncertainty need not all have the same number of RHP poles. This model is called stable-factor uncertainty, and can be described as follows: Let P_0 be the nominal plant model, and factor $P_0(s)$ as $N_0(s)[D_0(s)]^{-1}$

* Received 14 May 1984; revised 18 June 1985. The original version of this paper was presented at the 9th IFAC World Congress on A Bridge Between Control Science and Technology which was held in Budapest, Hungary during July 1984. The Published Proceedings of this IFAC Meeting may be ordered from Pergamon Press Limited, Headington Hill Hall, Oxford OX3 0BW, U.K. This paper was recommended for publication in revised form by Associate Editor J. Ackermann under the direction of Editor H. Kwakernaak. This research was supported by the Natural Sciences and Engineering Research Council of Canada under grant No. A-1240.

† Department of Electrical Engineering, University of Waterloo, Waterloo, Ontario, Canada N2L 3G1.

‡ Department of Control Engineering, Osaka University, Toyonaka, Osaka, Japan.

where N_0, D_0 are right-coprime stable rational matrices [see Desoer *et al.* (1980) and Vidyasagar *et al.* (1982)]. The class $S(N_0, D_0, r)$ consists of all plants P that satisfy $P(s) = N(s)[D(s)]^{-1}$ for some stable rational matrices N, D such that

$$\left\| \begin{bmatrix} N - N_0 \\ D - D_0 \end{bmatrix} (s) \right\| < |r(s)| \forall s \text{ with } Re\, s \geq 0, \quad (1.4)$$

where r is a specified stable rational function. Thus the class $S(N_0, D_0, r)$ depends not just on P_0 and r but on a particular right-coprime factorization of P_0. In Section 4, necessary and sufficient conditions that a controller must satisfy in order to stabilize all plants in the class $S(N_0, D_0, r)$ are derived.

Finally, in Section 5, the model of plant uncertainty is a ball centered at P_0. Thus it is assumed that the true plant is only known to lie within some distance r of the nominal plant P_0, where r is a positive real number, and the distance between P and P_0 is measured using the graph metric defined in Vidyasagar (1984). In this case the domain of uncertainty is

$$\mathbf{B}(P_0, r) = \{P : d(P, P_0) < r\}, \quad (1.5)$$

where d denotes the graph metric distance. This uncertainty model has the advantages that (i) P need not have the same number of RHP poles as P_0, and (ii) the domain of uncertainty $\mathbf{B}(P_0, r)$ depends only on P_0 and r, and not on any particular factorization of P_0. In Section 5, sufficient conditions are given for a controller to stabilize all plants in the class $\mathbf{B}(P_0, r)$. These sufficient conditions are readily extended to the case where both the plant and controller are perturbed.

As a preparation for the principal results, Section 2 contains a precis of known facts. Finally, Section 6 contains some concluding remarks.

2. NOTATION AND PRELIMINARIES

Throughout the paper, $\mathbf{S}$ denotes the set of proper stable rational functions with real coefficients. The symbol $\mathbf{M}(\mathbf{S})$ is a generic symbol denoting the set of all matrices (of whatever order) whose elements belong to $\mathbf{S}$. Thus, $A, B \in \mathbf{M}(\mathbf{S})$ does not imply that A and B have the same order. This notation is very useful because, almost always, the actual orders of the various matrices encountered in the discussion need not be displayed explicitly and can easily be determined should the need arise.

The set $\mathbf{S}$ is a subset of the space H_∞ of analytic functions bounded over the right half-plane. Specifically, H_∞ consists of all complex-valued analytic functions f over the open RHP with the property that

$$\limsup_{\sigma \to 0^+ Re\, s \geq \sigma} |f(s)| < \infty. \quad (2.1)$$

If $f \in H_\infty$, then the domain of definition of f can be extended to include the $j\omega$-axis, and the boundary function $\omega \to f(j\omega)$ is in $L_\infty(-\infty, \infty)$. Moreover, if the norm of $f \in H_\infty$ is defined as

$$\|f\|_\infty = \sup_{\sigma > 0} \sup_{Re\, s \geq \sigma} |f(s)|, \quad (2.2)$$

then actually

$$\|f\|_\infty = \operatorname*{ess\,sup}_\omega |f(j\omega)|. \quad (2.3)$$

A function $f \in H_\infty$ is symmetric if $f(s) = \bar{f}(\bar{s}) \forall s$ in the RHP, where the bar denotes complex conjugation. Note that $\mathbf{S}$ is precisely the set of symmetric rational functions in H_∞. For further basic facts about H_∞, see Duren (1970).

A function $f \in H_\infty$ is inner if $|f(j\omega)| = 1$ for almost all ω. The definition of an outer function is more technical, but a rational function $f \in H_\infty$ is outer if and only if $f(s) \neq 0$ whenever $Re\, s > 0$ (however, f can have zeros on the $j\omega$-axis or at infinity). For convenience, let us refer to the $j\omega$-axis plus the point at infinity as the extended $j\omega$-axis. If f is rational, outer, and also does not vanish at any point on the extended $j\omega$-axis, then f is a unit of H_∞, in that the function $s \to 1/f(s)$ also belongs to H_∞. Every rational $f \in H_\infty$ can be factored as $f_i f_o$, where f_i is inner and f_o is outer. In particular, if f is rational and has no zeros on the extended $j\omega$-axis, then its outer factor f_o is a unit of H_∞.

If $F \in \mathbf{M}(H_\infty)$, i.e. if all components of the matrix F are H_∞-functions, then we define

$$\|F\|_\infty = \operatorname*{ess\,sup}_\omega \bar{\sigma}(F(j\omega)), \quad (2.4)$$

where $\bar{\sigma}(\cdot)$ denotes the largest singular value of a matrix.

A rational matrix $F \in H_\infty^{m \times n}$ is inner if $F^*(j\omega)F(j\omega) = I \forall \omega$, where * denotes the conjugate transpose, and is outer if $F(s)$ has full row rank at all s in the open RHP. Note that if $F \in H_\infty^{m \times n}$ is inner (resp. outer), then $m \geq n$ (resp. $m \leq n$). Every rational matrix $F \in H_\infty^{m \times n}$ with $m \geq n$ can be factored as $F_i F_o$ where F_i is inner and F_o is square outer. Every rational matrix $F \in H_\infty^{m \times n}$ with $m \leq n$ can be factored as $G_i G_o$ where G_i is square inner and G_o is outer. Finally, if $F \in \mathbf{M}(H_\infty)$ is rational and square, then both types of factorization are possible. If F has full rank at all points on the extended $j\omega$-axis, then its outer factor is a unit matrix in $\mathbf{M}(H_\infty)$, i.e. its inverse also belongs to $\mathbf{M}(H_\infty)$. Note that if F is square inner, then its adjoint matrix F^{adj} and its determinant $|F|$ are also inner. Finally, multiplication (left or right) by an inner matrix preserves norms. Thus, if F, G are inner and $H \in \mathbf{M}(H_\infty)$, then

$\|FH\|_\infty = \|H\|_\infty, \|HG\|_\infty = \|H\|_\infty$. For further results concerning factorizations, see Sz-Nagy and Foias (1970).

Suppose $\lambda_1, \ldots, \lambda_n$ are distinct points in the open RHP, and $F_1, \ldots, F_n$ are complex matrices, all of the same order, with $\|F_i\| < 1 \; \forall i$. Then a classical result states that there exists an $F \in \mathrm{M}(H_\infty)$ such that $\|F\|_\infty \leq 1$ and $F(\lambda_i) = F_i \; \forall i$ if and only if the matrix

$$P = \begin{bmatrix} P_{11} & \cdots & P_{1n} \\ \vdots & \vdots & \vdots \\ P_{n1} & \cdots & P_{nn} \end{bmatrix}, \tag{2.5}$$

where

$$P_{ij} = \frac{I - F_i^* F_j}{\bar{\lambda}_i + \lambda_j}, \tag{2.6}$$

is nonnegative definite. See Walsh (1935) for the scalar case and Delsarte *et al.* (1979) for the matrix case.

Let $\mathbb{R}(s)$ denote the set of rational functions with real coefficients, and suppose a plant has the transfer matrix $P \in \mathrm{M}(\mathbb{R}(s))$. Suppose a feedback controller $C \in \mathrm{M}(\mathbb{R}(s))$ is applied to the plant P, resulting in the closed-loop transfer matrix

$$H(P, C) = \begin{bmatrix} (I + PC)^{-1} & -P(I + CP)^{-1} \\ C(I + PC)^{-1} & (I + CP)^{-1} \end{bmatrix} \tag{2.7}$$

We say that the pair (P, C) is *stable*, or that C *stabilizes* P if $H(P, C) \in \mathrm{M}(\mathrm{S})$. The symbol $S(P)$ denotes the set of all controllers that stabilize P.

Given a plant P, the set $S(P)$ can be explicitly parametrized using the concept of coprime factorizations over S defined in Vidyasagar (1975) and exploited in Desoer *et al.* (1980); see also Vidyasagar *et al.* (1982). A pair (N, D) is a right-coprime factorization (r.c.f.) of P if

(i) $N, D \in \mathrm{M}(\mathrm{S})$ and $P(s) = N(s)[D(s)]^{-1}$;

(ii) there exist $X, Y \in \mathrm{M}(\mathrm{S})$ such that $XN + YD = I$. A left-coprime factorization (l.c.f.) $(\tilde{D}, \tilde{N})$ of P is defined analogously.

Theorem 1. (Desoer *et al.*, 1980; Vidyasagar *et al.*, 1982.) Suppose $P \in \mathrm{M}(\mathbb{R}(s))$, and let (N, D), $(\tilde{D}, \tilde{N})$ be any r.c.f. and any l.c.f. of P. Let $X, Y, \tilde{X}, \tilde{Y}$ be solutions of

$$XN + YD = I, \qquad \tilde{N}\tilde{X} + \tilde{D}\tilde{Y} = I. \tag{2.8}$$

Then

$$S(P) = \{(Y - R\tilde{N})^{-1}(X + R\tilde{D}) : R \in \mathrm{M}(\mathrm{S}) \text{ and } |Y - R\tilde{N}| \neq 0\} \tag{2.9}$$

$$= \{(\tilde{X} + DR)(\tilde{Y} - NR)^{-1} : R \in \mathrm{M}(\mathrm{S}) \text{ and } |\tilde{Y} - NR| \neq 0\}. \tag{2.10}$$

Moreover, if P is strictly proper, the nonsingularity constraint can be dropped.

3. ADDITIVE AND MULTIPLICATIVE PERTURBATIONS

Suppose $P_0 \in \mathrm{M}(\mathbb{R}(s))$ is a nominal plant, $r \in \mathrm{S}$, and consider the classes $A(P_0, r)$, $M(P_0, r)$ defined in the introduction. Conditions for a controller $C \in S(P_0)$ to stabilize all plants in each of these classes are available in the literature.

Theorem 2. (Doyle and Stein, 1981; Chen and Desoer, 1982). A controller $C \in S(P_0)$ stabilizes all plants in the class $A(P_0, r)$ if and only if

$$\|C(I + P_0 C)^{-1} r\|_\infty \leq 1. \tag{3.1}$$

C stabilizes all plants in the class $M(P_0, r)$ if and only if

$$\|P_0 C(I + P_0 C)^{-1} r\|_\infty \leq 1. \tag{3.2}$$

It is natural to ask whether, given a nominal plant P_0 and a function r, there actually exists a C that satisfies (3.1) or (3.2). The purpose of this section is to answer these questions.

Theorem 3. Suppose a nominal plant P_0, free of extended $j\omega$-axis poles, and a function $r \in \mathrm{S}$, free of zeros on the extended $j\omega$-axis, are specified. Let (N, D), $(\tilde{D}, \tilde{N})$ be any r.c.f. and l.c.f. of P_0, and let $\tilde{X}$, $\tilde{Y} \in \mathrm{M}(\mathrm{S})$ be any particular solutions of the identity $\tilde{N}\tilde{X} + \tilde{D}\tilde{Y} = I$. Factor D, $\tilde{D}$ in the form

$$D = D_i D_o, \; \tilde{D} = \tilde{D}_o \tilde{D}_i, \tag{3.3}$$

where D_i, $\tilde{D}_i$ are inner and D_o, $\tilde{D}_o$ are outer, and assume without loss of generality that $|D_i| = |\tilde{D}_i|$. By Kailath (1980, p. 446), $|D|$ and $|\tilde{D}|$ are associates; so are $|D_o|$ and $|\tilde{D}_o|$, since both are units. Hence $|D_i|$ and $|\tilde{D}_i|$ are also associates. Since both are inner, we have $|D_i| = \pm|\tilde{D}_i|$ in any case. Finally, factor r as $r_i r_o$ where r_i is inner and r_o is a unit of S. Under these conditions, there exists a $C \in S(P_0)$ that stabilizes all plants in the class $A(P_0, r)$ if and only if

$$\min_{S \in \mathrm{M}(\mathrm{S})} \|D_i^{\mathrm{adj}} \tilde{X} \tilde{D}_o r_o + \delta S\|_\infty \leq 1, \tag{3.4}$$

where $\delta = |D_i| = |\tilde{D}_i|$.

If δ has only simple RHP zeros (or equivalently, if P_0 has only simple RHP poles), (3.4) can be stated in a more readily testable form.

Corollary 1.1. Suppose δ has only simple zeros, and let $\lambda_1, \ldots, \lambda_n$ denote these zeros. Let

$$F_j = (D_i^{\mathrm{adj}} \tilde{X} \tilde{D}_o r_o)(\lambda_j), \; j = 1, \ldots, n, \text{ and define}$$

$$Q_{jk} = (I - F_j^* F_k)/(\bar{\lambda}_j + \lambda_k), \qquad 1 \leq j, k \leq n. \tag{3.5}$$

$$Q = \begin{bmatrix} Q_{11} & \cdots & Q_{1n} \\ \vdots & \vdots & \vdots \\ Q_{n1} & \cdots & Q_{nn} \end{bmatrix}. \tag{3.6}$$

Then there exists a $C \in S(P_0)$ that stabilizes all plants in the class $A(P_0, r)$ if and only if Q is nonnegative definite.

The treatment of multiplicative perturbations is entirely similar to that of additive perturbations.

Theorem 4. Suppose a function $r \in \mathrm{S}$ and a nominal plant $P_0 \in \mathrm{M}(\mathbb{R}(s))$ are specified, and suppose in addition that P_0 has no poles on the extended $j\omega$-axis. Let (N, D), $(\tilde{D}, \tilde{N})$ be any r.c.f. and l.c.f. of P_0, and let $X, Y \in \mathrm{M(S)}$ be any particular solution of the identity $XN + YD = I$. Under these conditions, there exists a $C \in S(P_0)$ that stabilizes all plants in the class $M(P_0, r)$ if and only if

$$\min_{R \in \mathrm{M(S)}} \|N(X + R\tilde{D})\|_\infty \leq 1. \tag{3.7}$$

Suppose in addition that P_0 has at least as many inputs as outputs, and that r has no zeros on the extended $j\omega$-axis. Then the condition (3.7) can be further simplified as follows: factor $N, \tilde{D}$ in the form

$$N = N_i N_o, \qquad \tilde{D} = \tilde{D}_o \tilde{D}_i, \tag{3.8}$$

where $N_i, \tilde{D}_i$ are inner and $N_o, \tilde{D}_o$ are outer. Finally, factor r as $r_i r_o$ where r_i is inner and r_o is a unit of S. Under these conditions, there exists a $C \in S(P_0)$ that stabilizes all plants in the class $M(P_0, r)$ if and only if

$$\min_{R \in \mathrm{M(S)}} \|N_o X r_o \tilde{D}_i^{\mathrm{adj}} + \delta R\|_\infty \leq 1, \tag{3.9}$$

where $\delta = |\tilde{D}_i|$.

Proof of Theorem 3. By Theorem 1, the set of all $C \in S(P_0)$ is described by

$$S(P_0) = \{(\tilde{X} + DS)(\tilde{Y} - NS)^{-1} : S \in \mathrm{M(S)} \text{ and } |\tilde{Y} - NS| \neq 0\}, \tag{3.10}$$

where the nonsingularity constraint is automatically satisfied if P is strictly proper. Moreover, if $C = (\tilde{X} + DS)(\tilde{Y} - NS)^{-1}$, then

$$C(I + P_0 C)^{-1} = (\tilde{X} + DS)\tilde{D}. \tag{3.11}$$

Hence there exists a $C \in S(P_0)$ satisfying (3.1) if and only if there exists an $S \in \mathrm{M(S)}$ such that $\|\tilde{X} + DS)\tilde{D}r\|_\infty \leq 1$, i.e. if and only if

$$\min_{S \in \mathrm{M(S)}} \|\tilde{X} + DS)\tilde{D}r\|_\infty \leq 1. \tag{3.12}$$

A few simple manipulations bring (3.12) into the form (3.4). Note that

$$\begin{aligned} \|\tilde{X}\tilde{D}r + DS\tilde{D}r\|_\infty &= \|\tilde{X}\tilde{D}_o\tilde{D}_i r_o r_i + D_i D_o S \tilde{D}_o \tilde{D}_i r_o r_i\|_\infty \\ &= \|D_i^{\mathrm{adj}} \tilde{X} \tilde{D}_o r_o r_i \delta + \delta^2 r_i D_o S \tilde{D}_o r_o\|_\infty \end{aligned}$$

after left multiplication by D_i^{adj} and right multiplication by $\tilde{D}_i^{\mathrm{adj}}$

$$= \|\delta r_i (D_i^{\mathrm{adj}} \tilde{X} \tilde{D}_o r_o + \delta S_1)\|_\infty$$

where $S_1 = D_o S \tilde{D}_o r_o$ is a new free parameter

$$= \|D_i^{\mathrm{adj}} \tilde{X} \tilde{D}_o r_o + \delta S_1\|_\infty$$

since δ and r_i are inner. (3.13)

Hence (3.12) and (3.4) are equivalent.

Proof of Corollary 1.1. Let F denote $D_i^{\mathrm{adj}} \tilde{X} \tilde{D}_o r_o$, and note that $F \in \mathrm{M}(H_\infty)$. Now, a matrix $G \in \mathrm{M}(H_\infty)$ is of the form $F + \delta S$ for some $S \in \mathrm{M}(H_\infty)$ if and only if $G(\lambda_j) = F(\lambda_j) = F_j \forall j$. Thus (3.4) holds if and only if there exists a matrix $G \in \mathrm{M}(H_\infty)$ with $\|G\|_\infty \leq 1$ such that $G(\lambda_j) = F_j \forall j$. Now apply (2.5) and (2.6).

In (3.4), $\tilde{X}$ is any particular solution of the identity $\tilde{N}\tilde{X} + \tilde{D}\tilde{Y} = I$. Hence it is nice to know that the test matrix Q is the same no matter which $\tilde{X}$ is used. To see this, let $\tilde{X}_1, \tilde{Y}_1$ be another set of matrices in M(S) satisfying $\tilde{N}\tilde{X}_1 + \tilde{D}\tilde{Y}_1 = I$. Then $\tilde{X}_1 = \tilde{X} + DR$ for some $R \in \mathrm{M(S)}$. Hence

$$\begin{aligned} D_i^{\mathrm{adj}} \tilde{X}_1 \tilde{D}_o r_o &= D_i^{\mathrm{adj}} (\tilde{X} + DR) \tilde{D}_o r_o \\ &= D_i^{\mathrm{adj}} \tilde{X} \tilde{D}_o r_o + \delta D_o R \tilde{D}_o \end{aligned} \tag{3.14}$$

since $D = D_i D_o$. Now the second term on the right side of (3.14) vanishes at all zeros of δ. Hence $F_1, \ldots, F_n$ are independent of which particular solution $\tilde{X}$ is used to compute them. Similar remarks apply even in the case where δ has repeated zeros.

If the plant P_0 is scalar, then the expression for F_j (or f_j) in the scalar case is more elegant. In this case, $d_i^{\mathrm{adj}} = 1$ (since by convention the adjoint of a 1×1 matrix is 1), and $f_j = (xd_o r_o)(\lambda_j)$. Now, in the scalar case $\delta = \delta_i$; hence it follows from $d = d_i d_o = \delta d_o$ and $nx + dy = 1$ that, at any zero λ_j of δ, we have $n(\lambda_j) x(\lambda_j) = 1$. Hence

$$x(\lambda_j) d_o(\lambda_j) = \frac{d_o(\lambda_j)}{n(\lambda_j)}. \tag{3.15}$$

With the plant p_0, associate the stable plant $q_0 = p_0 \delta = n/d_o$. Then

$$f_j = \frac{d_o(\lambda_j) r_o(\lambda_j)}{n(\lambda_j)} = \frac{r_o(\lambda_j)}{q_o(\lambda_j)}. \tag{3.16}$$

The above result was obtained earlier by Kimura

(1983) using quite different methods. He also extended these results to the case where both the unperturbed and perturbed plants had a simple pole at $s = 0$.

The treatment of multiplicative perturbations is entirely similar to that of additive perturbations.

Proof of Theorem 4. From Theorem 1, every $C \in S(P_0)$ is of the form $(Y - R\tilde{N})^{-1}(X + R\tilde{D})$ for some $R \in M(S)$. Moreover,

$$P_0C(I + P_0C)^{-1} = P_0(I + CP_0)^{-1}C = N(X + R\tilde{D}). \tag{3.17}$$

Hence (3.2) holds for some $C \in S(P_0)$ if and only if there exists an $R \in M(S)$ such that (3.7) holds. This proves the first part of the theorem.

To prove the second part, let $\gamma = |N_i|$ and note that γ is inner. Using by now familiar manipulations, we arrive at

$$\begin{aligned}\|NXr + NR\tilde{D}r\|_\infty &= \|N_iN_oXr_or_i + N_iN_oR\tilde{D}_or_o\tilde{D}_ir_i\|_\infty \\ &= \|\gamma r_i(N_oXr_o\tilde{D}_i^{adj} + \delta S)\|_\infty \\ &\quad \text{where } S = N_oR\tilde{D}_or_o \\ &= \|N_oXr_o\tilde{D}_i^{adj} + \delta S\|_\infty. \end{aligned} \tag{3.18}$$

Now (3.9) follows readily.

If the plant P_0 is scalar, one can again obtain a simple expression for the quantity $N_oXr_o\tilde{D}_i^{adj}$ evaluated at the zeros of δ. Suppose $nx + dy = 1$, and suppose $\delta(\lambda) = 0$ at some point λ in the RHP. Then $d(\lambda) = 0$, $(nx)(\lambda) = 1$, and $(N_oXr_o\tilde{D}_i^{adj})(\lambda) = \dfrac{r_o(\lambda)}{n_i(\lambda)}$.

The foregoing results also lead very naturally to the notion of optimally robust compensators. Suppose a nominal plant P_0 is given, together with a function $f \in S$, which represents an uncertainty profile. Consider the class $A(P_0, r)$ consisting of all plants satisfying (1.1), where $r = \lambda f$. It is now reasonable to ask: what is the largest value of the parameter λ for which the class $A(P_0, r)$ is robustly stabilizable, and what is a corresponding robustly stabilizing compensator? This problem can be solved very easily using the results derived so far. Factor f as f_if_o where f_i, f_o are respectively inner and outer, and define

$$\gamma = \min_{S \in M(S)} \|D_i^{adj}\tilde{X}\tilde{D}_of_o + \delta S\|_\infty, \tag{3.19}$$

where all symbols are as in Theorem 3. Now, by applying Theorem 3 with $r = \lambda f$, it follows that there exists a single compensator that stabilizes all the plants in the class $A(P_0, r)$ if and only if $\gamma\lambda \le 1$. Hence the largest value of λ for which there exists a robustly stabilizing compensator is given by $\lambda = 1/\gamma$. Further, if S is any matrix which attains the minimum in (3.19), then the compensator $C = (\tilde{X} + DS)(\tilde{Y} - NS)^{-1}$ is an optimally robust compensator. The case of multiplicative perturbations is entirely similar and is left to the reader.

If the plant P_0 has more rows than columns, then the condition (3.7) for robust stabilizability cannot be simply tested in terms of the nonnegative definiteness of a test matrix of the form (3.6). In this case one has to use the iterative method of Doyle (1983) to test the condition (3.7). If the plant P_0 has at least as many columns as rows, then the simpler condition (3.9) applies, which can be more readily tested.

4. STABLE FACTOR PERTURBATIONS

In the previous section, we studied the case where the perturbed and unperturbed plants had the same number of RHP poles, and neither had poles on the $j\omega$-axis. However, this assumption is not satisfied in certain applications such as large flexible spacecraft. Thus it is desirable to develop a theory that removes this assumption by considering a different class of plant perturbations. The class of stable factor perturbations has this feature. The conditions for the robust stabilizability in the case of stable factor perturbations are given in the next theorem.

Theorem 5. Suppose $P_0 \in M(\mathbb{R}(s))$. Suppose an r.c.f. (N_0, D_0) of P_0 and a function $r \in S$ are specified, and define the class $S(N_0, D_0, r)$ as in (1.4). Suppose $C \in S(P_0)$, and select an l.c.f. $(\tilde{D}_c, \tilde{N}_c)$ of C such that $\tilde{D}_cD_0 + \tilde{N}_cN_0 = I$. Then C stabilizes all P in the class $S(N_0, D_0, r)$ if and only if

$$\|[\tilde{D}_c \;\; \tilde{N}_c]r\|_\infty \le 1. \tag{4.1}$$

Proof. For convenience, define

$$A_0 = \begin{bmatrix} D_0 \\ N_0 \end{bmatrix}, \quad A = \begin{bmatrix} D \\ N \end{bmatrix}, \quad \tilde{A}_c = [\tilde{D}_c \;\; \tilde{N}_c]. \tag{4.2}$$

"if": Suppose (4.1) holds, and suppose P is an arbitrary plant in the class $S(N_0, D_0, r)$. Then P has an r.c.f. (N, D) such that $\|A(s) - A_0(s)\| \le |r(s)|$ $\forall s \in C_{+e}$. Now consider the return difference matrix $\tilde{D}_cD + \tilde{N}_cN = \tilde{A}_cA$. Since $\tilde{A}_cA_0 = I$, it follows that $\tilde{A}_cA = I + A_c(A - A_0)$. However, from (4.1), we get $\|[\tilde{A}_c(A - A_0)](s)\| \le \|\tilde{A}_c(s)\| \; \|[A - A_0](s)\| < 1$ $\forall s \in C_{+e}$. This shows that $|\tilde{A}_cA(s)| \ne 0 \; \forall s \in C_{+e}$, so that $\tilde{A}_cA$ is unimodular. Hence C stabilizes P.

"only if": Suppose (4.1) is false; we will construct a plant $P \in S(N_0, D_0, r)$ that is not stabilized by C. Since (4.1) is false, there exists a ω_0 such that $\|\tilde{A}_c(j\omega_0)\| \; |r(j\omega_0)| > 1$. Select unitary matrices U, V

such that*

$$U\tilde{A}_c(j\omega_0)V = \begin{bmatrix} \sigma_1 & \dots & 0 & 0 \\ \vdots & \vdots & \vdots & \vdots \\ 0 & \dots & \sigma_m & 0 \end{bmatrix}, \quad (4.3)$$

where $\sigma_1 \geq \cdots \geq \sigma_m$ are the singular values of $A_c(j\omega_0)$. Since $A_c(j\omega_0)A(j\omega_0) = I$, it follows that $A(j\omega_0)$ must be of the form

$$A_0(j\omega_0) = V \begin{bmatrix} 1/\sigma_1 & \dots & 0 \\ \vdots & \vdots & \vdots \\ 0 & \dots & 1/\sigma_m \\ & T & \end{bmatrix} U \quad (4.4)$$

where T is some matrix. Now, by assumption, $\sigma_1 |r(j\omega_0)| > 1$. Define

$$Q(s) = -V \begin{bmatrix} 1/\sigma_1 & \dots & 0 \\ \vdots & \vdots & \vdots \\ 0 & \dots & 0 \\ q & 0 & 0 \end{bmatrix} Ur(s)/r(j\omega_0). \quad (4.5)$$

where q is a column vector chosen such that (i) $\|Q(j\omega_0)\| < |r(j\omega_0)|$, and (ii) the first column of $A_0(j\omega_0) + Q(j\omega_0)$ is not identically zero.† Let P be the plant ND^{-1} where $[D' \ N']' = A = A_0 + Q$. Then P belongs to the class $S(N_0, D_0, r)$, since $\|Q(s)\|$ $|r(s)| < 1 \ \forall s \in C_{+e}$. [The matrix Q may not be a real rational matrix, but this is easily fixed; see the discussion following the proof.] However, since

$$A(j\omega_0) = A_0(j\omega_0) + Q(j\omega_0)$$

$$= \begin{bmatrix} 0 & 0 & \dots & 0 \\ 0 & 1/\sigma_2 & & \vdots \\ \vdots & \vdots & \vdots & 1/\sigma_m \\ \times & \times & \times & \times \end{bmatrix} U \quad (4.6)$$

we see that $|\tilde{A}_c A(j\omega_0)| = 0$, so that $A_c A$ is not unimodular.

The question is: does the singularity of $|\tilde{D}_c D + \tilde{N}_c N|$ at $j\omega_0$ imply that C fails to stabilize P? We can conclude that this is so provided we can show that this singularity does not come about as a result of a nontrivial common right divisor between N and D. To amplify this argument, suppose $N = N_1 B$, $D = D_1 B$, where B is a greatest common right divisor of N and D, and N_1 and D_1 are right-coprime. Then $[\tilde{D}_c D + \tilde{N}_c N| = |\tilde{D}_c D_1 + \tilde{N}_c N_1| \cdot |B|$.

* Note that $A_c(s)$ is a "fat" matrix and has full row rank at all $s \in C_{+e}$.

† If the first column of the matrix T in (4.4) is nonzero, simply choose $q = 0$. Otherwise, choose q to be any nonzero vector of sufficiently small norm that $\|Q(j\omega_0)\| < |r(j\omega_0)|$. Since the norm of the top part of $Q(j\omega_0)$ is $1/\sigma < |r(j\omega_0)|$, such a q can always be found.

Hence, if it can be shown that $|B(j\omega_0)| \neq 0$, then $|\tilde{D}_c D_1 + \tilde{N}_c N_1|$ vanishes at $j\omega_0$, and C does not stabilize P. Thus the proof is complete if it can be shown that $|B(j\omega_0)| \neq 0$. For this it is enough to show that $A_0 + Q = [D' \ N']'$ has full column rank at $j\omega_0$. But this last fact is immediate from (4.6), since the first column of T is nonzero.

Let us digress briefly to consider the possibility that the matrix Q defined in (4.5) may not be a real rational matrix. If $\omega_0 = 0$ or ∞, then $Q(s)$ is clearly real rational. If ω_0 is nonzero and finite, proceed as follows: rewrite (4.5) in the form $Q(s) = r(s)M$ where M is a constant matrix whose definition is self-evident. Now M is a rank one, possibly complex, matrix, and $\|M\| < 1$ since $\sigma_1 |r(j\omega_0)| > 1$ by assumption. It is easy to see that the argument in the proof of Theorem 5 is unaffected if the constant matrix M is replaced by a function $T(s)$ so long as (i) $\|T(\cdot)\|_\infty < 1$, and (ii) $T(j\omega_0) = M$. Thus it is shown that, given *any* complex rank one matrix M and any nonzero finite number ω_0, it is possible to construct a real rational function $T(s)$ such that $T(j\omega_0) = M$, and $\|T(\cdot)\|_\infty = \|M\|$. Factor the rank one matrix M in dyadic form as xy', and suppose without loss of generality that ω_0 is greater than zero. Express the vectors x, y in the form

$$x_i = \alpha_i \exp(j\phi_i), \ y_i = \beta_i \exp(j\theta_i), \quad (4.7)$$

where the α_i, β_i are all real and $\phi_i, \theta_i \in (-\pi, 0] \ \forall i$. The idea is to generate a collection of all-pass functions such that at the frequency ω_0 they have the right values. This is done by defining

$$a_i(s) = \alpha_i \frac{s - \gamma_i}{s + \gamma_i}, \qquad b_i(s) = \beta_i \frac{s - \delta_i}{s + \delta_i}, \quad (4.8)$$

and adjusting the constants γ_i, δ_i such that $a_i(j\omega_0) = x_i$, $b_i(j\omega_0) = y_i$. The matrix $T(s) = a(s)[b(s)]'$, where a, b are the vectors of the a_i, b_i, has the required properties. As this construction is quite general, it follows that in robustness studies one can as well use complex rational matrices instead of real rational matrices, as the proofs are much more transparent in this case.

There is an interesting anomaly associated with the robustness condition (4.1). Comparing the contents of Doyle and Stein (1981) and Chen and Desoer (1982), we see that it makes very little difference, in the case of additive and multiplicative perturbations, whether the class of perturbations is defined with "<" or "≤". If the class of perturbations is defined with a strict inequality, the robustness condition has a nonstrict inequality, and vice versa. But this is not so in the case of stable factor perturbations. Define the class $\bar{S}(N_0, D_0, r)$ by

$\bar{S}(N_0, D_0, r) =$

$$\left\{P = ND^{-1}: \left\|\begin{bmatrix}(N - N_0)(s)\\(D - D_0)(s)\end{bmatrix}\right\| \le |r(s)| \forall s \in C_{+e}\right\}. \tag{4.9}$$

One might then be tempted to conjecture the following result: $C \in S(P_0)$ stabilizes all P in the class $\bar{S}(N_0, D_0, r)$ if and only if

$$\sup_{\omega} \|\tilde{A}_c(j\omega)\| \, |r(j\omega)| < 1. \tag{4.10}$$

But this is false: (4.10) is certainly sufficient for robust stability, but not necessary, as the next example shows.

Example 1. Consider

$$p_0(s) = \frac{4}{s-3} = \frac{n_0(s)}{d_0(s)},$$

where

$$n_0(s) = \frac{4(s+7)}{(s+1)^2}, \qquad d_0(s) = \frac{(s-3)(s+7)}{(s+1)^2}.$$

Let $c(s) = 4/(s+5)$. Then c stabilizes p_0. Moreover, $d_c d_0 + n_c n_0 = 1$, where

$$d_c(s) = \frac{s+5}{s+7}, \qquad n_c(s) = \frac{4}{s+7}.$$

Now consider the class $\bar{S}(n_0, d_0, r)$ with $r = 1$. Since

$$\left\|\begin{bmatrix}d_c(\infty)\\n_c(\infty)\end{bmatrix}\right\| = \left\|\begin{bmatrix}1\\0\end{bmatrix}\right\| = 1,$$

(4.10) does not hold. Nevertheless, c stabilizes every p in the class $\bar{S}(n_0, d_0, r)$, which shows that (4.8) is not always necessary for robust stability.

To show this, suppose that $p \in \bar{S}(n_0, d_0, r)$. Thus

$$p(s) = \frac{n(s)}{d(s)} \text{ where } \|[n - n_0 \;\; d - d_0]'(s)\| \le 1 \forall s \in C_{+e}.$$

Let $Q = [q_1 \; q_2]'$ denote $[d - d_0 \;\; n - n_0]'$. Then there are two cases to consider, namely: (i) $q_1(\infty) + 1 \ne 0$, and (ii) $q_1(\infty) + 1 = 0$. In either case we have $\|[d_c(s) \;\; n_c(s)]\| < 1$ for all (finite) s in the RHP, so that

$$(d_c d + n_c n)(s) = 1 + [d_c \;\; n_c]Q\big|_s \ne 0 \forall s \in C_{+}.$$

On the other hand, $[d_c(\infty) \;\; n_c(\infty)] = [1 \;\; 0]$ so that

$$(d_c d + n_c n)(\infty) = 1 + q_1(\infty).$$

Hence, if $q_1(\infty) \ne 1$, then the return difference $d_c d + n_c n$ has no zeros in C_{+e}, and is thus a unit of S. Therefore c stabilizes p.

It only remains to show that c stabilizes p even if $q_1(\infty) = -1$. In this case $q_2(\infty) = 0$ since $\|Q(\infty)\| \le 1$. Hence $n(\infty) = d(\infty) = 0$, and $\alpha = 1/(s+1)$ is a common divisor of n and d in the ring S. Further, since $(d_c d + n_c n)(s) \ne 0$ for all finite s in the RHP, α, its powers and associates are the only possible common divisors of n and d. Now it is claimed that, whatever be q_1, the function $d = d_0 + q_1$ can have only a simple zero at infinity. Let us accept this claim for a moment; then α is a greatest common divisor of n and d, since d/α does not vanish at infinity. Let $n_1 = n/\alpha$, $d_1 = d/\alpha$. Then (n_1, d_1) is a coprime factorization of p. Now

$$(d_c d_1 + n_c n_1)(\infty) = d_1(\infty)$$

$\ne 0$ since $d_1 = d/\alpha$ and d has only a simple zero at infinity. On the other hand, since it has already been established that $(d_c d + n_c n)(s) \ne 0$ for all finite s in the RHP, it follows that

$$(d_c d_1 + n_c n_1)(s) \ne 0 \forall s \in C_{+e}.$$

This shows that c stabilizes p.

Thus the example is complete if the claim can be established. This is most easily done using the bilinear transformation $z = (s-1)/(s+1)$, which sends the function d_0 into

$$a_0(z) = d_0((1+z)/(1-z)) = (4-3z)/(2z-1)$$

and sends $q_1(s)$ into an associated rational function $t_1(z)$. In this context, the claim is that $a_0 + t_1$ has only a simple zero at $z = 1$ whenever $t_1(1) = -1$ and $\|t_1\|_\infty = 1$. We prove the contrapositive, namely, if g is a rational H_∞-function such that $g(1) = 1$ and $a_0 - g$ has a double zero at $z = 1$, then $\|g\|_\infty > 1$. Expand g in a power series around $z = 1$, as

$$g(z) = \sum_{i=0}^{\infty} g_i (z-1)^i.$$

If $a_0 - g$ has (at least) a double zero at $z = 1$, then

$$g(1) = a_0(1) = 1, g'(1) = a_0'(1) = -1.$$

Hence, for $z < 1$ and sufficiently close to 1, we have $g(z) > 1$. Therefore $\|g\| > 1$.

This completes the example.

Theorem 5 provides a necessary and sufficient condition for a $C \in S(P_0)$ to stabilize all plants in the class $S(N_0, D_0, r)$. The issue of whether such a C exists can be formulated in terms of an H_∞-norm minimization problem.

Theorem 6. Suppose a function $r \in$ S and a

nominal plant $P_0 \in \mathbf{M}(\mathbf{S})$ are specified, together with an r.c.f. (N_0, D_0) and an l.c.f. $(\tilde{D}, \tilde{N})$ of P_0. Let X, $Y \in \mathbf{M}(\mathbf{S})$ satisfy $XN_0 + YD_0 = I$. Then there exists a $C \in S(P_0)$ that stabilizes all plants in the class $S(N_0, D_0, r)$ if and only if

$$\inf_{R \in \mathbf{M}(\mathbf{S})} \|A - RB\| \leq 1, \tag{4.11}$$

where

$$A = [Y \ X]r, \qquad B = [-\tilde{N}_0 \ \tilde{D}_0]r. \tag{4.12}$$

The proof is obvious and is left to the reader. Note that the matrix B in (4.11) always has more columns than rows. Hence the condition (4.12) can only be verified using the iterative method of Doyle (1983).

5. GRAPH METRIC PERTURBATIONS

In this section, we consider the case where the uncertainty in the plant is characterized by a graph metric perturbation. It turns out that simultaneous perturbations in both the plant and the controller can be very easily handled in this framework.

We begin with a brief description of the graph metric, which is defined in Vidyasagar (1984). An r.c.f. (N, D) of a plant P is said to be normalized if

$$[D'(-s)N'(-s)]\begin{bmatrix} D(s) \\ N(s) \end{bmatrix} = I, \forall s. \tag{5.1}$$

It can be shown that every P has a normalized r.c.f., which is unique to within right multiplication by an orthogonal matrix. If (N, D) is a normalized r.c.f. of P, then $A = [D' \ N']'$ is inner, and as a result multiplication by A is an isometry. Now suppose P_1, P_2 are two plants with normalized r.c.f.s (N_1, D_1), (N_2, D_2), respectively. To define the graph metric distance $d(P_1, P_2)$ between the two plants, let

$$A_i = \begin{bmatrix} D_i \\ N_i \end{bmatrix}, \qquad \text{for } i = 1, 2. \tag{5.2}$$

$$\delta(P_1, P_2) = \inf_{U \in \mathbf{M}(\mathbf{S}),\ \|U\|_\infty \leq 1} \|A_1 - A_2 U\|_\infty, \tag{5.3}$$

$$d(P_1, P_2) = \max\{\delta(P_1, P_2), \delta(P_2, P_1)\}. \tag{5.4}$$

Then d is a metric on $\mathbf{M}(\mathbb{R}(s))$, taking values in the interval $[0, 1]$.

The presentation of the results in this section is made clearer by a bit of notation. Given P, $C \in \mathbf{M}(\mathbb{R}(s))$, define

$$T(P, C) = H(P, C) - \begin{bmatrix} I & 0 \\ 0 & 0 \end{bmatrix}, \tag{5.5}$$

where $H(P, C')$ is defined in (2.7). Using this last equation, one can derive a few other useful representations of $T(P, C)$. For instance.

$$T(P, C') = \begin{bmatrix} -PC(I + PC)^{-1} & -P(I + CP)^{-1} \\ C(I + PC)^{-1} & (I + CP)^{-1} \end{bmatrix}$$
$$= \begin{bmatrix} -P \\ I \end{bmatrix}(I + CP)^{-1}[C \ I]. \tag{5.6}$$

Theorem 7. Suppose the pair (P_0, C_0) is stable, and that P_0, C_0 are perturbed to P, C, respectively. Then the pair (P, C') is stable provided

$$d(P, P_0)\|T(P_0, C_0)\|_\infty + d(C, C_0)\|T(C_0, P_0)\|_\infty < 1. \tag{5.7}$$

Corollary 5.1. Under the hypotheses of Theorem 7, the pair (P, C_0) is stable provided $d(P, P_0) < 1/\|T(P_0, C_0)\|_\infty$.

The significance of Theorem 7 and Corollary 5.1 is as follows: the type of plant perturbations studied in these two results is the most unstructured one considered in this paper, in that (i) one is permitted to perturb simultaneously both the plant and the controller, (ii) there is no restriction on the number of RHP poles of the perturbed and unperturbed plant being the same, and (iii) the perturbations are not couched in terms of a particular coprime factorization of the plant. The stability condition (5.7) is interesting in that the effects of the perturbations in the plant and controller enter additively. When only the plant is perturbed, the stability condition given in Corollary 5.1 is reminiscent of the small gain theorem [see e.g. Desoer and Vidyasagar (1975)]. These results also serve to bring out the significance of the concept of the graph metric introduced in Vidyasagar (1984).

The proof requires the following easily proved result.

Lemma 5.1. Suppose (N_p, D_p), (N_c, D_c) are r.c.f.s of P, C, respectively. Then the pair (P, C) is stable if and only if the matrix

$$U = \begin{bmatrix} D_p & -N_c \\ N_p & D_c \end{bmatrix} \tag{5.8}$$

is unimodular.

Proof. Define

$$G = \begin{bmatrix} C & 0 \\ 0 & P \end{bmatrix}, \quad N_g = \begin{bmatrix} N_c & 0 \\ 0 & N_p \end{bmatrix}, \quad D_g = \begin{bmatrix} D_c & 0 \\ 0 & D_p \end{bmatrix}, \tag{5.9}$$

$$F = \begin{bmatrix} 0 & -I \\ I & 0 \end{bmatrix}. \tag{5.10}$$

Then it is easy to verify that (N_g, D_g) is an r.c.f. of G, and that the transfer matrix $H(P, C)$ equals

$(I + FG)^{-1}$. Since F is a constant matrix, an r.c.f. of $H(P, C)$ is given by the pair $(D_g, D_g + FN_g)$. Hence $H(P, C) \in \mathbf{M}(\mathbf{S})$ if and only if its "denominator" matrix $D_g + FN_g$ is unimodular. Now routine computation shows that this matrix is precisely U.

Proof of Theorem 7. Let (N_0, D_0), $(\tilde{X}_0, \tilde{Y}_0)$ be normalized r.c.f.s of P_0, C_0, respectively. Then from the stability of the pair (P_0, C_0), the matrix

$$U_0 = \begin{bmatrix} D_0 & -\tilde{X}_0 \\ N_0 & \tilde{Y}_0 \end{bmatrix} \tag{5.11}$$

is unimodular. Let $V_0 = U_0^{-1}$, and partition V_0 as

$$V_0 = \begin{bmatrix} Y_0 & X_0 \\ -\tilde{N}_0 & \tilde{D}_0 \end{bmatrix}. \tag{5.12}$$

Then $(\tilde{D}_0, \tilde{N}_0)$, (Y_0, X_0) are l.c.f.s of P_0, C_0, respectively.

Now select real numbers $\delta_p > d(P, P_0)$, $\delta_c > d(C, C_0)$ such that

$$\delta_p \|T(P_0, C_0)\|_\infty + \delta_c \|T(C_0, P_0)\|_\infty < 1. \tag{5.13}$$

This is possible in view of (5.7). Let (N_p, D_p), (N_c, D_c) be any normalized r.c.f.s of P, C respectively. Then, from the definition of the graph metric, there exist matrices W_p, $W_c \in \mathbf{M}(\mathbf{S})$ with $\|W_p\|_\infty \le 1$, $\|W_c\|_\infty \le 1$ such that

$$\left\| \begin{bmatrix} D_0 - D_p W_p \\ N_0 - N_p W_p \end{bmatrix} \right\|_\infty \le \delta_p, \tag{5.14}$$

$$\left\| \begin{bmatrix} \tilde{Y}_0 - D_c W_c \\ \tilde{X}_0 - N_c W_c \end{bmatrix} \right\|_\infty \le \delta_c. \tag{5.15}$$

Define F as in (5.11) and observe that F is inner so that multiplication by F is norm-preserving. Thus (5.15) implies that

$$\left\| \begin{bmatrix} -(\tilde{X}_0 - N_c W_c) \\ \tilde{Y}_0 - D_c W_c \end{bmatrix} \right\|_\infty \le \delta_c. \tag{5.16}$$

Next, define

$$D_g = \begin{bmatrix} D_c & 0 \\ 0 & D_p \end{bmatrix},$$

$$V = \begin{bmatrix} D_p W_p & -N_c W_c \\ N_p W_p & D_c W_c \end{bmatrix}$$

$$= \begin{bmatrix} D_p & -N_c \\ N_p & D_c \end{bmatrix} \begin{bmatrix} W_p & 0 \\ 0 & W_c \end{bmatrix} \tag{5.17}$$

It is now shown that (5.13) implies the unimodularity of V. This will show, *a fortiori*, that W_p and W_c are both unimodular, and that the pair (P, C) is stable, the latter conclusion following from Lemma 5.1. Note that $V - U_0 = [A \;\; B]$ where

$$A = \begin{bmatrix} D_0 - D_p W_p \\ N_0 - N_p W_p \end{bmatrix}, \qquad B = \begin{bmatrix} -(\tilde{X}_0 - N_c W_c) \\ \tilde{Y}_0 - D_c W_c \end{bmatrix} \tag{5.18}$$

Now, if $\|[A \;\; B] U_0^{-1}\|_\infty < 1$, then $\|(V - U_0) U_0^{-1}\|_\infty < 1$, which implies that V is unimodular. From (5.12),

$$[A \;\; B] U_0^{-1} = [A \;\; B] V_0 = A[Y_0 \;\; X_0] + B[-\tilde{N}_0 \;\; \tilde{D}_0]. \tag{5.19}$$

$$\begin{aligned} \|[A \;\; B] U_0^{-1}\|_\infty &\le \|A[Y_0 \;\; X_0]\|_\infty + \|B[-\tilde{N}_0 \;\; \tilde{D}_0]\|_\infty \\ &\le \delta_p \|[Y_0 \;\; X_0]\|_\infty + \delta_c \|[-\tilde{N}_0 \;\; \tilde{D}_0]\|_\infty \end{aligned} \tag{5.20}$$

where the last step follows from (5.14) and (5.16). The proof is completed by showing that

$$\begin{aligned} \|[Y_0 \;\; X_0]\|_\infty &= \|T(P_0, C_0)\|_\infty, \\ \|[-\tilde{N}_0 \;\; \tilde{D}_0]\|_\infty &= \|T(C_0, P_0)\|_\infty. \end{aligned} \tag{5.21}$$

Then (5.13) and (5.20) will imply the unimodularity of V and the stability of (P, C).

To prove the first part of (5.21), recall that (N_0, D_0) is a normalized r.c.f. of P_0, and (Y_0, X_0) is the corresponding l.c.f. of C_0 such that $Y_0 D_0 + X_0 N_0 = I$. Hence, from (5.6),

$$\begin{aligned} T(P_0, C_0) &= \begin{bmatrix} -N_0 X_0 & -N_0 Y_0 \\ D_0 X_0 & D_0 Y_0 \end{bmatrix} \\ &= \begin{bmatrix} -N_0 \\ D_0 \end{bmatrix} [X_0 \;\; Y_0]; \end{aligned} \tag{5.22}$$

$$\begin{aligned} \|T(P_0, C_0)\|_\infty &= \|[X_0 \;\; Y_0]\|_\infty \\ &\quad \text{since } \begin{bmatrix} -N_0 \\ D_0 \end{bmatrix} \text{ is an isometry} \\ &= \|[Y_0 \;\; X_0]\|_\infty. \end{aligned} \tag{5.23}$$

The proof of the second half of (5.21) follows essentially by symmetry arguments, after noting that $\|[-\tilde{N}_0 \;\; \tilde{D}_0]\|_\infty = \|[\tilde{N}_0 \;\; \tilde{D}_0]\|_\infty$.

It is not known at present how close the conditions of Theorem 7 and Corollary 5.1 are to being necessary. In particular, it is not known whether the condition $\|T(P_0, C)\|_\infty < r^{-1}$ is necessary for a controller C to stabilize all plants within a distance of r from the plant P_0. Nevertheless, it is reasonable to seek, among all stabilizing controllers for P_0, an "optimally robust" controller C_0 for which $\|T(P_0, C_0)\|$ is as small as possible. The problem of finding such a C_0 can be formulated as an H_∞-norm minimization problem. As C varies over $S(P_0)$, the corresponding $T(P_0, C)$ vary over all matrices of the form

$$T_1(R) = \begin{bmatrix} -N_0(X_0 + R\tilde{D}_0) & -N_0(Y_0 - R\tilde{N}_0) \\ D_0(X_0 + R\tilde{D}_0) & D_0(Y_0 - R\tilde{N}_0) \end{bmatrix} \tag{5.24}$$

where (N_0, D_0), $(\tilde{D}_0, \tilde{N}_0)$ are any r.c.f. and l.c.f. of P_0, X_0, $Y_0 \in \mathbf{M}(\mathbf{S})$ satisfy $X_0 N_0 + Y_0 D_0 = I$, and $R \in \mathbf{M}(\mathbf{S})$ is a free parameter. Thus, minimizing $\|T(P_0, C)\|_\infty$ over all $C \in S(P_0)$ is equivalent to the unconstrained minimization of $\|T_1(R)\|_\infty$ as R varies over $\mathbf{M}(\mathbf{S})$. Now note that $T_1(R)$ is of the form $U - VRW$, where

$$U = \begin{bmatrix} -N_0 \\ D_0 \end{bmatrix} [X_0 \;\; Y_0], \qquad V = \begin{bmatrix} -N_0 \\ D_0 \end{bmatrix}, \tag{5.25}$$

$$W = [\tilde{D}_0 \;\; -\tilde{N}_0].$$

However, since V has more rows than columns and W has more columns than rows, the quantity

$$\inf_{C \in S(P_0)} \|T(P_0, C)\|_\infty = \inf_{R \in \mathbf{M}(\mathbf{S})} \|U - VRW\|_\infty \tag{5.26}$$

can only be computed iteratively using the method of Doyle (1983).

6. CONCLUSIONS

This paper contains three distinct contributions: (i) in the case of additive and multiplicative perturbations, necessary and sufficient conditions are given for the existence of a robustly stabilizing controller; (ii) in the case of stable-factor perturbations, necessary and sufficient conditions are given for a controller to stabilize all plants within the specified class; (iii) in the case of graph metric perturbations, which also includes simultaneous perturbations in the plant as well as controller, sufficient conditions for robust stabilization are given. In the latter two cases, conditions for the existence of a robustly stabilizing controller are given in the form of the value of a certain H_∞-norm minimization problem being less than one.

Acknowledgement—The first author thanks John Doyle for his comments on Theorem 5.

REFERENCES

Chang, B.-C. and J. B. Pearson (1982). Optimal disturbance reduction in linear multivariable systems. *IEEE Trans. Aut. Control*, **AC-29**, 880.

Chen, M. J. and C. A. Desoer (1982). Necessary and sufficient conditions for robust stability of linear distributed feedback systems. *Int. J. Control*, **35**, 255.

Delsarte, P. R., Y. Genin and Y. Kamp (1979). The Nevanlinna–Pick problem for matrix-valued functions. *SIAM J. appl. Math.*, **36**, 47.

Desoer, C. A., R.-W. Liu, J. Murray and R. Saeks (1980). Feedback system design: the fractional representation approach to analysis and synthesis. *IEEE Trans. Aut. Control*, **AC-25**, 399.

Desoer, C. A. and M. Vidyasagar (1975). *Feedback Systems: Input–Output Properties*. Academic Press, New York.

Doyle, J. C. (1983). Synthesis of robust controllers and filters. *Proc. IEEE Conf. Decision and Control*, San Antonio, Dec. 1983, pp. 109–114.

Doyle, J. C. and G. Stein (1981). Multivariable feedback design: concepts for a classical modern synthesis. *IEEE Trans. Aut. Control*, **AC-26**, 4.

Duren, P. L. (1970). *The Theory of H^p-Spaces*. Academic Press, New York.

Kailath, T. (1980). *Linear Systems*. Prentice-Hall, Engelwood Cliffs, New Jersey.

Kimura, H. (1983). Robust stabilizability for a class of transfer functions. *IEEE Trans. Aut. Control*, **AC-29**, 788.

Sz-Nagy, B. and C. Foias (1970). *Harmonic Analysis of Operators on Hilbert Space*. Elsevier, New York.

Vidyasagar, M. (1975). Coprime factorizations and the stability of multivariable distributed feedback systems. *SIAM J. Control*, **13**, 1144.

Vidyasagar, M. (1984). The graph metric for unstable plants and robustness estimates for feedback stability. *IEEE Trans. Aut. Control*, **AC-29**, 403.

Vidyasagar, M., H. Schneider and B. A. Francis (1982). Algebraic and topological aspects of feedback stabilization. *IEEE Trans. Aut. Control*, **AC-27**, 880.

Walsh, J. L. (1935). *Interpolation and Approximation by Rational Functions in the Complex Domain*. AMS Colloqium, Providence, R.I.

Part III
H^2 and H^∞ Sensitivity Optimization

Minimum Sensitivity Design of Linear Multivariable Feedback Control Systems by Matrix Spectral Factorization

JOSEPH J. BONGIORNO, JR., MEMBER, IEEE

Abstract—**A scalar measure of system sensitivity to plant parameter variations is employed in the design of linear lumped stationary multivariable feedback control systems. The plant parameters are treated as random variables, and design formulas are derived which lead to systems with the smallest expected value for the chosen scalar sensitivity measure. The design formulas give physically realizable feedback and tandem compensation network transfer function matrices provided the overall system transfer function matrix is properly specified. The solution of the minimum sensitivity design problem is obtained by first solving the multivariable semi-free-configuration Wiener problem.**

Introduction

THE RESULTS of an earlier effort [1] are extended to linear lumped stationary multivariable control systems in this paper. The system considered is shown in Fig. 1. The plant is represented by the rational transfer function matrix $G_p(s,\alpha)$. It is assumed that the plant is asymptotically stable. (When the plant is not asymptotically stable, but is completely controllable, it can always be made asymptotically stable with state variable feedback [2] or with output feedback through a compatible observer [3].) The N-dimensional column vector $\boldsymbol{\alpha}$ represents the mean or expected value of the plant parameters, and any deviation from the mean is denoted by $\delta\boldsymbol{\alpha}$. Thus,

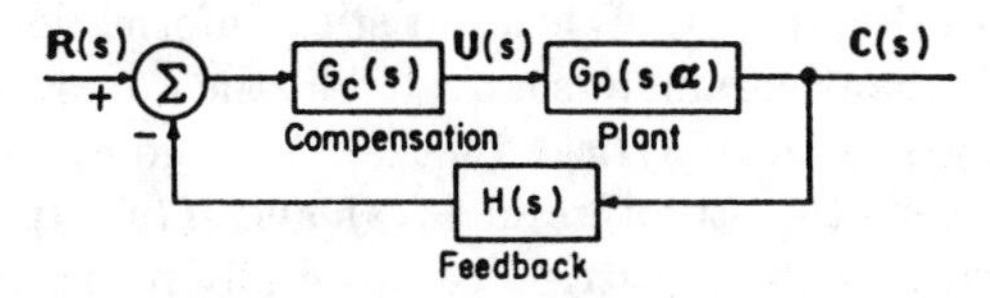

Fig. 1. System.

$$E\{\delta\alpha_i\} = 0, \quad i = 1,2,\cdots,N \tag{1}$$

where $E\{\cdot\}$ denotes the expected value, and $\delta\alpha_i$ is the element in the ith row of $\delta\boldsymbol{\alpha}$. It is assumed that the covariance matrix (the prime denotes the transpose)

$$\Sigma = E\{\delta\alpha\delta\alpha'\} = [\sigma_{ij}], \sigma_{ji} = \sigma_{ij} = E\{\delta\alpha_i\delta\alpha_j\} \tag{2}$$

is known, and that the variations $\delta\alpha_i$ are small and independent of the signals in the system. The input $\boldsymbol{R}$ is generated by a stationary stochastic process with known power spectral density matrix.

The rational transfer function matrices $G_c(s)$ and $H(s)$ represent, respectively, the tandem compensation network

Manuscript received February 3, 1969. This work was supported by NASA under Grant NGR 33-006-042.
The author is with the Department of Electrical Engineering, Polytechnic Institute of Brooklyn, Brooklyn, N. Y. 11201.

Reprinted from *IEEE Trans. Automat. Contr.*, vol. AC-14, no. 6, pp. 665–673, Dec. 1969.

and the feedback network. The bilateral Laplace transform is used exclusively and attention is restricted to only those cases in which the strip of convergence for all transforms includes the imaginary axis of the complex s plane. In this setting, a transfer function matrix is physically realizable (i.e., the impulse response matrix is causal) if, and only if, all of its elements are analytic in $\text{Re}\, s \geq 0$.

The objective is the determination of physically realizable transfer function matrices $G_c(s)$ and $H(s)$ for which

$$W(s,\alpha) = G_p(s,\alpha)G_c(s)[1_n + H(s)G_p(s,\alpha)G_c(s)]^{-1} \quad (3)$$

(1_n denotes the $n \times n$ identity matrix and n is the dimension of $\boldsymbol{R}$) satisfies the dynamic performance requirements placed on the system and for which the scalar sensitivity measure

$$S = E\{(\boldsymbol{c}_a - \boldsymbol{c})'Q(\boldsymbol{c}_a - \boldsymbol{c})\} \quad (4)$$

is a minimum. The square matrix Q is real, symmetric, constant, and nonnegative definite. The response $\boldsymbol{c}_a(t)$ is the output response $\boldsymbol{c}(t)$ when instead of $\boldsymbol{\alpha}$ the parameter vector takes on the value $\boldsymbol{\alpha} + \delta\boldsymbol{\alpha}$.

For the case of single-input–output systems, the sensitivity index (4) reduces to one similar to that employed by Mazer [13]. Here, however, the expectation is taken over the random plant parameters as well as the stochastic inputs. The sensitivity measure (4), except for taking the expected value, is also identical to the one considered by Perkins and Cruz [4], [5]. These same authors in collaboration with Gonzales [6] recently treated the design of the system shown in Fig. 1 from a minimax parameter optimization point of view. Using a computational algorithm they obtain the values of parameters which determine $G_c(s)$ and $H(s)$. When statistical information on the variable plant parameters is available, the approach taken here leads to an analytical solution. Also, no constraint is imposed on the structure of $G_c(s)$ and $H(s)$. It is only required that these matrices be physically realizable. This freedom in the choice of $G_c(s)$ and $H(s)$, however, leads to many designs in which differentiators are required. These differentiators, then, must be approximated with practical circuits.

The solution of the problem posed here is accomplished by first solving the multivariable semi-free-configuration Wiener problem.[1] The solution of this Wiener problem is in itself of some theoretical interest. It was first treated by Hsieh and Leondes [7]. They reduced the problem to the solution of a system of algebraic equations, but never proved that this system of equations has a solution. Indeed, for the free-configuration problem Davis [8] states that their method fails in the case of a predictor. The solution for the semi-free-configuration problem is achieved here using the idea of matrix spectral factorization. The conditions under which a matrix can be spectrally factored were first derived by Youla [9]. A computer program for factoring those square rational matrices which can be factored has recently been developed by Tuel [10].

The notation used in this paper is now summarized for easy reference. For an arbitrary matrix A the transpose, the complex conjugate, the adjoint (the complex conjugate transpose), the inverse, the trace, and the determinant of A are denoted by A', $\bar{A}$, A^*, A^{-1}, $\text{tr}[A]$, and $|A|$, respectively. A diagonal matrix Λ with diagonal elements $\lambda_1,\lambda_2,\cdots,\lambda_n$ is written as $\Lambda = \text{diag}[\lambda_1,\lambda_2,\cdots,\lambda_n]$. Column vectors are represented by $\boldsymbol{x}$, $\boldsymbol{y}$, etc., or in the alternative fashion $\boldsymbol{x} = (x_1,x_2,\cdots,x_n)'$ whenever it is desirable to indicate the components explicitly. The $n \times n$ identity matrix, the n-dimensional zero vector, and the $n \times m$ zero matrix are denoted by 1_n, $\boldsymbol{o}_n$, and O_{nm}, respectively. The n-dimensional column vector with unity in the ith row and all other elements equal to zero is denoted by $\boldsymbol{e}_i$. The right inverse of a $p \times q$ matrix A is the $q \times p$ matrix A^{-1} which has the property $AA^{-1} = 1_p$.

A matrix $A(s)$ is rational when each of its elements is rational. The matrix $A(s)$ is analytic in a region when each of its elements is analytic in the region. $A(s)$ is said to be real if $\bar{A}(s) = A(\bar{s})$. When for the matrix $A(s)$ there exists one minor of order ν which does not vanish identically, and when all minors of order greater than ν vanish identically, then $A(s)$ is said to be a matrix with normal rank ν. A point s_0 is a pole of $A(s)$ if some element of $A(s)$ has a pole at $s = s_0$. It is also convenient to introduce the notation

$$A_*(s) = A^*(-\bar{s}) \quad (5)$$

which for real matrices—the only kind of interest here—reduces to

$$A_*(s) = A'(-s). \quad (6)$$

Preliminary Analysis

When the number of plant outputs exceeds the normal rank of the plant transfer function matrix, one can always restrict attention to the n independent outputs. Since the number of independent outputs is always less than or equal to the number of inputs, it is always possible to choose n inputs to control the plant. One can choose for the n inputs those associated with the n columns of any nonzero minor of $G_p(s,\boldsymbol{\alpha})$ of order n. Once attention is restricted to $n \times n$ plant transfer function matrices, it immediately follows for input vectors $\boldsymbol{r}(t)$ of dimension n that both $G_c(s)$ and $H(s)$ are $n \times n$ matrices. In the sequel, therefore, all transfer function matrices are square and of order n, and the normal rank of the plant transfer function matrix is n.

The sensitivity index (4) is equivalent to

$$S = E\left\{\int_0^\infty \int_0^\infty \boldsymbol{r}'(t-\tau_1)\delta W'(\tau_1)Q\delta W(\tau_2)\boldsymbol{r}(t-\tau_2)\, d\tau_1\, d\tau_2\right\} \quad (7)$$

where in terms of impulse-response matrices

$$\delta W(t) = W(t,\boldsymbol{\alpha} + \delta\boldsymbol{\alpha}) - W(t,\boldsymbol{\alpha}). \quad (8)$$

[1] A reviewer has indicated to the author that a recent paper by Loo [14] also contains a solution of the semi-free-configuration Wiener problem.

Using the fact that the $\delta\alpha_i$ and $r_i(t)$ are independent, denoting the expectation with respect to the $\delta\alpha_i$ by $E_{\delta\alpha}$ and the expectation with respect to the $r_i(t)$ by E_r, and recognizing

$$\operatorname{tr}[AB] = \operatorname{tr}[BA] \quad (9)$$

whenever the indicated matrix products are defined, one easily obtains (provided the double integral exists)

$$S = E_{\delta\alpha}\left\{\operatorname{tr}\left[\int_0^\infty \int_0^\infty \Phi_r(\tau_2 - \tau_1)\Psi(\tau_1,\tau_2)\, d\tau_1\, d\tau_2\right]\right\} \quad (10)$$

where

$$\Phi_r(\tau) = E_r\{r(t)r'(t+\tau)\} \quad (11)$$

$$\Psi(\tau_1,\tau_2) = \delta W'(\tau_1)Q\delta W(\tau_2). \quad (12)$$

Attention is restricted here to those cases in which the elements of both $\Phi_r(\tau)$ and $\delta W(\tau)$ are bounded by a decaying exponential function of τ over the internal $\tau \geq 0$. The existence of the double integral in (10) is guaranteed, then, and Parseval's relationship leads to

$$S = E_{\delta\alpha}\left\{\operatorname{tr}\left[\frac{1}{2\pi j}\int_{-j\infty}^{j\infty} \Phi_r(s)\delta W'(s)Q\delta W(-s)\, ds\right]\right\} \quad (13)$$

where $\Phi_r(s)$ and $\delta W(s)$ are the bilateral Laplace transforms of $\Phi_r(\tau)$ and $\delta W(\tau)$, respectively. The matrix $\Phi_r(s)$ is the power spectral density matrix for the process generating the inputs $r(t)$.

The approximation

$$\delta W(s) = \sum_{l=1}^{N} [\partial W(s,\alpha)/\partial\alpha_l]\delta\alpha_l \quad (14)$$

is now made since the $\delta\alpha_i$ are small. Substituting (14) into (13), recalling (2), and interchanging the expectation, trace, and integral operations yields

$$S = \frac{1}{2\pi j}\int_{-j\infty}^{j\infty} \operatorname{tr}\{\sum_{l=1}^{N}\sum_{m=1}^{N} \sigma_{lm}\Phi_r(s) \cdot[\partial W'(s,\alpha)/\partial\alpha_l]Q[\partial W(-s,\alpha)/\partial\alpha_m]\}\, ds. \quad (15)$$

Since Q is a symmetric nonnegative definite matrix, it can always be written as

$$Q = \tilde{Q}'\tilde{Q} \quad (16)$$

where the rank of Q is equal to the number of rows of $\tilde{Q}$. This being the case, one can take $Q = 1_n$ without any loss in generality: formulas for the case $Q \neq 1_n$ are simply obtained from the formulas derived in the sequel by replacing W with $\tilde{Q}W$. With $Q = 1_n$ in (15) one obtains after substituting

$$\partial W/\partial\alpha_k = (1_n - WH)(\partial G_p/\partial\alpha_k)G_p^{-1}W \quad (17)$$

and defining

$$\Phi = \sum_{l=1}^{N}\sum_{m=1}^{N} \sigma_{lm}A_{m*}W_*'\Phi_r W'A_l \quad (18)$$

where

$$A_k = [(\partial G_p/\partial\alpha_k)G_p^{-1}]' \quad (19)$$

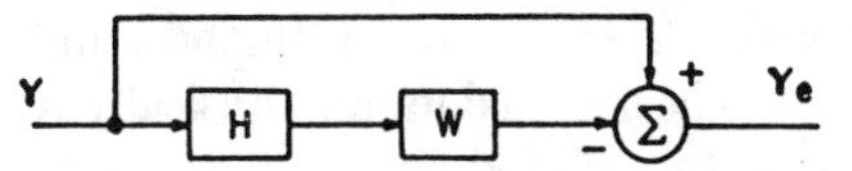

Fig. 2. System defining optimum $H(s)$.

the relationship[2]

$$S = \frac{1}{2\pi j}\int_{-j\infty}^{j\infty} \operatorname{tr}[(1_n - WH)'(1_n - WH)_*'\Phi]\, ds. \quad (20)$$

Since the normal rank of G_p is n, the existence of G_p^{-1} is guaranteed almost everywhere in the complex s plane. In (17)–(20) and in the rest of this paper, the dependence of Φ, G_p, W, and A_k on s and α, and the dependence of Φ_r, G_c, and H on s is not shown explicitly unless necessary for clarity.

Multivariable Semi-Free-Configuration Wiener Problem

It is not difficult to show, and interesting to note, that the mean-square error for the system shown in Fig. 2,

$$S = E\{y_e'(t)y_e(t)\} \quad (21)$$

is given by the value of S in (20) when the power spectral density matrix for the process generating the input $y(t)$ is Φ. Finding the physically realizable H which minimizes S given W is, therefore, equivalent to solving the multivariable semi-free-configuration Wiener problem for the system shown in Fig. 2. This is done here using in part the techniques in Sections 4-2 and 4-3 of [11].

Replacing H by $H + \epsilon H_1$ in (20), one obtains

$$S = S_0 - 2\epsilon S_1 + \epsilon^2 S_2 \quad (22)$$

where

$$S_0 = \frac{1}{2\pi j}\int_{-j\infty}^{j\infty} \operatorname{tr}[(1_n - WH)'(1_n - WH)_*'\Phi]\, ds \quad (23)$$

$$S_1 = \frac{1}{2\pi j}\int_{-j\infty}^{j\infty} \operatorname{tr}[(1_n - WH)'W_*'H_{1*}'\Phi]\, ds \quad (24)$$

$$S_2 = \frac{1}{2\pi j}\int_{-j\infty}^{j\infty} \operatorname{tr}[(WH_1)'(WH_1)_*'\Phi]\, ds. \quad (25)$$

When $S_1 = 0$ and $S_2 \geq 0$ for all physically realizable H_1, then there is no physically realizable choice for the feedback network transfer function matrix other than H which gives a smaller value for the sensitivity index S.

The condition

$$S_1 = \frac{1}{2\pi j}\int_{-j\infty}^{j\infty} \operatorname{tr}[\Phi(1_n - WH)'W_*'H_{1*}']\, ds = 0 \quad (26)$$

[2] Equation (20) can also be derived by making use of the comparison sensitivity matrix of Cruz and Perkins [4]. Indeed, it is not difficult to verify that $1_n - WH$ is an expression for the Cruz–Perkins sensitivity matrix. Also, Φ is identical with the power spectral density matrix for the nominally equivalent open-loop system error.

is necessary for H to be the optimum transfer function matrix. It must be satisfied for all physically realizable H_1. Hence, it must be satisfied for

$$H_1(s) = h(s)[e_i e_j'] \tag{27}$$

where the scalar function $h(s)$ is analytic in Re $s \geq 0$ and satisfies

$$\lim_{s\to\infty} sh(s) \operatorname{tr}[\Phi(1_n - WH)'W_*'(e_i e_j')] = 0. \tag{28}$$

Substituting (27) into (26) and using (28) gives

$$\lim_{\rho\to\infty} \int_{C_\rho} [x_{ji}(s)]h(-s)\, ds = 0 \tag{29}$$

where $x_{ji}(s)$ is the element in the jth row, ith column of

$$X \equiv \Phi(1_n - WH)'W_*' \tag{30}$$

and C_ρ is the contour in the complex s plane consisting of the imaginary axis for $|\omega| \leq \rho$ and the semicircle $s = \rho \exp(j\theta)$, $\pi/2 \leq \theta \leq 3\pi/2$. It follows from (29) and the fact that $h(-s)$ is analytic in Re $s \leq 0$, that $x_{ji}(s), j,i = 1,2,\cdots,n$ must be analytic everywhere in the half-plane Re $s \leq 0$. That is, H must satisfy (30) where the matrix X is analytic in Re $s \leq 0$, but is otherwise arbitrary.

Conditions are discussed in the next section under which it is possible to write

$$\Phi = \Delta_* \Delta \tag{31}$$

$$\Omega \equiv W_* W = \Gamma_* \Gamma \tag{32}$$

where the $n \times n$ square matrices Δ and Γ together with their inverses are analytic in Re $s \geq 0$. Assuming these conditions are met, one obtains from (30) after transposing and making the substitutions (31) and (32)

$$\Gamma_*^{-1}(W_* - \Gamma_*\Gamma H)\Delta' = \Gamma_*^{-1}X'(\Delta_*')^{-1} \equiv \tilde{X} \tag{33}$$

where $\tilde{X}$ is analytic in Re $s \leq 0$. Now one can write

$$\Gamma_*^{-1}W_*\Delta' = \{\Gamma_*^{-1}W_*\Delta'\}_+ + \{\Gamma_*^{-1}W_*\Delta'\}_- \tag{34}$$

where

$$\{\Gamma_*^{-1}W_*\Delta'\}_+ = \int_0^\infty \epsilon^{-st}\left[\frac{1}{2\pi j}\int_{-j\infty}^{j\infty} \Gamma_*^{-1}W_*\Delta'\epsilon^{st}\, ds\right] dt \tag{35}$$

is analytic in Re $s \geq 0$ and

$$\{\Gamma_*^{-1}W_*\Delta'\}_- = \int_{-\infty}^0 \epsilon^{-st}\left[\frac{1}{2\pi j}\int_{-j\infty}^{j\infty} \Gamma_*^{-1}W_*\Delta'\epsilon^{st}\, ds\right] dt \tag{36}$$

is analytic in Re $s \leq 0$. It therefore follows from (33) that

$$\{\Gamma_*^{-1}W_*\Delta'\}_+ - \Gamma H\Delta' = \tilde{X} - \{\Gamma_*^{-1}W_*\Delta'\}_-. \tag{37}$$

The left-hand side of (37) is analytic everywhere in the half-plane Re $s \geq 0$ and the right-hand side of (37) is analytic everywhere in the half-plane Re $s \leq 0$. Thus, the left-hand side of (37) must be analytic everywhere. This is the case if and only if

$$\{\Gamma_*^{-1}W_*\Delta'\}_+ - \Gamma H\Delta' = K \tag{38}$$

where K is an arbitrary polynomial matrix. Solving for H and recognizing that (32) implies $\Gamma_*^{-1}W_* = \Gamma W^{-1}$ yields ($|W| \not\equiv 0$ is assumed)

$$H = H_0 - \Gamma^{-1}K(\Delta')^{-1} \tag{39}$$

where

$$H_0 = \Gamma^{-1}\{\Gamma W^{-1}\Delta'\}_+(\Delta')^{-1}. \tag{40}$$

Equation (39) is arrived at after consideration of a special class of variations for H_1. It can not be stated, therefore, that $S_1 = 0$ for all physically realizable H_1 when H is given by (39). It can be stated, however, that if an optimum H exists it must be included among the family of functions defined by (39). It is now shown that H_0 is the optimum H. Substituting (39) into (20) one obtains

$$S = \tilde{S}_0 + 2\tilde{S}_1 + \tilde{S}_2 \tag{41}$$

where

$$\tilde{S}_0 = \frac{1}{2\pi j}\int_{-j\infty}^{j\infty} \operatorname{tr}[(1_n - WH_0)'(1_n - WH_0)_*'\Phi]\, ds \tag{42}$$

$$\tilde{S}_1 = \frac{1}{2\pi j}\int_{-j\infty}^{j\infty} \operatorname{tr}[(1_n - WH_0)'W_*'(\Gamma_*')^{-1}K_*'\Delta_*^{-1}\Phi]\, ds \tag{43}$$

$$\tilde{S}_2 = \frac{1}{2\pi j}\int_{-j\infty}^{j\infty} \operatorname{tr}[\Delta^{-1}K'(\Gamma')^{-1}W'W_*'(\Gamma_*')^{-1}K_*'\Delta_*^{-1}\Phi]\, ds. \tag{44}$$

[Equations (41)–(44) can also be arrived at by substituting $\epsilon = 1$, $H = H_0$, and $H_1 = -\Gamma^{-1}K(\Delta')^{-1}$ in (22)–(25).] From (40) it follows that

$$(1_n - WH_0) = W\Gamma^{-1}(\Gamma W^{-1}\Delta' - \{\Gamma W^{-1}\Delta'\}_+)(\Delta')^{-1} \tag{45}$$

or

$$(1_n - WH_0) = W\Gamma^{-1}\{\Gamma W^{-1}\Delta'\}_-(\Delta')^{-1}. \tag{46}$$

Using (31), (32), and (46) in (42)–(44) gives

$$\tilde{S}_0 = \frac{1}{2\pi j}\int_{-j\infty}^{j\infty} \operatorname{tr}[(\{\Gamma W^{-1}\Delta'\}_-)_*(\{\Gamma W^{-1}\Delta'\}_-)]\, ds \tag{47}$$

$$\tilde{S}_1 = \frac{1}{2\pi j}\int_{-j\infty}^{j\infty} \operatorname{tr}[(\{\Gamma W^{-1}\Delta'\}_-)K_*]\, ds \tag{48}$$

$$\tilde{S}_2 = \frac{1}{2\pi j}\int_{-j\infty}^{j\infty} \operatorname{tr}[KK_*]\, ds. \tag{49}$$

The above results are arrived at with the aid of (9) and the additional fact that a matrix and its transpose have the same trace.

It follows from (41) and (47)–(49) that if either

$$K = O_{nn} \tag{50}$$

or

$$K \to -\{\Gamma W^{-1}\Delta\}_-, \quad |s| \to \infty \tag{51}$$

then, and only then, can the sensitivity index be finite.[3] The condition (51) is never satisfied unless W^{-1} has no poles in Re $s \geq 0$: in this case $\{\Gamma W^{-1}\Delta'\}_- = O_{nn}$ and the best result possible, $S = 0$, is obtained. When (50) and (51) are not satisfied it follows that

$$\operatorname{tr}[KK_*] = q(s^2) \not\equiv 0 \tag{52}$$

where $q(s^2)$ is an arbitrary polynomial in s^2. This is the case because

$$\operatorname{tr}[KK_*] = \sum_{i=1}^{n} k_i' k_{i*}' = 0 \tag{53}$$

if and only if $k_i = o_n$, $i = 1,2,\cdots,n$, where k_i' is the ith row of K. Substituting (52) into (49) immediately leads to $\tilde{S}_2$ and therefore S being infinite. The only possible choice for H in (39) is then $H = H_0$.

It is now verified that $S_2 \geq 0$ for all physically realizable H_1, and therefore that $H = H_0$ is the optimum choice for the feedback network transfer function matrix. From (25) it suffices to show that

$$\begin{aligned} I(j\omega) &= [(WH_1)_*'\Phi(WH_1)']\,|_{s=j\omega} \\ &= [(WH_1)_*'\Delta_*\Delta(WH_1)']\,|_{s=j\omega} \end{aligned} \tag{54}$$

is a nonnegative definite Hermitian matrix. Applying the definition (5) with $s = j\omega$ to (54) gives

$$I(j\omega) = Z^*(j\omega)Z(j\omega) \tag{55}$$

where

$$Z(j\omega) = [\Delta(WH_1)']\,|_{s=j\omega}. \tag{56}$$

Clearly, $I(j\omega)$ is a nonnegative definite Hermitian matrix. Thus, $H = H_0$ is optimum, and (47) is a compact formula for the minimum value of S:

$$\min S = S\,|_{H=H_0} = \tilde{S}_0 \tag{57}$$

The results of this section are conveniently summarized in the following theorem.

Theorem

The physically realizable transfer function matrix

$$H = \Gamma^{-1}\{\Gamma W^{-1}\Delta'\}_+(\Delta')^{-1}$$

is the one for which the mean-square error of the system shown in Fig. 2 is a minimum. The minimum value of the mean-square error is given by

$$\min E\{y_e'(t)y_e(t)\}$$

$$= \frac{1}{2\pi j}\int_{-j\infty}^{j\infty} \operatorname{tr}[(\{\Gamma W^{-1}\Delta'\}_-)_*(\{\Gamma W^{-1}\Delta'\}_-)]\,ds.$$

On the Spectral Factorizations

The fundamental theorem regarding the spectral factorization of rational matrices is contained in [9]. An abbreviated statement of the theorem suitable for the problem being treated here is the following theorem.

[3] Note that for $K = -\{\Gamma W^{-1}\Delta'\}_-$ the integrands in (47)–(49) sum to zero.

Theorem 1

When the $n \times n$ real rational matrix $A(s)$ satisfies

a_1) $A(s) = A'(-s) = A_*(s)$,
a_2) $A(s)$ is analytic on the finite $s = j\omega$ axis, and
a_3) $A(j\omega)$ is positive definite for all finite ω,

then there exists an $n \times n$ real rational matrix $B(s)$ such that

b_1) $A(s) = B'(-s)B(s) = B_*(s)B(s)$, and
b_2) $B(s)$ and $B^{-1}(s)$ are both analytic in Re $s \geq 0$.

An immediate consequence of Theorem 1 is Theorem 2.

Theorem 2

Sufficient conditions for the $n \times n$ real rational matrix $\Omega = W_*W$ to have the spectral factorization $\Omega = \Gamma_*\Gamma$ where Γ and Γ^{-1} are both $n \times n$ real rational matrices analytic in Re $s \geq 0$ are

c_1) $W(s)$ be physically realizable and
c_2) $|W(j\omega)| \neq 0$, for all finite ω, or
c_3) $W^{-1}(s)$ be analytic on the finite $s = j\omega$ axis.

Proof: Clearly, $\Omega(s) = W'(-s)W(s) = \Omega'(-s)$ and condition a_1) is satisfied. Since $W(s)$ is physically realizable, $W(s)$ is analytic in Re $s \geq 0$ and $W'(-s)$ is analytic in Re $s \leq 0$. Hence, $\Omega(s)$ is analytic for $s = j\omega$, and condition a_2) is met. Finally, $|W(j\omega)| \neq 0$ guarantees that the Hermitian matrix $\Omega(j\omega)$ is positive definite, and a_3) is satisfied. The equivalence of c_2) and c_3) follows immediately from $W(s)W^{-1}(s) = 1_n$. Wherever $W^{-1}(s)$ is analytic on the $s = j\omega$ axis, $|W^{-1}(j\omega)|$ is finite and

$$|W(j\omega)|\,|W^{-1}(j\omega)| = 1$$

leads to $|W(j\omega)| \neq 0$. On the other hand, wherever $|W(j\omega)| \neq 0$, $|W^{-1}(j\omega)|$ must be finite. This is the case for finite ω only if $W^{-1}(s)$ is analytic on the finite $s = j\omega$ axis.

The spectral factorization of the matrix Φ defined by (18) is now considered. The conditions under which it is possible to accomplish the required spectral factorization are embodied in

Theorem 3

The $n \times n$ real rational matrix Φ has the spectral factorization $\Phi = \Delta_*\Delta$ where Δ and Δ^{-1} are both analytic in Re $s \geq 0$ whenever all of the following conditions are satisfied:

d_1) Φ_r satisfies a_1) through a_3),
d_2) W satisfies c_1) and c_2),
d_3) G_p^{-1} is analytic on the finite $s = j\omega$ axis,
d_4) the $N \times N$ covariance matrix $\Sigma = [\sigma_{ij}]$ is positive definite,
d_5) the rank of the $n \times nN$ matrix

$$\nabla = [\partial G_p/\partial\alpha_1 \mid \partial G_p/\partial\alpha_2 \mid \cdots \mid \partial G_p/\partial\alpha_N]$$

is n everywhere on the finite $s = j\omega$ axis.

Proof: It is not difficult to establish from d_1) and d_2) that the representation

$$W_*'\Phi_r W' = V_* V \quad (58)$$

is possible where the real rational $n \times n$ matrices V and V^{-1} are analytic in Re $s \geq 0$. Equation (18) is, therefore, of the form

$$\Phi = \sum_{l=1}^{N} \sum_{m=1}^{N} \sigma_{lm} M_{m*} M_l \quad (59)$$

where

$$M_k = VA_k = V[(\partial G_p/\alpha_k)G_p^{-1}]'. \quad (60)$$

Since G_p is real and rational it follows that the M_k are also. Hence, Φ is an $n \times n$ real rational matrix. Moreover,

$$\Phi_* = \sum_{l=1}^{N} \sum_{m=1}^{N} \sigma_{lm} M_{l*} M_m = \sum_{m=1}^{N} \sum_{l=1}^{N} \sigma_{ml} M_{l*} M_m = \Phi \quad (61)$$

and condition a_1) is satisfied.

Because G_p is physically realizable and rational it is true that the $\partial G_p/\partial \alpha_k$ are analytic in Re $s \geq 0$. Condition d_3) insures, therefore, that the M_k are analytic on the finite $s = j\omega$ axis. It immediately follows that Φ is analytic on the finite $s = j\omega$ axis and condition a_2) is satisfied. It only remains to show that Φ satisfies condition a_3).

Equation (59) is equivalent to

$$\Phi = M_* \Sigma M \quad (62)$$

where

$$M = [M_1' \mid M_2' \mid \cdots \mid M_N']' \quad (63)$$

$$\tilde{\Sigma} = \Sigma \times 1_n = [\sigma_{ij} 1_n]. \quad (64)$$

The matrix $\tilde{\Sigma}$ is a Kronecker product (see [12, p. 227]). Since both Σ and 1_n are positive definite or, equivalently, have only positive eigenvalues, it follows that all eigenvalues of $\tilde{\Sigma}$ are positive. Thus, $\Phi(j\omega)$ is positive definite for every finite ω if and only if there exist no n-dimensional nonzero column vector $\boldsymbol{a}$ and no finite ω for which $M(j\omega)\boldsymbol{a} = \boldsymbol{o}_{(nN)}$. This is the case if and only if the rank of $M'(j\omega)$ is n for all finite ω. Now

$$M' = \nabla \operatorname{diag}[G_p^{-1}, G_p^{-1}, \cdots, G_p^{-1}] \cdot \operatorname{diag}[V', V', \cdots, V'] \quad (65)$$

and it immediately follows from d_5) that rank $M'(j\omega)$ is indeed n for all finite ω provided V and G_p are nonsingular in the finite $s = j\omega$ axis. That $V(j\omega)$ is nonsingular follows from the fact that the right-hand side of (58) is positive definite on the finite $s = j\omega$ axis. Arguments identical with those used to establish the equivalence of c_2) and c_3) can be used with d_3) to establish that $G_p(j\omega)$ is nonsingular for all finite ω.

It is not difficult to verify that the plant transfer function matrix has the form

$$G_p(s,\boldsymbol{\alpha}) = G(s) \operatorname{diag}[\alpha_1, \alpha_2, \cdots, \alpha_n] \quad (66)$$

when the plant is described by the vector differential equation

$$\dot{x} = Ax + B \operatorname{diag}[\alpha_1, \alpha_2, \cdots, \alpha_n]u$$
$$c = Dx. \quad (67)$$

The $\delta\alpha_i$ then represent variations in control effort gains. It is now shown for the case in which G^{-1} is analytic on the finite $s = j\omega$ axis and $\alpha_i \neq 0$, for each $i = 1, 2, \cdots, n$, that condition d_5) is satisfied. The α_i are nonzero in keeping with the fact that only plant transfer function matrices with normal rank n need be considered. The analyticity of G^{-1} on the finite $s = j\omega$ axis assures the satisfaction of d_3).

When the jth column of G is denoted by the column vector $\boldsymbol{g}_j$, then one can write

$$\partial G_p/\partial \alpha_j = \boldsymbol{g}_j \boldsymbol{e}_j'. \quad (68)$$

Substituting (68) into the matrix ∇ defined in d_5) establishes that the only nonzero columns of ∇ are the columns $\boldsymbol{g}_j$, $j = 1, 2, \cdots, n$. Since G^{-1} is analytic on the finite $s = j\omega$ axis, it follows that the minor of order n formed from these n columns of ∇ is nonzero on the finite $s = j\omega$ axis. Therefore the rank of ∇ on the finite $s = j\omega$ axis is n, and condition d_5) is satisfied.

Physical Realizability of $G_c(s)$

The preceding developments are concerned with the determination of H_0 once W is specified. Attention is now turned to the computation of G_c, and the determination of conditions which guarantee that G_c is physically realizable. Solving (3) with $H = H_0$ for G_c and using (46) in the result yields

$$G_c = G_p^{-1}(1_n - WH_0)^{-1}W = G_p^{-1}\Delta' L^{-1}\Gamma \quad (69)$$

where

$$L = \{\Gamma W^{-1}\Delta'\}_-. \quad (70)$$

It is clear from (69) that G_c is not generally physically realizable. The matrices G_p^{-1} and L^{-1} can have poles in Re $s \geq 0$. The problem facing the designer and the one discussed here is the specification of W so that the G_c given by (69) is physically realizable.

Any W satisfying conditions c_1) and c_2) or c_3) can be written in the form

$$W = (q/q_*)\tilde{W} \quad (71)$$

where $q = q(s)$ is a monic polynomial with zeros in Re $s > 0$ only. The zeros of q include any zeros in Re $s > 0$ common to every element of W and also any other possible poles of W^{-1} in Re $s > 0$. From c_1), c_3), and the definition of q, it follows that $\tilde{W}^{-1}$ is analytic in Re $s \geq 0$. Substituting (71) into (70), recalling that Γ and Δ are analytic in Re $s \geq 0$, and collecting terms in the partial fraction expansion of $\Gamma W^{-1}\Delta'$ associated with the zeros of q, one obtains

$$L = \tilde{L}/q \quad (72)$$

where $\tilde{L}$ is a polynomial matrix. When, for example, the zeros of q are all simple and denoted by s_i, $i = 1,2,\cdots,l$, then

$$\tilde{L} = \sum_{i=1}^{l} \{[(s - s_i)q_*\Gamma\tilde{W}^{-1}\Delta']/q|_{s=s_i}\}[q/(s - s_i)]. \quad (73)$$

From (69) and (72) it follows that

$$G_c = qG_p^{-1}\Delta'\tilde{L}^{-1}\Gamma = q\tilde{G}_c \quad (74)$$

is physically realizable if and only if the poles of $\tilde{G}_c$ in $\operatorname{Re} s \geq 0$ are canceled by zeros of q.

It is of interest to examine the case in which G_p^{-1} has only one pole in $\operatorname{Re} s \geq 0$ at $s = \sigma_0 > 0$, σ_0 real. When it is possible to meet the dynamic performance requirements placed on the system with

$$W = -[(s - \sigma_0)/(s + \sigma_0)]\tilde{W} \quad (75)$$

where $\tilde{W}$ and $\tilde{W}^{-1}$ are analytic in $\operatorname{Re} s \geq 0$, then one can choose $\Gamma = -\tilde{W}$ and obtain from (73)

$$\tilde{L} = 2\sigma_0\Delta'(\sigma_0). \quad (76)$$

It now follows from (74) that

$$G_c = -(1/2\sigma_0)(s - \sigma_0)G_p^{-1}\Delta'[\Delta'(\sigma_0)]^{-1}\tilde{W}. \quad (77)$$

Since Δ^{-1} is analytic in $\operatorname{Re} s \geq 0$, the matrix $[\Delta'(\sigma_0)]^{-1}$ is finite. Moreover, $(s - \sigma_0)G_p^{-1}$, Δ, and $\tilde{W}$ are analytic in $\operatorname{Re} s \geq 0$, and G_c is therefore physically realizable.

Another interesting case for which general conclusions can be drawn occurs when G_p^{-1} has no poles in $\operatorname{Re} s \geq 0$, and the dynamic performance requirements of the system are met by a physically realizable W whose inverse is analytic in $\operatorname{Re} s \geq 0$. Under these conditions

$$\{\Gamma W^{-1}\Delta'\}_+ = \Gamma W^{-1}\Delta' \quad (78)$$

$$\{\Gamma W^{-1}\Delta'\}_- = O_{nn}. \quad (79)$$

Hence (40) reduces to $H_0 = W^{-1}$, and (57) leads to $\min S = 0$. Substituting $H_0 = W^{-1}$ into (69) indicates that

$$G_c = G_p^{-1}KW \quad (80)$$

which, aside from the fact that K is a constant matrix each of whose elements are infinitely large, is physically realizable. That is, the sensitivity can be made arbitrarily small at the expense of high gains in the tandem compensation network. This is simply a generalization to the multivariable case of the well-known result for single-input-output systems that overall system sensitivity to plant parameter variations can be made arbitrarily small when the specified plant and overall system transfer functions are minimum phase.

A Numerical Example

A plant with nominal transfer function matrix

$$G_p(s,\alpha) = G_p(s,\alpha) = (\alpha/s)\begin{bmatrix} 1/(s+1) & 1/(s+2) \\ 1/(s+3) & 1/(s+5) \end{bmatrix} \quad (81)$$

is considered. The usual practice of taking the pole at the origin to be located at $s = -\epsilon$, $\epsilon > 0$, and arbitrarily small, is followed here. A plant of this type is encountered when in (67) $n = 2$, $\alpha_1 = \alpha_2 = \alpha$, and $\delta\alpha_1 = \delta\alpha_2 = \delta\alpha$.

The input power spectral density matrix is

$$\Phi_r = \Delta_{r*}\Delta_r \quad (82)$$

where

$$\Delta_r = \begin{bmatrix} 1/(s+1) & 1/(s+10) \\ 1/(s+10) & 1/(s+2) \end{bmatrix}. \quad (83)$$

Both Δ_r and Δ_r^{-1} are analytic in $\operatorname{Re} s \geq 0$.

It is required that the overall transfer function matrix be a diagonal one (a noninteracting system is desired) and that the bandwidth of each diagonal element be equal to the bandwidth of the corresponding diagonal element of Φ_r. It is also required that the dc gain of the diagonal elements be unity. A transfer function matrix which closely meets these requirements is

$$W = -\left(\frac{s-1}{s+1}\right)\begin{bmatrix} 1/(s^2+\sqrt{2}s+1) & 0 \\ 0 & 4/(s^2+2\sqrt{2}s+4) \end{bmatrix}. \quad (84)$$

The damping ratio for each pair of complex conjugate poles is 0.707. The zero at $s = 1$ guarantees the physical realizability of G_c: the inverse of the given plant transfer function matrix has a pole at $s = 1$.

Using the given data, one obtains from (18)

$$\Phi = (\sigma/\alpha)^2(\tilde{W}_*\Delta_{r*})(\Delta_r\tilde{W}). \quad (85)$$

Since Δ_r^{-1} and $\tilde{W}^{-1}$ are analytic in $\operatorname{Re} s \geq 0$, a suitable choice for Δ is

$$\Delta = (\sigma/\alpha)\Delta_r\tilde{W}. \quad (86)$$

Also, $W_*W = \tilde{W}_*\tilde{W}$ and one can make the choice $\Gamma = -\tilde{W}$. Equation (77) is applicable in this example and a straightforward computation leads to

$$G_c = \begin{bmatrix} g_{c11} & g_{c12} \\ g_{c21} & g_{c22} \end{bmatrix} \quad (87)$$

where

$$g_{c11} = \frac{-7.84s(s+3)(s^5+1.8s^4-28s^3-97.3s^2-138s-78.0)}{\alpha(s+2)(s+10)(s^2+\sqrt{2}s+1)^2(s^2+2\sqrt{2}s+4)} \quad (88)$$

$$g_{c12} = \frac{-30.3s(s+3)(s^5+23.7s^4+132s^3+243s^2+216s+84.0)}{\alpha(s+2)(s+10)(s^2+\sqrt{2}s+1)(s^2+2\sqrt{2}s+4)^2} \quad (89)$$

$$g_{c21} = \frac{7.84s(s+5)(s^4 - 1.5s^3 - 23.7s^2 - 49.0s - 39.7)}{\alpha(s+10)(s^2+\sqrt{2}s+1)^2(s^2+2\sqrt{2}s+4)} \tag{90}$$

$$g_{c22} = \frac{30.3s(s+5)(s^4 + 20.3s^3 + 74.7s^2 + 87.0s + 50.0)}{\alpha(s+10)(s^2+\sqrt{2}s+1)(s^2+2\sqrt{2}s+4)^2}. \tag{91}$$

The optimum feedback network transfer function matrix can be computed using (40). It is more convenient to use the equivalent expression

$$H_0 = \Gamma^{-1}(\Gamma W^{-1}\Delta' - \{\Gamma W^{-1}\Delta'\}_-)(\Delta')^{-1} \tag{92}$$

in this example, however. The computation gives

$$H_0 = \begin{bmatrix} h_{011} & h_{012} \\ h_{021} & h_{022} \end{bmatrix} \tag{93}$$

where

$$h_{011} = \frac{1.41(10)^{-2}(s+1)(s^2+\sqrt{2}s+1)(s^3+24.2s^2+174s+291)}{(s+5.76)} \tag{94}$$

$$h_{012} = \frac{-3.53(10)^{-3}(s+2)(s+10)(s^2+2\sqrt{2}s+4)(s^2+1.40s+0.991)}{(s+5.76)} \tag{95}$$

$$h_{021} = \frac{-3.64(10)^{-3}(s+1)(s+10)(s+\sqrt{2}s+1)(s^2+2.83s+4.62)}{(s+5.76)} \tag{96}$$

$$h_{022} = \frac{9.10(10)^{-4}(s^2+2\sqrt{2}s+4)(s^4+29.3s^3+286s^2+902s+512)}{(s+5.76)}. \tag{97}$$

Clearly, repeated differentiations—which can only be approximated in practice—are required in the feedback network.

It is not difficult to verify that

$$\{\Gamma W^{-1}\Delta'\}_- = \frac{(\sigma/\alpha)}{(s-1)} K \tag{98}$$

where the constant matrix K is given by

$$K = \begin{bmatrix} \dfrac{1}{2+\sqrt{2}} & \dfrac{2}{11(2+\sqrt{2})} \\ \dfrac{8}{11(5+2\sqrt{2})} & \dfrac{8}{3(5+2\sqrt{2})} \end{bmatrix}. \tag{99}$$

Substituting (98) into (47) immediately yields

$$\min S = \tilde{S}_0 = \tfrac{1}{2}(\sigma/\alpha)^2 \operatorname{tr}[K'K] = 0.107(\sigma/\alpha)^2. \tag{100}$$

The minimum value of S is in and of itself not very informative. The real usefulness of (100) is for comparison purposes with other designs of a more practical nature. For example, the success of any instrumentation which approximates the repeated differentiations required in the feedback network can be gauged by comparing the value of S for the practical system with the one given by (100).

Conclusions

The approach taken in this paper is significant in that it leads to the analytical design of minimum sensitivity feedback systems. The method is applicable when the plant is a linear, time-invariant, lumped, finite-dimensional dynamical system and the uncertain plant parameters can be viewed as random variables. A part of the development important in its own right is the solution of the semi-free-configuration Wiener problem for the multivariable case.

Acknowledgment

The author wishes to thank Prof. D. C. Youla for his interest and help.

References

[1] J. J. Bongiorno, Jr., "Minimum sensitivity designs," *1967 IEEE Internatl. Conv. Rec.*, vol. 15, pt. 3, pp. 129–135.
[2] M. W. Wonham, "On pole assignment in multi-input controllable linear systems," *IEEE Trans. Automatic Control*, vol. AC-12, pp. 660–665, December 1967.
[3] J. J. Bongiorno, Jr., and D. C. Youla, "On observers in multivariable control systems," *Internatl. J. Control*, vol. 8, pp. 221–243, September 1968.
[4] J. B. Cruz, Jr., and W. R. Perkins, "A new approach to the sensitivity problem in multivariable feedback system design," *IEEE Trans. Automatic Control*, vol. AC-9, pp. 216–223, July 1964.
[5] ——, "The parameter variation problem in state feedback control systems," *Trans. ASME, J. Basic Engrg.*, vol. 87, ser. D, pp. 120–124, March 1965.

[6] W. R. Perkins, J. B. Cruz, Jr., and R. L. Gonzales, "Design of minimum sensitivity systems," *IEEE Trans. Automatic Control*, vol. AC-13, pp. 159–167, April 1968.

[7] H. C. Hsieh and C. T. Leondes, "Techniques for the optimum synthesis of multipole control systems with random processes as inputs," *IRE Trans. Automatic Control*, vol. AC-4, pp. 212–231, December 1959.

[8] M. C. Davis, "Factoring the spectral matrix," *IEEE Trans. Automatic Control*, vol. AC-8, pp. 296–305, October 1963.

[9] D. C. Youla, "On the factorization of rational matrices," *IRE Trans. Information Theory*, vol. IT-7, pp. 172–189, July 1961.

[10] W. G. Tuel, Jr., "Computer algorithm for spectral factorization of rational matrices," *IBM J. Res. Develop.*, vol. 12, pp. 163–170, March 1968.

[11] S. S. L. Chang, *Synthesis of Optimal Control Systems.* New York: McGraw-Hill, 1961.

[12] R. Bellman, *Introduction to Matrix Analysis.* New York: McGraw-Hill, 1960.

[13] W. M. Mazer, "Specification of the linear feedback system sensitivity function," *IRE Trans. Automatic Control*, vol. AC-5, pp. 85–93, June 1960.

[14] S. G. Loo, "Applications of Wiener–Hopf theory to linear multivariable systems with randomly sampled signals," *IFAC Symp. Papers on Automatic Control in Basic Industries* (Sidney, Australia, 1968), pp. 45–50.

Feedback, Minimax Sensitivity, and Optimal Robustness

GEORGE ZAMES, FELLOW, IEEE, AND BRUCE A. FRANCIS, MEMBER, IEEE

Abstract—In this paper, we look for feedbacks that minimize the sensitivity function of a linear single-variable feedback system represented by its frequency responses. Sensitivity to disturbances and robustness under plant perturbations are measured in a weighted H^∞ norm.

In an earlier paper, Zames proposed an approach to feedback design involving the measurement of sensitivity by "multiplicative seminorms," which have certain advantages over the widely used quadratic norm in problems where there is plant uncertainty, or where signal power-spectra are not fixed, but belong to sets. The problem was studied in a general setting, and some H^∞ examples were solved.

Here, a detailed study of the single-variable case is undertaken. The results are extended to unstable plants, and explicit formulas for the general situation of a finite number of right half-plane (RHP) plant zeros or poles are provided. The Q or "approximate-inverse" parametrization of feedbacks that maintain closed-loop stability is extended to the case of unstable plants. The H^∞ and Wiener–Hopf approaches are compared.

I. Introduction

In this paper we study the servomechanism problem, and seek a feedback F (see Fig. 1) to attenuate disturbances d or plant perturbations ΔP in a linear single-variable plant P characterized by its frequency response. We view this problem from a classical perspective, but would like to replace the heuristic aspect of classical design by an explicit mathematical theory. Our approach depends upon the minimization of a weighted H^∞ norm of the sensitivity function.

The main features of this approach were outlined in Zames [1]. It may be worth recalling the reasons for the interest in H^∞ norms and frequency domain models.

A. Motivation

At present, mathematical theories of servodesign are based mainly on quadratic minimization of the Wiener–Hopf–Kalman type, usually applied to state-space models. However, despite the academic success of these methods, classical frequency response techniques relying on "lead-lag compensators" to reduce sensitivity have con-

Manuscript received August 11, 1981; revised May 5, 1982 and June 9, 1982. Paper recommended by E. W. Kamen, Past Chairman of the Linear Systems Committee. This work was supported by the National Sciences and Engineering Research Council of Canada and the National Science Foundation under Grant ECS-80-12-565.

G. Zames is with the Department of Electrical Engineering, McGill University, Montreal, P.Q., Canada.

B. A. Francis is with the Department of Electrical Engineering, University of Waterloo, Waterloo, Ont., Canada.

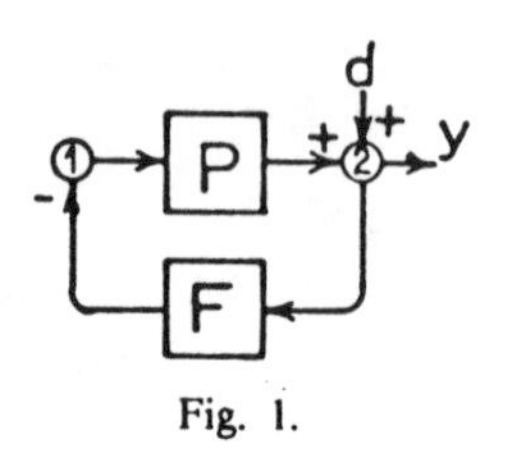

Fig. 1.

tinued to dominate industrial servodesign. One reason is that quadratic designs tend to have poor sensitivity. Various weighting strategies have been introduced to improve sensitivity, and may give good results in practice. However, weighting strategies are implicit, as the weighting depends upon the feedback sought, and leave unresolved the issue of what that feedback should be.

A more basic reason is related to the ability of systems to perform without exact knowledge of plant models or signal power-spectra. This ability is a distinguishing feature of the systems loosely called "servomechanisms," and is emphasized in classical "cybernetics," but has been difficult to capture in the quadratic[1]-state-space framework. The following discussion suggests that the difficulties may be intrinsic, and fresh approaches to the servoproblem may be worth seeking.

Variations in Power-Spectra: In quadratic methods the *integral-squared* sensitivity is minimized for disturbances d having a single *fixed* power-spectrum. If d is assumed to be deterministic for simplicity, its power spectrum coincides with the magnitude $|d(j\omega)|$. Typically, disturbance power is assumed to be small in any small frequency interval $(\omega_o - \frac{1}{2}\delta, \omega_o + \frac{1}{2}\delta)$, and there is no *a priori* reason for sensitivity to be small in that interval.

The assumption of a fixed spectrum is reasonable in those filtering problems in which the disturbance source is steady, e.g., 60 Hz noise in power supplies. However, most servos must respond to *large classes* of spectra, including steps, pulses, and narrow-band signals of various frequencies. In the case of a narrow-band spectrum whose power is concentrated near one frequency ω_o, large sensitivity at ω_o is of course not acceptable.

It would appear, then, that a plausible disturbance class for servos should contain a variety of spectra, including

[1]The term "quadratic" here refers to the norm on system transfer functions, and not the norm on inputs or outputs. The latter norm will always be assumed to be quadratic.

Reprinted from *IEEE Trans. Automat. Contr.*, vol. AC-28, no. 5, pp. 585–601, May 1983.

narrow-band disturbances in particular, subject to restrictions on how disturbance power depends upon frequency. Here, frequency dependence will be described by a (continuous bounded) weighting function $W(j\omega)$, and a class of spectra will be considered consisting of those $|d(j\omega)|$ in $L^2(-\infty,\infty)$ satisfying the inequality

$$\int_{-\infty}^{\infty} |d(j\omega)|^2 |W(j\omega)|^{-2} d\omega < 1. \tag{1.1}$$

$|W(j\omega)|$ can be viewed (see Remark 1) in Section II-D) as the maximal power of those disturbances whose spectra in $\mathfrak{D}$ are narrow-band and concentrated near ω.

As any spectrum may be persistent, servos are usually designed to perform dependably for all members of the class, e.g., classical specs such as phase margin or maximal magnitude are worst-case specs. The weighted H^∞ methods employed here in effect minimize maximum response power over all disturbances in the class. In this respect, the H^∞ methods resemble the classical ones. On the other hand, quadratic methods ensure good sensitivity for only one disturbance spectrum (normalized to have unit integral-squared), e.g., the average, and therefore cannot be relied upon for dependable performance under large spectral variations.

Plant Uncertainty: A related issue (which is considered at greater length in [1]) is that quadratic norms are not multiplicative, and therefore not well suited to plant perturbation studies. The effect of a plant perturbation ΔP on closed-loop behavior has the form of a product $E\Delta P$, E being a "return-difference" operator. It is therefore desirable to employ norms in which the norm of the product $\|AB\|$ of any two systems A, B, bears a simple relation to their separate norms, e.g., $\|AB\| \leqslant \|A\| \cdot \|B\|$. The weighted H^∞ norm employed here has such multiplicative[2] properties, whereas pure quadratic norms do not.

Frequency Response Descriptions: The revived interest in these partly follows the observation that certain problems of plant approximation/identification, which are easily described in terms of frequency responses, become virtually untractable in a state-space framework. For example, suppose that the inputs and outputs of two plants P and P_a lie in $L_2(0,\infty)$, and we measure how well P_a approximates the input–output behavior of P by the resulting induced norm of their difference $\|P_a - P\|$. Then, P_a approximates P to a tolerance ϵ in input–output behavior if and only if their frequency responses approximate each other, i.e., $|P_a(j\omega) - P(j\omega)| \leqslant \epsilon$ for all ω. However, even though ϵ is small, the pole-zero configurations of P and P_a may be quite different, and so may be their $[A,B,C,D]$ state descriptions. Consequently, in identification experiments, if we obtain a sequence of approximants P_n, $n=1,2,\cdots$, to the input–output behavior of a system, their $[A_n, B_n, C_n, D_n]$ descriptions may grow in complexity and fail to converge as $n \to \infty$. On the other hand, their frequency response descriptions $P_n(j\omega)$ must converge uniformly in ω. In some modeling problems, therefore, the input–output behavior is viewed as the primary mathematical object, with which it may be impossible to associate a unique state description.

[2] i.e., H^∞ is a normed algebra.

B. Relation to Previous Works

In [1] a fresh approach to the servoproblem was outlined in the general setting of an algebra of input–output maps. It was shown that many design issues can be reduced to the minimization of a norm of the sensitivity operator, that this norm must differ from the primary norm of the algebra, and that multiplicative norms are desirable in sensitivity studies. The weighted-H^∞ norm was shown to be an example of an appropriate multiplicative norm, and applied to multivariable systems without RHP (right half-plane) zeros, and single variable systems with a single RHP zero, with emphasis on plants that are stable or have been stabilized. Here, we undertake a detailed study of the single-variable case, extend the results to unstable plants, and provide explicit formulas for the general case of a finite number of RHP plant zeros or poles.

In [1] the feedbacks that maintain closed-loop stability were parametrized by a stable system-variable Q, which was interpreted as an approximate inverse. Here, we extend the parametrization to unstable plants, and interpret Q as a constrained approximate inverse.

The H^∞ approach is compared to the Wiener–Hopf method in an example involving a representative servomechanism.

Our results are an application of the theory of optimal interpolation developed early in the century by Cartheodory, Schur, Pick, Nevanlinna, and others (see Walsh [2, ch. 10]). An exposition of optimal H^∞ interpolation, involving the functional-analytic concept of duality, was developed several decades later by Rogosinski and Shapiro [3], and is outlined in Duren [4] and Koosis [5]. We shall use a combination of these early and later approaches.

The Q-parametrization is related to the well-known results of Youla *et al.* [6], and the interesting generalizations recently obtained by Desoer *et al.* [7], for coprime factorable plants and feedbacks. The Q-parametrization is convenient here because it displays the relationship between sensitivity and invertibility.

Safonov and Athans [8] and Doyle and Stein [9] have noted the lack of robustness of purely quadratic methods, and tried to reconcile them with classical theory.

II. Problem Formulation

The feedback system of Fig. 1 consists of a plant and feedback "compensator," whose input–output behavior is represented by the frequency responses $P(s)$ and $F(s)$, respectively. The system graph has two nodes, which have been labeled 1 and 2. The CL (closed-loop) behavior is represented by a 2×2 matrix of frequency responses

$$\left[K_{ij}\right]_{i,j=1}^{2} \triangleq \begin{bmatrix} (1+FP)^{-1} & P(1+FP)^{-1} \\ F(1+PF)^{-1} & (1+PF)^{-1} \end{bmatrix} \tag{2.1}$$

where K_{ij} is the response from the ith to the jth nodes.

In particular, the response of the output y to a disturbance d is $K_{22} = (1+PF)^{-1}$. This term appears in most expressions for sensitivity, whether to additive disturbances such as d, or to perturbations in the plant P. We shall call K_{22} the *sensitivity function*.[3] Our objective will be to optimize certain measures of this sensitivity function while keeping all the K_{ij} stable.

A *frequency response*, for our purposes, is any function $G(s)$ of a complex variable $s \triangleq \sigma + j\omega$, which is meromorphic in the open RHP $\mathrm{Re}(s) > 0$, i.e., is analytic in $\mathrm{Re}(s) > 0$ except, perhaps, at a countable number of poles.

A frequency response $G(s)$ is *proper* iff the limit $\lim_{R\to\infty} \sup_{|s|>R,\, \mathrm{Re}(s)>0} |G(s)|$ is finite, and *strictly proper* iff the limit is zero. $G(s)$ is *stable* iff it is bounded, i.e., $|G(s)| \leqslant$ constant, and analytic in $\mathrm{Re}(s) > 0$.

If G_1 and G_2 are stable frequency responses and c is any complex constant, then the sum $G_1 + G_2$ and products G_1G_2 and cG_2 are all stable frequency responses, i.e., stable frequency responses form an algebra over the field of complex numbers. The norm of any stable frequency response G is defined to be $\|G\| = \sup_{\mathrm{Re}(s)>0} |G(s)|$. Under this norm, the algebra of stable frequency responses is called the Hardy space of order ∞, and denoted by H^∞. It can be shown that the H^∞ norm can be found from the magnitude of $G(s)$ along the $j\omega$-axis, i.e., $\|G\| = \sup_\omega |G(j\omega)|$. The strictly proper frequency responses in H^∞ form a subalgebra denoted by H_o^∞.

It will be assumed throughout that $P(s)$ and $F(s)$ are frequency responses such that $P(s)F(s)$ is not identically equal to -1 (which will certainly be true if $P(s)$ and $F(s)$ are strictly proper). This assumption ensures that all the inverses in (2.1) are well defined frequency responses. Furthermore, *it will be assumed that $P(s)$ is the sum of an H^∞ function and a proper rational function (so that $P(s)$ is proper) and that $P(s)$ has at most a finite number of poles and zeros in $Re(s) \geqslant 0$.*

A. Fractional Transformation

We shall employ a transformation which serves to parametrize the feedbacks that keep the CL system stable, and thereby ensures a stable design. This transformation was introduced in [1] for stable plants, and is extended here to unstable plants. Let Q be a new variable defined by the fractional transformation

$$Q = F(1+PF)^{-1}. \tag{2.2a}$$

As $P(s)F(s) \not\equiv -1$ by hypothesis, (2.2a) is well defined. Moreover, $P(s)Q(s) \not\equiv 1$, and $F(s)$ can be recovered from $Q(s)$ by the reverse transformation

$$F = Q(1-PQ)^{-1} \tag{2.2b}$$

and vice versa. Observe that as $P(s)$ is proper, (2.2a), (2.2b) map any strictly proper $F(s)$ into a strictly proper $Q(s)$, and vice versa. We shall view $Q(s)$ as a representation of the feedback $F(s)$, which is useful for design purposes. In terms of Q, (2.1) assumes the form

$$[K_{ij}] = \begin{bmatrix} (1-QP) & P(1-QP) \\ Q & (1-PQ) \end{bmatrix}. \tag{2.3}$$

If the matrix (2.3) is stable, then Q is stable. Conversely, any choice of stable Q makes (2.3) stable provided that Q satisfies the following constraints.

C1: PQ and $(1-PQ)P$ are both stable. By using Q, subject to these constraints, as a design variable instead of F, we replace a potentially unstable design variable by a stable one and, more importantly, *automatically ensure closed-loop stability*, without further recourse to stability tests.

If P is stable to begin with then C1 is satisfied automatically for stable Q. Otherwise, C1 is a condition for Q to represent a stabilizing feedback for P. For stable Q, C1 is equivalent to the following condition.

C2: At any pole a of $P(s)$ of order m in $Re(s) \geqslant 0$, $(1-P(s)Q(s))$ must have m or more zeros.

That C1 implies C2 is clear. Conversely, suppose that C2 is true. Then, as Q is bounded in $\mathrm{Re}(s) > 0$, neither $(1-PQ)P$ nor $(1-PQ)$ can have any poles in $\mathrm{Re}(s) \geqslant 0$, except possibly at the poles of P, where in fact both are now finite. Both $(1-PQ)P$ and $(1-PQ)$ must therefore be bounded and analytic in $\mathrm{Re}(s) > 0$, and therefore stable, i.e., C1 is true.

It can be verified that under C2, Q has precisely m zeros at a.

B. Behavior Near ∞: Weightings

Some notation for limiting the growth or decay of frequency responses as $s \to \infty$ will be adopted.

Notation: A frequency response G is of inferior order $0(-l)$ at ∞, $l \geqslant 0$, iff for some $R > 0$,

$$|G(s)| \geqslant \mathrm{const}|s|^{-l} \tag{2.4}$$

for $|s| > R$, $Re(s) \geqslant 0$. The statement that G is in $(1+s)^l H^\infty$ means that $(1+s)^{-l}G(s)$ is in H^∞.

The H^∞ norm by itself is not a useful measure of sensitivity because if P is a strictly proper plant, $P(j\omega)$ approaches 0 as $\omega \to \infty$, with the consequence that if Q is in H^∞, the sensitivity function, whose expression in terms of Q is $(1-PQ)$, approaches 1. It follows that $\|1-PQ\| \geqslant 1$. It is impossible to make the H^∞ norm of $(1-PQ)$ small.[4] It is possible to design Q so as to make the sensitivity function small over some given finite frequency interval, but then it turns out that whenever $P(s)$ has RHP zeros, the sensitivity becomes very large over the remaining frequencies. The problem, in that case, is to achieve some compromise between the two frequency behaviors. One way of rationalizing such a compromise (see [1]) is to employ a frequency weighting function W.

[3] $(1+PF)^{-1}$ is often called the reciprocal return difference in engineering texts.

[4] For a more general discussion of these issues, see [1].

C. The Problem

The main objective of this paper is to solve the following problem. Suppose that the plant $P(s)$ and a weighting $W(s)$ are fixed proper frequency responses, not identically equal to zero, and

1) $P(s)$ is the sum of an H^∞ function continuous in $\mathrm{Re}(s) \geqslant 0$, and a proper rational function; $P(s)$ has q poles, $a_1, \cdots, a_q$, and r zeros, $b_1, \cdots, b_r$, in $\mathrm{Re}(s) > 0$, (q, r, finite), each enumerated according to its multiplicity; has no poles or zeros on the $j\omega$-axis; and $P(s)$ has inferior order $0(-k)$ at ∞.

2) $W(s)$ is in H^∞, is continuous in $\mathrm{Re}(s) \geqslant 0$, has norm $\|W\| \leqslant 1$, has no zeros in $\mathrm{Re}(s) \geqslant 0$, and is of inferior order $0(-l)$ at ∞.

Problem 1: Find a strictly proper[5] *feedback frequency response $F(s)$ to minimize*

$$\|W(1+PF)^{-1}\| \tag{2.5}$$

subject to the constraint that the CL matrix (2.1) is stable. As the value of (2.5) is identical to that of

$$\|W(1-PQ)\| \tag{2.6}$$

an equivalent problem, which is easier to solve because (2.6) is affine in Q, whereas (2.5) is nonlinear in F, is the following.

Problem 2: Find a stable strictly proper[5] *Q in H_o^∞ to minimize (2.6), subject to the constraint that Q stabilize the CL system, i.e., that conditions C1 or C2 hold.*

The expression (2.5), (2.6) will be called the *weighted sensitivity*. The *optimal weighted sensitivity*, denoted by $\mu(P)$, is defined by either one of the equivalent expressions

$$\mu(P) = \inf_F \|W(1+PF)^{-1}\| \tag{2.7}$$

$$= \inf_Q \|W(1-PQ)\| \tag{2.8}$$

where the infima are taken over all strictly proper frequency responses F, or over all Q in H_o^∞ satisfying the constraints C1 or C2.

D. Remarks

1) The weighted sensitivity represents the power output maximized over all disturbances whose spectra $|d(j\omega)|$ are in $L^2(-\infty,\infty)$ and satisfy (1.1), i.e., belong to the class

$$\mathfrak{D} = \{|d(\cdot)| \in L^2 \colon \|W^{-1}d\|_{L^2} \leqslant 1\}.$$

In other words,

$$\sup_{|d| \in \mathfrak{D}} \|(1+PF)^{-1}d\|_{L^2} = \|W(1+PF)^{-1}\|_{H^\infty}.$$

The weighting function $|W(j\omega)|$ can be interpreted as giving the maximal power of those disturbance spectra in $\mathfrak{D}$ that are narrow-band and concentrated at ω. More precisely, let $\chi(\omega_o, \delta; \omega)$, $0 < \delta < \omega_o$, denote the function of unit norm in $L^2(-\infty,\infty)$ satisfying the equations

$$(\omega_o, \delta; \omega) = \begin{cases} \dfrac{1}{\sqrt{\delta}} & \text{for } \omega \in \left[\omega_o - \dfrac{\delta}{2}, \omega_o + \dfrac{\delta}{2}\right] \\ 0 & \text{elsewhere.} \end{cases}$$

$\mathfrak{D}$ certainly contains each narrow-band spectrum of the type $|d(j\omega)| = \lambda\chi(\omega_o, \delta; \omega)$, centered at some ω_o, provided that

$$\lambda \leqslant \sup\{|W(j\omega)| \colon \omega \in [\omega_o - \tfrac{1}{2}\delta, \omega_o + \tfrac{1}{2}\delta]\}. \tag{2.8a}$$

The parameter λ determines narrow-band power, and we have

$$|W(j\omega_o)| = \lim_{\delta \to 0} \sup\{\lambda \colon \lambda\chi[\omega_o, \delta; \cdot] \in \mathfrak{D}\}.$$

$\mathfrak{D}$ also contains all convex combinations (see Luenberger [10, pp. 18, 43] of narrow-band spectra satisfying (2.8a), among other functions).

2) W generates an auxiliary norm $\|\cdot\|_W$ on the algebra of stable frequency responses, defined by $\|G\|_W \triangleq \|WG\|$, and our objective therefore is to minimize this auxiliary norm of the sensitivity function $\|(1+PF)^{-1}\|_W$. The auxiliary norm has the "multiplicative" properties

$$\|G_1G_2\|_W \leqslant \|G_1\| \cdot \|G_2\|_W, \qquad \|G_1G_2\|_W \leqslant \|G_1\|_W \|G_2\|. \tag{2.9}$$

It was pointed out in [1] that inequalities (2.9) are important in the study of plant perturbations or plant uncertainty, but are not valid for the quadratic norms used in LQG methods. One of our main reasons for pursuing the present weighted-H^∞ approach is that it can be extended to problems of plant uncertainty, along the lines of [1].

3) It is clear from (2.8) that the optimal sensitivity depends upon how close PQ can be made to 1 under the W weighting, i.e., on how invertible P is. For stable plants, $\mu(P)$ is called [1] the *measure of singularity* of P in H^∞. It is shown in [1] that $0 \leqslant \mu(P) \leqslant 1$, and that $\mu(P) \geqslant W(b_i)$, from which it follows that small sensitivity cannot be achieved if zeros are present in any heavily weighted part of the RHP. It will be shown that these notions can be extended to the unstable case.

4) As the norm of any H^∞ function depends only upon its $j\omega$-axis magnitude, all weightings W having a common magnitude on the $j\omega$-axis are equivalent. It follows that the assumption that W in H^∞ has no RHP zeros involves no loss of generality. Indeed, if $W_1(s)$ is any H^∞ function continuous in $\mathrm{Re}(s) \geqslant 0$, and $W_1(s)$ has a finite number of RHP zeros $c_1, \cdots, c_m$, there exists an equivalent weighting W_1 in H^∞ without RHP zeros, defined by the equation

$$W(s) = W_1(s) \prod_{i=1}^{m} (c_i - s)^{-1}(\bar{c}_i + s)$$

which satisfies $|W(j\omega)| = |W_1(j\omega)|$.

[5] As $P(s)$ is proper, the strict propriety of $F(s)$ or $Q(s)$ ensures that $P(s)Q(s) \not\equiv -1$ and the problem is well posed.

5) Each frequency response G in H^∞ determines a bounded, causal, linear, time-invariant operator $\boldsymbol{G}$ from the space $L^2(0,\infty)$ of (square-integrable) inputs into the similar space of outputs; conversely, each such operator $\boldsymbol{G}$ determines a G in H^∞. The H^∞ norm of G coincides with the operator norm of $\boldsymbol{G}$, i.e., $\|G\| = \sup_{0 \neq u \in L^2} \{\|\boldsymbol{G}u\|_{L^2}/\|u\|_{L^2}\}$. In view of this 1:1 correspondence between operators and frequency responses, the problem of minimizing the weighted operator norm $\|\boldsymbol{W}(\boldsymbol{I}+\boldsymbol{PF})^{-1}\|$ of the sensitivity operator can be done entirely in the frequency domain.

E. Relaxation of Propriety

The optimization problem is simplified if the propriety constraints on Q and the other CL responses of the matrix (2.2) are relaxed. It will be convenient initially to solve the simpler problem (in Section IV), and later (in Section V) to modify the improper solution obtained for Q, by inserting some high frequency attenuation, and so obtain a family of strictly proper solutions to the original problem. The relaxed problem is defined as follows.

Let H^B denote the algebra of functions $G(s)$ which are analytic in $\mathrm{Re}(s)>0$ and bounded on every half-disk $|s|<R$, $\mathrm{Re}(s)>0$ of finite radius $R>0$. H^∞ consists precisely of those functions in H^B that are proper.

Problem 2^0: Find a (not necessarily proper) Q in H^B to minimize (2.6) subject to the constraint C3: that $(1-PQ)P$ is[6] in H^B.

This problem is divorced from any consideration of F.

III. Allowable Sensitivities and Feedbacks: Interpolation Constraints

The optimization Problems 2 and 2^0 will be transformed here into optimal interpolation problems. In particular, the constraints C2, C3 on Q will be expressed as interpolation constraints.

Suppose that P and W are fixed as in Section II-C and $Q(s)$ is in H^B. We shall wish to characterize those $Q(s)$ for which $P(s)Q(s)$ is in H^B, and observe that there are three equivalent ways of characterizing $Q(s)$.

1) $Q(s)$ has zeros at the poles of $P(s)$ in $\mathrm{Re}(s)>0$, taking into account their multiplicities.

2) $Q(s)$ satisfies the interpolation constraints

$$Q(s)=0,\quad dQ/ds=0,\cdots,\quad (d/ds)^{m-1}Q=0 \tag{3.1}$$

at each distinct pole of $P(s)$ of multiplicity m in $\mathrm{Re}(s)>0$.

3) The ratio $Q(s)/B_p(s)$ is in H^B, where

$$B_p(s)=\prod_{i=1}^{q}\left(\frac{a_i-s}{\bar{a}_i+s}\right) \tag{3.2}$$

[recall that $a_1,\cdots,a_q$ are the $\mathrm{Re}(s)>0$ poles of $P(s)$.]

A rational function of the form (3.2) is called a *Blaschke product*. In engineering terminology, a frequency response is called *allpass* if its magnitude is constant on the $j\omega$-axis. As $|B_p(j\omega)|=1$, Blaschke products are allpass.[7]

A. Constraints Induced by RHP Plant Poles

A (possibly improper) frequency response $Q(s)$ in H^B satisfies the condition C3 for the CL system to be in H^B iff $[1-P(s)Q(s)]$ has zeros at the RHP poles of $P(s)$, taking into account their multiplicities. There are, again, several equivalent ways of expressing C3, namely, the following.

1) Q satisfies the interpolation constraints

$$\left.\begin{aligned}[1-P(s)Q(s)]&=0\\ &\vdots\\ (d/ds)^{m-1}[1-PQ]&=0\end{aligned}\right|_{\text{at } s=a} \tag{3.3a}$$

at each RHP pole a of $P(s)$ of multiplicity m.

2) A compact way of expressing (3.3a) is to say that B_p divides $(1-PQ)$ in H^B, i.e.,

$$(1-PQ)B_p^{-1}\in H^B \tag{3.3b}$$

or that

3)

$$W(1-PQ)B_p^{-1}\in H^B. \tag{3.3c}$$

Equation (3.3c) is equivalent to (3.3b) because W and W^{-1} are in H^B, the latter because $W(s)$ has no zeros in $\mathrm{Re}(s)\geqslant 0$.

The conditions for a stable $Q(s)$ to stabilize $P(s)$, namely C1, C2, are equivalent to the requirement that the expression in (3.3b) be in H^∞ or that (3.3a) hold.

From the fact that the expressions in (3.3a)–(3.3c) are all affine in Q it is easy to deduce the following characterization of the functions that stabilize $P(s)$. (See Appendix I for Proof.)

Proposition 3A1: Suppose there exists a function $Q_o(s)$ in H_o^∞ which stabilizes $P(s)$; then

a) $Q_o(s)$ can be expressed in the form $Q_o(s)=Q_{oA}(s)B_p(s)$, where Q_{oA} is in H_o^∞.

b) The functions Q in H_o^∞ which stabilize $P(s)$ are those and only those which have the form

$$Q=B_pQ_{oA}+B_p^2Q_1 \tag{3.4}$$

where Q_1 is in H_o^∞.

c) More generally, the functions $Q(s)$ in H^B for which $(1-PQ)B_p^{-1}$ is in H^B (or H^∞) are those and only those which can be expressed in the form (3.4), with $Q_1(s)$ in H^B (and such that $P(s)Q(s)$ is proper).

Any of the available constructions for finding a stabilizing feedback could now be used to determine a suitable $Q_o(s)$, but our characterization itself suggests a natural method for finding a stabilizing feedback. Let the RHP poles of $P(s)$ be distinct for simplicity, and let ν_i be the

[6]Note: $Q\in H^B$ and $(1-PQ)P\in H^B$ implies $PQ\in H^B$.

[7]Allpass functions of unit norm are also called inner.

residue of $P(s)$ at a_i, $i=1,\cdots,q$. By (3.3), the stabilizing $Q(s)\in H_o^\infty$ are those functions which invert $P(s)$ at its RHP poles in the sense that $P(s)Q(s)=1$ at a_i. Equivalently, $Q_{oA}(s)$ can be any H_o^∞ function satisfying the q interpolation constraints

$$Q_{oA}(a_i)=\nu_i^{-1}, \qquad i=1,\cdots,q. \tag{3.4a}$$

An easy way of finding such a stabilizing function makes use of certain interpolating functions $B_{pj}(s)$ in H_o^∞, which assume the value zero at each RHP pole a_i, $i\neq j$, of $P(s)$, and the value unity at each pole a_j. These $B_{pj}(s)$ are defined by the equations

$$\tilde{B}_{pj}(s)\triangleq\frac{1}{(s+1)}\prod_{\substack{i=1\\ i\neq j}}^{q}\frac{(a_i-s)}{(\bar{a}_i+s)}, \qquad B_{pj}(s)\triangleq\frac{\tilde{B}_{pj}(s)}{\tilde{B}_{pj}(a_j)}.$$

The desired $Q_{oA}(s)$ can now be defined as the following linear combination of the $B_{pj}(s)$:

$$Q_{oA}(s)\triangleq\sum_{i=1}^{q}\nu_i^{-1}B_{pj}(s). \tag{3.4b}$$

Clearly, $Q_{oA}(s)$ satisfies (3.4a). Therefore, the function $Q_o(s)\triangleq B_p(s)Q_{oA}(s)$ inverts $P(s)$ at its RHP poles, and stabilizes $P(s)$.

B. Constraints Added by RHP Plant Zeros

We are interested in optimizing the weighted sensitivity function $W(1-PQ)$, which will be denoted by X and is related to Q by the equations

$$X=W(1-PQ) \tag{3.5}$$

$$Q=P^{-1}(1-W^{-1}X). \tag{3.6}$$

Let us see how the zeros of $P(s)$ affect $X(s)$, when $Q(s)$ satisfies the pole constraints (3.3). Let Ω denote the set of those $Q(s)$ in H^B that satisfy the pole constraints (3.3) and for which WPQ is proper. If Q is in Ω and X is given by (3.5), then it follows from (3.3) and the propriety of WPQ that both X and XB_p^{-1} are in H^∞. From (3.5) it is clear that X must satisfy the following plant-zero interpolation constraints:

$$\begin{aligned} X(b)&=W(b)\\ &\vdots\\ (d/ds)^{m-1}X&=(d/ds)^{m-1}W \qquad \text{at } s=b \end{aligned} \tag{3.7}$$

at each distinct plant zero b of multiplicity m. The total of such constraints is r.

We would like to show that Q can be optimized by first minimizing $\|X\|$ subject to (3.7), and then using (3.6) to calculate Q. The following proposition, proved in Appendix I, provides the basis for this approach.

Proposition 3B1:

1) Equations (3.5), (3.6) establish a 1:1 *correspondence between functions $Q\in\Omega$, and those weighted sensitivity functions $X\in H^\infty$ for which $XB_p^{-1}\in H^\infty$ and which satisfy the plant-zero constraints (3.7).*

2) *Ω is a subset of $(1+s)^{k+l}H^\infty$.*

C. Remarks

1) Stabilization and Desensitization Versus Inversion: To achieve a small sensitivity $W(1-PQ)$, PQ must be close to 1, i.e., Q must act as an approximate inverse of P subject to the plant-pole constraints (3.3). Also, we observe that (3.3a) means that in order to stabilize $P(s)$, $Q(s)$ must invert $P(s)$ *exactly* at its RHP poles. Here the problems of sensitivity reduction and stabilization appear as two aspects of the problem of inversion, the former approximate, the latter exact at the RHP poles. This interpretation is constructive to the extent that it suggests ways of constructing stabilizers [see, for example, (3.4b)] and compensators namely in terms of certain "approximate inverses." Stabilization becomes a matter of achieving zero sensitivity at the RHP poles, and can be viewed as an extreme case of desensitization.

The minimal sensitivity $\mu(P)$ can be interpreted as a measure of the constrained singularity of P.

2) Coprime Parametrizations: There is a connection between Proposition 3A1 and [7, Theorem 3]. The function $N\triangleq PB_p$ is in H^∞. It can be shown that there exists H^∞-functions Y and Z such that

$$B_pY+NZ=1. \tag{3.7a}$$

On taking $Q_{oA}=Z$, substituting (3.4) into (2.2b), using (3.4a), and cancelling B_p, we get the expression

$$F=(Q_{oA}+B_pQ_1)(Y-NQ_1)^{-1}. \tag{3.7b}$$

This is the parametrization of strictly proper F's that stabilize P of Youla *et al.* [6] and Desoer *et al.* [7], which is based on the existence of a coprime factorization of $P(s)$.

The assumption that "a coprime factorization exists" roughly amounts to the assumption that a stabilizing feedback exists, and by itself offers little indication as to how to find such a feedback. On the other hand, the interpretation of Q as an approximate inverse satisfying the interpolation constraints (3.1) is constructive in the sense mentioned in Remark 1) of Section III-C, and offers a guide to finding (possibly suboptimal) compensators. Observe that the characterization of Q via interpolation constraints, as in (3.1), avoids any notion of coprime factorization. Note that (3.4) is affine in Q_1, whereas (3.7b) is nonlinear in Q_1.

IV. Improper Minimization

We turn now to the optimization Problem 2^0, of finding the minimum

$$\min\{\|X\|: X=W(1-PQ)\} \tag{4.1}$$

under the relaxed constraint that $Q(s)$ is a possibly improper function in H^B which satisfies the plant-pole constraints (3.3). (Equivalently, Q is in Ω.) Here, $P(s)$ and $W(s)$ are the fixed functions defined in Section II, and *it will be assumed that $P(s)$ is either strictly proper or has at least one RHP zero*, so as to exclude the trivial case in which $P(s)$ has a stable proper inverse.

In view of the correspondence between X and Q established in Proposition 3B1, X can be minimized first, subject to (3.7) and the condition that $XB_p^{-1} \in H^\infty$, and the optimal Q calculated second by (3.6). Furthermore, as $\|X\| = \|XB_p^{-1}\|$, the minimization of X can be accomplished by minimizing XB_p^{-1} and multiplying the result by B_p.

Our first major Lemma characterizes the optimal weighted sensitivity function $\tilde{X}$ (and will enable us to calculate $\tilde{X}$ and $\tilde{Q}$ in Sections VI and VII). The Lemma is an application of the results of Rogosinski and Shapiro [3], which are based on the concept of duality. An exposition of these results and proof of the Lemma, drawn from Duren [4] and Koosis [5], are in Appendix II.

Lemma 1:

a) There is a unique weighted sensitivity function $\tilde{X}$ in H^∞ which attains the minimum norm (4.1). The necessary and sufficient conditions for $\tilde{X}$ to be that function are that $\tilde{X}$ satisfies the interpolation constraints (3.7), and has the (allpass) form

$$\tilde{X}(s) = D \prod_{i=1}^{m} \left(\frac{c_i - s}{\bar{c}_i + s} \right) \prod_{j=1}^{q} \left(\frac{a_j - s}{\bar{a}_j + s} \right) \tag{4.2}$$

in which[8] *$Re(c_i) \geqslant 0$, D is a constant*[9] *satisfying $|D| = \|\tilde{X}\|$, and $m = r - 1$.*

b) The optimal $\tilde{Q}$ is $\tilde{Q} = P^{-1}(1 - W^{-1}\tilde{X})$ and belongs to $(1+s)^{k+l}H^\infty$.

c) If the plant $P(s)$ and weighting $W(s)$ have conjugate symmetry, then so do $\tilde{X}$ and $\tilde{Q}$, and the coefficients in (4.2) are real or occur in conjugate pairs.

The term $\|W(1 - PQ)\|$ can be interpreted as the weighted distance from PQ to the identity. The characterization of the optimal sensitivity as an allpass function is analogous to the characterization of the shortest distance from a vector to a subspace as the perpendicular, which is widely used in quadratic optimization theory. The analogy is made quite specific in duality theory (see Luenberger [10]).

V. Minimization Over Strictly Proper Q

As the minimal weighted sensitivity function $\tilde{X}$ is allpass, no strictly proper Q can attain it. The optimal $\tilde{Q}$ is improper and can not be realized by physical feedback. The most we can hope for is to find a sequence of strictly proper Q_n in H_o^∞ to give sensitivities approaching $\|\tilde{X}\|$ [which will be shown to equal $\mu(P)$].

Two procedures for constructing such a sequence will be given. The simpler procedure, valid for strictly proper weightings W, involves the modification of $\tilde{Q}$ by some high-frequency attenuation. The more general procedure, valid for nonstrictly proper weightings, involves the modification of $\tilde{X}$.

Suppose that $P(s)$ and $W(s)$ are fixed as in Section II-C, but will be *assumed to have conjugate symmetry.* $\tilde{Q}(s)$ and $\tilde{X}(s)$ are the extrema specified in Lemma 1, and $Q_o(s)$ is any function in H_o^∞ representing a strictly proper feedback that stabilizes $P(s)$. Let $Q_n(s)$, $n = 1, 2, \cdots$, be defined in one of the following two ways.

Case 1: If $W(s)$ is strictly proper, then

$$\tilde{Q}_n(s) \triangleq Q_o(s) + (\tilde{Q} - Q_o)(s)\left[n(s+n)^{-1}\right]^{k+l+1}. \tag{5.1}$$

Case 2: If $W(s)$ is proper, and $\lim_{|s| \to \infty, Re(s) \geqslant 0} |W(s)| \triangleq |W(\infty)| \leqslant |\tilde{X}(j\omega)|$, then

$$\tilde{X}_{1n}(s) \triangleq \tilde{X}(s)\left[1 - B_z(s)\frac{\lambda s}{n+s}\right], \qquad \lambda = \frac{\tilde{X}(\infty) - W(\infty)}{B_z(\infty)\tilde{X}(\infty)} \tag{5.2a}$$

$$\tilde{Q}_{1n}(s) \triangleq P^{-1}(s)\left[1 - \tilde{X}_{1n}(s)W^{-1}(s)\right] \tag{5.2b}$$

$$\tilde{Q}_n(s) \triangleq Q_o(s) + \left[\tilde{Q}_{1n}(s) - Q_o(s)\right]\left[m_n(s+m_n)^{-1}\right]^{k+l} \tag{5.2c}$$

where, for any $n > 0$, $m_n > 0$ is a sufficiently large integer, and $B_z(s)$ is the Blaschke product formed with the plant zeros.

In either case, the corresponding sequence of feedbacks is given by the fractional transformation

$$\tilde{F}_n(s) = \tilde{Q}_n(s)\left[1 - P(s)\tilde{Q}_n(s)\right]^{-1}. \tag{5.3}$$

If the numerator and denominator of (5.3) are divided by $B_p(s)$, (5.3) can also be written as

$$\tilde{F}_n(s) = \tilde{Q}_n(s)B_p^{-1}(s)\left\{\left[1 - P(s)\tilde{Q}_n(s)\right]B_p^{-1}(s)\right\}^{-1}. \tag{5.4}$$

Theorem 1: The functions $\tilde{Q}_n(s)$ defined by (5.1) in Case 1, or (5.2) in Case 2, are (strictly proper) in H_o^∞, stabilize $P(s)$, and are optimal in the sense of producing sensitivities approaching the infimal value $\mu(P)$, i.e.,

$$\lim_{n \to \infty} \|W(1 - P\tilde{Q}_n)\| = \mu(P). \tag{5.5}$$

Similarly the feedbacks $\tilde{F}_n(s)$ are in H_o^∞ and optimal, i.e.,

$$\lim_{n \to \infty} \|W(1 + \tilde{F}_n P)^{-1}\| = \mu(P). \tag{5.6}$$

Moreover, $\mu(P) = \|\tilde{X}\|$.

Proof: See Appendix I.

Remarks 5.1:

1) The numerator and denominator of the expression (5.3) for the optimal feedback $F_n(s)$ are not relatively prime, for both are divisible by $B_p(s)$. Common RHP zeros have been eliminated from (5.4).

[8] Whenever c_i is imaginary and $\tilde{X}(s)$ has conjugate symmetry, the numerator and denomenator of (4.2) have common factors, and $\tilde{X}$ has order $< (r-1)$.

[9] It will be shown in Section V that $\|\tilde{X}\| = \mu(P)$.

Common zeros are admissible in an input–output description, at least as long as no attempt is made to employ separate approximations to the numerator and denominator. Even then, common LHP zeros are admissible.

2) The stabilizing feedback $F_o \triangleq Q_o(1-P_oQ_o)^{-1}$ can be selected at will, using any of the several methods now available, e.g., by (3.4b). Indeed, as the expressions for $\tilde{Q}_n$ are affine in Q_o, the problem of stabilization can be isolated from desensitization, at least in principle.

3) If $|W(\infty)| > |\tilde{X}(j\omega)|$, the theorem is false, for the infimal sensitivity $\mu(P)$ produced by strictly proper compensators must exceed $\|\tilde{X}\|$. Indeed, if Q is strictly proper, $\|X\|$ satisfies the inequalities $\|X\| = \|W(1-PQ)\| \geqslant |W(\infty)|$ from which it follows that $\mu(P) \geqslant |W(\infty)| > |\tilde{X}(j\omega)| = \|\tilde{X}\|$.

A. Bounds on the Minimal Sensitivity

Proposition 5A1:
a)

$$\mu(P) \geqslant \max_i |W(b_i)B_p^{-1}(b_i)| \tag{5.7}$$

$$> \max_i |W(b_i)| \tag{5.8}$$

assuming P is unstable, i.e., $B_p \not\equiv 1$, in (5.8).

b) If P_1 and P_2 are plants with identical RHP zeros, P_1 is stable, and P_2 has poles in the RHP, then $\mu(P_2) > \mu(P_1)$, provided[10] $|W(\infty)| \leqslant \|\tilde{X}_1\|$.

Proof: See Appendix I.

Remark: It follows from (5.8), (5.7) that small sensitivity cannot be achieved if there are zeros in any heavily weighted part of the RHP, or poles anywhere near these zeros [since $B_p(b_i)$ is smallest near the RHP poles of $P(s)$].

Proposition 5A1b shows that the insertion of unstable poles in $\mathrm{Re}(s) > 0$ into an otherwise stable system always deteriorates the achievable optimal sensitivity.

B. Determination of $\tilde{X}$ and $\tilde{Q}$

The plant $P(s)$ can be expressed uniquely as the product

$$P(s) = B_z(s)B_p^{-1}(s)P_1(s) \tag{5.9}$$

consisting of the two Blaschke products $B_z(s)$ and $B_p(s)$ determined by the $\mathrm{Re}(s) > 0$ plant zeros and poles, respectively, and a factor $P_1(s)$ which is always in H^∞. The optimal weighted sensitivity function $\tilde{X}$ specified in (4.2) is not affected by $P_1(s)$, which is simply inverted by $\tilde{Q}$. The dependence of $\tilde{X}$ on the plant poles is explicitly given by $B_p(s)$. The burden of the optimization problem therefore falls on the calculation of the remaining factor of (4.2), namely,

$$\frac{\tilde{X}(s)}{B_p(s)} = D\prod_{i=1}^{r-1}\left(\frac{c_i - s}{\bar{c}_i + s}\right) \tag{5.10}$$

whose coefficients are not known *a priori*.

[10] i.e., Cases 1 or 2 apply.

Assumption: (For simplicity) suppose the plant zeros b_i are distinct.

By Lemma 1, XB_p^{-1} is unique, and by Proposition 3B1, is determined by the interpolation constraints (3.7), which take the form of r equations

$$\frac{\tilde{X}(b_j)}{B_p(b_j)} \triangleq D\prod_{i=1}^{r-1}\left(\frac{c_i - b_j}{\bar{c}_i + b_j}\right) = \theta_j \tag{5.11}$$

where $\theta_j \triangleq W(b_j)/B_p(b_j)$ are complex constants depending upon the values of the weighting $W(s)$ at the RHP plant zeros, and the location of the plant poles in relation to these zeros.

The simultaneous equations (5.11) are nonlinear. A general procedure for their solution is given in Section VII. However, when r is small, it is more efficient to solve (5.11) by a simple elimination scheme, which we proceed to do next.

VI. Explicit Formulas: Few RHP Zeros

A. One RHP Plant Zero

Here the expression for the plant is $P(s) = (b-s)(b+s)^{-1}P_1(s)B_p^{-1}(s)$, $b > 0$, $r = 1$, and conjugate symmetry is assumed. By Lemma 1a, $\tilde{X}B_p^{-1}$ is a constant, which is

$$\tilde{X}(s)B_p^{-1}(s) = \theta_1 = W(b)B_p^{-1}(b)$$

by (5.11). By Lemma 1b,

$$\tilde{Q}(s) = (b+s)\left[\frac{1 - \dfrac{W(b)}{W(s)}\dfrac{B_p(s)}{B_p(b)}}{(b-s)}\right]P_1^{-1}(s)B_p(s).$$

If we concentrate on stable plants ($B_p(s) = 1$), and consider the weighting $W(s) = (\kappa + \xi s)(\kappa + s)^{-1}$, $\kappa > 0$, $0 \leqslant \xi < W(b)$, which emphasizes low frequencies and deemphasizes high ones, then the optimal sensitivity is $\mu(P) = (\kappa + \xi b)(\kappa + b)^{-1}$. (For $\xi = 0$, this is the result of [1].) $\mu(P)$ increases with the parameter κ, which determines the "bandwidth" of disturbances that have to be accommodated. The optimal Q is

$$\tilde{Q}(s) = \frac{(b+s)}{(\kappa + \xi s)}\lambda P_1^{-1}(s), \qquad \lambda = \frac{\kappa(1-\xi)}{\kappa + b}.$$

If $P_1(s) = 1$, then $\tilde{Q}(s)$ is a lead filter, which reduces the lagging phase of the plant $(b-s)(b+s)^{-1}$ at all frequencies, but $\tilde{Q}(s)$ does this at the cost of increased deviation from 1 of the plant magnitude (which, to begin with is $|(b-s)(b+s)^{-1}| = 1$), especially at high frequencies.

Suppose that $P(s)$ has one zero at $s = \infty$, ($k = 1$), that Q_o is the representation of any stabilizing feedback for P, and that $\xi = 0$ so that $W(s)$ has one zero at $s = \infty$, ($l = 1$). Then, an optimal sequence of strictly proper functions $\tilde{Q}_n$ in H_o^∞ is given by the equation

$$\tilde{Q}_n(s) = Q_o(s) + [\tilde{Q}(s) - Q_o(s)]n^3(s+n)^{-3}$$

[by (5.1)], and an optimal sequence of strictly proper feedbacks by $\tilde{F}_n = Q_n(1 - PQ_n)^{-1}$.

Remark: If W assigns some minimal nonzero weighting to high frequencies, then the unweighted sensitivity $(1 - P\tilde{Q})$ is bounded at $s = \infty$.

B. Two RHP Plant Zeros

In this case,

$$P(s) = \frac{(b_1 - s)(b_2 - s)}{(b_1 + s)(b_2 + s)} P_1(s) B_p^{-1}(s)$$

where b_i, $i = 1,2$, are real or occur in conjugate pairs, and $\mathrm{Re}(b_i) > 0$. Here $r = 2$. By Lemma 1, the minimal weighted sensitivity function $\tilde{X}$ must have the form

$$\tilde{X}(s) B_p^{-1}(s) = D(c - s)(c + s)^{-1} \tag{6.1}$$

and the interpolation constraints are

$$D\frac{(c - b_i)}{(c + b_i)} = \theta_i \triangleq W(b_i) B_p^{-1}(b_i), \qquad i = 1,2. \tag{6.2}$$

If D is eliminated from the pair of equations (6.2), a quadratic equation is obtained for c, only one of whose solutions lies in $\mathrm{Re}(s) \geqslant 0$, namely,

$$c = -\frac{(b_2 - b_1)}{2}\left(\frac{\theta_2 + \theta_1}{\theta_2 - \theta_1}\right) + \sqrt{\frac{(b_2 - b_1)^2}{4}\left(\frac{\theta_2 + \theta_1}{\theta_2 - \theta_1}\right)^2 + b_2 b_1}\,. \tag{6.3}$$

If D is expressed in terms of c using (6.2) and (6.3), there results

$$D = -\left(\frac{\theta_2 - \theta_1}{b_2 - b_1}\right)\frac{(b_2 + b_1)}{2} - \mathrm{sgn}\left(\frac{\theta_2 - \theta_1}{b_2 - b_1}\right) \cdot \sqrt{\frac{(\theta_2 + \theta_1)^2}{4} + \left(\frac{\theta_2 - \theta_1}{b_2 - b_1}\right)^2 b_1 b_2} \tag{6.4}$$

from which it follows that since $\mu(P) = |D|$, the minimal sensitivity is

$$\mu(P) = \left|\frac{\theta_2 - \theta_1}{b_2 - b_1}\right| \cdot \frac{|b_2 + b_1|}{2} + \sqrt{\frac{|\theta_1 + \theta_2|^2}{4} + \left|\frac{\theta_2 - \theta_1}{b_2 - b_1}\right|^2 \cdot |b_1 b_2|}. \tag{6.5}$$

In the language of [1], (6.5) is the measure of singularity of P.

It can be deduced from (6.5), after some manipulations, that $\mu(P) \geqslant \max(|\theta_1|, |\theta_2|)$ in conformity with Proposition 5A1a.

If we assume that $B_p(s) = 1$, then the optimal improper $\tilde{Q}$ is

$$\tilde{Q} = \left[1 - \frac{D(c - s)}{W(s)(c + s)}\right] \cdot \frac{(b_1 + s)(b_2 + s)}{(b_1 - s)(b_2 - s)}.$$

If $W(s)$ is the "low-pass" weighting $\kappa(s + \kappa)^{-1}$, $\kappa > 0$, then

$$\tilde{Q}(s) = \frac{D(b_1 + s)(b_2 + s)}{\kappa(c + s)}$$

which, again, is a lead-type function with an extra "break." The optimal sequences of strictly proper feedbacks $\tilde{Q}_n$ and $\tilde{F}_n$ can be computed as for a simple zero.

VII. Explicit Formulas: Many RHP Zeros

The r simultaneous equations (5.11) for the coefficients c_i can be solved by a method based on the Schur–Pick–Nevanlinna theory. (See Walsh [2, ch. 10] for an exposition.) That theory, which predated duality by several decades, provided the original existence-uniqueness proof for H^∞ interpolation, although here we rely on it mainly for its algorithm.

The theory reduces the r equations (5.11) to a single algebraic equation for the constant $\|\tilde{X}\|$. It then provides explicit formulas for the c_i.

There are two cases. If all the constraints $\theta_1, \cdots, \theta_r$ are equal, or if $r = 1$, we have the *degenerate case*, in which the minimal weighted sensitivity function is a constant $\tilde{X}(s) \equiv \theta_i$. $\tilde{X}(s)$ certainly satisfies the constraints; by the maximum modulus principle, no smaller H^∞ function can do so, and the minimum is unique.

In the *nondegenerate case*, $r \geqslant 2$ and the constraints are not all equal. Here, $\tilde{X}(s)$ cannot be a constant, and $|\theta_i| \leqslant \|\tilde{X}\|$, as a nonconstant H^∞ function cannot attain its maximum modulus inside $\mathrm{Re}(s) > 0$. In this case, the theory depends on a special transformation, which reduces the number of interpolation constraints by one in each application. The starting point is the invertible mapping $u \leftrightarrow x$

$$u = \frac{M^2(x - \theta)}{M^2 - \bar{\theta}x}, \quad x = \frac{M^2(u + \theta)}{M^2 + \bar{\theta}u}, \qquad |\theta| < |M|$$

where M is some constant. It follows from elementary complex variables that this mapping is analytic in x, and establishes a 1:1 correspondence between points in the disks $|x| \leqslant |M|$ and $|u| \leqslant |M|$ of the complex u and x planes, respectively. Furthermore, points on the disk boundary are mapped into points on the boundary, and points in the interior into points in the interior, i.e.,

$$|x| < |M| \leftrightarrow |u| < |M|, \text{ and } |x| = |M| \leftrightarrow |u| = |M|.$$

Let M be fixed, and for any $|\theta| < |M|$; let U_θ: $H^\infty \to H^\infty$ denote the mapping $x(\cdot) \mapsto u(\cdot)$ satisfying the equation

$$u(s) = \frac{M^2[x(s) - \theta]}{M^2 - \bar{\theta}x(s)}. \tag{7.1}$$

Equation (7.1) maps the (solid) ball of radius $|M|$ of H^∞ into itself, and the set of allpass functions of norm $|M|$ into itself. Moreover, if $x(\cdot)$ in H^∞ satisfies an interpolation constraint $x(b)=\theta$, then $u(s)$ has a zero at b.

Equation (7.1) is now modified by division by $(b-s)(\bar{b}+s)^{-1}$, thereby removing the zero at b, but keeping $u(\cdot)$ in H^∞ to obtain the transformation $U_{\theta,b}$: $H^\infty \to H^\infty$, $x(\cdot) \mapsto x_1(\cdot)$ which satisfies

$$x_1(s)=\frac{M^2[x(s)-\theta]}{M^2-\bar{\theta}x(s)}\,\frac{(\bar{b}+s)}{(b-s)}. \tag{7.2}$$

The division taking (7.1) into (7.2) leaves $j\omega$-axis magnitudes and H^∞ norms invariant (although it may increase magnitudes in $\mathrm{Re}(s)>0$). Therefore, (7.2) still maps the $|M|$-ball of H^∞ into itself. Equation (7.2) also establishes a 1:1 correspondence between the set of all allpass functions $x(s)$ of norm $|M|$ that assumes the value θ at b, and the set of all allpass functions $x_1(s)$ of norm $|M|$ (whose value at b is arbitrary). This property of (7.2), together with the result of Lemma 1 that the smallest H^∞ function satisfying an interpolation constraint is a unique allpass function, i.e., a Blaschke product, immediately yield the following result. Suppose that $\theta_i < M$, and $\mathrm{Re}(b_i)>0$, $i=1,\cdots,r$.

Proposition 7.1: If the smallest H^∞ function assuming the values $(\theta_1,\cdots,\theta_r)$ at the $r \geqslant 2$ points $(b_1,\cdots,b_r)$ is $X(s)$ and has norm $|M|$, then the smallest H^∞ function assuming the transformed values $[U_1(\theta_2), U_1(\theta_3),\cdots,U_1(\theta_r)]$, $U_1 \triangleq U_{\theta_1,b_1}$ at the $(r-1)$ points $(b_2,\cdots,b_r)$ also has norm $|M|$ and is uniquely $(U_1X)(s)$.

Proposition 7.1 offers the possibility of reducing any H^∞ minimization problem involving $r \geqslant 2$ constraints to a problem involving $(r-1)$ *new* constraints. This reduction can be repeated, either $(r-1)$ times until one constraint remains, or less than $(r-1)$ times, until several equal constraints remain. At this point we have the degenerate case, for which the smallest H^∞ interpolating function has been shown to be a constant of norm $|M|$. Let us detail this procedure.

Here, $r \geqslant 2$ and the constraint values are nondegenerate. Recall that $\tilde{X}B_p^{-1}$ is the function of smallest H^∞ norm assuming the constraint values $\theta_1,\cdots,\theta_r$ at the points $b_1,\cdots,b_r$ in $\mathrm{Re}(s)>0$. Let M be a complex constant of magnitude $\|\tilde{X}\|$, and angle to be determined. Consider an iteration on the constraint set (illustrated in Fig. 2), as follows. Denote the initial constraints by $\{x_i^{(o)}\}$, i.e.,

$$\{x_1^{(o)},\cdots,x_r^{(o)}\}=\{\theta_1,\cdots,\theta_r\}.$$

For any $i \geqslant 1$, if the $(i-1)$st constraint set is nondegenerate and has the form $\{x_i^{(i-1)}, x_{i+1}^{(i-1)},\cdots,x_r^{(i-1)}\}$, let the ith constraint set be

$$\{x_{i+1}^{(i)}, x_{i+2}^{(i)},\cdots,x_r^{(i)}\} = \{T^{(i)}x_{i+1}^{(i-1)}, T^{(i)}x_{i+2}^{(i-1)},\cdots,T^{(i)}x_r^{(i-1)}\}$$

where $T^{(i)}$ denotes the transformation

$$[T^{(i)}x](s) \triangleq \frac{M^2[x(s)-x_i^{(i-1)}]}{M^2-\overline{x_i^{(i-1)}}x(s)}\,\frac{(\bar{b}_i+s)}{(b_i-s)}.$$

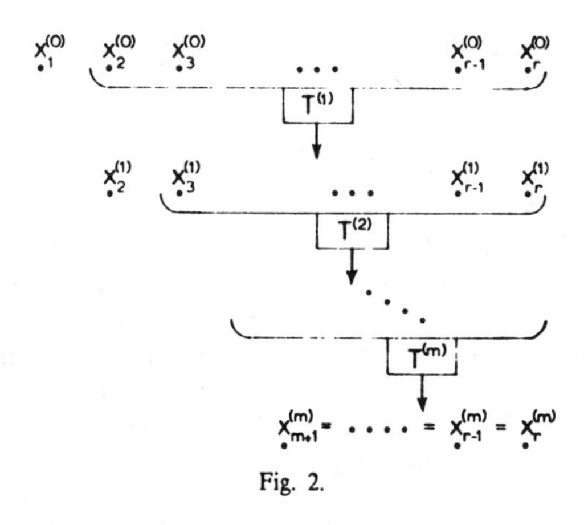

Fig. 2.

If the ith constraint set is degenerate, let $i=m$. Certainly, $m \leqslant (r-1)$. We have the following main result.

Theorem 2:

a) M satisfies the equation

$$[T^{(m)}T^{(m-1)}\cdots T^{(1)}x^{(o)}](b_r)=M \tag{7.3}$$

(which, after evaluation, yields an algebraic equation in the single variable M).

b) The optimal sensitivity function $\tilde{X}$ is determined explicitly by the equation

$$\tilde{X}(s)=B_p(s)[T^{(1)}]^{-1}[T^{(2)}]^{-1}\cdots[T^{(m)}]^{-1}(M)$$

where $[T^{(i)}]^{-1}$ satisfies

$$\{[T^{(i)}]^{-1}x\}(s)=\frac{M^2[x(s)+x_i^{(i-1)}]}{[M^2-\overline{x_i^{(i-1)}}x(s)]}\,\frac{(b_i-s)}{(\bar{b}_i+s)},\quad i=1,2,\cdots,m.$$

For a plant with conjugate symmetry and one or two RHP zeros, (7.3) obviously gives the same results as in Section VI, but with the complication that intermediate constraint sets obtained during the iteration may lack conjugate symmetry.

VIII. Comparison to Wiener–Hopf Quadratic Minimization

The well-known Wiener–Hopf quadratic minimization method, e.g., in Youla *et al.* [6], provides the optimal feedback for disturbances of *fixed* power spectrum.

In Fig. 1, let d be a disturbance obtained by passing unit variance white noise through the filter $W(s)$. The mean-square value of the output y is

$$\frac{1}{2\pi}\int_{-\infty}^{\infty}|X(j\omega)|^2\,d\omega \tag{8.1}$$

and the Wiener–Hopf objective is to minimize (8.1) subject to the constraint that F stabilizes the CL system.

Introduce the Hardy space H^2 of functions $G(s)$ which are analytic in $\mathrm{Re}(s)>0$ and for which the norm is

$$\|G\|_2 \triangleq \sup_{\sigma>0}\left[\int_{-\infty}^{\infty}|G(\sigma+j\omega)|^2\,d\omega\right]^{1/2}.$$

This norm can be shown to be computable on the $j\omega$-axis, and

$$\|G\|_2 = \left[\int_{-\infty}^{\infty} |G(j\omega)|^2 \, d\omega\right]^{1/2}.$$

The Wiener–Hopf problem is equivalent to the problem of minimizing the H^2 norm of the weighted sensitivity function X, with the same constraints as in Proposition 3B1, and can be solved by the same method as the H^∞ problem. The result is that there is a unique weighted sensitivity function $\tilde{\tilde{X}}$ in H^2 of minimum norm, which is characterized by the formula

$$\tilde{\tilde{X}}(s) = \frac{\phi(s)}{\prod_{i=i}^{r}(b_i + s)} \prod_{j=1}^{q} \left(\frac{a_j - s}{a_j + s}\right) \tag{8.2}$$

for plants with conjugate symmetry, where $\phi(s)$ is a polynomial of degree $< r$, which is uniquely determined by the plant-zero interpolation constraints (37). $\phi(s)$ can be computed, e.g., by the Lagrange interpolation formula. (Equation (8.2) is equivalent to [6, (41)].)

A. An Example

Let us consider a representative servo problem involving a single RHP plant zero, and compare the H^∞ and H^2 solutions.

As RHP plant poles affect both solutions similarly [cf. (8.2) and (4.2)], the plant will be assumed stable. Suppose that

$$P(s) = (b - s)(b + s)^{-1} P_1(s), \qquad b > 0, \; P_1 \in H^\infty$$

and that disturbance power-spectra can be arbitrary subject to the "weighted-power" constraint $\|W^{-1}d\|_{H^2} \leqslant 1$. The optimal weighted sensitivity function, obtained in Section VI-A, is $\tilde{X}(s) = W(b)$.

Although this problem has no completely satisfactory solution in H^2, a widely used current approach is to design X as though $|d(j\omega)|$ were a fixed power-spectrum equal to $|W(j\omega)|$. Strictly speaking, this is possible if $W(j\omega) \in H^2$, where we get the result that $\tilde{\tilde{X}}(s) = 2b(s + b)^{-1} W(b)$. The sensitivity ratios, weighted or unweighted, are

$$\tilde{\tilde{X}}(s)/\tilde{X}(s) = \left(1 - P\tilde{\tilde{Q}}\right)/\left(1 - P\tilde{Q}\right) = 2b(s + b)^{-1}$$

and are independent of the weighting $W(s)$. *The peak optimal weighted quadratic sensitivity* $\sup_\omega |H(j\omega)|$ *is always double the* H^∞ *one.*

Let us now concentrate on a special case. Typically, servomechanisms must be able to respond to disturbances having not one, but a variety of power spectra. Low frequency disturbances, which must be attenuated, occur most often. However, even very high frequency disturbances appear occasionally, and response to them must be limited, even though attenuation may be impractical. This situation is tailor-made for a description in H^∞, subject to a weighting which remains finite as $s \to \infty$.

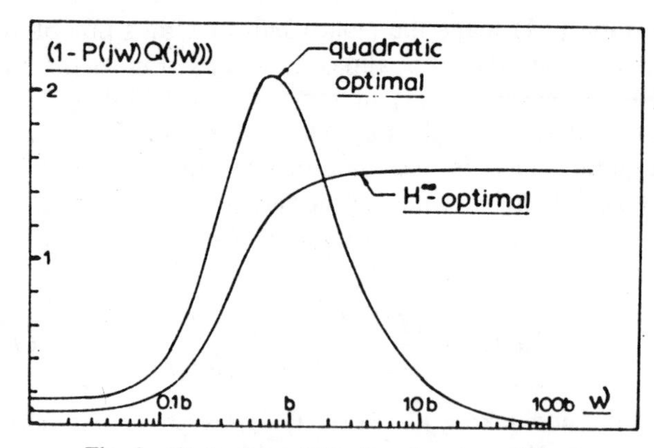

Fig. 3. Optimal sensitivity functions (unweighted).

Consider, e.g., the weighting

$$W(s) = \left(\frac{0.1b + \xi s}{0.1b + s}\right)^3, \qquad \xi = \left(\frac{1}{20}\right)^{1/3}$$

which is substantial for frequencies smaller than b, ($|W(j\omega)| > 0.707$ for $|\Omega| \in [0, 0.1b]$) then drops off quite rapidly, but never falls below the lower bound $\xi^3 = 1/20$. The optimal unweighted H^∞ sensitivity function is

$$\left(1 - P\tilde{Q}\right) = W(b)/W(s) = 0.077\left(\frac{0.1b + s}{0.1b + \xi s}\right)^3$$

and is illustrated in Fig. 3.

$W(s)$ is in H^2 only if $\xi = 0$. In that case $r = 1$ in (8.2), $\phi(s)$ is a constant, and on matching interpolation constraints, the result

$$\left(1 - P\tilde{\tilde{Q}}\right) = \left[W(b)/W(s)\right]\frac{2b}{s + b} = \left(1 - P\tilde{Q}\right)\frac{2b}{s + b} \tag{8.3}$$

is obtained. *The unweighted sensitivity function is greater by a factor of nearly two than in the* H^∞ *case over all frequencies of importance.* If $\xi \neq 0$, $W(s)$ can be approximated in a distributional sense by a sequence of H^2 functions, and the result (8.3) and conclusion still hold in the limit, as shown in Fig. 3.

Remarks: The doubling of peak weighted sensitivity under the H^2 method is inconsequential under the H^2 scenario, as the power spectrum $|d(j\omega)|$ is fixed, and the fraction of disturbance power subjected to near-peak sensitivities is small. However, under the H^∞-servomechanism scenario, narrow-band disturbances occur whose power is concentrated near the peak frequency. Even then, an increase in peak unweighted sensitivity by a factor of < 2 is not necessarily serious. However, in a feedback configuration, the attendant decrease in "stability margin" (i.e., sensitivity to large plant perturbations; see [1]) is usually large and undesirable.

The H^2 method forces the integral-squared value of the sensitivity function $(1 - PQ)$ to be small over the high-frequency interval $\omega \in (b, \infty)$, where in fact, under the H^∞ scenario, the integral is of no consequence (indeed, in practice cannot be finite, as $P\tilde{Q}$ is strictly proper), and where it is enough to maintain an upper-bound on $(1 - PQ)$.

This gratuitous reduction comes at the expense of the sensitivity at all frequencies of importance in H^∞.

IX. Minimization of Plant Perturbations

A major reason for our interest in weighted H^∞ optimization is that it yields a method for minimizing the effects of certain plant uncertainties or perturbations.

Let $P(s)$ be the plant and $W(s)$ be the weighting postulated in Theorem 1. Consider an uncertain frequency response $\underline{P}(s)$ whose deviation from $P(s)$ is the perturbation $\Delta P(s) \triangleq \underline{P}(s) - P(s)$. For any $\epsilon > 0$, the possible plant ϵ-perturbations are those for which the inequality $|\Delta P(j\omega)/P(j\omega)| \leq \epsilon |W(j\omega)|$ is satisfied, and for which $P + \Delta P$ continues to satisfy the hypotheses assumed in Theorem 1 for P; in particular, those for which $\underline{P}$ has the same number of poles and zeros in $\mathrm{Re}(s) > 0$ as P.

For any feedback F which stabilizes $P(s)$ as in Theorem 1, $K_{12}(s)$ is in H^∞. Also, the closed-loop frequency response $\underline{K}_{12}$ produced by $\underline{P}$, and $\Delta K_{12} \triangleq \underline{K}_{12} - K_{12}$, both belong to H^∞ provided that $\sup_\omega |\Delta P F(1+PF)^{-1}(j\omega)| < 1$, by the small gain theorem. In particular, $\underline{K}_{12}$ is in H^∞ for all possible ϵ-perturbations if

$$\epsilon \leq \|X - W\|^{-1} \tag{8.4}$$

since $F(1+PF)^{-1} = Q$, and $|\Delta P(j\omega) Q(j\omega)| \leq |W(j\omega) P(j\omega) Q(j\omega)| = \epsilon |X(j\omega) - W(j\omega)|$.

For any $\epsilon > 0$ satisfying (8.4), the sensitivity to plant ϵ-perturbations is defined to be $\eta_\epsilon(F) = (1/\epsilon)\sup_{\Delta P}\sup_\omega |\Delta K_{12}(j\omega)/K_{12}(j\omega)|$, where ΔP ranges over all possible ϵ-perturbations. For any $\epsilon > 0$, the *optimal sensitivity to plant ϵ-perturbations* is the infimum $\underline{\eta}_\epsilon = \inf_F[\eta_\epsilon(F)]$ over all feedbacks F that stabilize $P(s)$ and satisfy (8.4).

Let $\tilde{F}_n$ be an optimal sequence of feedbacks for the attenuation of disturbances, as in Theorem 1, but let $d = 0$.

Corollary: For any $\epsilon < [\mu(P) + \|W\|]^{-1}$, the optimal sensitivity to plant ϵ-perturbations $\underline{\eta}_\epsilon$ satisfies the inequalities

$$\mu(P) \leq \underline{\eta}_\epsilon \leq \underline{\lim}_{n \to \infty} \eta_\epsilon(\tilde{F}_n) \leq \frac{\mu(P)}{1 - \epsilon[\mu(P) + \|W\|]} \tag{8.5}$$

and approaches $\mu(P)$ when $\epsilon \to 0$, as do all the bounds in (8.5).

In this sense, the sequence $\tilde{F}_n$ approaches optimality for the attenuation of plant ϵ-perturbations.

IX. Conclusions

When there are no plant uncertainties and the disturbance power-spectrum exactly equals $|W(j\omega)|$, quadratic optimization gives the best sensitivity function. However, this sensitivity function can be quite poor for other spectra, even within the passband of $W(j\omega)$. For example, it is poor for a disturbance whose power is concentrated in a narrow-band around the frequency of the peak in Fig. 3.

H^∞ optimization, on the other hand, gives a sensitivity function optimized for a *set* of disturbance power-spectra, namely, those described in Remark 1) of Section II-D.

Of course, neither the H^∞ nor the quadratic solutions presented here consider practical issues of bandwidth cost, power cost, etc. However, the H^∞ formulation appears better able to cope with those systems that are required to tolerate large classes of disturbance power spectra, as feedback servomechanisms are (and as opposed, say, to single-source communications systems).

Problems involving additive perturbations or uncertainties in the plant frequency response appear to be untractable under a quadratic norm, as multiplicative inequalities of the type $\|X\Delta P\| \leq \|X\| \cdot \|\Delta P\|$ are lacking. On the other hand, a viable and even optimal sensitivity theory for plant perturbations is possible in H^∞.

Appendix I
Proofs

Proof of Proposition 3A1:

a) Follows from the reasoning preceding (3.2) and the strict propriety of Q_o.

c) Suppose Q_o stabilizes P, so $(1 - PQ_o)B_p^{-1} \in H^\infty$. If, now, Q is any other function in H^B the following statements are equivalent:

i) $(1 - PQ)B_p^{-1} \in H^\infty$;

ii) the difference $P(Q - Q_o)B_p^{-1}$ is in H^∞.

We express that difference as

$$P(Q - Q_o)B_p^{-1} = (PB_p)Q_1 \tag{A1.1}$$

where $Q_1 \triangleq B_p^{-2}(Q - Q_o)$, and recall that any function is in H^∞ iff it is in H^B and proper. Therefore, ii) is equivalent to the following.

iii) $(PB_p)Q_1$ is in H^B and proper.

Now $(PB_p)Q_1 \in H^B$ iff $Q_1 \in H^B$. For, $Q_1 \in H^B$ implies $(PB_p)Q_1 \in H^B$. Conversely, if $(PB_p)Q_1 \in H^B$, then any RHP poles of Q_1 are contained in the zeros of PB_p; but the identity $Q_1 = B_p^{-2}(Q - Q_o)$ implies that any RHP poles of Q_1 are contained among the a_i, which is a contradiction; therefore Q_1 had no such poles and belongs to H^B. It follows that iii) is equivalent to the following.

iv) Q_1 is in H^B and PQ is proper.

We recall now that $Q = Q_o + B_p^2 Q_1$ by definition of Q_1, and $Q_o = B_p Q_{oA}$, so iv) means that Q has the stipulated form (3.4).

Similarly, $(1 - PQ)B_p^{-1} \in H^\infty$ iff (3.4) holds with $Q_1 \in H^\infty$, and c) is true.

b) follows from c) and the observation that if Q is in H_o^∞, then PQ is proper, and Q_1 in (3.4) is strictly proper (because Q_o is strictly proper by hypothesis, and $|B_p(s)| \to 1$ as $s \to \infty$). Q.E.D.

Proof of Proposition 3B1:

a) Equations (3.5), (3.6) certainly establish a 1 : 1 invertible correspondence $Q \leftrightarrow X$ between functions meromorphic in $\mathrm{Re}(s) > 0$. If Q is in Ω, then it is in H^B and satisfies the pole constraints (3.3). Therefore, $(1 - PQ)B_p^{-1}$ is in H^B. It follows that XB_p^{-1} is in H^B and so is X. As WPQ is proper, by (3.5) so is X, and therefore XB_p^{-1} is proper. By the reasoning of Section III-B applied to (3.5), it now follows that X satisfies the plant-zero constraints.

Conversely, if $X \in H^\infty$ and Q satisfies (3.6), then as W has no zeros in $\mathrm{Re}(s) \geqslant 0$, $Q(s)$ can have no poles in $\mathrm{Re}(s) \geqslant 0$ except possibly at the zeros of $P(s)$. However, since X satisfies the zero constraints (3.7), such poles cannot occur in Q. To show that $Q \in H^B$ it is enough now to establish its continuity on the $j\omega$-axis. By (3.6), this follows from the hypotheses that W and P are continuous and nonzero on the $j\omega$-axis. Finally, if XB_p^{-1} is in H^∞ as well, then Q satisfies the pole constraints (3.3a), and therefore Q is in Ω; i.e., part a) is true.

b) If Q is in Ω, then X defined by (3.5) is in H^∞ by part a). Therefore, Q is related to X by (3.6), from which it is clear that the growth of $Q(s)$ as $s \to \infty$ in $\mathrm{Re}(s) > 0$ is bounded by const.$|P^{-1}(s)W^{-1}(s)|$, which in turn is bounded by const. $|1+s|^{k+l}$. As $Q(s)$ is analytic and bounded on every bounded subset of $\mathrm{Re}(s) > 0$, part b) is true. Q.E.D.

Proof of Theorem 1: Let J_n denote the high-frequency attenuation function in H_o^∞, $J_n(s) \triangleq n(s+n)^{-1}$.

Case 1:

Step 1: It will be shown that $\tilde{Q}_n$ stabilizes P by showing that $\tilde{Q}_n$ satisfies the pole constraints (3.3a), and then that $\tilde{Q}_n$ is strictly proper, and hence in H_o^∞.

By construction of $\tilde{Q}$ (see Lemma 1), $W(1-P\tilde{Q})B_p^{-1}$ is in H^∞. As $W(s)$ is continuous and has no zeros in $\mathrm{Re}(s) \geqslant 0$, $W^{-1}(s)$ is in H^B, and therefore $(1-P\tilde{Q})B_p^{-1}$ is in H^B, i.e., $\tilde{Q}$ satisfies the pole constraints, and must therefore have the form (3.4) stipulated in Proposition 3A1c. Therefore $(\tilde{Q}-Q_o)B_p^{-2}$ is in H^B. It follows that $\tilde{Q}_n$ defined by (5.1) still has the form (3.4), with Q_1 now equal to $(\tilde{Q}-Q_o)B_p^{-2}J_n^{k+l+1}$, and therefore $\tilde{Q}_n$ satisfies the pole constraints (3.3a), or equivalently, that $(1-P\tilde{Q}_n)B_p^{-1} \in H^B$, as claimed.

Now $\tilde{Q}_n$ must be in H_o^∞ because $Q_o(s)$ is in H_o^∞, and we shall show that $[\tilde{Q}-Q_o]J_n^{k+l+1}$ is also in H_o^∞. The last assertion follows from the fact that $[\tilde{Q}-Q_o]$ is in $(1+s)^{k+l}H^\infty$ by Lemma 1, and J_n^{k+l+1} therefore has enough zeros at infinity of s to cancel the poles of $[\tilde{Q}-Q_o]$ and produce at least one net zero at $s=\infty$, i.e., $\tilde{Q}_n$ is strictly proper.

Step 2, Optimality of $\tilde{Q}_n$: By definition, $\mu(P)$ is an infimum over those Q in H_o^∞ that stabilize P, i.e., that satisfy the pole constraints (3.3a). $\|\tilde{X}\|$ is an infimum over Q in Ω. Since ω contains all $Q \in H_o^\infty$ that satisfy (3.3a), $\mu(P) \geqslant \|\tilde{X}\|$. It will be shown that

$$\lim_{n\to\infty} \|W(1-P\tilde{Q}_n)\| \triangleq \|\tilde{X}\| \tag{A1.2}$$

which implies that $\mu(P) \leqslant \|\tilde{X}\|$. It will follow that $\mu(P) = \|\tilde{X}\|$, and that $\tilde{Q}_n$ is an optimal sequence.

For any n and $m \triangleq k+l+1$ we have

$$\begin{aligned}
&\|W(1-P\tilde{Q}_n)\| \\
&\quad = \|W(1-P\tilde{Q}J_n^m) - WPQ_o(1-J_n^m)\| \quad \text{[by (5.1)]} \\
&\quad = \|W(1-P\tilde{Q})J_n^m + (W-WPQ_o)(1-J_n^m)\| \\
&\quad \leqslant \|W(1-P\tilde{Q})\| + \|(W-WPQ_o)(1-J_n^m)\| \\
&\qquad (\text{as } \|J_n^m\| = 1) \\
&\quad = \|\tilde{X}\| + \|(W-WPQ_o)(1-J_n^m)\|.
\end{aligned} \tag{A1.3}$$

Now,

$$\lim_{n\to\infty} \|(W-WPQ_o)(1-J_n^m)\| = 0 \tag{A1.4}$$

because W and WPQ_o are both in H_o^∞, so the magnitude of $(W-WPQ_o)$ is bounded on the $j\omega$-axis, and approaches 0 as $|\omega| \to \infty$. Also, $|1-J_n^m(\omega)| \leqslant$ const. for all ω and on any finite interval $(-\omega_1, \omega_1)$, $|1-J_n^m(\omega)| \to 0$ uniformly as $n \to \infty$. Therefore, (A1.4) is true, and (A1.2) follows by (A1.3). Therefore, $\tilde{Q}_n$ is an optimal sequence and (5.5) holds. Equation (5.6) is now true because $(1-PQ_n) = (1+F_nP)^{-1}$, by a property of the fractional transformation. Q.E.D.

Case 2:

Step 1: $\tilde{X}$ divided by B_p is in H^∞, as in Case 1. Since the other factor on the RHS of (5.2a) is also in H^∞, $\tilde{X}_{1n}B_p^{-1}$ must be in H^∞. Since $B_z(b_i) = 0$ by construction, $\tilde{X}_{1n}(b_i) = \tilde{X}(b_i)$ by (5.2a), and therefore $\tilde{X}_{1n}(s)$ satisfies (3.7). It follows that $\tilde{Q}_{1n}$ satisfies the pole constraints (3.3a) and is in H^B. By an argument similar to that of Case 1, Step 1, Q_n [given by (5.2c)] stabilizes P, and has been modified by enough zeros at ∞ to ensure that $Q_n \in H_o^\infty$.

Step 2: From Case 1, we know that $\mu(P) \geqslant \|\tilde{X}\|$, and will establish the converse inequality.

Let us show, first that given any $\epsilon > 0$, $\|\tilde{X}_{1n}\| \leqslant \|\tilde{X}\| + \epsilon$ for n large enough. If $\lambda = 0$ this is obvious. Assume $\lambda \neq 0$. Certainly, for ω inside any finite interval $[-\omega_o, \omega_o]$, we have

$$\begin{aligned}
\tilde{X}_{1n}(j\omega)| &\leqslant |\tilde{X}(j\omega)| + \epsilon \\
&= \|\tilde{X}\| + \epsilon
\end{aligned} \tag{A1.5}$$

for n large enough, since $|\lambda B_z(j\omega)| = |\lambda|$, and $(j\omega)(n+j\omega)^{-1} \to 0$ as $n \to \infty$. It is enough, therefore, to show that there exists such an interval outside of which (A1.5) holds for all n.

Let $\delta_X(j\omega) \triangleq \tilde{X}(j\omega) - \tilde{X}(\infty)$ and $\delta_B(j\omega) \triangleq B_z(j\omega) - B_z(\infty)$. Since $\delta_X(j\omega)$ and $\delta_B(j\omega)$ approach 0 as $\omega \to \infty$, there is a number ω_o with the property that $|\delta_X(j\omega)| < \frac{1}{2}\epsilon$ and $|\delta_B(j\omega)| < \frac{1}{2}\epsilon|\lambda|(|\tilde{X}(\infty)| + \frac{1}{2}\epsilon)^{-1}$ for all $|\omega| \geqslant \omega_o$. By (5.2a), we now have, for any $n > 0$

$$\begin{aligned}
&|\tilde{X}_{1n}(j\omega)| \\
&\quad = |\tilde{X}(\infty) + \delta_x(j\omega)| \cdot \left| 1 - \frac{\lambda[B_z(\infty) + \delta_B(j\omega)]j\omega}{n+j\omega} \right| \\
&\quad \leqslant \left(\|\tilde{X}\| + \frac{1}{2}\epsilon\right) \cdot \left| \frac{n + (1-\lambda B_z(\infty))j\omega - \lambda\delta_B(j\omega)j\omega}{n+j\omega} \right| \\
&\qquad \left(\text{since } |\delta_X(j\omega)| < \frac{1}{2}\epsilon\right) \\
&\quad \leqslant \left(\|\tilde{X}\| + \frac{1}{2}\epsilon\right) \left\{ \left| \frac{n + W(\infty)\tilde{X}^{-1}(\infty)j\omega}{n+j\omega} \right| + |\lambda\delta_B(j\omega)| \right\}
\end{aligned} \tag{A1.6}$$

since $1 - \lambda B_z(\infty) = W(\infty)\tilde{X}^{-1}(\infty)$ by choice of λ, and $|j\omega(n+j\omega)^{-1}| < 1$. Since $|W(\infty)\tilde{X}^{-1}(\infty)| \leqslant 1$ by hypothesis, the underscored magnitude on the right-hand side of

(A1.6) must be $\leqslant 1$. Therefore,

$$|\tilde{X}_{1n}(j\omega)| \leqslant \left(\|\tilde{X}\| + \frac{1}{2}\epsilon\right)[1+|\lambda|\cdot|\delta_B(j\omega)|] \leqslant \|\tilde{X}\| + \epsilon$$

and (A1.5) must be true for $|\omega| > |\omega_o|$ by choice of ω_o.

Next, we have the inequalities

$$\begin{aligned}
&\|W(1-P\tilde{Q}_n)\| \\
&= \|W - WP\tilde{Q}_{1n}J_{m_n}^{k+l} - WPQ_o(1-J_{m_n}^{k+l})\| \quad \text{[by (5.2c)]} \\
&= \|W - WP\tilde{Q}_{1n} + WP\tilde{Q}_{1n}(1-J_{m_n}^{k+l}) - WPQ_o(1-J_{m_n}^{k+l})\| \\
&= \|\tilde{X}_{1n} + (W-\tilde{X}_{1n})(1-J_{m_n}^{k+l}) - WPQ_o(1-J_{m_n}^{k+l})\| \\
&\qquad (\text{since by (5.2b), } \tilde{X}_{1n} = W - WP\tilde{Q}_{1n}) \\
&\leqslant \|\tilde{X}_{1n}\| + \|(W-\tilde{X}_{1n})(1-J_{m_n}^{k+l})\| + \|WPQ_o(1-J_{m_n}^{k+l})\|.
\end{aligned} \tag{A1.7}$$

Now for any n, as $m_n \to \infty$, the last term on the right-hand side of (A1.7) approaches zero (by the reasoning of Case 1, Step 2) since WPQ_o is in H_o^∞; the middle term will be shown below to approach zero; and the first term is bounded by $\|\tilde{X}\| + \epsilon$. Therefore, $\lim_{n\to\infty}\|W(1-P\tilde{Q}_n)\| \leqslant \|\tilde{X}\| + \epsilon$, which implies that $\mu(P) = \|\tilde{X}\|$ as ϵ was arbitrary, and therefore the sequence (5.2) is optimal.

The convergence of the middle term on the right-hand side of (A1.7) to zero is established by the same reasoning as was used in (A1.4), from the fact that by construction of $\tilde{X}_{1n}$, $(\tilde{X}_{1n} - W)$ is in H_o^∞, since $[\tilde{X}_{1n}(s) - W(s)] \to W(\infty) - W(\infty)$ as $s \to \infty$ in $\text{Re}(s) \geqslant 0$. Q.E.D.

Proof of Proposition 5A1:

a) We have, $\mu(P) \geqslant \|\tilde{X}\|$ by Theorem 1, and $\|\tilde{X}\| = \|\tilde{X}B_p^{-1}\|$ as $|B(j\omega)| = 1$. Now, $\tilde{X}B_p^{-1}$ is analytic in $\text{Re}(s) \geqslant 0$, approaches a constant at $s = \infty$, and assumes the values $W(b_i)B_p^{-1}(b_i)$ at the plant zeros, by (3.7). By the maximum modulus principle, therefore, (5.7) is true. Equation (5.8) follows from (5.7) because any Blaschke product analytic in $\text{Re}(s) \geqslant 0$ has a magnitude < 1 in $\text{Re}(s) > 0$.

b) Let $\tilde{X}_1$ and $\tilde{X}_2$ be the optimal sensitivity functions for P_1 and P_2, respectively. $\tilde{X}_1$ and $\tilde{X}_2$ must both assume the value $W(b_i)$ at each RHP zero of P_1 or P_2. Since $\tilde{X}_1$ is the minimal function in H^∞ satisfying these constraints and the minimum is unique, by Lemma 1, either $\tilde{X}_1$ coincides with $\tilde{X}_2$ or

$$\|\tilde{X}_1\| < \|\tilde{X}_2\|. \tag{A1.8}$$

Since $\tilde{X}_1$ has fewer zeros than $\tilde{X}_2$, (A1.8) and the conclusion of b) must be true. Q.E.D.

Proof of Corollary: For any feedback F which stabilizes P, and all possible ϵ-perturbations satisfying (8.4), the identity $\Delta K_{12} = (1+PF)^{-1}(1+\underline{P}F)^{-1}\Delta P$ can be established by algebraic operations (see, e.g., [1]). Consequently,

$$\begin{aligned}
|\Delta K_{12}(j\omega)/K_{12}(j\omega)| &= |(1+\underline{P}F)^{-1}(\Delta P/P)(j\omega)| \\
&\leqslant \epsilon|(1+\underline{P}F)^{-1}W(j\omega)| \\
&= \epsilon|W(1+PF+\Delta PF)^{-1}(j\omega)| \\
&\leqslant \epsilon|W(1+PF)^{-1}(1+\Delta PQ)^{-1}(j\omega)|
\end{aligned}$$

where the identity $F(1+PF)^{-1} = Q$ has been used. Since

$$|\Delta P(j\omega)Q(j\omega)| \leqslant \epsilon|X(j\omega) - W(j\omega)| \leqslant \epsilon(\|X\| + \|W\|),$$

we have that

$$|\Delta K_{12}(j\omega)/K_{12}(j\omega)| \leqslant \epsilon\|X\|[1-\epsilon(\|X\|+\|W\|)]^{-1}$$

whenever $\epsilon(\|X\|+\|W\|) < 1$. If we now identify F with $\tilde{F}_n$ of Theorem 1, and observe that $\lim_{n\to\infty}\|X_n\| = \mu(P)$, the upper bound on η_ϵ in (8.5) follows. The lower bound in (8.5) is obtained by noting that for any $\epsilon > 0$, and any feedback F satisfying $\epsilon(\|X\|+\|W\|)^{-1} < 1$, the ϵ-permutation $\Delta P_1 \triangleq \epsilon WPB_p \in H^\infty$ is a possible perturbation by the small gain theorem, and yields

$$|\Delta K_{12}(j\omega)/K_{12}(j\omega)| = \epsilon|W(1+PF)^{-1}(j\omega)| \geqslant \epsilon\mu(P),$$

which implies that $\eta_\epsilon(F) \geqslant \mu(P)$, and that $\eta/\epsilon \geqslant \mu(P)$. Q.E.D.

Appendix II
Derivation of the Optimal Sensitivity

This derivation is adapted from Duren [4] and Koosis [5]. Some details will be omitted.

H^1 denotes the Hardy space of functions G analytic in $\text{Re}(s) > 0$ under the norm

$$\|G\|_1 \triangleq \sup_{\sigma > 0}\int_{-\infty}^{\infty}|G(\sigma + j\omega)|\,d\omega < \infty.$$

The L^p spaces, $p = 1$ or ∞, will be viewed as consisting of functions defined on the imaginary axis, and each H^p will be considered to be a subspace of L^p. This is possible if each function in H^p is represented by its limit function on the imaginary axis. All norms in this Appendix are L^p norms.

C_o is the subspace of L^∞ consisting of continuous functions which approach 0 as $|j\omega| \to \infty$. A_o^∞ is the subspace of H_o^∞ consisting of functions continuous in $\text{Re}(s) \geqslant 0$, i.e., $A_o^\infty = C_o \cap H_o^\infty$.

The basis for Lemma 1 is a min–max identity whose derivation, involving the theory of duality and the Hahn–Banach theorem, is sketched out in Appendix II-B.

Proposition A2.1 A Min–Max Identity: If $Y_1 \in C_o$, then

a)

$$\max_{G \in H^1,\, \|G\|_1 \leqslant 1}\left|\int_{\infty}^{\infty} Y_1(j\omega)G(j\omega)\,d\omega\right| = \min_{V \in H^\infty}\|Y_1 - V\|_\infty \triangleq \mu \tag{A2.1}$$

b) There exists $\tilde{G} \in H^1$ and a *unique* $\tilde{V} \in H^\infty$ [which attain the extrema in (A2.1)] such that

$$\int_{-\infty}^{\infty} Y_1(j\omega)\tilde{G}(j\omega)\,d\omega = \mu = \|Y_1 - \tilde{V}\|_\infty.$$

c) If $\tilde{Y} \triangleq Y_1 - \tilde{V}$, then $\tilde{Y} \in L^\infty$ is unique, $\tilde{Y}(j\omega)\tilde{G}(j\omega)$ is nonnegative real, and $\tilde{Y}(j\omega) = \mu|\tilde{G}(j\omega)|/\tilde{G}(j\omega)$ a.a.ω.

A. Proof of Lemma 1

By Lemma 3B1, the weighted sensitivity functions are those functions $X \in H^\infty$ which satisfy the pole-zero constraints (3.3),(3.7). There exists at least one such function in A_o^∞. Indeed, if $W(\infty)=0$, let $X_1 \triangleq W(1-PQ_o)$, where Q_o is defined in the paragraph following (3.4b); if $W(\infty) \neq 0$, let

$$X_1(s) = W(s)[1-P(s)Q_o(s)]\left[1-B_z(s)\frac{\lambda s}{s+1}\right],$$

λ given by (5.2a). In either case, $X_1 \in H^\infty$ satisfies the pole-zero constraints and approaches 0 as $|j\omega| \to \infty$ by construction, is continuous on the imaginary axis by the hypotheses on P and W, and is therefore in A_o^∞.

Now *every* $X \in H^\infty$ satisfying the pole-zero constraints must be expressible in the form $X = X_1 + B_p B_z V$ for some $V \in H^\infty$. The extremal problem of Lemma 1 is therefore equivalent to the problem

$$\inf_{V \in H^\infty} \|X_1 - B_p B_z V\|_\infty. \tag{A2.2}$$

Let $Y_1 = B_p^{-1} B_z^{-1} X_1$. As multiplication by $B_p B_z$ is norm preserving and invertible, (A2.2) is equivalent to

$$\inf_{V \in H^\infty} \|Y_1 - V\|_\infty \tag{A2.3}$$

where $Y_1 \in C_o$. By Proposition A2.1, there exists a unique $V \in H^\infty$ which attains the inf in (A2.3), and if we write $\tilde{Y} \triangleq Y_1 - \tilde{V}$, $\|\tilde{Y}\| \triangleq \mu$, then $\tilde{Y} \in L^\infty$ and $|\tilde{Y}(j\omega)| = \mu$ a.a.ω. The inf in (A2.2) is therefore attained uniquely by $\tilde{X} \triangleq X_1 + B_p B_z \tilde{V}$, where $\tilde{X} \in H^\infty$ and $|\tilde{X}(j\omega)| = \mu$ a.a.ω.

Let us show that $\tilde{X}(s)$ has the rational form (4.2). Let $\tilde{G} \in H^1$ be the extremal function which exists by Proposition A2.1, and define $R(s) \triangleq \tilde{Y}(s)\tilde{G}(s)$. Now as $R(s) = \tilde{X}(s)B_p^{-1}(s)B_z^{-1}(s)\tilde{G}(s)$, and as $\tilde{X}B_z^{-1} \in H^\infty$ by the pole constraints, $R(s)$ must be analytic in $\mathrm{Re}(s) > 0$ except at the poles b_j. By Proposition A2.1c, $R(j\omega)$ is nonnegative real. Therefore, $R(j\omega)$ must have a continuation to the entire plane, with poles at b_j and $-\bar{b}_j$, by the Schwartz reflection principle, as the continuous condition

$$\lim_{\sigma \to 0} \int_{\omega_2}^{\omega_1} |R(\sigma + j\omega) - R(j\omega)|\, d\omega \to 0$$

is fulfilled for all ω_1, ω_2. Therefore, $R(s)$ is rational, and has the form

$$R(s) = C\frac{\prod_{i=1}^{m}(s-c_j)(s+\bar{c}_j)}{\prod_{j=1}^{r}(s-b_j)(s+\bar{b}_j)}, \qquad \mathrm{Re}(c_j) \geqslant 0,\ c > 0 \tag{A2.4}$$

where $m \leqslant (r-1)$ because $R \in L^1$.

Let us deduce the expression (4.2) for $\tilde{X}(s)$ from (A2.4). As $\tilde{X}\tilde{G}$ is in H^1 it has a unique factorization into, at most, an outer (i.e., minimum phase) factor and an inner (i.e., of unit magnitude a.e. on the imaginary axis) factor. The inner factor has a further unique factorization into, at most, a Blaschke product, a pure delay, and a singular part. Now, as $R(s)$ is rational, $\tilde{X}(s)\tilde{G}(s)$ is also rational and has neither singular nor delay terms, and therefore $\tilde{X}(s)$ has no such terms. Since $|\tilde{X}(j\omega)| = \mu$ a.a.ω., $\tilde{X}$ is inner and, more particularly, a Blaschke product. The zeros of $\tilde{X}(s)$ must be among the c_j in (A2.4), and therefore part a) is true.

Part b) of Lemma 1 follows from (3.6), Proposition 3B1, and the assumed behavior of $P(s)$ and $W(s)$ as $|s| \to \infty$.

Part c) is proved as follows. Since P and W have conjugate symmetry, the poles and zeros of $P(s)$ are real or occur in conjugate pairs, and the interpolation constraints at the singularities of $P(s)$ have conjugate symmetry. For example, if $\tilde{X}(s)$ is constrained to have the value θ_i at $s = b_i$, then it is also constrained to have the value $\bar{\theta}_i$ at $s = \bar{b}_i$. Consequently, the function $\bar{\tilde{X}}(\bar{s})$, which is in H^∞ and has the same norm as $X(s)$, satisfies the same interpolation constraints as $X(s)$. But the H^∞ function with these properties is unique by part a). Consequently, $\bar{\tilde{X}}(\bar{s}) = \tilde{X}(s)$, $\tilde{X}$ has conjugate symmetry, and so does $\tilde{Q}$ by (3.6).

B. Duality

Let us briefly outline the relationship between Proposition A2.1 and standard results on duality. If A is any normed linear space, its dual space A^* is the linear space of all bounded linear functionals ϕ mapping A into the complex numbers under the norm $\|\phi\|_{A^*} = \sup\{|\phi(x)|: x \in A, \|x\| \leqslant 1\}$. A^* is a Banach space. The orthogonal complement of any subspace A_1 of A is the subspace $A_1^\perp$ of A^*, $A_1 \triangleq \{a^* \in A^*: a^*(a) = 0 \text{ for all } a \in A_1\}$.

Let A and B be Banach spaces, and for $a \in A$, $b \in B$, let $\langle a, b\rangle$ denote a bilinear form mapping $A \times B$ into the complex numbers, and assume the property (P) that for each $a \in A$, $\langle a, (\cdot)\rangle$ is a bounded linear functional on B, and hence belongs to B^*, and that similarly for each $b \in B$, $\langle (\cdot), b\rangle$ is in A^*. With each element $a \in A$ we shall associate the functional $\langle a, (\cdot)\rangle \in B^*$ produced by a, and so view A as a subspace of B^*. Similarly, B will be viewed as a subspace of A^*.

Our starting point is an adaptation of a result based on the Hahn–Banach theorem (see Duren [4, Theorems 7.1, 7.2] or Luenberger [10, ch. 5.8, Theorems 1 and 2]). This result, which we state without proof, extends the Euclidean notion that the shortest distance from a vector to a subspace equals the length of the longest projection of the vector onto the orthogonal complement of the subspace.

Let A_1 and B_1 be subspaces of the Banach spaces A and B, respectively.

Proposition A2.2: If $b \in B$ is fixed, and either $A_1 = B_1^\perp$ or $B_1 = A_1^\perp$, then the following occurs.

a) There is a constant $d \geqslant 0$ such that

$$\sup_{\substack{a_1 \in A_1 \\ \|a_1\| = 1}} |\langle a_1, b\rangle| = \inf_{b_1 \in B_1} \|b - b_1\| \triangleq d. \tag{A2.5}$$

b) If $B_1 = A_1^\perp$, the inf on the right is attained.

c) If $A_1 = B_1^\perp$, the sup on the left is attained, and there exists a (not necessarily unique) $a_1 \in A_1$ such that $\langle a_1, b\rangle = d$.

Proposition A2.1 is an application of Proposition A2.2 to the bilinear form

$$\langle Y_1, G\rangle \triangleq \int_{-\infty}^{\infty} Y_1(j\omega)G(j\omega)\,d(\omega) \tag{A2.6}$$

which is well defined on $L^\infty \times L^1$, and has the property (P) specified above. We shall require the following two facts from analytic function theory.

1) Each bounded linear functional on L^1 has the integral representation (A2.6) for some $Y_1 \in L^\infty$, and the functionals which vanish for all $G \in H^1$ are precisely those for which $Y_1 \in H^\infty$; i.e. (Koosis [5, p. 193]), $(L^1/H^1)^* = H^\infty$ and

$$(H^1)^\perp = H^\infty. \tag{A2.7}$$

2) Although the representation (A2.6) is too restrictive to represent all linear functionals on L^∞, it can be used to represent those on the subspace C_o. Indeed, each bounded linear functional in $(C_o)^*$ has the form (A2.6) for some $G \in L^1$, and the functionals which vanish for all Y_1 in the subspace A_o^∞ of C_o are precisely those for which $G \in H^1$; i.e. (Koosis [5, p. 193]), $(C_o/A_o^\infty)^* = H^1$ and

$$(A_o^\infty)^\perp = H^1. \tag{A2.8}$$

Proof of Proposition A2.1:

Part a): By (A2.7) and Proposition A2.2b applied to (A2.6), there is a $\tilde{V} \in H^\infty$ such that

$$\sup_{\substack{G\in H^1\\ \|G\|_1\leq 1}} \int_{-\infty}^{\infty} Y_1(j\omega)G(j\omega)\,d\omega = \min_{V\in H^\infty} \|Y_1 - V\|_\infty = \|Y_1 - \tilde{V}\|_\infty \triangleq \mu. \tag{A2.9}$$

Since $Y_1 \in C_o$, we can employ Proposition A2.2c and (A2.8) to conclude that there is a $\tilde{G} \in H^1$, $\|\tilde{G}\|_1 = 1$, which satisfies $\langle Y_1, \tilde{G}\rangle = \inf_{V\in A_o^\infty}\|Y_1 - V\|_\infty$. Therefore, we have

$$\min_{V\in H^\infty}\|Y_1 - V\|_\infty \leq \inf_{V\in A_o^\infty}\|Y_1 - V\|_\infty = \langle Y_1, \tilde{G}\rangle \tag{A2.10}$$

as $A_o^\infty \subset H^\infty$. But

$$\langle Y_1, \tilde{G}\rangle \leq \sup_{\substack{G\in H^1\\ \|G\|_1 = 1}} |\langle Y_1, G\rangle| = \min_{V\in H^\infty}\|Y_1 - V\|_\infty \tag{A2.11}$$

by (A2.9). But (A2.10) and (A2.11) can be valid only if identity holds throughout both expressions, and part a) of Proposition A2.1 is true.

Parts b), c) Uniqueness of $\tilde{V}$: We have shown that there exists (at least) one $G \in H^1$, which we have denoted by $\tilde{G}$, for which A2.1 holds. Since $\tilde{V} \in H^\infty = (H^1)^\perp$, we have

$$\int_{-\infty}^{\infty} [Y_1(j\omega) - \tilde{V}(j\omega)]\tilde{G}(j\omega)\,d\omega = \|Y_1 - \tilde{V}\|_\infty \|\tilde{G}\|_1 = \mu, \tag{A2.12}$$

i.e., an instance of Holder's inequality with identity. This can happen only if the integrand in (A2.12) is nonnegative real, and if

$$[Y_1(j\omega) - \tilde{V}(j\omega)]\tilde{G}(j\omega) = \mu|\tilde{G}(j\omega)| \qquad \text{a.a.}\omega.$$

If $\tilde{V}_1$ also attains the min in (A2.1) then

$$[Y_1(j\omega) - \tilde{V}_1(j\omega)]\tilde{G}(j\omega) = \mu|\tilde{G}(j\omega)| \qquad \text{a.a.}\omega.$$

and by subtraction, $[\tilde{V}(j\omega) - \tilde{V}_1(j\omega)]\tilde{G}(j\omega) = 0$ a.a.ω. But as $\tilde{G} \in H^1$, $\tilde{G}(j\omega) \neq 0$ a.a.ω. Consequently, $\tilde{V} = \tilde{V}_1$, and $\tilde{V}$ is unique. Part b) is true, and part c) follows immediately. [Note that the Hahn–Banach theorem establishes the existence of *some* $\tilde{G} \in H^1$. Uniqueness depends upon additional properties of the integral representation and $\tilde{G}_1(j\omega)$.]

Q.E.D.

ACKNOWLEDGMENT

The authors are indebted to J. W. Helton for introducing them to the literature on the Schur–Pick algorithm.

REFERENCES

[1] G. Zames, "Feedback and optimal sensitivity: Model reference transformations, multiplicative seminorms, and approximate inverses," in *Proc. 17th Allerton Conf.*, Oct. 1979, pp. 744–752; also in *IEEE Trans. Automat. Contr.*, vol. AC-26, pp. 301–320, Apr. 1981.

[2] J. L. Walsh, *Interpolation and Approximation by Rational Functions in the Complex Domain*. Providence, RI: AMS Colloq. Pub., 1935.

[3] W. W. Rogosinski and H. S. Shapiro, "On certain extremum problems for analytic functions," *Acta Math.*, vol. 90, pp. 287–318, 1953.

[4] W. L. Duren, *Theory of H^P Spaces*. New York: Academic, 1970.

[5] P. J. Koosis, *Introduction to H^P Spaces*, (London Math. Soc. Lecture Note Series No. 40). Cambridge, MA: Cambridge Univ. Press, 1980.

[6] D. C. Youla, H. Jabr, and J. J. Bongiorno, Jr., "Modern Wiener-Hopf design of optimal controllers—Part I," *IEEE Trans. Automat. Contr.*, vol. AC-21, pp. 3–13, Feb. 1967; and "Modern Wiener-Hopf design of optimal controllers—Part II," *IEEE Trans. Automat. Contr.*, vol. AC-21, pp. 319–338, June 1967.

[7] C. A. Desoer, R. W. Liu, J. Murray, and R. Saeks, "Feedback system design: The fractional representation approach to analysis and synthesis," *IEEE Trans. Automat. Contr.*, vol. AC-25, pp. 399–412, 1980.

[8] M. G. Safonov and M. Athans, "Gain and phase margin of multiloop LQG regulators," *IEEE Trans. Automat. Contr.*, vol. AC-22, pp. 173–178, Apr. 1977.

[9] J. C. Doyle and G. Stein, "Multivariable feedback design: Concepts for a classical/modern synthesis," *IEEE Trans. Automat. Contr.*, vol. AC-26, pp. 4–16, Feb. 1981.

[10] D. G. Luenberger, *Optimization by Vector Space Methods*. New York: Wiley, 1969.

SYNTHESIS OF ROBUST CONTROLLERS AND FILTERS

John C. Doyle

Honeywell, Inc., Systems and Research Center, Minneapolis, Minnesota
and
Department of Mathematics, University of California, Berkeley

ABSTRACT

This paper outlines a general framework for analysis and synthesis of linear control systems and reports on a new solution to a very general L_∞/H_∞ optimal control problem.

I. Introduction

This paper outlines a general framework for analysis and synthesis of linear control systems that unifies and extends many existing methods. These include covariance, singular value [1], and structured singular value ([2], [3]) analysis. The synthesis methods include the Wiener-Hopf-Kalman (WHK) approaches (e.g., Wiener and Kalman filtering, LQG, etc.) for time-invariant systems and the newer L_∞/H_∞ methods (see [4] for a review of recent work on this subject). The main result reported in this paper is a new solution to a very general L_∞/H_∞ optimal control problem. The practical significance of this result is enhanced by the fact that the H_∞-optimal controller can be computed using standard real matrix operations (i.e., solving Lyapunov and Riccati equations, QR and SV decompositions, etc.) on state-space representations.

This paper will present a rather bare outline of these results. The final sections rely heavily on the theory of Ball and Helton [5]. A tutorial introduction to the key issues in this theory is presented in this proceedings [4].

III. Analysis

Various modeling assumptions will be considered and the impact of these assumptions on analysis and synthesis methods will be explored. Referring to Figure 1a, the nominal model is assumed throughout to be a Linear, Time-Invariant Ordinary Differential Equation (LTIODE). The uncertain inputs are assumed to be either filtered white noise or weighted Lp-norm bounded signals. The plant uncertainty is modelled as perturbations (not necessarily small) to the nominal. Performance is measured in terms of either the weighted error covariance or the weighted Lp-norm of the error.

These performance measures are intended to reflect engineering issues such as good command response or small errors in regulation or estimation. Perturbations typically arise in an attempt to model changes and uncertainty in operating conditions and plant characteristics as well as unmodelled dynamics. Uncertain inputs model disturbances, noises and commands. The analysis and synthesis framework used in this paper includes all the standard linear time-invariant filtering and control problems, including the so-called two-degree-of-freedom control problem. This last problem is obtained when commands are modelled in the usual way as uncertain input signals.

Since the focus of this section is on analysis, the controller can be viewed as just another system component. Thus for analysis purposes, Figure 1a may be reduced to Figure 1b. Here P is a 2x2 block transfer function matrix providing connections from external inputs and perturbations to outputs and perturbations. Note than any interconnection of inputs and outputs with components and perturbations may be rearranged into this form. Then the output can be written as

$$e = (P_{22} + P_{21}\Delta(I-P_{11}\Delta)^{-1}P_{12})u. \tag{3.1}$$

It is assumed that stability is always a performance requirement and thus that P has all its poles in the open left-half plane.

The standard modelling assumptions and the resulting analysis methods are summarized in Table I. The first option is that uncertainty is modelled as white noise and performance is measured in terms of error covariance. It is well-known that the error covariance can be evaluated in terms of the L_2-norm of P_{22}. This model is appealing in that many physical noises and disturbances have existing, accepted models as filtered white noise and that computation of $\|P_{22}\|_2$ is quite easy using Lyapunov equations. Furthermore, synthesis (the so-called Wiener-Hopf-Kalman (WHK) theory) in this context involves linear approximation in a Hilbert space, also computationally appealing. Unfortunately, few physical systems are adequately modelled with additive white noise as the only uncertainty.

Figure 1b. Analysis Model

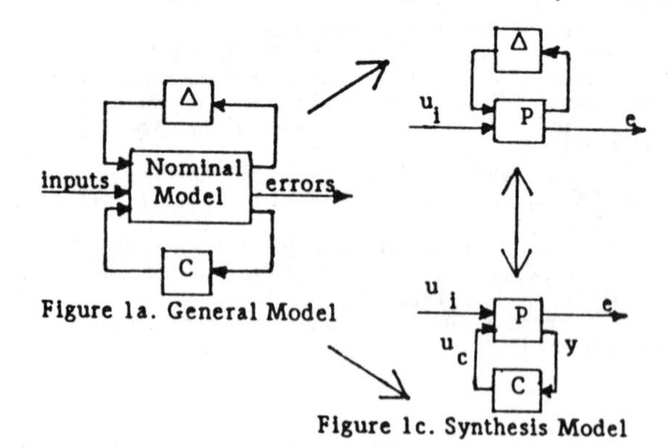

Figure 1a. General Model

Figure 1c. Synthesis Model

Case	Modeling Assumptions: Input	Modeling Assumptions: Perturbation	Performance Requirement	Analysis Test
1	White Noise Unit Covariance	$\Delta=0$	Covariance ≤ 1	$\frac{1}{2\pi}\|P_{22}\|_2 \le 1$
2	$\|u_i\|_2 \le 1$	$\Delta=0$	$\|e\|_2 \le 1$	$\|P_{22}\|_\infty \le 1$
3	-	$\|\Delta\|_\infty < 11$	BIBO Stable	$\|P_{11}\|_\infty \le 1$
4	-	$\Delta=$	BIBO Stable	$\|P_{11}\|_\mu \le 1$
5	$\|u_i\|_2 \le 1$	diag($\Delta 1, \Delta 2, \ldots, \Delta_n$) (structured uncertainty)	$\|e\|_2 \le 1$	$\|P\|_\mu \le 1$

Table 1. Analysis Summary

* This work has been supported by Honeywell Internal Research and Development Funding, The Office of Naval Research under ONR Research Grant N00014-82-C-0157, and the U.S. Air Force of Scientific Research Grant F49620-82-C-0090.

Reprinted from *IEEE Proc. 22nd Conf. Decision Contr.*, pp. 109–114, Dec. 1983.

Some alternatives to the white noise view of uncertainty are summarized in the remainder of Table I. These options may be thought of as being separate cases of the following general form of a performance/robustness theorem:

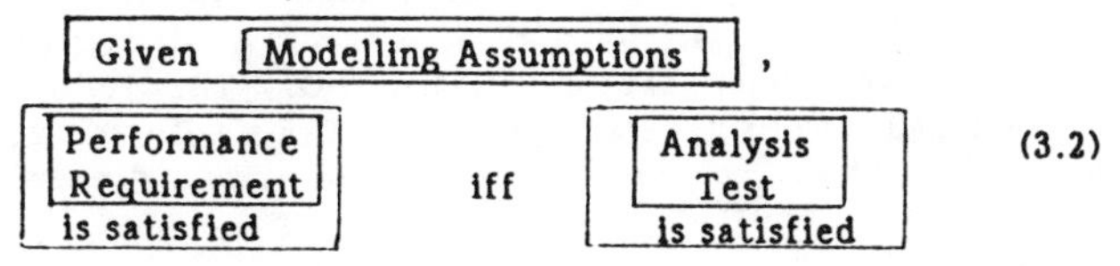

(3.2)

The first alternative (case 2) involves modelling inputs as unknown-but-bounded (in an L_2-sense) and requiring that the output remain bounded (in L_2) below a specified level for all such inputs. The resulting analysis test involves the L_∞-norm of P_{22}, the same norm which appears in the robust stability test of Case 3. The chief advantage of these assumptions over Case 1 is that both uncertain inputs and plant perturbations are handled with the same $\|\bullet\|_\infty$ test. A less compelling reason is that designs for unknown-but-bounded L_2 inputs and outputs can be given a minimax energy interpretation.

Note that any induced operator norm would provide an analysis test that would handle both uncertain norm-bounded inputs and (induced) norm-bounded perturbations. For example, modeling signals as unknown-but-bounded in magnitude (and using the resulting induced norm for convolution operators on L_∞ for analysis) has obvious advantages in applications where signal magnitude is a more natural notion than energy. On the other hand, there are some important reasons for choosing L_2 signal models, $\|\bullet\|_\infty$ perturbation bounds, and $\|\bullet\|_\infty$ analysis tests over, say, other L_p spaces:

1. The induced convolution-operator norm on L_2 (i.e., $\|\bullet\|_\infty$ on transfer functions) is the only induced norm which yields necessary as well as sufficient robust stability tests.

2. Perturbation models of this type are currently the most easily obtained.

3. An optimal synthesis theory analogous to that of WHK is now available. (i.e., the main result of this paper)

4. Engineers have developed substantial experience with these methods through the use of Bode plots and more recently, their singular value generalizations.

Clearly, these reasons are not entirely independent. The $\|\bullet\|_\infty$ norm on transfer functions is reasonably easily computed, but it does involve a search over one frequency variable.

It should be noted that in practice the use of weights on signals and perturbations is essential, since both vary with direction and frequency. This is true independent of the particular assumptions being made. By absorbing any weightings into the interconnection function P, the weighted case can be reduced to that considered in Figure 1a and Table I. This is one advantage of the framework proposed here over less general ones in that any interconnection of signals, systems and perturbations, including weights, can be rearranged to fit the framework.

While case 2 and 3 provide a single framework in which to analyze performance and robustness (of stability), the $\|\bullet\|_\infty$ norm alone provides no systematic, reliable method for analyzing robust performance. Furthermore, $\|\bullet\|_\infty$ analyzes robustness with respect to purely unstructured uncertainty. A more sophisticated tool that treats robust performance with respect to structured uncertainty involves the structured singular value, μ, and $\|\bullet\|_\mu$ [2], [3]. Although μ is not a norm, we will abuse notation and let $\|P\|_\mu \equiv \operatorname{ess\,sup}_\omega \mu(P(j\omega))$.

Cases 4 and 5 of Table I summarize the two basic applications of μ to analysis. Case 4 gives a structured version of Case 3 by characterizing robust stability with respect to block-diagonal perturbations. This is quite general since any interconnection of perturbations can be rearranged to fit the structure of Figure 1b with a block diagonal Δ. Case 5 generalizes cases 2, 3, and 4 by characterizing the performance (in an L_2-bounded sense) for systems with structured uncertainty. This is currently the only available method for systematically analyzing the performance of complex systems with plant perturbations.

We have seen that the standard analysis tools of linear control theory plus a new, more powerful method based on μ can be viewed as special cases of the general framework outlined in this section and summarized in Figure 1b, Table I and (3.2). In this framework analysis of system performance and robustness reduces to computing $\|P\|_\alpha$ for $\alpha=2$, ∞, or μ for some transfer function P. The goal of the remainder of the paper is to develop a similar framework for the synthesis of controllers to meet specifications expressed in terms of these analysis methods.

IV. Synthesis Framework

From the previous section on analysis, we know that evaluating the performance/robustness of the control system in Figure 1a can be reduced to the computation of $\|P(C)\|_\alpha$ for $\alpha=2$, ∞ or μ. Here $P(C) = P_{11} + P_{12}C(I-P_{22}C)^{-1}P_{21}$ as shown in Figure 1c (the P_{ij}'s here are in general different from the P_{ij}'s in Figure 1b). A natural approach to synthesis would be to solve

$$\min_C \|P(C)\|_\alpha \qquad \alpha=2,\ \infty \text{ or } \mu \tag{4.1}$$

The rest of this paper is primarily concerned with a synthesis framework that provides solutions for the $\alpha=2$ and ∞ cases. The $\alpha=\mu$ case will be considered briefly at the end of this paper. Until then, α will be used to denote those cases when either $\alpha=2$ or $\alpha=\infty$ apply equally well.

The first step in the solution is to use the stabilizing controller parameterization [6] to turn $\min\|P(C)\|$ into a problem affine in a stable parameter Q. For simplicity, suppose throughout that P is open-loop stable (the P_{ij} have all their poles in the open left-half plane). Relaxing this assumption complicates the parameterization but does not affect the synthesis methods. If the $\{P_{ij}\}$ are open-loop stable, P(C) is stable iff $Q=C(I-P_{22}C)^{-1}$ is stable. Using this parameterization, the synthesis problem becomes

$$\min_{Q\in RH_\alpha} \|P_{11} + P_{12}QP_{21}\|_\alpha \tag{4.2}$$

where the prefix R denotes real-rational. It greatly simplifies the discussion to drop the requirement that Q be real-rational. It is a fortunate consequence of the theory that the optimal Q is in fact real-rational when the P_{ij} are real-rational.

The two cases $\alpha=2$ and $\alpha=\infty$ can be developed in a parallel fashion. Figure 2 gives a flowchart outlining the steps involved in solving the synthesis problem with labels indicating the technique and section relevant to the step. Note that in each case the general problem is reduced to finding the nearest H_α approximation to a function in L_α. The first step in this reduction involves inner-outer factorization of rational matrices.

V. Inner-Outer and Spectral Factorizations

This section will develop purely state-space methods for performing the factorizations needed in the remaining sections. By using standard algorithms involving only real matrix algebra, these methods should prove computationally reliable. The key idea is to reduce each factorization to solving the standard Algebraic Riccati Equation

$$F^TX + XF - XWX + H=0 \tag{5.1}$$

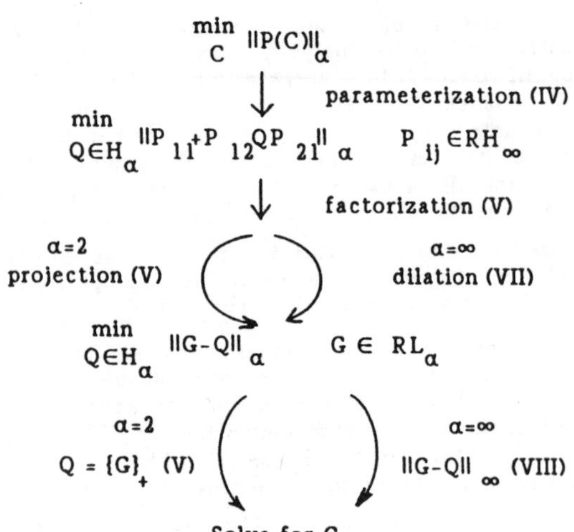

Figure 2. Synthesis Summary

The following theorem gives an algebraic relation between solutions of (5.1) and spectral factorizations (SF) [7], [8], [9].

Theorem (SF) Solving (5.1) for X with $F=A-BR^{-1}S^T$, $W=BR^{-1}B^T$, $H=P-SR^{-1}S^T$, $P=P^T$, $R=R^T>0$, and letting $K=R^{-1}(S^T+B^TX)$ yields the factorization

$$[B^T(-sI-A^T)^{-1}\ \ I]\begin{bmatrix}P & S\\ S^T & R\end{bmatrix}\begin{bmatrix}(sI-A)^{-1}B\\ I\end{bmatrix} = M^* RM$$

where $M = I+K(sI-A)^{-1}B$.

The proof is straightforward algebra and well-known. The different solutions of (5.1) correspond to alternative zero patterns for M. If (A,B) is controllable, there exists a unique solution $X=X^T \geq 0$ so that M has only lhp zeros [9], [10]. This theorem will be required in Section VII to form $(\gamma^2 I-G^*G)^{1/2}$ for $\gamma>\|G\|_\infty$. Here $(\ )^{1/2}$ denotes a spectral factor with all poles and zeros in the open left half plane. Note that if G has rhp poles the theorem must be applied twice, the second time to M^{-1} to reflect the rhp poles into the lhp.

Suppose $G(s) = D+C(sI-A)^{-1}B \in RH_\infty^{nxm}$, $(n \geq m)$ with rank (D) = m, [C,A,B] minimal, and no zeros with zero real part. An inner-outer factorization (IOF) of G is G(s) = Θ(s) M(s) where

1) $M(s) \in RH_\infty^{mxm}$, $[M(s)]^{-1} \in RH_\infty^{mxm}$ (outer)

2) $\Theta(s) \in RH_\infty^{nxm}$ and $\Theta^*\Theta=I$ $\quad (\Theta^* \equiv \Theta^T(-s))$ (inner)

Suppose wolog that D is already factored so that $D^TD=I$. Let $D_\perp$ be such that $[D\ D_\perp]$ is an orthogonal matrix. If n=m, then $D_\perp$ does not exist. Under these conditions, the following theorem holds:

Theorem (IOF) G has an IOF with

$$\Theta(s) = D + (D_\perp D_\perp^T C-DB^TX)(sI-F+BB^TX)^{-1}B$$

$$M(s) = I + (D^TC+B^TX)(sI-A)^{-1}B \text{ where } X=X^T \geq 0 \text{ solves}$$

(5.1) with $F=A-BD^TC$, $G=BB^T$, $H=C^TD_\perp D_\perp^TC$.

Proof: Obtain M(s) directly from SF theorem with

$$\begin{bmatrix}P & S\\ S^T & R\end{bmatrix} = \begin{bmatrix}C^T\\ D^T\end{bmatrix}[C\ D].$$ The Θ(s) is obtained from $\Theta(s) = G(s)\,[M(s)]^{-1}$ with a little algebra. Note that if G has rhp poles everything applies except that M will also have the same rhp poles.□

The inner-outer factorization is sufficient to solve the H_2 optimization problem, but the H_∞ problem is simplified by the introduction of an additional factorization. Suppose G is as before (except possibly unstable) and n>m. If Θ(s) is nx(n-m) and inner and $\Theta^*G=0$ then Θ will be called a complementary inner factor (CIF) of G. The following theorem and corollary relate particular CIF's with solutions to (5.1). When no confusion should arise, dimension superscripts are omitted..

Theorem (stable CIF) An RH_∞ CIF for G is

$$\Theta = [D_\perp+(DB^T-D_\perp D_\perp^T CX)(sI-F+C^TD_\perp D_\perp^T X)^{-1}C^TD_\perp]$$

where $F = -A^T + C^TDB^T$ and $X=X^T \geq 0$ solves

$$F^TX + XF - XC^TD_\perp D_\perp^T CX + BB^T = 0$$

Corollary (unstable CIF) An $R\bar{H}_\infty$ CIF Θ (all poles in rhp, $\Theta^* \epsilon RH_\infty$) may be obtained by replacing X in the stable CIF with the negative semidefinite solution to the same Riccati equation.

Proof: Write $G(s) = D+C(sI-A)^{-1}B$

$$= [D_\perp D]\left(\begin{bmatrix}0\\ I\end{bmatrix} + \begin{bmatrix}D_\perp^T\\ D^T\end{bmatrix} C(sI-A)^{-1}B\right) = [D_\perp D]\begin{bmatrix}G_1(s)\\ G_2(s)\end{bmatrix}$$

Let $G_3 = [D_\perp D]\begin{bmatrix}I\\ -[G_1G_2^{-1}]^*\end{bmatrix}$. Then $G_3^*G=0$.

The RH_∞ CIF is obtain as the inner factor Θ of G_3 from the IOF Theorem. The RH_∞ CIF is obtained in the same way with s replaced by -s. Note that if $(D_\perp^TC,A)$ is not observable, the Riccati equation may not have a positive semidefinite solution. In this case, simply proceed using a reduced, minimal realization. This will result in a similar formula for Θ but with new state coordinates. □

Combining these results yields a factorization $G(s)=\Theta(s)\begin{bmatrix}M(s)\\ 0\end{bmatrix}$ where $\Theta(s) = [\Theta_1\Theta_2]$ and $\Theta^*\Theta=I$. A similar result may be obtained for cases where n<m simply by taking transposes throughout.

These two factorizations play a central role in both the L_2/H_2 and L_∞/H_∞ solutions because both $\|\bullet\|_2$ and $\|\bullet\|_\infty$ are invariant under unitary (i.e., inner or allpass) transformations. In particular, suppose P_{12} and P_{21} factor as

$$P_{12} = \Theta\begin{bmatrix}M_{12}\\ 0\end{bmatrix} \text{ and } P_{21} = [M_{21}\ 0]\Psi$$

with $\Theta^*\Theta=I$, $\Psi\Psi^*=I$ and the M's and their inverses in RH_∞. Then for both α=2 and ∞

$$\|P_{11}+P_{12}QP_{21}\|_\alpha = \|\Theta^*(P_{11}+P_{12}QP_{21})\Psi^*\|_\alpha$$

$$= \left\|\begin{bmatrix}\Theta_1^*P_{11}\Psi_1^* + M_{12}QM_{21} & \Theta_1^*P_{11}\Psi_2^*\\ \Theta_2^*P_{11}\Psi_1^* & \Theta_2^*P_{11}\Psi_2^*\end{bmatrix}\right\|_\alpha \quad (5.5)$$

Since L_2 is a Hilbert space the optimal Q for α=2 is obtained immediately by orthogonal projection from L_2 onto H_2 as

$$Q_{opt} = -M_{12}^{-1}\{\Theta_1^*P_{11}\Psi_1^*\}_+ M_{21}^{-1} \quad (5.6)$$

where $\{\ \}_+$ indicates projection onto H_2 (e.g. by partial fractions expansion).

This is the classical Wiener-Hopf solution. The simplicity of the L_2/H_2 optimal solution is very appealing. The L_∞/H_∞ problem cannot be solved quite so simply because L_∞ is not a Hilbert space and therefore there is no notion of orthogonal projection from L_∞ onto H_∞. Nevertheless, (5.5) can be reduced to a problem of approximating L_∞ function by functions in H_∞. An algorithm for performing this reduction is developed in Section VII and an algorithm for obtaining an optimal RH_∞^{mxn} approximation to an RL_∞^{mxn} function will be developed in Section VIII. The next section (VI) uses the factorizations of this section and the H_2-optimal Q of (5.6) to provide a simple solution to the standard "LQG" control problem.

Note that the factorization theorems require the D term to be maximal rank. This implies that the synthesis methods in this paper require that P_{12} and P_{21}, though nonsquare, must have maximal rank D terms. This assumption is quite reasonable from an engineering point of view, but does exclude certain singular problems (e.g., no penalty on the control signal or no sensor noise) which may still be of some interest. Most of these problems can be treated by using some simple modifications and will not be studied in detail here.

VI. L_2/H_2 Optimal Controllers and the Standard LQG Problem

In this section, we will briefly digress to look at the special case of the standard LQG problem using the more general L_2/H_2 theory from the last section. It is hoped that this will provide readers well-versed in the LQG problem with a familiar reference point.

Consider the standard problem

$$x = Ax + Bu_c + Gd \quad , \quad y = Cx + Nn$$

$$e = \begin{bmatrix} r \\ v \end{bmatrix} = \begin{bmatrix} Hx \\ Ru_c \end{bmatrix}, \; u_1 = \begin{bmatrix} d \\ n \end{bmatrix} \tag{6.1}$$

where x is the state, u_c the control, d and n white noise with identity covariance, and y the measured output. The control objective is to design a linear controller L(s) (i.e., $u_c(s)=L(s)y(s)$ that minimizes $E(|e|^2)$, the steady state "error" covariance. All variables are vector quantities of compatible, but otherwise arbitrary finite dimension.

This fits naturally into the synthesis framework of the last section and can be solved using inner-outer factorization and the L_2/H_2 optimal Q of (5.6). The interconnection structure is

$$\begin{bmatrix} e \\ y \end{bmatrix} = \begin{bmatrix} r \\ v \\ y \end{bmatrix} = \begin{bmatrix} P_{11} & P_{12} \\ P_{21} & P_{22} \end{bmatrix} \begin{bmatrix} u_1 \\ u_c \end{bmatrix}$$

$$= \begin{bmatrix} H\phi G & 0 & H\phi B \\ 0 & 0 & R \\ C\phi G & N & C\phi B \end{bmatrix} \begin{bmatrix} d \\ n \\ u_c \end{bmatrix}$$

$$u_c(s) = L(s)y(s) \qquad \phi = (sI-A)^{-1}$$

The parameterization $Q=L(I-P_{22}L)^{-1}$ leads to the optimization problem

$$\min_{Q \in H_2} \|P_{11} + P_{12}QP_{21}\|_2$$

and the solution is given by (5.6). It simply remains to perform the algebra to obtain a simpler description of Q_{opt} and L_{opt}. The steps are as follows:

1)

Factor $P_{12} = \begin{bmatrix} H\phi B \\ R \end{bmatrix} = \begin{bmatrix} H\phi BM_{12}^{-1} \\ RM_{12}^{-1} \end{bmatrix} M_{12} = \Theta_1 M_{12}$

From the IOF theorem (5.3) $M_{12}(s) = (R^TR)^{1/2}\,[I+K\phi B]$

where $K = (R^TR)^{-1}B^TX$ and $X=X^T \geq 0$ solves

$$A^TX + XA - XB(R^TR)^{-1}B^TX + H^TH = 0$$

This is the standard (state-feedback) controller Riccati equation.

2) $P_{21} = [C\phi G \;\; N] = M_{21}[M_{21}^{-1}C\phi G \quad M_{21}^{-1}M] = M_{21}\Psi_1$

where $M_{21} = [I+C\phi F](NN^T)^{1/2}$, $F = \Sigma C^T(NN^T)^{-1}$,

and $\Sigma = \Sigma^T \geq 0$ solves the standard filter Riccati equation.

$$A\Sigma+\Sigma A^T - \Sigma C^T(NN^T)^{-1}C\Sigma + GG^T = 0.$$

3) From (5.6),

$$Q_{opt} = -M_{12}^{-1}\{\Theta_1^* P_{11}\Psi_1^*\}_+ M_{21}^{-1}$$

$$= -M_{12}^{-1}\{[H\phi BM_{12}^{-1}]^* H\phi G\,[M_{21}^{-1}C\phi G]^*\}_+ M_{21}^{-1}$$

A little algebra yields:

$$Q_{opt} = -[I+K(sI-A)^{-1}B]^{-1}K(sI-A)^{-1}F[I+C(sI-A)^{-1}F]^{-1}$$

4) $L_{opt} = Q(I+C\phi BQ)^{-1}$ which after a little more algebra reduces to $L_{opt} = -K(sI-A+BK+FC)^{-1}F$.

This is the well-known formula for the optimal LQG controller for the special case considered here.

VII. L_∞/H_∞: Reduction to min‖G-Q‖

In this section we will develop an algorithm for reducing the general L_∞/H_∞ synthesis problem in (4.2) (for $\alpha=\infty$) to the problem of approximating an L_∞ function with one in H_∞. This is similar to the L_2/H_2 case and involves similar factorizations, but L_∞ lacks the Hilbert space structure so the algorithm will be more complicated. The approach taken here is closely related to that of Davis, et al [11], who characterize the contraction dilations of a contraction. The interested reader should compare the algorithm (7.4) in this section with (5.5) (which isn't formed explicitly) and with the central problem treated in [11].

Two simple facts are needed before proceeding. Suppose

$$X \in L_\infty^{mxn}, \; Y \in L_\infty^{kxn}, \; W \in L_\infty^{mxp}, \text{ and } \|X\| < \gamma.$$

For this section $\|\bullet\|$ will denote $\|\bullet\|_\infty$. Then the following two facts are easily verified.

$$1) \quad \left\| \begin{bmatrix} Y \\ X \end{bmatrix} \right\| \leq \gamma \quad \text{iff} \quad \|Y(\gamma^2 I - X^*X)^{-1/2}\| \leq 1 \tag{7.1}$$

$$2) \quad \|[W \;\; X]\| \leq \gamma \quad \text{iff} \quad \|(\gamma^2 I - XX^*)^{-1/2}W\| \leq 1 \tag{7.2}$$

The main result of this section will use these two facts and the factorizations from Section V to reduce the general L_∞/H_∞ problem to a simple approximation problem.

This reduction will be expressed as a series of equivalent statements starting with the most general L_∞/H_∞ problem in (4.2) and ending with a ‖G-Q‖ problem, which is solved in Section VIII. Recall that the most general L_∞/H_∞ synthesis problem involves solving

$$\min_{Q_o \in H_\infty^{mxn}} \|G_o - H_oQ_oK_o\|_\infty \tag{7.3}$$

$$G_o \in RH_\infty^{kxp}, \; H_o \in RH_\infty^{kxm}, \; K_o \in RH_\infty^{nxp}.$$

or nontriviality, suppose $k \geq m$ and $p \geq n$ and further suppose that both H_o and K_o have maximal rank D terms. Previous results in this area have required k=m and p=n. With these assumptions and $\gamma \in R$, $\gamma > o$ the following holds.

$$\|G_o - H_o Q_o K_o\| \leq \gamma \tag{7.4a}$$

$$\text{iff } \|G_o - H_o Q_o M_1 \Psi_1\| \leq \gamma \tag{7.4b}$$

$$\text{where } K_o = M_1 \Psi_1 = [M_1 \ 0] \begin{bmatrix} \Psi_1 \\ \Psi_2 \end{bmatrix}$$

$$\text{iff } \|G_o \Psi^* - H_o [Q_1 \ 0]\| \leq \gamma \text{ where } Q_1 = Q_o M_1 \tag{7.4c}$$

$$\text{iff } \|[G_o \Psi_1^* - H_o Q_1 \quad G_o \Psi_2^*]\| \leq \gamma \tag{7.4d}$$

$$\text{iff } \|M_2 (G_o \Psi_1^* - H_o Q_1)\| \leq 1 \tag{7.4e}$$

$$\text{by (7.2) with } M_2 = (\gamma^2 I - G_o \Psi_2^* \Psi_2 G_o^*)^{-1/2}$$

$$\text{iff } \|G_1 - H_1 Q_1\| \leq 1 \tag{7.4f}$$

$$\text{where } G_1 = M_2 G_o \Psi_1^* \text{ and } H_1 = M_2 H_o$$

$$\text{iff } \|G_1 - \Theta_1 M_3 Q_1\| \leq 1 \tag{7.4g}$$

$$\text{where } H_1 = \Theta_1 M_3 = [\Theta_1 \Theta_2] \begin{bmatrix} M_3 \\ 0 \end{bmatrix}$$

$$\text{iff } \|\Theta^* G_1 - \begin{bmatrix} Q_2 \\ 0 \end{bmatrix}\| \leq 1 \text{ where } Q_2 = M_3 Q_1 \tag{7.4h}$$

$$\text{iff } \left\| \begin{matrix} \Theta_1^* G_1 - Q_2 \\ \Theta_2^* G_2 \end{matrix} \right\| \leq 1 \tag{7.4i}$$

$$\text{iff } \|(\Theta_1^* G_1 - Q_2) M_4\| \leq 1 \tag{7.4j}$$

$$\text{by (7.1) with } M_4 = (I - G_1^* \Theta_2 \Theta_2^* G_1)^{-1/2}$$

$$\text{iff } \|G_2 - Q_3\| \leq 1 \tag{7.4k}$$

$$\text{where } G_2 = \Theta_1^* G_1 M_4 \text{ and } Q_3 = Q_2 M_4$$

If there exists a Q_3 such that $\|G_2 - Q_3\| \leq 1$, then letting $Q_o = M_3^{-1} Q_3 M_4^{-1} M_1^{-1}$ yields a Q_o such that $\|G_o - H_o Q_o K_o\| \leq \gamma$. The computation of the optimal Q_3 will be treated in the next section.

The equations in (7.4) can be turned into an algorithm for solving (7.3) by guessing a γ and computing (7.4b)-(7.4k) successively. If γ is too small, either (7.4e) or (7.4j) will fail or $\min\|G_2 - Q_3\| > 1$. If γ is too large then $\min\|G_2 - Q_3\| < 1$. Note that, just as in the case of analysis using $\|\bullet\|_\infty$, the synthesis problem involves a one-parameter search (over γ) to find the optimal norm. A solution arbitrarily close to the optimal can be found in a finite number of iterations of (7.4). By using the factorizations from Section V each iteration involves fairly routine computations involving real matrix operations on the state-space representations.

Although not required for this paper, it is possible to relax the conditions on G_o, H_o and K_o to RL_∞ without altering (7.4). An extra factorization must be performed in steps c and g to insure the M_1 and M_3 are in H_∞. Note also that the steps in (7.4) apply equally well when considering $Q_o \in H_{k,\infty}^{mxn}$, in which case $Q_3 \in H_{k,\infty}^{mxn}$ ($H_{k,\infty}$ allows k rhp poles).

VIII. Solution of $\min\|G-Q\|_\infty$

This section will outline a method to solve

$$\min_{Q \in H_\infty^{mxn}} \|G-Q\|_\infty, \quad G \in R\bar{H}_\infty^{mxn}. \tag{8.1}$$

In the last section, the more general problem (7.3) was reduced to (8.1) with $G \in RL_\infty^{mxn}$. By partial fractions expansion, the H_∞^{mxn} part of G may be absorbed into Q, leaving (8.1). The solution to (8.1) will exploit Silverman and Bettayeb's [12] treatment of Hankel operators using Moore's balanced realization to provide a computational scheme based on the theory of Ball and Helton ([4], [5]).

To use these results, we must transform the right half-plane onto the unit disc, by say, taking $s=(\lambda+z)/(\lambda-z)$ $s=(\lambda+z)/(\lambda>o$ and $\lambda \notin$ spectrum (A). Suppose, then, that G(z) is given as

$$G(z) = C(zI-A)^{-1}B = \sum_{k=1}^{\infty} G_k z^{-k}, \quad G_k = CA^{k-1}B \tag{8.2}$$

with the associated infinite Hankel matrix

$$H = \begin{bmatrix} G_1 & G_2 & G_3 & \cdots \\ G_2 & G_3 & \cdots & \\ G_3 & & & \\ & & & \end{bmatrix} \begin{bmatrix} C \\ CA \\ CA^2 \\ \end{bmatrix} [B \ AB \ A^2B \ \cdots] = OR. \tag{8.3}$$

The controllability and observability grammians $W=RR^T$ and $M=O^TO$ can be computed as the unique solutions of the Lyapunov equations

$$\begin{aligned} AWA^T - W &= -BB^T \\ A^TMA - M &= -C^TC. \end{aligned} \tag{8.4}$$

Assume that (A,B,C) is a balanced realization so that $M=W=\Sigma=\mathrm{diag}(\sigma_1,\sigma_2,\ldots,\sigma_r)$ with $\sigma_1 \geq \sigma_2 \geq \ldots \geq \sigma_r$. Then the following hold

1) $H^T H x = \lambda x \rightarrow \Sigma^2(Gx) = \lambda(Gx)$

2) $H^T H(R^T e_i) = \sigma_i^2 (R^T e_i)$ (8.5)

3) $HH^T(Oe_i) = \sigma_i^2(Oe_i)$

This yields the singular value decomposition of H as

$$H = U_H \Sigma V_H^T = (O\Sigma^{-1/2})\Sigma(\Sigma^{-1/2} G) \tag{8.6}$$

A corresponding SVD of the Hankel operator in terms of its symbol G is

$$\{G(z)U(z)\}_- = V(z)\Sigma \tag{8.7}$$

where $V=C(zI-A)^{-1}\Sigma^{-1/2}$ and $U=B^T(I-zA^T)^{-1}\Sigma^{-1/2}$. It is well known that the minimum in (8.1) is equal to σ_1. Suppose that G has been normalized by a constant so that $\sigma_1=1$.

Let <F,G> for $F \in L_\infty^{mxj}$, $G \in L_\infty^{mxk}$ denote the unique matrix in C^{nxk} that solves $\langle Fx,Gy\rangle = x^* \langle F,G\rangle y$ for every $x \in C^j$ and $y \in C^k$. Then <V,V>=I and <U,U>=I, and it is easily verified that

$$\langle U,zU\rangle = \Sigma^{-1/2} A \Sigma^{1/2} \text{ and } \langle V,zV\rangle = \Sigma^{1/2} A \Sigma^{-1/2} \tag{8.8}$$

The Ball-Helton theory requires the introduction of several spaces. Let $U=UH_2^r$ and $V=VH_2^r$ and let the ambient Krein space be

$$X = \begin{bmatrix} V \\ 0 \end{bmatrix} + \begin{bmatrix} H_2^m \\ H_2^n \end{bmatrix} \tag{8.9}$$

with the usual indefinite inner product denoted by [,]. Define the subspace $M \subset X$ as

$$M \overset{\Delta}{=} \begin{bmatrix} G \\ I \end{bmatrix} H_2^n + \begin{bmatrix} I \\ 0 \end{bmatrix} H_2^m$$

$$= \begin{bmatrix} V\Sigma \\ U \end{bmatrix} C^r + \begin{bmatrix} 0 \\ I \end{bmatrix} U_\perp + \begin{bmatrix} I \\ 0 \end{bmatrix} H_2^n . \qquad (8.10)$$

As in ([4], [5]), we want to find a maximal negative, shift-invariant subspace of M and the operator whose graph is this subspace. This will solve (8.1). The key step is to obtain an explicit representation for the wandering subspace $L = M \ominus (SM)'$, where S denotes shift (multiplication by z). As a first step, note that

$$M' = \begin{bmatrix} V \\ U\Sigma \end{bmatrix} C^r \qquad (8.11)$$

$$SM = \begin{bmatrix} zV\Sigma \\ zU \end{bmatrix} C^r + \begin{bmatrix} 0 \\ zI \end{bmatrix} U_\perp + \begin{bmatrix} zI \\ 0 \end{bmatrix} H_2^m$$

$$(SM)' = \begin{bmatrix} zY & 0 & C_\perp \\ zU\Sigma & I & 0 \end{bmatrix} \begin{bmatrix} C^r \\ C^n \\ C^p \end{bmatrix} = \Theta(z)\, C^{r+n+p}$$

where $C_\perp$ spans the orthogonal complement of the range of C. Usually, C is onto and p=0.

Let [F,G] for $F \in L_\infty^{(m+n)xj}$ and $G \in L_\infty^{(m+n)xk}$ denote the matrix in C^{jxk} that solves $[Fx,Gy] = x^* [F,G]y$ for every $x \in C^j$, $y \in C^k$. Then a basis for $L = M \cap (SM)'$ can be found from (8.11) by letting $P \epsilon C^{(r+n+p)xq}$ span the kernel of

$$\Gamma = \left[\left[\begin{smallmatrix} V \\ U\Sigma \end{smallmatrix} \right], \Theta(z) \right]$$

$$= [\ \Sigma^{1/2} A \Sigma^{-1/2} - \Sigma^{1/2} A \Sigma^{3/2} \quad -\Sigma^{1/2} B \quad 0] \qquad (8.12)$$

Then $L = \Theta(z) P C^q$. Equation (8.12) is obtained by taking the Krein inner product of the representations for M' and (SM)' from (8.11) using (8.8).

The next step is to extract a maximal negative subspace of L. To this end let

$$M_1 = [\Theta,\Theta] = \begin{bmatrix} I-\Sigma^2 & 0 & 0 \\ 0 & -I & 0 \\ 0 & 0 & C_\perp^T C_\perp \end{bmatrix}, \qquad M_2 = P^T M_1 P,$$

and find Y such that $M_2 = Y^T \Lambda Y$ with

$$\Lambda = \begin{bmatrix} 0 & 0 & 0 \\ 0 & -I & 0 \\ 0 & 0 & I \end{bmatrix} \quad \text{a signature matrix.}$$

Partition $Y = [Y_0\, Y_-\, Y_+]$ to match Λ and let $L_-(z) = \Theta(z) P [Y_0\, Y_-]$. This gives a Krein-orthogonal basis for a maximal negative subspace $L_- C^{q-}$. There is, in general, an infinite family of maximal negative subspaces but this is a natural choice as it includes no part of the positive subspace from the Krein-orthogonal decomposition of L.

By shift variance, $L_- H_2^{q-}$ is a maximal negative shift-invariant subspace of M. Partition

$$L_- = \begin{matrix} L_1(z) \\ L_2(z) \end{matrix} \qquad L_1 \in RL_\infty^{mxq-} \text{ and } L_2 \in RL_\infty^{nxq-}.$$ By the

unique correspondence between maximal negative sift-invariant subspaces and graphs of contractions, there is a unique solution $Q_{opt} \in RH_\infty^{mxn}$ to

$$(G - Q_{opt}) L_2 = L_1 . \qquad (8.13)$$

The simplest way to solve (8.13) is to transform the unit disk back into the rhp. Generically, this will result in full rank D terms for the transfer functions in (8.13). Then simple formulas will yield a nonminimal realization for Q_{opt} which can be reduced to obtain the optimal controller. Note that all the operations performed to obtain the optimal Q can be done using standard computations on real matrices.

The algorithm for finding the optimal $Q \in H_\infty^{mxn}$ is easily generalized to handle $Q \in H_{k,\infty}^{mxn}$ by simply dividing through by σ_{k+1} initially instead of σ_1. The rest of the solution goes through unchanged. This is a further consequence of the Ball-Helton theory [5]. The $Q \in H_{k,\infty}^{mxn}$ problem arises in the optimal Hankel-norm model reduction problem, and the computational scheme outlined above may prove useful there.

IX. Synthesis for Plants With Structured Uncertainty

We have seen that L_∞/H_∞ optimal control theory can be generalized to handle as rich a class of problems as the L_2/H_2 theory. The advantage of the L_∞/H_∞ framework is it is potentially more relevant to practical engineering problems since it handles both uncertain inputs and uncertain plants. The price is increased conceptual and computational complexity.

While the results reported in this paper are encouraging, they are just one more step towards a truly practical, systematic synthesis method for linear systems. The next important step would be to synthesize optimal controllers for performance/robustness expressed in terms of μ by solving

$$\min_{Q \in RH_\infty} \| P_{11} + P_{12} Q P_{21} \|_\mu \qquad (9.1)$$

An appealing approach to this problem is to "solve"

$$\min_{Q,D} \| D(P_{11} + P_{12} Q P_{21}) D^{-1} \|_\infty \qquad (9.2)$$

by iteratively minimizing over Q and D. Here $D \epsilon RH_\infty$ is taken to be of the form appropriate for the uncertainty structure of the problem ([2], [3]). Each minimization over D or Q with the other fixed is a convex problem and a global solution can be found. Unfortunately, (9.2) is not convex jointly in D and Q. It is a reasonable conjecture, however, that the global solution to (9.2) is approached with such a scheme. Lacking a proof of this conjecture, this scheme remains ad hoc.

Acknowledgements

Many people contributed to the results outlined in this paper. My colleagues at Honeywell Systems and Research Center have greatly influenced this work. I would particularly like to thank Drs. Wall, Stein, Lehtomaki and Prof. Safonov. The formalism presented here was also inspired by the work of Professors Zames, Francis and Youla. I would also like to thank Prof. Sarason, my advisor, and Prof. Helton for guiding my study of the H_∞ theory, and Prof. Kahan for many useful comments.

References

[1] J.C. Doyle and G. Stein, IEEE Trans. 1981, AC-26, pp. 4-16.
[2] J.C. Doyle, IEE Proc. Vol. 12a, Pt.D, No. 6, November 1982, pp. 242-250.
[3] J.C. Doyle, J.E. Wall and G. Stein, 1982, CDC Proceedings, Orlando, Florida, pp. 629-636.
[4] B.A. Francis and G. Zames, this proceedings.
[5] J.A. Ball and J.W. Helton, J. Op. Theory 9, 1983, pp. 107-142.
[6] D.C. Youla, H. Jabr, and J. Bongiorno, Jr., IEEE Trans., 1976, AC-21, pp. 319-338.
[7] D.C. Youla, IRE Trans. Info. Theory IT-7, 1961, pp. 172-189.
[8] B.D.O. Anderson, IEEE Trans., AC-16, 1971, pp. 621-634.
[9] Jan C. Willems, IEEE Trans., AC-16, 1971, pp. 621-634.
[10] V. Kucera, IEEE Trans. AC-17, 1972, pp. 344-347.
[11] C. Davis, W.M. Kahan, & H.F. Weinberger, SIAM J. Numer. Anal., Vol. 19, No. 3, June 1982.
[12] L. Silverman and M. Bettayeb, 1980 JACC.

L^∞-compensation with mixed sensitivity as a broadband matching problem

M. VERMA and E. JONCKHEERE

Department of Electrical Engineering Systems, University of Southern California, Los Angeles, CA 90089-0781, USA

Received 19 February 1984

A simplified approach to the design of a mixed-sensitivity, L^∞-optimal compensator is developed. Mixed-sensitivity L^∞-compensation is reduced, in a rather straightforward way, to Helton's broadband matching problem, which has itself received an elegant operator-theoretic solution. The simplicity of the approach developed herein allows a clarification of some issues which have emerged in less transparent mathematical settings.

Keywords: Feedback, Robustness, Sensitivity, Frequency-domain design, Broadband matching.

1. Introduction

Among the attributes of the compensator C in the single-channel servomechanism of Figure 1 is its ability to maintain the error signal e within some bounds without sacrificing too much control effort u despite the presence of an arbitrary, but bandlimited, reference signal r.

Requiring the compensator C to operate properly under the 'worst frequency-domain condition' [13] leads to the design criterion

$$\inf_{C} \sup_{\|r\|_2 \leq 1} \left(\|W_1 e\|_2^2 + \|W_2 u\|_2^2\right). \tag{1}$$

In the above, $\|\cdot\|_2$ denotes the L^2-norm, i.e.,

$$\|f\|_2^2 = (1/2\pi)\int_{-\infty}^{+\infty} \|f(j\omega)\|^2 d\omega.$$

The frequency-domain worst case is achieved by taking the supremum over all reference signals r with total energy $\|r\|_2^2$ not exceeding 1. The infimum is taken over all stabilizing compensators C. W_1 and W_2 are H^∞ frequency-dependent

This research was supported by NSF Grant ECS-8212479.

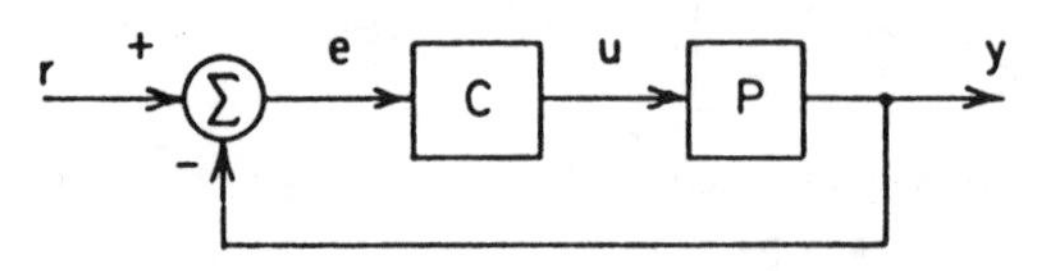

Fig. 1.

weights which are used to make the fine tuning of the design.

The design objective (1) can be further manipulated by observing that

$$e = Sr \quad \text{and} \quad u = P^{-1}(1 - S)r;$$

S is the inverse of the return difference:

$$S := (1 + PC)^{-1}, \tag{2a}$$

that is, the *sensitivity* function; also, remember that the 'size' of

$$1 - S = \left(1 + (PC)^{-1}\right)^{-1} \tag{2b}$$

is inversely proportional to the stability margin. Then it is easily seen that the design objective (1) can be rewritten in the L^∞-optimal, mixed-sensitivity format:

$$\inf_{C} \sup_{\omega \in \mathbf{R}} \left(|W_1(j\omega)S(j\omega)|^2 + |W_2(j\omega)P^{-1}(j\omega)(1 - S(j\omega))|^2\right). \tag{3}$$

The L^∞-optimal compensation problem was initiated by Zames [3] and solved by Zames and Francis [8] in the SISO, single-sensitivity (S) case using the Nevalinna–Pick theory. The multivariable S-sensitivity optimization problem was solved by Francis, Helton, and Zames [7] using the Ball–Helton theory [9], by Safonov and Verma [6] using Krein's Hankel theory [1], and by Chang and Pearson [10] using interpolation theory. A fairly general SISO, mixed-sensitivity, L^∞-optimal compensator design procedure was developed by Kwakernaak [14] using a polynomial approach. The most general solution to date is Doyle's [12]

Reprinted with permission from *Systems and Control Letters,* vol. 4, pp. 125–129, May 1984.

which uses the geometric theory of Ball and Helton [9]. Another L^∞-theory close to our point of view is due to Helton [13]. Here, we develop what is believed to be a simplified and straightforward approach to the SISO, mixed-sensitivity, L^∞-problem. The specific feature of this approach is that it reduces mixed-sensitivity, L^∞-compensation to broadband matching [4].

2. Reduction of mixed-sensitivity L^∞-compensation to broadband matching

First, we parameterize the set of stabilizing compensators. Following [2], and assuming P has no poles/zeros on the $j\omega$-axis, let

$$P = ND^{-1}, \quad N, D \in H^\infty, \tag{4}$$

be the rational coprime factorization of the plant P. Define

$$X, Y \in H^\infty \tag{5a}$$

as the solution of the Bezout identity

$$NX + DY = 1. \tag{5b}$$

Then the compensator C is known to be stabilizing iff

$$C = (X + DZ)/(Y - NZ) \tag{6a}$$

for some

$$Z \in H^\infty. \tag{6b}$$

Observing that

$$S = D(Y - NZ) \quad \text{and} \quad (1 - S) = N(X + DZ)$$

and defining

$$G_1 = W_1 DY, \qquad G_2 = W_2 DX, \tag{7a}$$

$$H_1 = W_1 DN, \qquad H_2 = -W_2 DD, \tag{7b}$$

we get

$$\begin{aligned} &W_1^* W_1 S^* S \\ &\quad + W_2^* W_2 (P^{-1})^* (P^{-1})(1 - S)^* (1 - S) \\ &\quad = G_1^* G_1 + G_2^* G_2 - (G_1^* H_1 + G_2^* H_2) Z \\ &\qquad - (G_1 H_1^* + G_2 H_2^*) Z^* \\ &\qquad + (H_1^* H_1 + H_2^* H_2) Z^* Z. \end{aligned} \tag{8}$$

Here, we invoke a result which Francis [5, Section 3] utilized in a solution to Youla's problem – namely, there exists a solution

$$H, G \in H^\infty \tag{9}$$

to the set of equations

$$H^* H = H_1^* H_1 + H_2^* H_2, \tag{10a}$$

$$GH^* = G_1 H_1^* + G_2 H_2^*. \tag{10b}$$

Then completing the squares in the right-hand side of (8) yields

$$\begin{aligned} &W_1^* W_1 S^* S \\ &\quad + W_2^* W_2 (P^{-1})^* (P^{-1})(1 - S)^* (1 - S) \\ &\quad = (G - HZ)^* (G - HZ) \\ &\qquad + G_1^* G_1 + G_2^* G_2 - G^* G. \end{aligned} \tag{11}$$

Now we prove the following.

Lemma. $G_1^* G_1 + G_2^* G_2 - G^* G \geq 0$.

Proof.

$$\begin{aligned} &G_1^* G_1 + G_2^* G_2 - G^* G \\ &\quad = G_1^* G_1 + G_2^* G_2 \\ &\qquad - \frac{G_1^* H_1 + G_2^* H_2}{H} \frac{G_1 H_1^* + G_2 H_2^*}{H^*} \\ &\quad = \{(G_1^* G_1 + G_2^* G_2)(H_1^* H_1 + H_2^* H_2) \\ &\qquad - (G_1^* H_1 + G_2^* H_2)(G_1 H_1^* + G_2 H_2^*)\}/HH^* \\ &\quad = \left(\frac{G_1 H_2 - G_2 H_1}{H}\right)^* \left(\frac{G_1 H_2 - G_2 H_1}{H}\right). \qquad \square \end{aligned}$$

Hence we can define F as the solution of the factorization equation

$$G_1^* G_1 + G_2^* G_2 - G^* G = F^* F, \quad F \in H^\infty. \tag{12}$$

Now the performance criterion (11) becomes

$$\begin{aligned} &W_1^* W_1 S^* S \\ &\quad + W_2^* W_2 (P^{-1})(P^{-1})^* (1 - S)^* (1 - S) \\ &\quad = (G - HZ)(G - HZ)^* + FF^*. \end{aligned} \tag{13}$$

Define

$$Q(j\omega) = H(j\omega) H(-j\omega) \quad (\geqslant 0), \tag{14}$$

$$K = GH^{-1} \in L^\infty. \tag{15}$$

Then the performance criterion takes its final form

$$W_1^* W_1 S^* S + W_2^* W_2 (P^{-1})(P^{-1})^*(1-S)^*(1-S) = (K-Z)Q(K-Z)^* + FF^*. \tag{16}$$

Observe that K, F, and Q are completely specified from the problem data (P, W_1, W_2), while Z is a stable (H^∞) transfer function which parameterizes the set of stabilizing compensators.

In this approach, the final twist is as follows: The L^∞-norm of the performance criterion (16) must be minimized. Assume that the optimal performance is ε, i.e.,

$$|Q(j\omega)||K(j\omega) - Z_0(j\omega)|^2 + |F(j\omega)|^2 \le \varepsilon \quad \forall\omega, \tag{17}$$

for some optimal (stabilizing) compensator, that is, for some $Z_0 \in H^\infty$. Observe that an a priori lower bound on the optimal performance is

$$\varepsilon \ge \|F\|_\infty^2. \tag{18}$$

The inequality (17) can be manipulated to

$$|K(j\omega) - Z_0(j\omega)|^2 \le \frac{\varepsilon - |F(j\omega)|^2}{|Q(j\omega)|} \quad \forall\omega. \tag{19}$$

In other words, if ε is the optimal performance, there must exist an H^∞-function Z_0 (i.e., a stabilizing compensator) which is within some tolerance close to the L^∞-center K. *This is Helton's broadband matching problem* [4]. An explicit condition to determine whether or not an H^∞-function exists and a recipe to compute it if it exists are available in [4, Lemmas I, II]. Therefore, the mixed-sensitivity L^∞-optimization problem would consist of (a) an initial bet on ε (for example, $\varepsilon = \|F\|_\infty^2$); (b) check whether there exists a Z in H^∞ verifying (19); if not, increase ε and go back to (b); if yes, compute Z.

To be more precise, we write out the whole algorithm:

(i) The problem data is (P, W_1, W_2).

(ii) Compute Q, K, and F by making use of (4), (5), (7), (9), (10), (12), (14), and (15).

(iii) Apply the bilinear transformation

$$j\omega = (1 - e^{j\theta})/(1 + e^{j\theta})$$

to $K(j\omega)$, $F(j\omega)$, and $Q(j\omega)$, and get their unit circle equivalents: $k(e^{j\theta})$, $f(e^{j\theta})$, and $q(e^{j\theta})$.

(iv) Choose $\varepsilon \ge \|F\|_\infty^2$.

(v) Compute

$$r(e^{j\theta}) := \left(\varepsilon - |f(e^{j\theta})|^2\right)/|q(e^{j\theta})|.$$

(vi) Compute the largest (generalized) eigenvalue τ^2 of the operator-theoretic problem

$$\mathcal{H}_k \mathcal{H}_k^* w = \tau^2 \mathcal{T}_r w.$$

($\mathcal{H}_k$ is the *Hankel* operator associated with k, while $\mathcal{T}_r$ is the *Toeplitz* operator associated with r.)

If $\tau > 1$, there does not exist a solution to the broadbanding problem; the performance ε has been taken too small and cannot be reached; hence, increase ε and go back to (v). If $\tau \le 1$, there exists a solution to the broadbanding problem; the performance ε can be achieved. If $\tau < 1$, ε is not the optimal performance; hence, decrease ε and go to (v). If $\tau = 1$, the performance ε is the optimal one; hence, set $\varepsilon = \varepsilon_0$ (and $\tau = \tau_0 = 1$), and go to (vii).

(vii) Compute the generalized eigenvector w corresponding to $\tau = \tau_0 = 1$.

(viii) Spectral factor r as

$$r(e^{j\theta}) = \alpha(e^{j\theta})\alpha(e^{-j\theta}),$$

with $\alpha, \alpha^{-1} \in H^\infty$.

(ix) Define $x = \mathcal{T}_\alpha^* w$.

(x) The solution to the broadbanding problem is

$$z_0(e^{j\theta}) = \tau_0 \alpha(e^{j\theta}) \left(\frac{k(e^{j\theta})}{\tau_0 \alpha(e^{j\theta})} - e^{-j\theta} \frac{x(e^{-j\theta})}{x(e^{j\theta})} \right).$$

(xi) Trace back to the continuous-time domain:

$$Z_0(j\omega) = z_0\left(\frac{1 - j\omega}{1 + j\omega}\right).$$

(xii) Compute the optimal compensator via (6).

The problem with the above algorithm is that it requires an iterative search of the optimal performance ε_0. The best one can do is to choose an ε slightly greater than the optimal one and then resolve the broadbanding problem. This will, of course, lead to a slightly suboptimal compensator. Similar difficulties have arisen in the solution to L^∞-compensation [11, Section 4.3] using the

Ball–Helton theory in the pseudo-regular case [9, Theorem 2.2].

Nevertheless, as a consequence of Lemma I of the broadbanding theory [4], the optimal performance ε_0 can be determined *exactly* as follows:

Theorem. *The optimal performance ε_0 is the largest eigenvalue of*

$$\mathscr{H}_k(\mathscr{T}_{q^{-1}})^{-1}\mathscr{H}_k^* + \mathscr{T}_{ff^*}.$$

Proof. Combine (19) with Lemma I of [4]. □

Finally, it also follows from the broadbanding theory that the optimal mixed sensitivity is all pass:

Theorem. *For all ω,*

$$|W_1(j\omega)S_0(j\omega)|^2 + |W_2(j\omega)P^{-1}(j\omega)(1 - S_0(j\omega))|^2 = \varepsilon_0.$$

Proof. From (16) and the bilinear transformation

$$\begin{aligned} &|W_1(j\omega)S_0(j\omega)|^2 + |W_2(j\omega)P^{-1}(j\omega)(1 - S_0(j\omega))|^2 \\ &= |K(j\omega) - Z_0(j\omega)|^2|Q(j\omega)| + |F(j\omega)|^2 \\ &= |q(e^{j\theta})|\left(|k(e^{j\theta}) - z_0(e^{j\theta})|^2 + \frac{|f(e^{j\theta})|^2}{|q(e^{j\theta})|}\right). \end{aligned}$$

From the broadbanding theory [4, Lemma I], we have

$$|k(e^{j\theta}) - z_0(e^{j\theta})|^2 = \tau_0^2|r(e^{j\theta})|.$$

Hence

$$\begin{aligned} &|W_1(j\omega)S_0(j\omega)|^2 + |W_2(j\omega)P^{-1}(j\omega)(1 - S_0(j\omega))|^2 \\ &= |q(e^{j\theta})|\left(|r(e^{j\theta})| + \frac{|f(e^{j\theta})|^2}{|q(e^{j\theta})|}\right) \\ &= |q(e^{j\theta})|\left(\frac{\varepsilon_0 - |f(e^{j\theta})|^2}{|q(e^{j\theta})|} + \frac{|f(e^{j\theta})|^2}{|q(e^{j\theta})|}\right) \\ &= \varepsilon_0. \quad \square \end{aligned}$$

3. Conclusion

A simple and transparent approach to SISO, mixed-sensitivity, L^∞-optimal compensation has been developed. The broadbanding theory, which this paper is based upon, is much likely to be amenable to a multivariable generalization, probably along the lines of Krein's 'extension' idea [1]. Therefore, it is most likely that the ideas developed here can be extended to the multivariate situation. Finally, we are currently under process to try out and evaluate this particular compensation scheme on an aerospace example.

Acknowledgment

Some useful discussions on this subject matter with Profs. H. Kwakernaak and J.W. Helton are gratefully acknowledged.

References

[1] V.M. Adamjan, D.Z. Arov, and M.G. Krein, Infinite Hankel block matrices and related extension problems, *American Math. Society Translations* **111** (1978) 133–156.

[2] C.A. Desoer, R.W. Liu, J. Murray, and R. Saeks, Feedback system design: The fractional representation approach to analysis and synthesis, *IEEE Trans. Automat. Control* **25** (1980) 399–412.

[3] G. Zames, Feedback and optimal sensitivity: model reference transformations, multiplicative seminorms, and approximate inverses, *IEEE Trans. Automat. Control* **26** (1981) 301–320.

[4] J.W. Helton, Broadbanding: gain equalization directly from data, *IEEE Trans. Circuits and Systems* **28** (1981) 1125–1137.

[5] B.A. Francis, On the Wiener-Hopf approach to optimal feedback design, *Systems & Control Letters* **2** (1982) 197–201.

[6] M.G. Safonov and M.S. Verma, Multivariable L^∞ sensitivity optimization and Hankel approximation, in: *Proc. American Control Conference*, San Francisco, California (1983).

[7] B. Francis, J.W. Helton, and G. Zames, H^∞-optimal feedback controllers for linear multivariable systems, *International Symposium on the Mathematical Theory of Networks and Systems*, Beer-Sheva, Israel (1983).

[8] G. Zames and B.A. Francis, Feedback, minimax sensitivity, and optimal robustness, *IEEE Trans. Automat. Control* **28** (1983) 585–601.

[9] J.A. Ball and J.W. Helton, A Beurling-Lax theorem for the Lie group $U(m,n)$ which contains most classical interpolation theory, *J. Operator Theory* **9** (1983) 107–142.

[10] B.-C. Chang and J.B. Pearson, Optimal disturbance reduc-

tion in linear multivariable systems, in: *Proc. IEEE Conference on Decision and Control*, San Antonio, Texas (1983) pp. 91–96.

[11] B.A. Francis and G. Zames, Design of H^∞-optimal multivariable feedback systems, in: *Proc. IEEE Conf. on Decision and Control*, San Antonio, Texas (1983) pp. 103–108.

[12] J.C. Doyle, Synthesis of robust controllers and filters, in: *Proc. IEEE Conf. Decision and Control*, San Antonio, Texas (1983) pp. 109–114.

[13] J.W. Helton, An H^∞ approach to control, in: *Proc. IEEE Conf. on Decision and Control*, San Antonio, Texas (1983) pp. 607–611.

[14] H. Kwakernaak, Robustness optimization of linear feedback systems, in: *Proc. IEEE Conf. on Decision and Control*, San Antonio, Texas (1983) pp. 618–624.

Optimal Disturbance Reduction in Linear Multivariable Systems

BOR-CHIN CHANG, MEMBER, IEEE, AND J. BOYD PEARSON, JR., FELLOW, IEEE

Abstract —**This paper presents a computational solution to an important optimization problem arising in optimal sensitivity theory. The approach is to treat the multivariable problem exactly as the scalar problem in that stability constraints are handled via interpolation. The resulting computations are easily implemented using existing methods.**

I. INTRODUCTION

IN this paper, a computational method is presented for solving a general optimization problem arising in optimal sensitivity theory [1]–[4]. This theory furnishes an alternative approach for the design of multivariable systems to the linear-quadratic-Gaussian (LQG) approach that has achieved some popularity over the past several years.

In a typical control system design problem, we are given a dynamic system to control (the plant), a description of the types of inputs to be encountered, and specifications concerning acceptable values of system errors, i.e., the difference between desired and actual responses. We are then required to design a controller so that the plant–controller combination satisfies the specifications. If we know the inputs precisely, e.g., steps, ramps, exponentials in the deterministic case or random processes with known means and covariances in the stochastic case and we can translate our specifications into a quadratic performance index, then LQG theory will furnish a useful design tool.

On the other hand, if the system inputs are not known precisely but can only be assigned to a certain class of signals, then it is possible to develop a design tool based upon minimization of the maximum error that can occur subject to all inputs belonging to this particular class. This is the approach of Zames *et al.* [1]–[4] and leads to the minimization of the norm of a weighted transfer function matrix in the following manner.

For the transfer function matrix

$$\Phi(s) = U(s)K(s)V(s) + H(s) \tag{1-1}$$

where $U(s) \in (RH^\infty)^{p\times p}$ and $V(s) \in (RH^\infty)^{q\times q}$ are inner matrices[1] and $H(s) \in (RH^\infty)^{p\times q}$, find a $K(s) \in (RH^\infty)^{p\times q}$ such that $\|\Phi\|_\infty$ is minimized, where $\|\Phi\|_\infty$ is defined as

$$\|\Phi\|_\infty := \sup_\omega \bar{\sigma}[\Phi(j\omega)] \tag{1-2}$$

and $\bar{\sigma}(A)$ is the maximum singular value of the matrix A, i.e., the square root of the maximum eigenvalue of A^*A, $(RH^\infty)^{p\times q}$ is the set of rational $p \times q$ matrices with real coefficients and with entries in H^∞, which is the space of all functions that are analytic and bounded in the right half plane.

Manuscript received December 16, 1982; revised July 22, 1983. Paper recommended by B. R. Barmish, Past Chairman of the Optimal Systems Committee. This work was supported by the National Science Foundation under Grant ECS 81-02895.

B.-C. Chang is with the Department of Electrical Engineering, Bradley University, Peoria, IL 61625.

J. B. Pearson, Jr. is with the Department of Electrical Engineering, Rice University, Houston, TX 77251.

[1]An inner matrix in $(RH^\infty)^{p\times q}$ is a proper rational matrix which is analytic in the closed right half plane and is unitary on the imaginary axis. In control problems, these matrices arise from right half plane system zeros.

Zames and Francis [2] and Francis and Zames [3] have solved the above problem in the scalar case (i.e., $p = q = 1$). Francis, Helton, and Zames [4] solved the multivariable case of the problem with $V(s) = I$ and Φ being square by using the recently developed Ball–Helton theory [5].

The problem considered here is more general than those in [1]–[4]. In the control problem, the measured outputs are not necessarily the controlled outputs, and the number of exogenous inputs is not necessarily the same as the number of controlled outputs. Therefore, $V(s)$ is not necessarily the identity and Φ is not necessarily square. The control problem is discussed in detail in Section V.

Our approach for solving this problem is quite different from that in [4] since here we use Nevanlinna–Pick theory to solve the problem. The paper is organized as follows. In Section II, several transformations of the optimization problem are made. We transform the problem in (1-1) to an equivalent problem

$$\hat{\Phi}(s) = \phi(s)K(s) + \hat{H}(s) \tag{1-3}$$

where $\|\hat{\Phi}\|_\infty = \|\Phi\|_\infty$ and $\phi(s)$ is a scalar inner function. The stability constraint on $K(s)$ can be replaced by interpolation constraints on $\hat{\Phi}(s)$. The problem then becomes that of finding a $\hat{\Phi}(s) \in (RH^\infty)^{p\times q}$ subject to these interpolation constraints such that $\|\hat{\Phi}\|_\infty$ is minimized. We divide the interpolation problem into two parts: first, computation of the minimal $\|\hat{\Phi}\|_\infty$ and second, the construction of an optimal $\hat{\Phi}$. Sarason's theory [6] is used for simplifying the first part of the problem. In Section III the computation of the minimal $\|\hat{\Phi}\|_\infty$ is discussed, and we will see it can be reduced to a generalized eigenvalue problem or a singular-value problem. In Section IV we will construct a $\hat{\Phi}$ subject to the interpolation constraints such that $\|\hat{\Phi}\|_\infty$ equals the minimal $\|\hat{\Phi}\|_\infty$ just computed in Section III. A matrix Nevanlinna algorithm [7] is used for this purpose.

The $\Phi(s)$ [or $\hat{\Phi}(s)$] in our problem is not necessarily square. The algorithm in Section III is only applicable to square matrices. This problem is solved by adding some zero row vectors or column vectors to $\hat{H}(s)$ in (1-3) so that (1-3) becomes

$$\hat{\Phi}_a(s) = \phi(s)K_a(s) + \hat{H}_a(s) \tag{1-4}$$

where $\hat{\Phi}_a(s)$, $K_a(s)$, and $\hat{H}_a(s)$ are square. We will see that the minimal $\|\hat{\Phi}_a\|_\infty$ and the minimal $\|\hat{\Phi}\|_\infty$ are the same.

In Section V, we discuss the application of the theory to a general control problem and state the necessary assumptions and restrictions. We will work our problem in both the s-plane (in Sections II and III) and z-plane (in Section IV). For notational convenience, we will drop the indeterminates s or z when it is clear which is appropriate.

Reprinted from *IEEE Trans. Automat. Contr.*, vol. AC-29, no. 10, pp. 880–887, Oct. 1984.

II. Equivalent Problems

Consider (1-1) and define

$$\phi = \phi_1 \phi_2 \tag{2-1}$$

where $\phi_1 = \det U$ and $\phi_2 = \det V$. Then ϕ is an inner function and $\phi_1 U^{-1}$ and $\phi_2 V^{-1}$ are inner matrices [15]. Rewrite (1-1) as

$$\phi U^{-1} \Phi V^{-1} = \phi K + \phi U^{-1} H V^{-1}. \tag{2-2}$$

Define

$$\hat{\Phi} = \phi U^{-1} \Phi V^{-1} \tag{2-3}$$

$$\hat{H} = \phi U^{-1} H V^{-1}. \tag{2-4}$$

Then (2-2) becomes

$$\hat{\Phi} = \phi K + \hat{H}. \tag{2-5}$$

It is easy to see that $\hat{H} \in (RH^\infty)^{p \times q}$ and if $K \in (RH^\infty)^{p \times q}$ then $\Phi, \hat{\Phi} \in (RH^\infty)^{p \times q}$. Since inner matrices are isometric, the following holds.

Proposition 2-1:

$$\|\hat{\Phi}\|_\infty = \|\Phi\|_\infty. \tag{2-6}$$

By this proposition, once a $K \in (RH^\infty)^{p \times q}$ minimizes $\|\hat{\Phi}\|_\infty$, it also minimizes $\|\Phi\|_\infty$. Therefore, the optimization problem in (1-1) is equivalent to finding a $K \in (RH^\infty)^{p \times q}$ in (2-5) such that $\|\hat{\Phi}\|_\infty$ is minimized.

If Φ (or $\hat{\Phi}$) is not square, we can add some zero row vectors or column vectors so that (2-5) becomes

$$\hat{\Phi}_a = \phi K_a + \hat{H}_a \tag{2.7}$$

with $\hat{\Phi}_a$, K_a, $\hat{H}_a$ square and $\hat{H}_a = [\hat{H} \;\; 0]$ or $\begin{bmatrix} \hat{H} \\ 0 \end{bmatrix}$. Let $r = \max\{p, q\}$, and define

$$\mu := \inf\left\{ \|\hat{\Phi}\|_\infty : K \in (RH^\infty)^{p \times q} \right\} \tag{2-8}$$

$$\mu_a := \inf\left\{ \|\hat{\Phi}_a\|_\infty : K_a \in (RH^\infty)^{r \times r} \right\}. \tag{2-9}$$

We can choose K_a of the form $[K \;\; 0]$ or $\begin{bmatrix} K \\ 0 \end{bmatrix}$, thus $\mu_a \leqslant \mu$. It is easy to show that

$$\bar{\sigma}\begin{bmatrix} A \\ B \end{bmatrix} \geqslant \bar{\sigma}(A) \quad \text{and} \quad \bar{\sigma}[A B] \geqslant \bar{\sigma}(A).$$

Therefore, $\|\hat{\Phi}_a\|_\infty \geqslant \|\hat{\Phi}\|_\infty$, so that $\mu_a \geqslant \mu$. Hence, $\mu_a = \mu$.

Next, in Proposition 2-2 we will see that the stability constraint on K_a is equivalent to some interpolation constraints on $\hat{\Phi}_a$.

Proposition 2-2: Suppose $\hat{H}_a \in (RH^\infty)^{r \times r}$ and all the zeros $\{s_i, i = 1, 2 \cdots, n\}$ of ϕ are distinct. Then $K_a \in (RH^\infty)^{r \times r}$ if and only if $\hat{\Phi}_a \in (RH^\infty)^{r \times r}$ and

$$\hat{\Phi}_a(s_i) = \hat{H}_a(s_i) \qquad i = 1, 2, \cdots, n. \tag{2-10}$$

Proof: If $K_a \in (RH^\infty)^{r \times r}$, then by (2-7), $\hat{\Phi}_a \in (RH^\infty)^{r \times r}$, and by (2-1), we have (2-10). Conversely, if $\hat{\Phi}_a \in (RH^\infty)^{r \times r}$, then by (2-7) $\phi K_a \in (RH^\infty)^{r \times r}$. The only possibility that $K_a \notin (RH^\infty)^{r \times r}$ is K_a has some some poles cancelled by the zeros of ϕ. In particular, suppose K_a has a pole at s_k, then $\phi K_{a|s=s_k} \neq 0$ which contradicts (2-10). Therefore,

$$K_a \in (RH^\infty)^{r \times r}.$$

Q.E.D.

The problem has now become a generalized interpolation problem.

$$\begin{aligned} &\text{Find a } \hat{\Phi}_a \in (RH^\infty)^{r \times r} \text{ such that} \\ &\|\hat{\Phi}_a\|_\infty \text{ is minimized} \\ &\text{subject to } \hat{\Phi}_a(s_i) = \hat{H}_a(s_i) \qquad i = 1, 2, \cdots, n. \end{aligned} \tag{2-11}$$

For the time being, we relax our problem to that of finding a $K_a \in (H^\infty)^{r \times r}$ (or equivalently, a $\hat{\Phi}_a \in (H^\infty)^{r \times r}$ subject to $\hat{\Phi}_a(s_i) = \hat{H}_a(s_i)$ $i = 1, \cdots, n$) such that $\|\hat{\Phi}_a\|_\infty$ is minimized and in Section IV we will see that an optimal K_a (or $\hat{\Phi}_a$) can be found in $(RH^\infty)^{r \times r}$. Define the coset

$$[\hat{H}_a] := \left\{ \hat{H}_a + \phi K_a : K_a \in (H^\infty)^{r \times r} \right\}. \tag{2-12}$$

Note that $[\hat{H}_a] \in (H^\infty)^{r \times r} / \phi (H^\infty)^{r \times r}$. Define

$$\mu_o := \|[\hat{H}_a]\| := \inf\left\{ \|\hat{\Phi}_a\|_\infty : \hat{\Phi}_a \in [\hat{H}_a] \right\}. \tag{2-13}$$

In the rest of the section we will use Sarason's theory to transform the problem of finding μ_o in (2-13) to a simpler equivalent problem. Here we are motivated by [12].

The space $(L^2)^{r \times r}$ is a Hilbert space with inner product

$$\langle G_1, G_2 \rangle := \frac{1}{2\pi} \int_{-\infty}^{\infty} \operatorname{tr}\left[G_1(j\omega)^* G_2(j\omega) \right] d\omega \tag{2-14}$$

where $G_1(j\omega)^*$ is the conjugate transpose of $G_1(j\omega)$. Let $\boldsymbol{K}$ be the orthogonal compliment of $\phi (H^2)^{r \times r}$ in $(H^2)^{r \times r}$ and let

$$P : (H^2)^{r \times r} \to \boldsymbol{K} \tag{2-15}$$

be the orthogonal projection. Each $\hat{\Phi}_a$ in $(H^\infty)^{r \times r}$ determines a linear operator $T_{\hat{\Phi}_a} : \boldsymbol{K} \to \boldsymbol{K}$ as follows:

$$T_{\hat{\Phi}_a} G = P(\hat{\Phi}_a G), \qquad G \in \boldsymbol{K}. \tag{2-16}$$

The kernel of the linear map $\hat{\Phi}_a | \to T_{\hat{\Phi}_a}$ is $\phi (H^\infty)^{r \times r}$, so the map $R : [\hat{\Phi}_a]| \to T_{\hat{\Phi}_a}$ is a linear bijection.

By Sarason [6], the mapping R

$$(H^\infty)^{r \times r} / \phi (H^\infty)^{r \times r} \to \left\{ T_{\hat{\Phi}_a} : \hat{\Phi}_a \in (H^\infty)^{r \times r} \right\} \tag{2-17}$$

is norm preserving where $\{T_{\hat{\Phi}_a} : \hat{\Phi}_a \in (H^\infty)^{r \times r}\}$ is equipped with the operator norm. That is,

$$\mu_o = \|[\hat{H}_a]\| = \|T_{\hat{H}_a}\|. \tag{2-18}$$

Therefore, we have

$$\mu_o = \|T_{\hat{H}_a}\| := \sup\left\{ \|T_{\hat{H}_a} G\| : G \in \boldsymbol{K}, \|G\| = 1 \right\}, \tag{2-19a}$$

i.e.,

$$\mu_o = \sup_{G \in \boldsymbol{K}} \frac{\|T_{\hat{H}_a} G\|}{\|G\|}, \qquad \|G\| \neq 0. \tag{2-19b}$$

Denote the adjoint of $T_{\hat{H}_a}$ by $T^*_{\hat{H}_a}$, then by the property of the adjoint operator

$$\|T^*_{\hat{H}_a}\| = \|T_{\hat{H}_a}\| \tag{2-20}$$

we have

$$\mu_o = \sup\left\{ \|T^*_{\hat{H}_a} G\| : G \in \boldsymbol{K}, \|G\| = 1 \right\}, \tag{2-21a}$$

i.e.,

$$\mu_o = \sup_{G \in K} \frac{\|T^*_{H_a} G\|}{\|G\|}, \qquad \|G\| \neq 0. \tag{2-21b}$$

III. Computation of Minimal $\|\Phi\|_\infty$

Let E_{ij} be the $r \times r$ matrix with a "1" in the ijth position and zeros elsewhere. If $\{s_k, k = 1, \cdots, n\}$ are distinct, then a basis for K consists of the $r \times r$ matrix-valued functions

$$F^k_{ij}(s) = \frac{1}{s + \bar{s}_k} \cdot E_{ij} \qquad k = 1, \cdots, n. \tag{3-1}$$

To see this, note that if $G \in (H^2)^{r \times r}$, then

$$\langle F^k_{ij}, G \rangle = \operatorname{tr}\left[E'_{ij} G(s_k) \right] = g_{ij}(s_k) \qquad k = 1, \cdots, n. \tag{3-2}$$

So, if $G \in \phi(H^2)^{r \times r}$, then $\langle F^k_{ij}, G \rangle = 0$ for all i, j, k. Conversely, if $\langle F^k_{ij}, G \rangle = 0$ for all i, j, k, then each component of G vanishes at $s_k, k = 1, \cdots, n$, so that $G \in \phi(H^2)^{r \times r}$.

Therefore, each G in K has a representation

$$G(s) = \sum_{k=1}^{n} \sum_{i=1}^{r} \sum_{j=1}^{r} \alpha_{ijk} F^k_{ij}(s) \tag{3-3}$$

with $\alpha_{ijk} \in C$. Define

$$G^k(s) := \frac{1}{s + \bar{s}_k} \cdot I_r \qquad k = 1, \cdots, n. \tag{3-4}$$

Then (3-3) can be written as

$$G(s) = \sum_{k=1}^{n} X^k G^k(s) \tag{3-5}$$

with $X^k \in C^{r \times r}$ for each k.

If G is represented as (3-5), then

$$\|G\|^2 := \langle G, G \rangle = \frac{1}{2\pi} \int_{-\infty}^{\infty} \operatorname{tr}\left[G(j\omega)^* G(j\omega) \right] d\omega$$

$$= \operatorname{tr}(X^* V X) \tag{3-6}$$

where

$$X^* = (X^{1*} X^{2*} \cdots X^{n*}) \tag{3-7}$$

and

$$V = [V_{ij}] \qquad i, j = 1, \cdots, n \tag{3-8}$$

with

$$V_{ij} = G^j(s_i).$$

For any $\Phi \in (H^\infty)^{r \times r}$ and any $G \in K$, $T_\Phi G$ can be represented by

$$P(\Phi G) = \sum_{i=1}^{n} Y^i G^i \tag{3-9}$$

with $Y^i \in C^{r \times r}$ for each i. If $\Phi(s_i) = W^i$, $i = 1, \cdots, n$, then by (3-5) and (3-9), we have

$$Y = V^{-1} W V X \tag{3-10}$$

where

$$Y^* = [Y^{1*} Y^{2*} \cdots Y^{n*}] \tag{3-11}$$

and

$$W = \text{block diag}\,[W^1 W^2 \cdots W^n]. \tag{3-12}$$

Proposition 3-1: If T_Φ is the operator on K defined by

$$T^*_\Phi G^i = W^{i*} G^i \qquad i = 1, \cdots, n \tag{3-13}$$

then $T_\Phi G = P(\Phi G)\ \forall G \in K$ if and only if

$$\Phi(s_i) = W^i \qquad i = 1, \cdots, n. \tag{3-14}$$

Proof: By hypothesis,

$$T^*_\Phi G^i = W^{i*} G^i. \tag{3-15}$$

If

$$T_\Phi G = P(\Phi G), \qquad G \in K. \tag{3-16}$$

Then by the property of adjoint operator

$$\left\langle T_\Phi \sum_{i=1}^{n} X^i G^i, \sum_{i=1}^{n} Z^i G^i \right\rangle = \left\langle \sum_{i=1}^{n} X^i G^i, T^*_\Phi \sum_{i=1}^{n} Z^i G^i \right\rangle, \tag{3-17}$$

i.e.,

$$\frac{1}{2\pi} \int_{-\infty}^{\infty} \operatorname{tr}\left\{ \left[P\left(\Phi \sum_{i=1}^{n} X^i G^i \right) \right]^* \cdot \sum_{i=1}^{n} Z^i G^i \right\} d\omega$$

$$= \frac{1}{2\pi} \int_{-\infty}^{\infty} \operatorname{tr}\left\{ \left(\sum_{i=1}^{n} X^i G^i \right)^* \cdot \sum_{i=1}^{n} W^{i*} Z^i G^i \right\} d\omega. \tag{3-18}$$

Suppose $\Phi(s_i) = \hat{W}^i$ $i = 1, \cdots, n$, and let

$$P\left(\Phi \sum_{i=1}^{n} X^i G^i \right) = \sum_{i=1}^{n} \hat{Y}^i G^i. \tag{3-19}$$

Then by (3-10), we have

$$\hat{Y} = V^{-1} \hat{W} V X \tag{3-20}$$

where

$$\hat{Y}^* = [\hat{Y}^{1*}, \hat{Y}^{2*}, \cdots, \hat{Y}^{n*}] \tag{3-21}$$

and

$$\hat{W} = \text{block diag}\,[\hat{W}^1, \hat{W}^2, \cdots, \hat{W}^n]. \tag{3-22}$$

Taking the conjugate transpose of (3-20)

$$\hat{Y}^* = X^* V^* \hat{W}^* V^{*-1} = X^* V \hat{W}^* V^{-1} \tag{3-23}$$

since V is Hermitian. By (3-18) and (3-19) we have

$$\operatorname{tr}[\hat{Y}^* V Z] = \operatorname{tr}[X^* V W^* Z] \tag{3-24}$$

and by (3-23), we have

$$\operatorname{tr}[X^* V \hat{W}^* Z] = \operatorname{tr}[X^* V W^* Z] \qquad \forall X^* \text{ and } Z. \tag{3-25}$$

Thus, $\hat{W} = W$, and therefore (3-14) holds. Tracing back the proof, we can prove that (3-11) implies $T_\Phi G = P(\Phi G)$. Q.E.D.

Recall (2-21b)

$$\mu_o = \max_{G \in K} \frac{\|T^*_{H_a} G\|}{\|G\|}, \qquad \|G\| \neq 0. \tag{3-26}$$

Now, suppose

$$\hat{H}_a(s_i) = W^i \qquad i=1,\cdots,n \tag{3-27}$$

and let

$$G = \sum_{i=1}^{n} Y^i G^i \tag{3-28}$$

then by Proposition 3-1, we have

$$T^*_{\hat{H}_a} G = T^*_{\hat{H}_a} \sum_{i=1}^{n} Y^i G^i = \sum_{i=1}^{n} W^{i*} Y^i G^i. \tag{3-29}$$

Therefore,

$$\mu_o^2 = \max \frac{\operatorname{tr}[Y^* W V W^* Y]}{\operatorname{tr}[Y^* V Y]}. \tag{3-30}$$

Then we have the following proposition.

Proposition 3-2: μ_o is the square root of the maximum λ such that

$$\det(WVW^* - \lambda V) = 0 \tag{3-31}$$

where W, V are given in (3-12) and (3-8), respectively, and

$$W^i = \hat{H}_a(s_i) \qquad i=1,\cdots,n. \tag{3-32}$$

Corollary 3-1: Define $V = UU^*$. Then μ_o is the maximum singular value of $U^{-1}WU$.

Now, the problem of finding μ_o is reduced to a generalized eigenvalue problem or a singular value problem. These reduced problems can be easily solved by using EISPACK routines [8] or LINPACK routines [9].

IV. Nevanlinna Algorithm

An optimal $\Phi \in (RH^\infty)^{p\times q}$ with $\|\Phi\|_\infty = \mu_o$ will be constructed. In this section we use the transformation

$$z = \frac{1-s}{1+s}. \tag{4-1}$$

Under this transformation, the closed right half of the s-plane maps into a unit disk centered at the origin of the z-plane.

Define

$$\Psi(z) := \frac{1}{\mu_o} \cdot \hat{\Phi}(s) \Big|_{s=(1-z)/(1+z)} \tag{4-2}$$

and

$$W_i := \frac{1}{\mu_o} \hat{H}(s_i) \qquad i=1,2,\cdots,n. \tag{4-3}$$

Now the problem becomes the following.

Find a $\Psi(z)$ which is analytic in the unit disk such that $\|\Psi\|_\infty \leq 1$ and

$$\Psi(z_i) = W_i \qquad i=1,\cdots,n \tag{4-4}$$

where

$$\|\Psi\|_\infty := \sup_\theta \bar{\sigma}[\Psi(e^{j\theta})] \tag{4-5}$$

and

$$z_i = \frac{1-s_i}{1+s_i}. \tag{4-6}$$

Note that $\|\Psi\|_\infty = 1/\mu_o \|\hat{\Phi}\|_\infty$.

Following Delsarte, Genin, and Kamp [7], define

$$L(E) := \begin{bmatrix} A & B \\ C & D \end{bmatrix} := \begin{bmatrix} (I-EE^*)^{1/2} & -(I-EE^*)^{-1/2}E \\ -(I-E^*E)^{-1/2}E^* & (I-E^*E)^{-1/2} \end{bmatrix} \tag{4-7}$$

where E is a constant matrix with $\bar{\sigma}(E) < 1$ and $M^{1/2}$ is the Hermitian square root of M, i.e., if $M = RR^*$, we denote R by $M^{1/2}$.

Define the linear fractional transformation

$$\Gamma_{L(E)}: X \mapsto (AX+B)(CX+D)^{-1} \tag{4-8}$$

$$y(\xi,z) := \begin{cases} \dfrac{\xi - z}{1-\bar{\xi}z} \cdot \dfrac{|\xi|}{\xi} & \text{if } \xi \neq 0 \\ -z & \text{if } \xi = 0 \end{cases} \tag{4-9}$$

and

$$\Gamma_{L(E),y(\xi,z)}: X \mapsto Y = y(\xi,z)^{-1}(AX+B)(CX+D)^{-1} \tag{4-10}$$

then

$$\Gamma^{-1}_{L(E),y(\xi,z)}: Y \mapsto X = [y(\xi,z)\cdot Y \cdot C - A]^{-1}[B - y(\xi,z)YD]. \tag{4-11}$$

Nevanlinna Algorithm [7]

1) Denote $\Psi(z)$ by $\Psi_1(z)$ and let $W_i^1 := W_i$, $i=1,2,\cdots,n$. Set

$$\Psi_2(z) = \Gamma_{L(W_1^1),y(z_1,z)}[\Psi_1(z)].$$

Note that if $\bar{\sigma}(W_1^1) < 1$, then $\Psi_2(z)$ is of class S if and only if $\Psi_1(z)$ is of class S and $\Psi_1(z_1) = W_1^1$. ($\Psi(z)$ is of class S (Schur) if it is analytic and satisfies $\bar{\sigma}(\Psi(z)) \leq 1$ in the domain $|z| < 1$.)

Define $W_i^2 = \Psi_2(z_i)$, $i=2,3,\cdots,n$. Now the problem is reduced to that of finding a $\Psi_2(z)$ in class S such that

$$\Psi_2(z_i) = W_i^2 \qquad i=2,3,\cdots,n.$$

2) Set

$$\Psi_3(z) = \Gamma_{L(W_2^2),y(z_2,z)}[\Psi_2(z)]$$

and define $W_i^3 = \Psi_3(z_i)$, $i=3,4,\cdots,n$.

3) Continue the iteration

$$\Psi_k(z) = \Gamma_{L(W_{k-1}^{k-1}),y(z_{k-1},z)}[\Psi_{k-1}(z)]$$

and

$$W_i^k = \Psi_k(z_i), \qquad i=k,\cdots,n$$

4) until

$$\Psi_n(z) = \Gamma_{L(W_{n-1}^{n-1}),y(z_{n-1},z)}[\Psi_{n-1}(z)]$$

$$W_n^n = \Psi_n(z_n).$$

Now the problem is reduced to that of finding a $\Psi_n(z)$ in class S such that $\Psi_n(z_n) = W_n^n$. There are infinitely many solutions to this problem, one of which is $\Psi_n(z) = W_n^n$.

5) Going backwards, we can find $\Psi_1(z)$ as follows:

$$\Psi_{n-1}(z)=\Gamma^{-1}_{L(W^{n-1}_{n-1}),y(z_{n-1},z)}[\Psi_n(z)]$$

$$\vdots$$

$$\Psi_{k-1}(z)=\Gamma^{-1}_{L(W^{k-1}_{k-1}),y(z_{k-1},z)}[\Psi_k(z)]$$

$$\vdots$$

$$\Psi_1(z)=\Gamma^{-1}_{L(W^1_1),y(z_1,z)}[\Psi_2(z)]. \tag{4-12}$$

Note that the inverse transformation can be implemented by (4-11). For example,

$$\Psi_1(z)=[y(z_1,z)\Psi_2(z)C-A]^{-1}[B-y(z_1,z)\Psi_2(z)D]$$

with A, B, C, and D evaluated at $E=W^1_1$.

Remark: In the above algorithm, for simplicity we assume $\|W^i_i\|<1$, $i=1,\cdots,n-1$. If $\|W^{k-1}_{k-1}\|=1$, steps 3) and 5) should be modified as follows. Do the singular value decomposition of W^{k-1}_{k-1} as

$$U^*_{k-1}\cdot W^{k-1}_{k-1}\cdot V_{k-1}=\begin{bmatrix} I & 0 \\ 0 & \hat{W}^{k-1}_{k-1}\end{bmatrix}$$

and

$$U^*_{k-1}\cdot W^{k-1}_i\cdot V_{k-1}=\begin{bmatrix} I & 0 \\ 0 & \hat{W}^{k-1}_i\end{bmatrix} \qquad i=k,\cdots,n$$

and define $\hat{\Psi}_{k-1}(z)$ as

$$U^*_{k-1}\cdot\Psi_{k-1}(z)\cdot V_{k-1}=\begin{bmatrix} I & 0 \\ 0 & \hat{\Psi}_{k-1}(z)\end{bmatrix}$$

where $\hat{\Psi}_{k-1}(z_i)=\hat{W}^{k-1}_i$ $i=k-1,\cdots,n$. Then set

$$\Psi_k(z)=\Gamma_{L(\hat{W}^{k-1}_{k-1}),y(z_{k-1},z)}[\hat{\Psi}_{k-1}(z)]$$

and

$$W^k_i=\Psi_k(z_i) \qquad i=k,\cdots,n.$$

In the backward step [i.e., step 5)], (4-12) is replaced by

$$\hat{\Psi}_{k-1}(z)=\Gamma^{-1}_{L(\hat{W}^{k-1}_{k-1}),y(z_{k-1},z)}[\Psi_k(z)]$$

and

$$\Psi_{k-1}(z)=U_{k-1}\cdot\begin{bmatrix} I & 0 \\ 0 & \hat{\Psi}_{k-1}(z)\end{bmatrix}\cdot V^*_{k-1}.$$

Once $\Psi(z):=\Psi_1(z)$ is constructed, by (4-2), $\hat{\Phi}(s)$ can be obtained as

$$\hat{\Phi}(s)=\mu_o\Psi(z)|_{z=(1-s)/(1+s)}.$$

Now, we have constructed an optimal $\hat{\Phi}$ which guarantees $K\in(RH^\infty)^{p\times q}$ with $\|\hat{\Phi}\|_\infty=\mu_o$. Then by (2-5), K is obtained.

The implementation of the Nevanlinna algorithm on the computer is quite easy. From step 1) to 4), it is nothing more than constant matrix manipulations. For step 5), we must compute

$$\Psi_{k-1}(z)=[y(z_{k-1},z)\cdot\Psi_k(z)\cdot C-A]^{-1}\cdot[B-y(z_{k-1},z)\Psi_k(z)D].$$

Denote $y(z_{k-1},z)$ and $\Psi_k(z)$ by $a(z)^{-1}b(z)$ and $Q(z)^{-1}P(z)$, respectively. Then

$$\Psi_{k-1}(z)=[b(z)P(z)C-a(z)Q(z)A]^{-1}\cdot[a(z)Q(z)B-b(z)P(z)D].$$

The multiplication and addition of polynomial matrices is straightforward. Furthermore, by the algorithm in [10], [11], we have

$$\Psi_{k-1}(z)=\hat{Q}^{-1}(z)\hat{P}(z)$$

with $\hat{Q}(z)$, $\hat{P}(z)$ left coprime.

V. Example

In this section we show how the control problem fits into the framework developed in the previous four sections and discuss the necessary assumptions and restrictions in terms of a specific example.

We consider the system

$$\begin{aligned} y&=-Nu+G_1e \\ z&=Pu+G_2e \end{aligned} \tag{5-1}$$

with

$$N=\begin{bmatrix}\frac{s-1}{s-2} & 0 \\ \frac{1}{s-2} & \frac{1}{s+1}\end{bmatrix},\qquad G_1=\begin{bmatrix}1 & 0 \\ 1 & 1\end{bmatrix}$$

$$P=\begin{bmatrix}\frac{s-1}{s-2} & 0\end{bmatrix},\qquad G_2=[1\quad 1].$$

This model (5-1) has been discussed in [13], [17] and consists of the measured output y, the regulated output z, the control input u, and the exogenous input e. The model is sufficiently general to include tracking and regulation problems where the regulated and measured outputs are not necessarily the same.

In this example, we will model the exogenous input as $e=Wd$ for some vector d where

$$W=\begin{bmatrix}\frac{10}{s+10} & 0 \\ 0 & \frac{10}{s+10}\end{bmatrix}.$$

This matrix can be chosen to furnish bounds on the frequency spectra of the exogenous signals and is discussed in [1] and [2]. We will always choose W to be stable and to have no closed right half plane zeros.

Define Φ as the transfer function matrix from d to z, i.e.,

$$z=\Phi d. \tag{5-2}$$

The problem is to find a proper controller C such that the closed-loop system is internally stable and $\|\Phi\|_\infty$ is minimized. It turns out that optimal controllers are not proper, in general. We therefore relax the properness constraint and obtain improper controllers that are optimal. We then modify the problem by introducing a sequence of proper controllers that are asymptotically optimal. This is the approach used in the work of Zames *et al.* [1]–[4].

Solution

It is easy to check that the McMillan degree of the unstable poles of N is the same as that of

$$\begin{bmatrix}-N & G_1 \\ P & G_2\end{bmatrix}$$

and therefore, the system is admissible, i.e., there exists a controller such that the closed-loop system is internally stable. This is the condition equivalent to detectability at y and stabilizability from u in a state space formulation or to the assumption of no unstable cancellation in certain polynomial matrices as stated in assumptions A1, A2, and A3 in the frequency domain formulation given in [17].

Let A_1, B_1, A_2, B_2, X_1, Y_1, X_2, and Y_2 be stable, proper, real-rational matrices such that

$$\begin{bmatrix} A_2 & B_2 \\ -Y_1 & X_1 \end{bmatrix}\begin{bmatrix} X_2 & -B_1 \\ Y_2 & A_1 \end{bmatrix} = \begin{bmatrix} I & 0 \\ 0 & I \end{bmatrix} \tag{5-3}$$

and

$$A_2^{-1}B_2 = B_1A_1^{-1} = N. \tag{5-4}$$

For computing (5-3), we use the transformation [16]

$$\lambda = \frac{1}{s+1} \tag{5-5}$$

then

$$\tilde{N}(\lambda) = N(s)|_{s=(1-\lambda)/\lambda} = \begin{bmatrix} \dfrac{1-2\lambda}{1-3\lambda} & 0 \\ \dfrac{\lambda}{1-3\lambda} & \lambda \end{bmatrix}. \tag{5-6}$$

By [10], [11], we have

$$\begin{bmatrix} \tilde{A}_2 & \tilde{B}_2 \\ -\tilde{Y}_1 & \tilde{X}_1 \end{bmatrix}\begin{bmatrix} \tilde{X}_2 & -\tilde{B}_1 \\ \tilde{Y}_2 & \tilde{A}_1 \end{bmatrix} = \begin{bmatrix} I & 0 \\ 0 & I \end{bmatrix} \tag{5-7}$$

and

$$\tilde{A}_2^{-1}\tilde{B}_2 = \tilde{N} \tag{5-8}$$

where

$$\tilde{A}_2 = \begin{bmatrix} 1 & -1 \\ \lambda - \frac{1}{3} & 0 \end{bmatrix}, \quad \tilde{B}_2 = \begin{bmatrix} 1 & -\lambda \\ \frac{2}{3}\lambda - \frac{1}{3} & 0 \end{bmatrix}$$

$$\tilde{A}_1 = \begin{bmatrix} 0 & \lambda - \frac{1}{3} \\ -1 & 0 \end{bmatrix}, \quad \tilde{B}_1 = \begin{bmatrix} 0 & \frac{2}{3}\lambda - \frac{1}{3} \\ -\lambda & -\frac{1}{3}\lambda \end{bmatrix}$$

$$\tilde{X}_2 = \begin{bmatrix} 0 & 6 \\ -1 & -3 \end{bmatrix}, \quad \tilde{Y}_2 = \begin{bmatrix} 0 & -9 \\ 0 & 0 \end{bmatrix}$$

$$\tilde{X}_1 = \begin{bmatrix} 0 & -1 \\ 6 & 0 \end{bmatrix}, \quad \tilde{Y}_1 = \begin{bmatrix} 0 & 0 \\ -9 & 0 \end{bmatrix}.$$

Then transform them back to s-domain

$$A_2 = \begin{bmatrix} 1 & -1 \\ \dfrac{-(s-2)}{3(s+1)} & 0 \end{bmatrix}, \quad B_2 = \begin{bmatrix} 1 & -\dfrac{1}{s+1} \\ \dfrac{-(s-1)}{3(s+1)} & 0 \end{bmatrix}$$

$$A_1 = \begin{bmatrix} 0 & \dfrac{-(s-2)}{3(s+1)} \\ -1 & 0 \end{bmatrix}, \quad B_1 = \begin{bmatrix} 0 & \dfrac{-(s-1)}{3(s+1)} \\ -\dfrac{1}{s+1} & -\dfrac{1}{3(s+1)} \end{bmatrix}$$

$$X_2 = \begin{bmatrix} 0 & 6 \\ -1 & -3 \end{bmatrix}, \quad Y_2 = \begin{bmatrix} 0 & -9 \\ 0 & 0 \end{bmatrix}$$

$$X_1 = \begin{bmatrix} 0 & -1 \\ 6 & 0 \end{bmatrix}, \quad Y_1 = \begin{bmatrix} 0 & 0 \\ -9 & 0 \end{bmatrix}.$$

By [14], the controller can be parameterized as

$$C = (Y_2 + A_1K)(X_2 - B_1K)^{-1} \tag{5-9}$$

with K stable.

The transfer function matrix Φ from d to z is

$$\Phi = (PA_1KA_2G_1 + PY_2A_2G_1 + G_2)W. \tag{5-10}$$

From the results of [17], it follows that PA_1, A_2G_1, and $PY_2A_2G_1 + G_2$ are proper and stable. In order to proceed, we require PA_1 to have full row rank and A_2G_1 full column rank. This assumption means that, in the control problem, we must have at least as many control inputs as regulated outputs and at least as many measured outputs as exogenous inputs. There is one other assumption required for the inner-outer factorization that follows. PA_1 and A_2G_1 must have no imaginary axis zeros. Define

$$B := PA_1 := B_iB_o \tag{5-11}$$

$$A := A_2G_1W := A_oA_i \tag{5-12}$$

$$H := (PY_2A_2G_1 + G_2)W \tag{5-13}$$

where A_i, B_i are inner matrices. B_o has a stable right inverse and A_o a stable left inverse. These are computed according to [18]. Let

$$\bar{K} := B_oKA_o. \tag{5-14}$$

Then (5-10) becomes

$$\Phi = B_i\bar{K}A_i + H. \tag{5-15}$$

Now we have

$$B_i = \frac{s-1}{s+1}, \quad B_o = \begin{bmatrix} 0 & -\frac{1}{3} \end{bmatrix}$$

$$A_o = \begin{bmatrix} 0 & -\dfrac{10}{s+10} \\ \dfrac{-10(s+2)}{3(s+1)(s+10)} & 0 \end{bmatrix}, \quad A_i = \begin{bmatrix} \dfrac{s-2}{s+2} & 0 \\ 0 & 1 \end{bmatrix}$$

$$H = \begin{bmatrix} \dfrac{10(4s-2)}{(s+1)(s+10)} & \dfrac{10}{s+10} \end{bmatrix}.$$

Define the inner function

$$\phi = \frac{s-1}{s+1} \cdot \frac{s-2}{s+2} \tag{5-16}$$

and define

$$\hat{\Phi} = \phi B_i^{-1}\Phi A_i^{-1} \tag{5-17}$$

$$\hat{H} = \phi B_i^{-1}HA_i^{-1} \tag{5-18}$$

then (5-15) becomes

$$\hat{\Phi} = \phi \bar{K} + \hat{H} \tag{5-19}$$

with

$$\hat{H} = \left[\frac{10(4s-2)}{(s+1)(s+10)} \quad \frac{10(s-2)}{(s+10)(s+2)} \right].$$

Add zero row vectors to (5-19) so that

$$\hat{\Phi}_a = \phi \bar{K}_a + \hat{H}_a \tag{5-20}$$

where $\hat{H}_a = \begin{bmatrix} \hat{H} \\ 0 \end{bmatrix}$.

$\bar{K}_a$ is stable if

$$\hat{\Phi}_a(1) = \hat{H}_a(1) = \begin{bmatrix} \frac{10}{11} & -\frac{10}{33} \\ 0 & 0 \end{bmatrix} := W^1$$

$$\hat{\Phi}_a(2) = \hat{H}_a(2) = \begin{bmatrix} \frac{5}{3} & 0 \\ 0 & 0 \end{bmatrix} := W^2.$$

Define

$$W = \begin{bmatrix} W^1 & 0 \\ 0 & W^2 \end{bmatrix}$$

$$V = \begin{bmatrix} \frac{1}{2} & 0 & \frac{1}{3} & 0 \\ 0 & \frac{1}{2} & 0 & \frac{1}{3} \\ \frac{1}{3} & 0 & \frac{1}{4} & 0 \\ 0 & \frac{1}{3} & 0 & \frac{1}{4} \end{bmatrix}.$$

The maximum γ such that

$$\det(WVW^* - \gamma V) = 0$$

is 8.480644801. Therefore,

$$\mu_o = 2.912154666.$$

Next, we will use the Nevanlinna algorithm to construct $\hat{\Phi}$. Define

$$\Psi_1(z) = \left. \frac{1}{\mu_o \hat{\Phi}(s)} \right|_{s=(1-z)/(1+z)}$$

and

$$W_i^1 = \frac{1}{\mu_o} \hat{H}(s_i) \qquad i = 1,2$$

where

$$\hat{H}(s_1) = \left[\frac{10}{11} \quad \frac{-10}{33} \right], \qquad \hat{H}(s_2) = \left[\frac{5}{3} \quad 0 \right].$$

Then by the forward steps of the Nevanlinna algorithm, we have

$$W_2^2 = [0.94447184 \quad 0.3285923].$$

Choose

$$\Psi_2(z) = W_2^2$$

then by the backward steps of the Nevanlinna algorithm, we have

$$\hat{\Phi}(s) = \frac{1}{\Delta(s)} [\phi_{11}(s) \phi_{12}(s)]$$

where

$$\Delta(s) = 1.3312071s + 0.78674165$$

$$\phi_{11}(s) = 3.823185s - 1.8977771$$

$$\phi_{12}(s) = 0.64180263s - 1.2836053.$$

By (5-19)

$$\bar{K} = \phi^{-1}(\hat{\Phi} - \hat{H}) = [\bar{k}_1(s) \bar{k}_2(s)]$$

where

$$\bar{k}_1(s) = \frac{(s+2)(3.823185s - 1.62147)}{(s+10)\Delta(s)}$$

$$\bar{k}_2(s) = \frac{(s+1)(0.6418026s - 4.968634)}{(s+10)\Delta(s)}.$$

By (5-14)

$$\bar{K} = B_o K A_o.$$

We can solve for K as

$$K(s) = \begin{bmatrix} 0 & 0 \\ k_{21}(s) & k_{22}(s) \end{bmatrix}$$

where

$$k_{21}(s) = 3 \cdot \frac{s+10}{10} \bar{k}_2(s)$$

$$k_{22}(s) = \frac{9(s+1)(s+10)}{10(s+2)} \bar{k}_1(s),$$

i.e.,

$$k_{21}(s) = \frac{3}{10} \cdot \frac{(s+1)(0.6418026s - 4.968634)}{\Delta(s)}$$

$$k_{22}(s) = \frac{9}{10} \cdot \frac{(s+1)(3.823185s - 1.62147)}{\Delta(s)}.$$

As previously noted, K is stable but not proper. By (5-9), the controller C is proper if K is strictly proper. We can use Zames' identity sequence multiplication technique [1] to make K strictly proper. That is, let

$$K_n(s) = K(s) \cdot \left[\frac{n}{s+n} \right]^2$$

then $K_n(s)$ is strictly proper and, therefore, the controller

$$C_n = (Y_2 + A_1 K_n)(X_2 - B_1 K_n)^{-1}$$

is proper. This sequence $\{C_n\}$ of proper controllers yields optimality as $n \to \infty$.

Acknowledgment

The authors are grateful to Prof. B. A. Francis for bringing [7] to our attention, to Prof. J. W. Helton for the preliminary version of [5], and to Prof. M. Vidyasagar for several valuable comments on Sections III and IV.

References

[1] G. Zames, "Feedback and optimal sensitivity: Model reference transformations, multiplicative seminorms, and approximate inverses," *IEEE Trans. Automat. Contr.*, vol. AC-26, pp. 301–320, Apr. 1981.

[2] G. Zames and B. A. Francis, "A new approach to classical frequency methods: Feedback and minimax sensitivity," in *Proc. IEEE Conf. Decision Contr.*, 1981, pp. 867–874.

[3] B. A. Francis and G. Zames, "On optimal sensitivity theory for SISO feedback systems," Dep. Elec. Eng., Univ. Waterloo, Waterloo, Ont., Canada, Research Rep., 1982.

[4] B. A. Francis, J. W. Helton, and G. Zames, "$\mathscr{H}^\infty$ optimal feedback controllers for linear multivariable systems," *IEEE Trans. Automat. Contr.*, this issue, pp. 888–900.

[5] J. A. Ball and J. W. Helton, "A Beurling–Lax theorem for the lie group $U(m, n)$ which contains most classical interpolation theory," *J. Operator Theory*, vol. 9, pp. 107–142, 1983.

[6] D. Sarason, "Generalized interpolation in H^∞," *Trans. AMS*, vol. 127, pp. 179–203, 1967.

[7] P. H. Delsarte, Y. Genin, and Y. Kamp, "The Nevanlinna–Pick problem for matrix-valued functions," *SIAM J. Appl. Math.*, vol. 36, pp. 47–61, Feb. 1979.

[8] B. S. Garbow, J. M. Boyle, J. J. Dongarra, and C. B. Moler, *Matrix Eigensystem Routines — EISPACK Guide Extension*. New York: Springer-Verlag, 1977.

[9] J. J. Dongarra, C. B. Moler, J. R. Bunch, and G. W. Stewart, "LINPACK user's guide," *SIAM*, 1979.

[10] B-C Chang and J. B. Pearson, "Algorithms for the solution of polynomial equations arising in multivariable control theory," Dep. Elec. Eng., Rice University, Houston, TX, Tech. Rep. 8208, May 1982.

[11] B-C Chang, "Programs for solving polynomial equations," Dep. Elec. Eng., Rice Univ., Houston, TX, Tech. Rep. 8209, May 1982.

[12] B. A. Francis, "Notes on optimal sensitivity theory: The single-input/single-output case," Dep. Elec. Eng., University of Waterloo, Waterloo, Ont., Canada, Tech. Rep. 81-08, Dec. 1981.

[13] L. Pernebo, "An algebraic theory for the design of controllers for linear multivariable systems—Parts I and II," *IEEE Trans. Automat. Contr.*, vol. AC-26, pp. 171–194, Feb. 1981.

[14] D. C. Youla, H. A. Jabr, and J. J. Bongiorno, "Modern Wiener–Hopf design of optimal controllers—Part II: The multivariable case," *IEEE Trans. Automat. Contr.*, vol. AC-21, pp. 319–338, June 1976.

[15] P. A. Fuhrmann, *Linear Systems and Operators in Hilbert Space*. New York: McGraw-Hill, 1981.

[16] M. Vidyasagar, "On the use of right-coprime factorizations in distributed feedback systems containing unstable subsystems," *IEEE Trans. Circuits Syst.*, vol. 25, no. 11, pp. 916–921, 1978.

[17] L. Cheng and J. B. Pearson, "Synthesis of linear multivariable regulators," *IEEE Trans. Automat. Contr.*, vol. AC-26, pp. 194–202, Feb. 1981.

[18] B-C. Chang and J. B. Pearson, "Inner-outer factorization of rational matrices," Rice Univ., Houston, TX, Tech. Rep. 8216, Nov. 1982.

$\mathscr{H}^\infty$-Optimal Feedback Controllers for Linear Multivariable Systems

BRUCE A. FRANCIS, MEMBER, IEEE, J. WILLIAM HELTON, MEMBER, IEEE, AND GEORGE ZAMES, FELLOW, IEEE

Abstract—**This paper treats the problem of designing, for a linear multivariable plant, a feedback controller which minimizes the $\mathscr{H}^\infty$-norm of a weighted sensitivity matrix. There exists a family of optimal improper feedbacks. This family is determined by application of a theory of Ball and Helton. A method for computing optimal feedbacks is described in detail and a numerical example is included. It is shown that an optimal improper feedback can be approximated by a proper one under certain conditions on the weighting matrices.**

I. Introduction

THIS paper treats the following basic control problem: design a feedback controller to minimize the effect of a disturbance on the plant output, subject to the constraint of internal stability of the closed-loop system. In this problem the disturbance is assumed to belong to a prespecified class of signals, and the energy of the plant output is to be minimized for the worst disturbance in the class. The distinction between this minimax type problem and that of the Wiener–Hopf approach [1] is that in the latter the disturbance class is a singleton, e.g., standard white noise in the stochastic formulation.

The plant is modeled by its $n \times m$ transfer matrix P and the feedback controller by its $m \times n$ transfer matrix F; P and F are real-rational matrices in the Laplace variable s. The corresponding sensitivity matrix is $S := (I + PF)^{-1}$. A weighted sensitivity matrix $X := W_1 S W_2$ is defined, where W_1 and W_2 are square real-rational matrices. Thus, X can be viewed as the transfer matrix from v to y in Fig. 1.

The class of disturbances is taken to be the set of signals

$$\{d : d = W_2 v, \qquad \text{energy of } v \text{ is } \leqslant 1\}.$$

The filter W_1 is chosen to reflect the variables to be attenuated. (Alternatively, W_1 and W_2 may be viewed as design parameters selected, perhaps iteratively, to "shape" the sensitivity matrix $S(j\omega)$.) The energy of y is to be minimized for the worst d in the above class. This is equivalent to minimizing the $\mathscr{H}^\infty$-norm of X, i.e., the supremum over ω of the largest singular value of $X(j\omega)$. By way of contrast, the Wiener–Hopf approach is to minimize the $\mathscr{H}^2$-norm of X.

Background and complementary work on the subject of this paper can be briefly summarized as follows. The $\mathscr{H}^\infty$-approach to feedback design was introduced by Zames [2], and the single-

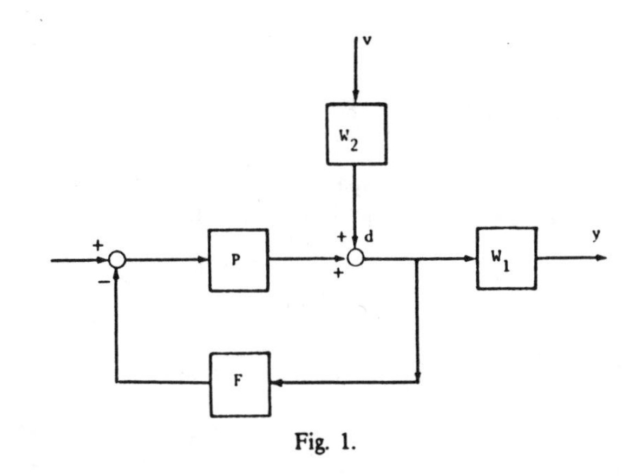

Fig. 1.

input, single-output (SISO) case of the problem posed above was solved in [3], [4]. In [3] the Nevanlinna–Pick theory was applied to the determination of the optimal X, which is a unique all-pass function in the SISO case. In addition, [2] and [3] contain a more complete motivation for the $\mathscr{H}^\infty$-approach, as well as a detailed comparison to the Wiener–Hopf approach. In [4] Sarason's theory [5] was applied to the determination of the optimal X and its norm. Kwakernaak [6] has obtained results on the SISO problem with a more general optimization criterion, and Kimura [7] has solved the related problem of robust stabilization (SISO case) using the Nevanlinna–Pick theory.

In the multiinput, multioutput (MIMO) problem solved in this paper, there is a family of optimal F's. Roughly speaking, minimization of the largest singular value of $X(j\omega)$ permits latitude in the other singular values. In this paper a theory of Ball and Helton [8] is applied to the determination of the family of optimal X's. Chang and Pearson [9] have solved the same problem using the matrix Nevanlinna–Pick theory of Delsarte, Genin, and Kamp [10]. Also, Safonov and Chen [11] solved a special MIMO problem which was equivalent to several independent SISO problems.

The next section summarizes the mathematical tools used in the paper. Section III contains the problem setup and a reformulation to make it amenable to the theory of [8], an account of which is given in Section IV. Section V shows in detail how the optimal X's can be computed, a numerical example being included. Section VI presents an asymptotically optimal sequence of proper feedbacks.

II. Preliminaries

The purpose of this section is to fix notation and to collect some definitions and facts.

The paper deals with both the right half-plane $\operatorname{Re} s \geqslant 0$, and the unit disk $|z| \leqslant 1$. A rational function of s is *stable* if it is analytic in $\operatorname{Re} s \geqslant 0$. A rational function with real coefficients is said to be *real-rational*.

Manuscript received February 10, 1983; revised July 27, 1983. Paper recommended by B. R. Barmish, Associate Editor for Optimal Systems. This work was supported by the Natural Sciences and Engineering Research Council of Canada and by the National Science Foundation.

B. A. Francis is with the Department of Electrical Engineering, University of Toronto, Toronto, Ont., Canada.

J. W. Helton is with the Department of Mathematics, University of California at San Diego, La Jolla, CA 92093.

G. Zames is with the Department of Electrical Engineering, McGill University, Montreal, P.Q., Canada.

Reprinted from *IEEE Trans. Automat. Contr.*, vol. AC-29, no. 10, pp. 888–899, Oct. 1984.

Two Hardy spaces for the unit disk are introduced next [12]. The space $\mathscr{H}^2$ is the space of complex-valued functions g which are analytic in $|z|<1$ and satisfy

$$\sup_{0\leqslant r<1}\int_0^{2\pi}|g(re^{j\theta})|^2\,d\theta<\infty.$$

The domain of each such g can be expanded to yield a boundary function defined on the unit circle. Then $\mathscr{H}^2$ is a Hilbert space with inner product

$$\langle g,h\rangle=\frac{1}{2\pi}\int_0^{2\pi}\overline{g(e^{j\theta})}\,h(e^{j\theta})\,d\theta$$

and norm

$$\|g\|_2=\left[\frac{1}{2\pi}\int_0^{2\pi}|g(e^{j\theta})|^2\,d\theta\right]^{1/2}.$$

The space $\mathscr{H}^\infty$ is the subspace of $\mathscr{H}^2$ of functions which are bounded and analytic in $|z|<1$. The $\mathscr{H}^\infty$-norm is

$$\|g\|_\infty=\operatorname*{ess\,sup}_\theta|g(e^{j\theta})|.$$

The subspace of real-rational functions in $\mathscr{H}^\infty$ will be denoted by $\mathscr{R}\mathscr{H}^\infty$, the space of real-rational functions having no poles in $|z|\leqslant 1$.

The notions of inner and outer functions are general, but we shall need them only as applied to rational functions. A rational function g in $\mathscr{H}^\infty$ is *inner* if $|g(e^{j\theta})|=1$ for all θ and *outer* if g has no zeros in $|z|<1$.

Hardy spaces of vectors and matrices are defined next [13]. The space $(\mathscr{H}^2)^k$ consists of k-dimensional vectors with entries from $\mathscr{H}^2$. This space is also a Hilbert space, with inner product

$$\langle g,h\rangle=\frac{1}{2\pi}\int_0^{2\pi}g(e^{j\theta})^*h(e^{j\theta})\,d\theta$$

where * denotes complex-conjugate transpose. Similarly, $(\mathscr{H}^2)^{k\times l}$ consists of $k\times l$ matrices with entries from $\mathscr{H}^2$, the inner product being

$$\langle G,H\rangle=\frac{1}{2\pi}\int_0^{2\pi}\operatorname{tr}G(e^{j\theta})^*H(e^{j\theta})\,d\theta.$$

The subspaces $(\mathscr{H}^\infty)^{k\times l}$ and $(\mathscr{R}\mathscr{H}^\infty)^{k\times l}$ are defined in the obvious way. The norm of G in $(\mathscr{H}^\infty)^{k\times l}$ is

$$\|G\|_\infty=\left\{\operatorname*{ess\,sup}_\theta\rho\left[G(e^{j\theta})^*G(e^{j\theta})\right]\right\}^{1/2}$$

where ρ denotes spectral radius (largest eigenvalue). It can be proved that

$$\|G\|_\infty=\sup\left\{\|Gh\|_2:h\in(\mathscr{H}^2)^l,\|h\|_2=1\right\}.$$

Let T be a linear operator $(\mathscr{H}^2)^l\to(\mathscr{H}^2)^k$. The graph of T is a subspace of the product space $(\mathscr{H}^2)^k\times(\mathscr{H}^2)^l$, which will be identified with $(\mathscr{H}^2)^{k+l}$. Thus, the *graph* of T is the space of vectors

$$\left\{\begin{bmatrix}Th\\h\end{bmatrix}:h\in(\mathscr{H}^2)^l\right\}.$$

If T is a *multiplication operator*, i.e., there is a matrix G in $(\mathscr{H}^\infty)^{k\times l}$ such that $Th=Gh$, then the graph of T has the representation

$$\begin{bmatrix}G\\I\end{bmatrix}(\mathscr{H}^2)^l.$$

A linear operator $(\mathscr{H}^2)^k\to(\mathscr{H}^2)^l$ of norm less than or equal to one is a *contraction*.

The concepts of inner and outer also apply to matrices [13]. A rational matrix G in $(\mathscr{H}^\infty)^{k\times l}$ is *inner* if

$$\|Gh\|_2=\|h\|_2\qquad\text{for all } h \text{ in } (\mathscr{H}^2)^l$$

or equivalently

$$G(e^{j\theta})^*G(e^{j\theta})=I\qquad\text{for all }\theta;$$

this requires $k\geqslant l$. It can be checked that if G is a square inner matrix, then $\det G$ is an inner function and $(\det G)G^{-1}$ is an inner matrix. A rational matrix G in $(\mathscr{H}^\infty)^{k\times l}$ is *outer* if

$$\operatorname{rank}G(z)=k\qquad\text{for all }|z|<1;$$

this requires $k\leqslant l$.

Every rational matrix G in $(\mathscr{H}^\infty)^{k\times l}$ has a factorization of the form $G=G_iG_o$ where G_i is inner and G_o is outer; G_i and G_o are unique up to multiplication by a constant unitary matrix.

Introduce the *shift operator*, multiplication by z:

$$\sigma:\mathscr{H}^2\to\mathscr{H}^2,$$
$$(\sigma g)(z)=zg(z).$$

Then σ can be extended in the obvious way to an operator $(\mathscr{H}^2)^k\to(\mathscr{H}^2)^k$. A subspace $\mathscr{S}$ of $(\mathscr{H}^2)^k$ is *shift-invariant* if $\sigma\mathscr{S}\subset\mathscr{S}$.

We shall not need the concept of Hardy space for the right half-plane. However, we shall need the definition of the $\mathscr{H}^\infty$-norm of a proper stable rational matrix $G(s)$, namely,

$$\|G\|_\infty=\{\sup\rho[G(j\omega)^*G(j\omega)]\}^{1/2}.$$

Finally, some miscellaneous notation: (a_{ij}) denotes a matrix with elements a_{ij}; if $\{\mathscr{S}_i\}$ is a countable family of subspaces of $(\mathscr{H}^2)^k$, then

$$\bigvee_{i=0}^{\infty}\mathscr{S}_i$$

denotes the smallest closed subspace containing all $\mathscr{S}_i$.

III. Problem Formulation

The following assumptions are made throughout the paper:

1) W_1, W_2, and P are real-rational matrices,[1]

2) P is strictly proper,

3) P has no poles on the imaginary axis and $\operatorname{rank}P(j\omega)=n$ for all ω,

4) W_1 and W_2 are nonsingular, proper, and stable, and they have stable inverses.

The rank assumption in 3) requires *a fortiori* that $n\leqslant m$, i.e., the number of outputs is less than or equal to the number of inputs. Condition 3) is the matrix generalization of the condition that P has neither poles nor zeros on the imaginary axis: a way to handle such poles and zeros is given in [4] for the scalar case.

The minimization of the $\mathscr{H}^\infty$-norm of

$$X:=W_1(I+PF)^{-1}W_2\tag{1}$$

begins with a parametrization of all real-rational proper F's which achieve internal stability. This approach based on coprime factorizations originated with Youla *et al.* [1]. It is convenient to use the modification due to Desoer *et al.* [14]. Thus, let

[1] The assumption that P is rational can be relaxed as in [3]. This relaxation can be significant in perturbation studies but nothing important is lost here by using the simpler assumption.

$A_i, B_i, Y_i, Z_i, (i=1,2)$, be stable, proper, real-rational matrices such that

$$P = A_1 B_1^{-1} = B_2^{-1} A_2 \tag{2}$$

$$Y_1 A_1 + Z_1 B_1 = I$$

$$A_2 Y_2 + B_2 Z_2 = I; \tag{3}$$

such matrices are known to exist when P is real-rational [15, Theorem 1]. Then the set of all proper, real-rational F's achieving internal stability is parametrized as [16, Lemma 3.2]

$$F = (Y_2 + B_1 Q)(Z_2 - A_1 Q)^{-1}$$

$$Q \text{ stable, proper, real-rational.} \tag{4}$$

Substitution of (2) and (4) into (1) yields

$$X = W_1 \left[I + B_2^{-1} A_2 (Y_2 + B_1 Q)(Z_2 - A_1 Q)^{-1} \right]^{-1} W_2$$

$$= W_1 (Z_2 - A_1 Q) [B_2 (Z_2 - A_1 Q) + A_2 (Y_2 + B_1 Q)]^{-1} B_2 W_2.$$

The quantity in the latter square brackets equals the identity matrix by virtue of (3) and the fact that $B_2 A_1 = A_2 B_1$ from (2). Thus,

$$X = W_1 (Z_2 - A_1 Q) B_2 W_2. \tag{5}$$

Hence, X is an affine function of the parameter Q.

The expression (5) leads to the definition

$$\mu := \inf \{ \| W_1 (Z_2 - A_1 Q) B_2 W_2 \|_\infty :$$

$$Q \text{ is stable, proper, real-rational} \} \tag{6}$$

as the *optimal weighted sensitivity*. This infimum is not usually achieved because of the constraint that Q be proper. Relax this constraint temporarily and define

$$\mu_1 := \inf \{ \| W_1 (Z_2 - A_1 Q) B_2 W_2 \|_\infty :$$

$$Q \text{ is stable, real-rational} \}. \tag{7}$$

It can be proved that this infimum is achieved; the problem is to determine the Q's that achieve it.

Since the results in [8] are developed for the unit disk, it is convenient at this point to map the right half-plane onto the unit disk via the mapping

$$s \mapsto z = (s-1)/(s+1).$$

The matrices $W_1(s)$, etc., on the right-hand side of (7) are then transformed into functions of z, which, by abuse of notation, will be denoted by $W_1(z)$, etc. Unless otherwise indicated, from here until the end of Section V all complex functions are functions of z.

The minimization problem (7) can be reduced to a problem of the form

$$\inf \left\{ \| G - UH \|_\infty : H \in (\mathcal{RH}^\infty)^{n \times n} \right\}$$

where G and U are fixed matrices in $(\mathcal{RH}^\infty)^{n \times n}$ with U inner. To do this reduction, first write

$$W_1 A_1 = (W_1 A_1)_i (W_1 A_1)_o \tag{8}$$

where $(W_1 A_1)_i$ is inner and $(W_1 A_1)_o$ is outer. Similarly, being square, $B_2 W_2$ can be factored as

$$B_2 W_2 = (B_2 W_2)_o (B_2 W_2)_i \tag{9}$$

where $(B_2 W_2)_o$ is outer and $(B_2 W_2)_i$ is inner. Define

$$\delta := \det (B_2 W_2)_i.$$

As observed in Section II δ is an inner function and $\delta (B_2 W_2)_i^{-1}$ is an inner matrix. It follows from (8) and (9) that

$$[W_1 (Z_2 - A_1 Q) B_2 W_2] \delta (B_2 W_2)_i^{-1} = G - UH \tag{10}$$

where

$$G := (W_1 Z_2 B_2 W_2) \delta (B_2 W_2)_i^{-1} \tag{11}$$

$$U := \delta (W_1 A_1)_i \tag{12}$$

$$H := (W_1 A_1)_o Q (B_2 W_2)_o. \tag{13}$$

From (7) and (10) and the fact that $\delta (B_2 W_2)_i^{-1}$ is a square inner matrix and multiplication by it is norm-preserving, we have

$$\mu_1 = \inf \left\{ \| G - UH \|_\infty : H \in (\mathcal{RH}^\infty)^{n \times n} \right\}. \tag{14}$$

In the next two sections it is shown how all H's which achieve the infimum in (14) can be found. From each such matrix H, a corresponding matrix Q can be obtained from (13). By virtue of the standing assumptions, the matrices $(W_1 A_1)_o(z)$ and $(B_2 W_2)_o(z)$ have rank n for all $|z| \leqslant 1$ except possibly at $z=1$ (i.e., $s=\infty$). It follows that (13) can be solved for $Q(z)$ having no poles in $|z| \leqslant 1$ except possibly at $z=1$. The corresponding $Q(s)$ is stable, but not necessarily proper; $Q(s)$ may have a pole at $s=\infty$. In this way one can obtain all $Q(s)$'s achieving the infimum in (7).

IV. Minimization of $\| G - UH \|_\infty$

The purpose of this section is to obtain a parametrization of all H's in $(\mathcal{RH}^\infty)^{n \times n}$ such that $\| G - UH \|_\infty$ equals its minimum μ_1, where G and U are fixed matrices in $(\mathcal{RH}^\infty)^{n \times n}$ and U is inner. The theory for this parametrization is due to Sarason [5], Adamjan *et al.* [17], [18], and Ball and Helton [8]. As the theory involves several concepts not widely known in the control community, an account of it is given here.

Some general remarks may make it easier to follow this section. First, it turns out that the infimum in (14) is not reduced if H is allowed to be a matrix in $(\mathcal{H}^\infty)^{n \times n}$ instead of $(\mathcal{RH}^\infty)^{n \times n}$. Thus, one can parametrize the set of all H's in $(\mathcal{H}^\infty)^{n \times n}$ such that $\| G - UH \|_\infty = \mu_1$, and then simply select the real-rational H's in this parametrization. Now the set of matrices

$$\left\{ G - UH : H \in (\mathcal{H}^\infty)^{n \times n} \right\}$$

is by definition the coset $[G]$ of G in the factor space $(\mathcal{H}^\infty)^{n \times n} / U (\mathcal{H}^\infty)^{n \times n}$. Thus, parametrizing the set of all H's in $(\mathcal{H}^\infty)^{n \times n}$ that minimize $\| G - UH \|_\infty$ is equivalent to parametrizing the set of all matrices in $[G]$ of minimum norm μ_1; this in turn is equivalent to parametrizing the set

$$\{ K \in [G_1] : \| K \|_\infty = 1 \} \tag{15}$$

where $G_1 := \mu_1^{-1} G$. The H's are determined from the K's via the equation

$$\mu_1 K = G - UH. \tag{16}$$

For each matrix K in the set (15), multiplication by K is a contraction on $(\mathcal{H}^2)^n$ whose graph is

$$\mathcal{K} := \begin{bmatrix} K \\ I \end{bmatrix} (\mathcal{H}^2)^n.$$

In Section IV-A it is shown that this graph contains a certain fixed subspace $\mathcal{N}$, that is, $\mathcal{N} \subset \mathcal{K}$ for all K in (15). In a special case, this containment condition in fact uniquely determines the graph $\mathscr{K}$, and hence the matrix K. For example, K is unique when $n=1$, a case solved by Sarason [5] and Adamjan *et al.* [18].

The geometric approach used in [8], an account of which is given in Section IV-B, is first to characterize the graphs of all K's in the set (15). It turns out that these graphs are characterized by three properties: they are shift-invariant, they are contained in a fixed subspace $\mathscr{M}$, and they are all maximal-negative relative to a certain Hermitian form. A parametrization of all subspaces with these three properties is obtained by a generalization of the Beurling–Lax theorem. (Beurling's theorem [19] is a characterization of all closed shift-invariant subspaces of $\mathscr{H}^2$ and Lax's extension [20] is to $(\mathscr{H}^2)^n$.)

The main result, Theorem 2, is a parametrization of the set (15) in terms of a fractional representation.

A. A Special Case: Unique K

Consider the operators on $(\mathscr{H}^2)^\kappa$ which are multiplication by matrices in the coset $[G_1]$. These operators all have a common part which depends only on G and U. To see what this common part is, let $\mathscr{U}$ denote the subspace $U(\mathscr{H}^2)^n$ and let $\mathscr{U}^\perp$ denote its orthogonal complement in $(\mathscr{H}^2)^n$

$$(\mathscr{H}^2)^n = \mathscr{U} \oplus \mathscr{U}^\perp.$$

Let $K \in [G_1]$, so that K satisfies (16) for some H. Then for each f in $(\mathscr{H}^2)^n$, the vectors $\mu_1 Kf$ and Gf have the same projection onto $\mathscr{U}^\perp$. This leads to the definition of the bounded linear operator Γ

$$\Gamma: (\mathscr{H}^2)^n \to \mathscr{U}^\perp$$
$$\Gamma f = \text{projection of } Gf \text{ onto } \mathscr{U}^\perp.$$

It will be seen in Section V-B that, since U is square and rational, $\mathscr{U}^\perp$ is finite-dimensional. Consequently, the self-adjoint operator

$$\Gamma\Gamma^*: \mathscr{U}^\perp \to \mathscr{U}^\perp$$

has a finite number of eigenvalues, which are real and nonnegative; let ν^2 denote the largest one and let v_1 be a corresponding eigenvector of unit norm. Observe that an alternative definition of ν is

$$\nu = \|\Gamma\| := \sup\left\{ \|\Gamma f\|_2 : f \in (\mathscr{H}^2)^n, \|f\|_2 = 1 \right\}.$$

Define v_2 in $\mathscr{U}^\perp$ via the equation

$$\Gamma^* v_1 = \nu v_2. \tag{17}$$

Then

$$\begin{aligned} \Gamma v_2 &= \nu^{-1}\Gamma\Gamma^* v_1 \\ &= \nu v_1. \end{aligned} \tag{18}$$

It can readily be shown that v_2 has unit norm too.

To see that $\mu_1 \geqslant \nu$, recall that for any K in $[G_1]$ the projection of $\mu_1 K v_2$ onto $\mathscr{U}^\perp$ equals Γv_2, and this latter vector equals νv_1 from (18). Thus,

$$\mu_1 K v_2 = \nu v_1 + (\text{a vector in } \mathscr{U}). \tag{19}$$

It follows that

$$\mu_1 \|K v_2\|_2 \geqslant \nu \|v_1\|_2 = \nu,$$

and hence that $\mu_1 \|K\|_\infty \geqslant \nu$. Since $\inf\{\|K\|_\infty : K \in [G_1]\} = 1$, we get $\mu_1 \geqslant \nu$. In fact, the equality holds.

Lemma 1 (A Corollary of [13, Theorem II.2.3]): The infimum

$$\mu_1 := \inf\left\{ \|G - UH\|_\infty : H \in (\mathscr{H}^\infty)^{n\times n} \right\}$$

is achieved, and $\mu_1 = \nu := \|\Gamma\|$.

Under certain circumstances there is a unique matrix in $[G_1]$ of unit (minimum) norm. To see when this happens, let $K \in [G_1]$, $\|K\|_\infty = 1$. From (19) and the fact that $\mu_1 = \nu$ we have

$$K v_2 = v_1 + (\text{a vector in } \mathscr{U}).$$

Since v_1 and v_2 have unit norms, this equation is consistent with the assumption that $\|K\|_\infty = 1$ only if $K v_2 = v_1$. Thus,

$$\begin{bmatrix} v_1 \\ v_2 \end{bmatrix} \in (\text{graph of } K) =: \mathscr{K}.$$

From (17) this implies that

$$\begin{bmatrix} I \\ \Gamma_1^* \end{bmatrix} v_1 \in \mathscr{K}$$

where $\Gamma_1 := \mu_1^{-1}\Gamma$. Since v_1 is an arbitrary element of the eigenspace $\ker(\mu_1^2 I - \Gamma\Gamma^*)$, we get

$$\begin{bmatrix} I \\ \Gamma_1^* \end{bmatrix} \ker(\mu_1^2 I - \Gamma\Gamma^*) \subset \mathscr{K}.$$

Finally, since $\mathscr{K}$ is a closed, shift-invariant subspace, we conclude that

$$\mathscr{N} \subset \mathscr{K} \tag{20}$$

where

$$\mathscr{N} := \bigvee_{i=0}^{\infty} \sigma^i \begin{bmatrix} I \\ \Gamma_1^* \end{bmatrix} \ker(\mu_1^2 I - \Gamma\Gamma^*). \tag{21}$$

To recap, *every* K *in* $[G_1]$ *of* (minimum) *norm one contains* $\mathscr{N}$ *in its graph.*

Define

$$p := \dim_{\mathbf{C}} \ker(\mu_1^2 I - \Gamma\Gamma^*)$$

and let N be a matrix in $(\mathscr{H}^2)^{2n\times p}$ whose columns form a basis (over $\mathbf{C}$) for

$$\begin{bmatrix} I \\ \Gamma_1^* \end{bmatrix} \ker(\mu_1^2 I - \Gamma\Gamma^*).$$

Because G and U are rational, it follows (see Section V-C) that N can be taken to be rational too. Thus, $N \in (\mathscr{H}^\infty)^{2n\times p}$. Hence, from (21) and the two equalities

$$(\mathscr{H}^2)^p = \bigvee_{i=0}^{\infty} \sigma^i \mathbf{C}^p$$

$$\begin{bmatrix} I \\ \Gamma_1^* \end{bmatrix} \ker(\mu_1^2 I - \Gamma\Gamma^*) = N\mathbf{C}^p$$

we get that

$$\mathscr{N} = \text{closure of } N(\mathscr{H}^2)^p. \tag{22}$$

Finally, partition N as

$$N = \begin{bmatrix} N_1 \\ N_2 \end{bmatrix} \tag{23}$$

where each N_i is $n \times p$.

Theorem 1: Each K *in* $[G_1]$ *of* (minimum) *norm one satisfies the equation* $KN_2 = N_1$. *This equation determines* K *uniquely iff* rank $N_2(z) = n$ *for some* (hence, almost all) z.

Proof: If $K \in [G_1]$ and $\|K\|_\infty = 1$, then (20) holds and so

$$\begin{bmatrix} N_1 \\ N_2 \end{bmatrix} (\mathcal{H}^2)^p \subset \begin{bmatrix} K \\ I \end{bmatrix} (\mathcal{H}^2)^n.$$

Multiplication of both sides by $[I, -K]$ yields $KN_2 = N_1$. The second statement of the theorem is standard linear algebra. ■

B. The General Case[2]

It was observed in Section IV-A that the graph of each K in $[G_1]$ of (minimum) norm one contains the subspace $\mathcal{N}$. However, this condition only partially constrains the graph in general. For a complete characterization we require three preliminary results, which are set out as Lemmas 2–4 below.

The first lemma characterizes multiplication operators in terms of their graphs.

Lemma 2: A bounded linear operator $(\mathcal{H}^2)^l \to (\mathcal{H}^2)^k$ *is a multiplication operator iff its graph is shift-invariant.*

The next lemma characterizes contractions in terms of their graphs. To motivate this characterization, consider a contraction $T: (\mathcal{H}^2)^l \to (\mathcal{H}^2)^k$. Let f be a vector in the graph of T. Thus,

$$f = \begin{bmatrix} f_1 \\ f_2 \end{bmatrix}$$

where $f_1 \in (\mathcal{H}^2)^k, f_2 \in (\mathcal{H}^2)^l, f_1 = Tf_2$. Since T is a contraction, we have that $\|f_1\|_2 \leqslant \|f_2\|_2$, or equivalently

$$\langle f_1, f_1 \rangle - \langle f_2, f_2 \rangle \leqslant 0. \tag{24}$$

It will be convenient to express this inequality in terms of an indefinite inner product on $(\mathcal{H}^2)^k \times (\mathcal{H}^2)^l$, defined as follows. For f and g in $(\mathcal{H}^2)^k \times (\mathcal{H}^2)^l$, with

$$f = \begin{bmatrix} f_1 \\ f_2 \end{bmatrix}, g = \begin{bmatrix} g_1 \\ g_2 \end{bmatrix}$$

define

$$[f, g] := \langle f_1, g_1 \rangle - \langle f_2, g_2 \rangle.$$

Then (24) assumes the form $[f, f] \leqslant 0$.

The space $(\mathcal{H}^2)^k \times (\mathcal{H}^2)^l$ equipped with the indefinite inner product $[,]$ will be called a *Krein space*.[3] A vector f in a Krein space is *positive*, *neutral*, or *negative* depending on whether $[f, f] \geqslant 0$, $= 0$, or $\leqslant 0$, respectively. The conclusion of the last paragraph is that every vector in the graph of a contraction is negative.

A subspace $\mathcal{S}$ of a Krein space is *negative* if each vector in $\mathcal{S}$ is negative and *maximal-negative* if, in addition, $\mathcal{S}$ is not properly contained in any negative subspace.

Lemma 3: A subspace of $(\mathcal{H}^2)^k \times (\mathcal{H}^2)^l$ *is maximal-negative iff it is the graph of a contraction* $(\mathcal{H}^2)^l \to (\mathcal{H}^2)^k$.

The third lemma characterizes matrices in $[G_1]$ in terms of the graphs of the corresponding multiplication operators. Define the Krein space

$$\mathcal{X} := (\mathcal{H}^2)^n \times (\mathcal{H}^2)^n,$$

the $2n \times 2n$ matrix

$$M := \begin{bmatrix} G_1 & U \\ I & 0 \end{bmatrix},$$

and the shift-invariant subspace $\mathcal{M} := M\mathcal{X}$ of $\mathcal{X}$. Every K in $[G_1]$ has the form $K = G_1 - UH$ for some H in $(\mathcal{H}^\infty)^{n \times n}$. Thus,

$$\begin{bmatrix} K \\ I \end{bmatrix} = M \begin{bmatrix} I \\ -H \end{bmatrix}$$

so that the graph of the operator $f \mapsto Kf$ on $(\mathcal{H}^2)^n$ is contained in $\mathcal{M}$. The converse is also true.

Lemma 4: Let $K \in (\mathcal{H}^\infty)^{n \times n}$ *and define the graph*

$$\mathcal{K} := \begin{bmatrix} K \\ I \end{bmatrix} (\mathcal{H}^2)^n.$$

Then $K \in [G_1]$ *iff* $\mathcal{K} \subset \mathcal{M}$.

The preceding three lemmas are now combined to yield immediately a characterization of each K in $[G_1]$ of minimum norm in terms of the graph of the corresponding multiplication operator.

Proposition 1: The mapping

$$K \mapsto \begin{bmatrix} K \\ I \end{bmatrix} (\mathcal{H}^2)^n$$

is a bijection from the set of all K *in* $[G_1]$ *of unit norm onto the set of all shift-invariant maximal-negative subspaces of* $\mathcal{M}$.

This proposition motivates us to characterize all shift-invariant maximal-negative subspaces of $\mathcal{M}$. This characterization uses the analog in a Krein space of Halmos' wandering subspace in the Hilbert space $(\mathcal{H}^2)^n$ [22]. (Let $\mathcal{R}$ be a closed shift-invariant subspace of $(\mathcal{H}^2)^n$. Then $\sigma\mathcal{R}$ is a closed subspace of $\mathcal{R}$. Let $\mathcal{S}$ denote the orthogonal complement of $\sigma\mathcal{R}$ in $\mathcal{R}$, i.e., $\mathcal{S} := \mathcal{R} \cap (\sigma\mathcal{R})^\perp$. Then it can be proved that the subspaces $\mathcal{S}, \sigma\mathcal{S}, \sigma^2\mathcal{S}, \cdots$ are pairwise orthogonal and they span $\mathcal{R}$, i.e.,

$$\mathcal{R} = \bigvee_{i=0}^{\infty} \sigma^i \mathcal{S}.$$

The subspace $\mathcal{S}$ is said to be a wandering subspace which generates $\mathcal{R}$.)

The concept of orthogonality in a Krein space is central in what follows. Vectors f and g in a Krein space are *orthogonal* (relative to $[,]$) if $[f, g] = 0$. Observe that if f and g are both negative, then $f + g$ is negative if f and g are orthogonal. This fact is useful in constructing a basis for a negative subspace. If $\mathcal{S}$ is a subspace of a Krein space, the *orthogonal companion* $\mathcal{S}'$ of $\mathcal{S}$ is the subspace of vectors orthogonal to all vectors in $\mathcal{S}$. Observe that each vector in $\mathcal{S} \cap \mathcal{S}'$ is neutral. Subspaces $\mathcal{S}$ and $\mathcal{T}$ are *orthogonal* if each vector in $\mathcal{S}$ is orthogonal to every vector in $\mathcal{T}$.

The wandering subspace analog is

$$\mathcal{L} := \mathcal{M} \cap (\sigma\mathcal{M})'. \tag{25}$$

It has the property that $\sigma^k\mathcal{L}$ and $\sigma^l\mathcal{L}$ are orthogonal for nonnegative integers $k \neq l$. (Proof: Let $f \in \mathcal{L}, g \in \mathcal{L}, 0 \leqslant k < l$. Then

$$[\sigma^k f, \sigma^l g] = [f, \sigma^{l-k} g].$$

But, $\sigma^{l-k} g \in \sigma\mathcal{M}$ and $f \in (\sigma\mathcal{M})'$. Thus, $\sigma^k f$ and $\sigma^l g$ are orthogonal.) It also turns out that $\mathcal{L}$ is finite-dimensional and its elements are bounded, i.e., $\mathcal{L} \subset (\mathcal{H}^\infty)^{2n}$. Moreover, as will be seen in the next section, it is possible to decompose $\mathcal{L}$ into a direct sum

$$\mathcal{L} = \mathcal{L}^+ \oplus \mathcal{L}^- \oplus \mathcal{L} \cap \mathcal{L}' \tag{26}$$

where $\mathcal{L}^+$ is strictly positive, i.e.,

$$[f, f] > 0, 0 \neq f \in \mathcal{L}^+$$

[2] Proofs of the results in this subsection are collected in the Appendix.

[3] Actually, $(\mathcal{H}^2)^k \times (\mathcal{H}^2)^l$ is a concrete example of the abstract concept of Krein space [21].

and $\mathcal{L}^-$ is strictly negative, i.e.,

$$[f,f]<0, 0\neq f\in\mathcal{L}^-.$$

Furthermore, the subspaces $\mathcal{L}^+, \mathcal{L}^-$, and $\mathcal{L}\cap\mathcal{L}'$ are pairwise orthogonal. It will further be shown that there is a basis $\{h_i^+\}$ for $\mathcal{L}^+$ which is orthonormal relative to [,], i.e.,

$$[h_i^+, h_j^+] = \begin{cases} 1, & i=j \\ 0, & i\neq j \end{cases}.$$

Similarly, there is a basis $\{h_i^-\}$ for $\mathcal{L}^-$ which is orthonormal relative to $-[,]$, i.e.,

$$[h_i^-, h_j^-] = \begin{cases} -1, & i=j \\ 0, & i\neq j \end{cases}.$$

Let L^+ (respectively, L^-) be the matrix whose columns are $\{h_i^+\}$ (respectively, $\{h_i^-\}$) and define

$$L := [L^+, L^-]. \tag{27}$$

The dimensions of L are $2n\times(q_+ + q_-)$ where

$$q_+ := \dim_{\mathbf{C}}\mathcal{L}^+, q_- := \dim_{\mathbf{C}}\mathcal{L}^-.$$

Introduce another Krein space

$$\mathcal{Y} := (\mathcal{H}^2)^{q_+}\times(\mathcal{H}^2)^{q_-}.$$

Multiplication by L is a linear operator $\mathcal{Y}\to\mathcal{X}$ which can be shown to be an isometry relative to [,], i.e.,

$$[f,g] = [Lf, Lg], f, g\in\mathcal{Y}. \tag{28}$$

This fact can be used to prove the following characterization of all shift-invariant maximal-negative subspaces of $\mathcal{M}$.

Proposition 2: The mapping $\mathcal{T}\mapsto L\mathcal{T}+\mathcal{N}$ is a bijection from the set of all shift-invariant maximal-negative subspaces of $\mathcal{Y}$ onto the set of all shift-invariant maximal-negative subspaces of $\mathcal{M}$.

This proposition is useful because it is easy to characterize the shift-invariant maximal-negative subspaces of $\mathcal{Y}$: from Lemmas 2 and 3 these subspaces are precisely those of the form

$$\mathcal{T} = \begin{bmatrix} T \\ I \end{bmatrix}(\mathcal{H}^2)^{q_-}$$

where $T\in(\mathcal{H}^\infty)^{q_+\times q_-}$, $\|T\|_\infty\leq 1$.

Propositions 1 and 2 are now combined to yield the main result. Recall the definition of N_1 and N_2 in (23). Also, partition L^+ and L^- as

$$L^+ = \begin{bmatrix} L_1^+ \\ L_2^+ \end{bmatrix}, \quad L^- = \begin{bmatrix} L_1^- \\ L_2^- \end{bmatrix}. \tag{29}$$

Theorem 2: Let $T\in(\mathcal{H}^\infty)^{q_+\times q_-}$, $\|T\|_\infty\leq 1$, and define

$$K_1 = [L_1^+T + L_1^-, N_1], K_2 = [L_2^+T + L_2^-, N_2]. \tag{30}$$

Then the equation

$$KK_2 = K_1 \tag{31}$$

has a unique solution K and $K\in[G_1]$, $\|K\|_\infty = 1$. Moreover, the mapping $T\mapsto K$ is a bijection from

$$\{T\in(\mathcal{H}^\infty)^{q_+\times q_-} : \|T\|_\infty\leq 1\} \tag{32}$$

onto

$$\{K\in[G_1] : \|K\|_\infty = 1\}. \tag{33}$$

Adamjan *et al.* [17, Theorem 6.1] obtained an analogous result for a similar problem.

The parameter matrix T in Theorem 2 varies over the unit ball of $(\mathcal{H}^\infty)^{q_+\times q_-}$ and the resulting K's lie in $(\mathcal{H}^\infty)^{n\times n}$. To obtain real-rational K's, T must satisfy an additional constraint which is determined next.

Each f in $\mathcal{H}^2$ has a power series representation $\Sigma c_i z^i$. Let $\check{f}$ denote the function whose power series is $\Sigma\bar{c}_i z^i$. Thus, if f is rational, it is real-rational iff $f=\check{f}$. The same notation will be used for rational vectors and matrices.

It will be seen in the next section that the matrices L^+, L^-, and N are rational. Moreover, the columns of L^+ occur in conjugate pairs, this being a consequence of the real-rationality of G and U; similarly for L^- and N. It follows that there are permutation matrices P_1, P_2, and P_3 such that

$$\check{L}^+P_1 = L^+ \tag{34a}$$

$$\check{L}^-P_2 = L^- \tag{34b}$$

$$\check{N}P_3 = N. \tag{34c}$$

Corollary 1: The set of all real-rational K's in $[G_1]$ with $\|K\|_\infty = 1$ is parametrized as follows:

$$KK_2 = K_1 \text{ where}$$
$$K_1 = [L_1^+T + L_1^-, N_1], K_2 = [L_2^+T + L_2^-, N_2],$$
$$T\in(\mathcal{H}^\infty)^{q_+\times q_-}, \|T\|_\infty\leq 1,$$
$$T \text{ rational}, P_1^{-1}\check{T}P_2 = T.$$

V. Details of Computation

The purpose of this section is to show how the previous theory can be implemented algorithmically, specifically, to show how the optimal K's (given in Corollary 1) can be computed starting from G and U in (11) and (12). This will be done by presenting a conceptual high-level algorithm and then showing how the steps in the algorithm can be carried out.

A. High-Level Algorithm

The theory of the last section leads directly to the following procedure. The input data are the matrices G and U in $(\mathcal{RH}^\infty)^{n\times n}$ with U inner.

Step 1: Define $\mathcal{U} = U(\mathcal{H}^2)^n$ and $\Gamma : (\mathcal{H}^2)^n\to\mathcal{U}^\perp$

$$\Gamma f = \text{projection of } Gf \text{ onto } \mathcal{U}^\perp.$$

Compute $\|\Gamma\|$.

Step 2: Define

$$\mu_1 = \|\Gamma\|, G_1 = \mu_1^{-1}G, \Gamma_1 = \mu_1^{-1}\Gamma$$

$$M = \begin{bmatrix} G_1 & U \\ I & 0 \end{bmatrix}$$

$$\mathcal{M} = M(\mathcal{H}^2)^{2n}, \mathcal{L} = \mathcal{M}\cap(\sigma\mathcal{M})'$$

$$q_+ = \dim\mathcal{L}^+, q_- = \dim\mathcal{L}^-$$

$$p = \dim\ker(\mu_1^2 I - \Gamma\Gamma^*).$$

Find a matrix L^+ (respectively, L^-) whose columns form a basis for $\mathcal{L}^+$ (respectively, $\mathcal{L}^-$) which is orthonormal relative to [,] (respectively, $-[,]$). Find a matrix N whose columns form a basis for

$$\begin{bmatrix} I \\ \Gamma_1^* \end{bmatrix}\ker(\mu_1^2 I - \Gamma\Gamma^*).$$

Partition L^+, L^-, and N as follows:

$$L^+ = \begin{bmatrix} L_1^+ \\ L_2^+ \end{bmatrix}, L^- = \begin{bmatrix} L_1^- \\ L_2^- \end{bmatrix}, N = \begin{bmatrix} N_1 \\ N_2 \end{bmatrix}$$

where each L_i^+, L_i^-, N_i has n rows.

The optimal K's are then given by Corollary 1. In summary, to be computed are $\|\Gamma\|$, N, L^+, and L^-.

B. Computation of $\|\Gamma\|$

The computation of $\|\Gamma\|$ requires the determination of a basis for $\mathscr{U}^\perp$. Assume first that $\det U(z)$ has only nonrepeated zeros in $|z|<1$ and denote them by $\{z_i\}$. Let a_i be a nonzero vector in $\mathbf{C}^n$ such that

$$a_i^* U(z_i) = 0.$$

Then it follows easily from the residue theorem that a basis for $\mathscr{U}^\perp$ is

$$\{(1-\bar{z}_i z)^{-1} a_i\}.$$

The case where $\det U$ has repeated zeros in $|z|<1$ is more complicated to describe. Let $\{z_i\}$ denote the distinct zeros. Then basis vectors for $\mathscr{U}^\perp$ take the form

$$\phi a_1 + \phi^{(1)} a_2 + \cdots + \phi^{(k-1)} a_k$$

where $\phi(z) = (1-\bar{z}_i z)^{-1}$, $\phi^{(i)}$ denotes the ith derivative of ϕ, and $a_i \in \mathbf{C}^n$. The details are omitted.

Let $\{f_i : i = 1,\cdots,r\}$ be a basis for $\mathscr{U}^\perp$, constructed as above, and define the $r \times r$ matrices

$$\Phi_1 = (\langle \Gamma^* f_i, \Gamma^* f_j \rangle)$$
$$\Phi_2 = (\langle f_i, f_j \rangle).$$

It is claimed that $\|\Gamma\|^2$ *equals the maximum* λ *such that*

$$\det(\Phi_1 - \lambda \Phi_2) = 0.$$

To prove this, observe that every element of $\mathscr{U}^\perp$ has the form

$$f = \sum c_i f_i$$

for some c_i in $\mathbf{C}$. Defining the vector $c = (c_i)$, we have

$$\begin{aligned} \|\Gamma\|^2 &= \|\Gamma^*\|^2 \\ &= \max\{\langle \Gamma^* f, \Gamma^* f \rangle : \langle f, f \rangle = 1\} \\ &= \max\{c^* \Phi_1 c : c^* \Phi_2 c = 1\}. \end{aligned}$$

The claim follows by application of the method of Lagrange multipliers to the latter maximization.

The entries in the matrix Φ_2 are routine to evaluate using residues. Similarly, the entries in Φ_1 can be evaluated once the vectors $\Gamma^* f_i$ are determined. To do this, let $G^-(z)$ denote the matrix $G(z^{-1})^T$. Then for any g in $(\mathscr{H}^2)^n$

$$\begin{aligned} \langle \Gamma^* f_i, g \rangle &= \langle f_i, \Gamma g \rangle \\ &= \langle f_i, G g \rangle \\ &= \langle G^- f_i, g \rangle. \end{aligned}$$

Thus, $\Gamma^* f_i =$ the projection of $G^- f_i$ onto $(\mathscr{H}^2)^n$. This projection can be determined by partial fraction expansion.

C. Computation of N

The computation of N requires the following lemma, which is proved in the Appendix.

Lemma 5:

$$\mathscr{M}' = \begin{bmatrix} I \\ \Gamma_1^* \end{bmatrix} \mathscr{U}^\perp \tag{35}$$

and

$$\mathscr{M} \cap \mathscr{M}' = \begin{bmatrix} I \\ \Gamma_1^* \end{bmatrix} \ker(\mu_1^2 I - \Gamma\Gamma^*). \tag{36}$$

From the basis $\{f_i : i = 1,\cdots,r\}$ for $\mathscr{U}^\perp$ can be obtained a basis $\{g_i : i = 1,\cdots,r\}$ for $\mathscr{M}'$ via (35), namely,

$$g_i = \begin{bmatrix} f_i \\ \Gamma_1^* f_i \end{bmatrix}. \tag{37}$$

Since $\mathscr{M} = (\mathscr{M}')'$, every element g of $\mathscr{M} \cap \mathscr{M}'$ can be expressed uniquely as

$$g = \sum c_i g_i \tag{38}$$

where $c_i \in \mathbf{C}$ and

$$[g_i, g] = 0 \qquad \text{for all } i. \tag{39}$$

Defining the $r \times r$ matrix

$$\Psi = ([g_i, g_j]) \tag{40}$$

and the vector $c = (c_i)$ and substituting (38) into (39) yields $\Psi c = 0$. Thus, $\mathscr{M} \cap \mathscr{M}'$ is isomorphic to $\ker \Psi$, so a basis for $\ker \Psi$ yields a basis for $\mathscr{M} \cap \mathscr{M}'$ via (38). This latter basis then becomes the columns of N in view of (36).

D. Computation of L^+ and L^-

The subspace

$$\mathscr{L} := \mathscr{M} \cap (\sigma\mathscr{M})'$$

can also be expressed as

$$\mathscr{L} = \mathscr{M} \cap (\mathbf{C}^{2n} \oplus \sigma\mathscr{M}').$$

(*Proof:* Each f in $(\mathscr{H}^2)^{2n}$ has a power series expansion

$$f = f_0 + \sigma f_1 + \sigma^2 f_2 + \cdots, f_i \in \mathbf{C}^{2n}.$$

Define $g := f_1 + \sigma f_2 + \cdots$. Then $f \in (\sigma\mathscr{M})'$ iff for every h in $\mathscr{M}$

$$\begin{aligned} 0 &= [f, \sigma h] \\ &= [f_0 + \sigma g, \sigma h] \\ &= [g, h], \end{aligned}$$

i.e., $g \in \mathscr{M}'$. Thus,

$$(\sigma\mathscr{M})' = \mathbf{C}^{2n} \oplus \sigma\mathscr{M}'.)$$

Let $\{e_i : i = 1,\cdots,2n\}$ denote the standard basis for $\mathbf{C}^{2n}$. Then a basis for

$$\mathbf{C}^{2n} \oplus \sigma \mathscr{M}'$$

is $\{h_i : i = 1, \cdots, 2n + r\}$ where

$$h_i = e_i, \qquad i = 1, \cdots, 2n \tag{41a}$$

$$h_{2n+i} = \sigma g_i, \qquad i = 1, \cdots, r. \tag{41b}$$

Thus, each element h of $\mathscr{L}$ can be written uniquely as

$$h = \Sigma c_i h_i$$

where $c_i \in \mathbf{C}$ and

$$[g_i, h] = 0 \qquad \text{for all } i. \tag{42}$$

Defining the $r \times (2n + r)$ matrix $\Lambda := ([g_i, h_j])$, we find that (42) is equivalent to $\Lambda c = 0$ where $c = (c_i)$. Thus, $\mathscr{L}$ is isomorphic to $\ker \Lambda$.

Let $q := \dim \ker \Lambda$ and let Ξ be a matrix whose columns form a basis for $\ker \Lambda$. Then the following map is an isomorphism:

$$x \mapsto h : \mathbf{C}^q \to \mathscr{L}$$

$$\text{where } h = \sum c_i h_i,$$

$$(c_i) = c = \Xi x. \tag{43}$$

The Hermitian form $[,]$ on $\mathscr{L} \times \mathscr{L}$ is mapped by the isomorphism (43) into a Hermitian form on $\mathbf{C}^q \times \mathbf{C}^q$, namely

$$(x, y) \mapsto x^* \Xi^* \P \Xi y \tag{44}$$

where $\P$ is the $(2n + r) \times (2n + r)$ matrix

$$\P := ([h_i, h_j]).$$

From (40) and (41) we find that

$$\P = \begin{bmatrix} I_n & 0 & 0 \\ 0 & -I_n & 0 \\ 0 & 0 & \Psi \end{bmatrix}. \tag{45}$$

The final step is to find a basis for $\mathbf{C}^q$ which is orthonormal relative to the Hermitian form (44) and then to transform this basis back to $\mathscr{L}$ via (43). Thus, choose a nonsingular matrix Δ such that $\Delta^* \Xi^* \P \Xi \Delta$ is a signature matrix. Recalling that $q_+ = \dim \mathscr{L}^+$ and $q_- = \dim \mathscr{L}^-$, we must have

$$\Delta^* \Xi^* \P \Xi \Delta = \begin{bmatrix} I_{q_+} & 0 & 0 \\ 0 & I_{q_-} & 0 \\ 0 & 0 & 0 \end{bmatrix}.$$

Then suitable L^+ and L^- are given by

$$[L^+ L^-] = [h_1 \cdots h_{2n+r}] \Xi \Delta \begin{bmatrix} I_{q_+} & 0 \\ 0 & -I_{q_-} \\ 0 & 0 \end{bmatrix}. \tag{46}$$

E. Example

The purpose of this example is simply to illustrate the calculations. Take

$$P(s) = \frac{1}{s+1} \begin{bmatrix} \frac{s-1}{s+1} & 0 \\ 0 & \frac{s-3}{s+3} \end{bmatrix}$$

$$W_1(s) = I, W_2(s) = \begin{bmatrix} -\frac{3s+1}{s+3} & 2 \\ -2\frac{s+1}{s+2} & 0 \end{bmatrix}.$$

Then we may take

$$A_1 = A_2 = P, \qquad B_1 = B_2 = I$$

$$Y_1 = Y_2 = 0, \qquad Z_1 = Z_2 = I$$

in (2) and (3). Then G and U in (11) and (13) are

$$G(z) = \begin{bmatrix} \frac{z+2}{z-2} & 2 \\ \frac{4}{z-3} & 0 \end{bmatrix}, \qquad U(z) = \begin{bmatrix} z & 0 \\ 0 & \frac{2z-1}{2-z} \end{bmatrix}.$$

By the method of Section V-B a basis for $\mathscr{U}^\perp$ is $\{f_1, f_2\}$ where

$$f_1(z) = \begin{bmatrix} 1 \\ 0 \end{bmatrix}, \qquad f_2(z) = (1 - 0.5z)^{-1} \begin{bmatrix} 0 \\ 1 \end{bmatrix}. \tag{47}$$

The matrix Φ_2 is the Gram matrix of these two vectors:

$$\Phi_2 = \begin{bmatrix} 1 & 0 \\ 0 & 4/3 \end{bmatrix}.$$

The matrix $G^\sim$ is

$$G^\sim(z) = \begin{bmatrix} \frac{1+2z}{1-2z} & \frac{4z}{1-3z} \\ 2 & 0 \end{bmatrix}.$$

Thus,

$$(G^\sim f_1)(z) = \begin{bmatrix} \frac{1+2z}{1-2z} \\ 2 \end{bmatrix}$$

$$(G^\sim f_2)(z) = \begin{bmatrix} \frac{4z}{(1-3z)(1-0.5z)} \\ 0 \end{bmatrix}.$$

Taking the $\mathscr{H}^2$-components of these vectors gives

$$(\Gamma^* f_1)(z) = \begin{bmatrix} -1 \\ 2 \end{bmatrix} \tag{48a}$$

$$(\Gamma^* f_2)(z) = (1 - 0.5z)^{-1} \begin{bmatrix} -8/5 \\ 0 \end{bmatrix}. \tag{48b}$$

Then Φ_1, the Gram matrix of this pair, is

$$\Phi_1 = \begin{bmatrix} 5 & 8/5 \\ 8/5 & 256/75 \end{bmatrix}.$$

The maximum λ such that

$$\det(\Phi_1 - \lambda \Phi_2) = 0$$

is $\lambda = 5.62619$. Thus,

$$\mu_1 = \|\Gamma\| = \sqrt{\lambda} = 2.37196.$$

A basis $\{g_1, g_2\}$ for $\mathscr{M}'$ is obtained from (37), (47), and (48). The corresponding $[,]$-Gram matrix is

$$\Psi = \begin{bmatrix} 0.111298 & -0.284385 \\ -0.284385 & 0.726646 \end{bmatrix}. \quad (49)$$

Ker Ψ is one-dimensional and the vector

$$\begin{bmatrix} 1 \\ 0.391366 \end{bmatrix}$$

lies in it. Thus, $\mathcal{M} \cap \mathcal{M}'$ is spanned by the vector $g_1 + 0.391366\, g_2$. Taking this vector as N gives

$$N(z) = \begin{bmatrix} 1 \\ 0.391366(1-0.5z)^{-1} \\ -0.421593 - 0.263995(1-0.5z)^{-1} \\ 0.843185 \end{bmatrix}. \quad (50)$$

For the calculation of L^+ and L^-, the matrices Λ and Ξ are as follows:

$$\Lambda = \begin{bmatrix} 1 & 0 & 0.421593 & -0.843185 & 0 & 0 \\ 0 & 1 & 0.674548 & 0 & -0.142192 & 0.363323 \end{bmatrix}$$

$$\Xi = \begin{bmatrix} -0.421593 & 0.843185 & 0 & 0 \\ -0.674548 & 0 & 0.142192 & -0.363323 \\ 1 & 0 & 0 & 0 \\ 0 & 1 & 0 & 0 \\ 0 & 0 & 1 & 0 \\ 0 & 0 & 0 & 1 \end{bmatrix}.$$

The matrix ¶ is given by (45) and (49). Then

$$\Xi^* ¶ \Xi = \begin{bmatrix} -0.367244 & -0.355481 & -0.095916 & 0.245079 \\ -0.355481 & -0.289039 & 0 & 0 \\ -0.095916 & 0 & 0.131517 & -0.336046 \\ 0.245079 & 0 & -0.336046 & 0.858650 \end{bmatrix}.$$

The signature of this matrix is 1,1,2. A suitable matrix Δ such that

$$\Delta^* \Xi^* ¶ \Xi \Delta = \begin{bmatrix} 1 & 0 & 0 & 0 \\ 0 & -1 & 0 & 0 \\ 0 & 0 & 0 & 0 \\ 0 & 0 & 0 & 0 \end{bmatrix}$$

is

$$\Delta = \begin{bmatrix} -4.12536 & 1.65015 & 1.37117 & -3.50357 \\ 4.26188 & 0 & -1.68637 & 4.30894 \\ 0 & 0 & 1 & 0 \\ 0 & 0 & 0 & 1 \end{bmatrix}.$$

Finally, (46) gives

$$L^+(z) = \begin{bmatrix} 5.33278 \\ 2.78275 \\ -4.12537 \\ 4.26189 \end{bmatrix}, L^-(z) = \begin{bmatrix} -0.695690 \\ -1.11310 \\ 1.65015 \\ 0 \end{bmatrix}. \quad (51)$$

From (50) and (51) we have

$$L_1^+(z) = \begin{bmatrix} 5.33278 \\ 2.78275 \end{bmatrix}, \quad L_1^-(z) = \begin{bmatrix} -0.695690 \\ -1.11310 \end{bmatrix}$$

$$L_2^+(z) = \begin{bmatrix} -4.12537 \\ 4.26189 \end{bmatrix}, \quad L_2^-(z) = \begin{bmatrix} 1.65015 \\ 0 \end{bmatrix}$$

$$N_1(z) = \begin{bmatrix} 1 \\ 0.391366(1-0.5z)^{-1} \end{bmatrix}$$

$$N_2(z) = \begin{bmatrix} -0.421593 - 0.263995(1-0.5z)^{-1} \\ 0.843185 \end{bmatrix}.$$

For a specific optimal K, take $T = 0$ in (30) to get

$$K(z) = \begin{bmatrix} -0.421593 & 0.975183 - .131998(1-0.5z)^{-1} \\ -0.674548 & -0.337274 + 0.252956(1-0.5z)^{-1} \end{bmatrix}.$$

The corresponding optimal H is determined from (16):

$$H = U^{-1}(G - \mu_1 K),$$

$$H(z) = \begin{bmatrix} \frac{2}{z-2} & \frac{0.313093}{2-z} \\ 0.8\frac{2-z}{z-3} & -0.4 \end{bmatrix}.$$

VI. Proper Controllers

Once an optimal $K(z)$ is determined, an optimal $H(z)$ is obtained from (16) and then an optimal $Q(z)$ is obtained from (13). Transformation back to the s-domain yields a $Q(s)$ achieving the minimum in (7). This Q is not proper in general, so neither is the corresponding F in (4). In this section is constructed a sequence $\{F_k\}$ of proper controllers which yields optimality in the limit as $k \to \infty$.

For the remainder of this section, all rational matrices are functions of s. Let Q be any solution of the optimization problem (7), obtained as just described. Choose an integer κ so that $s^{-\kappa}Q(s)$ is proper. Now define a sequence $F_k, k \geq 1$, of feedbacks as follows:

$$F_k = (Y_2 + B_1 Q_k)(Z_2 - A_1 Q_k)^{-1}$$
$$Q_k = J_k Q$$
$$J_k(s) = [k/(s+k)]^\kappa.$$

Clearly, Q_k is stable and proper. It follows that F_k is proper and it achieves internal stability. For large enough k, Q_k is a low-frequency approximation of Q. Intuitively, F_k should therefore be nearly optimal if the weighting is zero at infinite frequency.

Theorem 3: If $W_1 W_2$ is strictly proper, then $\mu = \mu_1$ and

$$\lim_{k \to \infty} \|W_1(I + PF_k)^{-1} W_2\|_\infty = \mu.$$

Proof: As in (5), the weighted sensitivity matrix corresponding to F_k is

$$\begin{aligned} X_k &:= W_1(I + PF_k)^{-1} W_2 \\ &= W_1 Z_2 B_2 W_2 - W_1 A_1 Q_k B_2 W_2 \\ &= W_1 Z_2 B_2 W_2 - J_k W_1 A_1 Q B_2 W_2. \end{aligned} \quad (52)$$

From (5)

$$W_1 A_1 Q B_2 W_2 = W_1 Z_2 B_2 W_2 - X.$$

Substitution of this into (52) gives

$$X_k = J_k X + (1 - J_k) W_1 Z_2 B_2 W_2. \quad (53)$$

From (3) and the fact that N_2 is strictly proper, we have

$$(B_2 Z_2)(\infty) = I$$

and hence

$$(Z_2 B_2)(\infty) = I,$$
$$(W_1 Z_2 B_2 W_2)(\infty) = (W_1 W_2)(\infty) = 0.$$

Thus, $W_1Z_2B_2W_2$ is strictly proper. Since $(1-J_k)(j\omega)$ converges to zero uniformly on compact intervals as $k\to\infty$ [2, Lemma 6.1], it follows that

$$\lim_{k\to\infty}\|(1-J_k)W_1Z_2B_2W_2\|_\infty = 0. \tag{54}$$

From (53), (54), and the fact that $\|J_k\|_\infty = 1$, we get

$$\lim_{k\to\infty}\|X_k\|_\infty \leqslant \|X\|_\infty.$$

But the opposite inequality is also true because X is optimal. Thus,

$$\lim_{k\to\infty}\|X_k\|_\infty = \|X\|_\infty. \tag{55}$$

Now $\|X\|_\infty = \mu_1$ and $\mu_1 \leqslant \mu$ [from (6) and (7)]. Thus, (55) implies that

$$\lim_{k\to\infty}\|X_k\|_\infty = \mu_1 \leqslant \mu.$$

But

$$\lim_{k\to\infty}\|X_k\|_\infty \geqslant \mu$$

by definition of μ. Thus,

$$\lim_{k\to\infty}\|X_k\|_\infty = \mu_1 = \mu. \qquad \blacksquare$$

VII. Concluding Remark

In the problem treated in this paper, the objective was solely to optimize the sensitivity matrix. A more realistic design problem would include plant uncertainty, sensor noise, a bound on controller bandwidth, etc. The Wiener–Hopf approach [1] accommodates some of these factors for the quadratic ($\mathscr{H}^2$) criterion. Helton [23] has made progress in incorporating these factors into the $\mathscr{H}^\infty$-approach.

Appendix
Proofs for Sections IV-B and V-C

Proof of Lemma 2: It is proved in [24, problem 116] that a bounded linear operator $T:\mathscr{H}^2\to\mathscr{H}^2$ is a multiplication operator iff T and σ commute. This result extends to the case where $T:(\mathscr{H}^2)^l\to(\mathscr{H}^2)^k$. Then it is routine to verify that T and σ commute iff the graph of T is shift-invariant. ■

Proof of Lemma 3: a) Let $T:(\mathscr{H}^2)^l\to(\mathscr{H}^2)^k$ be a contraction and let $\mathscr{T}$ denote its graph. To see that $\mathscr{T}$ is negative, let

$$f\in\mathscr{T}, f=\begin{bmatrix} f_1\\ f_2\end{bmatrix},$$

$$f_1\in(\mathscr{H}^2)^k, f_2\in(\mathscr{H}^2)^l.$$

Then $f_1 = Tf_2$, so

$$\begin{aligned}[f,f] &= \|f_1\|_2^2 - \|f_2\|_2^2\\ &\leqslant (\|T\|^2-1)\|f_2\|_2^2\\ &\leqslant 0.\end{aligned}$$

To see that $\mathscr{T}$ is maximal-negative, suppose

$$\mathscr{T}\subset\mathscr{S}\subset(\mathscr{H}^2)^k\times(\mathscr{H}^2)^l, \mathscr{S} \text{ negative.}$$

To show that $\mathscr{T}=\mathscr{S}$, let

$$f\in\mathscr{S}, f=\begin{bmatrix} f_1\\ f_2\end{bmatrix}$$

$$f_1\in(\mathscr{H}^2)^k, \quad f_2\in(\mathscr{H}^2)^l.$$

Then

$$\begin{bmatrix} Tf_2\\ f_2\end{bmatrix}\in\mathscr{T}\subset\mathscr{S}$$

so that

$$g := \begin{bmatrix} f_1-Tf_2\\ 0\end{bmatrix} = \begin{bmatrix} f_1\\ f_2\end{bmatrix} - \begin{bmatrix} Tf_2\\ f_2\end{bmatrix}\in\mathscr{S}.$$

Since $\mathscr{S}$ is negative

$$0\geqslant[g,g] = \|f_1 - Tf_2\|_2^2.$$

Thus, $f_1 = Tf_2$ and so $f\in\mathscr{T}$. Thus, $\mathscr{T}=\mathscr{S}$.

b) Let $\mathscr{T}$ be a maximal-negative subspace of $(\mathscr{H}^2)^k\times(\mathscr{H}^2)^l$ and define the $l\times(k+l)$ projection matrix

$$\Pi_2 = [0, I].$$

It can be proved that $\Pi_2\mathscr{T}$ is closed. It is claimed that

$$\Pi_2\mathscr{T} = (\mathscr{H}^2)^l. \tag{56}$$

If not, then there exists a nonzero vector f_2 in the orthogonal complement of $\Pi_2\mathscr{T}$ in $(\mathscr{H}^2)^l$, i.e.,

$$\langle f_2, g_2\rangle = 0, g_2\in\Pi_2\mathscr{T}.$$

Let $f = \begin{bmatrix} 0\\ f_2\end{bmatrix}$. Then for any $g=\begin{bmatrix} g_1\\ g_2\end{bmatrix}$ in $\mathscr{T}$, $g_2\in\Pi_2\mathscr{T}$, and hence

$$[f,g] = -\langle f_2, g_2\rangle = 0.$$

It follows that $\mathscr{T}+\mathrm{span}(f)$ is a negative subspace which properly contains $\mathscr{T}$; this contradicts the maximal-negativity of $\mathscr{T}$. This proves (56).

Define an operator $T:(\mathscr{H}^2)^l\to(\mathscr{H}^2)^k$ as follows. For each f_2 in $(\mathscr{H}^2)^l$ there exists by (56) an f_1 in $(\mathscr{H}^2)^k$ such that

$$\begin{bmatrix} f_1\\ f_2\end{bmatrix}\in\mathscr{T}.$$

Define $Tf_2 = f_1$. The operator T is well-defined because f_2 uniquely determines f_1; indeed, if

$$\begin{bmatrix} f_1\\ f_2\end{bmatrix}, \begin{bmatrix} \bar f_1\\ f_2\end{bmatrix}\in\mathscr{T}$$

then

$$\begin{bmatrix} f_1-\bar f_1\\ 0\end{bmatrix}\in\mathscr{T}$$

and hence $f_1=\bar f_1$ since $\mathscr{T}$ is negative.

Finally, it is straightforward to verify that $\mathscr{T}$ is the graph of T and that T is a contraction because $\mathscr{T}$ is negative. ■

Proof of Lemma 4: Necessity was proved in Section IV-B. Sufficiency is proved by reversing the argument. ■

Proof of Lemma 5: It follows immediately from the definitions of $\mathscr{M}$ and Γ_1 that

$$\mathscr{M} = \begin{bmatrix} \Gamma_1\\ I\end{bmatrix}(\mathscr{H}^2)^n \oplus \begin{bmatrix} I\\ 0\end{bmatrix}\mathscr{U}. \tag{57}$$

Let $f_1, f_2\in(\mathscr{H}^2)^n$. Then from (57)

$$\begin{bmatrix} f_1 \\ f_2 \end{bmatrix} \in \mathscr{M}'$$

if and only if

$$\langle \Gamma_1 g + h, f_1 \rangle = \langle g, f_2 \rangle, g \in (\mathscr{H}^2)^n, h \in \mathscr{U}$$

or equivalently

$$f_1 \in \mathscr{U}^\perp, f_2 = \Gamma_1^* f_1.$$

This proves that

$$\mathscr{M}' = \begin{bmatrix} I \\ \Gamma_1^* \end{bmatrix} \mathscr{U}^\perp. \tag{58}$$

Finally, intersection of the right-hand sides of (57) and (58) and the fact that $\mathscr{U} \cap \mathscr{U}^\perp = \{0\}$ yield

$$\mathscr{M} \cap \mathscr{M}' = \begin{bmatrix} I \\ \Gamma_1^* \end{bmatrix} \ker(I - \Gamma_1 \Gamma_1^*)$$

which is equivalent to (36). ■

The proof of Proposition 2 requires a preliminary result.

Lemma 6:

$$\mathscr{M} \cap \mathscr{N}' = L\mathscr{Y} + \mathscr{N}. \tag{59}$$

Proof: The matrix M can be factored as

$$M := \begin{bmatrix} G_1 & U \\ I & 0 \end{bmatrix} = \begin{bmatrix} I & G_1 \\ 0 & I \end{bmatrix} \begin{bmatrix} 0 & U \\ I & 0 \end{bmatrix}.$$

Thus, $\mathscr{M}$ is a closed subspace of $\mathscr{X}$ since

$$\begin{bmatrix} 0 & U \\ I & 0 \end{bmatrix}$$

is an inner matrix and

$$\begin{bmatrix} I & G_1 \\ 0 & I \end{bmatrix}$$

is invertible over $\mathscr{H}^\infty$. Hence, $\mathscr{M}$ is a pseudoregular subspace in the sense of [8]. It is proved in [8, equation (2.7)] that

$$\mathscr{M} \cap \mathscr{N}' = \bigvee_{i=0}^{\infty} \sigma^i \mathscr{L}. \tag{60}$$

Next, it is claimed that

$$\mathscr{L} \cap \mathscr{L}' = \mathscr{M} \cap \mathscr{M}' + \sigma(\mathscr{M} \cap \mathscr{M}'). \tag{61}$$

This is proved as follows. Beginning with (25) we get

$$\begin{aligned} \mathscr{L} \cap \mathscr{L}' &= \mathscr{M} \cap (\sigma\mathscr{M})' \cap \mathscr{L}' \\ &= \mathscr{M} \cap (\sigma\mathscr{M} + \mathscr{L})' \\ &= \mathscr{M} \cap \left[\sigma\mathscr{M} + \mathscr{M} \cap (\sigma\mathscr{M})'\right]' \text{ from (25)} \\ &= \mathscr{M} \cap \left\{\left[\sigma\mathscr{M} + (\sigma\mathscr{M})'\right] \cap \mathscr{M}\right\}' \text{ since } \sigma\mathscr{M} \subset \mathscr{M} \\ &= \mathscr{M} \cap \left\{\mathscr{M}' + \left[\sigma\mathscr{M} + (\sigma\mathscr{M})'\right]'\right\} \\ &= \mathscr{M} \cap \left[\mathscr{M}' + (\sigma\mathscr{M})' \cap \sigma\mathscr{M}\right] \\ &= \mathscr{M} \cap \mathscr{M}' + (\sigma\mathscr{M})' \cap \sigma\mathscr{M} \\ &= \mathscr{M} \cap \mathscr{M}' + \sigma(\mathscr{M} \cap \mathscr{M}') \end{aligned}$$

since σ is an isometry.

Finally, to obtain (59), begin with (60):

$$\begin{aligned} \mathscr{M} \cap \mathscr{N}' &= \bigvee_{i=0}^{\infty} \sigma^i \mathscr{L} \\ &= \bigvee_{i=0}^{\infty} \sigma^i(\mathscr{L}^+ \oplus \mathscr{L}^-) + \bigvee_{i=0}^{\infty} \sigma^i(\mathscr{L} \cap \mathscr{L}') \text{ from (26)} \\ &= L\mathscr{Y} + \bigvee_{i=0}^{\infty} \sigma^i(\mathscr{M} \cap \mathscr{M}') \text{ from (27) and (61)} \\ &= L\mathscr{Y} + \mathscr{N}. \end{aligned}$$

The last equality follows from

$$\mathscr{N} = \bigvee_{i=0}^{\infty} \sigma i(\mathscr{M} \cap \mathscr{M}') \tag{62}$$

which in turn follows from (21) and (36). ■

Proof of Proposition 2: i) It is first shown that $\mathscr{N} \subset \mathscr{M} \cap \mathscr{N}'$. That $\mathscr{N} \subset \mathscr{M}$ follows from (62) and the fact that $\mathscr{M}$ is shift-invariant. Also, the subspaces $\sigma^k(\mathscr{M} \cap \mathscr{M}')$ and $\sigma^l(\mathscr{M} \cap \mathscr{M}')$ are orthogonal for any $k, l \geqslant 0$. Thus (62) implies that $\mathscr{N} \subset \mathscr{N}'$. ii) Let $\mathscr{T}$ be a shift-invariant maximal-negative subspace of $\mathscr{Y}$, and define $\mathscr{K} := L\mathscr{T} + \mathscr{N}$. Then $\mathscr{K}$ is shift-invariant because $L\mathscr{T}$ and $\mathscr{N}$ are, and $\mathscr{K} \subset \mathscr{M}$ from (59). To see that $\mathscr{K}$ is negative, let $g \in \mathscr{T}, h \in \mathscr{N}$, and set

$$f = Lg + h \in \mathscr{K}.$$

Then $Lg \in \mathscr{N}'$ from (59), and hence Lg and h are orthogonal. Thus,

$$\begin{aligned} [f, f] &= [Lg, Lg] \\ &= [g, g] \text{ from (28)} \\ &\leqslant 0 \text{ since } \mathscr{T} \text{ is negative.} \end{aligned}$$

The proof that $\mathscr{K}$ is maximal-negative involves three steps.

The first step is to show that $\mathscr{K}' \cap \mathscr{M}$ is positive. Suppose, to the contrary, that there is some f in $\mathscr{K}' \cap \mathscr{M}$, $[f, f] < 0$. Then f is orthogonal to $\mathscr{K}$, so $\mathscr{G} := \mathscr{K} + \operatorname{span}(f)$ is a negative subspace of $\mathscr{M}$. Also, $\mathscr{N} \subset \mathscr{G}$. Since $\mathscr{N} \subset \mathscr{K}$, it follows that $\mathscr{K}' \subset \mathscr{N}'$, so that $f \in \mathscr{N}'$. Since $\mathscr{K} \subset \mathscr{N}'$ by (59), we get that $\mathscr{G} \subset \mathscr{N}'$. We conclude that $\mathscr{G} \subset \mathscr{M} \cap \mathscr{N}'$. Hence, from (59)

$$\begin{aligned} \mathscr{G} &= (L\mathscr{Y} + \mathscr{N}) \cap \mathscr{G} \\ &= \mathscr{G} \cap L\mathscr{Y} + \mathscr{N}. \end{aligned}$$

Thus,

$$\mathscr{G} = L\mathscr{S} + \mathscr{N} \tag{63}$$

where

$$\mathscr{S} := L^{-1}\mathscr{G} := \{f \in \mathscr{Y} : Lf \in \mathscr{G}\}.$$

Now $\mathscr{K} \subset \mathscr{G}$ implies that

$$L\mathscr{T} \subset L\mathscr{S} + \mathscr{N}$$

which in turn implies that

$$\mathscr{T} \subset \mathscr{S} + L^{-1}\mathscr{N}.$$

However, $\mathscr{S} + L^{-1}\mathscr{N}$ is negative, so by maximal-negativity of $\mathscr{T}$

$$\mathscr{T} = \mathscr{S} + L^{-1}\mathscr{N}.$$

Thus,

$$\begin{aligned} \mathscr{K} &= L\mathscr{T} + \mathscr{N} \\ &= L(\mathscr{S} + L^{-1}\mathscr{N}) + \mathscr{N} \\ &= L\mathscr{S} + \mathscr{N} = \mathscr{G}. \end{aligned}$$

The equality $\mathscr{X} = \mathscr{G}$ contradicts the definition of $\mathscr{G}$.

The second step is to show that $\mathscr{X}'$ is positive. Since $\mathscr{N} \subset \mathscr{X}$ and $\mathscr{M} \cap \mathscr{M}' \subset \mathscr{N}$ [from (62)], we have successively

$$\begin{aligned} &\mathscr{M} \cap \mathscr{M}' \subset \mathscr{X}, \\ &\mathscr{X}' \subset \mathscr{M} + \mathscr{M}', \\ &\mathscr{X}' = (\mathscr{M} + \mathscr{M}') \cap \mathscr{X}' \\ &\quad = \mathscr{X}' \cap \mathscr{M} + \mathscr{M}'. \end{aligned} \tag{64}$$

Now it follows from (35) that $\mathscr{M}'$ is positive. Since $\mathscr{M}'$ and $\mathscr{X}' \cap \mathscr{M}$ are orthogonal, (64) shows that $\mathscr{X}'$ is positive.

The final step is to show that if $\mathscr{X}$ is negative and $\mathscr{X}'$ is positive, then $\mathscr{X}$ is maximal-negative. The proof of this is similar to part b) of the proof of Lemma 3. Positivity of $\mathscr{X}'$ implies that $\Pi_2 \mathscr{X} = (\mathscr{H}^2)^n$, for if f_2 is in the orthogonal complement of $\Pi_2 \mathscr{X}$ in $(\mathscr{H}^2)^n$, then $f = \begin{bmatrix} 0 \\ f_2 \end{bmatrix}$ is a negative vector in $\mathscr{X}'$. Define K in $(\mathscr{H}^\infty)^{n \times n}$ as follows. For each f_2 in $(\mathscr{H}^2)^n$ let $Kf_2 = f_1$ where f_1 is uniquely determined via $\begin{bmatrix} f_1 \\ f_2 \end{bmatrix} \in \mathscr{X}$. Then

$$\mathscr{X} = \begin{bmatrix} K \\ I \end{bmatrix} (\mathscr{H}^2)^n$$

and $\|K\|_\infty \leqslant 1$ since $\mathscr{X}$ is negative. Thus, $\mathscr{X}$ is maximal-negative by Lemma 3.

iii) To see that the mapping $\mathscr{T} \mapsto L\mathscr{T} + \mathscr{N}$ is surjective, let $\mathscr{X}$ be a shift-invariant maximal-negative subspace of $\mathscr{M}$. Then since $\mathscr{N} \subset \mathscr{N}'$, $\mathscr{X} + \mathscr{N}$ is negative too, so $\mathscr{N} \subset \mathscr{X}$ by maximal-negativity of $\mathscr{X}$. It follows from this that $\mathscr{X} \subset \mathscr{N}'$, and hence $\mathscr{X} \subset \mathscr{M} \cap \mathscr{N}'$. As in (63), we now obtain that

$$\mathscr{X} = L\mathscr{T} + \mathscr{N}$$

where $\mathscr{T} = L^{-1}\mathscr{X}$. The subspace $\mathscr{T}$ is shift-invariant since $\mathscr{X}$ is. To see that $\mathscr{T}$ is negative, let $f \in \mathscr{T}$. Then $Lf \in \mathscr{X}$ and so

$$\begin{aligned} [f, f] &= [Lf, Lf] \text{ from (28)} \\ &\leqslant 0 \text{ since } \mathscr{X} \text{ is negative.} \end{aligned}$$

To see that $\mathscr{T}$ is maximal-negative, suppose $\mathscr{T} \subset \mathscr{S} \subset \mathscr{Y}$, $\mathscr{S}$ negative. Then

$$\mathscr{X} = L\mathscr{T} + \mathscr{N} \subset L\mathscr{S} + \mathscr{N}$$

and hence

$$L\mathscr{T} + \mathscr{N} = L\mathscr{S} + \mathscr{N} \tag{65}$$

since $L\mathscr{S} + \mathscr{N}$ is negative and $\mathscr{X}$ is maximal-negative. Thus, from (65)

$$\mathscr{S} \subset \mathscr{T} + L^{-1}\mathscr{N}.$$

But the containment $\mathscr{N} \subset \mathscr{X}$ implies that

$$L^{-1}\mathscr{N} \subset L^{-1}\mathscr{X} = \mathscr{T}$$

and hence $\mathscr{S} = \mathscr{T}$.

iv) Finally, to show that the mapping $\mathscr{T} \mapsto L\mathscr{T} + \mathscr{N}$ is injective, suppose $\mathscr{T}_1, \mathscr{T}_2$ are maximal-negative subspaces of $\mathscr{Y}$ and

$$L\mathscr{T}_1 + \mathscr{N} = L\mathscr{T}_2 + \mathscr{N}.$$

Then

$$\mathscr{T}_1 \subset \mathscr{T}_2 + L^{-1}\mathscr{N}. \tag{66}$$

Since $\mathscr{T}_2 + L^{-1}\mathscr{N}$ is negative and $\mathscr{T}_2$ is maximal-negative, we get $L^{-1}\mathscr{N} \subset \mathscr{T}_2$, and hence $\mathscr{T}_1 \subset \mathscr{T}_2$ from (66). By symmetry $\mathscr{T}_2 \subset \mathscr{T}_1$. Thus, $\mathscr{T}_1 = \mathscr{T}_2$. ■

Proof of Theorem 2: Let $T \in (\mathscr{H}^\infty)^{q_+ \times q_-}$, $\|T\|_\infty \leqslant 1$, and define

$$\mathscr{T} = \begin{bmatrix} T \\ I \end{bmatrix} (\mathscr{H}^2)^{q_-}. \tag{67}$$

Then by Lemmas 2 and 3, $\mathscr{T}$ is a shift-invariant maximal-negative subspace of $\mathscr{Y}$. By Proposition 2

$$\mathscr{X} := L\mathscr{T} + \mathscr{N} \tag{68}$$

is a shift-invariant maximal-negative subspace of $\mathscr{M}$. So by Proposition 1, $\mathscr{X}$ has the form

$$\mathscr{X} = \begin{bmatrix} K \\ I \end{bmatrix} (\mathscr{H}^2)^n$$

for some (unique) K in $[G_1]$, $\|K\|_\infty = 1$. Now from (22), (27), and (67),

$$L\mathscr{T} + \mathscr{N} = [L^+T + L^-, N](\mathscr{H}^2)^{q_- + p}.$$

But from (23), (29), and (30)

$$[L^+T + L^-, N] = \begin{bmatrix} K_1 \\ K_2 \end{bmatrix}.$$

Thus, (68) is equivalent to

$$\begin{bmatrix} K \\ I \end{bmatrix} (\mathscr{H}^2)^n = \begin{bmatrix} K_1 \\ K_2 \end{bmatrix} (\mathscr{H}^2)^{q_1 + p}. \tag{69}$$

Multiplication of (69) by $[I, -K]$ yields

$$KK_2 = K_1 \tag{70}$$

and multiplication of (69) by $[0, I]$ implies that $K_2(\mathscr{H}^2)^{q_- + p} = (\mathscr{H}^2)^n$, which implies uniqueness of the solution of (70).

That the mapping $T \to K$ is a bijection follows from Proposition 2, i.e., the mapping $\mathscr{T} \to \mathscr{X}$ is a bijection. ■

Proof of Corollary 1: Regarding the bijection $T \mapsto K$ from (32) onto (33), it is required to show that K is rational iff T is rational and that $K = \check{K}$ iff $P_1^{-1}\check{T}P_2 = T$.

If T is rational, then so are K_1 and K_2 in (30). Thus, the solution K of (31) is rational. Substitution of (30) into (31) yields the following equation for T in terms of K:

$$(KL_2^+ - L_1^+)T = L_1^- - KL_2^-. \tag{71}$$

From Theorem 2, for each K in the set (33), (71) has a unique solution T in the set (32). Thus, if K is rational, so is T.

Finally, let T belong to (32) and let K be the corresponding matrix in (33). It is routine to verify using (30), (31), and (34) that the parameter matrix corresponding to $\check{K}$ is $P_1^{-1}\check{T}P_2$. Since the mapping $T \mapsto K$ is bijective, it follows that $K = \check{K}$ iff $T = P_1^{-1}\check{T}P_2$. ■

References

[1] D. C. Youla, H. A. Jabr, and J. J. Bongiorno, Jr., "Modern Wiener–Hopf design of optimal controllers: Part II," *IEEE Trans. Automat. Contr.*, vol. AC-21, pp. 319–338, 1976.

[2] G. Zames, "Feedback and optimal sensitivity: Model reference transformations, multiplicative seminorms, and approximate inverses," *IEEE Trans. Automat. Contr.*, vol. AC-26, pp. 301–320, 1981.

[3] G. Zames and B. A. Francis, "Feedback, minimax sensitivity, and optimal robustness," *IEEE Trans. Automat. Contr.*, vol. AC-28, pp. 585–601, 1983.

[4] B. A. Francis and G. Zames, "On optimal sensitivity theory for SISO feedback systems," *IEEE Trans. Automat. Contr.*, vol. AC-29, pp. 9–16, Jan. 1984.

[5] D. Sarason, "Generalized interpolation in $\mathscr{H}^\infty$," *Trans. AMS*, vol. 127, pp. 179–203, 1967.

[6] H. Kwakernaak, "Robustness optimization of linear feedback systems," Dep. Appl. Math., Twente Univ. Technol., Internal Rep.

[7] H. Kimura, "Robust stabilizability for a class of transfer functions," Dep. Contr. Eng., Osaka Univ., Osaka, Japan, Tech. Rep. 83-02.

[8] J. A. Ball and J. W. Helton, "A Beurling–Lax theorem for the lie group $U(m, n)$ which contains most classical interpolation theory," *J. Operator Theory*, vol. 9, no. 1, pp. 107–142, 1983.
[9] B.-C. Chang and J. B. Pearson, "Optimal disturbance reduction in linear multivariable systems," Dep. Elec. Eng., Rice Univ., Tech. Rep. 8214.
[10] P. H. Delsarte, Y. Genin, and Y. Kamp, "The Nevanlinna–Pick problem for matrix-valued functions," *SIAM J. Appl. Math.*, vol. 36, pp. 47–61, 1979.
[11] M. G. Safonov and B. S. Chen, "Multivariable stability margin optimization with decoupling and output regulation," *Proc. IEE*, part D, pp. 276–282, 1982.
[12] P. L. Duren, *Theory of $\mathscr{H}^p$ Spaces*. New York: Academic, 1970.
[13] B. Sz.-Nagy and C. Foias, *Harmonic Analysis of Operators on Hilbert Space*. New York: Elsevier, 1970.
[14] C. A. Desoer, R. W. Liu, J. Murray, and R. Saeks, "Feedback system design: The fractional representation approach," *IEEE Trans. Automat. Contr.*, vol. AC-25, pp. 399–412, 1980.
[15] M. Vidyasagar, "On the use of right-coprime factorizations in distributed feedback systems containing unstable subsystems," *IEEE Trans. Circuits Syst.*, vol. CAS-25, pp. 916–921, 1978.
[16] M. Vidyasagar, H. Schneider, and B. A. Francis, 'Algebraic and topological aspects of feedback stabilization," *IEEE Trans. Automat. Contr.*, vol. AC-27, pp. 880–894, 1982.
[17] V. M. Adamjan, D. Z. Arov, and M. G. Krein, "Infinite Hankel matrices and related extension problems," *AMS Transl.*, vol. 2, pp. 133–156, 1978.
[18] ——, "Analytic properties of Schmidt pairs for a Hankel operator and the generalized Schur–Takagi problem," *Mat. Sb.*, vol. 86, pp. 34–75, 1971.
[19] A. Beurling, "On two problems concerning linear transformations in Hilbert space," *Acta Math.*, vol. 81, pp. 239–255, 1949.
[20] P. D. Lax, "Translation invariant subspaces," *Acta Math.*, vol. 101, pp. 163–178, 1959.
[21] J. Bognar, *Indefinite Inner Product Spaces*. New York: Springer-Verlag, 1974.
[22] P. R. Halmos, "Shifts on Hilbert space," *J. reine angew. Math.*, vol. 208, pp. 102–112, 1961.
[23] J. W. Helton, "Worst case analysis in the frequency domain: The $\mathscr{H}^\infty$ approach to control," Dep. Math., Univ. California, San Diego, Internal Rep.
[24] P. R. Halmos, *A Hilbert Space Problem Book*. New York: Van Nostrand, 1967.

L^∞ Optimization and Hankel Approximation

MICHAEL G. SAFONOV AND MADANPAL S. VERMA

Abstract—**An important L^∞-optimization problem arising in feedback system design is shown to be equivalent to an optimal Hankel-norm approximation problem. This enables the problem to be solved using state-space methods based on balanced realizations.**

In this note we show that the following L^∞-optimization problem.

Problem 1:

$$\min_{X(s)\in H^\infty} \|A(s)+B(s)X(s)C(s)\|_\infty \tag{1}$$

where A, B, and $C \in H^\infty$ are given rational matrices of dimension $k \times p$, $k \times m (k \leq m)$, and $n \times p (n \geq p)$, respectively, which occurs in many feedback system design problems is equivalent to

Problem 2:

$$\min_{Y(s)\in H^\infty} \|G(s)-Y^*(s)\|_\infty \tag{2}$$

where $G \in H^\infty$ is obtained from A, B, and C and $Y^*(s) \triangleq Y^T(-s)$.□

Problem 2 can be solved by optimal Hankel-norm approximation methods. This enables Problem 1 to be solved using state-space algorithms based on balanced realizations which exist in the Hankel-norm approximation literature.

The instances of Problem 1 in feedback system design are the problems of sensitivity minimization [1]–[6], stability margin optimization [7], [8], robust stabilization [9], and more general performance specification optimization [10], [11]. The parametrization in (1) in terms of $X \in H^\infty$ is related to the realizability constraint on the set of closed-loop transfer functions and can in general be obtained using coprime fractional methods [12].

We reduce Problem 1 to Problem 2 in two steps. First we show that B and C in (1) can be assumed to be square and inner. Then we use the all-pass property of B and C to reduce (1) to (2). When $k < m$ and/or $p < n$, extra rows and/or columns can be added to A, B, and C to reduce to the case $k = m$ and $p = n$ without affecting the L^∞-norm in (1). Thus, without loss of generality, $k = m$ and $p = n$, and B and C can be factored as [11], [13]

$$B(s)=\theta_1(s)M_1(s) \tag{3}$$

$$C(s)=M_2(s)\theta_2(s) \tag{4}$$

where θ_1 and θ_2 are square and inner and M_1 and M_2 are square and outer. A rational matrix ψ in H^∞ is inner if

$$\psi^*(s)\psi(s)=I \tag{5}$$

or

$$\psi(s)\psi^*(s)=I. \tag{6}$$

A rational matrix W in H^∞ is outer if W has no r.h.p. zeros. Then

$$A+BXC=A+\theta_1M_1XM_2\theta_2$$
$$=A+\theta_1\tilde{X}\theta_2$$

where $\tilde{X} \triangleq M_1XM_2$. Thus,

$$\min_{X\in H^\infty}\|A+BXC\|_\infty=\min_{\tilde{X}\in H^\infty}\|A+\theta_1\tilde{X}\theta_2\|_\infty$$

which shows that B and C can be assumed to be square and inner.

Manuscript received January 23, 1984.

The authors are with the Department of Electrical Engineering-Systems, University of Southern California, Los Angeles, CA 90089-0781.

Assuming B and C in (1) to be square and inner, we get

$$\bar{\sigma}[A(j\omega)+B(j\omega)X(j\omega)C(j\omega)]=\bar{\sigma}[B^*(j\omega)A(j\omega)C^*(j\omega)+X(j\omega)],\ \forall\omega. \tag{7}$$

Hence,

$$\|A+BXC\|_\infty=\|B^*AC^*+X\|_\infty=\|CA^*B+X^*\|_\infty. \tag{8}$$

Let

$$G \triangleq [CA^*B]_+ \tag{9}$$

$$X_o^* \triangleq G-CA^*B \tag{10}$$

$$Y \triangleq X_o-X$$

where $[\cdot]_+$ denotes the stable part of a rational matrix in its partial fraction expansion. Clearly, G, $X_o \in H^\infty$ and

$$\|CA^*B+X^*\|_\infty=\|G-Y^*\|_\infty.$$

As $X \in H^\infty$ iff $Y \in H^\infty$ we have

$$\min_{X\in H^\infty}\|A+BXC\|_\infty=\min_{Y\in H^\infty}\|G-Y^*\|_\infty,$$

i.e., Problem 1 is equivalent to Problem 2. In the case when $k > m$ or $p > n$, Problem 1 can be reduced to a sequence of square problems having the form of Problem 2 via a technique developed by Doyle [11].

Problem 2 is nothing but the zeroth-order optimal Hankel-norm approximation problem. This link between L^∞ optimization and Hankel approximation was first reported by us in [5], [6]. The present correspondence gives a concise summary of these results.

In the scalar case the existence of a solution to Problem 2 was first established by Nehari [14] who also showed that the minimum attained in (2) equals the norm of the Hankel operator induced by G. Later, Adamjan, Arov, and Krein [15] showed that the optimal $G - Y^*$ is a unique all-pass function in the scalar case.

Based on these results, algorithms to compute the optimal solution and the minimal norm were obtained by Genin and Kung [16] using a polynomial formulation and by Bettayeb, Silverman, and Safonov [17], [18] using a state-space formulation based on balanced realizations.

The multivariable version of Problem 2 was first solved by Adamjan, Arov, and Krein [19] who showed that the minimum attained still equals the norm of the Hankel operator induced by G, but the solution in general is not unique. They characterize the family of optimal solutions and give conditions under which the solution is unique. Based on these results, Kung and Lin [20] give an algorithm for computing solutions arbitrarily close to the optimal using the singular vectors of the Hankel matrix of an augmented G and a polynomial matrix formulation. A result of Ball and Helton [21] and the recent state-space algorithm of Doyle [11] can also be used to solve Problem 2. Glover [22] gives a complete characterization of optimal solutions to Problem 2 and a computational algorithm based on state-space methods and balanced realizations.

The significance of the equivalence between Problem 1 and Problem 2 established in the present correspondence is that the L^∞-optimization Problem 1 can be solved using purely state-space algorithms. The major computations involved are doing inner–outer factorizations of B and C, doing a partial fraction expansion to obtain G and computing a zeroth order optimal Hankel approximant of G. A state-space algorithm for doing inner–outer factorizations is given by Doyle [11] and state-space algorithms based on balanced realizations for computing optimal Hankel approximants are given by Bettayeb *et al.* [17], [18] for the scalar case and by Glover [22] for the multivariable case.

REFERENCES

[1] G. Zames, "Feedback and optimal sensitivity: Model reference transformations, multiplicative seminorms, and approximate inverses," *IEEE Trans. Automat. Contr.*, vol. AC-26, pp. 301–320, Apr. 1981.

[2] G. Zames and B. A. Francis, "Feedback, minimax sensitivity and optimal

Reprinted from *IEEE Trans. Automat. Contr.*, vol. AC-30, no. 3, pp. 279–280, Mar. 1985.

robustness," *IEEE Trans. Automat. Contr.*, vol. AC-28, pp. 585-600, May 1983.

[3] B. Francis, J. W. Helt and G. Zames, "H^∞ optimal feedback controllers for linear multivariable systems," presented at the MTNS Symp., Beer-Sheva, Israel, June 1983; and *IEEE Trans. Automat. Contr.*, vol. AC-29, pp. 888-900, Oct. 1984.

[4] B. C. Chang and J. B. Pearson, "Optimal disturbance reduction in linear multivariable systems," Dep. Elec. Eng., Tech. Rep. 8214, Rice Univ., Houston, TX, Oct. 1982, and *IEEE Trans. Automat. Contr.*, vol. AC-29, pp. 880-887, Oct. 1984.

[5] M. G. Safonov, "Singular value response shaping and H^∞-optimization—The MIMO case," Honeywell Systems and Research Center, Minneapolis, MN, Aug. 1982.

[6] M. G. Safonov and M. Verma, "Multivariable L^∞-sensitivity optimization and Hankel approximation," in *Proc. Amer. Contr. Conf.*, San Francisco, CA, June 22-24, 1983.

[7] M. G. Safonov and B. S. Chen, "Multivariable stability margin optimization with decoupling and output regulation," *IEE Proc.*, vol. 129, pt. D, pp. 276-282, Nov. 1982.

[8] M. G. Safonov, "Stability margins of diagonally perturbed multivariable feedback systems," *IEE Proc.*, vol. 129, pt. D, pp. 251-256, Nov. 1982.

[9] H. Kimura, "Robust stabilizability for a class of transfer functions," in *Proc. IEEE Conf. Decision Contr.*, San Antonio, TX, Dec. 14-16, 1983.

[10] H. Kwakernaak, "Robustness optimization of linear feedback systems," in *Proc. IEEE Conf. on Decision and Contr.*, San Antonio, TX, Dec. 14-16, 1983.

[11] J. Doyle, "Synthesis of robust controllers and filters with structured plant uncertainty," in *Proc. IEEE Conf. Decision Contr.*, San Antonio, TX, Dec. 14-16, 1983.

[12] C. A. Desoer, R. W. Liu, J. Murray, and R. Saeks, "Feedback system design: The fractional representation approach to analysis and synthesis," *IEEE Trans. Automat. Contr.*, vol. AC-25, pp. 399-412, June 1980.

[13] D. C. Youla, "On the factorization of rational matrices," *IRE Trans. Inform. Theory*, pp. 172-189, July 1961.

[14] Z. Nehari, "On bounded bilinear forms," *Ann. Math*, vol. 65, pp. 153-162, 1957.

[15] V. M. Adamjan, D. Z. Arov, and M. G. Krein, "Analytic properties of schmidt pairs for a Hankel operator and the generalized Schur-Takagi problem," *Math USSR Sbornik*, vol. 15, pp. 31-73, 1971.

[16] Y. V. Genin and S. Y. Kung, "A two-variable approach to the model reduction problem with Hankel norm criterion," *IEEE Trans. Circuits Syst.*, vol. CAS-28, pp. 912-924, Sept. 1983.

[17] M. Bettayeb, L. M. Silverman, and M. G. Safonov, "Optimal approximation of continuous-time systems," *Proc. IEEE Conf. Decision Contr.*, Albuquerque, NM, Dec. 10-12, 1980.

[18] L. M. Silverman and M. Bettayeb, "Optimal approximation of linear systems," in *Proc. Joint Automat. Contr. Conf.*, San Francisco, CA, 1980.

[19] V. M. Adamjan, D. Z. Arov, and M. G. Krein, "Infinite block matrices and related extension problems," *AMS Trans.*, vol. 111, pp. 133-156, 1978.

[20] S. Y. Kung and D. W. Lin, "Optimal Hankel-norm model reduction: Multivariable systems," *IEEE Trans. Automat. Contr.*, vol. AC-26, pp. 832-852, Aug. 1981.

[21] J. A. Ball and J. W. Helton, "A Beurling-Lax theorem for the lie group $U(m, n)$ which contains most classical interpolation theory," *J. Oper. Theory*, vol. 9, pp. 107-142, 1983.

[22] K. Glover and D. J. N. Limebeer, "Robust multivariable control system design using optimal reduced order plant models," *Proc. Amer. Contr. Conf.*, San Francisco, CA, June 1983.

A Feedback Theory of Two-Degree-of-Freedom Optimal Wiener–Hopf Design

DANTE C. YOULA, FELLOW, IEEE, AND JOSEPH J. BONGIORNO, JR., FELLOW, IEEE

Abstract—**The design of linear two-degree-of-freedom stabilizing controllers is treated in a quadratic-cost setting. The class of all such controllers which give finite cost is established and the tradeoff possible between optimum performance, tracking-cost sensitivity, and stability margins is discussed.**

I. Introduction

In general, every control problem possesses two natural degrees of freedom corresponding to the availability of an exogenous set-point input u and a sensor output w (Fig. 1). However, in the conventional single-loop configuration of Fig. 2, the controller generates the plant input r either by processing only the difference $u - w(F = 1)$, or by processing u and w in a special interdependent manner $(F \neq 1)$.[1] Unfortunately, this restriction has made it difficult, if not impossible, to incorporate essential sensitivity and stability-margin requirements into a quadratic-cost setting. It is therefore necessary to evolve an optimization procedure that allows the controller to process u and w independently. This paper contains several new results which provide a substantial theoretical framework for the execution of this basic idea.

1) The class of all two-degree-of-freedom stabilizing controllers for a prescribed plant and feedback sensor pair is identified in terms of two *free* matrix parameters (Theorem 1).

2) In (42) an explicit expression is obtained for the percent change in the closed-loop characteristic polynomial, due to a change in the plant transfer matrix. This formula reveals that the percent change involves only one of the free parameters in 1). Thus, this parameter can be used to adjust various stability margins.

3) Three new measures, one absolute and two relative, are introduced to describe the effect of plant sensitivity on closed-loop performance. The latter two are used to augment the standard cost-functional and the optimization problem is viewed as one of minimizing cost with a prescribed bound on tracking-cost sensitivity.

4) The augmented cost-functional E is shown to decompose into a sum $E = E_u + E_w$, where E_u is determined solely by R_u, the closed-loop transfer-matrix from u to r, and E_w is determined solely by R_w, the closed-loop transfer-matrix from $-m$ to r (Fig. 1). A complete characterization of all admissible R_u's which yield finite E_u and all admissible R_w's which yield finite E_w is obtained in Theorems 2 and 3, respectively. Once again, the representations are in terms of two free rational matrix parameters $Z_u(s)$ and $Z_w(s)$. Moreover, the unique R_u which minimizes E_u is paired with $Z_u = 0$ while the unique R_w which minimizes E_w is paired with $Z_w = 0$.

5) An illustrative example worked out in detail in Section III serves to clarify both the scope and potential of the method, and the discussion in Section IV offers a comparison with some of the existing literature.

Manuscript received February 3, 1984; revised July 18, 1984. This paper is based on a prior submission of June 7, 1983. Paper recommended by Past Associate Editor, B. R. Barmish. This work was supported in part by the National Science Foundation under Grant ECS-8209748.

The authors are with the Department of Electrical Engineering and Computer Science, Polytechnic Institute of New York, Farmingdale, NY 11735.

[1] This interdependence is most subtle in the multivariable case.

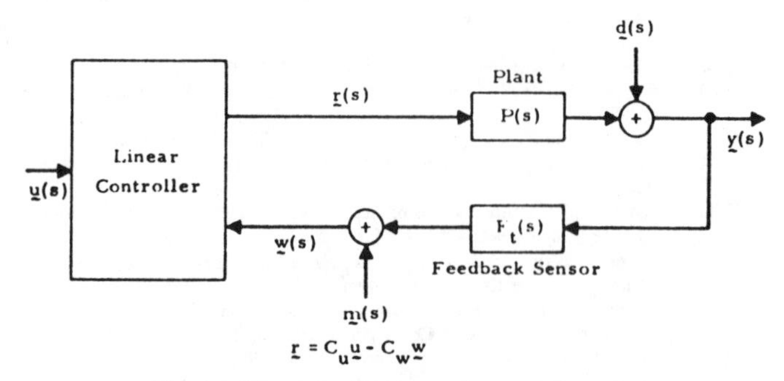

Fig. 1. The general servoregulator configuration.

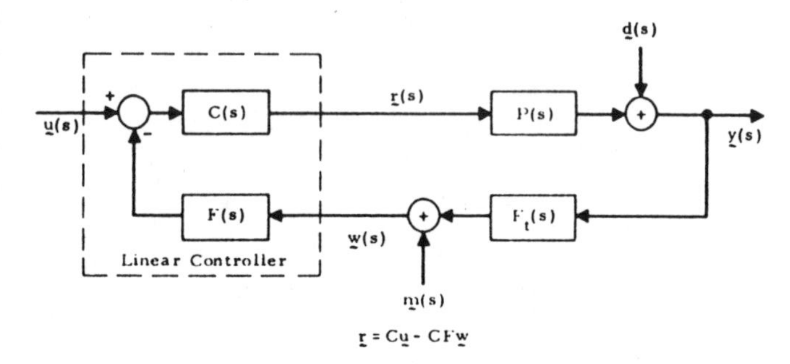

Fig. 2. The classical single-loop servoregulator configuration.

II. Problem Formulation and Results

Apart from all other considerations, the design of a linear controller has two clearly defined objectives. First,[2] transient acquisition of a prescribed set-point input $u(s)$ by the plant output $y(s)$ must be accomplished satisfactorily in the presence of measurement noise $m(s)$ and a saturation constraint on the plant input $r(s)$. Second, regulation in the face of load disturbance $d(s)$ entering through the plant must be acceptable. In the most general situation both requirements are present simultaneously and the problem is one of servoregulation. Available to the designer are two degrees of freedom embodied in the exogeneous input $u(s)$ and the feedback sensor output $w(s)$ and the overall configuration may therefore be depicted schematically as shown in Fig. 1.

To exploit both degrees of freedom fully, the controller must be permitted to generate an output

$$r(s) = C_u(s)u(s) - C_w(s)w(s) \tag{1}$$

in which the transfer matrices $C_u(s)$ and $C_w(s)$ are arbitrary

[2] We work exclusively in the complex $s = \sigma + j\omega$ – domain and all matrices $A(s)$ are assumed to be real and rational. We also omit arguments wherever convenient.

Reprinted from *IEEE Trans. Automat. Contr.*, vol. AC-30, no. 7, pp. 652–665, July 1985.

subject solely to the demand of closed-system asymptotic stability. But as has already been pointed out, the classical single-loop structure of Fig. 2 constrains the processing of $u(s)$ and $w(s)$ and the two inherent degrees of freedom are not completely utilized. Thus, it is not surprising to find that in a multivariable single-loop environment the two design objectives often conflict and even appear mutually contradictory. Moreover, to make matters worse, usually little or no design flexibility remains for the inclusion of any sensitivity or stability-margin considerations.[3] Our goal in this paper is to develop a rigorous theory of two-degree-of-freedom optimal controller design that subsumes these requirements analytically and also respects the notion of feedback from the outset.[4] Theorems 1, 2, and 3 constitute the core of this theory and their proofs are given in full in the Appendix.

The transfer matrices $P(s)$ of the plant and $F_t(s)$ of the sensor are assumed to be prescribed in advance. Clearly, according to (1)

$$T_c(s) = [C_u(s) | -C_w(s)] \tag{2}$$

is the controller transfer matrix from the pair of inputs $u(s)$, $w(s)$ to the output $r(s)$. A straightforward analysis of Fig. 1 yields

$$r = R_u u - R_w(F_t d + m) \tag{3}$$

where

$$R_u = (1 + C_w F_t P)^{-1} C_u \tag{4}$$

and

$$R_w = (1 + C_w F_t P)^{-1} C_w. \tag{5}$$

It is not difficult to see from (3) that $R_u(s)$, $R_w(s)$ and

$$T(s) = P(s) R_u(s) \tag{6}$$

are closed-system transfer matrices from u to r, $-m$ to r, and u to y, respectively. Of course, in any stable design, all three must be analytic in $\mathrm{Re}\, s \geq 0$.[5]

In principle, a meaningful cost-functional should incorporate those design factors that are judged to be of engineering significance. Although there exists considerable difference of opinion as to how they are to be assessed and rated in relative importance, it is nevertheless generally conceded that transient error, large rms plant input, and excessive closed-system sensitivity to plant uncertainty are all factors that deserve to be penalized.

Let $G_e(s)$ denote the spectral density of the error

$$e(s) = u(s) - y(s) \tag{7}$$

$G_r(s)$ that of the input $r(s)$ and let $Q(s)$ denote some suitably chosen full-rank para-Hermitian nonnegative-definite weight-matrix.[6] Then,[7] useful cost-functionals for tracking error and plant saturation are given by

$$E_t = \frac{1}{2\pi j} \int_{-j\infty}^{j\infty} \mathrm{tr}\, G_e(s)\, ds \tag{8}$$

and

$$E_s = \frac{1}{2\pi j} \int_{-j\infty}^{j\infty} \mathrm{tr}(Q G_r)\, ds, \tag{9}$$

respectively [3]. As usual, to permit tradeoff between E_t and E_s, it proves more convenient to work with the single cost

$$E_0 = E_t + k E_s \tag{10}$$

k an adjustable nonnegative constant. In terms of the individual spectral densities $G_u(s)$, $G_d(s)$, and $G_m(s)$ of $u(s)$, $d(s)$, and $m(s)$, we obtain the expanded form

$$\begin{aligned} E_0 = {} & \frac{1}{2\pi j} \int_{-j\infty}^{j\infty} \mathrm{tr}[(1 - PR_u) G_u (1 - PR_u)_* + kQR_u G_u R_{u*}]\, ds \\ & + \frac{1}{2\pi j} \int_{-j\infty}^{j\infty} \mathrm{tr}[(1 - PR_w F_t) G_d (1 - PR_w F_t)_* \\ & + kQR_w F_t G_d F_{t*} R_{w*}]\, ds \\ & + \frac{1}{2\pi j} \int_{-j\infty}^{j\infty} \mathrm{tr}[R_w G_m R_{w*} (P_* P + kQ)]\, ds. \end{aligned} \tag{11}$$

It follows readily from (3)–(5) that any uncertainty $\delta P(s)$ in $P(s)$ produces respective first-order perturbations

$$\delta e = -S(\delta P) r \tag{12}$$

and

$$\delta r = -R_w F_t (\delta P) r \tag{13}$$

in e and r, where

$$S(s) = 1 - P(s) R_w(s) F_t(s) \tag{14}$$

is the closed-system sensitivity matrix. Consequently, since (12) and (13) express δe and δr with respect to the *same* reference r, the matrix multipliers

$$S(\delta P) \tag{15}$$

and

$$R_w F_t (\delta P) \tag{16}$$

serve as feedback indicators of the relative importance of plant-uncertainty on tracking-error and plant-saturation, respectively.

It is therefore consistent with our weighting of e and r in E_0 to introduce the plant-uncertainty spectral density[8]

$$G_s(s) = \langle (\delta P)(\delta P)_* \rangle, \tag{17}$$

another adjustable nonnegative constant μ, and

$$\begin{aligned} E = E_0 & + \frac{\mu}{2\pi j} \int_{-j\infty}^{j\infty} \mathrm{tr}(S G_s S_*)\, ds \\ & + \frac{\mu}{2\pi j} \int_{-j\infty}^{j\infty} \mathrm{tr}(kQR_w F_t G_s F_{t*} R_{w*})\, ds \end{aligned} \tag{18}$$

as our final candidate for the quadratic measure of total performance.[9] Explicitly,

$$\begin{aligned} E = {} & \frac{1}{2\pi j} \int_{-j\infty}^{j\infty} \mathrm{tr}[(1 - PR_u) G_u (1 - PR_u)_* + kQR_u G_u R_{u*}]\, ds \\ & + \frac{1}{2\pi j} \int_{-j\infty}^{j\infty} \mathrm{tr}[(1 - PR_w F_t)(G_d + \mu G_s)(1 - PR_w F_t)_* \\ & + kQR_w F_t (G_d + \mu G_s) F_{t*} R_{w*}]\, ds \\ & + \frac{1}{2\pi j} \int_{-j\infty}^{j\infty} \mathrm{tr}[R_w G_m R_{w*} (P_* P + kQ)]\, ds. \end{aligned} \tag{19}$$

[3] The two early excellent texts by I. Horowitz [1] and S. Chang [2] are notable for their appreciation of this point.

[4] For example, if the plant and feedback-sensor are asymptotically stable and load disturbance is absent, the detrimental effects of sensor noise can be completely circumvented by using open-loop compensation. Thus, in this case there might be other reasons for insisting on feedback. In the design theory proposed in this paper, a feedback solution emerges naturally by incorporating two sensitivity measures in the cost functional.

[5] $\mathrm{Re}\, a \equiv$ "real part" of a.

[6] $Q(s) = Q_*(s) \equiv Q'(-s)$, $Q(j\omega) \geq 0$ for all real ω and $\det Q(s) \not\equiv 0$. (The prime denotes matrix transposition and $\det A \equiv$ determinant A.)

[7] $\mathrm{tr}\, A$ = trace A and all random processes are assumed to be second-order stationary with possible shape-deterministic mean values.

[8] The ensemble average in (17) is over all parameters that contribute to $\delta P(s)$.

[9] The use of spectral densities to characterize plant-uncertainty is not new [4].

Observe that (19) is obtained from (11) by replacing G_d with $G_d + \mu G_s$. Hence, it is possible to account for *some* of the feedback effects of plant-uncertainty by increasing load disturbance and, in a sense, this conclusion justifies an old conjecture.

There are many pairs (R_u, R_w) that yield both finite cost E and an asymptotically stable closed-system design. Included in these are absolutely optimal ones that minimize E absolutely and conditionally optimal ones that minimize E, subject to prescribed bounds on stability margins and *tracking-cost* sensitivity.[10] To delineate either of these sets, it is first necessary to solve the intermediate problem of finding all pairs (R_u, R_w) *acceptable* for the given datum (P, F_t).

Definition 1: The two real rational matrices $R_u(s)$ and $R_w(s)$ are said to be acceptable for the given data $P(s)$ and $F_t(s)$, if there exists a controller which realizes them as the designated closed-system transfer matrices of an internally asymptotically stable configuration of generic type shown in Fig. 1.

This definition of acceptability does not exclude nondynamical controllers from consideration, so that $T_c(s)$ (2) is not necessarily proper. It has been our experience that properness is almost always guaranteed in correctly posed optimization problems. In any case, the matter can always be investigated after the larger class has been identified.

Assumption 1: The plant and feedback sensor are both free of hidden poles in Re $s \geq 0$ and any finite pole of $F_t(s)P(s)$ in Re $s \geq 0$ has a McMillan degree which is equal to the sum of its McMillan degrees as a pole of $F_t(s)$ and as a pole of $P(s)$. (Any such pair is said to admissible.)

It should be obvious that all parts of Assumption 1 are necessary for the existence of a stabilizing controller. Theorem 1 characterizes the pairs (R_u, R_w) acceptable for a prescribed admissible datum pair (P, F_t) in terms of two arbitrary real rational matrix parameters $H_1(s)$ and $K_1(s)$ analytic in Re $s \geq 0$. Let

$$A^{-1}B = F_t P = B_1 A_1^{-1} \tag{20}$$

where (A, B) is any left and (B_1, A_1) is any right coprime pair of polynomial matrices. As is well known [3], [5], [6], there exist real polynomial matrices $X(s)$, $Y(s)$, $X_1(s)$, and $Y_1(s)$ such that $BA_1 = AB_1$, $A_1Y_1 = YA$,

$$AX + BY = 1 \tag{21}$$

$$X_1A_1 + Y_1B_1 = 1 \tag{22}$$

and

$$\det X \cdot \det X_1 \neq 0. \tag{23}$$

Theorem 1: Under Assumption 1, a pair (R_u, R_w) is acceptable for (P, F_t) iff

$$R_u = A_1 H_1 \tag{24}$$

and

$$R_w = A_1(Y_1 + K_1 A) \tag{25}$$

where $H_1(s)$ and $K_1(s)$ are *arbitrary* compatibly dimensioned real rational matrices analytic in Re $s \geq 0$ such that

$$\det (X_1 - K_1 B) \neq 0. \tag{26}$$

The associated controller transfer matrix $T_c(s)$ that realizes the pair (R_u, R_w) is then given by

$$T_c = (X_1 - K_1 B)^{-1}[H_1 | -(Y_1 + K_1 A)] \tag{27}$$

and any one of its realizations (minimal or otherwise) that is free of unstable hidden modes defines an asymptotically stable closed-system design. More precisely, if

$$K_1 = L_1^{-1} M_1 \tag{28}$$

is any left-coprime decomposition of $K_1(s)$, the resultant closed-system characteristic polynomial $\Delta(s)$ is given by

$$\Delta = \frac{\Delta_c}{\det (L_1 X_1 - M_1 B)} \cdot \frac{\Delta_t \Delta_p}{\det A_1} \cdot \det L_1 \tag{29}$$

where $\Delta_c(s)$, $\Delta_t(s)$, and $\Delta_p(s)$ are the characteristic polynomials of the controller, feedback sensor, and plant, respectively.

The first factor in (29) is strict-Hurwitz because the controller is realized without unstable hidden modes, the second because (P, F_t) is an admissible pair, and the third because $K_1(s)$ is analytic in Re $s \geq 0$. (The parametrizations of $R_u(s)$, $R_w(s)$, and $T_c(s)$ given in (24)–(27) are useful generalizations of those presented in [3] for one-degree-of-freedom controllers.)[11]

Although many different generic controller topologies can be employed for the realization of $T_c(s)$, it is nonetheless easily seen that the single-loop configuration of Fig. 2 is inadequate, even in the scalar case.[12] For, observe that if $C_u = C$ and $C_w = CF$, then (27) leads to the equations

$$C = (X_1 - K_1 B)^{-1} H_1 \tag{30}$$

and

$$CF = (X_1 - K_1 B)^{-1}(Y_1 + K_1 A) \tag{31}$$

for the determination of $C(s)$ and $F(s)$. Hence, any finite zero of $H_1(s)$ in Re $s \geq 0$ which is neither a zero of $X_1 - K_1 B$ nor $Y_1 + K_1 A$, is simultaneously a zero of $C(s)$ and a pole of $F(s)$, i.e., it is an unstable hidden mode in the realization of $T_c(s)$.

One possible remedy is to use a precompensator $C_f(s)$ outside the loop as shown in Fig. 3. Under these conditions (30) and (31) go into

$$CC_f = (X_1 + K_1 B)^{-1} H_1 \tag{32}$$

and

$$CF = (X_1 - K_1 B)^{-1}(Y_1 + K_1 A), \tag{33}$$

respectively, and the troublesome zeros of $H_1(s)$ in Re $s \geq 0$ can be transferred into $C_f(s)$. Two obvious disadvantages of the method are that the precompensator must be stable and dynamical because it lies external to the loop. When necessary, such difficulties should be avoided by the use of a different structure and in our opinion, the topic of controller topology deserves more attention.

The problem of constructing tractable stability-margin measures for single-input, single-output plants is, of course, classical [1], [2]. Nevertheless, it is only recently [9]–[18], because of renewed interest in frequency-domain controller design, that the multivariable case has come under careful scrutiny. Our own approach attempts to exploit the double-parameter representation of $T_c(s)$.

Let $P(s)$ and $F_t(s)$ denote nominal transfer matrices and let

$$T_c(s) = [C_u(s) | -C_w(s)] \tag{34}$$

[10] To be defined.

[11] By using results contained in a recent comprehensive paper by Desoer and Gustafson [44], it should be possible to extend the parametrizations (24)–(27) beyond the rational domain. The two excellent papers [7], [8] by Pernebo also solve several important problems involving two-degree-of-freedom dynamical controller design. However, the restriction to dynamical controllers precludes the use of derivative compensation and appears to lose the convenient parametrization (27). Moreover, performance, at least in the sense that we have defined it, is not an explicit issue. Moreover, it is also possible to arrive at Theorem 1 directly from the class of stabilizing controllers derived in [3] using results available in [39].

[12] Our argument is scalar oriented but has a direct multivariable analog.

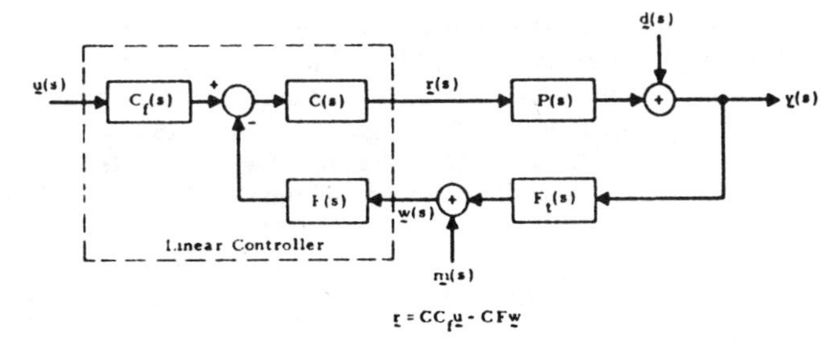

Fig. 3. The precompensated classical single-loop servoregulator configuration.

denote the transfer matrix of any companion stabilizing controller. Then, from Theorem 1,

$$C_w=(X_1-K_1B)^{-1}(Y_1+K_1A) \tag{35}$$

where $K_1(s)$ is any real rational matrix analytic in Re $s \geq 0$.

Consider a perturbation that modifies *only* the plant.[13] Under such a change, $F_t(s)$, $C_u(s)$, $C_w(s)$, $\Delta_t(s)$, and $\Delta_c(s)$ remain unaltered but the characteristic polynomials $\Delta_p(s)$ and $\Delta(s)$ can change. Let the transfer matrix and characteristic polynomial of the modified plant be denoted by $P(s)+\delta P(s)$ and $\Delta_{p'}(s)$, respectively. Concomitantly, let

$$F_t(P+\delta P)=(B_1+\delta B_1)(A_1+\delta A_1)^{-1} \tag{36}$$

where $B_1(s)+\delta B_1(s)$, $A_1(s)+\delta A_1(s)$ are right-coprime polynomial matrices. An explicit expression for the characteristic polynomial $\Delta_\delta(s)$ of the perturbed closed-system structure is easily derived.

From first principles [3],

$$\Delta_\delta=\Delta_c\Delta_t\Delta_{p'}\cdot \det\,(1+C_wF_t(P+\delta P)) \tag{37}$$

or, after some straightforward simplification with the aid of (22), (35), (36), $K_1 = L_1^{-1}M_1$ and the identity[14]

$$(L_1X_1-M_1B)A_1+(L_1Y_1+M_1A)B_1=L_1, \tag{38}$$

we obtain

$$\Delta_\delta=\frac{\Delta_c}{\det\,(L_1X_1-M_1B)}\cdot\frac{\Delta_t\Delta_{p'}}{\det\,(A_1+\delta A_1)}\cdot \det\,(L_1+\delta L_1) \tag{39}$$

where

$$\delta L_1=(L_1X_1-M_1B)\cdot\delta A_1+(L_1Y_1+M_1A)\cdot\delta B_1. \tag{40}$$

Note, that with no plant change, (39) reduces to (29) and $\Delta_\delta(s)$ equals $\Delta(s)$. Each of the three multiplicative factors in (39) has a specific bearing on stability.

The first is strict-Hurwitz because the controller for the nominal design is realized without unstable hidden modes. As for the second, the *a priori* admissibility of the pair (P, F_t) guarantees only the strict-Hurwitz character of $\Delta_t\Delta_p/\det A_1$ and not that of $\Delta_t\Delta_{p'}/\det\,(A_1+\delta A_1)$. To conclude that the latter is strict-Hurwitz it is necessary to assume admissibility of the perturbed pair $(P+\delta P, F_t)$.[15] If this is granted, there remains only the third factor $\det\,(L_1+\delta L_1)$ to consider.

Clearly,

$$\det\,(L_1+\delta L_1)=\det\,L_1\cdot\det\,(1+L_1^{-1}\delta L_1). \tag{41}$$

Since the nominal design is stable, $\det L_1(s)$ is a strict-Hurwitz polynomial which is invariant under the perturbation $\delta P(s)$.

Hence, the dependence of the zeros of

$$\det\,(1+L_1^{-1}\delta L_1)=\det\,(1+(X_1-K_1B)\cdot\delta A_1+(Y_1+K_1A)\cdot\delta B_1) \tag{42}$$

on the variations δA_1 and δB_1 reveals the tolerance of the nominal design to instability under the plant change. (We prove in the Appendix that these zeros are uniquely determined by $P(s)$ and $\delta P(s)$, so that the choice of coprime factors is immaterial.)

Ideally, to accommodate "large" excursions in $A_1(s)$ and $B_1(s)$, both $X_1(s)-K_1(s)B(s)$ and $Y_1(s)+K_1(s)A(s)$ must be made simultaneously "small" and unfortunately, in view of the identity

$$(X_1-K_1B)A_1+(Y_1+K_1A)B_1=1 \tag{43}$$

this is impossible to achieve. Nevertheless, preliminary analysis suggests that regardless of their eventual definitions, all stability margins will involve constraints imposed either directly or indirectly on the two matrices X_1-K_1B and Y_1+K_1A through the choice of free parameters $K_1(s)$.

The two terms that have been added to E_0 in (18) to give the total cost E constitute only relative penalties of δe and δr with respect to r as reference. However, r also depends on the particular nominal design and what we now need is one additional performance measure that refers exclusively to itself. From (8) and (12) it follows that the increment δe in e produced by plant uncertainty $\delta P(s)$ gives rise to an increment

$$\delta E_t=\frac{1}{2\pi j}\int_{-j\infty}^{j\infty}\text{tr}[S\langle(\delta P)rr_*(\delta P)_*\rangle S_*]ds \tag{44}$$

in the tracking-cost E_t.[16] Or, since $\langle rr_*\rangle = G_r$,

$$\delta E_t=\frac{1}{2\pi j}\int_{-j\infty}^{j\infty}\text{tr}[S\langle(\delta P)G_r(\delta P)_*\rangle S_*]ds \tag{45}$$

where

$$G_r=R_uG_uR_{u*}+R_w(F_tG_dF_{t*}+G_m)R_{w*}. \tag{46}$$

By definition,

$$\eta_t=\left(\frac{\delta E_t}{E_t}\right)^{1/2} \tag{47}$$

is the tracking-cost sensitivity.

The calculation of η_t involves $R_u = A_1H_1$, $R_w = A_1(Y_1+K_1A)$ and a rather difficult ensemble average $\langle(\delta P)G_r(\delta P)_*\rangle$.[17] In general, therefore, restrictions on η_t impose constraints on *both* of the free parameters $H_1(s)$ and $K_1(s)$. To summarize, our design procedure takes into explicit account performance, plant-saturation, and plant-uncertainty. The latter is penalized both relatively, by the addition of two extra terms to E_0, and absolutely by the imposition of bounds on η_t, the tracking-cost sensitivity. Moreover, a potential quantitative dependence of stability margin on the parameter $K_1(s)$ is also indicated.

The expression for E in (19) decomposes into the sum

$$E=E_u+E_w$$

in which

$$E_u=\frac{1}{2\pi j}\int_{-j\infty}^{j\infty}\text{tr}[(1-PR_u)G_u(1-PR_u)_*+kQR_uG_uR_{u*}]ds \tag{48}$$

[13] Thus far, the size of the perturbation is arbitrary.

[14] $BA_1 = AB_1$.

[15] For example, if $F_t(s) = 1$, admissibility is assured if the perturbed and unperturbed plants have identical hidden modes, etc.

[16] E_t is defined in (8) and $\langle\delta P(s)\rangle = 0$, all s.

[17] A convenient expression for this average in terms of G_r and $G_s = \langle(\delta P)(\delta P)_*\rangle$ is still not available.

includes all terms involving R_u and

$$E_w = \frac{1}{2\pi j}\int_{-j\infty}^{j\infty} \mathrm{tr}[(1-PR_wF_t)(G_d+\mu G_s)(1-PR_wF_t)_* + kQR_wF_t(G_d+\mu G_s)F_{t*}R_{w*}]ds + \frac{1}{2\pi j}\int_{-j\infty}^{j\infty} \mathrm{tr}[R_wG_mR_{w*}(P_*P+kQ)]ds \tag{49}$$

all those involving R_w. Consequently, the problem of finding all acceptable pairs (R_u, R_w) that yield finite E, is equivalent to finding all acceptable R_u's that yield finite E_u and all acceptable R_w's that yield finite E_w. This separation is of great practical and theoretical significance. Theorems 2 and 3 completely solve the problem and rest on six important additional assumptions, many of which have been extensively discussed in [3].

Assumption 2: The spectral density $G_u(s)$ vanishes at least as fast as $1/s^2$ as $s \to \infty$, i.e.,[18]

$$G_u(s) \leq 0(s^{-2}). \tag{50}$$

Assumption 3: $(F_t - 1)P$ and F_t are analytic on the finite part of the $s = j\omega$-axis.

Assumption 4: AG_uA_*, $A_{1*}(P_*P + kQ)A_1$ and Q are analytic and nonsingular on the finite part of the $s = j\omega$-axis.

Theorem 2: Suppose that Assumptions 1–4 are satisfied and let the matrices $\Lambda(s)$ and $\Omega_u(s)$ denote the Wiener–Hopf spectral solutions[19] of the two equations

$$A_{1*}(P_*P+kQ)A_1 = \Lambda_*\Lambda \tag{51}$$

and

$$G_u = \Omega_u\Omega_{u*}. \tag{52}$$

Let

$$\Gamma = \Lambda_*^{-1}A_{1*}P_*\Omega_u. \tag{53}$$

Then the following holds.

1) The set of all acceptable R_u's that yield finite E_u is generated by the formula[20]

$$R_u = A_1\Lambda^{-1}(Z_u + \{\Gamma\}_+)\Omega_u^{-1} \tag{54}$$

where $Z_u(s)$ is an arbitrary real rational matrix $\leq 0(s^{-1})$ and analytic in Re $s \geq 0$.

2) The acceptable R_u that minimizes E_u is given by

$$\tilde{R}_u = A_1\Lambda^{-1}\{\Gamma\}_+\Omega_u^{-1} \tag{55}$$

and corresponds to the choice $Z_u(s) = 0$.

3) Let the minimum cost E_u be denoted by $\tilde{E}_u$. Then, for any $Z_u(s)$

$$E_u = \tilde{E}_u + \frac{1}{2\pi j}\int_{-j\infty}^{j\infty} \mathrm{tr}(Z_uZ_{u*})ds \geq \tilde{E}_u. \tag{56}$$

4) Every R_u given by (54) preserves the steady-state tracking potential of the plant.

It is important to note that formulas (51)–(55) permit us to relate the poles of PR_u, which condition the transient response, to the choice of nonnegative constant k and weighting matrix Q. Thus, poor transient response can be predicted and avoided in advance. Of even greater practical significance is the fact that *no* coprime factorizations are needed if the plant and sensor transfer matrices are analytic in the *strict* right half-plane, Re $s > 0$!

Corollary: Let $P(s)$ and $F_t(s)$ be free of finite poles in Re $s > 0$ and let $\Lambda_r(s)$ denote the Wiener–Hopf solution of the "reduced" equation

$$P_*P + kQ = \Lambda_{r*}\Lambda_r. \tag{56a}$$

Then, under the conditions of Theorem 2, (54) and (55) simplify to

$$R_u = \Lambda_r^{-1}(Z_u + \{\Gamma\}_+)\Omega_u^{-1} \tag{56b}$$

and

$$\tilde{R}_u = \Lambda_r^{-1}\{\Gamma_r\}_+\Omega_u^{-1}, \tag{56c}$$

respectively, where

$$\Gamma_r = \Lambda_{r*}^{-1}P_*\Omega_u. \tag{56d}$$

Proof: Clearly, $F_t(s)P(s) = B_1(s)A_1^{-1}(s)$ is analytic in Re $s > 0$; hence, det $A_1(s) \neq 0$, Re $s > 0$. Consequently, from (51) and the special properties of the spectral factors $\Lambda_r(s)$ and $\Lambda(s)$, we obtain $\Lambda(s) = \Lambda_r(s)A_1(s)$. Lemma 1 now follows immediately by substitution into (54), (55), and (53), Q.E.D.

Assumption 5: The spectral densities $G_d(s)$ and $G_s(s)$ are both $\leq 0(s^{-2})$ and $G_m(s)$ is analytic on the finite part of the $s = j\omega$-axis.

Assumption 6: Let

$$G = G_m + F_t(G_d + \mu G_s)F_{t*}. \tag{56e}$$

The matrix AGA_* is analytic and nonsingular on the finite part of the $s = j\omega$-axis.[21]

Assumption 7: $F_t(s)$ is nonsingular on the finite part of the $s = j\omega$-axis.

Theorem 3: Suppose that Assumptions 1–7 are satisfied and let $\Omega(s)$ denote the Wiener–Hopf spectral soltuion of the equation

$$AGA_* = \Omega\Omega_*. \tag{56f}$$

Let

$$I = AF_t(G_d + \mu G_s)PA_1. \tag{56g}$$

Then the following holds.

1) The set of all acceptable R_w's that yield finite E_w is generated by the formula

$$R_w = A_1\Lambda^{-1}(\{\Lambda_*^{-1}I_*\Omega_*^{-1}\}_+ + \{\Lambda A_1^{-1}Y\Omega\}_- + Z_w)\Omega^{-1}A \tag{56h}$$

where $Z_w(s)$ is an arbitrary real rational matrix $\leq 0(s^{-1})$ and analytic in Re $s \geq 0$.

2) The acceptable R_w that minimizes E_w is given by

$$\tilde{R}_w = A_1\Lambda^{-1}(\{\Lambda_*^{-1}I_*\Omega_*^{-1}\}_+ + \{\Lambda A_1^{-1}Y\Omega\}_-)\Omega^{-1}A \tag{56i}$$

and corresponds to the choice $Z_w = 0$.

3) Let the minimum cost E_w be denoted by $\tilde{E}_w$. Then, for any $Z_w(s)$,

$$E_w = \tilde{E}_w + \frac{1}{2\pi j}\int_{-j\infty}^{j\infty} \mathrm{tr}(Z_wZ_{w*})ds \geq \tilde{E}_w. \tag{56j}$$

Corollary: Let $P(s)$ and $F_t(s)$ be free of finite poles in Re $s > 0$

[18] $A(s) \leq 0(s^r)$ means that no entry in $A(s)$ grows faster than s^r as $s \to \infty$.

[19] $\Lambda(s)$ and $\Omega_u(s)$ are square real rational matrices that are analytic together with their inverses in Re $s > 0$. They are unique up to constant orthogonal matrix multipliers [19].

[20] In the partial fraction expansion of $\Gamma(s)$, the contributions made by all its finite poles in Re $s \leq 0$, Re $s > 0$ and at $s = \infty$ are denoted by $\{\Gamma\}_+$, $\{\Gamma\}_-$, and $\{\Gamma\}_\infty$, respectively. Clearly, $\{\Gamma\}_+$ is analytic in Re $s > 0$, $\{\Gamma\}_-$ in Re $s \leq 0$, and both are $\leq 0(s^{-1})$.

[21] Assumptions 1–7 are not completely independent. For example, as is shown in the Appendix (A-39) the $j\omega$-analyticity of $(F_t - 1)P$ implies that of PA_1. In turn, the $j\omega$-analyticity of PA_1 and Q implies that of $A_{1*}(P_*P + kQ)A_1$.

and let $\Omega_r(s)$ denote the Wiener-Hopf solution of the reduced equation

$$G=\Omega_r\Omega_{r*}. \tag{56k}$$

Then, under the conditions of Theorem 3, (56h) and (56i) simplify to

$$R_w=\Lambda_r^{-1}(\{\Lambda_{r*}^{-1}I_{r*}\Omega_{r*}^{-1}\}_+ + Z_w)\Omega_r^{-1} \tag{56l}$$

and

$$\tilde{R}_w=\Lambda_r^{-1}\{\Lambda_{r*}^{-1}I_{r*}\Omega_{r*}^{-1}\}_+\Omega_r^{-1}, \tag{56m}$$

respectively, where

$$I_r=F_t(G_d+\mu G_s)P. \tag{56n}$$

Proof: By hypothesis, A_1 and A are nonsingular in Re $s>0$. Thus, $\Lambda=\Lambda_rA_1$, $\Omega=A\Omega_r$,

$$\{\Lambda A_1^{-1}Y\Omega)_-=0 \tag{56o}$$

and

$$\Omega^{-1}I\Lambda^{-1}=\Omega_r^{-1}A^{-1}AF_t(G_d+\mu G_s)PA_1A_1^{-1}\Lambda_r^{-1}\Omega_r^{-1}=I_r\Lambda_r^{-1}. \tag{56p}$$

The conclusions now follow immediately by direct substitution into (56h) and (56i). Q.E.D.

Comment: The importance of the corollaries to Theorems 2 and 3 cannot be overemphasized, since the majority of plants encountered in practice are modeled by transfer matrices that are analytic in Re $s>0$ but usually possess one or more $j\omega$-axis poles. Our design formulas (56a)–(56d) and (56k)–(56n) incorporate all such poles naturally, without any need to move them first into Re $s<0$.

III. Illustrative Example

For the pure servoproblem without measurement noise, both G_d and G_m are zero, $G_r=R_uG_uR_{u*}$,

$$E=\frac{1}{2\pi j}\int_{-j\infty}^{j\infty}\mathrm{tr}[(1-PR_u)G_u(1-PR_u)_* + kQR_uG_uR_{u*}]ds$$
$$+\frac{\mu}{2\pi j}\int_{-j\infty}^{j\infty}\mathrm{tr}[(1-PR_wF_t)G_s(1-PR_wF_t)_*$$
$$+kQR_wF_tG_sF_{t*}R_{w*}]ds \tag{57}$$

and

$$\eta_t^2=\left(\frac{\delta E_t}{E_t}\right)=\frac{\frac{1}{2\pi j}\int_{-j\infty}^{j\infty}\mathrm{tr}[(1-PR_wF_t)\langle(\delta P)R_uG_uR_{u*}(\delta P)_*\rangle(1-PR_wF_t)_*]ds}{\frac{1}{2\pi j}\int_{-j\infty}^{j\infty}\mathrm{tr}[(1-PR_u)G_u(1-PR_u)_*]ds}. \tag{58}$$

Clearly, in this case E_w reflects only plant-uncertainty so that $E_u=E_t+kE_s$ combines both the tracking and saturation costs. Note also, that because of the neglect of m, the choice of an acceptable R_w that yields finite E_w is independent of μ.

Consider a single-input, single-output plant with nominal transfer function given by

$$P(s)=\frac{s-1}{s(s-2)}. \tag{59}$$

Suppose also that

$$\delta P(s)=\frac{a_0s-a_1}{s(s-2)} \tag{60}$$

where $\langle a_0\rangle=\langle a_1\rangle=0$ and

$$\langle a_0^2\rangle=\sigma^2=\langle a_1^2\rangle. \tag{61}$$

Under the conditions $k=Q=F_t=1$, design a two-degree-of-freedom servo to track a step-input $u(s)$ whose spectral density $G_u(s)=-1/s^2$. Compare the results against the standard one-degree-of-freedom optimal design.

Solution: It is easily verified that Assumptions 1–7 are satisfied and that

$$A=s(s-2)=A_1,\qquad B=s-1=B_1 \tag{62}$$

$$\Lambda=s^2+\sqrt{7}s+1,\qquad \Omega_u=\frac{1}{s}. \tag{63}$$

Let us first select the acceptable pair $(\tilde{R}_u,\tilde{R}_w)$ that minimizes both E_u and E_w in (57). Then, with the aid of (55) and the results in Theorem 3, we readily obtain[22]

$$\tilde{R}_u=-\frac{s(s-2)}{s^2+\sqrt{7}s+1},\quad \tilde{R}_w=\frac{s(s-2)((8+3\sqrt{7})s-1)}{(s+1)(s^2+\sqrt{7}s+1)} \tag{64}$$

$$E_u=3.65,\ E_t=2.70,\ E_s=0.95,\ \eta_t=4.18\sigma. \tag{65}$$

The simple single-loop two-degree-of-freedom realization of the optimal pair $(\tilde{R}_u,\tilde{R}_w)$ that is shown in Fig. 4 corresponds to the controller transfer matrix

$$T_c(s)=\left[\frac{-(s+1)}{s-(5+2\sqrt{7})}\;\middle|\;\frac{(8+3\sqrt{7})s-1}{s-(5+2\sqrt{7})}\right]. \tag{66}$$

Under the assumption that $C(s)$, $P(s)$, and $F(s)$ are realized minimally,[23] $\Delta_f(s)=1$, $\Delta_p(s)=s(s-2)$, and

$$\Delta_c(s)=(s+1)(s-(5+2\sqrt{7})). \tag{67}$$

Furthermore, since $K_1=L_1^{-1}M_1$ and

$$\tilde{R}_w=A_1(Y_1+K_1A)=A_1L_1^{-1}(L_1Y_1+M_1A), \tag{68}$$

$L_1(s)$ is identified as the denominator of $\tilde{R}_w(s)$. Hence, from (64)

$$L_1(s)=(s+1)(s^2+\sqrt{7}s+1) \tag{69}$$

and (29) yields

$$\Delta(s)=(s+1)^2(s^2+\sqrt{7}s+1) \tag{70}$$

for the associated closed-system characteristic polynomial. [Note that an extra $s+1$ factor is due to the presence of a hidden mode in the realization of $T_c(s)$.]

It is not possible to realize $\tilde{R}_u(s)$ with a single-loop one-degree-of-freedom configuration.[24] Indeed, under this restriction, $F_t(s)=F(s)=1$ in Fig. 2,

$$\tilde{R}_u=(1+CP)^{-1}C, \tag{71}$$

[22] $G_s(s)=\dfrac{\sigma^2(1-s^2)}{s^2(s^2-4)}$.

[23] The McMillan degree of $T_c(s)$ equals one, but a minimal realization of the pair $C(s)$, $F(s)$ in Fig. 4 requires two integrators. Thus, $T_c(s)$ has not been realized minimally.

[24] Hence, LQG does not work either [3].

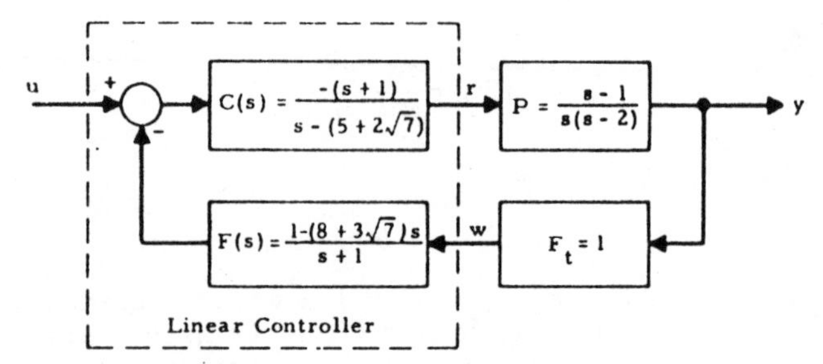

Fig. 4. Two-degree-of-freedom single-loop realization of the optimal pair $(\tilde{R}_u, \tilde{R}_w)$, (64).

$$C=\tilde{R}_u(1-P\tilde{R}_u)^{-1}=\frac{2-s}{s-(1+\sqrt{7})} \tag{72}$$

and closed-system stability is precluded by the $\text{Re}\, s > 0$ pole-zero cancellation between $P(s)$ and $C(s)$ at $s = 2$.

But if we insist on this topology, the smallest possible realizable cost is given by $E_u = 131$ and is achieved [3] by the design

$$R_u=\frac{s(s-2)((11+4\sqrt{7})s-2)}{(s+2)(s^2+\sqrt{7}s+1)}, \quad C=\frac{(11+4\sqrt{7})s-2}{s-(7+3\sqrt{7})} \tag{73}$$

$$E_t=42.35, \quad E_s=88.65, \quad \eta_t=3.65\sigma. \tag{74}$$

In this second design, the costs E_u, E_t, and E_s have been increased, respectively, by factors of 35.89, 15.69, and 93.32 for a mere 12.7 percent reduction in tracking-cost sensitivity! Actually, both $\eta_t = 4.18\sigma$ and $\eta_t = 3.65\sigma$ are totally unacceptable.[25] Yet, by exploiting the extra degree of freedom, it is very easy to reduce η_t to a value $\leq 0.25\sigma$ without any change in stability margins and at a total cost E_u approximately equal to E_t in the one-degree-of-freedom design (73), (74).

To preserve stability margins, the free parameter $K_1(s)$ must be left intact; hence, $R_w(s) = \tilde{R}_w(s)$. It follows then, in view of (58), that to reduce η_t we must vary $R_u(s)$ and this we accomplish by a suitable choice of $Z_u(s)$ in (54). In particular, for

$$Z_u(s)=\frac{1}{s+\omega_0}, \quad \omega_0=0.0126, \tag{75}$$

$$E_u=\tilde{E}_u+\frac{1}{2\pi j}\int_{-j\infty}^{j\infty} Z_u Z_{u*}\, ds=\tilde{E}_u+\frac{1}{2\omega_0}=\tilde{E}_u+0.0063 \tag{76}$$

and we obtain the third design

$$R_w=\tilde{R}_w, \quad R_u=\tilde{R}_u \cdot \frac{\omega_0}{s+\omega_0} \tag{77}$$

$$E_u=48.3333, \quad E_t=43.3038, \quad E_s=0.02874, \quad \eta_t=0.243\sigma. \tag{78}$$

The simple realization shown in Fig. 5 is suggested immediately by (77).[26]

The extremely small 3 dB bandwidth $\omega_0 = 0.0126$ rad/s of the precompensator $C_f(s)$ leads to a significant decrease in the bandwidth of the overall transfer function and a corresponding erosion in transient performance. This sacrifice of bandwidth for tracking-cost sensitivity explains why almost all of $E_u = 43.3333$ is now contributed by $E_t = 43.3038$. Nevertheless, and this a key point, except for a truly negligible increase in E_t, this suboptimal two-degree-of-freedom design (77) + (78) is, in almost every respect, vastly superior to the optimal one-degree-of-freedom design (73), (74).[27] Furthermore, the two designs have almost identical traditional gain and phase margins. In fact, in both cases the loop transfer function is given by

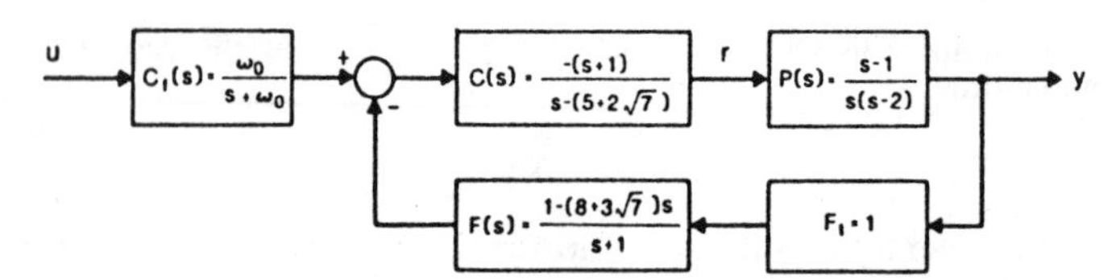

Fig. 5. Two-degree-of-freedom single-loop realization of the pair (R_u, R_w), (76).

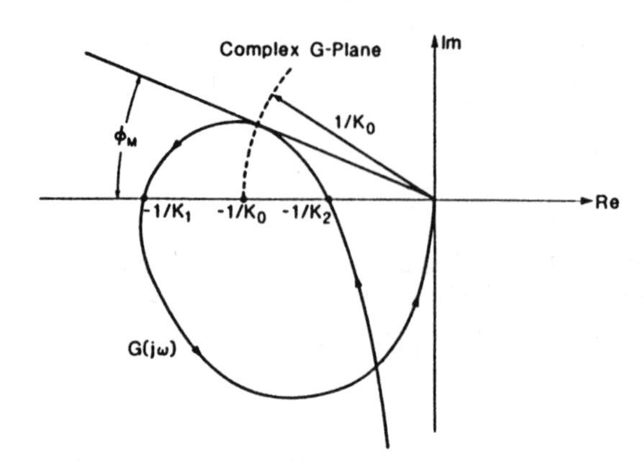

Fig. 6. Form of the Nyquist diagram for the examples.

$$KG(s)=\frac{K(s-1)(s-\sigma_0)}{s(s-2)(s-\sigma_p)} \tag{78a}$$

where σ_0 and σ_p are positive and K is a real number. It follows that the Nyquist plot of $G(j\omega)$ for $\omega > 0$ has the form shown in Fig. 6. Since the loop transfer function possesses poles in $\text{Re}\, s > 0$ at $s = 2$ and $s = \sigma_p$, the closed-loop system is stable iff the Nyquist diagram encircles the $-1/K$ point in the complex G-plane twice in a counterclockwise direction, i.e., iff

$$0<K_1<K<K_2. \tag{78b}$$

Thus, the one and two-degree-of-freedom optimal designs are stable iff

$$0.791<\frac{K}{K_0}<1.26 \tag{78.c}$$

and

$$0.780<\frac{K}{K_0}<1.21, \tag{78d}$$

respectively, where $K_0 = 11 + 4\sqrt{7}$ and $K_0 = 8 + 3\sqrt{7}$ are the corresponding nominal gains. Observe that the stability intervals in (78c) and (78d) are practically the same.

The angle ϕ_M shown in Fig. 6 is usually accepted as a conventional measure of stability phase margin and equals 10.4° for the first design and 9.5° for the second. Note that the agreement is again very close. In view of the difficulties associated with controlling this particular plant [47], [48], and [57], it is not at all clear that these phase margins can be significantly improved without accepting a serious deterioration in performance. In fact, we have explored a class of suboptimal two-degree-of-freedom designs with $Z_u(s) \equiv 0$ and

$$\frac{\sigma_o}{\sqrt{\mu}} \cdot Z_w(s)=\frac{\sigma\alpha}{s+\beta}, \quad \beta>0 \tag{78e}$$

and compared them using $\max_\omega |T(j\omega)|$ as a criterion, where $T = PR_wF_t$. (This stability margin measure is now receiving a great

[25] For a (reasonable) 10 percent variation in the numerator coefficients of $P(s)$, $\sigma \approx 0.1$ and $\eta_t = 3.65\sigma = 0.365$ represents a (most unreasonable) 36.5 percent change in tracking-cost about the nominal.

[26] The associated closed-loop characteristic polynomial $\Delta(s)$ is again given by (70).

[27] For example, a root mean-square deviation $\sigma = 0.1$ in the numerator coefficients of $P(s)$ now entails only a 2.43 percent change in tracking cost, etc.

deal of attention [9]–[13], [20]. For the optimal design ($\alpha = 0$), $\max_\omega |T(j\omega)| = 15.6$ dB which is only slightly larger than

$$\inf_{\substack{-\infty<\alpha<\infty \\ \beta>0}} \left(\sup_\omega |T(j\omega)|\right) = \left(\max_\omega |T(j\omega)|\right)\Big|_{\substack{\beta=4.28\alpha \\ \alpha\to\infty}} = 11.7 \text{ dB} . \tag{78f}$$

However, since

$$\frac{\sigma_\varrho^2}{\mu} \cdot (E_w - \tilde{E}_w) = \frac{(\sigma\alpha)^2}{2\beta} \tag{78g}$$

we see that this 3.9 dB improvement can only be obtained at the expense of an enormous increase in cost. Clearly then, the decision to retain the stability margins realized with the optimum two-degree-of-freedom design is justified.

The free parameter $Z_w(s)$ is expected to play a much more active design role after suitable margins have been precisely defined. In any case, we believe that our examples should dispel any remaining doubts about either the superiority or practical feasibility of a two-degree-of-freedom design theory of optimal controllers.

IV. Discussion and Suggestions for Future Research

The analytic design technique described in Sections II and III is quite comprehensive and takes performance, plant saturation, and plant sensitivity into account *explicitly*, with stability margins playing a passive background role. Naturally, its real significance can only be judged by its success in practical applications. In our opinion, its main features are the following.

1) Every component of the cost-functional has a clear physical significance.

2) The two available degrees of freedom are fully exploited.

3) Tradeoffs are easily recognized and included in the design.

4) There are no restrictions on the plant except that it be finite-dimensional and free of unstable hidden modes.

A comparison with some other published work in this area reveals many points of contact but also some differences in philosophy. Of the references we cite, it appears that only [9], [11], [12], and [18] consider the possibility of sacrificing cost to enhance feedback sensitivity and stability margins. In [18], state-variable feedback is used exclusively and stability margin is traded off against optimal regulator performance. In [12], essentially the same problem is considered and its LQG formulation is discussed briefly. However, as noted in [12], the unrealistic assumption is made that the observer corresponds to the perturbed plant and not the nominal. In both [9] and [11] a procedure for regulator design is outlined based on qualitative physical considerations similar to those employed by us. Namely, reasonable stability margin criteria are first translated into restrictions on the singular values of the two sensitivity matrices $S(j\omega)$ and $1 - S(j\omega)$ and a quadratic cost-functional is then used to penalize tracking error and saturation. If the optimal design obtained by minimizing the cost fails to satisfy the constraints imposed on the singular values, it is discarded, and the process is repeated with a different set of weighting matrices until a satisfactory solution is reached [11]. The similarities in the above approach and that taken in our paper should be evident, despite the fact that we bound tracking-cost instead of singular values, do not incorporate the effects of sensitivity in the same way and have not yet suggested precise quantitative stability margins. But in any case, since [9] and [11] also gauge performance in terms of a quadratic cost which must be finite to be meaningful, all acceptable R_w's are necessarily given by (56h). Hence, instead of changing weighting matrices in the cost, it may be easier and perhaps more efficient to go to a suboptimal design by choosing a nonzero $Z_w(s)$ and then to calculate the loss in performance by means of (56j).

A relatively new development is the use of nonquadratic norms to cope with the problem of sensitivity minimization [33]–[35], [46]. In [33], disturbance rejection and the problem of plant uncertainty are considered mainly in a one-degree-of-freedom setting. Unfortunately, the two-degree-of-freedom treatment of plant uncertainty neglects both measurement noise and disturbance inputs and places most of the emphasis on plants that are either stable or have been stabilized from the outset. In [34], [35], and [46] unstable single-input-output plants are admitted but attention is restricted to one-degree-of-freedom designs. Although it is probably true that minmax worst-case designs cannot be completely avoided if analytic models for plant uncertainty are not available, it also seems likely that a quadratic cost-functional must play some role in the evaluation of performance. The expressions (54) and (56h) for all acceptable R_u's and R_w's will then subsume all possible designs through the choice of free parameters $Z_u(s)$ and $Z_w(s)$. Clearly, only further study can determine whether a tractable comprehensive theory of optimal design based on nonquadratic norms is really feasible and advantageous. The growing number of papers on this subject [49]–[56] suggests that an answer may soon be forthcoming.

The treatment of stability margins for linear time-invariant finite-dimensional multivariable systems given in [9]–[12], [14], and [20] employs only feedback structures of the type shown in Fig. 3, and envisages either additive or multiplicative perturbations from a nominal loop transfer matrix $G_0(s)$, i.e.,

$$G(s) = G_0(s) + \delta G_0(s) \tag{79}$$

or

$$G(s) = [1 + \delta G_0(s)]G_0(s). \tag{80}$$

Correspondingly,

$$S_0 = [1 + G_0(s)]^{-1} \tag{81}$$

and

$$T_0 = G_0(s)[1 + G_0(s)]^{-1} \tag{82}$$

are the respective sensitivity matrices to be considered. But, as our work in Section III indicates, *both* $X_1 - K_1B$ and $Y_1 + K_1A$ must play an important role, so that it is probable that neither (79) nor (80), taken alone, suffices. Since $X_1 - K_1B$ and $Y_1 + K_1A$ cannot be made "too small," simultaneously, it is anticipated that some of the gain-bandwidth ideas described in [21] will prove helpful. In addition, there is reason to believe that the Re $s \geq 0$ analyticity of $L_1^{-1}\delta L_1$ in (42) can be exploited to develop sharp $j\omega$-axis stability-margin criteria.

In view of the computational complexities in multivariable system design, it is fortunate to have available computer algorithms for spectral factorization [22], [23] and the determination of matrix fraction descriptions [24]–[26]. Other important references [27], [40]–[42], [55] contain additional results which should ease the burden of numerical computation. As we have seen in the lemmas to Theorems 2 and 3, for plants with left-half plane and $j\omega$-axis poles only, it is possible to eliminate all coprime decompositions.

The polynomial approach advocated by Kučera [27], [28] dispenses with partial fraction expansions and deserves to be considered. In fact, Šebek [45] has applied this approach to single-input-output systems to obtain an optimal two-degree-of-freedom controller for the case $G_d = G_m = F_t = 1, E = E_0$. At this point, we are not certain that their method can be employed in our general formulation of the design problem but the matter is not closed.

One aspect of multivariable design that has not been addressed in this paper is that of robustness with respect to tracking and regulation [29], [30]. The reason for this omission is the following. In our development the only ramp-like signals that can be tracked and regulated are those whose Laplace transforms have poles that coincide with the finite $j\omega$-axis poles of $P(s)$. To attempt to track or regulate any other type of ramp requires the presence of such a ramp at the plant input and plant saturation is usually inevitable. Hence, the performance index (19) is infinite and design is meaningless.

Attempts to circumvent the difficulty by modifying the saturation part of the cost functional only avoid the issue. Clearly, the saturation constraint must be respected if design is based on a linearized model. At present, it appears to us that the inclusion of Re $s > 0$ poles in the controller to ensure robust asymptotic tracking and regulation can conflict with the requirement of no plant saturation. (Thus, the feedback problems treated in [32], [39], and [43] may need elaboration.)

Consequently, in our opinion, if plant saturation is a primary concern, robustness only makes sense if the plant already possesses the necessary $j\omega$-axis poles. But then, since such poles are usually structural in origin, they already exhibit great resistance to changes in plant parameters. For example, in the positioning of an inertial load J with viscous damping B, the transfer function relating angular displacement θ to applied torque λ is given by

$$\frac{\theta(s)}{\lambda(s)}=\frac{1}{s(Js+B)}. \tag{83}$$

Evidently, the pole at the origin is firmly placed and totally insensitive to variations in the parameters J and B. In this case, the plant is already able to track step inputs robustly and there is no need to include an additional pole in the controller. However, if plant saturation presents no problem the notion of robustness is meaningful and its inclusion is accomplished by simply setting $kQ = 0$ in the cost-functional (19).

Naturally, the selection of a suitable controller for the realization of an optimal pair (R_u, R_w) is an important problem. That one exists is assured, but many structures are possible. In [1, p. 241] eight different two-degree-of-freedom configurations are given which include, in particular, those shown in Figs. 2 and 3. Can they all be generalized to the multivariable case and, if so, is one of them preferable? It may be possible to single one out on the basis of system sensitivity to controller uncertainty, a consideration we have ignored. Qualitatively, it seems that the most desirable topology is the one shown in Fig. 2 with $F(s) \equiv 1$ because the sensitivity with respect to $C(s)$ is the same as that with respect to $P(s)$. Unfortunately, this is a one-degree-of-freedom unity-feedback structure and is only applicable if $R_u = R_w$. (If $R_u \neq R_w$, a two-degree-of-freedom configuration is compulsory.) References [7], [8], and [33]–[39] may prove helpful. The following is a list of some future research objectives.

1) Define physically meaningful stability margins and translate them into constraints on Z_w involving the plant uncertainties δA_1 and δB_1.

2) Determine, if possible, an acceptable pair (R_u, R_w) that minimizes the cost-functional E, maintains prescribed stability margins, and also keeps tracking-cost sensitivity within given bounds.

3) Investigate the potential role played by robustness in the context of our design philosophy.

4) Study the impact of sensitivity on the choice of controller structure used to realize an acceptable pair (R_u, R_w).

Appendix

Proof of Theorem 1

Necessity: Let $\Delta_c(s)$, $\Delta_t(s)$, and $\Delta_p(s)$ denote the characteristic polynomials of the controller, feedback sensor, and plant, respectively. Then [3]

$$\Delta=\Delta_c\Delta_t\Delta_p \cdot \det\,(1+C_wF_tP) \tag{A-1}$$

is the characteristic polynomial of the general two-degree-of-freedom structure shown in Fig. 1. Since $T_c = [C_u | -C_w]$,

$$1+C_wF_tP=1+T_c\left[\frac{0}{-1}\right]F_tP. \tag{A-2}$$

Thus, in terms of the coprime decompositions $F_tP = B_1A_1^{-1}$ and $T_c = A_c^{-1}B_c$,

$$\Delta=\frac{\Delta_c\Delta_t\Delta_p}{\det A_c \cdot \det A_1}\cdot \det\left(A_cA_1+B_c\left[\frac{0}{-1}\right]B_1\right). \tag{A-3}$$

Clearly, $\det A_c$ divides Δ_c and $\Delta_t\Delta_p/\det A_1$ is finite and $\neq 0$ in Re $s \geq 0$ because of the assumed admissibility of the pair (P, F_t). Consequently, the polynomial

$$g(s)=\det\left(A_cA_1+B_c\left[\frac{0}{-1}\right]B_1\right) \tag{A-4}$$

must be strict-Hurwitz if the structure is to be asymptotically stable.

It now follows that

$$R_u=(1+C_wF_tP)^{-1}C_u=\left(1+T_c\left[\frac{0}{-1}\right]F_tP\right)^{-1}T_c\left[\frac{1}{0}\right] \tag{A-5}$$

$$=A_1\left(A_cA_1+B_c\left[\frac{0}{-1}\right]B_1\right)^{-1}B_c\left[\frac{1}{0}\right]=A_1H_1 \tag{A-6}$$

where $H_1(s)$ is analytic in Re $s \geq 0$. Similarly,

$$R_w=(1+C_wF_tP)^{-1}C_w$$

$$=A_1\left(A_cA_1+B_c\left[\frac{0}{-1}\right]B_1\right)^{-1}B_c\left[\frac{0}{-1}\right] \tag{A-7}$$

$$=A_1(A_cA_1+\tilde{B}_cB_1)^{-1}\tilde{B}_c \tag{A-8}$$

in which

$$\tilde{B}_c=B_c\left[\frac{0}{-1}\right]. \tag{A-9}$$

Let us define $K_1(s)$ by means of the equation

$$Y_1+K_1A=(A_cA_1+\tilde{B}_cB_1)^{-1}\tilde{B}_c. \tag{A-10}$$

Then, $R_w = A_1(Y_1 + K_1A)$ and it remains to be shown that $K_1(s)$ is analytic in Re $s \geq 0$. (The Re $s \geq 0$ analyticity of $Y_1 + K_1A$ is obvious.)

From (A-10),

$$X_1-K_1B=X_1-((A_cA_1+\tilde{B}_cB_1)^{-1}\tilde{B}_c-Y_1)A^{-1}B. \tag{A-11}$$

Since $F_tP = A^{-1}B = B_1A_1^{-1}$ and $A_1^{-1} = X_1 + Y_1B_1A_1^{-1}$, (A-11) and (A-2) yield

$$X_1-K_1B=A_1^{-1}-(A_cA_1+\tilde{B}_cB_1)^{-1}\tilde{B}_cF_tP$$

$$=A_1^{-1}-A_1^{-1}(1+C_wF_tP)^{-1}C_wF_tP$$

$$=A_1^{-1}(1+C_wF_tP)^{-1}=(A_cA_1+\tilde{B}_cB_1)^{-1}A_c. \tag{A-12}$$

Therefore, $X_1 - K_1B$ is also analytic in Re $s \geq 0$ and the identity

$$(Y_1+K_1A)X-(X_1-K_1B)Y=Y_1X-X_1Y+K_1 \tag{A-13}$$

immediately establishes the Re $s \geq 0$ analyticity of $K_1(s)$. Note that

$$\det\,(X_1-K_1B)=\frac{\det A_c}{g}\neq 0 \tag{A-14}$$

is necessary.

Sufficiency: Let $R_u = A_1H_1$ and let $R_w = A_1(Y_1 + K_1A)$ where H_1 and K_1 are real rational matrices analytic in Re $s \geq 0$. Suppose also that det $(X_1 - K_1B) \not\equiv 0$. Our task is to prove that R_u, R_w are acceptable for the given admissible pair (P, F_t) in the sense of Definition 1.

From (5), $C_w = (1 + C_wF_tP)R_w$, hence,

$$\begin{aligned} C_w &= R_w(1-F_tPR_w)^{-1} \\ &= A_1(Y_1+K_1A)(1-B_1(Y_1+K_1A))^{-1} \\ &= A_1(1-(Y_1+K_1A)B_1)^{-1}(Y_1+K_1A) \\ &= A_1(X_1A_1-K_1BA_1)^{-1}(Y_1+K_1A) \\ &= (X_1-K_1B)^{-1}(Y_1+K_1A) \end{aligned} \tag{A-15}$$

since $X_1A_1 + Y_1B_1 = 1$. In the same way, from (4),

$$\begin{aligned} C_u &= (1+C_wF_tP)R_u \\ &= (X_1-K_1B)^{-1}(X_1-K_1B+(Y_1+K_1A)A^{-1}B)A_1H_1 \\ &= (X_1-K_1B)^{-1}(X_1+Y_1B_1A_1^{-1})A_1H_1 \\ &= (X_1-K_1B)^{-1}H_1. \end{aligned} \tag{A-16}$$

Thus,

$$T_c = [C_u| - C_w] = (X_1-K_1B)^{-1}[H_1| - (Y_1+K_1A)]. \tag{A-17}$$

Let $K_1 = L_1^{-1}M_1$ be any left-coprime decomposition of K_1. Then,

$$T_c = (L_1X_1-M_1B)^{-1}[L_1H_1| - (L_1Y_1+M_1A)]. \tag{A-18}$$

The pair $(L_1X_1 - M_1B)$, $(L_1Y_1 + M_1A)$ is left-coprime. Indeed, assume that for some fixed s there exists a vector $\boldsymbol{a}$ such that

$$\boldsymbol{a}'[L_1X_1-M_1B|L_1Y_1+M_1A] = O'. \tag{A-19}$$

Then, in view of the identity

$$(L_1X_1-M_1B)A_1+(L_1Y_1+M_1A)B_1 = L_1 \tag{A-20}$$

we have $\boldsymbol{a}'L_1 = O'$ and (A-19) reduces to

$$\boldsymbol{a}'M_1[-B|A] = O'. \tag{A-21}$$

But A, B is left-coprime, hence, $\boldsymbol{a}'M_1 = O'$ also. In short,

$$\boldsymbol{a}'[L_1|M_1] = O' \tag{A-22}$$

which implies $\boldsymbol{a} = O$ because L_1, M_1 is left-coprime and our assertion is established.

It is now easy to conclude from (A-18) that *every* finite pole of $T_c(s)$ in Re $s \geq 0$ of McMillan degree m, is a zero of det $(L_1Y_1 - M_1B)$ of multiplicity precisely equal to m. Consequently, with the aid of (A-15), we find that

$$\begin{aligned} \Delta &= \Delta_c\Delta_t\Delta_p \cdot \det(1+C_wF_tP) \\ &= \frac{\Delta_c\Delta_t\Delta_p}{\det(L_1X_1-M_1B)} \cdot \det(L_1X_1-M_1B+L_1Y_1B_1A_1^{-1}+M_1B) \\ &= \frac{\Delta_c}{\det(L_1X_1-M_1B)} \cdot \frac{\Delta_t\Delta_p}{\det A_1} \cdot \det L_1. \end{aligned} \tag{A-23}$$

The first factor in (A-23) is strict-Hurwitz because det $(L_1X_1 - M_1B)$ divides Δ_c and $T_c(s)$ is realized without unstable hidden modes, the second because the pair (P, F_t) is admissible, and the third because $K_1(s)$ is analytic in Re $s \geq 0$. Hence, the controller $T_c(s)$ yields an asymptotically stable design and the proof of Theorem 1 is complete. Q.E.D.

Proof of Theorem 2

Since $G_u = \Omega_u\Omega_{u*}$ and all matrices are rational, it is clear from (48) that E_u is finite iff $(1 - PR_u)\Omega_u \leq O(s^{-1})$, $QR_uG_uRu_* \leq O(s^{-2})$ and both are $j\omega$-analytic. But from Assumption 2, $\Omega_u \leq O(s^{-1})$; hence, $(1 - PR_u)\Omega_u \leq O(s^{-1})$ iff $PR_u\Omega_u \leq O(s^{-1})$ which implies $PR_u\Omega_u\Omega_{u*} = PR_uG_u \leq O(s^{-2})$. Let us rewrite E_u as

$$E_u = \frac{1}{2\pi j}\int_{-j\infty}^{j\infty} \operatorname{tr}\,[G_u - 2PR_uG_u + (P_*P+kQ)R_uG_uR_{u*}]\, ds. \tag{A-24}$$

Then, G_u and $PR_uG_u \leq O(s^{-2}) \rightarrow \operatorname{tr}\,[(P_*P + kQ)R_uG_uR_{u*}] \leq O(s^{-2})$. Equivalently, if we cycle the trace and use (51), (52), and (54),

$$\operatorname{tr}\,(\Omega_{u*}R_{u*}A_{1*}^{-1}\Lambda_*\Lambda A_1^{-1}R_u\Omega_u) \leq O(s^{-2}); \tag{A-25}$$

therefore,[28]

$$\Lambda A_1^{-1}R_u\Omega_u = Z + \{\Gamma\}_+ \leq O(s^{-1}). \tag{A-26}$$

But by definition, $\{\Gamma\}_+ \leq O(s^{-1})$ so that $Z \leq O(s^{-1})$. In brief, if the integrand ρ of E_u is $\leq O(s^{-2})$, then $Z \leq O(s^{-1})$. The converse is also true.

In fact, it suffices to verify that $Z \leq O(s^{-1})$ implies $PR_u\Omega_u \leq O(s^{-1})$. Observe first that

$$PR_u\Omega_u = PA_1\Lambda^{-1}(Z+\{\Gamma\}_+); \tag{A-27}$$

hence, $PA_1\Lambda^{-1}$ proper $\rightarrow PR_u\Omega_u \leq O(s^{-1})$. However, from (51),

$$1 = (PA_1\Lambda^{-1})_*(PA_1\Lambda^{-1}) + k(A_1\Lambda^{-1})_*Q(A_1\Lambda^{-1}), \tag{A-28}$$

the sum of two para-Hermitian nonnegative-definite matrices. Consequently, $PA_1\Lambda^{-1}$ must be bounded at $s = \infty$.

Let us say that a rational matrix is "good" if it is analytic on the finite part of the $s = j\omega$-axis. We will now prove that ρ is good iff Z is good. As we have already indicated, ρ is good iff both $(1 - PR_u)\Omega_u$ and $QR_uG_uR_{u*}$ are good. By Assumption 4, Q is good and nonsingular for all finite $s = j\omega$; thus, $QR_uG_uR_{u*}$ is good iff $R_u\Omega_u$ is good. In short, ρ is good iff $\Psi_1 \equiv (1 - PR_u)\Omega_u$ and $\Psi_2 \equiv R_u\Omega_u$ are good.

Let $R_u = (Y + A_1K)A$. Then, Ψ_1 and Ψ_2 are good iff K is good. Indeed, it follows easily from Assumption 4 that $A\Omega_u$ and Λ are good and nonsingular for all finite $s = j\omega$. Hence, Ψ_2 is good iff

$$A_1K = \Psi_2(A\Omega_u)^{-1} - Y \tag{A-29}$$

is good. Similarly, from the identity

$$\Psi_1 = (1-PR_u)\Omega_u = (1-F_tPR_u+(F_t-1)PR_u)\Omega_u \tag{A-30}$$

and the postulated goodness of $(F_t - 1)P$ in Assumption 3, it is evident, given that Ψ_1 is good, iff

$$(1-F_tPR_u)\Omega_u = (X-B_1K)A\Omega_u \tag{A-31}$$

is good. That is, iff B_1K is good. Consequently, ρ is good iff

$$\left[\frac{A_1}{B_1}\right]K \tag{A-32}$$

is good, i.e., iff K is good because the pair B_1, A_1 is right-coprime.

Let us substitute $R_u = (Y + A_1K)A$ into (54) and solve for Z to obtain

$$Z = \Lambda A_1^{-1}(Y+A_1K)A\Omega_u - \{\Gamma\}_+. \tag{A-33}$$

[28] The subscript u on Z_u has been dropped.

Since Λ^{-1} and $(A\Omega_u)^{-1}$ are good (Assumption 4), Z will be good iff

$$\Lambda^{-1}Z(A\Omega_u)^{-1}=A_1^{-1}Y+K-\Lambda^{-1}\{\Gamma\}_+(A\Omega_u)^{-1} \quad \text{(A-34)}$$

is good, i.e., iff

$$A_1^{-1}Y-\Lambda^{-1}\{\Gamma\}_+(A\Omega_u)^{-1}$$
$$=A_1^{-1}Y-\Lambda^{-1}(\Gamma-\{\Gamma\}_- -\{\Gamma\}_\infty)(A\Omega_u)^{-1} \quad \text{(A-35)}$$

is good.[29]

Clearly, by definition, $[\Gamma]_-$ and $[\Gamma]_\infty$ are good, so that Z is good iff

$$\Lambda^{-1}\Gamma(A\Omega_u)^{-1}-A_1^{-1}Y=(\Lambda_*\Lambda)^{-1}(PA_1)_*A^{-1}-A_1^{-1}Y \quad \text{(A-36)}$$

is good. Or, with the aid of (51), iff

$$(\Lambda_*\Lambda)^{-1}A_{1*}[P_*(A^{-1}-PY)-kQY] \quad \text{(A-37)}$$

is good. Since $A^{-1} = X + F_tPY$, the matrix in (A-37) is good iff

$$(\Lambda_*\Lambda)^{-1}[(PA_1)_*(X+(F_t-1)PY)-kA_{1*}QY] \quad \text{(A-38)}$$

is good and this is surely true if PA_1 is good. But by Assumption 3,

$$PA_1=F_tPA_1-(F_t-1)PA_1=B_1-(F_t-1)PA_1 \quad \text{(A-39)}$$

is obviously good and the implication ρ good $\rightarrow$ Z good has been established. Conversely, by retracing our steps we can also show that Z good $\rightarrow$ ρ good.

Thus, let Z be good. Then, from (A-33),

$$K=\Lambda^{-1}(Z+\{\Gamma\}_+)(A\Omega_u)^{-1}-A_1^{-1}Y \quad \text{(A-40)}$$

is evidently good because

$$\Lambda^{-1}\{\Gamma\}_+(A\Omega_u)^{-1}-A_1^{-1}Y \quad \text{(A-41)}$$

is good. Finally, K good implies Ψ_1 and Ψ_2 both good which implies ρ good.

Consider the collection of all matrices $R_u = A_1H_1$, where

$$H_1=\Lambda^{-1}(Z+\{\Gamma\}_+)\Omega_u^{-1} \quad \text{(A-42)}$$

and Z is analytic in Re $s \geq 0$. All such R_u's are acceptable. For clearly, due to the properties of Λ, Ω_u and $\{\Gamma\}_+$, H_1 is analytic in Re $s > 0$ iff Z is analytic in Re $s > 0$. Futhermore, since Z good implies K good and

$$H_1=KA+A_1^{-1}YA \quad \text{(A-43)}$$

Z good implies H_1 good if it can be shown that $A_1^{-1}YA$ is good. This we accomplish by noting that $X_1A_1+Y_1B_1 = 1$ yields

$$A_1^{-1}Y=X_1Y+Y_1B_1A_1^{-1}Y=XY_1+Y_1A^{-1}BY$$
$$=X_1Y+Y_1A^{-1}(1-AX)=X_1Y+Y_1A^{-1}-Y_1X; \quad \text{(A-44)}$$

hence,

$$A_1^{-1}YA=X_1YA+Y_1-Y_1XA, \quad \text{(A-45)}$$

a polynomial matrix. To summarize, R_u is acceptable and yields finite cost E_u iff it has the form $R_u = A_1H_1$, where H_1 is given in (A-42) and Z is $\leq O(s^{-1})$ and analytic in Re $s \geq 0$.

Let E_u and $\tilde{E}_u$ denote the costs associated with the choices Z and $Z = 0$, respectively. Then, after some tedious but straightforward algebra we obtain

$$2\pi j(E_u-\tilde{E}_u)=\int_{-j\infty}^{j\infty} \text{tr}\,(\Delta_*\Omega_{u*}^{-1}Z_*\Lambda_*^{-1})\,ds+\int_{-j\infty}^{j\infty} \text{tr}(ZZ_*)ds \quad \text{(A-46)}$$

where

$$\Delta_*=2(A_{1*}(P_*P+kQ)A_1\tilde{H}_1-A_{1*}P_*)G_u \quad \text{(A-47)}$$

and

$$\tilde{H}_1=\Lambda^{-1}\{\Gamma\}_+\Omega_u^{-1}. \quad \text{(A-48)}$$

It follows easily from (A-47) and (53) that

$$\Delta_*\Omega_{u*}^{-1}=\Lambda_*\{\Lambda_*^{-1}A_{1*}P_*\Omega_u\}_+-A_{1*}P_*\Omega_u$$
$$=-\Lambda_*(\{\Gamma\}_-+\{\Gamma\}_\infty). \quad \text{(A-49)}$$

Hence, $\Delta_*\Omega_{u*}^{-1}$ and $Z_*\Lambda_*^{-1}$ are free of poles in Re $s \leq 0$.

Now, we have already established that E_u and $\tilde{E}_u$ are both finite because they correspond to Z's that are $\leq O(s^{-1})$ and good. Since the second integral on the right side of (A-46) is clearly finite, the same must be true of the first one. Thus, its integrand is $\leq O(s^{-2})$ and analytic in Re $s \leq 0$. By Cauchy's theorem, this integral equals zero; hence,

$$E_u-\tilde{E}_u=\frac{1}{2\pi j}\int_{-j\infty}^{j\infty} \text{tr}\,(ZZ_*)\,ds\geq 0. \quad \text{(A-50)}$$

Therefore, $\tilde{E}_u$ is the minimum value of the cost E_u and is realized by the unique acceptable choice $\tilde{R}_u = A_1\tilde{H}_1$.

Lastly, it remains to prove that an optimal design preserves the steady-state tracking capability of the plant. Stated differently (see [3, pp. 322 and 323]) every finite $j\omega$-axis pole of $P(s)$ of McMillan degree ν is a zero of the determinant of the closed-system error transfer matrix of multiplicity at least equal to ν.

Since PR_u is the system transfer matrix from u to r, $1 - PR_u$ is the error matrix. Moreover,

$$1-PR_u=\Psi_1\Omega_u^{-1}=\Psi_1(A\Omega_u)^{-1}A \quad \text{(A-51)}$$

and

$$\det\,(1-PR_u)=\frac{\det\,\Psi_1\cdot\det\,A}{\det\,(A\Omega_u)}. \quad \text{(A-52)}$$

But for any acceptable R_u that yields finite E_u, Ψ_1 is good. Moreover, $A\Omega_u$ is nonsingular on the finite $s = j\omega$-axis. Consequently, every purely imaginary zero of det A is a zero of the left side of (A-52) of at least the same multiplicity. Q.E.D.

Proof of Theorem 3

Let ρ denote the integrand of E_w, (49). Then, E_w finite $\rightarrow \rho \leq O(s^{-2})$. In particular,

$$\text{tr}\,[(1-PR_wF_t)G_{ds}(1-PR_wF_t)_*]\leq O(s^{-2}) \quad \text{(A-53)}$$

where

$$G_{ds}=G_d+\mu G_s. \quad \text{(A-54)}$$

But by Assumption 5, $G_{ds} \leq O(s^{-2})$, hence, invoking (A-53),

$$PR_wF_tG_{ds}\leq O(s^{-2}). \quad \text{(A-55)}$$

In turn, (A-55), (56h) and the alternative expression[30]

$$\rho=\text{tr}\,G_{ds}-2\,\text{tr}\,(PR_wF_tG_{ds})$$
$$+\text{tr}\,[(F_tG_{ds}F_{t*}+G_m)R_{w*}(P_*P+kQ)R_w] \quad \text{(A-56)}$$

[29] Actually, Assumption 4 implies $\{\Gamma\}_\infty = 0$ but we do not need this result.

[30] $\int_{-j\infty}^{j\infty} \text{tr}\,(PR_wF_tG_{ds})\,ds=\int_{-j\infty}^{j\infty} \text{tr}\,(PR_wF_tG_{ds})_*\,ds.$

imply that

$$\text{tr } [(F_t G_{ds} F_{t*} + G_m) R_{w*} (P_* P + kQ) R_w]$$
$$= \text{tr } (A^{-1} \Omega \Omega_* A_*^{-1} R_{w*} A_{1*}^{-1} \Lambda_* \Lambda A_1^{-1} R_w)$$
$$= \text{tr } (\Sigma \Sigma_*) \le 0(s^{-2}) \quad \text{(A-58)}$$

in which[31]

$$\Sigma = \{\Lambda_*^{-1} I_* \Omega_*^{-1}\}_+ + \{\Lambda A_1^{-1} Y \Omega\}_- + Z. \quad \text{(A-59)}$$

Consequently, $\Sigma \le 0(s^{-1})$, i.e., $Z \le 0(s^{-1})$. Conversely, $Z \le 0(s^{-1}) \to \rho \le 0(s^{-2})$. In fact, it suffices to show that $Z \le 0(s^{-1}) \to \text{tr } (PR_w F_t G_{ds}) \le 0(s^{-2})$.

Clearly, since $R_w = A_1 \Lambda^{-1} \Sigma \Omega^{-1} A$,

$$PR_w F_t G_{ds} = PA_1 \Lambda^{-1} \Sigma \Omega^{-1} A F_t G_{ds}. \quad \text{(A-60)}$$

However, $AGA_* = \Omega \Omega_*$ gives

$$1 = (\Omega^{-1} A) G_m (\Omega^{-1} A)_* + (\Omega^{-1} A F_t) G_{ds} (\Omega^{-1} A F_t)_* \quad \text{(A-61)}$$

and a familiar argument permits us to infer that $\Omega^{-1} A F_t G_{ds} \le 0(s^{-1})$.[32] Furthermore, according to (A-28), $PA_1 \Lambda^{-1}$ is proper and since $Z \le 0(s^{-1}) \to \Sigma \le 0(s^{-1})$, the assertion follows immediately from (A-60).

R_w is acceptable, i.e.,

$$R_w = A_1 (Y_1 + K_1 A) \quad \text{(A-62)}$$

where $K_1(s)$ is analytic in Re $s \ge 0$, iff $Z(s)$ is analytic in Re $s \ge 0$. In fact, by solving (A-62) for K_1 we obtain, with the help of the identity $A_1 Y_1 = YA$ [3]

$$k_1 = k_1 + \Lambda^{-1} \{\Lambda A_1^{-1} Y \Omega\}_\infty \Omega^{-1} \quad \text{(A-63)}$$
$$= \Lambda^{-1} (\{\Lambda_*^{-1} I_* \Omega_*^{-1}\}_+ - \{\Lambda A_1^{-1} Y \Omega\}_+ + Z) \Omega^{-1} \quad \text{(A-64)}$$
$$= \Lambda^{-1} \{\Lambda_*^{-1} I_* \Omega_*^{-1} - \Lambda A_1^{-1} Y \Omega\}_+ \Omega^{-1} + \Lambda^{-1} Z \Omega^{-1}. \quad \text{(A-65)}$$

Clearly, K_1 analytic in Re $s \ge 0 \to Z$ analytic in Re $s > 0$ and, in addition, $\{\ \}_+$ good implies Z good.

Now, since Λ_*^{-1} and Ω_*^{-1} are good and nonsingular (Assumptions 4 and 6),

$$\{\ \}_+ = \{\Lambda_*^{-1} (I_* - \Lambda_* \Lambda A_1^{-1} Y \Omega \Omega_*) \Omega_*^{-1}\}_+$$
$$= \{\Lambda_*^{-1} (A_{1*} P_* G_{ds} F_{t*} A_* - A_{1*} (P_* P + kQ) Y A G A_*) \Omega_*^{-1}\}_+ \quad \text{(A-67)}$$

will be good iff the matrix function in round parentheses is good. That is, iff

$$(PA_1)_* F_t^{-1} (G - G_m) A_* - A_{1*} (P_* P + kQ) Y A G A_* \quad \text{(A-68)}$$
$$= (PA_1)_* F_t^{-1} (A^{-1} - F_t P Y)(AGA_*)$$
$$- (PA_1)_* F_t^{-1} G_m A_* - k A_{1*} Q Y (AGA_*)$$
$$= (PA_1)_* F_t^{-1} X (AGA_*) - (PA_1)_* F_t^{-1} G_m A_* - k A_{1*} Q (AGA_*) \quad \text{(A-69)}$$

is good. And this is obviously true because PA_1 is good and because Assumptions 3 and 7 imply that F_t^{-1} is good.

R_w acceptable implies ρ good. To see this, we must examine (49) carefully. It is not difficult to verify directly that ρ is good iff

$$(I - PR_w F_t) J, \ \text{tr } [QR_w G R_{w*}], \ \text{tr } (PR_w G_m R_{w*} P_*) \quad \text{(A-70)}$$

are all good ($G_{ds} = JJ_*$). Clearly, $A_1(Y_1 + K_1 A) = (Y + A_1 K_1) A$ follows from $A_1 Y_1 = YA$. Thus,

$$QR_w G R_{w*} = Q(Y + A_1 K_1)(AGA_*)(Y + A_1 K_1)_* \quad \text{(A-71)}$$

is good since Q, AGA_* and K_1 are good. Similarly, PA_1 good and G_m good (Assumption 5) imply $PR_w G_m R_{w*} P_*$ good. Lastly, AGA_* good $\to AF_t J$ good;[33] hence,

$$(1 - PR_w F_t) J = (1 - P(Y + A_1 K_1) A F_t) J \quad \text{(A-72)}$$

is good iff $(1 - PYAF_t)J = (1 - PYAF_t)(AF_t)^{-1}(AF_t J)$ is good. But

$$(1 - PYAF_t)(AF_t)^{-1}$$
$$= (AF_t)^{-1} (1 - AF_t PY) = (AF_t)^{-1} (1 - BY) = F_t^{-1} X \quad \text{(A-73)}$$

is good. To sum up, R_w is acceptable and yields finite cost E_w iff it has the form (56h), where $Z_w(s)$ is an arbitrary real rational matrix $\le 0(s^{-1})$ and analytic in Re $s \ge 0$.

Let E_w and $\tilde{E}_w$ denote the costs associated with the choices Z and $Z = 0$. Straightforward analysis yields

$$2\pi j (E_w - \tilde{E}_w) = -2 \int_{-j\infty}^{j\infty} \text{tr}[Z_* (\{\Lambda_*^{-1} I_* \Omega_*^{-1}\}_- + \{\Lambda_*^{-1} I_* \Omega_*^{-1}\}_\infty] ds + \int_{-j\infty}^{j\infty} \text{tr}(ZZ_*) ds. \quad \text{(A-74)}$$

From the finiteness of the second integral on the right side of (A-74) and the guaranteed finiteness of E_w and $\tilde{E}_w$, we easily conclude by Cauchy's theorem that the first integral vanishes. Therefore,

$$E_w = \tilde{E}_w + \frac{1}{2\pi j} \int_{-j\infty}^{j\infty} \text{tr}(ZZ_*) ds \ge \tilde{E}_w. \quad \text{(A-75)}$$

Q.E.D.

Finally, to conclude, we will show that the zeros of det $(1 + L_1^{-1} \delta L_1)$, (42), depend solely on $P(s)$, $\delta P(s)$ and the nominal design and not on the choice of coprime factorizations. From (29) and (39) we obtain

$$\frac{\Delta_\delta}{\Delta} = \frac{\Delta_{p'}}{\Delta_p} \cdot \frac{\det A_1}{\det (A_1 + \delta A_1)} \cdot \det (1 + L_1^{-1} \delta L_1). \quad \text{(A-76)}$$

Clearly, by their very definitions, Δ, Δ_δ, Δ_p and $\Delta_{p'}$, are uniquely specified by $P(s)$, $\delta P(s)$ and the nominal design for $T_c(s)$. However, A_1 and $A_1 + \delta A_1$, being right-coprime matrix denominator polynomials for $F_t P$ and $F_t(P + \delta P)$, respectively, are only determined up to right elementary polynomial matrix multipliers [6]. Thus, it follows immediately from (A-76) that det $(1 + L_1^{-1} \delta L_1)$ is uniquely determined by the data up to a nonzero multiplicative constant.

Q.E.D.

References

[1] I. M. Horowitz, *Synthesis of Feedback Systems*. New York: Academic, 1963.

[31] We also drop the subscript w on Z_w.

[32] Let $G_{ds} = JJ_*$. Then, $J \le 0(s^{-1})$ and $\Omega^{-1} A F_t J$ is proper, so that

$$\Omega^{-1} A F_t J J_* = \Omega^{-1} A F_t G_{ds} \le 0(s^{-1}).$$

[33] Due to the impossibility of any $j\omega$-axis pole cancellation in the sum

$$AGA_* = AF_t G_{ds} F_{t*} A_* + AG_m A_*$$

each term must be good if AGA_* is good.

[2] S. S. L. Chang, *Synthesis of Optimal Control Systems*. New York: McGraw-Hill, 1961.

[3] D. C. Youla, H. Jabr, and J. J. Bongiorno, Jr., "Modern Wiener-Hopf design of optimal controllers—Part II: The multivariable case," *IEEE Trans. Automat. Contr.*, vol. AC-21, pp. 319-338, June 1976.

[4] J. J. Bongiorno, Jr., "Minimum sensitivity design of linear multivariable feedback control systems by matrix spectral factorization," *IEEE Trans. Automat. Contr.*, vol. AC-14, pp. 665-673, Dec. 1969.

[5] H. R. Rosenbrock, *State-Space and Multivariable Theory*. New York: Wiley, 1970.

[6] T. Kailath, *Linear Systems*. Englewood Cliffs, NJ: Prentice-Hall, 1980.

[7] L. Pernebo, "An algebraic theory for the design of controllers for linear multivariable systems—Part I: Structure matrices and feedforward design," *IEEE Trans. Automat. Contr.*, vol. AC-26, pp. 171-182, Feb. 1981.

[8] L. Pernebo, "An algebraic theory for the design of controllers for linear multivariable systems—Part II: Feedback realizations and feedback design," *IEEE Trans. Automat. Contr.*, vol. AC-26, pp. 183-193, Feb. 1981.

[9] J. C. Doyle and G. Stein, "Multivariable feedback design: Concepts for a classical modern synthesis," *IEEE Trans. Automat. Contr.*, vol. AC-26, pp. 4-16, Feb. 1981.

[10] I. Postlethwaite, J. M. Edmunds, and A. G. J. MacFarlane, "Principal gains and principal phases in the analysis of linear multivariable feedback systems," *IEEE Trans. Automat. Contr.*, vol. AC-26, pp. 32-46, Feb. 1981.

[11] M. G. Safonov, A. J. Laub, and G. L. Hartmann, "Feedback properties of multivariable systems: The role and use of the return difference matrix," *IEEE Trans. Automat. Contr.*, vol. AC-26, pp. 47-65, Feb. 1981.

[12] N. A. Lehtomaki, N. R. Sandell, Jr., and M. Athans, "Robustness results in linear-quadratic Gaussian based multivariable control designs," *IEEE Trans. Automat. Contr.*, vol. AC-26, pp. 75-92, Feb. 1981.

[13] M. G. Safonov and M. Athans, "A multiloop generalization of the circle criterion for stability margin analysis," *IEEE Trans. Automat. Contr.*, vol. AC-26, pp. 415-421, Apr. 1981.

[14] M. G. Safonov, "Stability margins of diagonally perturbed multivariable feedback systems," in *Proc. 20th IEEE Conf. Decision Contr.*, San Diego, CA, Dec. 1981, pp. 1472-1478.

[15] N. R. Sandell, "Robust stability of systems with application to singular perturbations," *Automatica*, vol. 15, pp. 467-470, 1979.

[16] J. S. Freudenberg, D. P. Looze, and J. B. Cruz, Jr., "Robustness analysis using singular value sensitivities," in *Proc. 20th IEEE Conf. Decision Contr.*, San Diego, CA, Dec. 1981, pp. 1158-1166.

[17] E. Polak and D. Q. Mayne, "On the solution of singular value inequalities over a continuum of frequencies," *IEEE Trans. Automat. Contr.*, vol. AC-26, pp. 690-694, June 1981.

[18] M. E. Sezer and D. D. Šiljak, "Robustness of suboptimal control: Gain and phase margin," *IEEE Trans. Automat. Contr.*, vol. AC-26, pp. 907-910, Aug. 1981.

[19] D. C. Youla, "On the factorization of rational matrices," *IRE Trans. Inform. Theory*, vol. IT-7, pp. 172-189, July 1961.

[20] J. B. Cruz, Jr., J. S. Freudenberg, and D. P. Looze, "A relationship between sensitivity and stability of multivariable feedback systems," *IEEE Trans. Automat. Contr.*, vol. AC-26, pp. 66-74, Feb. 1981.

[21] D. C. Youla, "A gain-bandwidth approach to the design of multivariable controllers," Internal Polytechnic Rep., 1974.

[22] D. C. Youla and N. N. Kazanjian, "Bauer-type factorization of positive matrices and the theory of matrix polynomials orthogonal on the unit circle," *IEEE Trans. Circuits Syst.*, vol. CAS-25, pp. 57-69, Feb. 1978.

[23] J. Rissanen, "Algorithm for triangular decomposition of block Hankel and Toeplitz matrices with application to factoring positive matrix polynomials," *Math. Comput.*, vol. 27, pp. 147-154, 1973.

[24] M. K. Sain, "A free-modular algorithm for minimal design of linear multivariable systems," in *Proc. 6th Int. Fed. Automat. Contr. Congr.*, Boston, MA, Aug. 1975.

[25] S. Kung, T. Kailath, and M. Morf, "Fast and stable algorithms for minimal design problems," in *Proc. Int. Fed. Automat. Contr. Symp. on Multiv. Technol. Syst.*, D. P. Atherton, Ed. New York: Pergamon, 1977, pp. 97-104.

[26] R. V. Patel, "Computation of matrix fraction descriptions of linear time-invariant systems," *IEEE Trans. Automat. Contr.*, vol. AC-26, pp. 148-161, Feb. 1981.

[27] V. Kučera, *Discrete Linear Control: The Polynomial Equation Approach*. New York: Wiley, 1979.

[28] ——"Stochastic multivariable control: A polynomial equation approach," *IEEE Trans. Automat. Contr.*, vol. AC-25, pp. 913-918, Oct. 1980.

[29] E. J. Davison, "The robust control of a servomechanism problem for linear time-invariant multivariable systems," *IEEE Trans. Automat. Contr.*, vol. AC-21, pp. 25-34, Feb. 1976.

[30] E. J. Davison and I. J. Ferguson, "The design of controllers for the multivariable robust servomechanism problem using parameter optimization methods," *IEEE Trans. Automat. Contr.*, vol. AC-26, pp. 93-110, Feb. 1981.

[31] H. Kwakernaak and R. Sivan, *Linear Optimal Control Systems*. New York: Wiley, 1972.

[32] R. Saeks and J. Murray, "Feedback system design: The tracking and disturbance rejection problems," *IEEE Trans. Automat Contr.*, vol. AC-26, Feb. 1981.

[33] G. Zames, "Feedback and optimal sensitivity: Model reference transformations, multiplicative seminorms, and approximate inverses," *IEEE Trans. Automat. Contr.*, vol. AC-26, pp. 301-320, Apr. 1981.

[34] G. Zames and B. A. Francis, "A new approach to classical frequency methods: Feedback and minimax sensitivity," in *Proc. 20th IEEE Conf. Decision Contr.*, San Diego, CA, Dec. 1981, pp. 867-874.

[35] B. A. Francis and G. Zames, "On optimal min-max servos," in *Proc. 20th IEEE Conf. Decision Contr.*, San Diego, CA, Dec. 1984, pp. 188-189.

[36] G. Bengtsson, "Feedback realizations in linear multivariable systems," *IEEE Trans. Automat. Contr.*, vol. AC-22, pp. 576-585, Aug. 1977.

[37] M. K. Sain, P. J. Antsaklis, R. R. Gejji, B. F. Wyman, and J. L. Peczkowski, "The total synthesis problem of linear multivariable control—Part II: Unity feedback and the design morphism," in *Proc. 20th IEEE Conf. Decision Contr.*, San Diego, CA, Dec. 1981, pp. 875-884.

[38] C. A. Desoer and M. J. Chen, "Design of multivariable feedback systems with stable plant," *IEEE Trans. Automat. Contr.*, vol. AC-26, pp. 408-415, Apr. 1981.

[39] L. Cheng and J. B. Pearson, Jr., "Synthesis of linear multivariable regulators," *IEEE Trans. Automat. Contr.*, vol. AC-26, pp. 194-202, Feb. 1981.

[40] L. M. Silverman and P. Van Dooren, "A system theoretic approach for GCD extraction," in *Proc. 1978 IEEE Conf. Decision Contr.*, San Diego, CA, Jan. 1979, pp. 525-528.

[41] A. Kontos and J. B. Pearson, "Computation of a unimodular matrix," presented at the Joint Automat. Contr. Conf., San Francisco, CA, 1980.

[42] A. Kontos, "APL programs for polynomial matrix manipulations," Dep. Elec. Eng., Rice Univ., Houston, TX, Tech. Rep. 7913, Dec. 1979.

[43] W. A. Wolovich, "Multipurpose controllers for multivariable systems," *IEEE Trans. Automat. Contr.*, vol. AC-26, pp. 162-170, Feb. 1981.

[44] C. A. Desoer and C. L. Gustafson, "Algebraic theory of linear multivariable feedback systems," in *Proc. 1983 Amer. Contr. Conf.*, June 1983; also in Univ. Calif., Berkeley, Memo UCB/ERL 82/90, Dec. 1982.

[45] M. Šebek, "Polynomial design of stochastic tracking systems," *IEEE Trans. Automat. Contr.*, vol. AC-27, pp. 468-470, Apr. 1982.

[46] G. Zames and B. A. Francis, "Feedback, minimax sensitivity, and optimal robustness," *IEEE Trans. Automat. Contr.*, vol. AC-28, pp. 585-601, May 1983.

[47] D. C. Youla, J. J. Bongiorno, Jr., and C. N. Lu, "Single-loop feedback-stabilization of linear multivariable dynamical plants," *Automatica*, vol. 10, pp. 159-173, 1974.

[48] J. J. Bongiorno, Jr., and D. C. Youla, "Author's reply," *IEEE Trans. Automat. Contr.*, vol. AC-23, pp. 964-965, Oct. 1978.

[49] J. W. Helton, "An H^∞ approach to control," in *Proc. IEEE Conf. Decision Contr.*, San Antonio, TX, Dec. 1983, pp. 607-611.

[50] H. Kwakernaak, "Robustness optimization of linear feedback systems," in *Proc. IEEE Conf. Decision Contr.*, San Antonio, TX, Dec. 1983, pp. 607-611.

[51] J. S. Freudenberg and D. P. Looze, "Sensitivity reduction, nonminimum phase zeros, and design tradeoffs in single loop feedback systems," in *Proc. IEEE Conf. Decision Contr.*, San Antonio, TX, Dec. 1983, pp. 607-611.

[52] ——, "Some remarks on H^∞-sensitivity minimization," in *Proc. IEEE Conf. Decision Contr.*, San Antonio, TX, Dec. 1983, pp. 634-635.

[53] B-C. Chang and J. B. Pearson, "Optimal disturbance reduction in linear multivariable systems," in *Proc. IEEE Conf. Decision Contr.*, San Antonio, TX, Dec. 1983, pp. 91-96.

[54] B. A. Francis and G. Zames, "Design of H^∞-optimal multivariable feedback systems," in *Proc. IEEE Conf. Decision Contr.*, San Antonio, TX, Dec. 1983, pp. 103-108.

[55] J. C. Doyle, "Synthesis of robust controllers and filters," *Proc. IEEE Conf. Decision Contr.*, pp. 109-114.

[56] M. G. Safonov, "L^∞-optimal sensitivity vs. stability margin," in *Proc. IEEE Conf. Decision Contr.*, San Antonio, TX, Dec. 1983, pp. 115-118.

[57] J. J. Bongiorno, Jr., and D. C. Youla, "On the design of single-loop single-input-output feedback control systems in the complex frequency domain," *IEEE Trans. Automat. Contr.*, vol. AC-22, pp. 416-423, June 1977.

Minimax Frequency Domain Performance and Robustness Optimization of Linear Feedback Systems

HUIBERT KWAKERNAAK, FELLOW, IEEE

Abstract—**It is shown that feedback system design objectives, such as disturbance attenuation and rejection, power and bandwidth limitation, and robustness, may be expressed in terms of required bounds of the sensitivity function and its complement on the imaginary axis. This leads to a minimax frequency domain optimization problem, whose solution is reduced to the solution of a polynomial equation.**

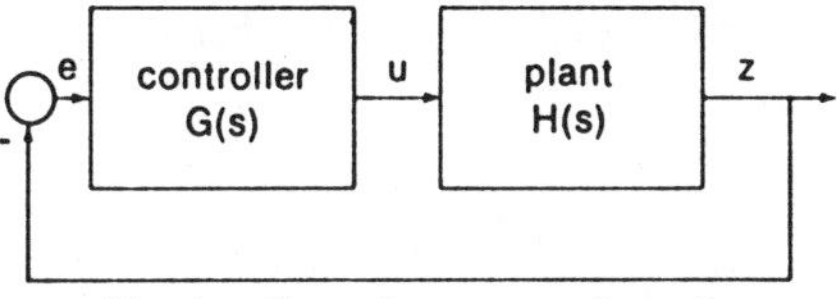

Fig. 1. Control system configuration.

I. Introduction

FEEDBACK systems are designed to fulfill a number of purposes. Among them are disturbance attenuation and rejection, good command response, and robustness. Restricted plant power and sensitivity to sensor noise form important limitations. Optimization is a powerful technique to deal with design problems with conflicting objectives. It therefore is little wonder that the optimization of control systems, particularly linear control systems, is a subject of long standing.

Inspired by the work of Wiener, much of the effort has been directed towards the minimization of quadratic integral-type criteria, either handled in the frequency domain such as in Wiener optimization, or in the time domain such as in linear optimal control theory. The principal reason for this emphasis on quadratic criteria is the existence of explicit solutions to the corresponding optimization problems and their accessibility for analysis.

Although linear optimal control theory with quadratic criteria is a powerful tool, it has limitations. One of its principal disadvantages is its inability to handle robustness problems. There seems to be no natural way to include robustness requirements as a design objective in quadratic integral criteria.

Much of the recent work on robustness has been focused on frequency domain characterizations [1]-[5]. The results mainly concern multiinput multioutput feedback systems in the configuration of Fig. 1, but the conclusions are of course also valid for single-input single-output systems, which are the subject of this paper. In the block diagram of Fig. 1, H is the plant transfer function and G the controller transfer function. The loop gain is denoted as $L(s) := H(s)G(s)$. From this recent work on robustness it is clear that the sensitivity function $S := 1/(1 + L)$ and its complement $T := 1 - S = L/(1 + L)$ play an important role in assessing the robustness of a closed-loop control system. It has also become evident that good robustness may be obtained if suitable bounds can be imposed on the behavior of S and T on the imaginary axis.

In this connection, an important field of study has been opened up by Zames [6]-[8]. Zames argues that under partial uncertainty about the nature of the disturbances it is necessary to impose bounds on the sensitivity function. This idea leads to an optimization problem that consists of the minimization of a criterion of the form $\sup_\omega|S(i\omega)V(i\omega)|$, where V is a suitable weighting function. The problem, as well as its multivariable version [9], [10], are solved by interpolation theory or functional analysis techniques. It is somewhat surprising that fairly explicit solutions exist and that no recourse needs be taken to nonlinear programming.

In the present paper it is attempted to take this work a little further and to show that many essential design objectives for linear feedback systems may be translated into required bounds on the sensitivity function S or its complement T, or both, on the imaginary axis. These design objectives include disturbance attenuation and rejection, power and bandwidth limitation, and robustness.

Section II of this paper is devoted to a discussion of these subjects. It is concluded that a balance between conflicting design objectives can be achieved by minimizing a criterion of the form

$$\sup_\omega \, [|V(i\omega)S(i\omega)|^2+|W(i\omega)T(i\omega)|^2] \tag{1.1}$$

where V and W are suitably chosen weighting functions. As in the work of Safonov, Laub, and Hartmann [11], the criterion involves both S and T, but it is a minimax and not an integral type criterion.

The actual solution of the minimax problem is undertaken in Section III. The solution technique differs from that of other authors and is related to the well-known fact in, for instance, statistical decision theory that "equalizers" yield minimax solutions. It is shown that the minimax problem may be reduced to the solution of a polynomial equation.

Section IV, which is quite brief, offers a few comments on the numerical solution of this polynomial equation. Section V deals with the specialization of the results to the minimization of $\sup_\omega|V(i\omega)S(i\omega)|$. Finally, in Section VI some examples are discussed. Three of them illustrate technical points. The fourth shows that application of the design technique to a stock example (the double integrator) leads to a classical design. The conclusions of the paper are summarized in Section VII.

Extension of the results of this paper to multiinput multioutput plants is feasible and is currently under investigation [12]. For single-input single-output systems, the results of the current paper may be generalized to a criterion of the form $\Sigma_k\,|V_k(i\omega)S(i\omega) + W_k(i\omega)T(i\omega)|^2$, with the V_k and W_k suitable weighting functions. These results are reported elsewhere [13]. Since the first version of the present paper was written, Verma and Jonckheere [14] published a solution of the problem of minimizing (1.1) based on Helton's work [15], while Francis [16] solved the multivariable version of the problem using functional analysis techniques.

We end this Introduction with some preliminaries. The plant transfer function H and the controller transfer function G are

Manuscript received October 19, 1983; revised August 21, 1984, November 21, 1984, and January 14, 1985. This paper is based on a prior submission of July 12, 1982. Paper recommended by Past Associate Editor, B. R. Barmish.

The author is with the Department of Applied Mathematics, Twente University of Technology, Enschede, The Netherlands.

Reprinted from *IEEE Trans. Automat. Contr.*, vol. AC-30, no. 10, pp. 994–1004, Oct. 1985.

written in the form

$$H(s)=\frac{\psi(s)}{\phi(s)}, \quad G(s)=\frac{\rho(s)}{\sigma(s)} \tag{1.2}$$

where ψ, ϕ, ρ, and σ are real polynomials such that ψ and ϕ have no common roots. The plant denominator polynomial ϕ has degree n and the numerator polynomial ψ degree m. G is not necessarily proper, but we assume that $H(s)$ and $L(s) = H(s)G(s)$ are. Under these assumptions the closed-loop system is well-defined as long as $L_\infty \neq -1$, where $L_\infty := \lim_{|s|\to\infty} L(s)$. When both ϕ and σ are monic, the closed-loop characteristic polynomial is given by

$$\chi_{ol}(s)\,\frac{1+L(s)}{1+L_\infty} \tag{1.3}$$

where $\chi_{ol}(s) := \phi(s)\sigma(s)$ is the open-loop characteristic polynomial. In the sequel it is convenient to work with the unnormalized characteristic polynomial

$$\chi_{cl}(s) := \chi_{ol}(s)(1+L(s)). \tag{1.4}$$

If the closed-loop system is well-defined, this polynomial equals the closed-loop characteristic polynomial within a nonzero constant factor, also when ϕ and σ are not monic. In terms of polynomials,

$$\chi_{cl}(s)=\phi(s)\sigma(s)+\psi(s)\rho(s). \tag{1.5}$$

The sensitivity function S and its complement T may be expressed as

$$S(s)=\frac{\phi(s)\sigma(s)}{\chi_{cl}(s)}, \quad T(s)=\frac{\psi(s)\rho(s)}{\chi_{cl}(s)}. \tag{1.6}$$

II. Performance and Robustness Optimization by Optimal Bounding

In this section it will first be shown that many essential control system design requirements can be translated into bounds on the behavior of the sensitivity function S and its complement T on the imaginary axis. The next step in the argument is to minimize these bounds. This leads to a minimax frequency domain optimization problem, whose solution is discussed in the next section. The design aspects considered in the present section are disturbance attenuation, disturbance rejection, power limitation—which is directly related to bandwidth limitation—robustness, and command response.

In the following, the discussion will be confined to closed-loop systems that have all their poles in the closed left-half plane. Controllers that result in control systems not satisfying this requirement are not admissible.

Disturbance Attenuation: For good disturbance attenuation, the sensitivity function S of the closed-loop system should have the property that $|S(i\omega)|$ is small over the frequency range of the disturbances. It has been convincingly argued by Zames [6] that this can be achieved by requiring that $|S(i\omega)V(i\omega)|$ is uniformly small for all ω, where V is a suitable weighting function reflecting the available information about the frequency content of the disturbances. Thus, for good disturbance attenuation we need to be ensured of a sufficiently small upper bound on $|S(i\omega)V(i\omega)|$.

Disturbance Rejection: Consider a disturbance (entering at the output of the plant) of the form $t^j e^{\lambda t}$, $j = 0, 1, \cdots, k-1$, with λ a given complex number such that Re $(\lambda) \geq 0$, and k a given integer. It is said that the closed-loop control system rejects this disturbance if the response of the controlled output to this disturbance eventually vanishes. A necessary and sufficient condition for disturbance rejection is that S has a zero of multiplicity k at $s = \lambda$.

It is mandatory to have rejection of disturbances that are plant-generated. These are disturbances that result from nonzero initial conditions of the plant, and are of the indicated type with λ any plant pole in the closed right-half plane and k its multiplicity. Since $S(s) = \phi(s)\sigma(s)/\chi_{cl}(s)$, S always has zeros, with the correct multiplicity, at the open-loop plant poles in the open right-half plane; by admissibility such poles can never cancel against roots of χ_{cl}. We can make sure that S has zeros, with the correct multiplicity, at the plant poles on the imaginary axis by requiring that $|S(i\omega)V(i\omega)|$ is bounded, where V has poles at all open-loop plant poles on the imaginary axis with corresponding multiplicity.

If V also has poles at other locations on the imaginary axis, rejection is achieved for the corresponding disturbances as well. This may be useful for vibration isolation (rejection of harmonic disturbances of a fixed frequency), but is particularly important for obtaining rejection of constant disturbances by letting V have a pole at $s = 0$. In particular, if V has a pole at $s = 0$ of multiplicity k, $|S(i\omega)V(i\omega)|$ can only be bounded if S has a zero at 0 of multiplicity k, which means that the closed-loop system is of "type-k."

Power and Bandwidth Limitation: In the closed-loop system, the transfer function from the disturbances, measurement noise, and command input to the plant input is $T(s)/H(s)$. Limitation of the input power of the plant can therefore be achieved by imposing a bound on $|W(i\omega)T(i\omega)|$, where W is a weighting function that is suitably chosen in relation to the plant transfer function H and the available information about the frequency content of disturbances, measurement noise, and command input.

We record some important considerations in the choice of the weighting function W. First, we note that if the plant has pole excess $e \geq 0$, $1/H(s) = 0(s^e)$ as $|s| \to \infty$. Therefore, in order to make sure that $|T(i\omega)/H(i\omega)|$ remains bounded as $\omega \to \infty$—which certainly is advisable—one should make sure that $W(s) = 0(s^e)$ as $|s| \to \infty$. If W has this property and $|W(i\omega)T(i\omega)|$ is bounded, T necessarily has pole excess e (or more). Since $L(s) = T(s)/(1 - T(s))$, the loop gain L has the same pole excess e as T. Consequently, since $L(s) = H(s)G(s)$, the controller transfer function G necessarily is proper. By letting $W(s) = 0(s^p)$ with $p \geq e$ we can in fact impose any desired pole excess $p - e$ on the controller transfer function G.

Next, assume that $W(s) = 0(s^p)$ for $|s| \to \infty$, with $p \geq e$. Then that frequency ω_1 from which the behavior of $W(i\omega)$ starts to be dominated by $(i\omega)^p$ is the frequency from which $T(i\omega)$ starts to decrease as $1/(i\omega)^p$; therefore, ω_1 is nothing else than the bandwidth of the closed-loop system. Thus, by a suitable choice of W it is possible to control the closed-loop bandwidth.

Finally, we note that $T(s)/H(s)$ will have poles at the plant zeros unless these are canceled by corresponding zeros of T. Because of power limitation, $T(s)/H(s)$ should have no poles on or to the right of the imaginary axis. Since $T(s)/H(s) = \phi(s)\sigma(s)/\chi_{cl}(s)$, poles to the right of the imaginary axis cannot occur by admissibility (see the definition of admissibility earlier in this section). We can make sure that T/H has no poles on the imaginary axis by letting W have poles at the plant zeros on the imaginary axis, with multiplicities equal to those of the zeros.

Robustness: From recent work [1]-[7] it has become clear that in questions of robustness, in particular stability robustness, both the sensitivity function and the complementary sensitivity function of the closed-loop system play an important role. We shall show here, in a variation on previous arguments and in a somewhat exemplary fashion, that S and T may simultaneously determine stability robustness. To this end, we assume that the plant transfer function is nominally $H(s) = \psi(s)/\phi(s)$, and that the nominal closed-loop system is asymptotically stable. Next we assume that the plant transfer function is perturbed to

$$H(s)=\frac{\psi'(s)}{\phi'(s)}\,h(s) \tag{2.1}$$

where ψ' is a polynomial of the same degree as ψ, ϕ' a polynomial of the same degree as ϕ, and where $h(s) := \bar{\psi}(s)/\bar{\phi}(s)$

is the parasitic (or singular) part of the perturbation. We assume $\bar{\psi}$ and $\bar{\phi}$ to be polynomial; the approach can be generalized to nonrational perturbations as well, however. We impose no restrictions on the relation of the roots of ψ' to those of ψ or on the roots of ϕ' to those of ϕ, but we require the roots of $\bar{\phi}$ to be in the open left-half complex plane. It is clear that (2.1) represents a large although not exhaustive class of perturbations. It includes both structured perturbations (induced by parameter variations), represented by ψ' and ϕ', and unstructured perturbations (induced by imperfect modeling), represented by h.

In the following the prime denotes the perturbed closed-loop system. With this convention we have $\chi_{cl} = \phi\sigma + \psi\rho$, $\chi'_{cl} = \phi'\bar{\phi}\sigma + \psi'\bar{\psi}\rho$ and it is easily established that

$$\frac{\chi'_{cl}}{\chi_{cl}\bar{\phi}} = \frac{\phi'}{\phi} S + \frac{\psi'}{\psi} hT. \tag{2.2}$$

By the principle of the argument [17] it is easily shown that under the assumption that χ_{cl} and $\bar{\phi}$ both have all their roots in the open left-half plane, χ'_{cl} has all its roots in the open left-half plane if and only if the image of $\chi'_{cl}(i\omega)/(\chi_{cl}(i\omega)\bar{\phi}(i\omega))$ does not encircle the origin as ω runs from $-\infty$ to $+\infty$. This means that the closed-loop system remains stable under perturbation as long as this condition is satisfied. Equivalently, the closed-loop system remains stable under perturbation as long as

$$\frac{\chi'_{cl}(i\omega)}{\chi_{cl}(i\omega)\bar{\phi}(i\omega)} - 1 = \left(\frac{\phi'(i\omega)}{\phi(i\omega)} - 1\right) S(i\omega) + \left(\frac{\psi'(i\omega)}{\psi(i\omega)} h(i\omega) - 1\right) T(i\omega) \tag{2.3}$$

does not encircle the point -1.

A sufficient condition for the right-hand side of (2.3) not to encircle the point -1 is that its modulus is less than 1 for all ω. Suppose that there exist rational functions V_1, W_1, and W_2 such that the various relative perturbations can be bounded as follows:

$$\left|\frac{\phi'(i\omega)}{\phi(i\omega)} - 1\right| \le |V_1(i\omega)|, \quad \left|\frac{\psi'(i\omega)}{\psi(i\omega)} - 1\right| \le |W_1(i\omega)|,$$

$$|h(i\omega) - 1| \le |W_2(i\omega)| \tag{2.4}$$

for all ω. Then since

$$\frac{\psi'}{\psi} h - 1 = \left(\frac{\psi'}{\psi} - 1\right)(h-1) + \left(\frac{\psi'}{\psi} - 1\right) + (h-1), \tag{2.5}$$

twofold application of the inequality

$$\left|\sum_{i=1}^{n} x_i\right|^2 \le n \sum_{i=1}^{n} |x_i|^2 \tag{2.6}$$

shows that the right-hand side of (2.3) may be bounded from above by $2(|V_1S|^2 + 3(|W_1W_2|^2 + |W_1|^2 + |W_2|^2)|T|^2)$, where the argument is $i\omega$ throughout. Thus, a sufficient condition for stability of the closed-loop system under perturbations satisfying (2.4) is that an expression of the form

$$|V(i\omega)S(i\omega)|^2 + |W(i\omega)T(i\omega)|^2 \tag{2.7}$$

is bounded by 1, $|V|^2 := 2|V_1|^2$ and $|W|^2 := 6(|W_1W_2|^2 + |W_1|^2 + |W_2|^2)$. The point that we wish to make is that given information about the possible relative perturbations in the form (2.4), robust stability can be ensured by bounding an expression of the form (2.7), which is a weighted combination of $|S(i\omega)|^2$ and $|T(i\omega)|^2$.

Command Response: In single-degree-of-freedom control systems the control system transfer function from the command input to the controlled output equals the complementary sensitivity function T. The design requirement that stipulates the sensitivity function to be small over the frequency range of the disturbance by implication requires $T = 1 - S$ to be close to 1 over that same frequency range. This in turn ensures good response to command signals in that frequency range. It is assumed that any corrections to the desired response to command signals can be accomplished by a suitable prefilter in a two-degree-of-freedom control configuration, and we impose no additional requirements on the loop properties.

Minimax Optimization: The conclusion from the previous considerations is that various control system design objectives, such as disturbance attenuation and rejection, power and bandwidth limitation and robustness, can be expressed in the form of required bounds on $|V(i\omega)S(i\omega)|$, $|W(i\omega)T(i\omega)|$ and $|V'(i\omega)S(i\omega)|^2 + |W'(i\omega)T(i\omega)|^2$, with V, W, V', and W' suitable weighting functions. Equivalently, the design objectives can be expressed in terms of required bounds on $|V(i\omega)S(i\omega)|^2$, $|W(i\omega)T(i\omega)|^2$ and $|V'(i\omega)S(i\omega)|^2 + |W'(i\omega)T(i\omega)|^2$. Since in each instance the smaller the bound is, the better the design objective is achieved, it is attractive to minimize the bounds in some way. We shall do this by minimizing the least upper bound of an expression of the form

$$a|V(i\omega)S(i\omega)|^2 + b|W(i\omega)T(i\omega)|^2 + c[|V'(i\omega)S(i\omega)|^2 + |W'(i\omega)T(i\omega)|^2] \tag{2.8}$$

where the nonnegative constants a, b, and c reflect the importance of each term. Given an upper bound on (2.8), corresponding upper bounds on each of the terms can immediately be ascertained. Simplifying the notation by combining terms and renaming coefficients, we are thus led to the problem of minimizing an expression of the form

$$\sup_{\omega} [|V(i\omega)S(i\omega)|^2 + |W(i\omega)T(i\omega)|^2] \tag{2.9}$$

with V and W suitable weighting functions. The solution of this problem is the subject of the next section.

III. Solution of the Minimax Problem

In this section we study the problem of minimizing the criterion (2.9).

Assumptions: We first introduce and explain a number of assumptions. To begin with, the weighting functions V and W are restricted to rational functions of the form

$$V(s) = \frac{\alpha_1(s)}{\beta_1(s)}, \quad W(s) = \frac{\alpha_2(s)}{\beta_2(s)} \tag{3.1}$$

where α_1, α_2, β_1, and β_2 are real polynomials such that deg $(\alpha_1) =: a_1$, deg $(\alpha_2) =: a_2$, deg $(\beta_1) =: b_1$, deg $(\beta_2) =: b_2$, satisfying the following hypotheses.

a) $a_1 \le b_1$ and $a_2 = b_2 + e$, with e a nonnegative integer.

b) α_1 and α_2 have all their roots in the closed left-half plane; β_1 and β_2 have all their roots in the closed left-half plane and have no common roots; α_1 and β_1 have no common roots; α_2 and β_2 have no common roots.

c) The polynomial $\gamma := \alpha_1\alpha_1^*\beta_2\beta_2^* + \alpha_2\alpha_2^*\beta_1\beta_1^*$ has no roots on the imaginary axis.

Here if f is any rational function, $f^*(s) := f(-s)$. Assumption a) ensures that $V(s) = O(1)$ and $W(s) = O(s^e)$ for $|s| \to \infty$, which is entirely acceptable in view of the discussion of the preceding section. The assumptions b) cause no loss of generality, except (very slightly) that β_1 and β_2 have no common roots. Assumption c) is needed to ensure that the problem has a solution at all.

Further to the assumptions at the end of Section I we introduce the factorizations

$$\phi = \phi_+\phi_o\phi_-, \quad \psi = \psi_+\psi_o\psi_- \tag{3.2}$$

where ϕ_+, ϕ_o, ϕ_-, ψ_+, ψ_o, and ψ_- are real polynomials with deg $(\phi_+) =: n_+$, deg $(\phi_o) =: n_o$, deg $(\phi_-) =: n_-$, deg $(\psi_+) =: m_+$, deg $(\psi_o) =: m_o$ and deg $(\psi_-) =: m_-$, such that ϕ_+ and ψ_+ have all their roots in the open right-half plane, ϕ_o and ψ_o have all their roots on the imaginary axis, and ϕ_- and ψ_- have all their roots in the open left-half plane.

Admissible Controllers: We restrict the admissible controllers to those that have a rational transfer function G such that the closed-loop system has all its poles in the closed left-half plane. In the preceding section we have seen that if $W(s) = 0(s^e)$ as $|s| \to \infty$, with e the plant pole excess, the controller transfer function is necessarily proper, provided (2.9) has a finite minimum at all. Later (Lemma 1) we shall see under what conditions on V and W the closed-loop system actually has all its poles in the *open* left-half plane.

Preliminaries: In this subsection we introduce the equations that will determine the solution of the minimax problem. First, we define the rational function

$$Z(s) := V(s)V(-s)S(s)S(-s) + W(s)W(-s)T(s)T(-s). \quad (3.3)$$

Thus, our problem is that of minimizing $\sup_\omega Z(i\omega)$. It will turn out that control systems that achieve this minimum have the property that

$$Z(s) = \lambda^2 \quad (3.4)$$

with λ a real constant, and we shall first consider which controllers from the admissible class achieve this. Suppose that the controller transfer function is $G = \rho/\sigma$, with ρ and σ real polynomials. Then

$$L = HG = \frac{\psi_+\psi_o\psi_-\rho}{\phi_+\phi_o\phi_-\sigma}. \quad (3.5)$$

Without loss of generality we can choose ρ and σ such that all factors in the numerator and denominator of L that have their roots in the closed left-half plane cancel. We, therefore, let ρ have a factor $\phi_-\phi_o$ and let σ have a factor $\psi_-\psi_o$. On the other hand, from (1.6) and (3.1)

$$VS = \frac{\alpha_1\phi\sigma}{\beta_1\chi_{cl}}, \quad WT = \frac{\alpha_2\psi\rho}{\beta_2\chi_{cl}}. \quad (3.6)$$

Since we want $Z(s) = \lambda^2$, VV^*SS^* cannot have poles at the roots of $\beta_1\beta_1^*$ and WW^*TT^* cannot have poles at the roots of $\beta_2\beta_2^*$. These requirements are satisfied when σ has a factor β_1 and ρ has a factor β_2.

In view of these observations, we let $\sigma = \psi_-\psi_o\beta_1\theta$ and $\rho = \phi_-\phi_o\beta_2\zeta$, with θ and ζ real polynomials to be determined. This means that the controller transfer function is chosen as

$$G = \frac{\phi_-\phi_o\beta_2\zeta}{\psi_-\psi_o\beta_1\theta}. \quad (3.7)$$

With this controller, the closed-loop characteristic polynomial is $\chi_{cl} = \phi_-\phi_o\psi_-\psi_o\chi$, where the polynomial χ is defined by

$$\chi := \beta_1\phi_+\theta + \beta_2\psi_+\zeta. \quad (3.8)$$

Correspondingly, the sensitivity function S and its complement T are given by

$$S = \frac{\beta_1\phi_+\theta}{\chi}, \quad T = \frac{\beta_2\psi_+\zeta}{\chi}. \quad (3.9)$$

Substitution of these formulas into (3.3) immediately shows that $Z(s) = \lambda^2$ is equivalent to

$$\alpha_1\alpha_1^*\phi_+\phi_+^*\theta\theta^* + \alpha_2\alpha_2^*\psi_+\psi_+^*\zeta\zeta^* = \lambda^2\chi\chi^*. \quad (3.10)$$

This equation constitutes a basic relation. Another crucial result follows by completing the square on the right-hand side of (3.3), and writing Z in the form

$$Z = \frac{(VV^*S - WW^*T)^*(VV^*S - WW^*T) + VV^*WW^*}{VV^* + WW^*}. \quad (3.11)$$

Substituting V and W as given by (3.1) and S and T as given by (3.9), it follows that

$$Z = \frac{(\alpha_1\alpha_1^*\beta_2^*\phi_+\theta - \alpha_2\alpha_2^*\beta_1^*\psi_+\zeta)(\alpha_1\alpha_1^*\beta_2^*\phi_+\theta - \alpha_2\alpha_2^*\beta_1^*\psi_+\zeta)^* + \alpha_1\alpha_1^*\alpha\ \alpha_2^*\chi\chi^*}{(\beta_1\beta_1^*\alpha_2\alpha_2^* + \beta_2\beta_2^*\alpha_1\alpha_1^*)\chi\chi^*}. \quad (3.12)$$

Therefore, $Z(s) = \lambda^2$ iff

$$(\alpha_1\alpha_1^*\beta_2^*\phi_+\theta - \alpha_2\alpha_2^*\beta_1^*\psi_+\zeta)(\alpha_1\alpha_1^*\beta_2^*\phi_+\theta - \alpha_2\alpha_2^*\beta_1^*\psi_+\zeta)^*$$
$$= \lambda^2\left(\gamma - \frac{1}{\lambda^2}\eta\right)\chi\chi^* \quad (3.13)$$

where

$$\gamma := \beta_1\beta_1^*\alpha_2\alpha_2^* + \beta_2\beta_2^*\alpha_1\alpha_1^*, \quad \eta := \alpha_1\alpha_1^*\alpha_2\alpha_2^*. \quad (3.14)$$

We now define the polynomial π_λ, for those real λ for which it exists, as the real polynomial of degree $b_1 + b_2 + e$ with nonnegative coefficients and all its roots in the closed left-half plane, such that

$$\pi_\lambda\pi_\lambda^* = \gamma - \frac{1}{\lambda^2}\eta. \quad (3.15)$$

The polynomial π_λ is well-defined for $|\lambda| \geq \lambda_o$, where $\lambda_o \geq 0$ is the first value of λ for which the right-hand side of (3.15) either loses degree or assumes a root on the imaginary axis as $|\lambda|$ is decreased from ∞. By assumption c), $\lambda_o < \infty$. Returning to (3.13), we observe that there are many ways of factoring this equation. It will turn out that the factorization we need is

$$\alpha_1\alpha_1^*\beta_2^*\phi_+\theta - \alpha_2\alpha_2^*\beta_1^*\psi_+\zeta = \lambda\pi_\lambda\chi^*. \quad (3.16)$$

Before proceeding, we summarize some properties of the controller (3.7).

Lemma 1: Consider the controller (3.7), and suppose that the polynomials θ and ζ satisfy the equation (3.16), with χ given by (3.8), for some real λ. Then the following holds. a) $Z(s) = \lambda^2$. b) If the polynomial χ has all its roots in the open left-half plane, the closed-loop system has all its poles in the closed left-half plane. c) If, in addition, β_1 has a factor ϕ_o and β_2 has a factor ψ_o, and these factors are canceled in the controller transfer function (3.7), the closed-loop system has all its closed-loop poles in the open left-half plane. □

Proof: a) That $Z(s) = \lambda^2$ follows by substituting (3.16) into (3.12). b) As we have seen following (3.7), the closed-loop polynomial is $\phi_-\phi_o\psi_-\psi_o\chi$, so that under the hypothesis the closed-loop poles are all in the closed left-half plane. c) If in the controller transfer function the factors ϕ_o and ψ_o are canceled, the closed-loop characteristic polynomial is $\phi_-\psi_-\chi$, which under the hypothesis has all its roots in the open left-half plane. □

The final step in this subsection is to establish the existence of a solution to (3.16). To this end, we first define the integers

$$x := n_+ + m_+ + b_1 + b_2 + e, \quad t := m_+ + b_2 + e, \quad z := n_+ + b_1. \quad (3.17)$$

Theorem 1: a) For $|\lambda| \geq \lambda_o$ the polynomial equation (3.16), with χ defined by (3.8), has a family of solutions χ_λ, θ_λ, ζ_λ with $\deg(\chi_\lambda) \leq x$, $\deg(\theta_\lambda) \leq t$, $\deg(\zeta_\lambda) \leq z$, unique within multiplication by a continuous function of λ, whose coefficients are continuous functions of λ, such that for $|\lambda|$ sufficiently large the polynomial χ_λ has degree x and has all its roots in the open left-half complex plane. b) Suppose that χ_{λ_o} or $\chi_{-\lambda_o}$ has a root in the open right-half plane. Then there exists a λ with $|\lambda| > \lambda_o$ such that $\deg(\chi_\lambda) \leq x - 1$, $\deg(\theta_\lambda) \leq t - 1$, $\deg(\zeta_\lambda) \leq z - 1$ and χ_λ has all its roots in the open left-half plane. □

In part b) of the theorem, it is to be understood that if a polynomial is assigned the degree -1, it is the zero polynomial. The proof of the theorem is given in Appendix A. That of part a) relies on first proving that (3.10) has a suitable solution (by the implicit function theorem), and then showing that this solution also solves (3.16). Part b) follows from the fact that under the hypothesis at least one root of χ_λ crosses over from the left- to the right-half plane for some $|\lambda| > \lambda_o$; this root can be canceled, resulting in a solution of reduced degree.

Solution of the Minimax Problem: Before presenting the details of the solution of the minimax problem, we establish an important property of the number λ_o previously defined.

Lemma 2: For any controller, $\sup_\omega Z(i\omega) \geq \lambda_o^2$. □

Proof: From (3.12) it is seen that $Z(i\omega) \geq \eta(i\omega)/\gamma(i\omega)$ for all real ω. It follows that $\sup_\omega Z(i\omega) \geq \sup_\omega[\eta(i\omega)/\gamma(i\omega)] = \inf\{\lambda^2 : \eta(i\omega)/\gamma(i\omega) \leq \lambda^2 \text{ for all } \omega\} = \inf\{\lambda^2 : \gamma(i\omega) - \eta(i\omega)/\lambda^2 \geq 0 \text{ for all } \omega\} = \lambda_o^2$. The last step is detailed in Appendix B. □

The following describes the solution to the problem of minimizing (2.9) in a somewhat exceptional situation.

Theorem 2: Suppose that χ_{λ_o} or $\chi_{-\lambda_o}$ has all its roots in the closed left-half plane. Then the controller (3.7), with θ replaced with θ_{λ_o}, respectively $\theta_{-\lambda_o}$, and ζ replaced with ζ_{λ_o}, respectively, $\zeta_{-\lambda_o}$, minimizes (2.9), and the minimal value is λ_o^2. □

Proof: Any roots of χ_{λ_o}, respectively $\chi_{-\lambda_o}$, on the imaginary axis may be canceled as in the proof of Theorem 1b). As the remaining roots are in the open left-half plane, by Lemma 1 the closed-loop system has all its poles in the closed left-half plane and achieves $Z(i\omega) = \lambda_o^2$. Then by Lemma 2 the control system minimizes (2.9). □

We briefly return to this situation in the Corollary to Theorem 3. In case Theorem 2 does not apply, we introduce the auxiliary problem of minimizing

$$\int_{-\infty}^{\infty} Z(i\omega)\Sigma(i\omega)\, d\omega \tag{3.18}$$

where $\Sigma(s)$ is a rational, strictly proper function of s with real coefficients such that $\Sigma(i\omega)$ is real and positive for $-\infty < \omega < \infty$. The following result explains our interest in controllers such that $Z(i\omega) = $ constant.

Lemma 3: Suppose that (for a given Σ) the auxiliary criterion (3.18) is minimized by an admissible controller that makes $Z(i\omega)$ a constant λ^2. Then this controller also minimizes $\sup_\omega Z(i\omega)$. □

Proof: This result is well known from statistical decision theory (see, e.g., [18]) and its proof is very simple. Suppose that there exists an admissible controller leading to a function $\tilde{Z}$ such that $\sup_\omega \tilde{Z}(i\omega) < \lambda^2$. Then necessarily $\tilde{Z}(i\omega) < \lambda^2 = Z(i\omega)$ for all ω. As a result, $\int_{-\infty}^{\infty} \tilde{Z}(i\omega)\Sigma(i\omega)d\omega < \int_{-\infty}^{\infty} Z(i\omega)\Sigma(i\omega)d\omega$, which contradicts the assumption that Z minimizes (3.18). □

A sufficient condition for the minimization of (3.18) is the following.

Lemma 4: A sufficient condition for the minimization of (3.18) is that $\Sigma\phi\psi(VV^*S^* - WW^*T^*)$ has all its poles in the closed left-half plane. If a minimizing solution exists, it is unique. □

A proof of this lemma is given in Appendix C. We are now able to state the solution to the problem of minimizing (2.9) in case Theorem 2 does not apply.

Theorem 3: Suppose that χ_{λ_o} or $\chi_{-\lambda_o}$ has at least one root in the open right-half plane. According to Theorem 1b) there exists a λ with $|\lambda| > \lambda_o$ such that $\deg(\chi_\lambda) \leq x - 1$, $\deg(\theta_\lambda) \leq t - 1$, $\deg(\zeta_\lambda) \leq z - 1$ and χ_λ has all its roots in the open left-half plane. Then the controller (3.7), with θ replaced with θ_λ and ζ replaced with ζ_λ, uniquely minimizes (2.9), and the minimal value is λ^2. □

Proof: We shall show that the solution χ_λ, θ_λ, ζ_λ of reduced degree minimizes an auxiliary criterion of the form (3.18), with Σ suitably chosen. With S and T given by (3.9), the optimality condition of Lemma 4 takes the form that the rational function

$$\Sigma\phi\psi \frac{\alpha_1\alpha_1^*\beta_2\phi_+^*\theta_\lambda^* - \alpha_2\alpha_2^*\beta_1\psi_+^*\zeta_\lambda^*}{\beta_1\beta_2\chi_\lambda^*} \tag{3.19}$$

should have all its poles in the closed left-half plane. For controllers that satisfy (3.16), this condition reduces to the requirement that

$$\Sigma\phi\psi \frac{\lambda\pi_\lambda^*\chi_\lambda}{\beta_1\beta_2\chi_\lambda^*} \tag{3.20}$$

have all its poles in the closed left-half plane. This requirement is indeed satisfied when

$$\Sigma = \frac{\chi_\lambda\chi_\lambda^*}{\phi_+\phi_+^*\psi_+\psi_+^*\pi_\lambda\pi_\lambda^*}. \tag{3.21}$$

Since for the particular solution of (3.16) that we are considering $\deg(\chi_\lambda) \leq x - 1$, and $\deg(\pi_\lambda) = b_1 + b_2 + e$ for $|\lambda| > \lambda_o$, Σ as given is strictly proper as required. Thus, by Lemma 4, the controller we consider minimizes the auxiliary criterion (3.18). Since by Lemma 1 for this controller $Z(i\omega) = \lambda^2 = $ constant, it follows from Lemma 3 that this controller minimizes (2.9). It remains to prove uniqueness. Suppose first that there exists a controller resulting in a Z such that $\sup_\omega Z(i\omega) = \lambda^2$ but $Z(i\omega) < \lambda^2$ over some nonzero interval. Then since Σ is strictly positive, this controller would achieve a smaller value for the auxiliary criterion (3.18), with Σ given by (3.21), than the controller considered so far. This contradicts the fact that the latter controller minimizes the auxiliary criterion. We conclude that for any controller that minimizes (2.9) necessarily $Z(i\omega) = \lambda^2$. Therefore, such a controller minimizes the auxiliary criterion (3.18) with Σ given by (3.21). Since by Lemma 4 the controller that minimizes the auxiliary criterion is uniquely determined, also the minimax controller is uniquely determined. □

Theorems 2 and 3 show that the solution of the minimax optimization problem may be reduced to determining suitable solutions to the polynomial equation (3.16), with χ defined by (3.8). Since solutions to (3.16) also satisfy (3.10), alternatively one may solve (3.10), with χ given by (3.8), but only solutions qualify that satisfy (3.16). For a further discussion see Section IV and Example 2 in Section VI.

Corollary: χ_{λ_o} has all its roots in the closed left-half plane if and only if $\chi_{-\lambda_o}$ has all its roots in the closed left-half plane.

Proof: Let χ_{λ_o} have all its roots in the closed left-half plane, but suppose that $\chi_{-\lambda_o}$ has a root in the open right-half plane. Then by Theorem 2 there exists an optimal controller that achieves $\sup_\omega Z(i\omega) = \lambda_o^2$, while by Theorem 3 there exists an optimal controller that achieves $\sup_\omega Z(i\omega) = \lambda^2 > \lambda_o^2$. This is a contradiction. The proof when $\chi_{-\lambda_o}$ has all its roots in the closed left-half plane is similar. □

In case Theorem 2 applies (which is a somewhat exceptional situation) the optimal controller may be nonunique. An example is given in Section VI (Example 1).

We conclude this section with a remark about the order of the optimal controller. Suppose that i) $e = n - m$, and ii) all roots of β_1 are also roots of $\phi_o\phi_-$ and all roots of β_2 are also roots of $\psi_o\psi_-$. Then it is easy to check that if Theorem 2 applies, the controller order is n, while in case Theorem 3 applies it is $n - 1$. Here n is the plant order. Assumption i) ensures that the controller is proper, while assumption ii) includes the situation where ϕ_o is a factor of β_1 and ψ_o is a factor of β_2, which guarantees the closed-loop system to be asymptotically stable. If i) and ii) do not hold, the controller order may be less or greater than n, respectively, $n - 1$. A controller order less than n, respectively, $n - 1$ is achieved only if the requirement that the controller be proper is dropped.

IV. Numerical Solution

Unless the various polynomials that are involved are of very low degree, it is not feasible to solve the polynomial equations that determine the solution of the minimax problem in closed form. Often it will be necessary to take recourse to numerical computation.

To check whether Theorem 2 applies, it is necessary to solve the (overdetermined) linear equation (3.16), with χ defined by (3.8), for $\lambda = \pm\lambda_o$. This is a routine problem from numerical linear algebra. Algorithms for the factorization (3.15) are well known (see, e.g., Kucera [19]).

If for $\lambda = \pm\lambda_o$ no solution is obtained such that χ has all its roots in the closed left-half plane it is useful to solve the equation (3.16) for a number of trial values of λ in order to obtain an estimate of the solution for which the first root of χ crosses over to the right-half plane as $|\lambda|$ is decreased from ∞.

To find the exact solution of reduced degree, it seems most convenient to solve χ, θ, ζ, and λ, with $\deg(\chi) = x - 1$, $\deg(\theta) = t - 1$, $\deg(\zeta) = z - 1$, from (3.10), with χ given by (3.8). Since these equations are homogeneous, it is necessary to add a normalization condition, for instance that χ is monic. This results in a set of nonlinear equations with as many unknowns as equations, which may be solved by a Newton–Raphson approach, using the approximate solution obtained as indicated in the preceding paragraph as a starting solution. Since solutions to (3.10), even if χ has all its roots in the open left-half plane, do not necessarily satisfy (3.16), the latter equation always has to be checked (see Example 2 in Section VI).

Computer codes (in Pascal) that implement these algorithms are available [20].

V. Specialization to the Case $W = 0$

In this section we consider the specialization of the minimax problem to the case where $W = 0$, which is the problem solved by Zames and Francis [7], [8]. To this end we take $\alpha_2 = 0$, $\beta_2 = 1$. It immediately follows that $\gamma = \alpha_1\alpha_1^*$, $\eta = 0$, $\lambda_o = 0$, and $\pi_\lambda = \alpha_1$, so that (3.8) and (3.16) reduce to $\chi = \beta_1\phi_+\theta + \psi_+\zeta$, $\alpha_1^*\phi_+\theta = \lambda\chi^*$. Elimination of χ leads to the equation

$$\frac{1}{\lambda}\alpha_1\phi_+^*\theta^* = \beta_1\phi_+\theta + \psi_+\zeta \tag{5.1}$$

where $\deg(\theta) = t = m_+$, $\deg(\zeta) = z = n_+ + b_1$. Note that although this equation does not appear in the work of Zames and Francis, it can immediately be obtained from their results.

We first check whether Theorem 2 applies. For $\lambda = \lambda_o = 0$ the equation (5.1) has the solution $\hat{\theta} = \psi_+^*$, $\zeta = \alpha_1\phi_+^*$, $\chi = \alpha_1\phi_+^*\psi_+$, where $\hat{\theta} := \theta/\lambda$. The polynomial χ has its roots in the open left-half plane if and only if $\psi_+ =$ constant (note that by assumption c) α_1 has all its poles in the open left-half plane). This solution corresponds to the well-known fact [8] that if the plant has no right-half plane zeros, one can make $|S(i\omega)|$ uniformly arbitrarily close to 0 by making the gain large enough.

When the plant possesses right-half plane zeros, Theorem 3 applies and we need to solve (5.1) with $\deg(\theta) = m_+ - 1$, $\deg(\zeta) = n_+ + b_1 - 1$. If $n_+ + b_1 = 0$, by assumption a) $a_1 = 0$, and (5.1) has the trivial solution $\lambda = \alpha_1/\beta_1$, $\theta = 1$, $\zeta = 0$. In all other cases, (5.1) can be reduced to a (generalized) eigenvalue problem as follows. Define $\boldsymbol{\theta}$ and $\boldsymbol{\zeta}$ as the m_+-, respectively, $(n_+ + b_1)$-dimensional column vectors whose elements are the coefficients of the polynomial θ, respectively, ζ, arranged in order of increasing power. Then in matrix form the polynomial equation (5.1) can be written as

$$\frac{1}{\lambda}T_{n_++m_++b_1,m_+}(\alpha_1\phi_+^*)J_{m_+}\boldsymbol{\theta} = T_{n_++m_++b_1,m_+}(\beta_1\phi_+)\boldsymbol{\theta} + T_{n_++m_++b_1,n_++b_1}(\psi_+)\boldsymbol{\zeta}. \tag{5.2}$$

Here if α is any polynomial, $T_{n,m}(\alpha)$ is the $n \times m$ lower triangular Toeplitz matrix formed from the coefficients of α (see Kailath [21]); furthermore, J_n is the $n \times n$ diagonal matrix $J_n := \mathrm{diag}(1, -1, 1, -1, \cdots)$. The equation (5.2) can be rearranged as

$$\left(\frac{1}{\lambda}A - B\right)\begin{bmatrix}\boldsymbol{\theta}\\ \boldsymbol{\zeta}\end{bmatrix} = 0 \tag{5.3}$$

where A and B are the $(n_+ + m_+ + b_1) \times (n_+ + m_+ + b_1)$ matrices

$$A := (T(\alpha_1\phi_+^*)J,\ 0),\quad B := (T(\beta_1\phi_+),\ T(\psi_+)). \tag{5.4}$$

Here for brevity we omitted the indexes. By the assumptions on β_1, ϕ_+, and ψ_+ the matrix B is nonsingular; hence (5.3) is equivalent to $(R/\lambda - I)\,\mathrm{col}(\boldsymbol{\theta}, \boldsymbol{\zeta}) = 0$, where $R := B^{-1}A$. It is seen that R is of the form

$$\begin{bmatrix}R_{11} & 0\\ R_{21} & 0\end{bmatrix} \tag{5.5}$$

where R_{11} is $m_+ \times m_+$ and R_{21} is $(n_+ + b_1) \times m_+$. As a result, we have

$$R_{11}\boldsymbol{\theta} = \lambda\boldsymbol{\theta},\quad \boldsymbol{\zeta} = \frac{1}{\lambda}R_{21}\boldsymbol{\theta}. \tag{5.6}$$

Clearly, λ is an eigenvalue of R_{11} and $\boldsymbol{\theta}$ a corresponding eigenvector. Now, from the proof of Theorem 1b) it follows that if there are several values of λ that yield a solution of (5.1) of reduced degree, the one we are looking for (namely that for which χ has all its roots in the open left-half plane) is that for which λ is the largest (in absolute value). Thus, λ is the largest (in absolute value) of the real eigenvalues of R_{11}.

The eigenvalue problem (5.6) that results from (5.1) may be routinely solved. This approach is offered as an alternative to the methods proposed by Zames and Francis [7], [8]. It leads of course to the same solutions. An example is given in Section VI (Example 3).

VI. Examples

Example 1: The first example serves to illustrate that there are cases where the lower bound λ_o^2 is achieved, and that in these cases the solution may be nonunique. Take $H = 1$, $V = (s + \epsilon)/(s + 1)$, $W = s$, where $0 < \epsilon < 1$. Then the polynomial π_λ is defined by $\pi_\lambda\pi_\lambda^* = (1 - 1/\lambda^2)s^4 + (-2 + \epsilon^2/\lambda^2)s^2 + \epsilon^2$. From this it may be verified that $\lambda_o = 1$ and $\pi_{\lambda_o} = (2 - \epsilon^2)^{1/2}s + \epsilon$. For $\lambda = \pm 1$ the equations (3.8) and (3.16) reduce to $\chi = (s + 1)\theta + \zeta$, $\pm((2-\epsilon^2)^{1/2}s + \epsilon)\chi^* = (-s^2 + \epsilon^2)\theta + s^2(-s + 1)\zeta$, where $\deg(\chi) = x = 2$, $\deg(\theta) = t = 1$ and $\deg(\zeta) = z = 1$. It is straightforward to solve the linear equations that result from expanding the polynomial equations in terms of their coefficients. It is found that for $\lambda = 1$ the solution may be written as $\theta = (\epsilon + 1) + ((2 - \epsilon^2)^{1/2} - 1)s$, $\zeta = (\epsilon^2 - 1)$, $\chi = \epsilon(1 + \epsilon) + (\epsilon + (2 - \epsilon^2)^{1/2})s + ((2 - \epsilon^2)^{1/2} - 1)s^2$, while for $\lambda = -1$ the solution is $\theta = (\epsilon - 1) + ((2 - \epsilon^2)^{1/2} + 1)s$, $\zeta = (1 - \epsilon^2)$, $\chi = \epsilon(1 - \epsilon) + (\epsilon + (2 - \epsilon^2)^{1/2})s + ((2 - \epsilon^2)^{1/2} + 1)s^2$. For $0 < \epsilon < 1$ in both cases χ has all its roots in the open left-half plane, and hence we have here two distinct asymptotically stable control systems that both achieve $Z(i\omega) = \lambda_o^2$.

Example 2: This example shows that solving (3.10) may yield nonoptimal solutions even if χ has all its roots in the open left-half plane. Let $H = 1$, $V = 1/(1 + \tau s)$, $W = s$. Then the equations (3.8) and (3.10) take the form

$$\chi = (1+\tau s)\theta + \zeta, \quad \lambda^2\chi\chi^* = \theta\theta^* - s^2\zeta\zeta^*. \tag{6.1}$$

For a solution of reduced degree, deg $(\chi) = 1$, deg $(\theta) = 0$, and deg $(\zeta) = 0$. Letting $\chi := s + \chi_0$, $\theta = \theta_0$, $\zeta = \zeta_0$, with χ_0, θ_0, and ζ_0 constants, it is not difficult to set up and solve the equations that follow from (6.1). For χ_0 four solutions are found, namely

$$\text{a) } \chi_0 = \frac{1}{2\tau}(1 \pm \sqrt{1+4\tau}), \quad \text{b) } \chi_0 = \frac{1}{2\tau}(1 \pm \sqrt{1-4\tau}). \tag{6.2}$$

In case a), for χ to have its roots in the left-half plane we need to replace the $\pm$ sign with $+$. In case b), to obtain real solutions we need $4\tau \leq 1$; if this is satisfied, χ has its roots in the left-half plane for both solutions. The corresponding values of λ are (for the cases where χ has left-half plane roots only)

$$\text{a) } \lambda^2 = \frac{1}{4\tau^2}(-1+\sqrt{1+4\tau})^2, \quad \text{b) } \lambda^2 = \frac{1}{4\tau^2}(-1 \pm \sqrt{1-4\tau})^2. \tag{6.3}$$

It can be verified that the values of λ^2 in case b) are always larger than that in case a); hence case b) cannot be optimal. Indeed, it turns out that the solution b) does not satisfy (3.16).

Example 3: To illustrate our results for the special case where $W = 0$, we consider an example also solved by Francis and Zames [8]. Let $H(s) = (s-1)^2/((s+1)^2(-s+2))$, $V(s) = (s+1)/(10s+1)$, so that $\psi_+ = (s-1)^2$, $\phi_+ = -s+2$, $\alpha_1 = s+1$, and $\beta_1 = 10s + 1$. Using the approach outlined in Section V, it is easy to determine that the matrices A and B are given by

$$A = \begin{bmatrix} 2 & 0 & 0 & 0 \\ 3 & -2 & 0 & 0 \\ 1 & -3 & 0 & 0 \\ 0 & -1 & 0 & 0 \end{bmatrix}, B = \begin{bmatrix} 2 & 0 & 1 & 0 \\ 19 & 2 & -2 & 1 \\ -10 & 19 & 1 & -2 \\ 0 & -10 & 0 & 1 \end{bmatrix}. \tag{6.4}$$

As a result,

$$R_{11} = \begin{bmatrix} 5/121 & 61/121 \\ 61/121 & -127/121 \end{bmatrix},$$

$$R_{21} = \begin{bmatrix} 232/121 & -122/121 \\ 610/121 & -1391/121 \end{bmatrix}. \tag{6.5}$$

It is easily found that R_{11} has the eigenvalues -1.24688 and 0.238613. It follows that $\lambda = -1.24688$. The corresponding eigenvector of R_{11} is $\theta = \text{col}(-0.391346, 1)$; hence $\theta = -0.391346 + s$. From $\zeta = -1/\lambda R_{21}\theta$ we find that $\zeta = 1.41041 + 10.8020\,s$. Given λ, θ, and ζ, the controller transfer function G and the resulting sensitivity function S are easily found.

Example 4—Double Integrator: As a final example we take the feedback control of a double integrator, with the purpose of illustrating the practical value of the optimization technique discussed in this paper. We assume that the (nominal) plant transfer function is given by

$$H(s) = k_o/s^2 \tag{6.6}$$

with k_o a constant. We could take this as a model of the transfer function of a d.c. motor. We first discuss the various considerations in the choice of the weighting functions.

Disturbance Attenuation and Rejection: The plant has a double pole at the origin. Therefore, to achieve rejection of plant-generated disturbances, V should have a double pole at the origin. A preliminary choice is $V = 1/(\tau_1 s)^2$, with the constant τ_1 to be determined. This weighting function ensures rejection of constant disturbances and attenuation of very low-frequency disturbances. To achieve disturbance attenuation over a wider frequency range, we modify $VV^* = 1/(\tau_1 s)^4$ to

$$VV^* = \frac{1}{(\tau_1 s)^4} + 1 = \frac{1+(\tau_1 s)^4}{(\tau_1 s)^4}. \tag{6.7}$$

Power and Bandwidth Limitation: Since $1/H(s) = s^2/k_o$, we choose $W(s) = (\tau_2 s)^2$, with the constant τ_2 to be determined. This choice of W makes sure that the controller transfer function will be proper.

Robustness: By way of illustration, suppose that the actual (perturbed) plant transfer function is

$$H(s) = \frac{k}{s(s+\alpha)} \tag{6.8}$$

where $k \neq k_o$ and $\alpha > 0$. This perturbation might be caused by gain changes and the occurrence of friction in the motor. In the notation of Section II, we have $\psi = k_o$, $\phi = s^2$, $\psi' = k$, $\phi' = s(s+\alpha)$, $h = 1$. Consequently,

$$\frac{\psi'-\psi}{\psi} = \frac{k-k_o}{k_o}, \quad \frac{\phi'-\phi}{\phi} = \frac{\alpha}{s}, \quad h-1=0. \tag{6.9}$$

These relative perturbations can be bounded as in (2.4), where

$$V_1 = \epsilon, \quad W_1 = \frac{\alpha_{max}}{s}, \quad W_2 = 0. \tag{6.10}$$

Here ϵ is the maximal relative variation of the gain k, and α_{max} the maximal value of α. Pursuing the argument following (2.4), we see that robust stability is ensured when $|VS|^2 + |WT|^2 < 1$ for all ω, where

$$V = \sqrt{2}\,\frac{\alpha_{max}}{s}, \quad W = \sqrt{2}\,\epsilon. \tag{6.11}$$

Inspection of the weighting functions shows that for the inequality to be satisfied, $|S|$ should be (very) small at low frequencies and $|T|$ should be sufficiently less than $1/(\epsilon\sqrt{2})$ at all frequencies. These requirements are fully compatible with the expected effects of the weighting functions that were selected for disturbance attenuation and power and bandwidth limitation. Consequently, there is no need for modification of V and W for robustness improvement.

Solution of the Optimization Problem: We discuss the minimization of sup $(|VS|^2 + |WT|^2)$ for the plant $H(s) = k_o/s^2$, with

$$VV^* = \frac{(1+(\tau_1 s)^4}{(\tau_1 s)^4}, \quad W = (\tau_2 s)^2. \tag{6.12}$$

It may be checked that all necessary assumptions are satisfied. Next, it may be found that the defining equation (3.15) for the polynomial π_λ and the equations (3.8), (3.16), and (3.10) take the form

$$\pi_\lambda\pi_\lambda^* = \tau_1^4\tau_2^4\left(1-\frac{1}{\lambda^2}\right)s^8 + \left(\tau_1^4 - \frac{1}{\lambda^2}\tau_2^4\right)s^4 + 1, \tag{6.13}$$

$$\chi = \tau_1^2 s^2\theta + \zeta, \tag{6.14}$$

$$\lambda\pi_\lambda\chi^* = (1+\tau_1^4 s^4)\theta - \tau_1^2\tau_2^4 s^6\zeta, \tag{6.15}$$

$$\lambda^2\chi\chi^* = (1+\tau_1^4 s^4)\theta\theta^* + \tau_2^4 s^4\zeta\zeta^* \tag{6.16}$$

where we took $\phi_+ = \psi_+ = \phi_- = \psi_- = 1$, $\phi_o = s^2$, $\psi_o = k_o$. The equations can be rewritten in terms of a single parameter $p := (\tau_2/\tau_1)^4$ by the substitution $\hat{s} = s\tau_1$. Immediately omitting the circumflex, we obtain

$$\pi_\lambda \pi_\lambda^* = \left(1 - \frac{1}{\lambda^2}\right) s^8 + \left(1 - \frac{p}{\lambda^2}\right) s^4 + 1, \tag{6.17}$$

$$\chi = s^2\theta + \zeta, \tag{6.18}$$

$$\lambda \pi_\lambda \chi^* = (1 + s^4)\theta - ps^6\zeta, \tag{6.19}$$

$$\lambda^2 \chi\chi^* = (1 + s^4)\theta\theta^* + ps^4\zeta\zeta^*. \tag{6.20}$$

From root locus considerations it follows that the lowest value λ_o of λ for which the factorization (6.17) is possible is assumed when the leading coefficient of the right-hand side vanishes; hence $\lambda_o = 1$. We conjecture, and indeed this is borne out by what follows, that the lower bound λ_o cannot be achieved, and immediately start looking for a solution of reduced degree. In the present case, $x = 4$, $t = 2$, $z = 2$; hence, we seek a solution with deg $(\chi) = 3$, deg $(\theta) = 1$, deg $(\zeta) = 1$.

We choose to solve the linear equation (6.18) together with the quadratic equation (6.20) for the polynomials $\chi := \chi_0 + \chi_1 s + \chi_2 s^2 + s^3$, $\theta := \theta_o + \theta_1 s$ and $\zeta := \zeta_0 + \zeta_1 s$ and for λ. It is a simple matter to set up the equations for the coefficients and reasonably straightforward to find that $\chi_0 = ((\lambda^4 - \lambda^2 - p)/(2p\lambda^3))^{1/2}$, $\chi_1 = ((\lambda^2 - 1)/p)^{1/2}$, $\chi_2 = \lambda\chi_0$, $\theta_0 = \lambda\chi_0$, $\theta_1 = 1$, $\zeta_0 = \chi_0$, $\zeta_1 = \chi_1$, while λ satisfies the equation

$$(\lambda^4 - \lambda^2 - p)^2 - 4p^{1/2}\lambda^5(\lambda^2 - 1)^{1/2} = 0. \tag{6.21}$$

In the process of obtaining these equations we used the fact that the coefficients of χ should all be positive.

Given p, (6.21) can be solved numerically to obtain λ, and, from this, the complete solution. In each case that was tried, it was found that there are two real values of λ satisfying the equations; only one of these results in real polynomials χ, θ, and ζ, however. Table I lists the numerical solutions for a few values of p. In Fig. 2, amplitude plots are given of the resulting sensitivity function S and complementary sensitivity function T. For p small (i.e., $\tau_2 \ll \tau_1$), the criterion is dominated by the term with S, and the modulus of the sensitivity function more or less behaves as $1/|V|$. As p increases, the term with T assumes more and more influence, pushing the bandwidth down and causing $|T|$ to fall off more steeply, at the cost of peaking of $|S|$ near the cutoff frequency. The controller transfer function in each case is

$$G(s) = \frac{\zeta_0 + \zeta_1 s}{k_o \tau_1^2(\theta_0 + \theta_1 s)} \tag{6.22}$$

(still in terms of the dimensionless frequency defined before). The controller thus is of a conventional lead-lag type.

It is instructive to obtain asymptotic formulas for p small. For $p \downarrow 0$, the desired root of (6.21) behaves as $\lambda^2 \sim 1 + 2^{4/3}p^{1/3}$. Correspondingly, $\chi \sim s^3 + 2^{1/6}p^{-1/3}s^2 + 2^{2/3}p^{-1/3}s + 2^{1/6}p^{-1/3}$, $\theta \sim s + 2^{1/6}p^{-1/3}$, $\zeta \sim 2^{2/3}p^{-1/3}s + 2^{1/6}p^{-1/3}$. Asymptotically, the polynomial χ, which in this case is precisely the closed-loop characteristic polynomial, has the roots $-2^{1/6}p^{-1/3}$ and $1/2\sqrt{2}(-1 \pm i)$. The latter pair of closed-loop poles is dominant and has a relative damping 1. Correspondingly, the closed-loop system bandwidth is 1 (in terms of the dimensionless frequency introduced before). The former closed-loop pole is "far-away" relative to the dominant pair; it is there because of our insistence that the controller be proper.

We observe that the design for the case where p is small is just the design that would have been obtained by classical frequency domain design methods. The advantage of the proposed method is that it can also be applied in situations (involving plants with several right-half plane poles and zeros, for instance) that cannot very easily be handled by classical techniques.

TABLE I
SOLUTIONS FOR THE DOUBLE INTEGRATOR

p	λ	χ_0	χ_1	χ_2
10^{-6}	1.01263	112.012	159.409	113.427
10^{-3}	1.12881	11.0055	16.5599	12.4232
1	2.39652	0.976371	2.17791	2.33990
10^3	11.9037	0.0749231	0.375100	0.891868

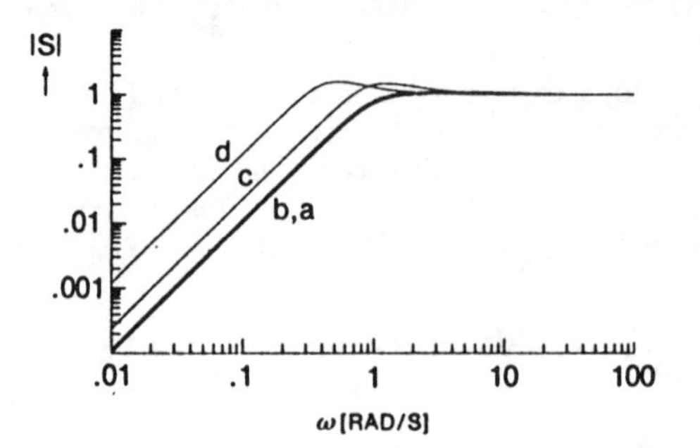

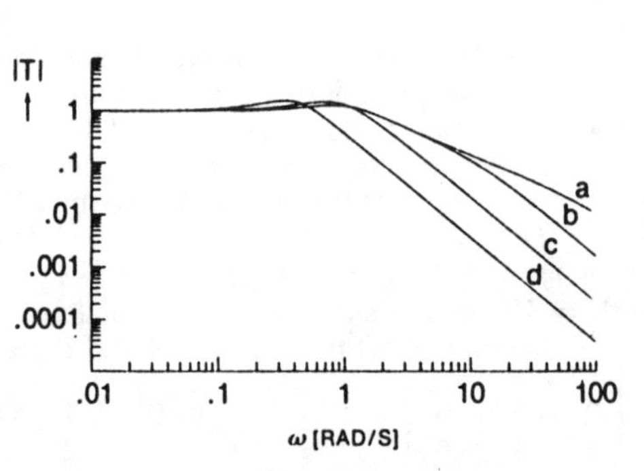

Fig. 2. Plots of optimal $|S|$ and $|T|$ for the double integrator: (a) $p = 10^{-6}$, (b) $p = 10^{-3}$, (c) $p = 1$, (d) $p = 10^3$.

VII. Conclusions

In this paper, various aspects of the performance of linear single-input single-output feedback systems are translated into required bounds on the sensitivity function S and its complement T on the imaginary axis. This leads to a frequency domain minimax optimization problem involving both these functions. The problem formulation is a generalization of that of Zames [6] and Zames and Francis [7]. It allows the inclusion of power and bandwidth constraints and permits handling plant poles and zeros on the imaginary axis.

The approach of the paper leads to a polynomial equation, whose solution yields the desired controller. Although it is clear that standard numerical techniques, such as the Newton-Raphson method, can be used for their solution, the question what is the best way of solving this equation for high-order systems needs further study. Another aspect that deserves more consideration is what flexibility there is in the choice of the weighting functions that occur in the optimization problem, and how modifications of the weighting functions affect the solution.

Appendix A

Proof of Theorem 1

Theorem 1 concerns the existence of a solution to the equation (3.16). It is convenient to rewrite the equation in terms of $\mu := 1/\lambda$. To this end, we define $\theta/\lambda =: \hat{\theta}$, $\zeta/\lambda =: \hat{\zeta}$, $\hat{\pi}_\mu := \pi_{1/\mu}$, replace χ with $\hat{\chi}$ (to emphasize that we now take μ as the variable rather than λ), and rewrite the relations (3.8), (3.10), and (3.16) as

$$\beta_1\phi_+\hat{\theta}+\beta_2\psi_+\hat{\zeta}=\mu\hat{\chi}, \tag{A.1}$$

$$\alpha_1\alpha_1^*\beta_2^*\phi_+\hat{\theta}-\alpha_2\alpha_2^*\beta_1^*\psi_+\hat{\zeta}=\hat{\pi}_\mu\hat{\chi}^*. \tag{A.2}$$

$$\alpha_1\alpha_1^*\phi_+\phi_+^*\hat{\theta}\hat{\theta}^*+\alpha_2\alpha_2^*\psi_+\psi_+^*\hat{\zeta}\hat{\zeta}^*=\hat{\chi}\hat{\chi}^*. \tag{A.3}$$

In the remainder of this Appendix for typographical simplicity all circumflexes are suppressed. We first consider the existence of a solution to (A.3) rather than (A.2).

Lemma A.1: For $|\mu|$ sufficiently small, the polynomial equations (A.3) and (A.1) have a unique family of solutions χ_μ, θ_μ, ζ_μ with $\deg(\chi_\mu) = x$, $\deg(\theta_\mu) \leq t$, $\deg(\zeta_\mu) \leq z$, whose coefficients depend continuously on μ, such that χ_μ is monic and has all its roots in the open left-half plane.

Proof: We first establish that the equation (A.3) has a solution as desired at $\mu = 0$ and then use the implicit function theorem to establish its existence for $|\mu|$ sufficiently small. For $\mu = 0$, (A.1) reduces to $\beta_1\phi_+\theta + \beta_2\psi_+\zeta = 0$. Since by the various assumptions $\beta_1\phi_+$ and $\beta_2\psi_+$ have no common roots, this equation has the general solution $\theta = \beta_2\psi_+a$, $\zeta = -\beta_1\phi_+a$, with a an arbitrary polynomial. Substitution into (A.3) yields

$$\begin{aligned}\chi\chi^* &= aa^*\phi_+\phi_+^*\psi_+\psi_+^*(\beta_1\beta_1^*\alpha_2\alpha_2^*+\beta_2\beta_2^*\alpha_1\alpha_1^*)\\ &= aa^*\phi_+\phi_+^*\psi_+\psi_+^*\pi_0\pi_0^*.\end{aligned} \tag{A.4}$$

We see that this equation indeed has a solution such that χ has degree x, has all its roots in the open left-half plane and is monic, given by $\chi_0 = a\phi_+^*\psi_+^*\pi_0$, where a is a constant such that χ_0 is monic. It follows that $\theta_0 = a\beta_2\psi_+$, $\zeta_0 = -a\beta_1\phi_+$.

We now apply the implicit function theorem [22] to the set of equations for the coefficients of the polynomials χ, θ, and ζ that results when the polynomial equations (A.1) and (A.3) are expanded in powers of s and the coefficients of like powers are equated. It is easily verified that (A.1) and (A.3) in this way yield the same number of equations as unknown coefficients if we look for a solution χ, θ, ζ of degrees, respectively x, t, and z, with χ monic. It follows from the implicit function theorem that these equations have a solution for $|\mu|$ sufficiently small, which is continuously dependent on μ, if the Jacobian of these equations with respect to the unknown coefficients is nonsingular at the solution that is obtained at $\mu = 0$. This Jacobian is nonsingular if and only if the homogeneous set of equations

$$\begin{aligned}&\beta_1\phi_+\tilde{\theta}+\beta_2\psi_+\tilde{\zeta}=0,\\ &\alpha_1\alpha_1^*\phi_+\phi_+^*(\theta_0\tilde{\theta}^*+\theta_0^*\tilde{\theta})+\alpha_2\alpha_2^*\psi_+\psi_+^*(\zeta_0\tilde{\zeta}^*+\zeta_0^*\tilde{\zeta})-(\chi_0\tilde{\chi}^*+\chi_0^*\tilde{\chi})\\ &=0,\end{aligned} \tag{A.5}$$

with $\deg(\tilde{\chi}) = x - 1$, $\deg(\tilde{\theta}) = t$, $\deg(\tilde{\zeta}) = z$, has the unique solution $\tilde{\chi} = 0$, $\tilde{\theta} = 0$, $\tilde{\zeta} = 0$. Here $\deg(\tilde{\chi}) = x - 1$ because we are looking for a solution such that χ is monic. The first of the equations (A.5) has the general solution $\tilde{\theta} = \beta_2\psi_+c$, $\tilde{\zeta} = -\beta_1\phi_+c$, with c an arbitrary polynomial. Since $\deg(\tilde{\zeta}) = z$ and also $\deg(\beta_1\phi_+) = z$, c is necessarily a constant. Substituting these solutions into the second of the equations (A.5), we obtain after rearrangement

$$\phi_+^*\psi_+^*\pi_0(\tilde{\chi}-c\phi_+^*\psi_+^*\pi_0)^*+\phi_+\psi_+\pi_0^*(\tilde{\chi}-c\phi_+^*\psi_+^*\pi_0)=0. \tag{A.6}$$

Now, if a is a polynomial that has no common roots with a^*, the polynomial equation $ax + a^*x^* = 0$ has the general solution $x = a^*\omega$, where ω is any polynomial such that $\omega^* = -\omega$. As a result it follows from (A.6) that $\tilde{\chi} - c\phi_+^*\psi_+^*\pi_0 = \phi_+^*\psi_+^*\pi_0\omega$, where ω is any polynomial such that $\omega^* = -\omega$. It follows that $\tilde{\chi} = (c + \omega)\phi_+^*\psi_+^*\pi_0$. Since $\deg(\tilde{\chi}) = x - 1$ and $\deg(\phi_+^*\psi_+^*\pi_0) = x$, necessarily $\tilde{\chi} = 0$, $c + \omega = 0$, so that $\omega = -c$. From $\omega^* = -\omega$ it follows that $c = 0$, $\omega = 0$, which proves that (A.5) has the unique solution $\tilde{\chi} = 0$, $\tilde{\theta} = 0$, $\tilde{\zeta} = 0$.

It follows that (A.3) has a solution for $|\mu|$ small enough such that χ_μ is monic. Since the coefficients of χ_μ vary continuously with μ, and χ_0 has all its roots in the open left-half plane, the roots of χ_μ remain in the open left-half plane for $|\mu|$ sufficiently small. This completes the proof of Lemma A.1. □

Proof of Theorem 1a): To prove part a) of Theorem 1 we shall first prove (A.2). This equation is obtained by factoring

$$\begin{aligned}&(\alpha_1\alpha_1^*\beta_2^*\phi_+\theta-\alpha_2\alpha_2^*\beta_1^*\psi_+\zeta)(\alpha_1\alpha_1^*\beta_2^*\phi_+\theta-\alpha_2\alpha_2^*\beta_1^*\psi_+\zeta)^*\\ &=\pi_\mu\pi_\mu^*\chi\chi^*\end{aligned} \tag{A.7}$$

[compare (3.13)], which is equivalent to $Z(s) = \lambda^2$. Thus, given a solution of (A.2), (A.3) holds. We first prove that (A.2) is the correct factorization of (A.7). To this end, define $q_\mu := \alpha_1\alpha_1^*\beta_2^*\phi_+\theta_\mu - \alpha_2\alpha_2^*\beta_1^*\psi_+\zeta_\mu$. Then (A.7) takes the form $qq^* = \pi_\mu\pi_\mu^*\chi\chi^*$, and (A.2) becomes $q = \pi_\mu\chi^*$. We thus have to prove that $q = \pi_\mu\chi^*$ is the correct factorization of $qq^* = \pi_\mu\pi_\mu^*\chi\chi^*$.

To this end, let the pair of polynomials g and h be any solution of the equation $\beta_1\phi_+g + \beta_2\psi_+h = 1$. By the various assumptions, $\beta_1\phi_+$ and $\beta_2\psi_+$ are relatively prime and a solution g, h exists. Then given $\mu\chi$, (A.1) has the general solution

$$\theta=g\mu\chi+\beta_2\psi_+\omega,\ \ \zeta=h\mu\chi-\beta_1\phi_+\omega \tag{A.8}$$

where ω is an arbitrary polynomial. Substitution of these expressions for θ and ζ into (A.3) yields an equation of the form

$$\begin{aligned}0=&-\chi^*\delta_{11}\chi+\mu\chi^*\phi_+\psi_+\delta_{12}^*\omega+\mu\chi\phi_+^*\psi_+^*\delta_{12}\omega^*\\ &+\omega^*\phi_+\phi_+^*\psi_+\psi_+^*\pi_0\pi_0^*\omega,\end{aligned} \tag{A.9}$$

where $\delta_{11} := 1 - \mu^2(\alpha_1\alpha_1^*\phi_+\phi_+^*gg^* + \alpha_2\alpha_2^*\psi_+\psi_+^*hh^*)$, $\delta_{12} := \alpha_1\alpha_1^*\phi_+\beta_2^*g - \alpha_2\alpha_2^*\psi_+\beta_1^*h$. Similarly, using (A.8) it follows from the definition of q that

$$q=\mu\delta_{12}\chi+\pi_0\pi_0^*\phi_+\psi_+\omega. \tag{A.10}$$

Comparison of (A.9) and (A.10) reveals that $0 = \chi^*(-\delta_{11}\chi + \mu\phi_+\psi_+\delta_{12}^*\omega) + q_\mu\phi_+^*\psi_+^*\omega^*$, which we rewrite in the form

$$q\phi_+^*\psi_+^*\omega^*=\chi^*\epsilon \tag{A.11}$$

where ϵ is the polynomial $\epsilon := \mu\phi_+\psi_+\delta_{12}^*\omega - \delta_{11}\chi$. Now, since χ is stable for μ sufficiently small, χ has no roots in common with ϕ_+ and ψ_+. Furthermore, for $\mu = 0$ we have $\omega_0 = a$, $\chi_0 = a\phi_+^*\psi_+^*\pi_0$ (see the proof of Lemma A.1) so that ω_0 and χ_0 have no common roots; therefore, χ and ω have no common roots for μ sufficiently small. Because as a result χ^* has no roots in common with $\phi_+^*\psi_+^*\omega^*$, it follows from (A.11) that any root of χ^* is a root of q, which means that q has a factor χ^*.

From this we conclude that $q = \bar{\pi}\chi^*$, with $\bar{\pi}$ a polynomial to be determined. From $qq^* = \pi_\mu\pi_\mu^*\chi\chi^*$ it follows that $\bar{\pi}\bar{\pi}^* = \bar{\pi}_\mu\bar{\pi}_\mu^*$. Now, from the definition of q it easily follows that $q_0 = a\pi_0\pi_0^*\phi_+\psi_+$, while from $q = \bar{\pi}\chi^*$ we obtain $q_0 = a\bar{\pi}_0\pi_0^*\phi_+\psi_+$, which shows that $\bar{\pi}_0 = \pi_0$. This implies that for $\mu = 0$, and hence for μ sufficiently small, $\bar{\pi}$ is stable. Therefore, the correct factorization of $\bar{\pi}\bar{\pi}^* = \pi_\mu\pi_\mu^*$ is $\bar{\pi} = \pi_\mu$, which proves that $q = \pi_\mu\chi^*$ as claimed. Having proved the validity of (A.2), we now consider (A.2) together with (A.1) as equations for the unknown polynomials θ, ζ, and χ. As has been shown, there exists a solution for μ small enough, which is unique within multiplication by a continuous function of μ.

We now observe that for μ given π_μ is known, and hence (A.2) and (A.1) form a set of homogeneous *linear* equations in the coefficients of the unknown polynomials χ, θ, and ζ. A coefficients-versus-equations count reveals that this linear set of equations is overdetermined (unless $b_1 = b_2 = e = 0$; then we have as many equations as unknowns). Let us arrange the unknown coefficients of the polynomials χ, θ, and ζ (not restricting χ to be monic) in a vector y; then the equations (A.2) and (A.1) can be put into the form $A_\mu y = 0$, where A_μ is a rectangular matrix. This equation has a nontrivial solution if and only if $A_\mu' A_\mu$ (with the prime denoting the transpose) is singular. Since a nontrivial solution exists for μ small enough, it follows that det $(A_\mu' A_\mu) = 0$ for μ small enough. Now the elements of A_μ are analytic functions of the coefficients of π_μ, and these coefficients are analytic functions of μ for $|\mu| < 1/\lambda_o$, so that det $(A_\mu' A_\mu)$ is an analytic function of μ for $|\mu| < 1/\lambda_o$. It follows that det $(A_\mu' A_\mu) = 0$ for $|\mu| \leq 1/\lambda_o$, and hence that (A.2) and (A.1) have a nontrivial solution with the properties stated in Theorem 1(a) for $|\mu| \leq 1/\lambda_o$. □

Proof of Theorem 1b): Suppose that χ_{1/λ_o} has a root in the open right-half plane. If χ_{-1/λ_o} has such a root, the proof follows similarly. We track the solution of (A.2) as μ is increased from 0 to $1/\lambda_o$. Since the coefficients of χ_μ vary continuously with μ, and for μ small enough all roots of χ_μ are in the open left-half plane but by hypothesis for $\mu = 1/\lambda_o$ at least one root of χ_μ is the open right-half plane, for some $\mu < 1/\lambda_o$ at least one root of χ_μ crosses over to the right-half plane, either via infinity or via the imaginary axis.

Let us consider what happens to the polynomials θ and ζ. Solving (A.1) and (A.2) for $\phi_+\theta$ and $\psi_+\zeta$ yields

$$\begin{bmatrix} \phi_+\theta \\ \psi_+\zeta \end{bmatrix} = \frac{1}{\pi_0\pi_0^*} \begin{bmatrix} \alpha_2\alpha_2^*\beta_1^* & \beta_2 \\ \alpha_1\alpha_1^*\beta_2^* & -\beta_1 \end{bmatrix} \begin{bmatrix} \mu\chi \\ \pi_\mu\chi^* \end{bmatrix}. \quad \text{(A.12)}$$

Suppose that a root of χ_μ crosses over via infinity, so that χ_μ loses degree. Then also χ_μ^* loses degree and (A.12) shows that θ_μ and ζ_μ simultaneously lose degree as well; clearly we have a solution as indicated in Theorem 1b). Suppose on the other hand that a root of χ_μ reaches the imaginary axis. Then by conjugate symmetry also χ_μ^* has this root, so that χ_μ and χ_μ^* have a common root on the imaginary axis. By (A.12), θ_μ and ζ_μ also have this root (note that by assumption c) π_0 has no roots on the imaginary axis so that it is impossible that the common root of θ_μ and ζ_μ is canceled by a root of π_0 or π_0^*). Thus, χ_μ, θ_μ, and ζ_μ have a common root on the imaginary axis, which can be canceled throughout in the equations (A.1) and (A.2), again leading to a solution of reduced degree. This completes the proof of Theorem 1b).

Appendix B

Proof of Lemma 2

In this Appendix the details of the last step of the proof of Lemma 2 are supplied. Suppose first that $\gamma - \eta/\lambda^2$ loses degree at λ_o and that $\gamma - \eta/\lambda^2$ has no roots on the imaginary axis for any $|\lambda| \geq \lambda_o$. Then $\gamma(i\omega) - \eta(i\omega)/\lambda^2 > 0$ for all $|\lambda| \geq \lambda_o$ and all ω. Since the leading coefficient of $\gamma - \eta/\lambda^2$ changes sign at λ_o, there exists a $|\lambda| < \lambda_o$ such that $\gamma(i\omega) - \eta(i\omega)/\lambda^2 < 0$ for ω large enough. It follows that inf $\{\lambda^2 : \gamma(i\omega) - \eta(i\omega)/\lambda^2 \geq 0$ for all $\omega\}$ $= \lambda_o^2$. Next, suppose that λ_o is the first value of λ such that $\gamma - \eta/\lambda^2$ has a root on the imaginary axis as λ is decreased from ∞, and suppose that this root is $i\omega_o$. Then $\gamma(i\omega) - \eta(i\omega)/\lambda^2 \geq 0$ for all $|\lambda| \geq \lambda_o$ and all ω. Now, the derivative of $\gamma - \eta/\gamma^2$ with respect to $1/\lambda^2$ at $s = i\omega_o$ equals $-\eta(i\omega_o)$. Since $\gamma(i\omega_o) - \eta(i\omega)/\lambda^2 = 0$ and [by assumption c)] $\gamma(i\omega_o) > 0$ it follows that $\eta(i\omega_o) > 0$; hence there exist a $|\lambda| < \lambda_o$ such that $\gamma(i\omega_o) - \eta(i\omega_o)/\lambda^2 < 0$. It follows again that inf $\{\lambda^2 : \gamma(i\omega) - \eta(i\omega)/\lambda^2 \geq 0$ for all $\omega\}$ $= \lambda_o^2$.

Appendix C

Proof of Lemma 4

The proof of this lemma is a standard application of Wiener optimization theory [23]–[25]. For a plant $H = \psi/\phi$ and a controller $G = \rho/\sigma$, the closed-loop characteristic polynomial is

$$\chi_{cl} = \phi\sigma + \psi\rho. \quad \text{(C.1)}$$

Let the polynomials ξ and η solve the diophantine equation $\phi\xi + \psi\eta = 1$ (since by assumption ϕ and ψ have no common roots a solution exists). Then for given χ_{cl}, (C.1) can be solved for σ and ρ in the form

$$\sigma = \xi\chi_{cl} + \psi\omega, \quad \rho = \eta\chi_{cl} - \phi\omega \quad \text{(C.2)}$$

with ω an arbitrary polynomial. Accordingly, we obtain for the sensitivity function and its complement

$$S = \frac{\phi\sigma}{\chi_{cl}} = \phi\xi + \phi\psi P, \quad T = \frac{\psi\rho}{\chi_{cl}} = \psi\eta - \phi\psi P \quad \text{(C.3)}$$

where $P := \omega/\chi_{cl}$. This is a well-known parametrization of S and T in terms of the rational function P. If the closed-loop system is required to have all its poles in the closed left-half plane, P can range over all rational functions having their poles in the closed left-half plane.

Define $\mathcal{K}$ as the set of rational functions P with all their poles in the closed left-half plane such that VS and WT are proper and have no poles on the imaginary axis. Without loss of generality we limit the class of admissible control systems to those for which $P \in \mathcal{K}$. Now, consider two control systems, respectively parametrized by $P \in \mathcal{K}$ and $\tilde{P} \in \mathcal{K}$. Then using (C.3) it is not difficult to find that one can write

$$\begin{aligned} \tilde{Z} = {} & VV^*SS^* + WW^*TT^* + (VV^*S - WW^*T)^*\phi\psi(\tilde{P} - P) \\ & + (VV^*S - WW^*T)\phi^*\psi^*(\tilde{P} - P)^* \\ & + (VV^* + WW^*)\phi\phi^*\psi\psi^*(\tilde{P} - P)(\tilde{P} - P)^*. \end{aligned} \quad \text{(C.4)}$$

Here, in an obvious notation, $\tilde{Z}$ corresponds to the control system parametrized by $\tilde{P}$, and S and T to that parametrized by P. A standard variational argument shows that if an optimal solution exists, the control system parametrized by P maximizes the auxiliary criterion (3.18) if and only if

$$\int_{-\infty}^{\infty} (\tilde{P} - P)\Sigma\phi\psi(|V|^2S - |W|^2T)^* \, d\omega = 0 \quad \text{(C.5)}$$

for all $\tilde{P} \in \mathcal{K}$. Since for P and $\tilde{P}$ in $\mathcal{K}$ the rational function $(\tilde{P} - P)\phi\psi(VV^*S^* - WW^*T^*)$ has no poles on the imaginary axis and is proper, a simple application of the residue theorem shows that a sufficient condition for (C.5) to hold is that $\Sigma\phi\psi(VV^*S^* - WW^*T^*)$ has all its poles in the closed left-half plane. To establish uniqueness, we observe by inspection of (C.4) using (C.5) that the control system parametrized by $\tilde{P}$ is optimal if and only if

$$\int_{\infty}^{\infty} \Sigma(|V|^2 + |W|^2)|\phi|^2|\psi|^2|\tilde{P} - P|^2 \, d\omega = 0. \quad \text{(C.6)}$$

Since by assumption Σ is positive for all ω this implies that $\tilde{P} = P$ and hence the optimal solution is unique.

This proof is due to H. Westdijk [26].

References

[1] J. C. Doyle, "Robustness of multiloop linear feedback systems," in *Proc. 17th IEEE Conf. Decision Contr.*, 1979, pp. 12–18.

[2] J. C. Doyle and G. Stein, "Multivariable feedback design: Concepts for a classical/modern synthesis," *IEEE Trans. Automat. Contr.*, vol. AC-26, pp. 4–16, 1981.

[3] J. C. Doyle, "Achievable performance in multivariable feedback systems," in *Proc. 18th IEEE Conf. Decision Contr.*, 1979, pp. 250–251.

[4] ——, "Limitations on achievable performance of multivariable feedback systems," in *Multivariable Analysis and Design-Techniques* (Agard Lecture Series 117) 1981.

[5] J. C. Doyle, J. E. Wall, and G. Stein, "Performance and robustness analysis for structured uncertainty," in *Proc. 21st IEEE Conf. Decision and Contr.*, 1982.

[6] G. Zames, "Feedback and optimal sensitivity: Model reference transformations, multiplicative seminorms, and approximate inverses," *IEEE Trans. Automat. Contr.*, vol. AC-26, pp. 301-320, 1981 (previously presented at the 1979 Allerton Conf.).

[7] G. Zames and B. A. Francis, "Feedback, minimax sensitivity, and optimal robustness," *IEEE Trans. Automat. Contr.*, vol. AC-28, pp. 585-601, 1983.

[8] B. Francis and G. Zames, "On $\mathcal{H}^\infty$-optimal sensitivity theory for SISO feedback systems," *IEEE Trans. Automat. Contr.*, vol. AC-29, pp. 9-16, 1984.

[9] B.-C. Chang and J. B. Pearson, "Optimal disturbance reduction in linear multivariable systems," in *Proc. IEEE Conf. Decision Contr.*, San Antonio, TX, Dec. 1983.

[10] B. A. Francis, J. W. Helton, and G. Zames, "H^∞-optimal feedback controllers for linear multivariable systems," *IEEE Trans. Automat. Contr.*, vol. AC-29, pp. 880-900, 1984.

[11] M. G. Safonov, A. J. Laub, and G. L. Hartmann, "Feedback properties of multivariable systems: The role and use of the return difference matrix," *IEEE Trans. Automat. Contr.*, vol. AC-26, pp. 47-65, 1981.

[12] H. Kwakernaak, "Minimax frequency domain optimization of multivariable linear feedback systems," in *Proc. 1984 IFAC World Congr.*, Budapest, Hungary, 1984.

[13] H. Kwakernaak, "Robustness optimization of linear feedback systems," in *Proc. 22nd IEEE Conf. Decision Contr.*, San Antonio, TX, Dec. 1983.

[14] M. Verma and E. Jonckheere, "L^∞-compensation with mixed sensitivity as a broadband matching problem," *Syst. Contr. Lett.*, vol. 4, pp. 125-129, 1984.

[15] J. W. Helton, "Worst case analysis in the frequency domain: the H^∞ approach to control," in *Proc. IEEE Conf. Decision Contr.*, San Antonio, TX, Dec. 1983.

[16] B. A. Francis, "Optimal disturbance attenuation with control weighting," in *Proc. Twente Workshop on Systems and Optimization*, LNCIS vol. 66, A. Bagchi and H. Th. Jongen, Eds. Berlin: Springer-Verlag, 1984.

[17] P. Henrici, *Applied and Computational Complex Analysis, Vol. I.* New York: Wiley, 1974.

[18] S. Zaks, *The Theory of Statistical Inference.* New York: Wiley, 1971.

[19] V. Kucera, *Discrete Linear Control.* New York: Wiley, 1979.

[20] S. Smulders, "Two computer programs for computing minimax frequency domain optimal controllers," Dep. Appl. Math., Twente Univ. Technol., Memo. 1984.

[21] Th. Kailath, *Linear Systems.* Englewood Cliffs, NJ: Prentice-Hall, 1980.

[22] J. Dieudonné, *Foundations of Modern Analysis.* New York: Academic, 1960.

[23] D. C. Youla, J. J. Bongiorno, and H. A. Jabr, "Modern Wiener-Hopf design of optimal controllers—Part I," *IEEE Trans. Automat. Contr.*, vol. AC-21, pp. 3-15, 1976.

[24] ——, "Modern Wiener-Hopf design of optimal controllers—Part II," *IEEE Trans. Automat. Contr.*, vol. AC-21, pp. 319-338, 1976.

[25] C. A. Desoer, R. W. Liu, J. Murray, and R. Saeks, "Feedback system design: The fractional representation approach to analysis and synthesis," *IEEE Trans. Automat. Contr.*, vol. AC-25, pp. 401-412, 1980.

[26] H. Westdijk, "Optimale robuustheid van lineaire regelsystemen," Dep. Appl. Math., Twente Univ. Technol., Enschede, The Netherlands, Student Project Rep., May 1983.

Part IV
Other Approaches to the Control of Uncertain Systems

OPTIMAL STATIONARY CONTROL OF A LINEAR SYSTEM WITH STATE-DEPENDENT NOISE*

W. M. WONHAM†

1. Introduction. Consider the linear control system described by the formal, vector stochastic differential equation

$$\dot{x} = Ax - Bu + C\dot{w}_1 + G(x)\dot{w}_2 . \tag{1.1}$$

In (1.1), u is the control and $\dot{w}_1$, $\dot{w}_2$ are independent Gaussian white noise disturbances.[1] The elements of the matrix G are assumed to be linear in x; and so the term $G(x)\dot{w}_2$ represents a disturbance of which the intensity is roughly proportional to the deviation of x from the origin $x = 0$. Equivalently, the disturbance can be regarded as a wideband random perturbation of the system matrix A.

Now consider the problem of choosing a feedback control $u = \phi(x)$ such that, in the steady state, the expected quadratic cost

$$\mathcal{E}\{x'Mx + u'Nu\} \tag{1.2}$$

is a minimum. If $G(x) \equiv 0$, the solution of this problem is well known [1], [2]. Under mild restrictions the optimal control always exists and is a linear function of x which is independent of the intensity of the additive disturbance $C\dot{w}_1$. In the present article it is shown that an optimal control exists for the more general system (1.1), provided the state-dependent noise $G(x)\dot{w}_2$ is sufficiently small. The optimal control is again linear, but is now rather critically dependent on the coefficients of G. Examples are provided to show that instability may result if this dependence is ignored.

The problem is stated precisely in §2; the proof of existence is given in §§3 and 4; and some examples studied in §§5 and 6. We conclude with some remarks on the interpretation of (1.1) and discuss alternative optimization problems which are closely related.

2. Statement of the problem. To make (1.1) precise we assume that x is an n-vector with stochastic differential

$$dx = Ax\,dt - Bu\,dt + C\,dw_1 + G(x)\,dw_2 . \tag{2.1}$$

* Received by the editors June 15, 1966, and in revised form November 7, 1966.

† Division of Applied Mathematics, Center for Dynamical Systems, Brown University, Providence, Rhode Island 02912. This research was supported in part by the National Aeronautics and Space Administration under Grant NGR 40-002-015, in part by the Office of Aerospace Research, United States Air Force, under Grant AF-AFOSR-693-66, and in part by the National Science Foundation under Grant GK-967.

This paper will also appear in Proceedings of the First All-Union Symposium on Statistical Problems in Technical Cybernetics (in Russian), Moscow, 1967.

[1] A precise interpretation of (1.1) is given in §2.

Here and below, all vectors and matrices have real-valued elements; A, B and C are constant matrices of dimensions $n \times n$, $n \times m$ and $n \times d_1$ respectively; $G(x)$ is an $n \times d_2$ matrix given by

$$G(x) = \sum_{k=1}^{n} x_k G_k , \tag{2.2}$$

where the $n \times d_2$ matrices G_k are constants. It is assumed that (A, B) is controllable, and that CC' is positive definite: that is, $d_1 \geqq n$ and C is of rank n. The latter assumption obviates fussy discussion about possible degeneracy of the ergodic measure introduced below. Finally, w_1 and w_2 are independent Wiener processes of dimension d_1, d_2 respectively.

In the following, E denotes Euclidean n-space; a prime ($'$), the transpose of a vector or matrix; and $|\cdot|$, the Euclidean norm: for a matrix F, $|F| = \max\{|Fx|:|x| = 1\}$.

In (2.1) let $u = \phi(x)$, where ϕ is defined on E and satisfies a uniform Lipschitz condition

$$|\phi(x) - \phi(y)| \leqq k\,|x - y|, \qquad x, y \in E. \tag{2.3}$$

With this choice of u, (2.1) becomes a stochastic differential equation of Itô's type [3]:

$$dx(t) = Ax(t)\,dt - B\phi[x(t)]\,dt + C\,dw_1(t) + G[x(t)]\,dw_2(t). \tag{2.4}$$

If $x(0)$ is a random variable independent of the w_1, w_2 increments, then (2.4), defined for $t \geqq 0$, determines a diffusion process

$$X_\phi = \{x_\phi(t) : t \geqq 0\}.$$

Diffusion processes are discussed extensively in [4]; a brief summary can be found in [5].

Of interest here is the case when X_ϕ is positive recurrent (for the definition of this term, see [5]). Under this condition it is known that there exists a unique ergodic probability measure μ_ϕ defined on the Borel sets of E: that is, if the distribution of $x(0)$ is μ_ϕ, then so is that of $x(t)$ for all $t > 0$. Let Φ be the class of *admissible* control functions ϕ, with the properties:

(i) ϕ satisfies (2.3) for some constant k;
(ii) X_ϕ is ergodic;
(iii) the corresponding ergodic measure μ_ϕ is such that

$$\mathcal{E}_\phi\{|x|^2\} \equiv \int_E |x|^2 \mu_\phi(dx) < \infty . \tag{2.5}$$

Now define

$$L(x, u) = x'Mx + u'Nu, \tag{2.6}$$

where M, N are constant symmetric positive definite matrices of dimensions $n \times n$, $m \times m$ respectively.

Reprinted with permission from *SIAM Journal on Control*, vol. 5, no. 3, pp. 486–500.

Our problem is the following: find a control $\phi^0 \in \Phi$ which is optimal in the sense that

$$\mathcal{E}_{\phi^0}\{L(x, \phi^0)\} = \min [\mathcal{E}_{\phi}\{L(x, \phi)\} : \phi \in \Phi].$$

3. Existence of an admissible control. In this section it will be shown that Φ is nonempty provided the matrices G_k of (2.2) are sufficiently small. This result will follow from the stability theorem stated below.

Let $V = V(x)$ be of class $C^{(2)}$ on E and let $\mathfrak{L}_u$ denote the elliptic operator

$$\mathfrak{L}_u V(x) \equiv \tfrac{1}{2} \operatorname{tr} \{C' V_{xx}(x) C + G(x)' V_{xx}(x) G(x)\} + (Ax - Bu)' V_x(x)' \tag{3.1}$$

In (3.1), tr denotes trace, V_x the vector $[\partial V/\partial x_i]$ and V_{xx} the matrix $[\partial^2 V/\partial x_i \partial x_j]$. The operator $\mathfrak{L}_\phi$, obtained by setting $u = \phi(x)$ in (3.1), is the *differential generator* of X_ϕ (see [4]).

The following theorem is an immediate consequence of (2.6) and Theorem 4.1 of [6].

THEOREM 3.1. *Let $\phi(x)$ satisfy (2.3). If there exist a function $V(x)$ of class $C^{(2)}$ on E, and a positive number λ such that*

$$V(x) \to \infty \quad as \quad |x| \to \infty, \tag{3.2}$$

and

$$\mathfrak{L}_\phi V(x) \leqq \lambda - L[x, \phi(x)], \qquad x \in E, \tag{3.3}$$

then $\phi \in \Phi$.

To apply the theorem, set

$$\phi(x) = Kx, \tag{3.4a}$$

$$V(x) = x'Px, \tag{3.4b}$$

where K, P are constant $m \times n$, $n \times n$ (respectively) matrices, to be determined so that

$$\mathfrak{L}_\phi V(x) = \lambda - L[x, \phi(x)], \qquad x \in E. \tag{3.5}$$

Let $\Gamma(P)$ be the symmetric $n \times n$ matrix with elements

$$[\Gamma(P)]_{kl} = \operatorname{tr}(G_k' P G_l). \tag{3.6}$$

Then a brief calculation shows that (3.4) determines a solution of (3.5) if and only if

$$\lambda = \operatorname{tr}(C'PC) \tag{3.7}$$

and

$$\Gamma(P) + (A - BK)'P + P(A - BK) + M + K'NK = 0. \tag{3.8}$$

By our assumption of controllability, K can be chosen so that all eigenvalues of the matrix $A - BK$ have negative real parts.[2] With K so chosen, the following lemma shows that (3.8) has a unique positive definite solution P provided $\sum_k |G_k|^2$ is sufficiently small. This together with Theorem 3.1 implies that $\phi \in \Phi$.

LEMMA 3.1. *If $Q > 0$ and A is stable, the equation*

$$\Gamma(P) + A'P + PA + Q = 0 \tag{3.9}$$

has a unique solution $P > 0$ provided

$$d_2\left(\sum_{k=1}^{n} |G_k|^2\right) \left| \int_0^\infty e^{tA'} e^{tA}\, dt \right| < 1. \tag{3.10}$$

Here and below $P > 0$ ($\geqq 0$) means P is positive definite (semi-definite); $P_1 > P_2$ means $P_1 - P_2 > 0$, etc.

Proof. Equation (3.9) is equivalent to the equation

$$P = R + T(P), \tag{3.11}$$

where

$$R = \int_0^\infty e^{tA'} Q e^{tA}\, dt$$

and

$$T(P) = \int_0^\infty e^{tA'} \Gamma(P) e^{tA}\, dt. \tag{3.12}$$

We observe that $\Gamma(P)$ is a linear function of P and $\Gamma(P) \geqq 0$ if $P \geqq 0$; it follows that $T(P)$ has the same properties. Define

$$P_1 = R, \quad P_{\nu+1} = R + T(P_\nu), \qquad \nu = 1, 2, \cdots.$$

The sequence P_ν is monotone nondecreasing; it is bounded if, for some $\theta \in (0, 1)$ and all $P \geqq 0$,

$$T(P) \leqq \theta\, |P|\, I. \tag{3.13}$$

If (3.13) holds, it follows by a result on positive operators (e.g., [7, Theorem 1, p. 189]) that the matrix

$$P = \lim_{\nu \to \infty} P_\nu$$

exists; and $P \geqq R > 0$. If (3.10) holds, T is actually a contraction; thus (3.13) holds, and P is unique. The proof is complete.

[2] This fact is easily proved by inspection of the canonical form for (A, B) obtained in [14].

Define

$$\kappa = \inf_K \left| \int_0^\infty e^{t(A-BK)'} e^{t(A-BK)} \, dt \right|.$$

Thus, for some K, (3.8) has a (unique) solution $P > 0$ provided

$$\sum_{k=1}^{n} |G_k|^2 < (\kappa d_2)^{-1}. \tag{3.14}$$

4. Existence of an optimal control. It will be shown that an optimal control ϕ^0 exists whenever (3.14) holds, and that ϕ^0 is linear. We use dynamic programming and the well-known method of approximation in policy space [8]. This approach was suggested by the work of Howard, who studied a similar problem for Markov chains [9]. The result depends on the following optimality theorem.

THEOREM 4.1. *Suppose there exist $\phi^0 \in \Phi$, a function $v(x)$ of class $C^{(2)}$ on E, and a positive number λ, with the following properties:*

$$\mathcal{E}_\phi\{|v(x)| + |x||v_x(x)| + |x|^2|v_{xx}(x)|\} < \infty \quad \text{for every} \quad \phi \in \Phi; \tag{4.1}$$

$$\mathcal{L}_{\phi^0} v(x) + L[x, \phi^0(x)] = \lambda, \qquad x \in E; \tag{4.2}$$

$$\mathcal{L}_u v(x) + L(x, u) \geqq \lambda \quad \text{for every } m\text{-vector} \quad u, \qquad x \in E. \tag{4.3}$$

Then ϕ^0 is optimal. Furthermore,

$$\lambda = \mathcal{E}_{\phi^0}\{L(x, \phi^0)\}. \tag{4.4}$$

Combining (4.2) and (4.3) we obtain the appropriate version of Bellman's equation:

$$\min_u \{\mathcal{L}_u v(x) + L(x, u)\} = \lambda. \tag{4.5}$$

To prove Theorem 4.1 we need the following lemma.

LEMMA 4.1. *Let X be a diffusion process determined by (2.4), with differential generator $\mathcal{L}$ and ergodic measure μ. If $v(x)$ is a function of class $C^{(2)}$ such that*

$$\mathcal{E}_\mu\{|v(x)| + |x||v_x(x)| + |x|^2|v_{xx}(x)|\} < \infty,$$

then

$$\mathcal{E}_\mu\{\mathcal{L}v(x)\} = 0.$$

A proof is given in Appendix 1.

To prove Theorem 4.1 observe that if $\phi \in \Phi$ then, by (4.2) and (4.3),

$$\lambda \leqq \mathcal{L}_\phi v(x) + L[x, \phi(x)], \qquad x \in E.$$

Taking expectations with respect to μ_ϕ on both sides, and applying Lemma 4.1, we obtain

$$\lambda \leqq \mathcal{E}_\phi\{L(x, \phi)\}.$$

Again by Lemma 4.1, (4.2) implies

$$\lambda = \mathcal{E}_{\phi^0}\{L(x, \phi^0)\},$$

and the result follows.

To compute an optimal control we seek a solution of Bellman's equation, in the form

$$v(x) = x'Px. \tag{4.6}$$

Substitution shows that (4.5) holds if and only if P satisfies (3.8), with

$$K = N^{-1}B'P. \tag{4.7}$$

The control determined by (4.5) is

$$\phi^0(x) = Kx. \tag{4.8}$$

We show next that (3.8) and (4.7) can be solved for a unique positive definite matrix P. For $\nu = 1, 2, \cdots$, let P_ν be a solution of (3.8) with $K = K_\nu$, and define

$$K_{\nu+1} = N^{-1}B'P_\nu. \tag{4.9}$$

By Lemma 3.1, we can choose K_1 so that P_1 exists. It will be shown that if K_2 is defined by (4.9), then P_2 exists and $0 < P_2 \leqq P_1$. Write $v_\nu(x) = x'P_\nu x$, $\phi_\nu = K_\nu x$ and $\mathcal{L}_\nu = \mathcal{L}_{\phi(\nu)}$. It can be verified directly that (4.9) is equivalent to the condition

$$\mathcal{L}_{\nu+1} v_\nu(x) + L[x, \phi_{\nu+1}(x)] \leqq \mathcal{L}_u v_\nu(x) + L(x, u), \qquad x \in E, \tag{4.10}$$

for all m-vectors u. That is, $\phi_{\nu+1}$ is determined by the minimizing operation (4.5) applied to v_ν. Setting $\nu = 1$ and $u = \phi_1(x)$ in (4.10), and using (3.8), we see that

$$\begin{aligned} -Q &\equiv \Gamma(P_1) + (A - BK_2)'P_1 + P_1(A - BK_2) + M + K_2'NK_2 \\ &\leqq 0. \end{aligned} \tag{4.11}$$

Write $A_2 = A - BK_2$. Since $P_1 > 0$ satisfies (4.11), it follows (by a standard Liapunov theorem) that A_2 is stable. Hence,

$$P_1 = \int_0^\infty e^{tA_2'}[M + K_2'NK_2 + \Gamma(P_1) + Q]e^{tA_2} \, dt. \tag{4.12}$$

Now P_2 is to be determined by (3.8) with $K = K_2$, or

$$P_2 = \int_0^\infty e^{tA_2'}[M + K_2'NK_2 + \Gamma(P_2)]e^{tA_2}\,dt. \tag{4.13}$$

As in the proof of Lemma 3.1, we solve (4.13) by successive approximations. Setting $P_2^{(1)} = 0$, we have

$$P_2^{(2)} = \int_0^\infty e^{tA_2'}(M + K_2'NK_2)e^{tA_2}\,dt \leqq P_1;$$

and similarly $P_2^{(\kappa)} \leqq P_1$, $\kappa = 2, 3, \cdots$. Since the $P_2^{(\kappa)}$ are nondecreasing and bounded,

$$P_2 \equiv \lim_{\kappa\to\infty} P_2^{(\kappa)} \tag{4.14}$$

exists and satisfies (4.13). Thus $P_2 \leqq P_1$, and $M > 0$ implies $P_2 > 0$.

It is not asserted that the solution of (4.13) is unique; however, we may now proceed by induction and define

$$P_\nu = \lim_{\kappa\to\infty} P_\nu^{(\kappa)}, \qquad \nu = 1, 2, \cdots.$$

In this way we obtain a sequence $\{P_\nu\}$ with $0 < P_{\nu+1} \leqq P_\nu$. Then

$$P = \lim_{\nu\to\infty} P_\nu, \qquad K = N^{-1}B'P \tag{4.15}$$

exist and satisfy (3.8) and (4.7).

Define

$$\phi^0(x) = Kx, \quad v(x) = x'Px, \quad \lambda = \mathrm{tr}\,(C'PC). \tag{4.16}$$

Theorem 4.1 will be applied to show that ϕ^0 is optimal. By construction, ϕ^0 satisfies (4.2) and (4.3). Furthermore, if $\phi \in \Phi$ then (2.5) and (4.16) imply the truth of (4.1). The existence of ϕ^0 is now established.

We observe that ϕ^0 is unique in the class of linear controls; for if ϕ is another optimal linear control and $\hat\lambda$, $\hat P$ are the corresponding quantities determined as before, then by (4.4), $\hat\lambda = \lambda$, and by (4.16),

$$\mathrm{tr}\,(C'\hat PC) = \mathrm{tr}\,(C'PC). \tag{4.17}$$

Since $\hat P$, P are independent of C, (4.17) holds for all C, and from this it easily follows that $\hat P = P$. Uniqueness of ϕ is a consequence of (4.7).

5. Example 1. The following artificial example is of interest because it illustrates the qualitative dependence of the control law on the intensity of the state-dependent noise. Let

$$dx_i = ax_i\,dt - bu_i\,dt + c\,dw_{1i} + g\,|x|\,dw_{2i}, \quad i = 1, \cdots, n, \tag{5.1}$$

and

$$L(x, u) = |x|^2 + |u|^2.$$

In (5.1) the matrix $G(x) = g\,|x|\,I$ is not linear in x (cf. (2.2)); nevertheless, because of the rotational symmetry, the methods used above apply equally well here, and (3.8), (4.7) become

$$g^2(\mathrm{tr}\,P)I + (aI - bK)'P + P(aI - bK) + I + K'K = 0,$$
$$K = bP.$$

This gives $P = pI$, $K = bpI$, and $\lambda = nc^2p$, where

$$p = (2b^2)^{-1}\{2a + ng^2 + [(2a + ng^2)^2 + 4b^2]^{1/2}\} \sim nb^{-2}g^2, \qquad g \to \infty. \tag{5.2}$$

For large g, $\phi^0(x) \sim nb^{-1}g^2x$, and the optimal control depends rather critically on noise intensity.

Now suppose that for some k, $u = \phi(x) = kx$ in (5.1). Solution of (3.5) and application of (4.7) yield

$$\lambda_\phi = \mathcal{E}_\phi\{|x|^2 + |u|^2\} = nc^2(1 + k^2)[2(bk - a) - ng^2]^{-1}$$

provided

$$bk - a > ng^2/2. \tag{5.3}$$

If this inequality fails (i.e., control is not sufficiently vigorous), then instability results, in the sense that either $\lambda_\phi = +\infty$ or λ_ϕ is not defined: that is, X_ϕ is no longer ergodic. Using the methods of [5] one can show that X_ϕ is ergodic (i.e., μ_ϕ exists) if and only if $bk - a > (n - 2)g^2/2$.

6. Example 2. State-dependent noise and instability. The example of this section illustrates the fact that an admissible linear control need not exist if the intensity of state-dependent noise is sufficiently large. Let

$$A = \begin{pmatrix} 0 & 1 \\ 0 & 0 \end{pmatrix}, \quad B = \begin{pmatrix} 0 \\ 1 \end{pmatrix}, \quad G(x) = \sqrt{\gamma}\begin{pmatrix} x_1 & x_2 \\ -x_2 & x_1 \end{pmatrix}, \tag{6.1}$$

where $\gamma > 0$ is a constant; thus $\Gamma(P) = \gamma(\mathrm{tr}\,P)I$.

We first show that, for the present example, (3.8) and (4.7) have a positive solution P if and only if $\gamma < 1$. Let $K = (k_1, k_2)$; then $A - BK$ is stable if and only if $k_1 > 0$, $k_2 > 0$. If (3.8) has a solution $P > 0$, then $A - BK$ is stable; thus

$$P = R + T(P), \tag{6.2}$$

where

$$R = \int_0^\infty e^{t(A-BK)'}(M + K'NK)e^{t(A-BK)}\,dt$$

and

$$T(P) = \gamma(\operatorname{tr} P)\int_0^\infty e^{t(A-BK)'}e^{t(A-BK)}\,dt. \tag{6.3}$$

Denote the integral in (6.3) by S. Computation yields

$$\inf\,(\operatorname{tr} S: k_1 > 0,\ k_2 > 0) = 1,$$

and the infimum is not attainable. Then

$$\begin{aligned} P \geqq T(P) &\geqq T^{(\nu)}(P) \\ &= \gamma^{\nu}(\operatorname{tr} P)(\operatorname{tr} S)^{\nu-1}S \\ &> \gamma^{\nu}(\operatorname{tr} P)S, \qquad \nu = 2, 3, \cdots, \end{aligned}$$

and necessarily $\gamma < 1$.

On the other hand, if $\gamma < 1$ there exist $k_1 > 0$, $k_2 > 0$ such that $\operatorname{tr} S < \gamma^{-1}$, and then

$$P = \sum_{\nu=0}^{\infty} T^{(\nu)}(R) > 0$$

exists and satisfies (3.8). Thus the construction used in §4 succeeds if and only if $\gamma < 1$.

Next we show that no admissible linear control exists if $\gamma > 1$. For this we need the following instability theorem, the proof of which is given in Appendix 2. Let X be a diffusion process determined by (2.4), with differential generator $\mathcal{L}$; let $L(x) \geqq 0$, and Hölder continuous, $x \in E$; and let $V_1(x)$, $V_2(x)$ be a pair of real-valued functions with the properties:

(i) for some $r < \infty$, V_1, V_2 are defined and of class $C^{(2)}$ for $|x| > r$;

(ii) there is a sequence $\{x_n\}$ with $|x_n| \to \infty$ such that $V_1(x_n) \to +\infty$;

(iii) $V_2(x) > 0, \quad |x| > r$;

(iv) $\displaystyle\overline{\lim_{\rho\to\infty}} \frac{\max\{V_1(x): |x| = \rho\}}{\min\{V_2(x): |x| = \rho\}} = 0$;

(v) $\mathcal{L}V_1(x) \geqq 0, \quad \mathcal{L}V_2(x) \leqq L(x), \quad |x| > r$.

THEOREM 6.1 (Instability). *Suppose X is positive recurrent, with ergodic probability measure μ. If there exist functions V_1, V_2 with properties* (i)–(v), *then*

$$\mathcal{E}_\mu\{L(x)\} = +\infty.$$

To apply the theorem in the present case, let $L(x) = |x|^2$ and suppose $\gamma > 1$. We choose $V_2(x) = \theta\,|x|^2$ for suitable θ, $0 < \theta < 1$; clearly this is possible. More critical is the choice of V_1: let

$$V_1(x) = [Q(x)]^q,$$

where $Q(x) = Q(x_1, x_2)$ is a positive definite quadratic form, and $0 < q < 1$. To find suitable Q and q, observe that

$$\mathcal{L}V(x) \equiv \tfrac{1}{2}\gamma\,|x|^2 \operatorname{tr} V_{xx} + \tfrac{1}{2}\operatorname{tr}(C'V_{xx}C) + x'(A - BK)'V_x.$$

Suppose first that $A - BK$ is stable. We choose Q so that

$$x'(A - BK)'Q_x(x) = -|x|^2.$$

Write $Q(1, 0) + Q(0, 1) = \operatorname{tr} Q$. Computation yields $\inf\{\operatorname{tr} Q: k_1 > 0, k_2 > 0\} = 1$, and

$$\begin{aligned}\mathcal{L}V_1(x) = q\,|x|^2 Q(x)^{q-1}\left[\frac{1}{2}\gamma(q-1)\frac{|Q_x(x)|^2}{Q(x)} + \gamma \operatorname{tr} Q - 1\right] \\ + O(|x|^{2q-2}), \qquad |x| \to \infty.\end{aligned}$$

Q being a positive quadratic form, we find

$$\sup_x \frac{|Q_x(x)|^2}{Q(x)} \leqq 4 \operatorname{tr} Q,$$

so that $\mathcal{L}V_1(x) \geqq 0$ for $|x|$ sufficiently large if

$$\gamma(\operatorname{tr} Q)\{2(q-1) + 1\} - 1 > 0.$$

Setting $q = (3 + \gamma^{-1})/4$, we have satisfied all the conditions of Theorem 6.1.

Finally, suppose $A - BK$ is not stable. Then integration of the stochastic differential equation

$$dx = (A - BK)x\,dt + C\,dw_1 + G(x)\,dw_2$$

yields

$$\mathcal{E}\{|x(t)|^2\} \geqq \operatorname{tr}\left\{\int_0^t e^{s(A-BK)}CC'e^{s(A-BK)'}\,ds\right\} \to \infty, \qquad t \to \infty;$$

and the control $\phi(x) = Kx$ cannot be admissible for any value of γ.

To conclude this section we remark that the boundary case $\gamma = 1$ apparently presents difficulty to application of Theorem 6.1, and will not be discussed at present. Our purpose has been to show that linearly state-dependent noise of sufficient intensity may make stabilization by linear control impossible, regardless of the choice of control parameters, even if the pair (A, B) is controllable in the usual sense. Although this result is not surprising, the situation deserves more detailed study than will be given here.

7. An alternative interpretation of (1.1). It is worth emphasizing that the choice of Itô's equation (2.1) as a precise version of (1.1) is somewhat arbitrary. We shall discuss briefly an alternative version of (1.1) which may be more appropriate in engineering applications. Equation (1.1) is a purely formal equation since the "derivatives" $\dot{w}_1, \dot{w}_2$ do not exist. In writing (1.1), we usually have in mind a physical system perturbed by noise with a power spectral density which is essentially constant within the frequency passband of the system. However, total noise power is presumably finite, and this fact is overlooked in adopting the precise model (2.1). Thus the question arises whether the diffusion process determined by (2.1) adequately reflects the properties of the physical random process of which (1.1) is a rough description. This question has been discussed in a precise fashion by Stratonovich [10], [11], and by Wong and Zakai [12]. It turns out that the proper Itô equation to associate with (1.1) will depend on what definition is adopted of the formal stochastic integral

$$J = \int_a^b G[x(t)]\dot{w}(t). \tag{7.1}$$

Let $\{t_\nu\}$ be a partition of the interval $[a, b]$. On the basis of results of [10]–[12] it is natural to adopt for (7.1) the definition

$$J = \text{l.i.m.} \sum_\nu G\left[\frac{x(t_\nu) + x(t_{\nu+1})}{2}\right][w(t_{\nu+1}) - w(t_\nu)]$$

as $\max_\nu (t_{\nu+1} - t_\nu) \to 0$. Let us now suppose that $x(t)$ has the Itô stochastic differential

$$dx(t) = f(x)\,dt + G(x)\,dw,$$

where $G(x) = [g_{ij}(x)]$. Then it can be shown [10] that

$$J = \frac{1}{2}\int_a^b G_x[x(t)]\cdot G[x(t)]\,dt + \int_a^b G[x(t)]\,dw(t), \tag{7.2}$$

where the second integral in (7.2) is an Itô stochastic integral, and $G_x \cdot G$ is the vector with ith component

$$\sum_{j,k} (\partial g_{ij}/\partial x_k) g_{kj}.$$

This result means that an alternative natural interpretation of (1.1) is that the process $x(t)$ has the Itô stochastic differential

$$dx = [Ax - Bu + \tfrac{1}{2}G_x(x)\cdot G(x)]\,dt + C\,dw_1 + G(x)\,dw_2. \tag{7.3}$$

Equation (7.3) differs from (2.1) by the presence of an additional drift term contributed by the coefficient of the state-dependent noise.

Suppose that $G(x)$ has the form (2.2). Then (7.3) can be written

$$dx = \hat{A}x\,dt - Bu\,dt + C\,dw_1 + G(x)\,dw_2,$$

where $\hat{A}$ is a modified system matrix with elements

$$\hat{a}_{ij} = a_{ij} + \tfrac{1}{2}\sum_{k,l} g_{ikl}\,g_{lkj}$$

and g_{ikl} is the (i, k)th element of G_l. After this modification the discussion of §§2–5 remains unchanged.

In light of this discussion consider again Example 1. Here $G(x) = g\,|x|\,I$, and

$$G_x(x)\cdot G(x) = g^2 x.$$

Thus $\hat{A} = aI + (g^2/2)I$, and the previous results hold with this replacement. With the new model,

$$\phi^0(x) \sim (n+1)b^{-1}g^2 x, \qquad g \to \infty;$$

that is, the optimal control gain is somewhat higher than previously. Suppose next that $u = \phi(x) = kx$. Then (cf. §5) $\lambda_\phi < \infty$ if and only if

$$bk - a > (n+1)g^2/2.$$

Comparing this result with (5.3) we see that the choice of mathematical model may be critical in an assessment of the stability properties of the physical system of interest.

8. Alternative problems. A variety of linear regulator problems with linearly state-dependent noise can be discussed by methods similar to the foregoing. If the index of performance is expectation of a quadratic functional, and if no a priori bound is placed on magnitude of the control vector, then in general the optimal control (when it exists) is linear in x and depends on noise intensity.

To mention one interesting variant, let

$$dx = Ax\,dt - Bu\,dt + G(x)\,dw, \tag{8.1}$$

and consider the problem of minimizing

$$\mathcal{E}_x\left\{\int_0^\infty [x(t)'Mx(t) + u(t)'Nu(t)]\,dt\right\}. \tag{8.2}$$

If $u = \phi(x)$ and $\phi(0) = 0$, then (8.1) admits the null solution $x(t) \equiv 0$ (see, e.g., [13]). The functional (8.2) is finite provided $x = 0$ is globally asymptotically stable in an appropriate sense. By a slight extension of the methods of [13], one can show that X_ϕ is stable if and only if a continuous function $V(x)$ exists such that:

(i) $V(x) > 0, x \neq 0, \quad V(0) = 0$;

(ii) $V(x) \to +\infty$ as $|x| \to \infty$;

(iii) $\mathcal{L}_\phi V(x) \leqq -|x|^2, x \neq 0$.

Call ϕ admissible if X_ϕ is stable. Just as in §3, we find that $\phi(x) = Kx$ is admissible if (3.8) has a positive solution P, and this is so whenever $G(x)$ is restricted by the inequality (3.14). Under these conditions the optimal linear control is determined exactly as before.

Appendix 1. Proof of Lemma 4.1. Let $\mathcal{E}_x$ denote expectation on the paths of X when $x(0) = x \in E$. Let $t > 0$ be fixed and write

$$w(x) = \mathcal{E}_x\{v[x(t)]\}.$$

We show first that w exists a.e. $[\mu]$ and

$$\mathcal{E}_\mu\{w\} = \mathcal{E}_\mu\{v\}. \tag{A1.1}$$

If v is a simple function, (A1.1) is obvious. If $v \geqq 0$ and v_n are simple functions with $0 \leqq v_n \uparrow v$, then

$$w_n(x) = \mathcal{E}_x\{v_n[x(t)]\}$$

is measurable and $w_n \uparrow w$. By monotone convergence,

$$\begin{aligned} \mathcal{E}_\mu\{w\} &= \mathcal{E}_\mu\{\lim w_n\} = \lim \mathcal{E}_\mu\{w_n\} \\ &= \lim \mathcal{E}_\mu\{v_n\} = \mathcal{E}_\mu\{v\}. \end{aligned}$$

The general result follows by applying the argument to the positive and negative parts of v.

Now let v be of class $C^{(2)}$ and of compact support. By the Itô-Dynkin formula [4],

$$\begin{aligned} \mathcal{E}_\mu\left\{\mathcal{E}_x\left\{\int_0^t \mathcal{L}v[x(s)]\,ds\right\}\right\} &= \mathcal{E}_\mu\{\mathcal{E}_x\{v[x(t)] - v(x)\}\} \\ &= 0. \end{aligned}$$

Since $\mathcal{L}v[x(s)]$ is bounded and almost surely continuous (in s) there follows, by dominated convergence,

$$\begin{aligned} \mathcal{E}_\mu\{\mathcal{L}v(x)\} &= \mathcal{E}_\mu\left\{\mathcal{E}_x\left\{\lim_{t \downarrow 0} t^{-1} \int_0^t \mathcal{L}v[x(s)]\,ds\right\}\right\} \\ &= \lim_{t \downarrow 0} \mathcal{E}_\mu\left\{\mathcal{E}_x\left\{t^{-1} \int_0^t \mathcal{L}v[x(s)]\,ds\right\}\right\} \\ &= 0. \end{aligned}$$

In general, suppose $v(x)$ satisfies the integrability condition of the hypothesis. Then for any $\epsilon > 0$, there exists a smooth function $\bar{v}(x)$ of compact support such that

$$|\mathcal{E}_\mu\{\mathcal{L}v(x)\} - \mathcal{E}_\mu\{\mathcal{L}\bar{v}(x)\}| < \epsilon;$$

that is, $|\mathcal{E}_\mu\{\mathcal{L}v(x)\}| < \epsilon$.

Appendix 2. Proof of Theorem 6.1. It is assumed that X is a diffusion process determined by the stochastic differential equation (2.4); in particular, X satisfies the hypotheses made in [5] and [6]. For brevity we use freely the methods and notation of [6].

LEMMA. *Let $D \subset E$ be a normal domain with boundary Γ and let τ_Γ be the first time the path $x(\cdot)$ hits Γ. A necessary and sufficient condition that*

$$\mathcal{E}_x\left\{\int_0^{\tau_\Gamma} L[x(t)]\,dt\right\} < \infty, \qquad x \in E - D,$$

is that the equation

$$\mathcal{L}[u(x)] = -L(x), \qquad x \in E - D,$$

have a smooth positive solution $u(x)$ in $E - D$.

Proof. The sufficiency was proved as Lemma 3.3 of [6]. To prove necessity let $\{D_n : n = 1, 2, \cdots\}$ be a sequence of domains as in [6, Lemma 3.3], and put

$$u_n(x) = \mathcal{E}_x\left\{\int_0^{\tau_n} L[x(t)]\,dt\right\}, \qquad x \in D_n - \bar{D},$$

where $\tau_n = \min\{t : x(t) \in \Gamma \cup \Gamma_n,\ x(0) \in \bar{D}_n - D\}$. Since $\tau_n \uparrow \tau_\Gamma$, $n \to \infty$, with probability 1, there follows

$$u_n(x) \uparrow u(x) = \mathcal{E}_x\left\{\int_0^{\tau_\Gamma} L[x(t)]\,dt\right\}, \qquad x \in E - \bar{D}.$$

Since the $u_n(x)$ satisfy $\mathcal{L}u_n(x) = -L(x)$, $x \in D_n - \bar{D}$, $u_n(x) = 0$, $x \in \Gamma \cup \Gamma_n$, it follows (see [6, Proof of Lemma 3.1]) that $u(x) \geqq 0$ has the required property.

Combining this result with that of [6, Lemma 3.2], we have that $\mathcal{E}_\mu\{L(x)\} < \infty$ if and only if there exists a smooth positive solution of

$$\mathcal{L}u(x) = -L(x), \qquad |x| > r, \tag{A2.1}$$

for some $r < \infty$. The proof of Theorem 6.1 is completed by showing that no positive solution of (A2.1) exists, and follows almost verbatim the proof of [5, Theorem 4].

Acknowledgment. I am indebted to Dr. J. G. Heller for useful discussion of this problem.

REFERENCES

[1] J. J. FLORENTIN, *Optimal control of continuous-time, Markov, stochastic systems*, J. Electronics Control, 10 (1961), pp. 473–488.

[2] W. M. Wonham, *Stochastic problems in optimal control*, Tech. Rep. 63-14, Research Institute for Advanced Studies, Baltimore, Maryland, 1963.
[3] K. Itô, *On stochastic differential equations*, Mem. Amer. Math. Soc., no. 4, 1951, 51 pp.
[4] E. B. Dynkin, *Markov Processes*, Academic Press, New York, 1965.
[5] W. M. Wonham, *Liapunov criteria for weak stochastic stability*, J. Differential Equations, 2 (1966), pp. 195–207.
[6] ———, *A Liapunov method for the estimation of statistical averages*, Ibid., 2 (1966), pp. 365–377.
[7] L. V. Kantorovich and G. P. Akilov, *Functional Analysis in Normed Spaces*, Macmillan, New York, 1964.
[8] R. Bellman, *Dynamic Programming*, Princeton University Press, Princeton, 1957.
[9] R. A. Howard, *Dynamic Programming and Markov Processes*, John Wiley, New York, 1960.
[10] R. L. Stratonovich, *A new form of representation of stochastic integrals and equations*, this Journal, 4 (1966), pp. 362–371.
[11] ———, *Conditional Markov Processes and their Applications in the Theory of Optimal Control*, Izd. Mosk. Univ., 1966.
[12] E. Wong and M. Zakai, *On the relation between ordinary and stochastic differential equations*, Internat. J. Engrg. Sci., 3 (1965), pp. 213–229.
[13] R. Z. Khasminskii, *On the stability of the trajectory of Markov processes*, J. Appl. Math. Mech., 26 (1962), pp. 1554–1565.
[14] C. E. Langenhop, *On the stabilization of linear systems*, Proc. Amer. Math. Soc., 15 (1964), pp. 735–742.

Adaptive Guaranteed Cost Control of Systems with Uncertain Parameters

SHELDON S. L. CHANG, FELLOW, IEEE, AND T. K. C. PENG, MEMBER, IEEE

***Abstract*—Guaranteed cost control is a method of synthesizing a closed-loop system in which the controlled plant has large parameter uncertainty. This paper gives the basic theoretical development of guaranteed cost control, and shows how it can be incorporated into an adaptive system. The uncertainty in system parameters is reduced first by either: 1) on-line measurement and evaluation, or 2) prior knowledge on the parametric dependence of a certain easily measured situation parameter. Guaranteed cost control is then used to take up the residual uncertainty.**

It is shown that the uncertainty in system parameters can be taken care of by an additional term in the Riccati equation. A Fortran program for computing the guaranteed cost matrix and control law is developed and applied to an airframe control problem with large parameter variations.

INTRODUCTION

A CONTROL problem can be represented as a system of differential equations characterized by a finite number of parameters, and system uncertainty is represented by the uncertainty of those parameters. The choice of control policy depends upon the optimization of some preassigned performance criterion that is usually subjective and has to be justified by the actual performance, i.e., the dynamics of the physical system or its simulation utilizing the control policy so determined. A useful control policy should also be simple to compute and realize while giving satisfactory performance. Complexity in control scheme usually imposes a cost in terms of reliability and limits its practicality.

The idea of guaranteed cost control was introduced in two previous papers [1], [2], as a special case of fuzzy dynamic programming. (The concept of fuzziness was introduced in [3].) When the uncertain parameters in the system equation of the control system vary within given bounds, it is useful to find a control law that guarantees an upper bound for the performance functional. It works very well in conjunction with adaptive control by making allowances for the inexactitude in the system identification processes. In some control problems, the system parameters vary with a certain easily measurable quantity or prominent parameter (e.g., the dynamic pressure in airplane and ship control) but are not completely determined by the latter. One can readily obtain a guaranteed cost control law as a simple function of the prominent parameter without any further need of the identification process.

Manuscript received January 18, 1971; revised February 14, 1972. Paper recommended by D. Sworder, Chairman of the IEEE S-CS Stochastic Control Committee. This research was sponsored by the Air Force Office of Scientific Research under Grant AFOSR-70-1890 and by the National Science Foundation under Grant GK-16017.

The authors are with the Department of Electrical Sciences, State University of New York, Stony Brook, N. Y. 11790.

In automatic control literature, the problem of systems with uncertain parameters has been treated in several categories according to different assumptions and approaches. The first is the stochastic approach where an *a priori* probability is assumed on the uncertain parameters, and the control is to minimize the expected value of some performance index, as in the case of stochastic optimal control [4]. If the statistical property of the uncertain parameters cannot be assumed beforehand but can be identified in the course of the dynamic process, the idea of adaptive or learning control is used [5]–[7]. The second is the sensitivity approach, which is relatively new but already showing promising success [8]. This approach is based on the assumption that the parameter uncertainty is small so that first-order perturbation equations can be obtained, from which a controller is designed to minimize the performance or trajectory sensitivity. If the uncertain parameters vary in a finite domain but the optimal control can be written as a Taylor series in the uncertain parameters, then the system is called optimally adaptive [9]; the first-order approximation of such a system is called the optimally sensitive system [10]. The third is the minimax approach where the only assumption on the uncertain parameters is the bound of their variation. Given any control law, there is a maximum cost over all possible parameter perturbation, and it is desired to find the minimum of this number over all admissible control laws [11], [12]. In general, a saddle point does not exist and usually the computation of the minimax control law is very complicated. In the case where the uncertain parameters are assumed to be an unknown constant, an algebraic minimax algorithm is obtained by Salmon [12].

The stochastic approach needs the knowledge of a distribution. After complicated computation, only the expected value of the performance index is minimized, leaving the possibility of occasional bad behavior of the system dynamics unchecked. Adaptive control gives good performance, but the procedure is usually complicated and expensive. The sensitivity approach is limited mainly by the assumption of small perturbation. The Taylor series method for large uncertainty is valid on the assumption that the control and some related functions are analytic in the uncertain parameters, which is not easy to justify in general. The minimax approach gives the minimum guaranteed cost but, except in very restricted conditions, it ends up with a nonlinear complicated

Reprinted from *IEEE Trans. Automat. Contr.*, vol. AC-17, no. 4, pp. 474–483, Aug. 1972.

control law even in a linear regulator problem. The algebraic minimax algorithm is effective but it can only be used when the uncertain parameter is a constant. Adaptive guaranteed cost as proposed in this paper makes use of all available information and results in a simple control law. The control law is linear if the controlled system is linear with time-varying uncertain parameters and a quadratic cost criterion.

Guaranteed Cost Control

System Equation

The controlled dynamic process is described by

$$\dot{x} = f(x, u, q, t) \tag{1}$$

where x is an n-dimensional state vector, u is an m-dimensional control vector, and q is a vector valued parameter that is an arbitrary measurable time function known to be in a closed and bounded region Ω:

$$q(t) \in \Omega.$$

The cost functional is given as

$$C[x(t_1), u, q] \triangleq h(x(t_2)) + \int_{t_1}^{t_2} g(x, u, q, t)\, dt \tag{2}$$

where $h(x(t_2)) \geq 0$ respresents the terminal cost and the integral from t_1 to t_2 represents accumulated cost along the path. A feedback control policy or control law is to express $u(t)$ as a function of the state variable x and t:

$$u(t) = \eta(x, t).$$

Guaranteed Cost and Guaranteed Cost Control

If there is a number V and a control u such that $C[x(t_1), u, q] \leq V$ for any allowed variation of q, then V is called a *guaranteed cost* for the system starting from $x(t_1)$ at time t_1, and u is called a *guaranteed cost control.*

Theorem 1: Let $V(x,t)$ be a scalar function with continuous first partial derivatives and let $\eta(x,t)$ be an m-dimensional vector function satisfying the following inequality:

$$g(x,\eta(x,t), q,t) + \frac{\partial V(x,t)}{\partial x} f(x,\eta(x,t), q,t) + \frac{\partial V(x,t)}{\partial t} \leq 0 \tag{3}$$

for all $q \in \Omega$, $t < t_2$, $x \in R^n$; and

$$V(x(t_2), t_2) = h(x(t_2)). \tag{4}$$

Then $V(x(t_1),t_1)$ is a guaranteed cost starting at $x(t_1)$ with any $t_1 < t_2$, and the feedback control policy $\eta(x,t)$, $t_1 \leq t \leq t_2$ is a guaranteed cost control.

Proof: Given any allowed function $\bar{q}$, let $\bar{x}(t)$ be the solution of (1) with $u(t) = \eta(\bar{x},t)$ and $q = \bar{q}$. It will be shown that

$$C[x(t_1), \eta, \bar{q}] \triangleq h(\bar{x}(t_2)) + \int_{t_1}^{t_2} g(\bar{x}, \eta(\bar{x},t), \bar{q}(t), t)\, dt \leq V(x(t_1),t_1). \tag{5}$$

The condition (3) can be written as

$$-\frac{dV(\bar{x},t)}{dt} \geq g(\bar{x}, \eta(\bar{x},t), \bar{q}(t), t). \tag{6}$$

Integrating (6) with respect to t gives

$$-V(\bar{x},t)\Big|_{t_1}^{t_2} \geq \int_{t_1}^{t_2} g(\bar{x}, \eta(\bar{x},t), \bar{q}(t), t)\, dt. \tag{7}$$

From (4),

$$-V(\bar{x},t)\Big|_{t_1}^{t_2} = V(x(t_1), t_1) - h(\bar{x}(t_2)). \tag{8}$$

Combining (7) and (8) and rearranging terms give (5).

Theorem 1 is a sufficient condition. The following corollary helps to locate an important class of guaranteed cost control.

Corollary: Let the function $F(V, x, u, q, t)$ be defined as

$$F(V, x, u, q, t) \triangleq g(x, u, q, t) + \frac{\partial V(x,t)}{\partial x} f(x, u, q, t) + \frac{\partial V(x,t)}{\partial t}. \tag{9}$$

Given $V(x,t)$, x, and t, let $\eta_0(x, t)$ denote the value of u that minimizes the expression

$$\max_{q \in \Omega} F(V, x, u, q, t). \tag{10}$$

Then if (3) is satisfied for all $q \in \Omega$ with $V(x,t)$ and *any* $\eta(x,t)$, it is satisfied with $\eta_0(x,t)$.

Proof: By definition of $\eta_0(x, t)$

$$\max_{q \in \Omega} F(V, x, \eta_0(x, t), q, t) \leq \max_{q \in \Omega} F(V, x, \eta(x, t), q, t) \leq 0.$$

The significance of the corollary is that, in the search for $V(x,t)$, one may limit the choice of $\eta(x,t)$ to the value of u that minimizes (10).

System Stability

In the control problem defined by (1) and (2), the system is said to be *autonomous in the wide sense* if the functions f and g do not depend explicitly on t:

$$f(x, u, q, t) = f(x, u, q) \tag{11a}$$

$$g(x, u, q, t) = g(x, u, q). \tag{11b}$$

It is assumed further that $t_2 \rightarrow \infty$. The functions $V(x, t)$ and $\eta(x, t)$ are then functions of x only:

$$V(x, t) = V(x) \tag{12}$$

$$\eta(x, t) = \eta(x). \tag{13}$$

With a given choice of control law, (1) becomes

$$\dot{x} = f(x, \eta(x), q). \tag{14}$$

Let Z denote the collection of (x, q) pairs for which $g(x, \eta(x), q) = 0$, $q \in \Omega$, and Z is said to be the *zero cost set*. The projection of Z on R^n (x space) is denoted Z_x. A subset I of Z_x is said to be *invariant in the wide sense* if, starting from any point $x_1(t_1) \in I$, there exist a $q(t)$ and solution $x(t)$ of (14) such that

1) $(x(t), q(t)) \in Z$, and $x(t) \in I$ for all $t \geq t_1$
2) $x(t_1) = x(t_2) = x_1(t_1)$.

A subset M of Z_x is said to be the *largest invariant set in the wide sense* if it contains *all* the invariant (in the wide sense) subsets I of Z_x.

Theorem 2: Let $V(x)$ be a scalar function with continuous first partial derivatives for all x, satisfying the following conditions for all $q \in \Omega$:

$$g(x, \eta(x), q) + \frac{\partial V(x)}{\partial x} f(x, \eta(x), q) \leq 0$$

$$V(x) \geq 0. \tag{15}$$

Let M be the largest invariant set in the wide sense contained in Z_x. Then all the solutions of (14) [for whatever $q(t)$] bounded for $t > 0$ tend to M as $t \to \infty$.

The proof of Theorem 2 is similar to the proof of Lyapunov theorem in [13, p. 66]. A slightly different version of this theorem was proved in [16], where an example of application was also given.

Linear Autonomous (in the Wide Sense) System with Uncertain Parameters

Based on Theorems 1 and 2, a general solution will be given for a subclass of linear systems:

$$\dot{x} = A(q)x + B(q)u \tag{16}$$

$$C[x(o), u, q] = \frac{1}{2} x'H_x + \frac{1}{2} \int_0^T (x'Px + u'Qu)\, dt \tag{17}$$

where H, P are symmetric, positive semidefinite matrices, Q is a symmetric, positive definite matrix, and

$$A(q) = A_0 + \sum_{i=1}^{i=n'} q_i(t) A_i \tag{18}$$

$$B(q) = q_b(t) B \tag{19}$$

$$-1 \leq q_i(t) \leq 1, \qquad i = 1,2,\cdots, n' \tag{20}$$

$$1 \leq q_b(t) \leq b. \tag{21}$$

One shall seek a guaranteed cost solution of the form

$$V(x,t) = \tfrac{1}{2} x'R(t)x \tag{22}$$

where $R(t)$ is taken to be symmetric, positive semidefinite, and will be called a guaranteed cost matrix. The condition (3) becomes

$$\frac{1}{2}(x'Px + \eta'Q\eta) + x'R(t)[A(q)x + B(q)\eta] + \frac{1}{2} x' \frac{dR(t)}{dt} x \leq 0 \tag{23}$$

where η stands for $\eta(x,t)$ and $R(T) = H$.

The left-hand side of (23) is the function F defined in the corollary of Theorem 1. Let η_0 denote the vector η that minimizes $\max_q F(V, x, \eta, q, t)$. To determine η_0, only the terms in F that involve η need to be considered:

$$F_1 \triangleq \tfrac{1}{2} \eta'Q\eta + x'R(t) q_b(t) B \eta.$$

Let $q_{b_0}(t)$ denote the value of $q_b(t)$ that maximizes F_1:

$$q_{b_0}(t) = b, \qquad \text{if } x'R(t)B\,\eta > 0$$
$$= 1, \qquad \text{if } x'R(t)B\,\eta < 0 \tag{24}$$

and F_1 is indifferent to $q_b(t)$ if $x'R(t)B\,\eta = 0$. Minimizing maximum F_1 gives

$$\eta_0 = -Q^{-1} q_{b_0}(t) B'R(t)x.$$

Since

$$x'R(t)B\eta_0 = -q_{b_0}(t) x'R(t)BQ^{-1}B'R(t)x \leq 0,$$

it follows from (24) that $q_{b_0}(t) = 1$ and

$$\eta_0(x,t) \triangleq \eta_0 = -Q^{-1}B'R(t)x. \tag{25}$$

It is to be noted that the same $\eta_0(x,t)$ minimizes the maximum value of $F(V,x,\eta,q,t)$, allowing all the uncertain parameters to take on their maximizing values. From the corollary, a necessary and sufficient condition for (23) to be satisfied with $R(t)$ and some $\eta(x,t)$ is that (23) is satisfied with $R(t)$ and $\eta_0(x,t)$. Substituting (25) into (23) gives[1]

$$-\frac{dR}{dt} \geq RA(q) + A'(q)R + P - RBQ^{-1}B'R \tag{26}$$

for all $q \in \Omega$. From (18), (26) is implied by

$$-\frac{dR}{dt} = P - RBQ^{-1}B'R + RA_0 + A_0'R + \mathcal{U}(R,A) \tag{27}$$

where $\mathcal{U}(R, A)$ is an upper bound of

$$\sum_{i=1}^{i=n'} q_i(RA_i + A_i'R)$$

in the sense that for all x and all $q_i \in [-1, 1]$

$$x'\mathcal{U}(R, A)x \geq x' \left[\sum_{i=1}^{i=n'} q_i(RA_i + A_i'R) \right] x. \tag{28}$$

Equation (27) is the well-known Riccati equation with the additional term $\mathcal{U}(R, A)$. One convenient form of $\mathcal{U}(R, A)$ is obtained as follows. Let S_i denote the orthonormal transformation that diagonalizes the symmetrical matrix $[RA_i + A_i'R]$:

$$S_i'(RA_i + A_i'R)S_i = \Lambda_i \tag{29}$$

where Λ_i is a diagonal matrix containing eigenvalues of $[RA_i + A_i'R]$. Then

$$RA_i + A_i'R = S_i \Lambda_i S_i'$$

and (28) is satisfied by a matrix $\mathcal{U}$ of the form

$$\mathcal{U}(R, A) = \sum_{i=1}^{i=n'} S_i |\Lambda_i| S_i' \tag{30}$$

where $|\Lambda_i|$ is a diagonal matrix with elements taking absolute values of elements in Λ_i. Substituting (30) into (27) gives

[1] For symmetric matrices A, B, $A > B$ ($A \geq B$) means that matrix $A - B$ is positive definite (semidefinite).

$$-\frac{dR}{dt} = P - RBQ^{-1}B'R + RA_0 + A_0'R + \sum_{i=1}^{i=n'} S_i|\Lambda_i|S_i'. \tag{31}$$

Equation (31) can be integrated backwards numerically with the initial condition

$$R(T) = H. \tag{32}$$

In control problems with $T \to \infty$, the matrix R is then a constant independent of T. It can be evaluated by a computer program quite similar to the RIAS Automatic Synthesis Program as developed by Kalman. A flow chart of the modified program is given in Fig. 1. Two problems then arise.

1) Under what condition is the reiterative process of Fig. 1 convergent?

2) Is the closed-loop system stable?

Because of the nonlinear, nonmonotone nature of the mapping $\mathfrak{U}$ in (30) (as a function of R when the A_i are given), the convergence proof is very difficult. But after careful study, the following was found true.

1) The integration of (31) is unique, symmetric, positive semidefinite, and has no finite escape time (see [14]).

2) If $[A(q), B]$ is of phase variable canonical form, $n' = 1$, and if $[A_0, D]$ is observable, where D is defined as

$$D'D \triangleq P, \tag{33}$$

then (31) has a stationary solution that is positive definite and symmetric (see [14]).

3) Assume that $[A_0, B]$ is controllable. The integration of (31) is bounded as $t \to -\infty$ if

$$\sum_{i=1}^{n'} \|A_i\| < \frac{1}{2}\left\{\inf_K \left\| \int_0^\infty \exp(A_0 + BK)'t \cdot \exp(A_0 + BK)t\, dt \right\|\right\}^{-1} \tag{34}$$

where $\|\cdot\|$ means the Euclidean norm. The proof is in Appendix I.

4) If there is a constant vector $q \in \Omega$ such that $[A(q), B]$ is not *stabilizable* (some unstable modes of $A(q)$ are not controllable, see [17]), but $[A(q), D]$ is *detectable*, then the integration of (31) goes to infinity at $t \to -\infty$.

The stability of the closed-loop uncertain system is, in general, decided after a stationary solution R of (31) is obtained. If $R > 0$ (positive definite), then the closed-loop system is stable in the sense of Lyapunov for any q variation. The asymptotic behavior is determined by Theorem 2. With $R > 0$ it is proved in [16] that all trajectories converge to the largest invariant set M (in the wide sense), and that set M is the origin if and only if the time-varying system $[A(q(t)) + BK, \bar{D}]$ is observable for any variation of $q(t)$, where $K = -Q^{-1}B'R$ and $\bar{D}$ is such that $\bar{D}'\bar{D} \triangleq D'D + K'QK$. Examples of closed-loop uncertain systems with this kind of observability can be found in [16] and in [14, lemma 3.1]. When $R > 0$ and M is the origin, the closed-loop system is globally asymptotically stable for any allowed parameter variation.

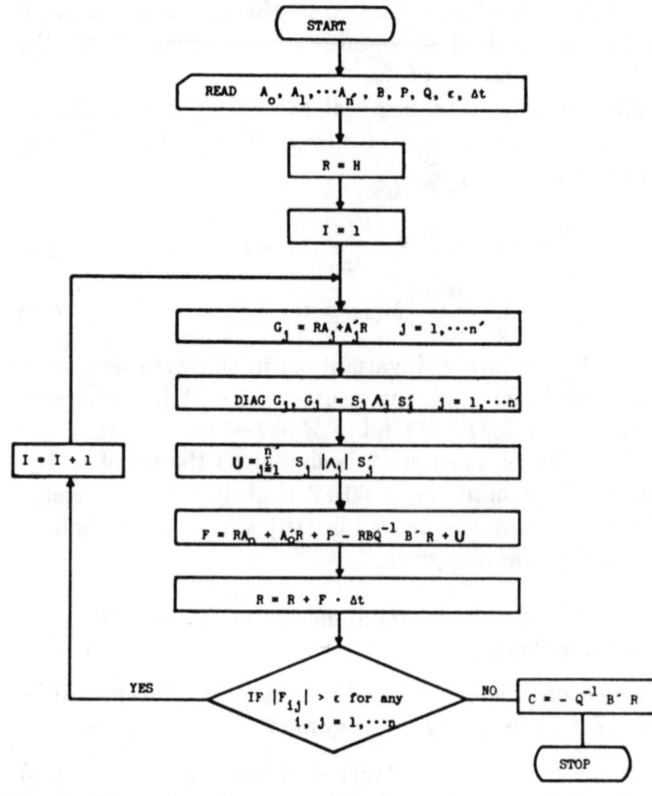

Fig. 1. Flow chart for computing the stationary solution R of (31) and the feedback gain C.

To illustrate the procedure and usefulness of the guaranteed cost approach, two examples are given.

Example 1: For the problem defined in (16)–(21), let $T \to \infty$, $n' = 1$,

$$A_0 = \begin{bmatrix} 0 & 1 & 0 \\ 0 & 0 & 1 \\ 0 & 0 & 0 \end{bmatrix}, \quad A_1 = \begin{bmatrix} 0 & 0 & 0 \\ 0 & 0 & 0 \\ 0 & 0 & 1.6 \end{bmatrix}, \quad B = \begin{bmatrix} 0 \\ 0 \\ 1 \end{bmatrix},$$

$$D = [1 \;\; 0 \;\; 0], \quad Q = 1,$$

$$-1 \le q_1(t) \le 1, \quad 1 \le q_b(t) \le 2.$$

The optimal cost matrix for the nominal system with $q_1(t) \equiv 0$, $q_b(t) \equiv 1$ is

$$R_0 = \begin{bmatrix} 2 & 2 & 1 \\ 2 & 3 & 2 \\ 1 & 2 & 2 \end{bmatrix}.$$

Using the nominal optimal control $K_0 \triangleq -Q^{-1}B'R_0$, the closed-loop system, when $q_b(t) \equiv 1$, is

$$A(q) + BK_0 = \begin{bmatrix} 0 & 1 & 0 \\ 0 & 0 & 1 \\ -1 & -2 & -2 + 1.6q_1(t) \end{bmatrix}$$

which is unstable when $q_1(t) \equiv 1$, as can be easily checked with the Hurwitz method, and the cost becomes infinity. However, if the procedure in Fig. 1 is used, we get a guaranteed cost matrix

$$R = \begin{bmatrix} 2.83 & 4.00 & 1.17 \\ 4.00 & 10.1 & 3.30 \\ 1.17 & 3.30 & 4.69 \end{bmatrix}.$$

Using guaranteed cost control $K = -Q^{-1}B'R$, the closed-loop system is globally asymptotically stable for any variation of q. The stability proof is the same as in the example in [16].

The following example shows that a nominal optimal control (or any stable control) is also a guaranteed cost control whenever the uncertainty is bounded in norm by a finite number. In this case, the closed-loop system is still asymptotically stable for any parameter variation.

Example 2: Let $A(q(t)) = A_0 + A_1(q(t))$ and $B(q(t)) = B$ in (16) and write $P = D'D$ in (17). Assume the nominal system $[A_0,B,D]$ to be controllable and observable so that the closed-loop system $A_0 + BK_0$ is stable (poles in left-half plane), where $K_0 \triangleq Q^{-1}B'R_0$ is the optimal control and $R_0 > 0$ is the optimal cost matrix for the nominal problem. If for all $q \in \Omega$

$$\|A_1(q)\| < \tfrac{1}{2}\,\|R_0 + R_I\|^{-1} \tag{35}$$

where R_I is calculated from

$$R_I(A_0 + BK_0) + (A_0 + BK_0)'R_I + I = 0, \tag{36}$$

then K_0 is also a guaranteed cost control for the uncertain system with $q(t) \in \Omega$, and the closed-loop system $A(q(t)) + BK_0$ is globally asymptotically stable for any allowed variations in $q(t)$. In this case, $R_0 + R_I$ is a guaranteed cost matrix. The proof is in Appendix II. The same analysis can be used to obtain some further theoretical results.

1) If $\dot{x} = Ax$ is stable (poles in left-half plane), the uncertain system $\dot{x} = [A + A_1(q(t))]x$, with $A_1(q)$ continuous and $q(t) \in \Omega$, is globally asymptotically stable for any variation in q if, for all $q \in \Omega$,

$$\|A_1(q)\| < \tfrac{1}{2}\,\|R_2\|^{-1} \tag{37}$$

where R_2 satisfies $R_2A + A'R_2 + I = 0$. This is because (37) implies $R_2(A + A_1(q)) + (A + A_1(q))'R_2 < 0$. By continuity of A_1 there exists $\delta > 0$ such that

$$R_2(A + A_1(q)) + (A + A_1(q))'R_2 \leq -\delta I < 0. \tag{38}$$

The result follows by taking $x'R_2x$ as a Lyapunov function.

2) If in (3) we allow $V(x,t) = \rho V_0(x,t)$, $\eta(x,t) = \eta_0(x,t)$, where $\rho > 1$, V_0, η_0 are the optimal return function and optimal control for the nominal system, and if we assume $g(x, u, q, t) = g(x, u, t)$, we obtain [18, condition (11)] as a special case. A weakness in [18] is in its (12) where matrix Q (corresponding to the matrix P in (17) of this paper) has to be nonsingular, which is rarely the case; otherwise, [18, condition (17)] may be vacuous. With guaranteed cost approach available, condition (17) can be written in a stronger form that allows Q to be singular. The detail is left in Appendix III.

Adaptive Guaranteed Cost Control of an Airframe

Description

When the ranges of the uncertain parameters are reduced by measurements or by *a priori* knowledge on their dependence on some readily measurable prominent parameter, only the reduced ranges are used in determining the guaranteed cost control law. The reduced ranges, and consequently the control law, vary from time to time and the control system becomes adaptive. The principle will now be discussed in the context of pitch control of an airframe.

On a conventional auto pilot, the closed-loop natural frequency and damping factor are kept within a narrow range known as the CAL Thumbprint. But one wonders, with on-line computers available, whether the CAL Thumbprint should still be our design goal. The speed of response of an airframe is a sharply rising function of its dynamic pressure. With essentially the same closed-loop ω and ζ at all flight conditions, the system is responding at its lowest capability corresponding to the lowest dynamic pressure. Yet in an emergency, the highest speed of response is called for. It seems that the augmented airframe should have ζ in the normally acceptable range of 0.6 to 1.0 but an ω as high as possible, consistent with structural limits and without causing excessive saturation. An approach in between time history computations (that would require an exceedingly fast computer) and the CAL Thumbprint appears desirable.

An on-line control system based on this concept is illustrated in Fig. 2. A fast control loop is represented by the heavy signal flow lines. At any time there are two possible modes of control: a slow mode with control law I conforming to the CAL Thumbprint, and a fast mode with control law II. The selection of one of the control modes is done by the pilot pushing a button. The control law is realized either by analog or digital means with a very high sampling rate at least an order of magnitude higher than the highest augmented frequency. An adaptive control loop is represented by the broken signal flow lines and is sampled at a considerably slower rate. From observed variables and other significant parameters that provide information on the system dynamics (e.g., the dynamic pressure), the computer evaluates the parameters of the state variable equation, computes both control laws I and II, and modifies the control laws in such a manner as not to cause transient disturbance.

An alternative approach that reduces the on-line computing requirements considerably is illustrated in Fig. 3. The prior knowledge of system parameters as a function of dynamic pressure is used to calculate the coefficients in the control laws. These coefficients are stored as a table in the on-line computer with overlapping ranges of dynamic pressure as the independent variable. As dynamic pressure varies, the precalculated coefficients are transferred to the processor for implementing the control law.

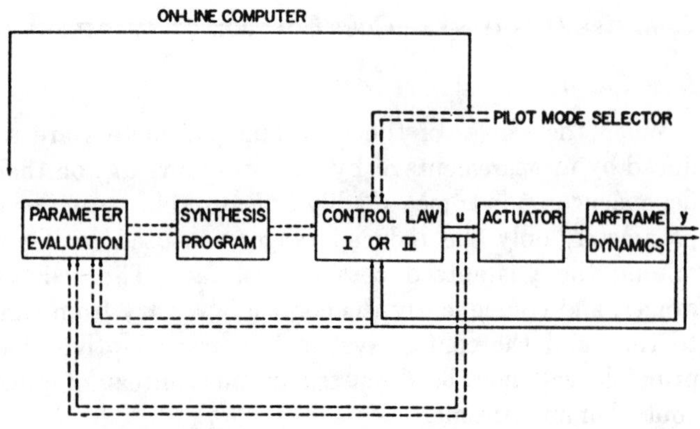

Fig. 2. On-line control of short period dynamics.

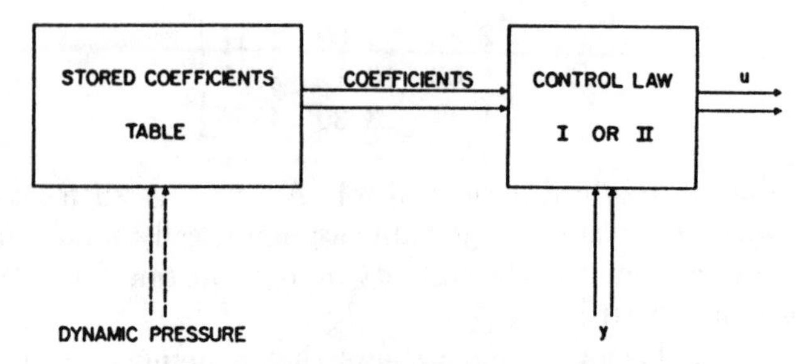

Fig. 3. Reduced on-line requirements.

In terms of required computer capacity, both Figs. 2 and 3 have the advantage that relatively little computation is required for the fast control loop. A linear control law can be executed by a modern computer in a matter of microseconds. While the computations required for the adaptive loop are involved, the time that can be allowed for such computations is of the order of a few seconds or more, since the control law rarely needs to be updated more frequently than that.

Synthesis of Control Law for Fast Mode

As a numerical example, the flow chart of Fig. 1 is applied to synthesize the control law for the fast mode. The coupling between pitch and lateral dynamics is quite negligible [15] and the two sets of state variables can be independently controlled.

1) Pitch Control: The longitudinal perturbations of an airplane in horizontal cruising flight are described in [15]:

$$\dot{\alpha} = -\frac{1}{\tau}\alpha + q \tag{39}$$

$$\dot{q} = -\omega_0^2(\alpha - Q\delta) \tag{40}$$

$$\dot{\theta} = q \tag{41}$$

$$J = \int_0^\infty \left\{\frac{\delta^2}{\delta_0^2} + \frac{(\theta - \alpha)^2}{\gamma_0^2}\right\} dt \tag{42}$$

where

α Perturbation from cruise angle of attack.

θ Perturbation from cruise pitch angle of the zero-lift axis.

$\omega_0 = \left(-\frac{M_\alpha}{I}\right)^{1/2}$ Undamped pitch natural frequency.

$Q = -\frac{M_\delta}{M_\alpha}$ Elevator effectiveness.

$\tau = \frac{MV}{L_\alpha}$ Lifting time constant.

δ Elevator deflection.

The constants M_α, M_δ, and L_α are approximately proportional to the dynamic pressure ρV^2. Let ρ_1, V_1 denote air density and forward velocity under some nominal cruising condition and let M_{α_1} denote the pitch moment to angle of attack coefficient under the same condition. Let ω_d denote the computed frequency

$$\omega_d \triangleq \left(-\frac{M_{\alpha_1}}{I}\cdot\frac{\rho V^2}{\rho_1 V_1^2}\right)^{1/2} \tag{43}$$

and t', q' the normalized time and pitch rate

$$t' \triangleq \omega_d t, \quad q' = q/\omega_d. \tag{44}$$

Equations (39)–(41) are replaced by

$$\frac{d\alpha}{dt'} = -\frac{1}{\omega_d\tau}\alpha + q' \tag{45}$$

$$\frac{dq'}{dt'} = -\left(\frac{\omega_0}{\omega_d}\right)^2(\alpha - Q\delta) \tag{46}$$

$$\frac{d\theta}{dt'} = q'. \tag{47}$$

The constants $(\omega_0/\omega_d)^2$ and Q are now approximately independent of the dynamic pressure, while $1/\omega_d\tau$ is proportional to $\sqrt{\rho}$. The ranges of variation of all these parameters are reduced. In (42), t is changed to t' and nothing else needs to be altered. Assuming the ranges of variation of: a) $1/\omega_d\tau$, b) $(\omega_0/\omega_d)^2$, c) $(\omega_0/\omega_d)^2 Q$, one then uses the flow chart in Fig. 1 to compute the feedback matrix $C = [C_1, C_2, C_3]$.

The resulting control law is

$$\delta = C_1\alpha + C_2 q' + C_3\theta$$
$$= C_1\alpha + \left(\frac{C_2}{\omega_d}\right)q + C_3\theta. \tag{48}$$

It can be realized by a very simple version of Fig. 3, as shown in Fig. 4.

In the present example, $\delta_0 = 1$, $\gamma_0 = 0.1$

$$0.06 \le \frac{1}{\omega_d\tau} \le 0.20$$

$$0.8 \le \left(\frac{\omega_0}{\omega_d}\right)^2 \le 1.20$$

$$0.4 \le \left(\frac{\omega_0}{\omega_d}\right)^2 Q \le 0.6.$$

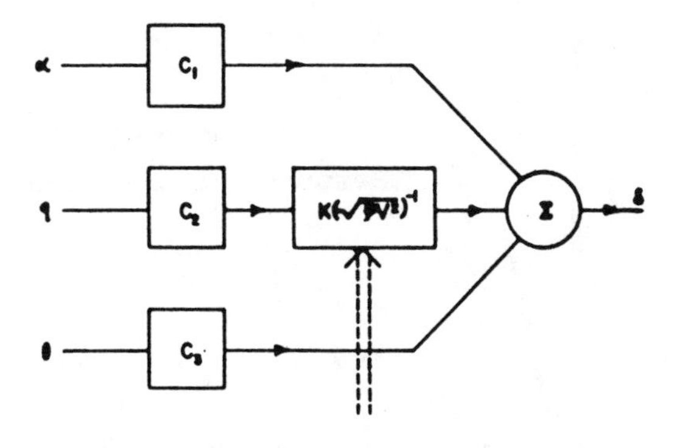

Fig. 4. On-line pitch control.

The computed guaranteed cost control (GCC) coefficients are

$$C_1 = 4.09, \quad C_2 = -6.59, \quad C_3 = -12.5.$$

Meanwhile the nominal optimal control (NOC) coefficients are

$$C_1 = 9.178, \quad C_2 = -1.824, \quad C_3 = -10.00.$$

Two examples of the pole locations of system (45)-(48) are shown in Fig. 5 as a function of different parameter settings: q_1, q_2, q_b represent the normalized variations of $1/\omega_{dT}$, $(\omega_0/\omega_d)^2$, $(\omega_0/\omega_d)^2 Q$, respectively, and 0, -1, $+1$ correspond to the nominal, the smallest and the largest values in the range of variation. For example, $(q_1, q_2, q_b) = (-1, +1, 0)$ means $(1/\omega_{dT}, (\omega_0/\omega_d)^2, (\omega_0/\omega_d)^2 Q) = (0.06, 1.20, 0.5)$. The locations of the complex poles show that the systems with guaranteed cost control are very stable under all operating conditions, and the locations of the real roots show that the same systems have longer response time. Although the left-half plane poles only imply stability when parameters are constant, it is clear from theory that the system with guaranteed cost control is stable even when the parameters are time varying. This is, in general, not true for nominal optimal systems.

2) Lateral Control: In this example of lateral control, our purpose is only to compare the sensitivity of guaranteed cost control and nominal optimal control. For this purpose, the adaptive method in the previous example for pitch control is not used here. The lateral perturbation of an airframe in cruising flight is a fifth-order system [15, p. 173].

$$\dot{\beta} + r = \frac{Y_\beta}{mV}\beta + \frac{g}{V}\phi \tag{49}$$

$$\dot{r} + \frac{Ixz}{Izz}\dot{p} = \frac{n_\beta}{Izz}\beta + \frac{n_r}{Izz}r + \frac{n_p}{Izz}p + \frac{n_{\delta r}}{Izz}\delta r \tag{50}$$

$$\dot{p} + \frac{Ixz}{Ixx}\dot{r} = \frac{l_p}{Ixx}p + \frac{l_\beta}{Ixx}\beta + \frac{l_r}{Ixx}r + \frac{l_{\delta a}}{Ixx}\delta a \tag{51}$$

$$\dot{\phi} = p \tag{52}$$

$$\dot{\psi} = r \tag{53}$$

where

- β Sideslip angle.
- ψ Yaw angle.

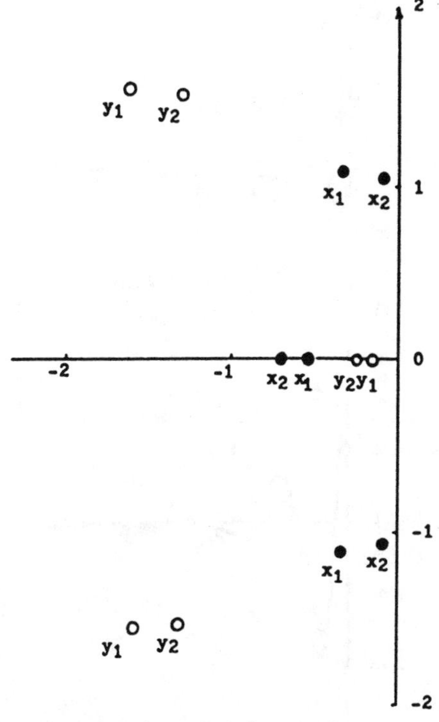

Fig. 5. Closed-loop locations of pitch control. x_1, x_2 represent NOC poles; y_1, y_2 represent GCC poles. Situation 1: $(q_1, q_2, q_b) = (0, 0, 0)$; situation 2: $(q_1, q_2, q_b) = (1, -1, -1)$.

- r Yaw angular velocity.
- ϕ Roll angle.
- p Roll angular velocity.
- δr Rudder deflection.
- δa Aileron deflection.

Use a quadratic cost criteria

$$J = \lim_{T\to\infty} \int_0^T \left(\frac{\delta a^2}{\delta a_0{}^2} + \frac{\delta r^2}{\delta r_0{}^2} + \frac{(\beta + \psi)^2}{\epsilon_0{}^2} + \frac{\phi^2}{\phi_0{}^2}\right) dt \tag{54}$$

to determine a gain matrix K such that

$$\begin{bmatrix}\delta r \\ \delta a\end{bmatrix} = K \begin{bmatrix}\beta \\ r \\ p \\ \phi \\ \psi\end{bmatrix}. \tag{55}$$

The numerical data in a specific cruising condition were taken from [15]. Let the parameter l_β vary in the range $[0.3\, l_{\beta_0}, 1.7\, l_{\beta_0}]$ with l_{β_0} as nominal value so that $l_\beta(t) = l_{\beta_0}(1 + 0.7\, q(t))$. The guaranteed cost control law was computed according to Fig. 1 and simulated on an analog computer. The results were then compared with that of a nominal optimal system. A typical outcome is shown in Fig. 6 where ϕ is the roll attitude with an initial disturbance of 0.1 in yaw rate and roll rate. The marked trajectories are with guaranteed cost control and the unmarked with nominal optimal control. Fig. 7 shows the effect of time varying parameter on δa, the aileron deflection, for an initial disturbance of 0.1 in all five state coordinates. These two figures exemplify the fact that guaranteed cost control systems are less sensitive to large parameter variations.

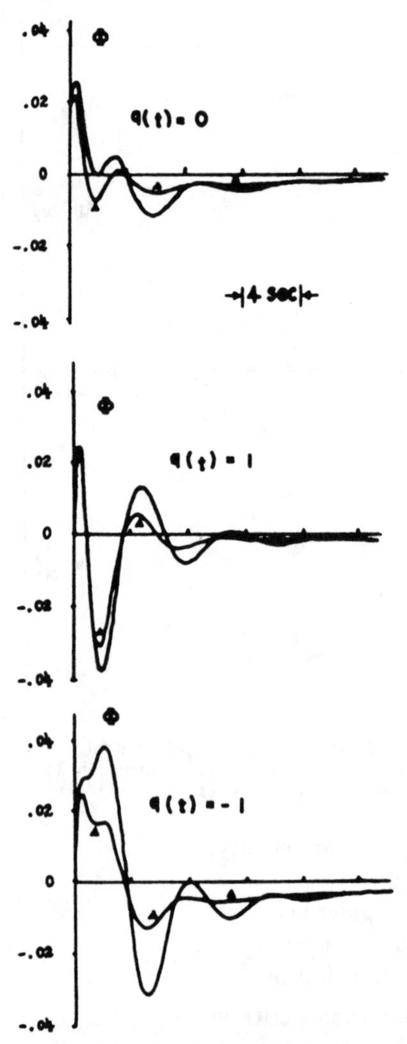

Fig. 6. Analog computer simulation of lateral motions for GCC (with Δ) and for NOC: constant parameters.

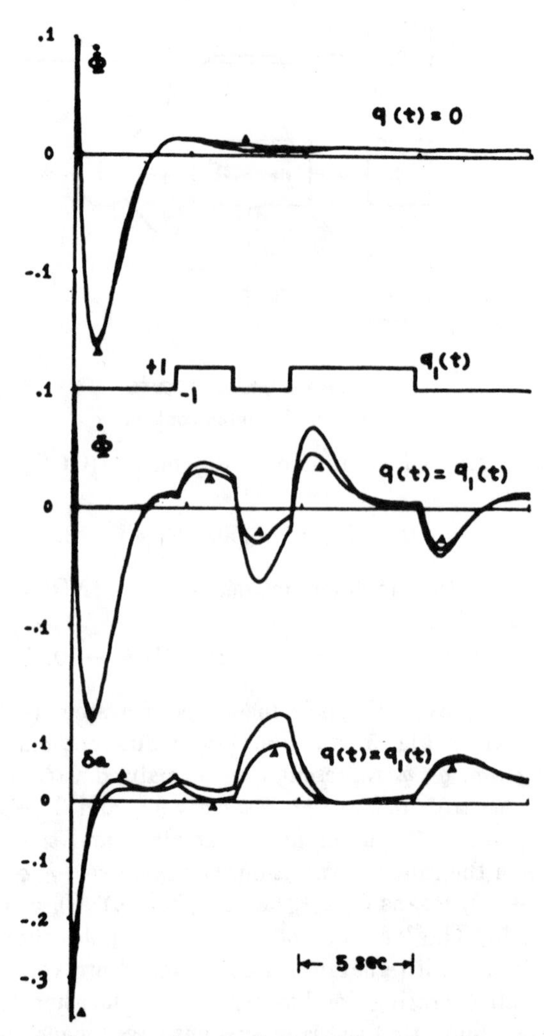

Fig. 7. Analog computer simulation of lateral motions for GCC (with Δ) and for NOC: time varying parameters.

Conclusion

In this paper the idea of guaranteed cost control is described. Its usefulness in stabilizing a system with large parameter variations is demonstrated. A Riccati equation with an added term, arising from the parameter uncertainty, is used to generate the guaranteed cost matrix and control law. A Fortran program is developed to compute these quantities accordingly.

Two types of adaptive controls are discussed. One is the on-line evaluation of system parameters; the other is the storage of the coefficients of the control law as a function of some important and easily updated situation parameters. The latter, combined with guaranteed cost control, is applied to an airframe pitch control system with large parameter variations. The control coefficients are constants, except one, which is modified by the dynamic pressure, an easily evaluated quantity. Pole locations of the closed-loop system show that longitudinal oscillations of the airframe are virtually eliminated. An analog computer plot for an airframe lateral autopilot shows that system response is less sensitive to large parameter variations with guaranteed cost control than a nominally optimum control.

Appendix I

We prove that (34) guarantees boundedness of integration of (31) as $t \to -\infty$.

It is proved in [14] that any finite time solution $R(t)$ of (31) and (32) is bounded by $\bar{R}(t)$, $R(t) \leq \bar{R}(t)$, where $\bar{R}(t)$ satisfies (31) and (32), but with matrix $\mathfrak{U}(R, A)$ in (31) replaced by $N(R, A) \triangleq 2(\sum_{i=1}^{n'} \|A_i\|)\|R\| I$. Notice that $N(R, A) \geq \mathfrak{U}(R, A)$. It is also proved in [14] (using the property that $N(R, A)$ is monotone in R) that $\bar{R}(t)$ is the minimum among solutions of the following equation for all K:

$$\frac{dR_K}{dt} + R_K(A_0 + BK) + (A_0 + BK)'R_K + P + K'QK + N(R_K, A) = 0, \qquad R_K(T) = H. \quad (56)$$

In other words, $\bar{R}(t) \leq R_K(t)$ for all $t \leq T$, all K. Choose K so that $A_0 + BK$ is stable. As $t \to -\infty$, the stationary solution of (56), if it exists, can be written as

$$R_K = \Phi_K(R_K) \triangleq \int_0^\infty \exp\,(A_0 + BK)'t[P + K'QK + N(R_K, A)] \exp\,(A_0 + BK)t\, dt. \quad (57)$$

Now $\Phi_K(R)$ is a *contraction mapping* for some K if (34) is satisfied, as can be observed from the following inequality:

$$\|\Phi_K(R_1) - \Phi_K(R_2)\|$$

$$\le \|N(R_1, A) - N(R_2, A)\|$$

$$\cdot \left\| \int_0^\infty \exp (A_0 + BK)'t \exp (A_0 + BK)t \, dt \right\|$$

$$\le 2 \sum_{i=1}^{n'} \|A_i\| \cdot \left\| \int_0^\infty \exp (A_0 + BK)'t \exp (A_0 + BK)t \right.$$

$$\left. \cdot dt \right\| \cdot \|R_1 - R_2\|.$$

Therefore (56) has a unique stationary solution R_K for certain K. Since $R(t) \le \bar{R}(t) \le R_K(t)$ for all $t \le T$, $R(t) \le \bar{R}(t) \le R_K$ as $t \to -\infty$.

Appendix II

We prove the assertion in Example 2.

Let $R = R_0 + (1 + \epsilon)R_I$ where $\epsilon > 0$ is any positive number. R will be a guaranteed cost matrix if, for all $q \in \Omega$,

$$R(A_0 + BK_0 + A_1(q)) + (A_0 + BK_0 + A_1(q))'R + D'D + K_0'QK_0 + \epsilon I \le 0. \quad (58)$$

This is true because (58) implies (23) with $dR/dt = 0$. Also, (58) implies asymptotic stability because with $R > 0$ and $D'D + K_0'QK_0 + \epsilon I > 0$, asymptotic stability follows from standard Lyapunov theory. Matrix R_0 satisfies the Riccati equation, which can be written in the form

$$R_0(A_0 + BK_0) + (A_0 + BK_0)'R_0 + D'D + K_0'QK_0 = 0. \quad (59)$$

This and (36), when substituted into (58), give

$$RA_1(q) + A_1'(q)R \le I, \quad (60)$$

which is satisfied if

$$\|A_1(q)\| \le \tfrac{1}{2} \|R_0 + (1 + \epsilon)R_I\|^{-1}. \quad (61)$$

Let $\epsilon \to 0$; then (35) implies (61) for all $q \in \Omega$. [Rigorously, this is true if $\|A_1(q)\|$ attains its maximum on Ω, which is the case when $\|A_1(q)\|$ is upper-semicontinuous in q. If $\|A_1(q)\|$ does not attain its maximum, but only a supremum on Ω, (61) should be used instead of (35).]

Appendix III

This is a stronger result than [18].

Referring to Example 2, let $\rho > 1$ be given. We can find a range of disturbance $\|A_1(q)\|$ so that ρR_0 is a guaranteed cost matrix, K_0 is a guaranteed cost control, and the closed-loop system is asymptotically stable. Let $R_\epsilon \triangleq (1 + \epsilon)R_I$ where $\epsilon > 0$ and R_I is defined in (36) ($R_I > 0$). Define $\beta \triangleq (\rho - 1)\lambda_{\min} \{R_\epsilon^{-1/2}R_0R_\epsilon^{-1/2}\}$, where $\lambda_{\min}\{\cdot\}$ means the minimum eigenvalue. Then $\beta I \le (\rho - 1)R_\epsilon^{-1/2}R_0R_\epsilon^{-1/2}$, and so $R_0 + \beta R_\epsilon \le \rho R_0$. Let

$$\|A_1(q)\| \le \frac{\beta}{2\rho\|R_0\|} \quad (62)$$

for all $q \in \Omega$, and let $R = R_0 + \beta R_\epsilon$. Then from (59)

$$R(A_0 + A_1(q) + BK_0) + (A_0 + A_1(q) + BK_0)'R + D'D + K_0'QK_0 + \beta \epsilon I$$

$$= RA_1(q) + A_1'(q)R - \beta I$$

$$\le (2\|A_1(q)\| \cdot \|R\| - \beta)I$$

$$\le \left(2 \cdot \frac{\beta}{2\rho\|R_0\|} \cdot \rho\|R_0\| - \beta\right)I = 0. \quad (63)$$

Therefore R is a guaranteed cost matrix, K_0 is a guaranteed cost control, and since $\beta\epsilon I > 0$, the closed-loop system is asymptotically stable for any variations in q. Now that $R \le \rho R_0$, ρR_0 is a guaranteed cost matrix by definition. Let $\epsilon \to 0$. Then $R_\epsilon \to R_I$ and (62) is implied by

$$\|A_1(q)\| < \left(\frac{\rho - 1}{2\rho}\right) \cdot \frac{\lambda_{\min}\{R_I^{-1/2}R_0R_I^{-1/2}\}}{\lambda_{\max}\{R_0\}} \quad (64)$$

for all $q \in \Omega$. Equation (64) is used under the same precaution as in the comment following (61); otherwise, (62) should be used. In any case, $\|A_1(q)\|$ always has a finite range as long as $\rho > 1$. In this analysis, the only assumption on P (corresponding to the Q in [18]) is that $[A_0, D]$ is observable, where $D'D \triangleq P$. P does not have to be nonsingular.

References

[1] S. S. L. Chang, "Fuzzy dynamical programming and approximate optimization of partially known systems," in *Proc. 2nd Hawaii Int. Conf. Syst. Sci.*, 1969.

[2] ——, "Fuzzy dynamical programming and decision making process," in *3rd Annu. Princeton Conf. Inform. Sci. Syst.*, 1969.

[3] L. A. Zadeh, "Fuzzy sets," *Inform. Contr.*, vol. 8, pp. 338–353, 1965.

[4] M. Aoki, *Optimization of Stochastic Systems.* New York: Academic, 1967.

[5] S. S. L. Chang, *Synthesis of Optimum Control Systems.* New York: McGraw-Hill, 1961, ch. 10, 11.

[6] K. S. Fu, "Learning system theory," in *System Theory*, L. A. Zadeh and E. Polak, Eds. New York: McGraw-Hill, 1969, ch. 11.

[7] S. S. L. Chang, "On the relative time of adaptive processes," *IEEE Trans. Automat. Contr.*, vol. AC-10, pp. 11–17, Jan. 1965.

[8] M. Sobral, Jr., "Sensitivity in optimal control systems," *Proc. IEEE*, vol. 56, pp. 1644–1652, Oct. 1968.

[9] R. A. Werner and J. B. Cruz, Jr., "Feedback control which preserves optimality for systems with unknown parameters," *IEEE Trans. Automat. Contr.*, vol. AC-13, pp. 621–629, Dec. 1968.

[10] P. V. Kokotović, J. B. Cruz, Jr., J. E. Heller, and P. Sannuti, "Synthesis of optimally sensitive systems," *Proc. IEEE*, vol. 56, pp. 1318–1324, Aug. 1968.

[11] H. S. Witsenhausen, "A minimax control problem for sampled linear systems," *IEEE Trans. Automat. Contr.*, vol. AC-13, pp. 5–21, Feb. 1968.

[12] D. Salmon, "Minimax controller design," *IEEE Trans. Automat. Contr.*, vol. AC-13, pp. 369–376, Aug. 1968.

[13] J. LaSalle and S. Lefschetz, *Stability by Liapunov's Direct Method with Applications.* New York: Academic, 1961.

[14] T. K. C. Peng, "On guaranteed cost controller and its application to uncertain regulator systems," Ph.D. dissertation, submitted to State Univ. New York at Stony Brook, Stony Brook, Jan. 1971.

[15] A. E. Bryson, Jr., and Y. C. Ho, *Applied Optimal Control.* Waltham, Mass: Blaisdell, 1969.

[16] T. K. C. Peng, "Invariance and stability for bounded uncertain systems," *SIAM J. Contr.*, to be published Nov. 1972.

[17] W. M. Wonham, "On a matrix Riccati equation of stochastic control," *SIAM J. Contr.*, vol. 6, no. 4, 1968.

[18] J. J. Rissanen, "Performance deterioration of optimal systems," *IEEE Trans. Automat. Contr.*, vol. AC-11, pp. 530–532, July 1966.

Sufficiently Informative Functions and the Minimax Feedback Control of Uncertain Dynamic Systems

DIMITRI P. BERTSEKAS AND IAN B. RHODES

***Abstract*—The problem of optimal feedback control of uncertain discrete-time dynamic systems is considered where the uncertain quantities do not have a stochastic description but instead are known to belong to given sets. The problem is converted to a sequential minimax problem and dynamic programming is suggested as a general method for its solution. The notion of a sufficiently informative function, which parallels the notion of a sufficient statistic of stochastic optimal control, is introduced, and conditions under which the optimal controller decomposes into an estimator and an actuator are identified. A limited class of problems for which this decomposition simplifies the computation and implementation of the optimal controller is delineated.**

I. Introduction

THIS PAPER is concerned with the optimal feedback control of a discrete-time dynamic system in the presence of uncertainty. The traditional treatment of this problem has been to assign probability distributions to the uncertain quantities and to formulate the optimization problem as one of minimizing the expected value of a suitable cost functional. In this paper, a nonprobabilistic description of the uncertainty is adopted, where, instead of being modeled as random vectors with given probability distributions, the uncertainties are considered to be unknown except for the fact that they belong to given subsets of appropriate vector spaces. The optimization problem is then cast as one of finding the feedback controller within a prescribed admissible class that minimizes the maximum value (over all possible values of the uncertain quantities) of a suitable cost functional. This worst case approach to the optimal control of uncertain dynamic systems is applicable to problems where a set-membership description of the uncertain quantities is more natural or more readily available than a probabilistic one, or when specified tolerances must be met with certainty.

The modeling of uncertainties as quantities that are unknown except that they belong to prescribed sets and the adoption of a worst case viewpoint in the context of the problem of feedback control of a dynamic system was first considered by Witsenhausen [1], [2] and received further attention in [4]–[10]. In this paper a general minimax feedback control problem which involves a

Manuscript received May 1, 1972; revised October 9, 1972. Paper recommended by D. Sworder, Chairman of the IEEE S-CS Stochastic Control Committee. This work was supported in part by the Air Force Office of Scientific Research under Grants AFOSR-69-1724 and F-44620-69-C-0116 and in part by NASA under Grant NGL-22-009 (124).

D. P. Bertsekas is with the Department of Engineering–Economic Systems, Stanford University, Stanford, Calif. 94305.

I. B. Rhodes is with the Laboratory of Control Systems Science and Engineering, Washington University, St. Louis, Mo. 63130.

Reprinted from *IEEE Trans. Automat. Contr.*, vol. AC-18, no. 2, pp. 117–123, Apr. 1973.

discrete-time dynamic system defined on a Euclidean space is formulated in Section II, and, in Section III, a dynamic programming algorithm is given for its solution. This algorithm is similar to one given earlier by Witsenhausen [1], but is somewhat more detailed and explicit than and somewhat different in its form from that in [1], although the same basic ideas are involved. Subsequently, in Section IV, an effort is made to identify conditions under which this algorithm can be simplified, and to deduce structural properties of the optimal controller. This is accomplished by introducing the notion of a sufficiently informative function, in analogy with the familiar notion of a sufficient statistic of stochastic optimal control. It is proved in Section IV that the decomposition of the optimal controller into an estimator and an actuator is possible. Finally, in Section V some special cases for which this decomposition is profitable are delineated.

II. Problem Formulation

Problem 1: Given is the discrete-time dynamic system

$$x_{k+1} = f_k(x_k,u_k,w_k), \qquad k = 0,1,\cdots,N-1 \tag{1}$$

where $x_k \in R^n$, $k = 0,1,\cdots,N$, is the state vector; $u_k \in R^m$, $k = 0,1,\cdots,N-1$, is the control vector; $w_k \in R^r$, $k = 0,1,\cdots,N-1$, is the input disturbance vector; and $f_k: R^n \times R^m \times R^r \to R^n$ is a known function for each $k = 0,1,\cdots,N-1$.

Available to the controller are measurements of the form

$$z_k = h_k(x_k,v_k), \qquad k = 1,2,\cdots,N-1 \tag{2}$$

where $z_k \in R^s$, $k = 1,2,\cdots,N-1$, is the measurement vector; $v_k \in R^p$, $k = 1,2,\cdots,N-1$, is the measurement noise vector; and $h_k: R^n \times R^p \to R^s$ is a known function for each $k = 0,1,\cdots,N-1$.

The uncertain quantities lumped in a vector $q \in R^{n+Nr+(N-1)p}$

$$q = (x_0',w_0',w_1',\cdots,w_{N-1}',v_1',v_2',\cdots,v_{N-1}')' \tag{3}$$

are known to belong to a given subset Q of $R^{n+Nr+(N-1)p}$.

Attention is restricted to control laws of the form

$$\mu_k: R^{k(s+m)} \to R^m, \qquad k = 0,1,\cdots,N-1 \tag{4}$$

taking values

$$u^k = \mu_k(z_1,z_2,\cdots,z_k,u_0,u_1,\cdots,u_{k-1}), \qquad k = 0,1,\cdots,N-1. \tag{5}$$

Because the control at time 0 depends only on *a priori* data, μ_0 may be interpreted as a constant vector.

It is required to find (if it exists) the control law in this class for which the cost functional

$$J(\mu_0,\mu_1,\cdots,\mu_{N-1}) = \sup_{q\in Q} F[x_1,x_2,\cdots,x_N,\mu_0,\mu_1(z_1,u_0), \cdots,\mu_{N-1}(z_1,\cdots,u_{N-1})] \tag{6}$$

is minimized subject to the system equation constraint, where $F: R^{N(n+m)} \to (-\infty,+\infty]$ is a given function.

It should be noted that this problem formulation implicitly includes the possible presence of state and control constraints, since we allow the function F to take the value $+\infty$. We need simply specify that F take the value $+\infty$ whenever some constraint is violated. Thus, for example, state and control constraints of the form $x_k \in X_k$ or $u_{k-1} \in U_{k-1}$, where X_k and U_{k-1}, $k = 1,2,\cdots,N$, are given sets, may be accounted for by additively including in F the function

$$\sum_{i=1}^{N} \{\delta(x_i|X_i) + \delta[\mu_{i-1}(z_1,\cdots,u_{i-2})|U_{i-1}]\}$$

where $(y|Y)$ denotes the indicator function of a set Y, viz.,

$$\delta(y|Y) = \begin{cases} 0, & y \in Y \\ \infty, & y \notin Y. \end{cases} \tag{7}$$

In the next section we present a dynamic programming algorithm for the solution of Problem 1. Using this algorithm we will then be able to reach some conclusions concerning the structure of the optimal control law.

III. Solution by Dynamic Programming

Consider the optimal value of the cost function (6)

$$\bar{J} = \inf_{\mu_k} \sup_{q\in Q} F(x_1,x_2,\cdots,x_N,u_0,u_1,\cdots,u_{N-1}), \qquad k = 0,1,\cdots,N-1. \tag{8}$$

The purpose of the dynamic programming algorithm is to convert the minimization problem indicated in the above equation to a sequence of simpler minimization problems by taking advantage of the sequential evolution of the system state and the information available to the controller according to (1) and (2). However, matters are somewhat complicated in the above problem by the presence of uncertainty, since in the process of generating the state and measurement vectors the disturbances are intermediately selected by, say, Nature with the objective of maximizing the value of the cost. For this reason the development of the dynamic programming algorithm requires a somewhat elaborate construction.

In order to simplify the notation we will make use of the vector $\zeta_k \in R^{k(s+m)}$, $k = 1,2,\cdots,N-1$, which consists of all the information available to the controller at time k, viz.,

$$\zeta_k = (z_1',z_2',\cdots,z_k',u_0',u_1',\cdots,u_{k-1}')'. \tag{9}$$

With this notation we write for the control law

$$\mu_k(z_1,z_2,\cdots,z_k,u_0,u_1,\cdots,u_{k-1}) = \mu_k(\zeta_k) = u_k. \tag{10}$$

Now consider the following definitions. Let $P(R^s)$ be the power set (the set of all subsets) of R^s and consider the following function:

$$\hat{Z}_k: R^{(k-1)s+km} \to P(R^s), \tag{11}$$

which assigns to the vectors ζ_{k-1},u_{k-1} the set $\hat{Z}_k(\zeta_{k-1}, u_{k-1}) \subset R^s$ of all measurement vectors z_k given by (2) which are consistent with the constraint set Q, the previous measurement vectors $z_1,z_2,\cdots,z_{k-1}$, and the previous control vectors $u_0,u_1,\cdots,u_{k-1}$. In other words, $z_k \in \hat{Z}_k(\zeta_{k-1},u_{k-1})$ if and only if there exists a vector $q = (x_0',w_0',w_1',\cdots,w_{N-1}',v_1',v_2',\cdots,v_{N-1}')' \in Q$ such that the vectors $x_0,w_0,\cdots,w_{k-1},v_1,\cdots,v_k,z_1,\cdots,z_k,u_0,\cdots,u_{k-1}$ together satisfy the system and measurement equations (1) and (2) for times $0,1,\cdots,k$.

We also define the function

$$\hat{Q}: R^{(N-1)s+Nm} \rightarrow P(R^{n+Nr+(N-1)p}), \tag{12}$$

which assigns to the vectors ζ_{N-1},u_{N-1} the set $\hat{Q}(\zeta_{N-1},u_{N-1}) \subset R^{n+Nr+(N-1)p}$ of all vectors $q \in Q$ [recall (3)] which are consistent with the measurements $z_1,z_2,\cdots,z_{N-1}$ and the control vectors $u_0,u_1,\cdots,u_{N-1}$. In other words a vector q belongs to the set $\hat{Q}(\zeta_{N-1},u_{N-1})$ if and only if $q \in Q$ and the vectors $x_0,w_0,\cdots,w_{N-1},v_1,\cdots,v_{N-1},z_1,\cdots,z_{N-1},u_0,\cdots,u_{N-1}$ together satisfy the system and measurement equations (1) and (2) for all k.

It should be noted that for some vectors ζ_{k-1} it is possible that the set $\hat{Z}_k(\zeta_{k-1},u_{k-1})$ or the set $\hat{Q}(\zeta_{N-1},u_{N-1})$ is empty for all $u_{k-1} \in R^m$, implying that the vector ζ_{k-1} is inconsistent with the constraint set Q and the system and measurement equations. Notice also that whether the set $\hat{Z}_k(\zeta_{k-1},u_{k-1})$ is empty or nonempty depends on the vector ζ_{k-1} alone and is entirely independent of u_{k-1}. From (2) the consistency of a vector ζ_k is equivalent to the existence of vectors x_k,v_k consistent with ζ_{k-1} and the constraint $q \in Q$ (for all u_{k-1}) and therefore is equivalent to the existence of a vector $z_k \in \hat{Z}_k(\zeta_{k-1},u_{k-1})$. In subsequent equations in which empty sets appear we will adopt the convention that the supremum of the empty set is $-\infty$ ($\sup \phi = -\infty$). Another possible approach would be to restrict the domain of definition of the functions $\hat{Z}_k,\hat{Q}$ to include only those vectors ζ_{k-1} for which the sets $\hat{Z}_k(\zeta_{k-1},u_{k-1})$, $\hat{Q}(\zeta_{N-1},u_{N-1})$ are nonempty. Since in any actual operation of the system these sets will always be nonempty, this restriction results in no loss of generality.

We are now ready to state and prove the following dynamic programming algorithm for the solution of Problem 1.

Proposition 1: Assume that for the functions H_k defined below we have $-\infty < H_k(\zeta_k)$, $k = 1,2,\cdots,N-2$, for all vectors ζ_k such that the set $\hat{Z}_{k+1}(\zeta_k,u_k)$ is nonempty (for all $u_k \in R^m$), and $-\infty < H_{N-1}(\zeta_{N-1})$ for all vectors ζ_{N-1} such that the set $\hat{Q}(\zeta_{N-1},u_{N-1})$ is nonempty. Then the optimal value $\bar{J}$ of the cost functional (6) is given by

$$\bar{J} = \inf_{u_0} E_1(u_0) \tag{13}$$

where the function $E_1: R^m \rightarrow (-\infty, +\infty]$ is given by the last step of the recursive algorithm

$$E_N(\zeta_{N-1},u_{N-1}) = \sup_{q \in \hat{Q}(\zeta_{N-1},u_{N-1})} F(x_1,x_2,\cdots,x_N,u_0,u_1,\cdots,u_{N-1}) \tag{14}$$

$$H_k(\zeta_k) = \inf_{u_k} E_{k+1}(\zeta_k,u_k), \qquad k = 1,2,\cdots,N-1 \tag{15}$$

$$E_{k+1}(\zeta_k,u_k) = \sup_{z_{k+1} \in \hat{Z}_{k+1}(\zeta_k,u_k)} H_{k+1}(\zeta_k,u_k,z_{k+1}) = \sup_{z_{k+1} \in \hat{Z}_{k+1}(\zeta_k,u_k)} H_{k+1}(\zeta_{k+1}), \qquad k = 0,1,\cdots,N-2. \tag{16}$$

The above proposition will not be proved here. Its proof involves standard but lengthy dynamic programming arguments and can be found in [4].

The dynamic programming algorithm of Proposition 1 can be profitably interpreted in terms of game theory, and in particular in terms of multistage games of perfect information [13]. The optimal value of the cost $\bar{J}$ can be viewed as the upper value (or min-max) of a game played by two opponents, the Controller selecting the control law $(\mu_0,\mu_1,\cdots,\mu_{N-1})$, and Nature selecting the uncertain quantities q from the set Q. The information, based on which the decision of the Controller is made, is fixed by the form of the functions μ_k, i.e., by the information vectors ζ_k. Since, however, only the upper value of the game is of interest here, a variety of equivalent methods of selections of the vector q and corresponding information patterns can be assigned to Nature. One such information pattern and method for selection of the components of the vector q corresponds to the following sequence of events: 1) Controller selects u_0; 2) Nature selects z_1 from the set $\hat{Z}_1(u_0)$; 3) Controller selects u_1; 4) Nature selects z_2 from the set $\hat{Z}_2(z_1,u_0,u_1),\cdots$, $2N-1$) Controller selects u_{N-1}; $2N$) Nature selects all the uncertain quantities $q = (x_0',w_0',w_1',\cdots,w_{N-1}',v_1',v_2',\cdots,v_{N-1}')'$ from the set $\hat{Q}(\zeta_{N-1},u_{N-1})$. Each selection by either Controller or Nature is made with full knowledge of the outcomes of previous selections.

This sequence of events is fictitious; however, it accurately reflects the sequence of events as viewed by the Controller whose only information concerning the course of the game at time k is the information ζ_k, i.e., all measurements and all control selections up to that time.

A moment's reflection shows that in fact the dynamic programming algorithm determines the (pure) value $\bar{J}$ of the game of perfect information described above. This value is the same as the optimal cost $\bar{J}$ of the Problem 1.

Finding the optimal cost $\bar{J}$ and the optimal control law from the dynamic programming algorithm of Proposition 1 is in general a very difficult task. Part of the difficulty stems from the fact that, loosely speaking, the objective of the Controller is dual in nature: first, to actuate the system in a favorable fashion, and, second, to try to improve the quality of his estimate of the uncertainty in the system. This is a familiar situation from stochastic optimal control, known as dual control problem [11], the formidable complexities of which have been widely discussed in the literature. In stochastic optimal control, insight into the structure of the optimal controller, and its dual function, can be obtained through the notion of a sufficient statistic [12]–[14]. Similar insight will be obtained for the minimax controller of this chapter by introducing in the next section the analogous concept of a sufficiently informative function.

IV. Sufficiently Informative Functions

Let us consider the following definition.

Definition 1: A function $\mathcal{S}_k$: $R^{k(s+m)} \to \Sigma_k$, where Σ_k is some space, will be called *sufficiently informative* with respect to Problem 1 if there exists a function $\bar{E}_{k+1}$: $\Sigma_k \times R^m \to (-\infty, +\infty]$ for all $k = 0,1,\cdots,N-1$ such that

$$\bar{E}_{k+1}[\mathcal{S}_k(\zeta_k),u_k] = E_{k+1}(\zeta_k,u_k) \tag{17}$$

where E_{k+1} is the function defined in (14) and (16) for $k = 0,1,\cdots,N-1$. The value of a sufficiently informative function at any point will be called *sufficient information.*

The clear consequence of the above definition is that, if $\mathcal{S}_k$ is a sufficiently informative function, then (15) may be rewritten as

$$\bar{H}_k(\mathcal{S}_k(\zeta_k)) = H_k(\zeta_k) = \inf_{u_k} \bar{E}_{k+1}(\mathcal{S}_k(\zeta_k),u_k);$$

so that we now seek the infimum over u_k of a function of u_k and the sufficient information $\mathcal{S}_k(\zeta_k)$. If this infimum is attained for all ζ_k, then there exists an optimal control law $\bar{\mu}_k$ that can also be written as

$$\bar{\mu}_k(\zeta_k) = \bar{\mu}_k{}^* \cdot \mathcal{S}_k(\zeta_k) \tag{18}$$

where $\bar{\mu}_k{}^*$ is a suitable function which can be determined by minimizing the function $\bar{E}_{k+1}$ of (17) with respect to u_k. As a result the control at any time need only depend on the sufficient information $\mathcal{S}_k(\zeta_k)$. If this sufficient information can be more easily generated or stored than the information vector ζ_k, and, furthermore, if it is easier to minimize the function $\bar{E}_{k+1}$ over u_k rather than the function E_{k+1}, then it is advantageous to compute and implement the control law in the form of (18).

Factorizations of the optimal control law into the composition of two functions, as in (18), have been widely considered in stochastic optimal control theory, and are commonly referred to as separation theorems whenever the function $\mathcal{S}_k$ can be interpreted as an estimator. In such problems the function $\mathcal{S}_k$ or its value is usually called a sufficient statistic. Particularly simple sufficient statistics have been found for problems involving a linear system, linear measurements, and Gaussian white input and measurement noises [12]. In other problems sufficient statistics of interest take the form of conditional probability distributions conditioned on the information available [12]. Such sufficient statistics imply the factorization of the optimal control law into an estimator $\mathcal{S}_k$ computing the conditional probability distribution of some quantities, which may differ depending on the problem given, and an actuator $\bar{\mu}_k{}^*$ applying a control input to the system. It has been demonstrated [3]–[5] that, in estimation problems which involve a set-membership description of the uncertainty, the set of possible states consistent with the measurements received plays a role analogous to that of conditional probability distributions in stochastic estimation problems. Thus it should not come as a surprise that for Problem 1 we shall be able to derive sufficiently informative functions that involve sets of possible system states (or other quantities) consistent with the measurements received. In what follows we obtain such sufficiently informative functions and further discuss the well-behaved case of a linear system and an energy constraint on the uncertain quantities for which, as was demonstrated in [4] and [7], the set of possible states can be characterized by a finite set of numbers. We first introduce the following notation.

We denote for all k by

$$S_k(x_1,\cdots,x_k,w_k,\cdots,w_{N-1},v_{k+1},\cdots,v_{N-1}|\zeta_k)$$

the subset of $R^{kn+(N-k)r+(N-k-1)p}$ which consists of all vectors $(x_1,\cdots,x_k,w_k,\cdots,w_{N-1},v_{k+1},\cdots,v_{N-1})$ that are consistent with the measurements $z_1,z_2,\cdots,z_k$, the control vectors $u_0,u_1,\cdots,u_{k-1}$, the system and measurement equations (1) and (2) and the constraint $q \in Q$. Similarly, we denote by $S_k(\cdot,\cdot,\cdots,\cdot|\zeta_k)$ the respective sets of all possible quantities within the parentheses that are consistent with the information vector ζ_k, the system and measurement equations, and the constraint $q \in Q$.

With the above notation we have the following proposition.

Proposition 2: A sufficiently informative function with respect to Problem 1 is the function

$$\mathcal{S}_k: R^{k(s+m)} \to P(R^{kn+(N-k)r+(N-k-1)p}) \times R^{km}$$

given for all ζ_k and k by

$$\mathcal{S}_k(\zeta_k) = [S_k(x_1,\cdots,x_k,w_k,\cdots,w_{N-1},v_{k+1},\cdots, v_{N-1}|\zeta_k),u_0,u_1,\cdots,u_{k-1}]. \tag{19}$$

Again, the proof of the above proposition is straightforward but tedious, and will not be presented here. It can be found in [4].

Proposition 2 shows that sufficient information in the case of Problem 1 is provided at each time by the set of past inputs together with the set of past and present states and future uncertainties that are consistent with the observed output sequence up to that time. In addition, Proposition 2 clearly illustrates the dual function of the optimal controller. By (18) the optimal control law is of the form

$$\bar{\mu}_k = \bar{\mu}_k{}^* \cdot \mathcal{S}_k, \tag{20}$$

i.e., it is the composition of the sufficiently informative function $\mathcal{S}_k$ and the function $\bar{\mu}_k{}^*$. The function $\mathcal{S}_k$ may be interpreted as an estimator, and the function $\bar{\mu}_k{}^*$ as an actuator. Alternatively, the optimal controller can be viewed as being composed of two cascaded parts. The first part produces an estimate set and the second part accepts as input this estimate set and produces a control vector. This control vector is stored and recalled in the future by the controller.

By adding additional structure to Problem 1, various important simplifications can be achieved in the sufficiently informative function (19). The additional structure takes the form of further assumptions on the form of the set Q in which the uncertainties lie, and the cost functional

F. The simplifications that these additional assumptions induce in the sufficiently informative function are reductions in the number of entities whose consistency with the past output measurements needs to be considered. For example, under certain conditions there is no need to include the past states $x_1,\cdots,x_{k-1}$ in the set of entities whose consistency with the output data is part of the sufficient information (19). Under other conditions there is no need to retain the future uncertainties in $S(\cdot|\zeta_k)$. In other cases, the controller need not recall past control inputs. Because the verification of each of these simplifications requires only a straightforward specialization of the proof of Proposition 2, we state them as a sequence of Corollaries to Proposition 2.

Corollary 1: If the cost functional F in (6) has the form

$$F(x_1,\cdots,x_N,u_0,\cdots,u_{N-1}) = f(x_N) + \sum_{i=0}^{N-1} g_i(u_i), \quad (21)$$

then the function

$$\mathcal{S}_k(\zeta_k) = S_k(x_k,w_k,w_{k+1},\cdots,w_{N-1},v_{k+1},v_{k+2},\cdots,v_{N-1}|\zeta_k) \quad (22)$$

is sufficiently informative.

Thus when F has the form (21) the past states $x_1,\cdots,x_{k-1}$ and controls $u_0,\cdots,u_{k-1}$ no longer appear in the sufficiently informative function. The dependence of the sufficient information on the future uncertainties can be removed if the constraint set Q for the uncertain quantities has a property implying that the set of values that any particular uncertain quantity can take is independent of the values of the other uncertain quantities.

Corollary 2: If the set Q has the form

$$Q = \{x_0,w_0,w_1,\cdots,w_{N-1},v_1,\cdots,v_{N-1}|x_0 \in X_0,$$
$$w_i \in w_i, \quad i = 0,1,\cdots,N-1,$$
$$v_k \in V_k, \quad k = 1,2,\cdots,N-1\} \quad (23)$$

where X_0, W_i, V_k are given subsets of the corresponding Euclidean spaces, then the function $\mathcal{S}_k$ given for all k by

$$\mathcal{S}_k(\zeta_k) = [S_k(x_1,x_2,\cdots,x_k|\zeta_k),\ u_0,u_1,\cdots,u_{N-1}]$$

is sufficiently informative.

The case where the constraint Q is of the form (23) should be considered analogous to the case of uncorrelated white input and measurement noises in the corresponding stochastic problem.

The natural combination of Corollaries 1 and 2 yields the following.

Corollary 3: If F has the form (21) and Q has the form (23), then the function

$$\mathcal{S}_k(\zeta_k) = S_k(x_k|\zeta_k)$$

is sufficiently informative.

In the stochastic problem analogous to Problem 1, important simplifications in the sufficient statistic result when the function F in (6) is additively separable [12], i.e., when

$$F(x_1,x_2,\cdots,x_N,u_0,u_1,\cdots,u_{N-1}) = \sum_{i=0}^{N-1} g_i(x_i,u_i) \quad (24)$$

where $g_i(\cdot,\cdot)$ are given real valued functions. In the minimax framework of interest here, simplification of the sufficiently informative function can be achieved only at the expense of adjoining to the system equation (1) an additional state $x_k^{(n+1)}$ defined by

$$x_{k+1}^{(n+1)} = x_k^{(n+1)} + g_k(x_k,u_k), \qquad x_0^{(n+1)} = 0$$

so that the function F given by (24) becomes simply $x_N^{(n+1)}$ and the simplification of the sufficient information afforded by Corollary 1 is applicable to the augmented system. The sufficiently informative function is reduced to

$$\mathcal{S}_k(\zeta_k) = S_k(\tilde{x}_k,w_k,\cdots,w_{N-1},v_{k+1},\cdots,v_{N-1}|\zeta_k)$$

where $\tilde{x}_k \triangleq (x_k,x_k^{(n+1)})$ is the augmented state. If, in addition, Q has the form of (23), we have

$$\mathcal{S}_k(\zeta_k) = S_k(\tilde{x}_k|\zeta_k).$$

The difference between the simplifications available for additively separable cost functionals in the stochastic setting and those available under the minimax formulation may be attributed to the fact that, whereas the expectation operation is linear and distributes over addition, the maximization operation is not.

Equation (20) demonstrates the structure of the optimal control law, provides an alternative conceptual framework for considering Problem 1, and can give insight concerning the complexity of the optimal control law. Furthermore, it can form the basis for the development and the analysis of suboptimal control schemes [15]. However, it appears that only for a limited class of problems is it profitable to implement the optimal control law in the form given by (20) since the estimator $\mathcal{S}_k$ is infinite dimensional in most cases. Some typical examples of this limited class of problems are discussed in the next section.

V. Some Special Cases

It is advantageous to use a specific sufficiently informative function $\mathcal{S}_k$ only if the sufficient information $\mathcal{S}(\zeta_k)$ has smaller dimension than ζ_k, i.e., $\mathcal{S}_k$ maps into a finite-dimensional space with dimension less than $k(s+m)$. Under these circumstances by using $\mathcal{S}_k$ the solution of the problem by dynamic programming will be implemented over a space of smaller dimension. Furthermore, the computation of the optimal control law $\bar{\mu}^*[\mathcal{S}_k(\zeta_k)]$, $k = 0,1,\cdots,N-1$, must be at least as easy as the computation of the optimal control law $\bar{\mu}(\zeta_k)$. Such situations occur in problems involving finite-state systems, one-dimensional problems with terminal cost where the set of possible states is an interval, and, of course, the case of perfect state information where system state is measured exactly and constitutes a sufficient information. Another situation in which the sufficient information has smaller dimension than ζ_k and leads to simplified computations is the case of a linear system with linear measurements and an energy constraint on the uncertain quantities. This case bears great similarity to the linear quadratic Gaussian case of stochastic control. We have

$$x_{k+1} = A_k x_k + B_k u_k + G_k w_k, \qquad k = 0,1,\cdots,N-1 \quad (25)$$

$$z_k = C_k x_k + v_k, \qquad k = 1,2,\cdots,N-1 \quad (26)$$

$$x_0'P^{-1}x_0 + \sum_{i=0}^{N-1} w_i'Q_i^{-1}w_i + \sum_{i=1}^{N-1} v_i'R_i^{-1}v_i \leq 1 \quad (27)$$

where P, Q_i, R_i are positive-definite symmetric matrices. It has been shown in [7] that under these circumstances the set $S(x_k,w_k,\cdots,w_{N-1},v_{k+1},\cdots,v_{N-1}\zeta_k)$ is the ellipsoid

$$\Big\{x_k,w_k,\cdots,w_{N-1},v_{k+1},\cdots,v_{N-1}\big|(x - \hat{x}_k)'\Sigma_{k|k}^{-1}(x - \hat{x}_k) + \sum_{i=k}^{N-1} w_i'Q_i^{-1}w_i + \sum_{i=k+1}^{N-1} v_i'R_i^{-1}v_i \leq 1 - \delta_k^2\Big\} \quad (28)$$

where the n-vector $\hat{x}_k$, the $n \times n$ matrix $\Sigma_{k|k}$, and the real number δ_k^2 are generated recursively by

$$\hat{x}_{i+1} = A_i\hat{x}_i + B_iu_i + \Sigma_{i+1|i+1}C_{i+1}'R_{i+1}^{-1} \cdot(z_{i+1} - C_{i+1}A_i\hat{x}_i - C_{i+1}B_iu_i) \quad (29)$$

$$\Sigma_{i|i} = [\Sigma_{i|i-1}^{-1} + C_i'R_i^{-1}C_i]^{-1} \quad (30)$$

$$\Sigma_{i+1|i} = A_i\Sigma_{i|i}A_i' + G_iQ_iG_i' \quad (31)$$

$$\delta_{i+1}^2 = \delta_i^2 + (z_{i+1} - C_{i+1}A_i\hat{x}_i - C_{i+1}B_iu_i)'(C_{i+1}\Sigma_{i+1|i} \cdot C_{i+1}' + R_{i+1})^{-1}(z_{i+1} - C_{i+1}A_i\hat{x}_i - C_{i+1}B_iu_i) \quad (32)$$

with initial conditions $\hat{x}_0 = 0$, $\Sigma_{0|0} = P$, $\delta_0^2 = 0$. Since the matrix $\Sigma_{k|k}$ is independent of the output z and precomputable from the problem data, the ellipsoid (28) is completely specified by the n-vector $\hat{x}_k$ and the scalar δ_k^2. If the function F has the form (21), combination of this result with Corollary 1 immediately yields the following result.

Corollary 4: For the system (25) and (26), the constraint set Q specified by (27), and the function F given by (21), the function S_k: $R^{k(s+m)} \to R^n \times [0, 1]$ defined by

$$S_k(\zeta) = \{\hat{x}_k, \delta_k^2\}$$

is sufficiently informative, where $\hat{x}_k$ and δ_k^2 are given above.

Thus, for the problem involving a linear system, linear measurements, an energy constraint on the uncertain quantities, and a cost functional involving a function F of the form (21), the estimator part of the optimal controller can be completely and efficiently characterized. Furthermore, the computational requirements of the dynamic programming algorithm leading to the calculation of the optimal controller are greatly reduced. This is due to the fact that the algorithm can be redefined over the space of the sufficient information as follows:

$$\bar{H}_{N-1}(\hat{x}_{N-1},\delta_{N-1}^2) = \inf_{u_{N-1}} \sup_{x\in X_{N|N-1}} \{f(x + B_{N-1}u_{N-1}) + g_{N-1}(u_{N-1})\} \quad (33)$$

where $X_{N|N-1}$ is the ellipsoid

$$X_{N|N-1} = \Big\{x\big|(x - A_{N-1}\hat{x}_{N-1})' \cdot\Sigma_{N|N-1}^{-1}(x - A_{N-1}\hat{x}_{N-1}) \leq 1 - \delta_{N-1}^2\Big\} \quad (34)$$

$$\bar{H}_k(\hat{x}_k,\delta_k^2) = \inf_{u_k} \sup_{z_{k+1}\in\hat{Z}_{k+1}} \{\bar{H}_{k+1}(\hat{x}_{k+1},\delta_{k+1}^2) + g_k(u_k)\}, \qquad k = 0,1,\cdots,N-2 \quad (35)$$

where $\hat{x}_{k+1}$, δ_{k+1}^2 are given in terms of $\hat{x}_k$, δ_k^2, u_k, and z_{k+1} by the estimator equations (29)–(32) and $\hat{Z}_{k+1}$ is the ellipsoid

$$\hat{Z}_{k+1} = \{z_{k+1}|(z_{k+1} - C_{k+1}A_k\hat{x}_k - C_{k+1}B_ku_k)' \cdot(C_{k+1}\Sigma_{k+1|k}C_{k+1}' + R_{k+1})^{-1} \cdot(z_{k+1} - C_{k+1}A_k\hat{x}_k - C_{k+1}B_ku_k) \leq 1 - \delta_k^2\}. \quad (36)$$

The optimal controller is of the form

$$u_k = \bar{\mu}_k^*(\hat{x}_k,\delta_k^2), \qquad k = 0,1,\cdots,N-1$$

where $\hat{x}_k,\delta_k^2$ are generated recursively by the estimator of (29)–(32). The function $\bar{\mu}_k^*$ is computed from the dynamic programming algorithm of (33)–(36). This algorithm is carried over a space of dimension $(n + 1)$—a substantial improvement over the case where the optimal control law is calculated as a function of ζ_k by means of the algorithm of Proposition 1.

A special case of the function (21) occurs in the target set reachability problem in which

$$F(x_1,\cdots,x_N,u_0,\cdots,u_{N-1}) = \delta(x_N|X_N) + \sum_{i=0}^{N-1} \delta(u_i|U_i)$$

where $\delta(y|Y)$ denotes the indicator function of a set Y given by (7). This problem has been examined in detail in [4].

As another representative example of the limited class of problems where the use of a suitable sufficiently informative function results in substantial reduction of computational and storage requirements, consider the following scalar system:

$$x_{k+1} = ax_k + u_k + w_k, \qquad k = 0,1,\cdots,N-1$$

where $a \geq 0$, with measurement equation

$$z_k = x_k + v_k,$$

and where the initial state and the input and measurement noises are known to belong to the following intervals:

$$x_0 \in [s_0^1, s_0^2], \qquad w_k \in [p^1, p^2], \qquad v_k \in [r^1, r^2].$$

It is required to find a control law μ_k, $k = 0,1,\cdots,N-1$ with $\mu_k(\zeta_k) \in U \subset R$, that minimizes a cost functional involving a function F of the form

$$F(x_1,\cdots,x_N,u_0,\cdots,u_{N-1}) = f(x_N) + \sum_{i=0}^{N-1} g(u_i)$$

where f, g are given real valued functions.

From Corollary 3 we have that the function

$$s_k(\zeta_k) = S(x_k|\zeta_k), \qquad k = 0,1,\cdots,N-1$$

is sufficiently informative where $S(x_k|\zeta_k)$ is the set of all states x_k consistent with the information vector $\zeta_k = (z_1,z_2,\cdots,z_k,u_0,u_1,\cdots,u_{k-1})$. The set $S(x_k|\zeta_k)$ is a closed interval for this problem and can be computed by using the recursive set algorithm given in [5]. We have

$$S(x_k|\zeta_k) = [s_k^1, s_k^2]$$

where s_k^1, s_k^2 are generated by the following estimator equations:

$$s_{k+1}^1 = \max\{as_k^1 + u_k + p^1, z_{k+1} - d^2\} \tag{37}$$

$$s_{k+1}^2 = \min\{as_k^2 + u_k + p^2, z_{k+1} - d^1\} \tag{38}$$

with initial conditions the endpoints s_0^1, s_0^2 of the interval of uncertainty for the initial state.

The dynamic programming algorithm in terms of the sufficiently informative function takes the form of

$$\bar{H}_{N-1}(s_{N-1}^1,s_{N-1}^2) = \inf_{u_{N-1}\in U} \sup_{\substack{w_{N-1}\in[p^1,p^2]\\ x_{N-1}\in[s_{N-1}^1,s_{N-1}^2]}} \{f(ax_{N-1} + u_{N-1} + w_{N-1}) + g(u_{N-1})\} \tag{39}$$

$$\bar{H}_{k-1}(s_{k-1}^1,s_{k-1}^2) = \inf_{u_{k-1}\in U} \sup_{z_k\in\hat{Z}_k(s_{k-1}^1,s_{k-1}^2,u_{k-1})} \{\bar{H}_k[\max\{as_{k-1}^1 + u_{k-1} + p^1, z_k - d^2\}, \min\{as_{k-1}^2 + u_{k-1} + p^2, z_k - d^1\}] + g(u_{k-1})\}, \qquad k = 1,2,\cdots,N \tag{40}$$

where the interval $\hat{Z}_k(s_{k-1}^1,s_{k-1}^2,u_{k-1})$ is given by

$$\hat{Z}_k(s_{k-1}^1,s_{k-1}^2,u_{k-1}) = a[s_{k-1}^1,s_{k-1}^2] + u_{k-1} + [p^1,p^2] + [r^1,r^2].$$

Thus for the above problem the optimal controller is of the form

$$u_k = \bar{\mu}_k^*(s_k^1,s_k^2), \qquad k = 0,1,\cdots,N-1$$

where s_k^1,s_k^2 are generated by the estimator of (37) and (38). The function $\bar{\mu}_k^*$, $k = 0,1,\cdots,N-1$, is computed from the dynamic programming algorithm (39), (40). This algorithm is carried over a two-dimensional half-space (recall that $s_k^1 \leq s_k^2$). It can be seen that the reduction in computational and storage requirements is substantial over the case where the optimal controller would be computed in the form $\mu_k(\zeta_k)$ by means of the algorithm of Proposition 1, since the dimension of ζ_k is $2k$ for this problem.

VI. Conclusions

In this paper we considered some general aspects of a minimax feedback control problem with imperfect state information. A dynamic programming algorithm was given for the solution of this problem, which in general must be implemented over the space of the information available to the controller. The notion of a sufficiently informative function, which parallels the notion of a sufficient statistic of stochastic control, was introduced with a twofold purpose. First, to provide an alternative conceptual framework for viewing the problem and to demonstrate the separation of the optimal controller into an estimator and an actuator. Second, to demonstrate the possibility of redefining and implementing the dynamic programming algorithm so that it is carried over the space of the sufficient information. For a limited class of problems it was shown that this alternate implementation is profitable. The results in this paper should not come as a surprise to anyone familiar with dynamic programming, sequential games, and stochastic control since they represent a formalization and extension of well-known concepts within the framework of the minimax problem. It is to be noted that, similar to the stochastic control case, the notion of a sufficiently informative function is useful for only a limited class of problems; however, this class does not include any special case with a solution as elegant as the case of a linear system with a quadratic cost in stochastic control.

References

[1] H. S. Witsenhausen, "Minimax control of uncertain systems," Mass. Inst. Technol., Cambridge, Rep. ESL-R-269, May 1966.

[2] ——,"A minimax control problem for sampled linear systems," *IEEE Trans. Automat. Contr.*, vol. AC-13, pp. 5–21, Feb. 1968.

[3] D. P. Bertsekas and I. B. Rhodes, "On the minimax reachability of target sets and target tubes," *Automatica*, vol. 7, pp. 233–247, 1971.

[4] D. P. Bertsekas, "Control of uncertain systems with a set-membership description of the uncertainty," Ph.D. dissertation, Dep. Elec. Eng., Mass. Inst. Technol., Cambridge, 1971.

[5] H. S. Witsenhausen, "Sets of possible states of linear systems given perturbed observations," *IEEE Trans. Automat. Contr.* (Short Papers), vol. AC-13, pp. 556–558, Oct. 1968; also in *Proc. 5th Annu. Allerton Conf. Circuit and System Theory*, 1967.

[6] F. C. Schweppe, "Recursive state estimation: Unknown but bounded errors and system inputs," *IEEE Trans. Automat. Contr.*, vol. AC-13, pp. 22–28, Feb. 1968.

[7] D. P. Bertsekas and I. B. Rhodes, "Recursive state estimation for a set-membership description of uncertainty," *IEEE Trans. Automat. Contr.*, vol. AC-16, pp. 117–128, Apr. 1971.

[8] D. P. Bertsekas, "Infinite-time reachability of state-space regions by using feedback control," *IEEE Trans. Automat. Contr.*, vol. AC-17, pp. 604–613, Oct. 1972.

[9] J. D. Glover and F. C. Schweppe, "Control of linear dynamic systems with set constrained disturbances," *IEEE Trans. Automat. Contr.*, vol. AC-16, pp. 411–423, Oct. 1971.

[10] D. P. Bertsekas and I. B. Rhodes, "On the minimax feedback control of uncertain dynamic systems," in *Proc. 1971 IEEE Conf. Decision and Control* (Miami, Fla.), pp. 451–455.

[11] A. A. Feldbaum, *Optimal Control Systems*. New York: Academic, 1965.

[12] C. T. Striebel, "Sufficient statistics in the optimal control of stochastic systems," *J. Math. Anal. Appl.*, vol. 12, pp. 576–592, 1965.

[13] D. Blackwell and M. A. Girshick, *Theory of Games and Statistical Decisions*. New York: Wiley, 1954.

[14] R. L. Stratonovich, "On the theory of optimal control: Sufficient coordinates," *Automat. Remote Contr.* (USSR), vol. 23, Jan. 1963.

[15] D. P. Bertsekas, "On the solution of some minimax problems," in *Proc. 1972 IEEE Conf. Decision and Control* (New Orleans, La.), 1972.

The Design of Controllers for the Multivariable Robust Servomechanism Problem Using Parameter Optimization Methods

EDWARD J. DAVISON, FELLOW, IEEE, AND IAN J. FERGUSON

***Abstract*—The problem of designing realistic multivariable controllers to solve the servomechanism problem is considered in this paper. Specifically, it is desired to find a controller for a plant to solve the robust servomechanism problem, so that closed-loop stability and asymptotic regulation occur, and also so that other desirable properties of the controlled system, such as fast response, low-interaction, integrity, tolerance to plant variations, etc., occur. The method of design is based on using state space methods via a two-stage process: 1) using theory, determine the existence of a solution and control structure required to solve the problem, and 2) using nonlinear programming methods, determine the unknown controller parameters so as to minimize a performance index for the system subject to certain constraint requirements. Numerous examples, varying from a single-input/single-output to a four-input/four-output system, are given to illustrate the design method, and the results obtained are compared with the results obtained by using other alternate design methods. In all cases, the controllers obtained have been highly competitive with controllers obtained by alternate design methods.**

I. INTRODUCTION

ONE of the most basic problems occurring in control engineering is the design problem associated with finding realistic controllers to solve the multivariable robust servomechanism problem. In this type of problem, a mathematical model of the plant to be controlled is assumed to be known to the designer, and various constraints, specific to the particular plant being studied must be considered in the problem formulation. In addition, decisions re tradeoffs between speed of response, integrity, tolerance to plant parameter variations, etc. must be made by the designer. These types of considerations imply that human judgment (and not mathematics) must ultimately be used in the final "shaping" of a satisfactory controller to solve the problem. The purpose of this paper is to develop a tool which can be used by the designer to give him guidance in the final shaping of such a controller. Hopefully, the tool will have the properties that the designer need not worry about any "intermediate" calculations required to determine the controller, but only about the final engineering properties of the controller itself, e.g., deciding on tradeoffs between speed of response and integrity of the controller say.

In the multivariable robust servomechanism problem to be considered, it is desired to find a controller so that:

1) the closed-loop system is asymptotically stable, with a desired "fast" transient behavior,

2) asymptotic tracking and regulation occur for all reference input signals of a given class, for all disturbance signals of a given class and such that "low interaction" occurs,

3) properties 1) and 2) hold in the presence of changes in the dynamics of the plant; in addition constraints on integrity, controller gains, damping of the system, etc. are to be satisfied.

The method of design adopted in this paper to form such a controller is based on state space methods using a two-stage process and is in the spirit of [1]–[4], [20], [21].

i) Using theory, determine the existence of a solution to the servomechanism problem, and the necessary controller structure to solve for it.

Manuscript received March 5, 1980; revised October 6, 1980. This work was supported by the National Research Council of Canada under Grant A4396 and by the Canada Council under a Killam Research Fellowship.

E. J. Davison is with the Department of Electrical Engineering, University of Toronto, Toronto, Ont., Canada M5S 1A4.

I. J. Ferguson was with the Department of Electrical Engineering, University of Toronto, Toronto, Ont., Canada M5S 1A4. He is now with Canadian General Electric, Peterborough, Ont., Canada.

Reprinted from *IEEE Trans. Automat. Contr.*, vol. AC-26, no. 1, pp. 93–110, Feb. 1981.

ii) Using parameter optimization methods, determine the controller parameters of the controller so as to minimize a performance index for the system, subject to certain constraint requirements.

This methodology differs from other design methods, e.g., frequency-domain design methods [5]–[9], graphical design methods [10], [11], where there is usually only an implicit assumption made that a solution exists, and where parts i) and ii) are generally combined together.

In developing the proposed design method, the following considerations are made—the method should automatically give a controller with *"fast speed"* of response and with *"low interaction"* occurring between channels, depending on what the fixed modes [12] of the system are, for the case of *both tracking and disturbance regulation*; the method should allow the designer to have control over the *positioning of the eigenvalues* of the resultant closed-loop system, e.g., damping; the method should allow the designer to specify the controller to be *tolerant to changes in certain specific plant parameters* of the system; the method should be able to guarantee the controller has certain desirable *fail-safe properties*, e.g., if a sensor failure occurs, the system should remain stable.

In part i) of the design procedure, conditions for the existence of a solution to the robust servomechanism problem are developed in which constraints on the damping factor, controller gain magnitude, tolerance, and integrity of the system are explicitly considered; these results then give guidance in the selection of a suitable control structure for the system. The key to the method for part ii) of the design procedure is to introduce a performance criterion which on minimization gives a system with low interaction and fast response; this is done by choosing a performance index which measures the deviations of certain outputs of the controller for the system, rather than one which directly measures error deviation in the system.

The paper is divided into six sections; Section II describes some preliminary results, Section III develops the controller structure results, Section IV describes the method to be used for controller parameter determination, Section V gives an illustration of the design method, and Section VI gives some design examples.

II. Preliminaries

In the development to follow, the plant to be controlled is assumed to be described by the following linear time-invariant model:

$$\begin{aligned} \dot{x} &= Ax + Bu + E\omega \\ y &= Cx + Du + F\omega \\ y_m &= C_m x + D_m u + F_m \omega \\ e &= y - y_{\text{ref}} \end{aligned} \tag{1}$$

where $x \in R^n$ is the state of the system, $u \in R^m$ are the inputs, $y \in R^r$ are the outputs to be regulated, $y_m \in R^{r_m}$ are the measurable outputs of the system, $\omega \in R^{\Omega}$ are the disturbances in the system, $y_{\text{ref}} \in R^r$ are the reference inputs, and $e \in R^r$ is the error in the system.

It is assumed that the disturbances ω belong to the following class of signals:

$$\begin{aligned} \dot{\eta}_1 &= \mathsf{A}_1\eta_1, \qquad \eta_1 \in R^{n_1} \\ \omega &= \mathsf{C}_1\eta_1 \end{aligned} \tag{2}$$

and that the reference inputs y_{ref} belong to the following class of signals:

$$\begin{aligned} y_{\text{ref}} &= G\sigma, \ \dot{\eta}_2 = \mathsf{A}_2\eta_2, \ \eta_2 \in R^{n_2} \\ \sigma &= \mathsf{C}_2\eta_2. \end{aligned} \tag{3}$$

This includes the case of most common type of disturbance/reference-inputs occurring in practice, e.g., polynomial, sinusoidal type inputs, etc.

It is assumed for nontriviality that $\sigma(\mathsf{A}_1) \subset C^+$, $\sigma(\mathsf{A}_2) \subset C^+$ where $\sigma(\cdot)$ denotes the eigenvalues of $(\cdot)$ and C^+ denotes the closed right-half complex plane. It is also assumed with no loss of generality that $(\mathsf{C}_1, \mathsf{A}_1)$, $(\mathsf{C}_2, \mathsf{A}_2)$ are observable, that rank $B = m$, rank $C = r$, and that rank $\binom{E}{F} =$ rank $\mathsf{C}_1 = \Omega$, rank $G =$ rank $\mathsf{C}_2 = \dim(\sigma)$.

The following controller will be used in characterizing a solution to the robust servomechanism controller for (1).

A. Stabilizing Compensator

Definition: Given the system

$$\begin{aligned} \dot{x} &= Ax + Bu \\ y_m &= C_m x + D_m u, \end{aligned}$$

which has no unstable fixed modes [12], a *stabilizing compensator* $\dot{\eta} = \Lambda_1\eta + \Lambda_2 y_m$, $u = K_1\eta + K_2 y_m$ is defined to be a controller which will stabilize the resultant closed-loop system.

A stabilizing compensator is not a unique device and may be constructed by using a number of different techniques, e.g., observer theory, linear optimal control, modal design methods, etc. The ability of the stabilizing compensator to achieve a "desired transient behavior" depends on the position of the fixed modes of (C_m, A, B), the transmission zeros [13] of (C_m, A, B, D_m), etc.

The following definitions are needed in characterizing the controller structure for a solution to the robust servomechanism problem.

Definition: Let the minimal polynomial of A_1, A_2 be denoted by $\Lambda_1(s)$, $\Lambda_2(s)$ respectively, and let the least common multiple of $\Lambda_1(s)$, $\Lambda_2(s)$ be given by $\Lambda(s)$. Let the zeros of $\Lambda(s)$ (multiplicities included) be given by

$$\lambda_1, \lambda_2, \cdots, \lambda_p. \tag{4}$$

Definition: The matrix $\mathsf{C} \in R^{p \times p}$ is defined as follows:

$$\mathsf{C} \triangleq \begin{bmatrix} 0 & 1 & 0 & \cdots & 0 \\ 0 & 0 & 1 & \cdots & 0 \\ \vdots & \vdots & \vdots & & \vdots \\ -\delta_1 & -\delta_2 & -\delta_3 & \cdots & -\delta_p \end{bmatrix} \tag{5}$$

where the coefficients δ_i, $i=1,\cdots,p$ are given by the coefficients of the polynomial $\Pi_{i=1}^{p}(\lambda-\lambda_i)$, where λ_i, $i=1,\cdots,p$ are given by (4), i.e.,

$$\lambda^p+\delta_p\lambda^{p-1}+\cdots+\delta_2\lambda+\delta_1 \triangleq \prod_{i=1}^{p}(\lambda-\lambda_i). \qquad (6)$$

The following compensator, called a general *servocompensator* is of fundamental importance in the design of robust controllers for (1).

B. Servocompensator [14]

Definition: A general servocompensator for (1) is a controller with input e and output $\eta \in R^{rp}$ given by

$$\dot{\eta}=\mathsf{C}^*\eta+\mathsf{B}^*e, \qquad e\triangleq(y-y_{\text{ref}}) \qquad (7)$$

where

$$\mathsf{C}^* \triangleq \mathsf{T}\ \text{block diag}\ \underbrace{(\mathsf{C},\mathsf{C},\cdots,\mathsf{C})}_{r\text{ matrices}}\mathsf{T}^{-1}$$

$$\mathsf{B}^* \triangleq \mathsf{TB} \qquad (8)$$

where T is a nonsingular real matrix, C is given by (5), and $\mathsf{B}\in R^{rp\times r}$ is a real matrix of rank r with the property that

$$\left\{\text{block diag}\underbrace{(\mathsf{C},\mathsf{C},\cdots,\mathsf{C})}_{r\text{ matrices}},\mathsf{B}\right\} \qquad (9)$$

is controllable.

III. Development (Controller Structure Results)

The following servomechanism problem is now posed.

A. Robust Servomechanism Problem

Find a controller for (1) (with input y_m, y_{ref} and output u) so that when applied to (1):

1) the resultant closed-loop system is asymptotically stable;

2) asymptotic tracking occurs, i.e., $\lim_{t\to\infty}e(t)=0$, $\forall x(0)\in R^n$, $\forall \eta_1(0)\in R^{n_1}$, $\forall\eta_2(0)\in R^{n_2}$ and for all initial conditions of the controller used;

3) condition 2) holds for any variations in the plant model (1) (including dynamic perturbations) which do not cause the resultant controlled system to become unstable.

B. Strong Robust Servomechanism Problem

Find a controller for (1) so that when applied to (1) conditions 1), 2), and 3) are true and in addition:

4) approximate error regulation occurs for any variations in the parameters of the controller used, the approximation becoming arbitrarily close to exact regulation as the controller parameter variation becomes smaller, i.e., let the controller parameters γ_i, $i=1,2,\cdots,N$ be perturbed to $\gamma_i+\delta\gamma_i$; then for any $\eta>0$, it is desired that there exist $\epsilon>0$, $T>0$ so that $|e(t)|<\eta$, $\forall t>T$, $\forall\delta\gamma_i\in\{\delta\gamma_i \| \delta\gamma_i|<\epsilon\}$, $i=1,2,\cdots,N$ for all bounded initial states of the system (1), (2), (3) and controller.

A solution to these problems are given as follows [14].

Lemma 1: There exists a solution to the robust servomechanism problem for (1) if and only if the following conditions all hold.

i) (C_m, A, B) has no unstable fixed modes [12].

ii) $m \geq r$.

iii) The transmission zeros [13] of (C, A, B, D) do not coincide with λ_i, $i=1,2,\cdots,p$ of (4).

iv) The outputs y are physically measurable [14].

There exists a solution to the strong robust servomechanism problem for (1) if and only if conditions i)–iv) hold and in addition:

v) the class of disturbance signals and reference-input signals is bounded, i.e., in (4), $\text{Re}(\lambda_i)=0$, $i=1,2,\cdots,p$ and $(\lambda_1,\lambda_2,\cdots,\lambda_p)$ are all distinct.

A controller for (1) which solves the robust servomechanism problem will now be described. In what follows, it will be assumed that $\text{Re}(\lambda_i)=0$, $i=1,2,\cdots,p$, but that λ_i are not necessarily distinct, i.e., polynomial-sinusoidal disturbances may be allowed in the problem formulation. (The controller obtained in this case will be robust but not necessarily strongly robust.)

C. Robust Servomechanism Controller

Consider the system (1) and assume that Lemma 1 holds; then any robust controller for (1) must have the following structure [14]:

$$u=K_0\hat{x}+K\eta \qquad (10)$$

where $\eta\in R^{rp}$ is the output of a *general servocompensator* (7) for (1) and where $\hat{x}$ is the output of a *stabilizing compensator* S^* with inputs y_m, y_{ref}, u, η where S^*, K_0, K are found to stabilize and give satisfactory transient behavior for the following augmented system:

$$\begin{pmatrix}\dot{x}\\ \dot{\eta}\end{pmatrix}=\begin{pmatrix}A & 0\\ \mathsf{B}^*C & \mathsf{C}^*\end{pmatrix}\begin{pmatrix}x\\ \eta\end{pmatrix}+\begin{pmatrix}B\\ \mathsf{B}^*D\end{pmatrix}u$$

$$\begin{pmatrix}y_m\\ \eta\end{pmatrix}=\begin{pmatrix}C_m & 0\\ 0 & I\end{pmatrix}\begin{pmatrix}x\\ \eta\end{pmatrix}+\begin{pmatrix}D_m\\ 0\end{pmatrix}u \qquad (11)$$

where the triple

$$\left\{\begin{pmatrix}C_m & 0\\ 0 & I\end{pmatrix},\begin{pmatrix}A & 0\\ \mathsf{B}^*C & \mathsf{C}^*\end{pmatrix},\begin{pmatrix}B\\ \mathsf{B}^*D\end{pmatrix}\right\}$$

has the same fixed modes as (C_m, A, B).

It is to be noted that the robust controller always has order $\geq rp$.

D. Special Classes of Stabilizing Compensators

Various special classes of stabilizing compensators which can be used to solve the problem are given as follows.

1) Observer-Stabilizing Compensator: Assume that Lemma 1 holds; then an *observer-stabilizing compensator* for (1) consists of the following device:

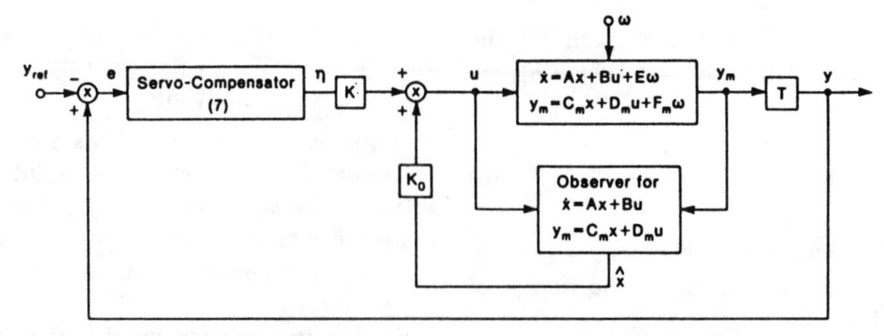

Fig. 1. Observer stabilizing compensator which can be used to solve the robust servomechanism problem. (Here y is physically measurable, i.e., there exists T so that $y = Ty_m$.)

$\hat{x}$ in (10) is the output of an observer of order $\hat{n}$ (either full order $\hat{n} \leq n$ or reduced order $\hat{n} \leq n-r$) with input y_m, u

$$\tag{12}$$

which estimates the state of the detectable system $\dot{x} = Ax + Bu$, $y_m = C_m x + D_m u$ and where K_0, K in (10) are found using standard methods (e.g., pole assignment, optimal control theory) to stabilize the following stabilizable system:

$$\begin{pmatrix} \dot{x} \\ \dot{\eta} \end{pmatrix} = \begin{pmatrix} A & 0 \\ \mathsf{B}^*C & \mathsf{C}^* \end{pmatrix}\begin{pmatrix} x \\ \eta \end{pmatrix} + \begin{pmatrix} B \\ \mathsf{B}^*D \end{pmatrix} u, \quad u = (K_0, K)\begin{pmatrix} x \\ \eta \end{pmatrix}$$

$$y_m = (C_m \quad 0)\begin{pmatrix} x \\ \eta \end{pmatrix} + D_m u \tag{13}$$

(see Fig. 1). This compensator has the following property.

Theorem 1: Consider the system (1) and assume that Lemma 1 holds; then the eigenvalues of the resultant closed-loop system obtained by applying the robust controller (10) to (1), where $\hat{x}$ in (10) is the output of the observer-stabilizing compensator (12), are equal to the eigenvalues of the observer (12) together with the eigenvalues of

$$\begin{bmatrix} A + BK_0 & BK \\ \mathsf{B}^*(C + DK_0) & \mathsf{C}^* + \mathsf{B}^*DK \end{bmatrix}.$$

Proof: For simplicity, assume the full order observer

$$\dot{\xi} = (A - \Lambda C_m)\xi + \Lambda \hat{y}_m + Bu, \quad \hat{y}_m \triangleq y_m - D_m u, \quad \hat{x} \triangleq \xi \tag{14}$$

is used for the observer-stabilizing compensator (12), where (C_m, A) is detectable and where Λ is the observer gain chosen so that $(A - \Lambda C_m)$ is stable. Then the closed-loop system obtained by applying the controller (10), (14) to (1) is given by

$$\begin{bmatrix} \dot{x} \\ \dot{\eta} \\ \dot{x} - \dot{\xi} \end{bmatrix} = \begin{bmatrix} A + BK_0 & BK & -BK_0 \\ \mathsf{B}^*(C + DK_0) & \mathsf{C}^* + \mathsf{B}^*DK & -\mathsf{B}^*DK_0 \\ 0 & 0 & A - \Lambda C_m \end{bmatrix}\begin{bmatrix} x \\ \eta \\ x - \xi \end{bmatrix} \tag{15}$$

from which the result follows. The corresponding result for a reduced order observer may be obtained in a similar way.

Remark 1: This observation implies that there exists a stabilizing compensator in (10) given by

$$\dot{\xi} = \Lambda_0 \xi + \Lambda_1 \hat{y}_m + \Lambda_2 \eta$$

$$\hat{y}_m \triangleq y_m - D_m u, \ \xi \in R^{\hat{n}}, \ \hat{n} \leq n - r$$

$$\hat{x} = \Lambda_3 \hat{y}_m + \Lambda_4 \xi \tag{16}$$

which solves the problem. It is also to be noted that arbitrary pole assignment can be achieved for the resultant closed loop system using (16) if (C_m, A, B) has no fixed modes [15].

In case the plant is open-loop stable, the following generalized three-term controller can be used.

2) Generalized Three-Term Servomechanism Controller:

Definition: Consider the system (1); then the following compensator

$$u = K\eta + K_0 \hat{y}_m + K_1 \tilde{\dot{y}}_m; \ \hat{y}_m \triangleq y_m - D_m u; \ \tilde{\dot{y}}_m \triangleq (\dot{\hat{y}}_m - C_m Bu) \tag{17}$$

where η is the output of the servocompensator (7) is called *a generalized three-term servomechanism controller for (1)* (see Fig. 2). This compensator has the following property.

Theorem 2: Consider the system (1) and assume that Lemma 1 holds and that A is asymptotically stable; then there exists gain matrices K, K_0, K_1 so that the resultant controlled system obtained by applying the generalized three-term controller (17) to (1) is asymptotically stable, i.e., there exists gain matrices K, K_0, K_1 so that

$$\begin{pmatrix} A + BK_0C_m + BK_1C_mA & BK \\ \mathsf{B}^*(C + DK_0C_m + DK_1C_mA) & \mathsf{C}^* + \mathsf{B}^*DK \end{pmatrix} \tag{18}$$

is asymptotically stable.

Proof: The proof follows directly from Lemma 2 (see the Appendix), where it is shown that there exists a K in (17) with $K_0 = 0$, $K_1 = 0$ to stabilize the system (18) provided that A is asymptotically stable and that $\mathrm{Re}(\lambda_i) = 0$, $i = 1, 2, \cdots, p$.

Remark 2: In this case, arbitrary pole assignment cannot necessarily be obtained using (17); however, it directly follows from [16] that K, K_0, K_1 in (17) can be found so that $(m + 2r + rp - 1)$ symmetric poles can be assigned to

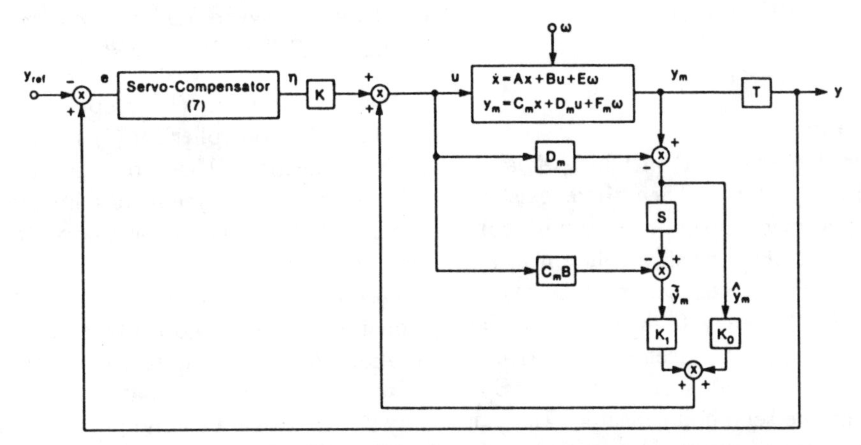

Fig. 2. Generalized three-term controller which can be used to solve the robust servomechanism problem when the plant is open-loop stable.

(18) for almost all (C_m, A, B) systems. This implies that if n is not too large, e.g., if $n \leq (m+2r-1)$ that complete pole assignment can be carried out (even if A is unstable) using (17).

In the servomechanism problem as originally stated, there were no constraints imposed on the closed-loop system with respect to eigenvalue position, integrity of the closed-loop system, etc. The following types of constraints will now be imposed.

1) Damping Factor Constraint: In order to prevent undesirable high-frequency oscillations occurring in the closed-loop system, it may be desirable to impose a constraint on the position of the closed-loop eigenvalues of the system.

2) Controller Gain Constraint: In order to prevent excessively large control signals occurring in the closed-loop system (which may cause "saturation effects" due to the nonlinearities in the system), it may be desirable to impose a limit on the magnitude of the controller gains in the system.

3) Integrity Constraint: In order to prevent "explosive instability" occurring in the closed-loop system when either an actuator or sensor failure occurs, it may be desirable to impose a constraint on the stability of the closed-loop system under certain types of failure conditions.

4) Tolerance Constraint: It may be desirable to impose a constraint on the stability of the closed-loop system, when certain specific plant parameters of the system change from their nominal values.

It is desired now to determine under what conditions the *existence of a solution* to the servomechanism problem can be guaranteed subject to the above constraints. The following results are obtained.

Definition: Given A asymptotically stable, let λ_i, $i=1, 2,\cdots,n$ denote the eigenvalues of A; then ζ_0 is *the damping factor* of A if $|\text{Re}(\lambda_i)| \geq \zeta_0 |\text{Im}(\lambda_i)|$, $i=1,2,\cdots,n$ and if $|\text{Re}(\lambda_j)| = \zeta_0 |\text{Im}(\lambda_j)|$ for some $j \in [1,\cdots,n]$.

Definition: Given $\zeta > 0$, $\epsilon_0 > 0$ and λ_i, $i=1,2,\cdots,p$ of (4), let

$$\Omega_{\zeta,\epsilon_0} = \left\{ \lambda \;\middle|\; \begin{array}{ll} 1) & \text{Re}(\lambda) < 0 \text{ and} \\ 2) & |\text{Re}(\lambda)| \geq \zeta\{|\text{Im}(\lambda)| - \max[|\text{Im}(\lambda_i)|, i=1,2,\cdots,p] - \epsilon_0\} \end{array} \right\}.$$

The following problem is now defined.

E. Robust Servomechanism Problem with Specified Damping ζ, ϵ_0

Assume that plant (1) is open-loop stable, i.e., A is asymptotically stable, with damping factor ζ_0. Then given ϵ_0, $\zeta < \zeta_0$ find a controller for (1) so that when applied to (1), 1)–3) of Section III-A are true and in addition the eigenvalues of the resultant controlled system are contained in $\Omega_{\zeta,\epsilon_0}$. The following result is obtained.

Theorem 3: Consider the system (1) in which A is assumed to be asymptotically stable; then there exists a solution to the robust servomechanism problem with specified damping ζ, ϵ_0 if and only if there exists a solution to the robust servomechanism problem (i.e., Lemma 1 holds).

Proof: The proof of sufficiency of the result will be by construction. Consider applying the control $u=K\eta$ to (1) where η is the output of the servocompensator (7). Then by Lemma 2 there exists $\hat{\epsilon}>0$ so that for any given ϵ, $0<\epsilon<\hat{\epsilon}$ a gain matrix $K=\Theta_\epsilon$ exists with the property that $\|\Theta_\epsilon\| < \epsilon$ and such that

$$\mathsf{A}_\epsilon = \begin{pmatrix} A & B\Theta_\epsilon \\ \mathsf{B}^*C & \mathsf{C}^* + \mathsf{B}^*D\Theta_\epsilon \end{pmatrix} \tag{19}$$

is asymptotically stable. On noting now that the eigenvalues of A_ϵ are continuous in the elements of Θ_ϵ, this implies that for a given ζ, ϵ_0 there always exists a scalar $\hat{\epsilon}_0 < \hat{\epsilon}$ so that the eigenvalues of the resultant closed-loop system $\mathsf{A}_{\hat{\epsilon}_0}$ are contained in $\Omega_{\zeta,\epsilon_0}$, which proves the result.

The following problem is now defined.

F. Robust Servomechanism Problem with Constrained Controller Gain c

Assume the plant (1) is open-loop stable. Then, given $c>0$, find a controller for (1) so that when applied to (1), 1)–3) of Section III-A are true and in addition the controller gains of the system (10) are bounded, i.e.,

$$\|K_0, K\| < c.$$

The following result is obtained.

Theorem 4: Consider the system (1) in which A is assumed to be asymptotically stable. Then there exists a solution to the robust servomechanism problem with constrained gain c if and only if there exists a solution to the robust servomechanism problem (i.e., Lemma 1 holds).

Proof: The proof of sufficiency follows in a similar way to the proof of Theorem 3, i.e., given c, there always exists by Lemma 2 a scalar $\hat{\epsilon}$ and a controller $u=\Theta_{\hat{\epsilon}}\eta$, where η is the output of the servocompensator (7), such that the resultant closed-loop system is asymptotically stable and such that $\|\Theta_{\hat{\epsilon}}\| < c$, which proves the result.

The following integrity problem is now considered.

Definition: Given $\epsilon^* > 0$, and λ_i, $i=1,2,\cdots,p$ of (4), let

$$\Omega_{\epsilon^*} = \left\{ \lambda \;\middle|\; \begin{array}{l} 1)\ \mathrm{Re}(\lambda) < 0 \text{ or} \\ 2)\ |\lambda - \lambda_i| < \epsilon^*, \quad i \in [1,2,\cdots,p] \end{array} \right\}. \tag{20}$$

Definition: Assume that the plant (1) is open-loop stable and consider the resultant system obtained by applying the controller (10), (7), (16), to (1). Let $u \triangleq (u_1, u_2, \cdots, u_m)'$, $y \triangleq (y_1, y_2, \cdots, y_r)'$. Then a *sensor j failure occurs in the controlled system (1), (10), (7), (16) if* $y_j(t) \equiv \bar{y}_j$ *(a constant), and an actuator j failure occurs if* $u_j(t) \equiv \bar{u}_j$ *(a constant).*

Definition (Single Sensor or Single Actuator Failure): Given $\epsilon^* > 0$, the system (1), (10), (7), (16) has *integrity with respect to a single-sensor failure if the resultant closed-loop system has eigenvalues contained in* Ω_{ϵ^*} *when a sensor j failure occurs,* $j=1,2,\cdots,r$.

The system (1), (10), (7), (16) *has integrity with respect to a single-actuator failure if the resultant closed-loop system has eigenvalues contained in* Ω_{ϵ^*} *when an actuator j failure occurs,* $j=1,2,\cdots,m$.

The following definition is made for multiple-sensor or/and multiple-actuator failure.

Definition (Multiple Sensor and Multiple Actuator Failure): Given $\epsilon^* > 0$, the system (1), (10), (7), (16) has *integrity with respect to a s-sensor failure and a t-actuator failure if the resultant closed-loop system has eigenvalues contained in* Ω_{ϵ^*} *for the case* when a sensor $i_1, i_2, \cdots, i_s$ failure occurs and an actuator $j_1, j_2, \cdots, j_t$ failure occurs $\forall i_k \in [0,1,\cdots,r]$, $k=1,2,\cdots,s$, $\forall j_l \in [0,1,\cdots,m]$, $l=1,2,\cdots,t$.

The following example gives an illustration of this definition.

Example: Let $r=3$, $s=2$; then given $\epsilon^* > 0$, a system has integrity with respect to a 2-sensor failure, if the resultant system always has eigenvalues $\in \Omega_{\epsilon^*}$ under the following failure conditions:

1) When a sensor 1, 2, or 3 failure occurs.

2) When a sensor 1 and sensor 2, or a sensor 1 and sensor 3, or a sensor 2 and sensor 3 simultaneously occur.

The following problem is now defined.

G. Robust Servomechanism Problem with s-Sensor and t-Actuator Failure Integrity ϵ^*

Assume the plant (1) is open-loop stable. Then given $\epsilon^* > 0$, find a controller for (1) so that when applied to (1), 1)–3) of Section III-A are true and in addition the resultant controlled system has integrity with respect to a s-sensor failure and a t-actuator failure. The following result is obtained.

Theorem 5: Consider the system (1) in which A is assumed to be asymptotically stable; then there exists a solution to the robust servomechanism problem with s-sensor failure and t-actuator failure integrity ϵ^* if and only if there exists a solution to the robust servomechanism problem (i.e., Lemma 1 holds).

Proof: The proof of sufficiency is similar to the proof of Theorem 3, i.e., by Lemma 2, there exists $\hat{\epsilon} > 0$ so that for any given ϵ, $0 < \epsilon < \hat{\epsilon}$ a controller $u = \Theta_\epsilon \eta$ exists, where η is the output of the servocompensator (7), such that $\|\Theta_\epsilon\| < \epsilon$ and such that the resultant closed-loop system matrix $\mathbf{A}_\epsilon$ given by (19) is asymptotically stable. On noting now that the eigenvalues of $\mathbf{A}_\epsilon$ are continuous in the elements of Θ_ϵ under all failure conditions, and that the eigenvalues of $\mathbf{A}_\epsilon$ with $\Theta_\epsilon = 0$ are contained in Ω_{ϵ^*} under all failure conditions, it is concluded that for a given ϵ^*, there exists $\hat{\epsilon}_0 < \hat{\epsilon}$ so that the eigenvalues of $\mathbf{A}_{\hat{\epsilon}_0}$ under all failure conditions are contained in Ω_{ϵ^*}, which proves the result.

Remark 3: If a system has integrity with respect to a s-sensor failure and a t-actuator failure, then in the case of sensor/actuator failure, the resultant failed system will not, in general, track/regulate, but will have the property that it will at the worst be only "slightly unstable" (depending on the choice of ϵ^*), i.e., the failed system will not be "explosively unstable."

The following problem is now considered.

Definition: In (1) assume that the plant model has the property that $A=A(\gamma)$, $B=B(\gamma)$, $C=C(\gamma)$, $C_m=C_m(\gamma)$, $D=D(\gamma)$ where γ is a vector of plant parameters, which under nominal operating conditions has the property that $\gamma=\gamma_0$, but depending on the operating conditions, may take on l different values $\gamma=\gamma_i$, $i=1,2,\cdots,l$. Then $\{C(\gamma), C_m(\gamma), A(\gamma), B(\gamma), D(\gamma)\}$, $\gamma=\gamma_i$, $i=0,1,\cdots,l$ is called *a plant with variable plant parameters* γ_i, $i=0,1,\cdots,l$.

The following problem is now defined.

H. Robust Servomechanism Problem with Tolerance (with Respect to γ_i, $i=0,1,\cdots,l$ *and* $\bar{\epsilon}$*)*

Assume that the plant (1) is a plant with variable plant parameters γ_i, $i=0,1,\cdots,l$, and assume that $A(\gamma)$ is asymptotically stable for $\gamma=\gamma_i$, $i=0,1,\cdots,l$. Then given $\bar{\epsilon}$, find a controller for (1) so that when applied to (1) with $\gamma=\gamma_0$, 1)–3) of Section III-A are true and in addition, the resultant controlled system has the property that the eigenvalues of the closed-loop system are contained in $\Omega_{\bar{\epsilon}}$ for all plant parameter perturbations γ_i, $i=1,2,\cdots,l$. The following result is obtained.

Theorem 6: Consider the system (1) which has variable

plant parameters γ_i, $i=0,1,\cdots,l$ and in which it is assumed that $A(\gamma_i)$ is asymptotically stable, $i=0,1,\cdots,l$; then there exists a solution to the robust servomechanism problem with tolerance with respect to γ_i, $i=0,1,\cdots,l$ and $\bar{\epsilon}$ if and only if there exists a solution to the robust servomechanism problem for (1) with $\gamma=\gamma_0$.

Proof: The proof is similar to the proof of Theorem 5 and so it is omitted.

Remark 4: If a solution to this problem exists, the system has the property that the resultant perturbed closed-loop system will at worst be only "slightly unstable" depending on the choice of $\bar{\epsilon}$.

Remark 5: The only assumption made re the perturbed plant is that it be open-loop stable for all perturbations (which may be arbitrarily large). Thus, the result obtained above is probably the "best" result that can be obtained without imposing more structure on the plant and on the type of perturbations allowed.

IV. Controller Design for the Robust Servomechanism Problem (Controller Parameter Determination)

So far, only the conditions for the existence of a solution and a characterization of the controller structure to solve the robust servomechanism problem have been discussed, i.e., Theorems 1 and 2 only give guidance re the structure of a stabilizing compensator which can be used and Theorems 3–6 only establish the existence of a solution to the various constrained servomechanism problems. It is desired now to determine the "optimal" values of the unspecified parameters of the controller so that the resultant controlled system has:

i) fast response, i.e., the transient errors of the controlled system should decay "as fast as possible"; and

ii) low interaction, i.e., when the reference signal at the jth output changes, the other outputs y_i, $i=1,2,\cdots,r$, $i\neq j$ should remain at their previous set-values "as much as possible."

A performance index will now be defined which reflects the speed of response and interaction in the servomechanism problem, and the unspecified controller parameters of the system will then be found to minimize this performance index.

Definition: Given $(\mathsf{C}^*,\mathsf{B}^*)$ of (8) let

$$\mathsf{D}\triangleq \text{block diag}\underbrace{(\alpha,\alpha,\cdots,\alpha)}_{r}\mathsf{T}^{-1} \tag{21}$$

where $\alpha\in R^{1\times p}$ is given by $\alpha=(1,0,\cdots,0)$.

A. Augmented System

Consider now the following augmented system, obtained by applying the servocompensator (7) to the plant (1):

$$\begin{pmatrix}\dot{x}\\ \dot{\eta}\end{pmatrix}=\begin{pmatrix}A & 0\\ \mathsf{B}^*C & \mathsf{C}^*\end{pmatrix}\begin{pmatrix}x\\ \eta\end{pmatrix}+\begin{pmatrix}B\\ \mathsf{B}^*D\end{pmatrix}u+\begin{pmatrix}E & 0\\ \mathsf{B}^*F & -\mathsf{B}^*\end{pmatrix}\begin{pmatrix}\omega\\ y_{\text{ref}}\end{pmatrix}$$

$$z=(0\ \ \mathsf{D})\begin{pmatrix}x\\ \eta\end{pmatrix}. \tag{22}$$

In this case z corresponds to the output of the servocompensator (7) rather than the error in the system. Assume that Lemma 1 holds; then this system has the following properties.

Proposition 1:

i) The system

$$\left\{(0\ \ \mathsf{D}),\begin{pmatrix}A & 0\\ \mathsf{B}^*C & \mathsf{C}^*\end{pmatrix},\begin{pmatrix}B\\ \mathsf{B}^*D\end{pmatrix}\right\}$$

is stabilizable and detectable and has the same fixed modes as (C,A,B).

ii) The transmission zeros of

$$\left\{(0\ \ \mathsf{D}),\begin{pmatrix}A & 0\\ \mathsf{B}^*C & \mathsf{C}^*\end{pmatrix},\begin{pmatrix}B\\ \mathsf{B}^*D\end{pmatrix}\right\}$$

are equal to the transmission zeros of (C,A,B,D).

Proof: For the proof of i) and ii), see Lemmas 5 and 6 of the Appendix, respectively.

Assume now that the following stabilizing controller is to be applied to the system (22).

B. Stabilizing Compensator

$$u=K\eta+K_0\hat{x},\ \dot{\xi}=\Lambda_0\xi+\Lambda_1\hat{y}_m+\Lambda_2\eta,\ \hat{y}_m\triangleq y_m-D_mu,\ \xi\in R^{\hat{n}}$$

$$\hat{x}=\Lambda_3\hat{y}_m+\Lambda_4\xi \tag{23}$$

and let the controller parameters be denoted by $\mathsf{K}\triangleq\{K,K_0;\Lambda_0,\Lambda_1,\Lambda_2;\Lambda_3,\Lambda_4\}$.

The controlled plant (22), (23) is then described by

$$\begin{aligned}\dot{\bar{x}}&=\mathsf{A}(\mathsf{K})\bar{x}+(B_1^*(\mathsf{K}),B_2^*)\begin{pmatrix}\omega\\ y_{\text{ref}}\end{pmatrix}\\ e&=C^*(\mathsf{K})\bar{x}-y_{\text{ref}}+(F+DK_0\Lambda_3F_m)\omega\\ z&=\mathsf{D}^*\bar{x}\\ u&=K^*(\mathsf{K})\bar{x}+K_0\Lambda_3F_m\omega\end{aligned} \tag{24}$$

where

$$\bar{x}\triangleq(x',\eta',\xi')'$$

and

$$\mathsf{A}(\mathsf{K})\triangleq\begin{bmatrix}A+BK_0\Lambda_3C_m & BK & BK_0\Lambda_4\\ \mathsf{B}^*(C+DK_0\Lambda_3C_m) & \mathsf{C}^*+\mathsf{B}^*DK & \mathsf{B}^*DK_0\Lambda_4\\ \Lambda_1C_m & \Lambda_2 & \Lambda_0\end{bmatrix},$$

$$B_1^*(\mathsf{K})\triangleq\begin{bmatrix}E+BK_0\Lambda_3F_m\\ \mathsf{B}^*F+\mathsf{B}^*DK_0\Lambda_3F_m\\ \Lambda_1F_m\end{bmatrix},\quad B_2^*\triangleq\begin{bmatrix}0\\ -\mathsf{B}^*\\ 0\end{bmatrix}$$

$$C^*(\mathsf{K})\triangleq(C+DK_0\Lambda_3C_m,DK,DK_0\Lambda_4),\quad \mathsf{D}^*\triangleq(0,\mathsf{D},0),$$

$$K^*(\mathsf{K})\triangleq(K_0\Lambda_3C_m,K,K_0\Lambda_4)$$

or alternatively on substituting for (2), (3) by

$$\begin{bmatrix}\dot{\bar{x}}\\ \dot{\eta}_1\\ \dot{\eta}_2\end{bmatrix}=\begin{bmatrix}\mathsf{A}(\mathsf{K}) & B_1^*(\mathsf{K})\mathsf{C}_1 & B_2^*G\mathsf{C}_2\\ 0 & \mathsf{A}_1 & 0\\ 0 & 0 & \mathsf{A}_2\end{bmatrix}\begin{bmatrix}\bar{x}\\ \eta_1\\ \eta_2\end{bmatrix}$$

$$e=\left[C^*(\mathsf{K}) \quad (F+DK_0\Lambda_3F_m)\mathsf{C}_1 \quad -G\mathsf{C}_2\right]\begin{bmatrix}\bar{x}\\ \eta_1\\ \eta_2\end{bmatrix}$$

$$z=\left[\mathsf{D}^* \quad 0 \quad 0\right]\begin{bmatrix}\bar{x}\\ \eta_1\\ \eta_2\end{bmatrix}$$

$$u=\left[K^*(\mathsf{K}) \quad K_0\Lambda_3F_m\mathsf{C}_1 \quad 0\right]\begin{bmatrix}\bar{x}\\ \eta_1\\ \eta_2\end{bmatrix}. \tag{25}$$

Assume now that the stabilizing compensator (23) and K have been chosen so that $\mathsf{A(K)}$ is asymptotically stable (always possible to do from Remark 1) and let

$$\tilde{x} \triangleq \bar{x}-\mathsf{T}_1(\mathsf{K})\eta_1-\mathsf{T}_2(\mathsf{K})\eta_2$$

in (25), where $\mathsf{T}_1(\mathsf{K})$, $\mathsf{T}_2(\mathsf{K})$ are given by the solution of

$$\mathsf{A(K)}(\mathsf{T}_1,\mathsf{T}_2)-(\mathsf{T}_1,\mathsf{T}_2)\begin{pmatrix}\mathsf{A}_1 & 0\\ 0 & \mathsf{A}_2\end{pmatrix}=-(B_1^*(\mathsf{K})\mathsf{C}_1, B_2^*G\mathsf{C}_2) \tag{26}$$

(since the eigenvalues of

$$\mathsf{A(K)} \text{ and } \begin{pmatrix}\mathsf{A}_1 & 0\\ 0 & \mathsf{A}_2\end{pmatrix}$$

are disjoint, a solution to (26) always exists) to obtain the following description of the controlled plant.

C. Controlled Plant

$$\left.\begin{aligned}\dot{\tilde{x}}&=\mathsf{A(K)}\tilde{x}\\ e&=C^*(\mathsf{K})\tilde{x}\\ \Delta z&=\mathsf{D}^*\tilde{x}\\ \Delta u&=K^*(\mathsf{K})\tilde{x}\end{aligned}\right\} \text{transient behavior of } e, \Delta z, \Delta u \tag{27}$$

where

$$\tilde{x}(0) \triangleq (I, -\mathsf{T}_1(\mathsf{K}), -\mathsf{T}_2(\mathsf{K}))\begin{bmatrix}\bar{x}(0)\\ \eta_1(0)\\ \eta_2(0)\end{bmatrix}$$

where

$$\Delta z \triangleq z-z_\infty, \quad \Delta u \triangleq u-u_\infty$$

and where

$$\left.\begin{aligned}z_\infty &\triangleq \mathsf{D}^*(\mathsf{T}_1(\mathsf{K})e^{\mathsf{A}_1t}\eta_1(0)+\mathsf{T}_2(\mathsf{K})e^{\mathsf{A}_2t}\eta_2(0))\\ u_\infty &\triangleq (K^*(\mathsf{K})\mathsf{T}_1(\mathsf{K})+K_0\Lambda_3F_m\mathsf{C}_1)e^{\mathsf{A}_1t}\eta_1(0)+K^*(\mathsf{K})\mathsf{T}_2(\mathsf{K})e^{\mathsf{A}_2t}\eta_2(0)\end{aligned}\right\} \text{steady-state behavior of } z, u$$

D. Performance Index for Controlled Plant

Define now the following performance index for (27),

$$J(\mathsf{K}) \triangleq E\int_0^\infty(\Delta z'Q\Delta z+\mu\Delta u'R\Delta u)\,d\tau$$

$$\triangleq \text{trace}\left\{(I,-\mathsf{T}_1(\mathsf{K}),-\mathsf{T}_2(\mathsf{K}))'\Gamma(\mathsf{K})(I,-\mathsf{T}_1(\mathsf{K}),-\mathsf{T}_2(\mathsf{K}))\right\} \tag{28}$$

where $Q>0$, $R>0$, $\mu>0$ and where $\Gamma(\cdot)$ is the solution of the matrix Lyapunov equation

$$\mathsf{A}'(\mathsf{K})\Gamma+\Gamma\mathsf{A}(\mathsf{K})=-\{\mathsf{D}^{*\prime}Q\mathsf{D}^*+\mu K^{*\prime}(\mathsf{K})RK^*(\mathsf{K})\} \tag{29}$$

where $\mu\to0$ and where K is constrained so that $\mathsf{A(K)}$ is asymptotically stable.

This performance index will be taken as a measure of the speed of response and as a measure of the noninteraction occurring in the controlled system, i.e., the term $\int_0^\infty\Delta z'Q\Delta z\,d\tau$ corresponds to the transient error associated with the tracking/regulator problem, which depends on the speed of response and the interaction in the system. An arbitrary small value of μ is chosen (say $\mu=10^{-8}$) to reflect the fact that the behavior of the controller inputs is not of direct concern (i.e., the so-called "cheap control" problem [17], [18] or "perfect control" problem [19]). Using this performance index measures the "average cost" of $\int_0^\infty(\Delta z'Q\Delta z+\mu\Delta u'R\Delta u)\,d\tau$ over all initial conditions $\bar{x}(0)$, $\eta_1(0)$, $\eta_2(0)$ uniformly distributed on a unit ball [20]–[22].

E. Optimal Stabilizing Compensator

It is of interest to know what type of structure an "optimal" stabilizing compensator (10) for (1) has. To determine this, assume Lemma 1 holds, $\omega\equiv0$, $y_{\text{ref}}\equiv0$ and consider the system

$$\dot{\bar{x}}=\mathsf{A}\bar{x}+\begin{pmatrix}B\\ \mathsf{B}^*D\end{pmatrix}u,\ \mathsf{A}\triangleq\begin{pmatrix}A & 0\\ \mathsf{B}^*C & \mathsf{C}^*\end{pmatrix},\ \bar{x}\triangleq\begin{pmatrix}x\\ \eta\end{pmatrix}$$

$$z=(0 \quad \mathsf{D})\bar{x}. \tag{30}$$

Then since the system

$$\left\{(0 \quad \mathsf{D}), \mathsf{A}, \begin{pmatrix}B\\ \mathsf{B}^*D\end{pmatrix}\right\}$$

is stabilizable and detectable from Proposition 1, there exists an optimal solution to the problem of

$$\min_u \int_0^\infty(z'Qz+\mu u'Ru)\,d\tau; \qquad Q>0, R>0, \mu>0 \tag{31}$$

for (30) given by $u=K_\mu^{\text{opt}}\bar{x}$. Thus the *optimal stabilizing compensator* in this case consists of *state feedback* and any other type of output dynamic stabilizing compensator will always result in a suboptimal controller. This optimal stabilizing compensator has the following property.

Definition: Given the system (30), assume Lemma 1 holds, and let the optimal stabilizing compensator for (30) be denoted by $u=K_\mu^{\text{opt}}\bar{x}$; then if this optimal controller has the property that $\int_0^\infty z'Qz\,d\tau\to0$ for all bounded $\bar{x}(0)\in R^{n+rp}$ as $\mu\to0$, the controller $u=K_\mu^{\text{opt}}\bar{x}$ is said to be a *perfect stabilizing compensator* for (30).

Definition: Given the system (1), the system (C, A, B, D) is said to be the *minimum phase* if its transmission zeros [13] are all contained in the open left-half complex plane.

The following result is obtained.

Theorem 7: Consider the system (1) and assume that Lemma 1 holds; then there exists a perfect stabilizing compensator for (30) if and only if the plant (C, A, B, D) is minimum phase and (A, B) is controllable.

Proof: The proof immediately follows from Proposition 1 and [17], [18] on noting that

$$\left\{(0 \quad \mathsf{D}), \mathsf{A}, \begin{pmatrix} B \\ \mathsf{B}^*D \end{pmatrix}\right\}$$

has no unstable transmission zeros if and only if (C, A, B, D) has no unstable transmission zeros, and that

$$\left\{\mathsf{A}, \begin{pmatrix} B \\ \mathsf{B}^*D \end{pmatrix}\right\}$$

is controllable if and only if (A, B) is controllable.

Remark 6: This result says that if a perfect stabilizing compensator is used in (30), then the performance criterion $\int_0^\infty z'Qzd\tau$ can be made arbitrarily small provided the plant is controllable and minimum phase. In this case, the resulting eigenvalues of the controlled system either approach the transmission zeros of the plant or approach negative infinity (in various Butterworth patterns) as $\mu \to 0$ [17], [18]. If the plant is nonminimum phase, this result emphasizes the inherent difficulty of trying to achieve "good regulation" in this case.

Remark 7: The motivation for using the performance index J given by (28) is that it has the property discussed previously, i.e., if a state feedback stabilizing compensator is applied to a minimum phase plant so as to minimize (28) with $E=0$, $F=0$, $F_m=0$, $G=0$, the eigenvalues of the controlled system will either approach negative infinity or the transmission zeros of the plant as $\mu \to 0$. This is not true with other choices of performance indexes, e.g., if a performance index like $J=E\int_0^\infty (e'Qe+\mu\Delta u'R\Delta u)d\tau$ is used, this implies (since the system

$$\left\{(C \quad 0), \mathsf{A}, \begin{pmatrix} B \\ \mathsf{B}^*D \end{pmatrix}, D\right\}$$

is nonminimum phase with transmission zeros on the imaginary axis) that some closed-loop eigenvalues of the resultant system will approach the imaginary axis as $\mu \to 0$, which is clearly undesirable.

F. Robust Servomechanism Problem with Fast Speed of Response/Low Interaction

Assume now that the stabilizing control structure (23) is specified, but that the controller parameters K are unknown and are to be determined. The following parameter optimization problem is now to be solved in order to maximize the speed of the response/minimize interaction effects of the resultant closed-loop system.

Parameter Optimization Problem: Given the system (1) and stabilizing compensator (23), find K so as to

$$\min_{\mathsf{K}} J(\mathsf{K}) \text{ defined by (28)} \tag{32}$$

subject to the constraint that $\mathsf{A(K)}$ be asymptotically stable.

Choice of Feasible Initial Starting Point: If Lemma 1 holds, then there always exists a feasible starting point for this optimization problem, provided the stabilizing compensator (23) has large enough order. In particular, by Theorem 1, there always exists a stabilizing compensator of order $\hat{n} < n-r$ for (23) by which there exists a K so that $\mathsf{A(K)}$ of (24) is asymptotically stable. If the plant (1) is open-loop asymptotically stable, the choice of $\Lambda_1=0$, $\Lambda_2=0$, $\Lambda_3=0$, $\Lambda_4=0$, $K_0=0$, $K=\Theta_\epsilon$, $\Lambda_0=-I$ in the stabilizing compensator (23) is a feasible starting point for any $\hat{n}$, where ϵ is chosen "small enough" and Θ_ϵ is obtained from Lemma 2 and [23].

Remark 8: In the special but important case of tracking/regulating constant set-points and constant disturbances, Θ_ϵ is given by $\Theta_\epsilon \triangleq -\epsilon(D-CA^{-1}B)^+$, $\epsilon>0$ [23].

Remark 9: In the case when output feedback is used in the stabilizing compensator (23), it is to be noted that the solution of (32) is not necessarily unique, i.e., (32) may have several local minima.

G. Robust Servomechanism Problem with Fast Speed of Response/Low-Interaction and Damping Factor, Gain, Integrity, and Tolerance Constraints

Assume now that given the stabilizing compensator structure (23), it is desired to determine the controller parameters K so as to achieve fast speed of response/low interaction, and also to satisfy the damping factor, constrained gain, integrity, and tolerance problem discussed previously. The following parameter optimization problem is now to be carried out.

Definition: Given that a matrix A is a function of a vector γ, let $sp[\mathsf{A}|_{\gamma=\gamma_0}]$ denote the eigenvalue spectrum of A with $\gamma=\gamma_0$.

Parameter Optimization Problem with General Constraints: Given the system (1) has variable plant parameters γ_i, $i=0, 1, \cdots, l$, assume that $A(\gamma_i)$ is asymptotically stable for $i=0,1,\cdots,l$ and that c, ϵ_0, ϵ^*, $\bar{\epsilon}$, and ζ are given, where $\zeta < \zeta_0$ and ζ_0 is the damping factor of $\mathsf{A}(\gamma_0)$. Then it is desired to find K so as to

$$\min_{\mathsf{K}} J(\mathsf{K}) \text{ defined by (28)} \tag{33}$$

subject to the constraints

1) $\mathsf{A(K)}\|_{\gamma=\gamma_0}$ is asymptotically stable	Stability constraint
2) $\\|K_0, K\\| < c$	Controller gain constraint
3) $sp[\mathsf{A(K)}\|_{\gamma=\gamma_0}] \subset \Omega_{\zeta,\epsilon_0}$	Damping factor constraint
4) $sp[\mathsf{A(K)}\|\gamma=\gamma_0 \quad] \subset \Omega_{\epsilon^*}$, $c_{i_1}=0,\cdots,c_{i_s}=0$, $b_{j_1}=0,\cdots,b_{j_t}=0$; $\forall i_k \in [0,1,\cdots,r], k=1,2,\cdots,s$; $\forall j_l \in [0,1,\cdots,m], l=1,2,\cdots,t$	Integrity constraint (s-sensor failure, t-actuator failure)
5) $sp[\mathsf{A(K)}\|_{\gamma=\gamma_1,\gamma_2,\cdots,\gamma_l}] \subset \Omega_{\bar{\epsilon}}$	Tolerance constraint

where

$$B \triangleq (b_1, b_2, \cdots, b_m),\ C \triangleq (c_1', c_2', \cdots, c'r)' \text{ in } (1).$$

Choice of Initial Feasible Starting Point: If Lemma 1 holds, it immediately follows from Theorems 3–6 that there exists a feasible starting point for this optimization problem. In particular, the choice of $\Lambda_1=0$, $\Lambda_2=0$, $\Lambda_3=0$, $\Lambda_4=0$, $K_0=0$, $K=\Theta_\epsilon$, $\Lambda_0=-I$ in the stabilizing compensator (23), is a feasible starting point, for any $\hat{n}$, where $\epsilon>0$ is chosen "small enough" depending on ϵ_0, ϵ^*, $\bar{\epsilon}$, c, ζ, and Θ_ϵ is obtained from [23].

Remark 10: It is clear that other constraints, which depend on the specific problem being considered, can also be easily imposed on the problem formulation. This will be illustrated in some of the design examples to follow.

The following is a summary of the suggested design method to solve the robust servomechanism problem.

H. Design Method for the Robust Servomechanism Controller Problem

Step 1: Given the plant (1), compute the transmission zeros of (C, A, B, D), the fixed modes of (C_m, A, B) and determine if a solution exists to the robust servomechanism problem (Lemma 1). If a solution exists go to Step 2.

Step 2: Synthesize the servocompensator (7) depending on the class of disturbances (2), reference inputs (3), and depending on the problem, specify the damping factor constraint, the controller gain constraint, the tolerance, and integrity constraint.

Step 3: Select the simplest possible stabilizing compensator which will solve the problem using the results of Theorems 1 and 2.

Step 4: Choose an appropriate value of Q, R, μ in (28) depending on the problem (typically $Q=I$, $R=I$, $\mu=10^{-8}$).

Step 5: Select an initial feasible starting point for the parameter optimization problem (32) or (33) using the method discussed in Sections IV-F or IV-G.

Step 6: Solve the parameter optimization problem (32) or (33).

Step 7: If the resulting controller obtained is not satisfactory, increase the complexity of the stabilizing controller [e.g., order $\hat{n}$ of (23)] and repeat Steps 5 and 6.

I. Computational Details

The following methods of solution for the above design procedure were used in all examples and have been found to be highly reliable.

1) The eigenvalues of the system were found by using a QR transformation method (as coded in EISPACK [24]).

2) The transmission zeros and fixed modes of the system were found by using eigenvalue methods as described in [25], [12], respectively.

3) The parameter optimization problem (33) was solved by using the algorithm given in [2], [26], [27]. This nonlinear programming method has the advantage that it searches orthogonal directions in the parameter space and thus is "robust," i.e., it does not easily fail on functions which have "ridge-like" contours or which are nondifferentiable. Alternatively other algorithms such as [28]–[31] could also be used.

4) The matrix Lyapunov equation (29) was solved by using [32] with a backup of [33] when the matrix A was nearly singular.

5) All simulations of the closed-loop system were carried out by using the algorithm [34].

All computations in the examples reported were done on an IBM 3033 digital computer using double precision arithmetic, with a value of $R=I$, $Q=I$, $\mu=10^{-8}$ in (28) being used. If an explicit value of E, F, F_m occurring in (1) was not known for a plant, the choice of $E=0$, $F=0$, $F_m=0$ was made in calculating the value of the performance index (28) in all design examples.

V. An Illustration of the Design Method—DC Motor

In order to demonstrate the design method, a relatively simple problem will initially be considered.

Problem: A dc motor modeled by the following equations [35]:

$$\dot{x}=\begin{bmatrix} -\dfrac{B}{J} & \dfrac{K_T}{J} \\ -\dfrac{K_e}{L} & -\dfrac{R}{L} \end{bmatrix} x+\begin{bmatrix} 0 \\ \dfrac{1}{L} \end{bmatrix} u+\begin{bmatrix} -\dfrac{1}{J} \\ 0 \end{bmatrix} \omega$$

$$y=(1 \quad 0)x,\ e \triangleq (y-y_{ref}) \tag{34}$$

where $B=0.0162$, $J=0.215$, $K_T=1.11$, $K_e=1.11$, $R=1.05$, $L=0.0053$ is to be regulated against constant load disturbances ω such that the output speed y tracks a constant set-point y_{ref}. The open-loop eigenvalues of the system are $(-5.70, -192.5)$.

Solution: The system has no fixed modes or transmission zeros, and therefore there exists a solution to the robust servomechanism problem from Lemma 1. The robust controller (10) for this problem is given by

$$u=K_0\hat{x}+K\eta, \qquad \dot{\eta}=e \tag{35}$$

where $\hat{x}$ is the output of a stabilizing compensator for the system (34). The following controllers are obtained from (35) on choosing stabilizing compensators of various complexity:

$$\text{I} \quad u=\frac{a_1}{s(1+a_4 s)}e(s)$$

$$\text{II} \quad u=\frac{a_1(1+a_2 s)}{s(1+a_4 s)}e(s)$$

$$\text{III} \quad u=\frac{a_1(1+a_2 s)(1+a_3 s)}{s(1+a_4 s)}e(s)$$

observer stabilizing compensator (12)

$$\text{IV} \quad u=(k_1, k_2)x(s)+\frac{K}{s}e(s)$$

state feedback stabilizing compensator

TABLE I
COMPARISON OF DIFFERENT CONTROLLERS OBTAINED FOR dc MOTOR PROBLEM

		Controller			
		I	II	III	IV
J_{opt} (28)		0.1244	10.7×10^{-3}	9.94×10^{-3}	9.93×10^{-3}
closed loop eigenvalues		$-2.3\pm j14$	$-51\pm j220$	$-119\pm j181$	$-115\pm j160$
		-194	-95	-192.5	-250
		-8.4×10^{7}	-2.2×10^{5}	-4.2×10^{3}	
controller parameters (optimal)	a_1	-38.4	-4970	-8770	
	a_2	--	1.23×10^{-2}	5.19×10^{-3}	
	a_3	--	--	5.19×10^{-3}	
	a_4	1.19×10^{-8}	4.56×10^{-6}	2.25×10^{-4}	
(k_1, k_2, K)		--	--	--	$(-98.0, -1.50, -10^{4})$

(a)

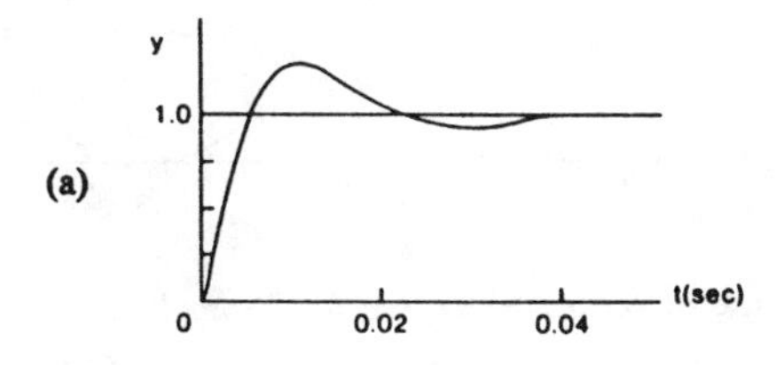

(b)

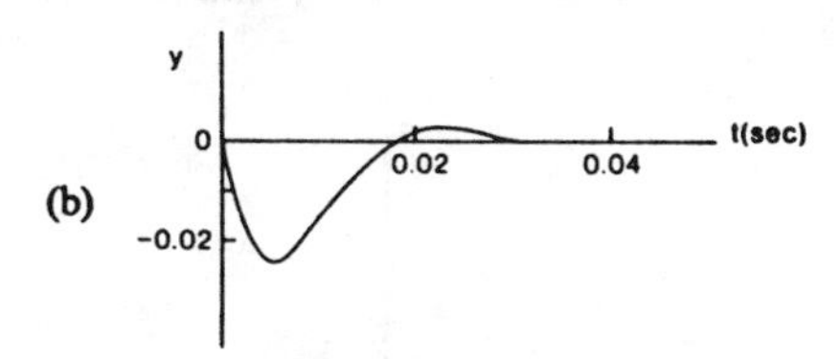

Fig. 3. Response of dc motor using proposed controller III. (a) $y_{ref}=1$, $\omega=0$. (b) $y_{ref}=0$, $\omega=1$.

and given the above controller structures, it is desired to find the optimal controller parameters to minimize (28) (with $Q=1$, $R=1$, $\mu=10^{-8}$, $E=0$, $F=0$). The results are obtained in Table I.

As expected it is seen that the performance of the system improves as the complexity of the stabilizing controller increases; it is interesting to note that the performance of controller II is only slightly worse than the optimal controller IV. A typical response of the controlled system using controller III is given in Fig. 3 for the case of a set point change in y_{ref} and for the case of a disturbance in ω for zero initial conditions. It is seen that excellent tracking/regulation occur as expected.

It is also to be observed that in the case of controller III (with an observer stabilizing compensator) given by

$$u(s) = -1050\frac{(s+192.5)(s+192.5)}{s(s+4450)}e(s)$$

the optimal controller parameters have been *automatically* found so that one of the compensator zeros cancels one of the plant's open-loop poles. This technique of choosing a compensator has been well established in classical single-input/single-output control theory, e.g., "an alternative way of choosing the compensator is to place its zeros where they will cancel some of the plant poles" [5, p. 83]. It is also to be noted that the other controllers I, II obtained are also consistent with classical design methods, e.g., the controller II obtained may be interpreted as being a "phase lead compensator" [5, p. 56] in terms of classical control theory.

A. Controller Gain Constraint

Suppose now that controller IV is used with a constraint on the controller gains (to account for saturation effects which might arise with excessively high controller gains). The following results are now obtained on minimizing (28) subject to the additional constraint $\sqrt{k_1^2+k_2^2+K^2} < 10$:

$$J_{opt} = 70.5\times10^{-3}$$

$$\text{closed-loop eigenvalues} = (-4.6\pm j28, -12)$$

$$(k_1, k_2, K)_{optimal} = (0.177, 0.937, -9.95).$$

In this case the magnitude constraint is satisfied and as expected, the performance of the closed-loop system is now slower compared to the case when this constraint is relaxed.

B. Integrity Type Constraint

Suppose now that controller IV is used with the constraint that the system be "fail-safe" with respect to the controller gains k_1, k_2. The following results are obtained on minimizing (28) subject to the additional constraint that the resultant closed-loop system remain stable when either $k_1\equiv0$ or $k_2\equiv0$:

$$J_{opt} = 22.3\times10^{-3}$$

$$\text{closed-loop eigenvalues} = (-41\pm j31, -2.1\times10^{4})$$

$$(k_1, k_2, K)_{optimal} = (-1800, -111.5, -5.84\times10^{4}).$$

In this case the constraints are satisfied, in particular, the eigenvalues of the resultant closed-loop system for the case of the active constraint $k_1\equiv0$ are given by $(-2.5\times10^{-9}\pm j52, -2.1\times10^{4})$. As expected the performance of the controlled system is now slower compared to when this constraint is relaxed. It is perhaps surprising to observe that the controller gains are now larger than in the unconstrained case.

C. Tolerance Constraint

1) Suppose now that controller IV is used with the constraint that when the plant parameters B, J, K_T, K_e, R, L are simultaneously allowed to vary by -50 percent or $+300$ percent, the resultant perturbed closed-loop system is to remain asymptotically stable. The following results are now obtained on minimizing (28) subject to the above constraints ($2^6+1=65$ constraints are now imposed):

$$J_{opt} = 9.94\times10^{-3}$$

$$\text{closed-loop eigenvalues} = (-118\pm j153, -257)$$

$$(k_1, k_2, K)_{optimal} = (-99.8, -1.567, -9907).$$

In this case all constraints are satisfied, in particular the eigenvalues of the two active constraints (occurring when B is decreased 50 percent, J is increased 300 percent, K_e is decreased 50 percent, R is decreased 50 percent, L is increased 300 percent, and when K_T is both decreased 50 percent and increased 300 percent) are given by $(-3.3\times 10^{-3}\pm j55, -99)$ and $(-4.6\times 10^{-4}\pm j156, -99)$, respectively.

2) Suppose now that controller IV is used with the constraint that when the above six plant parameters are simultaneously allowed to vary by ± 10 percent, the eigenvalues of the resultant closed-loop system are to remain in the region $\Omega=\{\lambda|\mathrm{Re}(\lambda)< -40 \text{ and } |\mathrm{Im}(\lambda)|/(|\mathrm{Re}(\lambda)|-40) < 2\}$. The following results are now obtained:

$$J_{\mathrm{opt}}=1.35\times 10^{-3}$$

$$\text{closed-loop eigenvalues}=(-79, -159\pm j136)$$

$$(k_1, k_2, K)_{\mathrm{optimal}}=(-69.7, -1.057, -3560).$$

In this case it may be verified that the constraints are satisfied; in particular, the eigenvalues of the active constraint $(-66, -139\pm j198)$ are a distance of 3×10^{-3} from the constraint boundary.

VI. Design Examples

A number of examples taken from the literature will now be considered.

Example 1 (Head Box): A pressurized head box described by the following equations [7]:

$$\dot{x}=\begin{pmatrix}-0.395 & 0.01145\\ -0.011 & 0\end{pmatrix}x+\begin{pmatrix}0.03362 & 1.038\\ 0.000966 & 0\end{pmatrix}u+E\omega$$

$$y=\begin{pmatrix}1 & 0\\ 0 & 1\end{pmatrix}x \tag{36}$$

is to be regulated against constant disturbances ω such that the output y tracks constant set points. The open-loop eigenvalues are $(-0.00032, -0.395)$.

Since the system has no transmission zeros and no fixed modes, there exists a solution to the robust servomechanism problem from Lemma 1. The following robust controller is obtained from (10) on assuming an output proportional term is used for the stabilizing compensator, i.e., see (17):

$$u=K_0 y+K_1\int_0^t (y-y_{\mathrm{ref}})d\tau \tag{37}$$

and it is desired to determine K_0, K_1 so as to minimize (28). The following results are obtained:

$$J_{\mathrm{opt}}=0.565$$

$$\text{closed-loop eigenvalues}=(-2.2\pm j2.2, -44\pm j62)$$

$$(K_0, K_1)_{\mathrm{optimal}}=\left(\begin{array}{cc|cc}28.3 & -4560 & 1750 & -10000\\ -85.3 & 150 & -5590 & 340\end{array}\right).$$

Some typical responses of the controlled system are given in Fig. 4 for the case of a set point change in y_{ref} and for

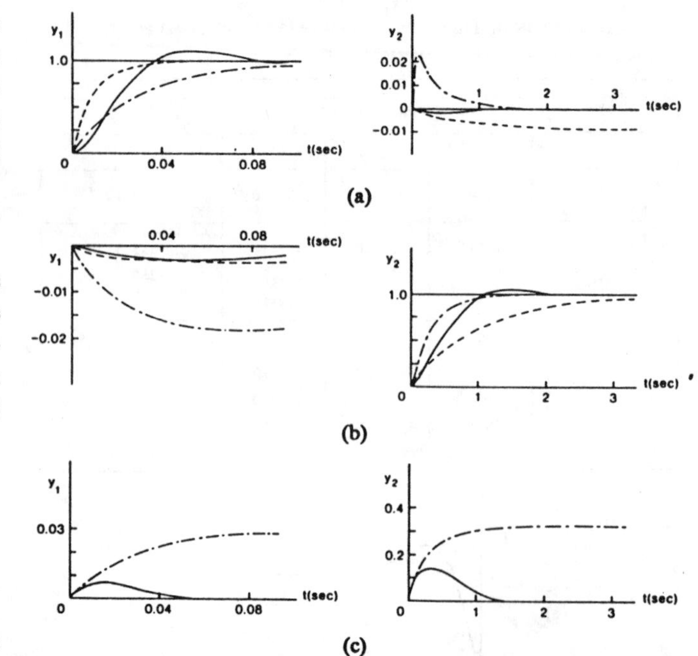

Fig. 4. Response of controlled system for Example 1. (a) $y_{\mathrm{ref}}=\begin{pmatrix}1\\0\end{pmatrix}$, $\omega=0$. (b) $y_{\mathrm{ref}}=\begin{pmatrix}0\\1\end{pmatrix}$, $\omega=0$. (c) Disturbance $E\omega=\begin{pmatrix}1\\1\end{pmatrix}$, $y_{\mathrm{ref}}=\begin{pmatrix}0\\0\end{pmatrix}$. Using proposed controller (37) ——. Using controller (38) ----. Using controller (39) ·—·—·—

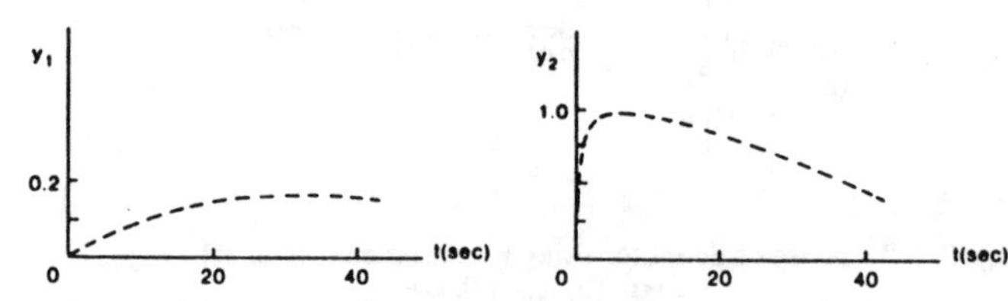

Fig. 5. Response of controlled system for Example 1 using controller (38) for case $y_{\mathrm{ref}}=\begin{pmatrix}0\\0\end{pmatrix}$, $E\omega=\begin{pmatrix}1\\1\end{pmatrix}$.

the case of an arbitrary constant disturbance $E\omega=\begin{pmatrix}1\\1\end{pmatrix}$ with zero initial conditions. It is seen that excellent tracking/regulation occur as expected. It is also to be noted that the above controller has integrity with respect to a sensor failure in either y_1 or y_2 (although an integrity constraint was not directly imposed on the problem).

It is of interest to compare this controller with controllers obtained by other design methods. The following controller is obtained by using the "characteristic locus design method" [7]

$$u=\begin{pmatrix}0 & -1030\\ -96.6 & 33.62\end{pmatrix}(y-y_{\mathrm{ref}})+\begin{pmatrix}-117.2 & 0\\ -0.12 & 1.1\end{pmatrix}\int_0^t (y-y_{\mathrm{ref}})d\tau \tag{38}$$

and produces closed-loop eigenvalues $(-0.02\pm j0.03, -1.0, -101)$. In this case some responses of the system using this controller are also given in Fig. 4—it may be observed that the eigenvalues $-0.02\pm j0.03$ will be excited for most disturbances occurring in the system and this will result in the system having a time-constant $100\times$ slower than the proposed controller. This effect is illustrated in Fig. 5.

The following controller is obtained by using a "triangular design method" [9]:

$$u=\begin{pmatrix} -1000 & -3064 \\ 0 & 100 \end{pmatrix}(y-y_{ref}) \tag{39}$$

which produces closed-loop eigenvalues $(-3.0, -34)$. In this case regulation will not occur as is illustrated in Fig. 4.

Example 2 (Unstable Batch Reactor): An unstable batch reactor described by the following equations [5]:

$$\dot{x}=\begin{bmatrix} 1.38 & -0.2077 & 6.715 & -5.676 \\ -0.5814 & -4.29 & 0 & 0.675 \\ 1.067 & 4.273 & -6.654 & 5.893 \\ 0.048 & 4.273 & 1.343 & -2.104 \end{bmatrix}x + \begin{bmatrix} 0 & 0 \\ 5.679 & 0 \\ 1.136 & -3.146 \\ 1.136 & 0 \end{bmatrix}u+E\omega$$

$$y=\begin{pmatrix} 1 & 0 & 1 & -1 \\ 0 & 1 & 0 & 0 \end{pmatrix}x \tag{40}$$

is to be regulated against constant disturbances such that the output y tracks constant set-points y_{ref}. The open-loop eigenvalues are $(1.99, 0.064, -5.057, -8.67)$.

Since the system has transmission zeros $(-1.192, -5.039)$ and no fixed modes, there exists a solution to the robust servomechanism problem from Lemma 1. The following robust controller is obtained from (17):

$$u=K_0 y+K_1\int_0^t (y-y_{ref})d\tau \tag{41}$$

and it is desired to determine K_0, K_1 so as to minimize (28) subject to the integrity constraint that the system's eigenvalues be not more than 10 percent "worse" than the system's open-loop unstable eigenvalues (i.e., $\mathrm{Re}(\lambda)<1.99 \times 1.1=2.19$) for a sensor failure in output y_1 or y_2. The following results are obtained:

$$J_{opt}=1.39\times 10^{-2}$$

closed-loop eigenvalues

$$=(-1.19, -5.04, -125\pm j125, -168\pm j169)$$

$$(K_0, K_1)_{optimal}=\left(\begin{array}{cc|cc} 0.95 & -58.4 & 100 & -1.0\times 10^4 \\ 79.3 & -0.72 & 1.0\times 10^4 & -100 \end{array}\right).$$

In this case the constraints are satisfied, in particular, the active constraint occurs when the output sensor for y_1 fails, and the resulting eigenvalues obtained in this case are given by $(2.15, 0, -1.01, -8.66, -168\pm j169)$. It is interesting to note that with this controller, approximate "pole-zero cancellation" occurs with respect to the eigenvalues -1.19, -5.04. This is to be expected from Remark 6. Some typical responses of the controlled system are given in Fig. 6 for the case of a set point change in y_{ref} and for the case of an arbitrary constant disturbance $E\omega=(1111)'$ for zero initial conditions. It is seen that excellent tracking and regulation occur as expected. It is interesting to compare this controller with the controller obtained by

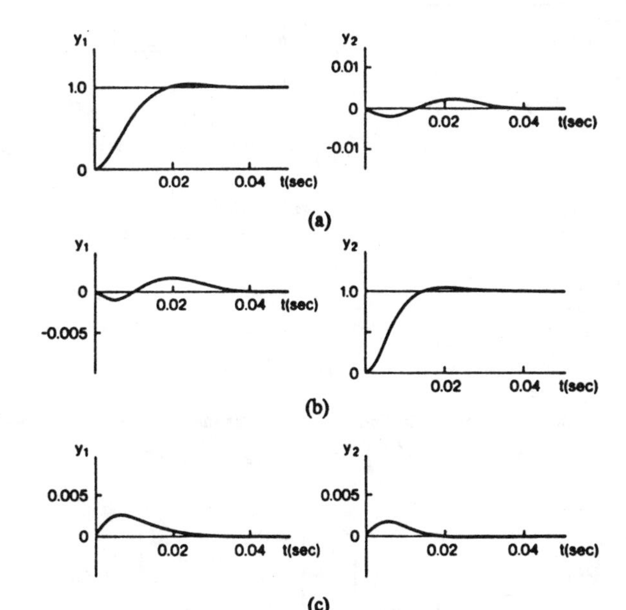

Fig. 6. Response of controlled system for Example 2 using proposed controller (41). (a) $y_{ref}=\begin{pmatrix}1\\0\end{pmatrix}$, $\omega=0$. (b) $y_{ref}=\begin{pmatrix}0\\1\end{pmatrix}$, $\omega=0$. (c) Disturbance $E\omega=\begin{bmatrix}1\\1\\1\\1\end{bmatrix}$, $y_{ref}=\begin{pmatrix}0\\0\end{pmatrix}$.

Rosenbrock [5]

$$u=K_0(y-y_{ref})+K_1\int_0^t (y-y_{ref})d\tau \tag{42}$$

where

$$K_0=\begin{pmatrix} 0 & -2 \\ 5 & 0 \end{pmatrix}$$

(K_1 is unspecified) which produces a time response some $100\times$ slower than the proposed controller (see Fig. 7). (It is to be noted, however, that this comparison is perhaps misleading since Rosenbrock [5] never suggested that his controller (42) was designed to be "fast.")

For interest, the following decentralized controller [36], [21],

$$u=\begin{pmatrix} k_{01} & 0 \\ 0 & k_{02} \end{pmatrix}y+\begin{pmatrix} k_{11} & 0 \\ 0 & k_{12} \end{pmatrix}\int_0^t (y-y_{ref})d\tau \tag{43}$$

was also considered to solve the problem, where the parameters k_{01}, k_{02}, k_{11}, k_{12} were found to minimize (28). The following results are obtained:

$$J_{opt}=39.0\times 10^{-2}$$

closed-loop eigenvalues

$$=(-0.06\pm j207, -1.16, -2.57, -2.79, -5.03)$$

$$(K_0, K_1)_{optimal}=\left(\begin{array}{cc|cc} -29.3 & 0 & -76.8 & 0 \\ 0 & -81.8 & 0 & -217 \end{array}\right).$$

In this case, the response of the system is considerably deteriorated as might be expected.

Example 3 (Rosenbrock Problem): The following system [6],

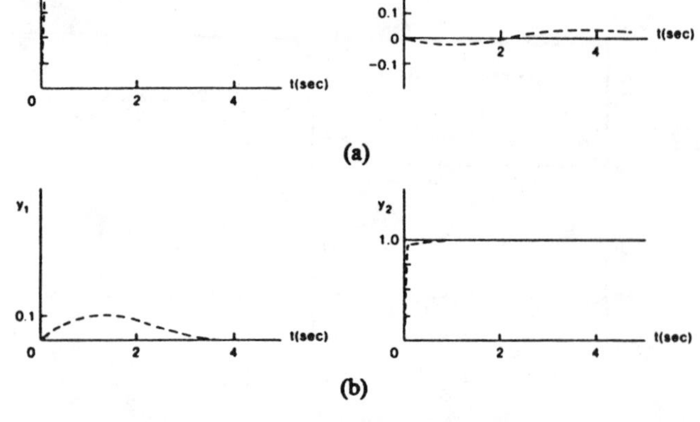

Fig. 7. Response of controlled system for Example 2 using controller (42). (a) $y_{ref}=\begin{pmatrix}1\\0\end{pmatrix}$, $\omega=0$. (b) $y_{ref}=\begin{pmatrix}0\\1\end{pmatrix}$, $\omega=0$.

$$\dot{x}=\begin{bmatrix}-1 & 1 & 0\\ 0 & -1 & 0\\ 0 & 0 & -1\end{bmatrix}x+\begin{bmatrix}-1/6 & 0\\ 2/3 & 1\\ 0 & 1/2\end{bmatrix}u+E\omega$$

$$y=\begin{pmatrix}3 & -3/4 & -1/2\\ 2 & -1 & 0\end{pmatrix}x, \qquad e=y-y_{ref} \tag{44}$$

is to be regulated against constant disturbances ω such that the output y tracks constant set points y_{ref}.

Since the system has no transmission zeros and no fixed modes, there exists a solution to the robust servomechanism problem from Lemma 1. The following robust controller is obtained from (17):

$$u=K_0y+K_1\int_0^t(y-y_{ref})d\tau \tag{45}$$

and it is desired to determine K_0, K_1 to minimize (28) subject to the integrity constraint that the system remain stable for a sensor failure in output y_1 or y_2 (i.e., one-sensor failure integrity with $\epsilon^*=10^{-8}$). The following results are obtained:

$$J_{opt}=0.254$$

$$\text{closed-loop eigenvalues}=(-2.8\pm j11, -4.9, -16\pm j21)$$

$$(K_0, K_1)_{optimal}=\left(\begin{array}{cc:cc}606 & -87 & 4960 & 2690\\ -595 & 117 & -4720 & -2280\end{array}\right).$$

In this case the constraints are satisfied, in particular, the eigenvalues of the resultant closed-loop system when the output sensor y_1, y_2 fails are given by $(0, -1.0, -7.2\pm j5.5, -17)$, $(0, -1.0, -3.4\pm j28, -5.7)$, respectively. Some typical responses of the closed-loop system are given in Fig. 8 for the case of a set point change in y_{ref} and for the case of an arbitrary constant disturbance $E\omega=(111)'$ with zero initial conditions. It is seen that excellent tracking/regulation occur.

It is of interest to compare this controller with the following controller obtained by Rosenbrock [6]:

$$u=\begin{pmatrix}100 & 300\\ -75 & -150\end{pmatrix}(y-y_{ref}) \tag{46}$$

which produces closed-loop eigenvalues $(-0.75\pm j3.8, -177)$. In this case robust regulation will not occur as is illustrated in Fig. 8.

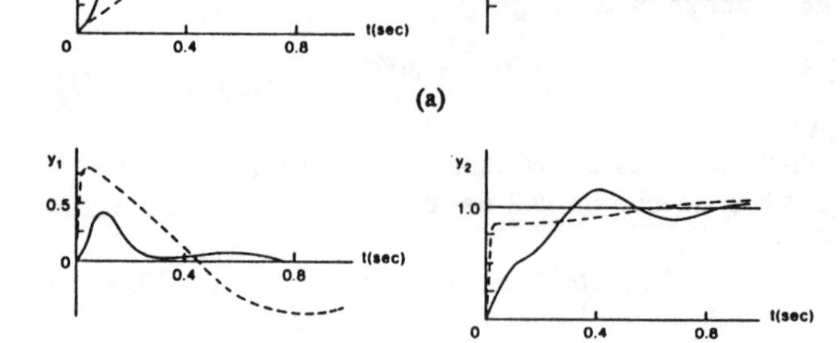

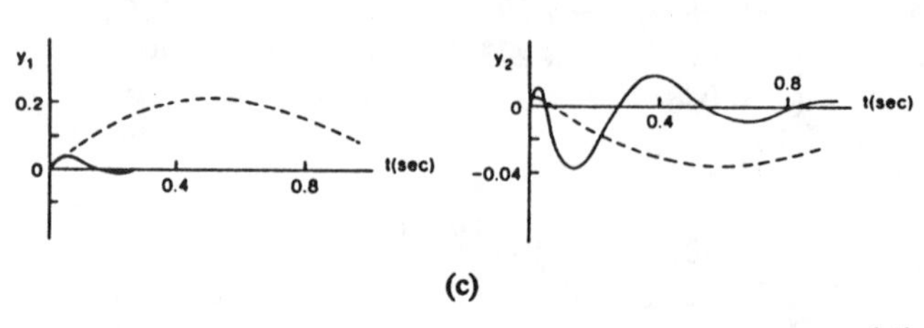

(c)

Fig. 8. Response of controlled system for Example 3. (a) $y_{ref}=\begin{pmatrix}1\\0\end{pmatrix}$, $\omega=0$. (b) $y_{ref}=\begin{pmatrix}0\\1\end{pmatrix}$, $\omega=0$. (c) $y_{ref}=\begin{pmatrix}0\\0\end{pmatrix}$, $E\omega=\begin{pmatrix}1\\1\\1\end{pmatrix}$. Using proposed controller (45) ——. Using controller (46) ----.

For interest the same problem is now considered with a constraint on the controller gains of the system, i.e., it is desired to determine K_0, K_1 to minimize (28) subject to the constraint that all elements of the controller gains be less than 100 in magnitude, i.e., $|k_{ij}|<100$, $\forall k_{ij}\in\{K_0, K_1\}$ in (45). The following results are obtained (nine constraints are now imposed):

$$J_{opt}=0.825$$

closed-loop eigenvalues

$$=(-1.23\pm j3.21, -2.00, -2.67\pm j6.67)$$

$$(K_0, K_1)_{optimal}=\left(\begin{array}{cc:cc}48.0 & 10.6 & 97.9 & 99.9995\\ -48.9 & -2.79 & -99.998 & -64.6\end{array}\right).$$

In this case the constraints are satisfied; in particular there are two active constraints occurring in the problem. As expected, the response of the system is now much slower as compared to the previous case.

Example 4 (Boiler Furnace): A boiler furnace with the following transfer function matrix [5]:

$$Y(s)=\begin{bmatrix}\frac{1}{1+4s} & \frac{0.7}{1+5s} & \frac{0.3}{1+5s} & \frac{0.2}{1+5s}\\ \frac{0.6}{1+5s} & \frac{1}{1+4s} & \frac{0.4}{1+5s} & \frac{0.35}{1+5s}\\ \frac{0.35}{1+5s} & \frac{0.4}{1+5s} & \frac{1}{1+4s} & \frac{0.6}{1+5s}\\ \frac{0.2}{1+5s} & \frac{0.3}{1+5s} & \frac{0.7}{1+5s} & \frac{1}{1+4s}\end{bmatrix}U(s) \tag{47}$$

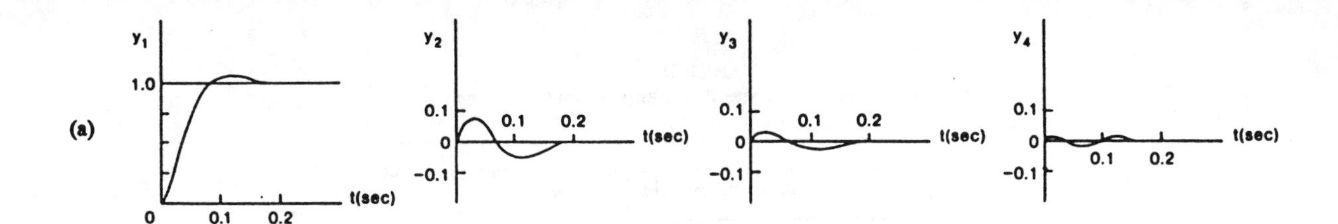

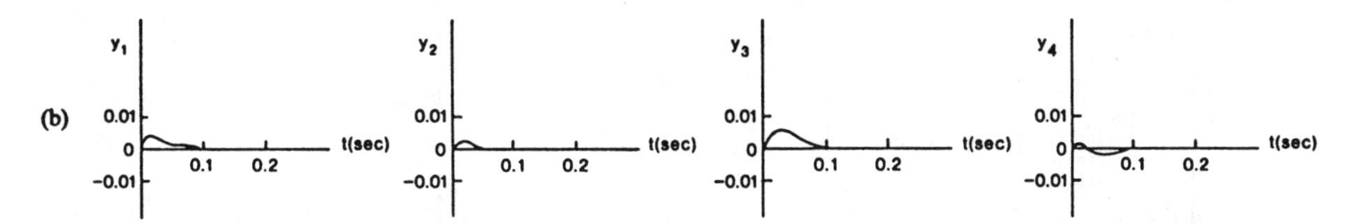

Fig. 9. Response of controlled system using proposed controller (49) for Example 4. (a) $y_{ref} = \begin{bmatrix} 1 \\ 0 \\ 0 \\ 0 \end{bmatrix}$, $\omega = 0$. (b) $y_{ref} = \begin{bmatrix} 0 \\ 0 \\ 0 \\ 0 \end{bmatrix}$, $E\omega = (1 \; 1 \; 1 \; 1 \; 1 \; 1 \; 1 \; 1)'$.

is to be regulated against constant disturbances such that the output y tracks constant set-points y_{ref}. In this case, a minimal realization of (47) is obtained [12] as follows:

$$\dot{x} = Ax + Bu + E\omega$$
$$y = Cx \tag{48}$$

where $x \in R^8$, $u \in R^4$, $y \in R^4$, and (C, A, B) are given in [12]. Since this system has transmission zeros $-0.145 \pm j0.0016$, -0.202, -0.225 and no fixed modes, there exists a solution to the robust servomechanism problem from Lemma 1. The open-loop eigenvalues of the system (48) are given by $(-0.2, -0.2, -0.2, -0.2, -0.25, -0.25, -0.25, -0.25)$. The following controller which has the simplest possible structure is considered for (48):

$$u = k_0 y + k_1 \int_0^t (y - y_{ref}) d\tau \tag{49}$$

where k_0, k_1 are scalars (this controller is a decentralized controller [36], [21] for (48) containing identical gains in all four channels). It is desired to determine k_0, k_1 so as to minimize (28) subject to the integrity constraint that the system remain stable for a sensor failure in either y_1, y_2, y_3, or y_4 (i.e., one-sensor failure integrity with $\epsilon^* = 10^{-8}$). The following results are obtained:

$$J_{opt} = 0.145$$

closed-loop eigenvalues

$$= \begin{Bmatrix} -0.14, -0.14, -0.20, -0.23, -21 \pm j28, -21 \pm j28, \\ -38, -45 \pm j25, -136 \end{Bmatrix}$$

$$(k_0, k_1)_{optimal} = (-346, -10300).$$

In this case the constraints are satisfied, e.g., when the sensor y_1 fails the eigenvalues of the resultant closed-loop system are given by $(0, -0.15, -0.17, -0.20, -0.22, -0.25, -24 \pm j27, -18 \pm j26, -32 \pm j28)$. Some representative responses of the controlled system are given in Fig. 9 for the case of a set point change in y_{ref} and for the case of an arbitrary constant disturbance $E\omega = (1\;1\;1\;1\;1\;1\;1\;1)'$ with zero initial conditions. It is seen that satisfactory tracking and regulation occur.

Remark 11: Some representative CPU times required to solve the above examples are given in Table II. Some additional examples of the design method applied to other problems may be found in [41].

VII. Conclusions

The problem of designing a controller for a plant to solve the robust servomechanism problem so that closed-loop stability and asymptotic regulation occur, and also so that other desirable properties of the controlled system result such as fast response, low interaction, integrity, tolerance to plant variations, etc. occur, is considered in this paper. The method of design is based on using state space methods via a two-stage process: 1) using theory determine the existence of a solution and controller structure required to solve the problem, and 2) using nonlinear programming methods, determine the unknown controller parameters so as to minimize a performance index for the system.

It is of interest to compare the proposed design method with the frequency domain design methods, e.g., [5]–[9] or graphical design methods, e.g., [10], [11].

1) The proposed design method guarantees that regulation/tracking occur for all perturbations of the plant, provided only that the resultant system remains stable. This differs from other design methods in which the servocompensator (7) is not considered to be an essential part of the compensator.

2) The proposed design method concentrates not only on the tracking problem, but also on the regulation problem. In contrast, the frequency-domain methods concentrate primarily on the tracking part of the servomechanism problem.

3) Existence results for the solvability of the problem are emphasized in the proposed design method. In other

TABLE II
REPRESENTATIVE CPU TIMES REQUIRED TO SOLVE EXAMPLES

Example	Order of Closed Loop Plant	No. of Parameters To Be Determined	No. of Constraints Imposed	CPU Time Required
DC Motor (Controller I)	4	2	1	2 sec
DC Motor (Controller IV)	3	3	1	4 sec
DC Motor (Controller IV with Controller Gain Constraint)	3	3	2	5 sec
DC Motor (Controller IV with Tolerance Constraint)	3	3	65	160 sec
Head Box	4	8	1	23 sec
Boiler Furnace	12	2	5	80 sec

design methods, the existence of a controller is only implicitly assumed.

4) In the proposed design method, constraints specified on the problem can be directly considered, whereas in the other design methods, there is no direct way of handling constraints.

5) The proposed design method considers disturbances/reference-input signals which are of a larger class than constant signals, unlike other design methods.

6) The sensitivity of the controller with respect to specific plant parameter perturbations (the tolerance problem) can be directly incorporated using the proposed design method; other sensitivity measures, e.g., [37]–[40], could also be used, but these measures do not reflect specific plant parameter perturbations, and as a result may tend to produce excessively conservative results.

Numerous examples are given to illustrate the design procedure, and comparisons made with other design methods indicate that the design procedure appears to be very attractive.

APPENDIX

The following lemmas are used in the development of the proof of the existence of a controller structure to satisfy the problem statements and in the development of the performance criterion for the parameter optimization problem.

Lemma 2 [23]: Given (C, A, B, D) of (1) and $(\mathsf{C}^*, \mathsf{B}^*)$ defined by (8), assume A is asymptotically stable; then there exists $\epsilon_0 > 0$, so that for any given ϵ, $0 < \epsilon < \epsilon_0$ a matrix $\Theta_\epsilon \in R^{m \times rp}$ exists with $\|\Theta_\epsilon\| < \epsilon$ such that

$$\begin{pmatrix} A & B\Theta_\epsilon \\ \mathsf{B}^* C & \mathsf{C}^* + \mathsf{B}^* D \Theta_\epsilon \end{pmatrix}$$

is asymptotically stable.

Lemma 3: Given $(\mathsf{C}^*, \mathsf{B}^*)$ defined by (8), and D defined by (21), then

$$\operatorname{rank}\begin{bmatrix} \mathsf{B}^* & \mathsf{C}^* - \lambda I \\ 0 & \mathsf{D} \end{bmatrix} = rp + r, \qquad \forall \lambda \in C \tag{1a}$$

Proof: By assumption $(\mathsf{C}^*, \mathsf{B}^*)$ is controllable, i.e.,

$$\operatorname{rank}\Big[\mathsf{B}, \text{block diag}\,(\underbrace{\mathsf{C} - \lambda I, \mathsf{C} - \lambda I, \cdots, \mathsf{C} - \lambda I}_{r})\Big] = rp, \qquad \forall \lambda \in C \tag{2a}$$

where B, C are defined by (8), (5), respectively. It now immediately follows that

$$\operatorname{rank}\begin{bmatrix} \mathsf{B} & \text{block diag}\,(\mathsf{C} - \lambda I, \mathsf{C} - \lambda I, \cdots, \mathsf{C} - \lambda I) \\ 0 & \text{block diag}\,(\alpha, \alpha, \cdots, \alpha) \end{bmatrix} = rp + r, \qquad \forall \lambda \in C \tag{3a}$$

from which it is concluded that (1a) is true.

Lemma 4: Given $(\mathsf{C}^*, \mathsf{B}^*)$ defined by (8) and D defined by (21), the pair

$$\left\{ (0 \;\; \mathsf{D}), \begin{pmatrix} A & 0 \\ \mathsf{B}^* C & \mathsf{C}^* \end{pmatrix} \right\}$$

has the same fixed modes as (C, A).

Proof: The proof immediately follows on noting that

$$\operatorname{rank}\begin{bmatrix} A - \lambda I & 0 \\ \mathsf{B}^* C & \mathsf{C}^* - \lambda I \\ 0 & \mathsf{D} \end{bmatrix} = \operatorname{rank}\left\{ \begin{bmatrix} I_n & 0 & 0 \\ 0 & \mathsf{B}^* & \mathsf{C}^* - \lambda I \\ 0 & 0 & \mathsf{D} \end{bmatrix} \begin{bmatrix} A - \lambda I & 0 \\ C & 0 \\ 0 & I_{rp} \end{bmatrix} \right\} \tag{4a}$$

and since from Lemma 3, it has been established that

$$\begin{pmatrix} \mathsf{B}^* & \mathsf{C}^*-\lambda I \\ 0 & \mathsf{D} \end{pmatrix}$$

is nonsingular $\forall \lambda \in C$, it is concluded that

$$\operatorname{rank}\begin{bmatrix} A-\lambda I & 0 \\ \mathsf{B}^*C & \mathsf{C}^*-\lambda I \\ 0 & \mathsf{D} \end{bmatrix} = \operatorname{rank}\begin{bmatrix} A-\lambda I & 0 \\ C & 0 \\ 0 & I_{rp} \end{bmatrix}$$

$$= rp + \operatorname{rank}\begin{pmatrix} A-\lambda I \\ C \end{pmatrix}, \qquad (5a)$$

i.e.,

$$\left\{(0 \quad \mathsf{D}), \begin{pmatrix} A & 0 \\ \mathsf{B}^*C & \mathsf{C}^* \end{pmatrix}\right\}$$

has the same fixed modes as (C, A).

Given (C, A, B, D), $(\mathsf{C}^*, \mathsf{B}^*)$, D defined previously, the following two lemmas now follow.

Lemma 5: Assume that Lemma 1 holds; then the system

$$\left\{(0,\mathsf{D}), \begin{pmatrix} A & 0 \\ \mathsf{B}^*C & \mathsf{C}^* \end{pmatrix}, \begin{pmatrix} B \\ \mathsf{B}^*D \end{pmatrix}\right\}$$

is stabilizable and detectable and has the same fixed modes as the system (C, A, B).

Proof: The proof immediately follows from Lemma 1 (see Section III-C) and Lemma 4.

Lemma 6: The transmission zeros of

$$\left\{(0,\mathsf{D}), \begin{pmatrix} A & 0 \\ \mathsf{B}^*C & \mathsf{C}^* \end{pmatrix}, \begin{pmatrix} B \\ \mathsf{B}^*D \end{pmatrix}\right\}$$

are equal to the transmission zeros of (C, A, B, D).

Proof: By definition, the transmission zeros are given by the set of λ which satisfy

$$\operatorname{rank}\begin{bmatrix} A-\lambda I & 0 & B \\ \mathsf{B}^*C & \mathsf{C}^*-\lambda I & \mathsf{B}^*D \\ 0 & \mathsf{D} & 0 \end{bmatrix} < n + rp + \min(r, m). \qquad (6a)$$

Now, on noting that

$$\operatorname{rank}\begin{bmatrix} A-\lambda I & 0 & B \\ \mathsf{B}^*C & \mathsf{C}^*-\lambda I & \mathsf{B}^*D \\ 0 & \mathsf{D} & 0 \end{bmatrix}$$

$$= \operatorname{rank}\left\{\begin{bmatrix} I_n & 0 & 0 \\ 0 & \mathsf{B}^* & \mathsf{C}^*-\lambda I \\ 0 & 0 & \mathsf{D} \end{bmatrix}\begin{bmatrix} A-\lambda I & 0 & B \\ C & 0 & D \\ 0 & I_{rp} & 0 \end{bmatrix}\right\} \qquad (7a)$$

it follows from Lemma 3 that

$$\begin{pmatrix} \mathsf{B}^* & \mathsf{C}^*-\lambda I \\ 0 & \mathsf{D} \end{pmatrix}$$

is nonsingular $\forall \lambda \in C$ and hence that

$$\operatorname{rank}\begin{bmatrix} A-\lambda I & 0 & B \\ \mathsf{B}^*C & \mathsf{C}^*-\lambda I & \mathsf{B}^*D \\ 0 & \mathsf{D} & 0 \end{bmatrix} = \operatorname{rank}\begin{bmatrix} A-\lambda I & 0 & B \\ C & 0 & D \\ 0 & I_{rp} & 0 \end{bmatrix}$$

$$= rp + \operatorname{rank}\begin{pmatrix} A-\lambda I & B \\ C & D \end{pmatrix} \qquad (8a)$$

from which it follows that the transmission zeros of

$$\left\{(0,\mathsf{D}), \begin{pmatrix} A & 0 \\ \mathsf{B}^*C & \mathsf{C}^* \end{pmatrix}, \begin{pmatrix} B \\ \mathsf{B}^*D \end{pmatrix}\right\}$$

are given by the transmission zeros of (C, A, B, D).

References

[1] E. J. Davison and W. Gesing, "A systematic design procedure for the multivariable servomechanism problem," in *Alternatives for Linear Multivariable Control*, M. K. Sain, J. L. Peczkowski, and J. L. Mesa, Eds. Chicago, IL: Nat. Eng. Consort., Inc., pp. 257–300.

[2] W. Gesing and E. J. Davison, "An exact penalty function algorithm for solving general constrained parameter optimization problems," *Automatica*, vol. 15, pp. 175–188, 1979.

[3] E. Polak and R. Trahan, "An algorithm for computer aided design problems," in *Proc. IEEE Contr. Decision Conf.*, 1976, pp. 537–542.

[4] D. Q. Mayne, "Computer-aided design of control systems via optimization," in *Proc. 1979 Joint Automat. Contr. Conf.*, pp. 371–374.

[5] H. H. Rosenbrock, *Computer-Aided Control System Design*. New York: Academic, 1974.

[6] ——, "Design of multivariable control systems using the inverse Nyquist array," *Proc. IEE*, vol. 116, pp. 1929–1936, 1969.

[7] A. G. J. MacFarlane and J. J. Belletrutti, "The characteristic locus design method," *Automatica*, vol. 9, pp. 575–588, 1973.

[8] D. Q. Mayne, "The design of linear multivariable systems," *Automatica*, vol. 9, pp. 201–208, 1973.

[9] N. T. Hung and B. D. O. Anderson, "Triangularization technique for the design of multivariable control systems," *IEEE Trans. Automat. Contr.*, vol. AC-24, pp. 455–460, June 1979.

[10] I. M. Horowitz, "A synthesis theory for linear time-varying feedback systems with plant uncertainty," *IEEE Trans. Automat. Contr.*, vol. AC-20, pp. 454–464, Aug. 1975.

[11] J. E. Ackermann, "A robust control system design," in *Proc. 1979 Joint Automat. Contr. Conf.*, pp. 877–883.

[12] E. J. Davison, W. Gesing, and S. H. Wang, "An algorithm for obtaining the minimal realization of a linear time-invariant system & determining if a system is stabilizable-detectable," *IEEE Trans. Automat. Contr.*, vol. AC-23, pp. 1048–1054, Dec. 1978.

[13] E. J. Davison and S. H. Wang, "Properties and calculation of transmission zeros of linear multivariable time-invariant systems," *Automatica*, vol. 10, pp. 643–658, 1974.

[14] E. J. Davison and A. Goldenberg, "The robust control of a general servomechanism problem: The servo compensator," *Automatica*, vol. 11, pp. 461–471, 1975.

[15] F. M. Brasch and J. B. Pearson, "Pole placement using dynamic compensators," *IEEE Trans. Automat. Contr.*, vol. AC-15, pp. 34–43, Feb. 1970.

[16] E. J. Davison and S. H. Wang, "On pole assignment in linear multivariable systems using output feedback," *IEEE Trans. Automat. Contr.*, vol. AC-20, pp. 516–518, Aug. 1975.

[17] H. Kwakernaak and R. Sivan, "The maximally achievable accuracy of linear optimal regulators and linear optimal filters," *IEEE Trans. Automat. Contr.*, vol. AC-17, pp. 79–86, Feb. 1972.

[18] B. A. Francis, "Perfect regulation and feedforward control of linear multivariable systems," in *Proc. 1977 Conf. Decision Contr.*, pp 760–762.

[19] E. J. Davison and S. G. Chow, "Perfect control in linear time-invariant multivariable systems: The control inequality principle," in *Control System Design by Pole-Zero Assignment*, F. Fallside, Ed. New York: Academic, 1975, pp. 1–15.

[20] E. J. Davison, N. S. Rau, and F. V. Palmy, "The optimal decentralized control of a power system consisting of a number of interconnected synchronous machines," *Int. J. Contr.*, vol. 6, pp. 1313–1328, 1973.

[21] E. J. Davison and W. Gesing, "Sequential stability & optimization of large scale decentralized systems," *Automatica*, vol. 15, pp. 307–324, 1979.

[22] W. S. Levine and M. Athans, "On the determination of the optimal constant output feedback gains for linear multivariable systems," *IEEE Trans. Automat. Contr.*, vol. AC-15, pp. 44–48, Feb. 1970.

[23] E. J. Davison, "Multivariable tuning regulators: The feedforward and robust control of a general servomechanism problem," *IEEE Trans. Automat. Contr.*, vol. AC-21, pp. 35–47, Feb. 1976.

[24] B. T. Smith *et al.*, *Matrix Eigensystem Routines—EISPACK Guide*. Berlin: Springer-Verlag, 1976.

[25] E. J. Davison and S. H. Wang, "An algorithm for the calculation of transmission zeros of the system (C, A, B, D) using high gain output feedback," *IEEE Trans. Automat. Contr.*, vol. AC-23, pp. 738–741, 1978.

[26] E. J. Davison and P. Wong, "A robust conjugate gradient algorithm which minimizes L functions," *Automatica*, vol. 11, pp. 297–308, 1975.

[27] W. Gesing and E. J. Davison, "Improvements on a robust conjugate-gradient algorithm which minimizes L functions," *Automatica*, vol. 14, pp. 515–516, 1978.

[28] E. Polak and D. Q. Mayne, "An algorithm for optimization problems with functional inequality constraints," *IEEE Trans. Automat. Contr.*, vol. AC-21, pp. 184–193, Apr. 1976.

[29] W. I. Zangwill, "Nonlinear programming via penalty functions," *Management Sci.*, vol. 13, pp. 344–358, 1967.

[30] F. A. Lootsma, "A survey of methods for solving constrained minimization problems via unconstrained minimization," in *Numerical Methods for Nonlinear Optimization*, F. A. Lootsma, Ed. New York: Academic, 1972, pp. 313–347.

[31] E. Polak, "Algorithms for a class of computer-aided design problems: A review," *Automatica*, vol. 15, pp. 531–538, 1979.

[32] R. H. Bartels and G. H. Stewart, "Algorithm 432, solution of the matrix equation $AX+XB=C$," *Commun. Ass. Comput. Mach.*, vol. 15, pp. 820–826, 1972.

[33] E. J. Davison and F. T. Man, "The numerical solution of $A'Q+QA=-C$," *IEEE Trans. Automat. Contr.*, vol. AC-13, pp. 448–449, Aug. 1968.

[34] E. J. Davison, "An algorithm for the computer simulation of very large dynamic systems," *Automatica*, vol. 9, pp. 665–675, 1973.

[35] H. W. Smith and E. J. Davison, "Design of industrial regulators: Integral feedback and feedforward control," *Proc. IEE*, vol. 119, no. 8, pp. 1210–1216, 1972.

[36] E. J. Davison, "The robust decentralized control of a general servomechanism problem," *IEEE Trans. Automat. Contr.*, vol. AC-21, pp. 14–24, Feb. 1976.

[37] A. Laub, "An inequality and some computations related to the robust stability of linear dynamic systems," *IEEE Trans. Automat. Contr.*, vol. AC-24, pp. 318–320, Apr. 1979.

[38] G. Stein and J. Doyle, "Singular values and feedback: Design examples," presented at the *16th Ann. Allerton Conf. Commun., Contr. Comput.*, Monticello, IL, 1978.

[39] N. R. Sandell, "Robust stability of linear dynamic systems with application to singular perturbation theory," *Automatica*, vol. 15, no. 4, pp. 467–470, 1979.

[40] R. V. Patel, M. Toda, and B. Sridhar, "Robustness of linear quadratic state feedback designs in the presence of system uncertainty," *IEEE Trans. Automat. Contr.*, vol. AC-22, pp. 945–949, Dec. 1977.

[41] E. J. Davison and I. Ferguson, "Design of controllers for the multivariable robust servomechanism problem using parameter optimization methods—Some case studies," *2nd IFAC Workshop on Control Applications of Nonlinear Programming and Optimization*, Munich, Germany, Sept. 15–17, 1980.

Quantitative feedback theory

Prof. Isaac Horowitz

Indexing terms: *Control theory, Feedback*

Abstract: In quantitative feedback theory, plant parameter and disturbance uncertainty are the reasons for using feedback. They are defined by means of a set $\mathscr{P} = \{P\}$ of plant operators and a set $\mathscr{D} = \{D\}$ of disturbances. The desired system performance is defined by sets of acceptable outputs $\mathscr{A}_u$ in response to an input, u, to be achieved for all $P \in \mathscr{P}$. If any design freedom remains in the achievement of the design specifications, it is used to minimise the effect of sensor noise at the plant input. Rigorous, exact quantitative synthesis theories have been established to a fair extent for highly uncertain linear, nonlinear and time-varying single-input single-output, single-loop and some multiple-loop structures; also for multiple-input multiple-output plants with output feedback and with internal variable feedback, both linear and nonlinear. There have been many design examples vindicating the theory. Frequency-response methods have been found to be especially useful and transparent, enabling the designer to see the trade-off between conflicting design factors. The key tool in dealing with uncertain nonlinear and multiple-input multiple-output plants is their conversion into equivalent uncertain linear time-invariant single-input single-output plants. Schauder's fixed-point theorem justifies the equivalence. Modern control theory, in particular singular-value theory, is examined and judged to be comparatively inadequate for dealing with plant parameter uncertainties.

List of principal symbols and abbreviations

$\mathscr{A}_u$	=	set of acceptable outputs in response to input u
$a, a(\omega)$	=	lower bound
B_h	=	universal high-frequency boundary of acceptable $L_0(j\omega)$, applies for all $\omega \geqslant \omega_h$
$b, b(\omega)$	=	upper bound
$\mathscr{D} = \{D\}$	=	set of disturbances
dB	=	decibels, $20 \log_{10}$
$L(j\omega), L_0$	=	loop transmission and its nominal value
LQR	=	linear quadratic regulator
LQG	=	linear quadratic Gaussian
LTI	=	linear time invariant
MIMO	=	multiple-input multiple-output
$\mathscr{P} = \{P\}$	=	set of LTI plant operators, or plant matrices
$P = [p_{ij}]$	=	plant matrix
$Q = [q_{ij}]$		with $P^{-1} = [1/Q_{ij}]$, P a matrix
QFT	=	quantitative feedback theory
RC	=	resistor-capacitor
SISO	=	single-input single-output
SVT	=	singular-value theory
$\mathscr{T} = \{T\}$	=	set of acceptable system response functions (matrices in MIMO systems) to command inputs
$\mathscr{T}_d = \{T_d\}$	=	as above, but for disturbance inputs
$\mathscr{T}_p(P(j\omega_1))$	=	plant template, set of complex numbers $\{P(j\omega_1)\}$
$\mathscr{W} = \{w\}$	=	set of nonlinear plant operators

1 Introduction

This paper is a survey of our work in feedback systems, denoted as quantitative feedback theory (QFT), and its comparsion with the modern control approach, in particular with singular-value theory. It is our view that feedback around the constrained 'plant' is mandatory only because of uncertainty in its parameters and/or in disturbances entering the plant. Fig. 1A depicts one of the simplest problems: a linear time-invariant (LTI) plant operator p with transfer function $P(s)$, whose output is the system output and can be measured, disturbance d not measurable, and command input $r(t)$ measurable. Suppose that p and d are precisely known, that the output y_r due to r is to be $y_r = r * h(t)$ ($*$ denotes convolution), and that the disturbance component $|y_d| \leqslant m(t)$ is given. Then simply insert a prefilter f_1 between r and p, with $p * f_1 = h$ or transform $F_1(s) = H(s)/P(s)$, and inject a signal z in Fig. 1A such that $|(z + d) * p| \leqslant m(t)$. This is so even if p is unstable, for a little thought shows that uncertainty in p and/or d must be invoked to render this method invalid. Uncertainty necessitates feedback around the plant, and a suitable canonic two-degrees-of-freedom structure (see section 6.1 of Reference 1) is shown in Fig. 1B.

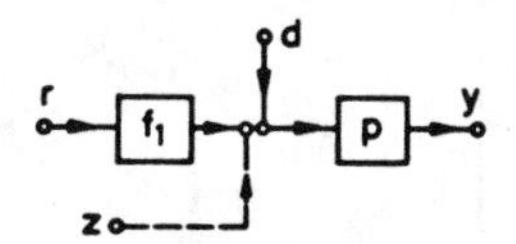

Fig. 1A *No feedback needed in absence of uncertainty*

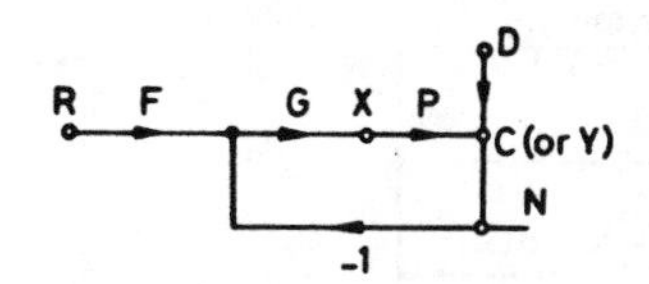

Fig. 1B *Canonic feedback structure*

$C = TR$
$T = (I + PG)^{-1}PGF$
$L = PG$

Most of the feedback control literature, both classical and modern, concentrates on realising a desired input-output relation under the constraint of a feedback structure around the plant, as if feedback is a tool in filter synthesis. If so, then feedback theory is just a branch of active network synthesis, and its merits should be compared with other techniques such as active RC etc., active RC elements against the transducers and the constant-gain infinite-bandwidth amplifiers needed in modern control theory. Such comparisons have never appeared.

The true importance of feedback is in 'achieving desired performance despite uncertainty'. If so, then obviously the actual design and the 'cost of feedback' should be closely related to the extent of the uncertainty and to the narrowness of the performance tolerances. In short, it should be *quantitative*. But feedback theory has not been quantitative: one hardly finds in the voluminous feedback literature any quantitative design techniques, or any quantitative problem

Paper 2206D, first received 2nd June and in revised form 2nd September 1982

The author is with the Department of Applied Mathematics, Weizmann Institute of Science, Rehovot, Israel, and the Department of Electrical Engineering, University of Colorado, Boulder, CO80309, USA

Reprinted with permission from *IEE Proc.*, vol. 129, Pt. D, no. 6, pp. 215–226, Nov. 1982.

statements. This is a fantastic phenomenon. so much teaching and research effort, such a huge literature, but the heart of the problem is almost ignored. Even the graduate Ph.D. often does not know the real reason why feedback is used in control.

2 Linear time-invariant single-input single-output feedback systems

In quantitative feedback theory, the uncertainties formulated as a set of plants $\mathcal{P} = \{P\}$ and a set of disturbances $\mathcal{D} = \{D\}$. There are specified a set of acceptable command response transfer functions $\mathcal{T} = \{T\}$ and a set of acceptable disturbance

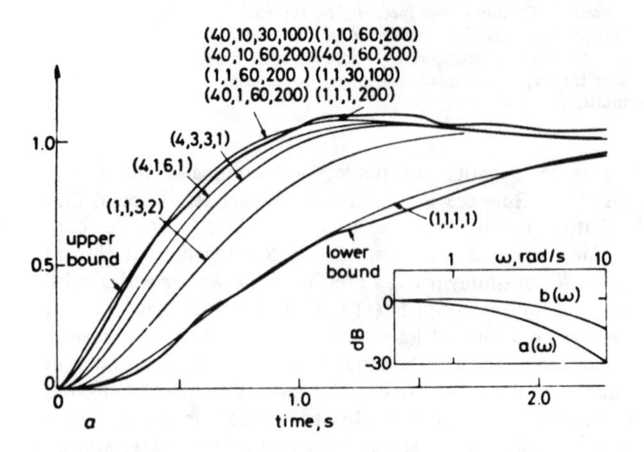

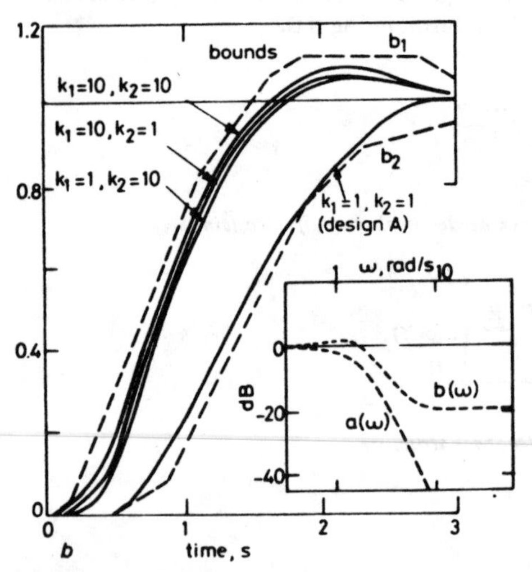

Fig. 2 *Time envelopes and their ω equivalents*

response functions $\mathcal{T}_d = \{T_d\}$, to be achieved for all $P \in \mathcal{P}$. Any freedom in doing this is used to minimise in some sense the effect of sensor noise at the plant input. QFT was first developed for linear time-invariant (LTI) single-input single-output (SISO) single-loop systems with output feedback only [2, 3]. Significant improvements in design execution have been made by East and Longdon [4–6]. This single-loop LTI theory is of key importance in solving the nonlinear and multiple-input multiple-output uncertainty problems, because these are rigorously converted into equivalent LTI SISO problems (see Sections 3 and 4); thus it is now summarised.

2.1 *Single-input single-output single-loop LTI system (Fig. 1B)*

The first step is to translate the tolerances on $\mathcal{T}$, if in the time domain, into ω-domain tolerances [3, 7]. If all $P \in \mathcal{P}$ are minimum phase (this forbids only right-half-plane zeros; uncertain right-half-plane poles are allowed), then bounds on $|T(j\omega)|$ suffice, in the form

$$a(\omega) \leqslant |T(j\omega)| \leqslant b(\omega) \tag{1}$$

Two examples are shown in Figs. 2*a* and *b*, where the final simulation results fill the envelope (but not when the disturbance response specifications are more severe than the command response specifications [3]). At any $\omega = \omega_1$, the set of points $\{P(j\omega_1)\}$ is a region in the complex plane, called the plant template $\mathcal{T}_p(P(j\omega_1))$. The first step is to find these $\mathcal{T}_p$ for a reasonable number of ω values; see Reference 3 for examples. In Fig. 1B,

$$T = \frac{FL}{1+L} \tag{2a}$$

$$\Delta \log T = \Delta \log \frac{L}{1+L} \tag{2b}$$

because there is negligible uncertainty in F (and G). Since $L = GP$, the variation in $P(j\omega)$ generates via eqn. 2*b* a variation in log $T(j\omega)$. The function of G in $L = GP$ is to guarantee that the *variation* in log T is within the amount allowed by the specifications. Let $L_0 = GP_0$ be a nominal loop transmission at a nominal plant P_0. It is convenient to find the bounds on L_0 in the Nichols chart which achieves this. Fig. 3 is an example of some bounds on $L_0(j\omega)$ for the plant shown and specifications of Fig. 2*a*:

$$P_i = k_i/s \qquad i = 1, 2, b \tag{3}$$

$$P_c = k_c/s^3 \qquad k_j \in [a_j, b_j] \qquad a_j = 20{,}50{,}1{,}1000$$
$$b_j = 800{,}500{,}60{,}200000$$
$$\text{for } i = 1, 2, b, c \tag{4}$$

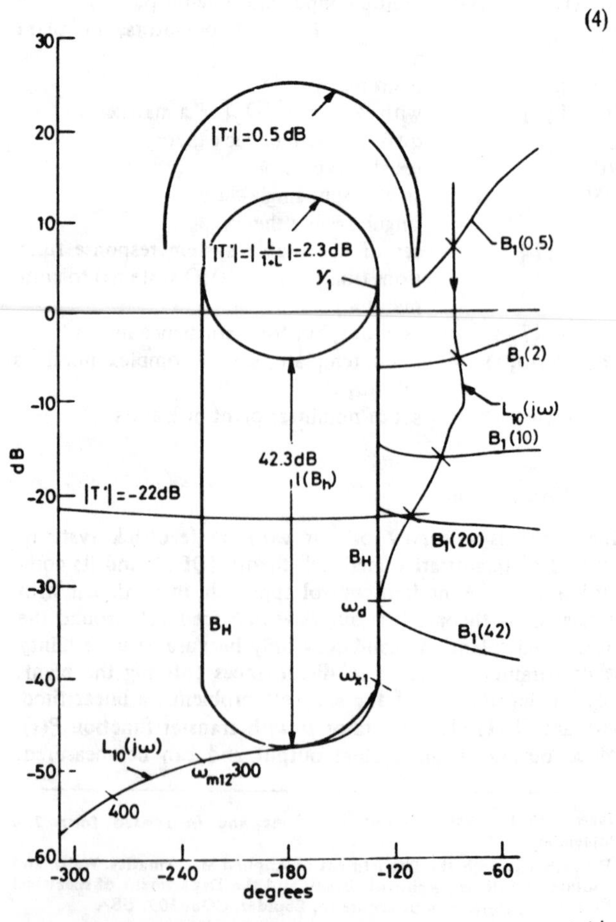

Fig. 3 *Bounds B(ω) on nominal outer-loop transmission $L_0(j\omega)$ in the Nichols chart*

In Fig. 3, it is treated as a single plant k/s^3, $k \in [2,24200]\ 10^3$. The next step is to find a rational $L_0(s)$ with sufficient excess of poles over zeros [so $G(s)$ is practical] which satisfies these bounds. It has been proved [8, 9] that the optimum, L_0 lies on its bound at each ω, so the designer can see how far he is from optimum, and judge whether the addition of more poles and zeros to L_0 is justified by the resulting decrease in bandwidth of L_0. This is the property of *transparency*, wherein the designer sees directly the trade-offs between the important system parameters, such as the narrowness of response tolerances, the extent of plant uncertainty, the complexity of the compensation (number of its poles and zeros), the resulting loop bandwidth needs and the effect of sensor noise. In this technique, the designer works *directly* with these parameters and easily sees the trade-offs. This is in sharp contrast with the 'modern' control technique of minimising a quadratic cost function at nominal plant values, and trying to control the above important system trade-offs by varying the various weights.

The final step in the design is to find $F(s)$ in Fig. 1B. For example, if the specifications permit $-10\,\text{dB} \leqslant |T(j\omega_1)| \leqslant -2\,\text{dB}$, and $L_0(j\omega_1)$ has been chosen so that $-6\,\text{dB} \leqslant |L(j\omega_1)/(1+L(j\omega_1))| \leqslant 0\,\text{dB}$, then $|F(j\omega_1)|$ may have any

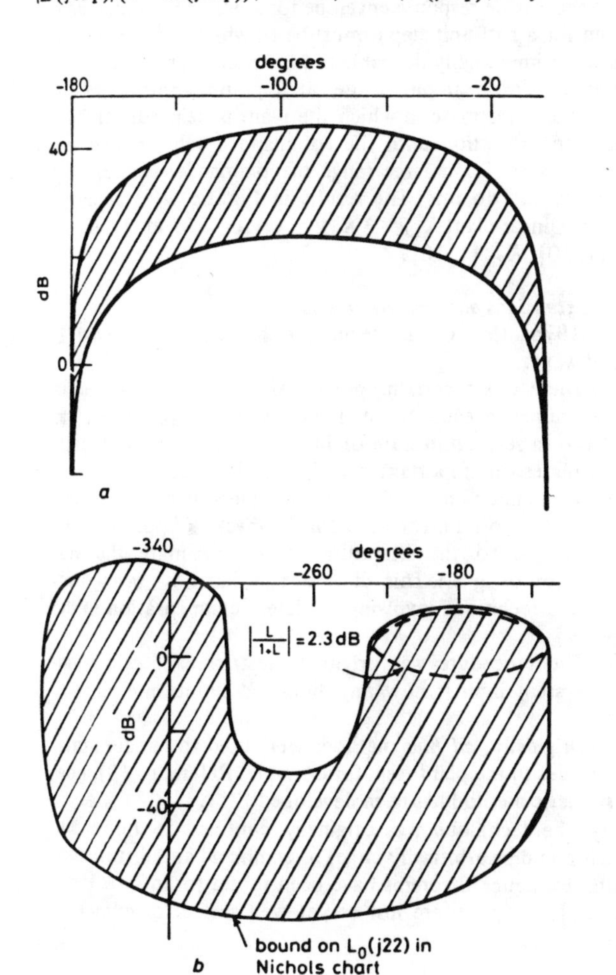

Fig. 4 *Plant template at $\omega = 22$ and bound on $L_0(j22)$ due to higher-order modes*

a Template of

$$\frac{k(22)\,10^4}{s(s^2+0.02s+B)(s^2+0.04s+4B)}$$

at $\omega = 22$ for k: [1, 10], B: [400, 600]

b Forbidden region for $L_0(j22)$ in Nichols chart

value between $-10-(-6) = -4\,\text{dB}$ and $-2-0 = -2\,\text{dB}$ in order to satisfy the bounds on $T(j\omega_1)$. In this way, bounds on $|F(j\omega)|$ are found, and $F(s)$ is found to satisfy these bounds.

2.2 Bending modes

Higher-order modes of any number and of any extent of uncertainty are easily and naturally incorporated into this design technique. One simply finds the templates $\mathcal{T}_p(P(j\omega))$ at these higher ω values in exactly the same manner as in the above and proceeds in exactly the same way at every step. Without them, $P(s) \to k/s^e$ at large s, the template of P becomes a vertical line of length k_{max}/k_{min}, and there emerges a 'universal high-ω boundary', such as B_h in Fig. 3 which applies for all $\omega >$ some ω_h value (~ 250 in Fig. 3).

But suppose that in the range of the higher-order modes

$$P(s) \simeq \frac{k(22)10^4}{s(s^2+0.02s+B)(s^2+0.04s+4B)} \tag{5}$$

with uncertainties $1 \leqslant k \leqslant 10$, $400 \leqslant B \leqslant 600$.

The resulting template of $P(j22)$, for example, (i.e. the set $\{P(j22)\}$) is shown in Fig. 4*a*. In this ω range, the dominating specification [3] is $|L/(1+L)| \leqslant$ some value, say 2.3 dB. If the nominal $k_0 = 1$ and $B_0 = 400$, then the resulting bound on $L_0(j22)$ is easily found to be that shown in Fig. 4*b*. Similar templates and bounds are determined for a discrete number of ω values, and then $L_0(j\omega)$ is shaped to satisfy these bounds.

Basically, the same technique is used for disturbance attenuation; see Reference 3 for details.

By now, many detailed design examples have been done. The techniques have been taught in graduate and undergraduate junior engineering courses, and no particular difficulty has been encountered. Modern control theory specialists, however, seem to find it difficult, perhaps owing to their estrangement from frequency-domain concepts.

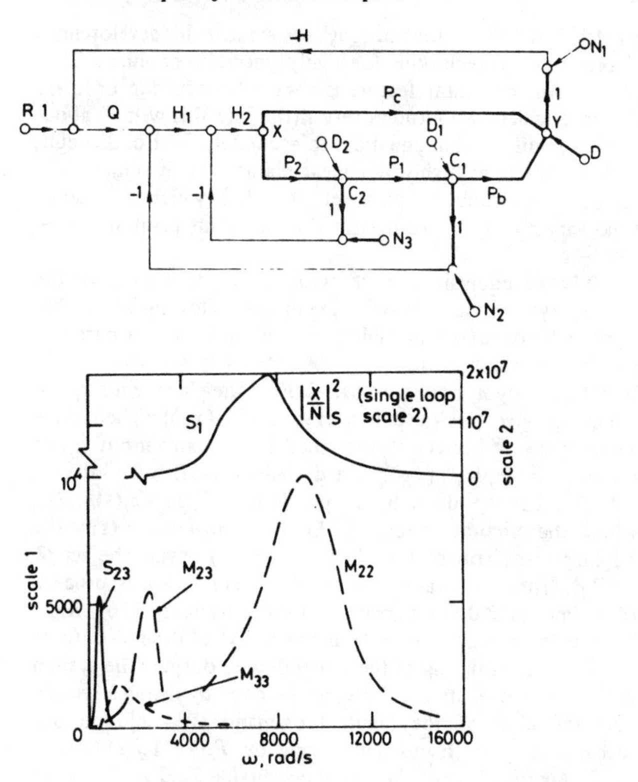

Fig. 5 *Tremendous reduction in noise effects (at X) of sensors due to multiple-loop feedback*

(Fig. 2*a*: P_i uncertainty factors: 40, 10, 60, 200)

2.3 *Extension to SISO multiple-loop systems [10–13]*

The technique has been extended to a number of multiple-loop SISO structures, wherein internal variables can be served and processed for feedback purposes. The great potential advantage is the possible vastly reduced effect of sensor noise. An example is shown in Fig. 5 for the problem [11] of Fig. 2*a*. If only outer-loop feedback (y alone) is used, the amplification of noise from sensor 1 *at the plant input* is given by curve S_1 in Fig. 5, for which scale 2 applies. If feedback from C_1 is also allowed and properly used, then the amplification of noise from sensor y is now given by curve S_{23} with scale 1, which is much less than S_1. But now there is sensor 2, and its noise amplification is curve M_{22} (scale 1). The latter can be drastically reduced by using a third sensor at C_2, and then the sensor-2 noise amplification is curve M_{23} (scale 1). Sensor-3 noise is amplified by curve M_{33}. Of course, this example has fantastically large uncertainty, and multiple-loop feedback would be essential to achieve a practical design.

A technique called 'design perspective' has been developed [12], whereby a detailed multiple-loop design is *not* needed in order to see the improvements possible to obtain this information fairly accurately, by means of rapid approximate calculations. The designer can thus decide early in the game the number of sensors to be used, and, if there are options available between sensor quality and cost, he has the information for intelligent decision.

2.4 *Uncertain nonminimum-phase plants*

In this case, bounds on both $|T(j\omega)|$ and $\arg T(j\omega)$ are needed. The same procedure is then followed, resulting, as before, in bounds on a nominal loop transmission $L_0(j\omega)$. However, it may be impossible to satisfy these bounds, and tests for determining this have been given. If this is so, the specifications must be relaxed [9].

3 Design for uncertain nonlinear SISO plants [14, 15]

In 1975–1976, a breakthrough was made in developing a rigorous design technique for highly uncertain nonlinear SISO plants. An important feature is that it permits the ordinary design engineer who knows very little (like this writer) about nonlinear differential equations to accurately control difficult, highly uncertain, highly nonlinear plants. This is achieved by replacing the nonlinear plant set $\mathscr{W} = \{w\}$ (which can also be time varying) by an equivalent time-invariant plant set $\mathscr{P} = \{P\}$.

$\mathscr{P}$ is the equivalent to $\mathscr{W}$ with respect to the set of the desired system outputs $\mathscr{A} = \{a\}$ in the following sense. Imagine a barrel of all possible plants w in $\mathscr{W}$ and a barrel of all desired plant outputs a of $\mathscr{A}$. Pick any w_i from the $\mathscr{W}$ barrel and any a_j of the $\mathscr{A}$ barrel. Find the plant *input* x_{ij} for which w_i gives an output a_j [i.e. $a_j = w_i(x_{ij})$]. Then there exists in the $\mathscr{P}$ barrel a P_{ij} such that, for the same input x_{ij}, its output is a_j [$a_j = P_{ij}(x_{ij})$]. The designer must first find the set $\mathscr{P}$. This can be done by simply letting $P_{ij}(s) = \hat{a}_j(s)/\hat{x}_{ij}(s)$, where the circumflex refers to Laplace transform: $\hat{f}(s)$ = the Laplace transform of $f(t)$. Repeat over i, j, giving the set $\mathscr{P} = \{P_{ij}\}$. There are various ways of streamlining this process, sometimes even doing a great part of it analytically [16, 18].

It is important to include in the set $\mathscr{A}$ of desired outputs a reasonable sampling of the actual desired outputs the system is to deliver over its life. This can be done by listing a reasonable sampling of the inputs (command $\mathscr{R} = \{r\}$ and disturbance $\mathscr{D} = \{D\}$) and multiplying by $T_r(s)$, $T_d(s)$, where T_r, T_d are members of the sets of acceptable T_r, T_d.

Once $\mathscr{P}$, the 'equivalent linear time-invariant' (ELTI) plant set, is found, the designer can forget about the nonlinear plant set $\mathscr{W}$ and design for the ELTI plant set $\mathscr{P}$. If he can solve this ELTI problem, then it is *guaranteed* that the same compensations work for the original highly uncertain nonlinear problem. The same desirable properties previously listed (systematic procedure, transparency, ease of trade-offs, practicality of compensation) therefore apply to the nonlinear design. No separate effort is needed to guarantee stability of the nonlinear system; this is automatically included. The main design effort, beyond that needed for ordinary LTI design, is to solve the nonlinear equations *backwards* on the computer, for which software packages exist; i.e. given the output, find the input, which is often much easier than solving it forwards, for example $(\ddot{y})^3\,\dot{y} + A\dot{y}y + By^3 = x$. Given y, it is easy to find x.

By now, over a dozen such design problems have been done [14–16, 18–21] with large uncertainty, defined time-domain tolerances and uniformly excellent results; some of these were Master's theses. One was a significant nonlinear flight-control problem for AFFDL, WPAFB with $c*$ the controlled output [18].

3.1 *Nonlinear performance specifications*

The design technique can handle nonlinear performance specifications. Fig. 6*a* is an example of linear system tolerances. If the acceptable response envelope for a unit step command is I, then for a half-unit step it must be II, which is one half of I. It is sometimes highly desirable to have nonlinear performance tolerances; for example, those of Fig. 6*b* appropriate for optimal time response in which the plant is being driven hard almost to saturation until the commanded value is reached, and this is so for a large range of commanded values. The technique can do this. The results in Fig. 6*b* were achieved for a nonlinear plant $K_1y^2 + K_2\dot{y} = x$ with [38] uncertainties $K_1 : [1, 10], K_2 : [1, 10]$.

3.2 *Extensions and improvements*

Since 1976, this design technique has been improved in several ways:

(*a*) There was a certain amount of overdesign involved in the technique, because, even if there was only one nonlinear plant (not a set), an infinite ELTI set would normally emerge. This violates an important principle in feedback theory: 'If there is no uncertainty (of plant parameters or disturbances), then there is no inherent need for feedback'. (Application of this principle to the literature would invalidate the vast majority of designs.) This disadvantage has now been overcome by a technique involving nonlinear compensation in the loop [19].

(*b*) The procedure was originally restricted to minimum-phase system inputs; now any bounded inputs are allowed [20].

(*c*) Originally, *ad hoc* methods were used to handle initial state values in the plant (see, for example, Reference 18): now a systematic method has been developed [21].

(*d*) The technique was originally confined to finite sets (which could be arbitrarily large) of command and disturbance inputs. Existence theorems have been found for infinite input sets [22], but they are not as yet well suited for numerical design.

4 Uncertain multiple-input multiple-output plants [17, 23–29]

Fig. 1B can be used to represent an $n \times n$ multiple-input multiple-output (MIMO) system if F, G, P, T are each $n \times n$ matrices, and $\mathscr{P} = \{P\}$ a set of matrices due to plant uncertainty. There are n^2 closed-loop system transfer functions t_{ij} relating output i to input j transforms: $\hat{y}_i(s) = t_{ij}(s)\hat{r}_j(s)$. In a quantitative problem statement, there are given tolerance bounds on each t_{ij}, giving n^2 sets $\mathscr{A}_{ij} = a_{ij}$ of acceptable t_{ij}.

To appreciate the difficulty of this problem, note the very complex expression for t_{11} below, for $n = 3$, even when simplified by letting the compensation matrix G in Fig. 1B be diagonal, $G = [g_i]$:

$$
\begin{aligned}
t_{11} = \{ & [p_{11}f_{11}g_1 + p_{12}f_{21}g_2 + p_{13}f_{31}g_3] \\
& [(1 + p_{22}g_2)(1 + p_{33}g_3) - p_{23}p_{32}g_2g_3] \\
& - [p_{21}f_{11}g_1 + p_{22}f_{21}g_2 + p_{23}f_{31}g_3] \\
& [p_{12}g_2(1 + p_{33}g_3) - p_{32}p_{13}g_2g_3] \\
& + [p_{31}f_{11}g_1 + p_{32}f_{21}g_2 + p_{33}f_{31}g_3] \\
& [p_{23}p_{12}g_2g_3 - (1 + p_{22}g_2)p_{13}g_3] \} / \\
& \{(1 + p_{11}g_1)[(1 + p_{22}g_2)(1 + p_{33}g_3) \\
& - p_{23}p_{32}g_2g_3] - p_{21}g_1[p_{12}g_2(1 + p_{33}g_3) \\
& - p_{32}p_{13}g_2g_3] + p_{31}g_1[p_{12}p_{23}g_2g_3 \\
& - p_{13}g_3(1 + p_{22}g_2)] \}
\end{aligned} \tag{6}
$$

There are $n^2 = 9$ such expressions (all have the same denominator), and there may be considerable uncertainty in the nine plant p_{ij} elements. The objective is to find nine f_{ij} and three g_i such that t_{ij} stays within its acceptable set $\mathscr{A}_{ij}$, no matter how the p_{ij} may vary. Clearly, this is a horrendous problem.

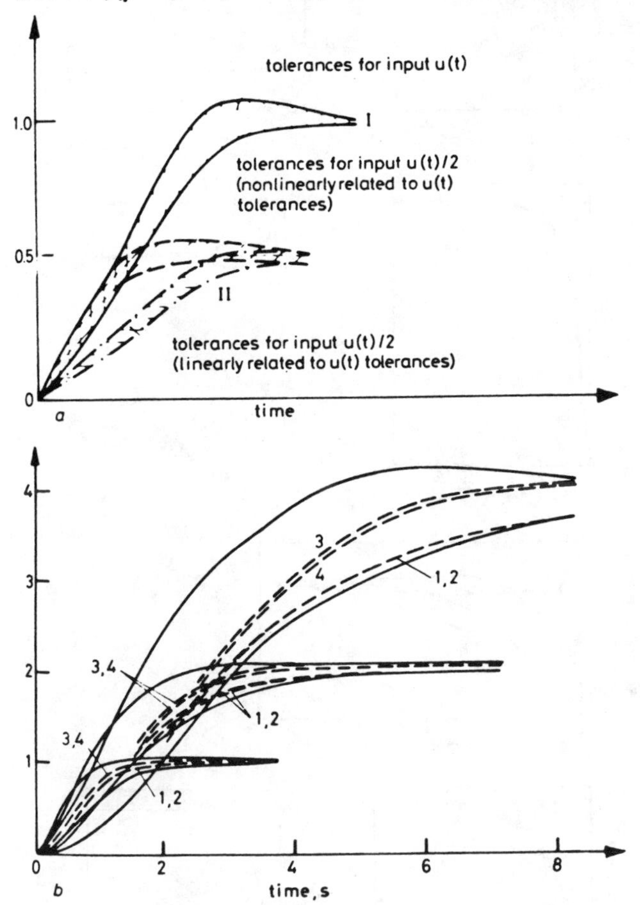

Fig. 6 *Design example of nonlinear tolerances $K_1 y^2 + K_2 \dot{y} = x$ and simulation results*

a Linear and nonlinear tolerances on step response

b

Case	K_1	K_2
1	10	10
2	10	1
3	1	1
4	1	10

Even the stability problem alone, ensuring the characteristic polynomial denominator of eqn. 6 has no right-half-plane zeros for all possible p_{ij}, is extremely difficult.

4.1 Single-loop equivalents

In this technique, justified by Schauder's fixed-point theorem [23], it is never necessary to consider the characteristic polynomial. Simply replace the $n \times n$ MIMO problem by n single loops and n^2 prefilters see Fig. 7 for $n = 3$. In Fig. 7, $P^{-1} = [1/Q_{uv}]$, and the uncertainty in P generates sets $\mathscr{Q}_{uv} = \{Q_{uv}\}$;

$$d_{11} = -\left(\frac{t_{21}}{Q_{12}} + \frac{t_{31}}{Q_{13}}\right) \tag{7}$$

is any member of the set $\mathscr{D}_{11}$ generated by the t_{j1} in $\mathscr{A}_{j1}$ and the Q_{1j} in $\mathscr{Q}_{1j}, j = 2, 3$.

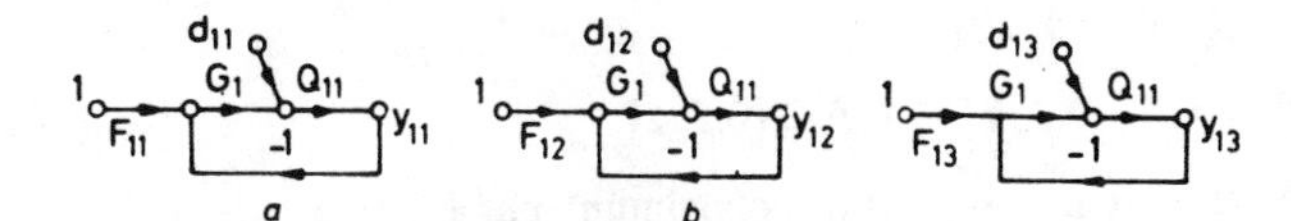

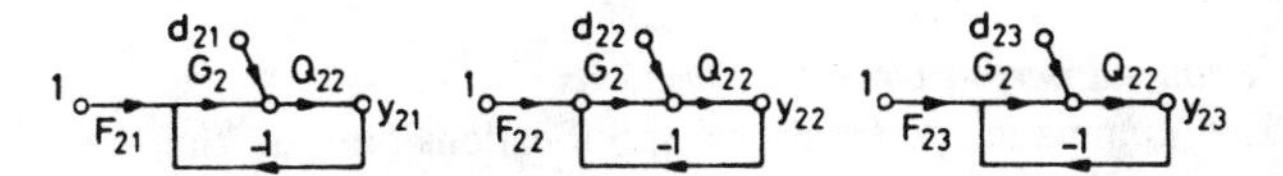

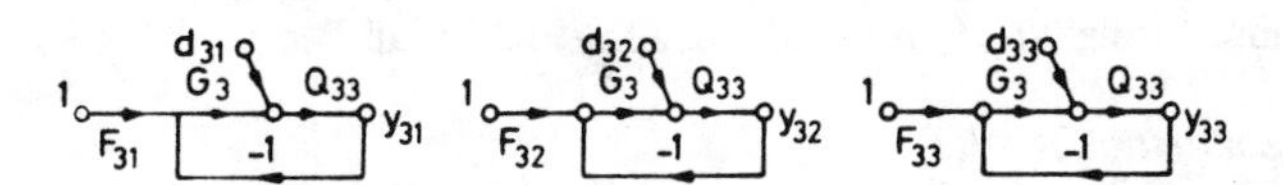

Fig. 7 *Final equivalent single-loop feedback structures which replace the 3 × 3 MIMO problem*

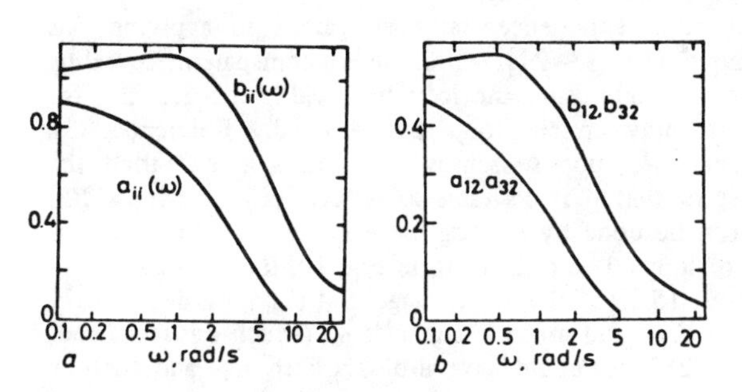

Fig. 8 *ω-domain bounds*

a For t_{ii}

b For t_{12}, t_{32} 3 × 3 MIMO example

In Fig. 7*a*, the SISO design problem is to find $L_1 = g_1 Q_{11}$, f_{11} such that the output is a member of the set $\mathscr{A}_{11}$ for all Q_{11} in $\mathscr{Q}_{11}$ and for all d_{11} in $\mathscr{D}$. Similarly, in Fig. 7*b*, find $L_1 = g_1 Q_{11}$, f_{12} so that its output is in $\mathscr{A}_{12}$ for all Q_{11} in $\mathscr{Q}_{11}$ and all d_{12} in $\mathscr{D}_{12}$ etc. Note that L_1 is the *same* for all the SISO structures in the first row of Fig. 7 etc. In each of these three structures, the uncertainty problem (due to the *sets* $\mathscr{Q}_{1i}$, $\mathscr{D}_{1i}$) gives *bounds* on the level of feedback L_1 needed, and so the toughest of these bounds must be satisfied by L_1.

If the designer designs these SISO systems to satisfy their above stated specifications, then it is guaranteed that these same f_{uv}, g_u satisfy the MIMO uncertainty problem. It is not necessary to consider the highly complex system characteristic equation (denominator of eqn. 6) with its uncertain p_{ij} plant parameters. System stability (and much more than that) for all P in $\mathscr{P}$ is automatically guaranteed.

4.2 Design example

The power of the technique is illustrated by presenting the results of Reference 25, done as a Master's thesis by a typical average graduate student. Here the plant and uncertainties are

$$p_{ij} = \frac{sA_{ij} + B_{ij}}{s^2 + Es + F} \tag{8}$$

$$\left.\begin{array}{l}
A_{11} \in [2, 4] \quad A_{12} \in [0.5, 1.1] \quad A_{22} \in [5, 10] \\
A_{31} \in [-0.8, -1.8] \quad \text{all other } A_{ij} = 0 \\
B_{11} \in [-0.15, 1] \quad B_{12} \in [-1, -2] \\
B_{13} \in [1, 4] \\
B_{21} \in [1, 2] \quad B_{22} \in [5, 10] \quad B_{23} \in [-1, -4] \\
B_{31} \in [-1, -2] \quad B_{32} \in [15, 25] \\
B_{33} \in [10, 20] \\
E \in [-0.2, 2] \quad F \in [0.5, 2]
\end{array}\right\} \tag{9}$$

Note that p_{12} is always nonminimum phase and p_{11} is so far part of the plant parameter range. Also, the plant is unstable for part of the parameter range.

Command performance specifications

The command performance specifications are in the frequency domain, shown in Fig. 8*a* for the three diagonal t_{ii} ($i = 1 - 3$) and in Fig. 8*b* for t_{12}, t_{32}. The other off-diagonal elements t_{13}, t_{21}, t_{23}, t_{31} are to be 'basically non-interacting', $|t_{uv}(j\omega)| \leqslant 0.1$ for all ω and for all P in $\mathscr{P}$.

Simulation results

The design details are given in Reference 25. Typical and extreme step responses are shown in Fig. 9 over the set $\mathscr{P}$.

Transparency and trade-off

Considerable experience has been gained in applying this technique [17, 23–27]. It is highly transparent, revealing means for minimising the loop bandwidths needed and for trade-off between the loops (see especially Reference 25). For example, suppose sensor 3 is much noisier than the others, so that it is desirable to reduce its loop bandwidth. This can be done by shifting more of the burden to one or both of loops 1 and 2. Thus, in Fig. 10, $|L_{30}| > |L_{10}|$, $|L_{20}|$ by about 15 dB in the low ω range. But $|L_{30}|$ was deliberately made $\simeq |L_{20}|$ and much less than $|L_{10}|$ (which was sacrificed) for $\omega > 250$. We could have also sacrificed L_{20} and thereby improved L_{30} even more, or made an even greater sacrifice of L_{10}.

4.3 Recent advances

The above dealt with $n \times n$ MIMO plants in which only the n outputs are available for feedback purposes. The technique has since been extended to $n \times n$ plants in which internal states are also available for feedback. The result is to replace the uncertain $n \times n$ system by a number of single-input single-output uncertain *multiple-loop* systems. The solutions of the latter SISO problem are guaranteed to solve the MIMO problem [28].

The technique requires a fixed-point theorem for its rigorous justification. Also, it involves a certain amount of inherent overdesign because, for example in Fig. 7*a*, there is some correlation between the t_{ul} appearing in d_{11} of eqn. 7 and the elements Q_{lu} in eqn. 7. This correlation is not normally used (Reference 25 suggests a method for doing so). Finally, a certain diagonal dominance condition as $s \to \infty$ must be satisfied for the technique to be applicable. For the 2 x 2 system with plant matrix $P = [p_{ij}]$, the condition is that

$$\lim_{\omega \to \infty} |p_{11}p_{22}| > |p_{12}p_{21}| \quad \text{or vice versa for all } p \text{ in } \mathscr{P} \tag{10}$$

(In Rosenbrock's technique [30], this condition must be satisfied for *all* ω.) It has been shown [29] that expr. 10 is inherent and necessary for *any* technique that may ever be invented, but only if arbitrary small sensitivity over arbitrary

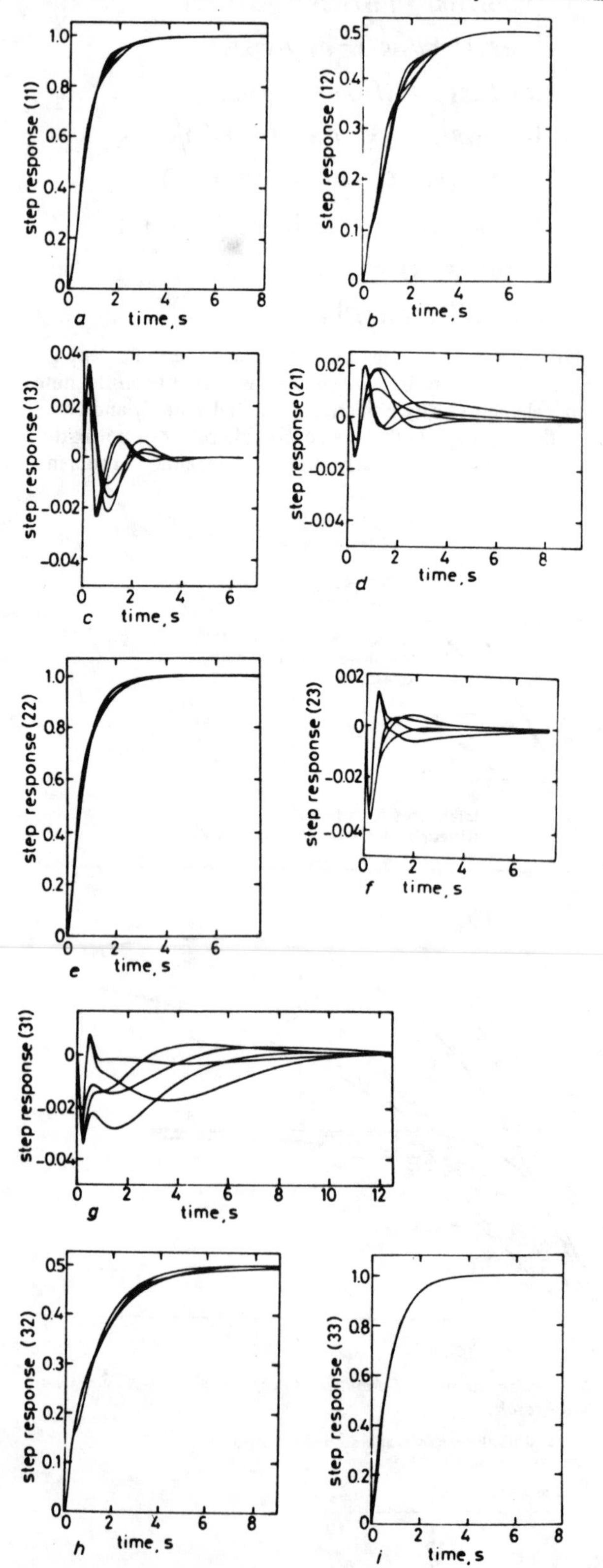

Fig. 9 *Representative time-domain step responses*

large bandwidth is desired. However, the technique of Section 4.1 *always* requires this, even for less demanding designs, and is therefore overdemanding in that respect. However, the newest extension [29] requires it also only for arbitrary small sensitivity over arbitrary large bandwidth and so in this respect is much better. Another advantage of this newest extension is that fixed-point theory is not needed for its justification; simple logical arguments suffice. A third advantage is that there is considerably less overdesign in the new technique.

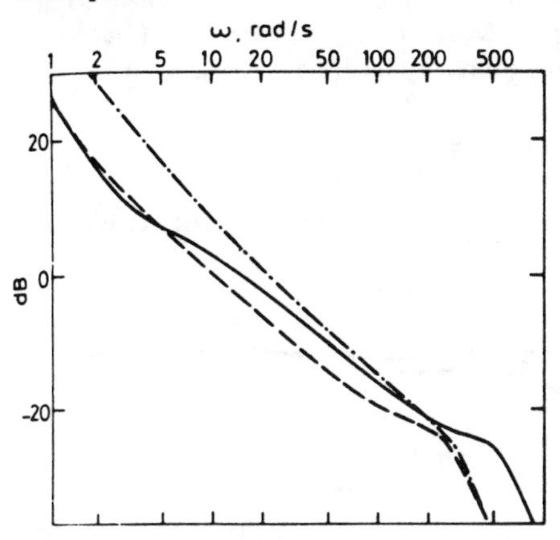

Fig. 10 *Trade-off between loops in 3 × 3 design*

——— $|L_{10}|$
– – – $|L_{20}|$
– · – · $|L_{30}|$

Disturbance attenuation

The above has emphasised response to commands, but the same approach is used for disturbance attenuation. The set of acceptable T_d system response functions (n^2 of them) to disturbances must be formulated, and then the d_{uv} in Fig. 7 contain the actual disturbance components as well. Design is easier because there is then no command input, only the disturbance input d_{uv}.

4.4 Digital compensators

The modern tendency is to use microprocessors as compensators in the feedback loops, which are, of course, essentially digital networks. Our design theory is particularly suited to consider the precautions that must be taken because of the *nonminimum-phase* property of the digital network. The latter property is seen by means of the w transformation, $w = u + jv = (z-1)/(z+1)$ with $z = \exp sT$, where T is the sampling period. The unit circle in the z-plane transforms into the imaginary v-axis in the w-plane, and so one can use it as the new frequency domain in exactly the same manner as $j\omega$ in the s-plane, i.e. frequency-domain compensation, plant uncertainty, and the design techniques of the preceding Sections. It is easy to show that every practical sampled device has a w transfer with a zero at $w = 1$ (see sections 11.11 and 11.15 of Reference 1) and so is nonminimum phase. This means 90° extra phase lag at $v = 1$ (which corresponds to $\omega = \omega_s/4$, where $\omega_s = 2\pi/T$, the sampling frequency).

5 Uncertain nonlinear MIMO plants

The nonlinear technique of Section 3 and the MIMO technique of Section 4 have been combined, giving a design technique for uncertain nonlinear $n \times n$ MIMO plants. The set $\mathscr{W}$ of the $n \times n$ nonlinear plants is replaced by a set $\mathscr{P} = \{p\}$ of $n \times n$ transfer-function matrices, and then the technique of Section 4 is used to solve this $n \times n$ LTI MIMO problem. This set $\mathscr{P}$ is equivalent to the nonlinear set $\mathscr{W}$ with respect to the entire set $\mathscr{A} = \{a\}$ of desired n system outputs. An element a of $\mathscr{A}$ consists of n time functions, corresponding to an n-vector of acceptable outputs in response to an n-vector of command inputs or disturbances. The procedure is illustrated by two design examples [17] with the same plant but different performance specifications.

5.1 Design example 1

Plant (2 × 2)

$$\dot{y}_1 + Ay_1^3 + B(y_1 + 1)y_2 = k_1 x_1$$
$$\dot{y}_2(1 + Cy_1) + Ey_1^2 + dy_2 = k_2 x_2 \qquad (11)$$

Uncertainties

$A \in [0.04, 0.05]$ $\quad B, C \in [0.08, 0.12]$

$D \in [0.8, 1.2]$ $\quad E \in [0.8, 1.5]$ $\quad k_1 k_2 \in [0.5, 2.5]$

all independently

Performance tolerances

In the first problem, the inputs are step commands $Mu(t)$, *only one at a time*, $M \in [-5, 5]$. In response to $r_i = Mu(t)$, it is required that $y_i \in [0.4M, 0.6M]$ within one second, $\geqslant 0.9M$ for all $t \geqslant 3.5$, and overshoot must not exceed 5%; $|y_k(t)|_{max} \leqslant 0.1M$ for $k \neq i$ (see Fig. 13).

Design execution

Analytic models were chosen for y_1, y_2 in response to $r_1 = Mu(t)$, and (and vice versa):

$$\left.\begin{aligned} y_1 &= M[1 - \lambda\exp(-\sigma t) - (1-\lambda)\exp(-\tau t)] \\ y_2 &= Nt\exp(-\beta t) \end{aligned}\right\} \qquad (12)$$

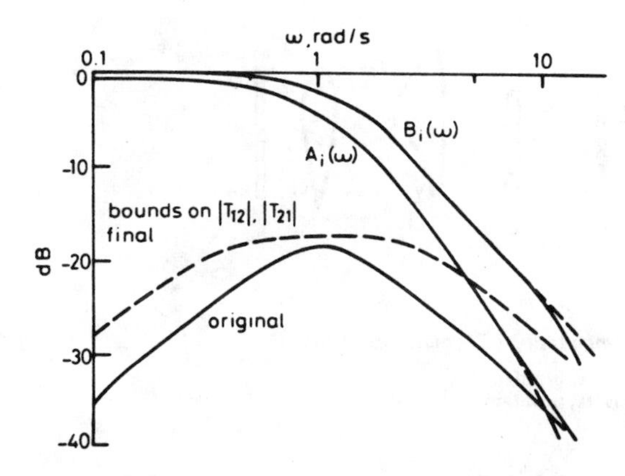

Fig. 11 *Design example: ω-domain bounds*

The parameters M, λ, σ, τ, N, β were chosen compatible with the tolerances. Those that passed were used to generate ω-domain trial bounds on $|T_{ii}(j\omega)|$ and $|T_{ik}(j\omega)|$, as shown as solid lines in Fig. 11. Then response functions which violate these trial bounds were tried, resulting in the dashed line enlargements in Fig. 11. These forms chosen for y_1, y_2 with so few free parameters certainly do not appear a good sampling of acceptable plant outputs. This is precisely the rigour-defying kind of short cut which very often works. The nonlinear equations appear to be well behaved, suggesting that smooth time responses of the form of eqn. 12 can be achieved. If, in practice, the resulting design gives $y(t)$ which significantly violates this assumption, then a more general form for y_i, y_k will be used, which permits a larger sampling. This was not found necessary.

LTI-equivalent plant templates and resulting design
The next step is to find the LTI-equivalent plant sets. Samples of y vectors were taken of the form of eqn. 12 and $x = W^{-1}y$ were obtained analytically from eqn. 11. 2×2 X, Y matrices were formed given the LTI equivalent $P^{-1} = XY^{-1}$. At any $\omega = \omega_1$, $\{Q_{uu}(j\omega_1)\}$ occupies a region in the complex plane, denoted as its template $\mathscr{T}_{P_{uu}}(\omega)$; the larger $\mathscr{T}_{P_{uu}}$, the larger the uncertainty. Some of these are shown in Fig. 12. The design technique detailed in Reference 25 is followed to complete the design. Representative and extreme simulation responses are shown in Fig. 13. The assigned tolerances were very nicely satisfied.

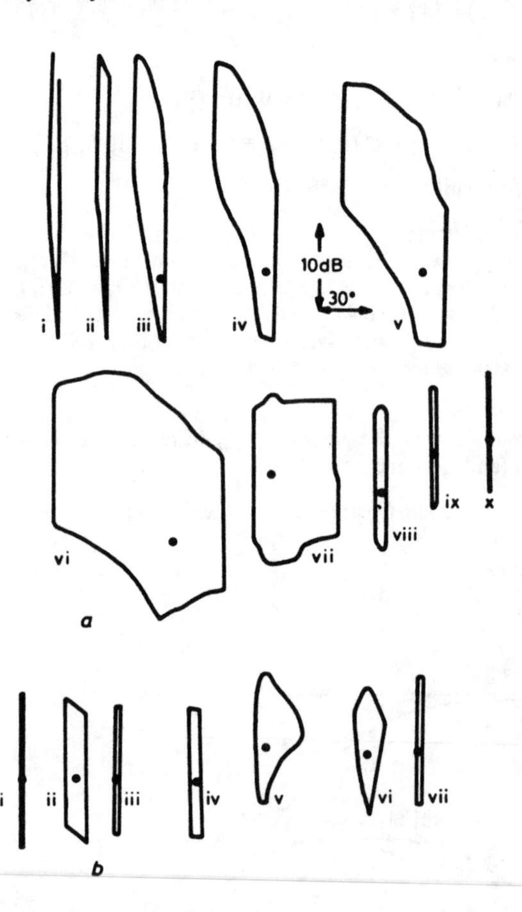

Fig. 12 *Templates of LTI-equivalent plants*
a Q_{11} templates, problem 1
b Q_{22} templates, problem 1

5.2 *Design example 2*
It is important to recall that this design was made only for command inputs r_1, r_2 *one at a time*. Several runs were made nevertheless, to see the results with both simultaneously applied $r_1 = M_1u(t)$, $r_2 = M_2u(t)$. These are shown in Fig. 14. It is seen that they are definitely not the superposition of the results obtained when r_1, r_2 were applied one at a time. To explain this divergence from superposition, templates of Q_{11}, Q_{22} were calculated on the assumption of superposition of outputs. The templates are shown in Fig. 15 for $\omega = 1$. Comparing with Fig. 12, it is seen that they are exceedingly larger, and so it is not at all surprising that a design made on the basis of the templates of Fig. 12 should be woefully inadequate for the templates of Fig. 15. The design technique is guaranteed only for the set of system inputs for which the design is executed. Thus a different design must be made if the command inputs can occur simultaneously. We shall also use here the nonlinear compensation technique of Reference 19.

Now the command input is

$$\left.\begin{aligned} r_1 &= Mu(t)\cos\theta \\ r_2 &= Mu(t)\sin\theta \qquad \theta: [0, 360^\circ] \\ &\qquad\qquad\qquad M: [2, 5] \end{aligned}\right\} \tag{13}$$

There is inserted in front of the plant (Fig. 16) ($W: y = Wx$) a nonlinear network (Λ: $x = \Lambda v$) such that $W\Lambda$: $y = W\Lambda v$ is LTI for all inputs v at a nominal $W = W_0$.

This method, previously applied only to SISO plants, is used here as follows. We want LTI response $y = H * v$ ($*$ denotes convolution), and for simplicity let H be diagonal with

$$y_1 = \int_0^t v_1\,dz \qquad y_2 = \int_0^t v_2\,dz$$

The equation $H * v = W_0x$ defines (see Fig. 16) Λ: $x = \Lambda v$.

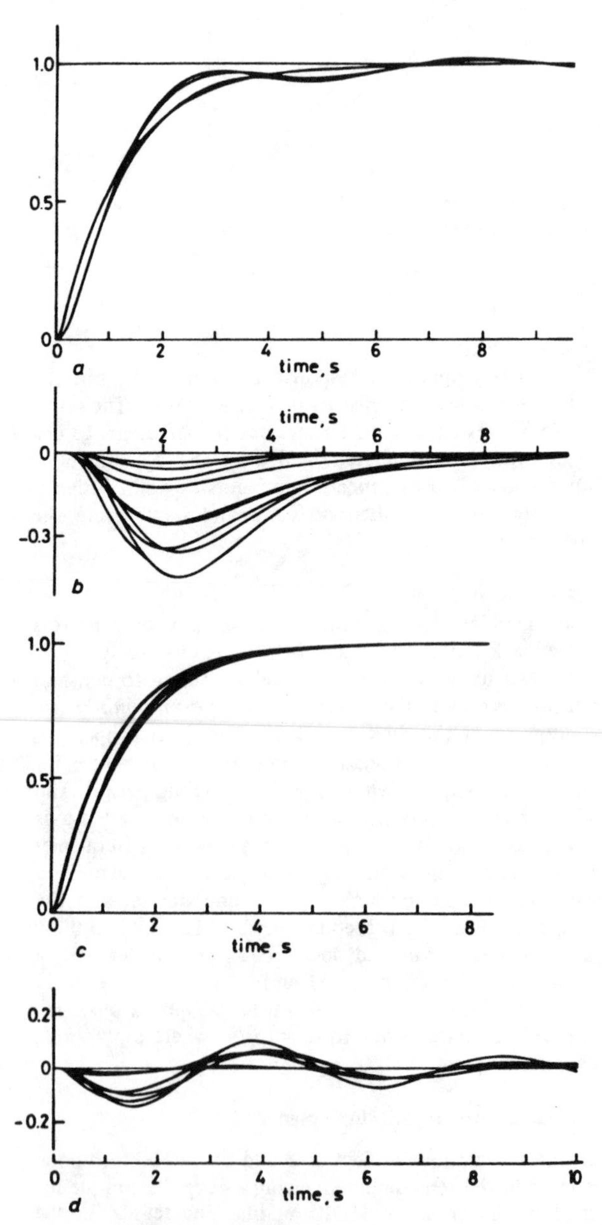

Fig. 13 *Step responses, design 1*
Representative normalised step responses
a y_1, due to r_1
b $10y_2$, due to r_1
c y_2, due to r_2
d $10y_1$, due to r_2

Eqn. 11 with $\dot{y}_1 = v_1, \dot{y}_2 = v_2$ becomes

$$\begin{aligned} v_1 + A_0\left[\int v_1\, dz\right]^3 + B_0\left[\int v_1\, dz + 1\right]\int v_2\, dz &= k_{10}x_1 \\ v_2\left[1 + C_0\int v_1\, dz\right] + D_0\int v_2\, dz + E_0\left[\int v_1\, dz\right]^2 &= k_{20}x_2 \end{aligned} \qquad (14)$$

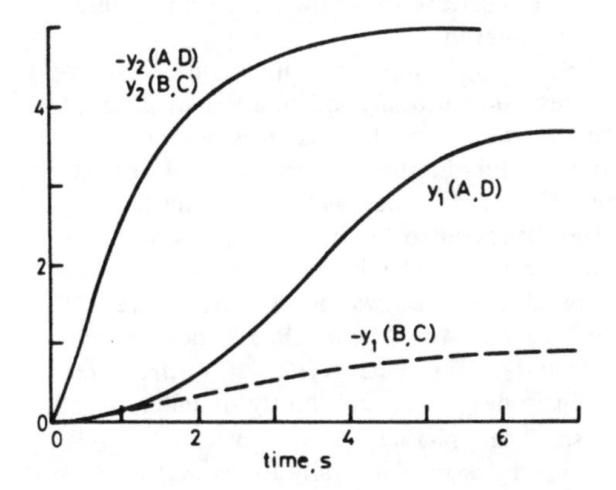

Fig. 14 *Outputs due to simultaneous step inputs, four runs*

	M_1	M_2
A	5	−5
B	5	5
C	−5	5
D	−5	−5

$r_1 = M_1 u(t)$
$r_2 = M_2 u(t)$

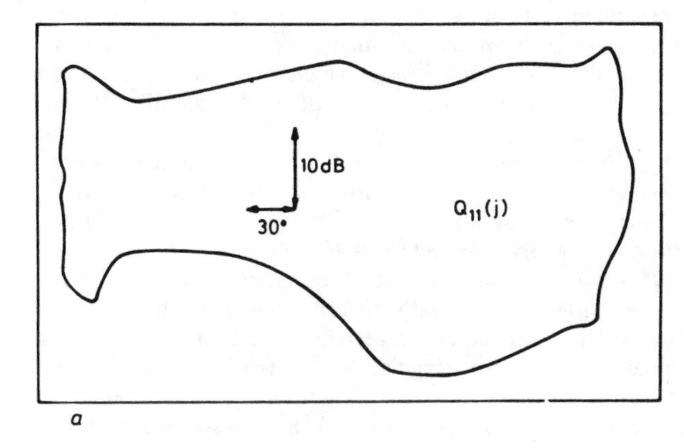

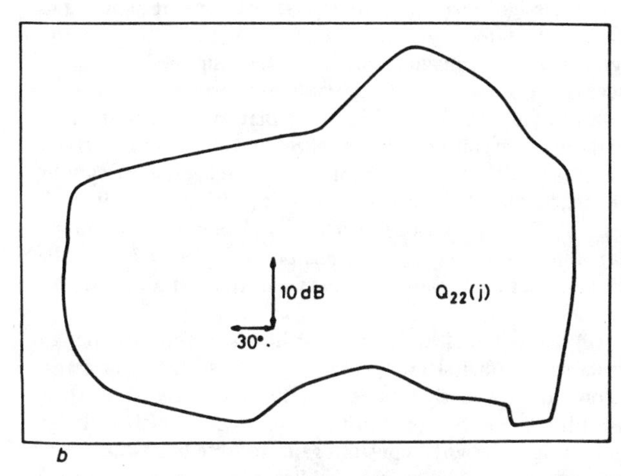

Fig. 15 *Plant templates due to simultaneous step inputs, no cancellation design*

Design problem 2, no cancellation

defining Λ, with the nominals used the same as before. The new effective LTI plant is $W\Lambda$, in place of W. Following the same procedure as in design 1, the new LTI equivalent plant is derived with the resulting plant templates shown in Fig. 16. Compare $\omega = 1$ in Fig. 16 with Fig. 15 to see the enormous decrease in equivalent uncertainty, achieved by use of the nonlinear Λ network. The reason is that, without Λ, nonlinearity by itself (even with no uncertainty) generates an uncertain linear equivalent set. The nonlinear Λ removes most of the uncertainty due to the nonlinearity, and so what remains is the actual uncertainty. There is therefore a much more economical design in terms of the loop gain and bandwidth needed.

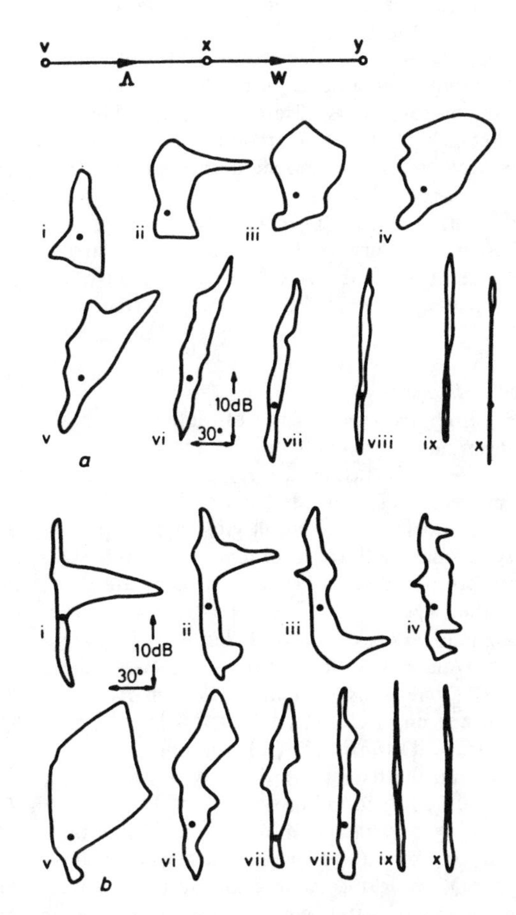

Fig. 16 *Plant templates due to simultaneous step inputs, cancellation design*

Design problem 2, cancellation design
a Q_{11} : ω = 0.1, 0.2, 0.5, 1, 2, 5, 10, 20, 50, 100
b Q_{22}

The next steps are the same as in design 1. Simulation results over a range of M, θ for $r_1 = M\cos\theta$, $r_2 = M\sin\theta$, $M \in [2, 5]$, are shown in Fig. 17.

Disturbance attenuation and command response
There is no basic difference in the design technique for disturbance attenuation alone or with simultaneous commands. Consider the actual system inputs and combinations of system inputs the system is subjected to in its actual real-life operating environment. Formulate the desired plant outputs in response to these inputs, but be sure that the plant is physically capable of delivering such outputs. In this way, obtain a representative set of the actual plant outputs over the life of the system: call it set $\mathscr{A} = \{a(t)\}$. Find $\mathscr{P}$, the LTI equivalent of the nonlinear plant set $\mathscr{W}$ with respect to the set $\mathscr{A}$. Thereafter $\mathscr{P}$ replaces

compensators, which proves that an unrealistic problem had been formulated. As explained by Athans [31], the answer to uncertainty lies the in stochastic LQG (linear quadratic Gaussian) approach, wherein plant and other uncertainties are somehow accounted for by means of Gaussian random zero-mean inputs with suitable covariance matrices. It is a matter of *faith* that uncertainty can be accounted for in this way, because no proof has been offered. The reason for this approach is simple. Optimal LQR theory cannot cope directly with parameter uncertainty and emerge with its 'elegant' Ricatti equation and constant gain compensation, but it can handle added inputs, and so one tries to *learn* somehow (by experience, iteration etc.) to handle the real problem, that of uncertainty, via suitable random inputs and weighting matrices.

For a while, the answer lay in *nominal* sensitivity and stability margins, nominal because LQG design is always at a fixed nominal set of parameters. There was much pride in the wonderful stability margins achieved by LQR theory (wonderful only because of the neglect of the bandwidth problem). This was easily shown as inadequate, because the sensitivity function is $S = (\partial T/T)/(\partial P/P)$, and so $\partial T/T = S\, \partial P/P$. Even if $|S|$ is small, one need only make $|\partial P/P|$ large by having poles or zeros near the $j\omega$-axis. The obvious answer is that sensitivity must be related to the *extent* of uncertainty.

6.1 Singular-value theory (SVT)

The latest and *definitely improved* answer of modern control theory to the uncertainty problem is to examine the actual uncertainty, and then to somehow *incorporate* it into the LQR (or LQG) procedure. The singular values of a matrix A are scalar functions which are an overall gross matrix parameter, used to try to simulate the transfer function of an SISO device [32]. They are not easy to calculate, as they are the square root of the eigenvalues of A^*A versus ω. The maximum and minimum values $\sigma_{max}[A]$ and $\sigma_{min}[A]$ are used. In order to avoid having to calculate them over the uncertainty set $\mathcal{P}$, one deals with *nominal* plant P_0 and obtains a bound on the uncertainty by letting $P = [I + V]P_0(j\omega)$ and finding $\sigma_{max}[V(j\omega)] < l_m(\omega)$, the upper bound over the plant variation matrix $V(j\omega)$. Stability is assured if $\sigma_{max}[L/(I + L)] < 1/l_m(\omega)$ for all ω (eqn. 17 of Reference 32). This gives stability constraints on loop matrix L and on $|L|$ in the low ω range, from σ_{min}. These constraints on L are in turn related to LQG weighting factors, and so constraints are thereby imposed on the latter, giving a first crack at an LQG design. Experimentation with weighting functions by checking against time-domain simulations is usually necessary, because LQG minimises a single scalar index, and so there is no direct control over time responses.

Comments on SVT design

(i) Phase information is lost in singular values as they are positive real functions of ω, and so the stability constraints tend to be very conservative. For example, let a scalar $P = k/(s + a)$, with uncertainties k: $[1, k_x]$, a: $[1, a_x]$. After some effort, it was found that the optimum nominals are $k_0 = (1 + k_x)/2$, $a_0 = (1 + a_x)/2$, giving $l_m(0) = (\beta - 1)/(\beta + 1)$, $\beta = k_x a_x$. The stability constraint gives $|L(j0)/(1 + L(j0))| < (\beta + 1)/(\beta - 1)$. At larger ω, this tends to $|L(j\omega)/[1 + L(j\omega)]| < (\lambda + 1)/(\lambda - 1)$, $\lambda = k_x$. Suppose $\lambda = 10$, $\beta = 100$; then, at larger ω, $|L/(1 + L)| < 1.74$ dB, which corresponds to an approximately 55° phase margin. If $\lambda = 40$ (possible in flight control), then at larger ω the bound is 0.43 dB, forcing a phase margin of about 73°, which is much higher than needed. This means that $|L|$ must decrease slower than inherently necessary. Modern control theorists boast about these large phase margins, but do not realise that the resulting slow decrease of $|L|$ gives a much larger sensor noise effect without getting anything needed in return. Our experience has been that SISO constraints and costs of feedback are accentuated in MIMO designs. Consider the 2 x 2 plant $P = [k_{ij}/s]$. Taking k_{ii}: [1, 5], k_{ij}: [0.2, 0.8] for the uncertainties and the nominal values as the average values, the result is $l_m = 1.6$, so that $|L_i/(1 + L_i)| < 0.62$ for all ω, which is a ridiculous constraint as it would force $|L_i|$ to decrease very slowly until the required gain margin has been achieved.

(ii) Doyle and Stein [33] emphasise 'unstructured' uncertainties, which correspond to *totally uncertain* plant phase at large ω. While this is true at *very* large ω, it is definitely not so in the crossover region in the vast majority of realistic control problems. They argue that well *before* the bending modes (in, say, the flight-control problem), there is no phase information about the plant. This is certainly untrue in the flight-control system. Their motivation is obvious, as the above unduly large margins would be justified if there really is total phase uncertainty. The true region of unstructured uncertainty in realistic designs occurs mostly at much larger ω, where $|L|$ is so small that phase no longer matters.

(iii) A good general design philosophy for MIMO systems should approach a good design method when applied to SISO systems. However, the singular-value approach does not: it retains all its disadvantages.

(iv) A good design technique should not be sensitive to the choice of nominal plant values. This is so in the QFT techniques previously presented: it matters not what values are chosen as nominal. However, l_m above is very sensitive to the choice of nominal values. Thus, in the above scalar example of comment (i), if $k_0 = 1$ is chosen, then $l_m = 4$ and $|L/(1 + L)| < 0.25$, which is a ridiculous constraint. It is not so simple to find the nominal plant which minimises l_m, even in the scalar case; it appears very difficult to do so in the MIMO case.

(v) Even at best, the singular value is a global matrix parameter. Thus there is no control of the individual matrix elements, nor trade-off between the individual loop transmissions as is easily achieved in QFT.

(vi) The relation between uncertainty and the actual time responses of the individual n^2 transmissions is very indirect. Note the steps: uncertainty → search for optimal nominal → search for singular values (with loss of phase information) → constraints on L for stability etc. → constraints on weighting functions in LQG → design → full state recovery design → simulate to see actual time response and uncertainty effects → weighting functions iterate. Many judgments must be made in course of the design [32–34].

(vii) SVT does not apply to nonlinear uncertain plants.

(viii) Doyle and Stein [33] note that 'a single worst-case uncertainty magnitude (is) applicable to all channels. If substantially different levels of uncertainty exist in various channels, it may be necessary to scale the input-output variables . . .'. This is a serious criticism, and they offer no proof that scaling is effective, nor any *ad hoc* examples. One might be sceptical whether it can help, since sensitivity is a normalised function.

(ix) It was noted in Section 4.4 how a quantitative design is suitable for digital compensation, because the loop transmissions are the design tools and the emphasis is on their bandwidth economy. In constrast, in LQR (which includes singular-value design), constant-gain infinite-bandwidth elements emerge, and there is no inkling as to the ω range over which the so-called 'constant' gain must be maintained.

It is urged that modern researchers in the uncertainty problem free themselves from the straitjacket of LQR theory.

$\mathscr{W}$, and so one is dealing with LTI problems. If these LTI problems are solved, then it is guaranteed that the nonlinear problems are solved, but only for those system inputs which give $\mathscr{A}$. The equivalence of $\mathscr{P}$ to $\mathscr{W}$ is only with respect to $\mathscr{A}$.

6 Discussion and comparison with modern control methods

Modern control theory ignored the uncertainty problem for a very long time. It first concentrated on eigenvalue realisation (pole placement) and emerged with the fantastic (from the feedback point of view) result of infinite-bandwidth amplifiers as compensators, and was even proud of this, because no 'dynamics' (poles and zeros) were needed in the compensation. This is comparable with parastic capacitance being neglected in a study of wideband amplifiers, or noise being neglected in a study of high-gain amplifiers. The practical designer may sometimes emerge with compensation which is finite as $s \to \infty$, but he knows in what ω range the inevitable poles are allowed to occur. In as much as the cost of feedback is precisely in the bandwidths of the compensation, there is no excuse for this in a presumably serious study of the feedback problem.

Originally, all the states had to be measured. After a few years of this, modern control theorists discovered that not all the states could usually be measured, which ushered in the 'observer' years of estimating the unavailable states from the measurable ones. After a few more years of this, modern control theorists became aware of the uncertainty problem, which brings us to the present 'robustness' period.

The solution from modern control theorists to the uncertainty problem is more patching of Kalman's linear quadratic regulator (LQR) theory, which is the foundation of the modern design approach to LTI MIMO systems. LQR theory is mathematically impeccable as a solution to a *nonengineering* feedback problem, because it emerges with infinite-bandwidth

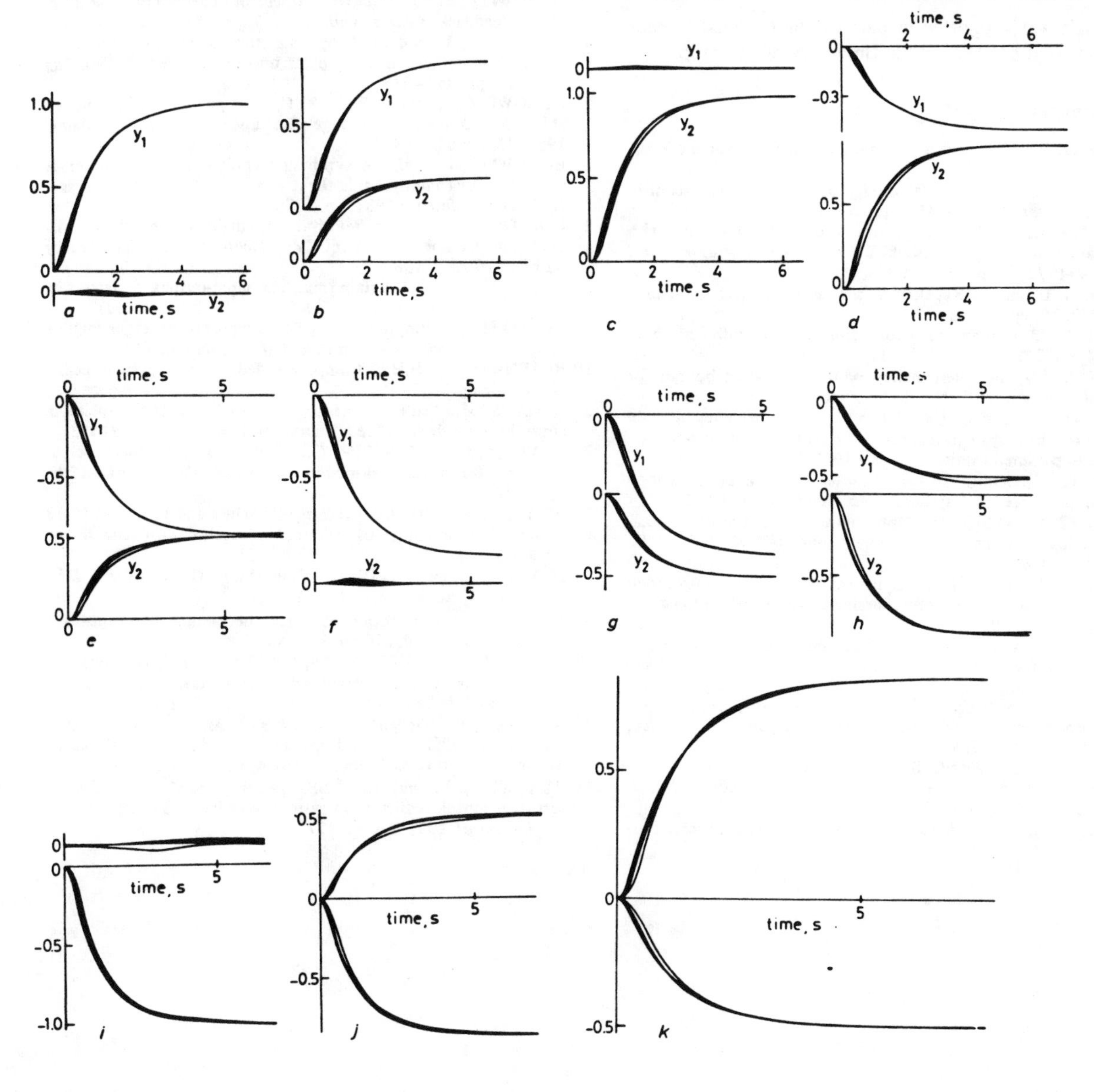

Fig. 17 *Step responses, design 2*

Normalised step responses
$r_1 = M \cos\theta$
$r_2 = M \sin\theta$

a $\theta = 0°$	b $\theta = 30°$	c $\theta = 90°$
d $\theta = 120°$	e $\theta = 150°$	f $\theta = 180°$
g $\theta = 210°$	h $\theta = 240°$	i $\theta = 270°$
j $\theta = 300°$	k $\theta = 330°$	

Its mathematics, its scalar figure of merit at a fixed nominal parameter set, its assumption of all states measurable are all woefully unsuited to quantitative feedback synthesis. For example, in the SISO problem of Fig. 1*b* with only two degrees of freedom, is it not much simpler to optimise on a two-product analytic function space? LQR has not been shown to cope properly even with this simple uncertainty problem.

6.2 Numerical methods

Numerical methods which exploit the power of modern computers have been developed [35–37]. They can be quite useful to the practical designer, but do not by themselves constitute a theory. They are best used in conjunction with a quantitative theory which provides a good initial first try.

7 Acknowledgment

This research was supported in part by the National Science Foundation grant 8101958 at the University of Colorado.

8 References

1 HOROWITZ, I.: 'Synthesis of feedback systems', (Academic Press, 1963)
2 HOROWITZ, I.: 'Fundamental theory of linear feedback control systems', *Trans. IRE*, 1959, AC-4, pp. 5–19
3 HOROWITZ, I., and SIDI, M.: 'Synthesis of feedback systems with large plant ignorance for prescribed time domain tolerances', *Int. J. Control*, 1972, **16**, pp. 287–309
4 LONGDON, L., and EAST, D.J.: 'A simple geometrical technique', *ibid.*, 1979, **30**, pp. 153–8
5 EAST, D.J.: 'A new approach to optimum loop synthesis', *ibid.*, 1981, **34**, pp. 731–748
6 EAST, D.J.: 'On the determination of plant variation bounds for optimum loop synthesis', *ibid.* (to be published)
7 KRISHNAN, K., and CRUICKSHANKS, A.: 'Frequency domain design of feedback systems for specified insensitivity of time-domain response to parameter variation', *ibid.*, 1977, 25, pp. 609–620
8 HOROWITZ, I.: 'Optimum loop transfer function in single-loop minimum-phase feedback systems', *ibid.*, 1973, **18**, pp. 97–113
9 HOROWITZ, I., and SIDI, M.: 'Optimum synthesis of nonminimum-phase feedback systems with parameter uncertainty', *ibid.*, 1978, **27**, pp. 361–386
10 HOROWITZ, I., and SIDI, M.: 'Synthesis of cascaded multiple-loop feedback systems with large plant parameter ignorance', *Automatica*, 1973, 9, pp. 589–600
11 HOROWITZ, I., and WANG, T.S.: 'Synthesis of a class of uncertain multiple-loop feedback systems', *Int. J. Control*, 1979, **29**, pp. 645–668
12 HOROWITZ, I., and WANG, T.S.: 'Quantitative synthesis of multiple-loop feedback systems with large uncertainty', *Int. J. Syst. Sci.*, 1979, **10**, pp. 1235–1268
13 HOROWITZ, I., and WANG, B.C.: 'Quantitative synthesis of uncertain cascade feedback systems with plant modification', *Int. J. Control*, 1979, **30**, pp. 837–862
14 HOROWITZ, I.: 'A synthesis theory for linear time-varying feedback systems with plant uncertainty', *IEEE Trans.*, 1975, **AC-20**, pp. 454–463
15 HOROWITZ, I.: 'Synthesis of feedback systems with non-linear time-varying uncertain plants to satisfy quantitative performance specifications', *IEEE Proc.*, 1976, **64**, pp. 123–130
16 HOROWITZ, I., and SHUR, D.: 'Control of a highly uncertain Van der Pol plant', *Int. J. Control*, 1980, **32**, pp. 199–219
17 HOROWITZ, I., and BREINER, M.: 'Quantitative synthesis of feedback systems with uncertain nonlinear multivariable plants', *Int. J. Syst. Sci.*, 1981, **12**, pp. 539–563
18 HOROWITZ, I. *et al.*: 'Research in advanced flight control design', AFFDL TR–79–3120, Jan. 1980
19 HOROWITZ, I.: 'Improvement in quantitative nonlinear feedback design by cancellation', *Int. J. Control*, 1981, **34**, pp. 547–560
20 HOROWITZ, I.: 'Quantitative synthesis of uncertain nonlinear feedback systems with nonminimum-phase inputs', *Int. J. Syst. Sci.*, 1981, **12**, pp. 55–76
21 HOROWITZ, I.: 'Nonlinear uncertain feedback systems with initial state values', *Int. J. Control*, 1981, **34**, pp. 749–764
22 HOROWITZ, I.: 'Feedback systems with nonlinear uncertain plants', *Int. J. Control*, 1982, **36**, pp. 155–171
23 HOROWITZ, I.: 'Quantitative synthesis of uncertain multiple input-output feedback systems', *ibid.*, 1979, **30**, pp. 81–106
24 HOROWITZ, I., and SIDI, M.: 'Practical design of multivariable feedback systems with large plant uncertainty', *Int. J. Syst. Sci.*, 1980, **11**, pp. 851–875
25 HOROWITZ, I., and LOECHER, C.: 'Design of a 3 × 3 multivariable feedback system with large plant uncertainty', *Int. J. Control*, 1981, **33**, pp. 677–699
26 HOROWITZ, I. *et al.*: 'A synthesis technique for highly uncertain and interacting multivariable flight control systems'. Proceedings of NAECON conference 1981, pp. 1276–1283
27 HOROWITZ, I. *et al.*: 'Multivariable flight control design with uncertain parameters (YF16CCV)'. Report to be published by AFFDL, WPAFB, 1982
28 HOROWITZ, I.: 'Uncertain multiple input-output systems with internal variable feedback', *Int. J. Control.* (to be published)
29 HOROWITZ, I.: 'Improved design technique for uncertain multiple input-output feedback systems', *ibid.*, (to be published)
30 ROSENBROCK, H.H.: 'Computer-aided control system design' (Academic Press, 1974)
31 ATHANS, M.: 'Role and use of the stochastic LQG problem in control system design', *IEEE Trans.*, 1971, **AC-16**, pp. 529–552
32 DOYLE, J.C., and STEIN, G.: 'Multivariable feedback design: Concepts for a classical/modern synthesis', *ibid.*, 1981, **AC-26**, pp. 4–16
33 STEIN, G., and DOYLE, C.: 'Singular values and feedback: Design examples'. Proceedings of Allerton conference on circuit theory, 1978, pp. 461–470
34 DOYLE, J.C., and STEIN, G.: 'Robustness with observers', *IEEE Trans.*, 1979, **AC-24**, pp. 607–611
35 ZAKIAN, V.: 'New formulation for the method of inequalities', *Proc. IEE*, 1979, **126**, (6), pp. 579–584
36 DAVISON, E., and FERGUSON, L.: 'The design of controllers for the multivariable robust servomechanism problem using parameter optimisation', *IEEE Trans.*, 1981, **AC-26**, pp. 93–110
37 GOLUBEV, B.: 'Design of feedback systems with large parameter uncertainty'. Ph.D. thesis, Department of Applied Mathematics, Weizmann Institute of Science, Rehovot, Israel, 1982
38 HOROWITZ, I., and ROSENBAUM, P.: 'Nonlinear design for cost of feedback reduction in systems with large plant uncertainty', *Int. J. Control*, 1975, **21**, pp. 977–1001

A NEW CLASS OF STABILIZING CONTROLLERS FOR UNCERTAIN DYNAMICAL SYSTEMS*

B. R. BARMISH†, M. CORLESS‡ AND G. LEITMANN‡

Abstract. This paper is concerned with the problem of designing a stabilizing controller for a class of uncertain dynamical systems. The vector of uncertain parameters $q(\cdot)$ is time-varying, and its values $q(t)$ lie within a prespecified bounding set Q in R^p. Furthermore, no statistical description of $q(\cdot)$ is assumed, and the controller is shown to render the closed loop system "practically stable" in a so-called *guaranteed sense*; that is, the desired stability properties are assured no matter what admissible uncertainty $q(\cdot)$ is realized. Within the perspective of previous research in this area, this paper contains one salient feature: the class of stabilizing controllers which we characterize is shown to include linear controllers when the nominal system happens to be linear and time-invariant. In contrast, in much of the previous literature (see, for example, [1], [2], [7], and [9]), a linear system is stabilized via nonlinear control. Another feature of this paper is the fact that the methods of analysis and design do not rely on transforming the system into a more convenient canonical form; e.g., see [3]. It is also interesting to note that a linear stabilizing controller can sometimes be constructed even when the system dynamics are nonlinear. This is illustrated with an example.

Key words. stability, uncertain dynamical systems, guaranteed performance

1. Introduction. During recent years, a number of papers have appeared which deal with the design of stabilizing controllers for uncertain dynamical systems; e.g., see [1]–[7]. In these papers the uncertain quantities are described only in terms of bounds on their possible sizes; that is, no statistical description is assumed. Within this framework, the objective is to find a class of controllers which *guarantee* "stable" operation for *all* possible variations of the uncertain quantities.

Roughly speaking, the results to date fall into two categories. There are those results which might appropriately be termed *structural* in nature; e.g., see [1]–[3], [6]. By this we mean that the uncertainty cannot enter arbitrarily into the state equations; certain preconditions must be met regarding the locations of the uncertainty within the system description. Such conditions are often referred to as *matching assumptions*. We note that in this situation uncertainties can be tolerated with an arbitrarily large prescribed bound. A second body of results might appropriately be termed *nonstructural* in nature; e.g., see [4] and [5]. Instead of imposing matching assumptions on the system, these authors permit more general uncertainties at the expense of "sufficient smallness" assumptions on the allowable sizes of the uncertainties.

This work falls within the class of structural results mentioned above. Our motivation comes from a simple observation. Namely, given a theory which yields stabilizing controllers for a class of uncertain nonlinear systems, it is often desirable for this theory to have the following property: upon specializing the "recipe" for controller construction from nonlinear to linear systems, one of the possible stabilizing control laws should be linear in form. It is of importance to note that existing results do *not* have this property. Upon specialization to the linear case, one typically obtains controllers of the discontinuous "bang-bang" variety; e.g., see [1] and [2]. One can often approximate these controllers using a so-called saturation nonlinearity; e.g., see

* Received by the editors March 4, 1981, and in revised form January 15, 1982.

† Department of Electrical Engineering, University of Rochester, Rochester, New York 14627. The work of this author was supported by the U.S. Department of Energy under contract no. ET-78-S-01-3390.

‡ Department of Mechanical Engineering, University of California at Berkeley, Berkeley, California 94720. The work of these authors was supported by the National Science Foundation under grant ENG 78-13931.

Reprinted with permission from *SIAM Journal on Control and Optimization*, vol. 21, no. 2, pp. 246–255.

[7]. Such an approach leads to uniform ultimate boundedness of the state to an arbitrarily small neighborhood of the origin; this type of behavior might be termed *practical stability*.[1]

Our desire in this paper is to develop a controller which is linear when the system dynamics are linear. By taking known results (such as in [3]) which were developed exclusively for linear systems, one encounters a fundamental difficulty when attempting to generalize[2] to a class of nonlinear systems; namely, it is no longer possible to transform the system dynamics to a more convenient canonical form. The subsequent analysis is free of such transformations.

2. Systems, assumptions and the concept of practical stability. We consider an uncertain dynamical system described by the state equation

$$\begin{aligned}\dot{x}(t) &= f(x(t), t) + \Delta f(x(t), q(t), t)\\ &\quad + [B(x(t), t) + \Delta B(x(t), q(t), t)]u(t),\end{aligned} \tag{2.1}$$

where $x(t) \in R^n$ is the *state*, $u(t) \in R^m$ is the *control*, $q(t) \in R^p$ is the *uncertainty* and $f(x, t)$, $\Delta f(x, q, t)$, $B(x, t)$ and $\Delta B(x, q, t)$ are matrices of appropriate dimensions which depend on the structure of the system. Furthermore, it is assumed that the uncertainty, $q(\cdot): R \to R^p$, is Lebesgue measurable and its values $q(t)$ lie within a prespecified *bounding set* $Q \subset R^p$ for all $t \in R$. We denote this by writing $q(\cdot) \in M(Q)$.

As mentioned in the introduction, given that "stabilization" is the goal, we must impose additional conditions on the manner in which $q(t)$ enters structurally into the state equations. We refer to such conditions as *matching assumptions*.

Assumption 1. There are mappings

$$h(\cdot): R^n \times R^p \times R \to R^m \quad \text{and} \quad E(\cdot): R^n \times R^p \times R \to R^{m \times m}$$

such that

$$\begin{aligned}&\Delta f(x, q, t) = B(x, t)h(x, q, t),\\ &\Delta B(x, q, t) = B(x, t)E(x, q, t),\\ &\|E(x, q, t)\| < 1\end{aligned}$$

for all $x \in R^n$, $q \in Q$ and $t \in R$.

We note that this assumption can sometimes be weakened. For example, in [9] a certain *measure of mis-match* is introduced and results are obtained under the proviso that this measure does not exceed a certain critical level termed the *mis-match threshold*.

Our second assumption reflects the fact that the uncertainties must be bounded in order to permit one to guarantee stability.

Assumption 2. The set $Q \subset R^p$ is compact.

Our next assumption is introduced to guarantee the existence of solutions of the state equations.

Assumption 3. The mappings $f(\cdot): R^n \times R \to R^n$, $B(\cdot): R^n \times R \to R^{m \times n}$, $h(\cdot)$ and $E(\cdot)$ (see Assumption 1) are continuous.[3]

[1] This notion is not to be interpreted in the sense of Lasalle and Lefschetz [12] but as defined subsequently.

[2] That is, one begins with a linear control law for a linear system and generalizes the controller in such a way that is applies to a class of nonlinear systems.

[3] In fact, one can modify the analysis to follow so as to allow mappings which are Carathéodory and satisfy certain integrability conditions. See, for example, Corless and Leitmann [7]. All the results of this paper still hold under this weakening of hypotheses.

In order to satisfy our final assumption, one may need to "precompensate" the so-called nominal system, that is, the system with $\Delta f(x, q, t) \equiv 0$ and $\Delta B(x, q, t) \equiv 0$; e.g., see [2]. Thus, prior to controlling the effects of the uncertainty, it may be necessary to employ a portion of the control to obtain an *uncontrolled nominal system*

$$\text{(UC)} \qquad \dot{x}(t) = f(x(t), t)$$

that has certain stability properties embodied in the next assumption.

Assumption 4. $f(0, t) = 0$ for all $t \in R$ and, moreover, there exist a C^1 function $V(\cdot): R^n \times R \to [0, \infty)$ and strictly increasing continuous functions $\gamma_1(\cdot)$, $\gamma_2(\cdot)$, $\gamma_3(\cdot): [0, \infty) \to [0, \infty)$ satisfying[4] $\gamma_1(0) = \gamma_2(0) = \gamma_3(0) = 0$ and $\lim_{r \to \infty} \gamma_1(r) = \lim_{r \to \infty} \gamma_2(r) = \lim_{r \to \infty} \gamma_3(r) = \infty$, such that for all $(x, t) \in R^n \times R$,

$$\gamma_1(\|x\|) \leqq V(x, t) \leqq \gamma_2(\|x\|). \tag{2.2}$$

Moreover, defining the Lyapunov derivative $\mathcal{L}_0(\cdot): R^n \times R \to R$ by

$$\mathcal{L}_0(x, t) \triangleq \frac{\partial V(x, t)}{\partial t} + \nabla_x' V(x, t) f(x, t), \tag{2.3}$$

where ∇_x' denotes the transpose of the gradient operation, we also require that

$$\mathcal{L}_0(x, t) \leqq -\gamma_3(\|x\|)$$

for all pairs $(x, t) \in R^n \times R$. This assumption, in effect, asserts the existence of a Lyapunov function for the uncontrolled nominal system (UC). Consequently, the origin, $x = 0$, is a uniformly asymptotically stable equilibrium point for the uncontrolled nominal system (UC).

The stability concept employed in this paper differs slightly from the traditional Lyapunov-type stability. To motivate this change of definition, consider the following very simple example of a system satisfying (2.1) and the associated assumptions: $\dot{x}(t) = x(t) + q(t) + u(t)$, with initial condition $x(t_0) = 1$ and uncertainty $q(\cdot)$ such that $|q(t)| \leqq 1$. Furthermore, suppose the control is a linear feedback of the form $u(t) = kx(t)$, with $k < -1$. Then, if a state $x(t) < -1/(1+k)$ is reached, an admissible uncertainty $q(t) \equiv 1$ results in motion of the state away from zero. Hence, although we cannot guarantee uniform asymptotic stability (using a finite gain), we can nevertheless drive the state to an arbitrarily small neighborhood of the origin.[5] The following uniform ultimate boundedness-type definition captures this notion.

DEFINITION 1. The uncertain dynamical system (2.1) is said to be *practically stabilizable* if, given any $\underline{d} > 0$, there is a control law $p_d(\cdot): R^n \times R \to R^m$ for which, given any admissible uncertainty $q(\cdot) \in M(Q)$, any initial time $t_0 \in R$ and any initial state $x_0 \in R^n$, the following conditions hold:

(i) The closed loop system

$$\begin{aligned} \dot{x}(t) &= f(x(t), t) + \Delta f(x(t), q(t), t) \\ &\quad + [B(x(t), t) + \Delta B(x(t), q(t), t)] p_d(x(t), t) \end{aligned} \tag{2.4}$$

possesses a solution $x(\cdot): [t_0, t_1] \to R^n$, $x(t_0) = x_0$.

[4] The limit condition on $\gamma_3(\cdot)$ can in fact be removed at the expense of a somewhat more technical development; e.g., see [7].

[5] This is not to be confused with Lyapunov stability, because the required gain k depends on the size of the neighborhood to which we wish to drive the state.

(ii) Given any $r>0$ and any solution $x(\cdot):[t_0, t_1]\to R^n$, $x(t_0)=x_0$, of (2.4) with $\|x_0\|\leqq r$, there is a constant $d(r)>0$ such that

$$\|x(t)\|\leqq d(r) \quad \text{for all} \quad t\in[t_0, t_1].$$

(iii) Every solution $x(\cdot):[t_0, t_1]\to R^n$ can be continued over $[t_0, \infty)$.

(iv) Given any $\bar{d}\geqq \underline{d}$, any $r>0$ and any solution $x(\cdot):[t_0, \infty)\to R^n$, $x(t_0)=x_0$, of (2.4) with $\|x_0\|\leqq r$, there exists a finite time $T(\bar{d}, r)<\infty$, possibly dependent on r but not on t_0, such that $\|x(t)\|\leqq \bar{d}$ for all $t\geqq t_0+T(\bar{d}, r)$.

(v) Given any $\bar{d}\geqq \underline{d}$ and any solution $x(\cdot):[t_0, \infty)\to R^n$, $x(t_0)=x_0$, of (2.4), there is a constant $\delta(\bar{d})>0$ such that $\|x_0\|\leqq \delta(\bar{d})$ implies that

$$\|x(t)\|\leqq \bar{d} \quad \text{for all} \quad t\geqq t_0.$$

3. Controller construction. We take $\underline{d}>0$ as given and proceed to construct a control law $p_{\underline{d}}(\cdot)$ which will later be shown to satisfy conditions (i)–(v) in the definition of practical stabilizability.

Construction of $p_{\underline{d}}(\cdot)$. The first step is to select functions $\Delta_1(\cdot)$ and $\Delta_2(\cdot):R^n\times R\to R$ satisfying

$$\Delta_1(x, t)\geqq \max_{q\in Q} \|h(x, q, t)\|, \tag{3.1}$$

$$1>\Delta_2(x, t)\geqq \max_{q\in Q} \|E(x, q, t)\|. \tag{3.2}$$

The standing Assumptions 1–4 assure that there is a $\Delta_2(x, t)$ such that

1) $\Delta_2(x, t)<1$ can be satisfied for all $(x, t)\in R^n\times R$;

2) $\Delta_1(\cdot)$ and $\Delta_2(\cdot)$ can be chosen to be continuous; e.g., see [10, p. 116].

Now, one simply selects any continuous function $\gamma(\cdot):R^n\times R\to[0, \infty)$ satisfying

$$\gamma(x, t)\geqq \frac{\Delta_1^2(x, t)}{4[1-\Delta_2(x, t)][C_2-C_1\mathcal{L}_0(x, t)]}, \tag{3.3}$$

where C_1 and C_2 are any (designer chosen) nonnegative constants such that

(3.4)

a) $C_1<1$;

b) either $C_1\neq 0$ or $C_2\neq 0$;

c) $C_2\neq 0$ whenever $\lim_{x\to 0}[\Delta_1^2(x, t)/\mathcal{L}_0(x, t)]$ does not exist;

d) $$\frac{C_2}{1-C_1}<(\gamma_3\circ\gamma_2^{-1}\circ\gamma_1)(\underline{d}).$$

Note that these conditions can indeed be satisfied because of continuity of the $\gamma_i(\cdot)$ and the fact that $\lim_{r\to 0}\gamma_i(r)=0$ for $i=1, 2, 3$.

This construction then enables one to let

$$p_{\underline{d}}(x, t)\triangleq -\gamma(x, t)B'(x, t)\nabla_x V(x, t). \tag{3.5}$$

Remark. In fact, (3.3) and (3.5) describe a class of controllers yielding practical stability. It will be shown in § 5 that this class includes linear controllers when the nominal system happens to be linear and time-invariant.

4. Main result and stability estimates. The theorem below and its proof differ from existing results (see [1], [2] and [6]) in one fundamental way: The control $p_{\underline{d}}(\cdot)$ which leads to the satisfaction of the conditions for practical stabilizability degenerates into a linear controller whenever the *nominal* system, obtained by setting $\Delta f(x(t), q(t), t) \equiv 0$ and $\Delta B(x(t), q(t), t) \equiv 0$ in (2.1), is linear and time-invariant. This will be demonstrated in the sequel. In fact, even for certain nonlinear nominal systems, the controller turns out to be linear. This phenomenon will be illustrated with an example of a nonlinear pendulum. Central to the proof of the theorem below is one fundamental concept: a system satisfying Assumptions 1–4 admits a control such that the Lyapunov function for the nominal system (UC) is also a Lyapunov function for the uncertain system (2.1).

THEOREM 1. *Subject to Assumptions* 1–4, *the uncertain dynamical system* (2.1) *is practically stabilizable.*

Proof. For a given $\underline{d} > 0$ and a given uncertainty $q(\cdot) \in M(Q)$, the Lyapunov derivative $\mathcal{L}(\cdot): R^n \times R \to R$ for the closed loop system obtained with the feedback control (3.5) is given by

$$\mathcal{L}(x, t) \triangleq \mathcal{L}_0(x, t) + \nabla_x' V(x, t)\{\Delta f(x, q(t), t) + [B(x, t) + \Delta B(x, q(t), t)]p_{\underline{d}}(x, t)\}. \tag{4.1}$$

By using the matching assumptions in conjunction with (3.5), (4.1) becomes

$$\begin{aligned}\mathcal{L}(x, t) = \mathcal{L}_0(x, t) - \gamma(x, t)\|B'(x, t)\nabla_x V(x, t)\|^2 &+ \nabla_x' V(x, t)B(x, t)[h(x, q(t), t) \\ &- \gamma(x, t)E(x, q(t), t)B'(x, t)\nabla_x V(x, t)].\end{aligned}$$

Letting $\phi(\cdot): R^n \times R \to R^m$ be given by

$$\phi(x, t) \triangleq B'(x, t)\nabla_x V(x, t),$$

and recalling the definition of $\Delta_1(\cdot)$ and $\Delta_2(\cdot)$, a straightforward computation yields

$$\mathcal{L}(x, t) \leqq \mathcal{L}_0(x, t) - [1 - \Delta_2(x, t)]\gamma(x, t)\|\phi(x, t)\|^2 + \Delta_1(x, t)\|\phi(x, t)\|.$$

Now there are two cases to consider.

Case 1. The pair (x, t) is such that $\Delta_1(x, t) = 0$. It then follows from the preceding inequality that

$$\mathcal{L}(x, t) \leqq \mathcal{L}_0(x, t).$$

Case 2. The pair (x, t) is such that $\Delta_1(x, t) \neq 0$. Then it follows from (3.3) that $\gamma(x, t) > 0$. Moreover, in view of (3.3) and the conditions on the C_i,

$$\begin{aligned}\mathcal{L}(x, t) &\leqq \mathcal{L}_0(x, t) - [1 - \Delta_2(x, t)]\gamma(x, t)\|\phi(x, t)\|^2 + \Delta_1(x, t)\|\phi(x, t)\| \\ &= \mathcal{L}_0(x, t) + \frac{\Delta_1^2(x, t)}{4\gamma(x, t)(1 - \Delta_2(x, t))} \\ &\quad - \frac{(1 - \Delta_2(x, t))\gamma(x, t)}{\Delta_1^2(x, t)}\left[\Delta_1(x, t)\|\phi(x, t)\| - \frac{\Delta_1^2(x, t)}{2(1 - \Delta_2(x, t))\gamma(x, t)}\right]^2 \\ &\leqq \mathcal{L}_0(x, t) + \frac{\Delta_1^2(x, t)}{4\gamma(x, t)(1 - \Delta_2(x, t))} \\ &\leqq (1 - C_1)\mathcal{L}_0(x, t) + C_2.\end{aligned}$$

Combining Cases 1 and 2, and noting that $C_1<1$, we conclude (as a consequence of Assumption 4) that

$$\mathscr{L}(x,t)\leqq(1-C_1)\mathscr{L}_0(x,t)+C_2\leqq-(1-C_1)\gamma_3(\|x\|)+C_2 \tag{4.2}$$

for all $(x,t)\in R^n\times R$. Having this inequality available now enables one to apply directly the results of [7]. That is, the closed loop system (2.4) possesses a solution $x(\cdot):[t_0,t_1]\to R^n$, $x(t_0)=x_0$, which is required by condition (i) in the definition of practical stabilizability. Moreover, in accordance with [7], if $\|x_0\|\leqq r$, one can satisfy the uniform boundedness requirement (ii) by selecting

$$d(r)\triangleq\begin{cases}(\gamma_1^{-1}\circ\gamma_2)(R) & \text{if } r\leqq R,\\ (\gamma_1^{-1}\circ\gamma_2)(r) & \text{if } r>R,\end{cases}$$

where

$$R\triangleq\gamma_3^{-1}(C_2/1-C_1).$$

It now follows that there is no finite escape time so that the solution is continuable over $[t_0,\infty)$ and hence condition (iii) holds. Again for $\bar{d}\geqq\underline{d}$, using the estimates provided in [7], one can define

$$T(\bar{d},r)\triangleq\begin{cases}0 \quad \text{if } r\leqq(\gamma_2^{-1}\circ\gamma_1)(\bar{d}),\\[2ex] \dfrac{\gamma_2(r)-(\gamma_1\circ\gamma_2^{-1}\circ\gamma_1)(\bar{d})}{(1-C_1)(\gamma_3\circ\gamma_2^{-1}\circ\gamma_1)(\bar{d})-C_2} \quad \text{otherwise,}\end{cases} \tag{4.3}$$

and in accordance with [7], the desired uniform ultimate boundedness condition (iv) holds with the proviso that

$$(1-C_1)(\gamma_3\circ\gamma_2^{-1}\circ\gamma_1)(\bar{d})-C_2>0. \tag{4.4}$$

Note that this requirement is implied by the satisfaction of condition (d) of (3.4) which entered into the construction of the controller.

Finally, to complete the proof, it remains to establish the desired uniform stability property. Indeed, let $\bar{d}\geqq\underline{d}$ be specified and notice that if $\delta(\bar{d})=R$, the following property will hold: Given any solution $x(\cdot):[t_0,\infty)\to R^n$, $x(t_0)=x_0$ of (2.4) with $\|x_0\|\leqq\delta(\bar{d})$, it follows (in view of the uniform boundedness property (ii) and the requirements on the C_i) that $\|x(t)\|\leqq d(R)\leqq\bar{d}$ for all $t\geqq t_0$. □

5. Specialization to linear systems. The objective of this section is to show that the flexibility permitted in choosing $\gamma(x,t)$ (see (3.3)) can be exploited in a "nice way" when the nominal system dynamics happen to be linear and time-invariant; that is, the control $p_d(\cdot)$ in (3.5) can be selected to be a linear time-invariant feedback of the state. We consider the special case when

$$\begin{aligned}\dot{x}(t)&=[A+\Delta A(q(t))]x(t)+[B+\Delta B(q(t))]u(t)+w(q(t)),\\ x(t_0)&=x_0,\end{aligned} \tag{5.1}$$

where A, $\Delta A(q(t))$, B and $\Delta B(q(t))$ are matrices of appropriate dimensions and $w(q(t))$ is an n-dimensional vector. In light of Assumptions 1 and 3 given in § 2, it follows that for all $q\in Q$

$$\begin{aligned}\Delta A(q)&=BD(q),\\ \Delta B(q)&=BE(q),\\ w(q)&=Bv(q),\\ E(q)\|&<1\end{aligned} \tag{5.2}$$

where $D(\cdot)$, $E(\cdot)$ and $v(\cdot)$ have appropriate dimensions and depend continuously on their arguments. In accordance with Assumption 4, the matrix A must be asymptotically stable. To obtain a Lyapunov function for the uncontrolled nominal system, we select simply an $n \times n$ positive-definite symmetric matrix H and solve the equation

$$A'P + PA = -H \tag{5.3}$$

for P which is positive-definite; see [11]. Then we have

$$V(x, t) = x'Px \tag{5.4}$$

and

$$\mathscr{L}_0(x, t) = -x'Hx. \tag{5.5}$$

It is clear from (5.4) and (5.5) that one can take the bounding functions $\gamma_i(\cdot)$ to be

$$\gamma_1(r) \triangleq \lambda_{\min}[P]r^2, \quad \gamma_2(r) \triangleq \lambda_{\max}[P]r^2, \quad \gamma_3(r) \triangleq \lambda_{\min}[H]r^2, \tag{5.6}$$

where $\lambda_{\max(\min)}[\cdot]$ denotes the operation of taking the largest (smallest) eigenvalue.

Construction of the controller. We take $\underline{d} > 0$ as prescribed and construct the controller $p_{\underline{d}}(\cdot)$ given in § 3. Using the notation above, we define first[6]

$$\rho_D \triangleq \max_{q \in Q} \|D(q)\|, \quad \rho_E \triangleq \max_{q \in Q} \|E(q)\| < 1, \quad \rho_v \triangleq \max_{q \in Q} \|v(q)\|. \tag{5.7}$$

Then, in agreement with (3.1) and (3.2), we may take

$$\Delta_1(x, t) = \rho_D\|x\| + \rho_v, \qquad \Delta_2(x, t) = \rho_E. \tag{5.8}$$

Using these choices in (3.3) and the fact that $\mathscr{L}_0(x, t) \leqq -\lambda_{\min}[H]\|x\|^2$, one can select $\gamma(\cdot)$ such that

$$\gamma(x, t) \geqq \frac{(\rho_D\|x\| + \rho_v)^2}{4(1-\rho_E)(C_1\lambda_{\min}[H]\|x\|^2 + C_2)}, \tag{5.9}$$

with the constants C_1 and C_2 yet to be specified. We shall examine three possible cases and see that in all instances one can take $\gamma(x, t) \equiv$ constant. Of course, this implies that the control $p_{\underline{d}}(\cdot)$ is a linear time-invariant feedback; that is,

$$p_{\underline{d}}(x, t) = -2\gamma_0 B'Px, \tag{5.10}$$

where γ_0 is the constant value of $\gamma(\cdot)$, which will be specified.

Case 1. $\rho_D > 0$, $\rho_v = 0$. In this case, we may select $C_2 = 0$ and $C_1 \in (0, 1)$. Consequently, we can satisfy (5.9) by choosing

$$\gamma(x, t) \equiv \gamma_0 \geqq \frac{\rho_D^2}{4(1-\rho_E)C_1\lambda_{\min}[H]}. \tag{5.11}$$

Case 2. $\rho_D = 0$, $\rho_v > 0$. Clearly, it suffices to take $C_1 = 0$ and

$$\gamma(x, t) \equiv \gamma_0 \geqq \frac{\rho_v^2}{4(1-\rho_E)C_2}, \tag{5.12}$$

where C_2 is required to satisfy condition d) of (3.4). Using the descriptions of the $\gamma_i(\cdot)$ given in (5.6), this amounts to restricting C_2 by

$$\frac{C_2}{1-C_1} < \frac{\lambda_{\min}[P]}{\lambda_{\max}[P]}\lambda_{\min}[H]\underline{d}^2 \tag{5.13}$$

with $C_1 = 0$ in the above.

[6] One can in fact use overestimates $\hat{\rho}_D$ and $\hat{\rho}_E$ for ρ_D and ρ_E as long as the inequality $\hat{\rho}_E < 1$ is satisfied.

Case 3. $\rho_D > 0$, $\rho_v > 0$. Now, in order to satisfy (5.9), we select $C_1 \in (0, 1)$, C_2 satisfying (5.13) and

$$\gamma(x, t) \equiv \gamma_0 \geqq \max_{r \geqq 0} \left\{ \frac{(\rho_D r + \rho_v)^2}{4(1-\rho_E)[C_1 \lambda_{\min}[H] r^2 + C_2]} \right\}. \tag{5.14}$$

Letting $f(r)$ denote the bracketed quantity in (5.14) above, a straightforward but lengthy differentiation yields

$$\max_{r \geqq 0} f(r) = \frac{1}{4(1-\rho_E)} \left\{ \frac{\rho_D^2}{C_1 \lambda_{\min}[H]} + \frac{\rho_v^2}{C_2} \right\}. \tag{5.15}$$

Hence, any γ_0 equal to or exceeding this maximum value will be appropriate in (5.14).

6. Illustrative example. We consider now the simple pendulum which was analyzed in [7]. However, here it will be shown that the desired practical stability can actually be achieved via a linear control. This may seem somewhat surprising in light of the fact that the nominal system dynamics are nonlinear. A pendulum of length l is subjected to a control moment $u(\cdot)$ (per unit mass). The point of support is subject to an uncertain acceleration $q(\cdot)$, with $|q(t)| \leqq \hat{\rho} l \equiv$ constant. Letting x_1 denote the angle between the pendulum's arm and a vertical reference line, one obtains the state equations

$$\begin{aligned} \dot{x}_1(t) &= x_2(t), \\ \dot{x}_2(t) &= -a \sin x_1(t) + u(t) - \frac{q(t) \cos x_1(t)}{l}, \end{aligned} \tag{6.1}$$

where $a > 0$ is a given constant. In order to satisfy the assumptions of § 2 one must assure a uniformly asymptotically stable equilibrium for (UC), the uncontrolled nominal system. Hence, for a given $\underline{d} > 0$, we propose a controller of the form

$$u(t) = -bx_1(t) - cx_2(t) + p_{\underline{d}}(x(t), t), \tag{6.2}$$

where b and c are positive constants and $p_{\underline{d}}(\cdot)$ will be specified later in accordance with the results of § 3. The linear portion of the controller (6.2) is used to obtain a stable nominal system. Substitution of (6.2) into (6.1) now yields the state equation

$$\dot{x}(t) = f(x(t), t) + B[p_{\underline{d}}(x(t), t) + h(x(t), q(t), t)], \tag{6.3}$$

where

$$\begin{aligned} B &= \begin{bmatrix} 0 \\ 1 \end{bmatrix}, \quad f(x, t) = \begin{bmatrix} x_2 \\ -bx_1 - cx_2 - a \sin x_1 \end{bmatrix}, \\ h(x, q, t) &= \frac{-q \cos x_1}{l}. \end{aligned} \tag{6.4}$$

A suitable Lyapunov function for the uncontrolled nominal system (with $x = 0$ as equilibrium) is

$$V(x, t) = (b + \tfrac{1}{2}c^2)x_1^2 + cx_1x_2 + x_2^2 + 2a(1 - \cos x_1), \tag{6.5}$$

and, provided b is sufficiently large, the associated $\gamma_i(\cdot)$ are given by

$$\begin{aligned} \gamma_1(r) &= \lambda_1 r^2, \\ \gamma_2(r) &= \begin{cases} \lambda_2 r^2 + 2a(1 - \cos r) & \text{if } r \leqq \pi, \\ \lambda_2 r^2 + 4a & \text{if } r > \pi, \end{cases} \\ \gamma_3(r) &= \lambda_3 r^2, \end{aligned} \tag{6.6}$$

(6.7) $$\lambda_1 \triangleq \lambda_{\min}[P], \quad \lambda_2 \triangleq \lambda_{\max}[P], \quad \lambda_3 \triangleq \min\{\bar{b}c, c\},$$

(6.8) $$P = \begin{bmatrix} b + \frac{1}{2}c^2 & \frac{1}{2}c \\ \frac{1}{2}c & 1 \end{bmatrix}, \qquad \bar{b} \triangleq b + a \min_{x_1}\left(\frac{\sin x_1}{x_1}\right) > 0.$$

Following the procedure described in § 3 for the construction of the controller $p_d(\cdot)$, we select first

(6.9) $$\Delta_1(x, t) = \hat{\rho}|\cos x_1|, \qquad \Delta_2(x, t) = 0.$$

Inequality (3.3) can then be assured by requiring

(6.10) $$\gamma(x, t) \geqq \frac{\hat{\rho}^2 \cos^2 x_1}{4[C_2 + C_1\lambda_3\|x\|^2]}.$$

Given our desire for a linear feedback, one can select $C_1 = 0$ and satisfy (6.10) by choosing

(6.11) $$\gamma(x, t) \equiv \gamma_0 \geqq \frac{\hat{\rho}^2}{4C_2}.$$

To complete the design, C_2 must be selected to satisfy condition d) of (3.4). The analysis must account for two cases, depending on the size of the given radius $\underline{d} > 0$.

Case 1. $\lambda_1\underline{d}^2 > \lambda_2\pi^2 + 4a$. The required conditions on C_2 are

(6.12) $$0 < C_2 < (\gamma_3 \circ \gamma_2^{-1} \circ \gamma_1)(\underline{d}) = \frac{\lambda_3}{\lambda_2}(\lambda_1\underline{d}^2 - 4a).$$

Case 2. $\lambda_1\underline{d}^2 \leqq \lambda_2\pi^2 + 4a$. In this case, the constraints imposed by (6.12) are met if $C_2 > 0$ is chosen sufficiently small so that

(6.13) $$\frac{\lambda_2 C_2}{\lambda_1\lambda_3} + \frac{2a}{\lambda_1}\left(1 - \cos\sqrt{\frac{C_2}{\lambda_3}}\right) < \underline{d}^2.$$

Having now selected C_2, the controller is specified by (3.5) in conjunction with (6.11); that is,

(6.14) $$p_d(x, t) = -\gamma_0 B'(x, t)\nabla_x V(x, t) = -\gamma_0(cx_1 + 2x_2),$$

with the proviso that $\gamma_0 \geqq \hat{\rho}^2/4C_2$. It is interesting to note that there is an obvious tradeoff between the required gain constant γ_0 and the given radius $\underline{d} > 0$. As the radius $\underline{d}$ decreases, C_2 decreases, which in turn implies that γ_0 increases. In contrast, the nonlinear saturation controller of [7] remains bounded by the bound of the uncertainty, and the radius $\underline{d}$ can be decreased by increasing the nonlinear gain; i.e., by approaching a discontinuous control.

7. Conclusion. This paper addresses the so-called problem of practical stabilizability for a class of uncertain dynamical systems. In contrast to previous work on problems of this sort, the main emphasis here is on the structure of the controller. It is shown that by choosing the function $\gamma(\cdot)$ in a special way, the resultant control law can often be realized as a linear time-invariant feedback.

REFERENCES

[1] G. LEITMANN, *Guaranteed asymptotic stability for some linear systems with bounded uncertainties*, J. Dynamic Systems, Meaurement and Control, 101 (1979), pp. 212–216.

[2] S. GUTMAN, *Uncertain dynamical systems, a Lyapunov min-max approach*, IEEE Trans. Automat. Control, AC-24 (1979), pp. 437–443; correction, 25 (1980), p. 613.

[3] J. S. THORP AND B. R. BARMISH, *On Guaranteed Stability of Uncertain Linear Systems via Linear Control*, J. Optim. Theory and Applications, in press.
[4] S. S. L. CHANG AND T. K. C. PENG, *Adaptive Guaranteed Cost Control of Systems with Uncertain Parameters*, IEEE Transactions on Automatic Control, AC-17 (1972), pp. 474–483.
[5] A. VINKLER AND I. J. WOOD, *Multistep Guaranteed Cost Control of Linear Systems with Uncertain Parameters*, J. Guidance and Control, 2, 6 (1980), pp. 449–456.
[6] P. MOLANDER, *Stabilization of Uncertain Systems*, Report LUTFD2/(TRFT-1020)/1-111/(1979), Lund Institute of Technology, August 1979.
[7] M. CORLESS AND G. LEITMANN, *Continuous State Feedback Guaranteeing Uniform Ultimate Boundedness for Uncertain Dynamic Systems*, IEEE Trans. Automat. Control, AC-26, 5 (1981), pp. 1139–1144.
[8] A. VINKLER AND I. J. WOOD, *A Comparison of Several Techniques for Designing Controllers of Uncertain Dynamic Systems*, Proceedings of the IEEE Conference on Decision and Control, San Diego CA, 1979.
[9] B. R. BARMISH AND G. LEITMANN, *On Ultimate Boundedness Control of Uncertain Systems in the Absence of Matching Conditions*, IEEE Transactions on Automatic Control, AC-27 (1982), pp. 153–157.
[10] C. BERGE, *Topological Spaces*, Oliver and Boyd, London, (1963).
[11] N. N. KRASOVSKII, *Stability of Motion*, Stanford University Press, Stanford, CA, (1963).
[12] J. P. LASALLE AND S. LEFSCHETZ, *Stability by Liapunov's Direct Method with Applications*, Academic Press, New York, (1961).

Part V
Related Theory and Numerical Computation

Single-Loop Feedback-Stabilization of Linear Multivariable Dynamical Plants*†

Stabilisation par Rétroaction à Boucle Simple d'Installations Dynamiques Multivariables Linéaires

Einschleifige Rückführungs-Stabilisierung linearer multivariabler dynamischer Anlagen

Одноконтурная стабилизация с обратной связью линейных многопараметрических динамических установок

D. C. YOULA, J. J. BONGIORNO, JR. and C. N. LU‡

It is possible to derive the complex s-plane restrictions that a general nonminimum-phase, unstable, multivariable plant must satisfy so that it can be stabilized with single-loop feedback employing an asymptotically stable controller and feedback sensor.

Summary—This paper derives the necessary and sufficient conditions for a multivariable plant $P(s)$ with asymptotically stable hidden modes to be stabilizable by means of single-loop feedback employing an asymptotically stable controller and feedback sensor. These conditions are completely general and therefore encompass unstable, nonminimum-phase plants as well. For single-input–output plants with zero gain at infinite frequency the conditions reduce to the sole requirement that no plant zeros on the nonnegative real axis of the complex s-plane lie to the left of an odd number of real plant poles, multiple poles counted according to their multiplicities.

It has also been possible to derive simple necessary and sufficient conditions for a closed-loop transfer function $T(s)$ to be realizable by asymptotically stable compensation around a *prescribed* single-input–output plant $P(s)$. It is expected that this latter result can be suitably generalized to the multivariable case.

In a real sense, this paper constitutes a continuation of some earlier unpublished work [8] by the first author.

1. INTRODUCTION

BY WAY of introducing the problems posed and solved in this paper, consider the standard feedback configuration shown in Fig. 1 with the included components satisfying the following assumptions:

(A_1) The plant, controller and feedback sensor are finite-dimensional, linear time-invariant dynamical systems.

(A_2) Any hidden modes of the plant are asymptotically stable.

(A_3) The controller and feedback sensor are completely observable and controllable.

(A_4) The closed loop is dynamical.

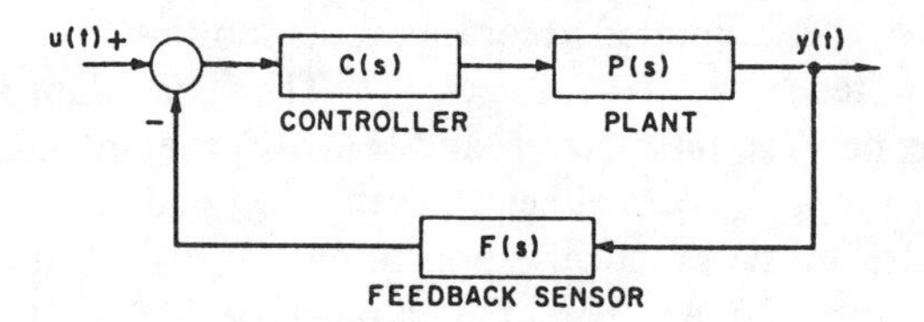

FIG. 1. Single-loop feedback systems.

Then [8], the plant, controller, feedback sensor, and closed loop are described fully by their 1×1 input–output transfer functions $P(s)$, $C(s)$, $F(s)$ and $T(s)$, respectively. Moreover [8, 9], (A_4) is equivalent to the inequality

$$1+P(\infty)C(\infty)F(\infty)\neq 0.$$

Recently, many papers have appeared dealing with the design of feedback systems which realize some or all of a prescribed set of closed-loop eigenvalues when the state of the plant is not completely available [1–6]. In Refs. 1–3, attention is restricted exclusively to non-dynamic compensators and it is found that in general, arbitrary

* Received 26 April 1973; revised 24 August 1973. The original version of this paper was not presented at any IFAC meeting. It was recommended for publication in revised form by Associate Editor B. D. O. Anderson.

† This work was supported by the National Science Foundation under Grant GK-31959.

‡ The authors are with the Department of Electrical Engineering and Electrophysics, Polytechnic Institute of New York, Long Island Center, Route 110, Farmingdale, New York 11735.

placement of all closed-loop eigenvalues is impossible and the concomitant question of closed-loop stability is therefore left unresolved. When dynamic compensation is employed in the manner suggested in [4–6], closed-loop stability is always achievable. However, the necessary compensators can be unstable and it has been shown [7] that, in certain cases, this instability appears to result in poor overall system sensitivity to variations in plant parameters. For these reasons, we have undertaken the solution to the following problem: Subject to assumptions (A_1)–(A_4) and *given* $P(s)$, find necessary and sufficient conditions to impose on $P(s)$ insuring the existence of an *asymptotically stable* compensator and feedback sensor yielding a dynamical asymptotically stable closed-loop standard feedback configuration. For short, such a plant is said to be strongly stabilizable (*ss*). Since the topology of Fig. 1 is quite special, it comes as no surprise to discover that not all plants are strongly stabilizable but what is really unexpected is the simplicity of the final answer given in the Theorems below. Note that full proofs are given in Section 2.

Theorem 1

Consider a dynamical plant satisfying (A_2)* and possessing the transfer function $P(s)$. Let the *distinct real* zeros of $P(s)$ in $Re\, s \geq 0$ (infinity included) be denoted by $\sigma_1, \sigma_2, \ldots, \sigma_l$ and let the total number of *real* poles of $P(s)$ to the right of σ_i, *each* counted according to its multiplicity, be denoted by ν_i, $(i=1, 2, \ldots, l)$. Then, the plant is strongly stabilizable if and only if the integers $\nu_1, \nu_2, \ldots, \nu_l$ are either *all* even or *all* odd.

In most practical applications $P(\infty)=0$ and because the number of real poles of $P(s)$ to the right of $+\infty$ is zero, an even integer, we obtain the following very important corollary.

Corollary 1. A dynamical plant satisfying (A_2) and $P(\infty)=0$ is strongly stabilizable if and only if every *real* zero of $P(s)$ in $Re\, s \geq 0$ lies to the left of an *even* number of *real* poles of $P(s)$, the latter counted according to their multiplicities.

The theorem reveals the crucial role played by the nonminimum-phase property of an unstable plant in determining if the plant is *ss*. The authors find it truly remarkable that only the parity interlacing property of the *real* zeros and poles of $P(s)$ in $Re s \geq 0$ need be taken into account. Indeed, we venture the opinion that the theorem and its corollary are probably most surprising to those accustomed to thinking in terms of root-locus. According to the corollary, if $P(\infty)=0$, plants with the pole-zero patterns shown in Fig. 3 are not *ss* whereas those with the patterns shown in Fig. 2 are, irrespective of the positions of the remaining complex zeros and poles in $Re\, s \geq 0$!

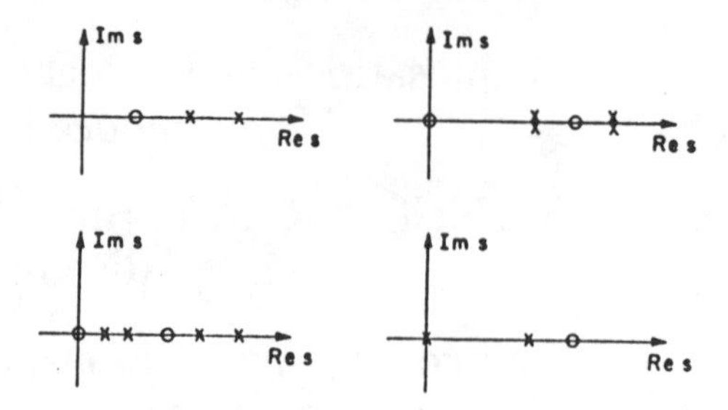

FIG. 2. Pole–zero patterns of stabilizable plants.

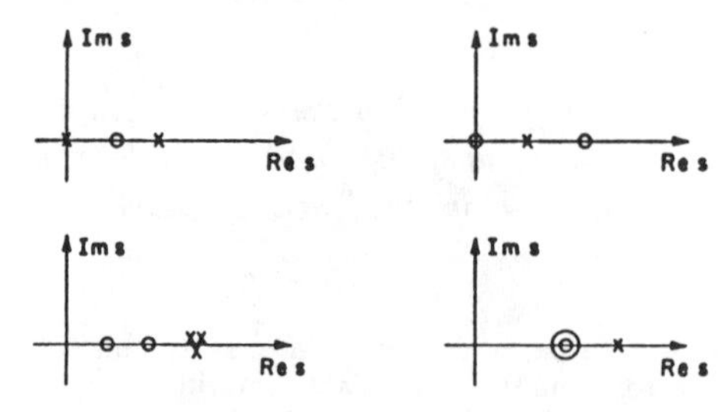

FIG. 3. Pole-zero patterns of nonstabilizable plants.

In the multivariable case $C(s)$, $P(s)$ and $F(s)$ are three proper,* real rational matrices of compatible sizes $m \times r$, $n \times m$ and $r \times n$, respectively. Figure 1 now represents a standard closed-loop interconnection of three multi-input, multi-output dynamical systems and the design objective is to drive the (n-dimensional) plant output $y(t)$ to the (r-dimensional) "set point" $u(t)$. Assuming again that (A_1)–(A_3) are satisfied, (A_4) is equivalent to the determinantal inequality [8, 9]†,

$$|1_n + P(\infty)C(\infty)F(\infty)| \neq 0. \tag{1}$$

Before stating theorem 2, it is necessary to recall several definitions. The point $s=s_o$ is a zero of $P(s)$ if it is a zero of *every* entry in $P(s)$; i.e. if $P(s_o) = O_{n,m}$. The McMillan degree, $\delta(P; s_o)$, of $s=s_o$ as a pole of $P(s)$ is the greatest multiplicity it possesses as a pole of any minor of $P(s)$ [14, 15]. The McMillan degree, $\delta(P)$, of $P(s)$ is the sum of the McMillan degrees of its distinct poles. Let the distinct (finite) poles of $P(s)$ be denoted by s_i and their associated McMillan degrees by δ_i, $(i=1, 2, \ldots, \mu)$. The monic polynomial

$$\psi_p(s) = \prod_{i=1}^{\mu} (s-s_i)^{\delta_i}$$

* Without this assumption, it is impossible to stabilize the plant if access is limited solely to its input and output terminals [8, 13].

* A proper rational matrix or transfer function is finite at infinity.

† $|A|$=det A=determinant A and $O_{r,k}$, O_r, 1_k denote, in the same order, the $r \times k$ zero matrix, the $r \times r$ zero matrix and the $k \times k$ identity matrix.

is the *characteristic denominator* of $P(s)$. It is easily shown [14] that $\psi_p(s)$ is also the monic least common multiple of all denominators of all minors of $P(s)$, each minor assumed expressed as the ratio of two relatively prime polynomials.

Theorem 2

Consider a dynamical plant satisfying (A_2) and possessing the transfer matrix $P(s)$. Let the *distinct real* zeros of $P(s)$ in $Re\, s \geq 0$ (infinity included) be denoted by $\sigma_1, \sigma_2, \ldots, \sigma_l$ and let the total number of real poles of $P(s)$ to the right of σ_i, *each counted according to its McMillan degree*, be denoted by ν_i, $(i=1, 2, \ldots, l)$. Then, the plant is strongly stabilizable if and only if the integers $\nu_1, \nu_2, \ldots, \nu_l$ are either *all* even or *all* odd. If such is the case, the choice r=normal rank $P(s)$ always permits compensation to be achieved under the constraint

$$|F(s)P(s)C(s)| \not\equiv 0.$$

A proper rational matrix $P(s)$ meeting the requirements of theorem 2 is said to have the p.i.p. (parity interlacing property).

Obviously, if a plant is *ss* the choice of $C(s)$ and $F(s)$ is not unique and in general, each possible closed-loop configuration defines a different overall transfer function $T(s)$. Let us say that $T(s)$ is strongly realizable for $P(s)$ if the plant is strongly stabilizable with a closed-loop configuration possessing the transfer function $T(s)$. The following corollary to Theorem 1 completely defines the class of T's strongly realizable for a single-input–output plant.

Theorem 1, *Corollary* 3

Consider a single-input–output plant satisfying (A_2) and possessing the transfer function $P(s)$. Then, the real rational function $T(s)$ is strongly realizable for $P(s)$ if and only if

(1) It is proper and analytic in $Re\, s \geq 0$.
(2) Every zero of $P(s)$ in $Re\, s \geq 0$ (infinity included) is a zero of $T(s)$ of at least the same multiplicity.
(3) No pole of $P(s)$ in $Re\, s \geq 0$ is a zero of $T(s)$.
(4) The real zeros of $T(s)$ in $Re\, s \geq 0$ (infinity included) and the real poles of $P(s)$ in $Re\, s \geq 0$ possess the parity interlacing property.

In Section 3 four examples are worked out to illustrate the several theorems and expose some of the design complications that can arise in their application. In addition, the potential difficulties inherent in the use of observer theory to arrive at controller and feedback sensor designs are revealed in Example 2. All of these difficulties appear to be inextricably linked with the question of sensitivity.

2. ANALYSIS

In the scalar case the sensitivity and closed-loop transfer functions are, respectively,

$$S(s) = \frac{1}{1+P(s)L(s)} \tag{2}$$

and

$$T(s) = \frac{P(s)C(s)}{1+P(s)L(s)}, \tag{3}$$

where

$$L(s) = C(s)F(s). \tag{4}$$

Before quoting a fundamental stability lemma, it is necessary to introduce some notation. Let

$$P(s) = \frac{n_p(s)}{d_p(s)}, \tag{5}$$

$$C(s) = \frac{n_c(s)}{d_c(s)}, \tag{6}$$

and

$$F(s) = \frac{n_f(s)}{d_f(s)}, \tag{7}$$

where the n's and d's are real polynomials in s. Because of the assumption that all components are dynamical, the degrees of the numerators do not exceed those of their respective relatively prime denominator mates:

$$\delta(n_p) \leq \delta(d_p); \quad \delta(n_c) \leq \delta(d_c); \quad \delta(n_f) \leq \delta(d_f). \tag{8}$$

Stability Lemma [8, 9]

Subject to assumptions (A_1)–(A_4) the closed-loop configuration of Fig. 1 is asymptotically stable if and only if the polynomial

$$\varphi(s) = d_p(s)d_c(s)d_f(s) + n_p(s)n_c(s)n_f(s) \tag{9}$$

is strict Hurwitz.* We can now proceed with the proof.

Proof of Theorem 1. According to (4) and (9),

$$S(s) = d_p(s) \cdot \frac{d_c(s)d_f(s)}{\varphi(s)} = \frac{1}{1+P(s)L(s)}. \tag{10}$$

Since by prescription, the compensator and feedback sensor are to be chosen asymptotically stable,

* That is, $\varphi(s)$ is free of zeros in $Re\, s \geq 0$.

$d_c(s)$ and $d_f(s)$ must be strict Hurwitz polynomials. Coupling this observation with the inequality

$$1+P(\infty)C(\infty)F(\infty)\neq 0 \tag{11}$$

and the fact that $\varphi(s)$ is also strict Hurwitz, we reach the following conclusions: If $P(s)$ is *ss*, then

(C_1) $S(s)$ is analytic in $Re\, s\geq 0$, $s=\infty$ included.

(C_2) The zeros of $S(s)$ in $Re\, s\geq 0$ are *precisely* the poles of $P(s)$ in $Re\, s\geq 0$, multiplicities included.

(C_3) Any zero of $P(s)$ in $Re\, s\geq 0$ of multiplicity k (infinity included) is a zero of $1-S(s)$ of *at least* multiplcity k. This may be seen by noting that

$$1-S(s)=\frac{n_p(s)}{d_p(s)}\cdot\frac{n_c(s)n_f(s)}{d_c(s)d_f(s)}\cdot S(s). \tag{12}$$

Conversely, suppose we succeed in constructing a real rational function $L(s)=n(s)/d(s)$, $n(s)$ and $d(s)$ relatively prime, such that

$$S(s)=\frac{1}{1+P(s)L(s)} \tag{13}$$

possesses the three properties enumerated in (C_1)–(C_3). We claim that (1) $\delta(n)\leq\delta(d)$; (2) $d(s)$ is strict Hurwitz; (3) the polynomial

$$\theta(s)=d_p(s)d(s)+n_p(s)n(s) \tag{14}$$

is strict Hurwitz and (4) $1+P(\infty)L(\infty)\neq 0$.

Proofs.

(1) Suppose $\delta(n)>\delta(d)$. Then, $L(\infty)=\infty$. If $P(\infty)\neq 0$, $S(\infty)=0$ and since $s=\infty$ is not a pole of $P(s)$ this violates (C_2). If $P(\infty)=0$, the order of $s=\infty$ as a zero of $1-S(s)$ is less than its order as a zero of $P(s)$ which contradicts (C_3).

(2) If $d(s)$ is not strict Hurwitz, either $S(s)$ possesses a zero in $Re\, s\geq 0$ which is not a pole of $P(s)$ or $S(s)$ possesses a zero in $Re\, s\geq 0$ of order greater than its multiplicity as a pole of $P(s)$ or some zero of $P(s)$ in $Re\, s\geq 0$ of multiplicity $k>0$ is a zero of $1-S(s)$ of multiplicity less than k and in every case either (C_2) or (C_3) is contradicted.

(3) From

$$S(s)=\frac{d_p(s)d(s)}{d_p(s)d(s)+n_p(s)n(s)}=\frac{d_p(s)d(s)}{\theta(s)} \tag{15}$$

and the analyticity of $S(s)$ in $Re\, s\geq 0$, any zero s_o of $\theta(s)$ in $Re\, s\geq 0$ must be a zero of $d_p(s)d(s)$ and therefore a zero of $n_p(s)n(s)$. Invoking (2), $d(s_o)\neq 0$ which implies that s_o is a zero of both $d_p(s)$ and $n(s)$ noting that $d_p(s)$ and $n_p(s)$ are relatively prime. But this means that s_o is a pole of $P(s)$ of multiplicity greater than its order as a zero of $S(s)$, another contradiction, Q.E.D.

Consequently, in view of the stability lemma, we may now choose $C(s)$ and $F(s)$ in any manner whatsoever consistent with the constraint

$$L(s)=C(s)F(s) \tag{16}$$

and the requirement that both be analytic in $Re\, s\geq 0$, $s=\infty$ included. In fact, under the compatibility condition (16),

$$n_c(s)n_f(s)=\lambda(s)n(s), \tag{17}$$

$$d_c(s)d_f(s)=\lambda(s)d(s) \tag{18}$$

where $\lambda(s)$ is a strict Hurwitz polynomial. Thus,

$$\varphi=\lambda(n_pn+d_pd)=\lambda\theta \tag{19}$$

is a product of two strict Hurwitz polynomials and the corresponding closed-loop configuration is asymptotically stable.* To sum up, the problem of ascertaining whether or not the plant is *ss* is completely equivalent to finding conditions making possible the construction of an $L(s)$ which defines via (13) an $S(s)$ satisfying (C_1)–(C_3). The first step is to verify that the parity interlacing property of the real zeros and poles of $P(s)$ in $Re\, s\geq 0$ is necessary.

Necessity

Suppose $S(s)$ satisfies (C_1)–(C_3) and write

$$d_p(s)=d_p^-(s)d_p^+(s) \tag{20}$$

where $d_p^-(s)$ and $d_p^+(s)$ are real polynomials, the former being strict Hurwitz and the latter monic.† Clearly, $d_p^+(s)$ is constructed from *all* the zeros of $d_p(s)$ in $Re\, s\geq 0$ which also represent the totality of poles of $P(s)$ in $Re\, s\geq 0$. To satisfy (C_1) and (C_2), $S(s)$ must be expressible in the form

$$S(s)=\frac{\varepsilon d_p^+(s)h(s)}{g(s)} \tag{21}$$

* We omit arguments whenever and wherever desirable. As is well known, to satisfy (A_3), $C(s)$ and $F(s)$ must be realized minimally [13].

† The coefficient of the highest power of s is unity.

where $\varepsilon=\pm 1$, $h(s)$ and $g(s)$ are two *strict* Hurwitz polynomials with positive coefficients and

$$\delta(g)=\delta(d_p^+)+\delta(h). \tag{22}$$

Let $\sigma_i \geq 0$ (finite or infinite) be any real non-negative zero of $P(s)$. According to (C_3), $S(\sigma_i)=1$ and using (21),

$$d_p^+(\sigma_i)=\varepsilon \frac{g(s)}{h(s)}\bigg|_{s=\sigma_i}. \tag{23}$$

However,

$$\frac{g(s)}{h(s)}\bigg|_{s=\sigma_i}>0$$

and $d_p^+(s)$ must assume the same sign when evaluated at any real zero of $P(s)$ in $Re\, s \geq 0$. Evidently,

$$d_p^+(s)=d_{p1}^+(s)d_{p2}^+(s) \tag{24}$$

where $d_{p1}^+(s)$ and $d_{p2}^+(s)$ are monic, the former absorbing the real zeros and $d_{p2}^+(s)$ the complex zeros of $d_p^+(s)$. Since $d_{p2}^+(\sigma)>0$ for all $\sigma \geq 0$, the l nonzero numbers $d_{p1}^+(\sigma_i)$, $(i=1, 2, \ldots, l)$, have the same signs.* Observing that sign $d_{p1}^+(\infty)=+1$ and that a sign change in $d_{p1}^+(\sigma)$ is produced whenever an odd number of zeros of $d_{p1}^+(s)$ is encountered in a traversal of the positive σ-axis from right to left, we immediately infer the correctness of the interlacing property.

Sufficiency

This part is constructive. By assumption the rea zeros and poles of $P(s)$ in $Re\, s \geq 0$ have the parity interlacing property and we can therefore choose $\varepsilon(\pm 1)$ so that

$$\varepsilon d_p^+(\sigma_i)>0, \quad i=1, 2, \ldots, l. \tag{25}$$

Using this ε, our initial choice for the sensitivity function is given by

$$S(s)=\frac{\varepsilon d_p^+(s)h(s)}{g(s)} \tag{26}$$

where $h(s)$ and $g(s)$ are any two strict-Hurwitz polynomials with positive coefficients meeting the degree requirement

$$\delta(g)=\delta(d_p^+)+\delta(h). \tag{27}$$

Conditions (C_1) and (C_2) are automatically satisfied and we now describe an inductive procedure which in a finite number of steps yields a sensitivity function also satisfying (C_3). Let the first line of the tableau

$$\begin{matrix} \xi_1, \xi_2, \ldots, \xi_\mu, & \sigma_1, \sigma_2, \ldots, \sigma_l(=+\infty) \\ j_1, j_2, \ldots, j_\mu, & r_1, r_2, \ldots, r_l \end{matrix} \tag{28}$$

denote the distinct zeros of $P(s)$ in $Re\, s \geq 0$ and the second, their related multiplicities. The σ's are the real zeros and the ξ's the complex ones. Of course, if $s=\infty$ is not a zero of $P(s)$, $r_l=0$. Starting with (26) we exhibit a modification of $S(s)$, $S_a(s)$ say, which is also of the form (26) and

(R_1) Satisfies (C_1), (C_2).

(R_2) Any zero of $1-S(s)$ in $Re\, s \geq 0$ is a zero of $1-S_a(s)$ of at least the same multiplicity.

(R_3) Given any particular s_o appearing in row 1 of (28), its multiplicity as a root of $1-S_a(s)$ may be made to exceed its multiplicity as a root of $1-S(s)$. Obviously, after a finite number of steps, we reach an $S(s)$ satisfying (C_1)–(C_3).

Let the monic polynomial $\eta(s)$ be constructed from all the finite zeros of $1-S(s)$ in $Re\, s \geq 0$, multiplicities included, and let the order of the zero at $s=\infty$ be denoted by ρ.

Case 1. $s_o=\sigma_i$. Choose any strict Hurwitz polynomial $q(s)$ with positive coefficients such that

$$\delta(q)=\rho+\delta(\eta) \tag{29}$$

and the function

$$\Omega(s)=\frac{\eta(s)}{q(s)} \tag{30}$$

has magnitude ≤ 1 in $Re\, s \geq 0$:*

$$|\Omega(s)| \leq 1; \quad Re\, s \geq 0.$$

Define $S_a(s)$ by

$$S_a(s)=(1+K\Omega(s))^m \cdot S(s) \tag{31}$$

where the real constant K and the positive integer m remain to be determined. Expanding (31) by the Binomial theorem, we get†

$$1-S_a=(1-S)-S \cdot \sum_{r=1}^{m} {}^mC_r(K\Omega)^r, \tag{32}$$

* Recall that $\sigma_1, \sigma_2, \ldots, \sigma_l$ are the distinct real zeros of $P(s)$ in $Re\, s \geqq 0$, infinity included, while the zeros of $d_{pl}(s)$ are *all* the *real* poles of $P(s)$ in $Re\, s \geqq 0$. Obviously, if $P(\infty)=0$, i.e. if $S(\infty)=1$, $\varepsilon=+1$ and all signs are positive.

* This is always possible.

† ${}^mC_r$ is the usual binomial coefficient.

from which it is seen immediately that every zero of $\Omega(s)$ in $Re\, s \geq 0$, finite or infinite, is a zero of $1-S_a(s)$ of at least the same multiplicity. Suppose first that $\Omega(\sigma_i) \neq 0$; then $S(\sigma_i) \neq 1$ and in view of (31), $s=\sigma_i$ is a zero of $1-S_a(s)$ if and only if

$$1=(1+K\Omega(\sigma_i))^m S(\sigma_i). \tag{33}$$

Since $\varepsilon d_p^+(\sigma_i)>0$,

$$S(\sigma_i)=\left.\frac{\varepsilon d_p^+(s)h(s)}{g(s)}\right|_{s=\sigma_i}>0$$

and the mth-principal root

$$\alpha_i=\sqrt[m]{S(\sigma_i)}$$

is positive and unequal to 1. (Observe that $P(\sigma_i)=0$ implies $d_p^+(\sigma_i) \neq 0$ and $S(\sigma_i) \neq 0$.) From (33),

$$K=\frac{\alpha_i^{-1}-1}{\Omega(\sigma_i)} \neq 0 \tag{34}$$

and by choosing m sufficiently large, the real number K can be made as small as desired. In particular, we can assume that $|K|<1$. With m and K so selected,

$$|K\Omega(s)|<1, \quad Re\, s \geq 0$$

whence

$$1+K\Omega(s) \neq 0, \quad Re\, s \geq 0$$

and $S_a(s)$ must have the form

$$S_a(s)=\frac{\varepsilon d_p^+(s)h_a(s)}{g_a(s)}; \tag{35}$$

$h_a(s)$ and $g_a(s)$ are strict Hurwitz, have positive coefficients and satisfy

$$\delta(g_a)=\delta(d_p^+)+\delta(h_a).$$

Suppose that $\Omega(\sigma_i)=0$. Then $s=\sigma_i$ is a root of $1-S(s)$ of order $\mu_i \geq 1$ say, and a root of $1-S_a(s)$ of order at least μ_i. To insure that σ_i is a zero of $1-S_a(s)$ of multiplicity $\geq \mu_i+1$, it suffices that the μ_ith-derivative

$$\left.S_a^{(\mu_i)}(s)\right|_{s=\sigma_i}=0.$$

Since

$$\left.\Omega^{(k-1)}(s)\right|_{s=\sigma_i}=0, \quad (k=1, 2, \ldots, \mu_i),$$

and

$$\left.S^{(k-1)}(s)\right|_{s=\sigma_i}=0, \quad (k=2, 3, \ldots, \mu_i),$$

but

$$\left.\Omega^{(\mu_i)}(s)\right|_{s=\sigma_i} \neq 0, \quad \left.S^{(\mu_i)}(s)\right|_{s=\sigma_i} \neq 0,$$

the Leibnitz rule applied to (31) yields easily,

$$K=-\frac{1}{m}\cdot\left.\frac{S^{(\mu_i)}(s)}{\Omega^{(\mu_i)}(s)}\right|_{s=\sigma_i} \neq 0. \tag{36}$$

Thus, by choosing m sufficiently large, $|K|<1$ and (35) is again valid.

Case 2. $s_o=\xi_i$. Let

$$S_a(s)=\left(1+K\frac{s-\beta}{s+\beta}\Omega(s)\right)^m \cdot S(s) \tag{37}$$

where K is a real constant, β a nonnegative constant, m a positive integer and all three remain to be determined. As before, suppose first that $\Omega(\xi_i) \neq 0$. Then $S(\xi_i) \neq 1$. Moreover, $\text{Im}\,\xi_i \neq 0$ implies that

$$\operatorname*{Inf}_{\beta \geq 0}\left|\frac{\xi_i-\beta}{\xi_i+\beta}\right|=\gamma>0. \tag{38}$$

In order that $S_a(\xi_i)=1$,

$$K\left(\frac{\xi_i-\beta}{\xi_i+\beta}\right)=\frac{\alpha_i^{-1}-1}{\Omega(\xi_i)} \neq 0 \tag{39}$$

in which the non-zero number

$$\alpha_i=\sqrt[m]{S(\xi_i)} \neq 1$$

is the mth-principal root of $S(\xi_i)$. (Again, $P(\xi_i)=0$ implies $d_p^+(\xi_i) \neq 0$ and $S(\xi_i) \neq 0$.) Evidently, independent of the choice of $\beta \geq 0$, it is always possible to select m large enough so that

$$\frac{1}{\gamma}\left|\frac{\alpha_i^{-1}-1}{\Omega(\xi_i)}\right|<1. \tag{40}$$

Thus, since

$$\arg\left(\frac{\xi_i-\beta}{\xi_i+\beta}\right)$$

varies from 0 to π or from 0 to $-\pi$ as β ranges from 0 to ∞, there exists a $\beta \geq 0$ such that

$$K=\frac{\xi_i+\beta}{\xi_i-\beta}\cdot\frac{\alpha_i^{-1}-1}{\Omega(\xi_i)} \tag{41}$$

is real. From (38) and (40), $|K|<1$ and making use of the obvious inequality

$$\left|\frac{s-\beta}{s+\beta}\right| \leq 1, \quad Re\, s \geq 0, \quad \beta \geq 0,$$

it is seen that the factor

$$1+K\frac{s-\beta}{s+\beta}\Omega(s) \tag{42}$$

is a real rational function analytic and devoid of zeros in $Re\, s \geq 0$, $s=\infty$ included. This means that $S_a(s)$ is again of the form (35). Finally, if $\Omega(\xi_i)=0$, ξ_i must be a zero of both $\Omega(s)$ and $1-S(s)$ of multiplicity $\mu_i \geq 1$, say, and reasoning as in Case 1,

$$K=-\frac{1}{m}\cdot\frac{\xi_i+\beta}{\xi_i-\beta}\cdot\frac{S^{(\mu_i)}(s)}{\Omega^{(\mu_i)}(s)}\bigg|_{s=\xi_i}. \tag{43}$$

By a proper choice of positive integer m and $\beta \geq 0$, K is determined as a real number of magnitude less than unity and it follows once more that (42) is a real rational function analytic and free of zeros in $Re\, s \geq 0$, $s=\infty$ included. Since there are no other cases to consider, the proof of the theorem is complete, Q.E.D.

A rational function $I(s)$ is said to make minimal contact k_o with the complex number c_o at $s=s_o$ if $s=s_o$ is a zero of $I(s)-c_o$ of multiplicity at least k_o. A review of the proof of sufficiency in the above theorem uncovers the following useful purely mathematical interpolatory result.

Corollary 2

Let $\gamma^+(s)$ be an arbitrary real polynomial whose zeros lie exclusively in $Re\, s \geq 0$. Consider a finite collection of triplets (s_i, c_i, k_i), $(i=1, 2, \ldots, r)$, where (1) the s_i's, finite or infinite, are distinct and lie in the closed first quadrant $Re\, s \geq 0$, $\mathrm{Im}\, s \geq 0$; (2) $\gamma^+(s_i) \neq 0$; (3) the c_i's are real or complex numbers and (4) the k_i's are arbitrary positive integers. Let $s_1', s_2', \ldots, s_l'$ denote the real s_i's and $c_1', c_2', \ldots, c_l'$ the corresponding c_i's. Then, there exist two strict Hurwitz polynomials $h(s)$ and $g(s)$ such that the rational function

$$I(s)=\frac{\gamma^+(s)h(s)}{g(s)} \tag{44}$$

makes minimal contact k_i with c_i at $s=s_i$, $(i=1, 2, \ldots, r)$, if and only if

$$0 \neq \prod_{i=1}^{r} c_i$$

and the l numbers

$$c_1'\gamma^+(s_1'),\ c_2'\gamma^+(s_2'),\ \ldots,\ c_l'\gamma^+(s_l')$$

are real and have the same signs. It is then always possible to affect the interpolation under the degree constraint

$$\delta(g)=\delta(\gamma^+)+\delta(h). \tag{45}$$

Obviously, if a plant is *ss* the choice of $C(s)$ and $F(s)$ is not unique and in general, each possible closed-loop configuration defines a different overall transfer function $T(s)$. Let us say that $T(s)$ is strongly realizable for $P(s)$ if the plant is strongly stabilizable with a closed-loop configuration possessing the transfer function $T(s)$.

Corollary 3

Consider a plant satisfying (A_2) and possessing the transfer function $P(s)$. Then, the real rational function $T(s)$ is strongly realizable for $P(s)$ if and only if

(1) It is proper and analytic in $Re\, s \geq 0$.

(2) Every zero of $P(s)$ in $Re\, s \geq 0$ (infinity included) is a zero of $T(s)$ of at least the same multiplicity.

(3) No pole of $P(s)$ in $Re\, s \geq 0$ is a zero of $T(s)$.

(4) The real zeros of $T(s)$ (infinity included) and the real poles of $P(s)$ in $Re\, s \geq 0$ possess the parity interlacing property.

Proof. The necessity of (1) and (2) follows immediately from the formula

$$T(s)=P(s)C(s)S(s)=\frac{n_p(s)n_c(s)d_f(s)}{\varphi(s)} \tag{46}$$

and the properties of $C(s)$ and $S(s)$ demanded by strong realizability. According to the stability lemma, the closed loop is asymptotically stable if and only if

$$\varphi(s)=d_p(s)d_c(s)d_f(s)+n_p(s)n_c(s)n_f(s) \tag{47}$$

is strict Hurwitz. Thus, $n_c(s)$ and $d_p(s)$ must be relatively prime in $Re\, s \geq 0$ and the necessity of (3) also follows from (46) since $d_f(s)$ is strict Hurwitz. Furthermore, the zeros of $S(s)$ in $Re\, s \geq 0$ are exactly cancelled by the poles of $P(s)$ and the zeros of $T(s)$ in $Re\, s \geq 0$ are therefore precisely those of the product $P(s)C(s)$. At every such zero s_i, $S(s_i)=1$ and the parity interlacing requirement (4) drops out immediately.

To establish sufficiency let $s_1, s_2, \ldots, s_r$ be the distinct, finite or infinite, zeros of $T(s)$ in $Re\, s \geq 0$ and let $k_1, k_2, \ldots, k_r$ be the associated multiplicities. Invoking (2), (3) and corollary 2, it is possible to choose $\varepsilon = \pm 1$ and to construct two strict Hurwitz polynomials $h(s)$ and $g(s)$ with positive coefficients such that $\delta(g)=\delta(d_p^+)+\delta(h)$ and

$$S(s)=\frac{\varepsilon d_p^+(s)h(s)}{g(s)}=\frac{1}{1+P(s)L(s)} \tag{48}$$

makes minimal contact k_i with $c_i=1$ at $s=s_i$, $(i=1, 2, \ldots, r)$. In view of (1), (2) and the structure of $S(s)$,

$$C(s)=\frac{T(s)}{P(s)S(s)} \tag{49}$$

is proper and analytic in $Re\, s\geq 0$. Since $L(s)=C(s)F(s)$, $F(s)$ must be determined from the formula

$$F(s)=\frac{L(s)}{C(s)}. \tag{50}$$

From (49), the (finite or infinite) zeros of $C(s)$ in $Re\, s\geq 0$ are all those zeros of $T(s)$ in $Re\, s\geq 0$ not supplied by $P(s)$. However,

$$1-S(s)=\frac{P(s)L(s)}{1+P(s)L(s)}$$

and by construction, all these zeros of $T(s)$ in $Re\, s\geq 0$ not supplied by $P(s)$ are zeros of $L(s)$ of at least the same multiplicities. Hence, $F(s)$ is analytic in $Re\, s\geq 0$, infinity included and the corresponding closed-loop configuration strongly stabilizes* $P(s)$ and realizes $T(s)$, Q.E.D.

As mentioned in the introduction, in the multi-input, multi-output case $C(s)$, $P(s)$ and $F(s)$ are three proper,† real rational matrices of respective sizes $m\times r$, $n\times m$ and $r\times n$. The proof of Theorem 2 depends on a correct generalization of the scalar stability lemma.

Multivariable stability lemma [8, 9]

Let $\psi_c(s)$, $\psi_p(s)$ and $\psi_f(s)$ denote the characteristic denominators of $C(s)$, $P(s)$ and $F(s)$, respectively. Then, subject to assumptions (A_1)–(A_4), the closed-loop configuration of Fig. 1 is asymptotically stable if and only if the polynomial

$$\varphi(s)=\psi_p(s)\psi_c(s)\psi_f(s)\cdot|1_r+F(s)P(s)C(s)| \tag{51}$$

is strict Hurwitz.‡

Thus, in view of the above lemma, a plant satisfying (A_2) is *ss* iff there exist two compatible proper real rational matrices $C(s)$ and $F(s)$ analytic in $Re\, s\geq 0$, such that the polynomial $\varphi(s)$ defined by (51) is strict Hurwitz. However, to avoid the technical inconvenience of having to distinguish between s finite and s infinite, the main part of the proof of Theorem 2 is carred out in the unit circle, $|z|\leq 1$. Corollary 2a is obtained by making the change of variable

$$z=\frac{s-1}{s+1} \tag{52}$$

and represents a slightly generalized but straightforward z-analogue of Theorem 1+Corollary 2 which takes advantage of the compactness of the domain $|z|\leq 1$.

Corollary 2a, *Theorem* 1

Consider any irreducible real rational function

$$t(z)=\frac{e_1(z)}{\chi(z)}. \tag{53}$$

There exists a real rational function $l(z)$ analytic in $|z|\leq 1$ and two strict Hurwitz polynomials $a(z)$, $b(z)$ such that

$$\frac{1}{1+t(z)l(z)}=\frac{\chi^+(z)b(z)}{a(z)}, \tag{54}$$

iff $t(z)$ has thep .i.p.; i.e. iff the zeros and poles o $t(z)$ lying in the closed interval $-1\leq z\leq 1$ have the parity interlacing property.*

Proof of Theorem 2. Suppose $P(s)$ is strongly stabilizable by means of the admissible pair $F(s)$, $C(s)$. Then, from (51),

$$\frac{1}{|1_r+F(s)P(s)C(s)|}=\frac{\psi_p(s)\psi_c(s)\psi_f(s)}{\varphi(s)}. \tag{55}$$

Invoking (1) and the strict Hurwitz character of the three polynomials $\psi_c(s)$, $\psi_f(s)$, $\varphi(s)$,

$$\frac{1}{|1_r+F(s)P(s)C(s)|}=\frac{\varepsilon\psi_p^+(s)h(s)}{g(s)} \tag{56}$$

where $\varepsilon=\pm 1$, $h(s)$, $g(s)$ are strict Hurwitz polynomials with positive coefficients and

$$\delta(g)=\delta(\psi_p^+)+\delta(h). \tag{57}$$

Let s_o be any zero (finite or infinite) of $P(s)$ in $Re\, s\geq 0$. Then $P(s_o)=0_{n,\, m}$. Since $F(s)$ and $C(s)$ are proper and analytic in $Re\, s\geq 0$, $F(s_o)P(s_o)C(s_o)=0_r$ and (56) yields

$$1=\left.\frac{\varepsilon\psi_p^+(s)h(s)}{g(s)}\right|_{s=s_o} \tag{58}$$

* By earlier reasoning, it is known that $L(s)$ is proper and analytic in $Re\, s\geq 0$.

† $A(s)$ is proper if $A(\infty)$ is finite and strictly proper if $A(\infty)=0$, the zero matrix.

‡ Note that $|1_r+FPC|=|1_n+PCF|=|1_m+CFP|$.

* The monic polynomial $\chi^+(z)$ absorbs all the zeros of $\chi(z)$ in $|z|\leq 1$. A function $f(z)$ is strict Hurwitz if it is devoid of zeros in $|z|\leq 1$ and proper if it is regular at $z=1$.

from which the necessity of the stipulated parity interlacing property follows immediately as in Theorem 1. To establish sufficiency, we describe a constructive procedure initialized around the SMITH–MCMILLAN canonic form of $P(s)$ [14, 15]. The underlying idea is to demonstrate the existence of an admissible pair $F(s)$, $C(s)$ satisfying (56). Under the mapping (52) the closed half-plane $Re\, s \geq 0$ is transformed into the closed unit circle $|z| \leq 1$ with $s=\infty$ going into $z=1$. Moreover, the entire nonnegative axis $0 \leq s \leq \infty$ is imaged into the diameter $-1 \leq z \leq 1$ in a strict monotonically increasing manner.

Let

$$W(z) \equiv P\left(\frac{1+z}{1-z}\right). \tag{59}$$

Since the bilinear transformation (52) leaves MCMILLAN degrees intact, $\delta_s(P)=\delta_z(W)$ and $P(s)$ has the p.i.p. iff $W(z)$ does. Assume therefore that $W(z)$ has the p.i.p., that normal rank $P(s)=k \geq 1$ and let*

$$\Omega(z)=\text{diag}\left[\frac{e_1(z)}{\psi_1(z)}, \frac{e_2(z)}{\psi_2(z)}, \ldots, \frac{e_k(z)}{\psi_k(z)}\right] \tag{60}$$

be the canonic form of $W(z)$. Then (1) the e's and ψ's are real monic polynomials uniquely determined by $W(z)$; (2) each $e_i(z)$ is relatively prime to its mate $\psi_i(z)$; $e_i(z)$ divides $e_{i+1}(z)$ and $\psi_{i+1}(z)$ divides $\psi_i(z)$, $(i=1, 2, \ldots, k-1)$; (4) the distinct zeros and poles of $W(z)$ in $|z| \leq 1$ are identical, respectively, with the distinct zeros of $e_1(z)$ and $\psi_1(z)$ in $|z| \leq 1$; (5) the MCMILLAN degree of any pole in $W(z)$ in $|z| \leq 1$ equals its multiplicity as a root of the polynomial

$$\chi(z)=\prod_{i=1}^{k} \psi_i(z). \tag{61}$$

Consequently, the scalar function

$$t(z)=\frac{e_1(z)}{\chi(z)} \tag{62}$$

has the p.i.p. and the obvious approach is to set $r=k$ and to define two real rational matrices $A(z)$ and $B(z)$ analytic in $|z| \leq 1$ such that

$$|1_k+B(z)W(z)A(z)|=1+t(z)l(z); \tag{63}$$

$l(z)$ is then constructed as described in Corollary 2a. Having accomplished this, a return to the s-plane transforms (63) into

$$\frac{1}{|1_k+F(s)P(s)C(s)|}=\frac{\varepsilon\psi_p^+(s)h(s)}{g(s)}$$

* A $k \times k$ diagonal matrix with diagonal elements $a_1, a_2, \ldots, a_k$ is written diag$[a_1, a_2, \ldots, a_k]$.

where the compensators

$$F(s) \equiv B\left(\frac{s-1}{s+1}\right) \tag{64}$$

and

$$C(s) \equiv A\left(\frac{s-1}{s+1}\right) \tag{65}$$

not only strongly stabilize $P(s)$ but are also proper and analytic in $Re\, s \geq 0$. The reader should have no trouble convincing himself that

$$\left.\frac{\chi^+(z)b(z)}{a(z)}\right|_{z=(s-1)/(s+1)}=\frac{\varepsilon\psi_p^+(s)h(s)}{g(s)}; \tag{66}$$

$\varepsilon=\pm 1$, $h(s)$ and $g(s)$ are two real strict Hurwitz polynomials and $\delta(g)=\delta(\psi_p^+)+\delta(h)$.

From the SMITH–MCMILLAN theorem [14, 15],

$$W(z)=U(z)(\Omega(z) \dotplus O_{n-k,\, m-k})V(z), \tag{67}$$

where $U(z)$ and $V(z)$ are two real square elementary polynomial matrices.* Choosing

$$B(z)=[B_1(z)|O_{k,\, n-k}]U^{-1}(z) \tag{68}$$

and

$$A(z)=V^{-1}(z)\left[\frac{1_k}{O_{m-k,\, k}}\right]A_1(z), \tag{69}$$

$$|1_k+B(z)W(z)A(z)|=|1_k+B_1(z)\Omega(z)A_1(z)|. \tag{70}$$

The two $k \times k$ matrices $B_1(z)$ and $A_1(z)$ remain to be determined. Obviously because of the polynomial character of $U^{-1}(z)$ and $V^{-1}(z)$, $B(z)$ and $A(z)$ are analytic in $|z| \leq 1$ iff $B_1(z)$ and $A_1(z)$ are analytic in $|z| \leq 1$. If $k=1$, $\chi(z)=\psi_1(z)$ and

$$|1_1+B(z)W(z)A(z)|=1+t(z)l(z), \tag{71}$$

$$l(z)=B_1(z)A_1(z). \tag{72}$$

We now invoke Corollary 2a and choose the functions $B_1(z)$, $A_1(z)$ analytic in $|z| \leq 1$ and consistent with (72). If $k \geq 2$ the argument is slightly more involved and relies on a simple but important observation. Namely, for $k \geq 2$, the k polynomials

$$\begin{aligned} \theta_1 &= \psi_2\psi_3\psi_4 \ldots \psi_k \\ \theta_2 &= e_2\psi_3\psi_4 \ldots \psi_k \\ \theta_3 &= e_2 e_3\psi_4 \ldots \psi_k \\ &\vdots \\ \theta_k &= e_2 e_3 e_4 \ldots e_k \end{aligned} \tag{73}$$

* $A \dotplus B$ is the "direct sum" of A and B. A square polynomial matrix $A(z)$ is elementary if $|A(z)|=\text{constant} \neq 0$.

are relatively prime. The easy proof is by induction and exploits the up-and-down divisibility properties of the e's and ψ's. By a well known theorem of algebra [16], there exist k real polynomials $b_1(z)$, $b_2(z), \ldots, b_k(z)$ such that

$$1=\sum_{i=1}^{k} b_i(z)\theta_i(z) \tag{74}$$

and

$$b_k(z) \not\equiv 0. \tag{75}$$

In order not to obscure the main idea with unessential notation we now carry out the explicit details for $k=3$, the generalization to arbitrary $k \geq 2$ being obvious. Put

$$B_1(z)=\begin{bmatrix} 0 & 0 & -1 \\ 0 & -1 & 0 \\ 1 & 0 & 0 \end{bmatrix}, \tag{76}$$

$$A_1(z)=\begin{bmatrix} lb_3 & lb_2 & lb_1 \\ 0 & 0 & 1 \\ 0 & 1 & 0 \end{bmatrix} \tag{77}$$

and

$$t_i(z)=\frac{e_i(z)}{\psi_i(z)}, \quad (i=1, 2, 3), \tag{78}$$

where $l(z)$ is an arbitrary rational function. Then,

$$B_1(z)\Omega(z)A_1(z)=\begin{bmatrix} 0 & -t_3 & 0 \\ 0 & 0 & -t_2 \\ lb_3t_1 & lb_2t_1 & lb_1t_1 \end{bmatrix} \tag{79}$$

and

$$\begin{aligned} |1_3+B_1(z)\Omega(z)A_1(z)| &= 1+l(b_3t_3t_2t_1+b_2t_2t_1 \\ &\quad +b_1t_1) \\ &= 1+\frac{e_1}{\chi}\cdot l\cdot(b_3\theta_3+b_2\theta_2 \\ &\quad +b_1\theta_1) \\ &= 1+t(z)l(z). \end{aligned} \tag{80}$$

In reaching (80) we have used (74) and (78). Finally, a last appeal to Corollary 2a determines $l(z)$ and completes the definitions of $B(z)$ and $A(z)$. Observe that $b_3(z) \not\equiv 0$ implies that

$$|B(z)W(z)A(z)|=l(z)b_3(z)t_3(z)t_2(z)t_1(z) \not\equiv 0$$

whence, $r=k=3$ and $|FPC| \not\equiv 0$. This concludes the proof of Theorem 2, Q.E.D.

Corollary 1

A dynamical plant which satisfies (A_2) and possesses a transfer matrix $P(s)$ devoid of real finite zeros in $Re\, s \geq 0$ is strongly stabilizable.

Proof. Since there are no poles of $P(s)$ to the right of $s=+\infty$, $P(s)$ automatically has the parity interlacing property. The importance of Corollary 1 stems from the fact that most multivariable plants encountered in practice have no finite zeros in $Re\, s \geq 0$.

It should be noted that at the present writing, the conditions to be satisfied by a transfer matrix $T(s)$ guaranteeing its strong realizability for a prescribed $P(s)$ appear too complicated to be of practical value and are therefore not quoted. The next section illustrates the theorems by means of several worked examples which highlight some unexpected conceptual and numerical difficulties.

3. EXAMPLES

The examples in this section have been chosen to (1) demonstrate the direct application of the theorems to actual design, (2) to expose some of the difficulties that are encountered when observer theory is used to derive the compensation and (3) to pinpoint some of the subtleties that arise in the stabilization of plants whose pole–zero patterns in $Re\, s \geq 0$ are "critical".

Example 1

We choose as our first example a plant with transfer function

$$P(s)=\frac{(1-s)(s+2)}{s(s-3)(s-4)}. \tag{81}$$

It is easily verified that

$$T(s)=\frac{(1-s)}{s^2+s+1} \tag{82}$$

satisfies all the conditions of Theorem 1, Corollary 3 and is therefore strongly realizable for $P(s)$. The first step in the determination of an admissible pair $C(s)$, $F(s)$ is the selection of a suitable sensitivity function. Once this is accomplished, the compensators are computed from the formulas

$$C(s)=\frac{T(s)}{P(s)S(s)} \tag{83}$$

and

$$F(s)=\frac{1-S(s)}{T(s)}. \tag{84}$$

Since $P(\infty)=0$, it follows from (21) that the sensitivity function must have the form

$$S(s)=\frac{s(s-3)(s-4)h(s)}{g(s)}. \tag{85}$$

The strict Hurwitz polynomials $h(s)$ and $g(s)$ are constructed so that the zeros of $1-S(s)$ in $Re\, s\geq 0$, counting multiplicities, include those of $T(s)$. Thus

$$1=S(\infty)=S(1). \tag{86}$$

The constraint $S(\infty)=1$ is met by choosing two strict Hurwitz monic polynomials $h(s)$ and $g(s)$ meeting the degree requirement (27). Obviously, it is desirable that the resultant compensators defined via (83) and (84) have the lowest possible degrees. It is precisely at this stage that the designer's judgment plays a key role because unfortunately, a systematic procedure for generating such an optimal design is not yet at hand. An obvious choice for the sensitivity function is

$$S(s)=\frac{s(s-3)(s-4)}{(s+1)(s^2+s+1)} \tag{87}$$

and the substitution of (82), (87) into (83) and (84), respectively, yields

$$C(s)=\frac{s+1}{s+2}, \tag{88}$$

$$F(s)=\frac{1-9s}{s+1}. \tag{89}$$

Observe that $C(s)$ and $F(s)$ are asymptotically-stable, proper, rational transfer functions and hence constitute an admissible pair.

Example 2

As pointed out in the first example, it is important to seek the simplest compensation. It is easily shown that the simplest compensation, i.e. the product $L(s)=C(s)F(s)$ of lowest degree, corresponds to a choice of admissible sensitivity function of minimum degree. From (2), (5) and (21),

$$L(s)=\frac{1-S(s)}{P(s)S(s)}=\frac{d_p(g-\varepsilon d_p^+ h)}{\varepsilon n_p d_p^+ h}. \tag{90}$$

Let $n_p=n_p^+ n_p^-$. (As usual, the monic polynomial n_p^+ contains all the finite plant zeros in $Re\, s\geq 0$ counted according to their multiplicities.) By design, $L(s)$ is analytic in $Re\, s\geq 0$. Thus, with $\hat{h}$ denoting a polynomial,

$$g-\varepsilon d_p^+ h=n_p^+ \hat{h} \tag{91}$$

and (90) reduces to

$$L(s)=\frac{d_p^- \hat{h}}{\varepsilon n_p^- h}. \tag{92}$$

Now, in view of (22),

$$\delta(L)=\delta(n_p^-)+\delta(h)=\delta(n_p^-)+\delta(S)-\delta(d_p^+) \tag{93}$$

since $L(s)$ and $S(s)$ are proper rational functions. The degrees of n_p^- and of d_p^+ are fixed by the plant and it is clear from (93) that $\delta(L)$ is a minimum when an acceptable $S(s)$ of least degree is employed.

According to Theorem 1, Corollary 3,

$$T(s)=\frac{5(s-1)(s-2)}{(s+10)(s^2+s+1)} \tag{94}$$

is strongly realizable for the plant

$$P(s)=\frac{5(s-1)(s-2)}{s(s-3)(s-4)} \tag{95}$$

and any associated admissible sensitivity function $S(s)$ must satisfy

$$S(1)=S(2)=S(\infty)=1. \tag{96}$$

From among the many compatible sensitivity functions that we have been able to generate, one which illustrates certain subtleties is*

$$S(s)=\frac{s(s-3)(s-4)(s+279{\cdot}5)}{(s+10)(s+50)(s^2+s+1)}\left[1+\frac{(39562{\cdot}5)(s-1)}{(s+200)^2}\right]^2. \tag{97}$$

Clearly, $\delta(S)=8$ and it follows from (93) that $\delta(L)$ is five. Since pole-zero cancellations in $Re\, s<0$ between $C(s)$ and $F(s)$ are allowed, it follows that

$$\delta(C)+\delta(F)\geq\delta(L) \tag{98}$$

and the sum of the degrees of the compensator and feedback sensor is at least five.

Another approach that can sometimes lead to lower-degree compensation involves the use of a

* It is possible to construct a compatible sensitivity function of degree seven but the difficulties persist.

LUENBERGER observer [10]–[12]. If the plant chosen in this example is assumed to be completely controllable and observable,* the desired transfer function can be realized using a dynamic observer of order $n-1=2$ and the resultant configuration is the one shown in Fig. 4. Both $F_1(s)$ and $F_2(s)$ have degree two and making the identifications

$$F(s)=-F_1(s), \tag{99}$$

$$C(s)=[1-F_2(s)]^{-1}, \tag{100}$$

it is clear that the possibility exists for achieving $\delta(C)+\delta(F)=4$ in the standard feedback configuration. However, it is shown in the sequel, that there exists no asymptotically stable observer for which the $C(s)$ given by (100) is analytic in $Re\, s\geq 0$. That is, the desired $T(s)$ cannot be strongly realized for the given $P(s)$ by means of the observer design technique.

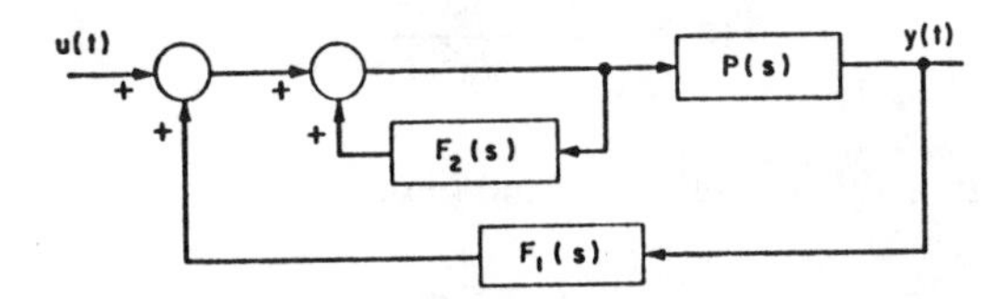

FIG. 4. Frequency domain representation of observer configuration.

A convenient starting point for design with an observer is a minimal state-variable representation for the plant transfer function. By straightforward calculation we find that for the dynamical pair

$$\dot{\mathbf{x}}=A\mathbf{x}+\mathbf{b}u \tag{101}$$

$$y=\mathbf{c}'\mathbf{x},$$

$$\frac{5(s-1)(s-2)}{s(s-3)(s-4)}=P(s)=\mathbf{c}'(s1_3-A)^{-1}\mathbf{b}$$

provided

$$A=\begin{bmatrix} 0 & 1 & 0 \\ 0 & 0 & 1 \\ 0 & -12 & 7 \end{bmatrix}, \tag{102}$$

$$\mathbf{b}'=(0 \quad 0 \quad 1) \tag{103}$$

and

$$\mathbf{c}'=5(2 \quad -3 \quad 1). \tag{104}$$

Transfer functions are computed with zero initial conditions and since under these conditions the observer gives a perfect estimate of the plant state vector, it follows that the closed-loop system transfer function is given by

$$T(s)=\mathbf{c}'(s1_3-A+\mathbf{b}\mathbf{k}')^{-1}\mathbf{b} \tag{105}$$

when $-\mathbf{k}'\hat{\mathbf{x}}$ is fed back to the plant input and $\hat{\mathbf{x}}$ is the estimate of the plant state vector generated by the observer. Using well-known matrix identities one can show that (105) is equivalent to [13]

$$T(s)=\frac{n_p(s)}{\det(s1_3-A+\mathbf{b}\mathbf{k}')}. \tag{106}$$

Setting the denominator of (106) equal to the denominator of (94) and solving for $\mathbf{k}'$ yields

$$\mathbf{k}'=(10 \quad -1 \quad 18). \tag{107}$$

The next step is the design of the observer. It is not difficult to establish with the aid of the formulas found in Section 3 of [11] that

$$F_1(s)=-\mathbf{k}'[M+N(s1_2-F)^{-1}G] \tag{108}$$

and

$$F_2(s)=-\mathbf{k}'N(s1_2-F)^{-1}H. \tag{109}$$

The matrices M, N, F, G and H are fixed [11] by the choice of matrices W_{22} and

$$W_{21}=\tfrac{1}{5}(w_1 \quad w_2)'. \tag{110}$$

It is easily shown that $F_1(s)$ and $F_2(s)$ are independent of the particular selection of nonsingular W_{22} and we therefore assume, without loss of generality, that $W_{22}=1_2$. In this case

$$F=\begin{bmatrix} 10w_1 & (1-4w_1) \\ (10w_2-12) & (7-4w_2) \end{bmatrix}, \tag{111}$$

$$N=\begin{bmatrix} \frac{1}{2} & -\frac{1}{2} \\ 1 & 0 \\ 0 & 1 \end{bmatrix} \tag{112}$$

and

$$H=(-w_1 \quad 1-w_2)'. \tag{113}$$

Now using (100), (107), (109) and (111)–(113) it is found that

$$|s1_2-F|=s^2+(4w_2-10w_1-7)s+22w_1-10w_2+12 \tag{114}$$

* Actually, this assumption although convenient is unnecessary [13].

and

$$C(s)=\frac{|s1_2-F|}{s^2+(6-24w_1+9w_2)s+(90w_1-24w_2+26)}. \tag{115}$$

In (114) and (115), the parameters w_1 and w_2 represent real numbers which are at the disposal of the designer. For closed-loop stability the observer eigenvalues must have negative real parts and the coefficients of the polynomial in (114) must be positive. This corresponds to the choice of pairs (w_1, w_2) falling in the shaded region marked A in Fig. 5. If the compensator is asymptotically stable, $C(s)$ is analytic in $Re\, s\geq 0$ and the coefficients of its denominator polynomial must also be positive. The appropriate region in the w_1w_2-plane is the one shown shaded and marked B in Fig. 5. Clearly, the shaded regions A and B have no points in common and it is therefore impossible to realize an asymptotically stable compensator $C(s)$ with observer eigenvalues having negative real parts.

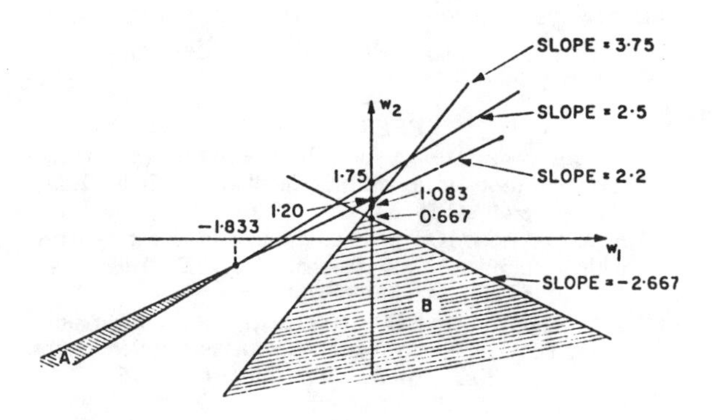

FIG. 5. Regions of stable observer and stable compensator.

Example 3

Having established that the observer design technique fails to yield asymptotically stable compensators for the plant (95) and the closed-loop transfer function (94), we complete the design with the aid of the sensitivity function (97). Substituting (95), (94) and (97) into (83) gives

$$C(s)=\frac{(s+50)(s+200)^4}{(s+279{\cdot}5)(s^2+39962{\cdot}5s+437{\cdot}5)^2} \tag{116}$$

and using (84),

$$F(s)=\frac{\frac{1}{5}\sum_{i=0}^{5}F_is^i}{(s+50)(s+200)^4}, \tag{117}$$

where

$$\left.\begin{array}{ll} F_0=3{\cdot}99997(10)^{11} & F_3=-4{\cdot}39872(10)^{11} \\ F_1=1{\cdot}05568(10)^{12} & F_4=-1{\cdot}61873(10)^{9} \\ F_2=1{\cdot}78288(10)^{12} & F_5=-79336{\cdot}5 \end{array}\right\}. \tag{118}$$

Evidently, $C(s)$ and $F(s)$ are both asymptotically stable and are therefore admissible. The actual transfer function $T_a(s)$ realized by compensation around the prescribed $P(s)$ with the pair (116), (117) is given by

$$T_a(s)=\frac{5(s-1)(s-2)(s+50)(s+200)^4}{\sum_{i=0}^{8}H_is^i}, \tag{119}$$

where

$$\left.\begin{array}{ll} H_0=7{\cdot}99994(10)^{11} & H_5=47\,087\,616 \\ H_1=9{\cdot}12006(10)^{11} & H_6=289\,360 \\ H_2=9{\cdot}15640(10)^{11} & H_7=861 \\ H_3=1{\cdot}15687(10)^{11} & H_8=1 \\ H_4=3{\cdot}68712(10)^{9} & \end{array}\right\}. \tag{120}$$

A Routh check shows that the denominator of (119) is strict Hurwitz and thus $T_a(s)$ is strictly proper and analytic in $Re\, s\geq 0$. With exact arithmetic the factors $(s+50)$ and $(s+200)^4$ would also be factors of the denominator polynomial but because of computer round-off this cancellation is not exact. To ascertain how close $T_a(s)$ approximates the desired transfer function $T(s)$ in (94), we decided to compare the two corresponding amplitude and phase characteristics on the $j\omega$-axis instead of calculating pole locations. Over the frequency range 0–2 Hz the per cent error in both amplitude and phase is less than 0·001 per cent. As mentioned in the introduction, one of the reasons for employing asymptotically stable controllers and feedback sensors was to avoid poor sensitivity [7]. But unstable compensation does not appear to be the principal cause of this overall system degradation. In fact, by rounding the denominator polynomials in (116) and the numerator in (117) to three places we obtain an unstable $T_a(s)$! It appears therefore that the whole question of sensitivity is more deeply rooted in the intrinsic structure of $P(s)$ than first anticipated. Since we intend to pursue the entire subject in greater depth in another paper we close this discussion with a simple rule-of-thumb observation: if asymptotically stable compensation is achievable "without excessive difficulty" by the use of Observer technique, the plant is strongly stabilizable and the closed-loop system sensitivity can be made to lie within bounds that are meaningful from an engineering design standpoint.

Example 4

Consider the 2×2 diagonal plant

$$P(s)=\left[\begin{array}{c|c} \dfrac{s-1}{s(s-4)} & 0 \\ \hline 0 & \dfrac{s-2}{s-3} \end{array}\right]. \tag{121}$$

Note that $P(\infty)\neq 0_2$. Although $P_{22}(s)$ is strongly stabilizable, $P_{11}(s)$ is not and individual *ss* single-loop compensation is not possible. Nevertheless, according to Theorem 2, $P(s)$ is *ss* but $C(s)$ and $F(s)$ cannot both be diagonal. From this example it is easy to understand the limitations inherent in a non-interacting approach to multivariable control problems.

In practice, multivariable plants rarely possess finite zeros. This observation taken in conjunction with Theorem 2, Corollary 1, implies that the ultimate impediment to strong closed-loop stability must be transportation lag.

One last point is worth making. Observe that in (121),

$$\hat{P}_{11}(s)\equiv \frac{1}{s-2}\cdot P_{11}(s)=C_1(s)P_{11}(s)$$

is strongly stabilizable. Thus, there exists an asymptotically stable compensator $C_2(s)$ and feedback sensor $F(s)$ which enclose $\hat{P}_{11}(s)$ in an asymptotically stable loop. This means that the closed-loop stabilization of $P(s)$ is achieved with an unstable $C(s)=C_2(s)C_1(s)$. We have therefore uncovered a simple proof of the well known fact that Observer theory can always succeed in stabilizing an arbitrary plant with the aid of a possibly unstable compensator $C(s)$.

4. CONCLUSIONS

A simple necessary and sufficient test has been derived for ascertaining whether or not a dynamical plant with transfer matrix $P(s)$ can be stabilized in a single-loop feedback configuration employing an asymptotically stable controller and feedback sensor. The test involves only the interlacing properties of the real poles and zeros of the transfer matrix in $Re\, s\geq 0$. A plant that can be so stabilized is called strongly stabilizable.

Not only has a test been described but a constructive procedure has also been provided for determining the controller and feedback sensor transfer matrices. Moreover, this procedure is completely scalar oriented and our prime objective is to eventually develop design techniques for multivariable plants based on single-input–output methodology. The design criteria sought are those leading to acceptable closed-loop system performance both in the frequency and time domain.

The initial motivation for using asymptotically stable controllers and feedback sensors was to attempt to improve sensitivity but the results of Example 3 reveal that this is not necessarily the case. Indeed, we have been able to substantially relax the sensitivity problem by using an unstable controller. It appears therefore that the whole question of sensitivity is more deeply rooted in the instinct structure of $P(s)$ than first anticipated. We intend to pursue this subject in greater depth in a future publication.

One final point warrants mention. Suppose $P(s)$ is not strongly stabilizable. Then one can select an unstable precompensator $C_1(s)$ so that $\hat{P}(s)=P(s)C_1(s)$ is strongly stabilizable. That is, there exists an asymptotically stable controller $C_2(s)$ and feedback sensor $F(s)$ which enclose $\hat{P}(s)$ in an asymptotically stable loop. This means that the closed-loop stabilization of $P(s)$ is achieved with an unstable $C(s)=C_1(s)C_2(s)$. The above idea when suitably generalized constitutes another proof of the well known fact that Observer theory can always succeed in stabilizing any dynamical plant with asymptotically stable hidden modes.

REFERENCES

[1] A JAMESON: Design of a single-input system for specified roots using output feedback. *IEEE Trans. Aut. Control* **AC-15**, 345–348 (1970).

[2] E. J. DAVISON: On pole assignment in linear systems with incomplete state feedback. *IEEE Trans. Aut. Control* **AC-15**, 348-351 (1970).

[3] E. J. DAVISON and D. CHATTERJEE: A note on pole assignment in linear systems with incomplete state feedback. *IEEE Trans. Aut. Control* **AC-16**, 98–99 (1971).

[4] F. M. BRASCH, JR. and J. B. PEARSON: Pole placement using dynamic compensators. *IEEE Trans. Aut. Control* **AC-15**, 34–43 (1970).

[5] C. T. CHEN: *Design of Pole-Placement Compensators for Multivariable Systems*, pp. 497–503. Proceedings of the 1970 Joint Automatic Control Conference.

[6] C. T. CHEN and C. H. HSU: *Design of Dynamic Compensators for Multivariable Systems.* Proceedings of the 1971 Joint Automatic Control Conference.

[7] L. SHAW: Pole placement: stability and sensitivity of dynamic compensators. *IEEE Trans. Aut. Control* **AC-16**, 210 (1971).

[8] D. C. YOULA: Modern Classical Multivariable Feedback Control Theory: Part 1, Rome Air Development Center Technical Report, RADC-TR-70-98. Griffiss Air Force Base, New York (1970).

[9] C. T. CHEN: *Introduction to Linear System Theory.* Holt, Rinehart & Winston, New York (1970).

[10] D. G. LUENBERGER: An introduction to observers. *IEEE Trans. Aut. Control* **AC-16**, 596–602 (1971).

[11] Y. Ö. YÜKSEL and J. J. BONGIORNO, JR.: Observers for linear multivariable systems with applications. *IEEE Trans. Aut. Control* **AC-16**, 603–613 (1971).

[12] W. A. WOLOVICH: *A Frequency Domain Approach to State Feedback and Estimation.* 1971 IEEE Decision and Control Conference, Miami Beach, Florida (1971).

[13] D. C. YOULA and J. J. BONGIORNO, JR.: On the complex *s*-plane design of multivariable feedback control systems. Rome Air Development Center Technical Report, RADC-TR-72-10 (1972).

[14] H. H. ROSENBROCK: *State-Space and Multivariable Theory*. Wiley, New York (1970).

[15] D. C. YOULA: On the factorization of rational matrices. *IRE Trans. Inform. Theory* IT-7, 172–189 (1961).

[16] K. HOFFMAN and R. KUNZE: *Linear Algebra*. Prentice-Hall, New Jersey (1961).

Résumé—Ce texte déduit les conditions nécessaires et suffisantes pour stabiliser une installation P (s) multivariable avec des modes cachés asymptotiquement stables grâce à une rétroaction à boucle simple utilisant un régulateur asymptotiquement stable et un senseur à rétroaction. Ces conditions sont entièrement générales et par conséquent comprennent également les installations à phase non minimum et instable. Pour les installations à débit-alimentation simple avec gain nul à fréquence infinie, les conditions sont réduites à la seule demande qu'aucun zéro d'installation sur l'axe réel non-négatif du plan complexe S ne soit à gauche d'un nombre impair de poles d'installations réelles, les poles multiples étant comptés selon leurs multiplicités.

Il a également été possible de déduire des conditions nécessaires et suffisantes simples pour réaliser une fonction de transfert T(s) à boucle fermée par une compensation stable asymptotique autour d'une installation P(s) à débit-alimentation simple *prescrite*. Il est prévuque ce dernier résultat peut être convenablement généralisé au cas multvariable.

Dans le sens réel, ce texte constitue une continuation de travaux précédents non publiés (8) par le premier auteur.

Zusammenfassung—Abgeleitet werden die notwendigen und hinreichenden Bedingungen für eine mehrvariable Anlage $P(s)$ mit asymptotisch stabilen verborgenen Moden, die mittels einschleifiger Rückführung unter Verwendung asymptotisch stabiler Regler und Rückführungs-Fühler stabilisierbar sind. Diese Bedingungen sind ganz allgemein und umfassen daher auch unstabile Anlagen mit nicht minimaler Phase. Für Anlagen mit einem Eingang und einem Ausgang mit dem Gewinn Null bei der Frequenz unendlich reduzieren sich die Bedingungen auf das einzige Erfordernis, daß keine Nullstellen der Anlage auf der nichtnegativen reellen Achse der komplexen s-Ebene zur Linken einer ungeraden Zahl von reellen Anlagenpolen liegen, wobei vielfache Pole ihrer Vielfachheit entsprechend gewählt werden.

Es gelang auch, einfache notwendige und hinreichende Bedingungen für die Übertragungsfunktion $T(s)$ der geschlossenen Schleife abzuleiten, die durch asymptotisch stabile Kompensation bei der *vorgeschriebenen* Anlage mit einem Eingang und einem Ausgang realisierbar ist. Erwartet wird, daß dieses letztere Resultat entsprechend auf den multivariablen Fall verallgemeinert werden kann.

In gewissem Sinne stellt diese Arbeit Fortserzung einer früheren unveröffentlichten Arbeit des ersten Autors (8) dar.

Резюме—В статье выводятся необходимые и достаточные условия стабилизируемости многопараметрической установки $P(s)$ с ассимтотически устойчивыми скрытыми формами стабилизуемости. Стабилизация проверяется с помощью одноконтурной обратной связи, использующей ассимтотически устойчивый регулятор и чувствительный элемент в цепи обратной связи. Эти условия полностью общие и поэтому пригодны также и для неустойчивой установки с неминимальными фазами. Для установпк с одним входом и одним выходом с нулевым коэффициентом усиления при бесконечной частоте эти условия уменьшаются до единственного требования, чтобы ни один из нулей комплексной плоскости-s на неотрицательной действительной оси не лежал слева от четного числа действительных полюсов плоскости, кратных полюсов, подсчитанных соответственно их кратности. Осазалось также не сложным вывести необходимые и достаточные условия для передаточной функции замкнутого контура $T(s)$. Эти условия реализуются с помощью ассимтотически устойчивой стабилизации относительно предписанной установки $P(s)$ с одним входом и одним выходом. Ожидается, что этот последний результат может быть распространен на многопараметрические случаи. В известном смысле, данная работа язляется продолжением ранее неопубликованной работы(8) первого автора.

Modern Wiener–Hopf Design of Optimal Controllers
Part I: The Single-Input-Output Case

DANTE C. YOULA, FELLOW, IEEE, JOSEPH J. BONGIORNO, JR., MEMBER, IEEE,
AND HAMID A. JABR, STUDENT MEMBER, IEEE

***Abstract*—An analytical feedback design technique is presented here for single-input-output processes which are characterized by their rational transfer functions. The design procedure accounts for the topological structure of the feedback system ensuring asymptotic stability for the closed-loop configuration. The plant or process being controlled can be unstable and/or nonminimum phase. The treatment of feedback sensor noise, disturbance inputs, and process saturation is another major contribution of this work.**

The cornerstone in the development is the selection of a performance index based on sound engineering considerations. It is these considerations, in fact, which ensure the existence of an optimal compensator for the system and make the performance index a natural one for the problem at hand.

I. Introduction

AN ANALYTICAL feedback design technique is presented for single-input-output processes which are characterized by their rational transfer functions. The design procedure accounts for the topological structure of the feedback system *and ensures the* asymptotic stability of the closed-loop configuration. The plant or process being controlled can be unstable and/or nonminimum phase. The treatment of feedback sensor noise, disturbance inputs, and process saturation is another major contribution of this work. The cornerstone of the development is the selection of a performance index based on sound engineering considerations. It is these considerations in fact which ensure the existence of an optimal compensator for the system and make the performance index a natural one for the problem at hand.

The classical treatment of the analytical feedback design problem by Newton is described in [1]. With his approach, which is inherently open loop, it is first necessary to find the transfer function $W_c(s)$ analytic in Re $s \geqslant 0$ of the optimal equivalent cascade compensator. The transfer function $C(s)$ of the corresponding controller for the feedback loop is then calculated by means of the formula

$$C(s) = W_c(s)/[1 - F(s)P(s)W_c(s)].$$

$F(s)$ and $P(s)$ denote the transfer functions of the feedback sensor and plant, respectively. Unfortunately, this procedure is flawed because it can, and often does, yield a computed $C(s)$ which possesses a zero in Re $s \geqslant 0$ coinciding with either a pole of the plant or feedback sensor. Clearly, if $C(s)$ possesses such a zero, the closed-loop system is unstable and the design is worthless. To exclude such a possibility Newton restricts the plant and feedback sensor to be asymptotically stable from the outset. In fact, several extensions of this idea to the multivariable case have already been made by Bongiorno and Weston [2], [3].

The earliest researchers to recognize the difficulty with right-half plane pole-zero cancellations within a feedback loop worked with sampled data systems [4]. The analogous treatment for continuous-time systems was presented by Bigelow [5].[1] His argument for ruling out pole-zero cancellations in Re $s \geqslant 0$ is based on the fallacious reasoning that exact cancellation cannot be achieved in practice. Although the observation concerning what can be achieved in practice is of course true, it is also true that even if perfect cancellation were possible the system would nevertheless still possess unstable "hidden" modes. Despite the error in physical reasoning these two papers succeeded in focusing attention on several meaningful engineering problems.

The frequency-domain optimization procedure described herein is the first one to correctly account for the asymptotic stability of the closed-loop system and to correctly treat plants which are not asymptotically stable. It also supplies significant insight into the essential role played by the classical sensitivity function in feedback system design. Although confined to single-input-output systems, these ideas can be extended to the multivariable situation. This extension is nontrivial and is the subject of Part II. Just as in [6], the scalar solution provided the necessary insight and impetus required to effect the breakthrough in the multivariable case. It is, therefore appropriate that both cases be presented in the literature. Moreover, it is only in the single-input-output case that the unique role of the sensitivity function manifests itself so clearly.

The limitations imposed by feedback sensor noise have been known for some time. Horowitz [7] has proposed a design philosophy for single-input-output minimum-phase

Manuscript received January 9, 1975; revised October 16, 1975. Paper recommended by J. B. Pearson, Chairman of the IEEE S-CS Linear Systems Committee. This work was supported by the National Science Foundation under Grant ENG 74-13054 and is taken in part from a Ph.D. dissertation submitted by H. A. Jabr to the Faculty of the Polytechnic Institute of New York.

D. C. Youla and J. J. Bongiorno, Jr. are with the Department of Electrical Engineering and Electrophysics, Polytechnic Institute of New York, Long Island Center, Farmingdale, NY 11735.

H. A. Jabr was with the Department of Electrical Engineering and Electrophysics, Polytechnic Institute of New York, Long Island Center, Farmingdale, NY 11735. He is now with the University of Petroleum and Minerals, Dhaharan, Saudi Arabia.

[1]The paper by Bigelow was kindly brought to the attention of the authors by P. Sarachik.

Reprinted from *IEEE Trans. Automat. Contr.*, vol. AC-21, no. 1, pp. 3–13, Feb. 1976.

stable plants which is quite imaginative but appears limited since it is modeled around a Bode two-terminal interstage equalization scheme. On the other hand, our approach takes the lumped character of the controller as an explicit constraint from the outset and nonminimum-phase and/or unstable plants offer no special obstacles.

A discussion of the relationship of our frequency-domain design procedure and some of the more popular state-variable techniques [14] is certainly in order and will be given in Part II. For now, we merely observe that the methods of this paper obviate the need to find state-variable representations and can handle stochastic inputs which are non-Gaussian and colored, as well as step and ramp-type disturbances. In addition, it permits the modeling and incorporation of feedback transducers such as tachometers, rate gyros, and accelerometers with nondynamical transfer functions.

II. Problem Statement and Preliminary Results

In this paper attention is restricted exclusively to the design of controllers for single-input-output finite-dimensional linear time-invariant plants embedded in an equivalent single-loop configuration shown in Fig. 1.[2] Suppose $y_d(s)$, the *desired* closed-loop output is related to $u_i(s)$, the *actual* input set-point signal in the linear fashion

$$y_d(s)=T_d(s)u_i(s)$$

via the *ideal* transfer function $T_d(s)$. The *prefilter* $H(s)$ can be selected in advance once and for all, but irrespective of the particular choice of criterion that is employed.[3]

$$u=H(u_i+n)$$

is the best available linear version of $y_d(s)$. Any reasonable performance measure must be based on the difference

$$e(s)=u(s)-y(s) \tag{1}$$

between the actual plant output $y(s)$ and the actual smoothed input $u(s)$ driving the loop. For a given plant and overall sensor $F(s)$ the design of the controller $C(s)$ should evolve from an appropriate minimization procedure subject to a power-like constraint on $r(s)$ to avoid plant saturation.

Plant disturbance $d(s)$ and measurement noise $m(s)$ are modeled in a perfectly general way by assuming that

$$y(s)=P(s)r(s)+P_0(s)d(s) \tag{2}$$

[2]To avoid proliferating symbols, all quantities are Laplace transforms, deterministic or otherwise, all stochastic processes are zero-mean second-order stationary with rational spectral densities and $\langle,\rangle$ denotes ensemble average.

[3]Function arguments are omitted wherever convenient.

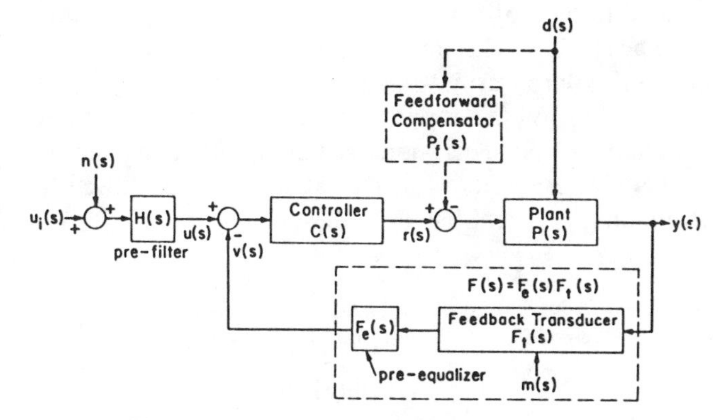

Fig. 1. Basic single-loop feedback configuration.

and

$$v(s)=F(s)y(s)+F_0(s)m(s) \tag{3}$$

where $P(s)$, $P_0(s)$, $F(s)$, and $F_0(s)$ are four real rational functions in the complex variable $s=\sigma+j\omega$. Moreover, by changing $P_0(s)$ into

$$P_0(s)-P(s)P_f(s) \tag{4}$$

it is also possible to envisage any desirable feedforward compensation $P_f(s)$.

Straightforward analysis yields

$$y=\frac{1-S}{F}(u-F_0m)+SP_0d, \tag{5}$$

$$r=\frac{1-S}{PF}(u-F_0m-FP_0d) \tag{6}$$

and

$$e=\left(\frac{F-1+S}{F}\right)u-SP_0d+\left(\frac{1-S}{F}\right)F_0m \tag{7}$$

where

$$S(s)=\frac{1}{1+F(s)P(s)C(s)} \tag{8}$$

is the closed-loop sensitivity function. In process control, the actual choice of a reliable feedback transducer $F_t(s)$ is more or less dictated by the problem at hand. However, as explained in greater detail later, some low-power-level preequalization $F_e(s)$ is almost always necessary to model delay in the feedback path, to improve stability margin and to assure zero steady-state error. In other words

$$F(s)=F_e(s)F_t(s). \tag{9}$$

We therefore assume that $P_0(s)$, $F_0(s)$, $F(s)$, and $P(s)$ are prescribed in advance. Equations (5)–(7) reveal the possibilities for tradeoff in the various frequency bands. Observe, that with unity feedback ($F=1$), $e(s)$ is the sum of the two errors

$$e_1(s)=S(u-P_0d) \tag{10}$$

and

$$e_2(s)=(1-S)F_0m \tag{11}$$

whose different origins are betrayed by the prefactors $S(s)$ and $1-S(s)$. The impossibility of making both $S(s)$ and $1-S(s)$ arbitrarily "small" over any frequency band is partly intrinsic and partly conditioned by the plant restrictions [8], [9]. This fundamental conflict is inevitable and largely responsible for a great deal of the difficulty surrounding practical feedback design.

Let

$$P(s)=\frac{n_p(s)}{d_p(s)} \tag{12}$$

$$F(s)=\frac{n_f(s)}{d_f(s)} \tag{13}$$

and

$$C(s)=\frac{n_c(s)}{d_c(s)} \tag{14}$$

where each numerator polynomial is relatively prime to its respective denominator mate. It is well known [10]–[12] that if the plant, controller and feedback sensor are free of unstable hidden modes, the closed loop of Fig. 1 is asymptotically stable iff the "reduced" characteristic polynomial

$$\varphi(s)=d_f(s)d_p(s)d_c(s)+n_f(s)n_p(s)n_c(s) \tag{15}$$

is strict Hurwitz; i.e., iff $\varphi(s)$ has no zeros in $\mathrm{Re}\,s\geq 0$. Hence, the pair $d_f(s)$, $n_p(s)$ as well as the pair $d_p(s)$, $n_f(s)$ must be devoid of common zeros in $\mathrm{Re}\,s\geq 0$ in which case $P(s)$ and $F(s)$ are said to be *admissible*.[4] Observe that once $H(s)$, $P(s)$, $F(s)$, $P_0(s)$, $F_0(s)$ and the statistics of $u_i(s)$, $n(s)$, $d(s)$, and $m(s)$ are specified, $y(s)$, $r(s)$, and $e(s)$ are uniquely determined by the choice of sensitivity function $S(s)$. Consequently, the following definition and its accompanying lemma have an obvious importance and are fundamental to our entire approach.

Definition 1: $S(s)$ is said to be *realizable* for an admissible pair $P(s)$, $F(s)$ if the closed-loop structure of Fig. 1 is asymptotically stable for some choice of controller $C(s)$ and possesses the sensitivity function $S(s)$.

Lemma 1 (Appendix): The function $S(s)\not\equiv 0$ is realizable for the admissible pair $P(s)$, $F(s)$ iff

1) $S(s)$ is analytic in $\mathrm{Re}\,s\geq 0$;
2) Every zero of the polynomial $d_f(s)d_p(s)$ in $\mathrm{Re}\,s\geq 0$ is a zero of $S(s)$ of *at least* the same multiplicity;
3) Every zero of the polynomial $n_f(s)n_p(s)$ in $\mathrm{Re}\,s\geq 0$ is a zero of $1-S(s)$ of *at least* the same multiplicity.

Let $G_i(-s^2)$, $G_n(-s^2)$, $G_u(-s^2)$, $G_d(-s^2)$, and $G_m(-s^2)$ denote the rational spectral densities of $u_i(s)$, $n(s)$, $u(s)$, $d(s)$, and $m(s)$, respectively. Setting aside for the moment all questions of convergence,[5]

$$2\pi jE_t=\int_{-j\infty}^{j\infty}\langle e_*(s)e(s)\rangle ds \tag{16}$$

is the usual quadratic measure of steady-state response. Similarly,[6] if $\boldsymbol{P}_s(s)$ represents the column-vector transfer matrix coupling the plant input $r(s)$ to those "sensitive" plant modes which must be especially protected against excessive dynamic excursions,

$$2\pi jE_s=\int_{-j\infty}^{j\infty}\langle r_*(s)\boldsymbol{P}_{s*}(s)\boldsymbol{P}_s(s)r(s)\rangle ds \tag{17}$$

is a proven useful penalty functional for saturation [1]. More explicitly,

$$2\pi jE_s=\int_{-j\infty}^{j\infty}Q(-s^2)\langle r_*(s)r(s)\rangle ds \tag{18}$$

where

$$Q(-s^2)=\boldsymbol{P}_{s*}(s)\boldsymbol{P}_s(s). \tag{19}$$

Thus,

$$E=E_t+kE_s. \tag{20}$$

k, a positive constant, serves as a weighted cost combining both factors. Using (6) and (7) and assuming all processes to be independent, a simple calculation yields the compact four-term expression

$$2\pi jE=\alpha+2\int_{-j\infty}^{j\infty}\frac{(F-1)S_*}{FF_*}G_u ds + \int_{-j\infty}^{j\infty}SS_*G_a ds+\int_{-j\infty}^{j\infty}(1-S)(1-S)_*G_b ds \tag{21}$$

where

$$\alpha=\int_{-j\infty}^{j\infty}\frac{(F-1)(F-1)_*}{FF_*}G_u ds, \tag{22}$$

$$G_a=\frac{G_u}{FF_*}+P_0P_{0*}G_d,\quad G_u=HH_*(G_i+G_n) \tag{23}$$

$$G_b=\frac{F_0F_{0*}}{FF_*}G_m+\frac{kQ}{PP_*}\left(P_0P_{0*}G_d+\frac{G_u+F_0F_{0*}G_m}{FF_*}\right). \tag{24}$$

Our entire physical discussion revolves around the implications of (21), and our assumptions are as follows.

Assumption 1: Rate gyros and tachometers are examples of practical sensing devices which are not modeled as dynamical systems.[7] Yet almost invariably, sensors are stable and their associated transfer functions $F_t(s)$ are analytic in $\mathrm{Re}\,s\geq 0$. For our purposes it suffices to restrict

[4]When $F_0(s)$, $H(s)$, $P_f(s)$, and $P_0(s)$ represent distinct physical blocks, these blocks must be stable: their transfer functions must be analytic in $\mathrm{Re}\,s\geq 0$. On the other hand if $F_0(s)$, $H(s)$, $P_f(s)$, and $P_0(s)$ are merely part of the paper modeling it is often possible to relax the analyticity requirements.

[5]If $A(s)$ is a real rational (or meromorphic) matrix in s, $A_*(s)\equiv A'(-s)$, the transpose of $A(-s)$.

[6]Column-vectors are written $\boldsymbol{a}$, $\boldsymbol{b}$, etc., and det A, A', $\bar{A}$, A^* $(\equiv\bar{A}')$ denote the determinant, transpose, complex conjugate, and adjoint of the matrix A, respectively. Note that for $A(s)$ real and meromorphic, $A_*(j\omega)=A^*(j\omega)$, ω real.

[7]A system with transfer function $A(s)$ is dynamical if $A(s)$ is proper; i.e., if $A(\infty)$ is finite.

$F(s)$ to be analytic on the finite $j\omega$-axis and to insist that the component of the cost α be finite. In particular, the integrand in (22) must be analytic on the $j\omega$-axis and $0(1/\omega^2)$ for $\omega^2 \to \infty$.

Suppose that parameter variations induce a change $\Delta\varphi(s)$ in the characteristic polynomial $\varphi(s)$. Clearly, if the nominal design is stable and structural changes are precluded

$$\eta(s)=\frac{\Delta\varphi(s)}{\varphi(s)}$$

is proper and analytic in $\mathrm{Re}\, s \geq 0$. Invoking the standard Nyquist argument it is immediately concluded that $\varphi(s)+\Delta\varphi(s)$, the reduced characteristic polynomial of the perturbed closed loop, is strict Hurwitz iff the normal plot of $\eta(j\omega)$ does not encircle the point $-1+j0$ in a clockwise direction. It is imperative therefore that at the nominal setting $|\varphi(j\omega)|$ be comparably large over those frequency ranges where $|\Delta\varphi(j\omega)|$ is expected to be large. Unfortunately, it does not appear possible to translate any nontrivial stability-margin criteria directly into manageable integral restrictions reconcilable with E. However, once the formula for the optimal $S(s)$ is available, the role played by $F(s)$ in securing adequate stability margin will be clarified and further discussion along these lines is postponed until the next section.

Assumption 2: A pole of $P(s)$ in $\mathrm{Re}\, s>0$ reveals true plant instability but a pole on the $j\omega$-axis is usually present because of intentional preconditioning and is not accidental. For example, with unity feedback ($F=1$) and $d(s)=m(s)=0$, a stable loop enclosing a plant whose transfer function possesses a pole of order ν at the origin will track any causal linear combination of the inputs $1, t,\cdots,t^{\nu-1}$ with zero steady-state error. Similarly, if $s=j\omega_0$, ω_0 real, is a pole of order ν of $P(s)$, a unity-feedback stable loop will track any linear combination of $e^{j\omega_0 t}$, $te^{j\omega_0 t},\cdots,t^{\nu-1}e^{j\omega_0 t}$ with zero steady-state error. These generalized ramp-modulated sinusoids constitute an important class of shape-deterministic information-bearing signals and play a key role in industrial applications. In a nonunity-feedback loop this perfect accuracy capability is lost unless $F_t(s)$ is also preconditioned compatibly. From (7) with $d(s)=m(s)=0$,

$$e=\left(\frac{F-1+S}{F}\right)u. \tag{25}$$

Now according to Lemma 1, a finite pole $s=j\omega_0$ of $P(s)$ of multiplicity ν must be a *zero* of $S(s)$ of order at least ν. Thus, if $s=j\omega_0$ is *also* a zero of $F(s)-1$ of order ν or greater, (25) shows that the loop is again capable of acquiring any linear combination of the inputs $e^{j\omega_0 t}$, $te^{j\omega_0 t},\cdots,t^{\nu-1}e^{j\omega_0 t}$ with zero steady-state error. By setting $u(s)=m(s)=0$ in (7) we obtain

$$e=-SP_0d \tag{26}$$

the loop error under load disturbance $d(s)$. In many areas, such as process control, the recovery of steady state under load changes is a requirement of paramount importance. As is seen from (26), if the shape deterministic component of $P_0(s)d(s)$ is envisaged to be the transform of a sum of ramp-modulated sinusoids, bounded zero steady-state error is possible iff $e(s)$ vanishes at infinity and is analytic in $\mathrm{Re}\, s \geq 0$. Assuming SP_0 proper and SP_0d analytic in $\mathrm{Re}\, s \geq 0$ is evidently sufficient. In particular, reasoning as above, the $j\omega$-axis poles of $P_0(s)d(s)$, multiplicities included, must be contained in those of $P(s)$. Summing up, $(F-1)P$,

$$d_p d_{p*}\langle uu_*\rangle = d_p G_u d_{p*} \tag{27}$$

and

$$d_p d_{p*}\langle P_0 dd_* P_{0*}\rangle = d_p P_0 G_d P_{0*} d_{p*} \tag{28}$$

must be $j\omega$-analytic. Equivalently, in view of Assumption 1 and (23),

$$d_p G_a d_{p*} \tag{29}$$

is analytic on the finite $j\omega$-axis.

Assumption 3: In general, the effects of parameter uncertainty on $P(s)$ and $F(s)$ are more pronounced as ω increases and closed-loop sensitivity is an important consideration. This sensitivity is usually expressed in terms of the percentage change in the loop transfer function

$$T(s)=\frac{P(s)C(s)}{1+F(s)P(s)C(s)}. \tag{30}$$

A straightforward calculation yields

$$\frac{\delta T}{T}=\frac{\delta(PC)}{PC}\cdot S-\frac{\delta F}{F}\cdot(1-S) \tag{31}$$

and once again $S(j\omega)$ and $1-S(j\omega)$ emerge as the pertinent gain functions for the forward and return links, respectively. Clearly then, to combat the adverse effects of high-frequency uncertainty in the modeling of $F(j\omega)$ and $P(j\omega)$ it is sound engineering practice to design $S(j\omega)$ proper and equal to 1 at $\omega=\infty$. This requirement is easily introduced into the analytic framework by imposing the restrictions $G_b(-s^2)\not\equiv 0$ and

$$G_b(-s^2)=0(\omega^{2l}), \qquad l>0 \tag{32}$$

for large ω^2.

Our final assumptions are fashioned for the express purpose of excluding from consideration certain mathematically possible but physically meaningless degeneracies. They are also motivated by Lemma 1, the structure of (21), and the requirement of finite cost.

Assumption 4: For large ω^2,

$$G_a(\omega^2)=0(1/\omega^{2\nu}), \qquad \nu>1. \tag{33}$$

Assumption 5: Q is analytic on the finite $j\omega$-axis, has no purely imaginary zeros in common with n_p, and the constant k is positive.

Assumption 6: Let

$$G=P_0G_dP_{0*}+\frac{G_u+F_0G_mF_{0*}}{FF_*}; \tag{34}$$

then

$$(d_pn_f)G(d_pn_f)_*$$

is analytic and nonzero on the finite $s=j\omega$-axis.

For later reference we record the useful formula

$$G_a+G_b=\left(1+\frac{kQ}{PP_*}\right)G \tag{35}$$

which drops out of (23), (24), and (34).

III. The Wiener–Hopf Solution

Recall that any rational function $A(s)$ possesses a Laurent expansion constructed from all its poles, finite or infinite and as is customary, $\{A(s)\}_+$ denotes that part of the expansion associated with all the *finite* poles of $A(s)$ in $\mathrm{Re}\, s<0$. Thus, $\{A(s)\}_+$ is analytic in $\mathrm{Re}\, s\geq 0$ and vanishes for $s=\infty$. The remainder of the expansion is written $\{A(s)\}_-$ and of course,

$$A(s)=\{A(s)\}_+ +\{A(s)\}_-.$$

Theorem 1 (Appendix): Let

$$d(s)=d_f(s)d_p(s) \tag{36}$$

$$n(s)=n_f(s)n_p(s) \tag{37}$$

$$\chi(s)=d(s)n(s) \tag{38}$$

and write

$$\chi(s)=\chi_l(s)\chi_r(s). \tag{39}$$

The polynomial $\chi_l(s)$ absorbs all the zeros of $\chi(s)$ in $\mathrm{Re}\, s<0$ and $\chi_r(s)$ all those in $\mathrm{Re}\, s\geq 0$. Perform the spectral factorization

$$\chi_r\chi_{r*}(G_a+G_b)=\Omega\Omega_* \tag{40}$$

where $\Omega(s)$ is free of zeros and poles in $\mathrm{Re}\, s>0$.

1) Under Assumptions 1–6, the optimal closed-loop sensitivity function $S_0(s)$ associated with any admissible pair $P(s)$, $F(s)$ is given by

$$S_0=\frac{\left\{\frac{\chi_r\chi_{r*}}{\Omega_*}\left(G_b-\frac{F-1}{FF_*}G_u\right)\right\}_+ +f}{\Omega}. \tag{41}$$

$f(s)$ a real polynomial. The requirements $E<\infty$ (finite cost) and $S_0(s)$ realizable for $P(s)$, $F(s)$ determine $f(s)$ *uniquely*.

2) The optimal controller $C_0(s)$ which realizes $S_0(s)$ for the pair $P(s),F(s)$ is obtained from the formula

$$C_0(s)=\frac{1-S_0(s)}{P(s)F(s)S_0(s)} \tag{42}$$

and can be improper, unstable or both.[8] Nevertheless, the closed-loop structure is always asymptotically stable and $S_0(s)$ is proper and analytic in $\mathrm{Re}\, s\geq 0$. (Assumptions 1–6 actually force $\Omega(s)$ to be free of zeros in $\mathrm{Re}\, s\geq 0$.)

With exact arithmetic the finite zeros and poles of $P(s)F(s)$ in $\mathrm{Re}\, s\geq 0$ are cancelled exactly by the zeros of $S_0(s)$ and $1-S_0(s)$, respectively. Thus, in any computer implementation of (41) and (42) it is necessary that all these exact arithmetic cancellations in $\mathrm{Re}\, s\geq 0$ be effected automatically by suitable preparation. Failure to do so will result in a nonstrict-Hurwitz stability polynomial $\varphi(s)$ and a corresponding unstable closed-loop design.

An examination of (41) reveals that the zeros of $\Omega(s)$ and the poles of

$$\left\{\frac{\chi_r\chi_{r*}}{\Omega_*}\left(G_b-\frac{F-1}{FF_*}G_u\right)\right\}_+ \tag{43}$$

constitute the poles of $S_0(s)$. Since the poles of $S_0(s)$ are all zeros of $\varphi(s)$, the stability margin of the optimal design is ascertainable *in advance*. This important feature cannot be overemphasized. From the formula

$$\Omega\Omega_*=\chi_r\chi_{r*}\left(1+\frac{kQ}{PP_*}\right)G \tag{44}$$

it is seen that the zeros of $\chi_r\chi_{r*}G$ and $1+kQ/PP_*$ in $\mathrm{Re}\, s<0$ emerge as poles of $S_0(s)$. The locations of these zeros depend on the choice of $F(s)$, the spectral density $G(-s^2)$ and the value of k. Changing k means compromising saturation (and accuracy). A more detailed analysis shows generally that the *negative images* of the right-half plane poles of $P(s)$ and $F(s)$ are zeros of $\chi_r\chi_{r*}$ and therefore poles of $S_0(s)$ unless $G(-s^2)$ is properly preconditioned. If some of these poles lie close to the $s=j\omega$-axis, it may be impossible to attain adequate stability margin. This difficulty can be circumvented by simply incorporating the offending poles into $G(-s^2)$. Hence, the rule, any pole of $P(s)F(s)$ in $\mathrm{Re}\, s>0$ which lies "too close" to the imaginary axis must be made a pole of $G(-s^2)$ of exactly twice the multiplicity. Last, we mention that delay τ in the feedback path can be simulated by introducing right-half plane zeros into $F_e(s)$ through one of the many available rational function approximations to $e^{-s\tau}$.

IV. Example

The theory developed in the preceding sections is now used to design the controller $C(s)$ for the system shown in Fig. 2. Since the theory is based on rational transfer

[8] $C(s)$ is proper if the integer l in Assumption 3 equals the order of the zero of $F(s)P(s)$ at infinity. This is often the case.

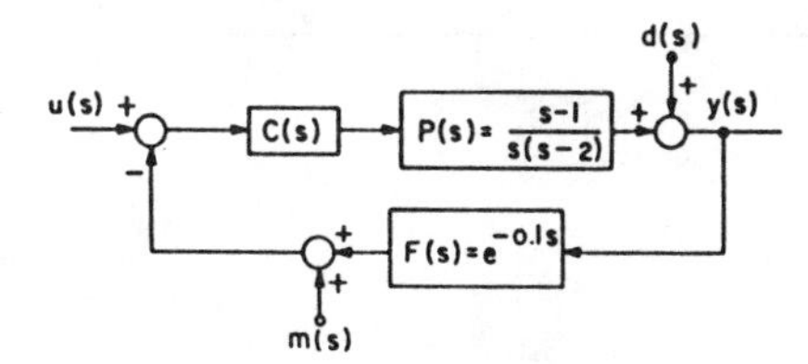

Fig. 2. Example.

functions, the first step is the selection of a suitable approximation for the ideal delay represented by $F(s) = e^{-0.1s}$. Highly satisfactory results are obtained with the second-order Padé approximation

$$F(s)=\frac{\frac{12}{(0.1)^2}-\frac{6}{(0.1)}s+s^2}{\frac{12}{(0.1)^2}+\frac{6}{(0.1)}s+s^2}\equiv\frac{p_f(-s)}{p_f(s)}. \tag{45}$$

Note that $F(s)$ and $P(s)$ are an admissible pair and (45) satisfies the condition $F-1=0$ at $s=0$, the only pole of $P(s)$ on the imaginary axis.

Because of the plant pole at the origin and the choice of rational approximation for $F(s)$, the closed loop is capable of following step inputs u with zero steady-state error when $m=d=0$. The simplest example calls for

$$G_u=-\frac{1}{s^2}. \tag{46}$$

For the remaining spectral densities we choose

$$G_d=\frac{1}{100-s^2} \tag{47}$$

and

$$G_m=1. \tag{48}$$

We also assume that the plant input is the signal most likely to cause saturation and put $Q=1$. The only remaining quantities needed for the calculation of the optimal controller are k, F_0, and P_0. Comparing Figs. 1 and 2, it is seen that $F_0=P_0=1$. With regard to k, we note that the performance index E is actually an auxilary cost function. The design objective is to minimize E_t subject to the constraint $E_s < N_s$, N_s a specified bound. Thus, k is a Lagrangian multiplier chosen to meet the design objectives.

The first step in the design of the optimal controller $C(s)$ is the determination of S_0 given by (41). For the determination of the optimal sensitivity function we need the quantities χ_r, Ω, G_a, and G_b. From (45) and Fig. 2 it follows that

$$\chi=n_pd_pn_fd_f=s(1-s)(2-s)p_f(s)p_f(-s) \tag{49}$$

and

$$\chi_r=s(1-s)(2-s)p_f(-s). \tag{50}$$

Substitution of the given data into (23), (24), (34), and (35) yields

$$G_a=G_u+G_d=\frac{-2(50-s^2)}{s^2(100-s^2)} \tag{51}$$

$$G_b=G_m+\frac{k}{PP_*}(G_m+G_u+G_d)=1+\frac{k(4-s^2)p_1(s)}{(1-s^2)(100-s^2)} \tag{52}$$

and

$$G_a+G_b=\frac{-kp_1(s)p_2(s)}{s^2(100-s^2)(1-s^2)} \tag{53}$$

where

$$p_1(s)=100-102s^2+s^4 \tag{54}$$

and

$$p_2(s)=\frac{1}{k}-\left(\frac{1}{k}+4\right)s^2+s^4. \tag{55}$$

It now follows from (40) that

$$\Omega=\frac{\sqrt{k}\;p_1^+p_2^+(2+s)p_f(s)}{10+s} \tag{56}$$

where

$$p_1^+=10+\sqrt{122}\;s+s^2 \tag{57}$$

and

$$p_2^+=\frac{1}{\sqrt{k}}+\sqrt{\frac{1}{k}+\frac{2}{\sqrt{k}}+4}\;s+s^2 \tag{58}$$

are the factors containing all the zeros of $p_1(s)$ and $p_2(s)$ in $\operatorname{Re} s<0$, respectively.

In addition,

$$\frac{(F-1)G_u}{FF_*}=\frac{120}{sp_f(s)} \tag{59}$$

where

$$\left\{\frac{\chi_r\chi_{r*}}{\Omega_*}\left(G_b-\frac{F-1}{FF_*}G_u\right)\right\}_+=\frac{k_0}{10+s} \tag{60}$$

$$k_0=-\left.\frac{\sqrt{k}\;s^2(2+s)(4-s^2)p_f(s)p_1^+(s)}{p_2^+(-s)}\right|_{s=-10} \tag{61}$$

We now find that

$$S_0=\frac{k_0+(10+s)f(s)}{\sqrt{k}\;(2+s)p_f(s)p_1^+(s)p_2^+(s)}\equiv\frac{h(s)}{g(s)}. \tag{62}$$

in which the sixth-degree polynomial $f(s)$ is uniquely determined by the interpolatory conditions

$$S_0(0)=S_0(2)=0 \tag{63}$$

$$S_0(1)=S_0(z_0)=S_0(\bar{z}_0)=1 \tag{64}$$

$$1-S_0(s)=0\left(\frac{1}{s^2}\right), \quad |s| \text{ large}. \tag{65}$$

In (64), z_0 and $\bar{z}_0$ are the zeros of $p_f(-s)$.

Due to roundoff error, the computed polynomial $f(s)$ will, in general, not satisfy conditions (63)–(65) exactly. However, we know from the theory that the poles of $F(s)$ and $P(s)$ in Re $s>0$ must be zeros of $S_0(s)$ and the zeros of $F(s)$ and $P(s)$ in Re $s>0$ must be zeros of $1-S_0(s)$. The computations are therefore conditioned so that

$$h(s)=s(s-2)h_1(s) \tag{66}$$

and

$$g(s)-h(s)=(s-1)(s-z_0)(s-\bar{z}_0)h_2(s). \tag{67}$$

This is accomplished by dividing the computed $h(s)$ by $s(s-2)$, setting $h_1(s)$ equal to the quotient and ignoring the remainder.[9] The same procedure is followed in obtaining (67), and (42) then yields

$$C_0(s)=\frac{p_f(s)h_2(s)}{h_1(s)} \tag{68}$$

for the optimal controller.

The design described above depends parametrically on k. The values of E_t and E_s have been computed for various values of k and the results are shown in Fig. 3. It is clear from the curves that the choice $k=4$ leads to a suitable compromise between the desire to minimize E_t while limiting E_s. (Note that all transfer functions in the frequency domain and all signals in the time domain are taken to be dimensionless quantities but time is measured in seconds. It follows that E_t and E_s have the dimensions of seconds.) The transient response of the optimally designed system to a unit step input has also been investigated. The error and plant input responses for several values of k are shown in Fig. 4 and Fig. 5. Since these responses are obtained with $d=m=0$ they do not, and should not, reflect the optimality of the design. They do show, however, that reasonable transient performance is obtained with the choice $k=4$.

It has been pointed out in Assumption 1 that $|\phi(j\omega)|$ should be large or, equivalently, its reciprocal should be small for good stability margin. We have in fact computed and plotted $\phi^{-1}(j\omega)$ for several values of ω in the range zero to infinity. The results are shown in Fig. 6 and are highly satisfactory. With $k=4$, $|\phi^{-1}(j\omega)|<10^{-4}$ and the system remains stable no matter what the phase of $\Delta\phi(j\omega)$

[9]The accuracy of the computations on a digital computer is such that this remainder is quite small (coefficients less than 10^{-5} in this example).

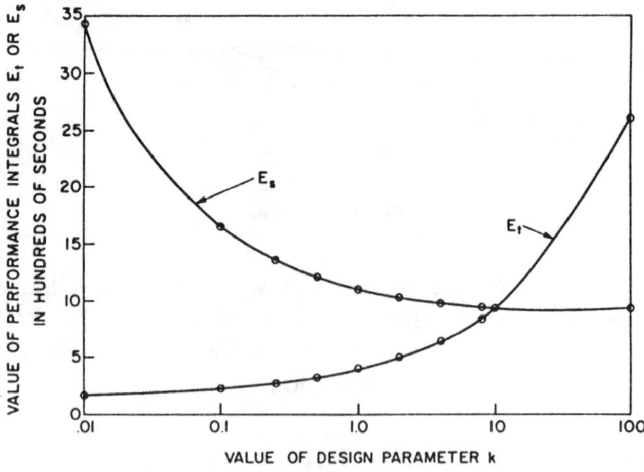

Fig. 3. Variation of performance integrals.

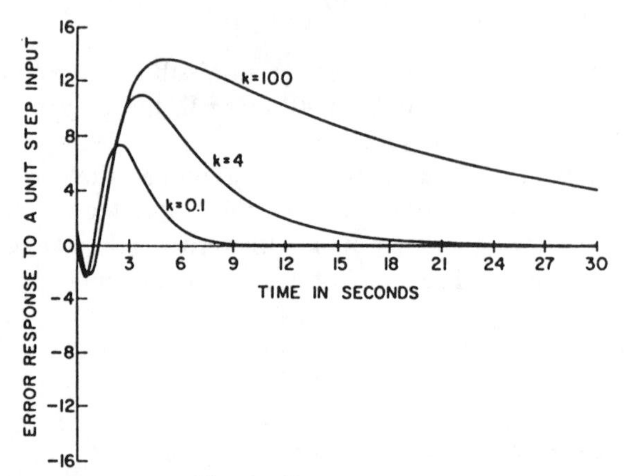

Fig. 4. Error responses.

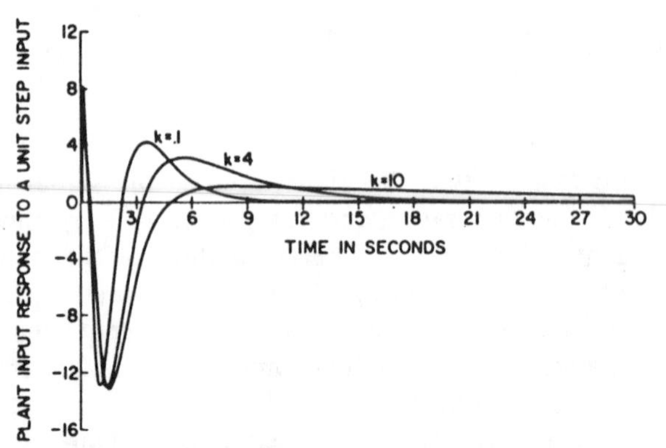

Fig. 5. Plant input responses.

provided only that $|\Delta\phi(j\omega)|<10\ 000$. For completeness, plots are also shown in Fig. 7 of $|S_0(j\omega)|$ versus ω for several choices of k.

Evidently, in the light of these observations, the choice $k=4$ makes engineering sense and the final step in the design is to compute the optimal controller transfer function with $k=4$. Using (68) we get

$$C_0(s)=\frac{K(s-\sigma_1)(s-\sigma_2)(s-s_0)(s-\bar{s}_0)}{(s-p_1)(s-p_2)(s-p_3)(s-s_p)(s-\bar{s}_p)}, \tag{69}$$

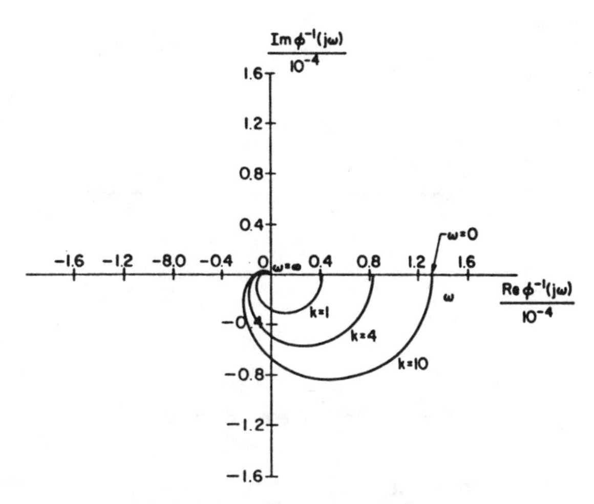

Fig. 6. Nyquist plot of $\phi^{-1}(j\omega)$.

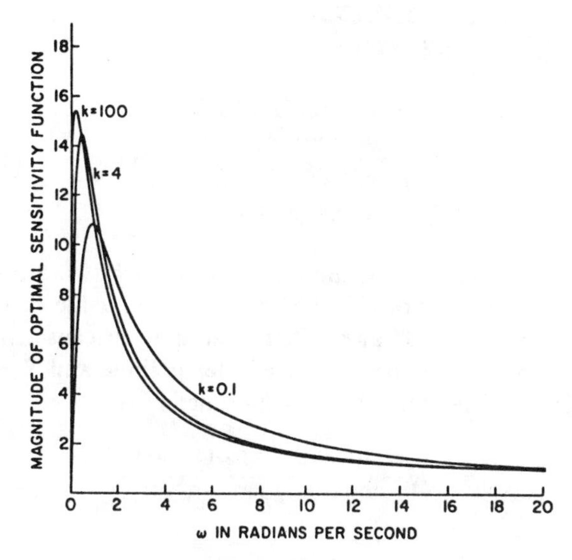

Fig. 7. Sensitivity function amplitude response.

where

$$\left.\begin{aligned} K &= 67.228808647 \\ \sigma_1 &= 0.014874634 \\ \sigma_2 &= -9.9999638 \\ \rho_1 &= 2.413271030575 \\ \rho_2 &= -9.9806403944 \\ \rho_3 &= -33.65463165144 \\ s_0 &= -30+j17.320508076 \\ s_p &= -18.05732390209+j14.991623794. \end{aligned}\right\} \tag{70}$$

The reader has probably noticed that σ_2 and ρ_2 are very nearly equal. Although we have in fact verified theoretically that these two quantities cannot be equal, it is nevertheless natural to inquire whether any significant deterioration in performance results by putting $\sigma_2=\rho_2$ and using the suboptimal controller

$$C(s)=\frac{67.2(s-0.015)(s^2+60s+1200)}{(s-2.4)(s+33.7)(s^2+36s+549)}. \tag{71}$$

Another aspect of the design which should be clarified is the use of the Padé approximation (45) for the delay $e^{-0.1s}$ in the feedback loop. These points have been taken up and the results are presented in Table I. A comparison of the first two columns in the table reveals that the use of the Padé approximation is certainly satisfactory while a comparison of the last two shows that the use of the suboptimal controller is justified. This is gratifying since no analog controller can be designed to the accuracy demanded by the values in (70).

We have also studied the stability margin with respect to variations in the delay $\tau=0.1$ s. With the suboptimal controller the feedback loop remains stable for $0\leq\tau\leq 0.155$ s. This stability margin is clearly satisfactory and the example indicates that the design procedure is a practical one.

One final point. It is quite obvious that the design equations are substantially simpler if $F(s)=1$ is used instead of the Padé approximation and the delay is ignored. However, when the corresponding controller is employed it is found that the system remains stable only for $0\leq\tau\leq 0.08$ s. Thus, with an actual delay $\tau=0.1$ s the system designed optimally with $F=1$ would be unstable. This facility to incorporate delay is of significant practical value.

TABLE I
COMPARISON OF RESULTS

	Optimal $C(s)$ Padé		Suboptimal $C(s)$
	Approximation for $F(s)$	$F(s)=e^{-0.1s}$	$F(s)=e^{-0.1s}$
E_t	646.9	646.1	676.8
E_s	986.7	957.6	952.2

V. Discussion

It appears from Fig. 4 that the transient performance of the optimally compensated loop in our example is poor. In fact, for $k=4$ a peak error response of 10.7 is obtained. Is this poor transient performance a consequence of the design procedure, inherent limitations imposed by the plant, or both? To answer this question the example described in the previous section is considered once again but with $G_m=G_d=k=0$. (Note that although the conditions $k>0$ and $G_b(-s^2)\not\equiv 0$ are violated, it is still possible to obtain an optimal solution. Optimal solutions can exist for cases which do not satisfy our assumptions on the data. Assumptions 1–6 are *sufficient* to guarantee the existence of an optimal controller and they hold in most cases of interest.)

The optimal solution obtained for $G_m=G_d=k=0$ is the one which minimizes the integral square error with a unit step input. The optimal controller in this case is

$$C_0=\frac{-\hat{K}(s-\sigma_3)(s-s_0)(s-\bar{s}_0)}{(s-\rho_4)(s-\rho_5)} \tag{72}$$

where s_0 is given in (70) and

$$\left.\begin{aligned}\hat{K}&=1.136452161649\\ \sigma_3&=0.240139\\ \rho_4&=56.64885134963\\ \rho_5&=9.55199217963.\end{aligned}\right\} \tag{73}$$

The error response to a unit step input with this controller in the system is shown in Fig. 8. The initial value of the error is $e(0)=-7.33$ and differs from unity because with $G_m=G_d=k=0$ the performance index is finite and $S(\infty)\neq 1$. In fact, the optimal sensitivity function is

$$S_0=-\tilde{K}\frac{s(s-2)(s-\rho_4)(s-\rho_5)}{(s+1)(s+2)(s-s_0)(s-\bar{s}_0)} \tag{74}$$

where

$$\tilde{K}=-7.328575728765. \tag{75}$$

It is clear from Fig. 8 that even in the best of circumstances, no disturbance inputs, no measurement noise, and no plant saturation constraints, the best possible transient performance is poor. The reason is that this particular nonminimum-phase unstable plant is one of the most difficult to control irrespective of whether the policy is optimal or suboptimal. Indeed, since it is impossible to stabilize this plant $P(s)=(s-1)/s(s-2)$ by means of any dynamical stable compensation whatsoever [6], lead-lag methods are futile. In our opinion, a design methodology which can accomodate disturbance inputs, feedback sensor noise, rms restrictions on plant inputs, and also yield results as encouraging as those shown in Figs. 4 and 5, is a valuable engineering tool.

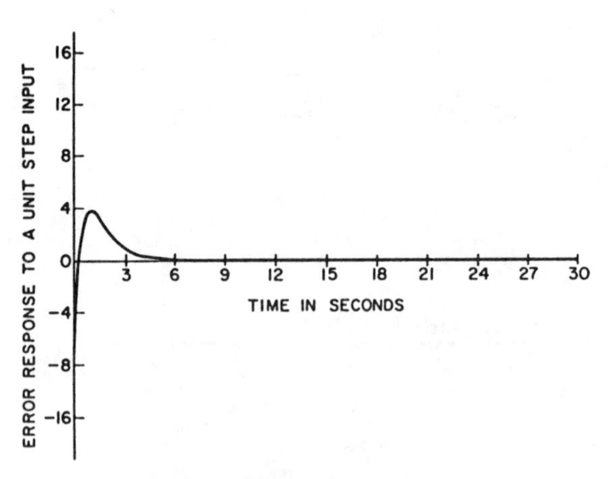

Fig. 8. Optimal error response.

APPENDIX

Proof of Lemma 1: The controller $C(s)=n_c(s)/d_c(s)$ is determined from the formula

$$S(s)=\frac{1}{1+P(s)F(s)C(s)}=\frac{d_f(s)d_p(s)d_c(s)}{\varphi(s)} \tag{76}$$

$$\varphi(s)=d_f(s)d_p(s)d_c(s)+n_f(s)n_p(s)n_c(s). \tag{77}$$

What must be shown is that $\varphi(s)$ is strict Hurwitz. According to Assumption 1, $S(s)$ is analytic in Re $s\geqslant 0$ and any zero s_0 of $\varphi(s)$ in Re $s\geqslant 0$ must be a zero of $d_f(s)d_p(s)d_c(s)$ and therefore of $n_f(s)n_p(s)n_c(s)$. If it is a zero of $d_f(s)d_p(s)$, it cannot be a zero of $n_f(s)n_p(s)$ because $P(s)$ and $F(s)$ are assumed to be admissible. Thus, s_0 must be a zero of $n_c(s)$ and consequently not one of $d_c(s)$ which is relatively prime to $n_c(s)$. But this means that the multiplicity of $s=s_0$ as a zero of $S(s)$ is less than its multiplicity as a zero of $d_f(s)d_p(s)$ which contradicts Assumption 2. If instead s_0 is assumed to be a zero of $d_c(s)$, it then follows that it is a zero of $n_f(s)n_p(s)$, but not a zero of $n_c(s)$. However, the expression

$$1-S(s)=\frac{n_f(s)n_p(s)n_c(s)}{\varphi(s)} \tag{78}$$

then reveals that the multiplicity of $s=s_0$ as a zero of $1-S(s)$ is less than its multiplicity as a zero of $n_f(s)n_p(s)$, a contradiction with Assumption 3. Q.E.D.[10]

Proof of Theorem 1: In (21), the candidate functions $S(s)$ must all be realizable for the prescribed admissible pair $P(s)$, $F(s)$, Hence, if $S_0(s)$ is minimizing and ϵ is any real number,

$$S(s)=S_0(s)+\epsilon\chi_r(s)\delta(s)$$

is a legitimate competitor provided $\delta(s)$ is analytic in Re$s\geqslant 0$ and the resulting cost is finite. The proof of this assertion is simple. Both $S_0(s)$ and $S(s)$ must include all the zeros of $d_f(s)d_p(s)$ in Re $s\geqslant 0$ and so must the difference $S-S_0$. Again, all the zeros of $n_f(s)n_p(s)$ in Re$s\geqslant 0$ must be zeros of $1-S_0(s)$ and $1-S(s)$ and therefore of $(1-S_0)-(1-S)=S-S_0$. But $d_f(s)d_p(s)$ and $n_f(s)n_p(s)$ are relatively prime in Re$s\geqslant 0$ and it follows that $S(s)-S_0(s)$ is divisible by

$$(d_fd_p)_r\cdot(n_fn_p)_r=\chi_r.$$

This quotient $\epsilon\delta(s)$ is analytic in Re$s\geqslant 0$ and the order of its zero at $s=\infty$ must be sufficiently high to guarantee the finiteness of E. To exploit the optimality of $S_0(s)$, we set

$$\left.\frac{dE}{d\epsilon}\right|_{\epsilon=0}=0 \tag{79}$$

and use the standard Wiener–Hopf variational argument [3] to obtain

$$\chi_{r*}S_0(G_a+G_b)-\chi_{r*}\left(G_b-\frac{F-1}{FF_*}G_u\right)=X_*(s) \tag{80}$$

$X(s)$ analytic in Re$s\geqslant 0$. Performing the spectral factorization

$$\chi_r\chi_{r*}(G_a+G_b)=\Omega\Omega_* \tag{81}$$

where $\Omega(s)$ is free of zeros and poles in Re$s>0$ and dividing both sides of (80) by Ω_*/χ_r gives, after re-

[10]The proof of necessity follows immediately from (76)+(78) and is trivial.

arrangement,

$$\Omega S_0-\left\{\frac{\chi_r\chi_{r*}}{\Omega_*}\left(G_b-\frac{F-1}{FF_*}G_u\right)\right\}_+=\frac{\chi_r X_*}{\Omega_*}+\left\{\frac{\chi_r\chi_{r*}}{\Omega_*}\left(G_b-\frac{F-1}{FF_*}G_u\right)\right\}_- . \quad (82)$$

Using (35) and (38) it is seen that

$$\chi\chi_*(G_a+G_b)=(n_pn_{p*}+kQd_pd_{p*})\cdot(d_pn_f)G(d_pn_f)_*\cdot(d_fd_{f*}) \quad (82a)$$

and invoking Assumptions 1, 5, and 6 we conclude that $\Omega(s)$ is actually free of zeros and poles in $\mathrm{Re}\,s\geqslant 0$. Thus, $S_0(s)\Omega(s)$ must be analytic in $\mathrm{Re}\,s\geqslant 0$. But then the left-hand side of (82) is also analytic in $\mathrm{Re}\,s\geqslant 0$ and equals the right-hand side which is analytic in $\mathrm{Re}\,s<0$. Being analytic in the entire finite s-plane,

$$\Omega S_0-\left\{\frac{\chi_r\chi_{r*}}{\Omega_*}\left(G_b-\frac{F-1}{FF_*}G_u\right)\right\}_+=f(s) \quad (83)$$

$f(s)$ a real polynomial; or

$$S_0=\frac{\left\{\frac{\chi_r\chi_{r*}}{\Omega_*}\left(G_b-\frac{F-1}{FF_*}G_u\right)\right\}_++f}{\Omega}, \quad (84)$$

which is (41).

Clearly, $S_0(s)$ is analytic in $\mathrm{Re}\,s\geqslant 0$. Since $G_b(\omega^2)=0(\omega^{2l})$, $l\geqslant 0$, the convergence of (21) forces

$$1-S_0(j\omega)=0(1/\omega^{l+1}). \quad (85)$$

Write

$$\left\{\frac{\chi_r\chi_{r*}}{\Omega_*}\left(G_b-\frac{F-1}{FF_*}G_u\right)\right\}_+=\frac{h(s)}{g(s)} \quad (86)$$

and

$$\Omega(s)=\frac{a(s)}{b(s)}. \quad (87)$$

$h(s)$, $g(s)$, $a(s)$, and $b(s)$ are four real polynomials. Then, as is easily checked,

$$\text{degree } g\geqslant \text{degree } h \quad (88)$$

$$\text{degree } a=\text{degree } b+\text{degree } \chi_r+l \quad (89)$$

and

$$S_0=\frac{(h+gf)b}{ag}. \quad (90)$$

To insure a proper $S_0(s)$, we must impose the degree restriction[11]

$$\delta(f)\leqslant\delta(a)-\delta(b) \quad (91)$$

or, using (89),

$$\delta(f)\leqslant\delta(\chi_r)+l. \quad (92)$$

To guarantee that $S_0(s)$ be realizable for $P(s)$, $F(s)$ every zero of $(d_fd_p)_r$ must be a zero of $S_0(s)$ and every zero of $(n_fn_p)_r$ must be a zero of $1-S_0(s)$. Coupling this with (85) we get a total of $\delta(\chi_r)+l+1$ interpolatory constraints on $f(s)$ which is one more than its permitted maximum degree, $\delta(\chi_r)+l$. Thus, $f(s)$ is the unique *Lagrange interpolation* polynomial [13] satisfying these conditions and if a minimizing $S_0(s)$ exists, it must be the one given by (84) because the cost functional is quadratic in $S(s)$.

Now part 2 is obviously correct and to complete the proof of Theorem 1 it suffices to show that $X_*(s)$, as given by (84), is actually analytic in $\mathrm{Re}\,s\leqslant 0$. Rearranging (80) with the aid of (81) and (84) leads to

$$\frac{\Omega\Omega_*}{\chi_r}\left(\frac{\{\Psi\}_++f}{\Omega}\right)-\frac{\Omega_*}{\chi_r}\Psi=\frac{\Omega_*}{\chi_r}(f-\{\Psi\}_-) \quad (93)$$

where

$$\Psi=\frac{\chi_r\chi_{r*}}{\Omega_*}\left(G_b-\frac{F-1}{FF_*}Gu\right). \quad (94)$$

It is apparent that (93) is analytic in $\mathrm{Re}\,s<0$ and it only remains to show that the same is true on the $s=j\omega$-axis. We observe first that the analyticity of $\{\Psi\}_-$ for $s=j\omega$ is implied by Assumption 2, (82a), and Assumption 6. Thus, (93) is analytic on the $j\omega$-axis if the purely imaginary zeros of χ_r are also zeros of $f-\{\Psi\}_-$ of at least the same multiplicities. But this is automatic whenever f is chosen to make $S_0(s)$ realizable for the pair $F(s)$, $P(s)$. For suppose that $s=j\omega_0$ is a zero of χ_r of order ν. Then, invoking Assumption 1, it is either a pole of P or a zero of FP of order ν.[12] Suppose it is a pole of P. Since the only $j\omega$-poles of G_b are either zeros of F or zeros of P, (94) shows that $s=j\omega_0$ is a zero of Ψ of order at least ν. It now follows from the identity

$$\{\Psi\}_++f=\Psi+(f-\{\Psi\}_-) \quad (95)$$

that $s=j\omega_0$ is a zero of $f-\{\Psi\}_-$ of multiplicity ν or higher [see (84)].

Suppose instead that $s=j\omega_0$ is a zero of χ_r which is a zero of FP of order ν. A direct calculation yields

$$1-S_0=\left(1-\frac{\Psi}{\Omega}\right)-\frac{1}{\Omega}(f-\{\Psi\}_-) \quad (96)$$

and from (81) and (94) we obtain

$$1-\frac{\Psi}{\Omega}=\frac{G_a+\frac{F-1}{FF_*}G_u}{G_a+G_b}. \quad (97)$$

With the aid of (82a) and Assumption 6 it is seen that

[11] $\delta(\cdot)\equiv$ degree $(\cdot)$.

[12] According to Assumption 1, F is $j\omega$-analytic.

$s=j\omega_0$ is a zero of the right-hand side of (97) of order ν or more. Since f has been constructed to make S_0 realizable for F and P, $s=j\omega_0$ is a zero of $1-S_0$ of order at least ν and therefore, in view of (96), a zero of $f-\{\Psi\}_-$ of multiplicity ν or greater. Q.E.D.

REFERENCES

[1] G. C. Newton, Jr., L.A. Gould, and J. F. Kaiser, *Analytical Design of Linear Feedback Controls*. New York: Wiley, 1957.

[2] J. J. Bongiorno, Jr., "Minimum sensitivity design of linear multivariable feedback control systems by matrix spectral factorization," *IEEE Trans. Automat. Contr.*, vol. AC-14, pp. 665–673, Dec. 1969.

[3] J. E. Weston and J. J. Bongiorno, Jr., "Extension of analytical design techniques to multivariable feedback control systems," *IEEE Trans. Automat. Contr.*, vol. AC-17, pp. 613–620, Oct. 1972.

[4] J. R. Ragazzini and G. F. Franklin, *Sampled-Data Control Systems*. New York: McGraw-Hill, 1958, pp. 155–158.

[5] S. C. Bigelow, "The design of analog computer compensated control systems," *AIEE Trans.* (Appl. Ind.), vol. 77, pp. 409–415, Nov. 1958.

[6] D. C. Youla, J. J. Bongiorno, Jr., and C.N. Lu, "Single-loop feedback-stabilization of linear multivariable dynamical plants," *Automatica*, vol. 10, pp. 159–173, Mar. 1974.

[7] I. Horowitz, "Optimum loop transfer function in single-loop minimum-phase feedback systems," *Int. J. Contr.* vol. 18, no. 1, pp. 97–113, 1973.

[8] H. W. Bode, *Network Analysis and Feedback Amplifier Design*. New York: Van Nostrand, 1945.

[9] D. C. Youla, "The modern design of optimal multivariable controllers via classical techniques," proposal to Nat. Science Foundation, Dec. 1973.

[10] ——, "Modern classical feedback control theory: Part I," Rome Air Development Center, Griffiss Air Force Base, NY, Tech. Rep. RADC-TR-70-98, June 1970.

[11] C. T. Chen, *Introduction to Linear System Theory*. New York: Holt, Rinehart and Winston, 1970.

[12] H. H. Rosenbrock, *State-Space and Multivariable Theory*. New York: Wiley Interscience, 1970.

[13] F. R. Gantmacher, *The Theory of Matrices*, vol. I. New York: Chelsea, 1960.

[14] C. D. Johnson, "Accommodation of external disturbances in linear regulator and servomechanism problems," *IEEE Trans. Automat. Contr.* (Special Issue on Linear-Quadratic-Gaussian Problem), vol. AC-16, pp. 635–644, Dec. 1971.

Modern Wiener–Hopf Design of Optimal Controllers — Part II: The Multivariable Case

DANTE C. YOULA, FELLOW, IEEE, HAMID A. JABR, MEMBER, IEEE,
AND JOSEPH J. BONGIORNO, JR., MEMBER, IEEE

Abstract—**In many modern-day control problems encountered in the fluid, petroleum, power, gas and paper industries, cross coupling (interaction) between controlled and manipulated variables can be so severe that any attempt to employ single-loop controllers results in unacceptable performance. In all these situations, any workable control strategy must take into account the true multivariable nature of the plant and address itself directly to the design of a compatible multivariable controller. Any practical design technique must be able to cope with load disturbance, plant saturation, measurement noise, process lag, sensitivity and also incorporate suitable criteria delimiting transient behavior and steady-state performance. These difficulties, when compounded by the fact that many plants (such as chemical reactors) are inherently open-loop unstable have hindered the development of an inclusive frequency-domain analytic design methodology. However, a solution based on a least-square Wiener–Hopf minimization of an appropriately chosen cost functional is now available. The optimal controller obtained by this method guarantees an asymptotically stable and dynamical closed-loop configuration irrespective of whether or not the plant is proper, stable, or minimum-phase and also permits the stability margin of the optimal design to be ascertained in advance. The main purpose of this paper is to lay bare the physical assumptions underlying the choice of model and to present an explicit formula for the optimal controller.**

I. Introduction

In many modern-day control problems encountered in the fluid, petroleum, power, gas, and paper industries, cross coupling (interaction) between controlled and manipulated variables can be so severe that any attempt to employ single-loop controllers results in unacceptable performance. In all these situations, any workable control strategy must take into account the true multivariable nature of the plant and address itself directly to the design of a compatible multivariable controller. Any practical design technique must be able to cope with load disturbance, plant saturation, measurement noise, process lag, sensitivity and also incorporate suitable criteria delimiting transient behavior and steady-state performance. These difficulties, when compounded by the fact that many plants (such as chemical reactors) are inherently open-loop unstable have hindered the development of an inclusive frequency-domain analytic design methodology. However, these obstacles have been overcome and a solution based on a least-squares Wiener–Hopf minimization of an appropriately chosen cost functional E is now available. This solution, which is the natural culmination of earlier work [1]–[4], offers the following concrete accomplishments:

1) There are no restrictions on the plant transfer matrix. It can be rectangular, unstable, improper,[1] and nonminimum phase.

2) The design incorporates input noise, load disturbance, measurement noise, and feedforward compensation. The noise can be colored.

3) The optimal controller minimizing E is proper and guarantees a dynamical asymptotically stable closed-loop design possessing proper sensitivity matrices equal to the identity matrix at $s=\infty$.

Manuscript received June 5, 1975; revised February 11, 1976. Paper recommended by J. B. Pearson, Chairman of the IEEE S-CS Linear Systems Committee. This work was supported by the National Science Foundation under Grant ENG 74-13054 and is taken in part from a Ph.D. dissertation submitted by H. A. Jabr to the Faculty of the Polytechnic Institute of New York.

D. C. Youla and J. J. Bongiorno, Jr., are with the Department of Electrical Engineering and Electrophysics, Polytechnic Institute of New York, Long Island Center, Farmingdale, NY 11735.

H. A. Jabr was with the Department of Electrical Engineering and Electrophysics, Polytechnic Institute of New York, Long Island Center, Farmingdale, NY 11735. He is now with the University of Jordan, Amman, Jordan.

[1]A transfer matrix $A(s)$ is proper if $A(\infty)$ is finite and strictly proper if $A(\infty)=O$, the zero matrix. Otherwise it is improper.

Reprinted from *IEEE Trans. Automat. Contr.*, vol. AC-21, no. 3, pp. 319–338, June 1976.

4) The loop can track ramp-type inputs and recover from step-type disturbances of the correct order with zero steady-state error.

5) Transient response (system accuracy) can be traded off against linear operation.

6) The stability margin of the optimal design is ascertainable in advance.

7) The sensor transfer matrices are absorbed directly into the cost and various delays can be simulated by suitable preequalization.

The primary purpose of this paper is to lay bare the physical assumptions underlying the choice of model and to derive an explicit expression for the optimal controller. To achieve this objective it is first necessary to solve several difficult intermediate problems of the "model matching" variety and, for the sake of continuity and clarity, some of the more involved details have been relegated to two Appendixes. Finally, to help place the contributions of the present work in perspective we offer a comparison with the linear quadratic Gaussian (LQG) approach [17]. It is pointed out that the problem addressed and solved by LQG is quite different from the one considered in this paper. Nevertheless, there is a common class of problems that can be treated by both methods and for any such problem the optimal controller is the same. However, we show by actual example that forcing the optimal controller to be realized via the Kalman structure is not always possible. This limitation is inherent in LQG, but not in ours.

II. THE MODEL

We focus our attention exclusively on the design of optimal controllers for multi-input–output finite-dimensional linear time-invariant plants imbedded in a multivariable single-loop configuration of generic type shown in Fig. 1.[2] Suppose $y_d(s)$, the *desired* closed-loop output is related to $u_i(s)$, the *actual* input set-point signal in the linear fashion

$$y_d(s)=T_d(s)u_i(s) \tag{1}$$

via the *ideal* transfer matrix $T_d(s)$. The *prefilter* $W(s)$ is selected in advance, but once chosen,[3]

$$u=W(u_i+n) \tag{2}$$

must be considered the best available linear version of $y_d(s)$. Any reasonable performance measure should be based on the vector error difference

$$e(s)=u(s)-y(s) \tag{3}$$

between the actual smoothed input $u(s)$ driving the loop and the plant output $y(s)$. If plant delays are excessive the suppression of load disturbance by means of feedback alone may not suffice and it is usually advisable to incorporate feedforward compensation $L(s)$ as an integral part of the design. For given choices of overall sensors $F(s)$ and $L(s)$, the design of the controller $C(s)$ evolves from an appropriate minimization procedure subject to a power-like constraint on $r(s)$ to avoid plant saturation and to extend the linear range. (In nonlinear applications the constraint on $r(s)$ is imposed to avoid permanent departures from the neighborhood of a desired equilibrium state.) Plant disturbance $d(s)$ and instrument noise $m(s)$, $l(s)$ are modeled in a perfectly general way by assuming that

$$y(s)=P(s)r(s)+P_o(s)d(s), \tag{4}$$

$$v(s)=F(s)y(s)+F_o(s)m(s) \tag{5}$$

and

$$z(s)=L(s)d(s)+L_o(s)l(s) \tag{6}$$

where $P(s)$, $P_o(s)$, $F(s)$, $F_o(s)$, $L(s)$, and $L_o(s)$ are real rational matrices.

In Fig. 1, $P(s)$ is $n\times m$, $F(s)$ is $n\times n$, and $C(s)$ is $m\times n$. Hence FPC is $n\times n$ and it is assumed of course that all other matrices are dimensioned compatible.[4] Straightforward analysis yields

$$y=PR(u-F_om-L_ol)+(P_o-PRP_d)d, \tag{7}$$

$$r=R(u-F_om-L_ol-P_dd), \tag{8}$$

$$e=(1_n-PR)u+PR(F_om+L_ol)-(P_o-PRP_d)d \tag{9}$$

where

$$R=CS, \tag{10}$$

$$S=(1_n+FPC)^{-1}, \tag{11}$$

$$P_d=FP_o+L. \tag{12}$$

In the absence of measurement noise and load disturbance, $y=(PR)u$. Thus,

$$T(s)=P(s)R(s) \tag{13}$$

is the closed-loop transfer matrix and $S(s)$ is the sensitivity matrix. In most industrial applications the available choices of physical sensing devices $L_t(s)$ and $F_t(s)$ are severely restricted and more or less dictated by the problem at hand. However, as explained later, low-power preequalizers $L_e(s)$ and $F_e(s)$ can and in many cases should be employed to improve stability margin, to assure

[2]To avoid proliferating symbols, all quantities are Laplace transforms, deterministic or otherwise. All stochastic processes are either zero-mean second-order stationary or shape-deterministic or a sum of both with rational spectral densities. For example, η/s, η a random variable, is the transform of a random step with spectral density

$$\langle(\eta/s)(\eta/s)_*\rangle=-\frac{\sigma^2}{s^2}$$

where $\sigma^2=\langle|\eta|^2\rangle$ and $\langle,\rangle$ denotes ensemble average.

[3]Function arguments are omitted wherever convenient and for any matrix A, A', $\bar{A}$, $A^*(\equiv\bar{A}')$ and det A denote the transpose, complex-conjugate, ajoint and determinant of A, respectively. Column vectors are written a, x, etc., or as $x=(x_1,x_2,\cdots,x_n)'$ to exhibit the components explicitly. Last, for any real rational matrix $A(s)$ of the complex frequency variable $s=\sigma+j\omega$, $A_*(s)\equiv A'(-s)$. Note that for $s=j\omega$, ω real, $A_*(j\omega)=A^*(j\omega)$.

[4]1_n is the $n\times n$ identity matrix and O_n, $O_{n,m}$, O_n denote the n-dimensional zero vector, the $n\times m$ and $n\times n$ zero matrices, respectively.

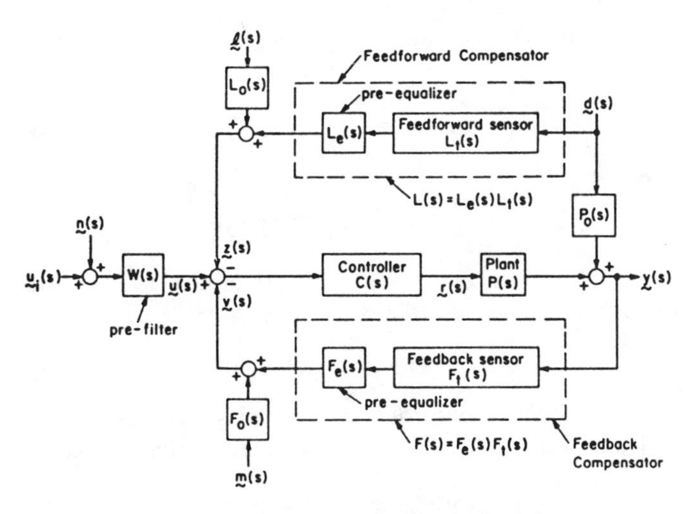

Fig. 1. Multivariable single-loop configuration.

zero steady-state error and to simulate delay in the feedback link. From this point on it is assumed that the data $P_o(s)$, $P(s)$, $L_o(s)$, $F_t(s)$, $L_t(s)$,

$$F(s)=F_e(s)F_t(s) \tag{14}$$

and

$$L(s)=L_e(s)L_t(s) \tag{15}$$

are prescribed in advance.

An interpretation of (9) with $F=1_n$ (unity feedback) and $z=O_n$ (no feedforward compensation) reveals most clearly the potential for tradeoff in the various frequency bands. In fact, for this case $1_n-PR=S$ and

$$e=S(u-P_od)+(1_n-S)F_om \tag{16}$$

is composed of two contributions $e_1(s)$ and $e_2(s)$. The first,

$$e_1=S(u-P_od), \tag{17}$$

subsumes steady-state error, accuracy, and load disturbance while the second,

$$e_2=(1_n-S)F_om, \tag{18}$$

is the error produced by measurement noise. The impossibility of making both $S(j\omega)$ and $1_n-S(j\omega)$ arbitrarily "small" over any frequency band is partly intrinsic and partly conditioned by the plant restrictions [5], [6]. This fundamental conflict is inevitable and responsible for a great deal of the difficulty surrounding practical analytic feedback design. According to (8), the spectral amplification from $m(j\omega)$ to $r(j\omega)$ must depend on the "size" of the matrix $R(j\omega)$. Since (with $F=1_n$)

$$P(j\omega)R(j\omega)=1_n-S(j\omega) \tag{19}$$

and plant bandwidth is usually confined to some low-frequency interval $0\leq\omega\leq\omega_o$, good transient response and sensitivity require

$$S(j\omega)\approx O_n, \qquad 0\leq\omega\leq\omega_o; \tag{20}$$

or, qualitatively,

$$C(j\omega)\approx\infty, \qquad 0\leq\omega\leq\omega_o. \tag{21}$$

Hence, even if $1_n-S(j\omega)$ approaches O_n as $\omega\to\infty$, there usually exists an intermediate high-frequency band $\omega_1\leq\omega\leq\omega_2$, $\omega_2-\omega_1\gg\omega_o$, over which $1_n-S(j\omega)\approx 1_n$ and $P(j\omega)\approx O_{n,m}$. In view of (19), this implies[5]

$$R(j\omega)\approx\infty, \qquad \omega_1\leq\omega\leq\omega_2, \tag{22}$$

and the rms value of $r(j\omega)$ can easily exceed the saturation level of the plant because of the extremely wide-band nature of $m(j\omega)$. If we assume the feedback sensors to be as noise free as possible the only remaining remedy is to concentrate on a "best" choice for $C(s)$ subject to the constraint that the closed loop be asymptotically stable. A successful quantitative reformulation of this latter requirement has enabled us to achieve a least-squares solution for $C(s)$ which takes into account all the pertinent performance criteria and is applicable to arbitrary open-loop unstable nonminimum-phase plants.

The McMillan degree, $\delta(A;s_o)$, of $s=s_o$ (finite or infinite) as a pole of the rational matrix $A(s)$ is the largest multiplicity it possesses as a pole of any minor of $A(s)$. The McMillan degree, $\delta(A)$, of $A(s)$ is the sum of the McMillan degrees of its distinct poles. Let the distinct *finite* poles of $A(s)$ be denoted by s_i and their associated McMillan degrees by δ_i, $i=1\to\mu$. The monic polynomial

$$\psi_A(s)=\prod_{i=1}^{\mu}(s-s_i)^{\delta_i} \tag{23}$$

is the *characteristic* denominator of $A(s)$. It is easily shown that $\psi_A(s)$ is also the monic least common multiple of all denominators of all minors of $A(s)$, each minor assumed expressed as the ratio of two relatively prime polynomials.

Lemma 1 (Appendix A): If the plant, feedback compensator, and controller (dynamical or otherwise) are free of unstable hidden modes, the closed loop of Fig. 1 is asymptotically stable iff

$$\varphi(s)\equiv\frac{\psi_P(s)\psi_C(s)\psi_F(s)}{\det S(s)} \tag{24}$$

is a strict Hurwitz polynomial.[6]

By assumption, $P(s)$ and $F(s)$ are given, but even granting that the plant and feedback compensator are free of unstable hidden modes, it may still be impossible to find a controller $C(s)$ which stabilizes the closed loop. Such is the case iff for some finite $s=s_o$, $\operatorname{Re}s_o\geq 0$,

$$\delta(FP;s_o)<\delta(P;s_o)+\delta(F;s_o). \tag{25}$$

This means that there exists at the output of the plant or feedback compensator an exponential ramp-modulated

[5]At this stage these arguments are intentionally informal. Incidentally, for nonminimum-phase unstable plants it is also possible to have $1_n-S(j\omega)\approx\infty$ over $\omega_1\leq\omega\leq\omega_2$. See the example in [6, p. 12] and the accompanying table in Fig. 5.

[6]A polynomial free of zeros in $\operatorname{Re}s\geq 0$ is said to be strict Hurwitz and "iff" abbreviates "if and only if." A system with transfer matrix $A(s)$ is dynamical if $A(s)$ is proper, i.e., if $A(\infty)$ is finite.

sinusoid whose growth in time exceeds that of the signal at the corresponding input and the necessary corrective action to effect stabilization is lacking.

Definition 1: The plant and feedback compensator form an admissible pair if each is individually free of unstable hidden modes and

$$\psi_{FP}^{+}(s)=\phi_{F}^{+}(s)\psi_{P}^{+}(s). \tag{26}$$

(The monic polynomials $\psi^{+}(s)$ and $\psi^{-}(s)$ absorb all the zeros of $\psi(s)$ in $\mathrm{Re}\,s\geq 0$ and $\mathrm{Re}\,s<0$, respectively. Thus, up to a multiplicative constant, $\psi=\psi^{+}\psi^{-}$.)

Lemma 2 (Appendix A): There exists a controller stabilizing the given plant and feedback compensator in the closed-loop configuration of Fig. 1 iff the pair is admissible.

Let the spectral densities of $\boldsymbol{u}(s)$, $\boldsymbol{d}(s)$, $\boldsymbol{l}(s)$, and $\boldsymbol{m}(s)$ be denoted by $G_u(s)$, $G_d(s)$, $G_l(s)$, and $G_m(s)$, respectively. Setting aside for the moment all questions of convergence,[7]

$$2\pi j E_t=\mathrm{Tr}\int_{-j\infty}^{j\infty}\langle \boldsymbol{e}(s)\boldsymbol{e}_*(s)\rangle ds \tag{27}$$

is the usual quadratic measure of steady-state response. Similarly, if $P_s(s)$ represents the transfer matrix coupling the plant input $\boldsymbol{r}(s)$ to those "sensitive" plant modes which must be especially guarded against excessive dynamic excursions,

$$2\pi j E_s=\mathrm{Tr}\int_{-j\infty}^{j\infty}\langle P_s(s)\boldsymbol{r}(s)\boldsymbol{r}_*(s)P_{s*}(s)\rangle ds \tag{28}$$

is a proven useful penalty function for saturation [7]. Hence,

$$E=E_t+kE_s, \tag{29}$$

k an adjustable positive constant, serves as a weighted cost combining both factors and the optimal controller is chosen to minimize E. Now referring to (8) and (9) it is seen that $R(s)$ determines $\boldsymbol{r}(s)$, $\boldsymbol{e}(s)$, and E. Consequently $R(s)$ embodies all the design freedom and the next lemma plays an obvious and indispensable role.

Lemma 3 (Appendix A): Let the given plant and feedback compensator form an admissible pair with transfer matrix descriptions $P(s)$, $F(s)$. Let

$$F(s)P(s)=A^{-1}(s)B(s)=B_1(s)A_1^{-1}(s) \tag{30}$$

where the pairs $A(s)$, $B(s)$ and $B_1(s)$, $A_1(s)$ constitute any left–right coprime polynomial decompositions of $F(s)P(s)$, respectively. Select polynomial matrices $X(s)$ and $Y(s)$ such that[8]

$$A(s)X(s)+B(s)Y(s)=1_n. \tag{31}$$

[7] $\mathrm{Tr}A=\mathrm{trace}A$. In (27) and (28), $s=j\omega$, ω real.

[8] The existence of real polynomial matrices $X(s)$, $Y(s)$ satisfying (31) is guaranteed by the left-coprimeness of $A(s)$ and $B(s)$ (Appendix A). These polynomials need not be unique.

Then, 1) the closed-loop of Fig. 1 is asymptotically stable iff

$$R(s)=H(s)A(s) \tag{32}$$

where

$$H(s)=Y(s)+A_1(s)K(s)$$

and $K(s)$ is any $m\times n$ real rational matrix analytic in $\mathrm{Re}\,s\geq 0$ which satisfies the constraint

$$\det(X(s)-B_1(s)K(s))\not\equiv 0. \tag{33}$$

2) The stabilizing controller associated with a particular choice of admissible $K(s)$ possesses the transfer matrix

$$C=(Y+A_1K)(X-B_1K)^{-1}. \tag{34}$$

(From $AB_1=BA_1$ and (31) we deduce that

$$A(X-B_1K)+B(Y+A_1K)=1_n, \tag{35}$$

a useful relationship.)

In view of this lemma, the natural way to attack the problem of minimizing E is to vary over all $m\times n$ real rational matrices $K(s)$ analytic in $\mathrm{Re}\,s\geq 0$ which satisfy restriction (33). We are now ready to discuss in detail the assumptions which justify the entire optimization scheme.

1) Rate gyros and tachometers are examples of practical sensing devices which are not modeled as dynamical systems. Yet almost invariably sensors are stable and their transfer matrices are analytic in $\mathrm{Re}\,s\geq 0$. However, for our purposes it suffices to assume that the feedforward cascade is stable and that $F(s)$ is analytic on the finite $s=j\omega$-axis. In particular, $L(s)$ is analytic in $\mathrm{Re}\,s\geq 0$. If $P_o(s)$ and $F_o(s)$ represent distinct physical blocks, these blocks must be stable and both $P_o(s)$ and $F_o(s)$ are analytic in $\mathrm{Re}\,s\geq 0$. On the other hand, if $P_o(s)$ and $F_o(s)$ are merely part of the paper modeling, it is possible to relax the analyticity requirements.

2) A pole of $F(s)P(s)$ in $\mathrm{Re}\,s>0$ reveals true open-loop instability whereas a finite pole on the $s=j\omega$-axis is usually present because of intentional high-gain preconditioning. Recall that in the absence of load disturbance and measurement noise, a unity-feedback single-input–output loop enclosing a plant whose transfer function possesses a pole of order ν at $s=j\omega_o$ will track any causal linear combination of $e^{j\omega_o t}\cdot t^{k-1}$, $k=1\rightarrow\nu$, with zero steady-state error. The correct generalization to the multivariable case is easy to find. Setting $\boldsymbol{d}=\boldsymbol{O}$, $\boldsymbol{l}=\boldsymbol{O}$, and $\boldsymbol{m}=\boldsymbol{O}$ in (7) and (9) we obtain

$$\boldsymbol{y}=T\boldsymbol{u} \tag{36}$$

and

$$\boldsymbol{e}=(1_n-T)\boldsymbol{u}. \tag{37}$$

For a stable configuration $T(s)$ is analytic in $\mathrm{Re}\,s\geq 0$, but not necessarily proper (Appendix A). Nevertheless, there are cogent reasons for insisting on a dynamical closed-loop design. Consider the conditions that must prevail if the loop is to track any one-sided input of the form

$$e^{j\omega_o t}\cdot \sum_{k=1}^{\nu} t^{k-1} d_{\nu-k}, \tag{38}$$

d_k a constant vector, $k=1\to\nu$, with zero steady-state error. Clearly,[9]

$$e(s)=\sum_{k=1}^{\nu} \Gamma(k)\cdot \frac{1_n - T(s)}{(s-j\omega_o)^k}\cdot d_{\nu-k} \tag{39}$$

is the transform of a bounded time function which vanishes as $t\to\infty$ iff it is analytic in $\mathrm{Re}\, s \geqslant 0$ and $e(\infty) = O$. As necessary consequences, $T(s)$ must be proper and

$$(1_n - T(j\omega_o))d_o = O_n. \tag{40}$$

As is well known, (40) possesses a nontrivial solution d_o iff

$$\det(1_n - T(j\omega_o))=0. \tag{41}$$

Conversely, any ω_o satisfying (41) generates a generalized ramp-modulated sinusoid (38) capable of being tracked with zero steady-state error. These inputs and their finite linear combinations constitute a most important class of shape-deterministic information-bearing signals and play a key role in industrial applications. The set of all such possible "infinite gain" frequencies ω_o coincides with the totality of real solutions of (41).

In view of the arguments presented in the initial paragraph of 2), poles of $P(s)$ on the $s=j\omega$-axis enable the loop to track certain inputs with zero steady-state error. With unity feedback, $1_n - T = S$ and all plant poles in $\mathrm{Re}\, s \geqslant 0$, counted according to their McMillan degrees, are indeed zeros of $\det S$ (Appendix A). However, if $F \neq 1_n$, this perfect tracking capability is lost unless $F(s)$ is also conditioned suitably. Employing the easily derived formulas

$$S=(X-B_1K)A \tag{42}$$

and

$$FPR=1_n - S, \tag{43}$$

it is seen that

$$\begin{aligned} 1_n - T &= S+(F-1_n)PR \\ &= (X-B_1K+(F-1_n)P(Y+A_1K))A. \end{aligned} \tag{44}$$

Since $P(s)$ and $F(s)$ form an admissible pair,

$$\det{}^+ A = \psi_{FP}^+ = \psi_F^+ \psi_P^+ \tag{45}$$

and (44) shows that the purely imaginary zeros of $\psi_P^+(s)$ will surely be zeros of $\det(1_n - T)$ provided

$$Z=X-B_1K+(F-1_n)P(Y+A_1K) \tag{46}$$

is designed to be analytic on the finite $s=j\omega$-axis.[10] This analyticity precludes any possibility of cancellation and is achieved iff $(F-1_n)P$ is analytic on the finite $s=j\omega$-axis. The proof of this assertion is somewhat tedious but because of its great importance we supply it in detail.

In 4) it is shown that $P(s)A_1(s)$ is automatically analytic for all finite $s=j\omega$ and it follows from (46) that the same is true of $Z(s)$ iff $(F-1_n)PY$ is analytic for all finite $s=j\omega$. Multiplication of both sides of (31) on the left by $A^{-1}(s)$ yields

$$X+FPY=A^{-1} \tag{47}$$

since $FP=A^{-1}B$. By assumption, $F(s)$ is analytic for $s=j\omega$ whence, by admissibility and (47), any finite purely imaginary pole $s_o=j\omega_o$ of $P(s)$ of McMillan degree ν_o must also be a pole of $P(s)Y(s)$ of the same degree. Write

$$PY=A_2^{-1}B_2 \tag{48}$$

and

$$P=A_p^{-1}B_p \tag{49}$$

where the polynomial pairs (B_2, A_2) and (B_p, A_p) are both left-coprime. By hypothesis,

$$(F-1_n)PY=(F-1_n)A_2^{-1}B_2 \tag{50}$$

is analytic for finite $s=j\omega$. But (X_2 and Y_2 are polynomial)

$$A_2X_2+B_2Y_2=1_n$$

implies

$$(F-1_n)X_2+(F-1_n)A_2^{-1}B_2Y_2=(F-1_n)A_2^{-1} \tag{51}$$

and $(F-1_n)A_2^{-1}$ is also analytic on the finite $s=j\omega$-axis. Substituting (49) into (48) we get

$$B_pY=A_pA_2^{-1}B_2 \tag{52}$$

and by an argument similar to the above, the analyticity of B_pY for finite $s=j\omega$ forces that of $A_pA_2^{-1}$. In other words,

$$A_2^{-1}=A_p^{-1}\mathcal{P} \tag{53}$$

where $\mathcal{P}(s)$ is analytic on the finite $s=j\omega$-axis. However, because the finite $j\omega$-axis poles of PY agree with those of P, McMillan degrees included, $\det A_2(s)$ and $\det A_p(s)$ possess the same $j\omega$-axis zeros, multiplicities included. Thus,

$$\det \mathcal{P}(s)\neq 0, \qquad s=j\omega \tag{54}$$

and $\mathcal{P}^{-1}(s)$ and $(F-1_n)A_2^{-1}$ are, therefore, both analytic on the finite $s=j\omega$-axis. It is now clear that

$$(F-1_n)P=(F-1_n)A_p^{-1}B_p=(F-1_n)A_2^{-1}\mathcal{P}^{-1}B_p \tag{55}$$

is analytic for all finite $s=j\omega$. Q.E.D.

This constraint is of decisive importance and replaces the usual unity-feedback desideratum $F=1_n$ which due to

[9] $\Gamma(k)$ is the Gamma function of argument k.

[10] The multiplicity of any zero of $\psi_P(s)$ equals its McMillan degree as a pole of $P(s)$. Hence, the purely imaginary zeros of $\psi_P^+(s)$ constitute the totality of finite $j\omega$-axis poles of $P(s)$.

ever-present delays and transducer inertia is never realizable. In the actual design the constraint is met by a correct choice of preequalizer $F_e(s)$ and the two degrees of freedom inherent in the problem are exploited to maximum advantage.

3) In process control the recovery of steady state under load disturbance $d(s)$ is a requirement of paramount importance. From (9), with u, m, and l set equal to O,

$$e=(P_0-PRP_d)d \tag{56}$$

$$=(1_n-PRF)P_o d-(PRL)d$$

$$=S_1P_o d-TLd=(S_1P_o-TL)d \tag{57}$$

where

$$S_1=(1_n+PCF)^{-1}. \tag{58}$$

Again as in 2), the shape-deterministic component of $d(s)$ is envisaged to be the transform of a sum of generalized ramp-modulated sinusoids and for bounded zero steady-state error, $e(s)$ must vanish at infinity and be analytic in $\mathrm{Re}\, s \geqslant 0$. Assuming S_1P_o—TL proper and $S_1P_o d$ and Ld analytic in $\mathrm{Re}\, s \geqslant 0$ is evidently sufficient. Invoking closed-loop stability it can be shown (Appendix A) that

$$S_1(s)=\mathcal{P}_1(s)A_p(s), \tag{59}$$

$\mathcal{P}_1(s)$ analytic in $\mathrm{Re}\, s \geqslant 0$. Hence, $S_1P_o d$ analytic in $\mathrm{Re}\, s \geqslant 0$ can be replaced by $A_pP_o d$ analytic in $\mathrm{Re}\, s \geqslant 0$ and once again the $j\omega$-axis poles of the plant are brought into evidence through $A_p(s)$. Observe that the $j\omega$-axis analyticity of $A_p(P_oG_dP_{o*})A_{p*}$, LG_dL_*, and $LG_dP_{o*}A_{p*}$ is a corollary.

4) Let

$$P=B_{p1}A_{p1}^{-1} \tag{60}$$

be any right-coprime decomposition of $P(s)$. Then,

$$B_1A_1^{-1}=FP=FB_{p1}A_{p1}^{-1} \tag{61}$$

and it follows from the assumed analyticity of $F(s)$ on the finite $j\omega$-axis that

$$B_1A_1^{-1}A_{p1}=FB_{p1} \tag{62}$$

is also analytic for all finite $s=j\omega$. Hence, reasoning as in 2),

$$A_{p1}(s)=A_1(s)\mathcal{P}_2(s), \tag{63}$$

$\mathcal{P}_2(s)$ analytic and nonsingular for all finite $s=j\omega$. Consequently,

$$PA_1=B_{p1}A_{p1}^{-1}A_1=B_{p1}\mathcal{P}_2^{-1} \tag{64}$$

is analytic for all finite $s=j\omega$. Q.E.D.

5) From (8), (9) and the definitions (28), (27),[11]

$$2\pi jE_s=\mathrm{Tr}\int_{-j\infty}^{j\infty} QR(G_u+G_{ml}+P_dG_dP_{d*})R_*\,ds \tag{65}$$

and

$$2\pi jE_t=\mathrm{Tr}\int_{-j\infty}^{j\infty}((1_n-PR)G_u(1_n-PR)_*+(PR)G_{ml}(PR)_*$$

$$+(P_o-PRP_d)G_d(P_o-PRP_d)_*)\,ds \tag{66}$$

where

$$Q(s)=P_{s*}(s)P_s(s) \tag{67}$$

and

$$G_{ml}=F_oG_mF_{o*}+L_oG_lL_{o*}. \tag{68}$$

In terms of

$$G=G_u+G_{ml}+P_dG_dP_{d*} \tag{69}$$

and H,

$$2\pi jE_s=\mathrm{Tr}\int_{-j\infty}^{j\infty} QH(AGA_*)H_*\,ds. \tag{70}$$

The nonnegative parahermetian matrices $G_{ml}(s)$ and $Q(s)$ are assumed to be free of finite $j\omega$-axis poles. (There is no physical reason for doing otherwise.) Since a stable closed-loop design forces $H(s)$ to be analytic in $\mathrm{Re}\, s \geqslant 0$ (Lemma 3), the integrand of (70) will be devoid of finite $j\omega$-axis poles if AGA_* is analytic on the $j\omega$-axis. Consider first AG_uA_* and the equality

$$P+(F-1_n)P=FP=A^{-1}B. \tag{71}$$

Write, as before, $P=A_p^{-1}B_p$ where A_p, B_p is a left-coprime pair and substitute into (71). Bearing in mind that $(F-1_n)P$ is assumed to be analytic on the finite $j\omega$-axis, familiar reasoning[12] permits us to conclude that $A(s)=\mathcal{P}_3(s)A_p(s)$, $\mathcal{P}_3(s)$ analytic and nonsingular for all finite $s=j\omega$. Hence, the analyticity of $A_pG_uA_{p*}$ on the finite $s=j\omega$-axis guarantees that of AG_uA_*. This $j\omega$-axis analyticity is in accord with our previous reasoning. Namely, the deterministic part of $u(s)$ is the transform of a sum of generalized ramp-modulated sinusoids whose resonant frequencies coincide with the $j\omega$-axis poles of $P(s)$. These poles *and only these poles* should appear as $j\omega$-axis poles of $G_u(s)$. But these poles are also imbedded in the Smith canonic structure of $A_p(s)$ and the $j\omega$-axis analyticity of $A_pG_uA_{p*}$ is merely a succinct formulation of one design objective. Regarding $A(P_dG_dP_{d*})A_*$, its $j\omega$-axis analyticity follows from the assumptions in 3) and the readily deduced relation

$$AF=\mathcal{P}_4A_p, \tag{72}$$

$\mathcal{P}_4(s)$ analytic for $s=j\omega$.

Let us now examine the $j\omega$-axis analyticity of the individual terms making up the integrand of (66). First,

[11] All random processes are assumed to be independent. Note $\mathrm{Tr}(L_1L_2)=\mathrm{Tr}(L_2L_1)$.

[12] Both the admissibility of the pair $F(s)$, $P(s)$ and the $j\omega$-axis analyticity of $F(s)$ must be invoked.

$$(1_n - PR)G_u(1_n - PR)_*$$

$$= (1_n - PR)A^{-1}(AG_uA_*)A_*^{-1}(1_n - PR)_* \quad (73)$$

$$= (A^{-1} - PH)(AG_uA_*)(A^{-1} - PH)_* \quad (74)$$

and it suffices to prove that $A^{-1} - PH$ is analytic on the $j\omega$-axis. This is clear because K, PA_1, and $(F - 1_n)P$ are analytic on $j\omega$, $A^{-1} = X + FPY$ and

$$A^{-1} - PH = A^{-1} - P(Y + A_1K) = A^{-1} - PY - PA_1K$$

$$= X + (F - 1_n)PY - (PA_1)K. \quad (75)$$

The $j\omega$-axis analyticity of the second and third terms in (66) follows from that of $G_{ml}(s)$ and 3). In order to exclude meaningless, but mathematically allowed physical degeneracies, we must also impose the restriction

$$\det(AGA_*)\cdot\det(A_{1*}(P_*P + kQ)A_1) \neq 0, \qquad s = j\omega. \quad (76)$$

This inequality is essential (Appendix B). It is also shown in Appendix B[13] that

$$G_u(j\omega) \leq O(1/\omega^2) \quad (77)$$

and

$$P_oG_dP_{o*} \leq O(1/\omega^2) \quad (78)$$

are suggested naturally by the requirement of finite cost. Furthermore, if

$$(P_*P + kQ)G \approx \omega^{2\mu}1_m, \quad (79)$$

$$G_d(j\omega) \approx \omega^{-2i}1 \quad (80)$$

and

$$P(s) = O(s^\nu), \quad (81)$$

the inequalities

$$\mu \geq \nu - 1 \quad (82)$$

and

$$i \leq 1 \quad (83)$$

assure the properness of T and $S_1P_o - TL$, respectively. In most applications load disturbance contains a step-component and (80) is satisfied with $i = 1$. (The integers μ and ν can be negative.)

6) In general, the effects of parameter uncertainty on P and F are more pronounced as ω increases and closed-loop sensitivity is an important consideration. Let F, PC, S, S_1, and T undergo changes from $(F)_a$, $(PC)_a$, $(S)_a$, $(S_1)_a$, and $(T)_a$ to $(F)_b$, $(PC)_b$, $(S)_b$, $(S_1)_b$, and $(T)_b$ at a fixed ω. Noting that $T = PCS = S_1PC$,

$$(T)_b - (T)_a = (S_1)_b(PC)_b - (PC)_a(S)_a$$

$$= (S_1)_b[(PC)_b(1_n + (F)_a(PC)_a) - (1_n + (PC)_b(F)_b)(PC)_a](S_a)$$

$$= (S_1)_b[(PC)_b - (PC)_a - (PC)_b \cdot ((F)_b - (F)_a)(PC)_a](S)_a.$$

Thus

$$\Delta T = (S_1)_b\cdot\Delta(PC)\cdot(S)_a - (T)_b\cdot(\Delta F)\cdot(T)_a, \quad (84)$$

an exact formula valid for arbitrary increments ΔF, $\Delta(PC)$. To first order,

$$\delta T = S_1\cdot\delta(PC)\cdot S - T(\delta F)T \quad (85)$$

and we recover the classical differential version of (84). If $\det(FPC) \not\equiv 0$, (85) may also be rewritten as

$$T^{-1}\cdot\delta T = (PC)^{-1}\cdot\delta(PC)\cdot S - (\delta F)\cdot F^{-1}\cdot(1_n - S). \quad (86)$$

In words, at frequency ω,

left percent change in T = (left percent change in PC) $\cdot S(j\omega)$ − (right percent change in F)$\cdot(1_n - S(j\omega))$ (87)

and again $S(j\omega)$ and $1_n - S(j\omega)$ emerge as the pertinent matrix gain functions for the forward and return links, respectively. Clearly then, to combat the adverse effects of high-frequency uncertainty in the modeling of $F(j\omega)$ and the plant matrix $P(j\omega)$, it is sound engineering practice to insist on a design with $S(j\omega)$ proper and equal to 1_n at $\omega = \infty$. This feature is easily introduced into the analytic framework by means of the constraint

$$O(P) + O(F) \leq \mu \quad (88)$$

which simultaneously ensures that the Wiener–Hopf controller defined by (34) makes sense and is proper if $\mu \geq -1$ (Appendix B). Furthermore, (88) also forces $S_1 = (1_n + PCF)^{-1} \to 1_n$ as $\omega \to \infty$ which is consistent with the engineering interpretation of the right-percentage formula for T,

$$(\delta T)\cdot T^{-1} = S_1\cdot\delta(PC)\cdot(PC)^{-1} - (1_n - S_1)\cdot F^{-1}\cdot\delta F. \quad (89)$$

We should like to emphasize that the cost E already imposes a weighted penalty on the choice of forward and return-link sensitivities through the (somewhat disguised) presence of S and $1_n - S$ in the error $\boldsymbol{e}$. (Equation (16) for $F = 1_n$ illustrates the point.) All this is in accord with a basic tenet of the classical theory which states that good immunity to load disturbance and good forward-link sensitivity usually go hand in hand.

III. The Optimal Controller

By way of recapitulation we shall collate the major working assumptions.

[13] $A(s) \leq O(s^r)$ means that no entry in $A(s)$ grows faster than s^r as $s \to \infty$. The order of $A(s)$ equals r, i.e., $A(s) = O(s^r)$ if 1) $A(s) \leq O(s^r)$ and 2) *at least one* entry grows exactly like s^r. For $A(s)$ square, $A(s) \approx s^r 1$ abbreviates

$$\lim_{s\to\infty} s^{-r}A(s) = A_\infty,$$

A_∞ *a constant nonsingular matrix*. Note $A(s) \approx s^r 1$ implies $A(s) = O(s^r)$, but not conversely.

Assumption 1: The plant and feedback compensator form an admissible pair (Definition 1), the feedforward compensator is asymptotically stable and the respective transfer matrices, $P(s)$, $F(s)$, $L(s)$ are prescribed in advance. (Note, in particular, that $L(s)$ is analytic in Re $s \geqslant 0$.)

Assumption 2: $P_o(s)$, $F_o(s)$, $L_o(s)$, $Q(s) = P_{s*}(s)P_s(s)$ and the spectral densities $G_u(s)$, $G_d(s)$, $G_m(s)$, $G_l(s)$ are given. Any block outside the loop which represents an actual physical component must be asymptotically stable and its transfer matrix is therefore analytic in Re $s \geqslant 0$. (On the other hand, if any such block is merely part of the paper modeling the analyticity requirement can be relaxed.) The input signal, load disturbance and measurement noises are stochastically independent.

Assumption 3: Let $P = A_p^{-1}B_p$ be any left-coprime factorization of $P(s)$ and let

$$G_{ml} = F_o G_m F_{o*} + L_o G_l L_{o*}. \tag{90}$$

The matrices Q, F, $(F - 1_n)P$, $A_p G_u A_{p*}$, $A_p(P_o G_d P_{o*})A_{p*}$, $LG_d L_*$, and G_{ml} are analytic on the finite $s = j\omega$-axis.

Assumption 4: Let k be any positive constant,

$$G = G_u + P_d G_d P_{d*} + G_{ml} \tag{91}$$

and

$$P_d = FP_o + L. \tag{92}$$

The matrices AGA_* and $A_{1*}(P_*P + kQ)A_1$ are nonsingular on the finite $s = j\omega$-axis. (Their $j\omega$-analyticity is ensured by the above assumptions.)

Assumption 5: The data satisfy the order relations[14]

$$G_u \leqslant O(1/s^2); \qquad P_o G_d P_{o*} \leqslant O(1/s^2)$$

$$G_d \approx s^{-2i}1; \qquad P = O(s^{\nu}) \tag{93}$$

$$O(P) + O(F) \leqslant \mu \tag{94}$$

and

$$(P_*P + kQ)G \approx s^{2\mu}1_m \tag{95}$$

where

$$i \leqslant 1; \qquad \mu \geqslant \max(\nu - 1, -1). \tag{96}$$

We are now in a position to state the master result.

Theorem 1 (Appendix B): Under Assumptions 1–5 the optimal design is carried out in the following manner.

1) Construct two square real rational matrices $\Lambda(s)$, $\Omega(s)$ *analytic together with their inverses in Res* $\geqslant 0$ such that

$$A_{1*}(P_*P + kQ)A_1 = \Lambda_*\Lambda \tag{97}$$

and

$$AGA_* = \Omega\Omega_*. \tag{98}$$

[14]Refer to footnote 13 for an explanation of the notation.

2) Let

$$I = A_{1*}P_*(G_u + P_o G_d P_{d*})A_* \tag{99}$$

and choose any two real polynomial matrices $X(s)$, $Y(s)$ such that

$$A(s)X(s) + B(s)Y(s) = 1_n. \tag{100}$$

3) The transfer matrix of the optimal controller is given by

$$C = (Y + A_1 K)(X - B_1 K)^{-1} \tag{101}$$

where[15]

$$K = \Lambda^{-1}\left(\{\Lambda_*^{-1} I \Omega_*^{-1}\}_+ + \{\Lambda A_1^{-1} Y \Omega\}_-\right)\Omega^{-1} - A_1^{-1}Y; \tag{102}$$

or, in a form more suitable for numerical implementation,

$$C = H_o(A^{-1}\Omega - FPH_o)^{-1}, \tag{103}$$

$$H_o = A_1\Lambda^{-1}\left(\{\Lambda_*^{-1} I \Omega_*^{-1}\}_+ + \{\Lambda A_1^{-1} Y \Omega\}_-\right). \tag{104}$$

The (nonhidden) poles of the optimally compensated loop are *precisely* the *zeros* of the strict Hurwitz polynomial

$$\theta(s) = \frac{\psi_F^-(s)\psi_P^-(s)}{\psi_{FP}^-(s)} \tag{105}$$

plus the finite *poles* of $K(s)$, each of these poles counted according to its McMillan degree. Both $H_o(s)$ and $K(s)$ are analytic in Re $s \geqslant 0$ and the *distinct* finite poles of $K(s)$ are included in those of the primary data

$$FP,\ (A_{1*}(P_*P + kQ)A_1)^{-1},\ (AGA_*)^{-1},$$
$$A_{1*}P_*(G_u + P_o G_d P_{d*})A_* \tag{106}$$

located in Re$s < 0$. Thus stability margin is ascertainable in advance.

Several comments are in order. First, there exist effective computer algorithms for the realization of the canonic factors $\Lambda(s)$, $\Omega(s)$ [8], [9]. Second, the combination of plant and feedback compensator is said to be nonminimum-phase if the polynomial matrix $B(s)$ appearing in the left-coprime decomposition $F(s)P(s) = A^{-1}(s)B(s)$ has rank *less* than row-rank for some finite $s = s_o$ in Re$s > 0$.[16] As is shown in Appendix A, any such s_o is also a zero of $\det(1_n - S)$. Now for any choice of nonzero constant $\eta \neq 1$ the zeros of the stability polynomial

$$\varphi(s,\eta) = \psi_F\psi_P\psi_C \cdot \det(1_n + \eta FPC)$$
$$= \psi_F\psi_P\psi_C \cdot \det(1_n + FPC + (\eta - 1)FPC) \tag{107}$$

[15]In the partial fraction expansion $\{\ \}_\infty + \{\ \}_+ + \{\ \}_-$ of any rational matrix, $\{\ \}_\infty$ is the part associated with the pole at infinity and $\{\ \}_+$, $\{\ \}_-$ the parts associated with all the finite poles in Re$s < 0$ and Re$s \geqslant 0$, respectively. Clearly, $\{\ \}_+$ is analytic in Re$s \geqslant 0$, $\{\ \}_-$ in Re$s < 0$ and both vanish at infinity.

[16]Although this definition is the most natural generalization of the one accepted in the scalar case, other definitions also make physical sense when examined in the context of the standard control problem [3].

coincide with those of[17]

$$\varphi(s,1)\cdot\det\left(\frac{1}{\eta-1}1_n+1_n-S\right). \tag{108}$$

Thus, by continuity, at least one of these zeros tends to $s=s_o$ as $|\eta|\to\infty$ and all attempts to decrease transient error to zero by a simple constant-gain modification of some already predetermined controller $C(s)$ must, therefore, fail.

Corollary 1: Suppose $F(s)P(s)$ is analytic in $\mathrm{Re}\,s\geqslant 0$. Then

$$C=H_o(\Omega_r-FPH_o)^{-1} \tag{109}$$

where

$$H_o=\Lambda_r^{-1}\{\Lambda_{r*}^{-1}I_r\Omega_{r*}^{-1}\}_+, \tag{110}$$

$$(P_*P+kQ)=\Lambda_{r*}\Lambda_r, \tag{111}$$

$$G=\Omega_r\Omega_{r*}, \tag{112}$$

$$I_r=P_*(G_u+P_oG_dP_{d*}) \tag{113}$$

and $\Lambda_r(s)$, $\Omega_r(s)$ are square, real rational matrices analytic together with their inverses in $\mathrm{Re}\,s\geqslant 0$. ($\Lambda_r,\Omega_r,I_r$ are "reduced" quantities.)

Proof: The analyticity of $F(s)P(s)$ in $\mathrm{Re}\,s\geqslant 0$ implies $\det^+A(s)=\det^+A_1(s)=\psi_{FP}^+(s)=1$. Thus $A\Omega_r=\Omega$, $\Lambda_rA_1=\Lambda$, $\{\Lambda A_1^{-1}Y\Omega\}_-=O$ and the rest follows by direct substitution. Q.E.D.

Under the conditions of the corollary, the feedback sensor and plant are asymptotically stable and the resulting simplification, as evidenced in (109)–(113), is striking. Note in particular that the polynomial factors $A(s)$, $A_1(s)$, and $Y(s)$ are no longer needed!

The general formula (103) for the optimal controller transfer matrix is excellently conditioned. In fact, it is easily shown (Appendix B) that in exact arithmetic

$$H_o=(Y+A_1K)\Omega \tag{114}$$

and

$$A^{-1}\Omega-FPH_o=(X-B_1K)\Omega. \tag{115}$$

Thus, in exact arithmetic both H_o and $A^{-1}\Omega-FPH$ are analytic in $\mathrm{Re}\,s\geqslant 0$. Consequently, if the numerical scheme employed to compute (103) automatically ensures the closed right half-plane analyticity of H_o and $A^{-1}\Omega-FPH_o$, the corresponding exact arithmetic K is such that A_1K and B_1K are also analytic in $\mathrm{Re}\,s\geqslant 0$.[18] But then,

$$X_1A_1K+Y_1B_1K=K \tag{116}$$

reveals that $K(s)$ is analytic in $\mathrm{Re}\,s\geqslant 0$ *and the closed-loop structure realized with the computed C(s) is asymptotically stable.*

[17] $1_n-S=(1_n+FPC)^{-1}FPC$.
[18] $\Omega^{-1}(s)$ is analytic in $\mathrm{Re}\,s\geqslant 0$.

Corollary 2: Let[19]

$$a=\Lambda_*^{-1}I\Omega_*^{-1}, \tag{117}$$

$$b=\Lambda A_1^{-1}Y\Omega, \tag{118}$$

$$c=\{a-b\}_- \tag{119}$$

and

$$\rho=G_u+P_oG_dP_{o*}-a_*a+c_*c. \tag{120}$$

Then, under the assumption[20]

$$a_\infty(s)=O, \tag{121}$$

the *minimum* cost $E_{\min}$ is given by

$$2\pi jE_{\min}=\mathrm{Tr}\int_{-j\infty}^{j\infty}\rho(s)\,ds. \tag{122}$$

In particular, if $F(s)P(s)$ is analytic in $\mathrm{Re}\,s\geqslant 0$ (the stable case) we can choose

$$\rho=G_u+P_oG_dP_{o*}-a_{+*}a_+. \tag{123}$$

Proof: From (B2), Appendix B,

$$a-b-\Lambda K\Omega=\Lambda_*^{-1}\Delta_*\Omega_*^{-1} \tag{124}$$

and it follows immediately that $c=\{a-b\}_-$ is analytic in $\mathrm{Re}\,s\leqslant 0$ (which includes the $j\omega$-axis). Now by combining (B19), (B22), and (B24) and exploiting the closed left half-plane analyticity of $c(s)$ with the aid of Cauchy's theorem, we easily reach (120)+(122). If $F(s)P(s)$ is analytic in $\mathrm{Re}\,s\geqslant 0$, $b_-(s)\equiv O$ and

$$\rho=G_u+P_oG_dP_{o*}-a_{+*}a_+-a_{+*}a_--a_{-*}a_+.$$

However, since $a_{+*}a_-$ is $O(1/\omega^2)$ and analytic in $\mathrm{Re}\,s\leqslant 0$, contour integration yields

$$\mathrm{Tr}\int_{-j\infty}^{j\infty}a_{+*}a_-\,ds=\mathrm{Tr}\int_{-j\infty}^{j\infty}a_{-*}a_+\,ds=0.$$

Thus

$$2\pi jE_{\min}=\mathrm{Tr}\int_{-j\infty}^{j\infty}(G_u+P_oG_dP_{o*}-a_{+*}a_+)\,ds.$$

Q.E.D.

In the stable case $G_u+P_oG_dP_{o*}$ and a_*a are both individually $j\omega$-analytic, but, in general, it is only the combination $G_u+P_oG_dP_{o*}-a_*a$ which is devoid of purely imaginary poles.

Corollary 3:[21] Let $P(s)$ be square and analytic *together with its inverse* in $\mathrm{Re}\,s\geqslant 0$, let $F=1$ (unity feedback), let $k=0$ (no saturation constraint) and assume feedforward compensation is not employed (L and G_l are zero). Then, if G and $G_u+P_oG_dP_{o*}$ are diagonal matrices, the optimal controller $C(s)$ satisfies the noninteraction condition

[19] Here $a_+\equiv\{a\}_+, b_-\equiv\{b\}_-, a_\infty\equiv\{a\}_\infty$, etc.
[20] Quite usual.
[21] Suggested some years ago by I. M. Horowitz [19].

$$P(s)C(s) = \text{diagonal matrix}. \tag{125}$$

Proof: Clearly, from (109)–(113) and the stipulated assumptions, it follows that $\Lambda_r = P$, Ω_r is diagonal and

$$I_r = \Lambda_{r*}(G_u + P_o G_d P_{o*}). \tag{126}$$

Thus

$$\Lambda_{r*}^{-1} I_r \Omega_{r*}^{-1} = (G_u + P_o G_d P_{o*})\Omega_{r*}^{-1}, \tag{127}$$

$$PH_o = \Lambda_r H_o = \{(G_u + P_o G_d P_{o*})\Omega_{r*}^{-1}\}_+ \tag{128}$$

and

$$PC = PH_o(\Omega_r - PH_o)^{-1} \tag{129}$$

are also diagonal. Q.E.D.

Let us mention some obvious generalizations. First, suppose the integrand in (27) is also weighted so that

$$2\pi j E_t = \int_{-j\infty}^{j\infty} \langle e_*(s) Q_t e(s)\rangle ds$$
$$= \text{Tr}\int_{-j\infty}^{j\infty} Q_t \langle e(s) e_*(s)\rangle ds, \tag{130}$$

Q_t an arbitrary real, constant, symmetric nonnegative-definite matrix. Then we simply make the substitutions

$$(P_*P + kQ) \rightarrow P_* Q_t P + kQ, \tag{130a}$$

$$I \rightarrow A_{1*} P_* Q_t (G_u + P_o G_d P_{d*}) A_* \tag{130b}$$

and

$$\rho \rightarrow Q_t(G_u + P_o G_d P_{o*}) - a_* a + c_* c \tag{130c}$$

and continue to use the same formulas as before. In particular, the canonic factor $\Lambda(s)$ is found from the decomposition

$$\Lambda_* \Lambda = A_{1*}(P_* Q_t P + kQ)A_1 \tag{130d}$$

and (76) is altered to read

$$\det(A_{1*}(P_* Q_t P + kQ)A_1)\cdot\det(AGA_*) \neq 0, \qquad s = j\omega. \tag{130e}$$

Second, we have assumed all processes to be zero-mean. This is always true for the measurement noises $\boldsymbol{m}$ and $\boldsymbol{l}$ and almost invariably true for $\boldsymbol{u}$. In any case, if at least one of the means $\langle \boldsymbol{u}\rangle$ or $\langle \boldsymbol{d}\rangle$ vanishes, all formulas remain intact. Otherwise, $\boldsymbol{e}$ in the integrand of E_t and $\boldsymbol{r}$ in the integrand of E_s must be replaced by $\boldsymbol{e} - \langle \boldsymbol{e}\rangle$ and $\boldsymbol{r} - \langle \boldsymbol{r}\rangle$, respectively. This then entails identifying G_u with the spectral density of $\boldsymbol{u} - \langle \boldsymbol{u}\rangle$ and G_d with that of $\boldsymbol{d} - \langle \boldsymbol{d}\rangle$. The optimal controller now minimizes the steady-state rms error fluctuation subject to a steady-state rms constraint on the fluctuation of the plant input.

IV. Discussion and Conclusions

It would be superfluous to list the numerical problems which beset algorithms involving the factorization and manipulation of rational matrices. Nevertheless, work now in progress leads us to believe that a feasible computer implementation of the optimal controller is within reach. The availability of such an algorithm will undoubtedly suggest related simpler suboptimal strategies.

Although many of the physical ideas propounded in this paper have already been touched upon in the literature by several authors [14], [15], the various attempts to evolve an inclusive frequency-domain least-squares approach to multivariable controller design have by and large been unsuccessful because of an imprecise grasp of the full implications of closed-loop stability and a failure to recognize at the outset the need to condition the cost functional in a manner compatible with the physical constraints introduced by the given data $F(s)$, $P(s)$. As this paper clearly shows, within a linear framework a correct treatment depends essentially on an in-depth analytic characterization of all those engineering factors which figure meaningfully in any practical design scheme and a successful parametric solution of the concomitant "model-matching" problems associated with the matrix Wiener–Hopf equation.[22]

In LQG [17], [18] the objective is to optimally reset the state of the plant to a fixed known equalibrium state in the face of both Gaussian white background noise and Gaussian white measurement noise given the measured output. However, in the problem solved in this paper the "set point" excursion $\boldsymbol{u}(t)$ is both variable and stochastic and the functions of time to be reset are subsumed as components of an output vector $\boldsymbol{y}(t)$. The task of the optimal controller is to optimally reset $\boldsymbol{y}(t)$ to a new level dictated by a shape-deterministic or second-order stationary $\boldsymbol{u}(t)$. This must be accomplished despite the presence of shape deterministic or second-order stationary load disturbance, measurement noise, nonideal sensor dynamics, a zero steady-state error requirement, etc. Accordingly, that part of the cost E_t reflecting loop accuracy has been imposed directly on $\boldsymbol{u}(t) - \boldsymbol{y}(t)$ and does not necessarily involve all the state variables. Nevertheless, these other variables are kept within bounds by an appropriate weighting of the saturation constraint E_s. It appears, therefore, that underlying our design philosophy is the assumption that any variable which is to be reset to a time-varying stochastic set-point must be available as a measured output. This attitude is of course consistent with the classical viewpoint and its true merit can only be judged after sufficient experience with applications of the optimal controller formula (103) to problems of industrial importance has been obtained.[23]

In LQG the optimal regulator structure is prescribed in advance in a manner which identifies the separate roles played by state estimation, Kalman-weighting (via the

[22] The Ph.D. dissertation of H. A. Jabr [16] contains some nontrivial fully worked numerical examples and also includes a transfer-matrix description of the stirred-tank chemical reactor linearized about its unstable equilibrium state.

[23] One possible way to enlarge LQG to encompass a special variable set-point problem is described in [18] but it appears to us that the proposed method can be very sensitive to parameter variations and we prefer the infinite-gain plant preconditioning scheme discussed in 2), Section II.

innovative input) and noise-free optimal deterministic state feedback [17]. However, it is easily shown that the entire configuration is simply a special case of Fig. 1 in which $u=O$, feedforward compensation is absent, $F(s)=1$, $F_o(s)=1$, $G_d(s)=\Sigma$ and $G_m(s)=\Theta$; Σ is real, constant, symmetric nonnegative-definite and Θ is real, constant, symmetric positive-definite. In the frequency-domain approach the objective is to find the optimal controller transfer matrix $C(s)$ and its mode of realization is based on other considerations. This shifting of the emphasis to $C(s)$ is really a restatement of the problem in invariantive fashion and has some distinct advantages.

For example, consider a time-invariant single-input, single-output plant with the constant-coefficient state-variable description

$$\dot{x}=\begin{bmatrix}0 & 1\\ 0 & 2\end{bmatrix}x+\begin{bmatrix}0\\ 1\end{bmatrix}u_o+\xi, \tag{131}$$

$$y=\begin{bmatrix}-1 & 1\end{bmatrix}x; \qquad z=y+\theta. \tag{132}$$

Here $u_o(t)$ is the plant input, $y(t)$ the output, and $z(t)$ the measured output. The noise processes $\xi(t)$ and $\theta(t)$ are both white Gaussian with respective covariance matrices

$$\langle\xi(t)\xi(\tau)\rangle=\begin{bmatrix}\sigma^2 & 0\\ 0 & 0\end{bmatrix}\cdot\delta(t-\tau)=\Sigma\cdot\delta(t-\tau) \tag{133}$$

and

$$\langle\theta(t)\theta(\tau)\rangle=\sigma_m^2\cdot\delta(t-\tau)=\Theta\cdot\delta(t-\tau). \tag{134}$$

Clearly,

$$P(s)=\frac{s-1}{s(s-2)}; \qquad P_o(s)=\left[-\frac{1}{s}\,\middle|\,\frac{s-1}{s(s-2)}\right] \tag{135}$$

$$G_d(s)=\Sigma; \qquad G_m(s)=\sigma_m^2. \tag{136}$$

Subject to the choices $k>0$ and

$$Q_o=\begin{bmatrix}1 & -1\\ -1 & 1\end{bmatrix}, \tag{137}$$

it is a consequence of the LQG solution that the cost functional

$$J=\lim_{T\to\infty}\frac{1}{2T}\int_{-T}^{T}(x'Q_ox+ku_o^2)\,dt \tag{138}$$

is minimized by choosing the control law

$$u_o=-\frac{1}{k}\left[\sqrt{k}\;|2k+\sqrt{4k^2+2k\sqrt{k}+k}\;\right]\hat{x} \tag{139}$$

and placing the Kalman column-vector gain

$$W=\begin{bmatrix}\dfrac{3\sigma}{\sigma_m}+4\\ \dfrac{4\sigma}{\sigma_m}+8\end{bmatrix} \tag{140}$$

at the innovative input.[24] The observer of course provides the state estimate $\hat{x}$. Since $x'Q_ox=y^2$, this problem also falls within the scope of our solution (with $u=O$) and we obtain[25]

$$C(s)=\frac{[(3+4c_1+4c_2+8c_1c_2)\sigma+4(1+2c_1+2c_2+4c_1c_2)\sigma_m]s-2\sigma}{\sigma_mc_1c_2s^2+[c_1c_2\sigma+(c_1+c_2+4c_1c_2)\sigma_m]s-[(3+3c_1+3c_2+4c_1c_2)\sigma+(3+4c_1+4c_2+8c_1c_2)\sigma_m]} \tag{141}$$

where

$$\sqrt{2}\,c_{1,2}=\sqrt{(4k+1)\pm\sqrt{(4k+1)^2-4k}}\;. \tag{142}$$

Now in view of (139) and (140), the LQG design fails if either $k=0$ or $\sigma_m=0$. Nevertheless, $C(s)$ is perfectly well-defined and, in fact, setting $k=0$ in (141) yields

$$C(s)=\frac{\left(\dfrac{7\sigma}{\sigma_m}+12\right)s-\dfrac{2\sigma}{\sigma_m}}{s-\left(\dfrac{6\sigma}{\sigma_m}+7\right)}; \qquad k=0. \tag{143}$$

If, in addition, σ_m also equals zero,

$$C(s)=\frac{1}{3}-\frac{7}{6}s; \qquad k=\sigma_m=0. \tag{144}$$

Thus, the LQG technique misses these extremely simple practical controllers. On the other hand, if $k\sigma_m\neq 0$, it is straightforward to show that (139) and (140) lead to an overall controller with transfer function given precisely by (141). It is interesting to note that in this limiting case ($k=\sigma_m=0$),

$$O(P)+O(F)=-1>-2=\mu$$

and since inequality (94) is violated it is not surprising to find that $C(s)$ is improper and $S(\infty)=-6\neq 1$.

The other Kalman gain alluded to in the previous footnote is given by

$$\hat{W}=\begin{bmatrix}-\dfrac{\sigma}{\sigma_m}\\ 0\end{bmatrix} \tag{145}$$

and its associated cost

$$\hat{J}=\sigma^2+\sigma\sigma_m+\sigma^2\sqrt{4k+2\sqrt{k}+1} \tag{146}$$

is obviously *less* than the cost

[24]Actually, there is a second Kalman column-vector gain which yields a smaller cost than (139) but leads to an unstable closed-loop design. We address this point later on.

[25]All details are omitted.

$$J=\left(9+9\sqrt{4k+2\sqrt{k}+1}+24\sqrt{k}+32k\right.$$
$$+16\sqrt{4k^2+2k\sqrt{k}+k}\,)\sigma^2+$$
$$+\left(25+24\sqrt{4k+2\sqrt{k}+1}+80\sqrt{k}+128k\right.$$
$$+64\sqrt{4k^2+2k\sqrt{k}+k}\,)\sigma\sigma_m+$$
$$+\left(20+16\sqrt{4k+2\sqrt{k}+1}+64\sqrt{k}+128k\right.$$
$$\left.+64\sqrt{4k^2+2k\sqrt{k}+k}\,\right)\sigma_m^2 \qquad (147)$$

induced by W, (140). The two gains W and $\hat{W}$ are generated by the two distinct solutions

$$Z_R=\left[\begin{array}{c|c} 4\sigma^2+9\sigma\sigma_m+4\sigma_m^2 & 4\sigma^2+12\sigma\sigma_m+8\sigma_m^2 \\ \hline 4\sigma^2+12\sigma\sigma_m+8\sigma_m^2 & 4\sigma^2+16\sigma\sigma_m+16\sigma_m^2 \end{array}\right] \qquad (148)$$

and

$$\hat{Z}_R=\left[\begin{array}{c|c} \sigma\sigma_m & 0 \\ \hline 0 & 0 \end{array}\right], \qquad (149)$$

respectively, of the pertinent matrix Riccati equation. Although Z_R is positive-definite, $\hat{Z}_R$ is only semipositive-definite. Observe, that with our choice of Σ in (133), the plant described by (131) is not controllable from the equivalent scalar disturbance input and the various theorems relating to the uniqueness of the solution of the Riccati equation do not apply [18, p. 36].[26]

The controller[27]

$$\hat{C}(s)=\frac{\frac{\sigma}{\sigma_m}(2-s)}{ks^2+\left(\frac{k\sigma}{\sigma_m}+\sqrt{4k^2+2k\sqrt{k}+k}\right)s+\sqrt{k}\left(1+\frac{\sigma}{\sigma_m}\right)+\frac{\sigma}{\sigma_m}\sqrt{4k^2+2k\sqrt{k}+k}} \qquad (150)$$

paired with the choice $\hat{Z}_R$ possesses a zero at $s=2$ coincident with a pole of $P(s)$, and this fact makes the instability of the LQG design immediately apparent. It appears, therefore, that the optimal stabilizing controller (141) yields a relative and not an absolute minimum for the cost functional J. This observation suggests the following question. Since LQG prejudges the structure of the controller and does not invoke closed-loop stability as an *a priori* constraint, is it really clear that the optimal stabilizing Z_R is always included in its several solutions? In any reasonable topology the collection of stabilizing controllers for a given plant-feedback sensor combination should form an open set and the answer is probably yes, but in our opinion the conjecture is in need of strict proof. On the other hand, the frequency-domain solution advanced in this paper not only absorbs many important practical factors easily and naturally, but also succeeds in completely circumventing the above difficulty.

[26]The existence of this second solution $\hat{Z}_R$ was kindly brought to the authors' attention by Dr. J. Boyd Pearson of Rice University, Houston, TX, who also supplied some interesting insights regarding its implications for LQG. We gratefully acknowledge his comments and helpful editorial suggestions.

[27]All details are omitted.

Appendix A

For sound practical reasons the components in the loop of Fig. 1 must not be restricted to be dynamical and a stability criterion must be general enough to encompass this case. Let the zero-state Laplace transform descriptions of the feedback compensator, controller, and plant be given by

$$F_i(s)\boldsymbol{x}_i(s)=G_i(s)\boldsymbol{u}_i(s), \qquad (A1)$$
$$\boldsymbol{y}_i(s)=J_i(s)\boldsymbol{u}_i(s)+H_i(s)\boldsymbol{x}_i(s), \qquad (A2)$$

$i=2\rightarrow4$, respectively. All coefficient matrices are real and polynomial, all F_i's are square and as usual $\boldsymbol{x}_i$, $\boldsymbol{u}_i$, $\boldsymbol{y}_i$ denote, in the same order, the internal state, the input, and the output. Physical degeneracies are excluded by imposing the determinantal condition,

$$\prod_{i=2}^{4}\det F_i(s)\not\equiv 0. \qquad (A3)$$

Clearly then

$$P_i(s)=J_i(s)+H_i(s)F_i^{-1}(s)G_i(s) \qquad (A4)$$

is the transfer matrix of system no. i, $i=2\rightarrow4$. As is well known [10], [11], system number i is asymptotically stable iff the scalar polynomial

$$\Delta_i(s)\equiv\det F_i(s) \qquad (A5)$$

has all its zeros in $\mathrm{Re}\,s<0$, $i=2\rightarrow4$. In the present notation,

$$\Delta_F(s)=\Delta_2(s); \qquad F(s)=P_2(s) \qquad (A6)$$
$$\Delta_C(s)=\Delta_3(s); \qquad C(s)=P_3(s) \qquad (A7)$$
$$\Delta_P(s)=\Delta_4(s); \qquad P(s)=P_4(s). \qquad (A8)$$

In Fig. 2 the three systems are shown interconnected through a linear, time-invariant frequency-insensitive grid, and it is assumed that the inputs to this grid uniquely determine its outputs. Hence, there exist real constant matrices M_a and M_b such that

$$\boldsymbol{u}_t(s)=M_a\boldsymbol{y}_t(s)+M_b\boldsymbol{u}(s) \qquad (A9)$$

where

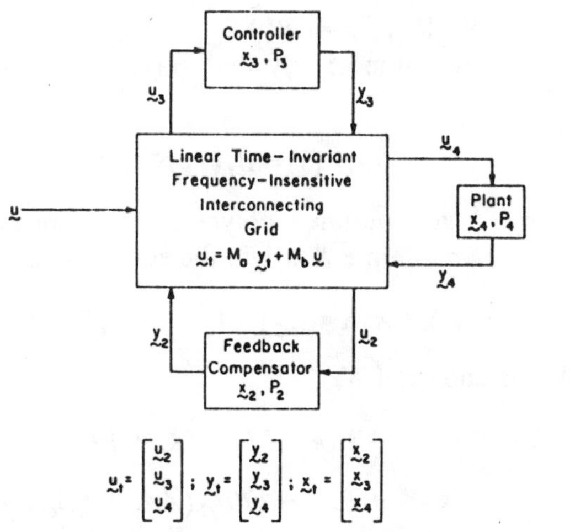

Fig. 2. Generalized interconnection scheme.

$$u_t(s) = \begin{bmatrix} u_2 \\ u_3 \\ u_4 \end{bmatrix}; \; y_t(s) = \begin{bmatrix} y_2 \\ y_3 \\ y_4 \end{bmatrix}; \; x_t = \begin{bmatrix} x_2 \\ x_3 \\ x_4 \end{bmatrix}. \tag{A10}$$

Eliminating u_t in (A1) and (A2) with the help of (A9) and (A10) we obtain

$$\left[\begin{array}{c|c} F & -GM_a \\ \hline -H & 1-JM_a \end{array}\right] \left[\begin{array}{c} x_t \\ \hline y_t \end{array}\right] = \left[\begin{array}{c} G \\ \hline J \end{array}\right] M_b u \tag{A11}$$

in which[28]

$$\begin{aligned} F &= F_2 \dotplus F_3 \dotplus F_4, \\ G &= G_2 \dotplus G_3 \dotplus G_4, \\ H &= H_2 \dotplus H_3 \dotplus H_4, \\ J &= J_2 \dotplus J_3 \dotplus J_4. \end{aligned} \tag{A12}$$

Consequently [10], [11], the interconnected system is asymptotically stable iff the determinant $\Delta(s)$ of the coefficient matrix on the left-hand side of (A11) has all its roots in $\operatorname{Re} s < 0$. A straightforward row operation yields

$$\Delta(s) = \det(1 - P_t(s)M_a) \cdot \prod_{i=2}^{4} \Delta_i(s), \tag{A13}$$

$$P_t(s) = P_2(s) \dotplus P_3(s) \dotplus P_4(s). \tag{A14}$$

The interconnection is nondegenerate iff

$$\det(1 - P_t(s)M_a) \not\equiv 0 \tag{A15}$$

which is exactly the necessary and sufficient condition for the existence of an overall transfer matrix description $T_t(s)$. In fact if $y_t = T_t u$,

$$T_t(s) = (1 - P_t(s)M_a)^{-1} P_t(s) M_b. \tag{A16}$$

(The easy derivation is left to the reader.)

For the topology depicted in Fig. 1,

$$\begin{bmatrix} u_2 \\ u_3 \\ u_4 \end{bmatrix} = \begin{bmatrix} O & O & 1 \\ -1 & O & O \\ O & 1 & O \end{bmatrix} \begin{bmatrix} y_2 \\ y_3 \\ y_4 \end{bmatrix} + \begin{bmatrix} O \\ 1 \\ O \end{bmatrix} u. \tag{A17}$$

Thus,

$$1 - P_t M_a = 1 - (F \dotplus C \dotplus P) \begin{bmatrix} O & O & 1 \\ -1 & O & O \\ O & 1 & O \end{bmatrix} \tag{A18}$$

$$= \begin{bmatrix} 1 & O & -F \\ C & 1 & O \\ O & -P & 1 \end{bmatrix} \tag{A19}$$

and

$$\det(1 - P_t M_a) = \det(1_n + FPC) = 1/\det S. \tag{A20}$$

Expression (A13) for $\Delta(s)$ now assumes the form

$$\Delta(s) = \frac{\Delta_P(s)\Delta_C(s)\Delta_F(s)}{\det S(s)}. \tag{A21}$$

If all components have asymptotically stable hidden modes,

$$\begin{aligned} \Delta_P(s) &= h_P(s)\psi_P(s), \\ \Delta_C(s) &= h_C(s)\psi_C(s), \\ \Delta_F(s) &= h_F(s)\psi_F(s) \end{aligned} \tag{A22}$$

where the h's are strict Hurwitz and ψ_P, ψ_C, ψ_F are the characteristic denominators of plant, controller, and feedback compensator, respectively [10], [11].[29] Thus, the loop is asymptotically stable iff

$$\varphi(s) = \frac{\psi_P(s)\psi_C(s)\psi_F(s)}{\det S(s)} \tag{A23}$$

is strict Hurwitz which is precisely the assertion of Lemma 1. Q.E.D.

Let

$$F(s)P(s) = A^{-1}(s)B(s) \tag{A24}$$

be any left-coprime polynomial decomposition of $F(s)P(s)$ and

$$C(s) = B_C(s)A_C^{-1}(s) \tag{A25}$$

any right-coprime decomposition of $C(s)$.

Then [10], [12],

$$\psi_{FP}(s) = \det A(s) \tag{A26}$$

and

$$\psi_C(s) = \det A_C(s). \tag{A27}$$

[28] $A \dotplus B$ is the "direct sum" of matrices A and B.

[29] Equation (A22) is also obvious from (A4).

Evidently,

$$S=(1_n+FPC)^{-1}=A_C(AA_C+BB_C)^{-1}A \quad \text{(A28)}$$

and

$$\det S=\frac{\psi_C\psi_{FP}}{g(s)}\not\equiv 0 \quad \text{(A29)}$$

where

$$g(s)=\det(AA_C+BB_C) \quad \text{(A30)}$$

is a polynomial. Substituting into (A23),

$$\varphi=g\cdot\frac{\psi_F\psi_P}{\psi_{FP}}. \quad \text{(A31)}$$

Since the McMillan degree of any pole of $F(s)P(s)$ cannot exceed the sum of its degrees as a pole of $F(s)$ and $P(s)$, ψ_{FP} must divide $\psi_F\psi_P$ without remainder and the quotient $\psi_F\psi_P/\psi_{FP}$ is polynomial. Thus, if the loop is asymptotically stable $g(s)$ is necessarily strict Hurwitz. Furthermore, any zero of the product $\psi_F^+\psi_P^+$ must be cancelled by a zero of ψ_{FP}^+ whence

$$\psi_{FP}^+(s)=\psi_F^+(s)\psi_P^+(s) \quad \text{(A32)}$$

is also necessary for closed-loop stability.

Suppose now that (A32) is satisfied and let the real polynomial matrices $X(s)$ and $Y(s)$ be chosen so that[30]

$$A(s)X(s)+B(s)Y(s)=1_n \quad \text{(A33)}$$

and

$$\det X(s)\not\equiv 0. \quad \text{(A34)}$$

Select any controller with asymptotically stable hidden modes and with transfer matrix

$$C(s)=Y(s)X^{-1}(s). \quad \text{(A35)}$$

According to (A33), the pair (Y,X) is right-coprime and

$$AA_C+BB_C=AX+BY=1_n. \quad \text{(A36)}$$

Thus $g(s)=1$ and the associated stability polynomial $\varphi(s)$ is given by

$$\varphi=\frac{\psi_F\psi_P}{\psi_{FP}}=\frac{\psi_F^-\psi_P^-}{\psi_{FP}^-} \quad \text{(A37)}$$

which is devoid of zeros in $\mathrm{Re}\,s>0$. Consequently, the closed-loop structure is asymptotically stable and Lemma 2 is established. Q.E.D.

Recall from (10) that $R=CS$ or, in terms of the polynomial factors A, B, B_C, and A_C,

$$R=B_C(AA_C+BB_C)^{-1}A=HA, \quad \text{(A38)}$$

$$H=B_C(AA_C+BB_C)^{-1}. \quad \text{(A39)}$$

[30] The left-coprimeness of the pair $A(s)$, $B(s)$ guarantees that such a choice is always possible [10], [12].

For a stable loop, $g=\det(AA_C+BB_C)$ is strict Hurwitz and it follows immediately that $H(s)$ is analytic in $\mathrm{Re}\,s>0$. Let

$$F(s)P(s)=B_1(s)A_1^{-1}(s) \quad \text{(A40)}$$

be any right-coprime polynomial factorization of $F(s)\,P(s)$ and define $K(s)$ via the equation

$$Y+A_1K=B_C(AA_C+BB_C)^{-1}. \quad \text{(A41)}$$

With this choice of K,

$$\begin{aligned}X-B_1K&=X-B_1A_1^{-1}\left(B_C(AA_C+BB_C)^{-1}-Y\right)\\&=X+FPY-FPB_C(AA_C+BB_C)^{-1}\\&=A^{-1}\left(1_n-BB_C(AA_C+BB_C)^{-1}\right)\\&=A_C(AA_C+BB_C)^{-1}=SA^{-1}.\end{aligned} \quad \text{(A42)}$$

Hence, if the polynomial matrices $X_1(s)$ and $Y_1(s)$ are constructed to satisfy

$$X_1A_1+Y_1B_1=1_m, \quad \text{(A43)}$$

(A41) and (A42) combine to give

$$K=(X_1B_C-Y_1A_C)(AA_C+BB_C)^{-1}+Y_1X-X_1Y \quad \text{(A44)}$$

which is obviously analytic in $\mathrm{Re}\,s>0$. Conversely, let $K(s)$ be any real rational matrix analytic in $\mathrm{Re}\,s>0$ such that $\det(X-B_1K)\not\equiv 0$ and select any controller with stable hidden modes and transfer matrix $C(s)$ given by

$$C=(Y+A_1K)(X-B_1K)^{-1}. \quad \text{(A45)}$$

Reasoning exactly as in Appendix B in the derivation of (B64) we find that

$$\varphi(s)=\frac{\psi_F^-(s)\psi_P^-(s)}{\psi_{FP}^-(s)}\cdot\psi_K(s). \quad \text{(A46)}$$

Since $K(s)$ is analytic in $\mathrm{Re}\,s>0$, $\psi_K(s)$ and therefore $\varphi(s)$ are both strict Hurwitz and the closed-loop is asymptotically stable. This completes the proof of Lemma 3. Q.E.D.

The closed-loop transfer matrix

$$T=PR=PCS=(1_n+PCF)^{-1}PC. \quad \text{(A47)}$$

Let $PC=A_5^{-1}B_5$ and $F=B_FA_F^{-1}$ be left-and right-coprime polynomial factorizations. Evidently,

$$S_1=(1_n+PCF)^{-1}=A_F(A_5A_F+B_5B_F)^{-1}A_5 \quad \text{(A48)}$$

and

$$\det S_1=\det S=\frac{\psi_F\psi_{PC}}{g_1(s)}, \quad \text{(A49)}$$

$$g_1=\det(A_5A_F+B_5B_F). \quad \text{(A50)}$$

Hence

$$\varphi(s)=g_1(s)\cdot\frac{\psi_P(s)\psi_C(s)}{\psi_{PC}(s)} \quad \text{(A51)}$$

and it follows as before that the conditions $g_1(s)$ strict Hurwitz and $\psi_P^+\psi_C^+=\psi_{PC}^+$ are both necessary for closed-loop stability. In particular,

$$T=PR=PCS=S_1PC=A_F(A_5A_F+B_5B_F)^{-1}B_5 \quad \text{(A52)}$$

is analytic in $\text{Re}\, s \geqslant 0$.

From (A42),

$$\det S=\det(X-B_1K)\cdot\det A=\det(X-B_1K)\psi_{FP} \quad \text{(A53)}$$

and because of the analyticity of $K(s)$ in $\text{Re}\, s \geqslant 0$, all zeros of $\psi_{FP}^+(s)$ are zeros of det $S(s)$, multiplicities included.

Introducing the left-right coprime polynomial decompositions $P=A_p^{-1}B_p$ and $CF=B_6A_6^{-1}$ into (A48) gives

$$S_1=A_6(A_pA_6+B_pB_6)^{-1}A_p=\mathcal{P}_1A_p \quad \text{(A54)}$$

and closed-loop stability forces $\mathcal{P}_1(s)$ to be analytic in $\text{Re}\, s \geqslant 0$. We have now justified the three comments preceeding (38), (42) and (59). Q.E.D.

Also, let us remark that for given polynomial matrices $A_7(s)$, $B_7(s)$, the existence of polynomial matrices $X_7(s)$, $Y_7(s)$ such that

$$X_7(s)A_7(s)+Y_7(s)B_7(s)=1 \quad \text{(A55)}$$

is possible iff for every fixed finite s the homogeneous pair

$$A_7(s)\boldsymbol{a}=\boldsymbol{O}, \quad \text{(A56)}$$

$$B_7(s)\boldsymbol{a}=\boldsymbol{O} \quad \text{(A57)}$$

admits only the trivial solution $\boldsymbol{a}=\boldsymbol{O}$. Necessity is trivial and sufficiency is easily established by actually constructing a solution pair $X_7(s)$, $Y_7(s)$ with the help of the Smith–McMillan theorem. The idea underlying the construction is very simple to grasp. Let

$$\Omega_c(s)=\text{diag}\left[\frac{e_1(s)}{\psi_1(s)},\frac{e_2(s)}{\psi_2(s)},\cdots,\frac{e_k(s)}{\psi_k(s)}\right] \quad \text{(A58)}$$

be the canonic form of $F(s)P(s)$. Then [8], 1) $k=$normal rank $F(s)P(s)$; 2) the e's and ψ's are real monic polynomials uniquely determined by $F(s)P(s)$; 3) each $e_i(s)$ is relatively prime to its mate $\psi_i(s)$, $i=1\rightarrow k$; 4) $e_i(s)$ divides $e_{i+1}(s)$ and $\psi_{i+1}(s)$ divides $\psi_i(s)$, $i=1\rightarrow k-1$; 5) the distinct finite zeros and poles of $F(s)P(s)$ are identical, respectively, with the distinct zeros of $e_1(s)$ and $\psi_1(s)$; 6) the McMillan degree of any finite pole of $F(s)P(s)$ equals its multiplicity as a root of the characteristic denominator

$$\psi_{FP}(s)=\prod_{i=1}^{k}\psi_i(s). \quad \text{(A59)}$$

From the Smith–McMillan theorem [8],

$$FP=U(\Omega_c\dot{+}O_{n-k,m-k})V \quad \text{(A60)}$$

where $U(s)$ and $V(s)$ are square, real elementary polynomial matrices.[31] Since $e_i(s)$ is relatively prime to $\psi_i(s)$ there exist [13] two real polynomials $\alpha_i(s)$, $\beta_i(s)$ such that $\beta_i(s)\not\equiv 0$ and

$$\alpha_i(s)e_i(s)+\beta_i(s)\psi_i(s)=1, \qquad i=1\rightarrow k. \quad \text{(A61)}$$

Let[32]

$$\epsilon=\text{diag}[e_1,e_2,\cdots,e_k], \quad \text{(A62)}$$

$$\chi=\text{diag}[\psi_1,\psi_2,\cdots,\psi_k], \quad \text{(A63)}$$

$$\alpha=\text{diag}[\alpha_1,\alpha_2,\cdots,\alpha_k], \quad \text{(A64)}$$

$$\beta=\text{diag}[\beta_1,\beta_2,\cdots,\beta_k]. \quad \text{(A65)}$$

Then, putting

$$A=U(\chi\dot{+}1_{n-k})U^{-1}, \quad \text{(A66)}$$

$$B_1=B=U(\epsilon\dot{+}O_{n-k,m-k})V, \quad \text{(A67)}$$

$$A_1=V^{-1}(\chi\dot{+}1_{m-k})V, \quad \text{(A68)}$$

$$X=U(\beta\dot{+}1_{n-k})U^{-1}, \quad \text{(A69)}$$

$$Y_1=Y=V^{-1}(\alpha\dot{+}O_{m-k,n-k})U^{-1} \quad \text{(A70)}$$

and

$$X_1=V^{-1}(\beta\dot{+}1_{m-k})V, \quad \text{(A71)}$$

we verify by inspection that

$$\alpha\epsilon+\beta\chi=1_k, \quad \text{(A72)}$$

$$AX+BY=1_n, \quad \text{(A73)}$$

$$X_1A_1+Y_1B_1=1_m, \quad \text{(A74)}$$

$$XA=AX=U(\chi\beta\dot{+}1_{n-k})U^{-1}, \quad \text{(A75)}$$

$$A^{-1}B=BA_1^{-1}=FP, \quad \text{(A76)}$$

$$XB=BX_1=U(\beta\epsilon\dot{+}O_{n-k,m-k})V, \quad \text{(A77)}$$

$$Y_1X=X_1Y=V^{-1}(\alpha\beta\dot{+}O_{m-k,n-k})U^{-1}. \quad \text{(A78)}$$

Of course, other decompositions may not possess all the symmetry properties enumerated in (A66)–(A78).

According to (A29), any zero of the characteristic denominator $\psi_{FP}(s)$ in $\text{Re}\, s \geqslant 0$ of multiplicity μ is a zero of det S of at least the same multiplicity. Define

$$e_{FP}(s)=\prod_{i=1}^{k}e_i(s) \quad \text{(A79)}$$

to be the characteristic *numerator* of $F(s)P(s)$. Suppose $\det(1_n-S)\not\equiv 0$. Then, any zero of $e_{FP}(s)$ in $\text{Re}\, s \geqslant 0$ of multiplicity μ is a zero of $\det(1_n-S)$ of multiplicity at least μ. For the proof, note that

$$1_n-S=1_n-(1_n+FPC)^{-1}=(1_n+FPC)^{-1}FPC;$$

or, using (A24), (A25), and (A30),

$$1_n-S=A_C(AA_C+BB_C)^{-1}\cdot(BB_C)A_C^{-1}. \quad \text{(A80)}$$

[31] Det $U(s)$ and det $V(s)$ equal nonzero constants.

[32] A square matrix A whose only nonzero elements are its main diagonal elements $a_1,a_2,\cdots,a_k$ is written $A=\text{diag}[a_1,a_2,\cdots,a_k]$.

$$\therefore \det(1_n - S) = \frac{\det(BB_C)}{g(s)} \tag{A81}$$

and it is clear from (A62) and (A67) with $n=k$ that $\det(BB_C)$ is divisible by $e_{FP}(s)$. Since $g(s)$ is a strict Hurwitz polynomial, the assertion follows. Q.E.D.

Appendix B

Adding (66) to $k\times(65)$ we obtain $2\pi j(E_t+kE_s)=2\pi jE$. Since $R=(Y+A_1K)A$,

$$\delta R = A_1(\delta K)A \tag{B1}$$

with $\delta K(s)$ analytic in $\operatorname{Re} s \geq 0$. Use of the standard variational argument [1] to examine the increment in E produced by the perturbation (B1) leads directly to the Wiener–Hopf equation

$$\Phi - A_{1*}(P_*P+kQ)A_1K(AGA_*)=\Delta_* \tag{B2}$$

where

$$\Phi = A_{1*}P_*(G_u+P_oG_dP_{d*})A_* - A_{1*}(P_*P+kQ)Y(AGA_*) \tag{B3}$$

and $\Delta(s)$ is analytic in $\operatorname{Re} s > 0$. If (B2) possesses a real rational matrix solution $K(s)$ analytic in $\operatorname{Re} s > 0$ which satisfies (33) and has a finite associated cost E, then this $K(s)$ is optimal. According to 5) and (76), $A_{1*}(P_*P+kQ)A_1$ and AGA_* are analytic for all $s=j\omega$ and the existence of a $K(s)$ with the desired properties implies the $j\omega$-axis analyticity of $\Phi(s)$. Since the latter is a unique construct from the prescribed data it is important to verify at the outset that this is indeed the case.

Using (69),

$$\Phi = A_{1*}P_*(G+(P_o-P_d)G_dP_{d*}-G_{ml})A_* - A_{1*}(P_*P+kQ)Y(AGA_*) \tag{B4}$$

and its $j\omega$-analyticity follows from that of[33]

$$A_{1*}P_*(P_o-P_d)G_dP_{d*}A_* + \left(A_{1*}P_*A^{-1}-A_{1*}(P_*P+kQ)Y\right)AGA_*$$

which in turn follows from that of

$$A_{1*}P_*(P_o-P_d)G_dP_{d*}A_* \tag{B5}$$

and

$$A_{1*}P_*A^{-1}-A_{1*}(P_*P+kQ)Y \equiv \beta. \tag{B6}$$

Expanding (B5),

$$\begin{aligned}(P_o-P_d)G_dP_{d*}A_* &= ((1_n-F)P_o-L)G_d(FP_o+L)_*A_*\\ &= (1_n-F)P_oG_dP_{o*}(AF)_* - LG_dL_*A_*\\ &\quad + (1_n-F)P_oG_dL_*A_* - LG_dP_{o*}(AF)_*.\end{aligned} \tag{B7}$$

[33] PA_1 and G_{ml} are $j\omega$-analytic.

All four terms are $j\omega$-analytic. First, $AF=\mathcal{P}_4A_p$, $\mathcal{P}_4(s)$ $j\omega$-analytic. Second, the analyticity of

$$(1_n-F)P_oG_dP_{o*}A_{p*}=(1_n-F)A_p^{-1}A_pP_oG_dP_{o*}A_{p*}, \tag{B8}$$

$$(1_n-F)P_oG_dL_*=(1_n-F)A_p^{-1}A_pP_oG_dL_*, \tag{B9}$$

LG_dL_* and $LG_dP_{o*}A_{p*}$ is implied by that of $(F-1_n)P$ and the assumptions introduced in 3).

With regard to (B6), replacing A^{-1} by $X+FPY$ transforms it into

$$A_{1*}P_*X+A_{1*}P_*(F-1_n)PY-kA_{1*}QY \tag{B10}$$

which is eveidently analytic on the $j\omega$-axis since Q, PA_1 and $(F-1_n)P$ are $j\omega$-analytic.

The solution of (B2) is now routine. Construct[34] two square real rational matrices $\Lambda(s)$, $\Omega(s)$ *analytic together with their inverses in* $\operatorname{Re} s \geq 0$ such that

$$A_{1*}(P_*P+kQ)A_1=\Lambda_*\Lambda \tag{B11}$$

and

$$AGA_*=\Omega\Omega_*. \tag{B12}$$

From (B2),

$$\Lambda_*^{-1}\Phi\Omega_*^{-1}-\Lambda K\Omega=\Lambda_*^{-1}\Delta_*\Omega_*^{-1}. \tag{B13}$$

Effect the partial fraction decomposition

$$\Lambda_*^{-1}\Phi\Omega_*^{-1}=\left\{\Lambda_*^{-1}\Phi\Omega_*^{-1}\right\}_\infty + \left\{\Lambda_*^{-1}\Phi\Omega_*^{-1}\right\}_+ + \left\{\Lambda_*^{-1}\Phi\Omega_*^{-1}\right\}_- \tag{B14}$$

where $\{\ \}_\infty$ is the polynomial part of the Laurent expansion of $\Lambda_*^{-1}\Phi\Omega_*^{-1}$ associated with the pole at infinity and $\{\ \}_+$, $\{\ \}_-$ the parts associated with all the poles in $\operatorname{Re} s<0$ and $\operatorname{Re} s \geq 0$, respectively. Clearly, since Φ is analytic on $j\omega$, $\{\ \}_+$ is analytic in $\operatorname{Re} s > 0$, $\{\ \}_-$ in $\operatorname{Re} s \leq 0$ and both vanish for $s=\infty$. The substitution of (B14) into (B13) yields

$$\left\{\Lambda_*^{-1}\Phi\Omega_*^{-1}\right\}_+ - \Lambda K\Omega = \Lambda_*^{-1}\Delta_*\Omega_*^{-1} - \left\{\Lambda_*^{-1}\Phi\Omega_*^{-1}\right\}_- - \left\{\Lambda_*^{-1}\Phi\Omega_*^{-1}\right\}_\infty. \tag{B15}$$

However, with $K(s)$ forced to be analytic in $\operatorname{Re} s > 0$, the left-hand side of (B15) is also analytic in $\operatorname{Re} s>0$ and equals the right-hand side which is analytic in $\operatorname{Re} s \leq 0$. Thus (B15) is polynomial and we obtain

$$K=\Lambda^{-1}J\Omega^{-1}+\Lambda^{-1}\left\{\Lambda_*^{-1}\Phi\Omega_*^{-1}\right\}_+\Omega^{-1}, \tag{B16}$$

$J(s)$ a real polynomial matrix to be determined by the requirement of finite cost. Observe that $K(s)$, as defined by (B16) is actually analytic in $\operatorname{Re} s \geq 0$ while

[34] Inequality (76) guarantees the analyticity of the factors $\Lambda^{-1}(s)$, $\Omega^{-1}(s)$ in $\operatorname{Re} s \geq 0$. Without (76) analyticity is assured only in $\operatorname{Re} s > 0$. It can be shown that the factors are unique up to real constant orthogonal multipliers [8], [9].

$$\Delta_* = \Lambda_*(\{\Lambda_*^{-1}\Phi\Omega_*^{-1}\}_- + \{\Lambda_*^{-1}\Phi\Omega_*^{-1}\}_\infty - J)\Omega_* \quad \text{(B17)}$$

is analytic in $\text{Re}s < 0$ (as it should be). In 5) we imposed conditions guaranteeing the $j\omega$-analyticity of all integrands in E_s and E_t and to study the convergence of the cost under the choice (B16) for K it suffices to examine the behavior of the integrand of E as $\omega \to \infty$. Denote this integrand by $\rho(s)$. Noting that $R = (Y + A_1K)A = HA$ and

$$\begin{aligned} \text{Tr}(H_*(P_*P+kQ)H(AGA_*)) &= \text{Tr}[kQRGR_* + (PR)G(PR)_*] \\ &= \text{Tr}(\Omega_*H_*(P_*P+kQ)H\Omega), \quad \text{(B18)} \end{aligned}$$

simple algebra yields

$$\begin{aligned} \rho = \text{Tr}(\Omega_*H_*(P_*P+kQ)H\Omega) + \text{Tr}\,G_u + \text{Tr}(P_oG_dP_{o*}) \\ -2\text{Tr}(PRG_u) - 2\text{Tr}(PRP_dG_dP_{o*}). \quad \text{(B19)} \end{aligned}$$

To evaluate the first term in ρ we need H. From (B16) and (B3),

$$\begin{aligned} K = \Lambda^{-1}J\Omega^{-1} + \Lambda^{-1}\{\Lambda_*^{-1}I\Omega_*^{-1}\}_+\Omega^{-1} \\ -\Lambda^{-1}\{\Lambda A_1^{-1}Y\Omega\}_+\Omega^{-1} \quad \text{(B20)} \end{aligned}$$

where

$$I = A_{1*}P_*(G_u + P_oG_dP_{d*})A_*. \quad \text{(B21)}$$

Multiplying (B20) on the left by A_1 and combining,

$$\Lambda A_1^{-1}H\Omega = J_1 + \{\Lambda_*^{-1}I\Omega_*^{-1}\}_+ + \{\Lambda A_1^{-1}Y\Omega\}_- \quad \text{(B22)}$$

where

$$J_1 = J + \{\Lambda A_1^{-1}Y\Omega\}_\infty \quad \text{(B23)}$$

is also polynomial. Since

$$\Omega_*H_*(P_*P+kQ)H\Omega = (\Lambda A_1^{-1}H\Omega)_*(\Lambda A_1^{-1}H\Omega), \quad \text{(B24)}$$

the integral of the first term in (B19) converges iff

$$\Lambda A_1^{-1}H\Omega \leqslant O(1/\omega), \qquad \omega \to \infty. \quad \text{(B25)}$$

Now both curly brackets in (B22) are already $\leqslant O(1/\omega)$ and, therefore, $J_1 \leqslant O(1/\omega)$. But being polynomial J_1 can only be $\leqslant O(1/\omega)$ for $\omega \to \infty$ if it is identically zero, whence

$$J = -\{\Lambda A_1^{-1}Y\Omega\}_\infty \quad \text{(B26)}$$

is identified. According to (B18) this convergence entails that of

$$\text{Tr}\int_{-j\infty}^{j\infty}(PR)G(PR)_*ds \quad \text{(B27)}$$

which entails that of

$$\text{Tr}\int_{-j\infty}^{j\infty}(PR)G_u(PR)_*ds, \quad \text{(B28)}$$

$$\text{Tr}\int_{-j\infty}^{j\infty}(PR)G_{ml}(PR)_*ds \quad \text{(B29)}$$

and

$$\text{Tr}\int_{-j\infty}^{j\infty}(PR)(P_dG_dP_{d*})(PR)_*ds. \quad \text{(B30)}$$

We can exploit the integrability of

$$\text{Tr}[H_*(P_*P+kQ)H(AGA_*)] = \text{Tr}((P_*P+kQ)RGR_*) \quad \text{(B31)}$$

to derive a sharp sufficient condition for $T(s) = PR$ to be proper. Let

$$G(j\omega) \approx \omega^{2l}1_n, \quad \text{(B32)}$$

$$(P_*P+kQ) \approx \omega^{2q}1_m \quad \text{(B33)}$$

and

$$R(j\omega) = O(\omega^r) \quad \text{(B34)}$$

for $\omega \to \infty$. Then[35]

$$(P_*P+kQ)RGR_* = O(\omega^{2l+2q+2r}) \quad \text{(B35)}$$

and invoking integrability, $l+q+r \leqslant -1$. Thus $r \leqslant -(1+l+q)$ and if $P(s) = O(s^\nu)$, order T = order $(PR) \leqslant \nu - (1+l+q)$. It follows that the constraint

$$\nu - 1 \leqslant l + q \quad \text{(B36)}$$

guarantees $T(s)$ proper. Stated differently, if

$$(P_*P+kQ)G \approx \omega^{2\mu}1_m, \quad \text{(B37)}$$

$$P(s) = O(s^\nu) \quad \text{(B38)}$$

and

$$\mu \geqslant \nu - 1, \quad \text{(B39)}$$

then $T(s)$ is proper. Irrespective of (B39), the assumptions

$$G_u(j\omega) \leqslant O(1/\omega^2) \quad \text{(B40)}$$

and

$$P_oG_dP_{o*} \leqslant O(1/\omega^2) \quad \text{(B41)}$$

plus the finiteness of (B28) and (B30) imply $E < \infty$. For, using Schwartz's inequality,[36]

$$\begin{aligned} |\text{Tr}\int(PRG_u)d\omega|^2 \leqslant \text{Tr}\int(PR)G_u(PR)d\omega \\ \cdot\text{Tr}\int G_ud\omega < \infty \quad \text{(B42)} \end{aligned}$$

[35]This conclusion is reached by making use of some properties of positive-definite matrices.

[36]$|\text{Tr}\int F_1F_2dx|^2 \leqslant \text{Tr}\int F_1F_1^*dx \cdot \text{Tr}\int F_2F_2^*dx$.

and

$$|\mathrm{Tr}\int PRP_dG_dP_{o*}d\omega|^2 \leq \mathrm{Tr}\int PR(P_dG_dP_{d*})(PR)_*d\omega \cdot \mathrm{Tr}\int P_oG_dP_{o*}d\omega < \infty. \quad \text{(B43)}$$

(The range of integration is over $|\omega| > \omega_o$, ω_o sufficiently large.) Writing $G_d = K_dK_{d*}$, it is seen that (B30) is finite iff

$$PRP_dK_d = ((1_n - S_1)P_o + TL)K_d \leq O(1/\omega). \quad \text{(B44)}$$

From (B41), $P_oK_d \leq O(1/\omega)$ and substituting into (B44) we obtain

$$(S_1P_o - TL)K_d \leq O(1/\omega). \quad \text{(B45)}$$

Obviously, if

$$G_d(j\omega) \approx \omega^{-2i}1, \qquad i \leq 1, \quad \text{(B46)}$$

then $K_d(j\omega) \approx \omega^{-i}1$ and (B45) forces $S_1P_o - TL$ to be proper.

From $FPR = 1_n - S$ and $PRF = 1_n - S_1$ it is clear that

$$O(P) + O(R) + O(F) \leq -1 \quad \text{(B47)}$$

is a sufficient condition for limit $S(j\omega) = \text{limit } S_1(j\omega) = 1_n$ as $\omega \to \infty$. Since $O(R) \leq -(1+\mu)$, (B47) is certainly valid if

$$O(P) + O(F) \leq \mu. \quad \text{(B48)}$$

It now follows from $R = CS$ that $O(C) = O(R)$ and, therefore, $\mu \geq -1$ guarantees $C(s)$ proper. Note that $S(j\omega) \to 1_n$ as $\omega \to \infty$ implies $\det S(j\omega) \not\equiv 0$ and in particular $\det(X - B_1K) \not\equiv 0$ because $S = (X - B_1K)A$. This means that $C(s)$, as defined by (34), makes sense.

Employing the formulas $R = CS = HA$, $S = 1_n - FPR$ and (B22) with $J_1 = O$ we obtain

$$C = R(1_n - FPR)^{-1} = H(A^{-1} - FPH)^{-1} = H(1_n - BH)^{-1}A \quad \text{(B49)}$$

where

$$H = A_1\Lambda^{-1}(\{\Lambda_*^{-1}I\Omega_*^{-1}\}_+ + \{\Lambda A_1^{-1}Y\Omega\}_-)\Omega^{-1}. \quad \text{(B50)}$$

The product $A_1\Lambda^{-1}\{\Lambda A_1^{-1}Y\Omega\}_-\Omega^{-1}$ is obviously analytic in $\mathrm{Re}\, s \geq 0$ and the closed right-half-plane analyticity of $H(s)$ is, therefore, apparent.[37]

According to Appendix A, to study the stability margin of the optimally compensated loop it is necessary to find the zeros of the associated polynomial[38]

$$\Delta(s) = \frac{\Delta_F(s)\Delta_P(s)\Delta_C(s)}{\det S(s)}. \quad \text{(B51)}$$

Granting that any hidden modes of the plant and feedback compensator are known or at least localizable and that $C(s)$ shall be realized minimally,[39] it suffices instead to locate the zeros of the polynomial

$$\varphi(s) = \frac{\psi_F(s)\psi_P(s)\psi_C(s)}{\det S(s)} \quad \text{(B52)}$$

where $\psi_F(s)$, $\psi_P(s)$ and $\psi_C(s)$ are the characteristic denominators of $F(s)$, $P(s)$, and $C(s)$, respectively. Since

$$\det S = \det(X - B_1K)\cdot\det A \quad \text{(B53)}$$

and $\psi_F^+\psi_P^+ = \psi_{FP}^+ = \det{}^+A$ (by admissibility),

$$\varphi = \frac{\psi_F^-\psi_P^-\psi_C}{\det(X - B_1K)\cdot\psi_{FP}^-}. \quad \text{(B54)}$$

Evidently, ψ_{FP}^- divides $\psi_F^-\psi_P^-$ and

$$\varphi = \frac{\psi_C}{\det(X - B_1K)}\cdot\theta \quad \text{(B55)}$$

where

$$\theta(s) = \frac{\psi_F^-(s)\psi_P^-(s)}{\psi_{FP}^-(s)} \quad \text{(B56)}$$

is a strict Hurwitz polynomial. To make further progress we must relate ψ_C to $\det(X - B_1K)$. Let $K = NM^{-1}$ be a right-coprime factorization of $K(s)$. Then

$$C = (Y + A_1K)(X - B_1K)^{-1} = (YM + A_1N)(XM - B_1N)^{-1} \quad \text{(B57)}$$

and the pair $(YM + A_1N,\ XM - B_1N)$ is right-coprime. For the proof it is necessary to show (Appendix A) that the equations

$$(YM + A_1N)a = O, \quad \text{(B58)}$$

$$(XM - B_1N)a = O \quad \text{(B59)}$$

possess only the trivial solution $a = O$ which we accomplish by using the identity

$$A(XM - B_1N) + B(YM + A_1N) = M. \quad \text{(B60)}$$

Clearly, in view of (B60) any a satisfying (B58) and (B59) must also satisfy

$$Ma = O, \quad \text{(B61)}$$

$$A_1Na = O; \quad B_1Na = O. \quad \text{(B62)}$$

Since the pair (A_1, B_1) is right-coprime, (B62) implies $Na = O$ and invoking (B61) and the right-coprimeness of (N, M), $a = O$. Q.E.D.

Hence, up to a multiplicative constant,

$$\psi_C = \det(XM - B_1N) \quad \text{(B63)}$$

and substituting into (B55),

$$\varphi(s) = \theta(s)\cdot\det M(s). \quad \text{(B64)}$$

[37]Unfortunately, the best numerical scheme for carrying out the computation (B50) is not so apparent.

[38]$\Delta(s)$ in (B51) has no connection with the $\Delta(s)$ appearing in (B2).

[39]If $C(s)$ is not realized minimally its hidden modes must also be localizable.

It is seen therefore that the (nonhidden) poles of the optimally compensated loop are *precisely* the zeros of $\theta(s)$ plus the finite poles of $K(s)$, each of the latter counted according to its McMillan degree.

An examination of the formula

$$K=\Lambda^{-1}\left(\left\{\Lambda_*^{-1}I\Omega_*^{-1}\right\}_+ +\left\{\Lambda A_1^{-1}Y\Omega\right\}_-\right)\Omega^{-1}-A_1^{-1}Y \tag{B65}$$

reveals immediately that the distinct finite poles of $K(s)$ are *included* in those of $A_1^{-1}(s)$, $\Lambda^{-1}(s)$, $\Omega^{-1}(s)$, and $I(s)$ in $\operatorname{Re}s<0$; or, in terms of primary data, in those of

$$FP,\ (A_{1*}(P_*P+kQ)A_1)^{-1},\ (AGA_*)^{-1},\qquad A_{1*}P_*(G_u+P_oG_dP_{d*})A_*. \tag{B66}$$

Finally, instead of (B49), experience indicates that the formula

$$C=H_o(A^{-1}\Omega-FPH_o)^{-1} \tag{B67}$$

where

$$H_o=H\Omega=A_1\Lambda^{-1}\left(\left\{\Lambda_*^{-1}I\Omega_*^{-1}\right\}_+ +\left\{\Lambda A_1^{-1}Y\Omega\right\}_-\right) \tag{B68}$$

is more suitable for computer implementation. To complete the proof of Theorem 1, Section III, it is finally necessary to prove that the controller defined by (B67) and (B68) provides a global minimum for the cost E from among the class of all admissible controllers.

Combining (B16) and (B26), it is seen that

$$K_o(s)=\Lambda^{-1}\left(\left\{\Lambda_*^{-1}\Phi\Omega_*^{-1}\right\}_+ -\left\{\Lambda A_1^{-1}Y\Omega\right\}_\infty\right)\Omega^{-1} \tag{B69}$$

is the Wiener–Hopf solution for $K(s)$. Clearly, $K_o(s)$ is analytic in $\operatorname{Re}s\geqslant 0$ and as we have already shown in great detail, the associated cost

$$E(K_o)=E_s(K_o)+kE_t(K_o) \tag{B70}$$

obtained by substituting $R_o=(Y+A_1K_o)A$ into (65) and (66) is finite. According to Lemma 3, any $R(s)$ corresponding to a stable closed-loop design must be of the form $R=(Y+A_1K)A$ where $K(s)$ is analytic in $\operatorname{Re}s\geqslant 0$. Hence, for our present purposes we say that $K(s)$ is admissible if it is analytic in $\operatorname{Re}s\geqslant 0$ and the associated cost $E(K)<\infty$.[40] Our objective is to prove that $E(K)\geqslant E(K_o)$ for any choice of admissible $K(s)$.

Let

$$(R_1,R_2)_s\equiv\operatorname{Tr}\int_{-\infty}^{\infty}QR_1GR_{2*}\,d\omega, \tag{B71}$$

$$(R_1,R_2)_u\equiv\operatorname{Tr}\int_{-\infty}^{\infty}(1_n-PR_1)G_u(1_n-PR_2)_*\,d\omega, \tag{B72}$$

[40]The argument that follows is independent of the assumption $\det(X-B_1K)\not\equiv 0$ in (33).

$$(R_1,R_2)_{ml}\equiv\operatorname{Tr}\int_{-\infty}^{\infty}(PR_1)G_{ml}(PR_2)_*\,d\omega \tag{B73}$$

and

$$(R_1,R_2)_d\equiv\operatorname{Tr}\int_{-\infty}^{\infty}(P_o-PR_1P_d)G_d(P_o-PR_2P_d)_*\,d\omega. \tag{B74}$$

In view of (66), (69), and (70),

$$2\pi E(K)=k(R,R)_s+(R,R)_u+(R,R)_{ml}+(R,R)_d. \tag{B75}$$

Moreover, since each of the four terms on the right-hand side of (B75) is nonnegative, $E(K)<\infty$ iff these terms are all finite.

Suppose R_1 and R_2 correspond to admissible choices K_1 and K_2, respectively. Then, $E(K_1)<\infty$, $E(K_2)<\infty$ and using the version of Schwartz's inequality given,[36] it is easily shown that

$$\begin{aligned}|(R_1,R_2)_s|^2&\leqslant(R_1,R_1)_s\cdot(R_2,R_2)_s<\infty,\\ |(R_1,R_2)_u|^2&\leqslant(R_1,R_1)_u\cdot(R_2,R_2)_u<\infty,\\ |(R_1,R_2)_{ml}|^2&\leqslant(R_1,R_1)_{ml}\cdot(R_2,R_2)_{ml}<\infty,\\ |(R_1,R_2)_d|^2&\leqslant(R_1,R_1)_d\cdot(R_2,R_2)_d<\infty.\end{aligned} \tag{B76}$$

For example, recalling that $Q=P_{s*}P_s$ and $G=\Omega\Omega_*$,

$$\begin{aligned}|(R_1,R_2)_s|^2&=\left|\operatorname{Tr}\int_{-\infty}^{\infty}(P_sR_1\Omega)(P_sR_2\Omega)_*\,d\omega\right|^2\\ &\leqslant\operatorname{Tr}\int_{-\infty}^{\infty}(P_sR_1\Omega)(P_sR_1\Omega)_*\,d\omega\\ &\quad\cdot\operatorname{Tr}\int_{-\infty}^{\infty}(P_sR_2\Omega)(P_sR_2\Omega)_*\,d\omega=\\ &=\operatorname{Tr}\int_{-\infty}^{\infty}QR_1GR_{1*}\,d\omega\cdot\operatorname{Tr}\int_{-\infty}^{\infty}QR_2GR_{2*}\,d\omega\\ &=(R_1,R_1)_s\cdot(R_2,R_2)_s<\infty.\end{aligned}$$

The other three inequalities are established in exactly the same way. (The result $\operatorname{Tr}AB=\operatorname{Tr}BA$ is used repeatedly.)

Identify R_1 with R_o and R_2 with any R defined by an admissible $K=K_o+\delta K$. Of course, $\delta K(s)$ is analytic in $\operatorname{Re}s\geqslant 0$ and $R=R_o+\delta R$ where $\delta R=A_1(\delta K)A$. Since

$$(R_o,R)_s=(R_o,R_o)_s+(R_o,\delta R)_s$$

is finite and $(R_o,R_o)_s<\infty$, it is also true that $\alpha_o=(R_o,\delta R)_s<\infty$. Similarly,

$$\begin{aligned}\alpha_1&=(R_o,\delta R)_{ml}<\infty,\\ \alpha_2&=\operatorname{Tr}\int_{-\infty}^{\infty}(1_n-PR_o)G_u(P\delta R)_*\,d\omega<\infty\end{aligned} \tag{B77}$$

and

$$\alpha_3=\operatorname{Tr}\int_{-\infty}^{\infty}(P_o-PR_oP_d)G_dP_{d*}(P\delta R)_*\,d\omega<\infty.$$

From

$$(R,R)_s=(R_o,R_o)_s+2(R_o,\delta R)_s+(\delta R,\delta R)_s$$

it now follows that $\beta_o=(\delta R,\delta R)_s<\infty$. In the same manner, exploiting the remaining inequalities in (B77), we get

$$\beta_1=\mathrm{Tr}\int_{-\infty}^{\infty}(P\delta R)G_{ml}(P\delta R)_*d\omega<\infty,$$

$$\beta_2=\mathrm{Tr}\int_{-\infty}^{\infty}(P\delta R)G_u(P\delta R)_*d\omega<\infty \qquad \text{(B78)}$$

and

$$\beta_3=\mathrm{Tr}\int_{-\infty}^{\infty}(P\delta R)P_dG_dP_{d*}(P\delta R)_*d\omega<\infty.$$

Clearly, all four β's are nonnegative.

Let $E(K)=E(K_o)+\delta E$. By a straightforward expansion of (B75),

$$2\pi(\delta E)=2(k\alpha_o+\alpha_1-\alpha_2-\alpha_3)+(k\beta_o+\beta_1+\beta_2+\beta_3). \qquad \text{(B79)}$$

However, it is readily verified by grouping terms that

$$-j(k\alpha_o+\alpha_1-\alpha_2-\alpha_3)=\mathrm{Tr}\int_{-j\infty}^{j\infty}\Delta_*(\delta K)_*ds \qquad \text{(B80)}$$

where $\Delta_*(s)$ is as defined in (B2) and (B3) and $K(s)$ replaced by $K_o(s)$. Now the Wiener–Hopf solution $K_o(s)$ guarantees the analyticity of $\Delta_*(s)$ in Re $s\leqslant 0$ and the finiteness of the α's implies that of the integral. The integrand

$$\mathrm{Tr}(\Delta_*(\delta K)_*)$$

is therefore analytic in Re $s\leqslant 0$ and $O(1/\omega^2)$ for large ω^2. By Cauchy's theorem the integral equals zero whence, $k\alpha_o+\alpha_1-\alpha_2-\alpha_3=0$ and

$$2\pi(\delta E)=k\beta_o+\beta_1+\beta_2+\beta_3\geqslant 0.$$

Consequently, $E(K)\geqslant E(K_o)$ for every admissible $K(s)$.

Q.E.D.

References

[1] J. E. Weston and J. J. Bongiorno, Jr., "Extension of analytical design techniques to multivariable feedback control systems," *IEEE Trans. Automat. Contr.*, vol. AC-17, pp. 613–620, Oct. 1972.

[2] D. C. Youla, "Modern classical multivariable feedback control theory: Part I," Rome Air Development Center, Griffiss Air Force Base, NY, Tech Rep. RADC-TR-70-98, June 1970.

[3] D. C. Youla, J. J. Bongiorno, Jr., and C. N. Lu, "Single-loop feedback stabilization of linear multivariable dynamical plants," *Automatica*, vol. 10, pp. 159–173, 1974.

[4] D. C. Youla, J. J. Bongiorno, Jr., and H. A. Jabr, "Modern Wiener–Hopf design of optimal controllers—Part I: The single-input-output case," *IEEE Trans. Automat. Contr.*, vol. AC-21, pp. 3–13, Feb. 1976.

[5] H. W. Bode, *Network Analysis and Feedback Design.* New York: Van Nostrand, 1945.

[6] D. C. Youla, "Network ideas applied to the design of optimal multivariable feedback controllers," Polytechnic Inst. of New York, Farmingdale, Memo POLY-EE/EP-75-146, Dec. 1973.

[7] G. C. Newton, Jr., L.A. Gould, and J. F. Kaiser, *Analytical Design of Linear Feedback Controls.* New York: Wiley, 1957.

[8] D. C. Youla, "On the factorization of rational matrices," *IRE Trans. Inform. Theory*, vol. IT-7, pp. 172–189, July 1961.

[9] W. G. Tuel, Jr., "Computer algorithm for spectral factorization of rational matrices," *IBM J. Res. Develop.*, vol. 12, pp. 163–170, Mar. 1968.

[10] H. H. Rosenbrock, *State-Space and Multivariable Theory.* New York: Wiley-Interscience, 1970.

[11] R. E. Kalman, "Irreducible realizations and the degree of a rational matrix," *J. Soc. Industrial and Applied Mathematics*, vol. 13, pp. 520–544, June 1965.

[12] H. H. Rosenbrock, "Relatively prime polynomial matrices," *Electron. Lett.*, no. 4, pp. 227–228, 1968.

[13] K. Hoffman and R. Kunze, *Linear Algebra.* Englewood Cliffs, NJ: Prentice-Hall, 1961.

[14] R. T. Yanushevskii, "Synthesis of closed-loop multivariable control systems for a certain class of plants with delay," *Automation and Remote Control*, pp. 1394–1402, Feb. 1975; Russian original, vol. 35, no. 9, pt.1, Sept. 1974.

[15] I. M. Horowitz and U. Shaked, "Superiority of transfer function over state-variable methods in linear time-invariant feedback system design," *IEEE Trans. Automat. Contr.*, vol. AC-20, pp. 84–97, Feb. 1975.

[16] H. A. Jabr, "Modern analytical design of optimal multivariable control systems," Ph.D. dissertation, Polytechnic Institute of New York, Farmingdale, 1975.

[17] *IEEE Trans. Automat. Contr. (Special Issue on Linear-Quadratic-Gaussian Problem)*, vol. AC-16, Dec. 1971.

[18] H. Kwakernaak and R. Sivan, *Linear Optimal Control Systems.* New York: Wiley-Interscience, 1972.

[19] I. M. Horowitz, *Synthesis of Feedback Systems.* New York: Academic, 1963.

Feedback System Design: The Fractional Representation Approach to Analysis and Synthesis

C. A. DESOER, FELLOW, IEEE, RUEY-WEN LIU, JOHN MURRAY, AND RICHARD SAEKS, FELLOW, IEEE

***Abstract*—The problem of designing a feedback system with prescribed properties is attacked via a fractional representation approach to feedback system analysis and synthesis. To this end we let H denote a ring of operators with the prescribed properties and model a given plant as the ratio of two operators in H. This, in turn, leads to a simplified test to determine whether or not a feedback system in which that plant is embedded has the prescribed properties and a complete characterization of those compensators which will "place" the feedback system in H. The theory is formulated axiomatically to permit its application in a wide variety of system design problems and is extremely elementary in nature requiring no more than addition, multiplication, subtraction, and inversion for its derivation even in the most general settings.**

I. Introduction

INTUITIVELY, the linear feedback system design process may be broken down into three steps: modeling, analysis, and synthesis; each of which may be carried out via a multiplicity of time and frequency domain techniques. In engineering practice, however, the three steps are loosely matched to one another. The purpose of the present paper is to use fractional representation models to the analysis and synthesis of feedback systems. Here, if one desires to design a system with prescribed properties the given plant is initially modeled as a quotient of two operators, each of which has the desired properties. Once such a model has been specified a similar model may be formulated for the feedback system constructed from that plant which, in turn, may be used to determine whether or not the feedback system has the desired properties. Moreover, the set of compensators which will cause the feedback system to have the prescribed properties may be completely characterized in terms of such a model. As such, by choosing a model for the plant which is matched to the design criteria the analysis and synthesis processes for a feedback system may be greatly simplified.

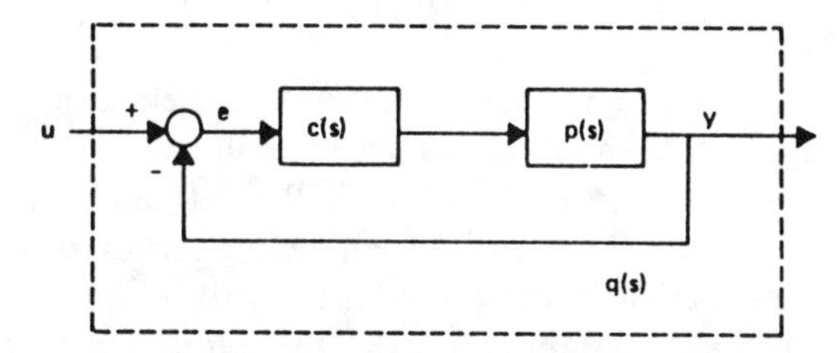

Fig. 1. Single-variate control system.

These ideas are illustrated by the following derivation of the set of stabilizing compensators for the single variate control system of Fig. 1.

We say that a transfer function $p(s)$ is *exponentially stable* (exp. stable) if $p(s)$ is a *proper rational* function with poles having *negative* real parts. Although the plant may naturally be modeled as a quotient of coprime polynomials [16],[19] $p(s)=a(s)/b(s)$ since our ultimate goal is a

Manuscript received April 2, 1979; revised October 15, 1979. Paper recommended by B. Francis, Chairman of the Linear Systems Committee. The work of C. A. Desoer was supported in part by the Joint Services Electronics Program at the University of California, Berkeley, under AFOSR Contract 76-C-0100. The work of R.-W. Liu was supported in part by ONR Contract 78-C-0444. The work of J. Murray and R. Saeks was supported in part by the Joint Services Electronics Program, Texas Tech University, under ONR Contract 76-C-1136.

C. A. Desoer is with the Department of Electrical Engineering and Computer Science, University of California, Berkeley, CA 94720.

R.-W. Liu is with the Department of Electrical Engineering, University of Notre Dame, Notre Dame, IN 46556.

J. Murray and R. Saeks are with the Department of Electrical Engineering, Texas Tech University, Lubbock, TX 79409.

Reprinted from *IEEE Trans. Automat. Contr.*, vol. AC-25, no. 3, pp. 399–412, June 1980.

stable system we prefer to model $p(s)$ as a quotient of exp. stable rational functions

$$p(s)=n(s)/d(s)=[a(s)/m(s)][b(s)/m(s)]^{-1} \quad (1.1)$$

where $m(s)$ is strictly Hurwitz polynomial of degree equal to the degree of $b(s)$. Moreover, since $a(s)$ and $b(s)$ are coprime, the rational functions $n(s)$ and $d(s)$ are coprime in the sense that there exist exp. stable rational functions $u(s)$ and $v(s)$ such that

$$u(s)n(s)+v(s)d(s)=1. \quad (1.2)$$

Similarly, we assume that our compensator is modeled as a quotient of exp. stable rational functions, $c(s)=x(s)/y(s)$, which are coprime in the above sense. Now, a little algebra will reveal that the closed-loop system transfer function from input u to output y is given by a ratio of exp. stable rational functions in the form

$$h_{yu}(s)=n(s)[y(s)d(s)+x(s)n(s)]^{-1}x(s). \quad (1.3)$$

Moreover, it can be shown[1] that $h_{yu}(s)$ will be stable if and only if

$$[y(s)d(s)+x(s)n(s)]=k(s) \quad (1.4)$$

has an exp. stable inverse. Since $k(s)$ is, itself, exp. stable this implies that the feedback system will be exp. stable if and only if $k(s)$ is nonzero for all $\mathrm{Re}\,s>0$, including ∞. An exp. stable function with these properties is called miniphase. As such, the problem of synthesizing an exp. stable feedback system reduces to the solution of (1.4) for exp. stable rational functions $x(s)$ and $y(s)$ given exp. stable functions $n(s)$ and $d(s)$ and a miniphase function $k(s)$.

By direct substitution one may verify that

$$y^h(s)=r(s)n(s) \quad \text{and} \quad x^h(s)=-r(s)d(s) \quad (1.5)$$

satisfy the homogeneous equation

$$y^h(s)d(s)+x^h(s)n(s)=0 \quad (1.6)$$

for all exp. stable rational functions $r(s)$. Moreover, since $n(s)$ and $d(s)$ are coprime it follows that all exp. stable rational solutions of (1.6) are of this form [15],[18]. On the other hand, a particular solution of (1.4) may be obtained by multiplying (1.2) by $k(s)$, which yields

$$y^p(s)=k(s)v(s) \quad \text{and} \quad x^p(s)=k(s)u(s). \quad (1.7)$$

As such, if we let $r(s)$ vary over the set of exp. stable rational functions and $k(s)$ vary over the set of miniphase functions we obtain a complete parameterization of the stabilizing compensators for our feedback system in the form

$$c(s)=\frac{x(s)}{y(s)}=\frac{[k(s)u(s)-r(s)d(s)]}{[k(s)v(s)+r(s)n(s)]}=\frac{[u(s)-w(s)d(s)]}{[v(s)+w(s)n(s)]} \quad (1.8)$$

where $w(s)=r(s)/k(s)$ ranges over the exp. stable rational functions.

A comparison of (1.8) with the class of stabilizing compensators derived by Youla, Bongiorno, and Jabr [24],[25],[29] will reveal that the two results differ only in that our $u(s)$, $v(s)$, $n(s)$, and $d(s)$ are exp. stable rational functions while theirs are polynomials.[2] Unlike their analytic derivation, however, the above result was obtained via elementary algebraic operations. Indeed, the only properties of the exp. stable rational functions employed are their closure under addition and multiplication together with the fact that the identity is an exp. stable rational function, i.e., the exp. stable rational functions form a ring with identity. As such, if the exp. stable rational functions of the above derivation were to be replaced by any prescribed ring of single-input single-output systems, (1.8) would yield a complete characterization of the compensators which would "place" the feedback system in that ring. If one works with a ring of rational functions with poles in a prescribed region a solution of the pole placement problem is obtained [18], whereas, if one chooses to work with stable transcendental functions a solution to the stabilization problem for distributed systems is obtained [7],[8] etc. Indeed, with minor modifications the derivation can be extended to noncommutative rings thereby including multivariate and time-varying systems. In each case, a simple solution to a fundamental problem of feedback system design is obtained by virtue of choosing a model for the given plant which is matched to the ultimate goal of the design problem. In particular, if we desire to design a feedback system which lies in a prescribed ring of operators we model the plant as a quotient of operators from that ring.

Consistent with the above philosophy the following section of the paper is devoted to the formulation of an axiomatic theory of fractional system representation. Here, a given system is modeled as a quotient of two operators lying in a prescribed ring H. The corresponding feedback system analysis and synthesis problems are then studied in the succeeding sections. In particular, Section III is devoted to the problem of determining whether or not a feedback system lies in H given that its plant is represented as a quotient of systems from H while Section IV is devoted to the problem of characterizing those compensators which will "place" the feedback system in H. The resultant axiomatic theory of feedback system design is applicable to multivariate, time-varying, distrib-

[1]See the axiomatic derivation of Section III for the details.

[2]From a computational point of view, it is more convenient to represent rational functions as ratios of polynomials, as per Youla *et al.*

TABLE I

EXAMPLES OF THE AXIOMATIC SYSTEM $\{G, H, I, J\}$

G	$R(s)$	$R_p(s)$	$R(s)^{n\times n}$	$R_p(s)^{n\times n}$	$\hat{B}(\sigma_0)$	$\hat{B}(\sigma)^{n\times n}$	$L_\infty(R)$	$B(H)$
H	$R[s]$	$R(\sigma_0)$	$R[s]^{n\times n}$	$R(\sigma_0)^{n\times n}$	$A_-(\sigma_0)$	$\hat{A}_-(\sigma_0)^{n\times n}$	$H_\infty(R)$	$C(H)$
I	$R[s] \neq 0$	$R^\infty(\sigma_0)$	$M \in R[s]^{n\times n}$ s.t. $\lvert M(s)\rvert \neq 0$	$M \in R(\sigma_0)^{n\times n}$ s.t. $\lvert M(s)\rvert \in R^\infty(\sigma_0)$	$\hat{A}_-^\infty(\sigma_0)$	$M \in \hat{A}_-(\sigma_0)^{n\times n}$ s.t. $\lvert M(s)\rvert \in \hat{A}_-^\infty(\sigma_0)$	$m \in H_\infty(R)$ s.t. $\inf_R \lVert m(j\omega)\rVert > 0$	$C_0(H)$
J	$m \in R[s]$ s.t. $m(s) = c \neq 0$	$m \in R^\infty(\sigma_0)$ s.t. $m(s) \neq 0$ for $s \in C_{\sigma_0^+}$	$M \in R[s]^{n\times n}$ s.t. $\lvert M(s)\rvert \neq 0$ for $s \in C_{\sigma_0^+}$	$M \in R(\sigma_0)^{n\times n}$ s.t. $\lvert M(s)\rvert \in R^\infty(\sigma_0)$ & $\lvert M(s)\rvert \neq 0$ for $s \in C_{\sigma_0^+}$	$m \in \hat{A}^\infty(\sigma_0)$ s.t. $m(s) \neq 0$ for $s \in C_{\sigma_0^+}$	$M \in \hat{A}_-(\sigma_0)^{n\times n}$ s.t. $\lvert M(s)\rvert \in \hat{A}_-^\infty(\sigma_0)$ $\lvert M(s)\rvert \neq 0$ for $s \in C_{\sigma_0^+}$	$m \in H_\infty(R)$ s.t. $\inf \lVert m(j\omega)\rVert > 0$ & m is outer	$CC(H)$
Ref.	19	31	19	15,18	4, 31	6,7,8	12	11,15

$R(s)$ = rational functions with real coeficients

$R_p(s)$ = proper rational functions with real coeficients

$X^{n\times n}$ = n by n matrices of elements in X.

A = distributions of the form $g(t) + \sum_{i=0} g_i \delta(t - t_i)$ where $g(t)$ is an integrable function s.t. $g(t) = 0$ for $t < 0$; g_i is a summable sequence and $0 = t_0 < t_1 < t_2 < \ldots$.

$\hat{A}_-(\sigma_0)$ = Laplace transforms of distributions g such that $g(t)e^{-\sigma_0 t}$ is in A for some $\sigma_1 < \sigma_0$

$\hat{A}_-^\infty(\sigma_0)$ = multiplicative subset of $\hat{A}_-(\sigma_0)$ consisting of elements bounded away from zero at ∞.

$\hat{B}(\sigma_0)$ = quotients of elements of the form m/n where $m \in A_-(\sigma_0)$ and $n \in A_-^\infty(\sigma_0)$.

$R[s]$ = polynomials with real coeficients

$C_{\sigma_0^+}$ = complex numbers with real part greater than or equal to σ_0

$R(\sigma_0)$ = proper rational functions with real coficients which are analytic in $C_{\sigma_0^+}$

$R^\infty(\sigma_0)$ = proper rational functions with real coefficients which are analytic in $C_{\sigma_0^+}$ and nonzero at ∞

$B(H)$ = bounded linear operators on a Hilbert Space H.

$C(H)$ = causal bounded linear operators on a Hilbert space H.

$C_0(H)$ = causal bounded linear operators with a bounded inverse on a Hilbert space H.

$CC(H)$ = causal bounded linear operators with a causal bounded inverse on a Hilbert space H.

$L_\infty(R)$ = essentially bounded Lebesque measurable functions defined on R.

$H_\infty(R)$ = the Hardy space of essentially bounded Lebesque measurable functions defined on R which have an analytic extension into $C_{0_0^+}$

uted, and some multidimensional systems and includes the stabilization, pole placement, and feedforward design problems. Several of these applications are illustrated by the examples of Section V. In the final section of the paper a partial generalization of the theory to nonlinear systems is described. This follows the algebraic pattern established in the linear case but is formulated in terms of a left-distributive ring to model the properties of a nonlinear system [23].

II. AXIOMATIC THEORY

Table I displays several examples of the axiomatic system developed below. Reference to it will help in visualizing the breadth and significance of the theory. Additional examples also appear in Section V.

Let G be a (not necessarily commutative) ring with identity and let H be a subring of G which includes the identity. The feedback system and its subsystems will be represented by operators which are elements of G. The compensator will be chosen so that the overall system will be represented by an operator in the subring H.

We define two multiplicative subsets [2], [27] of H,

$$I = \{h \in H \mid h^{-1} \in G\}, \tag{2.1}$$

i.e., I is the set of elements of H which have an inverse in G;

$$J = \{h \in H \mid h^{-1} \in H\}, \tag{2.2}$$

i.e., J is the *subgroup* of H consisting of all invertible elements of H. Note that

$$J \subset I \subset H \subset G. \tag{2.3}$$

Given the above structure we say that a system $g \in G$ has a *right fractional representation* in $\{G, H, I, J\}$ if there exist $n_r \in H$ and $d_r \in I$ such that $g = n_r d_r^{-1}$. Furthermore, we say that the pair $(n_r, d_r) \in H \times H$ is *right coprime* if there exist u_r and v_r in H such that

$$u_r n_r + v_r d_r = 1. \tag{2.4}$$

The right fractional representation $n_r d_r^{-1}$ in $\{G, H, I, J\}$ is said to be *right coprime* if the pair (n_r, d_r) is right coprime.

The relationship between our concept of coprimeness and the usual common factor criterion for coprimeness [28] is given by the following properties.

Property 1: Let the pair $(n_r, d_r) \in H \times H$ be right coprime. Let n_r and d_r have a common right factor $r \in H$, i.e., $n_r = x_r r$, $d_r = y_r r$ for some $x_r \in H$ and $y_r \in H$. Then r has a *left inverse* in H.

Proof: Substitute the assumed factorizations of n_r and d_r into (2.4) and obtain

$$u_r n_r + v_r d_r = (u_r x_r + v_r y_r) r = 1. \tag{2.5}$$

Since H is a ring, $u_r x_r + v_r y_r \in H$. From (2.5) it follows that $r^{-L} = u_r x_r + v_r y_r$ is a left-inverse of r. ■

Property 2: Let $g = n_r d_r^{-1}$ be a right coprime fractional representation of g in $\{G, H, I, J\}$. Let $g = x_r y_r^{-1}$ be a second (not necessarily coprime) right fractional representation of g in $\{G, H, I, J\}$. Then there exists an r in H such that

$$x_r = n_r r \quad \text{and} \quad y_r = d_r r. \tag{2.6}$$

Proof: Given the two factorizations of g, let $r = d_r^{-1}y_r$; hence $r \in G$. Then

$$y_r = d_r r \tag{2.7}$$

and, performing calculations in the ring G, we obtain

$$x_r = gy_r = (n_r d_r^{-1})y_r = n_r(d_r^{-1}y_r) = n_r r. \tag{2.8}$$

From (2.7) and (2.8), r is a common right factor of x_r and y_r. To show that $r \in H$, consider

$$r = d_r^{-1}y_r = (u_r n_r + v_r d_r)d_r^{-1}y_r = u_r n_r d_r^{-1}y_r + v_r y_r$$

$$= u_r g y_r + v_r y_r = u_r x_r + v_r y_r \in H \tag{2.9}$$

where we used the equality $g = x_r y_r^{-1} = n_r d_r^{-1}$ to derive (2.9). ■

Although G is, in general, a noncommutative ring, the entire theory developed above for right fractional representations can be replicated for left fractional representations. In particular, we say that $g \in G$ has a *left fractional representation* in $\{G,H,I,J\}$ if there exist $n_l \in H$ and $d_l \in I$ such that $g = d_l^{-1}n_l$. Furthermore we say that the pair $(n_l, d_l) \in H \times H$ is *left coprime* if there exist u_l and v_l in H such that

$$n_l u_l + d_l v_l = 1. \tag{2.10}$$

The left fractional representation $d_l^{-1}n_l$ is said to be *left coprime* if the pair (n_l, d_l) is left coprime. With these definitions the existence of a *common left factor* for a left fractional representations of g is characterized by the following properties.

Property 1': Let the pair (n_l, d_l) be left coprime. Let n_l and d_l have a common left factor l in H, i.e., $n_l = lx_l$, $d_l = ly_l$ for some $x_l \in H$ and $y_l \in H$. Then l has a *right inverse* $\in H$.

Property 2': Let $g = d_l^{-1}n_l$ be a left coprime fractional representation of g in $\{G,H,I,J\}$. Let $g = y_l^{-1}x_l$ be a second (not necessarily coprime) left fractional representation of g in $\{G,H,I,J\}$. Then there exists an l in H such that

$$x_l = ln_l \quad \text{and} \quad y_l = ld_l. \tag{2.11}$$

The above properties of a coprime fractional representation have all been derived under the assumption that such a representation exists. Of course, if G denotes the rational matrices and H denotes the polynomial matrices the existence of a coprime representation is implied by classical analysis [16],[19]. Indeed, the classical analysis readily extends to the case where H is taken to be the exp. stable rational matrices or the ring of proper rational matrices with poles in a prescribed region [18]. On the other hand for multidimensional [26], distributed [4],[8], and time-varying systems [11],[15] there is no assurance that an arbitrary $g \in G$ will admit a fractional representation nor even that the set of $g \in G$ which admit such a representation will be a linear space. Moreover, all g's which admit a fractional representation may not admit a coprime fractional representation [26]. In general, the set of $g \in G$ which admit a fractional representation in $\{G,H,I,J\}$ will form a subring of G if and only if the Ore condition[3] is satisfied while criteria for coprimeness have been formulated in various special cases though no general theory exists [1],[4],[26]. The standard condition for the existence of fractional representations which are coprime in the sense of (2.4) is that H be a right principal ideal domain.

Reference to Table I shows that in applications it is important to have conditions under which g will be in H and these conditions should be expressed in terms of its fractional representation.

Property 3: Let $g = n_r d_r^{-1}$ with $n_r \in H$ and $d_r \in I$.

a) If $d_r \in J$, then $g \in H$.

b) If $g = n_r d_r^{-1}$ is a *right coprime fractional representation* of g in $\{G,H,I,J\}$, then $g \in H$ implies that $d_r \in J$.

Proof:

a) We have $d_r \in J$; hence by (2.2), $d_r^{-1} \in H$ and thus $n_r d_r^{-1} = g \in H$.

b) We have $g \in H$. Furthermore, $n_r = gd_r$, $d_r = 1d_r$ implies that d_r is a right common factor of n_r and d_r; hence by Property 1, d_r has a left inverse in H. But $d_r \in I$ by assumption, so d_r^{-1} exists and is an element of G; thus $d_r^{-1} = d_r^{-L} \in H$; hence, by (2.2), $d_r \in J$. ■

Property 3': Let $g = d_l^{-1}n_l$ with $n_l \in H$ and $d_l \in I$.

a) If $d_l \in J$, then $g \in H$.

b) If $g = d_l^{-1}n_l$ is a *left coprime fractional representation* of g in $\{G,H,I,J\}$, then $g \in H$ implies that $d_l \in J$.

Property 4: Let $g = n_r d^{-1} n_l$ where $n_r, n_l \in H$, and $d \in I$.

a) If $d \in J$, then $g \in H$.

b) Let, in addition, $n_r d^{-1}$ be a right coprime fractional representation in $\{G,H,I,J\}$ and $d^{-1}n_l$ be a left coprime fractional representation in $\{G,H,I,J\}$; then $g \in H$ implies that $d \in J$.

Proof:

a) By assumption, $d \in J$; hence $d^{-1} \in H$. So $g = n_r d^{-1} n_l \in H$.

b) Since $d^{-1}n_l$ is a left coprime fractional representation there exist $u_l, v_l \in H$ such that

$$n_l u_l + dv_l = 1, \tag{2.12}$$

thus,

$$n_r d^{-1} = n_r d^{-1}(n_l u_l + dv_l) = n_r d^{-1} n_l u_l + n_r v_l = gu_l + n_r v_l. \tag{2.13}$$

Now $g \in H$ hence (2.13) gives $n_r d^{-1} \in H$. By Property 3, $n_r d^{-1} \in H$ together with the fact that the pair (n_r, d) is right coprime implies $d \in J$. ■

III. Analysis

To start with consider the feedback system Σ_p of Fig. 2. Suppose that the plant is described by a right coprime fractional representation $p = n_r d_r^{-1}$ in $\{G,H,I,J\}$. The

[3] $\{G,H,I,J\}$ satisfies the Ore condition for right fractional representations if, whenever $g \in G$ admits a left fractional representation it also admits a right fractional representation and vice versa.[2]

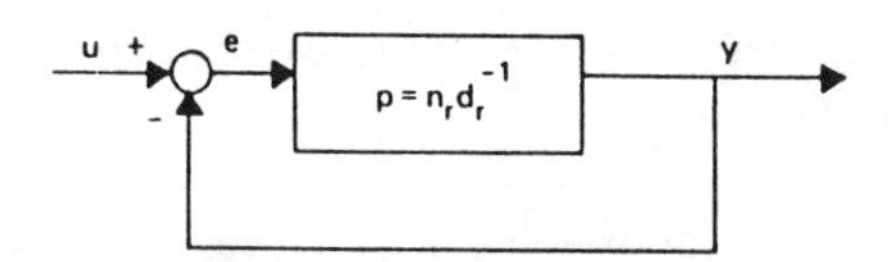

Fig. 2. Unity gain negative feedback system.

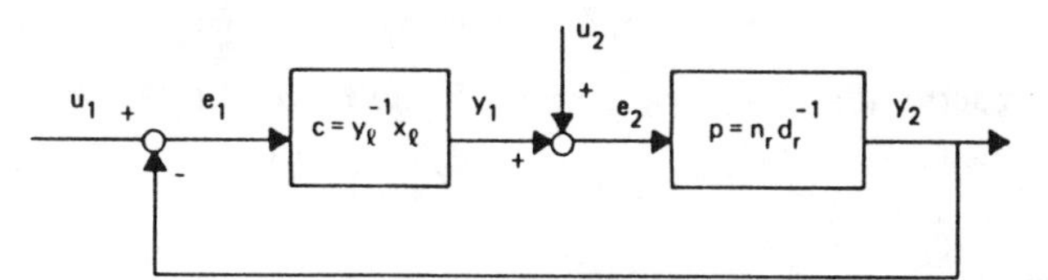

Fig. 3. Feedback system with plant and compensator.

closed-loop dynamics of Σ_p are described by the maps

$$h_{eu}: u \mapsto e\;; \qquad h_{eu}=(1+p)^{-1}=d_r(d_r+n_r)^{-1} \tag{3.1}$$

$$h_{yu}: u \mapsto y\;; \qquad h_{yu}=p(1+p)^{-1}=n_r(d_r+n_r)^{-1}. \tag{3.2}$$

Note that

$$h_{eu}+h_{yu}=1. \tag{3.3}$$

We say that Σ_p is *well defined in* G, (H, respectively), if $h_{eu}\in G$, (H, respectively).

Note that the pairs (n_r, d_r+n_r) and (d_r, d_r+n_r) are right coprime; indeed, the right coprimeness of (n_r, d_r) implies (2.4), hence

$$(u_r-v_r)n_r+v_r(d_r+n_r)=1 \tag{3.4}$$

while

$$(v_r-u_r)d_r+u_r(d_r+n_r)=1. \tag{3.5}$$

Theorem 1: Consider the feedback system Σ_p of Fig. 2.

a) Let $p=n_rd_r^{-1}$ be a fractional representation in $\{G,H,I,J\}$ of the element $p\in G$; then Σ_p *is well defined in* G if and only if $d_r+n_r\in I$.

b) Let $p=n_rd_r^{-1}$ be a *right coprime* fractional representation in $\{G,H,I,J\}$ of the element $p\in G$; then Σ_p *is well defined in* H if and only if $d_r+n_r\in J$.

Proof: a) $\Rightarrow$. $h_{eu}\in G$ and $d_r\in I$ imply

$$d_r^{-1}h_{eu}=d_r^{-1}(1+p)^{-1}=d_r^{-1}d_r(d_r+n_r)^{-1}=(d_r+n_r)^{-1}\in G. \tag{3.6}$$

Now $d_r\in I\subset H$ and $n_r\in H$, so $d_r+n_r\in H$. This together with (3.6) implies $d_r+n_r\in I$.

a) $\Leftarrow$. $d_r+n_r\in I$ implies $(d_r+n_r)^{-1}\in G$; hence $h_{eu}=d_r(d_r+n_r)^{-1}\in G$.

b) Follows from Property 3, together with (3.4) and (3.5). ■

Of course, a similar theorem holds for left factorizations.

We now consider the feedback system Σ of Fig. 3 where the plant p is preceded by a compensator c; p and c belong to G and are specified by their coprime fractional representation in $\{G,H,I,J\}$ $n_rd_r^{-1}$ and $y_l^{-1}x_l$, respectively.

To describe the feedback system Σ we consider the map $h_{eu}: (u_1,u_2)\mapsto(e_1,e_2)$. Simple calculations give

$$h_{eu}=\begin{bmatrix} h_{e_1u_1} & h_{e_1u_2} \\ h_{e_2u_1} & h_{e_2u_2}\end{bmatrix}=\begin{bmatrix}(1+pc)^{-1} & -p(1+cp)^{-1} \\ c(1+pc)^{-1} & (1+cp)^{-1}\end{bmatrix}. \tag{3.7}$$

Now let $h_{yu}: (u_1,u_2)\mapsto(y_1,y_2)$. Using the summing node equations it is easy to see that

$$h_{yu}=K(h_{eu}-1) \quad\text{and}\quad h_{eu}=1-Kh_{yu} \tag{3.8}$$

where K is the symplectic matrix

$$K=\begin{bmatrix}0 & 1\\ -1 & 0\end{bmatrix}. \tag{3.9}$$

It is well known that in the case of multivariable rational matrices, one has to consider the four submatrices of h_{eu} in (3.8) because examples show that any one of the submatrices may be unstable while the remaining ones are stable. (For detailed examples, see [30].) Let us calculate

$$\begin{aligned} h_{e_1u_1}&=(1+pc)^{-1}=1-pc(1+pc)^{-1}\\ &=1-p(1+cp)^{-1}c\\ &=1-p\big[y_l^{-1}(y_ld_r+x_ln_r)d_r^{-1}\big]^{-1}c\\ &=1-n_r(y_ld_r+x_ln_r)^{-1}x_l \end{aligned} \tag{3.10}$$

$$\begin{aligned} h_{e_2u_1}&=c(1+pc)^{-1}=(1+cp)^{-1}c\\ &=d_r(y_ld_r+x_ln_r)^{-1}x_l \end{aligned} \tag{3.11}$$

$$\begin{aligned} h_{e_2u_2}&=(1+cp)^{-1}=(1+y_l^{-1}x_ln_rd_r^{-1})^{-1}\\ &=\big[y_l^{-1}(y_ld_r+x_ln_r)d_r^{-1}\big]^{-1}\\ &=d_r(y_ld_r+x_ln_r)^{-1}y_l \end{aligned} \tag{3.12}$$

$$h_{e_1u_2}=-p(1+cp)^{-1}=-n_r(y_ld_r+x_ln_r)^{-1}y_l. \tag{3.13}$$

We say that Σ is *well defined in* G, (H, respectively) if and only if each entry of h_{eu} defined in (3.8) belongs to G, (H, respectively).

Theorem 2: Consider the feedback system Σ of Fig. 3. Let $n_rd_r^{-1}$ and $y_l^{-1}x_l$ be a right and left fractional representations of p and c in $\{G,H,I,J\}$.

a) If $y_ld_r+x_ln_r\in I$, then Σ is well defined in G.

b) If $y_ld_r+x_ln_r\in J$, then Σ is will defined in H.

c) If $h_{e_2u_2}\in G$, then $y_ld_r+x_ln_r\in I$ hence if Σ is well defined in G, then $y_ld_r+x_ln_r\in I$.

d) Assume, in addition, that $n_r(y_ld_r)^{-1}$ and $(y_ld_r)^{-1}x_l$ are right coprime and left coprime fractional representation, respectively; then $h_{e_1u_1}\in H$ implies that $y_ld_r+x_ln_r\in J$, and hence, if Σ is well defined in H, then $y_ld_r+x_ln_r\in J$.

Proof: a) and b). If $y_ld_r+x_ln_r\in I$, (J, respectively), then by the definition (2.1) of I, [(2.2) of J, respectively], the formulas (3.10)–(3.13), and the closure of the ring G, (H, respectively), the conclusion follows.

c) If $h_{e_2u_2}\in G$, then so is $d_r^{-1}h_{e_2u_2}y_l^{-1}$ since $d_r\in I$ and $y_l\in I$. Now,

$$d_r^{-1}h_{e_2u_2}y_l^{-1}=d_r^{-1}(1+cp)^{-1}y_l^{-1}$$
$$=d_r^{-1}(1+y_l^{-1}x_ln_rd_r^{-1})^{-1}y_l^{-1}$$
$$=d_r^{-1}[y_l^{-1}(y_ld_r+x_ln_r)d_r^{-1}]y_l^{-1}=(y_ld_r+x_ln_r)^{-1} \tag{3.14}$$

hence the fact that $h_{e_2u_2}\in G$ implies that $(y_ld_r+x_ln_r)^{-1}\in G$ and thus $(y_ld_r+x_ln_r)\in I$.

d) First we prove that the pair $(n_r,y_ld_r+x_ln_r)$ is right coprime. Since (n_r,y_ld_r) is right coprime, there exists $\tilde{u}_r$ and $\tilde{v}_r\in H$ such that

$$\tilde{u}_rn_r+\tilde{v}_ry_ld_r=1; \tag{3.15}$$

hence

$$(\tilde{u}_r-\tilde{v}_rx_l)n_r+\tilde{v}_r(y_ld_r+x_ln_r)=1 \tag{3.16}$$

and the claim is established. Similarly, we show that $(y_ld_r+x_ln_r,x_l)$ is left coprime. Now consider

$$h_{e_1u_1}=1-n_r(y_ld_r+x_ln_r)^{-1}x_l. \tag{3.17}$$

By assumption, $h_{e_1u_1}\in H$; then the special assumption of d) and Property 4 imply that $y_ld_r+x_ln_r\in J$. This completes the proof. ■

Note, the special assumptions used in d) to the effect that $n_r(y_ld_r)^{-1}$ is right coprime and $(y_ld_r)^{-1}x_l$ is left coprime, imply, in some sense, that p and c have no common factors. More precisely, since J serves as the group of units in our theory these conditions imply that any common factors of p and c must lie in J.

IV. Design

Consistent with our approach of matching the plant model to the goal of the given feedback system design problem the present section is devoted to the problem of characterizing the set of compensators which will "place" a feedback system in a prescribed ring H given that both the plant and compensator are modeled by fractional representations in $\{G,H,I,J,\}$.

Theorem 3: For the feedback system Σ of Fig. 3, let the plant p have a right coprime and a left coprime fractional representation $p=n_rd_r^{-1}=d_l^{-1}n_l$ in $\{G,H,I,J\}$. Let u_r and v_r both in H be such that (2.4) holds. Then for any $w\in H$ such that $wn_l+v_r\in I$, the compensator

$$c=(wn_l+v_r)^{-1}(-wd_l+u_r)\in G \tag{4.1}$$

results in a feedback system Σ well defined in H. For such a compensator, $h_{eu}\in H^{2\times 2}$ and

$$h_{eu}=\begin{bmatrix}1-n_r(-wd_l+u_r) & -n_r(wn_l+v_r)\\ d_r(-wd_l+u_r) & d_r(wn_l+v_r)\end{bmatrix}. \tag{4.2}$$

Conversely, if Σ is well defined in H and if the compensator $c=y_l^{-1}x_l$ is such that (n_r,y_ld_r) and (y_ld_r,x_l) are right coprime and left coprime respectively, then c is given by expression (4.1).

Proof:

Step 1: Choose any $k\in J$, (hence $k^{-1}\in H$), and solve for y_l and $x_l\in H$ the equation

$$y_ld_r+x_ln_r=k. \tag{4.3}$$

Observe that if (y_l,x_l) is *any solution in H* of (4.3), then

$$k^{-1}(y_ld_r)+k^{-1}(x_ln_r)=1 \tag{4.4}$$

and

$$(y_ld_r)k^{-1}+(x_ln_r)k^{-1}=1, \tag{4.5}$$

hence, (n_r,y_ld_r) is right coprime and (y_ld_r,x_l) is left coprime. Thus, the assumptions of Theorem 2, part d) holds for *any* solution of (4.3).

Step 2: Obtain all solutions of the homogeneous equation

$$y_l^hd_r+x_l^hn_r=0. \tag{4.6}$$

Since $p=n_rd_r^{-1}=d_l^{-1}n_l$, direct calculation shows that for any $r\in H$,

$$y_l^h=rn_l \qquad x_l^h=-rd_l \tag{4.7}$$

are solutions of (4.6).

It remains to show that all solutions of (4.6) are of the form (4.7); so we *assume* that y_l^h and $x_l^h\in H$ and satisfy (4.6). Let $r=-x_l^hd_l^{-1}$; hence

$$x_l^h=rd_l. \tag{4.8}$$

Now using (4.6)

$$y_l^h=y_l^hd_rd_r^{-1}=-x_l^hn_rd_r^{-1}=-x_l^hp$$
$$=-x_l^hd_l^{-1}n_l=rn_l. \tag{4.9}$$

Equations (4.8) and (4.9) show that any solution of (4.6) has the form of (4.7); it remains, however, to show that $r\in H$,

$$r=-x_l^hd_l^{-1}=-x_l^hd_l^{-1}(d_lv_l+n_lu_l)$$
$$=-x_l^hv_l-x_l^hd_l^{-1}n_lu_l=-x_l^hv_l+y_l^hu_l\in H. \tag{4.10}$$

Step 3: Obtain a particular solution of (4.3). From the right coprimeness condition for (n_r,d_r),

$$kv_rd_r+ku_rn_r=k \tag{4.11}$$

hence

$$y_l^p=kv_r, \quad x_l^p=ku_r. \tag{4.12}$$

Hence any solution of (4.3) is of the form

$$y_l=rn_l+kv_r$$
$$x_l=-rd_l+ku_r \qquad \text{for some } r\in H \tag{4.13}$$

and for any such solution (n_r,y_ld_r) is right coprime and (y_ld_r,x_l) is left coprime.

Step 4: Consider the condition

$$r\in H \text{ and } k\in J \qquad \text{such that } rn_l+kv_r\in I \quad (4.14)$$

or equivalently, if we set $w=k^{-1}r\in H$,

$$w\in H \qquad \text{such that } wn_l+v_r\in I. \quad (4.15)$$

If (4.15) holds,

$$c=(wn_l+v_r)^{-1}(-wd_l+u_r)\in G \quad (4.16)$$

is a compensator in G which can also be written as [see (4.13)]

$$c=(rn_l+kv_r)^{-1}(-rd_l+ku_r). \quad (4.17)$$

If we let $y_l=rn_l+kv_r$ and $x_l=-rd_l+ku_r$, then, by (4.17), $c=y_l^{-1}x$ and, by calculation, we verify that (4.3) holds. Thus for any such compensator, by Theorem 2, the feedback system Σ is well defined in H.

Step 5: Conversely consider a feedback system well defined in H with a compensator $c=y_l^{-1}x_l$ such that $(n_r, y_l d_r)$ and $(y_l d_r, x_l)$ are right coprime and left coprime, respectively. By Theorem 2, (4.3) holds for some $k\in J$, hence by the analysis above, c is also given by (4.1) for some $w\in H$ such that $wn_l+v_r\in I$. The proof is thus complete. ■

The theorem yields a complete parameterization of all possible controllers which will place a plant in H given the existence of:

1) right and left coprime fractional representations of p and

2) a w in H for which (wn_l+v_r) is in I.

In the multivariable case where p is a square matrix whose elements are proper rational functions it is well known that p has left and right coprime fractional representations [19]. In order to obtain a *proper* controller one has to choose w in (4.1) so that $\det[w(s)n_l(s)+v_r(s)]\neq 0$ at infinity. Methods for obtaining such a proper stabilizing controller have been reported in [32] and [33]. Alternatively, one can verify the existence of such a w in our algebraic setting by invoking the fact that n_l and d_r are right coprime and applying linear algebraic arguments thereto. Of course, these arguments apply to distributed systems as well as lumped systems using the formulation of [7] and [8].

In the most general ring theoretic setting neither right nor left coprime fractional representations of p, nor a w such that $(wn+v_r)$ is in I, are assured to exist. At present, the only known counterexample to the latter is, however, in the ring of integers which is of no system theoretic interest.

Conditions 1) and 2) have been conjectured to be both necessary and sufficient conditions for the existence of a compensator, c, which places the feedback system in H [3]. In fact, if c places the feedback system in H, then from (3.7) we obtain left and right fractional representations

$$p=(-h_{e_1u_2})(h_{e_2u_2})^{-1}=(h_{e_1u_1})^{-1}(-h_{e_1u_2}). \quad (4.18)$$

Note that there is no guarantee that these fractional representations are coprime. These representations are, however, coprime when the compensator is in H. Indeed, in that case they satisfy a stronger condition which completely characterizes those plants which can be placed in H by a compensator in H. For an early analogous result, see [10, pp. 85–87].

Corollary 1: For the feedback system Σ of Fig. 3 there exists a c in H which places the feedback system in H if and only if p admits left and right fractional representations $p=d_l^{-1}n_l=n_rd_r^{-1}$ such that n_r is a right factor of $1-d_r$ and n_l is a left factor of $1-d_l$.

Proof: If the feedback system is placed in H by a c in H it admits the fractional representations of (4.18). By calculation [see (3.7)]

$$h_{e_2u_2}-ch_{e_1u_2}=1 \quad (4.19)$$

and

$$h_{e_1u_1}-h_{e_1u_2}c=1 \quad (4.20)$$

which verifies their coprimeness since c is in H. Moreover, upon rearranging the terms in (4.19) and (4.20) the conditions of the corollary follow. Conversely, if fractional representations exist which satisfy the conditions of the corollary there exists u_r in H such that

$$u_rn_r=d_r=1 \quad (4.21)$$

(equivalently $p=n_rd_r^{-1}$ is a right coprime fractional representation with $v_r=1$). Now, by using this right fractional representation in (4.1) (with any left coprime fractional representation) and $w=0$ we obtain a compensator $c=u_r$ in H, which places the feedback system in H. ■

V. EXAMPLES

Example 1: A Single Variate Servomechanism Problem[4]

Here G is the ring of proper rational functions and H is subring of functions analytic in $\mathrm{Re}\, s > -1$. Consider the problem of designing a compensator for the unstable plant $p(s)=(s+1)/(s^2-4)$ which will simultaneously place the poles of the feedback system in the region, $\mathrm{Re}(s)<-1$, and cause the system to asymptotically track a step input. Since our transfer functions are commutative we may adopt common right and left fractional representation for $p(s)$. In particular,

$$p(s)=\frac{(s+1)}{(s^2-4)}=\left[\frac{(s+1)}{(s+2)^2}\right]\left[\frac{(s-2)}{(s+2)}\right]^{-1}=n(s)d(s)^{-1} \quad (5.1)$$

while

$$\left[\frac{16}{3}\right]\left[\frac{(s+1)}{(s+2)^2}\right]+\left[\frac{(s+2/3)}{(s+2)}\right]\left[\frac{(s-2)}{(s+2)}\right]=u(s)n(s)+v(s)d(s)=1. \quad (5.2)$$

[4]The purpose of this example is merely to give a simple illustration of the theory. In this situation, a much more highly developed theory is available in [29].

Here, each of the four rational functions, $n(s)$, $d(s)$, $u(s)$, and $v(s)$, lie in the ring of operators with poles in the region $\mathrm{Re}(s)<-1$ and hence the set of all compensators which will place the feedback system in this ring is given by Theorem 3 with $w(s)$ also in the ring. Moreover, for an arbitrary $w(s)$ the input–output mapping for the resultant feedback system will take the form

$$h_{y_2u_1}(s) = -\left[\frac{(s+1)(s-2)}{(s+2)^3}\right]w(s)+\left[\frac{16(s+1)}{3(s+2)^2}\right]$$
$$= -n(s)d(s)w(s)+n(s)u(s). \quad (5.3)$$

By the final value theorem the feedback system will asymptotically track a step input if and only if $h_{y_2u_1}(0)=1$ (equivalently $c(s)$ has a pole at zero). As such, to simultaneously place the poles of the feedback system in the region, $\mathrm{Re}(s)<-1$, and cause the feedback system to asymptotically track a step input we must find a $w(s)$ with poles in this region such that $h_{y_2u_1}(0)=1$. Evaluating (5.3) at $s=0$ and setting it equal to one yields

$$h_{y_2u_1}(0) = \frac{1}{4}w(0)+\frac{4}{3}=1, \quad (5.4)$$

implying that $w(0)=-4/3$. As such, the simplest $w(s)$ which will achieve our simultaneous goals is the constant $w(s)=-4/3$ whose poles are trivially in the prescribed region. Adopting this $w(s)$, a little algebra with the expressions of Theorem 3 will reveal that the required compensator takes the form

$$c(s)=\frac{(20s+24)(s+2)}{(3s+4)s} \quad (5.5)$$

while the input–output mapping for the feedback system takes the form

$$h_{y_2u_1} = \frac{(s+1)(20s+24)}{3(s+2)^3}. \quad (5.6)$$

Clearly, $c(s)$ has the required pole at zero (for $h_{y_2u_1}(0)=1$), although it is by no means obvious that this quasi-stable compensator will transfer the unstable poles of $p(s)$ to the prescribed region. Indeed, this illustrates the underlying power of the proposed design technique in that when one designs the system in terms of $w(s)$ rather than $c(s)$ the pole placement or stabilization process is automatically resolved by working with a $w(s)$ whose poles lie in the prescribed region while the remainder of the design process is simplified by the affine relationship between $w(s)$ and the matrices h_{eu} and h_{yu}. Finally, we note that $c(s)$ has a zero at $s=-2$ which may cancel with the pole of $p(s)$ at $s=-2$. This, however, does not contradict the coprimeness assumptions of Theorem 3 since the common factors involved lie in J which serves as the group of units in our theory. Fortunately, such common factors can never lead to an erroneous design since by assumption the poles and zeros of the rational functions in J lie in the prescribed region. As such, any cancellations which may take place are benign.

Since the previous compensator design was achieved with an especially simple $w(s)$ let us add an additional constraint to the problem by requiring that $h_{y_2u_1}(s)$ have zeros at $\pm j$ (so that the system will be insensitive to a noise source at that frequency). Now, from (5.4) it follows that the above design is the only compensator which will make $h_{y_2u_1}(0)=1$ with a constant $w(s)$; hence to satisfy this additional design constraint we will work with the first order $w(s)$ in the form

$$w(s)=\frac{as-4}{bs+3}. \quad (5.7)$$

Here, by specifying the zeroth-order coefficients of $w(s)$ we assure that $w(0)=-4/3$ while we are left with the parameters a and b to create the required zeros. Of course, to achieve our stability condition we must have $-3/b<-1$. Substituting the $w(s)$ of (5.7) into (5.3) yields

$$h_{y_2u_1}(s)=\frac{(s+1)\left[(16b-3a)s^2+(60+6a+32b)s+72\right]}{3(s+2)^3(bs+3)}. \quad (5.8)$$

To obtain the desired zeros at $s-\pm j$ the equation

$$\left[(16b-3a)s^2+(60+6a+32b)s+72\right]=k\left[s^2+1\right] \quad (5.9)$$

must be satisfied. Now, this represents three linear equations in three unknowns and has the unique solution

$$a=-17,\quad b=\frac{21}{16},\quad \text{and } k=72. \quad (5.10)$$

Moreover, $-3/b=-16/7<-1$; hence this choice of $w(s)$ will also assure the prescribed degree of stabilization. As such, we take

$$w(s)=\frac{-(17s+4)}{(21s/16+3)}=\frac{-(272s+64)}{(21s+48)} \quad (5.11)$$

which yields

$$c(s)=\frac{128(s+2)(s^2+1)}{(7s^2-56s-60)s} \quad (5.12)$$

and

$$h_{y_2u_1}(s)=\frac{384(s+1)(s^2+1)}{(s+2)^3(21s+48)} \quad (5.13)$$

satisfying all of our design criteria.

Example 2: A Multivariate Lumped-Distributed Decoupling Problem

Consider the multivariate, lumped-distributed plant

$$p(s)=\begin{bmatrix} \dfrac{e^{-1/s}}{(s+1)} & \dfrac{(s-1)}{(s+1)} \\ 0 & \dfrac{1}{(s-1)} \end{bmatrix} \quad (5.14)$$

which we desire to stabilize and simultaneously decouple by feedback. For most lumped-distributed systems one can take H to be a ring of matrices whose elements lie in the algebra $\hat{\mathcal{Q}}_-(\sigma_0)$ of stable transfer functions generated by lumped elements and delays while G is a ring of matrices whose elements lie in $\hat{\mathcal{B}}(\sigma_0)$, the algebra of quotients of elements in $\hat{\mathcal{Q}}_-(\sigma_0)$, as per Table I. In our case, however, although $e^{-1/s}$ is L_2-stable (since it is analytic on the right half-plane and bounded on the imaginary axis [10]) it has a "nasty" singularity at $s=0$ and hence does not lie in $\hat{\mathcal{Q}}(\sigma_0)$ for any $\sigma_0<0$. As such, we take H to be a ring of 2×2 matrices whose elements are transfer functions lying in the Hardy space $H_\infty(R)$ of functions which are (essentially) bounded on the $j\omega$ axis and admit an analytic extension into the right half-plane (thereby making them L_2-stable) [12]. Similarly, we let G be a ring of 2×2 matrices whose entries are transfer functions lying in the Lebesgue space $L_\infty(R)$ [12]. With this setup I becomes the set of H_∞ functions which are uniformly bounded below on the $j\omega$ axis while J is the set of H_∞ functions whose analytic extension is uniformly bounded below in the right half-plane [12]. Equivalently, J is the set of invertible outer functions in $H_\infty(R)$ [12].

Using these spaces a little algebra will reveal that $p(s)$ has the right and left coprime fractional representations in $\{G,H,I,J\}$ shown below:

$$p(s)=\begin{bmatrix}\frac{e^{-1/s}}{(s+1)} & \frac{(s-1)^2}{(s+1)^2}\\ 0 & \frac{1}{(s+1)}\end{bmatrix}\begin{bmatrix}1 & 0\\ 0 & \frac{(s-1)}{(s+1)}\end{bmatrix}^{-1}$$

$$=n_r(s)d_r(s)^{-1} \tag{5.15}$$

$$p(s)=\begin{bmatrix}1 & 0\\ 0 & \frac{(s-1)}{(s+1)}\end{bmatrix}^{-1}\begin{bmatrix}\frac{e^{-1/s}}{(s+1)} & \frac{(s-1)}{(s+1)}\\ 0 & \frac{1}{(s+1)}\end{bmatrix}$$

$$=d_l(s)^{-1}n_l(s) \tag{5.16}$$

where

$$\begin{bmatrix}0 & 0\\ 0 & 2\end{bmatrix}\begin{bmatrix}\frac{e^{-1/s}}{(s+1)} & \frac{(s-1)^2}{(s+1)^2}\\ 0 & \frac{1}{(s+1)}\end{bmatrix}+\begin{bmatrix}1 & 0\\ 0 & 1\end{bmatrix}\begin{bmatrix}1 & 0\\ 0 & \frac{(s-1)}{(s+1)}\end{bmatrix}$$

$$=u_r(s)n_r(s)+v_r(s)d_r(s)=1 \tag{5.17}$$

and

$$\begin{bmatrix}\frac{e^{-1/s}}{(s+1)} & \frac{(s-1)}{(s+1)}\\ 0 & \frac{1}{(s+1)}\end{bmatrix}\begin{bmatrix}0 & 0\\ 0 & 2\end{bmatrix}+\begin{bmatrix}1 & 0\\ 0 & \frac{(s-1)}{(s+1)}\end{bmatrix}\begin{bmatrix}1 & \frac{-2(s-1)}{(s+1)}\\ 0 & 1\end{bmatrix}=n_l(s)u_l(s)+d_l(s)v_l(s)=1. \tag{5.18}$$

Upon substitution of these matrices into the expression for $h_{y_2u_1}(s)$ from Theorem 3 one obtains

$$h_{y_2u_1}(s)=-\begin{bmatrix}\frac{e^{-1/s}}{(s+1)} & \frac{(s-1)^2}{(s+1)^2}\\ 0 & \frac{1}{(s+1)}\end{bmatrix}\begin{bmatrix}w_{11}(s) & w_{12}(s)\\ w_{21}(s) & w_{22}(s)\end{bmatrix}\cdot\begin{bmatrix}1 & 0\\ 0 & \frac{(s-1)}{(s+1)}\end{bmatrix}+\begin{bmatrix}0 & \frac{2(s-1)^2}{(s+1)^2}\\ 0 & \frac{2}{(s+1)}\end{bmatrix} \tag{5.19}$$

which will be stable if and only if the $w_{ij}(s)$ are stable.

Now, to decouple the system we require that

$$h^{12}_{y_2e_1}(s)=\frac{(s-1)e^{-1/s}}{(s+1)^2}w_{12}(s)+\frac{(s-1)^3}{(s+1)^3}w_{22}(s)+\frac{2(s-1)^2}{(s+1)^2}=0 \tag{5.20}$$

and

$$h^{21}_{y_2e_1}(s)=\frac{1}{(s+1)}w_{21}(s)=0. \tag{5.21}$$

Clearly, $w_{21}(s)=0$ solves (5.21). On the other hand (5.20) has numerous solutions none of which are, however, stable. As such, the system cannot be decoupled and stabilized simultaneously. Note, since our theory guarantees that all stable feedback systems with plant $p(s)$ take the form of (5.19) if we cannot find stable w's which decouple (5.19) we are assured that it is *impossible* to simultaneously stabilize and decouple $p(s)$ by feedback (using a compensator as specified in Theorem 3) and we need not consider other formulations.

Since we cannot simultaneously stablize and decouple $p(s)$ by feedback the best we can do is to try to stabilize $p(s)$ while preserving its triangularity (which will allow us to sequentially adjust its various outputs). Formally, this can be achieved by taking $w(s)=0$ which yields the input–output mapping

$$h_{y_2u_1}(s)=\begin{bmatrix}0 & \frac{2(s-1)^2}{(s+1)^2}\\ 0 & \frac{2}{(s+1)}\end{bmatrix}. \tag{5.22}$$

Unfortunately, the first input has been rendered useless by this compensator and hence the goal of being able to sequentially tune the outputs is not achieved. On the other hand, if we take

$$w(s)=\begin{bmatrix}1 & 0\\ 0 & 0\end{bmatrix}, \tag{5.23}$$

then

$$h_{y_2u_1}(s)=\begin{bmatrix}\dfrac{e^{-1/s}}{(s+1)} & \dfrac{2(s-1)^2}{(s+1)^2}\\ 0 & \dfrac{2}{(s+1)}\end{bmatrix} \tag{5.24}$$

which has the desired property is obtained. In particular, one can tune the second input to control the second output and then adjust the first input to simultaneously cancel out the effects of the second input on the first output and control the first output. Of course, since $w(s)$ is stable so is $h_{y_2u_1}(s)$.

Finally, we note that as we have formulated our theory one can deal only with square matrices (since rectangular matrices are not closed under multiplication). The extension to rectangular matrices is, however, straightforward [19] and yields an identical theory the details of which are left to the reader.

Example 3: A Multidimensional Image Restoration Problem

Let

$$p(z_1,z_2)=\frac{z_1+z_2}{z_1^2+z_1z_2+3} \tag{5.25}$$

denote the discrete two-dimensional transfer function for a device in a digital image processing system. Since this represents an IIR (infinite impulse response) transfer function the image processing device will tend to "smear" the image with the data observed at any one pixel distorting all other pixels at the output of the device. In an effort to reduce this "smearing" effect we would like to place the device in a feedback system whose input–output transfer function minimizes the "smearing" effect. In particular, that means that the input–output mapping for the feedback system should have an FIR (finite impulse response) transfer function with its "point-spread function" concentrated about a single point as closely as possible.

Since the FIR transfer functions are just the polynomials we let H be the ring of polynomials in two variables and G be the ring of rational functions in two variables [16]. Once again employing only a single fractional representation since these rings are commutative we obtain the coprime fractional representation

$$\begin{aligned}p(z_1,z_2)&=[z_1+z_2][z_1^2+z_1z_2+3]^{-1}\\ &=n(z_1,z_2)d(z_1,z_2)^{-1}\end{aligned} \tag{5.26}$$

where

$$\begin{aligned}&\left[-\frac{1}{3}z_1\right][z_1+z_2]+\left[\frac{1}{3}\right][z_1^2+z_1z_2+3]\\ &\quad=u(z_1,z_2)n(z_1,z_2)+v(z_1z_2)d(z_1,z_2)=1.\end{aligned} \tag{5.27}$$

As such, the set of all possible FIR transfer functions which can be obtained from $p(z_1,z_2)$ by feedback takes the form

$$\begin{aligned}h_{y_2u_1}(z_1,z_2)=&-[z_1^3+2z_1^2z_2+z_1z_2^2+3z_1+3z_2]\\ &\cdot w(z_1,z_2)-\frac{1}{3}[z_1^2+z_1z_2]\end{aligned} \tag{5.28}$$

where $w(z_1,z_2)$ is an arbitrary polynomial in two variables. Clearly, $w(z_1,z_2)$ should be low order to keep the "point-spread function" of $h_{y_2u_1}(z_1,z_2)$ as concentrated as possible. Indeed, if we take $w(z_1,z_2)=0$ we obtain

$$h_{y_2u_1}(z_1,z_2)=-\frac{1}{3}[z_1^2+z_1z_2] \tag{5.29}$$

in which the response from a given pixel effects only two adjacent pixels. Note that the fact that these pixels are not centered around the input point does not cause any difficulty since one can always shift the origin of the raster to compensate. Taking this $w(z_1,z_2)$ we obtain the simple compensator $c(z_1,z_2)=-z_1$ which represents a one directional shift and a 180° phase shift.

An alternative design which also yields a "point-spread function" which affects only two pixels, although it is shifted further from the origin, is obtained with $w(z_1,z_2)=-(1/9)z_1$. This yields

$$h_{y_2u_1}(z_1,z_2)=\frac{1}{9}[z_1^4+2z_1^3z_2] \tag{5.30}$$

and

$$c(z_1,z_2)=\frac{z_1^2(z_1+z_2)}{z_1^2+z_1z_2+3}. \tag{5.31}$$

Since two-thirds of the output energy in this design is concentrated at a single point whereas the energy is equally divided in the previous design it may be argued that this represents a superior design. On the other hand, the shift from the origin is greater and the compensator more complex in this case. Finally, since all FIR transfer functions are stable (in an appropriate sense) the feedback systems obtained via either choice of $w(z_1,z_2)$ are stable. Moreover, both compensators are, themselves, stable as is $p(z_1,z_2)$ [6].

Example 4: A Time-Varying Differential-Delay Stochastic Optimal Control Problem

Consider the feedback system of Fig. 4 where the plant represents a cascade of a time-varying function f with an ideal predictor e^s. The system is driven by a stochastic process a, which is derived from white noise by passing it through a miniphase filter with transfer function $(s+2)/(s+1)$. We desire to choose a compensator which will stabilize the system and minimize the performance measure

$$J=E\|b\|^2+E\|d\|^2 \tag{5.32}$$

under the constraint of stability. Here, d is the stochastic process observed at the output of the system, b is the

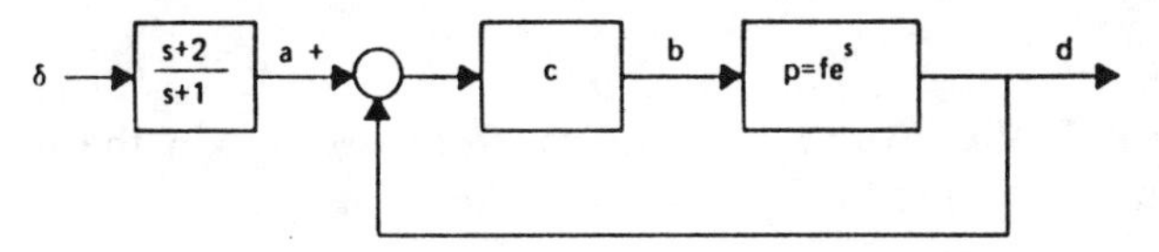

Fig. 4. Stochastic control system.

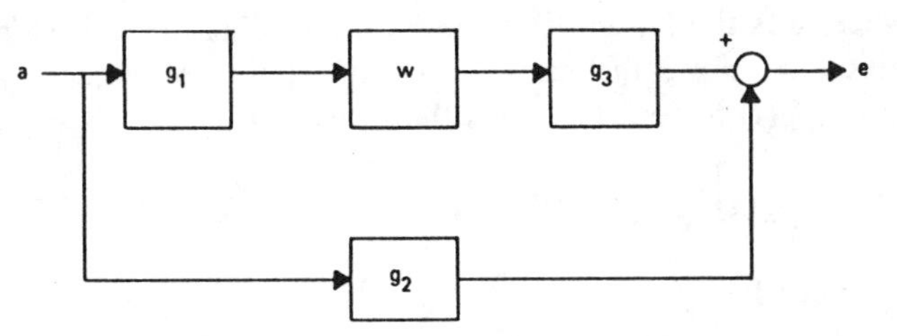

Fig. 5. Open-loop optimization problem.

stochastic process observed at the plant input, and E is the expected value operator.

Since we have a time-varying component, a rational component, and a delay component we formulate our theory in an abstract operator theoretic setting [20] with G taken to be the bounded operators on the Hilbert space $L_2(R)$ and H taken to be the causal bounded operators (which correspond to the stable systems in such a setting) [20],[23]. Note, in this setting we will denote the time-invariant operators by their transfer function and the time-varying multiplication operators by their characteristic function. Of course, one must be careful with such notation since the operational calculus associated with the time-invariant components is only partially valid in such a setting.

Since the inverse of a predictor is the ideal delay which is causal one immediately obtains the right and left coprime fractional representations for p in the form

$$p=[f][e^{-s}]^{-1}=[e^{-s}f^{-1}]^{-1}[1]=n_r d_r^{-1}=d_l^{-1}n_l \tag{5.33}$$

where

$$[f^{-1}][f]+[0][e^{-s}]=u_r n_r+v_r d_r=1 \tag{5.34}$$

and

$$[1][1]+[e^{-s}f^{-1}][0]=n_l u_l+d_l v_l=1. \tag{5.35}$$

Here, we have assumed that f^{-1} exists and is bounded (i.e., f is bounded away from zero) while f and f^{-1} are both causal since multiplication by a function of time is a memoryless operation [20]. From Theorem 3 it now follows that the input–output and input-plant input mappings for our feedback system with compensator defined by a causal operator w will take the form

$$h_{y_2u_1}=-[f]w[e^{-s}][f^{-1}]+1 \tag{5.36}$$

and

$$h_{e_2u_1}=-[e^{-s}]w[e^{-s}][f^{-1}]+[e^{-s}][f^{-1}]. \tag{5.37}$$

As such, our optimization problem reduces to choosing the causal w which minimizes the performance measure of (5.32) where $d=h_{y_2u_1}a$ and $b=h_{e_2u_1}a$.

It is significant to note that even though we are interested in designing an optimal closed-loop system by minimizing over the operator w rather than the compensator we have transformed the problem into the open-loop optimization problem of Fig. 5.

Here we desire to minimize $J=E\|e\|^2$ over all causal operators w, where g_1, g_2, and g_3 are arbitrarily specified bounded operators. In our case we take

$$g_1=[e^{-s}][f^{-1}] \tag{5.38}$$

$$g_2=\begin{bmatrix} -1 \\ \hdashline -[e^{-s}][f^{-1}] \end{bmatrix} \tag{5.39}$$

and

$$g_3=\begin{bmatrix} -[f] \\ \hdashline -[e^{-s}] \end{bmatrix} \tag{5.40}$$

in which case the output of the open-loop system is $e=(d,b)$ in the product space constructed from two copies of the (Hilbert) space on which the given system is defined. Now, if we take the a in our open-loop problem to coincide with the given a in the closed-loop optimization problem then the Pythogorean law (in Hilbert space) implies that

$$J=E\|e\|^2=E\|d\|^2+e\|b\|^2. \tag{5.41}$$

As such, our two optimization problems coincide.

Interestingly, an explicit solution has recently been given for the above open-loop optimization problem [9]. Indeed, the optimal causal w is given by

$$w_0=\lambda^{-1}[\lambda^{*-1}g_3^*g_2Q_ag_1^*\theta^{*-1}]_C\theta^{-1} \tag{5.42}$$

where λ and θ are causal, causally invertible operators such that

$$\lambda^*\lambda=g_3^*g_3 \quad \theta\theta^*=g_1Q_ag_1^*. \tag{5.43}$$

Q_a is the covariance for the stochastic processes a,$[\ ]_C$ denotes the causal part of an operator, and "$*$" denotes the adjoint operator. To apply this general theory to our example we represent the adjoint operation when applied to a transfer function by $g(s)^*=g(-s)$ which coincides with the classical adjoint on the $j\omega$ axis. Of course, the memoryless multiplication operators, $[f]$ and $[f^{-1}]$, are self adjoint. Finally, since a is the stochastic process generated from white noise by passing it through the filter $(s+2)/(s+1)$

$$Q_a=\left[\frac{(s+2)}{(s+1)}\right]\left[\frac{(s+2)}{(s+1)}\right]^*=\frac{(s+2)(s-2)}{(s+1)(s-1)}. \tag{5.44}$$

First, we calculate λ and θ via

$$\lambda^*\lambda=[-f \,\vdots\, -e^{s}]\begin{bmatrix} -f \\ \hdashline -e^{-s} \end{bmatrix}=f^2+1 \tag{5.45}$$

and

$$\theta\theta^* = [e^{-s}][f^{-1}]\left[\frac{(s+2)(s-2)}{(s+1)(s-1)}\right][f^{-1}][e^s]$$
$$= [f_{-1}^{-1}][e^{-s}]\left[\frac{(s+2)(s-2)}{(s+1)(s-1)}\right][e^s][f_{-1}^{-1}]$$
$$= [f_{-1}^{-1}]\frac{(s+2)(s-2)}{(s+1)(s-1)}[f_{-1}^{-1}]. \tag{5.46}$$

Here $f_{-1}(t)=f(t-1)$ and we have used the properties of the delay and predictor to obtain the equalities $[e^{-s}][f^{-1}] = [f_{-1}^{-1}][e^{-s}]$ and $[f^{-1}][e^s]=[e^s][f_{-1}^{-1}]$. Of course, the exponential transfer functions commute with the rational transfer functions allowing the cancellation of the exponential terms in (5.46). From (5.45) and (5.46) one may now readily obtain the required causal, causally invertible λ and θ operators in the form

$$\lambda=\lambda^*=\sqrt{f^2+1} \text{ and } \lambda^{-1}=\lambda^{*-1}=\frac{1}{\sqrt{f^2+1}} \tag{5.47}$$

while

$$\theta=[f_{-1}^{-1}]\left[\frac{(s+2)}{(s+1)}\right], \quad \theta^*=\left[\frac{(s-2)}{(s-1)}\right][f_{-1}^{-1}],$$
$$\theta^{-1}=\left[\frac{(s+1)}{(s+2)}\right][f_{-1}], \text{ and } \theta^{*-1}=[f_{-1}]\left[\frac{(s-1)}{(s-2)}\right]. \tag{5.48}$$

The next step in evaluating (5.42) is to compute the term in the bracket, i.e.,

$$\lambda^{*-1}g_3^*g_2Q_ag_1^*\theta^{*-1}$$
$$=\frac{1}{\sqrt{f^2+1}}[\,-f \;\vdots\; -e^s\,]\left[\frac{-1}{[-e^{-s}][f^{-1}]}\right]$$
$$\cdot\left[\frac{(s+2)(s-2)}{(s+1)(s-1)}\right][f^{-1}][e^s]\cdot[f_{-1}]\left[\frac{(s-1)}{(s-2)}\right]$$
$$=\frac{1}{\sqrt{f^2+1}}[f+f^{-1}]\left[\frac{(s+2)(s-2)}{(s+1)(s-1)}\right]$$
$$\cdot[f^{-1}][f]\frac{(s-1)}{(s-2)}[e^s]$$
$$=[f^{-1}]\sqrt{f^2+1}\left[\frac{(s+2)}{(s+1)}\right][e^s] \tag{5.49}$$

whose causal part must now be computed. Recalling that the memoryless term factors through the causal part bracket [9] it suffices to compute the causal part of the time-invariant system with transfer function

$$g(s)=\left[\frac{(s+2)}{(s+1)}e^s\right]. \tag{5.50}$$

Taking the inverse Laplace transform we obtain the impulse response of this system in the form

$$g(t)=\delta(t+1)+\bar{e}^{(t+1)}U(t+1) \tag{5.51}$$

where δ is the Dirac delta function and U is the unit step function. Now, the causal part of $g(t)$ is obtained by setting $g(t)$ to zero for t less than zero; hence

$$[g(t)]_C=g(t)U(t)=\bar{e}^{(t+1)}U(t)=\frac{1}{e}e^{-t}U(t) \tag{5.52}$$

or equivalently

$$[g(s)]_C=\frac{1}{e(s+1)}. \tag{5.53}$$

Multiplying through by the memoryless factor from (5.49) we then obtain

$$[\lambda^{*-1}g_3^*g_2Q_ag_1^*\theta^{*-1}]_C=[f^{-1}]\sqrt{f^2+1}\,\frac{1}{c(s+1)} \tag{5.54}$$

and finally

$$w_0=\lambda^{-1}[\lambda^{*-1}g_3^*g_2Q_ag_1^*\theta^{*-1}]_C\theta^{-1}$$
$$=\frac{1}{\sqrt{f^2+1}}[f^{-1}]\sqrt{f^2+1}$$
$$\cdot\frac{1}{e}\left[\frac{1}{(s+1)}\right]\left[\frac{(s+1)}{(s+2)}\right][f_{-1}]$$
$$=\frac{1}{e}[f^{-1}]\frac{1}{(s+2)}[f_{-1}] \tag{5.55}$$

which is surprisingly simple given the complexity of the derivation.

Substituting the expression of (5.55) into the formula of Theorem 3 now yields an expression for our optimal compensator and the input–output mapping for the resultant feedback system in the form

$$c=e[f_{-1}^{-1}](s+2)-1 \tag{5.56}$$

and

$$h_{y_2u_1}=\frac{-e^{-s}}{e(s+2)}+1. \tag{5.57}$$

Note that $h_{y_2u_1}$ is stable, as required, even though both p and c are unstable.

VI. Nonlinear Feedback Systems

From an algebraic point of view the fundamental difference between linear and nonlinear systems is the fact that nonlinear systems fail to satisfy the right-distributive property, $x(y+z)=xy+xz$. They do, however, satisfy all of the other axioms for a ring with identity including the left-distributive property $(y+z)x=yx+zx$. As such, one can attempt to extend the preceding development to nonlinear systems by carrying it out in left-distributive rings, G and H [23]. Indeed, if we define a right coprime fractional representation for a system g in a left-distributive ring G relative to $\{G,H,I,J\}$ precisely as we did in Section II the fundamental properties 1, 2, and 3 go through without modification.

Property 1N: Let $g=n_rd_r^{-1}$ be a right coprime fractional representation of g in $\{G,H,I,J\}$ where G and H are left-distributive rings with identity. Let n_r and d_r have a common right factor $r \in H$, i.e., $n_r=x_rr$, $d_r=y_rr$ for some $x_r \in H$ and $y_r \in H$. Then r has a left inverse in H.

Property 2N: Let $g=n_rd_r^{-1}$ be a right coprime fractional representation of g in $\{G,H,I,J\}$ where G and H are left-distributive rings with identity. Let $g=x_ry_r^{-1}$ be a second (not necessarily coprime) right fractional representation of g in $\{G,H,I,J\}$; then there exists r in H such that

$$x_r=n_rr \quad \text{and} \quad y_r=d_rr. \tag{6.1}$$

Property 3N: Let $g=n_rd_r^{-1}$ with $n_r \in H$ and $d_r \in I$ where G and H are left-distributive rings with identity.

a) If $d_r \in J$, then $g \in H$.

b) If $g=n_rd_r^{-1}$ is a *right coprime fractional representation* of g in $\{G,H,I,J\}$, then $g \in H$ implies $d_r \in J$.

With the aid of property 3N one can do a complete analysis of a nonlinear feedback system $h_{yu}=p(1+p)^{-1}=n_rd_r^{-1}$ where $n_rd_r^{-1}$ is a right coprime fractional representation of h_{yu}. Indeed, h_{yu} is well defined in G if and only if $d_r \in I$ and it is well defined in H if and only if $d_r \in J$. Note, however, that we cannot construct our fractional representation for h_{yu} from a fractional representation for p since the verification that such a representation is coprime appears to require right-distributivity [see (3.4) and (3.5)].

The right coprime fractional representation plays a special role in the *nonlinear* case because $h_{yu}=p(1+p)^{-1}$ holds, whereas $h_{yu}=(1+p)^{-1}p$ does not (even though the latter formula is true for the *linear* case). As such, those results on the analysis of feedback systems which assume a left coprime fractional representation theory fail as does the design theorem since it simultaneously employs both left and right coprime fractional representations. We believe, however, that these results should hold, at least in part, for nonlinear systems with an appropriate modification of the theory. In particular, since the rings G and H are asymmetric we believe that asymmetric concepts of left and right coprimeness will be required to achieve this end.

VII. Conclusions

Although several of our examples are characterized by a deep analytic structure the key to our fractional representation approach to feedback system design is the algebraic nature of the main results. Indeed, the entirety of our modeling, analysis, and synthesis theory was derived with no more sophisticated mathematics than addition, multiplication, subtraction, and inversion. As such, it applies to essentially any class of linear systems and by proper choice of the rings G and H the results are applicable to a variety of systems problems.

Although we believe that the present work represents the first attempt at the formulation of an axiomatic fractional representation theory for systems which may be matched to the feedback system analysis and synthesis problems of interest the work owes much to a number of recent results on the input–output theory of linear systems. The use of a fractional representation theory for multivariate systems, though implicit in a number of classical results, was popularized by Rosenbrock's polynomial matrix fractions [19]. Interestingly, however, Rosenbrock's goal was apparently to permit the powerful analytic and arithmetic theory available for polynomial matrices to be applied to rational matrices whereas the present fractional representation theory is motivated by the desire to formulate a representation theory for systems which is closed under inversion. Over the years numerous generalizations of the polynomial matrix fraction concept have been formulated for distributed systems [4], [5], [13], [21], and multidimensional systems [9],[24] while partial extensions to the time-varying and nonlinear cases have appeared in a number of unpublished reports [11],[22].

For any type of fractional representation theory to be meaningful it must be identified with an appropriate coprimeness concept. Indeed, the key to the present formulation is the use of the algebraic coprimeness concept of (2.4) in lieu of the more classical common factor criterion. Such a criterion has previously been applied by one of the authors in a study of fractional representations for distributed system [4] and was also shown to be the strongest of several possible coprimeness criteria for multidimensional systems by Youla and Gnavi [26]. Of course, it is well known as one of the several equivalent criteria for coprimeness in the polynomial matrix fraction theory [16],[19].

The feedback system analysis theorems of Section III are motivated by the now classical theorems for determining the stability of a multivariate feedback system in terms of its polynomial matrix fraction representation [10]. Moreover, the system synthesis theorem is an outgrowth of the feedback system stabilization theorem of Youla *et al.* [24],[25]. Indeed, the present work began with an attempt to give a simple proof of this most powerful analytic theorem and developed through several stages of generalization and simplification into its present form. Finally, the optimization theory used in Example 4 represents the generalization [9] to an operator theoretic setting of a result originally developed by Youla *et al.* in the frequency domain for use in conjunction with their stabilization theorem [24],[25].

Acknowledgment

The authors wish to thank an anonymous reviewer for numerous useful and perceptive comments.

REFERENCES

[1] W. Averson, "Interpolation problems in nest algebras," *J. Functional Analysis*, vol. 20, pp. 208–233, 1975.
[2] Bourbaki, *Commutative Algebra*. Paris: Hermann, 1972.
[3] C. I. Byrnes, private communication.
[4] F. M. Callier, and C. A. Desoer, "An algebra of transfer functions for distributed linear time-invariant systems," *IEEE Trans. Circuits Syst.*, vol. CAS-25, pp. 651–662, 1978.
[5] ——, "Open-loop unstable convolution feedback systems with dynamical feedback," *Automatica*, vol. 12, pp. 507–518, 1976.
[6] ——, "Dynamic output feedback stabilization of a control system," presented at Amer. Math. Soc. 761st Meeting, Charleston, Nov. 1978 (abstract appeared in *Notices AMS*, vol. 25, p. A-665, 1978).
[7] ——, "Stabilization, tracking and disturbance rejection in linear multivariable distributed systems," in *Proc. 17th IEEE Conf. Decision and Control*, San Diego, Jan. 1979, p. 513.
[8] ——, "Stabilization, tracking and disturbance rejection in multivariable convolution systems," Tech. Memo. UCB/ERL M78/83, Dec. 1978.
[9] R. M. DeSantis, R. Saeks, and J. J. Tung, "Basic optimal estimation and control problems in Hilbert space," *Math. Syst. Theory*, to be published.
[10] C. A. Desoer and M. Vidyasagar, *Feedback Systems: Input-Output Properties*. New York: Academic, 1975.
[11] C. A. Desoer, R. W. Liu, J. Murray, and R. Saeks, "On stabilization of feedback systems," unpublished notes, 1978.
[12] R. G. Douglas, *Banach Algebra Techniques in Operator Theory*. New York: Academic, 1972.
[13] B. Francis, "The multivariable servomechanism problem from the input–output viewpoint," *IEEE Trans. Automat. Contr.*, vol. AC-22, pp. 322–328, 1977.
[14] M. L. J. Hautus and M. Heymann, "Linear feedback—An algebraic approach," *SIAM J. Contr. Optimiz.*, vol. 16, pp. 83–105, 1978.
[15] N. T. Hung and B. D. O. Anderson, "Triangularization for the design of multivariable control systems," *IEEE Trans. Automat. Contr.*, to be published; also in *Proc.* 1978 *IEEE CDC Conf.*, pp. 795–800, Jan. 1973.
[16] E. I. Jury, *Inners and Stability of Dynamic Systems*. New York: Wiley, 1974.
[17] M. B. Morf, B. Levy, and S. Y. Kung, "New results in 2-D system theory, part I: 2-D polynomial matrices, factorization and coprimeness," *Proc. IEEE*, vol. 65, pp. 861–872, 1977.
[18] A. S. Morse, "System invariants under feedback and cascade control," in *Mathematical Systems Theory*, Udine, 1975, G. Marchesini and S. K. Mitter, Eds. Heidelberg: Springer-Verlag, 1976.
[19] H. H. Rosenbrock, *State-Space and Multivariable Theory*. London: Nelson-Wiley, 1970
[20] R. Saeks, *Resolution Space, Operators, and Systems*. Heidelberg: Springer-Verlag, 1973.
[21] M. Vidyasagar, "On the input–output stability of multivariable linear time-invariant feedback systems containing unstable systems," *IEEE Trans. Circuits Syst.*, to be published.
[22] ——, unpublished notes, 1978.
[23] J. C. Willems, *Analysis of Feedback Systems*. Cambridge, MA: MIT Press, 1971.
[24] D. C. Youla, J. J. Bongiorno, and H. A. Jabr, "Modern Wiener–Hopf design of optimal controllers—Part I," *IEEE Trans. Automat. Contr.*, vol. AC-21, pp. 3–15, 1976.
[25] ——, "Modern Wiener–Hopf design of optimal controllers—Part II," *IEEE Trans. Automat. Contr.*, vol. AC-21, pp. 319–338, 1976.
[26] D. C. Youla and G. Gnavi "Notes on *n*-dimensional system theory," unpublished notes, 1978.
[27] O. Zariski and P. Samuel, *Commutative Algebra*, vol. I. Princeton, NJ: Van Nostrand, 1958 p. 46, exp.
[28] S. MacLane and G. Birkhoff, *Algebra*. New York: MacMillan, 1971, p. 154.
[29] J. J. Bongiorno and D. C. Youla, "On the design of single-loop single-input single-output feedback control systems in the complex frequency domain," *IEEE Trans. Automat. Contr.*, vol. AC-22, pp. 416–423, 1977.
[30] C. A. Desoer and W. S. Chan, "The feedback interconnection of lumped linear time-invariant systems," *J. Franklin Inst.*, vol. 300, pp. 335–351, 1975.
[31] F. M. Callier and C. A. Desoer, "Simplification and clarification on the paper 'An algebra of transfer function for distributed linear time-invariant systems,'" *IEEE Trans. Circuits Syst.*, to be published.
[32] L. Cheng and J. B. Pearson, "Frequency domain synthesis of multivariable linear regulators," *IEEE Trans. Automat. Contr.*, vol. AC-23, pp. 3–15, 1978.
[33] P. J. Antsaklis and J. B. Pearson, "Stabilization and regulation in linear multivariable systems," *IEEE Trans. Automat. Contr.*, vol. AC-23, pp. 928–930, 1978.

ON THE ROLE OF THE NEVANLINNA–PICK PROBLEM IN CIRCUIT AND SYSTEM THEORY*

PH. DELSARTE, Y. GENIN AND Y. KAMP

Philips Research Laboratory Brussels, Av. Van Becelaere 2, Box 8, B-1170 Brussels, Belgium

SUMMARY

The paper is concerned with applications of the standard Nevanlinna–Pick problem in various technical domains, namely the following: interpolation by reflectance functions, polynomial stability checking, cascade synthesis of passive one-ports, and model reduction with a Hankel norm criterion. Some fundamental results on the Nevanlinna–Pick problem are shown to be of a definite interest in each of these subjects.

1. INTRODUCTION

The Nevanlinna–Pick interpolation problem has a long history in mathematics and can actually be traced back to the work of Pick[1] and Nevanlinna.[2] A strong interest has been brought to this problem, extending considerably the original setting and putting into light various relationships between *a priori* unrelated topics.[3–5] This research has produced a comprehensive bunch of mathematical techniques, some of them being of direct use in the field of circuit and system theory, as first observed by Fenyves.[6]

The aim of the present paper is precisely to discuss certain applications of the Nevanlinna–Pick problem in this field. Four different subjects will be successively considered, namely interpolation by reflectance functions,[6–9] polynomial stability checking,[10] cascade synthesis of passive one-ports,[11–13] and model reduction in Hankel norm.[14] A clear-cut connection will be established between each of these problems and a particular form of the Nevanlinna–Pick problem, which in turn will provide the reader with efficient mathematical methods of solution. Earlier results will thus appear to be simplified, unified and sometimes extended. It is true enough to say that the role played by the Nevanlinna–Pick problem in these questions was anticipated by several authors, in more or less explicit forms. However, there seems to be a definite need in the technical literature to present these facts in a forceful and cogent manner.

The Nevanlinna–Pick problem can actually be approached in four different but essentially equivalent ways, depending on whether one considers positive (impedance) or bounded (reflectance) functions in the right half plane (continuous time case) or in the unit disk (discrete time case). For the sake of clarity, a unique setting of the problem will be adopted in this paper; as is well known, the results can be carried over to any of the three other situations by the usual bilinear transformation argument.

One of the simplest forms of the Nevanlinna–Pick problem is examined in Section 2. A set of distinct points $z_1, \ldots, z_n$ with $|z_i| < 1$ and of complex values $u_1, \ldots, u_n$ being given, one is interested in obtaining a rational Schur function (i.e., a rational function of modulus not larger than unity in the closed unit disk), interpolating the prescribed value u_i at each point z_i. This type of problem has been considered to some extent in the technical literature,[6–9] mostly in connection with broadband matching and circuit modelling. The classical solvability criterion, due to Pick,[1] is first recalled. Next it is shown that the Nevanlinna recursion[2] provides a constructive algorithm yielding the unique solution in the 'degenerate case' and a parametrization of all minimum degree solutions in the 'nondegenerate case'. It is furthermore pointed out that the Nevanlinna recursion can be rephrased in terms of well-defined homographic transformations (see Reference 4). An interesting reformulation of the problem as a linear equation in three polynomial unknowns is finally introduced.

* Dedicated to V. Belevitch, on the occasion of his sixtieth birthday.

Section 3 is devoted to establishing a polynomial stability test entirely based on the numerical values assumed by the given polynomial at a certain number of points which can be arbitrarily chosen inside the unit circle. In particular, it is shown that the stability of a polynomial $q(z)$ of degree n is equivalent to the solvability of an associated Nevanlinna–Pick problem with n interpolation points z_i and the corresponding values $u_i = \hat{q}(z_i)/q(z_i)$. Applying the Nevanlinna algorithm to solve that problem directly produces a new stability test which appears naturally as a generalization of the classical Schur–Cohn criterion.[10]

The subject of Section 4 is the Darlington cascade synthesis problem for passive one-ports.[11-13] It turns out that the successive sections in such a synthesis can be identified with the successive homographic transformations occurring in the solution of the following Nevanlinna–Pick problem: to find a normalized minimum degree Schur function interpolating the reflectance of a given one-port at its transmission zeros. This result, rather obvious in the present context, was more than anticipated a decade ago. In fact, Youla and Saito[7] used essentially the Nevanlinna recursion to construct some particular solutions to an interpolation problem with positive real functions. Next, Belevitch[9] recognized that the transmission zeros of the reflectance function solving a given Nevanlinna–Pick problem coincide with the corresponding interpolation points. Later on, Belevitch and Genin[15] used the polynomial formulation of the Nevanlinna–Pick problem to derive the general cascade decomposition of lossless two-ports, thus unifying earlier work on the subject by Youla[16] and Fettweis.[17] More recently, Dewilde, Vieira and Kailath[18] established the fundamental equivalence of the Darlington synthesis with a particular form of the Nevanlinna recursion met in the theory of optimal linear filtering for stationary stochastic processes. This result was amplified and put forth in full generality by Dewilde and Dym.[19]

Section 5 is concerned with an entirely different subject. Given a stable discrete transfer function $h(z)$, the model reduction problem requires one to find a stable approximate model $h'(z)$ of lowest possible degree such that a well-defined norm of the difference $h(z)-h'(z)$ does not exceed a prescribed tolerance.[20] The so-called Hankel norm has recently received a considerable attention for that purpose (see Reference 14 and the papers quoted therein). The fundamental theory of bounded Hankel operators that underlies this particular model reduction problem was developed by Adamjan, Arov and Krein.[5] The final section of the present paper gives a brief account of part of this theory and shows that the rational approximation problem in Hankel norm is closely related to a generalized version of the Nevanlinna–Pick problem. In fact, it turns out that the model reduction problem in Hankel norm can be approached via an interpolation problem of the Nevanlinna–Pick type, with the only difference that the solution has no longer to be looked for in the class of Schur functions but in the class of rational functions of unit modulus on the unit circle. In spite of this difference, the mathematical techniques developed to solve the classical Nevanlinna–Pick problem can still be applied. As a result, a fairly good insight into this promising model reduction method can be gained by potential users without resorting to very sophisticated mathematics.

2. THE NEVANLINNA–PICK PROBLEM

Let n distinct complex numbers $z_1, z_2, \ldots, z_n$, with $|z_i|<1$, and n complex numbers $u_1, u_2, \ldots, u_n$ be given. The particular case of the *Nevanlinna–Pick problem* considered here consists in constructing a *rational Schur function* $s(z)$, i.e. a rational function of modulus not larger than unity in the closed unit disk $|z|\leq 1$, interpolating the prescribed value u_i at each point z_i, thus satisfying $s(z_i) = u_i$ for $i = 1, \ldots, n$. (Note that the problem can be extended in a natural way to the case where the interpolation points are not necessarily distinct, at the cost of some extra algebraic manipulations.)

Obviously, the interpolation problem above does not always admit a solution. Let us now quote a criterion for its solvability, in terms of the *Nevanlinna–Pick Hermitian matrix* $\mathbf{N}$ defined as

$$\mathbf{N} = \left[\frac{1-\bar{u}_i u_j}{1-\bar{z}_i z_j}: i, j = 1, 2, \ldots, n\right]. \tag{1}$$

It turns out that the Nevanlinna–Pick problem is solvable if and only if $\mathbf{N}$ is non-negative definite (see e.g. References 1 and 3).

Assuming a solution to exist, one is faced with the question of its actual computation. We shall now discuss in some detail an efficient iterative algorithm achieving this goal, and closely related to the cascade synthesis of passive networks as shown in Section 4.

Let $s_1(z)$ denote a solution to the Nevanlinna–Pick interpolation problem. For the present discussion we disregard the degenerate case where $s_1(z)$ is a constant of unit modulus. From the given Schur function $s_1(z)$ and the interpolation pair (z_1, u_1) define the rational function

$$s_2(z) = \frac{1-\bar{z}_1 z}{z-z_1}\,\frac{s_1(z)-u_1}{1-\bar{u}_1 s_1(z)} \tag{2}$$

From the fact that $s_1(z)$ is analytic in the closed unit disk $|z| \leq 1$ and satisfies $s_1(z_1) = u_1$ it follows that $s_2(z)$ is analytic in $|z| \leq 1$. Otherwise one would have $1-\bar{u}_1 s_1(\zeta) = 0$ for some point ζ with $|\zeta| \leq 1$, so that the assumptions $|u_1| \leq 1$ and $|s_1(\zeta)| \leq 1$ force $|s_1(\zeta)| = |u_1| = 1$, hence $s_1(z) \equiv u_1$ as a consequence of the maximum modulus theorem, in contradiction with our assumption of non-degeneracy. On the other hand, one readily deduces from (2) the identity

$$1-|s_2(e^{i\theta})|^2 = \frac{(1-|u_1|^2)(1-|s_1(e^{i\theta})|^2)}{|1-\bar{u}_1 s_1(e^{i\theta})|^2} \tag{3}$$

Hence $|s_2(e^{i\theta})| \leq 1$ for all θ, so that $s_2(z)$ appears to be a Schur function in view of the maximum modulus theorem.

By inverting (2) one further deduces

$$s_1(z) = \frac{u_1(1-\bar{z}_1 z) + (z-z_1)s_2(z)}{1-\bar{z}_1 z + \bar{u}_1(z-z_1)s_2(z)} \tag{4}$$

An argument similar to that indicated above shows that, given an arbitrary rational Schur function $s_2(z)$, the function $s_1(z)$ defined from (4) is a Schur function interpolating the value u_1 at the point z_1. As a result, the original Nevanlinna–Pick problem is transformed via (2) and (4) into an entirely similar problem with one less interpolation point. In fact, the constraints to be satisfied by $s_2(z)$ are found by use of (2) to be

$$s_2(z_i) = u_{i,2} = \frac{(1-\bar{z}_1 z_i)(u_i - u_1)}{(z_i - z_1)(1-\bar{u}_1 u_i)}, \qquad i = 2, \ldots, n \tag{5}$$

Iterating this procedure up to exhausting all interpolation points yields the *Nevanlinna recursion scheme*,[2,6] that can be described as follows.

(a) Compute the so-called Fenyves array of complex numbers $u_{i,k}$ by using the recurrence formula

$$u_{i,k} = \frac{(1-\bar{z}_{k-1} z_i)(u_{i,k-1} - u_{k-1,k-1})}{(z_i - z_{k-1})(1-\bar{u}_{k-1,k-1} u_{i,k-1})} \tag{6}$$

for $1 \leq k \leq i \leq n$, with the initialization $u_{i,1} = u_i$.

(b) From the diagonal entries $v_k = u_{k,k}$ of the Fenyves array compute iteratively the general solution $s_1(z)$ with the help of the formula

$$s_k(z) = \frac{v_k(1-\bar{z}_k z) + (z-z_k)s_{k+1}(z)}{1-\bar{z}_k z + \bar{v}_k(z-z_k)s_{k+1}(z)} \tag{7}$$

for $k = n, n-1, \ldots, 2, 1$, where $s_{n+1}(z)$ is an arbitrary rational Schur function.

In the description above it is implicitly understood that no degeneracy of the form $|v_k| = 1$ occurs, so that one has $|v_k| < 1$ for all k. Note that, in this non-degenerate situation, the Nevanlinna–Pick problem has infinitely many solutions $s_1(z)$, parametrized by $s_{n+1}(z)$. To complete the discussion let us now briefly show that the degenerate case where $|v_{r+1}| = 1$ for some $r < n$ (with $|v_1| < 1, \ldots, |v_r| < 1$) leads to a unique solution $s_1(z)$, which has degree r. As indicated earlier, $|v_{r+1}| = 1$ is only possible if $s_{r+1}(z)$ is identically equal to v_{r+1}. Thus equation (7) has to be iterated with the initialization $k = r$ and $s_{r+1}(z) = v_{r+1}$, so that the resulting Schur function $s_1(z)$ is uniquely determined and has degree r. In addition, successive application of (3) shows that

$s_1(e^{i\theta})$ has unit modulus, implying that $s_1(z)$ can be written as

$$s_1(z)=\hat{q}(z)/q(z) \tag{8}$$

where $q(z)$ is a polynomial of degree $\leq r$, devoid of zeros in the closed unit disk, and $\hat{q}(z)$ denotes its reciprocal of degree r, i.e. $\hat{q}(z)=z^r\bar{q}(1/\bar{z})$. It can be shown that this degenerate situation yielding a unique solution of degree r occurs precisely when the Nevanlinna–Pick matrix **N** is singular (non-negative definite) of rank r.

For future use let us now interpret the backward Nevanlinna recursion formula (7) as a homographic transformation associated with a **J**-lossless matrix.[4,18] Assuming $|v_k|<1$, with $v_k=u_{k,k}$ as above, define the 2×2 matrix function

$$\mathbf{L}_k(z)=(1-|v_k|^2)^{-\frac{1}{2}}\begin{bmatrix}1 & v_k\\ \bar{v}_k & 1\end{bmatrix}\begin{bmatrix}(z-z_k)/(1-\bar{z}_k z) & 0\\ 0 & 1\end{bmatrix} \tag{9}$$

Denoting by **J** the diagonal matrix diag $(1,-1)$ one easily verifies that $\mathbf{L}_k(z)$ is **J**-*lossless*, in the sense that it satisfies

$$\mathbf{L}_k(z)\mathbf{J}\tilde{\mathbf{L}}_k(z)\begin{cases}\leq \mathbf{J} \text{ for } |z|\leq 1\\ =\mathbf{J} \text{ for } |z|=1\end{cases} \tag{10}$$

(Here $\mathbf{A}\leq\mathbf{B}$ signifies that $\mathbf{B}-\mathbf{A}$ is non-negative definite, and $\tilde{\mathbf{X}}$ denotes the conjugate transpose of **X**.) Thus $\mathbf{L}_k(z)$ is **J**-*contractive* inside the unit circle and **J**-*unitary* on the boundary. Given any **J**-contractive matrix $\mathbf{L}=[l_{i,j}\colon i,j=1,2]$ let us then define the *homographic transform* $y=L[x]$ of any complex number x with $|x|\leq 1$ as $y=(l_{1,1}x+l_{1,2})/(l_{2,1}x+l_{2,2})$. Note that $\mathbf{LJ\tilde{L}}\leq\mathbf{J}$ and $|x|\leq 1$ imply $|y|\leq 1$. It is readily verified that (7) can be rewritten by use of (9) as the homographic transformation

$$s_k(z)=\mathbf{L}_k(z)[s_{k+1}(z)] \tag{11}$$

Next put the matrix $\mathbf{T}(z)=\mathbf{L}_1(z)\mathbf{L}_2(z)\ldots\mathbf{L}_n(z)$, which is **J**-lossless in view of (10). Using the multiplicative property of homographic transformations one immediately deduces from (11)

$$s_1(z)=\mathbf{T}(z)[s_{n+1}(z)] \tag{12}$$

As a result, in case **N** is positive definite the general solution $s_1(z)$ of the Nevanlinna–Pick problem can be expressed in compact form by (12), i.e. more explicitly, by

$$s_1(z)=\frac{t_{1,1}(z)s_{n+1}(z)+t_{1,2}(z)}{t_{2,1}(z)s_{n+1}(z)+t_{2,2}(z)} \tag{13}$$

where the $t_{i,j}(z)$ are well-defined polynomials of degree not exceeding n. Moreover, it turns out that the minimal degree solutions $s_1(z)$ result from choosing the Schur function $s_{n+1}(z)$ to be an arbitrary constant σ with $|\sigma|\leq 1$. In particular, from the fact that $\mathbf{T}(e^{i\theta})$ is **J**-unitary it follows that taking $|\sigma|=1$ yields a solution $s_1(z)=\mathbf{T}(z)[\sigma]$ with unit modulus on the unit circle; hence $s_1(z)$ has the form (8) where $q(z)$ is a polynomial of full degree n.

Let us finally reformulate the Nevanlinna–Pick problem so as to exhibit its close relationship with a very peculiar approximation problem. This theme will be further developed in Section 5, in a more general setting. From the data z_i and u_i let us introduce the polynomials $a(z)$ and $b(z)$, of degrees n and $n-1$, respectively, given by

$$a(z)=\prod_{i=1}^{n}(z-z_i),\qquad b(z)=\sum_{i=1}^{n}\frac{u_i\hat{a}(z_i)a(z)}{a'(z_i)(z-z_i)} \tag{14}$$

with $a'(z)$ the derivative of $a(z)$. By construction, one has $b(z_i)/\hat{a}(z_i)=u_i$ for $i=1,2,\ldots,n$, so that one can write

$$b(z)/\hat{a}(z)-s_1(z)=a(z)f(z) \tag{15}$$

for any solution $s_1(z)$ of the Nevanlinna–Pick problem, with $f(z)$ denoting a rational function analytic in $|z|\leq 1$. Writing $s_1(z)=p(z)/q(z)$ for some relatively prime polynomials $p(z)$ and $q(z)$ one deduces from (15)

$$b(z)q(z)-\hat{a}(z)p(z)=a(z)r(z) \tag{16}$$

where $r(z)=f(z)\hat{a}(z)q(z)$ is a well-defined polynomial. As a consequence, the Nevanlinna–Pick problem is equivalent to the determination of three polynomials $p(z)$, $q(z)$ and $r(z)$ satisfying (16) with the requirement that $s_1(z)=p(z)/q(z)$ is a Schur function (see Reference 15 in that respect). The particular solutions $s_1(z)$ with unit modulus on $|z|=1$ have the form (8), so that (16) can be written in this case as

$$\frac{b(z)}{a(z)}-\frac{r(z)}{q(z)}=\frac{\hat{a}(z)\hat{q}(z)}{a(z)q(z)} \tag{17}$$

This exhibits the difference between $b(z)/a(z)$ and $r(z)/q(z)$ as a function of unit modulus on $|z|=1$. Thus $r(z)/q(z)$ can be viewed as an approximation of $b(z)/a(z)$ with an error of unit L_∞-norm on $|z|=1$. In Section 5 this interpretation will be shown to be quite useful in the problem of the optimal Hankel approximation of a given rational function $b(z)/a(z)$ analytic in $|z|\geq 1$. As a matter of fact, the algebraic machinery described in the present section to solve the Nevanlinna–Pick problem will be shown to apply as well to this approximation problem, although the latter is generally not reducible to a classical Nevanlinna–Pick problem.

3. POLYNOMIAL STABILITY CHECKING

The problem of determining whether a given polynomial $q(z)$ is devoid of zeros in the closed unit disk $|z|\leq 1$ is standard in numerous applications of circuit and system theory concerned with stability problems. Most of the practical tests that have been devised to check that property can be viewed as different implementations of the same criterion, commonly referred to as the Schur–Cohn criterion.[10]

The main idea underlying this criterion can be presented as follows. If $q(z)$ does not vanish for $|z|\leq 1$, the rational function $s(z)=\hat{q}(z)/q(z)$ is irreducible, analytic in $|z|\leq 1$, and has unit modulus on $|z|=1$; hence $s(z)$ is a Schur function of type (8). Conversely, it is quite obvious that the denominator $q(z)$ of any irreducible rational Schur function of the form $s(z)=\hat{q}(z)/q(z)$ is devoid of zeros in $|z|\leq 1$. As a matter of fact, the *Schur–Cohn criterion* relative to a given polynomial $q(z)$ of degree n precisely amounts to checking whether $s(z)=\hat{q}(z)/q(z)$ is a Schur function of degree n by performing n steps of the classical Schur algorithm.[10]

An alternative way of doing the same thing consists of solving a particular Nevanlinna–Pick problem. Let indeed $(z_1, z_2, \ldots, z_n)$ be any n-tuple of distinct points in the unit disk $|z|<1$. Consider then the following problem: given a polynomial $q(z)$ of degree n, to find a rational Schur function $s_1(z)$ of degree n having unit modulus on $|z|=1$ and satisfying the interpolation constraints

$$s_1(z_i)=s(z_i)=\hat{q}(z_i)/q(z_i) \tag{18}$$

for $i=1,2,\ldots,n$. As shown in Section 2, the problem admits a solution if and only if the Nevanlinna–Pick matrix

$$\mathbf{N}=\left[\frac{1-\bar{s}(z_i)s(z_j)}{1-\bar{z}_i z_j}: i,j=1,\ldots,n\right] \tag{19}$$

is positive definite. In this case, $q(z)$ turns out to be devoid of zeros in $|z|\leq 1$. Indeed, a solution to the interpolation problem (18) can be computed by means of the Nevanlinna algorithm and can be expressed in the form (12) as $s_1(z)=\mathbf{T}(z)[\varepsilon]$ where ε is an arbitrary constant of unit modulus. Let us choose $\varepsilon=\mathbf{T}^{-1}(z_0)[s(z_0)]$ for any point z_0 with $|z_0|=1$ and $q(z_0)\neq 0$; this actually yields $|\varepsilon|=1$ as a consequence of $|s(z_0|=1$ because $\mathbf{T}(z_0)$ is $\mathbf{J}$-unitary. By construction, one has $s_1(z)=\hat{q}_1(z)/q_1(z)$ for a certain polynomial $q_1(z)$ of degree n devoid of zeros in $|z|\leq 1$. On the other hand, $s(z)$ and $s_1(z)$ coincide in the $2n+1$ points

$z_0, z_1, \ldots, z_n, 1/\bar{z}_1, \ldots, 1/\bar{z}_n$; this clearly implies $s(z) = s_1(z)$ and $q(z) = cq_1(z)$ for some constant c, which proves the assertion.

As a result, a stability criterion for the polynomial $q(z)$ appears to be the positive definiteness of the Nevanlinna–Pick matrix (19) or, equivalently, of the matrix

$$\mathbf{N}' = \left[\frac{\tilde{q}(z_i)q(z_j) - \tilde{\hat{q}}(z_i)\hat{q}(z_j)}{1 - \bar{z}_i z_j} : i, j = 1, \ldots, n\right] \tag{20}$$

Strangely enough, this test has received little attention in the technical literature. Of course, it requires the evaluation of $q(z)$ at the $2n$ points z_i and $1/\bar{z}_i$. However, there exist applications (typically, interpolation problems) where the required values are *a priori* known. On the other hand, one could take benefit of the freedom of choice for the points z_i in situations where the usual tests of the Schur–Cohn type are inappropriate because of an excessive numerical sensitivity, i.e. situations where some zeros of $q(z)$ are very close to the unit circle.

Let us now emphasize that the Nevanlinna–Pick test just discussed can be organized in terms of recursive formulae quite similar to the Schur–Cohn formulae. Indeed, taking into account the particular form $\hat{q}(z)/q(z)$ of the Schur function $s_1(z)$, one easily verifies from (9) and (11) that $s_k(z)$ can be written as $\hat{q}_k(z)/q_k(z)$ where $q_k(z)$ is a polynomial of degree $n - k + 1$ determined from the recursion

$$q_{k+1}(z) = \frac{1}{1 - \bar{z}_k z}[q_k(z) - \bar{v}_k \hat{q}_k(z)] \tag{21}$$

with $v_k = \hat{q}_k(z_k)/q_k(z_k)$ and the initialization $q_1(z) = q(z)$. As pointed out in Section 2, the positive definiteness of the matrix (20) is equivalent to the set of inequalities

$$|v_k| < 1, \qquad k = 1, 2, \ldots, n \tag{22}$$

It is worth mentioning that the recursive form (21) (22) of the Nevanlinna–Pick test degenerates into the classical Schur–Cohn test when all interpolation points z_k coalesce to the origin of the z-plane, as is apparent from putting $z_k = 0$ in (21). In general, the test (21), (22) can be viewed as a simple modification of the Schur–Cohn test, obtained from applying the homographic transformation $z \to (z + z_k)/(1 + \bar{z}_k z)$ at the kth step of the algorithm. (See Reference 18 in that respect.)

4. THE DARLINGTON SYNTHESIS

The present section is devoted to showing that the iterative solution to the Nevanlinna–Pick problem has an interesting connection with the *Darlington cascade synthesis* of passive lumped networks.[11-13]

Let $s(z)$ be a rational Schur function of degree n satisfying $|s(z)| < 1$ in the unit disk $|z| \leq 1$. Performing the bilinear transformation $z = (1 - p)/(1 + p)$ one can interpret $s(z)$ as the reflectance of a passive one-port network of degree n with no transmission zero on Re $p = 0$.

Given any n-tuple $(z_1, \ldots, z_n)$ of distinct points z_i satisfying $|z_i| < 1$, let us denote by $(u_1, \ldots, u_n)$ the n-tuple of values assumed by $s(z)$ at those points:

$$u_i = s(z_i), \qquad i = 1, 2, \ldots, n \tag{23}$$

We now consider the Nevanlinna–Pick problem with interpolation data (z_i, u_i) for the subclass of rational Schur functions $s_1(z)$ of degree n. In view of (12), the general solution has the form

$$s_1(z) = \mathbf{T}(z)[\sigma] \tag{24}$$

where σ is an arbitrary constant of modulus not larger than unity. In general, $s_1(z)$ does of course not coincide with $s(z)$. However, it is shown below how a particular choice of the interpolation points z_i and the parameter σ actually yields $s_1(z) = s(z)$.

Let $s(z)=p(z)/q(z)$ and $s_1(z)=p_1(z)/q_1(z)$ be the representations of $s(z)$ and $s_1(z)$ as irreducible fractions. By construction, one can write

$$\begin{bmatrix} p_1(z) \\ q_1(z) \end{bmatrix} = \prod_{k=1}^{n} (1-\bar{z}_k z) \, . \, \mathbf{T}(z) \begin{bmatrix} \sigma \\ 1 \end{bmatrix} \tag{25}$$

in view of (9) and (24). Using the fact that $\mathbf{T}(z)$ is $\mathbf{J}$-unitary on $|z|=1$ one easily deduces from (25) the identity

$$\hat{q}_1(z)q_1(z)-\hat{p}_1(z)p_1(z)=(1-|\sigma|^2)\prod_{k=1}^{n}(z-z_k)(1-\bar{z}_k z) \tag{26}$$

Let us now choose the interpolation points $z_1, \ldots, z_n$ as the transmission zeros of $s(z)$ inside the unit circle, i.e. the zeros of the polynomial $\hat{q}(z)q(z)-\hat{p}(z)p(z)$ in the region $|z|<1$. Thus we make the restrictive assumption that these zeros are distinct. (Note that there is no transmission zero on the unit circle, in view of the property $|s(z)|<1$ for $|z|\leq 1$.) Let us further normalize (24) by choosing $\sigma=\mathbf{T}^{-1}(z_0)[s(z_0)]$ where z_0 is any point with $|z_0|=1$, which implies $|\sigma|<1$ and $s_1(z_0)=s(z_0)$.

It is then easily seen that $s(z)$ and $s_1(z)$ coincide in $2n+1$ points and hence are necessarily identical. Indeed, for $k=0,1,\ldots,n$ one has $s_1(z_k)=s(z_k)$ by construction, while for $k=1,\ldots,n$ one has both $s_1(1/\bar{z}_k)=1/\bar{s}_1(z_k)$ in view of (26) and $s(1/\bar{z}_k)=1/\bar{s}(z_k)$ since z_k is a transmission zero of $s(z)$, which forces $s_1(1/\bar{z}_k)=s(1/\bar{z}_k)$ and thus proves the assertion. As a result, the given reflectance $s(z)$ can be written as $s(z)=\mathbf{T}(z)[\sigma]$ and hence admits a realization of the form depicted in Figure 1, i.e. a cascade of two-ports with transfer matrices $\mathbf{L}_k(z)$ of degree one, closed on the constant one-port with reflectance σ. Note that these two-ports are passive and lossless in view of (10) while the one-port is a complex resistance with positive real part since $|\sigma|<1$. Observe also that each transfer matrix $\mathbf{L}_k(z)$ depends on two parameters (z_k and v_k), so that the realization of Figure 1 is 'canonical'.

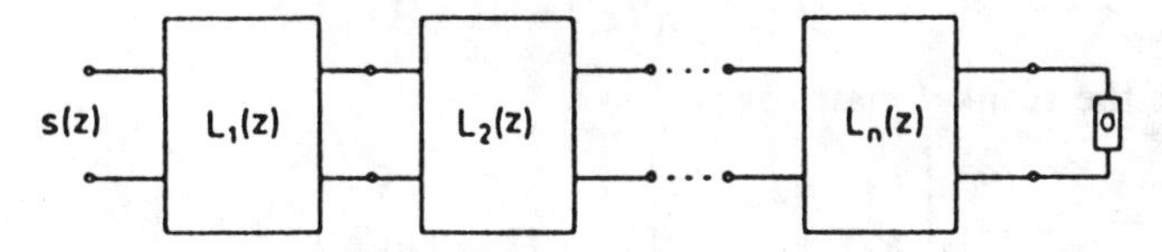

Figure 1. Cascade realization of $s(z)$

In fact, the Darlington synthesis of passive one-ports is easily recognized in the construction just described. Of course, some more algebra is needed to allow one to obtain the Youla sections (by combining the sections built on pairs of complex conjugate transmission zeros) and the Darlington sections (by increasing the degree so as to achieve a reciprocal realization). On the other hand, one should get rid of the assumption that the transmission zeros are distinct. All this can be done and has actually been known for decades, but does not really affect the following conclusion: *the Darlington synthesis procedure applied to the reflectance $s(z)$ is essentially the Nevanlinna recursive algorithm solving the Nevanlinna–Pick problem associated with $s(z)$ when the interpolation points are chosen as the transmission zeros of $s(z)$ inside the unit circle.*

5. RATIONAL APPROXIMATION IN HANKEL NORM

The discrete version of the *model reduction problem*, which is standard in system theory and applications, can be described as follows. Let $h(z)$ denote the transfer function of a stable discrete-time finite-dimensional linear system. By definition, $h(z)$ is a rational function analytic in the region $|z|\geq 1$. We make the additional assumption that $h(z)$ is proper, i.e. vanishes at infinity. The problem consists in deriving from $h(z)$ a stable proper rational approximation $h'(z)$ of degree as low as possible but still good enough with regard to the specific application considered. Then the reduced model of the system is defined via its transfer function $h'(z)$.

Of course, the problem makes sense only when a precise way of measuring the deviation of the approximant $h'(z)$ with respect to the original function $h(z)$ is given. This classically leads to introducing an error norm in the problem. The so-called Hankel norm criterion has recently received special attention and has been shown to be very attractive.[20] This criterion can be introduced as follows.

Any proper rational function $x(z)$ analytic in $|z|\geq 1$ is uniquely representable by a power series

$$x(z)=\sum_{i=1}^{\infty} x_i z^{-i} \tag{27}$$

convergent in the region $|z|\geq 1$. Let us now consider the infinite *Hankel matrix* $\mathbf{X}$ associated with $x(z)$, namely

$$\mathbf{X}=\begin{bmatrix} x_1 & x_2 & x_3 & \cdots \\ x_2 & x_3 & x_4 & \cdots \\ x_3 & x_4 & x_5 & \cdots \\ \vdots & \vdots & \vdots & \end{bmatrix} \tag{28}$$

The *Hankel norm* of $x(z)$ is then defined as the usual l_2-norm of the matrix $\mathbf{X}$, viewed as an operator acting on the space of square-summable sequences. We recall that the l_2-norm $\|\mathbf{X}\|_2$ equals the largest singular value of $\mathbf{X}$. Thus

$$\|x(z)\|=\|\mathbf{X}\|_2 \tag{29}$$

defines $\|x(z)\|$ as the Hankel norm of the function $x(z)$.

The deviation of the transfer function $h'(z)$ of the reduced model from the transfer function $h(z)$ of the original system is henceforth measured by the Hankel norm of the difference $h(z)-h'(z)$. Thus, in view of (29), this deviation is

$$\|h(z)-h'(z)\|=\|\mathbf{H}-\mathbf{H}'\|_2, \tag{30}$$

where $\mathbf{H}$ and $\mathbf{H}'$ denote the Hankel matrices

$$\mathbf{H}=\begin{bmatrix} h_1 & h_2 & \cdots \\ h_2 & h_3 & \cdots \\ \vdots & \vdots & \end{bmatrix}, \qquad \mathbf{H}'=\begin{bmatrix} h'_1 & h'_2 & \cdots \\ h'_2 & h'_3 & \cdots \\ \vdots & \vdots & \end{bmatrix} \tag{31}$$

built on the power series expansions

$$h(z)=\sum_{i=1}^{\infty} h_i z^{-i}, \qquad h'(z)=\sum_{i=1}^{\infty} h'_i z^{-i} \tag{32}$$

both convergent in $|z|\geq 1$. Note that the sequences $(h_1, h_2, \ldots)$ and $(h'_1, h'_2, \ldots)$ occurring in (32) are the impulse responses of the original system and of the reduced model, respectively.

Let us now mention three reasons motivating the current interest in the Hankel norm criterion.[14]

(a) With regard to intrinsic relevance: the Hankel norm criterion turns out to be a satisfactory trade-off between the popular least-squares and uniform error criteria; indeed, the Hankel norm is bounded from below by the Euclidean norm and from above by the Chebyshev norm, i.e.

$$\left(\sum_{i=1}^{\infty} |x_i|^2\right)^{\frac{1}{2}} \leq \|x(z)\| \leq \max_{0\leq\theta\leq 2\pi} |x(e^{i\theta})|. \tag{33}$$

(b) With regard to robustness: the Hankel norm is fairly stable with respect to numerical perturbations due, for example, to the effect of noise on the impulse response.

(c) With regard to practical implementation: a Hankel approximation of minimum degree with a preassigned error tolerance can be achieved at a moderate computational cost.

It turns out that the theory of rational approximation in Hankel norm is closely related to a generalized form of the Nevanlinna–Pick interpolation problem. We shall now explain this relationship by using a simplified version[14] of the fundamental theory of Hankel operators developed by Adamjan, Arov and Krein.[5]

Given a stable proper rational function $h(z)$ one is asked to find an approximation $h'(z)$ of $h(z)$ with a preassigned tolerance λ in the Hankel norm; thus one must have

$$\|h(z)-h'(z)\| \leq \lambda \tag{34}$$

for a given positive number λ. Suppose a rational function $f(z)$ has been determined under two requirements: $f(z)$ is analytic at each pole of $h(z)$, and $\lambda^{-1}[h(z)-f(z)]$ is an *all-pass function*, which means

$$|h(e^{i\theta})-f(e^{i\theta})| = \lambda \tag{35}$$

for all real θ. Define then the approximant $h'(z)$ to be the *stable proper part of* $f(z)$, written as

$$h'(z) = f_{st}(z) \tag{36}$$

i.e. the rational function obtained as the partial fraction expansion of $f(z)$ having all its poles inside the unit circle and vanishing at infinity.

To exhibit the main properties of the function $h'(z)$ given by (36) we need to consider the ordered sequence

$$\lambda_1 \geq \lambda_2 \geq \ldots \geq \lambda_n > 0 \tag{37}$$

formed by the n non-zero *singular values* λ_i of the Hankel matrix $\mathbf{H}$ built on $h(z)$. Note that λ_1 is the Hankel norm $\|h(z)\|$. (Let us recall that the squared singular values of $\mathbf{H}$ are the eigenvalues of $\mathbf{H}\tilde{\mathbf{H}}$.) By convention, we put $\lambda_0 = \infty$ and $\lambda_{n+1} = 0$. The most striking features of the construction above can be described as follows.

(a) The degree t of $h'(z)$ is uniquely defined from the location of the parameter λ with respect to the singular values of $\mathbf{H}$ by the inequalities

$$\lambda_t > \lambda \geq \lambda_{t+1} \tag{38}$$

(b) The function $h'(z)$ is a minimum-degree approximant of $h(z)$ with tolerance λ, in the sense that it satisfies the requirement (34) and has the lowest possible degree among all such approximants. In addition, one has $\|h(z)-h'(z)\| \geq \lambda_{t+1}$.

(c) If λ is chosen to be the singular value λ_{t+1}, then $h'(z)$ is uniquely determined as the minimum-error approximation of $h(z)$ with degree not exceeding t. Thus $h'(z)$ minimizes the Hankel norm of $h(z)-h'(z)$ in the class of all stable proper rational functions of degree $\leq t$. In this case the approximation error is given by $\|h(z)-h'(z)\| = \lambda_{t+1}$.

Our main objective in the sequel is to put into full clarity the equivalence between determining a suitable function $f(z)$ and solving a generalized version of the Nevanlinna–Pick problem. For the sake of simplicity, we make the restrictive assumption that all poles of the given function $h(z)$ are simple. Let us write $h(z) = b(z)/a(z)$ where $a(z)$ and $b(z)$ are relatively prime polynomials with $\deg b(z) < \deg a(z) = n$. By assumption, $a(z)$ has n distinct zeros $z_1, z_2, \ldots, z_n$ in the open unit disk $|z|<1$. Since all-pass rational functions have the general form $\hat{p}(z)/p(z)$, the condition (35) for the irreducible fraction $f(z) = r(z)/q(z)$, required to be analytic at the poles z_i of $h(z)$, clearly amounts to

$$\frac{b(z)}{a(z)} - \frac{r(z)}{q(z)} = \lambda \frac{\hat{a}(z)\hat{q}(z)}{a(z)q(z)} \tag{39}$$

Note that (39) is formally identical to (17), except for the presence of the parameter λ. Writing (39) as

$$\frac{b(z)}{\hat{a}(z)} - \lambda \frac{\hat{q}(z)}{q(z)} = \frac{r(z)a(z)}{q(z)\hat{a}(z)} \tag{40}$$

one immediately observes that the problem of determining $f(z) = r(z)/q(z)$ can be reformulated as follows: to find an all-pass rational function $s_1(z) = \hat{q}(z)/q(z)$ assuming the values

$$s_1(z_i) = b(z_i)/\lambda\hat{a}(z_i) \tag{41}$$

at the n zeros z_i of the polynomial $a(z)$. This generalized Nevanlinna–Pick interpolation problem is known to be solvable in the class of all-pass rational functions $s_1(z)$ of degree not exceeding n, except possibly when the parameter λ coincides with a singular value of **H**. But even in these exceptional cases the conclusions (a)(b)(c) remain valid for the function $h'(z)$ resulting via continuity from the general situation. Thus the solution to the 'parametric interpolation problem' always enjoys the properties quoted above. Note the essential difference with respect to the classical Nevanlinna–Pick problem: a solution has no longer to be found in the class of Schur functions but in the class of all-pass functions. (Thus the requirement of analyticity in the unit disk is dropped.)

It can be shown that the recursive algorithm described in Section 2 to solve the classical Nevanlinna–Pick interpolation problem still applies to the present generalized version, except for a finite number of values of the parameter λ. Let us examine the first step of the algorithm, which is based on (4). It appears from (3) that if $s_2(z)$ is an all-pass function so is $s_1(z)$ provided $|u_1| \neq 1$. Disregarding this case and, more generally, assuming none of the numbers $v_k = u_{k,k}$ in (6) to have unit modulus, one easily convinces oneself that the Nevanlinna recursive formulas (6) and (7) can still be used to generate the solution $s_1(z)$, which thus takes the form

$$s_1(z) = \mathbf{T}(z)[\varepsilon] \tag{42}$$

where $\mathbf{T}(z)$ is the product $\mathbf{L}_1(z)\mathbf{L}_2(z)\ldots\mathbf{L}_n(z)$ of the matrices (9), and ε is any constant of unit modulus. Note that $\mathbf{T}(e^{i\theta})$ is not always **J**-unitary as in the classical problem, because all numbers v_k need not have modulus smaller than unity. In fact $\mathbf{T}(e^{i\theta})$ satisfies $\mathbf{T}(e^{i\theta})\mathbf{J}\tilde{\mathbf{T}}(e^{i\theta}) = \tau\mathbf{J}$ with $\tau = \prod_{k=1}^{n} \operatorname{sgn}(1-|v_k|^2)$. Note also that v_k is a rational function of λ, so that the condition $|v_k| \neq 1$ excludes only a finite set of values of the parameter λ, containing all singular values λ_i.

It is interesting to mention how the singular values can be determined from the generalized Nevanlinna–Pick matrix **N** given by

$$\mathbf{N} = \left[\frac{\lambda^2 - \bar{w}_i w_j}{1 - \bar{z}_i z_j}: i, j = 1, \ldots, n\right] \tag{43}$$

where $w_i = b(z_i)/\hat{a}(z_i)$. It turns out that the eigenvalues $\lambda_1^2, \lambda_2^2, \ldots, \lambda_n^2$ of $\mathbf{H}\tilde{\mathbf{H}}$ are the zeros of the determinant of **N**. In fact, one can prove a stronger result:[14] the matrix **N** is reducible under a constant congruence transformation to the matrix diag $(\lambda^2-\lambda_1^2, \lambda^2-\lambda_2^2, \ldots, \lambda^2-\lambda_n^2)$. This property leads to a simple derivation of part of the result (a) as we now briefly explain. Define $\alpha_m = \prod_{k=1}^{m}(1-|v_k|^2)$ for $m = 1, 2, \ldots, n$. From (21) it is not difficult to show, by use of Rouché's theorem, that the number of zeros of $q(z)$ in $|z|<1$ equals the number of indices m such that $\alpha_m < 0$. On the other hand, it turns out that the matrix **N** is congruent to diag $(\alpha_1, \alpha_2, \ldots, \alpha_n)$. In view of Sylvester's law of inertia, this readily leads to the desired conclusion, namely that the number of zeros of $q(z)$ in $|z|<1$ is the integer t defined from $\lambda_t > \lambda > \lambda_{t+1}$.

REFERENCES

1. G. Pick, 'Uber die Beschränkungen analytischer Funktionen, welche durch vorgegebene Funktionswerte bewirkt werden', *Math. Ann.*, **77**, 7–23 (1916).
2. R. Nevanlinna, 'Uber beschrankte analytische Funktionen', *Ann. Acad. Sci. Fenn.*, **A32**, 1–75 (1929).
3. N. I. Akhiezer, *The Classical Moment Problem*, Oliver & Boyd, London, 1965.
4. V. P. Potapov, 'The multiplicative structure of J-contractive matrix functions', *Amer. Math. Soc. Transl.*, **15**, 131–243 (1960).
5. V. M. Adamjan, D. Z. Arov and M. G. Krein, 'Analytic properties of Schmidt pairs for a Hankel operator and the generalized Schur–Takagi problem', *Math. USSR Sbornik*, **15**, 31–73 (1971).
6. F. Fenyves, 'Beitrag zur Realisierung von Zweipolen mit vorgegebener Charakteristik', *Atheneum*, Budapest, 1938.
7. D. C. Youla and M. Saito, 'Interpolation with positive-real functions', *J. Franklin Inst.*, **284**, 77–108 (1967).
8. J. Gregor, 'O interpolaci ve tride realne positivnich funkci', *Acta Polytechnica Praha*, **4**, 23–28 (1969).
9. V. Belevitch, 'Interpolation matrices', *Philips Res. Repts.*, **25**, 337–369 (1970).
10. P. Henrici, *Applied and Computational Complex Analysis, Vol. I*, Wiley, New York, 1974.
11. S. Darlington, 'Synthesis of reactance 4-poles which produce prescribed insertion loss characteristics', *J. Math. Phys.*, **18**, 257–355 (1939).
12. D. C. Youla, 'A new thory of cascade synthesis', *IRE Trans. Circuit Theory*, **CT-8**, 244–260 (1961).

13. V. Belevitch, *Classical Network Theory*, Holden-Day, San Francisco, 1968.
14. Y. Genin and S. Y. Kung, 'A two-variable approach to the model reduction problem with Hankel norm criterion', *IEEE Trans. Circuits and Systems*, to appear.
15. V. Belevitch and Y. Genin, 'Cascade decomposition of lossless 2-ports', *Philips Res. Repts.*, **26**, 326–340 (1971).
16. D. C. Youla, 'On the cascade decomposition of lossless 2-ports', *Memo 1377*, Polytech. Inst. of Brooklyn, 1967.
17. A. Fettweis, 'Factorization of transfer matrices of lossless two-ports', *IEEE Trans. Circuit Theory*, **CT-17**, 86–94 (1970).
18. P. Dewilde, A. Vieira and T. Kailath, 'On a generalized Szegö–Levinson realization algorithm for optimal linear predictors based on a network synthesis approach', *IEEE Trans. Circuits and Systems*, **CAS-25**, 663–675 (1978).
19. P. Dewilde and H. Dym, 'Schur recursions, error formulas and convergence of rational estimators for stationary stochastic sequences', Delft University of Technology, *Report 92*, 1979.
20. S. Y. Kung, 'A new identification and model reduction algorithm via singular value decomposition', *Proc. 12th Annual Asilomar Conference on Circuits, Systems and Computers*, November 1978.

Numerical Algorithms for the Nevanlinna-Pick Problem

A.C. Allison[1] and N.J. Young[2]

[1] Department of Computing Science, University of Glasgow, Scotland
[2] Department of Mathematics, University of Glasgow, Scotland

Summary. Based on the formulation in [14], efficient algorithms are devised for two variants of the Nevanlinna-Pick problem. Their implementation is validated by several examples. The method depends on establishing the equivalence of two different operator-theoretic approaches to the solution of classical interpolation problems [1, 14].

Subject Classifications: AMS(MOS): 65E05, 30D50, 47B38; CR: G1.2.

1. Introduction

A problem which occupied several of the leading analysts in the first two decades of this century was that of interpolating by analytic functions in the open unit disc U subject to the requirement of a minimal or restricted norm. Among the fruits of their endeavours are criteria for the existence of functions with the desired properties and a recursive algorithm for the construction of such functions (Pick [12], Nevanlinna [11]). It was nearly 20 years later that these discoveries were first applied to problems of design engineering (Fenyves [6]) and now, after a further 40 years, there are several quite different design problems which can be reduced to the problem solved by Pick and Nevanlinna, or some variant thereof. The most important of these is probably that of model reduction - that is, approximating a rational transfer function to within a prescribed tolerance by a rational function of smaller degree, thereby reducing the complexity of the circuit needed to model the function ([4]). More spectacular mathematically is an application to the problem of broadband matching. This arises, for example, in the design of circuits for the amplification of speech; in a tour de force of analysis, Helton [8] reduced it to the Nevanlinna-Pick problem. Other instances of the problem are in digital filter design [7] and the cascade synthesis of networks [4].

Of course there is quite a shift in viewpoint from the pure mathematics of 1916 to the electronics of 1980, and though the solutions of the old masters have been used in the newer context, it is not surprising that we need to go beyond them. Present requirements are for stable and efficient numerical computation of solutions, as well as for interpolation by functions from the wider class of meromorphic functions. In this paper we describe new algorithms for two such interpolation problems and report on the very successful practical testing of the programs we have written to implement them. The method we use is based on the theory of operators on Hilbert space, and the algorithm which results seems to us to have substantial advantages over others which have been proposed.

The norm we shall be trying to minimise is the supremum norm on the unit circle ∂U. More precisely, we denote by L^∞ the space of equivalence classes of bounded measurable functions on ∂U, two functions being equivalent if they are equal almost everywhere with respect to Lebesgue measure. For $f \in L^\infty$ we define $\|f\|_\infty$ to be the essential supremum of $|f(z)|$ for $z \in \partial U$. We shall seek interpolating functions from the subset H_k^∞ of L^∞, $k=0,1,2,\ldots$, consisting of (restrictions to ∂U of) functions meromorphic in U and having at most k poles there, counting multiplicities. In the particular case $k=0$, H_0^∞ is the space of bounded analytic functions in U, normally written H^∞, which is the case treated by Pick and Nevanlinna. They solved the case $k=0$ of the following.

Problem I. Given distinct points $\alpha_1, \ldots, \alpha_n \in U$, complex numbers $w_1, \ldots, w_n$ and an integer k, $0 \leqq k < n$, find a function $f \in H_k^\infty$ such that $f(\alpha_i) = w_i$, $1 \leqq i \leqq n$, and $\|f\|_\infty$ is minimised.

It turns out that this minimum is always attained, and that there is a unique function in H_k^∞ satisfying the constraints and attaining the minimum norm; moreover this function is rational, with both numerator and denominator being of degree less than n.

We can re-formulate Problem I slightly differently. Suppose ψ is any bounded analytic function satisfying $\psi(\alpha_i) = w_i$, $1 \leqq i \leqq n$; then a function $f \in H_k^\infty$ satisfies the interpolation conditions if and only if $f - \psi$ vanishes at the α_i, or in other words, if $f - \psi$ is divisible by the polynomial

$$p(z) = (z-\alpha_1)(z-\alpha_2)\ldots(z-\alpha_n),$$

in the sense that the function

$$g = (f-\psi)/p$$

is analytic at the zeros of p. Thus Problem I is mathematically equivalent to:

Problem II. Given a polynomial p (of degree n) with all its zeros lying in U, a function $\psi \in H^\infty$ and an integer k, $0 \leqq k < n$, find a function $f \in H_k^\infty$ such that $f - \psi$ is divisible by p and $\|f\|_\infty$ is minimised.

From the point of view of computation problems I and II are rather different. To pass from I to II we have to construct an interpolating function ψ - not difficult but still a step of order n^3. Going in the opposite direction, assuming that the polynomial p is given in terms of its coefficients rather than its zeros, requires the solution of a polynomial equation. It is clearly desirable to have algorithms which apply directly to the two formulations of the problem. Our two programs depend on the same theorems about operators, which

Reprinted with permission from *Numerische Mathematik,* vol. 42, pp. 125-145, 1983.

yield concrete computational techniques through the choice of bases in certain vector spaces. By making different choices of basis appropriate to the two forms of data we obtain two algorithms of the desired types.

Some rather technical operator theory is needed to justify our algorithms, and this is presented in Sect. 2. Those who are willing to take this on trust should still be able to follow the rest of the paper: they should read our notation for the backward shift operator T on H^2 at the beginning of Sect. 2 and then turn to Theorem 2 at the end of the section, where the main result is re-stated in the matrix form which we actually use for computation. Section 3 discusses the bases of spaces of rational functions which we use in our algorithms, while Sect. 4 deals with some aspects of implementation and testing.

Let us recall some standard terminology from operator theory. By an *operator* we shall alway mean a bounded linear operator between Hilbert spaces. If $A: H \to K$ is an operator then A^* is the operator from K to H satisfying $(Ax, y) = (x, A^* y)$ for all $x \in H$, $y \in K$. If M is a matrix then M^* is the conjugate transpose of M. The connection is that if M is the matrix of A with respect to *orthonormal* bases in H, K then M^* is the matrix of A^* with respect to the same bases. We shall have to deal also with non-orthogonal bases, and so we point out that in general, to write down the matrix of A^* knowing that of A we must also know the *Gram matrices* of the chosen bases of H and K. The Gram matrix of the basis $u_1, \ldots, u_n$ is defined to be the $n \times n$ matrix $[(u_j, u_i)]$. More details are in Lemma 4 below.

We shall also come clean about the proper definition of H_k^∞. For the purposes of this paper the description above is adequate, since we only deal with rational functions having no poles on ∂U, and it is clear that these can be regarded as elements of L^∞. For a precise statement of Problems I and II, however, we need to be more careful. Any function $f \in H^\infty$ has a radial limit almost everywhere on ∂U and hence defines a function $f_1 \in L^\infty$. Since can be recovered from f_1 (its boundary values) by means of Poisson's integral formula, there is no harm in identifying f and f_1, especially as the maximum principle shows that the L^∞ norm agrees with the natural norm of H^∞:

$$\|f_1\|_{L^\infty} = \|f\|_{H^\infty} \stackrel{\text{def}}{=} \sup_{z \in U} |f(z)|.$$

Hence we may regard H^∞ as a subspace of L^∞. We may then define H_k^∞ to consist of all functions of the form $f(z)/(z-z_1)\ldots(z-z_k)$ when $z_1, \ldots, z_k \in U$ and $f \in H^\infty$. Note that H_k^∞ is not a subspace unless $k=0$. A full treatment of these questions can be found in [9] and [5]; incidentally, Chap. 8 of the latter reference contains a historical account of extremal problems including those discussed here.

2. Operators for Interpolation

In 1968 two seminal papers were published giving solutions of classical interpolation problems in operator-theoretic terms. The two approaches, due to Sarason [14] on the one hand and Adamjan et al. [1] on the other, look at first sight quite different, but since both papers deal in the spaces H^2 and L^2 of functions on the circle one might expect that there would be a close connection between the two methods. This is in fact so, and we have found it necessary for our present purpose to make this connection explicit. The reason is that our algorithm is based on Sarason's formulation, though we require a result going somewhat further than those in [14]. Sarason considered only interpolation by analytic functions – the case $k=0$ of problems I and II – whereas Adamjan, Arov and Krein treat the general case. It would be interesting to generalize Sarason's methods to derive Theorem 1 below directly, but as it was not clear to us how to do this we have deduced it from the AAK result of [1].

The calculation of the desired interpolating functions will be reduced to the determination of the singular values and vectors of certain operators on Hilbert space. Let us explain our terminology. If a bounded linear operator A: $H \to K$, H, K being Hilbert spaces, has finite rank r then there exist positive numbers $s_0 \geqq s_1 \geqq \ldots \geqq s_{r-1}$ and orthonormal sequences $e_0, \ldots, e_{r-1}$ in H and $f_0, \ldots, f_{r-1}$ in K such that

$$A = \sum_{j=0}^{r-1} s_j(\cdot, e_j) f_j \tag{1}$$

in the sense that, for all $x \in H$,

$$Ax = \sum_{j=0}^{r-1} s_j(x, e_j) f_j.$$

We make the convention that $s_j = 0$ for $j \geqq r$. The s_j are called the *singular values of* A and are unique, being the eigenvalues of $(A^* A)^{\frac{1}{2}}$, together with 0. e_j and f_j are called *singular vectors of* A. An ordered pair $\langle e, f \rangle \in H \times K$ is called a *Schmidt pair corresponding to s* if

$$Ae = sf \quad \text{and} \quad A^* f = se.$$

Thus $\langle e_j, f_j \rangle$ is a Schmidt pair corresponding to s_j. The relation (1) is called a *singular value decomposition* of A.

The Hilbert spaces we are concerned with are L^2 and H^2 and their subspaces. L^2 is the space of all square-integrable Lebesgue measurable functions on the circle ∂U, modulo the subspace of functions equal to zero almost everywhere, with inner product

$$(f, g) = \frac{1}{2\pi} \int_0^{2\pi} f(e^{i\theta}) g(e^{i\theta})^- d\theta.$$

H^2 is the space of functions $\left\{ \sum_0^\infty x_j z^j : x_j \in \mathbb{C}, \sum_0^\infty |x_j|^2 < \infty \right\}$, with inner product

$$(\sum x_j z^j, \sum y_j z^j) = \sum x_j \bar{y}_j.$$

We can think of elements of H^2 as functions on either U or ∂U at will. By the Fischer-Riesz theorem,

$$L^2 = \left\{ \sum_{-\infty}^{\infty} x_j z^j : \sum_{-\infty}^{\infty} |x_j|^2 < \infty \right\},$$

so that we can regard H^2 as a subspace of L^2. We can therefore introduce the orthogonal projection operator π_+: $L^2 \to H^2$. The key operator for Sarason's approach is the *backward shift operator* T on H^2: this is defined by

$$T(x_0+x_1 z+x_2 z^2+\ldots)=x_1+x_2 z+x_3 z^2+\ldots$$

or alternatively by

$$Tf(z)=\begin{cases}\frac{1}{z}(f(z)-f(0)) & \text{if } z\neq 0\\ f'(0) & \text{if } z=0\end{cases} \tag{2}$$

Consider any polynomial p. We write p^* for the conjugate polynomial, defined by $p^*(z)=p(\bar{z})^-$. The kernel of the operator $p^*(T)$ on H^2, which we denote by $\operatorname{Ker} p^*(T)$, is clearly invariant under T – that is,

$$T(\operatorname{Ker} p^*(T))\subseteq \operatorname{Ker} p^*(T).$$

We may therefore introduce the *restriction of T to* $\operatorname{Ker} p^*(T)$, by which we mean the operator T_p on $\operatorname{Ker} p^*(T)$ given by

$$T_p f=Tf, \quad f\in \operatorname{Ker} p^*(T).$$

Our algorithm is based on the following.

Theorem 1. *Let $\psi\in H^\infty$ and let p be any polynomial. The infimum M of $\|h\|_{L^\infty}$, over all functions $h\in H_k^\infty$ such that $h-\psi$ is divisible by p, is given by $M=s_k$, where $s_0\geqq s_1\geqq\ldots$ are the singular values of $\psi(T_p^*)$. Moreover there is a unique such function h for which this infimum is attained; it is a rational function given explicitly by*

$$h=\frac{\psi(T_p^*)g}{g},$$

g being any element of $\operatorname{Ker} p^*(T)$ *such that, for some f, $\langle g,f\rangle$ is a Schmidt pair of $\psi(T_p^*)$ corresponding to s_k.*

The case $k=0$ of this theorem is contained in Propositions 2.1 and 5.1 of [14]. Indeed, [14] allows p to be an H^∞ function, not just a polynomial, but for the purpose of computation we can only handle a finite number of interpolation points, so the extra generality is not relevant here.

Note that the shift operator T has operator norm 1 and has no eigenvalue of unit modulus. It follows that all eigenvalues of T_p lie in U, and hence $\psi(T_p^*)$ is well defined.

We now turn to the result of Adamjan, Arov and Krein from which we shall deduce Theorem 1. To state this we define the *Hankel operator* $\Gamma(f)$ *determined by* $f\in L^\infty$ to be the operator on H^2 whose matrix with respect to the standard orthonormal basis $1, z, z^2,\ldots$ is $[\hat{f}(-i-j+1)]_{i,j=1}^\infty$, where $\hat{f}(k)$ is the kth Fourier coefficient of f:

$$\hat{f}(k)=\frac{1}{2\pi}\int_0^{2\pi} f(e^{i\theta})e^{-ik\theta}\,d\theta.$$

The following is a slight specialization of Theorem 1.3 of [1].

Lemma 1. *If $f\in L^\infty$ is a rational function and $s_0\geqq s_1\geqq\ldots$ are the singular values of $\Gamma(f)$ then there is a unique $h\in H_k^\infty$, $k=0,1,2,\ldots$, such that*

$$\|f-h\|_{L^\infty}=s_k,$$

and this h is given explicitly by the formula

$$h=f-s_k(J\eta/\xi) \tag{3}$$

where $\langle\xi,\eta\rangle$ is any Schmidt pair of $\Gamma(f)$ corresponding to s_k and J is defined (as an operator from H^2 to L^2) by $J\eta(z)=\frac{1}{z}\eta\left(\frac{1}{z}\right)$.

Since f is rational, $\Gamma(f)$ has finite rank (this is a well known theorem of Kronecker) and [1, Theorem 1.3] applies directly. It is further shown in [1] that $\|f-h\|_{L^\infty}\geqq s_k$ for all $h\in H_k^\infty$, so the h obtained in (3) is the unique function in H_k^∞ which minimises $\|f-h\|_{L^\infty}$.

This result gives a solution of Problems I and II. Let us denote by ϕ the Blaschke product corresponding to p: that is,

$$\phi=\frac{p}{\hat{p}}, \quad \hat{p}(z)=z^n p(1/\bar{z})^-, \tag{4}$$

where n is the degree of p and we assume that all zeros of p lie in U. Thus, if $p(z)=(z-\alpha_1)(z-\alpha_2)\ldots(z-\alpha_n)$,

$$\phi(z)=\prod_{i=1}^{n}\frac{z-\alpha_i}{1-\bar{\alpha}_i z}.$$

Corollary. *In the notation of Theorem 1, $M=s_k$ where $s_0\geqq s_1\geqq\ldots$ are the singular values of $\Gamma(\psi/\phi)$, as long as the zeros of p belong to U. Moreover the unique minimising function $h\in H_k^\infty$ is given by*

$$h=s_k\phi(J\eta)/\xi$$

where $\langle\xi,\eta\rangle$ is any Schmidt pair for $\Gamma(\psi/\phi)$ corresponding to s_k.

Proof. Any interpolating function $h\in H_k^\infty$ is of the form $h=\psi-\phi g$, with $g\in H_k^\infty$. Such an h has minimal norm when

$$\|\psi-\phi g\|_{L^\infty}=\|\psi/\phi-g\|_{L^\infty}=\min$$

(note that ϕ is unimodular on ∂U).

By Lemma 1 this minimum is the singular value s_k of $\Gamma(\psi/\phi)$, and the minimum is attained when

$$g=\frac{\psi}{\phi}-s_k\frac{J\eta}{\xi}.$$

Then

$$h=\psi-\phi g=\psi-\phi\left(\frac{\psi}{\phi}-s_k\frac{J\eta}{\xi}\right)=s_k\phi\frac{J\eta}{\xi}.$$

Comparison of the Corollary with Theorem 1 suggests forcibly that the operators $\Gamma(\psi/\phi)$ and $\psi(T_p^*)$ must be somehow "the same": we give the precise relationship. Roughly speaking, $\Gamma(\psi/\phi)$ is the orthogonal direct sum of a zero operator and a multiple of $\psi(T_p^*)$ by a unitary. Observe that any bounded operator A is the orthogonal direct sum of the zero operator from Ker A to (Range $A)^\perp$ and the restriction of A mapping (Ker $A)^\perp$ to Range A. To apply this observation to $\Gamma(\psi/\phi)$ we need the following information. We say that ψ and p are relatively prime if their only common divisors in H^∞ are units (invertible elements), or equivalently, if they have no common zero in the open unit disc.

Lemma 2. *If ψ and p are relatively prime then the range of $\Gamma(\psi/\phi)$ is* Ker $p(T)$. *Moreover* Ker $p(T)$ *consists of all rational functions $g/\tilde{p}$ where g is a polynomial of degree less than n, n=degree of p, and*

$$\tilde{p}(z)=z^n p(1/z).$$

Proof. We first calculate Ker $\Gamma(f)$ for any $f\in L^\infty$. From the definition of Hankel operators,

$$\begin{aligned}\Gamma(f)z^j &= \hat{f}(-j-1)z^0+\hat{f}(-j-2)z^1+\dots\\ &=\pi_+ z^{-j-1}\sum_{r=-\infty}^{\infty}\hat{f}(-r)z^r\\ &=\pi_+ z^{-1}f(z^{-1})z^{-j}.\end{aligned}$$

Thus, for any $g\in H^2$,

$$\Gamma(f)g=\pi_+ z^{-1}f(z^{-1})g(z^{-1}). \tag{5}$$

It follows that $\Gamma(f)g=0$ if and only if $z^{-1}f(z^{-1})g(z^{-1})\perp H^2$, or equivalently, $fg\in H^2$. That is, Ker $\Gamma(f)=\{g\in H^2: fg\in H^2\}$.

Now observe that, since $\Gamma=\Gamma(\psi/\phi)$ has finite rank, Range Γ is closed and therefore equals (Ker $\Gamma^*)^\perp$. As $\Gamma^*=\bar{\Gamma}$ it is clear that Γ^* is the Hankel operator $\Gamma(f^*)$ corresponding to the function f^* whose Fourier coefficients are the complex conjugates of those of $f=\psi/\phi$: $f^*(z)=f(\bar{z})^-$. Thus

$$\text{Ker}\,\Gamma^*=\{g\in H^2: \psi^* g/\phi^*\in H^2\}.$$

Since ψ^* and p^* are relatively prime, $\psi^* g/\phi^*\in H^2$ if and only if g is divisible by ϕ^*, i.e. $g\in p^* H^2$.

The adjoint T^* of the shift operator T is the operation of multiplication by z on H^2 (the "forward shift operator"). Hence

$$\text{Ker}\,\Gamma^*=p^* H^2=\text{Range}\,p^*(T^*).$$

Taking orthogonal complements we obtain

$$\text{Range}\,\Gamma=\text{Ker}\,p(T)$$

as desired. To prove the second statement write $p(z)=(z-\alpha)p_1(z)$ for some polynomial of degree $n-1$ and use induction on n.

For future reference let us extract an observation from the above proof: for any $f\in L^\infty$,

$$\Gamma(f)=\pi_+ RM_f|H^2 \tag{6}$$

where M_f is the operation of multiplication by f on L^2 and R: $L^2\to L^2$ is defined by $Rf(z)=z^{-1}f(z^{-1})$. This is simply equation (5).

Corollary. *If ψ and p are relatively prime then*

$$(\text{Ker}\,\Gamma)^\perp=\text{Range}\,\Gamma^*=\text{Ker}\,p^*(T),$$

and Ker $p^*(T)$ *consists of all function $g/\hat{p}$ where g is a polynomial of degree less than n.*

This follows on applying Lemma 2 to $\Gamma^*=\Gamma(\psi^*/\phi^*)$.

Lemma 3. *Let p be a polynomial of degree n, all of whose zeros lie in the open unit disc and let $\psi\in H^\infty$ be relatively prime to p. The diagram*

$$\begin{array}{ccc} & \text{Ker}\,p(T) & \\ {\scriptstyle U}\nearrow & & \nwarrow{\scriptstyle \gamma} \\ \text{Ker}\,p^*(T) & \xleftarrow[\psi(T_p^*)]{} & \text{Ker}\,p^*(T)\end{array}$$

commutes, where γ is the restriction of $\Gamma(\psi/\phi)$ mapping Ker $p^*(T)$ *to* Ker $p(T)$ *and U is the unitary operator defined by*

$$\left(U\frac{g}{\hat{p}}\right)(z)=\frac{z^{n-1}g(\bar{z})}{\tilde{p}(z)},\qquad z\in\partial U.$$

Thus the Sarason operator $\psi(T_p^*)$ is a unitary multiple of the non-zero part of the Adamjan-Arov-Krein operator Γ.

Proof. U is unitary since, when $|z|=1$,

$$\begin{aligned}\left(U\frac{g}{\hat{p}}\right)(\bar{z}) &= \frac{(\bar{z})^{n-1}g(t)}{\tilde{p}(\bar{z})}=\frac{z\,g(z)}{p(z)}\\ &=\frac{z}{\phi(z)}\frac{g}{\hat{p}}(z),\end{aligned}$$

so that $|Uf(\bar{z})|=|f(z)|$ for $f\in\text{Ker}\,p^*(T)$. One easily verifies that U^{-1}: Ker $p(T)\to$ Ker $p^*(T)$ is given by

$$\left(U^{-1}\frac{g}{\tilde{p}}\right)(z)=\frac{z^{n-1}g(\bar{z})}{\hat{p}(z)},\qquad z\in\partial U$$

and of course $U^*=U^{-1}$.

We must prove that $\gamma=U\psi(T_p^*)$, or equivalently $\gamma^*=\psi^*(T_p)U^*$. For $g/\tilde{p}\in\text{Ker}\,p^*(T)$,

$$\psi^*(T_p)U^*\frac{g}{\tilde{p}}=\psi^*(T)\frac{z^{n-1}g(\bar{z})}{\hat{p}(z)}.$$

Now another way of writing the definition (2) of T is

$$Tf(z)=\pi_+(\bar{z}f(z)),$$

from which it follows that, for $h\in H^\infty$,

$$h(T)f=\pi_+(h(\bar{z})f(z)).$$

Thus

$$\begin{aligned}\psi^*(T_p)\,U^*\frac{g}{\tilde{p}}&=\pi_+\,\psi^*(\bar{z})\frac{z^{n-1}g(\bar{z})}{\hat{p}(z)}\\&=\pi_+\,R\,\bar{z}\frac{\psi^*(z)\,g(z)}{z^{n-1}\,\hat{p}(\bar{z})}\\&=\pi_+\,R\frac{\psi^*g}{p^*}\\&=\pi_+\,R\frac{\psi^*}{\phi^*}\frac{g}{\tilde{p}}\\&=\Gamma(\psi^*/\phi^*)\frac{g}{\tilde{p}}\qquad\text{(compare (6))}\\&=\Gamma^*\frac{g}{\tilde{p}}.\end{aligned}$$

Hence $\psi^*(T_p)\,U^*=\Gamma^*$ on $\operatorname{Ker} p(T)$, and so $\Gamma=U\psi(T_p^*)$ on $\operatorname{Ker} p^*(T)$ as required.

Proof. of Theorem 1. We simply use Lemma 3 to recast the solution of the interpolation problem given in the Corollary to Lemma 1 in terms of Sarason's operator $\psi(T_p^*)$.

Suppose that $\psi(T_p^*)$ has singular value decomposition

$$\psi(T_p^*)=\sum_{j=0}^{n-1}s_j(\cdot,g_j)f_j$$

where $s_0\geqq s_1\geqq\ldots\geqq s_{n-1}$. It follows from Lemma 3 that a singular value decomposition of Γ is

$$\Gamma=\sum_{j=0}^{n-1}s_j(\cdot,g_j)\,Uf_j.$$

By the Corollary to Lemma 1 the solution of our interpolation problem is $M=s_k$, the unique minimising function $h\in H_k^\infty$ being given by

$$h=s_k\,\phi\cdot\frac{JUf_k}{g_k}.\tag{7}$$

Now for any $f\in\operatorname{Ker} p^*(T)$, $\phi JUf=f$. For if $f=q/\hat{p}$, where q has degree less than n,

$$\begin{aligned}\phi JUf&=\phi J(z^{n-1}q(\bar{z})/\tilde{p}(z))\\&=\phi\,\bar{z}\,q(z)/z^{n-1}\,\tilde{p}(\bar{z})\\&=\phi\,q/p=q/\hat{p}.\end{aligned}$$

Hence we can write (7) as

$$h=s_kf_k/g_k,$$

and since $\psi(T_p^*)\,g_k=s_kf_k$, this becomes

$$h=\psi(T_p^*)\,g_k/g_k.$$

This minimising function h has a surprising property, established in [1]: $|h(z)|$ is constant (and so equal to s_k) on the unit circle. This affords a useful check of the computed solution.

For the sake of concrete calculation it is helpful to have a restatement of Theorem 1 in terms of matrices.

Theorem 2. *Suppose given $\psi\in H^\infty$, a polynomial p of degree n having all its zeros in the open unit disc and an integer k, $0\leqq k<n$. Let $u_1,\ldots,u_n$ be a basis of $\operatorname{Ker} p^*(T)$ and let T_p have matrix B with respect to $u_1,\ldots,u_n$. Let the Gram matrix of $u_1,\ldots,u_n$ be G and let U be an $n\times n$ matrix such that $U^*U=G^{-1}$. Then the infimum M of $\|h\|_{L^\infty}$, over all functions $h\in H_k^\infty$ such that $h-\psi$ is divisible by p, is given by*

$$M=s_k$$

where $s_0\geqq s_1\geqq\ldots$ are the singular values of the matrix

$$A=U\,\psi(B^*)\,U^{-1}.$$

The unique such function h for which this infimum is attained is given by

$$h=\frac{\eta_1u_1+\ldots+\eta_nu_n}{\xi_1u_1+\ldots+\xi_nu_n}$$

where $U^*x=(\xi_1,\ldots,\xi_n)$, $U^*Ax=(\eta_1,\ldots,\eta_n)$ and x is a right eigenvector of A^*A corresponding to the eigenvalue s_k^2.

The proof depends on standard facts from linear algebra which we summarize.

Lemma 4. *Let E be an n-dimensional Hilbert space and let A, Q be linear transformations on E having matrices A_u, Q_u respectively with respect to a basis $u_1,\ldots,u_n$.*

(i) If Q is invertible then the matrix of A with respect to $Qu_1,\ldots,Qu_n$ is $Q_u^{-1}A_uQ_u$.

(ii) $Qu_1,\ldots,Qu_n$ is an orthonormal basis of E if and only if $Q_uQ_u^*=G^{-1}$, where G is the Gram matrix $[(u_j,u_i)]_{i,j=1}^n$ of $u_1,\ldots,u_n$.

These facts are elementary and can be found in many texts – e.g. [10].

Proof of Theorem 2. Let Q be the operator on $\operatorname{Ker} p^*(T)$ whose matrix with respect to $u_1,\ldots,u_n$ is U^*. By Lemma 4, $Qu_1,\ldots,Qu_n$ is an orthonormal basis of $\operatorname{Ker} p^*(T)$ and the matrix of T_p with respect to this basis is $U^{*-1}BU^*$. Thus the matrix of $\psi^*(T_p)$ with respect to this basis is A^*. A^* therefore has the same singular values as $\psi^*(T_p)$, and the first statement of the theorem follows from Theorem 1. Now let $\psi(T_p^*)$ have singular value decomposition

$$\psi(T_p^*)=\sum s_j(\cdot, g_j) f_j,$$

so that

$$\psi^*(T_p)=\sum s_j(\cdot, f_j) g_j$$

and hence

$$\psi^*(T_p)\psi^*(T_p)^*=\sum s_j^2(\cdot, g_j) g_j.$$

We see that g_k is an eigenvector of $\psi^*(T_p)\psi^*(T_p)^*$ corresponding to the eigenvalue s_k^2, and hence $g_k=x_1 Q u_1+\ldots+x_n Q u_n$ where $x=(x_1,\ldots,x_n)$ is a right eigenvector of A^*A. Hence, if $U^*=[q_{ij}]$,

$$\begin{aligned} g_k &= \sum_{i,j} x_i q_{ji} u_j = \sum_j \Big(\sum_i q_{ji} x_i\Big) u_j \\ &= \sum_j (U^* x)_j u_j = \sum_j \xi_j u_j \end{aligned}$$

and

$$\begin{aligned} \psi(T_p^*) g_k &= \psi^*(T_p)^* g_k = \sum_i (Ax)_i Q u_i \\ &= \sum_i (U^* A x)_i u_i = \sum_i \eta_i u_i. \end{aligned}$$

It now follows from Theorem 1 that

$$h=\sum \eta_j u_j \Big/ \sum \xi_j u_j.$$

3. Three Bases

In using Theorem 2 to solve a Nevanlinna-Pick problem the first step is to choose a basis of $\operatorname{Ker} p^*(T)$. We must make our choice in such a way that both the Gram matrix of the basis and the matrix of T_p^* can be calculated economically. We discuss three natural alternatives.

Let p be given by

$$\begin{aligned} p(z) &= a_0 + a_1 z + \ldots + a_n z^n \\ &= (z-\alpha_1)(z-\alpha_2)\ldots(z-\alpha_n). \end{aligned} \tag{8}$$

Basis 1 is Everyman's basis and is available when $\alpha_1,\ldots,\alpha_n$ are distinct, for in this case $g_1,\ldots,g_n$ constitutes a basis, where

$$g_j(z)=(1-\bar{\alpha}_j z)^{-1}.$$

At first glance this looks particularly attractive, since the matrix of T_p with respect to the g_j's is diagonal:

$$T_p g_j = T g_j = \bar{\alpha}_j g_j,$$

and their Gram matrix is given by a simple formula:

$$\begin{aligned} (g_j, g_i) &= (1+\bar{\alpha}_j z+\bar{\alpha}_j^2 z^2+\ldots, 1+\bar{\alpha}_i z+\bar{\alpha}_i^2 z^2+\ldots) \\ &= 1+\bar{\alpha}_j \alpha_i+\bar{\alpha}_j^2 \alpha_i^2+\ldots \\ &= (1-\bar{\alpha}_j \alpha_i)^{-1}. \end{aligned}$$

It is thus quite easy to write down all the entities occurring in Theorem 2 for this basis. If we do so we can deduce that the solution of problem I is $M=\lambda_k$, where $\lambda_0 \geqq \lambda_1 \geqq \ldots$ are the generalized eigenvalues of the problem

$$\left[\frac{\lambda - w_i \bar{w}_j}{1-\alpha_i \bar{\alpha}_j}\right] x = 0.$$

This is a classical result due to Pick [12] (for $k=0$) and Akhiezer [2] in general.

The disadvantage of this choice of basis is that we have to know the zeros α_j of p, whereas in the principal applications we have in mind only the coefficients of p are available. One could of course solve for the α_j, but this is plainly a step to be avoided if possible. Luckily there is a choice of basis which enables us to do without the α_j.

Basis 2 of Ker $p^*(T)$ is $f_1,\ldots,f_n$ where f_j is the unique member of $\operatorname{Ker} p^*(T)$ having a power series expansion of the form

$$f_j(z)=z^{j-1}+O(z^n). \tag{9}$$

That is, $f_j=b_j/\hat{p}$, where b_j is a polynomial of degree less than n and $\hat{p}$ is defined by (4), as before, and f_j satisfies (9). The latter relation can be written

$$b_j(z)=z^{j-1}\hat{p}(z)+z^n \hat{p}(z) k(z) \tag{10}$$

for some $k \in H^2$. On referring to (4) we perceive that

$$\hat{p}(z)=\bar{a}_0 z^n+\bar{a}_1 z^{n-1}+\ldots+\bar{a}_n,$$

and hence (10) is satisfied precisely when

$$b_j(z)=\bar{a}_n z^{j-1}+\bar{a}_{n-1} z^j+\ldots+\bar{a}_j z^{n-1}.$$

That is,

$$f_j(z)=(\bar{a}_n z^{j-1}+\bar{a}_{n-1} z^j+\ldots+\bar{a}_j z^{n-1})/\hat{p}(z).$$

The matrix of T_p with respect to basis 2 is the *companion* (or *Frobenius*) matrix C_{p^*} of p^* (see, for example, Sect. 3 of [15]):

$$C_{p^*}=\begin{bmatrix} 0 & 1 & 0 & \ldots & 0 \\ 0 & 0 & 1 & \ldots & 0 \\ \cdot & \cdot & \cdot & \ldots & \cdot \\ 0 & 0 & 0 & \ldots & 1 \\ -\bar{a}_0/\bar{a}_n & -\bar{a}_1/\bar{a}_n & -\bar{a}_2/\bar{a}_n & \ldots & -\bar{a}_{n-1}/\bar{a}_n \end{bmatrix}.$$

This is one of the advantages of basis 2: the special form of C_{p^*} enables functions of C_{p^*} to be calculated economically. The second, and more remarkable, advantage is that there is a very simple formula for the inverse G^{-1} of the Gram matrix of basis 2:

$$G^{-1}=I-Q^*Q \tag{11}$$

where

$$Q=\begin{bmatrix} \beta_1 & \beta_2 & \beta_3 & \dots & \beta_n \\ 0 & \beta_1 & \beta_2 & \dots & \beta_{n-1} \\ \cdot & \cdot & \cdot & \dots & \cdot \\ 0 & 0 & 0 & \dots & \beta_1 \end{bmatrix}$$

and

$$\beta_1=\bar{a}_0/a_n,$$
$$\beta_i=(\bar{a}_{i-1}-a_{n-i+1}\beta_1-a_{n-i+2}\beta_2-\dots-a_{n-1}\beta_{i-1})/a_n, \quad 2\leqq i\leqq n. \tag{12}$$

A rather complicated and indirect proof of this formula is given in [16, Remark 2]. Since it is this which constitutes the theoretical innovation underlying our algorithm for Problem II, it is worth sketching a much shorter proof.

Let P: $H^2\to \operatorname{Ker} p^*(T)$ be the orthogonal projection operator; then P^*: $\operatorname{Ker} p^*(T)\to H^2$ is the natural injection and P^*P is therefore the operator on H^2 which maps each function onto its projection on $\operatorname{Ker} p^*(T)$. Now $\phi(T^*)$ is the operator on H^2 of multiplication by the unimodular function ϕ, and is hence an isometry, so that $\phi(T^*)^*\phi(T^*)$ is the identity. Hence $\phi(T^*)\phi(T^*)^*$ is the orthogonal projection operator on Range $\phi(T^*)$, and so $I-\phi(T^*)\phi(T^*)^*$ is the orthogonal projection on

$$\operatorname{Range}\phi(T^*)^\perp=\operatorname{Ker}\phi(T^*)^*=\operatorname{Ker} p^*(T).$$

Thus

$$P^*P=I-\phi(T^*)\phi(T^*)^* = I-\phi^*(T)^*\phi^*(T)$$

Now let N: $H^2\to \operatorname{Ker} T^n$ be the orthogonal projection operator, and let τ: $\operatorname{Ker} T^n\to \operatorname{Ker} p^*(T)$ be the linear mapping defined by

$$\tau(x_0+x_1z+\dots+x_{n-1}z^{n-1})=x_0f_1+x_1f_2+\dots+x_{n-1}f_n.$$

The defining property (9) of basis 2 can then be written

$$\tau^{-1}=NP^*$$

Hence we have

$$(\tau^*\tau)^{-1}=\tau^{-1}\tau^{*-1}=NP^*PN^* = N(I-\phi^*(T)^*\phi^*(T))N^* = I_n-\phi^*(T_n)^*\phi^*(T_n)$$

where I_n is the identity operator on Ker T^n and $T_n=NSN^*$ on Ker T^n. It is easy to calculate that, with respect to the natural basis $1,z,\dots,z^{n-1}$ of Ker T^n, T_n has matrix

$$\mathscr{S}=\begin{bmatrix} 0 & 1 & 0\dots 0 \\ 0 & 0 & 1\dots 0 \\ \cdot & \cdot & \cdot\dots\cdot \\ 0 & 0 & 0\dots 1 \\ 0 & 0 & 0\dots 0 \end{bmatrix}$$

while $\tau^*\tau$ has matrix G, the Gram matrix of $f_1,\dots,f_n$. Hence

$$G^{-1}=I-B^*B$$

where

$$B=\phi^*(\mathscr{S})=p^*(\mathscr{S})\tilde{p}(\mathscr{S})^{-1}.$$

The latter Toeplitz matrix is precisely the one given by relation (12).

We remark that it requires exactly n^2 multiplications and divisions to generate G^{-1} using (11) and (12).

Practical testing, which we describe below, confirms that basis 2 is indeed a good choice for the solution of Problem II. It is less appropriate for Problem I: knowledge of the α_j's puts more bases at our disposal. We could use basis 1 above, but this involves us with the Gram matrix $[(1-\alpha_i\bar{\alpha}_j)^{-1}]$, an unattractive beast which is singular if two α's coincide and is therefore likely to lead to ill-conditioning when some α's are close together. We have therefore preferred to put our trust in an alternative basis which has the great merit of being orthonormal (so that its Gram matrix is the identity). It is true we have to work a bit harder to get the matrix of $\psi(T_p^*)$, but a recursion method enables us to achieve this fairly economically.

Basis 3. An *orthonormal* basis of Ker $p^*(T)$ is $e_1,\dots,e_n$ where

$$e_1(z)=s_1/(1-\bar{\alpha}_1 z),$$
$$e_j(z)=\frac{z-\alpha_1}{1-\bar{\alpha}_1 z}\,\frac{z-\alpha_2}{1-\bar{\alpha}_2 z}\cdots\frac{z-\alpha_{j-1}}{1-\bar{\alpha}_{j-1} z}\,\frac{s_j}{1-\bar{\alpha}_j z}, \quad 2\leqq j\leqq n,$$

and $s_j=(1-|\alpha_j|^2)^{\frac{1}{2}}$. This basis is discussed in [13, Sect. 3]. In particular it is shown that the matrix of T_p^* with respect to basis 3 is

$$L_n=\begin{bmatrix} \alpha_1 & 0 & 0 & 0 & \dots \\ s_1s_2 & \alpha_2 & 0 & 0 & \dots \\ -s_1\bar{\alpha}_2s_3 & s_2s_3 & \alpha_3 & 0 & \dots \\ s_1\bar{\alpha}_2\bar{\alpha}_3s_4 & -s_2\bar{\alpha}_3s_4 & s_3s_4 & \alpha_4 & \dots \\ \cdot & \cdot & \cdot & \cdot & \dots \end{bmatrix}.$$

For the calculation of $\psi(L_n)$ we shall find the following foklore observation useful.

Lemma 4. *Let L_n be an $n\times n$ matrix which can be partitioned as follows:*

$$L_n=\begin{bmatrix} L_{n-1} & 0 \\ v_n & \alpha_n \end{bmatrix}$$

where L_{n-1} is $(n-1)$ square and v_n is of type $1 \times (n-1)$. If ψ is a function analytic at the eigenvalues of L_n then

$$\psi(L_n) = \begin{bmatrix} \psi(L_{n-1}) & 0 \\ v_n \psi_n(L_{n-1}) & \psi(\alpha_n) \end{bmatrix}$$

where

$$\psi_n(z) = \begin{cases} \dfrac{\psi(z)-\psi(\alpha_n)}{z-\alpha_n} & \text{if } z \neq \alpha_n \\ \psi'(\alpha_n) & \text{if } z = \alpha_n. \end{cases}$$

One may prove this by induction in the case $\psi(z)=z^k$ and deduce the case of $\psi \in H^\infty$, or alternatively one can prove it directly from the definition of $\psi(L_n)$ via Cauchy's integral formula.

4. Implementation and Testing

Two computer programs have been written to solve Problems I and II by Theorem 2 using bases 3 and 2 respectively. The basic algorithm for both problems is

```
begin
choose the basis for Ker p*(T);
calculate the inverse of the Gram matrix G for this basis;
find a matrix U such that U* U = G^-1;
generate the matrix B of T_p and form ψ(B*);
calculate the singular values and vectors of the
matrix U^-1 ψ(B*) U;
form the function h in its lowest terms;
check the characteristics of the solution;
end.
```

The programming language Pascal was chosen because of its elegant style, readability, availability and the claimed ease with which algorithms could be expressed in the language.

The main operations in the program are those which involve manipulation of polynomials with complex coefficients and of complex matrices.

Accordingly we defined new types of objects

```
complex = record realpart, imagpart: real end:
    and
polys   = record degree: integer;
                 coeff:  array [0...maxdegree] of complex end,
```

with the matrices being represented as two dimensional arrays of complex type.

We wrote a module, common to both programs, which consisted of procedures for all the standard arithmetic, vector and matrix operations on complex numbers. A standard routine to find the singular values and vectors of a *real* matrix was available from the N.A.G. subroutine library and we generated the linear algebra routines required to perform the necessary transformations of a complex matrix to a form where this routine could be used. We will comment only on a few of the more significant procedures in this module. In retrospect, we note that choice of another programming language (e.g. Algol 68) which allowed redefinition of the standard arithmetic operations (+, −, *, /) for complex operands would have produced more compact, more readable code.

One of the major problems which affects all manipulation of non-integer numbers by computer is the finite precision imposed by the number representation. For example, if one divides a polynomial by another polynomial which is known to be a factor of the first then only in the unlikely event of all coefficients being exactly represented in the computer will the terms of the remainder polynomial be zero. We have introduced a procedure called ADJUST DEGREE which reduces the degree of a polynomial by removal of its highest terms when their coefficients are negligibly small. The criterion we have adopted is

$$\frac{|a_n|^2}{|a_M|^2} < 10 * \varepsilon,$$

where a_M is the coefficient of largest modulus, a_n the coefficient of highest degree at some stage and ε is a parameter which is input to the program.

Problem II

Input: A rational function $\psi \in H^\infty$ presented as two polynomials, numerator (N) and denominator (D) together with the polynomial p of degree n. Additional parameters, including k and ε.

Output: A function $f \in H_k^\infty$ such that $\|f\|_\infty$ is minimised.

We have observed that the matrix of T_p with respect to basis 2 is the companion matrix C_{p^*} of p^*. The required function $\psi(C_{p^*}^*)$ may be calculated by first finding a polynomial g of degree $< n$ such that $\psi - g$ is divisible by p. The Euclidean algorithm for polynomials may be used to find a polynomial S such that

$$D \times S \equiv 1 \bmod p$$

whence

$$N \times S \equiv g \bmod p.$$

Then $\psi(C_{p^*}^*) = g(C_{p^*}^*)$ and the problem reduces to the calculation of a polynomial with matrix argument. If the degree of the polynomial is m then a direct evaluation would require mn^3 multiplications. We can do better by exploiting the special form of C_{p^*}. Note that the first $n-1$ rows of C_{p^*} constitute rows 2 to n of I_n. This means that, for any $n \times n$ matrix A the product $C_{p^*}A$ has, as its first $n-1$ rows, the last $n-1$ rows of A. Thus when we calculate $C_{p^*}^{k+1}$ from $C_{p^*}^k$ only the new nth row need be calculated

i.e. nth row of $C_{p^*}^{k+1} = (n\text{th row of } C_{p^*}^k)\, C_{p^*}$

$$= [r_1 r_2 \ldots r_n] \begin{bmatrix} 0 & 1 & 0 & \ldots & 0 \\ \cdot & \cdot & \cdot & \ldots & \cdot \\ c_0 & c_1 & & \ldots & c_{n-1} \end{bmatrix}$$

$$= [r_n c_0 \;\; r_1 + r_n c_1 \ldots r_{n-1} + r_n c_{m-1}]. \tag{13}$$

The process of calculating powers of C_{p^*} may be visualised by introducing an $\infty \times n$ matrix

$$\begin{bmatrix} R_0 \\ R_1 \\ \cdot \\ R_{n-1} \\ R_n \\ \cdot \\ \cdot \\ \cdot \end{bmatrix} = \begin{bmatrix} 1 & 0 & 0 & \dots & 0 \\ 0 & 1 & 0 & \dots & 0 \\ \cdot & \cdot & \cdot & \dots & \cdot \\ 0 & 0 & 0 & \dots & 1 \\ c_0 & c_1 & c_2 & \dots & c_{n-1} \\ \cdot & \cdot & \cdot & \dots & \cdot \\ \cdot & \cdot & \cdot & \dots & \cdot \\ \cdot & \cdot & \cdot & \dots & \cdot \end{bmatrix}$$

where successive rows are generated from the previous row using relation (13).

Then $C^i_{p^*}$ is obtained by taking rows $i,\ i+1 \dots i+n-1$ of this infinite matrix:

$$C^i_{p^*} = \begin{bmatrix} R_i \\ R_{i+1} \\ \vdots \\ R_{i+n-1} \end{bmatrix}.$$

Hence

$$g(C_{p^*}) = g_0 I + g_1 C_{p^*} \dots + g_{n-1} C^m_{p^*}$$

$$= g_0 \begin{bmatrix} R_0 \\ \vdots \\ R_{n-1} \end{bmatrix} + g_1 \begin{bmatrix} R_1 \\ \vdots \\ R_n \end{bmatrix} + \dots + g_m \begin{bmatrix} R_m \\ \vdots \\ R_{m+n-1} \end{bmatrix}.$$

Now $R_0 \dots R_{n-1}$ require no calculation so we can start by generating the matrix

$$\begin{bmatrix} g_0 & g_1 & g_2 & \dots & g_{n-1} \\ 0 & g_0 & g_1 & \dots & g_{n-2} \\ \cdot & \cdot & \cdot & \dots & \cdot \\ 0 & 0 & 0 & \dots & g_0 \end{bmatrix}.$$

We then successively calculate $R_n, R_{n+1} \dots R_{n+m+1}$ using (13) and add appropriate multiples of R_k to successive rows of G.

The algorithm is

for values of k s.t. $n \leqq k \leqq m+i-1$

for values of i s.t. $i \geqq \max(1, k-m+1)$

add $g_{k-i+1} R_k$ to row i.

The number of operations in this approach is $O(n^3)$ rather than mn^3 for the straightforward method mentioned earlier.

The inverse of the Gram matrix is then calculated from (11) and a Choleski decomposition yields the matrix U given by $U^* U^* = G^{-1}$. All the singular values and singular vectors of $U^{-1} \psi^*(C^*_{p^*}) U$ are calculated and the function $h \in H^\infty_k$ evaluated from the singular vector corresponding to the kth singular value.

Problem I

Input: A set of points $\alpha_1, \alpha_2 \dots \alpha_n$, $|\alpha_i| < 1$, $1 \leqq i \leqq n$ and a corresponding set of values $w_1, w_2 \dots w_n$.

Additional parameters, including k and ε.

Output: A function $f \in H^\infty_k$ such that $f(\alpha_i) = w_i$, $1 \leqq i \leqq n$ and $\|f\|_\infty$ is minimised.

Note that, in this formulation, we do not know the function ψ, only the values $\psi(\alpha_j) = w_j$. However, if we adopt basis 3 then the matrix of T^*_p with respect to this basis is L_n (defined in Sect. 3) and we require $\psi(L_n)$. The most obvious course to evaluate this function is to compute an interpolating polynomial ψ of degree n, and then evaluate $\psi(L_n)$ directly. However the latter step requires $O(n^4)$ operations and we can do better by invoking Lemma 4.

We can proceed recursively as follows – we have

$$\psi(L_1) = [\psi(\alpha_1)] = w_1,$$

$$\psi(L_j) = \begin{bmatrix} \psi(L_{j-1}) & 0 \\ v_j \psi_j(L_{j-1}) & w_j \end{bmatrix}, \quad 2 \leqq j \leqq n,$$

where (cf. Lemma 4),

$$v_j = [(-1)^j s_1 \bar{\alpha}_2 \dots \bar{\alpha}_{j-1} s_j \ \ (-1)^{j-1} s_2 \bar{\alpha}_3 \dots \bar{\alpha}_{j-1} s_j \dots -s_{j-2} \bar{\alpha}_{j-1} s_j \ \ s_{j-1} s_j]$$

and

$$\psi_j(L_{j-1}) = (\psi(L_{j-1}) - w_j I)(L_{j-1} - \alpha_j I)^{-1}.$$

Thus, if we suppose that $\alpha_1, \dots, \alpha_j$ are distinct, we have

$$\psi(L_j) = \begin{bmatrix} \psi(L_{j-1}) & 0 \\ u_j & w_j \end{bmatrix}$$

where u_j is the unique solution of the linear equation

$$u_j(L_{j-1} - \alpha_j I) = v_j \psi(L_{j-1}) - w_j v_j.$$

This algorithm requires $O(n^3)$ operations.

Since basis 3 is orthonormal the Gram matrix is the identity matrix so we evaluate all the singular values and vectors of $\psi(L_n)$ and generate the function $h \in H^\infty_k$ as before.

The algorithms for Problems II and I are now at the same stage and the comments in the rest of this section refer to both problems equally.

From the definition of the rational function h of Theorem 2 it can be seen that the degrees of both numerator and denominator are less than n. These degrees are usually exactly $n-1$ but there are singular cases in which the above methods can result in an extremal function which is not in its lowest terms. Typically this will be where the numerator and denominator have a common factor $z - \alpha$, and it can happen that $|\alpha| < 1$, resulting in a (removable) singularity in the unit disc. That this has occurred is shown by the required singular value being repeated, m times say, indicating that $m-1$ common linear factors may be removed. In practice we consider singular values to be repeated if the relative error between the required value and any other value is less than ε. As algorithms for highest common factors are not well suited to real arithmetic we adopted a different approach.

Using the notation of Theorem 2 we generated all vectors Ux with x being a right eigenvector of AA^* corresponding to one of the repeated singular values. $m-1$ Householder transformations are then applied to the corresponding denominator polynomials thus obtaining a linear combination of such polynomials which is of suitably reduced degree. The final numerator polynomial is obtained by applying the same linear combination to the product UA^*.

When designing any computer program it is important to build in as many checks as possible in order to give confidence in the results. We will describe two such.

Our extremal function $h \in H_k^\infty$, so the denominator must have at most k zeros within the unit circle. We have used a method due to Cohn ([3]) that finds the number of zeros within the unit circle of a given polynomial. This beautiful paper, published long before the advent of computers is a model of clear, concise algorithmic expression. The implementation is straightforward.

As mentioned in Chap. 2 the extremal function has the surprising property that $|h(z)|$ is constant on the unit circle. We print out this quantity at six equidistant points on the unit circle and expect the results to be constant.

Testing

We will describe three test cases from the many combinations of data that have run successfully for the solution of Problems I and II. In each case the input is generated in such a way that exactly the same extremal function is produced from both programs.

Case 1. If the extremal function is to be $h(z)=\dfrac{z(z-\frac{1}{2})}{1-\frac{1}{2}z}$ then a possible choice of input is:

(a) for Problem I, $k=0$

$$\alpha\text{'s: } 0 \quad \tfrac{1}{2} \quad -\tfrac{1}{2} \quad \tfrac{1}{2}i$$
$$w\text{'s: } 0 \quad 0 \quad \tfrac{2}{5} \quad -\tfrac{3}{17}-\tfrac{5}{17}i$$

(b) for Problem II, $k=0$

$$N\colon \tfrac{1}{255}\{(112+28i)z-136z^2-(176+112i)z^3\}$$
$$D\colon \tfrac{1}{15}\{-15-(16-2i)z-20z^2-(16+8i)z^3\}$$
$$p\colon \tfrac{1}{8}\{iz-2z^2-4iz^3+8z^4\}.$$

Note that since in (a) four points are specified while in (b) the polynomial p has degree 4, a straightforward approach would produce an extremal function with numerator and denominator of maximum degree 3.

Both programs do indeed generate the same function with a numerator of degree 3 and a denominator of degree 2. This function cannot be in its lowest terms and both programs correctly take the next step of removing a common linear factor from numerator and denominator to give the required function.

The runs were performed with two different precisions. On input the rational numbers shown in the input above were presented as decimal numbers

(i) to 8 decimal places with a corresponding value of $\varepsilon=5\times10^{-8}$,
(ii) to 3 decimal places with a corresponding value of $\varepsilon=5\times10^{-3}$.

In both cases all the coefficients, real and imaginary parts, in the calculated result differed from the exact value by $<\varepsilon$.

The Cohn algorithm shows that the denominator has no zeros in the unit disc and it is confirmed that the function is indeed unimodular.

Case 2. We must test the case $k\neq0$ and we choose as input:

(a) Problem I, $k=1$,

$$\alpha\text{'s: } 0 \quad \tfrac{1}{2} \quad -\tfrac{1}{2} \quad \tfrac{1}{2}i$$
$$w\text{'s: } 1 \quad 0 \quad \tfrac{3}{5} \quad \tfrac{19}{34}+\tfrac{9}{34}i$$

(b) Problem II, $k=1$

$$N\colon \tfrac{1}{85}\{-85-(20+124i)z+272z^2+(216+496i)z^3\}$$
$$D\colon \tfrac{1}{15}\{-15-(16-2i)z-20z^2-(16+8i)z^3\}$$
$$p\colon \tfrac{1}{8}\{iz-2z^2-4iz^3+8z^4\}.$$

The rational numbers were input as decimal numbers correct to 8 decimal places and $\varepsilon=5\times10^{-8}$ in both cases.

Once again, after performing very different operations, the programs produced the same singular values and vectors of the appropriate matrix. The singular values are, to within ε,

$$0,\quad 0.5,\quad 0.5,\quad 5.408\ldots.$$

Since $k=1$, this means that the required norm is 0.5 and the repeated singular value indicates that the calculated function is not in its lowest terms. A linear factor is extracted and the final result is

$$h(z)=\frac{(z-\frac{1}{2})(z-4)}{2(1-\frac{1}{2}z)(1-4z)}.$$

The Cohn algorithm correctly shows that the denominator has one zero inside the unit circle and the square of the modulus of the function is correctly evaluated to be 0.25 on the unit circle.

Case 3. A more realistic case involving polynomials of higher degree than used in cases 1 and 2 is necessary. To this end we have written a program that generates interpolation points from the expression

$$m\times\text{factor}\times\exp(2\pi i j/Q) \quad \text{with} \quad \begin{array}{l}0\leqq j\leqq Q-1,\\ 1\leqq m\leqq M.\end{array}$$

A minimising function of norm 1 is then constructed using the Schur algorithm and the corresponding values found. Finally the polynomial p is calculated as a product of linear factors,

$$p(z) = \prod_{j=1}^{n} (z - \alpha_j).$$

Thus data can be generated in a form suitable for Problem I and Problem II and both programs should return the identical minimising function.

For our test case we used twenty points defined by the choice $M = 2$, factor $= 0.4567$ and $Q = 10$. Hence n, the degree of p, was 20. The coefficients were input to 10 decimal places and ε was taken to be 5×10^{-10}. With this value of n there is a larger build-up of error but all the coefficients in the minimising functions produced by the two programs differ by at most 2×10^{-5}. Again the Cohn algorithm showed that there were no zeros inside the unit circle while the calculations of the square of the modulus of the function at six points on the unit circle returned a value of $1.0 \pm 3 \times 10^{-9}$. The execution time, for twenty points, was approximately 5 seconds for Problem I and 10 for Problem II on a ICL 2976 computer.

References

1. Adamjan, V.M., Arov, D.Z., Krein, M.G.: Analytic properties of Schmidt pairs and the generalised Schur-Takagi problem. Math. USSR Sbornik **15**, 31–73 (1971)
2. Akhiezer, N.I.: On a minimum problem in the theory of functions and on the number of roots of an algebraic equation which lie inside the unit circle. Izv. Akad. Nauk. SSSR Otd. Mat. **9**, 1169–1189 (1931)
3. Cohn, A.: Über die Anzahl der Wurzeln einer algebraischen Gleichung in einem Kreise. Math. Z. **14**, 110–148 (1922)
4. Delsarte, Ph., Genin, Y., Kamp, Y.: On the role of the Nevanlinna-Pick problem in circuit and system theory. Circuit Theory and Applications **9**, 177–187 (1981)
5. Duren, P.: Theory of H^p spaces. New York: Academic Press 1970
6. Fenyves, F.: Beitrag zur Realisierung von Zweipolen mit vorgegebener Charakteristik. Budapest: Atheneum 1938
7. Gutknecht, M.H., Trefethen, L.M.: Recursive digital filter design by the Caratheodory-Fejér Method. Numer. Anal. Proj. Manuscript NA-80-01 1980
8. Helton, J.W.: The distance of a function to H^∞ in the Poincaré metric: electrical power transfer. J. Functional Analysis **38**, 273–314 (1980)
9. Hoffman, K.: Banach Spaces of Analytic Functions. New Jersey: Prentice Hall 1962
10. Mal'cev, A.I.: Foundations of Linear Algebra. San Francisco: W.H. Freeman 1963
11. Nevanlinna, R.: Über beschränkte analytische Funktionen. Ann. Acad. Sci. Fenn **A 32**, 1–75 (1929)
12. Pick, G.: Über die Beschränkungen analytischer Funktionen, welche durch vorgegebene Funktionswerte bewirkt werden. Math. Ann. **77**, 7–23 (1916)
13. Pták, V., Young, N.J.: Functions of operators and the spectral radius. Linear Algebra and its Applications **29**, 357–392 (1980)
14. Sarason, D.: Generalised interpolation in H^∞. Trans. Amer. Math. Soc. **127**, 179–203 (1967)
15. Young, N.J.: The singular value decomposition of an infinite Hankel matrix. Linear Algebra and its Applications **50**, 639–656 (1983)
16. Young, N.J.: The solution of Lyapunov's equation by a geometric method. Proc. Roy. Soc. Edinburgh **86 A**, 347–354 (1980)

Received September 29, 1982 / April 15, 1983

All optimal Hankel-norm approximations of linear multivariable systems and their L^∞-error bounds†

KEITH GLOVER‡

The problem of approximating a multivariable transfer function $G(s)$ of McMillan degree n, by $\hat{G}(s)$ of McMillan degree k is considered. A complete characterization of all approximations that minimize the Hankel-norm $\|G(s)-\hat{G}(s)\|_H$ is derived. The solution involves a characterization of all rational functions $\hat{G}(s)+F(s)$ that minimize $\|G(j\omega)-\hat{G}(j\omega)-F(j\omega)\|_{L^\infty}$, where $\hat{G}(s)$ has McMillan degree k, and $F(s)$ is anticausal. The solution to the latter problem is via results on balanced realizations, all-pass functions and the inertia of matrices, all in terms of the solutions to Lyapunov equations. It is then shown that $\|G(j\omega)-\hat{G}(j\omega)\|_{L^\infty} \leqslant \sigma_{k+1}(G(s)) + \|F(j\omega)\|_{L^\infty}$ where $\sigma_{k+1}(G(s))$ is the $(k+1)$st Hankel singular value of $G(s)$ and $\|F(j\omega)\|_{L^\infty} \leqslant \sigma_{k+2}(G(s))+\ldots+\sigma_n(G(s))$ for one class of optimal Hankel-norm approximations. The method is not computationally demanding and is applied to a 12-state model.

Nomenclature

$\mathbb{R}^n$, $\mathbb{C}^n$	n-dimensional real and complex euclidean spaces
$\mathrm{Re}\,(z)$, $\bar{z}$, $\lvert z\rvert$	real part, complex conjugate and modulus of $z\in\mathbb{C}$
$\mathbb{C}^{n\times m}$	space of $n\times m$ complex matrices
I_n	$n\times n$ identity matrix
A', A^*, $A^\dagger$, $\mathrm{In}\,(A)$	transpose, complex conjugate transpose, pseudo-inverse of $A\in\mathbb{C}^{n\times m}$, $\mathrm{In}\,(A)=(\pi(A), \nu(A), \delta(A))$; Definition 3.1
$\lambda_i(A)$	ith eigenvalue of $A\in\mathbb{C}^{n\times n}$
$\lambda_{\max}(A)$	eigenvalue of $A\in\mathbb{C}^{n\times n}$ with maximum modulus
$\bar{\sigma}(A)$	largest singular value of $A\in\mathbb{C}^{n\times m}$ $(=\lambda_{\max}{}^{1/2}(A^*A))$
(A, B, C, D)	state-space realization; eqns. (2.1), (2.2)
$G(s)$	transfer function; eqn. (2.3)
$\hat{G}(s)$	reduced-order approximation to $G(s)$
$\mathrm{diag}\,(A_1, A_2, \ldots, A_m)$	block diagonal matrix with $A_i\in\mathbb{C}^{n_i\times n_i}$
$\sigma_i(G(s))$	ith Hankel singular value of $G(s)$; Definition 2.1
$\lVert G(s)\rVert_H$	Hankel-norm of $G(s)$; Definitions 2.2 and 2.3
L^2, L^∞	Definitions 6.2 and 6.1
$\langle\cdot,\cdot\rangle_{L^2}$	inner product in L_2; Definition 6.2
$H_+{}^2$, $H_-{}^2$	Hardy spaces; Definition 6.4
$H_+{}^\infty$, $H_-{}^\infty$	Hardy spaces of functions bounded and analytic in the right (resp. left) half-plane; Definition 6.3
$\exists$, $\Rightarrow$, $\Leftarrow$, $\Leftrightarrow$	there exists, implies, is implied by, if and only if

---

Received 3 January 1984.

† This research was completed and the paper written while the author was a Visiting Fellow with the Department of Systems Engineering, The Australian National University, Canberra, Australia.

‡ Control and Management Systems Division, Department of Engineering, Cambridge University, Mill Lane, Cambridge CB2 1RX, U.K.

Reprinted with permission from *Int. J. Control,* vol. 39, no. 6, pp. 1115–1193, June 1984.

1. Introduction

In many areas of engineering, high-order linear state-space models of dynamic systems are derived and it is desirable if they can be replaced by reduced-order models without incurring too much error. For example,

(i) Plant models can be derived by combining models of the separate components, hence forming a single model of high order. If a reduced-order model could be substituted then a subsequent control system design would be computationally less demanding and possibly numerically more reliable.

(ii) If a high-order dynamic compensator has been designed for a plant then a reduced-order model of the compensator would involve fewer components or computing resources in an implementation.

(iii) In filter design it is sometimes possible to simply produce satisfactory high-order filters and reduced-order filters would then save in implementation.

(iv) In modelling some distributed parameter systems high-order approximate models can be produced from finite element analysis or modal analysis.

A wide variety of methods for model reduction have been proposed over the years, with the earliest probably that due to de Prony (1795)†. Until recently methods have generally either been computationally very simple (for example, Padé approximations, modal approximation, or continued fraction expansions) which generally have no guaranteed performance, or computationally demanding non-linear programming exercises. In the latter case there is still no guarantee that a globally optimal solution has been found.

It is the author's opinion that two recent developments have dramatically changed the status of model reduction. These are the theories of balanced realizations and optimal Hankel-norm approximations (see § 2 for definitions and bibliography). With these techniques it is possible to calculate the achievable error between the frequency responses of the full-order model and any reduced-order model of McMillan degree k. It is also possible to put lower bounds on the same errors. This calculation is not computationally demanding taking the equivalent of about two symmetric eigenvalue problems and two non-symmetric eigenvalue problems, all of size equal to the state dimension.

It is the purpose of this paper to present a new characterization of all optimal Hankel-norm approximations to multivariable linear systems and derive the frequency response error bounds. The development will be restricted to rational transfer functions and generally uses only straightforward mathematics. It is also hoped that many of the intermediate results, required to prove the main results, are of independent interest. The paper has been made as self-contained as possible and hence contains a certain amount of tutorial material.

Section 2 contains the necessary background to formulate the problem and gives a detailed outline of the paper.

---

† See Hildebrand (1956).

2. Problem formulation and an outline of the main results

In this section some background material is given, the optimal Hankel-norm model-reduction problem is formulated and our development is outlined in some detail. It is also hoped that this section will motivate and give the rationale for the Hankel-norm approach as well as giving some physical insights.

2.1. *The Hankel singular values of a linear system*

Throughout this paper we will be concerned with the standard linear, time-invariant dynamical system

$$\dot{x}(t) = Ax(t) + Bu(t) \tag{2.1}$$

$$y(t) = Cx(t) + Du(t) \tag{2.2}$$

where for each t, $u(t) \in \mathbb{C}^m$, $x(t) \in \mathbb{C}^n$, $y(t) \in \mathbb{C}^p$ are the vectors of inputs, states and outputs, respectively. The matrices $A \in \mathbb{C}^{n \times n}$, $B \in \mathbb{C}^{n \times m}$, $C \in \mathbb{C}^{p \times n}$ and $D \in \mathbb{C}^{p \times m}$ are assumed to be constant. The transfer function of the system is

$$G(s) = D + C(sI - A)^{-1}B \tag{2.3}$$

The notions of controllability and observability are central to the state-space description of dynamical systems and if the eigenvalues of A are assumed to be strictly in the left half-plane then we can define the controllability gramian as

$$P \triangleq \int_0^\infty \exp(At)BB^* \exp(A^*t)\, dt \tag{2.4}$$

and the observability gramian as

$$Q \triangleq \int_0^\infty \exp(A^*t)C^*C \exp(At)\, dt \tag{2.5}$$

By considering the corresponding matrix differential equations it is easily verified that P and Q satisfy the following linear matrix equations (Lyapunov equations)

$$AP + PA^* + BB^* = 0 \tag{2.6}$$

$$A^*Q + QA + C^*C = 0 \tag{2.7}$$

One interpretation of the controllability gramian is illustrated by considering the following minimum energy problem

$$\min_{u \in L^2(-\infty, 0)} J(u) \quad \text{subject to } x(0) = x_0$$

where

$$J(u) = \int_{-\infty}^0 u^*(t)u(t)\, dt \tag{2.8}$$

Since

$$x(0) = \int_0^\infty \exp(At)Bu(-t)\, dt$$

is a linear functional of u the minimum-norm solution is given by the pseudo-inverse as (Luenberger 1969)

$$u_{opt}(t) = B^* \exp(-A^*t)P^{-1}x_0 \tag{2.9}$$

$$J(u_{opt}) = x_0^* P^{-1}x_0 \tag{2.10}$$

Hence, if P^{-1} is ' large ' there will be some states that can only be reached if a large input energy is used.

If the system is released from $x(0) = x_0$ with $u(t) = 0$, $t \geq 0$ then

$$\int_0^\infty y^*(t)y(t)\, dt = x_0^* Q x_0 \tag{2.11}$$

and it is seen that if the observability gramian Q is nearly singular then some initial conditions will have little effect on the output. A standard result is as follows.

Lemma 2.1

If Re $(\lambda_i(A)) < 0$ $\forall i$ then

(*a*) $P > 0$ if and only if (A, B) is completely controllable.

(*b*) $Q > 0$ if and only if (A, C) is completely observable. ■

If the state-space coordinates of the system are changed to $z \triangleq Tx$ for some non-singular T then $\dot{z} = TAT^{-1}z + TBu$, $y = CT^{-1}z + Du$. Furthermore, the controllability and observability gramians become TPT^* and $T^{*-1}QT^{-1}$, respectively. The gramians P and Q depend strongly on the state-space coordinates but their product $PQ \rightarrow TPQT^{-1}$ in the transformed coordinates. Therefore the eigenvalues of PQ are invariant under state-space transformations, and are hence input/output invariants.

Definition 2.1

Let Re $(\lambda_i(A)) < 0$ $\forall i$ then the Hankel singular values of $G(s)$ are defined as

$$\sigma_i(G(s)) \triangleq \{\lambda_i(PQ)\}^{1/2}$$

where by convention $\sigma_i(G(s)) \geq \sigma_{i+1}(G(s))$. $G(s)$, P, Q are defined by (2.3), (2.6), (2.7), respectively. ■

It will be a continuing claim in this paper that the Hankel singular values of a linear system are the fundamental invariants related to both gain and complexity. A useful state-space realization in this respect is the balanced realization where $P = Q = \text{diag}(\sigma_1, \sigma_2, \ldots, \sigma_n)$. These are described in § 3 and were proposed by Moore (1981).

2.2. *Discrete-time systems*

All the results given in this paper will be derived for continuous-time models, however, it is shown here that corresponding discrete-time results can be obtained via a bilinear transformation. This is a standard device used to relate functions analytic in a half-plane to those analytic in the unit disc (see Hoffman 1962)†. We will consider the discrete-time state-space model

$$x(t+1)=\tilde{A}x(t)+\tilde{B}u(t) \tag{2.12}$$

$$y(t)=\tilde{C}x(t)+\tilde{D}u(t) \tag{2.13}$$

If $|\lambda_i(\tilde{A})|<1$ $\forall i$ then the controllability and observability gramians will be, respectively

$$\tilde{P}=\sum_{0\leq k<\infty}\tilde{A}^k\tilde{B}\tilde{B}^*\tilde{A}^{*k} \tag{2.14}$$

$$\tilde{Q}=\sum_{0\leq k<\infty}\tilde{A}^{*k}\tilde{C}^*\tilde{C}\tilde{A}^k \tag{2.15}$$

and will satisfy the linear, matrix equations

$$\tilde{P}-\tilde{A}\tilde{P}\tilde{A}^*=\tilde{B}\tilde{B}^* \tag{2.16}$$

$$\tilde{Q}-\tilde{A}^*\tilde{Q}\tilde{A}=\tilde{C}^*\tilde{C} \tag{2.17}$$

The z-plane transfer function will be given by

$$F(z)=\tilde{D}+\tilde{C}(zI-\tilde{A})^{-1}\tilde{B} \tag{2.18}$$

and the Hankel singular values of $F(z)$ will be defined as

$$\sigma_i(F(z))\triangleq\{\lambda_i(\tilde{P}\tilde{Q})\}^{1/2} \tag{2.19}$$

Now consider the bilinear transformation

$$s=\frac{(z-1)}{(z+1)} \tag{2.20}$$

$$\Rightarrow z=\frac{(1+s)}{(1-s)} \tag{2.21}$$

Notice that $|z|<1\Leftrightarrow\text{Re }(s)<0$ and $|z|=1\Leftrightarrow\text{Re }(s)=0$. Let

$$G(s)=F\left(\frac{1+s}{1-s}\right) \tag{2.22}$$

then a straightforward calculation gives $G(s)=D+C(sI-A)^{-1}B$ where

$$A=(I+\tilde{A})^{-1}(\tilde{A}-I) \tag{2.23}$$

$$B=\sqrt{2}(I+\tilde{A})^{-1}\tilde{B} \tag{2.24}$$

$$C=\sqrt{2}\tilde{C}(I+\tilde{A})^{-1} \tag{2.25}$$

$$D=\tilde{D}-\tilde{C}(I+\tilde{A})^{-1}\tilde{B} \tag{2.26}$$

Substituting (2.23) and (2.24) gives

$$\begin{aligned}&(I+\tilde{A})\{A\tilde{P}+\tilde{P}A^*+BB^*\}(I+\tilde{A}^*)\\&=(\tilde{A}-I)\tilde{P}(I+\tilde{A}^*)+(I+\tilde{A})\tilde{P}(\tilde{A}^*-I)+2\tilde{B}\tilde{B}^*\\&=-2\{\tilde{P}-\tilde{A}\tilde{P}\tilde{A}^*-\tilde{B}\tilde{B}^*\}=0\quad\text{by (2.16)}\end{aligned}$$

† This transformation is used for the present problem by Adamjan *et al.* (1971) and by Lin and Kung (1982).

Hence, $A\tilde{P}+\tilde{P}A^*+BB^*=0$ and the controllability gramian of (A, B) is $\tilde{P}$. Similarly (2.23), (2.25) and (2.17) imply that $A^*\tilde{Q}+\tilde{Q}A+C^*C=0$ and hence the observability gramian of (A, C) is $\tilde{Q}$. Hence, the Hankel singular values of the discrete and continuous-time systems will be identical. Furthermore

$$F(\exp(j\theta))=G(j\tan\tfrac{1}{2}\theta)$$

and the McMillan degrees of F and G are clearly the same. Hence, if we can produce an approximation to $G(s)$, say $\hat{G}_k(s)$, with McMillan degree k, then $\hat{F}_k(z)$ given by $\hat{F}_k(z)=\hat{G}_k((z-1)/(z+1))$ will satisfy

$$\sigma_i(F(z)-\hat{F}_k(z))=\sigma_i(G(s)-\hat{G}_k(s))$$

and

$$F(\exp(j\theta))-\hat{F}_k(\exp(j\theta)=G(j\tan\tfrac{1}{2}\theta)-\hat{G}_k(j\tan\tfrac{1}{2}\theta)$$

Therefore, a solution to the equivalent continuous-time problem generates a solution to the discrete-time problem in the sense that the Hankel singular values of the errors will be the same, as will the frequency responses. The frequency axis will, however, be rescaled so that the L^p-norms of the errors will be different except when $p=\infty$. All the frequency response results of this paper will be in terms of L^∞-norms, i.e. $\sup_\omega \bar{\sigma}(E(j\omega))$.

2.3. *The Hankel operator and Hankel norm*

We will now relate the Hankel singular values to the Hankel operator or Hankel matrix in the discrete-time case. Firstly, consider the discrete-time system (2.12), (2.13) with Markov parameters $H_k=\tilde{C}\tilde{A}^{k-1}\tilde{B}$, $k=1, 2, 3, \ldots$ The (block) Hankel matrix H is then defined as the doubly infinite matrix whose (i, j)th block is H_{i+j-1}. H can be written as

$$H=W_oW_c \tag{2.27}$$

where

$$W_o=[C^*, A^*C^*, \ldots, A^{*k}C^*, \ldots]^* \tag{2.28}$$

$$W_c=[B, AB, \ldots, A^kB, \ldots] \tag{2.29}$$

Hence, rank $(H)\leqslant n$ and, moreover, a basic result from realization theory is that rank $(H)=$ McMillan degree of $F(z)$.

The singular values of H can be evaluated from

$$[i\text{th singular value of } H]^2=\lambda_i(H^*H)=\lambda_i(W_c^*\,W_o^*\,W_oW_c)$$
$$=\lambda_i(W_cW_c^*\,W_o^*\,W_o)=\lambda_i(PQ) \tag{2.30}$$

and hence the Hankel singular values of $F(z)$ are the singular values of H.

The Hankel matrix can be interpreted as the mapping from the past inputs to the future outputs via the state $x(0)$, at time $t=0$, as illustrated in Fig. 1. that is, if $u(t)=0$, $t\geqslant 0$ then

$$x(0)=W_c\begin{bmatrix}u(-1)\\ u(-2)\\ \vdots\end{bmatrix} \tag{2.31}$$

$$\begin{bmatrix}y(0)\\ y(1)\\ \vdots\end{bmatrix}=W_ox(0) \tag{2.32}$$

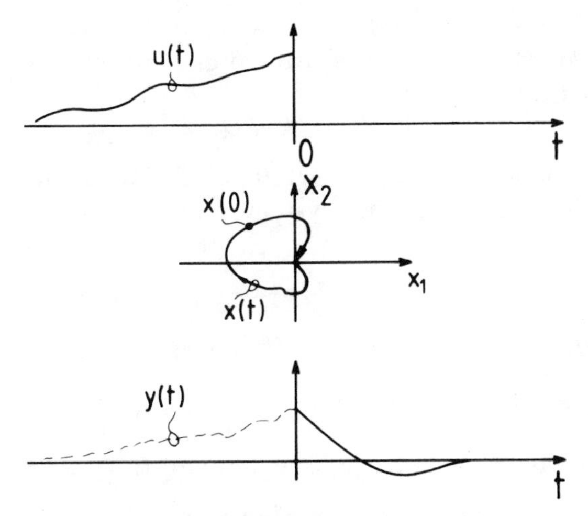

Figure 1. The Hankel operator.

The Hankel singular values are therefore the singular values of the mapping from the past inputs to the future outputs.

In continuous-time the Hankel operator can be defined analogously (see Fig. 1) as $\Gamma_G : L^2(0, \infty) \to L^2(0, \infty)$ where

$$(\Gamma_G v)(t) \triangleq \int_0^\infty C \exp [A(t+\tau)]Bv(\tau)\, d\tau \tag{2.33}$$

Notice that if the input $u(t) = v(-t)$ for $t < 0$ then the output for $t > 0$ will be $y(t) = (\Gamma_G v)(t)$. Similarly we have

$$(\Gamma_G^* y)(t) = \int_0^\infty B^* \exp [A^*(t+\tau)]C^* y(\tau)\, d\tau \tag{2.34}$$

To find the singular values of Γ_G, suppose that σ_i is a singular value with v the corresponding eigenvector of $\Gamma_G^* \Gamma_G$, i.e. $\Gamma_G^* \Gamma_G v = \sigma_i^2 v$. Let

$$y \triangleq \Gamma_G v = C \exp (At) x_0 \tag{2.35}$$

where

$$x_0 = \int_0^\infty \exp (A\tau) Bv(\tau)\, d\tau \tag{2.36}$$

then

$$\begin{aligned} \Gamma_G^* \Gamma_G v &= \Gamma_G^* y \\ &= B^* \exp (A^* t) \int_0^\infty \exp (A^*\tau) C^* C \exp (A\tau) x_0\, d\tau \\ &= B^* \exp (A^* t) Q x_0 \\ &= \sigma_i^2 v \end{aligned} \tag{2.37}$$

Hence

$$v(t) = B^* \exp (A^* t) Q x_0 \sigma_i^{-2} \tag{2.38}$$

and substituting (2.38) into (2.36) gives

$$PQx_0 = \sigma_i^2 x_0 \tag{2.39}$$

Hence, x_0 is an eigenvector of PQ with corresponding eigenvalue σ_i^2. Similarly if (2.39) is assumed, (2.38) is easily shown to give an eigenvector of $\Gamma_G^* \Gamma_G$. So we have verified that

$$\sigma_i^2(\Gamma_G) = \lambda_i(PQ) \tag{2.40}$$

The induced norm of $\Gamma_G : L^2(0, \infty) \to L^2(0, \infty)$ is given by its largest singular value and for stable, strictly proper systems this also gives a norm for $G(s)$ as follows.

Definition 2.2

Let $G(s) = C(sI - A)^{-1}B$ with $\mathrm{Re}\,(\lambda_i(A)) < 0 \ \forall i$ then the Hankel-norm of $G(s)$ is defined as

$$\|G(s)\|_{\mathrm{H}} \triangleq \bar{\sigma}(\Gamma_G) = \lambda_{\max}{}^{1/2}(PQ)$$

where Γ_G, P and Q are defined by (2.33), (2.6) and (2.7), respectively.

Similarly for discrete-time systems we have

Definition 2.3

Let $F(z) = \tilde{C}(zI - \tilde{A})^{-1}\tilde{B}$ with $|\lambda_i(\tilde{A})| < 1 \ \forall i$, then the Hankel-norm of $F(z)$ is defined as

$$\|F(z)\|_{\mathrm{H}} \triangleq \bar{\sigma}(H) = \lambda_{\max}{}^{1/2}(\tilde{P}\tilde{Q})$$

where H, $\tilde{P}$ and $\tilde{Q}$ are defined by (2.27), (2.16) and (2.17), respectively.

It will be observed that Γ_G, H and their norms are independent of the D or $\tilde{D}$ feed-through terms.

Finally, it is noted that if the input $u(t) = 0$ for $t \geqslant 0$ and the output is $y(t)$ then

$$\|G(s)\|_{\mathrm{H}} = \sup_{u \in L^2(-\infty, 0)} \frac{\|y\|_{L^2(0, \infty)}}{\|u\|_{L^2(-\infty, 0)}} \tag{2.41}$$

and we see that the Hankel-norm gives the L^2-gain from past inputs to future outputs.

2.4. *Hankel-norm approximations*

The model-reduction problem is to find $\hat{G}_k(s)$ of McMillan degree $k < n$ so as to minimize the norm of the error $\|G(s) - \hat{G}_k(s)\|$. The choice of norm is influenced by what norms can be minimized with reasonable computational effort and whether the chosen norm is an appropriate measure of error. It will turn out that the Hankel-norm is appropriate on both counts.

Consider first the discrete-time problem where we wish to find a Hankel matrix $\hat{H}$ of rank k such that $\|H - \hat{H}\|$ is minimized. The problem of approximating given matrices of lower rank was one of the earliest applications of the singular-value decomposition (Eckart and Young 1936 where the Frobenius norm was used). Mirsky (1960, Theorem 2) gives the following result.

Proposition 2.2

Let $X \in \mathbb{C}^{n \times m}$ have a singular value decomposition given by $X = UDV$, where $U \in \mathbb{C}^{n \times n}$ and $V \in \mathbb{C}^{m \times m}$ are unitary matrices and $D = \begin{pmatrix} D_r & 0 \\ 0 & 0 \end{pmatrix}$, $D_r =$

diag $(\alpha_1, \alpha_2, \ldots, \alpha_r)$, and $\alpha_1 \geqslant \alpha_2 \geqslant \ldots \geqslant \alpha_r > 0$ are the singular values of X. Let $\|\cdot\|$ denote any unitarily invariant matrix norm, i.e. $\|E\| = \|UE\| = \|EU\|$ for any unitary U. Then

$$\inf_{\text{rank}(\hat{X}) \leqslant k} \|X - \hat{X}\| = \|\text{diag}(\alpha_{k+1}, \alpha_{k+2}, \ldots, \alpha_r)\| \tag{2.42}$$

Further, the bound is achieved by

$$\hat{X} = U\hat{D}_k V \tag{2.43}$$

where

$$\hat{D}_k = \begin{pmatrix} D_k & 0 \\ 0 & 0 \end{pmatrix}, \quad D_k = \text{diag}(\alpha_1, \alpha_2, \ldots, \alpha_k)$$

■

Examples of unitarily invariant norms are :

(i) the largest singular value,
(ii) the sum of the singular values (= the trace norm),
(iii) the square root of the sum of the squared singular values (= the Frobenius norm).

Since the formula (2.43) minimizes all unitarily invariant norms it is tempting to use it for our problem. However, (2.43) cannot be applied because the resulting rank k approximation $\hat{H}$ to H will not generally be a Hankel matrix.

The remarkable result of Adamjan *et al.* (1971) is that for infinite-dimensional scalar Hankel matrices with $\|H - \hat{H}\| = \bar{\sigma}(H - \hat{H})$ the restriction that $\hat{H}$ needs to be a Hankel matrix does *not* affect the achievable error, i.e. there exists $\hat{H}$ of rank k such that $\bar{\sigma}(H - \hat{H}) = (k+1)$st singular value of $H = \sigma_{k+1}(F(z))$.

The significance of this result in systems theory was noted by de Wilde and is mentioned in Kailath (1980, p. 640). This work has been applied to the model-reduction problem by Bultheel and de Wilde (submitted for publication) Kung and Lin (1981 a, b) and Silverman and Bettayeb (1980), where explicit algorithms are derived for rational $F(z)$ or $G(s)$.

The present paper differs from others in several respects. Firstly, the development is entirely self-contained using little more than linear algebra. Secondly, *all* solutions to the optimal Hankel-norm approximation problem are derived and shown to be a simple function of a balanced realization. Thirdly, uniform error bounds on $\bar{\sigma}(G(j\omega) - \hat{G}(j\omega))$ are derived, i.e. there exists optimal Hankel-norm approximations that also match the frequency response of the system.

2.5. *Paper outline*

Since the development given here is, although straightforward, rather lengthy an outline of the route taken is now given. Section 3 gives some results determining the number of eigenvalues of A in the left half-plane from the number of eigenvalues of P in the right half-plane when they satisfy the Lyapunov equation (2.6). This result on the inertia of matrices (Theorem 3.3) will be central in determining the McMillan degree of the ‘ stable part ’ of certain transfer functions.

Section 4 gives several results on balanced realizations which will be the state-space description used in the development of the computational algorithms. Section 5 classifies all-pass transfer functions. It is shown that if $G(s)G^*(-\bar{s})=I$ then $PQ=I$ (Theorem 5.1). This is true for stable and unstable systems, and for stable systems gives a relationship between Hankel singular values and frequency response. The results of §§ 3, 4 and 5 are generally quite minor variations or extensions to existing results. For completeness they are all proved in detail.

In § 6 the problem of approximating $G(s)$ by a function $F(s)$ all of whose poles are in the right half-plane is considered ($F(s)$ can be thought of as an anti-causal transfer function). The norm used is $\sup_\omega \bar{\sigma}(G(j\omega)-F(j\omega))$. This problem was solved for scalar discrete-time systems by Nehari (1957) and in the multivariable case by Adamjan *et al.* (1978). The derivation given here is only for rational multivariable $G(s)$ but the approach is very explicit, straightforward and suitable for calculations. The main result is Theorem 6.1 which states that $\inf_F \sup_\omega \bar{\sigma}(G(j\omega)-F(j\omega))=\|G(s)\|_H$. The proof involves constructing a $F(s)$ such that $(G(s)-F(s))/\|G(s)\|_H$ is all-pass (Theorem 6.3). The construction is based on a balanced realization of $G(s)$ (§ 4), the pole positions are verified using Theorem 3.3 and the all-pass property is shown using Theorem 5.1.

In § 7 a class of solutions to the optimal Hankel-norm approximation problem is derived. Mirsky's result (Proposition 2.2) can be used to show that $\|G(s)-\hat{G}(s)\|_H \geq \sigma_{k+1}(G(s))$ for McMillan degree of $\hat{G}(s)=k$. If the bound is achieved for some $\hat{G}(s)$ then Theorem 6.1 will imply that there exists a $F(s)$ that is anticausal and such that $(G(s)-\hat{G}(s)-F(s))/\sigma_{k+1}(G(s))$ is all-pass. The construction of Theorem 6.3 can then be used to find such a $\hat{G}(s)+F(s)$. The above is for square systems. Non-square systems are treated in Corollary 7.3 by imbedding them in a $(p+m)\times(p+m)$ square system.

In § 8 the results of § 7 are extended to find all optimal Hankel-norm approximations. The derivation proceeds by characterizing all $\hat{G}(s)+F(s)$ such that $(G(s)-\hat{G}(s)-F(s))/\sigma_{k+1}(G(s))$ is all-pass. This is done for square systems in the state-space via Lyapunov equations in Theorem 8.4 and Lemma 8.5. These results are then interpreted in terms of transfer functions in Corollary 8.6. Finally, Theorem 8.7 solves the problem for the cases when $G(s)$ is not necessarily square by imbedding in a square system. The result is shown in Fig. 4 where $H(s)$ is a simple function of the balanced realization of $G(s)$ and $K(s)$ is an anticausal function with $K(j\omega)K^*(j\omega)\leq I$ $\forall\omega$ and $C_2+K(s)B_2{}^*=0$ (where C_2 and B_2 are parts of the C and B matrices in a balanced realization of $G(s)$).

Section 9 is concerned with frequency response bounds. It is shown first that there exists a D-matrix such that $\bar{\sigma}(G(j\omega)-D)\leq\sigma_1+\sigma_2+\ldots+\sigma_n$ $\forall\omega$. This result is proven via a representation of stable square systems as the sum of scaled all-pass functions (Theorem 9.2)

$$G(s)=D_0+\sigma_1E_1(s)+\sigma_2E_2(s)+\ldots+\sigma_nE_n(s)$$

where $E_k(s)$ are stable all-pass functions $\forall k$, and the partial sums $(D_0+\sigma_1E_1+\ldots+\sigma_kE_k)$ have McMillan degree k. Theorem 9.6 shows that approximations based on balanced realizations truncated to the first k states satisfy

$$\bar{\sigma}(G(j\omega)-\hat{G}(j\omega))\leq 2(\sigma_{k+1}+\sigma_{k+2}+\ldots+\sigma_n) \quad \forall\omega$$

Finally, Corollary 9.9 shows that for the optimal Hankel-norm approximations of Corollary 7.3 there exists D_0 such that

$$\bar{\sigma}(G(j\omega)-\hat{G}(j\omega)-D_0)\leqslant \sigma_{k+1}+\delta \quad \forall\omega$$

where

$$\delta=\sigma_1(F(-s))+\sigma_2(F(-s))+\ldots$$
$$\leqslant \sigma_{k+2}(G(s))+\ldots+\sigma_n(G(s))$$

Notice also that by Theorem 6.1 $\bar{\sigma}(G(j\omega)-\hat{G}(j\omega)-D_0)\geqslant\sigma_{k+1}$ for some ω and hence if δ is small this theoretical lower bound is nearly achieved.

These upper bounds on the error in the frequency response can be applied to give sufficient conditions for closed-loop stability (Glover and Limebeer 1983).

In § 10.1 the proposed algorithm is described in more detail and a particular method for choosing D_0 is outlined. This method has been implemented by the author and is used in § 10.2 to obtain reduced-order models for a 12th-order model of a 2-input 2-output automobile gas turbine.

3. Results on the inertia of matrices

The Hankel singular values of a system are determined by the solutions of the Lyapunov equations (2.6) and (2.7). These equations were introduced by Lyapunov in studying the stability of the equation $\dot{x}=Ax$ and his results can in fact be incorporated in a general result on the inertia of matrices. The purpose of this section is to state and prove Theorem 3.3 which will be required in the following sections.

Definition 3.1

The inertia of a general complex, square matrix A denoted In (A) is the triple $(\pi(A),\ \nu(A),\ \delta(A))$ where

$\pi(A)$ = number of eigenvalues of A in the open right half-plane,

$\nu(A)$ = number of eigenvalues of A in the open left half-plane,

$\delta(A)$ = number of eigenvalues of A on the imaginary axis. ■

A fundamental paper on the inertia of matrices is Ostrowski and Schneider (1962) with many of the results summarized in Barnett (1971, § 4.2). The following theorem shows the connection between the inertias of two matrices satisfying a Lyapunov equation.

Theorem 3.1 (Ostrowski and Schneider 1962)

Given a complex matrix A, there exists a unique hermitian matrix H satisfying

$$A^*H+HA=R \tag{3.1}$$

where $R>0$, if and only if $\lambda_i(A)+\bar{\lambda}_j(A)\neq 0\ \forall i,\ j$ and then In (A) = In (H). ■

Unfortunately in our application this very strong result cannot be applied directly since the Lyapunov equations considered will typically have R positive semidefinite and not necessarily positive definite. The following corollary can be used.

Corollary 3.2 (Ostrowski and Schneider 1962)

Given a complex matrix A such that $A+A^* \geqslant 0$ and $H=H^*$ then

$$\pi(AH) \leqslant \pi(H) \tag{3.2}$$

$$\nu(AH) \leqslant \nu(H) \tag{3.3}$$

■

It will be of central importance in the next sections to determine the exact number of stable poles for certain matrices satisfying Lyapunov equations, and the following theorem gives a number of results to assist in this determination.

Theorem 3.3

Given complex $n \times n$ and $n \times m$ matrices A and B, and hermitian matrix $P=P^*$ satisfying

$$AP+PA^*+BB^*=0 \tag{3.4}$$

then

(1) There is a unique solution to (3.4) if and only if $\lambda_i(A)+\bar{\lambda}_j(A) \neq 0$, $\forall i, j$.
(2) If $\delta(P)=0$ then $\pi(A) \leqslant \nu(P)$, $\nu(A) \leqslant \pi(P)$.
(3) If $\delta(A)=0$ then $\pi(P) \leqslant \nu(A)$, $\nu(P) \leqslant \pi(A)$.
(4)† If (A, B) is completely controllable then $\pi(A)=\nu(P)$, $\nu(A)=\pi(P)$, $\delta(A)=\delta(P)=0$.
(5) rank $P=\pi(P)+\nu(P) \geqslant \text{rank}\,(B, AB, \ldots, A^{n-1}B)$, and if $\lambda_i(A)+\bar{\lambda}_j(A) \neq 0$ $\forall i, j$ then the equality holds.
(6) If $\pi(A)=n$ then $\pi(P)=0$.
(7) If $\nu(A)=n$ then $\nu(P)=0$.

Proof

(1) See Barnett (1971, Theorem 1.9) or Gantmacher (1960).

(2) If $\delta(P)=0$ then observe that (3.4) implies that

$$(-AP)+(-AP)^* \geqslant 0 \tag{3.5}$$

and by Corollary 3.2 ($A \to -AP$, $H \to P^{-1}$)

$$\nu(A)=\pi(-APP^{-1}) \leqslant \pi(P^{-1})=\pi(P)$$
$$\pi(A)=\nu(-APP^{-1}) \leqslant \nu(P^{-1})=\nu(P)$$

(3) Assume $\delta(A)=0$ and that $P=U\begin{pmatrix} P_1 & 0 \\ 0 & 0 \end{pmatrix}U^*$ with $\delta(P_1)=0$, $U^*U=I$, and define

$$\tilde{A}=U^*AU=\begin{pmatrix} A_{11} & A_{12} \\ A_{21} & A_{22} \end{pmatrix}, \quad \tilde{B}=U^*B=\begin{pmatrix} B_1 \\ B_2 \end{pmatrix}$$

---

† It has been pointed out to the author by W. A. Coppel that point (4) is the same as the correct part of Proposition 4, p. 65 in Coppel (1978). It is also noted that his derivation is independent of Ostrowski and Schneider (1962). See also Wimmer (1974).

Then $U^*(3.4)U$ gives

$$\tilde{A}\begin{pmatrix} P_1 & 0 \\ 0 & 0 \end{pmatrix} + \begin{pmatrix} P_1 & 0 \\ 0 & 0 \end{pmatrix}\tilde{A}^* + \tilde{B}\tilde{B}^* = 0 \tag{3.6}$$

$$(3.6) \Rightarrow \qquad B_2 B_2^* = 0 \Rightarrow B_2 = 0 \tag{3.7}$$

$$(3.6), (3.7) \Rightarrow \qquad A_{21}P_1 = 0 \Rightarrow A_{21} = 0 \tag{3.8}$$

$$(3.6) \Rightarrow \qquad A_{11}P_1 + P_1 A_{11}^* + B_1 B_1^* = 0 \tag{3.9}$$

$$\Rightarrow (\text{by part (2)}) \quad \nu(A_{11}) \leqslant \pi(P_1)$$

$$\pi(A_{11}) \leqslant \nu(P_1)$$

but since $\delta(A_{11}) = \delta(P_1) = 0$

$$\pi(P_1) = \nu(A_{11}) \leqslant \nu(A)$$

$$\nu(P_1) = \pi(A_{11}) \leqslant \pi(A)$$

(4) We first show that $\delta(A) = 0$. Suppose the contrary, i.e. $\exists x$ such that $x^*A = \lambda x^*$ and $\lambda + \lambda^* = 0$. Then $x^*(3.4)x$ gives

$$(\lambda + \lambda^*)x^*Px + x^*BB^*x = 0$$

$$\Rightarrow x^*B = 0$$

But this gives a contradiction to the assumed complete controllability of (A, B) from the Popov–Belevitch–Hautus test (Kailath 1980, Theorem 2.4–8).

Now with $\delta(A) = 0$ we can apply part (3) whose proof showed that if $\delta(P) > 0$ then

$$U^*AU = \begin{pmatrix} A_{11} & A_{12} \\ 0 & A_{22} \end{pmatrix}, \quad U^*B = \begin{pmatrix} B_1 \\ 0 \end{pmatrix} \tag{3.10}$$

Since this is not completely controllable $\delta(P) = 0$. Now with $\delta(P) = \delta(A) = 0$ both parts (2) and (3) can be applied to give the result.

(5) The inequality is an immediate consequence of (3.7) and (3.8) in the proof of part (3). If (A, B) is completely controllable then the result follows immediately from part (4). If (A, B) is not completely controllable then without loss of generality assume that

$$A = \begin{pmatrix} A_{11} & A_{12} \\ 0 & A_{22} \end{pmatrix}, \quad B = \begin{pmatrix} B_1 \\ 0 \end{pmatrix}$$

where (A_{11}, B_1) is completely controllable. Now since $\lambda_i(A) + \bar{\lambda}_j(A) \neq 0$ $\forall i, j$ is assumed, by part (1) there is a unique P solving (3.4) and this will be given by

$$P = \begin{pmatrix} P_1 & 0 \\ 0 & 0 \end{pmatrix}$$

where P_1 satisfies (3.9). Now by part (4) and (3.9)

$$\begin{aligned}\text{rank } P &= \pi(P_1) + \nu(P_1) = \dim(A_{11}) \\ &= \text{rank } [B, AB, \ldots, A^{n-1}B]\end{aligned}$$

(6) $\pi(A) = n \Rightarrow \nu(A) = \delta(A) = 0 \Rightarrow \pi(P) \leqslant \nu(A) = 0$ by part (3).

(7) $\nu(A) = n \Rightarrow \pi(A) = \delta(A) = 0 \Rightarrow \nu(P) \leqslant \pi(A) = 0$ by part (3).

This completes the proof of Theorem 3.3. ■

Remark 3.1

Part (4) of the theorem gives the standard stability result that if (A, B) is completely controllable then A is asymptotically stable if and only if P satisfying (3.4) is positive definite. Similarly for asymptotically stable A, part (5) gives a test for controllability.

Remark 3.2

To show that the additional assumption in part (5) cannot be completely removed, let

$$A = \begin{pmatrix} 1 & 0 \\ 0 & -1 \end{pmatrix}, \quad B = \begin{pmatrix} 0 \\ 0 \end{pmatrix}, \quad P = \begin{pmatrix} 0 & 1 \\ 1 & 0 \end{pmatrix}$$

then $\pi(P) = \nu(P) = 1$, but the system is not controllable.

When (A, B, C) satisfies (2.6) and (2.7) then it can be shown that (A, B, C) cannot have any controllable and observable modes on the imaginary axis as follows.

Theorem 3.4

Given matrices (A, B, C) and hermitian matrices $P = P^*$, $Q = Q^*$ satisfying $AP + PA^* + BB^* = 0$, and $A^*Q + QA + C^*C = 0$, then

$$\begin{aligned}G(s) &\triangleq C(sI - A)^{-1}B \\ &= \hat{C}(sI - \hat{A})^{-1}\hat{B}\end{aligned}$$

for some $(\hat{A}, \hat{B}, \hat{C})$ such that $\delta(\hat{A}) = 0$. ■

This result is of peripheral interest and will only be required in Theorem 4.2, part (3), the proof is therefore relegated to Appendix A. Notice that no additional assumptions, apart from those explicitly stated, are needed, i.e. P and Q may be indefinite, A may have stable and unstable modes and (A, B, C) may be uncontrollable and unobservable.

Finally, a standard result in linear algebra that will be needed in the later sections is stated here for convenience.

Lemma 3.5

Given matrices $X \in \mathbb{C}^{n \times m}$ and $Y \in \mathbb{C}^{n \times r}$, $r \geqslant m$, such that $XX^* = YY^*$, then there exists $U \in \mathbb{C}^{m \times r}$ such that $UU^* = I$ and $Y = XU$.

Proof

Let X have a singular-value decomposition $X = U_1 \Sigma V_1^*$ where $U_1 \in \mathbb{C}^{n \times n}$, $V_1 \in \mathbb{C}^{m \times m}$, $\Sigma \in \mathbb{R}^{n \times m}$, $U_1 U_1^* = I$, $V_1 V_1^* = I$, $\Sigma = \begin{pmatrix} \Sigma_1 & 0 \\ 0 & 0 \end{pmatrix}$, $\Sigma_1 > 0$ and diagonal. Hence, $YY^* = XX^* = U_1 \Sigma \Sigma^* U_1^*$ so that Y has a singular-value decomposition that can be written as

$$Y = U_1[\Sigma, 0] \begin{pmatrix} V_2^* \\ \\ V_3^* \end{pmatrix}$$

where $V_2^* V_2 = I_m$. Therefore $Y = U_1 \Sigma V^*_2 = X V_1 V_2^*$ and setting $U = V_1 V_2^*$ gives the result since $UU^* = I$. ■

4. Balanced realizations

In this section some of the results for balanced realizations will be presented, and for completeness those that we will subsequently require will be proved. Most of these results have previously appeared in Moore (1981) or Pernebo and Silverman (1982). We will first derive a state-space transformation to a balanced realization (Laub 1980). Given a minimal stable (A, B, C), let P and Q be the hermitian solutions to

$$AP + PA^* + BB^* = 0 \tag{4.1}$$

$$A^*Q + QA + C^*C = 0 \tag{4.2}$$

Let Q have a Cholesky factorization

$$Q = R^*R \tag{4.3}$$

then RPR^* will be a positive-definite matrix and can be diagonalized as

$$RPR^* = U\Sigma^2 U^* \quad \text{with } U^*U = I \tag{4.4}$$

$$\Sigma = \text{diag}\,(\sigma_1, \sigma_2, \ldots, \sigma_n), \quad \sigma_1 \geqslant \sigma_2 \;\ldots\; \geqslant \sigma_n > 0 \tag{4.5}$$

Now a balancing transformation is given by

$$T = \Sigma^{-1/2} U^* R \tag{4.6}$$

since

$$TPT^* = \Sigma^{-1/2} U^* RPR^* U \Sigma^{-1/2}$$

$$= \Sigma \tag{4.7}$$

and

$$(T^*)^{-1} Q T^{-1} = \Sigma^{1/2} U^* (R^*)^{-1} (R^* R) R^{-1} U \Sigma^{1/2}$$

$$= \Sigma \tag{4.8}$$

That is the controllability and observability gramians are equal and diagonal. It is remarked that the author has had a similar method to the above algorithm programmed in the Cambridge Linear Analysis and Design Programs since 1979. The programs solve the Lyapunov equations using the Bartels–Stewart (Bartels and Stewart 1972) algorithm with only the upper Schur form of A required for both equations. The equation (4.3) is obtained via the less computationally efficient symmetric eigenvalue decomposition, but this gives accurate information on the observability properties of the original realization that is often useful in removing unobservable modes before continuing. Recently an ingenious new method for solving the Lyapunov equations has been derived by Hammarling (1982), that calculates a Cholesky factor of the solution directly and which will have superior numerical properties, particularly in the present application.

The balanced realization is not quite unique as is shown by the following result.

Lemma 4.1

Given an asymptotically stable, completely controllable and completely observable system (A, B, C) then it has a balanced realization that is unique up to an arbitrary transformation T such that $T\Sigma = \Sigma T$ and $T^*T = I$. In the case of distinct $\{\sigma_i\}$ and real T and (A, B, C) then T must be a sign matrix.

Proof

The balanced realization is characterized by the gramians $P = Q = \Sigma$. A state transformation T that preserves this relation must satisfy

$$T\Sigma T^* = (T^*)^{-1}\Sigma T^{-1} = \Sigma$$

$$\Rightarrow T\Sigma^2 T^{-1} = \Sigma^2$$

$$\Rightarrow \quad T\Sigma^2 = \Sigma^2 T$$

$$\Rightarrow \quad T\Sigma = \Sigma T \quad (\text{since } \sigma_i > 0)$$

$$\Rightarrow \quad \Sigma = T\Sigma T^* = \Sigma T T^*$$

$$\Rightarrow \quad TT^* = I = T^*T$$

Conversely, if $T\Sigma = \Sigma T$ and $T^*T = I$ then

$$T\Sigma T^* = \Sigma T T^* = \Sigma$$

and

$$(T^*)^{-1}\Sigma T^{-1} = T\Sigma T^{-1} = \Sigma$$

Finally, if σ_i are distinct then T must be a diagonal unitary matrix, which will be a sign matrix in the real case (i.e. the diagonal entries are ± 1). ■

An important result that determines the stability of truncations of a balanced realization is now given. Theorem 4.2 in fact states a generalization of a result in Silverman and Pernebo (1982) to the case of indefinite matrices Σ. Although this generalization is not of immediate application for the present problem, it is nevertheless included since with the results of Theorem 3.3 its derivation is no more involved than when $\Sigma > 0$.

Theorem 4.2

Let (A, B, C) be complex matrices satisfying

$$A\Sigma + \Sigma A^* + BB^* = 0 \tag{4.9}$$

$$A^*\Sigma + \Sigma A + C^*C = 0 \tag{4.10}$$

for

$$\Sigma = \Sigma^* = \begin{pmatrix} \Sigma_1 & 0 \\ 0 & \Sigma_2 \end{pmatrix} \quad \text{with } \delta(\Sigma) = 0$$

Partition (A, B, C) conformally with Σ as

$$A = \begin{pmatrix} A_{11} & A_{12} \\ A_{21} & A_{22} \end{pmatrix}, \quad B = \begin{pmatrix} B_1 \\ B_2 \end{pmatrix}, \quad C = (C_1, C_2)$$

Then

(1) If $\nu(\Sigma_1) = 0$ then $\pi(A_{11}) = 0$.

(2) If $\delta(A) = 0$ and $\lambda_i(\Sigma_1^2) \neq \lambda_j(\Sigma_2^2)$ $\forall i, j$ then $\text{In}(A_{11}) = \text{In}(-\Sigma_1)$ and $\text{In}(A_{22}) = \text{In}(-\Sigma_2)$.

(3) Any minimal realization of $C_1(sI - A_{11})^{-1}B_1$ given by $\hat{C}(sI - \hat{A})^{-1}\hat{B}$ satisfies $\delta(\hat{A}) = 0$.

Proof

(1) This was originally shown by Moore (1981) and is a direct consequence of Theorem 3.3, part (2) since (4.9) implies

$$A_{11}\Sigma_1 + \Sigma_1 A_{11}^* + B_1 B^*_1 = 0$$

and hence $0 = \nu(\Sigma_1) \geqslant \pi(A_{11}) = 0$.

(2) It is first shown that $\delta(A_{11}) = 0$. Assume the contrary; that $\exists x, \lambda$ such that $A_{11}x = \lambda x$ and $\lambda + \bar{\lambda} = 0$. It is shown in Pernebo and Silverman (1982), eqns. (3.15) to (3.29)) that if $\lambda_i(\Sigma_1^2) \neq \lambda_j(\Sigma_2^2)$ then $A_{21}x = 0$. (Their definiteness assumptions on Σ are not required for this step and since our derivation is identical it is omitted. It is interesting to note that the proof requires both (4.9) and (4.10) to be satisfied in order to show that $A_{21}x = 0$.) It is therefore true that $A \begin{pmatrix} x \\ 0 \end{pmatrix} = \lambda \begin{pmatrix} x \\ 0 \end{pmatrix}$ and hence $\delta(A) > 0$, but this is in contradiction to the assumption that $\delta(A) = 0$, and therefore $\delta(A_{11}) = 0$. $\delta(\Sigma_1) = 0$ is assumed and therefore Theorem 3.3, parts (2) and (3) can be applied to show $\text{In}(A_{11}) = \text{In}(-\Sigma_1)$. Similarly for $\text{In}(A_{22})$.

(3) Since the (1, 1) blocks of (4.9) and (4.10) are satisfied Theorem 3.4 can be applied to give the result immediately. ∎

Remark

For balanced realizations $\Sigma > 0$ and hence Theorem 4.2, part (2) implies that A_{11} will be asymptotically stable if $\lambda_i(\Sigma_1^2) \neq \lambda_j(\Sigma_2^2)$ $\forall i, j$, i.e. this truncation of the state variable description will be stable. Pernebo and Silverman (1982) also show that if $\lambda_i(\Sigma_1) = \lambda_j(\Sigma_2)$ then almost all of the equivalent balanced realizations, which are now not unique, will have A_{11} asymptotically stable. However, in this case Theorem 4.2, part (3) shows that a minimal realization of (A_{11}, B_1, C_1) will be asymptotically stable and therefore a reduced-order model based on (A_{11}, B_1, C_1) will be well-behaved.

Finally, the standard result on the decomposition of a state-space model into four components (controllable and observable, controllable and not observable, observable and not controllable, neither controllable nor observable) can be reproduced in terms of state-space realizations that have diagonal gramians.

Theorem 4.3

For any stable system (A, B, C) there exists T such that (TAT^{-1}, TB, CT^{-1}) has controllability and observability gramians given by

$$P = \text{diag}(\Sigma_1, \Sigma_2, 0, 0)$$

$$Q = \text{diag}(\Sigma_1, 0, \Sigma_3, 0)$$

respectively, with $\Sigma_1, \Sigma_2, \Sigma_3 > 0$ and diagonal. ■

The proof involves diagonalizing the product PQ and is given in Appendix B.

5. All-pass transfer functions

Relating the Hankel singular values of a system to the frequency response will be an important theme throughout this paper, and will also play a central role in some of the proofs. A fundamental identity occurs for the case of square all-pass transfer functions (also called square paraconjugate unitary rational matrices, see Youla (1961)), that is $G(s)$ such that $G(s)G^*(-\bar{s}) = I$. In this case Theorem 5.1 shows that all the Hankel singular values will be equal to unity. Theorem 5.1 also characterizes all-pass transfer functions without assuming the stability of the system ; in this case, the gramians are not defined but the Lyapunov equations will have solutions P and Q that will be indefinite but still satisfy $PQ = I$.

Theorem 5.1

Given a realization (A, B, C) (not necessarily stable) with $A \in \mathbb{C}^{n \times n}$, $B \in \mathbb{C}^{n \times m}$, $C \in \mathbb{C}^{m \times n}$, then

(1) If (A, B, C) is completely controllable and completely observable the following two statements are equivalent :

(*a*) $\exists\ D$ such that $G(s)G^*(-s) = \sigma^2 I\ \forall s$ where $G(s) \triangleq D + C(sI - A)^{-1}B$.

(*b*) $\exists\ P, Q \in \mathbb{C}^{n \times n}$ such that

(i) $P = P^*$, $Q = Q^*$

(ii) $AP + PA^* + BB^* = 0$

(iii) $A^*Q + QA + C^*C = 0$

(iv) $PQ = \sigma^2 I$.

(2) Given that part (1 *b*) is satisfied then $\exists\ D$ satisfying

$$D^*D = \sigma^2 I$$

$$D^*C + B^*Q = 0$$

and

$$DB^* + CP = 0$$

and any such D will satisfy part (1 *a*) (note, observability and controllability are not assumed).

Proof

Any system satisfying part (1 *a*) or (1 *b*) can be transformed to the case $\sigma=1$ by $\hat{B}=\sigma^{-1/2}B$, $\hat{C}=\sigma^{-1/2}C$, $\hat{D}=\sigma^{-1}D$, $\hat{P}=\sigma^{-1}P$, $\hat{Q}=\sigma^{-1}Q$. Hence, without loss of generality the proof will be given for the case $\sigma=1$ only.

(1 *a*)$\Rightarrow$(1 *b*)

This is proved by constructing P and Q to satisfy (1 *b*) as follows. Given (1 *a*), $G(\infty)=D \Rightarrow DD^*=I$. Also $G(s)G^*(-\bar{s})=I \Rightarrow G^*(-\bar{s})=G^{-1}(s)$, which can be evaluated by the matrix-inversion lemma as

$$
\begin{aligned}
G^{-1}(s) &= [D+C(sI-A)^{-1}B]^{-1} \\
&= D^*-D^*C[sI-A+BD^*C]^{-1}BD^* \\
&= G^*(-\bar{s}) \\
&= D^*+B^*(-sI-A^*)^{-1}C^*
\end{aligned}
$$

These two transfer functions are identical and both minimal (since (A, B, C) is assumed to be minimal), and hence $\exists$ a similarity transformation T relating the state-space descriptions, i.e.

$$-A^*=T(A-BD^*C)T^{-1} \tag{5.1}$$

$$C^*=TBD^* \tag{5.2}$$

$$B^*=D^*CT^{-1} \tag{5.3}$$

Further

(5.2) $\Rightarrow$ $$B^*=D^*CT^{*-1} \tag{5.4}$$

(5.3) $\Rightarrow$ $$C^*=T^*BD^* \tag{5.5}$$

(5.1) $\Rightarrow$ $$-A^*=-C^*DB^*+(T^{-1}A^*T)^*$$
$$=T^*(A-T^{*-1}C^*DB^*T^*)T^{*-1}$$

(5.2) and (5.3)$\Rightarrow$ $$=T^*(A-BD^*C)T^{*-1} \tag{5.6}$$

Hence, T and T^* satisfy identical equations, (5.1) to (5.3) and (5.4) to (5.6), and minimality implies these have a unique solution and hence $T=T^*$.

Now setting

$$Q=-T \tag{5.7}$$

$$P=-T^{-1} \tag{5.8}$$

clearly satisfies part (1 *b*), eqns. (i) and (iv). Further, (5.1) and (5.2) imply

$$TA+A^*T-C^*C=0 \tag{5.9}$$

which verifies (1 *b*), eqn. (iii). Also (5.9) implies

$$AT^{-1}+T^{-1}A^*-T^{-1}C^*CT^{-1}=0 \tag{5.10}$$

which together with (5.3) implies part (1 *b*), eqn. (ii).

(1 *b*)$\Rightarrow$(1 *a*)

This is proven by first constructing D according to part (2) and then verifying part (1 *a*) by calculation. Firstly note that since $Q=P^{-1}$, Q((1 *b*), eqn. (ii))Q gives

$$QA+A^*Q+QBB^*Q=0 \tag{5.11}$$

which together with part (1 b), eqn. (iii) implies that

$$QBB^*Q = C^*C \tag{5.12}$$

and hence by Lemma 3.5 $\exists$ D such that $D^*D = I$ and

$$DB^*Q = -C \tag{5.13}$$

$$DB^* = -CQ^{-1} = -CP \tag{5.14}$$

Equations (5.13) and (5.14) imply that the conditions of part (2) are satisfied. Now note that

$$\begin{aligned} BB^* &= (sI - A)P + P(-sI - A^*) \\ &\Rightarrow C(sI - A)^{-1}BB^*(-sI - A^*)^{-1}C^* \\ &= CP(-sI - A^*)^{-1}C^* + C(sI - A)^{-1}PC^* \\ \text{(by (5.14))} \quad &= -DB^*(-sI - A^*)^{-1}C^* - C(sI - A)^{-1}BD^* \end{aligned}$$

Hence, on expanding $G(s)G^*(-\bar{s})$ we get

$$\begin{aligned} G(s)G^*(-\bar{s}) &= (D + C(sI - A)^{-1}B)(D^* + B^*(-sI - A^*)^{-1}C^*) \\ &= I \end{aligned}$$

Part (2) follows immediately from the proof of (1 b)$\Rightarrow$(1 a) above. ■

Remark 5.1

Notice that this theorem does not assume $\lambda_i(A) + \lambda_j(A^*) \neq 0$ $\forall i, j$ which would imply that the Lyapunov equations have unique solutions. If $\lambda_i(A) + \lambda_j(A^*) = 0$ for some i and j then there will be an infinite number of solutions to part (1 b), eqns. (i), (ii) and (iii) and when part (1 a) is satisfied only some of these solutions also satisfy part (1 b), eqn. (iv). For example, let

$$A = \begin{pmatrix} -1 & 0 \\ 0 & 1 \end{pmatrix}, \quad B = \begin{pmatrix} 1 & 0 \\ 0 & 1 \end{pmatrix}, \quad C = \begin{pmatrix} 1 & 1 \\ -1 & 1 \end{pmatrix}$$

then

$$P = \begin{pmatrix} \frac{1}{2} & p \\ p & -\frac{1}{2} \end{pmatrix}, \quad Q = \begin{pmatrix} 1 & q \\ q & -1 \end{pmatrix}$$

satisfy part (1 b), eqns. (i), (ii) and (iii), and setting $q = 2p$, $\sigma^2 = \frac{1}{2} + 2p^2$ satisfies part (1 b), eqn. (iv). Hence P, Q and σ are not uniquely determined, but given any values satisfying part (1 b) D can be found to satisfy part (1 a).

Remark 5.2

It is interesting at this point to interpret one of the main results, Theorem 7.2, on approximating systems as it applies to all-pass transfer functions. This result states that the minimal Hankel-norm of the error is σ_{k+1} when a kth-order approximation is used. But for all-pass systems $\sigma_i = 1$ for all i and so the error will be unity, equaling the Hankel-norm of the original system. Therefore, in this sense, approximating all-pass transfer functions by lower-order models is futile. This can also be interpreted in the frequency domain for scalar systems as follows. The phase change in an nth-order all-pass transfer function is necessarily $n\pi$ with constant amplitude, whereas any kth-order model can have at most a phase change of $k\pi$ and this implies that the error necessarily becomes large at some frequency. This does not imply, however, that an all-pass system in series with another system cannot be adequately approximated, as is now illustrated.

Example 5.1

Let

$$g(s)=\left(\frac{1-0{\cdot}05s}{1+0{\cdot}05s}\right)^{10}\frac{1}{(1+sT)}$$

then the Hankel singular values of $g(s)$ are given in Fig. 2 as T varies. It is seen that all $\sigma_i \to 1$ as $T \to 0$ but as T increases and the high frequency behaviour of the all-pass term is attenuated then the smaller σ_is decrease significantly. Indeed for $T=10$ a second-order approximation would be quite good.

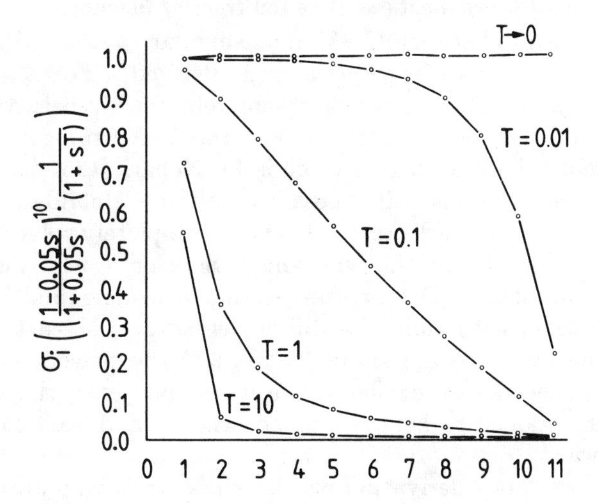

Figure 2. All-pass term in series with simple lag.

This section is concluded with a result on non-square systems such that $G(s)G^*(-\bar{s})\leqslant I$ for all $s=j\omega$. If all the poles of $G(s)$ are in the open left half-plane and $G(s)$ is real, then such $G(s)$ are referred to as bounded real rational functions (see Anderson and Vongpanitlerd (1973), Ch. 7). The following result is a slight extension to the bounded real lemma and the technique used in the proof mimics that given for the generalized positive real lemma outlined in Exercise 5.6.1 of Anderson and Vongpanitlerd (1973). Theorem 5.2 shows that any such bounded $G(s)$ can be imbedded in a square $(p+m)\times(p+m)$ all-pass function.

Theorem 5.2

Given a rational $p\times m$ transfer function $G(s)$ of McMillan degree n with a minimal realization (A, B, C, D) and such that

$$G(j\omega)G^*(j\omega)\leqslant I \quad \forall\omega \tag{5.15}$$

then there exists

$$\tilde{D} \triangleq \begin{pmatrix} D & D_{12} \\ D_{21} & D_{22} \end{pmatrix} \in \mathbb{C}^{(p+m\times(p+m)} \tag{5.16}$$

$$\tilde{B} \triangleq (B \;\; B_2) \in \mathbb{C}^{n\times(p+m)} \tag{5.17}$$

$$\tilde{C} \triangleq \begin{pmatrix} C \\ C_2 \end{pmatrix} \in \mathbb{C}^{(p+m)\times n} \tag{5.18}$$

such that

$$H(s) \triangleq \tilde{D} + \tilde{C}(sI - A)^{-1}\tilde{B} \tag{5.19}$$

is all-pass. That is $\|G(j\omega)\|_{L^\infty} \leqslant 1 \Rightarrow G(s)$ is a submatrix of an all-pass function with the same ' A-matrix '. ■

The proof is given in Appendix D.

6. Optimal anticausal approximations of causal transfer functions

In § 5 a relationship between the Hankel singular values and the frequency response of an all-pass transfer function was derived. For a general causal transfer function there exists a most surprising relationship between the Hankel singular values and the transfer function and this is the subject of the present section. The seminal work on this topic is by Nehari (1957) who considered scalar discrete-time systems with possibly infinite McMillan degree ; the mutivariable version of Nehari's result has been completely solved by Adamjan *et al.* (1978) who classify all the solutions ; Sarason (1968) also includes a solution to this problem. The present section considers multivariable continuous-time systems with finite McMillan degree and since the continuous and discrete time problems are equivalent (§ 2.2) the present derivation will be seen to be a special case of the above work, however, the techniques of proof and constructions are believed to be original and particularly explicit and straightforward.

The major part of our derivation will be via a straightforward calculation, however, in order to state the results rigorously we will first need to depart from our general intent of only using elementary mathematics and introduce H^p and L^p spaces (see Hoffman (1962), Duren (1970)).

Definition 6.1

Let $G(s): \mathbb{C} \to \mathbb{C}^{p\times m}$ then $G(j\omega) \in L^\infty(-\infty, \infty)$ if and only if $\exists M < \infty$ such that $\bar{\sigma}(G(j\omega)) \leqslant M \; \forall \omega \in \mathbb{R}$ and we denote $\|G(j\omega)\|_{L^\infty} \triangleq \sup_{\omega\in\mathbb{R}} \bar{\sigma}(G(j\omega))$.

Definition 6.2

Let $X(s): \mathbb{C} \to \mathbb{C}^n$ then $X(j\omega) \in L^2(-\infty, \infty)$ if and only if

$$\int_{-\infty}^{\infty} X^*(j\omega)X(j\omega)\, d\omega < \infty$$

and we denote

$$\|X(j\omega)\|_{L^2}^2 \triangleq \int_{-\infty}^{\infty} X^*(j\omega)X(j\omega)\, d\omega$$

$$\langle Y(j\omega), X(j\omega)\rangle_{L^2} \triangleq \int_{-\infty}^{\infty} Y^*(j\omega)X(j\omega)\, d\omega$$

We will be concerned with the Hardy spaces H^2 and H^∞ which will be defined as follows.

Definition 6.3

Let $G(s): \mathbb{C} \to \mathbb{C}^{p \times m}$ the $G \in H_+^\infty$ (respectively H_-^∞) if and only if G is analytic in the open right (respectively left) half-plane and bounded in the closed right (respectively left) half-plane. This definition implies that H_+^∞ and H_-^∞ are subspaces of L^∞.

Definition 6.4

Let $X(s): \mathbb{C} \to \mathbb{C}^n$ then $X \in H_+^2$ (respectively H_-^2) if and only if X is analytic in the open right (respectively left) half-plane and $\|X(\sigma + j\omega)\|_{L^2} \leqslant M < \infty$ for all $\sigma \geqslant 0$ (respectively $\sigma \leqslant 0$).

We see that $L^\infty(-\infty, \infty)$ corresponds to the transfer functions bounded on $s = j\omega$, whereas H_+^∞ corresponds to the causal bounded transfer functions (i.e. poles in the left half-plane) and H_-^∞ corresponds to the anticausal bounded transfer functions (i.e. poles in the right half-plane). All the results of this paper remain valid if H_-^∞ is replaced throughout by the space of rational transfer functions with poles in the open right half-plane, and then the proof of Lemma 6.2 can be simplified.

The main result of this section is now stated.

Theorem 6.1

Let $G(s)$ be a rational transfer function, analytic in the closed right half-plane, then

$$\sigma_1(G(s)) \triangleq \|G(s)\|_{\mathrm{H}} = \inf_{F \in H_-^\infty} \|G(j\omega) - F(j\omega)\|_{L^\infty} \qquad \blacksquare$$

This result is indeed surprising and states that if one wished to approximate the causal transfer function $G(j\omega)$ by an anticausal function $F(j\omega)$, then the smallest $L^\infty(-\infty, \infty)$ error that can be achieved is precisely the Hankel-norm of $G(s)$. The proof will require some auxillary results and proceeds by firstly showing that $\|G(s)\|_{\mathrm{H}} \leqslant \|G(j\omega) - F(j\omega)\|_{L^\infty}$ for any $F \in H_-^\infty$ (Lemma 6.2) and then explicitly constructing an $F \in H_-^\infty$ such that equality holds (Theorem 6.3).

Lemma 6.2

Given a rational transfer function $G(s)$ analytic in the right half-plane, then for any $F \in H_-^\infty$

$$\|G(s)\|_{\mathrm{H}} \leqslant \|G(j\omega) - F(j\omega)\|_{L^\infty}$$

Proof

We will use the characterization of the Hankel-norm given in § 2.3, eqn. (2.41) that is, let

$$u(t) \in L^2(-\infty, \infty) \quad \text{with } u(t) = 0 \quad \text{for } t \geqslant 0 \tag{6.1}$$

$$U(j\omega) = \int_{-\infty}^{\infty} \exp(-j\omega t) u(t)\, dt \tag{6.2}$$

$$Y(j\omega) = G(j\omega) U(j\omega) \tag{6.3}$$

$$y(t) = \frac{1}{2\pi} \int_{-\infty}^{\infty} \exp(j\omega t) Y(j\omega)\, d\omega \tag{6.4}$$

$$v(t) = \begin{cases} 0 & t \leqslant 0 \\ y(t) & t > 0 \end{cases} \tag{6.5}$$

$$V(j\omega) = \int_{-\infty}^{\infty} \exp(-j\omega t) v(t)\, dt \tag{6.6}$$

then

$$\| G(s) \|_{\mathrm{H}} = \sup_{u \in L^2} \frac{\| v \|_{L^2}}{\| u \|_{L^2}} \tag{6.7}$$

$$= \sup_{U \in L^2} \frac{\| V(j\omega) \|_{L^2}}{\| U(j\omega) \|_{L^2}} \tag{6.8}$$

Firstly, note that since $G(j\omega)$ is bounded then $Y(j\omega) \in L^2$ and hence $y(t) \in L^2$ and $v(t) \in L^2$ and $V(j\omega) \in L^2$. The transforms given by (6.2), (6.3), (6.4) and (6.6) are to be considered as the extension of the Fourier transform from $L^1 \cap L^2$ to L^2 (sometimes referred to as the Plancherel transform, see Rudin (1966), p. 187). We will now bound $\| V(j\omega) \|_{L^2}$ as follows. From (6.5) we obtain

$$\langle v(t), v(t) \rangle_{L^2} = \langle v(t), y(t) \rangle_{L^2} \tag{6.9}$$

and hence by Parseval equality and (6.3)

$$\| V(j\omega) \|_{L^2}^2 = \langle V(j\omega), Y(j\omega) \rangle_{L^2}$$

$$= \langle V(j\omega), G(j\omega) U(j\omega) \rangle_{L^2} \tag{6.10}$$

Now consider $\langle V(j\omega), F(j\omega) U(j\omega) \rangle_{L^2}$ for some $F(s) \in H_-^{\infty}$. Since $U(s) \in H_-^2$ and $F(s) \in H_-^{\infty}$, $F(s) U(s) \in H_-^2$. But, $V(s) \in H_+^2$ and therefore $\langle V(j\omega), F(j\omega) U(j\omega) \rangle_{L^2} = 0$ since by the Paley–Weiner theorem (see Hoffman (1962), p. 131, 135) H_+^2 and H_-^2 are orthogonal subspaces of L^2. Hence from (6.10) for any $F(s) \in H_-^{\infty}$

$$\| V(j\omega) \|_{L^2}^2 = \langle V(j\omega), (G(j\omega) - F(j\omega)) U(j\omega) \rangle_{L^2}$$

$$\leqslant \| V(j\omega) \|_{L^2} \| (G(j\omega) - F(j\omega)) U(j\omega) \|_{L^2} \tag{6.11}$$

$$\leqslant \| G(j\omega) - F(j\omega) \|_{L^\infty} \| V(j\omega) \|_{L^2} \| U(j\omega) \|_{L^2} \tag{6.12}$$

where the Schwartz and Holder inequalities have been used (Rudin 1966, p. 62). The final result then follows from (6.8) and (6.12) to give

$$\| G(s) \|_{\mathrm{H}} = \sup_{U} \frac{\| V(j\omega) \|_{L^2}}{\| U(j\omega) \|_{L^2}} \leqslant \| G(j\omega) - F(j\omega) \|_{L^\infty} \tag{6.13}$$

■

In order to show that the lower bound in Lemma 6.2 can in fact be achieved requires a construction of such an $F(s)$. This construction can be made very explicit when the balanced realization of $G(s)$ is used and is given by eqns. (6.19) to (6.22). The construction of Theorem 6.3 will be central to our development of the solution to the Hankel-norm approximation problems, as will be seen in § 7, and it is therefore stated in more generality than is required to prove Theorem 6.1. The required $F(s)$ will be given by $F(s) = \hat{D} + \hat{C}(sI - \hat{A})^{-1}\hat{B}$. Theorem 6.3 will show that $\sigma^{-1}(G(s) - F(s))$ is an all-pass function.

Theorem 6.3

Let (A, B, C, D) with $A \in \mathbb{C}^{n\times n}$, $B \in \mathbb{C}^{n\times m}$, $C \in \mathbb{C}^{m\times n}$, $D \in \mathbb{C}^{m\times m}$ satisfy

$$AP + PA^* + BB^* = 0 \tag{6.14}$$

$$A^*Q + QA + C^*C = 0 \tag{6.15}$$

for

$$P = P^* = \text{diag}\,(\Sigma_1, \sigma I_r) \tag{6.16}$$

$$Q = Q^* = \text{diag}\,(\Sigma_2, \sigma I_r) \tag{6.17}$$

with Σ_1 and Σ_2 diagonal, $\sigma \neq 0$ and $\delta(\Sigma_1\Sigma_2 - \sigma^2 I) = 0$.

Partition (A, B, C) conformally with P, as

$$A = \begin{pmatrix} A_{11} & A_{12} \\ A_{21} & A_{22} \end{pmatrix}, \quad B = \begin{pmatrix} B_1 \\ B_2 \end{pmatrix}, \quad C = (C_1, C_2) \tag{6.18}$$

and define

$$\hat{A} \triangleq \Gamma^{-1}(\sigma^2 A_{11}^* + \Sigma_2 A_{11} \Sigma_1 - \sigma C_1^*\, U B_1^*) \tag{6.19}$$

$$\hat{B} \triangleq \Gamma^{-1}(\Sigma_2 B_1 + \sigma C_1^*\, U) \tag{6.20}$$

$$\hat{C} \triangleq C_1 \Sigma_1 + \sigma U B_1^* \tag{6.21}$$

$$\hat{D} \triangleq D - \sigma U \tag{6.22}$$

where U is a unitary matrix satisfying

$$B_2 = -C_2^*\, U \tag{6.23}$$

and

$$\Gamma \triangleq \Sigma_1\Sigma_2 - \sigma^2 I \tag{6.24}$$

Also define the error system

$$A_e \triangleq \begin{pmatrix} A & 0 \\ 0 & \hat{A} \end{pmatrix}, \quad B_e \triangleq \begin{pmatrix} B \\ \hat{B} \end{pmatrix}, \quad C_e \triangleq (C, \; -\hat{C}), \quad D_e \triangleq (D \; - \hat{D}) \tag{6.25}$$

Then

(1) (A_e, B_e, C_e) satisfy

$$A_e P_e + P_e A_e^* + B_e B_e^* = 0 \tag{6.26}$$

$$A_e^*\, Q_e + Q_e A_e + C_e^*\, C_e = 0 \tag{6.27}$$

with

$$P_e = \begin{pmatrix} \Sigma_1 & 0 & I \\ 0 & \sigma I & 0 \\ I & 0 & \Sigma_2\Gamma^{-1} \end{pmatrix} \tag{6.28}$$

$$Q_e = \begin{pmatrix} \Sigma_2 & 0 & -\Gamma \\ 0 & \sigma I & 0 \\ -\Gamma & 0 & \Sigma_1\Gamma \end{pmatrix} \tag{6.29}$$

$$P_e Q_e = \sigma^2 I \tag{6.30}$$

(2) Defining $E(s) \triangleq D_e + C_e(sI - A_e)^{-1}B_e$ then $E(s)E^*(-\bar{s}) = \sigma^2 I$.

(3) If $\delta(A) = 0$ then

(a) $\delta(\hat{A}) = 0$

(b) If $\delta(\Sigma_1\Sigma_2) = 0$ then

$$\text{In}(\hat{A}) = \text{In}(-\Sigma_1\Gamma) = \text{In}(-\Sigma_2\Gamma)$$

(c) If $P > 0$, $Q > 0$ then the McMillan degree of the stable part of $(\hat{A}, \hat{B}, \hat{C})$ equals $\pi(\Sigma_1\Gamma) = \pi(\Sigma_2\Gamma)$.

(d) If either

(i) $\Sigma_1\Gamma > 0$ and $\Sigma_2\Gamma > 0$

or

(ii) $\Sigma_1\Gamma < 0$ and $\Sigma_2\Gamma < 0$ then $(\hat{A}, \hat{B}, \hat{C})$ is a minimal realization.

Proof

For notational convenience it will be assumed that $\sigma = 1$, and as with the proof of Theorem 5.1 this can be done without loss of generality since B, C and Σ can be simply rescaled to give $\sigma = 1$.

It is first necessary to verify that there exists a unitary matrix U satisfying (6.23). The (2, 2) blocks of (6.14) and (6.15) give

$$A_{22} + A_{22}^* + B_2B_2^* = 0 \tag{6.31}$$

$$A_{22}^* + A_{22} + C_2^* C_2 = 0 \tag{6.32}$$

and hence $B_2B_2^* = C_2^* C_2$ and by Lemma 3.5 there exists a unitary U satisfying (6.23).

(1) The proof of eqns. (6.26) to (6.30) is by a straightforward calculation, as follows. To verify (6.26) and (6.28) we need (6.14) which is assumed, together with

$$A_{11} + \hat{A}^* + B_1\hat{B}^* = 0 \tag{6.33}$$

$$A_{21} + B_2\hat{B}^* = 0 \tag{6.34}$$

$$\hat{A}\Sigma_2\Gamma^{-1} + \Sigma_2\Gamma^{-1}\hat{A}^* + \hat{B}\hat{B}^* = 0 \tag{6.35}$$

which will now be verified.

$$B_2\hat{B}^* = B_2(B_1^* \Sigma_2 + U^*C_1)\Gamma^{-1} \tag{6.36}$$

$$(6.23) \Rightarrow = (B_2B_1^* \Sigma_2 - C_2^* C_1)\Gamma^{-1} \tag{6.37}$$

$$= ((-A_{21}\Sigma_1 - A_{12}^*)\Sigma_2 + A_{12}^* \Sigma_2 + A_{21})\Gamma^{-1} \tag{6.38}$$

$$= -A_{21} \Rightarrow (6.34)$$

where $B_2B_1^*$ and $C_2^* C_1$ were substituted in (6.37) using the (2, 1) blocks of (6.14) and (6.15), respectively. To verify (6.33)

$$B_1\hat{B}^* = (B_1B_1^* \Sigma_2 + B_1U^*C_1)\Gamma^{-1} \tag{6.39}$$

$$= (-A_{11}\Sigma_1\Sigma_2 - \Sigma_1A_{11}^* \Sigma_2 + B_1U^*C_1)\Gamma^{-1} \tag{6.40}$$

$$= -A_{11} - \hat{A}^* \Rightarrow (6.33)$$

where $B_1B_1^*$ was substituted using the (1, 1) block of (6.14) and (6.19) substituted in (6.40). Finally to verify (6.35) consider

$$\begin{aligned}
&\Gamma\hat{A}\Sigma_2+\Sigma_2\hat{A}^*\Gamma \\
&\quad=(A_{11}^*+\Sigma_2A_{11}\Sigma_1-C_1^*\,UB_1^*)\Sigma_2+\Sigma_2(A_{11}+\Sigma_1A_{11}^*\,\Sigma_2-B_1U^*C_1) \quad (6.41)\\
&\quad=-(\Sigma_2B_1+C_1^*\,U)(B_1^*\,\Sigma_2+U^*C_1)+(A_{11}^*\,\Sigma_2+\Sigma_2A_{11}+C_1^*\,C_1)\\
&\qquad\qquad+\Sigma_2(A_{11}\Sigma_1+\Sigma_1A_{11}^*+B_1B_1^*)\Sigma_2 \quad (6.42)\\
&\quad=-\Gamma\hat{B}\hat{B}^*\Gamma\Rightarrow(6.35)
\end{aligned}$$

where (6.19)⇒(6.41) and (6.42) is a rearrangement of (6.41) and, finally, the (1, 1) blocks of (6.14) and (6.15) are used. Equations (6.26) and (6.28) are hence verified.

Similarly in order to verify (6.27) and (6.29) we need (6.15) which is assumed together with

$$A_{11}^*(-\Gamma)+(-\Gamma)\hat{A}-C_1^*\,\hat{C}=0 \quad (6.43)$$

$$A_{12}^*(-\Gamma)-C_2^*\,\hat{C}=0 \quad (6.44)$$

$$\hat{A}^*\Sigma_1\Gamma+\Sigma_1\Gamma\hat{A}+\hat{C}^*\hat{C}=0 \quad (6.45)$$

Equations (6.43) to (6.45) are now verified in an analogous manner to eqns. (6.33) to (6.35)

$$\begin{aligned}
C_2^*\,\hat{C}&=C_2^*(C_1\Sigma_1+UB_1^*)\\
&=(-A_{12}^*\,\Sigma_2-A_{21})\Sigma_1-B_2B_1^*\\
&=-A_{12}^*\,\Gamma\Rightarrow(6.44)\\
C_1^*\,\hat{C}&=C_1^*\,C_1\Sigma_1+C_1^*\,UB_1^*\\
&=-A_{11}^*\,\Sigma_2\Sigma_1-\Sigma_2A_{11}\Sigma_1+C_1^*\,UB_1^*\\
&=-A_{11}^*\,\Gamma-\Gamma\hat{A}\Rightarrow(6.43)
\end{aligned}$$

$$\begin{aligned}
&\hat{A}^*\Sigma_1\Gamma+\Sigma_1\Gamma\hat{A}\\
&\quad=(A_{11}+\Sigma_1A_{11}^*\,\Sigma_2-B_1U^*C_1)\Sigma_1+\Sigma_1(A_{11}^*+\Sigma_2A_{11}\Sigma_1-C_1^*\,UB_1)\\
&\quad=-(\Sigma_1C_1^*+B_1U^*)(C_1\Sigma_1+UB_1^*)\\
&\qquad+\Sigma_1(A_{11}^*\,\Sigma_2+\Sigma_2A_{11}+C_1^*\,C_1)\Sigma_1+(A_{11}\Sigma_1+\Sigma_1A_{11}^*+B_1B_1^*)\\
&\quad=-\hat{C}^*\hat{C}\Rightarrow(6.45)
\end{aligned}$$

Therefore, (6.27) and (6.29) have been verified, (6.30) is immediate, and the proof of part (1) is complete.

(2) Equations (6.26), (6.27) and (6.30) ensure the conditions of Theorem 5.1, part (1 *b*) are satisfied and Theorem 5.1, part (2) can be used to show that the D_e given in (6.25) makes $E(s)$ all-pass. (Note it is still assumed that $\sigma=1$.) We hence need to verify that

$$D_e^*\,D_e=I \quad (6.46)$$

$$D_e^*\,C_e+B_e^*\,Q_e=0 \quad (6.47)$$

$$D_eB_e^*+C_eP_e=0 \quad (6.48)$$

Equation (6.46) is immediate, (6.47) follows by substituting the definitions of $\hat{B}$, $\hat{C}$, D_e and Q, and (6.48) follows from $D_e(6.47)P_e$.

(3) (*a*) To show that $\delta(\hat{A})=0$ if $\delta(A)=0$ we will assume that there exists $x\in\mathbb{C}^n$ and $\lambda\in\mathbb{C}$ such that $\hat{A}x=\lambda x$ and $\lambda+\lambda^*=0$, and show that this implies $x=0$. From $x^*(6.45)x$,

$$(\lambda+\lambda^*)x^*\Sigma_1\Gamma x+x^*\hat{C}^*\hat{C}x=0 \tag{6.49}$$

$$\Rightarrow \hat{C}x=0 \tag{6.50}$$

Now (6.43)x gives

$$-A_{11}^*\,\Gamma x-\Gamma(\lambda x)+C_1^*\hat{C}x=0$$

$$\Rightarrow x^*\Gamma A_{11}=(-\lambda^*)x^*\Gamma \tag{6.51}$$

Also (6.44)x and (6.50) give

$$A_{12}^*\,\Gamma x=0 \tag{6.52}$$

Equations (6.51) and (6.52) imply that $(x^*\Gamma, 0)A=(-\lambda^*)(x^*\Gamma, 0)$ but since it is assumed that $\delta(A)=0$, $(\lambda+\lambda^*)=0$ and Γ^{-1} exists this implies that $x=0$ and $\delta(\hat{A})=0$ is proven.

(*b*) Since $\delta(\hat{A})=0$ has been proved and $\delta(\Sigma_1\Sigma_2)=0$ is asssumed ($\Rightarrow\delta(\Sigma_2\Gamma^{-1})=\delta(\Sigma_1\Gamma)=0$) Theorem 3.3, parts (2) and (3) can be applied since eqns. (6.35) and (6.45) have been verified. Hence

$$\text{In}\,(\hat{A})=\text{In}\,(-\Sigma_1\Gamma^{-1})=\text{In}\,(-\Sigma_1\Gamma)$$
$$=\text{In}\,(-\Sigma_2\Gamma)$$

(*c*) Assume that there exists $x\neq 0\in\mathbb{C}^n$ and $\lambda\in\mathbb{C}$ such that $\hat{A}x=\lambda x$ and $\hat{C}x=0$ (i.e. $(\hat{A}, \hat{C})$ is not completely observable). Then (6.43)x and (6.44)x give

$$-A_{11}^*\,\Gamma x-\Gamma\lambda x=0$$
$$-A_{12}^*\,\Gamma x=0$$

hence $(-\lambda^*)$ is an eigenvalue of A since $\Gamma x\neq 0$. However, since $P>0$ and $\delta(A)=0$ are assumed then $\text{In}\,(A)=(0, n, 0)$ and all the unobservable modes must be in the open right half-plane. Similarly, if it is assumed that $(\hat{A}, \hat{B})$ is not completely controllable then (6.33) and (6.34) will give the analogous conclusion and therefore all the modes in the left half-plane are controllable and observable, and the condition in (3 *b*) gives their number.

(*d*) (i) If $\Sigma_1\Gamma>0$ or $\Sigma_2\Gamma>0$ then by (3 *b*) $\text{In}\,(\hat{A})=(0, n-r, 0)$ and by (3 *c*) the McMillan degree of $(\hat{A}, \hat{B}, \hat{C})$ is $(n-r)$ and the result is proven.

(ii) Assume there exists x such that $\hat{A}x=\lambda x$ and $\hat{C}x=0$. Then $x^*(6.45)x$ gives

$$(\lambda+\lambda^*)x^*\Sigma_1\Gamma x=0$$

but $(\lambda+\lambda^*)\neq 0$ by (3 *a*) and $\Sigma_1\Gamma<0$ is assumed so that $x=0$. Hence $(\hat{A}, \hat{C})$ is completely observable. Similarly (6.35) gives $(\hat{A}, \hat{B})$ completely controllable. ■

Proof of Theorem 6.1

Firstly, if $p\neq m$ then augment $G(s)$ by rows or columns of zero to make it square. This does not change the Hankel singular values. Lemma 6.2 gives that

$$\|G(s)\|_H\leqslant\|G(j\omega)-F(j\omega)\|_{L^\infty}$$

for all $F \in H_-^\infty$, and we just need to exploit the construction of Theorem 6.3 to find a minimizing F. Since $G(s)$ is stable and rational it will have a balanced realization in which the Hankel singular values are $\sigma_1 = \sigma_2 \ldots = \sigma_r > \sigma_{r+1} \geqslant \sigma_{r+2} \ldots \geqslant \sigma_n > 0$. Now let

$$\Sigma = \begin{pmatrix} \Sigma_1 & 0 \\ 0 & \sigma_1 I_r \end{pmatrix}$$

where $\Sigma_1 = \text{diag}\,(\sigma_{r+1}, \sigma_{r+2}, \ldots, \sigma_n)$, and let (A, B, C) be the corresponding balanced realization of $G(s)$ satisfying (6.14) and (6.15) (with $\sigma = \sigma_1$, $\Sigma_2 = \Sigma_1$). Note that $(\Sigma_1^2 - \sigma_1^2 I) < 0$ so that Theorem 6.3, part (3 *b*) gives that $\text{In}\,(\hat{A}) = \text{In}\,(-\Sigma_1(\Sigma_1^2 - \sigma_1^2 I)) = (n-r, 0, 0)$, that is all the poles of $F(s) \triangleq \hat{D} + \hat{C}(sI - \hat{A})^{-1}\hat{B}$ are in the open right half-plane and hence $F \in H_-^\infty$. Finally, Theorem 6.3, part (2) gives that $\| G(j\omega) - F(j\omega) \|_{L^\infty} = \sigma_1$, and the result is proved. ■

Example 6.1

Take

$$G(s) = \frac{39s^2 + 105s + 250}{(s+2)(s+5)^2}$$

This has a balanced realization given by

$$A = \begin{pmatrix} -2 & 4 & -4 \\ -4 & -1 & -4 \\ -4 & 4 & -9 \end{pmatrix}, \quad B = \begin{pmatrix} 2 \\ 1 \\ 6 \end{pmatrix}, \quad C = (2, -1, 6), \quad \Sigma = \begin{pmatrix} 1 & 0 & 0 \\ 0 & \frac{1}{2} & 0 \\ 0 & 0 & 2 \end{pmatrix}$$

Now using the above construction with $\Sigma_1 = \begin{pmatrix} 1 & 0 \\ 0 & \frac{1}{2} \end{pmatrix}$, $\sigma = 2$ gives

$$\hat{A} = \tfrac{1}{3}\begin{pmatrix} 2 & 10 \\ -8 & 5 \end{pmatrix}, \quad \hat{B} = \tfrac{1}{3}\begin{pmatrix} 2 \\ -2 \end{pmatrix}, \quad \hat{C} = (-2, -5/2), \quad \hat{D} = 2$$

$$F(s) = \frac{6s^2 - 13s + 90}{3s^2 - 7s + 30}$$

$$G(s) - F(s) = \frac{2(-s+2)(-s+5)^2(3s^2 + 7s + 30)}{(s+2)(s+5)^2(3s^2 - 7s + 30)}$$

$F(s)$ is an optimal anticausal approximation to $G(s)$ with L^∞-error of 2.

Example 6.2

Let us also illustrate Theorem 6.3 when $(\Sigma_1^2 - \sigma^2 I)$ is indefinite. Take $G(s)$ as in Example 6.1 and permute the first and third states of the balanced realization so that $\Sigma = \text{diag}\,(2, \frac{1}{2}, 1)$, $\Sigma_1 = \text{diag}\,(2, \frac{1}{2})$, $\sigma = 1$. The construction of Theorem 6.3 now gives

$$\hat{A} = \begin{pmatrix} -3 & 2 \\ 8 & 3 \end{pmatrix}, \quad \hat{B} = \begin{pmatrix} 2 \\ -2 \end{pmatrix}, \quad \hat{C} = (6, -3/2), \quad \hat{D} = 1$$

Theorem 6.3, part (3 *b*) implies that

$$\text{In}(\hat{A}) = \text{In}(-\Sigma_1(\Sigma_1{}^2 - \sigma^2 I)) = \text{In}\begin{pmatrix} -6 & 0 \\ 0 & \frac{3}{8} \end{pmatrix} = (1, 1, 0)$$

which is verified by noting that $\hat{A}$ has eigenvalues of 5 and -5.

$$\hat{G}(s) = \hat{D} + \hat{C}(sI - \hat{A})^{-1}\hat{B} = \frac{s+20}{s+5}$$

and we note that the stable part of $\hat{G}(s)$ has McMillan degree 1 as predicted by Theorem 6.3, part (3 *c*). However, this example has been constructed to show that $(\hat{A}, \hat{B}, \hat{C})$ itself may not be minimal when the conditions of part (3 *d*) are not satisfied, and in this case the unstable pole at $+5$ is both uncontrollable and unobservable. $(s+20)/(s+5)$ is in fact an optimal Hankel-norm approximation to $G(s)$ of degree 1 and

$$E(s) = 1\,\frac{(-s+2)(-s+5)^2}{(s+2)(s+5)^2}$$

In general the error $E(j\omega)$ will have modulus equal to σ but $E(s)$ will contain unstable poles.

Example 6.3

Let us finally complete the analysis of this $G(s)$ by permuting the second and third states in the balanced realization of Example 6.2 to obtain $\Sigma_1 = \text{diag}(2, 1)$, $\sigma = \frac{1}{2}$. We will find

$$\hat{A} = \begin{pmatrix} -15 & -4 \\ -20 & -6 \end{pmatrix}, \quad \hat{B} = \begin{pmatrix} 4 \\ 4 \end{pmatrix}, \quad \hat{C} = (15, 3), \quad \hat{D} = -\tfrac{1}{2}$$

$$\hat{G}(s) = \hat{D} + \hat{C}(sI - \hat{A})^{-1}\hat{B} = \tfrac{1}{2}\,\frac{(-s^2 + 123s + 110)}{(s^2 + 21s + 10)}$$

$$E(s) = G(s) - \hat{G}(s) = -\tfrac{1}{2}\,\frac{(-s+2)(-s+5)^2(s^2 - 21s + 10)}{(s+2)(s+5)^2(s^2 + 21s + 10)}$$

Note that $-\Sigma_1\Gamma = \text{diag}(-15/2, -3/4)$ so that $\hat{A}$ is stable by Theorem 6.3, part (3 *b*). $|E(j\omega)| = \frac{1}{2}$ by Theorem 6.3, part (2), $(\hat{A}, \hat{B}, \hat{C})$ is minimal by Theorem 6.3, part (3 *d*). $\hat{G}(s)$ is in fact an optimal second-order Hankel-norm approximation to $G(s)$.

The Nyquist diagrams for $G(s)$ and the three approximations are given in Fig. 3. The fact that the difference between $G(s)$ and the approximation have constant moduli can be observed by noting the frequency scales on each plot.

7. One class of solutions to the optimal Hankel-norm approximation problem

Recall that the optimal Hankel-norm approximation problem is, 'given $G(s)$ of McMillan degree n, find $\hat{G}(s)$ of McMillan degree $k < n$ such that $\|G(s) - \hat{G}(s)\|_H$ is minimized'. One class of solutions to this problem can now be derived immediately using the results of § 6.

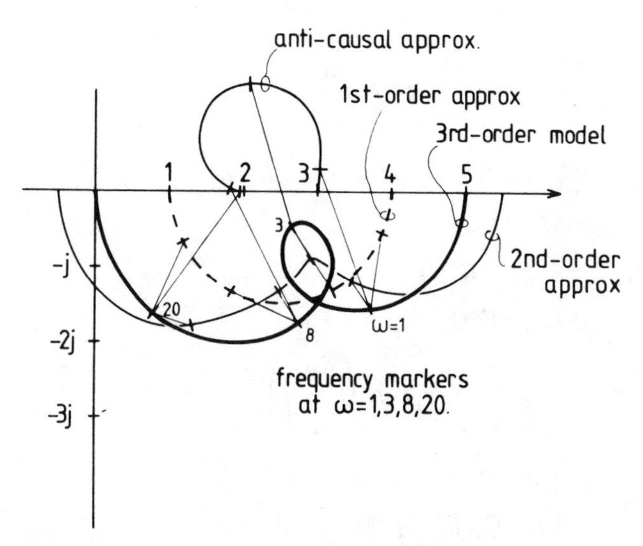

Figure 3. Nyquist diagrams for Examples 6.1 to 6.3.

The scalar discrete and continuous time problems were solved by Adamjan *et al.* (1971). These results have been made more explicit by de Wilde, Kung, Silverman and their co-workers. The results of Adamjan *et al.* (1978) on block Hankel matrices have been exploited by Kung and Lin (1981 a, b) to approach a solution to the multivariable discrete-time problem.

The derivation given here is for multivariable continuous-time systems and is self-contained, straightforward and explicit. Firstly, Lemma 7.1 gives a lower bound on the achievable Hankel-norm of the error $\|G(s)-\hat{G}(s)\|_H \geqslant \sigma_{k+1}(G)$, and then Theorem 7.2 shows that the construction of Theorem 6.3 can be used to achieve this lower bound.

Lemma 7.1

Given a stable, rational, $p \times m$, transfer-function matrix $G(s)$ with Hankel singular values $\sigma_1 \geqslant \sigma_2 \ldots \geqslant \sigma_k \geqslant \sigma_{k+1} \geqslant \sigma_{k+2} \ldots \geqslant \sigma_n > 0$, then for all $\hat{G}(s)$ stable and of McMillan degree $\leqslant k$

$$\|G(s)-\hat{G}(s)\|_H \geqslant \sigma_{k+1}(G(s)) \tag{7.1}$$

Proof

Let $(\hat{A}, \hat{B}, \hat{C})$ be a minimal state-space realization of $\hat{G}(s)$, then (A_e, B_e, C_e) given by (6.25) will be a state-space realization of $(G(s)-\hat{G}(s))$. Now let $P=P^*$ and $Q=Q^*$ satisfy (6.26) and (6.27) respectively (but not necessarily (6.28) and (6.29)) and write

$$P=\begin{pmatrix} P_{11} & P_{12} \\ P_{12}{}^* & P_{22} \end{pmatrix}, \quad Q=\begin{pmatrix} Q_{11} & Q_{12} \\ Q_{12}{}^* & Q_{22} \end{pmatrix} \tag{7.2}$$

Since $P \geqslant 0$ it can be factored as

$$P=RR^* \tag{7.3}$$

where

$$R = \begin{pmatrix} R_{11} & R_{12} \\ 0 & R_{22} \end{pmatrix} \tag{7.4}$$

with

$$R_{22} = P_{22}^{1/2}, \quad R_{12} = P_{12}P_{22}^{-1/2}, \quad R_{11}R_{11}^* = P_{11} - R_{12}R_{12}^* \tag{7.5}$$

$$\|G(s) - \hat{G}(s)\|_H^2 = \lambda_{\max}(PQ) \tag{7.6}$$

$$= \lambda_{\max}(RR^*Q) \tag{7.7}$$

$$= \lambda_{\max}(R^*QR) \tag{7.8}$$

$$\geqslant \lambda_{\max}\left([R_{11}^*, 0]Q\begin{pmatrix} R_{11} \\ 0 \end{pmatrix}\right) \tag{7.9}$$

$$= \lambda_{\max}(R_{11}^*\, Q_{11}R_{11}) \tag{7.10}$$

$$= \lambda_{\max}(Q_{11}R_{11}R_{11}^*) \tag{7.11}$$

$$= \lambda_{\max}(Q_{11}(P_{11} - R_{12}R_{12}^*)) \tag{7.12}$$

$$= \lambda_{\max}(Q_{11}^{1/2}\, P_{11}Q_{11}^{1/2} - XX^*) \quad \text{where } X = Q_{11}^{1/2}\, R_{12} \tag{7.13}$$

$$\geqslant \sigma_{k+1}^2\,(G(s)) \tag{7.14}$$

where (7.10) follows since $R_{11}^*\, Q_{11}R_{11} \geqslant 0$ is a submatrix of $R^*QR \geqslant 0$, and (7.14) follows from Mirsky's result (Proposition 2.2) since X is of dimension $n \times k$ ($\Rightarrow$rank $XX^* \leqslant k$) and $\lambda_i(Q_{11}^{1/2}\, P_{11}Q_{11}^{1/2}) = \sigma_i^2\,(G(s))$. ■

Note that this lemma is no more than a continuous-time version of Mirsky's result on approximating matrices by matrices of lower rank.

We can now give a solution to the optimal Hankel-norm approximation problem for square transfer functions.

Theorem 7.2

Given a stable, rational, $m \times m$, transfer function $G(s)$ then

(1) $\sigma_{k+1}(G(s)) = \inf_{F \in H_-^\infty, \hat{G}} \|G(j\omega) - \hat{G}(j\omega) - F(j\omega)\|_{L^\infty}$, McMillan degree $(\hat{G}) \leqslant k$ (7.15)

(2) If $G(s)$ has Hankel singular values $\sigma_1 \geqslant \sigma_2 \ldots \geqslant \sigma_k > \sigma_{k+1} = \sigma_{k+2} \ldots = \sigma_{k+r} > \sigma_{k+r+1} \geqslant \ldots \geqslant \sigma_n > 0$ then $\hat{G}(s)$ of McMillan degree k is an optimal Hankel-norm approximation to $G(s)$ if and only if there exists $F(s) \in H_-^\infty$ (whose McMillan degree can be chosen $\leqslant n + k - 1$) such that $E(s) \triangleq G(s) - \hat{G}(s) - F(s)$ satisfies

$$E(s)E^*(-\bar{s}) = \sigma_{k+1}^2\, I \tag{7.16}$$

In which case

$$\|G(s) - \hat{G}(s)\|_{\mathrm{H}} = \sigma_{k+1} \tag{7.17}$$

(3) Let $G(s)$ be as in (2) above, then an optimal Hankel-norm approximation of McMillan degree k, $\hat{G}(s)$, can be constructed as follows. Let (A, B, C) be a balanced realization of $G(s)$ with corresponding $\Sigma = \mathrm{diag}\,(\sigma_1, \sigma_2, \ldots, \sigma_k, \sigma_{k+r+1}, \ldots, \sigma_n, \sigma_{k+1}, \ldots, \sigma_{k+r})$, and define $(\hat{A}, \hat{B}, \hat{C}, \hat{D})$ from eqns. (6.18) to (6.23). Then

$$\hat{G}(s) + F(s) = \hat{D} + \hat{C}(sI - \hat{A})^{-1}\hat{B} \tag{7.18}$$

where $\hat{G}(s) \in H_+^\infty$ and $F \in H_-^\infty$.

Proof

From Lemma 6.2 and Lemma 7.1, for all $F \in H_-^\infty$, and $\hat{G}$ of McMillan degree k

$$\|G(j\omega)-\hat{G}(j\omega)-F(j\omega)\|_{L^\infty} \geqslant \|G(s)-\hat{G}(s)\|_H$$
$$\geqslant \sigma_{k+1}(G(s)) \tag{7.19}$$

Now define $\hat{G}$ and F via eqn. (7.18), then Theorem 6.3, part (2) implies that (7.16) holds and hence

$$\|E(j\omega)\|_{L^\infty} = \sigma_{k+1} \tag{7.20}$$

Also from Theorem 6.3, part (3 *b*)

$$\operatorname{In}(\hat{A}) = \operatorname{In}(-\Sigma_1(\Sigma_1{}^2 - \sigma_{k+1}{}^2 I)) = (n-k-r, k, 0) \tag{7.21}$$

Hence, $\hat{G}$ has McMillan degree k and is in the correct class, and therefore (7.20) implies that the inequalities in (7.19) become equalities, and part (1) is proven, as is part (3). Clearly the sufficiency of part (2) can be similarly verified by noting that (7.16) implies that (7.19) is satisfied with equality.

To show the necessity of part (2) suppose that $\hat{G}(s)$ is an optimal Hankel-norm approximation to $G(s)$ of McMillan degree k, i.e. eqn. (7.17) holds. Now Theorem 6.3 can be applied to $(G(s)-\hat{G}(s))$ to produce an optimal anticausal approximation $F(s)$, such that $(G(s)-\hat{G}(s)-F(s)/\sigma_{k+1}(G)$ is all-pass since $\sigma_{k+1}(G)=\sigma_1(G-\hat{G})$. Further, the McMillan degree of this $F(s)$ will be (the McMillan degree of $(G(s)-\hat{G}(s))$ minus the multiplicity of $\sigma_1(G-\hat{G}) \leqslant n+k-1$. ■

Remark 7.1

Note that part (1) shows the relation between the Hankel singular values and an L^∞ optimization problem. Part (2) gives a characterization of all optimal Hankel-norm approximations, but this is only implicit. Part (3) gives a particular optimal Hankel-norm approximation.

The class of optimal Hankel-norm approximations given in Theorem 7.2, part (3) can be enlarged by allowing the matrix U to satisfy $U^*U \leqslant I$ rather than being a unitary matrix. This will also give a solution for non-square systems as is now proven.

Corollary 7.3

With the same assumptions as for Theorem 7.2, part (3) except that $G(s)$ may be $p \times m$, let $(\hat{A}, \hat{B}, \hat{C}, \hat{D})$ be defined by (6.19) to (6.22) where U satisfies (6.23) and

$$U^*U \leqslant I \tag{7.22}$$

Then $\hat{G}(s)$ defined by (7.18) will give an optimal Hankel-norm approximation to $G(s)$ of McMillan degree k. Further, a particular admissible value of U is

$$U_0 = -C_2 B_2{}^{*\dagger} \tag{7.23}$$

where $B_2{}^\dagger$ is the Moore–Penrose pseudo-inverse of B_2 (see Barnett 1971, Ch. 6). In particular when rank $B_2 = r$ then $B_2{}^{*\dagger} = (B_2 B_2{}^*)^{-1} B_2$. ■

In order to prove Corollary 7.3 we first state an inequality that is an immediate consequence of Theorem 7.2, part (1) (since the L^∞ norm of a submatrix is less than or equal to that of the full matrix).

Lemma 7.4

Given a stable

$$G(s)=\begin{pmatrix} G_{11}(s) & G_{12}(s) \\ G_{21}(s) & G_{22}(s) \end{pmatrix}$$

then

$$\sigma_k(G(s)) \geq \sigma_k(G_{ij}(s)) \quad \forall i, j, k$$ ∎

Proof of Corollary 7.3

The proof will show that the given approximation $\hat{D}+\hat{C}(sI-\hat{A})^{-1}\hat{B}$ can be augmented by additional rows and columns to give an optimal Hankel-norm approximation to the augmented $(p+m)\times(p+m)$ system

$$G_a(s)=\begin{pmatrix} G(s) & 0 \\ 0 & 0 \end{pmatrix}$$

which will have a state-space representation

$$\left(A, [B, 0], \begin{pmatrix} C \\ 0 \end{pmatrix}, \begin{pmatrix} D & 0 \\ 0 & 0 \end{pmatrix}\right) \triangleq (A_a, B_a, C_a, D_a)$$

Let (A, B, C) be a balanced realization ordered and partitioned as in Theorem 7.2, part (3), then (A_a, B_a, C_a) will also be a balanced realization with the same Σ. Equations (6.23) and (7.22) imply that

$$\begin{pmatrix} C_2 \\ 0 \end{pmatrix} + V \begin{pmatrix} B_2{}^* \\ 0 \end{pmatrix} = 0$$

where

$$V=\begin{pmatrix} U & (I-UU^*)^{1/2} \\ (I-U^*U)^{1/2} & -U^* \end{pmatrix}$$

since

$$B_2(I-U^*U)B_2{}^* = B_2B_2{}^* - C_2{}^*C_2 = 0$$

(cf. eqns. (6.31) and (6.32)). Also $V^*V=I$ (this is easily verified on noting that $U^*(I-UU^*)^{1/2}=(I-U^*U)^{1/2}U^*$). Hence, Theorem 7.2, part (3) can be applied to produce an optimal Hankel-norm approximation to $G_a(s)$, as follows (eqns. (6.19) to (6.22))

$$\tilde{A} \triangleq \Gamma^{-1}\left(\sigma_{k+1}{}^2 A_{11}{}^* + \Sigma_1 A_{11}\Sigma_1 - \sigma_{k+1}[C_1{}^*, 0]V\begin{pmatrix} B_1{}^* \\ 0 \end{pmatrix}\right) = \hat{A}$$

$$\tilde{B} \triangleq \Gamma^{-1}(\Sigma_1[B_1, 0] + \sigma_{k+1}[C_1{}^*, 0]V) = (\hat{B}, \hat{B}_2)$$

$$\tilde{C} \triangleq \begin{pmatrix} C_1 \\ 0 \end{pmatrix}\Sigma_1 + \sigma_{k+1}V\begin{pmatrix} B_1{}^* \\ 0 \end{pmatrix} = \begin{pmatrix} \hat{C} \\ \hat{C}_2 \end{pmatrix}$$

$$\tilde{D} \triangleq D_a - \sigma_{k+1}V$$

Therefore, the system $\hat{D}+\hat{C}(sI-\hat{A})^{-1}\hat{B}$ is given by the first p rows and m columns of the system $\tilde{D}+\tilde{C}(sI-\tilde{A})^{-1}\tilde{B}$. Let $\hat{G}_a(s)$ and $\hat{G}(s)$ be the resulting approximations to $G_a(s)$ and $G(s)$, respectively. Then $\|G_a(s)-\hat{G}_a(s)\|_H=\sigma_{k+1}(G(s))$ but $(G(s)-\hat{G}(s))$ is a submatrix of $(G_a(s)-\hat{G}_a(s))$ and by Lemma 7.4 $\|G(s)-\hat{G}(s)\|_H \leqslant \|G_a(s)-\hat{G}_a(s)\|_H=\sigma_{k+1}(G(s))$. The optimality of the answer is hence proven via Lemma 7.1 which is also proven for $p\times m$ $G(s)$.

To show that U_0 is an admissible value for U, note that there exists a U such that $C_2+UB_2{}^*=0$. Hence, using properties of pseudo-inverses (Barnet 1971, p. 131)

$$\begin{aligned} C_2+U_0B_2{}^* &= C_2-C_2B_2{}^{*\dagger}\,B_2{}^* \\ &= -UB_2{}^*+UB_2{}^*\,B_2{}^{*\dagger}\,B_2{}^* \\ &= 0 \end{aligned}$$

Also

$$\begin{aligned} U_0{}^*\,U_0 &= B_2{}^\dagger\,C_2{}^*\,C_2B_2{}^{*\dagger} \\ &= B_2{}^\dagger\,B_2B_2{}^*\,B_2{}^{*\dagger} \\ &= I-(I-B_2{}^\dagger\,B_2)(I-B_2{}^*\,B_2{}^{*\dagger}) \\ &\leqslant I \end{aligned}$$

Hence, U_0 is an admissible value for U. ■

Remark 7.2

Corollary 7.3 now gives an explicit algorithm to approximate $G(s)$ based on its balanced realization and is given by finding the stable part of $\hat{C}(sI-\hat{A})^{-1}\hat{B}$.

The choice of the ' D-matrix ' for the approximation is arbitrary since the Hankel-norm does not depend on D. However, $\|G(j\omega)-\hat{G}(j\omega)\|_{L^\infty}$ depends on D and in § 9 and § 10.1 a particular choice of D is suggested. The choice of the matrix U generating the solution is also arbitrary subject to (6.23) and (7.22). It will also be shown later that the choice given by (7.23) can give favourable L^∞-errors.

It might be of interest at this point to re-examine Examples 6.1, 6.2 and 6.3. Note that since these examples are single-input and single-output eqn. (6.23) has a unique solution and in fact the given approximations are the only solutions.

8. Characterizations of all solutions to the optimal Hankel-norm approximation problem

In § 7 it was shown how one class of solutions to the optimal Hankel-norm approximation problem could be derived immediately from the results of § 6. For multivariable systems the set of all solutions to this problem is larger than that given in Corollary 7.3 and we may now solve the problem of characterizing all solutions to this problem. It was noted in Remark 7.1 that Theorem 7.2, part (2) gives characterization in terms of frequency response, namely $(G(s)-\hat{G}(s)-F(s))/\sigma_{k+1}(G)$ being all-pass. This is not explicit because the constraints on the McMillan degree of $\hat{G}(s)$ and the pole positions of $F(s)$ have not been simply incorporated. However, this frequency response relation can be re-written in state-space form using the result of Theorem 5.1 and Theorem

6.3, part (1). The constraints on the pole positions of $\hat{G}(s)+F(s)$ are then guaranteed by specifying the inertia of certain matrices that are solutions of Lyapunov equations. All solutions to these equations are then derived in Lemmas 8.2 and 8.3. The optimal Hankel-norm approximations are then characterized in state-space form in Theorem 8.4 and Lemma 8.5. These solutions can in fact be rewritten in the form of Fig. 4 where $H(s)$ is fixed and the degrees of freedom are contained in the transfer function $K(s)$ that satisfies the constraint $C_2+K(s)B_2^*=0$ and is either all-pass (Corollary 8.6) or $\|K(j\omega)\|_{L^\infty}\leqslant 1$ (Theorem 8.7). Also Theorem 8.7 includes a uniqueness result. Finally, an example is included whose solutions can be analytically determined.

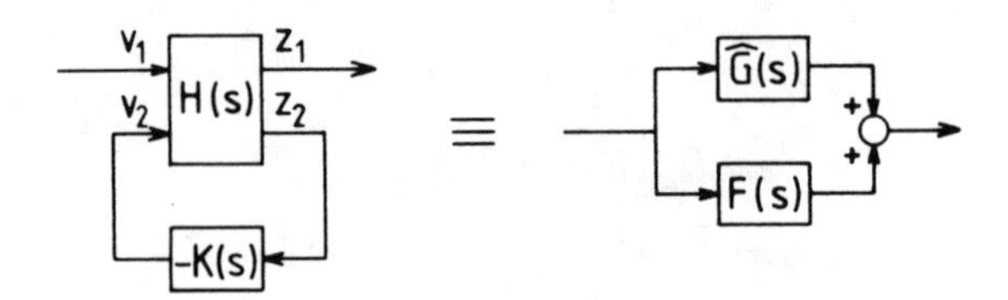

Figure 4. All optimal Hankel-norm approximations.

8.1. *State-space characterization*

We first give necessary and sufficient conditions for $\hat{G}(s)$ to be an optimal Hankel-norm approximation to $G(s)$ in terms of the matrices P_e and Q_e.

Lemma 8.1

Given a stable, rational, $m\times m$ transfer-function $G(s)=C(sI-A)^{-1}B$, with Hankel singular values

$$\sigma_1\geqslant\sigma_2\ldots\geqslant\sigma_k>\sigma_{k+1}=\sigma_{k+2}\ldots=\sigma_{k+r}>\sigma_{k+r+1}\geqslant\ldots\geqslant\sigma_n>0$$

and $\hat{G}(s)$ of McMillan degree $k\leqslant n-r$. Then $\hat{G}(s)$ is an optimal Hankel-norm approximation to $G(s)$ *if and only if* there exists $(\hat{A},\hat{B},\hat{C})$, P_e, Q_e such that

(*a*) $\hat{G}(s)$ is the stable part of $\hat{C}(sI-\hat{A})^{-1}\hat{B}$,

(*b*) eqns. (6.25), (6.26), (6.27) hold, and

$$P_eQ_e=\sigma_{k+1}^2 I \tag{8.1}$$

and

(*c*) for P_e and Q_e partitioned conformally with A_e in (6.25) as

$$P_e=\begin{pmatrix}P_{11} & P_{12}\\ P_{12}^* & P_{22}\end{pmatrix},\quad Q_e=\begin{pmatrix}Q_{11} & Q_{12}\\ Q_{12}^* & Q_{22}\end{pmatrix} \tag{8.2}$$

then

$$\text{In}\,(P_{22})=\text{In}\,(Q_{22})=(k,l,0) \tag{8.3}$$

Further, dim $(\hat{A})\triangleq k+l$ can be chosen $\leqslant n+2k-1$.

Proof

Sufficiency. Assuming conditions (*a*) to (*c*) hold we will show that $\hat{G}(s)$ satisfies part (2) of Theorem 7.2. Firstly note that we can write $\hat{G}(s)+F(s)=\hat{D}+\hat{C}(sI-\hat{A})^{-1}\hat{B}$ for some $F \in H_-^{\infty}$. Also $(A_e, B_e, C_e, -\hat{D})$ will be a state-space realization of $(G(s)-\hat{G}(s)-F(s))$. Now (8.1), (6.26), (6.27) and Theorem 5.1, part (2) imply there exists $\hat{D}$ such that $(G(s)-\hat{G}(s)-F(s))/\sigma_{k+1}$ is all-pass. Finally, assumption (8.3) together with the (2, 2) block of (6.26) or (6.27) and Theorem 3.3, part (2) show that $\nu(\hat{A}) \leqslant k$ and hence $\hat{G}(s)$ has McMillan degree $\leqslant k$. Part (2) of Theorem 7.2 has therefore been verified.

Necessity. Given that $\|G(s)-\hat{G}(s)\|_H=\sigma_{k+1}$ we need to show that there exist $(\hat{A}, \hat{B}, \hat{C})$, P_e, Q_e satisfying (*a*) to (*c*). Let $\hat{G}(s)=\hat{C}_1(sI-\hat{A}_1)^{-1}\hat{B}_1$ with $\dim(\hat{A}_1)=k$, then $(G(s)-\hat{G}(s))$ has the state-space realization

$$\begin{pmatrix} A & 0 \\ 0 & \hat{A}_1 \end{pmatrix}, \quad \begin{pmatrix} B \\ \hat{B}_1 \end{pmatrix}, \quad (C, \ -\hat{C}_1) \tag{8.4}$$

The construction of Theorem 6.3 can now be applied to (8.4) with $\sigma=\sigma_{k+1}(G)=\sigma_1(G-\hat{G})$. The construction would first find a realization of (8.4) with diagonal gramians P and Q, and this will be possible by Theorem 4.3 since A and $\hat{A}_1$ are assumed to be stable ($(G-\hat{G})$ may not have McMillan degree $(n+k)$ so that a balanced realization related to (8.4) via a similarity transformation may not exist).

The construction will then give an anticausal system $(\hat{A}_2, \hat{B}_2, \hat{C}_2, \hat{D}_2)$ of dimension $l \leqslant n+k-1$. Now defining

$$\hat{A}=\begin{pmatrix} \hat{A}_1 & 0 \\ 0 & \hat{A}_2 \end{pmatrix}, \quad \hat{B}=\begin{pmatrix} \hat{B}_1 \\ \hat{B}_2 \end{pmatrix}, \quad \hat{C}=(\hat{C}_1, \hat{C}_2), \quad \hat{D}=\hat{D}_2$$

gives (6.25), (6.26), (6.27) and (8.1).

Now to verify (8.3) we observe that since (A, B, C) is assumed stable and minimal then $P_{11}>0$ and $Q_{11}>0$. Further, by the Schur formula and (8.1)

$$\det(P_{22}-P_{12}^* P_{11}^{-1} P_{12})=\det(P_e)/\det(P_{11}) \neq 0$$

Further, (8.1) implies $Q_e=\sigma_{k+1}^2 P_e^{-1}$ so that by the matrix inversion lemma

$$Q_{22}=\sigma_{k+1}^2 [P_{22}-P_{12}^* P_{11}^{-1} P_{12}]^{-1}$$

and hence $\delta(Q_{22})=0$. Similarly $\delta(P_{22})=0$. By construction $\text{In}(\hat{A})=(l, k, 0)$ and hence by Theorem 3.3, parts (2) and (3) (8.3) holds. ■

Lemma 8.1, therefore, gives an explicit condition on the matrices P_e and Q_e such that $(\hat{A}, \hat{B}, \hat{C})$ generates an optimal Hankel-norm approximation. Let us assume that (A, B, C) is a balanced realization of $G(s)$ with

$$\begin{aligned} \Sigma &= \text{diag}(\sigma_1, \sigma_2, \ldots, \sigma_k, \sigma_{k+r+1}, \ldots, \sigma_n, \sigma_{k+1}, \ldots, \sigma_{k+r}) \\ &= \text{diag}(\Sigma_{11}, \Sigma_{12}, \sigma_{k+1} I_r) \\ &= \text{diag}(\Sigma_1, \sigma_{k+1} I_r) \end{aligned} \tag{8.5}$$

Examining the (1, 1) blocks of (6.26) and (6.27) we have in (8.2) that

$$P_{11}=\Sigma,\quad Q_{11}=\Sigma \tag{8.6}$$

and all the solutions to eqns. (8.1), (8.2), (8.3), (8.6) are given in the following lemma.

Lemma 8.2

Let $P_e=P_e^*$, $Q_e=Q_e^*$ be given by

$$P_e=\begin{pmatrix}\Sigma & M\\ M^* & R\end{pmatrix},\quad Q_e=\begin{pmatrix}\Sigma & N\\ N^* & S\end{pmatrix} \tag{8.7}$$

with Σ given by (8.5) and $R, S\in\mathbb{C}^{(k+l)\times(k+l)}$. Then *all* solutions to eqn. (8.1), subject to

$$\operatorname{In}(R)=\operatorname{In}(S)=(k, l, 0) \tag{8.8}$$

are given by

$$R=T\begin{pmatrix}\Sigma_1\Gamma^{-1} & 0\\ 0 & -\sigma_{k+1}I\end{pmatrix}T^* \tag{8.9}$$

$$S=T^{*-1}\begin{pmatrix}\Sigma_1\Gamma & 0\\ 0 & -\sigma_{k+1}I\end{pmatrix}T^{-1} \tag{8.10}$$

$$M=\begin{pmatrix}I_{n-r} & 0\\ 0 & 0\end{pmatrix}T^* \tag{8.11}$$

$$N=\begin{pmatrix}-\Gamma & 0\\ 0 & 0\end{pmatrix}T^{-1} \tag{8.12}$$

where

$$\Gamma\triangleq(\Sigma_1^2-\sigma_{k+1}^2 I)$$

and T is any $(k+l)\times(k+l)$ non-singular matrix. Further, $l\geqslant n-k-r$ and the zero-columns of MT^{*-1} and NT are only present if $l>n-k-r$. ■

Note that the transformation T only corresponds to a change of coordinates of the state-space realization $(\hat{A}, \hat{B}, \hat{C})$ and hence the solution for P_e and Q_e is essentially unique given the dimension l. The proof of Lemma 8.2 requires the following lemma.

Lemma 8.3

Given that $M\in\mathbb{C}^{(n\quad)\times(k+l)}$ satisfies

$$MS_2M^*=S_1 \tag{8.13}$$

where

$$S_1=\begin{pmatrix}I_k & 0 & 0\\ 0 & -I_{n-k-r} & 0\\ 0 & 0 & 0_{r,r}\end{pmatrix} \tag{8.14}$$

$$S_2=\begin{pmatrix}I_k & 0\\ 0 & -I_l\end{pmatrix} \tag{8.15}$$

then $l \geqslant n-k-r$ and $\exists\ T$ such that

$$T^{-1}S_2T^{*-1}=S_2 \tag{8.16}$$

and

$$MT=\begin{pmatrix} I_{n-r} & 0 \\ 0 & 0 \end{pmatrix} \tag{8.17}$$

■

The proof of Lemma 8.3 is a somewhat intricate but nevertheless straightforward exercise in matrix manipulation and is given in Appendix C.

Proof of Lemma 8.2

Firstly note that the solutions given by (8.9) to (8.12) indeed satisfy (8.1) and (8.8) and it remains to show that any solution (R, S, M, N) can be transformed to that given via an appropriate matrix T. We will construct such a matrix via three transformations and Lemma 8.3. Equation (8.1) is equivalent to, (defining $\sigma \triangleq \sigma_{k+1}$)

$$\Sigma^2+MN^*=\sigma^2 I \tag{8.18}$$

$$\Sigma N+MS=0 \tag{8.19}$$

$$M^*N+RS=\sigma^2 I \tag{8.20}$$

Substituting for N from (8.19) into (8.18) gives

$$MSM^*=\Sigma(\Sigma^2-\sigma^2 I) \tag{8.21}$$

Since In $(S)=(k, l, 0)$ there exists a congruence transformation T_1 such that

$$T_1^*\ ST_1=\text{diag}\ (I_k,\ -I_l)\triangleq S_2 \tag{8.22}$$

Also define

$$W\triangleq \text{diag}\ (\Sigma_{11}{}^{1/2}(\Sigma_{11}{}^2-\sigma^2 I)^{1/2},\ \Sigma_{12}{}^{1/2}(\sigma^2 I-\Sigma_{12}{}^2)^{1/2},\ I_r) \tag{8.23}$$

then (8.21), (8.22), (8.23) give

$$(W^{-1}MT_1{}^{*-1})S_2(W^{-1}MT_1{}^{*-1})^*=\text{diag}\ (I_k,\ -I_{n-k-r},\ 0) \tag{8.24}$$

and the assumptions of Lemma 8.3 are satisfied. Hence, there exists $T_2{}^{*-1}$ such that

$$(W^{-1}MT_1{}^{*-1})T_2{}^{*-1}=\begin{pmatrix} I_{n-r} & 0 \\ 0 & 0 \end{pmatrix} \tag{8.25}$$

and

$$T_2{}^*\ S_2T_2=S_2 \tag{8.26}$$

Now define

$$T_3\triangleq \text{diag}\ (\Sigma_{11}{}^{1/2}(\Sigma_{11}{}^2-\sigma^2 I)^{1/2},\ \Sigma_{12}{}^{1/2}(\sigma^2 I-\Sigma_{12}{}^2)^{1/2},\ \sigma^{1/2}I_{l+k+r-n}) \tag{8.27}$$

$$T\triangleq T_1T_2T_3 \tag{8.28}$$

Hence, (8.25) gives

$$MT^{*-1}=W\begin{pmatrix} I_{n-r} & 0 \\ 0 & 0 \end{pmatrix}T_3{}^{*-1}=\begin{pmatrix} I_{n-r} & 0 \\ 0 & 0 \end{pmatrix} \tag{8.29}$$

Further (8.22), (8.26) and (8.27) give

$$
\begin{aligned}
T^*ST &= T_3{}^* \, T_2{}^* \, T_1{}^* \, ST_1T_2T_3 \\
&= T_3{}^* \, T_2{}^* \, S_2T_2T_3 \\
&= T_3{}^* \, S_2T_3 \\
&= \begin{pmatrix} \Sigma_1(\Sigma_1{}^2 - \sigma^2 I) & 0 \\ 0 & -\sigma I \end{pmatrix}
\end{aligned}
\tag{8.30}
$$

Equations (8.19), (8.29) and (8.30) give

$$
\begin{aligned}
NT &= -\Sigma^{-1}MT^{*-1}T^*ST \\
&= \begin{pmatrix} -(\Sigma_1{}^2 - \sigma^2 I) & 0 \\ 0 & 0 \end{pmatrix}
\end{aligned}
\tag{8.31}
$$

Equations (8.20), (8.29), (8.30) and (8.31) give

$$
\begin{aligned}
T^{-1}RT^{*-1} &= T^{-1}(\sigma^2 I - M^*N)S^{-1}T^{*-1} \\
&= [\sigma^2 I - (MT^{*-1})^*(NT)](T^*ST)^{-1} \\
&= \begin{pmatrix} \Sigma_1(\Sigma_1{}^2 - \sigma^2 I)^{-1} & 0 \\ 0 & -\sigma I \end{pmatrix}
\end{aligned}
\tag{8.32}
$$

Hence eqns. (8.9) to (8.12) have been verified. ■

Lemmas 8.1 and 8.2 can now be combined to characterize all state-space solutions to the optimal Hankel-norm approximation problem as follows.

Theorem 8.4

Let $G(s)$ be a stable, rational, $m \times m$ transfer-function with Hankel singular values

$$\sigma_1 \geqslant \sigma_2 \ldots \geqslant \sigma_k > \sigma_{k+1} = \sigma_{k+2} = \sigma_{k+r} > \sigma_{k+r+1} \geqslant \ldots \geqslant \sigma_n > 0$$

and a balanced realization (A, B, C) ordered as in Theorem 7.2, part (3). Then $\hat{G}(s)$ is an optimal Hankel-norm approximation to $G(s)$ of McMillan degree k *if and only if* there exists $(\hat{A}, \hat{B}, \hat{C})$ such that

(*a*) $\hat{G}(s) =$ stable part of $(\hat{C}(sI - \hat{A})^{-1}\hat{B})$

and

(*b*) for (R, S, M, N) given by eqns. (8.9) to (8.12) with $T = I$, the following equations in $(\hat{A}, \hat{B}, \hat{C})$ hold

$$\hat{A}R + R\hat{A}^* + \hat{B}\hat{B}^* = 0 \tag{8.33}$$

$$\hat{A}^*S + S\hat{A} + \hat{C}^*\hat{C} = 0 \tag{8.34}$$

$$AM + M\hat{A}^* + B\hat{B}^* = 0 \tag{8.35}$$

$$A^*N + N\hat{A} - C^*\hat{C} = 0 \tag{8.36}$$

Further, dim $(\hat{A}) \geqslant n - r$ and may be chosen $\leqslant n + 2k - 1$.

Proof

Lemma 8.1 gives necessary and sufficient conditions for $\hat{G}(s)$ to be an optimal Hankel-norm approximation. All solutions to conditions (*b*) and (*c*) in Lemma 8.1 for P_e and Q_e have been given in Lemma 8.2. The matrix T in Lemma 8.2 only corresponds to a change of coordinates for the system ($\hat{A}$, $\hat{B}$, $\hat{C}$) and can hence, without loss of generality, be set to the identity matrix. Finally, eqns. (8.33) to (8.36) are equivalent to eqns. (6.26) and (6.27) for P_e and Q_e given by (8.7). ■

The characterization of the solutions given in Theorem 8.4 is not quite explicit. However, eqns. (8.33) to (8.36) can be solved explicitly for ($\hat{A}$, $\hat{B}$, $\hat{C}$), as in the following lemma, with a small number of remaining degrees of freedom in the results.

Lemma 8.5

All solutions to eqns. (8.33) to (8.36) in Theorem 8.4 can be written as

$$\hat{A}=\begin{pmatrix}\hat{A}_{11} & \hat{A}_{12}\\ \hat{A}_{21} & \hat{A}_{22}\end{pmatrix},\quad \hat{B}=\begin{pmatrix}\hat{B}_1\\ \hat{B}_2\end{pmatrix},\quad \hat{C}=(\hat{C}_1,\hat{C}_2) \tag{8.37}$$

where

$$\hat{A}_{11}=\Gamma^{-1}(\sigma_{k+1}^2 A_{11}^*+\Sigma_1 A_{11}\Sigma_1-\sigma_{k+1}C_1^*\,UB_1^*) \tag{8.38}$$

$$\hat{B}_1=\Gamma^{-1}(\Sigma_1 B_1+\sigma_{k+1}C_1^*\,U) \tag{8.39}$$

$$\hat{C}_1=C_1\Sigma_1+\sigma_{k+1}UB_1^* \tag{8.40}$$

$$B_2\hat{B}_2^*=0 \tag{8.41}$$

$$\hat{C}_2=-U\hat{B}_2^* \tag{8.42}$$

$$\hat{A}_{12}=-\Gamma^{-1}C_1^*\,\hat{C}_2 \tag{8.43}$$

$$\hat{A}_{21}=-\hat{B}_2B_1^* \tag{8.44}$$

$$\hat{A}_{22}+\hat{A}_{22}^*=\hat{B}_2\hat{B}_2^*/\sigma_{k+1} \tag{8.45}$$

where (A, B, C) is a balanced realization of $G(s)$ with corresponding Σ given by (8.5) and partitioned conformally with $\Sigma=\text{diag}\,(\Sigma_1,\sigma_{k+1}I_r)$ as in (6.18). Also U is a unitary matrix satisfying

$$C_2+UB_2^*=0 \tag{8.46}$$

Further if ($\hat{A}_{22}$, $\hat{B}_2$) is chosen to be completely controllable then In $(\hat{A})=$ (dim $(\hat{A})$ $-k$, k, 0).

Proof

The components of eqns. (8.33) to (8.36) can be written out as follows:
(8.33)⇔

$$\hat{A}_{11}\Sigma_1\Gamma^{-1}+\Sigma_1\Gamma^{-1}\hat{A}_{11}^*+\hat{B}_1\hat{B}_1^*=0 \tag{8.47}$$

$$-\sigma_{k+1}\hat{A}_{12}+\Sigma_1\Gamma^{-1}\hat{A}_{21}^*+\hat{B}_1\hat{B}_2^*=0 \tag{8.48}$$

$$-\sigma_{k+1}(\hat{A}_{22}+\hat{A}_{22}^*)+\hat{B}_2\hat{B}_2^*=0 \tag{8.49}$$

(8.34)$\Leftrightarrow$

$$\hat{A}_{11}^* \Sigma_1 \Gamma + \Sigma_1 \Gamma \hat{A}_{11} + \hat{C}_1^* \hat{C}_1 = 0 \tag{8.50}$$

$$-\sigma_{k+1} \hat{A}_{21}^* + \Sigma_1 \Gamma \hat{A}_{12} + \hat{C}_1^* \hat{C}_2 = 0 \tag{8.51}$$

$$-\sigma_{k+1}(\hat{A}_{22} + \hat{A}_{22}^*) + \hat{C}_2^* \hat{C}_2 = 0 \tag{8.52}$$

(8.35)$\Leftrightarrow$

$$A_{11} + \hat{A}_{11}^* + B_1 \hat{B}_1^* = 0 \tag{8.53}$$

$$\hat{A}_{21}^* + B_1 \hat{B}_2^* = 0 \tag{8.54}$$

$$A_{21} + B_2 \hat{B}_1^* = 0 \tag{8.55}$$

$$B_2 \hat{B}_2^* = 0 \tag{8.56}$$

(8.36)$\Leftrightarrow$

$$-A_{11}^* \Gamma - \Gamma \hat{A}_{11} - C_1^* \hat{C}_1 = 0 \tag{8.57}$$

$$-\Gamma \hat{A}_{12} - C_1^* \hat{C}_2 = 0 \tag{8.58}$$

$$-A_{12}^* \Gamma - C_2^* \hat{C}_1 = 0 \tag{8.59}$$

$$-C_2^* \hat{C}_2 = 0 \tag{8.60}$$

It is first necessary to verify that the solution given by eqns. (8.37) to (8.46) does indeed satisfy eqns. (8.47) to (8.60). Note that the formulae for $\hat{A}_{11}$, $\hat{B}_1$ and $\hat{C}_1$ are identical to those given in (6.19), (6.20) and (6.21) respectively. Therefore (8.53), (8.55), (8.47), (8.57), (8.59) and (8.50) are verified by identical arguments to those verifying (6.33), (6.34), (6.35), (6.43), (6.44) and (6.45), respectively. The remaining equations are immediate from (8.39) to (8.46) as follows: (8.39), (8.43) and (8.44)$\Rightarrow$(8.48); (8.45)$\Rightarrow$(8.49); (8.40), (8.43), and (8.44)$\Rightarrow$(8.51); (8.42) and (8.45)$\Rightarrow$(8.52); (8.44)$\Rightarrow$(8.54); (8.41)$\Rightarrow$(8.56); (8.43)$\Rightarrow$(8.58); (8.41), (8.42) and (8.46)$\Rightarrow$(8.60).

It is now necessary to show that every solution to eqns. (8.33) to (8.36) can be written in the form of eqn. (8.37) to (8.46). Since the solutions to the Lyapunov equations (6.26) and (6.27) satisfy $P_e Q_e = \sigma_{k+1}^2 I$ then by Theorem 5.1, part (2) there exists D_e such that

$$D_e^* D_e = \sigma_{k+1}^2 I \tag{8.61}$$

$$D_e B_e^* + C_e P_e = 0 \tag{8.62}$$

Recall that

$$B_e^* = (B_1^*, B_2^*, \hat{B}_1^*, \hat{B}_2^*) \tag{8.63}$$

$$C_e = (C_1, C_2, -\hat{C}_1, -\hat{C}_2) \tag{8.64}$$

and P_e is given by (8.7) with (8.9) and (8.11). Equation (8.61) implies that there exist a unitary matrix U such that

$$D_e = \sigma_{k+1} U \tag{8.65}$$

and the second block column of (8.62) gives (8.46). The first block column of (8.62) gives (8.40). The third block column of (8.62) together with (8.40) gives (8.39). The fourth block column of (8.62) gives (8.42). Equation (8.38) is implied by (8.53) together with (8.39) and substituting $B_1 B_1^* = -A_{11}\Sigma_1 - \Sigma_1 A_{11}^*$. Finally, the other equations are immediate: (8.56)$\Rightarrow$(8.41); (8.58)$\Rightarrow$(8.43); (8.54)$\Rightarrow$(8.44); (8.52)$\Rightarrow$(8.45). Hence we have shown that any solution to eqns. (8.47) to (8.60) or equivalently eqns. (8.33) to (8.36) can be written in the form of eqns. (8.37) to (8.46).

The inertia of $\hat{A}$ is determined by first noting that Theorem 3.3 and (8.33) imply that In $(\hat{A})=$ In $(-R)$ if $\delta(\hat{A})=0$. To verify that $\delta(\hat{A})=0$, assume that there exists $x\neq 0$, λ such that $x^*\hat{A}=\lambda x^*$ and $\lambda+\bar{\lambda}=0$. $x^*(8.33)x \Rightarrow x^*\hat{B}=0$ since $\lambda+\bar{\lambda}=0$. $(8.35)x \Rightarrow AMx+\bar{\lambda}Mx=0$ and since $\delta(A)=0$ is assumed this implies $Mx=0$. Writing $x^*=(x_1^*, x_2^*)$ and (8.11) gives that $x_1=0$. $x_2^*(8.45)x_2 \Rightarrow x_2^*\ \hat{B}_2=0$ and we also have that $x_2^*\ \hat{A}_{22}=\lambda x_2^*$ so that by the assumed controllability of $(\hat{A}_{22}, \hat{B}_2) \Rightarrow x_2=0$. That is a contradiction so that $\delta(\hat{A})=0$ and the result is proven. ■

Remark 8.1

Theorem 8.4 together with Lemma 8.5 therefore give a complete characterization of all solutions to the optimal Hankel-norm approximation problem in terms of the state-space representation $(\hat{A}, \hat{B}, \hat{C})$. The procedure to generate all solutions is first to find a unitary U satisfying (8.46) ; if $m>1$ and $r=1$ then there will be an infinite number of such U characterized by an arbitrary unitary matrix of dimension $m-1$. Then $\hat{B}_2$ is chosen arbitrarily to satisfy (8.41) and $\hat{A}_{22}$ to satisfy (8.45) and then $(\hat{A}, \hat{B}, \hat{C})$ is uniquely determined from U, $\hat{B}_2$ and $\hat{A}_{22}$. Note that a unitary change of basis on $(\hat{A}_{22}, \hat{B}_2, \hat{C}_2)$ will neither change the approximation $\hat{G}(s)$ nor P_e and Q_e and so the forms of $(\hat{A}_{22}, \hat{B}_2, \hat{C}_2)$ can be made more restrictive if desired. Finally, this procedure can be repeated for the dimension of $\hat{A}_{22}$ lying between 0 and $(n+k-1)$. This procedure may generate the same answer $\hat{G}(s)$ by a number of different $(\hat{A}, \hat{B}, \hat{C})$ choices.

8.2. *Transfer function characterization*

The solution given in Lemma 8.5 can be rewritten in terms of the block diagram of Fig. 4 where $K(s)$ is any all-pass transfer function satisfying $C_2+K(s)B_2=0\ \forall s$. This is proved in the following corollary.

Corollary 8.6

With the same assumptions and notation as Theorem 8.4, all optimal Hankel-norm approximations to $G(s)$ can be written as $\hat{G}(s)$ where for some $F(s)\in H_-^\infty$

$$\hat{G}(s)+F(s)=H_{11}(s)-H_{12}(s)(K^{-1}(s)+H_{22}(s))^{-1}H_{21}(s) \tag{8.66}$$

where

$$H(s)=\begin{pmatrix} H_{11}(s) & H_{12}(s) \\ H_{21}(s) & H_{22}(s) \end{pmatrix}$$

$$=\begin{pmatrix} 0 & I \\ \sigma_{k+1}I & 0 \end{pmatrix}+\begin{pmatrix} C_1\Sigma_1 \\ -\sigma_{k+1}B_1^* \end{pmatrix}(sI-\tilde{A})^{-1}[\Gamma^{-1}\Sigma_1B_1,\ -\Gamma^{-1}C_1^*] \tag{8.67}$$

$$\tilde{A}=\Gamma^{-1}(\sigma_{k+1}{}^2\ A_{11}^*+\Sigma_1A_{11}\Sigma_1) \tag{8.68}$$

and $K(s)\in H_-^\infty$ is any all-pass function satisfying

$$C_2+K(s)B_2^*=0\quad \forall s \tag{8.69}$$

Proof

We will represent the system $\hat{G}(s)+F(s)$ as shown in Fig. 4 via the partitioned state-space realization

$$\dot{x}_1=\hat{A}_{11}x_1+\hat{A}_{12}x_2+\hat{B}_1v_1 \tag{8.70}$$

$$\dot{x}_2=\hat{A}_{21}x_1+\hat{A}_{22}x_2+\hat{B}_2v_1 \tag{8.71}$$

$$z_1=\hat{C}_1x_1+\hat{C}_2x_2+\hat{D}v_1 \tag{8.72}$$

where the matrices are defined by eqns. (8.38) to (8.46). Define

$$z_2=-\sigma_{k+1}B_1{}^*\,x_1+\sigma_{k+1}v_1 \tag{8.73}$$

and substitute (8.44) and (8.73) into (8.71) to give

$$\dot{x}_2=\hat{A}_{22}x_2+\sigma_{k+1}{}^{-1}\,\hat{B}_2z_2 \tag{8.74}$$

Now define

$$v_2=\hat{C}_2x_2-Uz_2 \tag{8.75}$$

then (8.73) gives

$$\hat{C}_2x_2=v_2+U(-\sigma_{k+1}B_1{}^*\,x_1+\sigma_{k+1}v_1) \tag{8.76}$$

and (8.39), (8.43) and (8.70) give

$$\begin{aligned}\dot{x}_1&=\hat{A}_{11}x_1-\Gamma^{-1}C_1{}^*\,\hat{C}_2x_2+\Gamma^{-1}(\Sigma_1B_1+\sigma_{k+1}C_1{}^*\,U)v_1\\&=(\hat{A}_{11}+\Gamma^{-1}C_1{}^*\,U\sigma_{k+1}B_1{}^*)x_1+\Gamma^{-1}\Sigma_1B_1v_1-\Gamma^{-1}C_1{}^*\,v_2\\&=\tilde{A}x_1+\Gamma^{-1}\Sigma_1B_1v_1-\Gamma^{-1}C_1{}^*\,v_2\end{aligned} \tag{8.77}$$

Equations (8.72), (8.40), (8.76) together with $\hat{D}=-\sigma_{k+1}U$ give

$$z_1=C_1\Sigma_1x_1+v_2 \tag{8.78}$$

Now define

$$K(s)=U-\hat{C}_2(sI-\hat{A}_{22})^{-1}\hat{B}_2\sigma_{k+1}{}^{-1} \tag{8.79}$$

then (8.42), (8.45) and Theorem 5.1 imply that $K(s)$ is all-pass. Further (8.41) and (8.46) imply (8.69). Hence, the representation (8.66) to (8.69) is verified on noting that $V_2(s)=-K(s)Z_2(s)$.

It is now necessary to show that any representation given by eqns. (8.66) to (8.69) is in fact equivalent to eqns. (8.38) to (8.46). For this it is only necessary to verify that the constraints on $K(s)$ are sufficient. Suppose $K(s)=D_3+C_3(sI-A_3)^{-1}B_3$ is a minimal balanced realization of any anticausal all-pass transfer function (i.e. $P=Q=-I$) and satisfies (8.69). Since $K(s)$ is all-pass Theorem 5.1 implies that D_3 is unitary, $A_3+A_3{}^*=B_3B_3{}^*$, and $D_3B_3{}^*-C_3=0$. Further, (8.69) implies that $C_2+D_3B_2{}^*=0$ and that $B_3B_2{}^*=0$ since (A_3, C_3) is assumed to be observable. Hence (8.41), (8.42), (8.45) and (8.46) are verified with $U=D_3$, $\hat{C}_2=-\sigma_{k+1}{}^{1/2}C_3$, $\hat{B}_2=\sigma_{k+1}{}^{1/2}B_3$, $\hat{A}_2=A_3$. So eqns. (8.66) to (8.69) generate optimal Hankel-norm approximations and the proof is complete. ■

The characterization of all solutions given in Corollary 8.6 for square systems can be extended to non-square systems by augmenting the transfer function by zeros to make it square and then approximating this system.

Imbedding $G(s)$ in a larger square system

$$G_a(s)=\begin{pmatrix} G(s) & 0 \\ 0 & 0 \end{pmatrix}$$

can be used also to obtain an alternative characterization of the solutions. The following theorem gives a complete characterization of all solutions to the optimal Hankel-norm approximation problem for $G(s)$ a $p\times m$ transfer function. Further, it reproduces, in the rational case, the characterization in Adamjan *et al.* (1978) of all optimal anticausal approximations.

Theorem 8.7

Let $G(s)$ be a stable, rational, $p\times m$, transfer function with Hankel singular values

$$\sigma_1 \geqslant \sigma_2 \ldots \geqslant \sigma_k > \sigma_{k+1} = \sigma_{k+2} \ldots = \sigma_{k+r} > \sigma_{k+r+1} \ldots \geqslant \sigma_n > 0$$

Then

(1) All optimal Hankel-norm approximations of $G(s)$ of McMillan degree k can be written as $\hat{G}(s)$ where

$$\hat{G}(s)+F(s)=H_{11}(s)-H_{12}(s)K(s)(I+H_{22}(s)K(s))^{-1}H_{21}(s) \tag{8.80}$$

for some $F\in H_-^\infty$. Where $H_{ij}(s)$ and $K(s)$ are defined by eqns. (8.67) to (8.69) and $K(s)\in H_-^\infty$ such that

$$\|K(j\omega)\|_{L^\infty} \leqslant 1 \tag{8.81}$$

Further there is a unique solution if rank $C_2=p$ or rank $B_2=m$.

(2) All rational matrices $F(s)\in H_-^\infty$ and $\hat{G}(s)$ of McMillan degree k such that $\|G(j\omega)-\hat{G}(j\omega)-F(j\omega)\|_{L^\infty}\leqslant \sigma_{k+1}$ can be expressed as eqn. (8.80) subject to (8.67) to (8.69) and (8.81).

Proof

(1) Let the augmented $(p+m)\times(p+m)$ square system be

$$G_a(s)\triangleq\begin{pmatrix} G(s) & 0 \\ 0 & 0 \end{pmatrix} \tag{8.82}$$

The Hankel singular values of $G_a(s)$ will be identical to those of $G(s)$ and recall that (A, B, C) is an appropriately ordered balanced realization of $G(s)$. If $\hat{G}_a(s)$ is an optimal Hankel-norm approximation to $G_a(s)$ then $\|G_a(s)-\hat{G}_a(s)\|_H=\sigma_{k+1}(G)$, and defining $\hat{G}(s)=[I, 0]\hat{G}_a(s)\begin{pmatrix} I \\ 0 \end{pmatrix}$ gives $\|G(s)-\hat{G}(s)\|_H=\sigma_{k+1}(G)$ by Lemma 7.4 and Lemma 7.1. We will therefore characterize the approximations of $G(s)$ via those of $G_a(s)$ (truncated) which is square and hence Corollary 8.6 can be used. Note that a balanced realization of $G_a(s)$ is given by $\left(A, [B, 0], \begin{pmatrix} C \\ 0 \end{pmatrix}\right)$. Applying Corollary 8.6 with an obvious notation gives

$$\hat{G}+F=\tilde{H}_{11}-\tilde{H}_{12}K_a(I+\tilde{H}_{22}K_a)^{-1}\tilde{H}_{21} \tag{8.83}$$

where

$$\tilde{H}_{11}(s)=[I_p,\ 0]H_{11}{}^a(s)\begin{pmatrix} I_m \\ 0 \end{pmatrix} \tag{8.84}$$

$$\tilde{H}_{12}(s)=[I_p,\ 0]H_{12}{}^a(s) \tag{8.85}$$

$$\tilde{H}_{21}(s)=H_{21}{}^a(s)\begin{pmatrix} I_m \\ 0 \end{pmatrix} \tag{8.86}$$

$$\tilde{H}_{22}(s)=H_{22}{}^a(s) \tag{8.87}$$

Hence using the augmented values for B and C in eqn. (8.67) gives

$$\begin{aligned} \tilde{H}(s) &= \begin{pmatrix} 0 & I_p & 0 \\ \sigma_{k+1}I_m & 0 & 0 \\ 0 & 0 & 0 \end{pmatrix} + \begin{pmatrix} C_1\Sigma_1 \\ -\sigma_{k+1}B_1{}^* \\ 0 \end{pmatrix} (sI-\tilde{A})^{-1}[\Gamma^{-1}\Sigma_1 B_1,\ -\Gamma^{-1}C_1{}^*,\ 0] \\ &= \begin{pmatrix} H(s) & 0 \\ 0 & 0 \end{pmatrix} \end{aligned} \tag{8.88}$$

Further, $K_a(s)$ satisfies

$$\begin{pmatrix} C_2 \\ 0 \end{pmatrix} + K_a(s)\begin{pmatrix} B_2{}^* \\ 0 \end{pmatrix} = 0 \tag{8.89}$$

and writing

$$K_a(s)=\begin{pmatrix} K(s) & K_{12}(s) \\ K_{21}(s) & K_{22}(s) \end{pmatrix} \tag{8.90}$$

gives

$$C_2 + K(s)B_2{}^* = 0 \quad \forall s$$

and since $K_a(s)$ is all-pass (8.81) is satisfied by $K(s)$. Finally, (8.83) gives

$$\begin{aligned} \hat{G}(s)+F(s) &= H_{11}(s)-[H_{12}K,\ H_{12}K_{12}]\begin{pmatrix} I+H_{22}K & H_{22}K_{12} \\ 0 & I \end{pmatrix}^{-1}\begin{pmatrix} H_{21} \\ 0 \end{pmatrix} \\ &= H_{11}-H_{12}K(I+H_{22}K)^{-1}H_{21} \end{aligned} \tag{8.91}$$

Here, all solutions can be written in the given form where $K(s)$ is given by (8.89) and (8.90) and $K_a(s)$ is all-pass. Note that this implies that (8.69) and (8.81) are satisfied.

Hence it has been shown that all solutions can be written in the given form and it remains to show that the constraints on $K(s)$ are sufficient to ensure that it can be augmented as in (8.90) to an all-pass function $K_a(s)$ satisfying (8.89). Theorem 5.2 can be applied to show that $K(s)\in H_-{}^\infty$ can be written as a sub-matrix of an all-pass $K_a(s)\in H_-{}^\infty$.

(8.69) and (8.90) $\Rightarrow$

$$K_a(s)\begin{pmatrix} B_2{}^* \\ 0 \end{pmatrix} = \begin{pmatrix} -C_2 \\ K_{21}(s)B_2{}^* \end{pmatrix}$$

The term $K_{21}(s)B_2{}^*$ can be shown to be zero as follows.

$$(K_{21}(j\omega)B_2{}^*)^*(K_{21}(j\omega)B_2{}^*) = B_2(I - K^*(j\omega)K(j\omega))B_2{}^*$$
$$= B_2B_2{}^* - C_2{}^*C_2$$
$$= 0$$

where $K_a{}^*(j\omega)K_a(j\omega) = I$ and (8.69) have been used. Therefore, $K_{21}(j\omega)B_2{}^* = 0 \ \forall\omega \Rightarrow K_{21}(s)B_2{}^* = 0 \ \forall s$ and hence (8.89) is satisfied. Therefore the appropriate $K_a(s)$ can be constructed given that $K(s)$ satisfies (8.81) and (8.69) and the proof of the characterization is complete.

The uniqueness result is immediate since if rank $B_2 = m$ then (8.69) will have a unique solution for $K(s)$ that will in fact be independent of s. Also if rank $C_2 = p$ approximating $G^*(\bar{s})$ gives uniqueness.

(2) To show that the given $\hat{G}(s) + F(s)$ also characterize all solutions to the equation $E_{11}(s) \triangleq G(s) - \hat{G}(s) - F(s)$, with $\|E_{11}(j\omega)\|_{L^\infty} \leqslant \sigma_{k+1}$ we will again imbed this in an augmented system. Theorem 5.2 implies that there exists E_{ij} such that

$$E(s) \triangleq \sigma_{k+1}{}^{-1}\begin{pmatrix} E_{11}(s) & E_{12}(s) \\ E_{21}(s) & E_{22}(s) \end{pmatrix}$$

is all-pass. Note that no entry in $E(s)$ will have a pole on the imaginary axis since Theorem 5.2⇒Theorem 5.1, part (*a*)⇒Theorem 5.1, part (*b*)⇒the result by Theorem 3.4. Now let $U(s)$, $V(s) \in H_-{}^\infty$ be, respectively $p \times p$ and $m \times m$ all-pass functions such that $F_{21}(s) \triangleq -U(s)E_{21}(s) \in H_-{}^\infty$, $F_{12}(s) \triangleq -E_{12}(s)V(s) \in H_-{}^\infty$ and $F_{22}(s) \triangleq -U(s)E_{22}(s)V(s) \in H_-{}^\infty$. Such functions $U(s)$ and $V(s)$ can be constructed as follows. If any entry in $[E_{21}(s), E_{22}(s)]$ has a pole, λ say, with Re $(\lambda) > 0$ multiply the corresponding row of $[E_{21}(s), E_{22}(s)]$ by the all-pass function $(s - \lambda)/(s + \lambda)$ to move it to $-\lambda$. Repeat this procedure until $[E_{21}(s), E_{22}(s)]$ has no poles in the left half-plane and let $U(s)$ be the accumulation of the corresponding row operations. Similarly let $V(s)$ operate on the columns of

$$\begin{pmatrix} E_{12}(s) \\ E_{22}(s) \end{pmatrix}$$

Hence

$$\tilde{F}(s) \triangleq \begin{pmatrix} F(s) & F_{12}(s) \\ F_{21}(s) & F_{22}(s) \end{pmatrix} \in H_-{}^\infty$$

and

$$\sigma_{k+1}{}^{-1}[G_a(s) - \hat{G}_a(s) - \tilde{F}(s)]$$

is all-pass where

$$\hat{G}_a(s) = \begin{pmatrix} \hat{G}(s) & 0 \\ 0 & 0 \end{pmatrix}$$

Now the development leading up to Corollary 8.6 precisely characterizes all $\hat{G}_a(s) + \tilde{F}(s)$ such that $\sigma_{k+1}{}^{-1}[G_a(s) - \hat{G}_a(s) - \tilde{F}(s)]$ is all-pass and the proof is complete. ■

Remark 8.2

It is interesting to note that for square systems Theorem 8.7 and Corollary 8.6 give alternative characterizations of the family of solutions. Both solutions are represented by Fig. 4 and the degrees of freedom are contained in $K(s)$ which must satisfy (8.69) and in Corollary 8.6 is all-pass whereas in Theorem 8.7 $\|K(j\omega)\|_{L^\infty} \leqslant 1$ which is clearly a less restrictive condition. The same approximation $\hat{G}(s)$ could therefore be produced by two routes ; Corollary 8.6 constructs $\hat{G}+F_1$ such that $(G(s)-\hat{G}(s)-F_1(s))/\sigma_{k+1}$ is all-pass whereas Theorem 8.7 constructs $\hat{G}+F_2$ such that $\|G(j\omega)-\hat{G}(j\omega)-F_2(j\omega)\|_{L^\infty}=\sigma_{k+1}$. For a particular $\hat{G}(s)$ the required order of $K(s)$ in Corollary 8.6 may be larger than that in Theorem 8.7.

Remark 8.3

The degrees of freedom in the solutions can also be represented by a matrix $K_2(s) \in H_-^\infty$ satisfying $\|K_2(j\omega)\|_{L^\infty} \leqslant 1$ but otherwise unrestricted as follows. Let V and W be unitary matrices such that

$$VC_2 = \begin{pmatrix} E \\ 0 \end{pmatrix}, \quad -WB_2^* = \begin{pmatrix} E \\ 0 \end{pmatrix}$$

where E has a full row rank $=r'$, say. Then (8.69)$\Rightarrow$

$$VK(s)W^* = \begin{pmatrix} I & 0 \\ 0 & K_2(s) \end{pmatrix}$$

where $K_2(s)$ is $(p-r')\times(m-r')$ and arbitrary but (8.81)$\Rightarrow \|K_2(j\omega)\|_{L^\infty} \leqslant 1$. Hence, the family of $K_2(s)$ constitutes a ball in H_-^∞. The centre of the family of admissible $K(s)$ is hence given by $K_2(s)=0$. Therefore in a loose sense the centre of the family of solutions $\hat{G}(s)$ is given by $K_2(s)=0$. It is relatively straightforward to show that this in fact corresponds to the solution given by Corollary 7.3 with U_0, and may be an appropriate particular solution to choose. When $K(s)$ is chosen to be constant and such that $KK^*=I$ or $K^*K=I$ then in a sense these solutions correspond to boundary values of the set of solutions. However these latter solutions can also give good results in an L^∞-sense.

Remark 8.4

Suppose that one wished to characterize all approximations $\hat{G}(s)$ of McMillan degree k and such that $\|G(s)-\hat{G}(s)\|_{\rm H} \leqslant \rho$ for some ρ, $\sigma_{k+1} < \rho < \sigma_k$. This has in fact been the initial problem considered in some other approaches (Adamjan *et al.* 1978, Kung and Lin 1981). This problem can in fact be simply solved from the above results by considering the augmented system

$$G_\rho(s) = \begin{pmatrix} G(s) & 0 \\ 0 & g(s) \end{pmatrix}$$

where $g(s)=2\rho/(s+1)$, i.e. $\|g(s)\|_{\rm H}=\rho$. The $(k+1)$st Hankel singular value of $G_\rho(s)$ will then be ρ and hence for all optimal Hankel-norm approximations to $G_\rho(s)$, the Hankel-norm of the error will be less than or equal to ρ. Hence

$\hat{G}(s)$, the truncation of the approximation to the augmented system will also satisfy $\|G(s)-\hat{G}(s)\|_H \leqslant \rho$. Similarly any $\hat{G}(s)$ satisfying $\|G(s)-\hat{G}(s)\|_H \leqslant \rho$ can be augmented to

$$\begin{pmatrix} \hat{G}(s) & 0 \\ 0 & 0 \end{pmatrix}$$

to give an optimal Hankel-norm approximation to $G_\rho(s)$. It is then a simple exercise to apply Theorem 8.7 and truncate the result. The family of solutions is given by (8.80) when (A, B, C) and ρ are substituted for (A_{11}, B_1, C_1) and σ_{k+1} respectively, and $K(s)$ satisfies (8.81) but does not need to satisfy (8.69).

Example 8.1

This section is now concluded with a simple multivariable example where it is in fact possible to characterize all the solutions in an analytic form. Let

$$G(s)=\begin{pmatrix} g_1(s) & 0 \\ 0 & g_2(s) \end{pmatrix}$$

where

$$g_1(s)=\frac{(2s+0{\cdot}45)}{(s^2+1{\cdot}25s+0{\cdot}09)}$$

$$g_2(s)=\frac{1}{(s+0{\cdot}5)}$$

This has $\sigma_1=2$, $\sigma_2=1$, $\sigma_3=\frac{1}{2}$ and a balanced realization

$$A=\begin{pmatrix} -0{\cdot}25 & 0 & -0{\cdot}4 \\ 0 & -0{\cdot}5 & 0 \\ -0{\cdot}4 & 0 & -1 \end{pmatrix},\quad B=\begin{pmatrix} 1 & 0 \\ 0 & 1 \\ 1 & 0 \end{pmatrix},\quad C=\begin{pmatrix} 1 & 0 & 1 \\ 0 & 1 & 0 \end{pmatrix}$$

We will now use Theorem 8.7 to characterize all optimal Hankel-norm approximations of degree $2(k=2)$.

Using Corollary 8.6 we obtain

$$\tilde{A}=\Gamma^{-1}(\sigma_3{}^2\, A_{11}{}^*+\Sigma_1 A_{11}\Sigma_1)=-\begin{pmatrix} a_1 & 0 \\ 0 & a_2 \end{pmatrix}$$

where $a_1=17/60$, $a_2=5/6$

$$H_{11}(s)=C_1\Sigma_1(sI-\tilde{A})^{-1}\Gamma^{-1}\Sigma_1 B_1=\text{diag}\,(16/15(s+a_1),\ 4/3(s+a_2))$$

$$H_{12}(s)=I-C_1\Sigma_1(sI-\tilde{A})^{-1}\Gamma^{-1}C_1{}^*=\text{diag}\left(\frac{(s-\frac{1}{4})}{(s+a_1)},\ \frac{(s-\frac{1}{2})}{(s+a_2)}\right)$$

$$H_{21}(s)=\sigma_3 I-\sigma_3 B_1{}^*\,(sI-\tilde{A})^{-1}\Gamma^{-1}\Sigma_1 B_1=\text{diag}\left(\frac{(s-\frac{1}{4})}{2(s+a_1)},\ \frac{(s-\frac{1}{2})}{2(s+a_2)}\right)$$

$$K(s)B_2{}^*+C_2=0\Rightarrow K(s)=\begin{pmatrix} -1 & 0 \\ 0 & k(s) \end{pmatrix}$$

where $|k(j\omega)|\leqslant 1\ \forall\omega$. Hence using (8.66) gives, after some manipulation that

$$\hat{G}(s)+F(s)=\operatorname{diag}\,(\hat{g}_1(s)+f_1(s),\ \hat{g}_2(s)+f_2(s))$$

where

$$\hat{g}_1(s)=\tfrac{1}{2}+\frac{3}{5(s+0{\cdot}15)},\quad f_1(s)=0$$

$$\hat{g}_2(s)+f_2(s)=\frac{4}{3(s+a_2)}-\frac{(s-\frac{1}{2})^2}{2(s+a_2)(\frac{2}{3}+(s+a_2)k^{-1}(s))}$$

Now we know from Theorem 8.4 that $\hat{g}_2+f_2$ has precisely one stable pole and hence it is also easily shown that the residue of the pole at $-a_2$ is zero. Hence if

$$\hat{g}_2(s)=\frac{R}{s-\lambda}$$

then

$$\tfrac{2}{3}k(\lambda)+\lambda+a_2=0$$

$\Rightarrow -3/2\leqslant\lambda\leqslant -1/6$ since $|k(\lambda)|\leqslant 1$ for all $\lambda\in\mathbb{C}^-$ by the maximum modulus principle. Further, the residue R can be evaluated using partial fractions expansion and after some manipulation one obtains

$$R=\tfrac{3}{4}(\lambda-\tfrac{1}{2})^2\left(1+\tfrac{2}{3}\frac{dk(\lambda)}{ds}\right)^{-1}$$

Now an application of the Schwartz lemma (Rudin 1966, pp. 240–242) can be used to show that

$$\left|\frac{dk(\lambda)}{ds}\right|\leqslant\frac{1-|k(\lambda)|^2}{-(\lambda+\bar{\lambda})}$$

and in particular it is possible to achieve this bound with a first-order all-pass $k(s)$ and such that

$$\frac{dk(\lambda)}{ds}=\frac{\alpha(1-|k(\lambda)|^2)}{-(\lambda+\bar{\lambda})}\quad |\alpha|=1$$

In this example we have

$$k(\lambda)=-\tfrac{3}{2}(\lambda+\tfrac{5}{6})$$

and hence we can show that

$$R_l\leqslant R\leqslant R_u$$

where

$$R_u=-\lambda(\lambda-\tfrac{1}{2})^2/(\lambda^2+\tfrac{1}{2}\lambda+\tfrac{1}{4})$$

$$R_l=\lambda(\lambda-\tfrac{1}{2})^2/(\lambda^2+3\lambda+\tfrac{1}{4})$$

The region of solutions is given in Fig. 5.

When Corollary 8.6 is used then the extreme values of λ will necessarily correspond to a constant $k(s)$ due to the maximum modulus principle. Further, the extreme values of R given λ correspond to first-order $k(s)$ due to the Schwarz lemma. The interior then corresponds to second-order $k(s)$ and higher-order $k(s)$ do not give any additional solutions.

When Theorem 8.7 is used then constant $k(s)$ give the line indicated on the figure and first-order $k(s)$ can represent the other solutions. Note also that the solution given by Corollary 7.3 with U_0 is also marked ; as are contours for the Hankel singular values of $(g_2-\hat{g}_2)$.

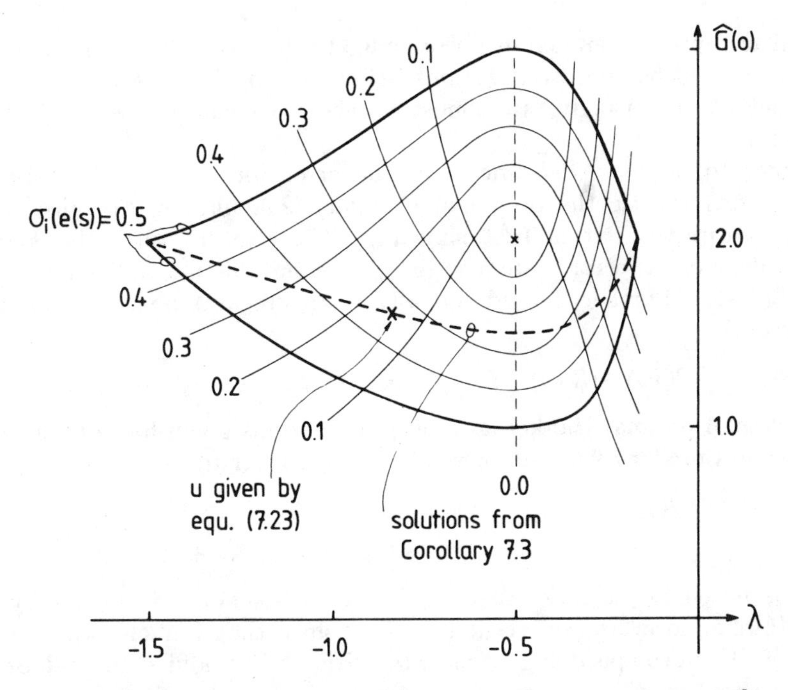

Figure 5. Hankel singular values for the system $e(s)=[1/(s+0{\cdot}5)-\hat{G}(0)/(1-s/\lambda)]$.

Since $G(s)$ is diagonal with $\sigma_1(g_1)=2$, $\sigma_2(g_1)=\frac{1}{2}$, $\sigma_1(g_2)=1$ the solution in fact has $\hat{g}_1$ as the unique optimal Hankel-norm approximation to g_1 with $\|g_1-\hat{g}_1\|_H=\frac{1}{2}$, and $\hat{g}_2$ can be any first-order function that such $\|g_2-\hat{g}_2\|_H \leqslant \frac{1}{2}$. The optimal choice in this very special case would presumably be to have $\hat{g}_2=g_2$ but it is not at all clear how this could be generalized.

9. Frequency response bounds

The natural question that arises now is, does the Hankel-norm being small imply that any other more familiar norms are also small? In this section a reasonably complete answer to this question is given for the L^∞-norm. In general the L^∞-norm cannot be bounded in terms of the Hankel-norm when $n\to\infty$, as has been shown by Jonckheere *et al.* (1981). However, given all the Hankel singular values it will be shown in Corollary 9.3 that $\|G(j\omega)\|_{L^\infty} \leqslant 2(\sigma_1+\sigma_2+\ldots+\sigma_n)$ and further that there exists a constant D such that $\|G(j\omega)+D\|_{L^\infty} \leqslant (\sigma_1+\sigma_2+\ldots+\sigma_n)$. The proof of this result is via a new representation of stable square transfer functions in Theorem 9.2. This shows that

$$G(s)=D_0+\sigma_1E_1(s)+\sigma_2E_2(s)+\ldots+\sigma_nE_n(s)$$

where $E_k(s)$ are all-pass functions and the partial sums $(D_0+\sigma_1E_1(s)+\ldots+\sigma_kE_k(s))$ have McMillan degrees k. This representation is obtained by reducing the order one dimension at a time via optimal Hankel-norm approximations. It also illustrates in a very explicit way our claim that the Hankel singular values give accurate information on both the gain and complexity of stable systems.

A similar approach is taken in Theorem 9.6 to show that the L^∞-norm of the error in truncated balanced realizations is bounded by $2(\sigma_{k+1}+\sigma_{k+2}+\ldots+\sigma_n)$. This bound is, surprisingly, an almost immediate consequence of Theorem 6.3, part (1).

In order to bound the L^∞-norm of the error for optimal Hankel-norm approximations we use the bound of Corollary 9.3, but on $F(s)$ instead of $(G(s)-\hat{G}(s))$ directly. It is first shown in Theorem 9.7 that the Hankel singular values of the error system are σ_{k+1} with suitable multiplicity together with $\sigma_i(F(-s))$. Further, bounds on $\sigma_i(F(-s))$ are obtained. It can then be shown that

$$\|G(j\omega)-\hat{G}(j\omega)-D_0\|_{L^\infty} \leqslant \sigma_{k+1}+\|F(j\omega)-D_0\|_{L^\infty}$$

For the class of optimal Hankel-norm approximations given by Corollary 7.3, it is shown in Corollary 9.9 that there exists D_0 such that

$$\begin{aligned}\|G(j\omega)-\hat{G}(j\omega)-D_0\|_{L^\infty} &\leqslant (\sigma_{k+1}+\delta)\\ &\leqslant (\sigma_{k+1}+\sigma_{k+2}+\ldots+\sigma_n)\end{aligned}$$

As was noted in Remark 7.2 the ' D-matrix ' is not determined by the optimal Hankel-norm approximation method since the Hankel-norm is independent of D. Also specifying $\hat{G}(s)+F(s)$ with $F\in H_-^\infty$ and $\hat{G}\in H_+^\infty$ does not determine $\hat{G}(\infty)$ or $F(\infty)$ unless, for example, $\hat{G}(s)$ or $F(s)$ is required to be strictly proper. The choices of D given in this section are not optimal in an L^∞-sense but do ensure that the above error bounds are met.

So far only stable systems have been considered and in § 9.4. a scheme for approximating systems with unstable poles is presented.

It is claimed that the L^∞-error bounds derived in this section give strong support for the utility of optimal Hankel-norm approximations in a variety of applications. They have been applied successfully by Glover and Limebeer (1983) to the approximation of plants and controllers in robust closed-loop control system design, where L^∞-norms are most natural.

9.1. *A representation of stable transfer functions*

The following lemma gives the properties of certain optimal Hankel-norm approximations when the degree is reduced by the multiplicity of σ_n. In this case some precise statements on the error and the approximating system can be made.

Lemma 9.1

Let $G(s)$ be a stable, rational $m\times m$, transfer function of McMillan degree n and such that $\sigma_n(G)$ has multiplicity r. Also let $\hat{G}(s)$ be an optimal Hankel-norm approximation of degree $n-r$ given by Theorem 7.2, part (3) (with $F(s)\equiv 0$) then

(1) $(G(s)-\hat{G}(s))/\sigma_n(G(s))$ is all-pass.

(2) $\sigma_i(\hat{G}(s))=\sigma_i(G(s))$, $i=1, 2, \ldots, n-r$

Proof

Theorem 7.2 gives that In $(\hat{A})=(0, n-r, 0)$ and hence $F(s)$ can be chosen to be zero (it could also have been a constant) and therefore $(G(s))-\hat{G}(s)/\sigma_n(G)$

is all-pass. The $\sigma_i(\hat{G}(s))$ are obtained by noting that in (8.33) and (8.34) for this case $RS = \Sigma_1^2$. ∎

Note that Lemma 9.1 will not necessarily be the case for the other optimal Hankel-norm approximations. However, those given in Corollary 7.3 will be embedded in a $(p+m) \times (p+m)$ system where the result holds.

Applying the above reduction procedure again on $\hat{G}(s)$ and repeating until $\hat{G}(s)$ has zero McMillan degree gives the following new representation of stable systems.

Theorem 9.2

Let $G(s)$ be a stable, rational $m \times m$, transfer function with Hankel singular values $\sigma_1 > \sigma_2 \ldots > \sigma_N$ where σ_i has multiplicity r_i and $r_1 + r_2 + \ldots + r_N = n$. Then there exists a representation of $G(s)$ as

$$G(s) = D_0 + \sigma_1 E_1(s) + \sigma_2 E_2(s) + \ldots + \sigma_N E_N(s) \tag{9.1}$$

where

(1) $E_k(s)$ are all-pass and stable for all k.

(2) For $k = 1, 2, \ldots, N$

$$\hat{G}_k(s) \triangleq D_0 + \sum_{1 \leqslant i \leqslant k} \sigma_i E_i(s)$$

has McMillan degree $r_1 + r_2 + \ldots + r_k$.

Proof

Let $\hat{G}_k(s)$ be the optimal Hankel-norm approximation to $\hat{G}_{k+1}(s)$ (given by Lemma 9.1) of degree $r_1 + r_2 + \ldots + r_k$, with $\hat{G}_N(s) \triangleq G(s)$. Lemma 9.1(2) applied at each step then gives that the Hankel singular values of $\hat{G}_k(s)$ will be $\sigma_1, \sigma_2, \ldots, \sigma_k$ with multiplicities $r_1, r_2, \ldots, r_k$, respectively. Hence Lemma 9.1(1) gives that $\hat{G}_k(s) - \hat{G}_{k-1}(s) = \sigma_k E_k(s)$ for some stable, all-pass $E_k(s)$. Note also that Theorem 6.3, part (3 d), relation (i) also ensures that each $\hat{G}_k(s)$ will have McMillan degree $r_1 + r_2 + \ldots + r_k$. Finally taking $D_0 = \hat{G}_0(s)$ which will be a constant and combining the steps gives the result. ∎

Remark 9.1

Note that the construction of Theorem 9.2 immediately gives an approximation algorithm that will satisfy $\|G(s) - \hat{G}(s)\|_{L^\infty} \leqslant (\sigma_{k+1} + \sigma_{k+2} + \ldots + \sigma_N)$. This will *not* be an optimal Hankel-norm approximation in general, but would involve less computation since the decomposition into $\hat{G}(s) + F(s)$ need not be done, and at each step a balanced realization of $\hat{G}_k(s)$ is given by $(\hat{A}_k, \hat{B}_k, \hat{C}_k)$ with a diagonal scaling.

9.2. *Bounds on the* L^∞*-norm of* $G(s)$

An upper-bound on the L^∞-norm of $G(s)$ is now obtained as an immediate consequence of Theorem 9.2.

Corollary 9.3

Let $G(s)$ be a stable, rational $p \times m$, transfer function with Hankel singular values $\sigma_1 > \sigma_2 \ldots > \sigma_N$, where each σ_i has multiplicity r_i, and such that $G(\infty) = 0$.

Then

(1) $\|G(j\omega)\|_{L^\infty} \leqslant 2(\sigma_1+\sigma_2+\ldots+\sigma_N)$

(2) there exists a constant D_0 such that

$$\|G(j\omega)-D_0\|_{L^\infty} \leqslant (\sigma_1+\sigma_2+\ldots+\sigma_N)$$

Proof

For $p=m$ consider the representation of $G(s)$ given by Theorem 9.2 then

$$\begin{aligned}\|G(j\omega)-D_0\|_{L^\infty} &= \|\sigma_1 E_1(j\omega)+\ldots+\sigma_N E_N(j\omega)\|_{L^\infty}\\ &\leqslant (\sigma_1+\sigma_2+\ldots+\sigma_N)\end{aligned}$$

since $E_k(s)$ are all-pass. Further setting $s=\infty$, since $G(\infty)=0$, gives

$$\bar{\sigma}(D_0) \leqslant (\sigma_1+\sigma_2+\ldots+\sigma_N)$$

$$\Rightarrow \|G(j\omega)\|_{L^\infty} \leqslant 2(\sigma_1+\sigma_2+\ldots+\sigma_N).$$

For the case $p \neq m$ just augment $G(s)$ by zero rows or columns to make it square, but will have the same L^∞-norm, then the above argument gives upper bounds on the L^∞-norm of this augmented system. ■

Remark 9.2

The above bound could also have been derived approximating $G(s)$ by an anticausal function $F_1(s)$ whose L^∞-error is σ_1 and such that $\{\sigma_i(F_1(-s))\} = \{\sigma_2(G), \ldots, \sigma_N(G)\}$. Then $F_1(s)$ can be approximated by a causal function $F_2(s)$ with error σ_2 and so on. That such a sequence of approximations exists is given in the following lemma.

Lemma 9.4

Let $G(s)$ be a stable, rational, $m \times m$, transfer function of McMillan degree n such that $\sigma_1(G)$ has multiplicity r_1. Also let $F(s)$ be an optimal anticausal approximation of degree $n-r_1$ given by the construction of Theorem 6.3. Then

(1) $(G(s)-F(s))/\sigma_1$ is all-pass.

(2) $\sigma_{i-r_1}(F(-s)) = \sigma_i(G(s))$, $i=r_1+1, \ldots, n$.

Proof

(1) is proved in Theorem 6.3, part (2). (2) is obtained from the forms of P_e and Q_e in Theorem 6.3, part (1). $F(-s)$ is used since it will be stable and have well-defined Hankel singular values. ■

9.3. L$^\infty$-*norm of the error in trucated balanced realizations*

Very similar techniques to those of Theorem 9.2 can be used to bound the error obtained by truncating a balanced realization†. We will first need a

† The author was informed by J. Doyle that the result of Theorem 9.6 had been shown by D. Enns of Stanford University ; the present derivation has since been obtained independently.

lemma that gives a perhaps surprising relationship between a truncated balanced realization of degree $(n-r_N)$ and an optimal Hankel-norm approximation of the same degree.

Lemma 9.5

Let (A, B, C) be a balanced realization of the stable, rational, $m \times m$ transfer function $G(s)$, and let

$$A=\begin{pmatrix} A_{11} & A_{12} \\ A_{21} & A_{22} \end{pmatrix}, \quad B=\begin{pmatrix} B_1 \\ B_2 \end{pmatrix}, \quad C=(C_1, C_2)$$

$$\Sigma=\begin{pmatrix} \Sigma_1 & 0 \\ 0 & \sigma I \end{pmatrix} \quad \text{with } \delta(\Sigma_1-\sigma I)=0$$

Let $(\hat{A}, \hat{B}, \hat{C}, \hat{D})$ be defined by eqns. (6.19) to (6.24) (where $\Sigma_2=\Sigma_1$) and define

$$G_b(s) \triangleq C_1(sI-A_{11})^{-1}B_1$$

$$G_h(s) \triangleq \hat{D}+\hat{C}(sI-\hat{A})^{-1}\hat{B}$$

then

(1) $(G_b(s)-G_h(s))/\sigma$ is all-pass.

(2) $\|G(s)-G_b(s)\|_{L^\infty} \leqslant 2\sigma$.

(3) If $\Sigma_1>\sigma I$ then $\|G(s)-G_b(s)\|_H \leqslant 2\sigma$.

Proof

(1) In order to prove that $(G_b(s)-G_h(s))/\sigma$ is all-pass we note that $G_b(s)-G_h(s)=\tilde{D}+\tilde{C}(sI-\tilde{A})^{-1}\tilde{B}$ where

$$\tilde{A}=\begin{pmatrix} A_{11} & 0 \\ 0 & \hat{A} \end{pmatrix}, \quad \tilde{B}=\begin{pmatrix} B_1 \\ \hat{B} \end{pmatrix}, \quad \tilde{C}=(C_1, -\hat{C}), \quad \tilde{D}=-\hat{D}$$

Now Theorem 6.3, part (1) gives that the solutions to the Lyapunov equations

$$\tilde{A}\tilde{P}+\tilde{P}\tilde{A}^*+\tilde{B}\tilde{B}^*=0 \tag{9.2}$$

$$\tilde{A}^*\tilde{Q}+\tilde{Q}\tilde{A}+\tilde{C}^*\tilde{C}=0 \tag{9.3}$$

are

$$\tilde{P}=\begin{pmatrix} \Sigma_1 & I \\ I & \Sigma_1\Gamma^{-1} \end{pmatrix}, \quad \tilde{Q}=\begin{pmatrix} \Sigma_1 & -\Gamma \\ -\Gamma & \Sigma_1\Gamma \end{pmatrix} \tag{9.4}$$

(This is verified by noting that the blocks of eqns. (9.2) and (9.3) are also blocks of eqns. (6.26) and (6.27) for P_e and Q_e.) Hence $\tilde{P}\tilde{Q}=\sigma^2 I$ and by Theorem 5.1 there exists $\tilde{D}$ such that $(G_b(s)-G_h(s))/\sigma$ is all-pass. That $\tilde{D}=-\hat{D}$ is an appropriate choice is verified from eqns. (6.46) to (6.48) and Theorem 5.1, part (2).

(2) $\sigma^{-1}(G(s)-G_b(s))=\sigma^{-1}(G(s)-G_h(s))+\sigma^{-1}(G_h(s)-G_b(s))$ but the first term on the right-hand side is all-pass by Theorem 6.3, part (2) and the second term is all-pass by part (1) above. Hence $\|G(j\omega)-G_b(j\omega)\|_{L^\infty} \leqslant 2\sigma$.

(3) Similarly using the fact that all-pass functions have unity Hankel-norms gives that

$$\|G(s)-G_{\mathrm{b}}(s)\|_{\mathrm{H}} \leqslant \|G(s)-G_{\mathrm{h}}(s)\|_{\mathrm{H}} + \|G_{\mathrm{h}}(s)-G_{\mathrm{b}}(s)\|$$
$$=2\sigma$$
■

Given the result of Lemma 9.5 bounds on the error in a truncated balanced realization are easily proved as follows.

Theorem 9.6

Let $G(s)$ be a stable, rational, $p\times m$, transfer function with Hankel singular values $\sigma_1>\sigma_2\ldots>\sigma_N$, where each σ_i has multiplicity r_i and let $\tilde{G}_k(s)$ be obtained by truncating the balanced realization of $G(s)$ to the first $(r_1+r_2+\ldots+r_k)$ states. Then

(1) $$\|G(j\omega)-\tilde{G}_k(j\omega)\|_{L^\infty} \leqslant 2(\sigma_{k+1}+\sigma_{k+2}+\ldots+\sigma_N)$$

(2) $$\|G(s)-\tilde{G}_k(s)\|_{\mathrm{H}} \leqslant 2(\sigma_{k+1}+\sigma_{k+2}+\ldots+\sigma_N)$$

Proof

If $p\neq m$ then augmenting B or C by zero columns or rows, respectively, will still give a balanced realization and the same argument is valid. Hence assume $p=m$. Notice that since truncation of balanced realization are also balanced, satisfying the truncated Lyapunov equations, the Hankel singular values of $\tilde{G}_i(s)$ will be $\sigma_1, \sigma_2, \ldots, \sigma_i$ with multiplicities $r_1, r_2, \ldots, r_i$, respectively. Also $\tilde{G}_i(s)$ can be obtained by truncating the balanced realization of $\tilde{G}_{i+1}(s)$ and hence $\|\tilde{G}_{i+1}(s)-\tilde{G}_i(s)\| \leqslant 2\sigma_{i+1}$ for both L^∞- and Hankel-norms. Hence $(G_N(s) \triangleq G(s))$

$$\|G(s)-\tilde{G}_k(s)\| = \left\| \sum_{k\leqslant i\leqslant N-1} (\tilde{G}_{i+1}(s)-\tilde{G}_i(s)) \right\|$$
$$\leqslant 2(\sigma_{k+1}+\sigma_{k+2}+\ldots+\sigma_N)$$

for both norms, and the proof is complete. ■

Remark 9.3

It is therefore seen that the L^∞-norm of the error in truncated balanced realizations can be bounded by ‘ twice the sum of the tail ’. The earlier papers on balanced realizations tended to state that if $\sigma_{k+1} \ll \sigma_k$ then the approximation would be good. However, generally, good results were also obtained when σ_{k+1} was close to σ_k, and this could be thought surprising since as $\sigma_{k+1}\to\sigma_k$ the poles of the truncation can get arbitrarily close to the imaginary axis. The above error bound and the result of Theorem 4.2, part (3) show that if a pole tends to the imaginary axis then it tends to become uncontrollable or unobservable and hence does not make the error unduly large. This error bound gives strong theoretical support to the observation that truncated balanced realizations give good results in practice.

9.4. *L^∞-norm of the error in optimal Hankel-norm approximations*

We are now in a position to bound the L^∞-norm of the error in optimal Hankel-norm approximation. The representation of Theorem 8.4 that determines the matrices P_e and Q_e is used in Theorem 9.7 together with the result of Lemma 9.8 to bound the Hankel singular values of the $(G(s)-\hat{G}(s))$ and $F(-s)$. The result of Corollary 9.3 can then be used to bound the L^∞-norm of $F(-s)$ and the approximation error. Finally Corollary 9.9 gives bounds in the L^∞-norm of the optimal Hankel-norm approximations given in Corollary 7.3.

Theorem 9.7

Let $\hat{G}(s)$ be an optimal Hankel-norm approximation of degree k to $G(s)$ as characterized by Theorem 8.4 and Lemma 8.5 or Corollary 8.6. Also let $F(s)\in H_-^\infty$ be as defined in Corollary 8.6, and let $\dim \hat{A}=k+l$ and assume $(\hat{A}_{22}, \hat{B}_2)$ is completely controllable (note that $n-k-r\leqslant l\leqslant n+k-1$ and equivalently in Corollary 8.6 the McMillan degree of $K(s)=l-n+k+r$). Then

(1) The Hankel singular values of the error are given by

$$\sigma_i(G(s)-\hat{G}(s))=\begin{cases}\sigma_{k+1}(G(s)), & i=1,2,\ldots,n+k-l\\ \sigma_{i+l-n-k}(F(-s)), & i=n+k+1-l,\ldots,n+k\end{cases}$$

(2) (*a*) If $l=n-k-r$ then

$$\sigma_i(F(-s))\leqslant\sigma_{i+k+r}(G(s)),\quad i=1,2,\ldots,l$$

(*b*) If $l>n-k-r$ then

$$\sigma_i(F(-s))\leqslant\begin{cases}\sigma_{k+1}(G(s)), & i=1,2,\ldots,l+k+r-n\\ \sigma_{i+n-l}(G(s)), & i=l+k+r-n+1,\ldots,l\end{cases}$$

(3) There exists D_0 such that

(*a*)
$$\delta\triangleq\|F(j\omega)-D_0\|_{L^\infty}\leqslant\sum_{1\leqslant i\leqslant l}\sigma_i(F(-s))$$

(*b*)
$$\|G(j\omega)-\hat{G}(j\omega)-D_0\|_{L^\infty}\leqslant\sigma_{k+1}(G(s))+\delta$$

(*c*) if $\delta<\sigma_{k+1}(G(s))$ then

$$\|[G(j\omega)-\hat{G}(j\omega)-D_0]^{-1}\|_{L^\infty}\leqslant[\sigma_{k+1}(G(s))-\delta]^{-1}$$ ■

Remark 9.4

The optimal Hankel-norm approximations just have to satisfy $\sigma_i(G(s)-\hat{G}(s))\leqslant\sigma_{k+1}(G(s))\ \forall i$. However, we see from Theorem 9.7, part (1) that in fact $\sigma_i(G(s)-\hat{G}(s))=\sigma_{k+1}(G(s))$ for $i=1,2,\ldots,n+k-l$. In the case of $p=m=1$ when the answer is unique and l can be taken as $n-k-r$, we see that the first $(2k+r)$ Hankel singular values of the error are necessarily equal to $\sigma_{k+1}(G)$. That is, in order to minimize $\sigma_1(G-\hat{G})$, the $(2k+r-1)$ subsequent $\sigma_i(G-\hat{G})$ have also to equal $\sigma_1(G-\hat{G})$. We see, therefore, that for single-input/single-output systems the Hankel-norm and the Hankel-constraint are remarkably well-matched. That is if the Hankel-norm is used without the Hankel-constraint there would be an infinite number of solutions, whereas for other norms with the Hankel-constraint the unconstrained minimum would not normally be achievable.

Proof of Theorem 9.7

(1) The distinguishing feature of the optimal Hankel-norm approximations characterized by Theorem 8.4 is that they are all generated via an error system (A_e, B_e, C_e, D_e) given by (6.25) and matrices P_e and Q_e solutions to (6.26) and (6.27) that satisfy

$$P_e Q_e = \sigma_{k+1}^2 (G) I \tag{9.5}$$

The error system will normally contain an anticausal function $F(s)$ that needs to be removed. Since $(\hat{A}_{22}, \hat{B}_2)$ is assumed to be completely controllable, $\delta(\hat{A}) = \delta(A_e) = 0$ by Lemma 8.5 so that there exists T_e such that

$$T_e A_e T_e^{-1} = \begin{pmatrix} A_{e1} & 0 \\ 0 & A_{e2} \end{pmatrix}, \quad T_e B_e = \begin{pmatrix} B_{e1} \\ B_{e2} \end{pmatrix}, \quad C_e T_e^{-1} = (C_{e1}, \ -C_{e2})$$

where In $(A_{e1}) = (0, n+k, 0)$, In $(A_{e2}) = (l, 0, 0)$.

Then $(G(s) - \hat{G}(s))$ has state-space realization (A_{e1}, B_{e1}, C_{e1}) and $F(s)$ has state-space realization (A_{e2}, B_{e2}, C_{e2}). Now let

$$T_e P_e T_e^* = \begin{pmatrix} P_{e1} & P_{e2} \\ P_{e2}^* & P_{e3} \end{pmatrix}, \quad T_e^{*-1} Q_e T_e^{-1} = \begin{pmatrix} Q_{e1} & Q_{e2} \\ Q_{e2}^* & Q_{e3} \end{pmatrix}$$

then the transformed Lyapunov equations give that $A_{e1} P_{e1} + P_{e1} A_{e1}^* + B_{e1} B_{e1}^* = 0$ and $A_{e1}^* Q_{e1} + Q_{e1} A_{e1} + C_{e1}^* C_{e1} = 0$, so that

$$\begin{aligned} \sigma_i^2(G(s) - \hat{G}(s)) &= \lambda_i(P_{e1} Q_{e1}) \\ &= \lambda_i(\sigma_{k+1}^2(G) I - P_{e2} Q_{e2}^*) \quad \text{by (9.5)} \end{aligned}$$

The rank of $P_{e2} Q_{e2}^* \leqslant \dim A_{e2} = l$ and hence $\sigma_i(G(s) - \hat{G}(s)) = \sigma_{k+1}(G)$ for $i = 1, 2, \ldots, n+k-l$

$$\begin{aligned} \sigma_{n+k-l+i}(G(s) - \hat{G}(s)) &= \lambda_i^{1/2}(\sigma_{k+1}^2(G) I - Q_{e2}^* P_{e2}) \\ &= \lambda_i^{1/2}(\sigma_{k+1}^2(G) I - P_{e2}^* Q_{e2}) \\ &= \lambda_i^{1/2}(P_{e3} Q_{e3}) \quad \text{by (9.5)} \\ &= \sigma_i(F(-s)) \quad \text{for } i = 1, 2, \ldots, l \end{aligned}$$

where the last step follows from the (2, 2) block of the transformed Lyapunov equations. Hence part (1) is proved.

(2) In order to bound $\sigma_i(F(-s))$ we need to examine the detailed structure of $(\hat{A}, \hat{B}, \hat{C})$ as given by (8.33) and (8.34) and use Lemma 9.8. Recall that R and S in (8.33) and (8.34) are given by

$$R = \begin{pmatrix} \Sigma_1 \Gamma^{-1} & 0 \\ 0 & -\sigma_{k+1} I_{l-n+k+r} \end{pmatrix}, \quad S = \begin{pmatrix} \Sigma_1 \Gamma & 0 \\ 0 & -\sigma_{k+1} I_{l-n+k+r} \end{pmatrix}$$

where

$$\Sigma_1 = \operatorname{diag}(\sigma_1, \sigma_2, \ldots, \sigma_k, \sigma_{k+r+1}, \ldots, \sigma_n)$$

$$\Gamma = (\Sigma_1^2 - \sigma_{k+1}^2 I)$$

Hence there exists a diagonal transformation L such that

$$LRL^* = \begin{pmatrix} \hat{\Sigma}_1 & 0 \\ 0 & -\hat{\Sigma}_2 \end{pmatrix} = L^{*-1}SL^{-1}$$

where

$$\hat{\Sigma}_1 = \text{diag}\,(\sigma_1, \sigma_2, \ldots, \sigma_k)$$

$$\hat{\Sigma}_2 = \text{diag}\,(\sigma_{k+r+1}, \ldots, \sigma_n, \sigma_{k+1}, \ldots, \sigma_{k+1})$$

The decomposition of $(\hat{A}, \hat{B}, \hat{C})$ into $\hat{G}(s) + F(s)$ can be accomplished via a transformation $\hat{T}$ such that (ignoring D-terms for the present)

$$\hat{T}L\hat{A}L^{-1}\hat{T}^{-1} = \begin{pmatrix} \hat{A}_1 & 0 \\ 0 & \hat{A}_2 \end{pmatrix}, \quad \hat{T}L\hat{B} = \begin{pmatrix} \hat{B}_1 \\ \hat{B}_2 \end{pmatrix}, \quad \hat{C}L^{-1}\hat{T}^{-1} = (\hat{C}_1, \hat{C}_2)$$

where $\text{In}\,(\hat{A}_2) = (l, 0, 0)$, $\text{In}\,(\hat{A}_1) = (0, k, 0)$ $F(s) = \hat{C}_2(sI - \hat{A}_2)^{-1}\hat{B}_2$. The transformed R and S matrices will be, respectively

$$\hat{T}\begin{pmatrix} \hat{\Sigma}_1 & 0 \\ 0 & -\hat{\Sigma}_2 \end{pmatrix}\hat{T}^* \triangleq \begin{pmatrix} P_1 & P_2 \\ P_2^* & P_3 \end{pmatrix}$$

$$\hat{T}^{*-1}\begin{pmatrix} \hat{\Sigma}_1 & 0 \\ 0 & -\hat{\Sigma}_2 \end{pmatrix}\hat{T}^{-1} \triangleq \begin{pmatrix} Q_1 & Q_2 \\ Q_2^* & Q_3 \end{pmatrix}$$

Note that

$$\hat{A}_2P_3 + P_3\hat{A}_2^* + \hat{B}_2\hat{B}_2^* = 0$$

$$\hat{A}_2^* Q_3 + Q_3\hat{A}_2 + \hat{C}_2^* \hat{C}_2 = 0$$

and hence $P_3 \leqslant 0$ and $Q_3 \leqslant 0$ by Theorem 3.3, part (7) since $\text{In}\,(\hat{A}_2) = (l, 0, 0)$. Also $\sigma_i^2(F(-s)) = \lambda_i(P_3Q_3)$. Therefore, Lemma 9.8 can be applied to give that $\sigma_i(F(-s)) \leqslant \lambda_i(\hat{\Sigma}_2)$ as desired. (Note that σ_{k+1} needs to be scaled to unity to apply Lemma 9.8 but this gives no problems.)

(3) Part (*a*) is an immediate consequence of Corollary 9.3. For parts (*b*) and (*c*) we note that

$$(G(s) - \hat{G}(s) - F(s)) = \sigma_{k+1}E(s)$$

where $E(s)$ is all-pass. Therefore

$$(G(s) - \hat{G}(s) - D_0) = \sigma_{k+1}E(s) + F(s) - D_0$$

and part (*b*) follows on taking norms. Part (*c*) follows from the standard singular-value inequality that

$$\bar{\sigma}[\sigma_{k+1}E(j\omega) + F(j\omega) - D_0]^{-1} = [\underline{\sigma}(\sigma_{k+1}E(j\omega) + F(j\omega) - D_0)]^{-1}$$

$$\leqslant [\sigma_{k+1} - \delta]^{-1} \quad \blacksquare$$

The proof of Theorem 9.7, part (2) used the following lemma in order to bound the Hankel singular values of $F(-s)$. The proof of this lemma involves an intricate sequence of inequalities and the result is crucial to the present development.

Lemma 9.8

Let

$$P \triangleq \begin{pmatrix} \hat{\Sigma}_1 & 0 \\ 0 & -\hat{\Sigma}_2 \end{pmatrix}, \quad Q \triangleq \begin{pmatrix} \hat{\Sigma}_1 & 0 \\ 0 & -\hat{\Sigma}_2 \end{pmatrix} \tag{9.6}$$

where $\hat{\Sigma}_1$ and $\hat{\Sigma}_2$ are real diagonal matrices such that

$$\hat{\Sigma}_1 > I \geqslant \hat{\Sigma}_2 > 0 \tag{9.7}$$

Let T be such that

$$TPT^* = \begin{pmatrix} P_1 & P_2 \\ P_2^* & P_3 \end{pmatrix}, \quad T^{*-1}QT^{-1} = \begin{pmatrix} Q_1 & Q_2 \\ Q_2^* & Q_3 \end{pmatrix} \tag{9.8}$$

where (9.8) is partitioned conformally with (9.6) and

$$P_3 \leqslant 0 \quad \text{and} \quad Q_3 \leqslant 0 \tag{9.9}$$

Then

$$\lambda_i(P_3Q_3) \leqslant \lambda_i(\hat{\Sigma}_2^2) \quad \forall i \tag{9.10}$$

Proof

Partition T and T^{-1} conformally with P and Q as

$$T = \begin{pmatrix} T_1 & T_2 \\ T_3 & T_4 \end{pmatrix}, \quad T^{-1} = \begin{pmatrix} S_1 & S_3 \\ S_2 & S_4 \end{pmatrix} \tag{9.11}$$

then (9.6), (9.8) and (9.9) imply

$$P_3 = T_3\hat{\Sigma}_1T_3^* - T_4\hat{\Sigma}_2T_4^* \leqslant 0 \tag{9.12}$$

$$Q_3 = S_3^*\,\hat{\Sigma}_1S_3 - S_4^*\,\hat{\Sigma}_2S_4 \leqslant 0 \tag{9.13}$$

It will now be shown that T_4 is non-singular. Assume there exists x such that $x^*T_4 = 0$, then $x^*(9.12)x \Rightarrow x^*T_3\hat{\Sigma}_1T_3^*\,x \leqslant 0 \Rightarrow x^*T_3 = 0$ since $\hat{\Sigma}_1 > 0$. Hence $[0, x^*]T = 0$ but T is assumed to be non-singular so that $x^* = 0$ and hence T_4 must be non-singular.

Define

$$Y \triangleq T_4^{-1}\,T_3 \tag{9.14}$$

$$Z \triangleq S_3T_4 \tag{9.15}$$

then (9.11) implies $T_3S_3 + T_4S_4 = I$ and hence

$$\begin{aligned} S_4T_4 &= I - T_4^{-1}\,T_3S_3T_4 \\ &= I - YZ \end{aligned} \tag{9.16}$$

Also define P_4 as follows and substitute (9.14) and (9.15) in (9.12)

$$P_4 \triangleq -T_4^{-1}\,P_3T_4^{*-1} = \hat{\Sigma}_2 - Y\hat{\Sigma}_1Y^* \geqslant 0 \tag{9.17}$$

Define Q_4 as follows and substitute (9.15) and (9.16) in (9.13)

$$Q_4 \triangleq -T_4^*\,Q_3T_4 = (I - YZ)^*\hat{\Sigma}_2(I - YZ) - Z^*\hat{\Sigma}_1Z \tag{9.18}$$

Since $\lambda_i(P_3Q_3)=\lambda_i(P_4Q_4)$ the approach now is to find upper bounds on P_4 and Q_4. Haynsworth's result (Barnett 1971, Theorem 4.14) gives that

$$\text{In}\begin{pmatrix} \hat{\Sigma}_1^{-1} & Y^* \\ Y & \hat{\Sigma}_2 \end{pmatrix} = \text{In}\,(\hat{\Sigma}_1^{-1}) + \text{In}\,(\hat{\Sigma}_2 - Y\hat{\Sigma}_1 Y^*)$$

$$= \text{In}\,(\hat{\Sigma}_2) + \text{In}\,(\hat{\Sigma}_1^{-1} - Y^*\hat{\Sigma}_2^{-1}Y)$$

and since $\hat{\Sigma}_1 > I$ and $\hat{\Sigma}_2 > 0$ (9.17)$\Rightarrow$

$$\hat{\Sigma}_1^{-1} - Y^*\hat{\Sigma}_2^{-1}Y \geqslant 0 \tag{9.19}$$

and assumption (9.7) now imply that

$$I - Y^*\hat{\Sigma}_2 Y > 0 \tag{9.20}$$

and

$$X \triangleq \hat{\Sigma}_1 - Y^*\hat{\Sigma}_2 Y > 0 \tag{9.21}$$

Relation (9.21) is required in order to 'complete the square' in (9.18) as follows

$$\begin{aligned} Q_4 &= \hat{\Sigma}_2 - (Z^* + \hat{\Sigma}_2 Y X^{-1})X(Z + X^{-1}Y^*\hat{\Sigma}_2) + \hat{\Sigma}_2 Y X^{-1} Y^* \hat{\Sigma}_2 \\ &\leqslant \hat{\Sigma}_2 + \hat{\Sigma}_2 Y X^{-1} Y^* \hat{\Sigma}_2 \\ &\leqslant \hat{\Sigma}_2 + \hat{\Sigma}_2 Y(I - Y^*\hat{\Sigma}_2 Y)^{-1} Y^* \hat{\Sigma}_2 \triangleq Q_5 \end{aligned} \tag{9.22}$$

Also from (9.17) and (9.7)

$$P_4 \leqslant \hat{\Sigma}_2 - YY^* \triangleq P_5 \tag{9.23}$$

Now (9.22) and (9.23)$\Rightarrow \lambda_i(P_3Q_3) \leqslant \lambda_i(P_5Q_5)$ since

$$\begin{aligned} \lambda_i(P_3Q_3) &= \lambda_i(P_4Q_4) = \lambda_i(P_4^{1/2}\,Q_4P_4^{1/2}) \\ &\leqslant \lambda_i(P_4^{1/2}\,Q_5P_4^{1/2}) = \lambda_i(P_4Q_5) \\ &\leqslant \lambda_i(Q_5^{1/2}\,P_4Q_5^{1/2}) \leqslant \lambda_i(Q_5^{1/2}\,P_5Q_5^{1/2}) \\ &= \lambda_i(P_5Q_5) \end{aligned}$$

Hence

$$\begin{aligned} \lambda_i(P_3Q_3) &\leqslant \lambda_i[(\hat{\Sigma}_2 - YY^*)\{\hat{\Sigma}_2 + \hat{\Sigma}_2 Y(I - Y^*\hat{\Sigma}_2 Y)^{-1}Y^*\hat{\Sigma}_2\}] \\ &= \lambda_i[\hat{\Sigma}_2^2 + \{-Y(I - Y^*\hat{\Sigma}_2 Y) + (\hat{\Sigma}_2 - YY^*)\hat{\Sigma}_2 Y\}(I - Y^*\hat{\Sigma}_2 Y)^{-1}Y^*\hat{\Sigma}_2] \\ &= \lambda_i[\hat{\Sigma}_2^2 - (I - \hat{\Sigma}_2^2)Y(I - Y^*\hat{\Sigma}_2 Y)^{-1}Y^*\hat{\Sigma}_2] \\ &= \lambda_i[\hat{\Sigma}_2^2 - (\hat{\Sigma}_2 - \hat{\Sigma}_2^3)^{1/2}Y(I - Y^*\hat{\Sigma}_2 Y)^{-1}Y^*(\hat{\Sigma}_2 - \hat{\Sigma}_2^3)^{1/2}] \\ &\leqslant \lambda_i(\hat{\Sigma}_2^2) \quad \text{by (9.20)} \end{aligned}$$

and the result is proved. ■

Remark 9.5

The error in non-square system approximation can be bounded by augmenting the systems by zeros to make it square and then applying Theorem 9.7. In this case the equalities of Theorem 9.7, part (1) need to be replaced by inequalities and the bound in Theorem 9.7, part (3 *c*) is not valid.

Remark 9.6

It will be seen that for $l > n-k-r$ the L^∞-error bounds of Theorem 9.7, part (3) will be generally worse than the case $l = n-k-r$ since the former involves $l-n+k+r$ terms near to σ_{k+1}. Therefore the $l = n-k-r$ case may be preferable. However, the class of solutions given by Corollary 7.3, in which $l > n-k-r$ normally, can be analysed by considering the augmented $(m+p) \times (m+p)$ system

$$G_a(s) = \begin{pmatrix} G(s) & 0 \\ 0 & 0 \end{pmatrix}$$

for which $l = n-k-r$ and the tighter bounds of Theorem 9.7 can be used. This is given in the following Corollary.

Corollary 9.9

Let $\hat{G}(s)$ be an optimal Hankel-norm approximation of degree k to the stable, rational $p \times m$ transfer function $G(s)$ given by Corollary 7.3. Also let $F(s) \in H_-^\infty$ be such that $\hat{G}(s) + F(s) = \hat{D} + \hat{C}(sI - \hat{A})^{-1}\hat{B}$. Then

(1)
$$\sigma_i(G(s) - \hat{G}(s)) \leqslant \begin{cases} \sigma_{k+1}(G(s)), & i = 1, 2, \ldots, 2k+r \\ \sigma_{i-k}(G(s)), & i = 2k+r+1, \ldots, n+k \end{cases}$$

(2)
$$\sigma_i(F(-s)) \leqslant \sigma_{i+k+r}(G(s)), \quad i = 1, 2, \ldots, n-k-r$$

(3) there exists D_0 such that

(*a*)
$$\delta \triangleq \| F(j\omega) - D_0 \|_{L^\infty} \leqslant \sum_{1 \leqslant i \leqslant n-k-r} \sigma_i(F(-s))$$

(*b*)
$$\| G(j\omega) - \hat{G}(j\omega) - D_0 \|_{L^\infty} \leqslant \sigma_{k+1}(G(s)) + \delta$$
$$\leqslant \sigma_{k+1}(G(s)) + \sum_{1 \leqslant i \leqslant n-k-r} \sigma_{i+k+r}(G(s))$$

Proof

The method of construction used in Corollary 7.3 was to imbed $G(s)$ in the $(p+m) \times (p+m)$ augmented system $G_a(s)$ and then we use the construction of Theorem 7.2, part (3) to obtain an optimal Hankel-norm approximation to $G_a(s)$. In the notation of Theorem 9.7 this approximation has $l = n-k-r$. Therefore, the approximation to $G_a(s)$ satisfies the bounds of Theorem 9.7, and the approximation to $G(s)$ is just obtained by deleting the final m rows and p columns of that of $G_a(s)$ and hence all its error bounds will be less than those given for $G_a(s)$. Note that $F(s)$ in the statement of the corollary will be a submatrix of the $F_a(s)$ of the construction. ■

Remark 9.7

Corollary 9.9, part (3) gives a bound on the L^∞-norm of the error in the approximation of Corollary 7.3 that is less than the 'sum of the tail'. In order to calculate this bound, δ needs to be found which requires a balanced realization of $F(-s)$, which is also necessary in order to find D_0 via Corollary 9.3. This requires slightly less computational effort than the original balancing of $G(s)$ and may be significant.

Remark 9.8

Recall that from Theorem 7.2, part (1) $\|G(j\omega)-\hat{G}(j\omega)\|_{L^\infty} \geqslant \sigma_{k+1}(G(s))$ for any $\hat{G}(s)$ of degree k, and we could only achieve this lower bound if we allowed the additional anticausal term $F(s)$. However, the bound of Corollary 9.9, part (3 b) may get quite close to this theoretical lower bound. This is because the $\sigma_i(G(s))$ often decrease fairly quickly and the inequality of Corollary 9.9, part (2) was obtained after a number of intermediate inequalities. Hence the term δ can be small relative to $\sigma_{k+1}(G)$. For very high-order systems where the $\sigma_i(G)$ do not decrease quickly (for example, behave as $O(1/i)$ or tend to a constant) then δ may be large even though $\sigma_{k+1}(G)$ is small. This may be because the bound is very weak, because the Hankel-norm approximation is not appropriate, or because no approximation can make the L^∞-norm of the error small. The first cause could be checked by evaluating the Bode diagram of the error system, but it is hard to distinguish between the other two possible causes.

Remark 9.9

It is interesting to consider whether the bounds of Theorem 9.7, part (3) for the case $l > n-k-r$ are excessively conservative (as is the case if they are applied to the method of Corollary 7.3 ; see Corollary 9.9). We will now outline an example where they are quite tight. Consider $p=m=2$ and let

$$G(s)=\begin{pmatrix} g_1(s) & 0 \\ 0 & g_2(s) \end{pmatrix}$$

where $\{\sigma_i(g_1(s))\}=\{\sigma_1, \sigma_2, \ldots, \sigma_k, \sigma_n\}$ and $\{\sigma_i(g_2(s)\}=\{\sigma_{k+1}, \sigma_{k+2}, \ldots, \sigma_{n-1}\}$. Now a $(n-1)$st-order optimal Hankel-norm approximation will be $\hat{G}(s)=\text{diag}\ (\hat{g}_1(s), \hat{g}_2(s))$ and will satisfy $\|g_1(s)-\hat{g}_1(s)\|_{\text{H}}=\sigma_n$ and $\|g_2(s)-\hat{g}_2(s)\|_{\text{H}} \leqslant \sigma_n$, where $\hat{g}_1(s)$ and $\hat{g}_2(s)$ have degrees k and $(n-k-1)$. $(g_2(s)-\hat{g}_2(s))$ could now be chosen so that its Hankel-norm is small but its L^∞-norm is large if n is large. Therefore, in an L^∞-sense certain optimal Hankel-norm approximations may be quite poor and worse than that obtained by a truncated balanced realization or by the simple method outlined in Remark 9.1 and based on Theorem 9.2. However the optimal Hankel-norm approximations of Corollary 7.3 will generally be superior to either of these methods in an L^∞-sense.

9.5. *Approximating systems with unstable poles*

All the results so far have been concerned with approximating a stable $G(s)$. We will now consider the case when

$$G(s)=G_1(s)+G_2(s)$$

where $G_1(s) \in H_+^\infty$ and $G_2(s) \in H_-^\infty$ of McMillan degrees n_1 and n_2 respectively (if there are any poles on the imaginary axis these should be retained unchanged). Suppose we are concerned with approximating a transfer function in a stable closed-loop configuration and have that the error is small on the imaginary axis. In this case stability of the closed-loop system with the approximate model can be deduced from that of full-order model if the number of right half-plane poles of the approximate model equals that of the full-order model and the L^∞-error is suitable small (see Glover and Limebeer (1983)). Hence we will consider approximations of $G(s)$ that have the same number of right half-plane poles as $G(s)$. That is

$$\hat{G}(s) = \hat{G}_1(s) \mp \hat{G}_2(s)$$

with $\hat{G}_1(s) \in H_+^\infty$, $\hat{G}_2(s) \in H_-^\infty$ of McMillan degrees $k < n_1$ and n_2 respectively.

One method of approximation is to make $\hat{G}_1(s)$ an optimal Hankel-norm approximation of $G_1(s)$ and let $\hat{G}_2(s) = G_2(s)$. The frequency response error bounds of Theorem 9.7 and Corollary 9.9 will then apply using $\sigma_i(G_1(s))$. Also a lower bound on the approximation error is given by Theorem 7.2, part (1) as

$$\|G(j\omega) - \hat{G}_1(j\omega) - \hat{G}_2(j\omega)\|_{L^\infty} \geqslant \sigma_{k+1}(G_1(s))$$

since $\hat{G}_2(s) \in H_-^\infty$ and $\hat{G}_1(s)$ has McMillan degree k.

An alternative method is to still make $\hat{G}_1(s)$ an optimal Hankel-norm approximation to $G_1(s)$ but to choose $\hat{G}_2(s)$ as follows. Let $F_1(s) \in H_-^\infty$ be such that

$$\|G_1(j\omega) - \hat{G}_1(j\omega) - F_1(j\omega)\|_{L^\infty} = \sigma_{k+1}(G_1(s))$$

(this is generated automatically in Corollary 7.3 and will have McMillan degree $\leqslant (n-k-r)$). Now consider

$$\|G(j\omega) - \hat{G}_1(j\omega) - \hat{G}_2(j\omega)\|_{L^\infty} \leqslant \|G_1(j\omega) - \hat{G}_1(j\omega) - F_1(j\omega)\|_{L^\infty} + \|G_2(j\omega) + F_1(j\omega) - \hat{G}_2(j\omega)\|_{L^\infty}$$

We can now choose $\hat{G}_2(-s)$ to be an optimal Hankel-norm approximation to $(G_2(-s) + F_1(-s))$ of McMillan degree n_2. It is a straightforward consequence of Theorem 7.2, part (1) that

$$\sigma_{n_2+i}(G_2(-s) + F_1(-s)) \leqslant \sigma_i(F_1(-s)) \quad \text{for } i \geqslant 1$$

and defining δ in Corollary 9.9 as

$$\delta \triangleq \sum_{1 \leqslant i \leqslant n-k-r} \sigma_{n_2+i}(G_2(-s) + F_1(-s))$$

we have that

$$\|G(j\omega) - \hat{G}_1(j\omega) - \hat{G}_2(j\omega)\|_{L^\infty} \leqslant \sigma_{k+1}(G(s)) + \delta$$

Hence, by slight changes in the unstable poles one may be able to decrease the L^∞-norm of the error further towards its theoretical lower bound of $\sigma_{k+1}(G_1(s))$.

10. A specific approximation algorithm

The results of the previous sections are now combined to give a model-reduction algorithm that is suitable for computer implementation. This algorithm is then applied to a 2-input, 2-output, 12-state model of an automobile gas turbine.

10.1. *A specific algorithm*

In this section the results of the previous sections are brought together to outline an optimal Hankel-norm approximation algorithm that the author has implemented and used on a variety of systems. The method is that of Corollary 7.3 with the choice of U given by (7.23), and the choice of $\hat{D}$ given by a minor variation of the results of § 9. It will be assumed that the system is real and stable and that $\sigma_k > \sigma_{k+1} > \sigma_{k+2}$. If the system is unstable the methods of § 9.5 can be applied. The equality of σ_{k+1} and σ_{k+2} would be difficult to detect numerically but, since $(\sigma_{k+1}{}^2 - \sigma_{k+2}{}^2)^{-1}$ is used, near equality of σ_{k+1} and σ_{k+2} may give numerical difficulties. The algorithm is now outlined.

(1) It is assumed that a state-space realization of $G(s)$, (A, B, C, D) has been supplied together with k equalling the required McMillan degree of the approximation.

(2) *Form a balanced realization* of $G(s)$ by the methods of § 4. If the realization is not minimal then minor modifications to these methods are required to retain only controllable and observable states. Let the Hankel singular values be $\sigma_1 \geqslant \sigma_2 \ldots \geqslant \sigma_k > \sigma_{k+1} > \sigma_{k+2} \ldots \geqslant \sigma_n > 0$, and let the balanced realization of state dimension n be reordered so that

$$A = \begin{pmatrix} A_{11} & A_{12} \\ A_{21} & A_{22} \end{pmatrix}, \quad B = \begin{pmatrix} B_1 \\ B_2 \end{pmatrix}, \quad C = (C_1, C_2)$$

$$\Sigma = \text{diag}\,(\sigma_1, \sigma_2, \ldots, \sigma_k, \sigma_{k+2}, \ldots, \sigma_n, \sigma_{k+1})$$

$$= \text{diag}\,(\Sigma_1, \sigma_{k+1})$$

with (A, B, C) partitioned conformally with Σ (i.e. A_{11} is $(n-1) \times (n-1)$).

(3) *Form* $\hat{G}(s) + F(s)$

$$U = -(C_2 B_2)/(B_2 B'_2)$$

$$\Gamma = (\Sigma_1{}^2 - \sigma_{k+1}{}^2 I)$$

$$\hat{A} = \Gamma^{-1}(\sigma_{k+1}{}^2 A_{11}{}^* + \Sigma_1 A_{11} \Sigma_1 - \sigma_{k+1} C_1{}^* U B_1{}^*)$$

$$\hat{B} = \Gamma^{-1}(\Sigma_1 B_1 + \sigma_{k+1} C_1{}^* U)$$

$$\hat{C} = C_1 \Sigma_1 + \sigma_{k+1} U B_1{}^*$$

$$\hat{D} = D - \sigma_{k+1} U$$

(4) *Block diagonalize* $\hat{A}$

(*a*) Reduce $\hat{A}$ to real upper Schur form, i.e. find V_1 such that $V_1' V_1 = I$ and $V_1' \hat{A} V_1$ is in upper Schur form.

(*b*) Find an orthogonal matrix V_2 such that

$$V'_2 V'_1 \hat{A} V_1 V_2 = \begin{pmatrix} \hat{A}_{11} & \hat{A}_{12} \\ 0 & \hat{A}_{22} \end{pmatrix}$$

where $\text{Re}\,(\lambda_i(\hat{A}_{11})) < 0$, $\text{Re}\,(\lambda_i(\hat{A}_{22})) > 0$, and note that $\hat{A}_{11} \in \mathbb{R}^{k \times k}$ (the details of this step are omitted).

(*c*) Find $X \in \mathbb{R}^{k \times (n-k-1)}$ such that

$$\hat{A}_{11} X - X \hat{A}_{22} + \hat{A}_{12} = 0$$

by the Bartels–Stewart algorithm.

(*d*) Let

$$T = V_1 V_2 \begin{pmatrix} I & X \\ 0 & I \end{pmatrix} = (T_1, T_2)$$

$$S = \begin{pmatrix} I & -X \\ 0 & I \end{pmatrix} V'_2 \, V'_1 = \begin{pmatrix} S_1 \\ S_2 \end{pmatrix}$$

(*e*) Let

$$\hat{B}_1 = S_1 \hat{B}$$
$$\hat{B}_2 = S_2 \hat{B}$$
$$\hat{C}_1 = \hat{C} T_1$$
$$\hat{C}_2 = \hat{C} T_2$$

(5) *Calculation of the D-matrix*

(*a*) Find a balanced realization of the system $(-\hat{A}_{22}, \hat{B}_2, \hat{C}_2, \hat{D})$ say $(A_3, B_3, C_3, \hat{D})$ with Hankel singular values $\mu_1 > \mu_2 \ldots > \mu_{n-k-1} \geqslant 0$.

(*b*) Let Z, $Y \in \mathbb{R}^{q \times (n-k-1)}$ be given by

$$Z = \begin{pmatrix} B'_3 \\ 0 \end{pmatrix}, \quad Y = \begin{pmatrix} C_3 \\ 0 \end{pmatrix}$$

and let z_i, y_i be the ith columns of Z and Y, respectively. (Note q is an integer greater than or equal to $(p+m)$.)

(*c*) For $i := 1$ to $n-k-1$

(i) Find Householder transformations (Stewart 1973, p. 233) such that

$$(I - \pi_1^{-1} W_1 W'_1) y_i = -(\alpha, 0, 0, \ldots, 0)'$$
$$(I - \pi_2^{-1} W_2 W'_2) z_i = -(\beta, 0, 0, \ldots, 0)'$$

(ii) Let

$$U := (I - \pi_1^{-1} W_1 W'_1) \begin{bmatrix} \frac{-\alpha}{\beta} & 0 & 0 & 0 \\ 0 & 0 & I_{p-1} & 0 \\ 0 & I_{m-1} & 0 & 0 \\ 0 & 0 & 0 & I_{q-p-m+1} \end{bmatrix} \times (I - \pi_2^{-1} W_2 W_2)$$

(iii) If $i < n-k-1$ then for $j := (i+1)$ to $(n-k-1)$

$$y := -(y_j \mu_j + U z_j \mu_i)(\mu_i^2 - \mu_j^2)^{-1/2}$$
$$z_j := (z_j \mu_j + U' y_j \mu_i)(\mu_i^2 - \mu_j^2)^{-1/2}$$
$$y_j := y$$

(iv)

$$\hat{D} := \hat{D} + (-1)^i \mu_i [I_p \;\; 0] U \begin{pmatrix} I_m \\ 0 \end{pmatrix}$$

(6)

$$\hat{G}(s) = \hat{D} + \hat{C}_1(sI - \hat{A}_{11})^{-1}\hat{B}_1$$

and

$$\|G(s) - \hat{G}(s)\|_{L^\infty} \leqslant \sigma_{k+1} + \mu_1 + \mu_2 + \dots + \mu_{n-k-1}$$ ■

Remark 10.1

The rationale for the $\hat{D}$-matrix is as follows. Remark 9.2 showed that a D-matrix could be found such that $\|F(j\omega) - D\|_{L^\infty} \leqslant \mu_1 + \mu_2 + \dots + \mu_{n-k-1}$. This was shown by constructing a sequence of anticausal and causal approximations. It appears that a good choice for an anticausal approximation is given by Theorem 6.3 but with U the minimum-norm solution to (6.23). Let $F_1(s)$ be such an approximation then $\|F(-j\omega) - F_1(j\omega)\|_{L^\infty} = \mu_1$ and $\sigma_i(F_1(-s)) \leqslant \mu_{i+1}$ for $i = 1, 2, \dots, n-k-2$. Now the procedure could be repeated on $F_1(-s)$ but this would require a balanced realization of $F_1(-s)$ and this is computationally demanding for modest gain. We therefore imbed the system in a $(q \times q)$ system and choose orthogonal matrices U. Now by Lemma 9.4 this ensures that the approximation $F_1(-s)$, satisfies $\sigma_i(F_1(-s)) = \mu_{i+1}$ and that a balanced realization of $F_1(-s)$ is simply obtained from a balanced realization of $F(-s)$. The particular form of U is so that the first one or two steps of the procedure actually correspond to taking the minimum-norm solution to (6.23). As has been remarked before the choice of $\hat{D}$ is somewhat arbitrary and that given here is acceptable in that the L^∞-error bound is satisfied.

That the μ_i are distinct is an assumption that could be violated, however μ_i close to μ_{i+1} does not appear to give difficulties.

Remark 10.2

As was observed in Theorem 8.7 in the multivariable case there are many degrees of freedom in the optimal Hankel-norm approximations. One feature of the present method, that is not necessarily true for other optimal Hankel-norm approximations, is that if $G(s)$ is block diagonal then so is $\hat{C}_1(sI - \hat{A}_{11})^{-1}\hat{B}_1$ although $\hat{D}$ may have some off-diagonal terms of order $\mu_2 + \mu_3 + \dots + \mu_{n-k-1}$. That this is true can be deduced by observing that U will be zero except for the block containing the σ_{k+1} as a Hankel singular value. This observation shows in a limited sense that this algorithm does not introduce approximation errors where it is not necessary.

Remark 10.3

An important practical consideration is that of the scaling of multivariable systems. The units of the different entries in a transfer function will generally be incompatible and the Hankel-norm or the L^∞-norm will combine these. It is therefore important that the $G(s)$ be scaled so that unit errors in each entry of $G(s)$ are of similar importance. Diagonal constant scaling factors are, therefore, normally introduced on the input and output to help achieve this ; which is also good practice in control system design. This step clearly requires knowledge of the physical significance of all the inputs and outputs and their likely amplitude variations which cannot be obtained by examining the algebraic form of $G(s)$ alone.

10.2. *A multivariable example*

In this section we illustrate the algorithm on a model of a 2-input, 2-output, automobile gas turbine. The full-order model has 12 states and is analysed in Hung and MacFarlane (1982, p. 164) and in Glover and Limebeer (1983). In the latter paper a compensator is designed based on a reduced-order plant model. Firstly, the balanced realization of $G(s)$ is formed from the given realization and the Hankel singular values are calculated to be

$$\sigma_1 = 7{\cdot}118, \quad \sigma_2 = 1{\cdot}488, \quad \sigma_3 = 0{\cdot}9225$$
$$\sigma_4 = 0{\cdot}5863, \quad \sigma_5 = 0{\cdot}4593, \quad \sigma_6 = 0{\cdot}2351$$
$$\sigma_7 = 0{\cdot}1605, \quad \sigma_8 = 0{\cdot}09334, \quad \sigma_9 = 0{\cdot}56 \times 10^{-3}$$
$$\sigma_{10} = 0{\cdot}205 \times 10^{-4}, \quad \sigma_{11} = 0{\cdot}21 \times 10^{-5}, \quad \sigma_{12} \simeq 0{\cdot}86 \times 10^{-7}$$

The σ_i are plotted in Fig. 6, and it is seen that $\sigma_1 \simeq 5\sigma_2$, then $\sigma_3, \ldots, \sigma_8$ decrease quite slowly until $\sigma_9, \ldots, \sigma_{12}$ which are almost negligible. Therefore an optimal eight-state model would be essentially indistinguishable from the 12-state model

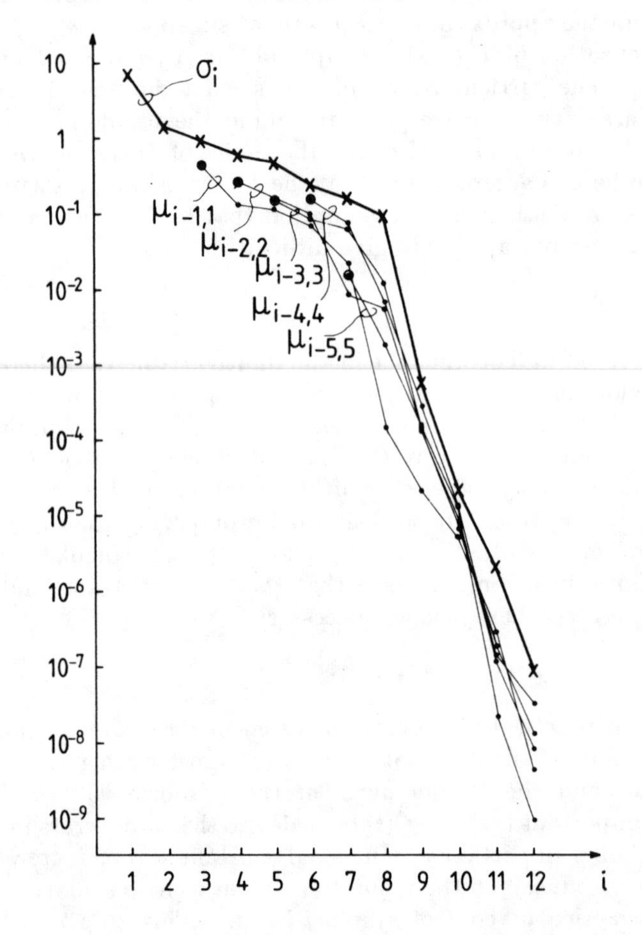

Figure 6. Hankel singular values for the example.

with an L^∞-error less than $\sigma_9 + \ldots + \sigma_{12} \simeq 0{\cdot}58 \times 10^{-3}$ (by Corollary 9.9). Bounds on the L^∞-errors of lower-order models are given by Corollary 9.9 as

$$\sigma_{k+1} \leqslant \| G(j\omega) - \hat{G}_k(j\omega) - D_0 \|_{L^\infty} \leqslant \sigma_{k+1} + \delta_k \triangleq M_k$$
$$\leqslant \sigma_{k+1} + \sigma_{k+2} + \ldots + \sigma_n \triangleq N_k$$

where the subscript k refers to the dimension of the reduced-order model. These bounds are evaluated below in the table.

k	σ_{k+1}	M_k	N_k
0	7·118	8·665	11·0636
1	1·488	2·284	3·9456
2	0·9225	1·474	2·4576
3	0·5863	0·902	1·5351
4	0·4593	0·697	0·9488
5	0·2351	0·253	0·4895
6	0·1605	0·172	0·2544
7	0·09334	0·0934	0·0939
8	$0{\cdot}56 \times 10^{-3}$	$0{\cdot}56 \times 10^{-3}$	$0{\cdot}58 \times 10^{-3}$

Note that σ_{k+1} and N_k are immediately available but that M_k which satisfies $\sigma_{k+1} \leqslant M_k \leqslant N_k$ needs $\mu_{ik} \triangleq \sigma_i(F_k(-s))$ to be evaluated. In order to determine what is an acceptable error one needs to consider the purpose of the reduced-order model. For example, open-loop simulations for different classes of relatively low frequency inputs would probably require less accurate models than if a closed-loop stability analysis were to be performed. In the latter case one approach would be to estimate the system gain $G(j\omega_0)$ at the desired ' cross-over frequency ' of the closed-loop system and then choose an error small compared to $\underline{\sigma}(G(j\omega_0))$. This estimate may well be possible from physical considerations. Alternatively an appropriate size for the error can be based on the prior confidence in the model which would be determined by the accuracy with which it was derived. Once an acceptable error has been determined the σ_{k+1} and N_k columns of the table then give the values of k which need to be considered. Suppose 0·3 is deemed to be an acceptable error then it is only necessary to consider $k = 5$ and 6. Now the optimal Hankel-norm approximation can be performed and M_k determined. In Fig. 6 the values for μ_{ik} are plotted for $k = 1, 2, 3, 4, 5$, and it is noted from the result of Corollary 9.9, part (2) that $\mu_{ik} \leqslant \sigma_{i+k+1}$ for $i = 1, 2, \ldots, n-k-1$, are satisfied. For $k = 5$ the 17 $(= n+k)$ Hankel singular values of $(G(s) - \hat{G}(s))$ have been calculated and are

0·23508	0·23505	0·23503	0·23502	0·23279
0·22750	0·22293	0·17888	0·17690	0·17369
0·17143	0·10112	$0{\cdot}343 \times 10^{-2}$	$0{\cdot}323 \times 10^{-3}$	$0{\cdot}159 \times 10^{-4}$
$0{\cdot}230 \times 10^{-7}$	$0{\cdot}971 \times 10^{-9}$			

Note that Corollary 9.9, part (1) is satisfied with $\sigma_i(G(s) - \hat{G}(s)) \leqslant \sigma_6(G(s))$ for $i = 1, 2, \ldots, 11$, and $\leqslant \sigma_{i-5}(G(s))$ for $i = 12, \ldots, 17$.

The family of solutions given by Corollary 7.3 are characterized by a 2×2 matrix U such that $U^*U \leqslant I$ and $C_2+UB_2^*=0$. Restricting ourselves to real matrices in the 2×2 case the solutions will be characterized by an arbitrary real number, u_2 say, with $-1 \leqslant u_2 \leqslant 1$. The solution given above corresponds to taking $u_2=0$; for $u_2=1, -1$ let the bounds corresponding to M_k be M'_k and M''_k, respectively. Then $M'_5=0{\cdot}270$, $M''_5=0{\cdot}334$ can be calculated and it is seen that $u_2=0$ gives smaller bounds in this case.

The two singular values of the error, $\underline{\sigma}(G(j\omega)-\hat{G}_5(j\omega)-D_0)$ and $\bar{\sigma}(G(j\omega)-\hat{G}_5(j\omega)-D_0)$ are plotted in Fig. 7 for $k=5$ and the error bound on the larger one is seen to be satisfied and to be very tight. Also the step responses of the system and $G_5(s)$ are plotted in Fig. 8 and the L^∞-norm of the step-response error is seen to be quite small. In fact this error can be bounded *a priori* in terms of M_k, n and k as will be shown in Glover (1984).

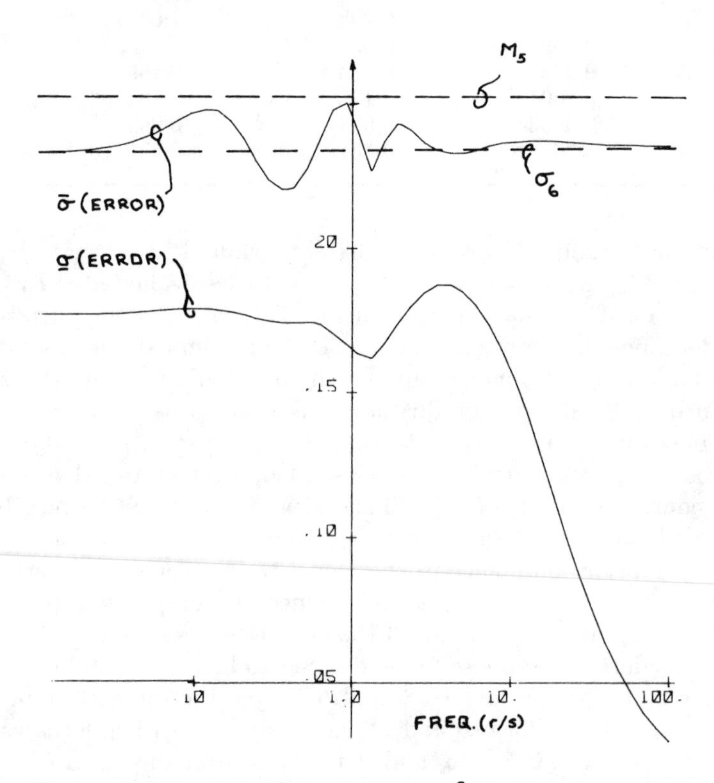

Figure 7. Singular values of $G(j\omega)-\hat{G}_5(j\omega)$ for the example.

11. Conclusions

A characterization of all optimal Hankel-norm approximations to a rational transfer function has been derived (Theorem 8.7). Bounds on the L^∞-norm of the error in the frequency response are calculated in Theorem 9.7. For the particular method of Corollary 7.3 it is shown in Corollary 9.9 that $\|G(j\omega)-\hat{G}(j\omega)-D_0\|_{L^\infty} \leqslant \sigma_{k+1}(G(s))+\delta$, where δ will often be small. In the case when δ is small the L^∞-error will be close to the lower bound of $\sigma_{k+1}(G(s))$, which can itself be only achieved if an anticausal term is included.

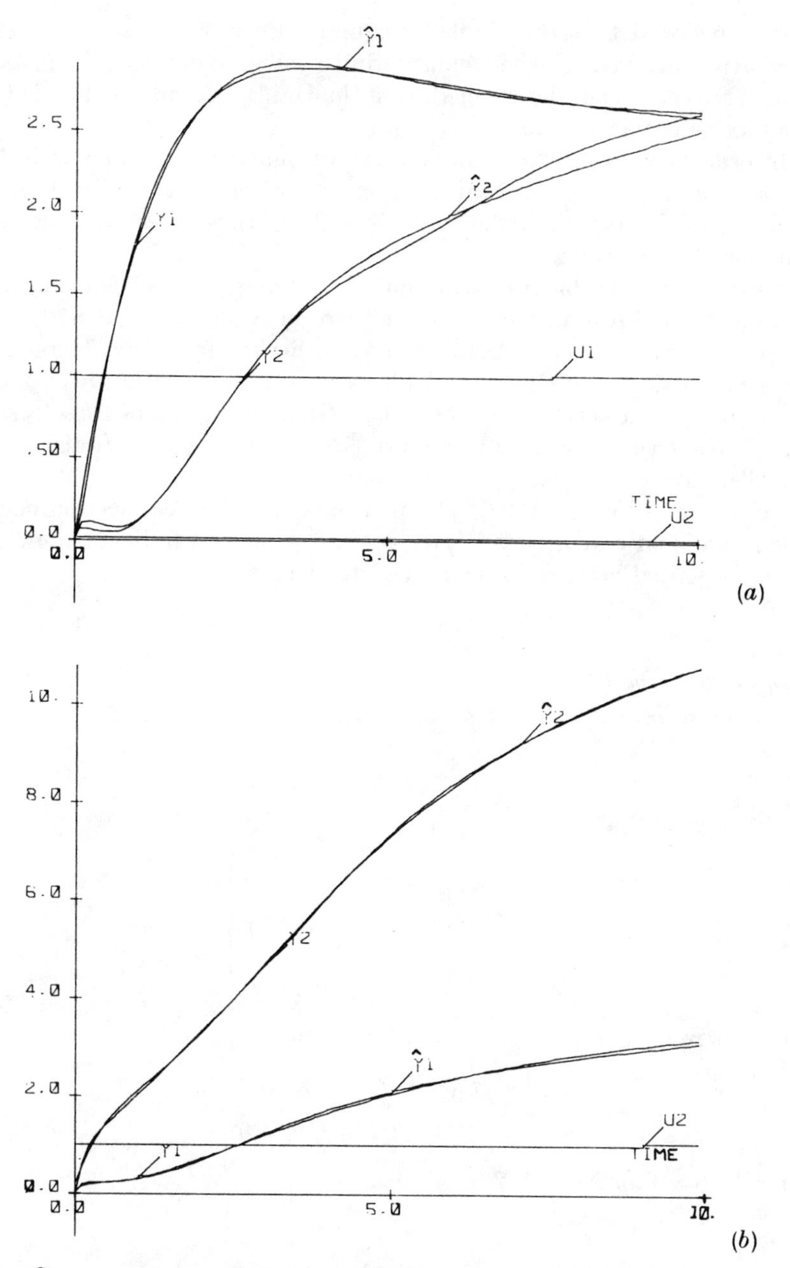

Figure 8. Step responses of system and 5th-order approximation. (*a*) input 1. (*b*) input 2.

The author has found the method described in § 10.1 to be most effective because it not only tells you what L^∞-errors can be achieved with this method but also what L^∞-errors cannot be achieved with any other approximation of the same McMillan degree. That is, once the Hankel singular values are calculated upper and lower bounds on the L^∞-error can be simply determined for each value of k.

It is believed that the Hankel singular values of a system are extremely informative invariants when considering system complexity and also gain. The feature that is not determined from the Hankel singular values is the distribution of gain over frequency.

Intermediate results on the inertia of matrices, characterizing all-pass functions and finding anti-causal approximations have also been derived which although very similar to other results in these areas are thought to be of independent interest.

An important problem is to weight the L^∞-norm by a frequency weighting function. Some results relating to this are in Anderson *et al.* (1984) and is a topic of current research. As mentioned earlier the frequency response bounds have been exploited in Glover and Limebeer (1983) to obtain results in closed-loop stability and performance bounds. Generalizations of these results have been developed in Curtain and Glover (1984) for designing finite dimensional controllers for a wide class of distributed parameter systems.

Safonov and Verma (1983) have pointed out a close connection between the present problem and L^∞-sensitivity optimization and indicates how the algorithms presented here could be applied to the latter problem.

Appendix A

Proof of Theorem 3.4

The proof of Theorem 3.4 proceeds via a lemma.

Lemma

Let

$$H = \begin{bmatrix} 0 & 1 & 0 & \dots & 0 \\ 0 & 0 & 1 & \dots & 0 \\ 0 & \cdot & \cdot & \dots & 1 \\ 0 & \cdot & \cdot & \dots & 0 \end{bmatrix}$$

then if there exist $P = P^*$ and $Q = Q^*$ such that

$$HP + PH^* + BB^* = 0 \tag{A 1}$$

$$H^*Q + QH + C^*C = 0 \tag{A 2}$$

then $CH^kB = 0$ for $k = 0, 1, 2, \dots$.

Proof

(1) It will be shown by induction that

$$H^{n-k}PH^{*k} = 0 \quad \text{and} \quad H^{n-k}B = 0 \quad \text{for } 0 \leqslant k \leqslant \frac{(n+1)}{2}$$

It is clearly true for $k = 0$ (note $n = \dim H$). Assume it is true for $k - 1$, then $H^{n-k}(14.1)H^{*n-k}$ gives for $n - 2k + 1 > 0$ that

$$(H^{n-(k-1)}PH^{*(k-1)})H^{*(n-2k+1)} + H^{n-2k+1}(H^{k-1}PH^{*n-(k-1)}) + H^{n-k}BB^*H^{*n-k} = 0$$

$$\Rightarrow H^{n-k}B = 0 \quad \text{by the inductive hypothesis.}$$

Now $H^{n-k}(\mathrm{A}\ 1)H^{*k-1}$ gives

$$\Rightarrow H^{n-k+1}PH^{*k-1}+H^{n-k}PH^{*k}+H^{n-k}BB^*H^{*k-1}=0$$
$$\Rightarrow H^{n-k}PH^{*k}=0$$

which proves the inductive hypothesis for $0\leqslant k\leqslant(n+1)/2$.

(2) Similarly it can be shown that

$$H^{*(n-k)}QH^k=0,\quad CH^{n-k}=0\quad \text{for } 0\leqslant k\leqslant\frac{(n+1)}{2}$$

(3) n *odd.* $(1)\Rightarrow H^{(n-1)/2}B=0\Rightarrow$

$$B=[b_1^*, b_2^*, \ldots, b_{(n-1/2)}^*, 0, \ldots, 0]^*$$

and $(2)\Rightarrow CH^{(n-1)/2}=0\Rightarrow$

$$C=[0, \ldots, 0, c_{(n+3/2)}, \ldots, c_n]$$
$$\Rightarrow CH^kB=0\quad \text{for all } k\geqslant 0.$$

(4) n *even.* (1) and $(2)\Rightarrow H^{n/2}B=0$, $CH^{n/2}=0\Rightarrow$

$$B=[b_1^*, \ldots, b_{n/2}^*, 0, \ldots, 0]^*,\quad C=[0, 0, \ldots, 0, c_{n/2+1}, \ldots, c_n]$$
$$\Rightarrow CH^kB=0\quad \text{for all } k\geqslant 0.$$

■

Proof of Theorem 3.4

Without loss of generality we will assume that A is in Jordan canonical form and combine all the Jordan blocks with eigenvalues not on the imaginary axis into A_1(i.e. $\delta(A_1)=0$) and $A_2, \ldots, A_l$ are the remaining Jordan blocks with for $i\geqslant 2$

$$A_i=\lambda_i I+H_i,\quad H_i=\begin{pmatrix} 0 & 1 & \cdots & 0 \\ \vdots & \ddots & \ddots & \\ \vdots & & \ddots & 1 \\ 0 & \cdots & \cdots & 0 \end{pmatrix},\quad \lambda_i+\bar{\lambda}_i=0$$

$$A=\operatorname{diag}(A_1, A_2, \ldots, A_l)$$

Now partition B, C, P and Q conformally with A with blocks B_i, C_i, P_{ij} and Q_{ij}. The (i, i)th block entries of the Lyapunov equations then give

$$A_iP_{ii}+P_{ii}A_i^*+B_iB_i^*=0$$
$$A_i^*Q_{ii}+Q_{ii}A_i+C_i^*C_i=0$$

$\Rightarrow$since $\lambda_i+\bar{\lambda}_i=0$, that for $i\geqslant 2$

$$H_iP_{ii}+P_{ii}H_i^*+B_iB_i^*=0$$
$$H_i^*Q_{ii}+Q_{ii}H_i+C_i^*C_i=0$$

Hence the lemma can be applied to give that for $i\geqslant 2$, $C_iH_i^kB_i=0$ for $k\geqslant 0$, $\Rightarrow C_iA_i^kB_i=0$ for $k\geqslant 0$. Therefore $CA^kB=C_1A_1^kB_1$ for $k\geqslant 0$ and since $\delta(A_1)=0$ the result is proven. ■

Appendix B

Proof of Theorem 4.3

Since A is assumed to be stable Theorem 3.3, part (7) implies that the gramians for (A, B, C) are $P = P^* \geqslant 0$, $Q = Q^* \geqslant 0$. Hence there exists a congruence transformation T_1 such that

$$T_1 P T_1^* = \begin{pmatrix} I & 0 \\ 0 & 0 \end{pmatrix}$$

Let

$$T_1^{*-1} Q T_1^{-1} = \begin{pmatrix} Q_{11} & Q_{12} \\ Q_{12}^* & Q_{22} \end{pmatrix}$$

then there exists a unitary U_1 such that

$$U_1^* Q_{11} U_1 = \begin{pmatrix} \Sigma_1^2 & 0 \\ 0 & 0 \end{pmatrix}, \quad \Sigma_1 > 0$$

Writing

$$T_2^{*-1} = \begin{pmatrix} U_1 & 0 \\ 0 & I \end{pmatrix}$$

gives

$$(T_2 T_1)^{*-1} Q (T_2 T_1)^{-1} = \begin{pmatrix} \Sigma_1^2 & 0 & Q_{13} \\ 0 & 0 & Q_{23} \\ Q_{13}^* & Q_{23}^* & Q_{33} \end{pmatrix}$$

But $Q \geqslant 0 \Rightarrow Q_{23} = 0$. Now let

$$T_3^{*-1} = \begin{pmatrix} I & 0 & 0 \\ 0 & I & 0 \\ -Q_{13}^* \Sigma_1^{-2} & 0 & I \end{pmatrix}$$

gives

$$(T_3 T_2 T_1)^{*-1} Q (T_3 T_2 T_1)^{-1} = \begin{pmatrix} \Sigma_1^2 & 0 & 0 \\ 0 & 0 & 0 \\ 0 & 0 & Q_{33} - Q_{13} \Sigma_1^{-2} Q_{13} \end{pmatrix}$$

Now find a unitary U_2 such that

$$U_2^* (Q_{33} - Q_{13}^* \Sigma_1^{-2} Q_{13}) U_2 = \begin{pmatrix} \Sigma_3 & 0 \\ 0 & 0 \end{pmatrix}$$

and defining

$$T_4^{*-1} = \begin{pmatrix} \Sigma_1^{-1/2} & 0 & 0 \\ 0 & I & 0 \\ 0 & 0 & U_2^* \end{pmatrix}$$

and letting $T = T_4 T_3 T_2 T_1$ gives the desired form for the transformed P and Q. ■

Appendix C

Proof of Lemma 8.3

Firstly note that rank inequalities on (8.13) give $k+l \geq \text{rank } M \geq \text{rank } (S_1) = n-r$ and hence $l \geq n-k-r$ as required. (In the derivation matrices will be written down assuming $l > n-k-r$, however in the case $l = n-k-r$ this just implies that certain submatrices are of zero dimension and the derivation is valid without modification.) Next partition M conformally with S_1 and S_2 as

$$M = \begin{pmatrix} \hat{M}_{11} & \hat{M}_{12} \\ \hat{M}_{21} & \hat{M}_{22} \\ \hat{M}_{31} & \hat{M}_{32} \end{pmatrix} \tag{C 1}$$

The (1, 1), (1, 3) and (3, 3) blocks of (8.13) give, respectively

$$\hat{M}_{11}\hat{M}_{11}^* = I + \hat{M}_{12}\hat{M}_{12}^* \tag{C 2}$$

$$\hat{M}_{11}\hat{M}_{31}^* = \hat{M}_{12}\hat{M}_{32}^* \tag{C 3}$$

$$\hat{M}_{31}\hat{M}_{31}^* = \hat{M}_{32}\hat{M}_{32}^* \tag{C 4}$$

(C 2)$\Rightarrow \hat{M}_{11}^{-1}$ exists and substituting for $\hat{M}_{31}$ from (C 3) into (C 4) gives

$$\hat{M}_{32}(I - \hat{M}_{12}^* \hat{M}_{11}^{*-1} \hat{M}_{11}^{-1} \hat{M}_{12})\hat{M}_{32}^* = 0 \tag{C 5}$$

The matrix inversion lemma applied to (C 5), and (C 2) then give that

$$\hat{M}_{32}(I + \hat{M}_{12}^*(\hat{M}_{11}\hat{M}_{11}^* - \hat{M}_{12}\hat{M}_{12}^*)^{-1}\hat{M}_{12})^{-1}\hat{M}_{32}^* = 0$$

$$\Rightarrow \hat{M}_{32}(I + \hat{M}_{12}^* \hat{M}_{12})^{-1}\hat{M}_{32}^* = 0 \tag{C 6}$$

and since $I + \hat{M}_{12}^* \hat{M}_{12} > 0$ (C 6)$\Rightarrow \hat{M}_{32} = 0 \Rightarrow \hat{M}_{31} = 0$ by (C 3). Now the (2, 2) block of (8.13) gives

$$\hat{M}_{22}\hat{M}_{22}^* = I + \hat{M}_{21}\hat{M}_{21}^* \tag{C 7}$$

Hence $\hat{M}_{22} \in \mathbb{C}^{(n-k-r)\times l}$ is of full row rank and there exists a unitary matrix V such that

$$\hat{M}_{22} V = [\hat{M}_{221} \;\; 0] \tag{C 8}$$

with $\hat{M}_{221} \in \mathbb{C}^{(n-k-r)\times(n-k-r)}$ invertible. Let

$$T_1 = \text{diag } (I_k, V) \tag{C 9}$$

then we can write

$$MT_1 = \begin{pmatrix} M_{11} & M_{12} \\ M_{21} & 0 \\ 0 & 0 \end{pmatrix} \tag{C 10}$$

where M_{11} is $k \times (n-r)$ and M_{21} is $(n-k-r) \times (n-r)$, and $T_1^{-1} S_2 T_1^{*-1} = S_2$. Equation (8.13) can now be written as

$$MT_1 \begin{pmatrix} S_3 & 0 \\ 0 & -I_{l-n+k+r} \end{pmatrix} (MT_1)^* = \begin{pmatrix} S_3 & 0 \\ 0 & 0 \end{pmatrix} \tag{C 11}$$

where $S_3 = \text{diag}(I_k, -I_{n-k-r})$. The (1, 1), (2, 1) and (2, 2) blocks of (C 11) now give, respectively

$$M_{11}S_3M_{11}^* - M_{12}M_{12}^* = I_k \tag{C 12}$$

$$M_{21}S_3M_{11}^* = 0 \tag{C 13}$$

$$M_{21}S_3M_{21}^* = -I_{n-k-r} \tag{C 14}$$

Now define T_2 as

$$T_2 \triangleq \begin{pmatrix} S_3M_{11}^* & -S_3M_{21}^* & -S_3M_{11}^* M_{12}X^{-1} \\ -M_{12}^* & 0 & X \end{pmatrix} \tag{C 15}$$

where

$$X \triangleq [I + M_{12}^* M_{12}]^{1/2} \tag{C 16}$$

and define

$$T \triangleq T_1T_2 \tag{C 17}$$

Now

$$[I_{n-r}, 0]MT$$

$$= \begin{pmatrix} M_{11} & M_{12} \\ M_{21} & 0 \end{pmatrix} T_2$$

$$= \begin{pmatrix} M_{11}S_3M_{11}^* - M_{12}M_{12}^* & -M_{11}S_3M_{21}^* & (-M_{11}S_3M_{11}^* M_{12} + M_{12}(I + M_{12}^* M_{12}))X^{-1} \\ M_{21}S_3M_{11}^* & -M_{21}S_3M_{21}^* & -M_{21}S_3M_{11}^* M_{12}X^{-1} \end{pmatrix}$$

$$= [I_{n-r}, 0_{n-r, l-n+k+r}] \tag{C 18}$$

where (C 18) follows using (C 12), (C 13) and (C 14). Hence (8.17) has been verified. It remains to verify (8.16). Define the matrix $R \in \mathbb{C}^{(k+l)\times(k+l)}$ as

$$R \triangleq \begin{pmatrix} M_{11} & M_{12} \\ M_{21} & 0 \\ 0 & I \end{pmatrix} \tag{C 19}$$

then from (C 18)

$$RT_2S_2T_2^* = \begin{pmatrix} I_{n-r} & 0 \\ [-M_{12}^*, 0] & X \end{pmatrix} S_2T_2^*$$

$$= RS_2 \tag{C 20}$$

after substituting for T_2 and S_2 and multiplying out. Equation (8.16) then follows if it is verified that R is non-singular. Assume there exists $x^* = (x_1^*, x_2^*, x_3^*)$ such that $x^*R = 0$, i.e.

$$x_1^* M_{11} + x_2^* M_{21} = 0 \tag{C 21}$$

$$x_1^* M_{12} + x_3^* = 0 \tag{C 22}$$

$[x_1^*\ (\text{C }12)+x_2^*\ (\text{C }13)]$ gives

$$(x_1^*\ M_{11}+x_2^*\ M_{21})S_3M_{11}^*-x_1^*\ M_{12}M_{12}^*=x_1^* \qquad (\text{C }23)$$

(C 21) and (C 23)$\Rightarrow x_1^*(I+M_{12}M_{12}^*)=0\Rightarrow x_1=0$ since $I+M_{12}M_{12}^*>0$. Hence from (C 22) $x_3^*=0$. Also [(C 13) x_1 + (C 14) x_2] gives

$$M_{21}S_3(M_{11}^*\ x_1+M_{21}^*\ x_2)=-x_2$$

$\Rightarrow x_2=0$, using (C 21). Hence $x=0$ and thus R is invertible. ■

Appendix D

Proof of Theorem 5.2

Since (A, C) is assumed to be completely observable there exists $K\in\mathbb{C}^{n\times p}$ such that $\lambda_i(A-KC)$ are in the open left half-plane. Define

$$X(s)\triangleq I-C(sI-A_k)^{-1}K$$

$$A_k\triangleq A-KC$$

then

$$X(s)G(s)=D+C(sI-A_k)^{-1}[-KD+(sI-A_k)(sI-A)^{-1}B-KC(sI-A)^{-1}B]$$
$$=D+C(sI-A_k)^{-1}(B-KD)$$

Hence (5.15)$\Rightarrow\forall s=j\omega$

$$\begin{aligned}0&\leqslant X(s)[I-G(s)G^*(-\bar{s})]X^*(-\bar{s})\\&=[I-C(sI-A_k)^{-1}K][I-K^*(-sI-A_k^*)^{-1}C^*]\\&\quad-[D+C(sI-A_k)^{-1}(B-KD)][D^*+(B^*-D^*K^*)\\&\quad\times(-sI-A_k^*)^{-1}C^*]\end{aligned} \qquad (\text{D }1)$$

Since Re $(\lambda_i(A_k))<0$, P_k can be defined as the unique solution to

$$A_kP_k+P_kA_k^*+(B-KD)(B-KD)^*-KK^*=0 \qquad (\text{D }2)$$

and defining

$$Z(s)\triangleq\tfrac{1}{2}(I-DD^*)+C(sI-A_k)^{-1}(-P_kC^*-K-BD^*+KDD^*) \qquad (\text{D }3)$$

then (D 1), (D 2) and (D 3) imply that $Z(s)+Z^*(-\bar{s})\geqslant 0\ \forall\text{Re}\,(s)>0$. Hence the positive real lemma can be applied (Ch. 5 in Anderson and Vongpanitlerd (1973), ex.5.2.2 since it is not known that (D 3) is a controllable realization of $Z(s)$). That is there exists $P_z=P_z^*\geqslant 0$, $L\in\mathbb{C}^{n\times p}$, $D_{12}\in\mathbb{C}^{p\times p}$ such that

$$A_kP_z+P_zA_k^*+LL^*=0 \qquad (\text{D }4)$$

$$P_zC^*=-P_kC^*-K-BD^*+KDD^*-LD_{12}^* \qquad (\text{D }5)$$

$$D_{12}D_{12}^*=I-DD^* \qquad (\text{D }6)$$

Now let $P\triangleq P_k+P_z$ then (D 5) and (D 6) implies that

$$\begin{aligned}PC^*&=-K-BD^*+K(I-D_{12}D_{12}^*)-LD_{12}^*\\&=-[B,\ B_2]\begin{pmatrix}D^*\\D_{12}^*\end{pmatrix}\end{aligned} \qquad (\text{D }7)$$

where $B_2 \triangleq L + KD_{12}$. Also (D 2)+(D 4) gives

$$A_kP + PA_k^* + (B-KD)(B-KD)^* - KK^* + LL^* = 0$$

$$\Rightarrow AP + PA^* - KCP - PC^*K^* + BB^* - KDB^* - BD^*K^* + KDD^*K^* - KK^* + LL^* = 0$$

and substituting for PC^* gives

$$AP + PA^* + BB^* + B_2B_2^* = 0 \qquad \text{(D 8)}$$

Now Theorem 3.3, part (4) implies that $\delta(P)=0$ since $(A, \tilde{B})$ is completely controllable ($\Leftarrow (A, B)$ completely controllable). Now let D_{21}, D_{22} be such that $\tilde{D}$ defined in (5.16) is unitary which is possible by (D 6) and by unitarily completing the basis for the rows of $\tilde{D}$. Define

$$\tilde{C} \triangleq -\tilde{D}\tilde{B}^*P^{-1}$$

which is of the form (5.18) due to (D 7). Then it is easily verified that $H(s)$ defined by (5.19) is all-pass (expand $H(s)H^*(-\bar{s})$ using (D 8) to obtain that $\tilde{B}\tilde{B}^* = (sI - A)P + P(-sI - A^*)$). ■

Acknowledgments

The author would like to thank the following: Mr. M. G. Hall for many helpful discussions in the early stages of this research; Mrs. F. H. Nex for programming the algorithms; Professor B. D. O. Anderson for several helpful suggestions; the Science and Engineering Research Council, U.K., and the Australian National University, Institute of Advanced Studies, for support of this work.

References

Adamjan, V. M., Arov, D. Z., and Krein, M. G., 1971, *Math. USSR Sbornik*, **15**, 31; 1978, *Am. Math. Soc. Transl.*, **111**, 133 (translation of *Iz. Akad. Nauk Armjan. SSR Ser. Mat.* 6 (1971), p. 87).

Anderson, B. D. O., and Latham, G. A., 1984, report under preparation.

Anderson, B. D. O., and Vongpanitlerd, S., 1973, *Network Analysis and Synthesis: A Modern Systems Theory Approach* (Englewood Cliffs: Prentice-Hall).

Barnett, S., 1971, *Matrices in Control Theory with Applications to Linear Programming* (Van Nostrand Reinhold).

Bartels, R. H., and Stewart, G. W., 1972, *Communs. Ass. comput. Mach.*, **415,** 820.

Bultheel, A., and Dewilde, P., *I.E.E.E. Trans. Circuits Syst.* (submitted for publication).

Bettayeb, M., Silverman, L. M., and Safonov, M. G., 1980, *Proc. I.E.E.E. Conf. on Decision and Control*, Albuquerque, New Mexico, p. 195.

Coppel, W. A., 1978, *Dichotomies in Stability Theory*, Lecture Notes in Mathematics No. 629 (Berlin: Springer).

Curtain, R. F., and Glover, K., 1984, report under preparation.

Duren, P. L., 1970, *Theory of* H^p *Spaces* (New York: Academic Press).

Eckart, C., and Young, G., 1936, *Psychometrika*, **1,** 211.

Gantmacher, F. R., 1966, *The Theory of Matrices*, Vols. I and II (Chelsea Publ. Co.).

Glover, K., 1984, report under preparation.

Glover, K., and Limebeer, D. J. N., 1983, *I.E.E.E. Trans. autom. Control* (submitted for publication; brief version appeared American Control Conf., June 1983, San Francisco, California).

Hammarling, S. J., 1982, *I.M.A. J. of Numer. Anal.*, **2,** 303.

HILDEBRAND, K., 1956, *Introduction to Numerical Analysis* (McGraw-Hill).
HOFFMAN, K., 1962, *Banach Spaces of Analytic Functions* (Englewood Cliffs: Prentice-Hall).
HUNG, Y. S. and MACFARLANE, 1982, *Multivariable Feedback: A Quasi-Classical Approach* (Springer Verlag).
JONCKHEERE, E. A., SAFONOV, M. G., and SILVERMAN, L. M., 1981, *Proc.* 20th *I.E.E.E. Conf. on Decision and Control*, p. 118.
KAILATH, T., 1980, *Linear Systems* (Englewood Cliffs: Prentice-Hall).
KUNG, S.-Y., and LIN, D. W., 1981 a, *I.E.E.E. Trans. autom. Control*, **26,** 832; 1981 b, *Ibid.*, **26,** 942.
LAUB, A. J., 1980, *Proc.* 1980 *Joint Automatic Control Conf.*, San Francisco, California.
LIN, D. W., and KUNG, S.-Y., 1982, *Circuits Syst. sig. Proc.*, **1,** 407.
LUENBERGER, D. G., 1969, *Optimization by Vector Space Methods* (New York: Wiley).
MIRSKY, L., 1960, *Q. Jl Math.* (2), **11,** 50.
MOORE, B. C., 1981, *I.E.E.E. Trans. autom. Control*, **26,** 17.
NEHARI, Z., 1957, *Ann. Math.*, **65,** 153.
OSTROWSKI, A. M., and SCHNEIDER, H., 1962, *J. math. Analysis Applic.*, **4,** 72.
PERNEBO, L., and SILVERMAN, L. M., 1982, *I.E.E.E. Trans. autom. Control*, **27,** 382.
DE PRONY, R., 1795, *J. Éc. polytech.* (*Paris*), **1,** 24.
RUDIN, W., 1966, *Real and Complex Analysis* (McGraw-Hill).
SAFONOV, M. G., and VERMA, M. S., 1983, *Proc. American Control Conf.*, San Francisco, California.
SARASON, D., 1967, *Am. Math. Soc. Transl.*, **127,** 180.
SILVERMAN, L. M., and BETTAYEB, M., 1980, *Proc. Joint Automatic Control Conf.*, San Francisco.
STEWART, G. W., 1973, *Introduction to Matrix Computation* (New York: Academic Press).
WIMMER, H. K., 1974, *Linear Algebra Applic.*, **8,** 337.
YOULA, D. C., 1961, *I.R.E. Trans. Inf. Theory*, **7,** 172.

A Connection Between State-Space and Doubly Coprime Fractional Representations

C. N. NETT, C. A. JACOBSON, AND M. J. BALAS

Abstract—**Explicit formulas for a doubly coprime fractional representation of the transfer function of a lumped LTI system are given in terms of a stabilizable and detectable state-space realization of the transfer function. These formulas allow existing computational algorithms to be utilized for the purpose of computing doubly coprime factorizations. Several additional implications of this result are briefly discussed, as are some immediate extensions and future research directions.**

I. Introduction

The fractional representation approach to linear feedback system analysis and synthesis has received a great deal of attention over the past few years (see [5] and the references therein). However, little has been done to develop numerically feasible algorithms to compute the coprime factorizations and the corresponding Bezout identity elements required for the application of the powerful theoretical results provided by this theory. The results presented here allow existing computational algorithms to be utilized for the purpose mentioned above and also provide a link to state-space methods.

II. Preliminaries

Let $\Omega \subset \mathbf{C}$ denote any closed superset of the closed right-half complex plane which is symmetric with respect to the real axis. A square real matrix is said to be Ω-Hurwitz if all its eigenvalues are contained in $\mathbf{C}$-Ω. A pair of real matrices $(A, B) \in R^{n \times n} \times R^{n \times m}$ is said to be Ω-stabilizable if there exists a $K \in R^{m \times n}$ such that $A - BK$ is Ω-Hurwitz. This is equivalent to all uncontrollable modes of (A, B) lying in $\mathbf{C}$-Ω. We define Ω-detectable in a similar manner.

Let H denote the ring of proper rational functions which are analytic in Ω. A rational matrix $X \in R(s)^{n \times m}$ is said to be Ω-stable if it has entries in H.

III. Main Result

The main result is contained in the following theorem. In the interest of brevity, we shall not discuss all its implications or extensions in detail. However, the remarks following the theorem should give an indication of the significance of the result and outline future research directions.

Theorem: Suppose $G(s) = C(sI - A)^{-1}B \in R_{p,o}(s)^{p \times m}$ *where* $A \in R^{n \times n}$, $B \in R^{n \times m}$, $C \in R^{p \times n}$, (C, A) *is* Ω*-detectable and* (A, B) *is* Ω*-stabilizable. Select* $K \in R^{m \times n}$, $F \in R^{n \times p}$ *such that* $A - BK$ *and* $A - FC$ *are* Ω*-Hurwitz. Define*

$$N(s) = C(sI - A + BK)^{-1}B \qquad D(s) = I - K(sI - A + BK)^{-1}B \quad (1)$$

$$U(s) = K(sI - A + FC)^{-1}F \qquad V(s) = I + K(sI - A + FC)^{-1}B \quad (2)$$

$$\bar{D}(s) = I - C(sI - A + FC)^{-1}F \quad \bar{N}(s) = C(sI - A + FC)^{-1}B \quad (3)$$

$$\bar{V}(s) = I + C(sI - A + BK)^{-1}F \quad \bar{U}(s) = K(sI - A + BK)^{-1}F. \quad (4)$$

Manuscript received November 3, 1983; revised December 20, 1983. This work was supported in part by the National Science Foundation under Grant ECS-80-16173, the National Aeronautic Space Administration under Grant NAG-1-171, and the Air Force Office of Scientific Research under Grant AFOSR-83-0124. The work of the first author was supported by the Office of Naval Research Graduate Fellowship Program. The work of the second author was supported by the Rockwell International Fellowship Program.

The authors are with the Department of Electrical, Computer, and Systems Engineering, Rensselaer Polytechnic Institute, Troy, NY 12181.

Then

i) all eight matrices described by (1)–(4) are Ω*-stable*

ii) $D(s)$ *and* $\bar{D}(s)$ *are nonsingular*

iii) $G(s) = N(s)D^{-1}(s) = \bar{D}^{-1}(s)\bar{N}(s)$

iv)

$$\begin{bmatrix} V(s) & U(s) \\ -\bar{N}(s) & \bar{D}(s) \end{bmatrix} \begin{bmatrix} D(s) & -\bar{U}(s) \\ N(s) & \bar{V}(s) \end{bmatrix} = \begin{bmatrix} I & 0 \\ 0 & I \end{bmatrix}.$$

Remark 1: In the terminology of [8], the above theorem provides a doubly coprime factorization of $G(s)$ with respect to the ring H. A weaker version of the above theorem, Lemma 6.1 of Vidyasagar [7], provides the coprime factorizations $N(s)D^{-1}(s), \bar{D}^{-1}(s)\bar{N}(s)$ of $G(s)$. However, in contrast to the above theorem, *no* expressions for the corresponding Bezout identity elements are given. This is a crucial extension in that the application of fractional representation theory depends critically on a knowledge of the Bezout identity elements.

Remark 2: The above theorem is readily extended to the case $G(s) = C(sI - A)^{-1}B + E$, where $E \in H^{p \times m}$, by 1) adding $ED(s)$ to the expression for $N(s)$, 2) adding $\bar{D}(s)E$ to the expression for $\bar{N}(s)$, 3) subtracting $U(s)E$ from the expression for $V(s)$, and 4) subtracting $E\bar{U}(s)$ from the expression for $\bar{V}(s)$.

Remark 3: Lumped LTI discrete-time systems may also be handled by replacing s by z and requiring that $\Omega \subset \mathbf{C}$ be any closed superset of the complement of the open unit disk which is symmetric with respect to the real axis.

Remark 4: The above theorem allows one to compute doubly coprime factorizations with *existing* computational algorithms. Indeed, to compute both right and left coprime factorizations and the corresponding Bezout identity elements of the transfer function of a lumped LTI dynamical system described by state equations one need only 1) solve two pole placement problems and 2) perform the algebraic operations given in (1)–(4). Of course, if $\Omega = \{s \in \mathbf{C}: \text{Re}\, s \geq -\sigma \text{ for some } \sigma \geq 0\}$, then 1) may be accomplished by solving Riccati equations, and in any case there exist a myriad of computational algorithms with which one can achieve the desired goal; 2) can be accomplished using the algorithm of Varga and Sima [6].

Remark 5: The theorem should also be useful in allowing one to establish connections between results obtained via state-space methods and those recently obtained via fractional representation theory (e.g., [5] and the references therein). For example, observe that knowledge of K and F in the above theorem is equivalent to knowledge of a single deterministic Kalman filter for the state-space realization corresponding to $G(s)$. Coupling this observation with the results in [8] parameterizing all stabilizing controllers for $G(s)$ given coprime factorizations and Bezout identity elements corresponding to $G(s)$, one sees that the knowledge of a single deterministic Kalman filter for a given linear plant provides one with an explicit parameterization of all linear controllers stabilizing the given plant.

Remark 6: Recently in [5] results were obtained linking the transfer function approach to distributed systems via fractional representation theory [1], [2] to the functional analytic approach to distributed systems via semi-groups and evolution equations [3]. Although here we restrict our attention to lumped systems, inspection of the proof of the above theorem shows it to be essentially algebraic in nature. Hence, modulo certain technical difficulties, the extension of the theorem to the class of linear distributed systems (described by evolution equations) discussed in [5, sects. 3 and 4] is straightforward. In order to do full justice to the technicalities in this case, a detailed treatment will be given elsewhere.

Proof: i) follows immediately from the definition of an Ω-Hurwitz matrix and the fact that for $X \in R^{n \times n}$ one has $(sI - X)^{-1} = [\text{adj}(sI - X)][\det(sI - X)]^{-1}$ where $\det(sI - X)$ is a monic polynomial in s of degree n and each entry of $\text{adj}(sI - X)$ is a polynomial in s of degree strictly less than n. From this it follows that $\lim_{s \to \infty} \det(D(s)) = \lim_{s \to \infty} \det(\bar{D}(s)) = 1$ since the determinant is a continuous function of

Reprinted from *IEEE Trans. Automat. Contr.*, vol. AC-29, no. 9, pp. 831–832, Sept. 1984.

its arguments. Hence, ii) is proven. To show iii), write

$$\begin{aligned}N(s)D^{-1}(s) &= C(sI-A+BK)^{-1}B\left[I-K(sI-A+BK)^{-1}B\right]^{-1}\\&= C(sI-A+BK)^{-1}\left[I-BK(sI-A+BK)^{-1}\right]^{-1}B\\&= C(sI-A+BK)^{-1}\left[(sI-A)(sI-A+BK)^{-1}\right]^{-1}B\\&= C(sI-A)^{-1}B = G(s).\end{aligned}$$

In a similar fashion, one shows $G(s)=\overline{D}^{-1}(s)\overline{N}(s)$. To show iv), we must verify three equalities only, the fourth being contained in iii). Now

$$\begin{aligned}I-V(s)D(s)&\\&= I-\left(I+K(sI-A+FC)^{-1}B\right)\left(I-K(sI-A+BK)^{-1}B\right)\\&= \left[I+K(sI-A+FC)^{-1}B\right]K(sI-A+BK)^{-1}B\\&\quad -K(sI-A+FC)^{-1}B\\&= K\left[I+(sI-A+FC)^{-1}BK\right](sI-A+BK)^{-1}B\\&\quad -K(sI-A+FC)^{-1}B\\&= K(sI-A+FC)^{-1}(sI-A+FC+BK)(sI-A+BK)^{-1}B\\&\quad -K(sI-A+FC)^{-1}B\\&= K(sI-A+FC)^{-1}\left[I+FC(sI-A+BK)^{-1}\right]B\\&\quad -K(sI-A+FC)^{-1}B\\&= K(sI-A+FC)^{-1}\left[I+FC(sI-A+BK)^{-1}-I\right]B\\&= \left[K(sI-A+FC)^{-1}F\right]\left[C(sI-A+BK)^{-1}B\right]\\&= U(s)N(s).\end{aligned}$$

Also,

$$\begin{aligned}V(s)\overline{U}(s)&\\&= \left[I+K(sI-A+FC)^{-1}B\right]K(sI-A+BK)^{-1}F\\&= K\left[I+(sI-A+FC)^{-1}BK\right](sI-A+BK)^{-1}F\\&= K(sI-A+FC)^{-1}(sI-A+FC+BK)(sI-A+BK)^{-1}F\\&= K(sI-A+FC)^{-1}\left[I+FC(sI-A+BK)^{-1}\right]F\\&= \left[K(sI-A+FC)^{-1}F\right]\left[I+C(sI-A+BK)^{-1}F\right]\\&= U(s)\overline{V}(s).\end{aligned}$$

Finally, one shows $I-\overline{D}(s)\overline{V}(s)=\overline{N}(s)\overline{U}(s)$ by manipulations similar to those above.

References

[1] F. M. Callier and C. A. Desoer, "An algebra of transfer functions for distributed linear time-invariant systems," *IEEE Trans. Circuits Syst.*, vol. CAS-25, pp. 651–663, Sept. 1978.
[2] —, "Stabilization, tracking and disturbance rejection in multivariable convolution systems," *Annales Soc. Scientif. de Bruxelles*, T. 94, I, pp. 7–51, 1980.
[3] R. F. Curtain and A. J. Pritchard, *Infinite Dimensional Linear Systems Theory*. New York: Springer-Verlag, 1978.
[4] B. A. Francis and M. Vidyasagar, "Algebraic and topological aspects of the regulator problem for lumped linear systems," *Automatica*, vol. 19, pp. 87–90, Jan. 1983.
[5] C. N. Nett, C. A. Jacobson, and M. J. Balas, "Fractional representation theory: Robustness results with applications to finite dimensional control of a class of linear distributed systems," in *Proc. 1983 IEEE Conf. Decision and Contr.*, to be published.
[6] A. Varga and V. Sima, "Numerically stable algorithm for transfer function matrix evaluation," *Int. J. Contr.*, vol. 33, pp. 1123–1133, 1981.
[7] M. Vidyasagar, "The graph metric for unstable plants and robustness estimates for feedback stability," *IEEE Trans. Automat. Contr.*, May 1984.
[8] M. Vidyasagar, H. Schneider, and B. A. Francis, "Algebraic and topological aspects of feedback stabilization," *IEEE Trans. Automat. Contr.*, vol. AC-27, pp. 880–894, Aug. 1982.

Bibliography

A. Books

[1] J. Ackermann, *Sampled-Data Control Systems, Analysis and Synthesis, Robust System Design*. Berlin and New York: Springer-Verlag, Inc., 1985.

[2] J. Ackermann, Ed., *Uncertainty and Control*. Berlin and New York: Springer-Verlag, Inc., 1985.

[3] B. D. O. Anderson and J. B. Moore, *Linear Optimal Control*. Englewood Cliffs, NJ: Prentice Hall, Inc., 1971.

[4] M. J. Ashworth, *Feedback Design of Systems with Significant Uncertainty*. Chichester: Research Studies Press (Division of John Wiley & Sons Ltd.), 1982.

[5] H. W. Bode, *Network Analysis and Feedback Amplifier Design*. Princeton, NJ: Van Nostrand, 1945.

[6] F. M. Callier and C. A. Desoer, *Multivariable Feedback Systems*. New York, NY: Springer-Verlag, 1982.

[7] J. B. Cruz, Ed., *System Sensitivity Analysis*. Stroudsburg, PA: Dowden, Hutchinson, and Ross, 1973.

[8] C. A. Desoer and M. Vidyasagar, *Feedback Systems: Input-Output Properties*. New York, NY: Academic Press, 1975.

[9] P. L. Duren, *Theory of H^p Spaces*. New York, NY: Academic Press, 1970.

[10] P. M. Frank, *Introduction to System Sensitivity Theory*. New York, NY: Academic Press, 1978.

[11] M. M. Gupta, Ed., *Adaptive Methods for Control System Design*. New York, NY: IEEE PRESS, 1986.

[12] I. Horowitz, *Synthesis of Feedback Systems*. New York, NY: Academic Press, 1963.

[13] Y. S. Hung and A. G. J. MacFarlane, *Multivariable Feedback: A Quasi-Classical Approach*. Berlin and New York: Springer-Verlag, Inc., 1982.

[14] P. J. Koosis, *Introduction to H^p Spaces*, London Math. Soc. Lecture Note Series No. 40, Cambridge, England: Cambridge Univ. Press, 1980.

[15] Y. D. Landau, *Adaptive Control: The Model Reference Approach*. New York, NY: Marcel Dekker, 1979.

[16] A. G. J. MacFarlane, Ed., *Frequency-Response Methods in Control Systems*. New York, NY: IEEE PRESS, 1979.

[17] A. G. J. MacFarlane, Ed., *Complex Variable Methods for Linear Multivariable Feedback Systems*. London, UK: Taylor and Francis Ltd., 1980.

[18] J. O'Reilly, *Observers for Linear Systems*. London, UK: Academic Press, 1983.

[19] S. C. Power, *Hankel Operators on Hilbert Space*. Boston, MA: Pitman Publishing Inc., 1982.

[20] H. H. Rosenbrock, *State Space and Multivariable Theory*. London, UK: Nelson, 1970.

[21] H. H. Rosenbrock, *Computer-Aided Control System Design*. New York, NY: Academic Press, 1974.

[22] M. G. Safonov, *Stability and Robustness of Multivariable Feedback Systems*. Cambridge, MA: MIT Press, 1980.

[23] F. C. Schweppe, *Uncertain Dynamic Systems*. Englewood Cliffs, NJ: Prentice-Hall, 1973.

[24] B. Sz.-Nagy and C. Foias, *Harmonic Analysis of Operators on Hilbert Space*. New York, NY: American Elsevier, 1970.

[25] S. G. Tzafestas, Ed., *Multivariable Control*. Boston, MA: D. Reidel, 1984.

[26] M. Vidyasagar, *Control System Synthesis: A Factorization Approach*. Cambridge, MA: MIT Press, 1985.

[27] J. Walsh, *Interpolation and Approximation by Rational Functions in the Complex Domain*, vol. 20, Fourth Edition. AMS Colloquium Publications, 1965.

[28] W. M. Wonham, *Linear Multivariable Control: A Geometric Approach*, 2nd Ed. New York, NY: Springer-Verlag, 1979.

B. Robust Analysis

[29] Y. Arkun, B. Manousiouthakis, and P. Putz, "Robust Nyquist array methodology: a new theoretical framework for analysis and design of robust multivariable feedback systems," *Int. J. Contr.*, vol. 40, pp. 603–629, Apr. 1984.

[30] M. Athans, "A tutorial on the LQG/LTR method," in *Proc. American Contr. Conf.*, Seattle, WA, June 1986, pp. 1289–1296.

[31] M. F. Barrett, "Conservatism with robustness tests for linear feedback control systems," in *Proc. 19th IEEE Conf. Decision Contr.*, Albuquerque, NM, Dec. 1980, pp. 885–890.

[32] J. D. Birdwell and A. J. Laub, "Balanced singular values for LQG/LTR design," in *Proc. American Contr. Conf.*, Seattle, WA, June 1986, pp. 409–414.

[33] H. S. Black, "Stabilized feedback amplifiers," U.S. Patent No. 2, 102, 671, 1927.

[34] S. Boyd, "A note on parametric and nonparametric uncertainties in control systems," *Proc. American Contr. Conf.*, Seattle, WA, June 1986, pp. 1847–1849.

[35] M. J. Chen and C. A. Desoer, "Necessary and sufficient conditions for robust stability of linear distributed feedback systems," *Int. J. Contr.*, vol. 35, pp. 255–267, Feb. 1982.

[36] V. H. L. Cheng and C. A. Desoer, "Limitations on the closed-loop transfer function due to right-half plant transmission zeros of the plant," *IEEE Trans. Automat. Contr.*, vol. AC-25, pp. 1218–1220, Dec. 1980.

[37] F. H. Clarke and P. D. Loewen, "Sensitivity analysis in optimal control," in *Proc. 23rd IEEE Conf. Decision Contr.*, Las Vegas, NV, Dec. 1984, pp. 1649–1654.

[38] J. D. Cobb, "Robust stabilization relative to the unweighted H^∞ norm is generically unattainable in the presence of singular plant perturbations," *IEEE Trans. Automat. Contr.*, vol. AC-32, pp. 51–53, Jan. 1987.

[39] J. B. Cruz, Jr., J. S. Freudenberg, and D. P. Looze, "A relationship between sensitivity and stability of multivariable feedback systems," *IEEE Trans. Automat. Contr.*, vol. AC-26, pp. 66–74, Feb. 1981.

[40] J. B. Cruz, Jr. and W. R. Perkins, "A new approach to the sensitivity problem in multivariable feedback systems," *IEEE Trans. Automat. Contr.*, vol. AC-9, pp. 216–223, July 1964.

[41] R. W. Daniel and B. Kouvaritakis, "A new robust stability criterion for linear and non-linear multivariable feedback systems," *Int. J. Contr.*, vol. 41, pp. 1349–1379, June 1985.

[42] C. A. Desoer, F. M. Callier, and W. S. Chan, "Robustness of stability conditions for linear time-invariant systems," *IEEE Trans. Automat. Contr.*, vol. AC-22, pp. 586–590, Aug. 1977.

[43] C. A. Desoer and Y. T. Wang, "Linear time-invariant robust servomechanism problem: A self-contained exposition," in *Control and Dynamical Systems*, vol. 6, C. T. Leondes, Ed. New York, NY: Academic Press, 1980.

[44] T. E. Djaferis, "Achieving robust stability by sensing additional system output," in *Proc. 25th IEEE Conf. Decision Contr.*, Athens, Greece, Dec. 1986, pp. 1451–1456.

[45] R. Doraiswami, "Measures of overshoot, speed of response and robustness," in *Proc. 1985 American Contr. Conf.*, Boston, MA, June 1985, pp. 757–761.

[46] J. C. Doyle, "Guaranteed margins for LQG regulators," *IEEE Trans. Automat. Contr.*, vol. AC-23, pp. 756–757, Aug. 1978.

[47] J. C. Doyle, "Robustness of multiloop linear feedback systems," in *Proc. 1978 IEEE Conf. Decision Contr.*, San Diego, CA, Jan. 1979, pp. 12–18.

[48] J. C. Doyle, "Analysis of feedback systems with structured uncertainties," *IEE Proc.-Part D*, vol. 129, pp. 242–250, Nov. 1982. Also in this volume, page 113.

[49] J. C. Doyle, "Structured uncertainty in control system design," in *Proc. 24th IEEE Conf. Decision Contr.*, Ft. Lauderdale, FL, Dec. 1985, pp. 260–265.

[50] J. C. Doyle and G. Stein, "Robustness with observers," *IEEE Trans. Automat. Contr.*, vol. AC-24, pp. 607–611, Aug. 1979.

[51] J. C. Doyle and G. Stein, "Multivariable feedback design: Concepts for a classical/modern synthesis," *IEEE Trans. Automat. Contr.*, vol. AC-26, pp. 4–16, Feb. 1981. Also this volume, page 50.

[52] J. C. Doyle, J. E. Wall, and G. Stein, "Performance and robustness analysis for structured uncertainty," in *Proc. 21st IEEE Conf. Decision Contr.*, Orlando, FL, Dec. 1982, pp. 629–636.

[53] A. K. El-Sakkary, "The gap metric: Robustness of stabilization of feedback systems," *IEEE Trans. Automat. Contr.*, vol. AC-30, pp. 240–247, Mar. 1985.

[54] A. K. El-Sakkary and N. E. Gough, "Effect of delay on robustness of feedback systems: The gap metric," in *Proc. 25th IEEE Conf. Decision Contr.*, Athens, Greece, Dec. 1986, pp. 905–909.

[55] Y. K. Foo and I. Postlethwaite, "Extensions of the small-μ test for robust stability," in *Proc. 23rd IEEE Conf. Decision Contr.*, Las Vegas, NV, Dec. 1984, pp. 324–329.

[56] J. S. Freudenberg, D. P. Looze, and J. B. Cruz, Jr., "Robustness analysis using singular value sensitivities," *Int. J. Contr.*, vol. 35, pp. 95–116, Jan. 1982.

[57] J. S. Freudenberg and D. P. Looze, "Relations between properties of multivariable feedback systems at different loop-breaking points: Part I," in *Proc. 24th IEEE Conf. Decision Contr.*, Ft. Lauderdale, FL, Dec. 1985, pp. 250–256.

[58] J. S. Freudenberg and D. P. Looze, "A generalization of Bode gain-phase relations to multiple-loop systems," in *Proc. 1985 American Contr. Conf.*, Boston, MA, June 1985, pp. 119–124.

[59] J. S. Freudenberg and D. P. Looze, "A sensitivity tradeoff for plants with time delay," in *Proc. 1985 American Contr. Conf.*, Boston, MA, June 1985, pp. 765–766.

[60] J. S. Freudenberg and D. P. Looze, "Right half-plane poles and zeros and design trade-offs in feedback systems," *IEEE Trans. Automat. Contr.*, vol. AC-30, pp. 555–565, June 1985. Also this volume, page 137.

[61] J. S. Freudenberg and D. P. Looze, "The relation between open-loop and closed-loop properties of multivariable feedback systems," *IEEE Trans. Automat. Contr.*, vol. AC-31, pp. 333–340, Apr. 1986.

[62] J. S. Freudenberg and D. P. Looze, "Relations between properties of multivariable feedback systems at different loop-breaking points: Part II," in *Proc. American Contr. Conf.*, Seattle, WA, June 1986, pp. 771–777.

[63] T. Fujii and N. Mizushima, "Robustness of the optimality property of an optimal regulator: multi-input case," *Int. J. Contr.*, vol. 39, pp. 441–453, Mar. 1984.

[64] R. E. de Gaston and M. G. Safonov, "Calculation of multiloop stability margin," in *Proc. American Contr. Conf.*, Seattle, WA, June 1986, pp. 761–770.

[65] C. V. Hollot, "Matrix uncertainty structures for robust stabilizability," in *Proc. 1985 American Contr. Conf.*, Boston, MA, June 1985, pp. 450–455.

[66] I. Horowitz and Y.-K. Liao, "Limitations of non-minimum phase feedback systems," *Int. J. Contr.*, vol. 40, pp. 1003–1013, May 1984.

[67] O. Huseyin, M. E. Sezer, and D. D. Siljak, "Robust decentralised control using output feedback," *IEE Proc.-Part D*, vol. 129, pp. 310–314, Nov. 1982.

[68] R. E. Kalman, "When is a linear control system optimal?" *Trans. ASME, Ser. D (J. Basic Engr.)*, vol. 86, pp. 51–60, Mar. 1964.

[69] J. C. Kantor and R. P. Andres, "Characterization of 'Allowable perturbations' for robust stability," *IEEE Trans. Automat. Contr.*, vol. AC-28, pp. 107–109, Jan. 1983.

[70] H. Kazerooni, P. K. Houpt, and T. B. Sheridan, "An approach to loop transfer recovery using eigenstructure assignment," in *Proc. 1985 American Contr. Conf.*, Boston, MA, June 1985, pp. 796–803.

[71] H. Kazerooni and P. K. Houpt, "On loop transfer recovery," *Int. J. Contr.*, vol. 43, pp. 981–996, Mar. 1986.

[72] H. K. Khalil, "On the robustness of output feedback control methods to modeling errors," *IEEE Trans. Automat. Contr.*, vol. AC-26, pp. 524–526, Apr. 1981.

[73] B. Kouvaritakis and I. Postlethwaite, "Principal gains and phases: insensitive robustness measures for assessing the closed-loop stability property," *IEE Proc.-Part D*, vol. 129, pp. 233–241, Nov. 1982.

[74] H. Kwakernaak, "A condition for robust stabilizability," *Syst. Contr. Lett.*, vol. 2, pp. 1–4, July 1982.

[75] N. A. Lehtomaki, D. A. Castanon, B. C. Levy, G. Stein, N. R. Sandell, Jr. and M. Athans, "Robustness and modeling error characterization," *IEEE Trans. Automat. Contr.*, vol. AC-29, pp. 212–220, Mar. 1984.

[76] N. A. Lehtomaki, N. R. Sandell, and M. Athans, "Robustness results in linear-quadratic Gaussian based multivariable designs," *IEEE Trans. Automat. Contr.*, vol. AC-26, pp. 75–92, Feb. 1981. Also this volume, page 89.

[77] G. Leitmann, E. P. Ryan, A. Steinberg, "Feedback control of uncertain systems: robustness with respect to neglected actuator and sensor dynamics," *Int. J. Contr.*, vol. 43, pp. 1243–1256, Apr. 1986.

[78] A. G. J. MacFarlane, "Return-difference and return-ratio matrices and their use in analysis and design of multivariable feedback systems," *Proc. IEE*, vol. 117, pp. 2037–2049, Oct. 1970.

[79] A. G. J. MacFarlane and I. Postlethwaite, "The generalized Nyquist stability criterion and multivariable root loci," *Int. J. Contr.*, vol. 25, pp. 81–127, Jan. 1977.

[80] A. N. Madiwale and D. E. Williams, "Some extensions of loop transfer recovery," in *Proc. 1985 American Contr. Conf.*, Boston, MA, June 1985, pp. 790–795.

[81] H. Maeda and M. Vidyasagar, "Infinite gain margin problem in multivariable feedback systems," *Automatica*, vol. 22, pp. 131–133, Jan. 1986.

[82] A. K. Mahalanabis and S. Purkayastha, "Frequency-domain criteria for stability of a class of nonlinear stochastic systems," *IEEE Trans. Automat. Contr.*, vol. AC-18, pp. 266–270, June 1973.

[83] J. B. Moore and L. Xia, "Loop recovery and robust state estimate feedback design," in *Proc. American Contr. Conf.*, Seattle, WA, June 1986, pp. 1807–1812.

[84] B. G. Morton and R. M. McAfoos, "A Mu-test for robustness analysis of a real-parameter variation problem," in *Proc. 1985 American Contr. Conf.*, Boston, MA, June 1985, pp. 135–138.

[85] K. Nordstrom, "On the trade-off between noise sensitivity and robustness for LQG-regulators," in *Proc. 1985 American Contr. Conf.*, Boston, MA, June 1985, pp. 767–771.

[86] K. Nordstrom, "A study of stability robustness tests," in *Proc. 25th IEEE Conf. Decision Contr.*, Athens, Greece, Dec. 1986, pp. 772–773.

[87] H. Nyquist, "Regeneration theory," *Bell Syst. Tech. J.*, vol. 11, pp. 126–147, 1932.

[88] H. P. Opitz, "Robustness properties of discrete-variable structure controllers," *Int. J. Contr.*, vol. 43, pp. 1003–1004, Mar. 1986.

[89] D. H. Owens, "The numerical range: A tool for robust stability studies?" *Syst. Control Lett.*, vol. 5, pp. 153–158, Dec. 1984.

[90] D. H. Owens, "Robust stability theory using both singular value and numerical range data," *IEEE Trans. Automat. Contr.*, vol. AC-31, pp. 348–349, Apr. 1986.

[91] D. H. Owens and A. Chotai, "On eigenvalues, eigenvectors and singular values in robust stability analysis," *Int. J. Contr.*, vol. 40, pp. 285–296, Feb. 1984.

[92] D. H. Owens and A. Chotai, "Robust sampled regulators for stable systems from plant step data," *Automatica*, vol. 20, pp. 465–469, July 1984.

[93] S. D. O'Young and B. A. Francis, "Sensitivity trade-offs for multivariable plants," *IEEE Trans. Automat. Contr.*, vol. AC-30, pp. 625–632, July 1985.

[94] R. V. Patel, M. Toda, and B. Sridhar, "Robustness of linear quadratic state feedback designs in the presence of system uncertainty," *IEEE Trans. Automat. Contr.*, vol. AC-22, pp. 945–949, Dec. 1977.

[95] R. V. Patel and M. Toda, "Quantitative measures of robustness for multivariable systems," in *Proc. Joint Automat. Contr. Conf.*, San Francisco, CA, June 1980, paper TD8-A.

[96] I. Postlethwaite, J. M. Edmunds, and A. G. J. MacFarlane, "Principal gains and phases in the analysis of linear multivariable feedback systems," *IEEE Trans. Automat. Contr.*, vol. AC-26, pp. 32–46, Feb. 1981.

[97] I. Postlethwaite and Y. K. Foo, "A robustness test for distributed feedback systems," *Int. J. Contr.*, vol. 41, pp. 973–980, Apr. 1985.

[98] I. Postlethwaite and Y. K. Foo, "Robustness with simultaneous pole and zero movement across the $j\omega$-axis," *Automatica*, vol. 21, pp. 433–443, July 1985.

[99] L. Qui and E. J. Davison, "New perturbation bounds for the robust stability of linear state space models," in *Proc. 25th IEEE Conf. Decision Contr.*, Athens, Greece, Dec. 1986, pp. 751–755.

[100] H. H. Rosenbrock, "The stability of multivariable systems," *IEEE Trans. Automat. Contr.*, vol. AC-17, pp. 105–107, Feb. 1972.

[101] M. Saeki, "A method of robust stability analysis with highly structured uncertainties," *IEEE Trans. Automat. Contr.*, vol. AC-31, pp. 935–940, Oct. 1986.

[102] M. G. Safonov, "Stability margins of diagonally perturbed multivariable feedback systems," *IEE Proc.-Part D*, vol. 129, pp. 251–256, Nov. 1982. Also this volume, page 107.

[103] M. G. Safonov, "Propagation of conic model uncertainty in hierarchical systems," *IEEE Trans. Automat. Contr.*, vol. AC-28, pp. 701–709, June 1983.

[104] M. G. Safonov, "Exact calculation of the multivariable structured-singular-value stability margin," in *Proc. 23rd IEEE Conf. Deci-

sion Contr., Las Vegas, NV, Dec. 1984, pp. 1224–1225.

[105] M. G. Safonov and M. Athans, "Gains and phase margin for multiloop LQG regulators," *IEEE Trans. Automat. Contr.*, vol. AC-22, pp. 173–179, Apr. 1977.

[106] M. G. Safonov and M. Athans, "A multiloop generalization of the circle criterion for stability margin analysis," *IEEE Trans. Automat. Contr.*, vol. AC-26, pp. 415–421, Apr. 1981. Also this volume, page 82. First presented at the Asilomar Conference on Circuits, Systems, and Computers, Nov. 1978.

[107] M. G. Safonov, A. J. Laub, and G. L. Hartmann, "Feedback properties of multivariable systems: the role and use of the return difference matrix," *IEEE Trans. Automat. Contr.*, vol. AC-26, pp. 47–65, Feb. 1981. Also this volume, page 63.

[108] M. E. Sezer and D. D. Siljak, "Robustness of suboptimal control: Gain and phase margin," *IEEE Trans. Automat. Contr.*, vol. AC-26, pp. 907–910, Aug. 1981.

[109] A. Sideris and R. R. E. de Gaston, "Multivariable stability margin calculation with uncertain correlated parameters," in *Proc. 25th IEEE Conf. Decision Contr.*, Athens, Greece, Dec. 1986.

[110] E. Soroka and V. Shaked, "On the robustness of LQ regulators," *IEEE Trans. Automat. Contr.*, vol. AC-29, pp. 664–665, July 1984.

[111] G. Stein and M. Athans, "The LQG/LTR procedure for multivariable feedback control design," *IEEE Trans. Automat. Contr.*, vol. AC-32, pp. 105–114, Feb. 1987.

[112] M. Vidyasagar, "The graph metric for unstable plants and robustness estimates for feedback stability," *IEEE Trans. Automat. Cont.*, vol. AC-29, pp. 403–418, May 1984. Also this volume, page 122.

[113] E. Yaz and N. Yildizbayrak, "Robustness of feedback-stabilized systems in the presence of non-linear and random perturbations," *Int. J. Contr.*, vol. 41, pp. 345–353, Feb. 1985.

[114] R. K. Yedavalli, "Perturbation bounds for robust stability in linear state space models," *Int. J. Contr.*, vol. 42, pp. 1507–1517, June 1985.

[115] R. K. Yedavalli, "Improved measures of stability robustness for linear state space models," *IEEE Trans. Automat. Contr.*, vol. AC-30, pp. 577–579, June 1985.

[116] R. K. Yedavalli, "Stability analysis of interval matrices: another sufficient condition," *Int. J. Contr.*, vol. 43, pp. 767–772, Mar. 1986.

[117] R. K. Yedavalli, S. S. Banda, and D. B. Ridgely, "Time-domain stability robustness measures for linear regulators," *AIAA J. Guidance, Contr. Dyn.*, vol. 8, pp. 520–524, July–Aug. 1985.

[118] R. K. Yedavalli and Z. Liang, "Reduced conservatism in stability robustness bounds by state transformations," *IEEE Trans. Automat. Contr.*, vol. AC-31, pp. 863–866, Sept. 1986.

[119] H.-H. Yen, S. S. Banda, and D. B. Ridgely, "Stability robustness measures utilizing structural information," *Int. J. Contr.*, vol. 41, pp. 265–387, Feb. 1985.

[120] G. Zames, "Functional analysis applied to nonlinear feedback systems," *IEEE Trans. Circuit Theory,* vol. CT-10, pp. 392–404, Sept. 1963. Also this volume, page 17.

[121] G. Zames, "On the input-output stability of nonlinear time-varying feedback systems, part I & II," *IEEE Trans. Automat. Contr.*, vol. AC-11, pp. 228–238 and 465–477, Apr. and July 1966.

[122] G. Zames, "Feedback and optimal sensitivity: Model reference transformations, multiplicative seminorms, and approximate inverses," *IEEE Trans. Automat. Contr.*, vol. AC-26, pp. 301–320, Apr. 1981. Also this volume, page 30. First presented at the 17th Allerton Conference, Oct. 1979.

[123] G. Zames and D. Bensousson, "Multivariable feedback, sensitivity and decentralized control," *IEEE Trans. Automat. Contr.*, vol. AC-28, pp. 1030–1035, Nov. 1983.

[124] G. Zames and A. El-Sakkary, "Unstable systems and feedback: the gap metric," in *Proc. Allerton Conf.*, 1980, pp. 380–385.

[125] G. Zames and B. A. Francis, "A new approach to classical frequency methods: Feedback and minimax sensitivity," in *Proc. 20th IEEE Conf. Decision Contr.*, Dec. 1981, pp. 867–874.

[126] D.-Z. Zheng, "Optimization of linear-quadratic regulator systems in the presence of parameter perturbations," *IEEE Trans. Automat. Contr.*, vol. AC-31, pp. 667–670, July 1986.

C. Robust Stabilization

[127] B.-S. Chen and C.-C. Chiang, "Robust stabilizer synthesis for feedback systems containing time-varying nonlinear perturbations," *IEEE Trans. Automat. Contr.*, vol. AC-31, pp. 768–771, Aug. 1986.

[128] M. J. Chen and C. A. Desoer, "Algebraic theory of robust stability of interconnected systems," *IEEE Trans. Automat. Contr.*, vol. AC-29, pp. 511–519, June 1984.

[129] R. F. Curtain and K. Glover, "Robust stabilization of infinite dimensional systems by finite dimensional controllers," *Syst. Control Lett.*, vol. 7, pp. 41–47, Feb. 1986.

[130] T. E. Djaferis, "Robust pole assignment and robust stabilization for systems with parameters," in *Proc. 1985 American Contr. Conf.*, Boston, MA, June 1985, pp. 784–789.

[131] P. Dorato and Yunzhi Li, "A modification of the classical Nevanlinna-Pick interpolation algorithm with applications to robust stabilization," *IEEE Trans. Automat. Contr.*, vol. 31, pp. 645–648, July 1986.

[132] B. K. Ghosh, "A robust reliable stabilization scheme for single input, single output systems using transcendental methods," *Syst. Control Lett.*, vol. 5, pp. 111–115, Nov. 1984.

[133] B. K. Ghosh and C. I. Byrnes, "Simultaneous stabilization and simultaneous pole-placement by non-switching dynamic compensation," *IEEE Trans. Automat. Contr.*, AC-28, pp. 735–741, June 1983.

[134] K. Glover, "Robust stabilization of linear multivariable systems: relations to approximation," *Int. J. Contr.*, vol. 43, pp. 741–766, Mar. 1986.

[135] J. Kautsky, N. K. Nichols and P. V. Dooren, "Robust pole assignment in linear state feedback," *Int. J. Contr.*, vol. 41, pp. 1129–1155, May 1985.

[136] P. P. Khargonekar, T. T. Georgiou, and A. Pascoal, "On the robust stabilizability of linear time-invariant plants with unstructured uncertainty," *IEEE Trans. Automat. Contr.*, vol. AC-32, pp. 201–207, Mar. 1987.

[137] P. P. Khargonekar and K. Poolla, "Robust stabilization of distributed systems," *Automatica,* vol. 22, pp. 77–84, Jan. 1986.

[138] P. P. Khargonekar and A. Tannenbaum, "Robust stabilization of systems with uncertain parameters," in *Proc. 23rd IEEE Conf. Decision Contr.*, Las Vegas, NV, Dec. 1984, pp. 1214–1218.

[139] P. P. Khargonekar and A. Tannenbaum, "Non-Euclidian metrics and the robust stabilization of systems with parameter uncertainty," *IEEE Trans. Automat. Contr.*, vol. AC-30, pp. 1005–1013, Oct. 1985. Also this volume, page 168.

[140] H. Kimura, "Robust stabilization for a class of transfer functions," *IEEE Trans. Automat. Contr.*, vol. AC-29, pp. 788–793, Sept. 1984. Also this volume, page 162.

[141] H. Maeda and M. Vidyasagar, "Some results on simultaneous stabilization," *Syst. Control Lett.*, vol. 5, pp. 205–208, Dec. 1984.

[142] K. D. Minto and M. Vidyasagar, "A state-space approach to simultaneous stabilization," *Contr. Theory Adv. Tech.*, vol. 2, pp. 39–64, Mar. 1986.

[143] C. N. Nett, C. A. Jacobson, and M. J. Balas, "Fractional representation theory: Robustness with applications to finite-dimensional control of a class of linear distributed systems," in *Proc. 22nd IEEE Conf. Decision Contr.*, San Antonio, TX, Dec. 1983, pp. 268–280.

[144] R. Saeks and J. Murray, "Fractional representation, algebraic geometry and the simultaneous stabilization problem," *IEEE Trans. Automat. Contr.*, vol. AC-27, pp. 895–903, Aug. 1982.

[145] M. G. Safonov and B. S. Chen, "Multivariable stability margin optimization with decoupling and output regulation," *IEE Proc.-Part D,* vol. 129, pp. 276–282, Nov. 1982.

[146] N. R. Sandell, "Robust stability of systems with application to singular perturbations," *Automatica,* vol. 15, pp. 467–470, July 1979.

[147] A. Tannenbaum, "Feedback stabilization of linear dynamical plants with uncertainty in the gain factor," *Int. J. Contr.*, vol. 32, pp. 1–16, Jan. 1980.

[148] A. Tannenbaum, "Modified Nevanlinna-Pick interpolation of linear plants with uncertainty in the gain factor," *Int. J. Contr.*, vol. 36, pp. 331–336, Feb. 1982.

[149] M. S. Verma, J. W. Helton, and E. A. Jonckeere, "Robust stabilization of a family of plants with varying number of right half plane poles," in *Proc. American Contr. Conf.*, Seattle, WA, June 1986, pp. 1827–1832.

[150] M. Vidyasagar, "Robust stabilization singularly perturbed systems," *Syst. Control Lett.*, vol. 5, pp. 413–418, May 1985.

[151] M. Vidyasagar and H. Kimura, "Robust controllers for uncertain linear multivariable systems," *Automatica,* vol. 22, pp. 85–94, Jan. 1986. Also this volume, page 177.

[152] M. Vidyasagar and N. Viswanadham, "Algebraic design techniques for reliable stabilization," *IEEE Trans. Automat. Contr.*, vol. AC-

27, pp. 1085–1095, Oct. 1982. Also this volume, page 151.

D. H^2 and H^∞ Sensitivity Optimization

[153] J. S. Bird and B. A. Francis, "On the robust disturbance attenuation problem," in *Proc. 25th IEEE Conf. Decision Contr.*, Athens, Greece, Dec. 1986, pp. 1804–1809.

[154] J. J. Bongiorno, Jr., "Minimum sensitivity design of linear multivariable feedback control systems by matrix spectral factorization," *IEEE Trans. Automat. Contr.*, vol. AC-14, pp. 665–673, Dec. 1969. Also this volume, page 189.

[155] B. C. Chang and J. B. Pearson, Jr., "Optimal disturbance reduction in linear multivariable systems," *IEEE Trans. Automat. Contr.*, vol. AC-29, pp. 880–887, Oct. 1984. Also this volume, page 225.

[156] B. S. Chen, "Controller synthesis of optimal sensitivity: multivariable case," *IEE Proc.-Part D*, vol. 131, pp. 47–51, Jan. 1984.

[157] B.-S. Chen and C.-C. Kung, "The robustness optimization of a multivariable feedback system in Hankel norm space," *Int. J. Contr.*, vol. 39, pp. 1211–1228, June 1984.

[158] M. A. Dahleh and J. B. Pearson, "L^1-optimal compensators for continuous-time systems," in *Proc. 25th IEEE Conf. Decision Contr.*, Athens, Greece, Dec. 1986, pp. 1826–1828.

[159] J. C. Doyle, "Synthesis of robust controllers and filters," in *Proc. 22nd IEEE Conf. Decision Contr.*, San Antonio, TX, Dec. 1983, pp. 109–114. Also this volume, page 214.

[160] A. Feintuch and B. A. Francis, "Uniformally optimal control of linear feedback systems," *Automatica*, vol. 21, pp. 563–574, Sept. 1985.

[161] A. Feintuch, P. Khargonekar, and A. Tannenbaum, "On the sensitivity minimization problem for linear time-varying periodic systems," *SIAM J. Contr.*, vol. 24, pp. 1076–1085, Sept. 1986.

[162] C. Foias, A. Tannenbaum, and G. Zames, "Weighted sensitivity minimization for delay systems," *IEEE Trans. Automat. Contr.*, vol. AC-31, pp. 763–766, Aug. 1986. First presented at the 24th IEEE Conference on Decision and Control, Dec. 1985.

[163] C. Foias, A. Tannenbaum, and G. Zames, "On decoupling the H^∞-optimal sensitivity problem for products of plants," in *Proc. American Contr. Conf.*, Seattle, WA, June 1986, pp. 1716–1721.

[164] Y. K. Foo and I. Postlethwaite, "An H^∞-minimax approach to the design of robust control systems," *Syst. Control Lett.*, vol. 5, pp. 81–88, Nov. 1984.

[165] B. A. Francis, "Optimal disturbance attenuation with control weighting," in *Proc. Twentieth Workshop on Systems and Optimization, LNCIS vol. 66*, A. Bagchi and H. Th. Jongen, Editors. Berlin: Springer-Verlag, 1984.

[166] B. A. Francis, J. W. Helton, and G. Zames, "H^∞-optimal feedback controllers for linear multivariable systems," *IEEE Trans. Automat. Contr.*, vol. AC-29, pp. 888–900, Oct. 1984. Also this volume, page 233.

[167] B. A. Francis and G. Zames, "On H^∞-optimal sensitivity theory for SISO feedback systems," *IEEE Trans. Automat. Contr.*, vol. AC-29, pp. 9–16, Jan. 1984.

[168] J. S. Freudenberg and D. P. Looze, "Some remarks on H^∞-sensitivity minimization," in *Proc. 22nd IEEE Conf. Decision Contr.*, San Antonio, TX, Dec. 1983, pp. 634–635.

[169] C. Ganesh and J. B. Pearson, "Design of optimal control systems with state feedback," in *Proc. 1986 American Contr. Conf.*, Seattle, WA, June 1986, pp. 1969–1973.

[170] T. T. Georgiou and P. P. Khargonekar, "A constructive algorithm for sensitivity minimization of periodic systems," *SIAM J. Contr.*, vol. 25, pp. 334–340, Mar. 1987.

[171] M. J. Grimble, "Optimal H_∞ robustness and the relationship to LQG design problems," *Int. J. Contr.*, vol. 43, pp. 351–372, Feb. 1986.

[172] M. J. Grimble and M. A. Johnson, "Robustness and optimality: A dual performance index," in *Preprints, IFAC Congress, Volume IX*, Budapest, Hungary, 1984, pp. 102–107.

[173] S. Hara and H. Katori, "On constrained H^∞ optimization for SISO systems," *IEEE Trans. Automat. Contr.*, vol. AC-31, pp. 856–858, Sept. 1986.

[174] P. P. Khargonekar and K. Zhou, "Weighted sensitivity minimization for delay systems," in *Proc. 25th IEEE Conf. Decision Contr.*, Athens, Greece, Dec. 1986, pp. 1950–1952.

[175] C.-C. Kung and B.-S. Chen, "LQG controller synthesis with plant and noise spectral uncertainties," *Int. J. Contr.*, vol. 43, pp. 705–715, Feb. 1986.

[176] H. Kwakernaak, "Minimax frequency domain performance and robustness optimization of linear feedback systems," *IEEE Trans. Automat. Contr.*, vol. AC-30, pp. 994–1004, Oct. 1985. Also this volume, page 261.

[177] D. J. N. Limebeer and G. Halikias, "A controller degree bound for H^∞ optimal control problems of the second kind," in *Proc. 25th IEEE Conf. Decision Contr.*, Athens, Greece, Dec. 1986, pp. 1802–1803.

[178] D. J. N. Limebeer and Y. S. Hung, "An analysis of the pole-zero cancellations in a class of H^∞ optimal control problems," in *Proc. American Contr. Conf.*, Seattle, WA, June 1986, pp. 1699–1740.

[179] S. D. O'Young and B. A. Francis, "Optimal performance and robust stabilization," *Automatica*, vol. 22, pp. 171–183, Mar. 1986. First presented at the 24th IEEE Conference on Decision and Control, Dec. 1985.

[180] L. Pandolfi and A. W. Olbrot, "On the minimization of sensitivity to additive disturbances for linear-distributed parameter MIMO feedback systems," *Int. J. Contr.*, vol. 43, pp. 389–399, Feb. 1986.

[181] M. G. Safonov, "L^∞-optimal sensitivity vs. stability margin," in *Proc. 22nd IEEE Conf. Decision Contr.*, San Antonio, TX, Dec. 1983, pp. 115–118.

[182] M. G. Safonov, "Optimal H^∞ synthesis of robust controllers for systems with structured uncertainty," in *Proc. 25th IEEE Conf. Decision Contr.*, Athens, Greece, Dec. 1986, pp. 1822–1825.

[183] M. G. Safonov, "Future directions in L^∞ robust control theory," in *Proc. 25th IEEE Conf. Decision Contr.*, Athens, Greece, Dec. 1986, pp. 375–377.

[184] M. G. Safonov and H. S. Verma, "L^∞ sensitivity optimization and Hankel approximation," *IEEE Trans. Automat. Contr.*, vol. AC-30, pp. 279–280, Mar. 1985. First presented at the 1983 American Control Conference.

[185] A. Sideris and M. G. Safonov, "A design algorithm for the robust synthesis of SISO feedback control systems using conformal maps and H^∞ theory," in *Proc. American Contr. Conf.*, Seattle, WA, June 1986, pp. 1710–1715.

[186] Y. C. Soh and R. J. Evans, "Robust multivariable regulator design—Special cases," in *Proc. 24th IEEE Conf. Decision Contr.*, Ft. Lauderdale, FL, Dec. 1985, pp. 1328–1332.

[187] M. Verma and E. Jonkheere, "L^∞ compensation with mixed sensitivity as a broadband matching problem," *Syst. Control Lett.*, vol. 4, pp. 125–130, May 1984. Also this volume, page 220.

[188] M. Vidyasagar, "Optimal rejection of persistent bounded disturbances," *IEEE Trans. Automat. Contr.*, vol. AC-31, pp. 527–534, June 1986.

[189] J. H. Xu and M. Mansour, "Design of H^∞-optimal robust controllers: stability, asymptotic regulation and disturbance rejection," in *Proc. 25th IEEE Conf. Decision Contr.*, Athens, Greece, Dec. 1986, pp. 1–6.

[190] D. C. Youla and J. J. Bongiorno, Jr., "A feedback theory of two-degree-of-freedom optimal Wiener-Hopf design," *IEEE Trans. Automat. Contr.*, vol. AC-30, pp. 652–665, July 1985. Also this volume, page 248.

[191] G. Zames, and B. A. Francis, "Feedback, minimax sensitivity and optimal robustness," *IEEE Trans. Automat. Contr.*, vol. AC-28, pp. 585–601, May 1983. Also this volume, page 198.

E. Other Approaches to the Control of Uncertain Systems

[192] J. Ackermann, "Parameter space design of robust control systems," *IEEE Trans. Automat. Contr.*, vol. AC-25, pp. 1058–1071, Dec. 1980.

[193] J. Ackermann, "Robustness against sensor failures," *Automatica*, vol. 20, pp. 211–215, Mar. 1984.

[194] B. R. Barmish, M. Corless, and G. Leitmann, "A new class of stabilizing controllers for uncertain dynamical systems," *SIAM J. Control and Optimization*, vol. 21, pp. 246–255, Mar. 1983. Also this volume, page 329.

[195] B. R. Barnish and S. Leitmann, "On ultimate boundedness control of uncertain systems in the absence of matching conditions," *IEEE Trans. Automat. Contr.*, vol. AC-27, pp. 153–158, Feb. 1982.

[196] D. S. Bernstein and S. W. Greeley, "Robust controller synthesis using the maximum entropy design equations," *IEEE Trans. Automat. Contr.*, vol. AC-31, pp. 362–364, Apr. 1986.

[197] D. S. Bernstein and S. W. Greeley, "Robust output-feedback stabilization: deterministic and stochastic perspectives," in *Proc. American Contr. Conf*, Seattle, WA, June 1986, pp. 1818–1826.

[198] D. P. Bertsekas and I. B. Rhodes, "Sufficiently informative functions and the minimax feedback control of uncertain dynamic systems,"

IEEE Trans. Automat. Contr., vol. AC-18, pp. 117–124, Apr. 1973. Also this volume, page 292.

[199] S. P. Bhattacharyya, A. C. Del Nero Gomes, and J. W. Howze, "The structure of robust disturbance rejection control," *IEEE Trans. Automat. Contr.*, vol. AC-28, pp. 874–881, Sept. 1983.

[200] N. K. Bose, E. I. Jury, and E. Zeheb, "On robust Hurwitz and Schur polynomials," in *Proc. 25th IEEE Conf. Decision Contr.*, Athens, Greece, Dec. 1986, pp. 739–744.

[201] D. Carlucci and F. Donati, "Control of norm uncertain systems," *IEEE Trans. Automat. Contr.*, vol. AC-20, pp. 792–795, Dec. 1975.

[202] S. S. Chan and M. B. Zarrop, "A suboptimal dual controller for stochastic systems with unknown parameters," *Int. J. Contr.*, vol. 41, pp. 507–524, Feb. 1985.

[203] S. S. L. Chang and T. K. C. Peng, "Adaptive guaranteed cost control of systems with uncertain parameters," *IEEE Trans. Automat. Contr.*, vol. AC-17, pp. 474–483, Aug. 1972. Also this volume, page 283.

[204] Y. H. Chen, "On the deterministic performance of uncertain dynamical systems," *Int. J. Contr.*, vol. 43, pp. 1557–1579, May 1986.

[205] M. J. Corless and G. Leitmann, "Continuous state feedback guaranteeing uniform ultimate boundedness for uncertain dynamical systems," *IEEE Trans. Automat. Contr.*, vol. AC-26, pp. 1139–1143, Oct. 1981.

[206] R. W. Daniel, "Frequency-response design of robust optimal controllers," *IEE Proc.-Part D*, vol. 129, pp. 257–262, Nov. 1982.

[207] E. J. Davison, "The output control of linear time-invariant multivariable systems with unmeasurable arbitrary disturbances," *IEEE Trans. Automat. Contr.*, vol. AC-17, pp. 621–629, Oct. 1972.

[208] E. J. Davison, "The systematic design of control systems for large multivariable linear time-invariant systems," *Automatica*, vol. 9, pp. 441–452, July 1973.

[209] E. J. Davison, "The robust control of a servomechanism problem for linear time-invariant multivariable systems," *IEEE Trans. Automat. Contr.*, vol. AC-21, pp. 25–34, Feb. 1976. First presented at the Allerton Conference, Oct. 1973.

[210] E. J. Davison, "The robust decentralized control of a general servomechanism problem," *IEEE Trans. Automat. Contr.*, vol. AC-21, pp. 14–24, Feb. 1976.

[211] E. J. Davison, "Multivariable tuning regulators: The feedforward and robust control of a general servomechanism problem," *IEEE Trans. Automat. Contr.*, vol. AC-21, pp. 35–47, Feb. 1976.

[212] E. J. Davison and B. Copeland, "Solutions of the robust servomechanism problem subject to specified gain margin and time lag tolerance constraints," *IEEE Trans. Automat. Contr.*, vol. AC-30, pp. 229–239, Mar. 1985.

[213] E. J. Davison and I. Ferguson, "Design of controllers for the multivariable robust servomechanism problem using parameter optimization methods," *IEEE Trans. Automat. Contr.*, vol. AC-26, pp. 93–110, Feb. 1981. Also this volume, page 299. First presented at the 19th IEEE Conference on Decision and Control, Dec. 1980.

[214] E. J. Davison and A. Goldenberg, "Robust control of a general servomechanism problem: The servo compensator," *Automatica*, vol. 11, pp. 461–471, Sept. 1975.

[215] T. E. Djaferis, "Robust observers and regulation for systems with parameters," in *Proc. 23rd IEEE Conf. Decision Contr.*, Las Vegas, NV, Dec. 1984, pp. 1234–1239.

[216] F. Donati and M. Vallauri, "Guaranteed control of 'almost-linear' plants," *IEEE Trans. Automat. Contr.*, vol. AC-29, pp. 34–41, Jan. 1984.

[217] R. Doraiswami, "Robust control strategy for a linear time-invariant multivariable sampled-data servomechanism problem," *IEE Proc.-Part D*, vol. 129, pp. 283–292, Nov. 1982.

[218] R. Doraiswami and A. Gulliver, "An interactive design procedure yielding robustness and sensitivity," in *Proc. 23rd IEEE Conf. Decision Contr.*, Las Vegas, NV, Dec. 1984, pp. 1228–1233.

[219] P. Dorato and A. Kestenbaum, "Applications of game theory to the sensitivity design of optimal systems," *IEEE Trans. Automat. Contr.*, vol. AC-12, pp. 85–87, Feb. 1967.

[220] P. Dorato and G. Menga, "Observer feedback for uncertain systems," in *Proc. 1974 IEEE Conf. Decision Contr.*, Phoenix, AZ, Nov. 1974, pp. 333–337.

[221] J. C. Doyle, "Quantitative feedback theory (QFT) and robust control," in *Proc. American Contr. Conf.*, Seattle, WA, June 1986, pp. 1691–1698.

[222] M. Eslami and D. L. Russell, "On stability with large parameter variations: stemming from the direct method of Lyapunov," *IEEE Trans. Automat. Contr.*, vol. AC-25, pp. 1231–1234, Dec. 1980.

[223] R. J. Evans and X. Xianya, "Robust regulator design," *Int. J. Contr.*, vol. 41, pp. 461–476, Feb. 1985.

[224] A. R. Galimidi, "A new class of uncertain linear systems stabilizable via output feedback," in *Proc. 25th IEEE Conf. Decision Contr.*, Athens, Greece, Dec. 1986, pp. 756–761.

[225] A. R. Galimidi and B. R. Barmish, "Dynamic compensation in robust stabilization using observers," in *Proc. 23rd IEEE Conf. Decision Contr.*, Las Vegas, NV, Dec. 1984, pp. 1560–1565.

[226] A. R. Galimidi and B. R. Barmish, "The constrained Lyapunov problem and its application to robust output feedback stabilization," *IEEE Trans. Automat. Contr.*, vol. AC-31, pp. 410–419, May 1986. First presented at the 1985 American Control Conference.

[227] B. F. Goldstein, "Minimax control of linear unknown systems using mismatched state observers," *Int. J. Contr.*, vol. 20, pp. 753–767, May 1974.

[228] M. J. Grimble, "Optimal control of linear uncertain multivariable stochastic systems," *IEE Proc.-Part D*, vol. 129, pp. 263–270, Nov. 1982.

[229] S. Gutman, "Uncertain dynamical systems, a Lyapunov minimax approach," *IEEE Trans. Automat. Contr.*, vol. AC-24, pp. 437–443, June 1979.

[230] C. V. Hollot, "Stabilizing uncertain systems via an observer," in *Proc. 1986 American Contr. Conf.*, Seattle, WA, June 1986, pp. 511–514.

[231] C. V. Hollott and A. R. Galimidi, "Stabilizing uncertain systems: recovering full state feedback performance via an observer," *IEEE Trans. Automat. Contr.*, vol. AC-31, pp. 1050–1053, Nov. 1986.

[232] I. Horowitz, "Quantitative feedback theory," *IEE Proc.-Part D.*, vol. 129, pp. 251–256. Nov. 1982. Also this volume, page 317.

[233] I. Horowitz and M. Breiner, "Quantitative synthesis of feedback systems with nonlinear multivariable plants," *Int. J. Syst. Sci.*, vol. 12, pp. 539–563, 1981.

[234] I. Horowitz, L. Neumann, and O. Yaniv, "Quantitative synthesis of uncertain cascaded multiple-input multiple-output feedback systems," *Int. J. Contr.*, vol. 42, pp. 273–303, Feb. 1985.

[235] I. M. Horowitz, S. Oldak, and O. Yaniv, "An important property of nonminimum phase multiple-input-multiple-output feedback systems," *Int. J. Contr.*, vol. 44, pp. 677–688, Sept. 1986.

[236] I. Horowitz and M. Sidi, "Optimum synthesis of nonminimum-phase feedback systems with parameter uncertainty," *Int. J. Contr.*, vol. 27, pp. 361–386, Mar. 1978.

[237] I. Horowitz and S.-H. Wang, "Weak and strong robustness in feedback control systems," in *Proc. 25th IEEE Conf. Decision Contr.*, Athens, Greece, Dec. 1986, pp. 1457–1458.

[238] I. Horowitz and T. S. Wang, "Quantitative synthesis of multiple-loop feedback systems with large uncertainty," *Int. J. Syst. Sci.*, vol. 10, pp. 1235–1268, 1979.

[239] I. Horowitz and O. Yaniv, "Quantitative cascaded multiple input-multiple output synthesis by an improved method," *Int. J. Contr.*, vol. 42, pp. 305–331, Feb. 1985.

[240] A. F. Hotz and R. E. Skelton, "Controller design for robust stability and performance," in *Proc. 1986 American Contr. Conf.*, Seattle, WA, June 1986, pp. 1813–1817.

[241] S. Jayasurija, M. J. Rabins, and R. D. Barnard, "Guaranteed tracking, behavior in the sense of input-output spheres for systems with uncertain parameters," *J. Dyn. Systems, Meas., Contr.*, vol. 106, pp. 273–279, Dec. 1984.

[242] L. H. Keel and S. P. Bhattacharyya, "Low order robust stabilizer design using Hurwitz conditions," in *Proc. 24th IEEE Conf. Decision Contr.*, Ft. Lauderdale, FL, Dec. 1985, pp. 693–696.

[243] L. H. Keel, J. W. Howze, and S. P. Bhattacharyya, "Robust compensation via pole assignment," in *Proc. 1985 American Contr. Conf.*, Boston, MA, June 1985, pp. 440–442.

[244] P. P. Khargonekar, K. Poolla, and A. Tannenbaum, "Robust control of linear time-invariant plants using periodic compensation," *IEEE Trans. Automat. Contr.*, vol. AC-30, pp. 1088–1096, Nov. 1985.

[245] G. Leitmann, "On the efficacy of nonlinear control in uncertain systems," *J. Dyn. Sys. Meas., Contr.*, vol. 10, pp. 95–102, June 1981.

[246] W.-S. Lu and K. S. P. Kumar, "A staircase model for unknown multivariable systems and design of regulators," *Automatica*, vol. 20, pp. 109–112, Jan. 1984.

[247] J. Lunze, "The design of robust feedback controllers in the time domain," *Int. J. Contr.*, vol. 39, pp. 1143–1260, June 1984.

[248] G. Menga and P. Dorato, "Observer-feedback design for linear

systems with parameter uncertainties," in *Proc. 1974 IEEE Conf. Decision Contr.*, Phoenix, AZ, Nov. 1974, pp. 872–878.

[249] T. Okada, M. Kihara, and H. Furinata, "Robust control system with observer," *Int. J. Contr.*, vol. 41, pp. 1207–1219, May 1985.

[250] D. H. Owens, A. Chotai, "Controller design for unknown multivariable systems using monotone modelling errors," *IEE Proc.-Part D*, vol. 129, pp. 57–69, Jan. 1982.

[251] D. H. Owens and A. Chotai, "Robust controller design for linear dynamic systems using approximate models," *IEE Proc.-Part D*, vol. 130, pp. 45–56, Mar. 1983.

[252] J. B. Pearson and P. W. Staats, Jr., "Robust controllers for linear regulators," *IEEE Trans. Automat. Contr.*, vol. AC-19, pp. 231–234, June 1974.

[253] I. R. Petersen, "A Riccati equation approach to the design of stabilizing controllers and observers for a class of uncertain linear systems," in *Proc. American Contr. Conf.*, Boston, MA, June 1985, pp. 772–777.

[254] I. R. Peterson and C. V. Hollot, "Using observers in the stabilization of uncertain linear systems and in disturbance rejection problems," in *Proc. 25th IEEE Conf. Decision Contr.*, Athens, Greece, Dec. 1986, pp. 1466–1471.

[255] B. Porter, "Design of robust digital set-point tracking controllers for linear multivariable plants with time-delayed inputs," *IEE Proc.-Part D*, vol. 129, pp. 293–297, Nov. 1982.

[256] R. K. Ragade and I. G. Sarma, "A game theoretic approach to optimal control in the presence of uncertainty," *IEEE Trans. Automat. Contr.*, vol. AC-12, pp. 395–402, Aug. 1967.

[257] D. Salmon, "Minimax controller design," *IEEE Trans. Automat. Contr.*, vol. AC-12, pp. 369–376, Aug. 1968.

[258] M. G. Safonov and M. Athans, "Robustness and computational aspects of nonlinear stochastic estimators and regulators," *IEEE Trans. Automat. Contr.*, vol. AC-23, pp. 717–725, Aug. 1978.

[259] W. E. Schmitendorf and T. H. Hopp, "Asymptotic tracking of uncertain systems in the absence of matching conditions," in *Proc. 1985 American Contr. Conf.*, Boston, MA, June 1985, pp. 423–426.

[260] Y. C. Soh and R. J. Evans, "Robust multivariable regulator design-general case," in *Proc. 24th. Conf. Decision Contr.*, Ft. Lauderdale, FL, Dec. 1985, pp. 1323–1327.

[261] D. Summers, "Lyapunov approximation of reachable sets for uncertain linear systems," *Int. J. Contr.*, vol. 41, pp. 1235–1243, May 1985.

[262] A. Swierniak, "State-inequalities approach to control systems with uncertainty," *IEE Proc.-Part D*, vol. 129, pp. 271–275, Nov. 1982.

[263] T. Tsai and T. Wang, "Optimal design of control systems with large plant uncertainty," *Int. J. Contr.*, vol. 43, pp. 1015–1028, Mar. 1986.

[264] A. Vinkler and L. J. Wood, "Multistep guaranteed cost control of linear systems with uncertain parameters," *J. Guidance Contr.*, vol. 2, pp. 449–456, Nov.–Dec. 1979.

[265] A. Vinkler and L. J. Wood, "A comparison of several techniques for designing controllers of uncertain dynamic systems," in *Proc. 1978 IEEE Conf. Decision Contr.*, San Diego, CA, Jan. 1979, pp. 31–38.

[266] K. H. Wei and B. R. Barmish, "On making a polynomial Hurwitz invariant by choice of feedback gains," in *Proc. 24th IEEE Conf. Decision Contr.*, Ft. Lauderdale, FL, Dec. 1985, pp. 679–685.

[267] J. L. Willems and J. C. Willems, "Robust stabilization of uncertain systems," *SIAM J. Contr. Opt.*, vol. 21, pp. 352–374, May 1983.

[268] H. S. Witsenhausen, "A minimax control problem for sampled-data systems," *IEEE Trans. Automat. Contr.*, vol. AC-13, pp. 5–21, Feb. 1968.

[269] W. M. Wonham, "Optimal stationary control of a linear system with state-dependent noise," *SIAM J. Control*, vol. 5, pp. 486–500, Aug. 1967. Also this volume, page 275.

[270] O. Yaniv and I. M. Horowitz, "A quantitative design method for MIMO linear feedback systems having uncertain plants," *Int. J. Contr.*, vol. 43, pp. 401–421, Feb. 1986.

[271] O. Yaniv and I. Horowitz, "Quantitative feedback theory—Reply to Doyle's criticisms," Accepted for publication in *Int. J. Contr.*, 1987.

[272] E. Yaz, "Further results on the stabilizing property of certainty equivalent controllers," *IEEE Trans. Automat. Contr.*, vol. AC-31, pp. 586–587, June 1986.

[273] R. K. Yedavalli, "Dynamic compensator design for robust stability of linear interval systems," in *Proc. 25th IEEE Conf. Decision Contr.*, Athens, Greece, Dec. 1986, pp. 34–36.

[274] T. Yoshikawa, T. Sugie, and H. Hanafusa, "On synthesis of robust servo systems," in *Preprints, IFAC Congress, Volume IX*, Budapest, Hungary, 1984, pp. 196–201.

[275] T. Yoshikawa, T. Sugie, and H. Hanafusa, "Synthesis of robust tracking systems with specified transfer matrices," *Int. J. Contr.*, vol. 43, pp. 1201–1214, Apr. 1986.

[276] E. Zeheb and D. Hertz, "Robust control of the characteristic values of systems with possible parameter variations," *Int. J. Contr.*, vol. 40, pp. 81–96, Jan. 1984.

[277] Y. Zhau and H. Kimura, "Dead-beat control with robustness," *Int. J. Contr.*, vol. 43, pp. 1427–1440, May 1980.

F. Related Theory and Numerical Computation

[278] V. M. Adamjan, O. Z. Arov, and M. G. Krein, "Analytic properties of Schmidt pairs for a Hankel operator and the generalized Schur-Takagi problem," *Mat. Sb.*, vol. 86, pp. 31–75, 1971.

[279] V. M. Adamjan, D. Z. Arov, and M. G. Krein, "Infinite Hankel matrices and related extension problems," *AMS Transl.*, vol. 2, pp. 133–156, 1978.

[280] A. A. Allison and N. J. Young, "Numerical algorithms for the Nevanlinna-Pick problem," *Numer. Math.*, vol. 24, pp. 125–145, 1983. Also this volume, page 412.

[281] B. D. O. Anderson, "An algebraic solution to the spectral factorization problem," *IEEE Trans. Automat. Contr.*, vol. AC-12, pp. 410–414, Aug. 1967.

[282] B. D. O. Anderson, "A note on the Youla-Bongiorno-Lu condition," *Automatica*, vol. 12, pp. 387–388, July 1976.

[283] B. D. O. Anderson, K. L. Hitz, and N. D. Diem, "Recursive algorithm for spectral factorization," *IEEE Trans. Circuit Syst.*, vol. CAS-21, pp. 742–750, 1974.

[284] J. A. Ball and J. W. Helton, "A Beurling-Lax theorem for the lie group $U(m, n)$ which contains most classical interpolation theory," *J. Operator Theory*, vol. 9, no. 1, pp. 107–142, 1983.

[285] J. Bence, J. W. Helton, and D. E. Marshall, "Optimization over H^∞," in *Proc. 25th IEEE Conf. Decision Contr.*, Athens, Greece, Dec. 1986, pp. 356–360.

[286] M. Bettayeb, L. M. Silverman, and M. G. Safonov, "Optimal approximation of continuous-time systems," in *Proc. 19th IEEE Conf. Decision Contr.*, Albuquerque, NM, Dec. 1980, pp. 195–198.

[287] A. Bhaya and C. A. Desoer, "Necessary and sufficient conditions on $Q(= c(1 + pc)^{-1})$ for stabilization of the linear feedback system $S(p, c)$," *Syst. Control Lett.*, vol. 7, pp. 35–38, Feb. 1986.

[288] S. Boyd and C. A. Desoer, "Subharmonic functions and performance bounds on linear time-invariant feedback systems," in *Proc. 23rd IEEE Conf. Decision Contr.*, Las Vegas, NV, Dec. 1984, pp. 311–312.

[289] B.-C. Chang and J. B. Pearson, "Iterative computation of minimal H^∞ norm," in *Proc. 24th IEEE Conf. Decision Contr.*, Ft. Lauderdale, FL, Dec. 1985, pp. 1307–1310.

[290] L. Cheng and J. B. Pearson, "Frequency domain synthesis of multivariable linear regulators," *IEEE Trans. Automat. Contr.*, vol. AC-23, pp. 3–15, Feb. 1978.

[291] R. Y. Chiang and M. G. Safonov, "The LINF computer program for L^∞ controller design," Rep. EECG-0785-1, Dept. Electrical Engineering Systems, University of Southern California, July 1986.

[292] C.-C. Chu and J. C. Doyle, "On inner-outer and spectral factorizations," in *Proc. 23rd IEEE Conf. Decision Contr.*, Las Vegas, NV, Dec. 1984, pp. 1764–1765.

[293] C.-C. Chu and J. C. Doyle, "The general distance problem in H^∞ synthesis," in *Proc. 24th IEEE Conf. Decision Contr.*, Ft. Lauderdale, FL, Dec. 1985, pp. 1311–1316.

[294] M. C. Davis, "Factoring the spectral matrix," *IEEE Trans. Automat. Contr.*, vol. AC-8, pp. 296–305, Oct. 1963.

[295] P. H. Delsarte, Y. Genin, and Y. Kamp, "The Nevanlinna-Pick problem for matrix-valued functions," *SIAM J. Appl. Math.*, vol. 36, pp. 47–61, Feb. 1979.

[296] P. Delsarte, Y. Genin, and Y. Kamp, "On the role of the Nevanlinna-Pick problem in circuit and system theory," *Int. J. Circuit Theory and Appl.*, vol. 9, pp. 177–187, 1981. Also this volume, page 401.

[297] C. A. Desoer and M. J. Chen, "Design of multivariable feedback systems with stable plant," *IEEE Trans. Automat. Contr.*, vol. AC-26, pp. 408–415, Apr. 1981.

[298] C. A. Desoer and C. L. Gustafson, "Algebraic theory of linear multivariable feedback systems," *IEEE Trans. Automat. Contr.*, AC-29, pp. 909–917, Oct. 1984.

[299] C. A. Desoer, R.-W. Liu, J. Murray, and R. Saeks, "Feedback system design: the fractional representation approach," *IEEE Trans. Automat. Contr.*, vol. AC-25, pp. 399–412, June 1980. Also this volume, page 387.

[300] J. C. Doyle and C.-C. Chu, "Matrix interpolation and H^∞ performance bounds," in *Proc. 1985 American Contr. Conf.*, Boston, MA, June 1985, pp. 129–134.
[301] B. A. Francis, "On the Wiener-Hopf approach to optimal feedback design," *Syst. Control Lett.*, vol. 2, pp. 197–201, Dec. 1982.
[302] B. A. Francis and M. Vidyasagar, "Algebraic and topological aspects of the regulator problem for lumped linear systems," *Automatica*, vol. 19, pp. 87–90, Jan. 1983.
[303] J. S. Freudenberg and D. P. Looze, "An analysis of H^∞ optimization methods," *IEEE Trans. Automat. Contr.*, vol. AC-31, pp. 194–200, Mar. 1986.
[304] T. T. Georgiou, "A topological view of the Nevanlinna-Pick problem: solution of dimension n," in *Proc. 23rd IEEE Conf. Decision Contr.*, Las Vegas, NV, Dec. 1984, pp. 315–319.
[305] T. T. Georgiou and P. P. Khargonekar, "Linear fractional transformations and spectral factorization," *IEEE Trans. Automat. Contr.*, vol. AC-31, pp. 345–347, Apr. 1986.
[306] T. T. Georgiou and P. P. Khargonekar, "Spectral factorization using analytic interpolation theory," in *Proc. 25th IEEE Conf. Decision Contr.*, Athens, Greece, Dec. 1986, pp. 7–11.
[307] T. T. Georgiou and P. P. Khargonekar, "Nevanlinna-Pick interpolation and spectral factorization," *SIAM J. Contr. Opt.*, vol. 25, pp. 754–756, May 1987.
[308] K. Glover, "All optimal Hankel-norm approximations of linear multivariable systems and their L^∞-error bounds," *Int. J. Control*, vol. 39, pp. 1115–1193, June 1984. Also this volume, page 423.
[309] K. Glover, J. Lam, and J. R. Partington, "Balanced realizations and Hankel-norm approximations of systems involving delays," in *Proc. 25th IEEE Conf. Decision Contr.*, Athens, Greece, Dec. 1986, pp. 1810–1815.
[310] C. L. Gustafson and C. A. Desoer, "A CAD methodology for linear multivariable feedback systems based on algebraic theory," *Int. J. Contr.*, vol. 41, pp. 653–675, Mar. 1985.
[311] J. W. Helton, "Operator theory and broadband matching," in *Proc. Allerton Conf. Circuits and Systems*, 1976, pp. 91–98.
[312] J. W. Helton, "Broadbanding: gain equalization directly from data," *IEEE Trans. Circuit Syst.*, vol. 28, pp. 1125–1137, Dec. 1981.
[313] J. W. Helton, "Non-Euclidean functional analysis and electronics," *Bull. AMS*, vol. 7, pp. 1–64, July 1982.
[314] J. W. Helton, "Worst case analysis in the frequency domain: The H^∞ approach to control," *IEEE Trans. Automat. Contr.*, vol. AC-30, pp. 1154–1170, Dec. 1985. Also presented at the 22nd IEEE Conference on Decision and Control, Dec. 1983.
[315] J. Jezek and V. Kucera, "Efficient algorithms for spectral factorization," in *Preprint Volume IX, IFAC Congress*, Budapest, Hungary, 1984, pp. 38–43.
[316] E. A. Jonckheere and J. C. Juang, " 'Toeplitz + Hankel' structures in H^∞-design and the avoidance of the 'ϵ-iteration'," in *Proc. American Contr. Conf.*, Seattle, WA, June 1986, pp. 1683–1689.
[317] E. A. Jonckheere and J. C. Juang, "A finite polynomial algorithm for computing the largest eigenvalue of the 'Toeplitz + Hankel' operator of the H^∞ problem," in *Proc. 25th IEEE Conf. Decision Contr.*, Athens, Greece, Dec. 1986, pp. 1816–1821.
[318] P. P. Khargonekar and K. R. Poolla, "Uniformally optimal control of linear time-invariant plants: nonlinear time-varying controllers," *Syst. Control Lett.*, vol. 6, pp. 303–308, Jan. 1986.
[319] P. P. Khargonekar and E. D. Sontag, "On the relation between stable matrix factorizations and regulable realizations of linear systems over rings," *IEEE Trans. Automat. Contr.*, vol. AC-27, pp. 627–638, June 1982.
[320] S. Y. Kung and D. W. Lin, "Optimal Hankel-norm model reduction: multivariable systems," *IEEE Trans. Automat. Contr.*, vol. AC-26, pp. 832–852, Aug. 1981.
[321] K. J. Kurman, "Comments on, 'On the design of single-loop single-input-output feedback systems in the complex-frequency domain,'" *IEEE Trans. Automat. Contr.*, vol. AC-23, pp. 963–964, Oct. 1978. Authors' reply, pp. 964–965.
[322] G. A. Latham and B. D. O. Anderson, "Frequency-weighted optimal Hankel-norm approximation of stable transfer functions," *Syst. Control Lett.*, vol. 5, pp. 229–236, Feb. 1985.
[323] A. J. Laub, "An inequality and some computations related to robust stability of linear dynamic systems," *IEEE Trans. Automat. Contr.*, vol. AC-24, pp. 318–320, Apr. 1979.
[324] Z. Nehari, "On bounded bilinear forms," *Annals of Math.*, vol. 65, no. 1, pp. 153–162, 1957.
[325] C. Nett, C. Jacobson, and M. Balas, "A connection between state space and doubly coprime fractional representations," *IEEE Trans. Automat. Contr.*, vol. AC-29, pp. 831–832, Sept. 1984. Also this volume, page 502.
[326] R. Nevanlinna, "Uber beschrankte funktionen die in gegebenen punkten vorgeschriebene werte annehmen," *Ann. Acad. Sci. Fenn.*, vol. 13, pp. 1–75, 1919.
[327] L. Pernebo, "An algebraic theory for the design of controllers for linear multivariable systems—Part I: Structure matrices and feedforward design," *IEEE Trans. Automat. Contr.*, vol. AC-26, pp. 171–182, Feb. 1981.
[328] L. Pernebo, "An algebraic theory for the design of controllers for linear multivariable systems—Part II: Feedback realizations and feedback design," *IEEE Trans. Automat. Contr.*, vol. AC-26, pp. 183–193, Feb. 1981.
[329] G. Pick, "Uber die beschrankungen analytischer funktionen, welche durch vorgegebenen werte annehmen," *Math. Ann.*, vol. 77, pp. 7–23, 1916.
[330] K. R. Poolla and P. P. Khargonekar, "Stabilizability and stable-proper factorizations for linear time-varying systems," *SIAM J. Contr. Opt.*, vol. 25, pp. 723–726, May 1987.
[331] M. Rosenblum and M. Rovnyak, "An operator theoretic approach to theorems of the Pick-Nevanlinna and Loewner types. I," *Integral Equations Op. Theory*, vol. 3, pp. 408–436, 1980.
[332] R. Saeks, "The factorization problem—A survey," *Proc. IEEE*, vol. 64, pp. 90–95, Jan. 1976.
[333] R. Saeks and J. Murray, "Feedback system design: The tracking and disturbance rejection problems," *IEEE Trans. Automat. Contr.*, vol. AC-26, Feb. 1981.
[334] M. G. Safonov, "Optimal diagonal scaling for infinity norm optimization," in *Proc. 1985 American Contr. Conf.*, Boston, MA, June 1985, pp. 125–128.
[335] D. Sarason, "Generalized interpolation in H^∞," *Transl. AMS*, vol. 127, pp. 180–203, 1967.
[336] L. M. Silverman and M. Bettayeb, "Optimal approximation of linear systems," in *Proc. Joint Automat. Contr. Conf.*, San Francisco, CA, Aug. 1980, Paper FA8-A.
[337] A. Tannenbaum, "H^∞-optimization theory for distributed systems," in *Proc. 25th IEEE Conf. Decision Contr.*, Athens, Greece, Dec. 1986, pp. 899–904.
[338] W. G. Tuel, Jr., "Computer algorithm for spectral factorization of rational matrices," *IBM J. Res. Develop.*, pp. 163–170, Mar. 1968.
[339] M. Vidyasagar, "On the use of right-coprime factorizations in distributed feedback systems containing unstable subsystems," *IEEE Trans. Circuit Syst.*, vol. 25, pp. 916–921, Nov. 1978.
[340] M. Vidyasagar and K. Dean Minto, "Computer-aided control design using the stable factorization approach," presented at 2nd Symposium on Computer-Aided Control System Design, Santa Barbara, CA, March 1985.
[341] M. Vidyasagar, H. Schneider, and B. A. Francis, "Algebraic and topological aspects of feedback stabilization," *IEEE Trans. Automat. Contr.*, vol. AC-27, pp. 880–894, Aug. 1982.
[342] W. M. Wonham and J. B. Pearson, Jr., "Regulation and internal stabilization in linear multivariable systems," *SIAM J. Contr.*, vol. 12, pp. 5–18, Feb. 1974.
[343] D. C. Youla, "On the factorization of rational matrices," *IRE Trans. Inform. Theory*, vol. IT-7, pp. 172–189, July 1961.
[344] D. C. Youla, J. J. Bongiorno, Jr., and C. N. Lu, "Single-loop feedback stabilization of linear multivariable dynamical plants," *Automatica*, vol. 10, pp. 159–173, March 1974. Also this volume, page 341.
[345] D. C. Youla, J. J. Bongiorno, Jr., and H. A. Jabr, "Modern Wiener-Hopf design of optimal controllers—Part I: The single-input-output case," *IEEE Trans. Automat. Contr.*, vol. AC-21, pp. 3–13, Feb. 1976. Also this volume, page 356.
[346] D. C. Youla, H. A. Jabr, and J. J. Bongiorno, Jr., "Modern Wiener-Hopf design of optimal controllers—Part II: The multivariable case," *IEEE Trans. Automat. Contr.*, vol. AC-21, pp. 75–93, Apr. 1976. Also this volume, page 367.
[347] D. C. Youla and M. Saito, "Interpolation with positive real functions," *J. Franklin Inst.*, vol. 284, pp. 77–108, Aug. 1967.
[348] N. J. Young, "The Nevanlinna-Pick problem for matrix-valued functions," *J. Operator Theory*, vol. 15, pp. 239–265, 1986.
[349] G. Zames and A. Tannenbaum, "Optimal H^∞ interpolation: A new approach," in *Proc. 25th IEEE Conf. Decision Contr.*, Athens, Greece, Dec. 1986, pp. 350–355.
[350] S.-Y. Zhang, "On solving the polynomial diophantine equation," in *Proc. 23rd IEEE Conf. Decision Contr.*, Las Vegas, NV, Dec. 1984, pp. 1768–1769.

G. Applications

[351] A. Arbel and N. K. Gupta, "Robust colocated control for large flexible space structures," in *Proc. Joint Automat. Contr. Conf.*, San Francisco, CA, Aug. 1980, Paper FP1-E.

[352] C. E. Baumgartner, H. P. Geering, C. H. Onder, and E. Shafai, "Robust multivariable idle speed control," in *Proc. 1986 American Contr. Conf.*, Seattle, WA, June 1986, pp. 258–265.

[353] J. A. Bossi, R. D. Jones, and U.-L. Ly, "Multivariable regulator design for robustness and performance. A realistic example," in *Proc. American Contr. Conf.*, Seattle, WA, June 1986, pp. 285–288.

[354] Y. H. Chen, "Stabilization of uncertain systems using state detection and application to a Maglev vehicle," in *Proc. 1986 American Contr. Conf.*, Seattle, WA, June 1986, pp. 521–525.

[355] G. W. M. Coppus, S. L. Shah, and R. K. Wood, "Robust multivariable control of a binary distillation column," *IEE Proc.-Part D*, vol. 130, pp. 201–208, Sept. 1983.

[356] V. Cvetkovic and M. Vukobratovic, "One robust, dynamic control algorithm for manipulation systems," *Int. J. Robotics*, vol. 1, pp. 15–28, Winter 1982.

[357] E. J. Davison, "Robust control for industrial systems," in *Proc. 25th IEEE Conf. Decision Contr.*, Athens, Greece, Dec. 1986, pp. 367–375.

[358] E. J. Davison and W. Gesing, "Robust attitude and shape control of third generation spacecraft," presented at AAS/AISS Astrodynamics Specialists Conf., Vail, CO, Aug. 1985.

[359] E. J. Davison, P. A. Taylor, and J. D. Wright, "On the application of tuning regulators to control a commercial heat exchanger," *IEEE Trans. Automat. Contr.*, vol. AC-25, pp. 361–375, June 1980.

[360] J. C. Doyle, K. Lenz, and A. Packard, "Design examples using μ-synthesis: Space shuttle lateral axis FCS during reentry," in *Proc. 25th IEEE Conf. Decision Contr.*, Athens, Greece, Dec. 1986, pp. 2218–2223.

[361] S. N. Franklin and J. Ackermann, "Robust flight control: a design example," *J. Guidance Contr., Dyn.*, vol. 4, pp. 597–605, Nov.–Dec. 1981.

[362] K. Furuta, K. Kosuge, O. Yamano, and K. Nosaki, "Robust control of a robot manipulator with nonlinearity," *Robotica*, vol. 2, pp. 75–81, 1984.

[363] H. P. Geering, "Robust multivariable idle speed control," in *Proc. 1986 American Contr. Conf.*, Seattle, WA, June 1986, pp. 258–265.

[364] M. R. Katebi, M. J. Grimble, and J. Byrne, "Disturbance rejection robustness of LQG controllers," in *Proc. 25th IEEE Conf. Decision Contr.*, Athens, Greece, Dec. 1986, pp. 361–366.

[365] G. J. Kissel and D. R. Hegg, "Stability enhancement for control of flexible space structures," *IEEE Contr. Syst. Mag.*, vol. 6, pp. 19–26, June 1986.

[366] R. L. Kosut, H. Salzwedel, and Abbas Emami-Naeini, "Robust control of flexible spacecraft," *J. Guidance Contr.*, vol. 6, pp. 103–111, Mar.–Apr. 1983.

[367] G.-L. Luo and G. N. Saridis, "Robust compensation of optimal control for manipulators," in *Proc. 21st IEEE Conf. Decision Contr.*, Orlando, FL, Dec. 1982, pp. 351–356.

[368] R. J. Martin, L. Valavani, and M. Athans, "Multivariable control of a submersible using LQG/LTR design methodology," in *Proc. American Contr. Conf.*, Seattle, WA, June 1986, pp. 1313–1324.

[369] L. D. Metz and B. Cyr, "Robust control performance of time-varying human controller models," *Automatica*, vol. 21, pp. 473–478, July 1985.

[370] J. K. Mills and A. Goldenberg, "Robust control of robotic manipulators with task space feedback," in *Proc. 25th IEEE Conf. Decision Contr.*, Athens, Greece, Dec. 1986, pp. 423–428.

[371] W. H. Pfeil, M. Athans, and H. A. Spang, III, "Multi-variable control of the GE T700 engine using the LQG/LTR design methodology," in *Proc. American Contr. Conf.*, Seattle, WA, June 1986, pp. 1297–1312.

[372] I. Postlethwaite, D. W. Gu, S. D. O'Young, and M. S. Tombs, "Industrial control system design using H^∞ optimization," in *Proc. 25th IEEE Conf. Decision Contr.*, Athens, Greece, Dec. 1986, pp. 12–13.

[373] A. A. Rodriguez and M. Athans, "Multivariable control of a twin left helicopter system using the LQG/LTR design methodology," in *Proc. American Contr. Conf.*, Seattle, WA, June 1986, pp. 1325–1332.

[374] M. G. Safonov and R. Y. Chiang, "CACSD using the state space L^∞ theory: A design example," in *Proc. IEEE Conf. on CACSD*, Washington, DC, Sept. 1986.

[375] C. Samson, "Robust non-linear control of robotic manipulators," in *Proc. 22nd IEEE Conf. Decision Contr.*, San Antonio, TX, Dec. 1983, pp. 1211–1216.

[376] S. N. Singh and A. A. R. Coelho, "Nonlinear control of mismatched uncertain linear systems and applications to control of aircraft," *Trans. ASME, J. Dynamical Sys. Measurement Contr.*, vol. 106, pp. 203–210, Sept. 1984.

[377] S. N. Singh and A. A. Schy, "Robust torque control of an elastic robotic arm based on invertibility and feedback stabilization," in *Proc. 24th IEEE Conf. Decision Contr.*, Ft. Lauderdale, FL, Dec. 1985, pp. 1317–1322.

[378] S. N. Singh and A. A. Schy, "Robust trajectory following control of robotic systems," *Trans. ASME J. Dynamical Sys. Measurement Contr.*, vol. 107, pp. 308–315, Dec. 1985.

[379] M. W. Spong and M. Vidyasagar, "Robust nonlinear control of robot manipulators," in *Proc. 24th Conf. Decision Contr.*, Ft. Lauderdale, FL, Dec. 1985, pp. 1767–1772.

[380] M. W. Spong, J. S. Thorp, and J. W. Kleinwaks, "The control of robot manipulators with bounded control. Part II: Robustness and disturbance rejection," in *Proc. 23rd IEEE Conf. Decision Contr.*, Las Vegas, NV, Dec. 1984, pp. 1047–1052.

[381] N. Sundararajan, S. M. Joshi, and E. S. Armstrong, "Robust controller synthesis for a large flexible space antenna," in *Proc. 23rd IEEE Conf. Decision Contr.*, Las Vegas, NV, Dec. 1984, pp. 202–208.

[382] O. Taiwo, "The design of robust control systems for plant with recycle," *Int. J. Contr.*, vol. 43, pp. 671–678, Feb. 1986.

[383] A. Vinkler and L. J. Wood, "Guaranteed cost control of linear systems with uncertain parameters—application to remotely piloted vehicle flight control systems," in *Proc. AIAA Guidance and Contr. Cong.*, Palo Alto, CA, Aug. 1978, pp. 226–233.

[384] G. S. West-Vukovich, E. J. Davison, and P. C. Hughes, "The decentralized control of large flexible space structures," *IEEE Trans. Automat. Contr.*, vol. AC-29, pp. 866–879, Oct. 1984.

[385] R. K. Yedavalli and Z. Liang, "Aircraft control design using improved time-domain stability robustness bounds," *AIAA J. Guidance Contr. Dyn.*, vol. 9, pp. 710–714, Nov.–Dec. 1986.

Author Index

Subject Index

T

U

W

Y

Editor's Biography

Peter Dorato (S'53, M'57, SM'67, F'77) is a native of New York, NY. He received his B.S.E.E. degree from the City College of New York in 1955, M.S.E.E. degree from Columbia University in 1956, and the D.E.E. degree from the Polytechnic University (formerly Polytechnic Institute of Brooklyn) in 1961. He was awarded a National Science Foundation faculty fellowship in 1960 to support his graduate studies at the Polytechnic University.

Dr. Dorato was a faculty member at the City College of New York from 1956 to 1957, and at the Polytechnic University from 1957 to 1972. In 1969–70, he spent a sabbatical year at the University of Colorado, Boulder. In 1972, he joined the University of Colorado at Colorado Springs as a professor and director of the Resource Systems Analysis program. In 1976, he joined the department of electrical engineering and computer science at the University of New Mexico as department chairman, where he served as chairman for eight years. He spent a sabbatical year, 1984–85, at the University of California at Santa Barbara, and is currently a professor of electrical and computer engineering at the University of New Mexico.

Dr. Dorato was a member of the Control Systems Society board of governors for the following periods of time: 1970–73 and 1984–89. He has been a past associate editor of *Automatica* (1969–83) and IEEE TRANSACTIONS ON AUTOMATIC CONTROL (1969–72, 1981–83), and was program chairman of the 1983 American Control Conference in San Francisco, CA. He is also a past Secretary/Administrator of the Control Systems Society (1985–86).

His areas of interest include robust control, optimization theory, and energy systems. Dr. Dorato is a Fellow of the IEEE, cited for his contributions to sensitivity analysis and design in automatic control systems, and a Distinguished Member of the Control Systems Society, cited for his service to the Society and the profession. He is currently chairman of the Control Systems Society Education Committee, and a registered professional engineer in the State of Colorado.